ORGANIC & MEDICINAL CHEMIS ...

Stages 1 & 2 of the MPharm Programme
Years 1, 2 & 3 of the BSc course in Pharmaceutical Science,
Technology and Business

Russell J Pearson, School of Pharmacy & Bioengineering,
Keele University

Oxford University Press, the world's leading academic publisher, offers you a tailored teaching and learning solution with this custom text which contains chapters selected from OUP texts.

This custom text has been compiled for Keele University.

Please bear in mind that any cross-references will apply to chapters in the original book rather than to chapters within this custom text. If you would like to follow up on any of these references, please consult the original work in which the chapter appears. See **www.oup.com** for the full catalogue of OUP books.

Each page in this custom text has two sets of page numbers for ease of use. The page numbering of the original books is included to enable correct citation and reference back to the book from which the chapter is taken. There is also consecutive page numbering throughout this custom text in a shaded band at the bottom of each page.

The authors included in this custom text are responsible only for the ideas expressed in their own writing and not for any views that might be expressed by authors of other chapters.

ORGANIC & MEDICINAL CHEMISTRY

for MPharm & Pharmaceutical Science Students

2nd Edition

COMPILED FROM

Organic Chemistry
2nd edition
By Jonathan Clayden, Nick Greeves, Stuart Warren

Solutions Manual to Accompany Organic Chemistry
2nd edition
By Jonathan Clayden and Stuart Warren

How to Succeed in Organic Chemistry
1st edition
By Mark C. Elliott

An Introduction to Medicinal Chemistry
6th edition
By Graham Patrick

OXFORD
UNIVERSITY PRESS

OXFORD
UNIVERSITY PRESS

Great Clarendon Street, Oxford, OX2 6DP,
United Kingdom

Oxford University Press is a department of the University of Oxford.
It furthers the University's objective of excellence in research, scholarship,
and education by publishing worldwide. Oxford is a registered trade mark of
Oxford University Press in the UK and in certain other countries

Published in the United States of America by Oxford University Press
198 Madison Avenue, New York, NY 10016, United States of America

British Library Cataloguing in Publication Data
Data available

ISBN 978-0-19-887056-2

Printed and bound by
CPI Group (UK) Ltd, Croydon, CR0 4YY

Compiled from:

Organic Chemistry, 2e
By Jonathan Clayden, Nick Greeves, Stuart Warren
ISBN: 9780199270293
© Jonathan Clayden, Nick Greeves, and Stuart Warren 2012

Solutions Manual to Accompany Organic Chemistry, 2e
By Jonathan Clayden and Stuart Warren
ISBN: 9780199663347
© Oxford University Press 2013

How to Succeed in Organic Chemistry, 1e
By Mark C. Elliott
ISBN: 9780198851295
© Mark C. Elliott 2020

An Introduction to Medicinal Chemistry, 6e
By Graham Patrick
ISBN: 9780198749691
© Graham L. Patrick 2017

CONTENTS

Acknowledgements

The author and Oxford University Press would like to thank the following people who have given advice on the various editions of this textbook:

Dr Lee Banting, School of Pharmacy and Biomedical Sciences, University of Portsmouth, UK

Dr Don Green, Department of Health and Human Sciences, London Metropolitan University, UK

Dr Mike Southern, Department of Chemistry, Trinity College, University of Dublin, Ireland

Dr Mikael Elofsson (Assistant Professor), Department of Chemistry, Umeå University, Sweden

Dr Ed Moret, Faculty of Pharmaceutical Sciences, Utrecht University, The Netherlands

Professor John Nielsen, Department of Natural Sciences, Royal Veterinary and Agricultural University, Denmark

Professor H. Timmerman, Department of Medicinal Chemistry, Vrije Universiteit, Amsterdam, The Netherlands

Professor Nouri Neamati, School of Pharmacy, University of Southern California, USA

Professor Kristina Luthman, Department of Chemistry, Gothenburg University, Sweden

Professor Taleb Altel, College of Pharmacy, University of Sarjah, United Arab Emirates

Professor Dirk Rijkers, Faculty of Pharmaceutical Sciences, Utrecht University, The Netherlands

Dr Sushama Dandekar, Department of Chemistry, University of North Texas, USA

Dr John Spencer, Department of Chemistry, School of Life Sciences, University of Sussex, UK

Dr Angeline Kanagasooriam, School of Physical Sciences, University of Kent at Canterbury, UK

Dr A. Ganesan, School of Chemistry, University of Southampton, UK

Dr Rachel Dickens, Department of Chemistry, University of Durham, UK

Dr Gerd Wagner, School of Chemical Sciences and Pharmacy, University of East Anglia, UK

Dr Colin Fishwick, School of Chemistry, University of Leeds, UK

Professor Paul O'Neil, Department of Chemistry, University of Liverpool, UK

Professor Trond Ulven, Department of Chemistry, University of Southern Denmark, Denmark

Professor Jennifer Powers, Department of Chemistry and Biochemistry, Kennesaw State University, USA

Professor Joanne Kehlbeck, Department of Chemistry, Union College, USA

Dr Robert Sindelar, Faculty of Pharmaceutical Sciences, University of British Columbia, Canada

Professor John Carran, Department of Chemistry, Queen's University, Canada

Professor Anne Johnson, Department of Chemistry and Biology, Ryerson University, Canada

Dr Jane Hanrahan, Faculty of Pharmacy, University of Sydney, Australia

Dr Ethel Forbes, School of Science, University of the West of Scotland, UK

Dr Zoe Waller, School of Pharmacy, University of East Anglia, UK

Dr Susan Matthews, School of Pharmacy, University of East Anglia, UK

Professor Ulf Nilsson, Organic Chemistry, Lund University, Sweden

Dr Russell Pearson, School of Physical and Geographical Sciences, Keele University, UK

Dr Rachel Codd, Sydney Medical School, The University of Sydney, Australia

Dr Marcus Durrant, Department of Chemical and Forensic Sciences, Northumbria University, UK

Dr Alison Hill, College of Life and Environmental Sciences, University of Exeter, UK

Dr Connie Locher, School of Biomedical, Biomolecular and Chemical Sciences, University of Western Australia, Australia

Associate Professor Jon Vabeno, Department of Pharmacy, University of Tromso, Norway

Dr Celine Cano, Northern Institute for Cancer Research, Newcastle University, UK

Professor Steven Bull, Department of Chemistry, University of Bath, UK

Professor John Marriott, School of Pharmacy, University of Birmingham, UK

Associate Professor Jonathan Watts, University of Southampton, UK

Associate Professor Alexander Zelikin, Aarhus University, Denmark

Prof. Dr Iwan de Esch, Division of Medicinal Chemistry, VU University Amsterdam, The Netherlands

Dr Patricia Ragazzon, School of Environment & Life Sciences, University of Salford, UK

Dr David Adams, School of Pharmacy and Biomedical Sciences, University of Central Lancashire, UK

The author would like to express his gratitude to Dr John Spencer of the University of Sussex for co-authoring Chapter 16, the preparation of several web articles, and for preparation of Journal Club to accompany the sixth edition. Much appreciation is due to Nahoum Anthony and Dr Rachel Clark of the Strathclyde Institute for Pharmaceutical and Biomedical Sciences at the University of Strathclyde, for their assistance with creating Figures 2.9, Box 8.2 Figures 1 and 3, Figures 17.9, 17.44, 20.15, 20.22, 20.54, and 20.55 from pdb files, some of which were obtained from the RSCB Protein Data Bank. Dr James Keeler of the Department of Chemistry, University of Cambridge, kindly generated the molecular models that appear on the book's Online Resource Centre. Thanks also to Dr Stephen Bromidge of GlaxoSmithKline for permitting the description of his work on selective 5-HT2C antagonists, and for providing many of the diagrams for that web article. Many thanks to Cambridge Scientific, Oxford Molecular, and Tripos for their advice and assistance in the writing of Chapter 17. Finally, thanks are due to Dr Des Nichol, Dr Jorge Chacon, Dr Ciaran Ewins, Dr Callum McHugh, and Dr Fiona Henriquez for their invaluable support at the University of the West of Scotland.

Abbreviations

Ac	Acetyl		DMS	Dimethyl sulfide
Acac	Acetylacetonate		DMSO	Dimethyl sulfoxide
AD	Asymmetric dihydroxylation		DNA	Deoxyribonucleic acid
ADP	Adenosine 52-diphosphate		E1	Unimolecular elimination
AE	Asymmetric epoxidation		E2	Bimolecular elimination
AIBN	Azobisisobutyronitrile		E_a	Activation energy
AO	Atomic orbital		EDTA	Ethylenediaminetetraacetic acid
Ar	Aryl		EPR	Electron paramagnetic resonance
ATP	Adenosine triphosphate		ESR	Electron spin resonance
9-BBN	9-Borabicyclo[3.3.1]nonane		Et	Ethyl
BHT	Butylated hydroxy toluene (2,6-di-*t*-butyl-4-methylphenol)		FGI	Functional group interconversion
			Fmoc	Fluorenylmethyloxycarbonyl
BINAP	Bis(diphenylphosphino)-1,1'-binaphthyl		GAC	General acid catalysis
			GBC	General base catalysis
Bn	Benzyl		HMPA	Hexamethylphosphoramide
Boc, BOC	*tert*-Butyloxycarbonyl		HMPT	Hexamethylphosphorous triamide
Bu	Butyl		HOBt	1-Hydroxybenzotriazole
s-Bu	*sec*-Butyl		HOMO	Highest occupied molecular orbital
t-Bu	*tert*-Butyl		HPLC	High performance liquid chromatography
Bz	Benzoyl			
Cbz	Carboxybenzyl		HIV	Human immunodeficiency virus
CDI	Carbonyldiimidazole		IR	Infrared
CI	Chemical ionization		KHMDS	Potassium hexamethyldisilazide
CoA	Coenzyme A		LCAO	Linear combination of atomic orbitals
COT	Cyclooctatetraene		LDA	Lithium diisopropylamide
Cp	Cyclopentadienyl		LHMDS	Lithium hexamethyldisilazide
DABCO	1,4-Diazabicyclo[2.2.2]octane		LICA	Lithium isopropylcyclohexylamide
DBE	Double bond equivalent		LTMP, LiTMP	Lithium 2,2,6,6-tetramethylpiperidide
DBN	1,5-Diazabicyclo[4.3.0]non-5-ene		LUMO	Lowest unoccupied molecular orbital
DBU	1,8-Diazabicyclo[5.4.0]undec-7-ene		*m*-CPBA	*meta*-Chloroperoxybenzoic acid
DCC	*N,N*-dicyclohexylcarbodiimide		Me	Methyl
DDQ	2,3-Dichloro-5,6-dicyano-1,4-benzoquinone		MO	Molecular orbital
			MOM	Methoxymethyl
DEAD	Diethyl azodicarboxylate		Ms	Methanesulfonyl (mesyl)
DIBAL	Diisobutylaluminum hydride		NAD	Nicotinamide adenine dinucleotide
DMAP	4-Dimethylaminopyridine		NADH	Reduced NAD
DME	1,2-Dimethoxyethane		NBS	*N*-Bromosuccinimide
DMF	*N,N*-Dimethylformamide		NIS	*N*-Iodosuccinimide
DMPU	1,3-Dimethyl-3,4,5,6-tetrahydro-2(1*H*)-pyrimidinone		NMO	*N*-Methylmorpholine-*N*-oxide

NMR	Nuclear magnetic resonance	**SOMO**	Singly occupied molecular orbital
NOE	Nuclear Overhauser effect	**STM**	Scanning tunnelling microscopy
PCC	Pyridinium chlorochromate	**TBDMS**	*Tert*-butyldimethylsilyl
PDC	Pyridinium dichromate	**TBDPS**	*Tert*-butyldiphenylsilyl
Ph	Phenyl	**Tf**	Trifluoromethanesulfonyl (triflyl)
PPA	Polyphosphoric acid	**THF**	Tetrahydrofuran
Pr	Propyl	**THP**	Tetrahydropyran
i-**Pr**	*iso*-Propyl	**TIPS**	Triisopropylsilyl
PTC	Phase transfer catalysis	**TMEDA**	N,N,N',N'-tetramethyl-1,2-ethylenediamine
PTSA	*p*-Toluenesulfonic acid		
Py	Pyridine	**TMP**	2,2,6,6-Tetramethylpiperidine
Red Al	Sodium *bis*(2-methoxyethoxy) aluminum hydride	**TMS**	Trimethylsilyl, tetramethylsilane
		TMSOTf	Trimethylsilyl triflate
RNA	Ribonucleic acid	**TPAP**	Tetra-N-propylammonium perruthenate
SAC	Specific acid catalysis		
SAM	*S*-Adenosyl methionine	**Tr**	Triphenylmethyl (trityl)
SBC	Specific base catalysis	**TS**	Transition state
S$_N$1	Unimolecular nucleophilic substitution	**Ts**	*p*-Toluenesulfonyl, tosyl
		UV	Ultraviolet
S$_N$2	Bimolecular nucleophilic substitution	**VSEPR**	Valence shell electron pair repulsion

SECTION 1

LAYING THE FOUNDATIONS

INTRODUCTION

There are certain key skills that you need to have in order to succeed at organic chemistry. You need to be able to draw good structural representations, and to see exactly what those structures mean. You need to know what a structure is called, and what structure to draw from a chemical name.

The eventual goal is that you see the name of a compound written, and you visualize the complete molecular structure.

In considering the various different structures with the same molecular formula, we will establish key concepts such as isomerism, and double bond equivalents.

Perhaps most importantly, we will use the convention of curly arrows to draw the 'mechanisms' of organic reactions.

Initially, we won't worry too much about what is happening at a molecular orbital level, but we will establish the rules that good curly arrow drawing must follow.

In this first section, we will establish many of the basic ideas that we will be building on in later sections. There will be quite a few 'Practice' chapters here, as this is how you get comfortable with the basics.

Don't just read this book. Draw the structures. Make notes from it. Have a go at the problems, and discuss your answers with your friends. If you are getting it wrong, someone will tell you. If you are getting it right, you'll be able to persuade someone else that you have the right answer.

Learning organic chemistry is not easy. There will be pain! That feeling of frustration as you draw yet another compound with a five-valent carbon, yet another curly arrow that doesn't make sense.

All of that is a necessary part of the journey. You didn't learn to ride a bicycle without falling off!

Gradually you will find that those mistakes disappear. Don't expect to not make the mistakes at all. Just feel confident that drawing a structure or a reaction mechanism again and again will internalize the knowledge. Drawing one or two extra examples will help it become instinctive.

If you don't master the basics, it will be impossible to draw the mechanism of a complex reaction. Even then though, you can use that mechanism as practice in order to 'fix' your knowledge of the basics.

Trust me. You have already learned to do many complicated things. Understand **how** you did it, and then stop worrying. Let it happen. There is a wonderful Japanese proverb. 'If you have one eye on your destination, you only have one eye left to find the way.' The exam is not the goal!

BASICS 1
Structures of Organic Compounds

Let's start with the absolute basics—drawing structures. We need to consider the different conventions that are in common usage, and what the benefits and disadvantages are of each.

Simply reading through this chapter will not make you good at drawing structures. Every reader will probably *understand* everything in this chapter. But some people will still make mistakes when they are under pressure. We are going to need to establish some habits and 'ground rules' for how you should approach your studies and this book.

NUMBER AND TYPE OF BONDS

The vast majority of organic compounds contain only the elements carbon, hydrogen, oxygen, and nitrogen. **In a stable, neutral molecule, carbon forms four bonds, hydrogen forms one bond, oxygen two bonds, and nitrogen three bonds.** Most organic chemistry textbooks begin with some of the theory of bonding in organic compounds. I want to take a slightly different approach. For now, let's simply state that we can have single bonds, double bonds, and triple bonds. We represent a single bond by a single line, a double bond by a double line, and a triple bond by a triple line.

ethane	ethene	ethyne

Clearly, since hydrogen can only form one bond, it cannot form a double bond! Carbon can form four bonds—a triple bond 'counts' as three bonds, so carbon can form single, double, or triple bonds. Nitrogen can form three bonds, so it can also form single, double, and triple bonds. Oxygen can only form two bonds, so it can only form single and double bonds. Here are examples with all permutations.

Most of the time, organic chemists will draw structures by habit. They do understand the bonding, but they don't actually need this understanding to draw the correct number of bonds. So, let's focus on building the habits. We will do this by looking at some factual information, and a little guidance. We will build this up over a couple of chapters, and then provide some examples to work through.

The more structures you draw, the better you will get, but I want to ensure that you undertake some reflection on what you have drawn.

If you are drawing structures with the wrong number of bonds to some atoms, you can already check this yourself. The act of checking will make it less likely for you to make the same mistake next time. You might want me to give you the answers to all of the problems. My preference is to provide you with the tools to allow you to check your own structures.

Hydrogen might seem like the simplest of all, and there is little to explain really. A hydrogen atom is the thing you add on to satisfy the valence of other atoms. However, don't get complacent about this simplicity. Most of the mistakes you make will be because you forgot where the hydrogen atoms were. Many important reactions feature either the addition or removal of a hydrogen atom (normally in the form of a proton). This is *really important!*

Everyone can learn these basic rules. You will make mistakes, and you will draw the occasional carbon atom with five bonds. The important thing is to keep looking for them, and eventually you will find you don't draw them anymore.

First of all, there are some conventions that we need to be aware of, and we will expand on these here.

REPRESENTATIONS OF ORGANIC STRUCTURES

We use several different representations for organic structures. It is important that you recognize all of them. For example, the following three structures are all representations of the same compound, *n*-butane (more about the names later).

1	2	3

4

Which of these is clearest? Most year 1 undergraduate students will feel that structure **3** is the clearest. All hydrogen atoms are explicitly drawn, along with all bonds. It is easy to work out the formula. Importantly, you'll never forget where a hydrogen atom is, or draw an incorrect mechanism as a result of this.

However, this structure can look quite cluttered. If you have even a moderately complex structure, this type of representation will look very messy, and you'll find that you don't get an overall feeling for the key features of the structure.

Structure **2** is almost as clear. I have explicitly drawn H_3C on the left, so it is clear that the bond to the next carbon is from the carbon. However, the next carbon is a CH_2 group, so something has to be on the left and something on the right. We would never think that because the H is on the right, the bond to the next carbon is from hydrogen. We will look at this convention at various points over the next few chapters.

For me, structure **1** is actually the clearest. I can see at a glance that this is a saturated (no double or triple bonds) compound with four carbon atoms.

I can only see this because of the conventions used.

- Each line represents a chemical bond.

- Where the lines join or terminate represents a carbon atom.

- If no label is used for the atom, then a carbon atom can be assumed. We are organic chemists after all!

For most organic chemistry textbooks, this is the most common sort of representation. This is the one you need to get used to. So, you need to very quickly dispense with representation **3** and practise drawing all structures in representation **1**.

1.1

*If you find yourself drawing structures in representation **3**, **STOP** and redraw in representation I. Perhaps draw representation **3** and representation I alongside one another for a while, or draw (part of) a structure in representation **3** if you cannot see what is going on. But every time you do this, redraw the same structure and mechanism in representation I.*

You probably read the above point and thought 'but I like representation **3**' or 'I'm comfortable with representation **3**'. Of course you are! It's easier to learn at first. But I can assure you that representation **1** can become just as comfortable. I'll go a little further than this. If you don't make the effort to become comfortable with representation **1**, you're going to be wasting time with the rest. You'll know when you are getting good at this, in the same way that you know you are getting good at playing a musical instrument. It will just feel right,[1] and you will instinctively know that you've drawn a nice clear correct structure. Don't take short cuts!

---

[1] Or sound right, in the case of a musical instrument!

EMPHASIZING VALENCE AND FURTHER CONVENTIONS

Carbon is always tetravalent.[2] It forms four bonds. Once we take this into account, structure **1** needs a total of 10 hydrogen atoms.

You should be careful to use consistent representations. You would never draw *n*-butane as in structure **4** or structure **5**.

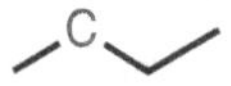

4 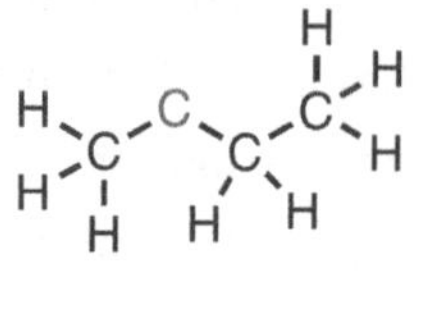   **5**

In both cases the second carbon atom from the left is explicitly drawn but without hydrogen atoms. This is just confusing. If you are going to explicitly write 'C' then you need to indicate the attached hydrogen atoms.

Conversely, if you see a structure with the atoms explicitly drawn, make sure you understand the bonding. Don't add other atoms to satisfy what you think the bonding is. Have a look at Common Error 1 for an example of this.

It may be that you use the less cluttered representation, but you need (perhaps when drawing a reaction mechanism) to show a particular hydrogen atom. This is fine. Just draw something like structure **6** below.

Just because you explicitly draw the H, you don't need to draw the C.

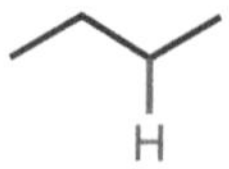

6

Again, there is little to understand. Every student who studies organic chemistry at university can 'get' all of this.

Let's add a bit more basic information.

CHARGES ON ATOMS

We aren't going to worry about positive and negative charges on carbon just yet. That will come up soon enough.

---

[2] There are exceptions. They are rare, and we don't need to concern ourselves with them at this point. If you develop the right habits, you'll be prepared when you encounter the exceptions, and they won't confuse you. For now, four bonds to carbon is where we are at.

We have established that oxygen forms two bonds. This is only the case if it is neutral.[3] An oxygen atom with one bond will normally have a negative charge and an oxygen atom with three bonds will normally have a positive charge. So, if I draw structure **7** below, then you would know that there was something wrong with the oxygen atom.

$$H_3C-O$$

7

You either need to add a hydrogen atom, as in **8**, or a negative charge, as in **9**. Which you do will depend on what the correct structure is, but it certainly isn't **7**!

$$H_3C-OH \qquad H_3C-O^{\ominus}$$

8 **9**

There is something else we haven't added to the oxygen atom—lone pairs. The oxygen in **8** has two lone pairs. We would show them as in **10**.

$$H_3C-\overset{..}{O}H$$

10

At a basic level, you know why this is. Oxygen is in group 6 of the periodic table, and therefore starts with six outer-shell electrons. It gains one extra electron from its bonding with carbon and one from hydrogen, giving an octet (eight electrons). Four of these are used in bonding, leaving four (two pairs) that are not bonding. A more rigorous discussion of bonding is coming in Basics 6.

But, to reiterate a point I made before, organic chemists will draw lone pairs on oxygen by habit rather than by thinking of the bonding. In my experience this happens whether you learn about the bonding first or not. So, let's make 'building habits' the major focus.

Remember that the main focus of this discussion is the conventions used to draw organic structures. I would be perfectly happy drawing structure **8** for methanol. But, if I needed to do something with it that required me to 'use' the lone pairs on oxygen, then I would draw structure **10**. The key point is that either representation is perfectly correct. There are always two lone pairs on the oxygen. We just don't always draw them. It is important to state this, because the sooner you start to get used to this, the faster you will learn. At no point can you say 'I forgot that there was a lone pair on that oxygen!'

A neutral nitrogen atom forms three bonds. It will have one lone pair. Again, we very often don't draw it unless we are going to use it in a reaction.

---

[3] If you are ever getting confused, think of a simple compound. For me, water is my 'go to' molecule with oxygen. It's important to have a 'go to' strategy.

MULTIPLE BONDS

We can have double bonds between carbon and carbon, or carbon and other atoms (most commonly oxygen). We can even have carbon–carbon triple bonds. We will look at the strengths of these bonds in due course, along with their bonding. Let's just look at these briefly for now in order to make two more points.

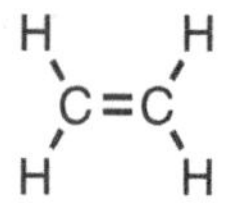

ethene

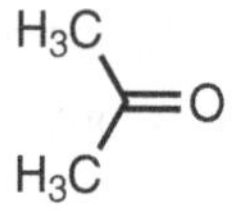

propanone
(trivial name acetone)

ethyne
(trivial name acetylene)

I have used two different representations above. For ethyne, I used the simplest skeletal representation, but I explicitly drew the hydrogen atoms in. I didn't need to do this. The representation below is perfectly correct.

ethyne

oct-4-yne

It is quite common in the early stages to get confused with alkynes and to count the number of carbon atoms incorrectly. The structure above right is oct-4-yne and has a total of eight carbon atoms. Count them! There is one at each end of the triple bond. This is easier to see than to explain. Look at it carefully for now, and we will include some examples in the exercises.

*The reason behind this problem is a tendency to draw the two bonds coming **from** the alkyne much shorter, and then to be confused about whether the two carbon atoms are **attached** to the alkyne, or are **actually** the alkyne carbons. This then leads to the assumption that this compound has only six carbon atoms. We will emphasize the importance of correct drawing in Habit 1.*

If you do make this mistake, you will never be able to draw a correct structure for any reaction product derived from the alkyne.

You don't need to be clever to get the number of carbon atoms right. You just need to have invested the effort at the right time to get used to it.

8

AROMATIC COMPOUNDS

When we put three double bonds in a six-membered ring, it gives it a special type of stability known as aromaticity. We will explain this later, but for now it is enough to mention the pattern. The simplest aromatic compound is benzene (**11**).

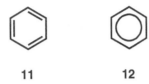

11   12

*Benzene does not have alternating single and double bonds. All of the carbon–carbon bonds are the same length. You have probably been told that structure **12** is a better representation, as it shows that all of the bonds are the same length. When we draw reaction mechanisms, we will find that representation **11** is better because it allows us to keep track of electrons, and it allows us to draw curly arrows (more about these later—much more!). When you get used to it, you will always recognize that benzene has six identical C–C bonds in structure **11**. Again, it simply needs to become a habit.*

Remember that not all rings are aromatic! Cyclohexane (**13**) is a perfectly good organic molecule. We use the prefix '*cyclo*' in the name to indicate that there is a ring.

13

Here we have two hydrogen atoms attached to each carbon. One of the most common mistakes I see is students drawing the hexagon, and then automatically drawing the circle that would turn it into benzene. Remember that cyclohexane is not a single lady. Even if you like it, don't put a ring on it![4]

CONCLUSION

I'm hoping by now that you're starting to realize why this book is different. My focus is on telling you how you should study, what mistakes you will make, and how to correct them.

---

[4] There will be quite a few bad jokes throughout this book. My book, my rules! If you don't get the joke, just Google 'Beyonce'. If you don't know what Google is, then the book has surpassed my wildest hopes with its longevity!

1

Don't beat yourself up about the mistakes. The only reason I know to highlight these mistakes is because I have seen them all, many times. Just be mindful of them, and you'll find it takes less time to spot the mistakes, right up to the point where you stop yourself before you actually draw an incorrect structure.

As we start to encounter various problems in organic chemistry, I will try to give you good problem-solving strategies.

You will also notice that this book is written in the first person. I am writing this for you, the reader. I want to use a writing style that makes it more personal—to achieve a direct connection.

HABIT 1
Always Draw Structures with Realistic Geometry

In the previous chapter, I mentioned the importance of forming good habits. I want to formalize this by including short chapters focused only on this one theme. Here is the first one!

REALISTIC GEOMETRY

Now we have looked at the different types of bond, we can consider the typical bond lengths and angles.

Here are the structures of ethane (**1**), ethene (**2**) and ethyne (**3**, nobody ever calls it ethyne—it is always acetylene). I've added the bond lengths in angstrom units (Å). There are two absolute key points to recognize at this level.

- Pretty much all chemical bonds in organic compounds are between 1 and 2 Å long.

- For a given type of bond, triple bonds are shorter than double bonds, which in turn are shorter than single bonds.

If you look closely at the numbers, you will notice some trends. Don't worry about them for now. We will add the details later.

And here are the bond angles. In ethane, all bond angles are essentially tetrahedral, 109.5°. For ethene, the atoms are all in a plane, with approximate angles of 120°. For acetylene (ethyne) the bonds are linear, 180°.

1.2

It is important to develop the habit of always trying to draw structures with realistic bond lengths and angles where possible. As you can see with ethane, this is not very easy! After a while, you will get used to the situations in which it is not possible to draw all of the bond lengths and angles realistically, and you will correctly interpret a given structure.

WEDGES AND DASHES—REPRESENTING SHAPE

We simply cannot draw methane on a flat page and have all bond angles be 109.5°.

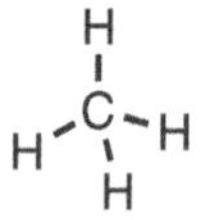

$$\begin{array}{c} H \\ | \\ H-C-H \\ | \\ H \end{array}$$

If we have one of the H−C−H angles close to 109.5°, all the others will look smaller. The key point in this representation is to 'internalize' the drawing of the structure, so that you instinctively recognize what the actual shape is, no matter how it is drawn.

When I see the structure above, I 'see' a tetrahedral structure. That's because, deep down, I know that it cannot be anything other than tetrahedral. Right now, you are reading a paragraph of text. You never stop to ask 'why is this the letter "a"?' You just 'know' it, and as a result, the words leap out of the page and into your head. You've done it so many times that it has become second nature. And you started doing that when you were three years old. So why on earth would you question your ability to do the same with molecular structures? All you need to do is practise enough.

We also have certain conventions that we can apply. We use a wedged bond to show that an atom is coming out of the page, and a dashed bond to show that an atom is going into the page.

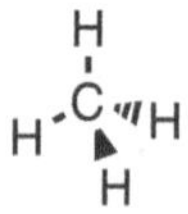

$$\begin{array}{c} H \\ | \\ H-C\text{\tiny\textbackslash}H \\ \blacktriangle \\ H \end{array}$$

There isn't much to 'understand' here. It is what it is! You need to get used to these representations, so that when you see it, you can visualize the molecule in three dimensions.

Use molecular models to help with this, particularly as the structures get more complex.

We aren't going to directly do much with stereochemistry until Section 3. However, in the various 'Practice' chapters, I will be drawing structures with wedged and dashed bonds. I will do that because the molecules are three-dimensional.

When you get to these chapters, don't ignore the stereochemistry. Start to 'absorb' it. Just don't stress about it. Once you've read Section 3, come back to these and have another go. You'll 'see' more in the structures.

BASICS 2
Functional Groups and 'R' Groups

INTRODUCTION

Functional groups are the parts of an organic compound that lead to its reactivity. There are many different functional groups. They all have names. The important thing is to know what each functional group is, what it is called, and not to confuse different functional groups.

In this section, I will list several important functional groups, and show how they are commonly drawn (and how they should **not** be drawn). Please read this section very carefully, so that you don't accidentally learn to draw them the wrong way.

Before we look at functional groups, we will look at the carbon chains and rings that they are attached to. Before that, though, a cautionary note.

If you associate a functional group with the name as a word, there is a risk that you will confuse it with another (sometimes very different) functional group with a similar name. Always try to associate the name with the structure. The best analogy I can give for this is learning a new language. At first, you will think of an object in your native language and ask 'what is the word for X?' Eventually you will see the object and the word will automatically come to you in your new second language.

ALKYL GROUPS—*PRIMARY*, *SECONDARY*, AND *TERTIARY*

We might draw a structure such as **1** to mean any aliphatic alcohol.

R – OH

1

The 'R' refers to a carbon chain in a very general sense. We need to look at these carbon chains. We refer to the 'R' group as an alkyl group. There are a great many of these, and we will look at how we name them, and their parent hydrocarbons, in Basics 3. For now,

though, we will look at a few of them and classify them according to the nature of the attached carbon atom. The ones we need are summarized in the table below:

Abbreviation	Name	Structure
Me	Methyl	CH_3-
Et	Ethyl	CH_3CH_2-
n-Pr	Propyl	$CH_3CH_2CH_2-$
i-Pr	Isopropyl	$(CH_3)_2CH-$
n-Bu	Butyl	$CH_3CH_2CH_2CH_2-$
t-Bu	*tert*-Butyl	$(CH_3)_3C-$

Note: in isopropyl and *tert*-butyl, we need to use the brackets to make it clear that the CH_3 groups are attached to the same C.

Here is the key distinction.

Ethyl, n-propyl, and n-butyl are primary alkyl groups. The point of attachment is a CH₂. Isopropyl is a secondary alkyl group. The point of attachment is a CH. tert-Butyl is a tertiary alkyl group. The point of attachment has no attached hydrogen atoms.[5]

 Can you work out what a *sec*-butyl group looks like? If you can, you will probably see why I didn't include it in the table.

AROMATIC RINGS

Benzene, **2**, is an aromatic compound. We will encounter the definition of aromaticity in Basics 10. For now, let's just focus on recognizing patterns.

2

If we have 'some sort of' aromatic ring attached to (for example) an amino group, we can refer to the aromatic ring as 'Ar', and we can abbreviate the structure as in **3** below.[6]

Ar$-NH_2$

3

---

[5] Is methyl a *primary* alkyl group? Methyl is methyl. If you had to classify it, you would say it was *primary*. Methyl is small and unhindered. It is better to understand the effect of different groups with different branching rather than to lump methyl in with all the other possible *primary* alkyl groups.

[6] You probably don't want to do this if you work on organoselenium chemistry!

14

Aromatic compounds have very different reactivity to aliphatic compounds, so it is important that within these 'generic' representations, we can make this important distinction.

We will look at the naming of aromatic compounds in Basics 3.

ALCOHOLS AND PHENOLS

Structure **4** is methanol, an alcohol. It has an OH group attached to carbon. Compound **5** is phenol. In this case, phenol is the name of the specific compound, but also the name of the functional group. All compounds with an OH group attached to a benzene ring are known as phenols. Perhaps the most commonly encountered phenol is 2,4,6-trichlorophenol (**6**), a fungicide/antiseptic.[7]

H_3C-OH

4

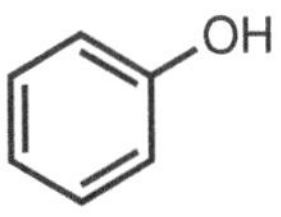

5

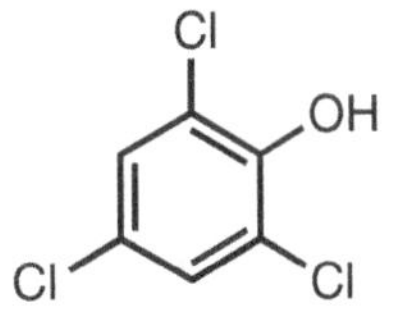

6

AMINES

Compounds **7**, **8** and **9** are all amines. They don't have to have a hydrogen atom attached to the nitrogen, but they can have.

H_3C-NH_2

$H_3C-NH \\ \ \ \ \ \ \ \ \ CH_3$

$H_3C-N \\ \ \ \ \ \ \ \ CH_3 \\ \ \ \ \ \ \ \ CH_3$

7, methylamine
a *primary* amine

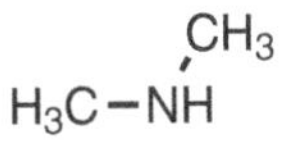

8, dimethylamine
a *secondary* amine

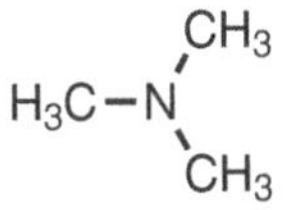

9, trimethylamine
a *tertiary* amine

PRIMARY, SECONDARY, AND *TERTIARY*—A NOMENCLATURE MINEFIELD

Before we get too much further, I want to illustrate the problems you face with chemical nomenclature—and the solutions.

---

[7] There seems to be somewhat of an urban myth that the antiseptic solution 'TCP' is so-called because it contains 2,4,6-trichlorophenol. In fact, TCP is an abbreviation for trichlorophenylmethyliodosalicyl, the active ingredient prior to the 1950s. This name makes no chemical sense, and as far as I know, nobody knows what the structure of the active ingredient actually was!

15

A *tertiary* alcohol has **one** *tertiary* alkyl group attached to the oxygen. A *tertiary* amine has three alkyl groups attached to nitrogen. These alkyl groups could be *primary, secondary*, or *tertiary*.

*Of course, since an alcohol has an OH group, and oxygen can only form two bonds, there can only ever be one alkyl group attached. If we refer to a tertiary alcohol, we **must** be talking about the alkyl group!*

What is the solution to this problem? At a simple level, just keep reading, looking at how people describe structures, and absorb the terminology.[8]

KEEP YOUR DATA CONNECTED!

Every alcohol has a hydrogen attached to oxygen. Not every amine has a hydrogen attached to nitrogen.

While I don't want to say too much about infrared spectroscopy, an N–H bond gives quite a characteristic peak. It is quite common for students to interpret an infrared spectrum and state that the compound concerned is an amine because they see an N–H peak. They will then draw a structure such as **9** which is indeed an amine but doesn't have the N–H bond that led to the conclusion in the first place. It is really important to keep track of those hydrogen atoms and what they mean.

ALKENES AND ALKYNES

A double bond between two carbon atoms is known as an alkene. A triple bond between two carbon atoms is known as an alkyne.

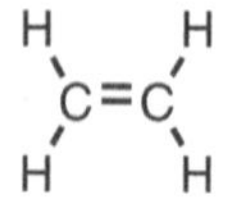

ethene
(an alkene)

ethyne
(an alkyne)

CARBONYL COMPOUNDS

A double bond between carbon and oxygen has the general name 'carbonyl', but there are so many different types of carbonyl compound depending on what else is attached.

---

[8] Here's a confession. I don't think I had really thought about this contradiction until I started writing this book. But I still used the terminology correctly. I had internalized the learning to the point that I didn't think there was anything to worry about. This is the best sort of learning!

They all have their own names. Structure **10** is an aldehyde. Structure **11** is a ketone. Structure **12** is an ester. Structure **13** is an amide. Again, it doesn't need to be NH_2 to be an amide. There can be other alkyl groups attached to the nitrogen atom.

| 10 | 11 | 12 | 13 |

We can abbreviate the aldehyde **10** as CH_3CHO. We would not normally abbreviate it as CH_3COH. This can be a bit confusing. The oxygen and hydrogen atoms are both bonded to carbon, so it isn't clear which should be first. However, in the second representation (RCOH) it looks as though the H is bonded to O. Basically, CH_3CHO is right and CH_3COH is wrong. Sometimes it is best to accept it and try to get used to it.

In some representations, we might want the aldehyde on the left, as in $OHCCH_3$. Again, you would not normally write $HOCCH_3$.

With an ester, the point of attachment becomes more important. If we were to abbreviate the above example (**12**), we would write $CH_3CO_2CH_2CH_3$. We would not write $CH_3OCOCH_2CH_3$. This would be a different compound, as it implies that the CH_3 group is directly bonded to oxygen.

This is a lot to take on board. There are many places you can make mistakes. However, if you practise, this will all become second nature. If you don't practise, you will always be trying to remember how to draw something or what to call it. Succeeding in organic chemistry is all about doing the right things at the right time.

Here is one final point, with one more functional group.

ETHERS

Compound **14** is an ether. It has two alkyl groups (or they could be aryl groups) attached to one oxygen atom. This one is diethyl ether, a common solvent in the lab.

$$H_3CH_2C-O-CH_2CH_3$$

14

ETHERS AND ESTERS—A POINT OF CONFUSION

So, here is a very important point. It is not uncommon for students to confuse an ester and an ether. After all, the names are very similar. Most of the time this doesn't cause any problems—they just use the wrong word, but they still know which functional group they are talking about.

But in other cases, a student may correctly interpret an infrared spectrum (don't worry if you don't know about infrared spectroscopy) as having an ester functional group due to the characteristic peak around 1740 cm^{-1}. They will then draw an ether as a possible structure for the compound.

This is because they are associating the data with the word rather than with the molecular feature.

It is important that you start, very early on, to rely on structures as your primary tool for all organic chemistry.

All I can do is caution you against these mistakes. The whole point is that you need to study correctly and internalize certain information and patterns. There is no way I can do this for you, but I can (and will) give you exercises for practice.

HISTORICAL PERSPECTIVE

There is another important point.

In the lab, you will use diethyl ether and petroleum ether as solvents. Diethyl ether is an ether. Petroleum ether is not an ether. It simply refers to the petroleum fraction distilling in a particular range (*e.g.* petroleum ether 40 – 60 is the fraction boiling in the range 40 to 60 °C. It is a hydrocarbon (alkane)).

The term 'ether' is a very old one, and it doesn't always refer to the specific functional group.

What a nightmare! And yet if you just relax and focus on understanding molecular structure, the rest of it will take care of itself in time. Don't get overwhelmed by any of this. Just come back to this chapter in a couple of weeks and have another read!

Now we have got a selection of functional groups covered, we can start to think about systematic nomenclature.

18

BASICS 3
Naming Organic Compounds

NOMENCLATURE OF ORGANIC COMPOUNDS

We have to have clear and understandable names for chemicals. We have already used some of these names. You will find that in most instances you learn the names of compounds as you progress through the book. We will reinforce this with practice.

Whenever you encounter a compound name, you should take a few minutes to ensure that you understand why this name is correct for a given molecular structure. It doesn't make any sense to try to learn how to name all possible compounds.[9] Understanding the names is important, but it's more important to understand what the compounds do and why they do it. Just 'absorb' nomenclature as you encounter it.

For now, we will simply start with some basic rules, and expand these as we look at various functional groups.

ALKANES

For simple saturated hydrocarbons, the name is made up of a component that refers to the number of carbon atoms, followed by 'ane'. 'Saturated' simply means that there are as many hydrogen atoms as the number of carbon atoms can take. We will come back to this point later.

Here is a table.

Number of Carbon Atoms	Formula	Structural Formula	Name
1	CH_4	CH_4	Methane
2	C_2H_6	CH_3CH_3	Ethane
3	C_3H_8	$CH_3CH_2CH_3$	Propane
4	C_4H_{10}	$CH_3CH_2CH_2CH_3$	Butane

[9] I've got a book that is just about naming of organic compounds! Does that make me sad?

3

Number of Carbon Atoms	Formula	Structural Formula	Name
5	C_5H_{12}	$CH_3(CH_2)_3CH_3$	Pentane
6	C_6H_{14}	$CH_3(CH_2)_4CH_3$	Hexane
7	C_7H_{16}	$CH_3(CH_2)_5CH_3$	Heptane
8	C_8H_{18}	$CH_3(CH_2)_6CH_3$	Octane
9	C_9H_{20}	$CH_3(CH_2)_7CH_3$	Nonane
10	$C_{10}H_{22}$	$CH_3(CH_2)_8CH_3$	Decane

You will notice two new aspects of structure representation in the above table. First of all, while we normally draw a single bond as a line between two atoms, we can 'write' the structure with no need for the lines.

We can actually represent alkenes and alkynes using this convention as well. Writing '$CH_3CHCHCH_3$' would give us but-2-ene (more about naming alkenes later in this chapter). There must be a double bond between the second and third carbon atoms, as we are explicitly writing the hydrogen atoms and there aren't enough for a saturated structure. Personally, I would still make it clearer by writing '$CH_3CH=CHCH_3$'.

You may notice that in the structural formula column, after butane I started to use brackets to indicate the total number of CH_2 groups rather than drawing them out. This is another convention that you should get used to.

If we have branched alkanes, we identify the longest carbon chain. This becomes the 'parent' name. Substituents along the chain are then identified by the type of substituent and their position along the chain. If there is more than one of the same type of substituent, we use prefixes such as 'di', 'tri', *etc*. This is best seen with examples.

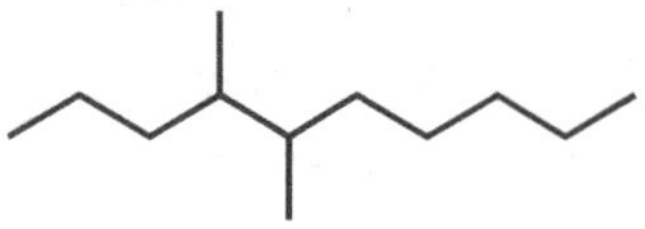

4,5-dimethyldecane

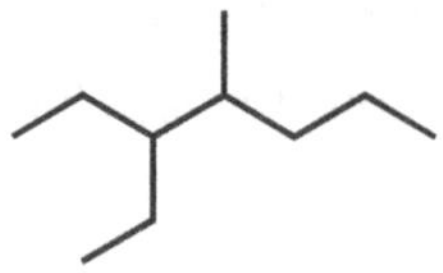

3-ethyl-4-methylheptane

Notice in the second example, there are actually two different 7-carbon chains that we could have considered. Actually, in this case it doesn't matter. The substituents are listed in alphabetical order, and numbering is done to give the lowest possible numbers (*i.e.* 3 and 4 rather than 4 and 5 if you counted from the other end).

PREFIXES FOR ALKYL CHAINS

We saw some simple alkyl substituents above. These are named from the parent alkane. Here they are again. I'm willing to bet that you can add more lines to this table.

20

Abbreviation	Name	Structure
Me	Methyl	CH_3-
Et	Ethyl	CH_3CH_2-
n-Pr	Propyl	$CH_3CH_2CH_2-$
i-Pr	Isopropyl	$(CH_3)_2CH-$
n-Bu	Butyl	$CH_3CH_2CH_2CH_2-$
t-Bu	*tert*-Butyl	$(CH_3)_3C-$

Note: in isopropyl and *tert*-butyl, we need to use the brackets to make it clear that the CH_3 groups are attached to the same C.

We defined the terms *primary, secondary,* and *tertiary* in Basics 2. We now need to consider how we would name compounds with various alkyl groups. Here is an example.

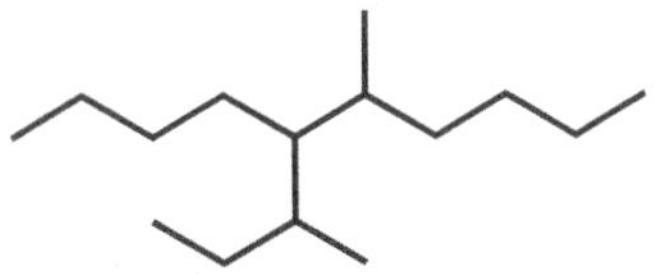

The longest alkyl chain has 10 carbon atoms. It is a derivative of decane. There is a methyl group on C5 or C6 depending on which end you start the numbering at. Then there is a *sec*-butyl group[10] on the other of these two carbon atoms. We could simply call it a *sec*-butyl group, or we could call it 'but-2-yl', where the '2-yl' bit makes it clear that the butyl group is attached via **its** 2-position.

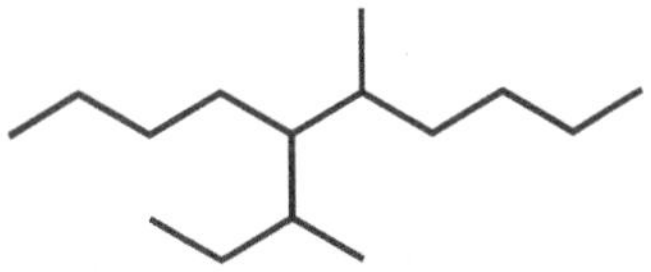

5-(but-2-yl)-6-methyldecane

There isn't much more to say about this. I could write out a dozen more examples, but you wouldn't learn anything new from them. I would much prefer to set them as problems, so you will be aware that you are getting better at naming compounds.

HALOALKANES

Most of the time, we simply consider the halogen to be a substituent. As above, we put substituents into alphabetical order. If there is only one place to put a substituent, there is no need for the number.

---

[10] Did you work it out correctly in Basics 2? Make sure you refer back. This constant reinforcement is how you will learn.

21

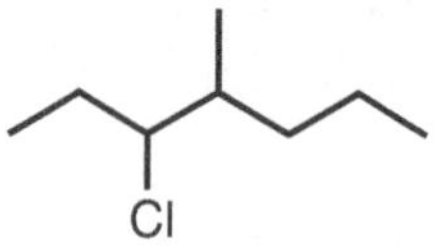

3-chloro-4-methylheptane 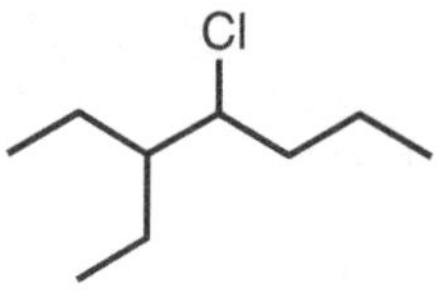 4-chloro-3-ethylheptane  bromoethane

Trivial names are commonly used. Most people would name bromoethane as ethyl bromide.

The main point about nomenclature is that if you write a compound name, someone reading it could only draw one possible structure. If you didn't put the substituent names in alphabetical order, this wouldn't be a massive problem. Sure, it wouldn't be strictly correct, but it would still do the job.

ALKENES

Here, the name of the compound ends with 'ene'. Since a double bond is always between two adjacent carbon atoms, you don't need to specify both carbon atoms.

In simple compounds, it is okay to put the number at the beginning. As names get more complicated, it is sometimes necessary to put the number directly before 'ene' to make sure it is clear that the number refers to the double bond position.

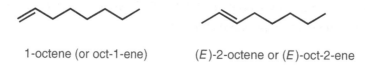

1-octene (or oct-1-ene) (*E*)-2-octene or (*E*)-oct-2-ene

2-Octene could exist as two double bond isomers, depending on whether the methyl and pentyl groups are on the same side of the double bond or on opposite sides. We use (*E*) and (*Z*) to describe these. (*E*) comes from the German word '*entgegen*', meaning 'against'. (*Z*) comes from the German word '*zusammen*' meaning 'together'. When you only have two substituents on a double bond it is clear which one to use. When you have three or four substituents it is less straightforward. We will come back to this in Habit 6.

ALKYNES

The naming of alkynes is similar to that of alkenes. Here are a couple of examples.

2-octyne 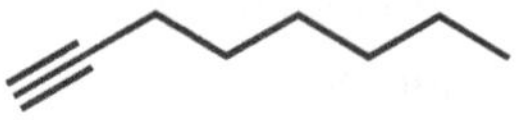 1-octyne

22

*As I mentioned in **Basics I**, the alkyne functional group confuses a lot of students. Take a moment to look at these structures and count the atoms. Make sure you understand why this representation shows eight carbon atoms in each case.*

AROMATIC COMPOUNDS

First of all, there is just one more abbreviation we need to add, Ph. Ph is short for phenyl, formula C_6H_5. A phenyl group is a benzene ring with one hydrogen atom removed to allow something else to be attached. For example, structures **1** and **2** represent the same structure.

$$Ph-NH_2$$

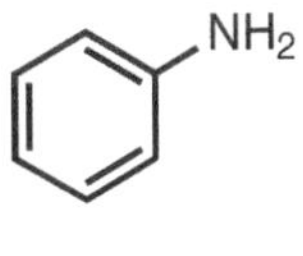

1   **2**

*There is one common point of confusion with phenyl. It sounds a bit like 'phenol'. I find that many students talk about 'the phenol group' when they mean 'the phenyl group'. Phenol is Ph-OH. I've tried to think what I can write to make this point any clearer. There isn't anything. Phenyl is phenyl, **and you just need to know this**. I will remind you of this lots, to make sure it becomes habit.*

Now we have got that out of the way, what happens if we have more than one substituent on a benzene ring? At that point, we can have **constitutional isomers** (Basics 4). Here are three phenol examples.[11]

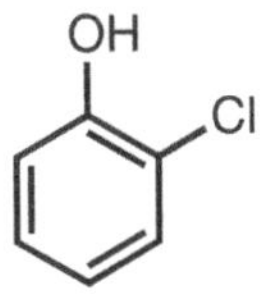

2-chlorophenol
o-chlorophenol

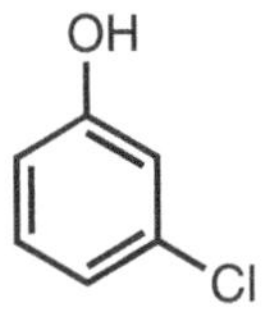

3-chlorophenol
m-chlorophenol

4-chlorophenol
p-chlorophenol

There are two conventions that we use. The first is numbering. In the compound on the left, the chlorine is in the 2-position relative to the OH group being on the 1-position. We don't need to specify both numbers. We can also use the terms *ortho, meta,* and *para* (usually abbreviated as *o-, m-,* and *p-* in the compound name) to indicate the relative placement of one substituent relative to another.

---

[11] These ones really are phenols!

23

This book doesn't cover reactions of aromatic compounds, but you still need to be familiar with the nomenclature.

ALCOHOLS

Depending on the compound, you can specify an alcohol with the suffix 'ol' or the substituent prefix 'hydroxy'. You would use the prefix if there is another substituent that is 'more important'. It takes a while to get used to which substituent will be more important, and you tend to learn by example rather than formally understanding. In the second example below, we have added an extra functional group.

heptan-1-ol (or 1-heptanol)

(*E*)-hept-4-en-1-ol
(or (*E*)-1-hydroxy-4-heptene)

In the second compound above, we only need to use one number to define the alkene position. If the alkene bond was between C3 and C4, then it would be a hept-3-ene, not a hept-4-ene.

ALDEHYDES

These are named with the suffix 'al'. Note that in most cases we don't need to specify a number for the aldehyde group. If it wasn't at the end, it wouldn't be an aldehyde. Everything else is numbered relative to the aldehyde being position '1'.

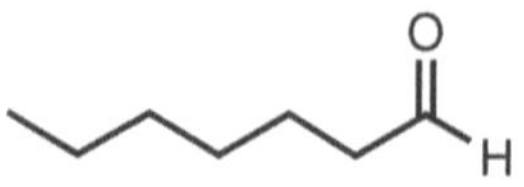

heptanal

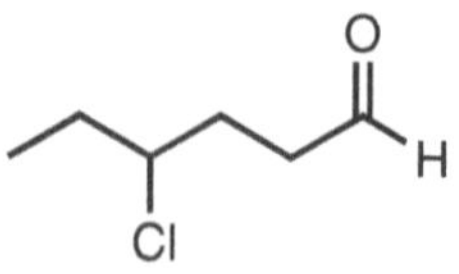

4-chlorohexanal

KETONES

Here you definitely do need to specify the position of the ketone group, unless it is symmetrical. Even then, it is better to specify the number (see heptan-4-one below) so it is clear that you didn't just forget!

24

heptan-2-one   heptan-4-one

3

SUMMARY

This discussion is really just to get you started. There are many more functional groups to consider, but they follow broadly similar rules. As with the structural representations, once you understand the principles and apply them with regular practice, you'll start to come up with plausible compound names without too much effort.

Fundamentally, a compound name needs to describe one compound and one compound only. Any name that is ambiguous **must** be wrong.

There's an easy way to check if you have named a compound sensibly. Take the name and draw a structure from it (or ask someone else to do this). If it is only possible to draw one structure, the name is good. We will use this approach in Practice 2.

PRACTICE 1
Drawing Structures from Chemical Names

THE PURPOSE OF THE EXERCISE

The point of this section is not really about drawing the correct structure from a chemical name. This is only one aspect of it. The main focus is to practise drawing good skeletal structures of organic compounds.

If you need to draw a structure that shows every hydrogen atom at first, that is fine. However, then redraw it as the skeletal structure. Draw bond angles as close to reality as you can, while still considering the hydrogen atoms that are not shown.

If it's a good structure, you will be able to clearly see what the compound is. If you get confused, perhaps you need to draw the structure more clearly.

THE EXERCISE

Draw structures for the following compounds:

1. 3-(Hex-2-yl)heptane
2. 4-Chlorohexanal
3. 4-Bromo-7-methylnonan-2-one
4. 5-Chloro-2-propylphenol
5. 3-Ethylhexanoic acid
6. 7-Chlorohept-3-yn-2-ol
7. Methyl propanoate
8. Ethyl benzoate
9. 3-Hydroxyacetophenone (this is a non-systematic name—get used to it!)
10. 2-Aminoheptan-4-ol

11. 4-Chloro-1-methylcyclohexan-2-ol

12. *N*-Ethyl-*N*-propyl-(2-hydroxypentylamine) (this one is tricky—what does the '*N*' mean?)[12]

Note: You will probably need to look up some compound names or discuss them. This is fine. **Basics 3** doesn't cover all aspects of naming compounds.

1.3

Above all, don't get stressed by this. Remember, once again, the main point is that the structures you draw should be neat and tidy, and should be the more economical skeletal structures. You should be able to draw them quickly and confidently.

Get your friends to draw the structures, and compare your structures with theirs. Discuss which ones are best, and why.

For the compounds you have drawn, identify the functional groups present. Your answers may depend on how you interpret the term 'functional group'. I don't normally think of a methyl group as a functional group, but if you have a reaction that leads to its functionalization, perhaps it is! Discuss and debate your answers with colleagues. Don't try to learn in a vacuum!

For now, you need to be looking at organic structures and taking the time to try to see what is in them.

FOLLOW-UP EXERCISE

Give the structures you have drawn to a friend and ask them to propose systematic names for the compounds. Don't worry about number 9, and number 12 might be quite difficult. Once they have had a go, discuss the names with them.

 Are the compound names in my list the best systematic names for all of the compounds?

Were there any that were particularly troublesome?

AND IN REVERSE!

If you can propose a structure from a chemical name, you should be able to propose a chemical name from a structure. Here are some structures of *relatively* simple hydrocarbons to practise with.

---

[12] This probably isn't a fair question, as I haven't explained this. But you can look at the rest of the name and make an educated guess. Also, I want you to get used to looking things up in a variety of sources.

27

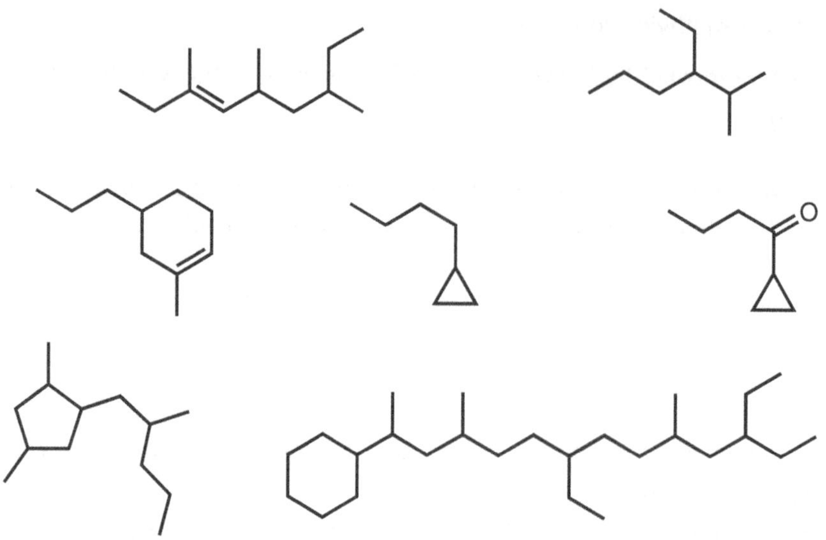

There are a few things to think about. In particular, compounds with rings can be challenging. Do you consider the ring to be the main feature, and everything else a substituent, or is it better to consider the ring itself to be a substituent? There isn't always a simple answer to this.[13]

Video 1.4 shows the worked answers to these problems. First of all, have a go and discuss your answers with friends.[14]

1.4

---

[13] Well, there are a set of definitive rules, but there is a lot to be said for giving the shortest possible **unambig-uous** name for a compound. Remember why we are doing this!

[14] Not too much though! I'd like you to keep your friends!

BASICS 4

Isomerism in Organic Chemistry—Constitutional Isomers

WHAT IS ISOMERISM?

The concept of isomerism is a very important one in all areas of chemistry. There are quite a few different types of isomer, so first of all, let's state clearly what we mean by isomers.

Isomers are compounds that have the same molecular formula, but are not the same compound.

Within this very broad definition, there are quite a few ways in which isomers could differ. For now, we will focus on one of these. Once you've got the hang of this, we will add another one.

CONSTITUTIONAL ISOMERS

Constitutional isomers, also known as **structural isomers**, are compounds that have the same molecular formula—they have the same number of each type of atom within the molecule—but the atoms are connected in a different way.

EXAMPLES

Let's have a look at some alkane structures, along with their names.

CH_4 /   ⌒

methane   ethane   propane

Note that I have used the most economical representation for each structure—lines to represent bonds. Where a line ends, or two lines join, there is a carbon atom. For methane, with only one carbon atom, we have to state 'CH_4'. We couldn't just draw a dot.

4

For these three compounds, we would find that if we were given a molecular formula, we could only draw these exact structures. There are no isomers. The molecular formula would be enough to fully define the structure.

If we have four carbon atoms, there are two possible structures. The first is where all four carbon atoms form a single chain. This is *n*-butane, or simply 'butane'. The second has a shorter carbon chain, with one carbon atom as a substituent. This is 2-methylpropane, because the longest carbon chain is three carbon atoms (propane). The methyl group is on the middle carbon, or the 2-position. A methyl group is 'CH₃'—never forget the attached hydrogen atoms. *n*-Butane and 2-methylpropane are constitutional isomers.

n-butane    2-methylpropane

The more carbon atoms we add, the more constitutional isomers there can be. For example, we have three isomers of pentane, although only one of them is actually called pentane.

n-pentane 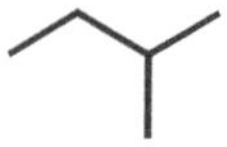   2-methylbutane    2,2-dimethylpropane

We often call this isomer '*n*-pentane'. The '*n*-' refers to 'normal' which relates to the straight chain (rather than branched).

Cyclohexane and 1-hexene are isomers. Don't take my word for it. Check! This sort of thing is good practice. The key point here is that one of these has a carbon–carbon double bond and the other does not. The molecular formula alone (work it out!) doesn't tell you anything about the functional groups present.

cyclohexane    1-hexene

Here is another example. The two compounds below are isomers, and again they have different functional groups.

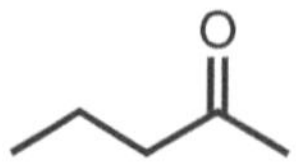

2-pentanone 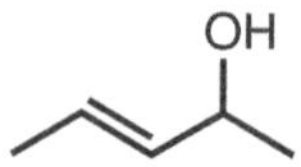   2-penten-4-ol

This brings us to the next topic. Some molecular formulae make sense, while others do not. Although you can't tell the full structure of a compound from the formula, you can get some information from it. We will look at this next, after a short 'Practice' chapter.

30

PRACTICE 2
Constitutional Isomers and Chemical Names

THE PURPOSE OF THE EXERCISE

It's really about getting more practice! Sometimes it's that simple. The added layer here is that it isn't always possible to draw a plausible structure for **any** given molecular formula. A good chemist will spot the impossible formulae quickly. There are several things to look out for. They are covered in the next chapter. But you already know enough to work them out for yourself.

QUESTION 1

How many different structures can you draw for the following molecular formulae?

1.5

1. $C_6H_{12}O$
2. $C_8H_{15}O$
3. $C_7H_{13}NO$
4. C_4H_7Cl
5. $C_6H_7NO_2$
6. $C_4H_{12}O$

 Can you actually draw a structure in every case?[15]

QUESTION 2

Propose a systematic chemical name for each of the structures you have drawn. Give the compound names to a friend and ask them to draw structures. They should draw the same structure you did, but it might not look exactly the same.

[15] You can bet the answer to this question is a resounding 'no'! The more important thing is learning to recognize the patterns that will allow you to spot an 'impossible' formula without even trying to draw a structure. Don't draw atoms with the wrong number of bonds just so that you have a molecule with the 'correct' formula.

Learning to spot two identical structures when they are drawn differently is a skill that cannot be overstated. No-one can teach you how to do this. You have to start with small structures and work up to the bigger structures.

You probably did 'spot the difference' puzzles as a child. After a while, you probably started to anticipate the parts of a picture where the artist would try to hide a difference. You would get better at solving the problem with experience. Funny, that!

QUESTION 3

For the structures you have drawn in **QUESTION 1**, identify all of the functional groups present. I cannot possibly give you all of the answers. You may need to look up some functional groups. It depends on how creative you have been!

I asked you to identify functional groups in Practice 1. Are you tempted to ignore the question here because you have already done it? Don't be tempted! If it is easy, it won't take you long. If it takes you too long, then you definitely need to do it. No short cuts!

32

HABIT 2
Identifying When a Formula is Possible

Let's start with something obvious. You cannot draw a structure for the formula 'CH$_6$'. This is because there are too many hydrogen atoms. Carbon can only form four bonds.

It's always good to reduce an idea to a simple and obvious example. Now let's develop the idea a bit further.

FORMULAE—WHAT IS POSSIBLE AND WHAT IS NOT?

If you have a compound with only carbon and hydrogen atoms, you must have an even number of hydrogen atoms. If the compound is completely saturated (as many hydrogen atoms as possible) and has no rings then the formula, for n carbon atoms, will be

$$C_nH_{2n+2}$$

You cannot add more hydrogen atoms than this.

Adding an oxygen atom doesn't change the maximum possible number of hydrogen atoms. Sometimes it is easiest to see this with structures.

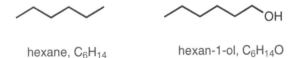

hexane, C_6H_{14} hexan-1-ol, $C_6H_{14}O$

As long as the compound is completely saturated, then adding an oxygen atom doesn't change the number of hydrogen atoms, and it doesn't change the fact that there need to be an even number of hydrogen atoms. We could insert an oxygen atom into any C−H bond or C−C bond and the rest of the formula would not change.

Adding a nitrogen atom does change things! If you have an odd number of nitrogen atoms, and only carbon, hydrogen, and oxygen otherwise, then you will have an odd number of hydrogen atoms.

Look at the following structure. It has a formula $C_6H_{15}N$. Make sure you know where all the hydrogen atoms are. In effect, you have taken hexane, removed one H and added an NH$_2$ group. The net effect is one additional hydrogen atom in the formula.

DOUBLE BOND EQUIVALENTS

When you make a compound, and you don't know the structure, the idea of 'double bond equivalents' is a very useful one. The only information you need is the formula. Take a look at the following structures.

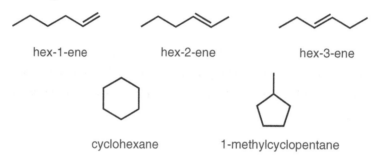

All of these structures have the formula C_6H_{12}. The saturated compound with six carbon atoms would be C_6H_{14}. You form a double bond by taking away two hydrogen atoms. You form a ring by taking away two hydrogen atoms and joining the ends. If we are given the information that a compound has the structure C_6H_{12}, we cannot tell whether it has a double bond or a ring. We just know that it must have one or the other.

There is a specific name for this lack of two hydrogen atoms in a formula. It is a **double bond equivalent**.

There is an equation that will allow you to calculate the number of double bond equivalents from the number of carbon, hydrogen, oxygen, and nitrogen atoms. I'm not going to show you this formula.

Don't look it up just yet!

I'd much rather you understand structure and be able to work it out. After all, the equation doesn't include any other elements, so you would need to be able to work out examples with things that are not in the equation.[16]

Let's see how this could work. As an example, we will use a formula C_6H_5NO. The simplest question you can ask is 'what would be the formula if there were no double bond equivalents?' That is, 'how many hydrogen atoms would it need?'

The easy way to answer this question is to draw a structure with the carbon, nitrogen, and oxygen atoms, then add the right amount of hydrogen. In this case, I would draw

HO⌒⌒⌒NH₂

---

[16] If you asked me to write down the equation right now, I'm not sure I could. I know what it looks like, but I will probably mix up the plus and the minus. I can take a bit of time, and I can work out the equation by using the principles (how many bonds each atom can form). But if I am going to do that, I might as well just skip the equation and apply the principles directly to a given molecular formula.

34

This has a formula $C_6H_{15}NO$. Therefore, C_6H_5NO is 'missing' 10 hydrogen atoms. Each double bond equivalent requires two hydrogen atoms, so that the formula has **5 double bond equivalents**. You can apply this method to any formula. Just make sure you draw a structure with no double bonds or rings, so your structure will have the maximum number of hydrogen atoms. You will find it doesn't matter where you put the oxygen or nitrogen!

TWO IMPORTANT POINTS

First up, if you are ever given a structure and asked to work out how many double bond equivalents it has, don't waste time determining the formula and then working it out.

Inspect the structure and identify the double bonds and/or rings. This shows that you actually understand what a double bond equivalent is.

Secondly, if you ever determine the number of double bond equivalents for a given structure or molecular formula, and your answer is half integral (something and a half!) then you have made a mistake. Either that, or the formula you have been given is nonsense. I do sometimes give my students nonsense formulae to check that they understand the principle.

PRACTICE 3
Double Bond Equivalents

THE PURPOSE OF THE EXERCISE

An organic chemist might have determined the molecular formula of something they have made, but they don't know the structure. Knowing how many double bonds or rings are present is useful information in determining the outcome of a reaction.

Beyond that, it is yet more practice and getting used to looking at increasingly complex molecules. The added layer of complexity is that the structures have wedged bonds and dashed bonds. We have looked at what these mean at the simplest level. We will sort out the detail in Section 3, but it's still worth getting used to seeing them now.

QUESTION 1

Here are the molecular formulae that we looked at in Practice 2. For each formula, determine the number of double bond equivalents present. Look back at your answers to Practice 2. Pick one or two of your structures and try to identify where the double bond equivalents are in the structure.

If the number of double bond equivalents looks 'wrong', what does this tell you about the structure? Were you able to draw any 'sensible' structures in those cases?

1. $C_6H_{12}O$
2. $C_8H_{15}O$
3. $C_7H_{13}NO$
4. C_4H_7Cl
5. $C_6H_7NO_2$
6. $C_4H_{12}O$

1.6

QUESTION 2

Now it's time to step it up a gear. Here are eight natural product structures. Identify the double bond equivalents.

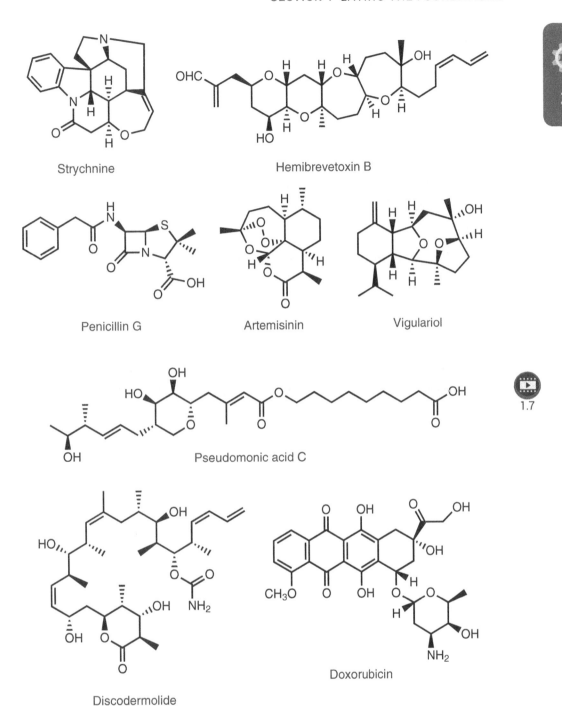

Strychnine

Hemibrevetoxin B

Penicillin G

Artemisinin

Vigulariol

Pseudomonic acid C

Discodermolide

Doxorubicin

Do this in each of the following three ways.

1. Work out the molecular formula of the compound. Then work out how many hydrogen atoms would be present if the molecule had no double bond equivalents.

2. Draw the structure. Identify which features correspond to double bond equivalents by inspection.

37

3. Look up the equation that I referred to in the previous chapter[17] and use it to calculate the number of double bond equivalents.

Which method is easiest? It probably depends on whether you are given a formula or a structure? Will you remember the equation? Can you be sure you will get it right?[18]

This is not easy! Look at the structure of artemisinin. You have a seven-membered ring bridged by two oxygen atoms. How many rings is this? We could have the seven-membered ring, and then two different rings that contain the bridge with two oxygen atoms.

My 'fallback' position is to 'break' a ring and add a couple of hydrogen atoms. Then see how many rings are left. You can do this as many times as you like!

A COMMENT!

If you want to check your answers, the formula of each compound can be readily found online! Did you get them all correct? If not, the equation is of no use to you!

What do these compounds do? You need to appreciate why organic compounds are important! These compounds, and many thousands of other compounds, save and enhance lives. And all of these compounds have been synthesized in the laboratory. That's why you are studying organic chemistry!

---

[17] I know it would be easier if I gave it to you. But I really don't like it!
[18] I can't! But then again, I have viable alternatives.

COMMON ERROR 1
Formulae, Functional Groups, and Double Bond Equivalents

Here is the structure of lysergic acid diethylamide, LSD.

? What is the molecular formula of this compound?

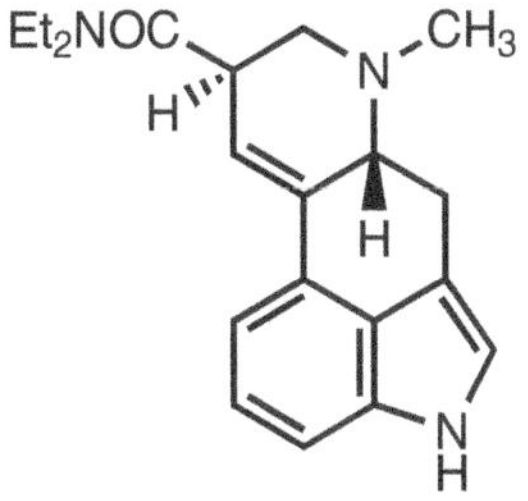

1.8

This is relatively straightforward, but there are a couple of bits that cause problems. The main one is the Et$_2$NOC substituent.

I have drawn it this way round so that it is clear that the carbon atom is joined to the ring.

There is a tendency to assume that the nitrogen atom is bonded to oxygen. After all, they are 'together' in the representation. This then leads to the following partial structure.

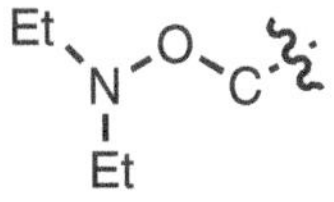

Note that I am using a wiggly line across the bond to indicate where it is attached to the rest of the molecule.

The problem here is that the carbon atom only has two bonds, not four.

How does this relate to the formula?

If you made this mistake, you might then add a couple of hydrogen atoms to the carbon to 'fix' the problem.

This doesn't fix anything.

This part of the molecule isn't a skeletal structure, so you cannot just add hydrogen atoms. You are explicitly told how many hydrogen atoms there are in this part of the structure.

Okay, so the nitrogen atom cannot be bonded to oxygen. It must be bonded to carbon. There is no other possibility. The oxygen must also be bonded to carbon, and to have two bonds to oxygen and four bonds to carbon, we need a double bond.

The next common error (at least in the very early stages) is very simple—not knowing that Et is C_2H_5 (or CH_3CH_2, which is the same thing). This error will disappear very quickly if you immerse yourself in molecular structures.

> *How many abbreviations do you need to know? You definitely need to know as far as butyl. We don't normally use two-letter abbreviations for alkyl chains longer than this.*

So, combining these two errors, you have the possibility of determining a molecular formula that isn't $C_{20}H_{25}N_3O$ (the correct answer!).

Everyone can make a mistake. Hopefully you will spot all three nitrogen atoms. There are an odd number, so there must be an odd number of hydrogen atoms. This gives you an opportunity to spot any mistakes.

? What functional groups are present in the structure?

As you might imagine, if you misidentified Et_2NOC then you won't realize that this is an amide. There is also an amine, and we won't worry about the heterocyclic (has a ring with an atom that is not carbon) ring system at the bottom of the structure for now.

? How many double bond equivalents are present?

We established that there is an equation that allows you to calculate this based on the number of carbon, hydrogen, and nitrogen atoms. I also told you that you will probably remember the formula incorrectly.

> *Some readers will have seen the equation before—you will already feel comfortable with it. If you determined the molecular formula incorrectly, the equation will give you the wrong answer. If you learned the equation incorrectly, it will also give you the wrong answer. It's just an opportunity to make mistakes.*

The structure has six double bonds (including the amide carbonyl that we discussed above). It has four rings. Each ring is a double bond equivalent. Six plus four equals ten. And that's it! Much easier than using an equation.

You're going to tell me that you wouldn't make these mistakes.

40

With enough practice, you probably won't! And then you will have three good ways to solve a problem of this type.

Until you have had enough practice, it is best to be aware of the pitfalls, and to know the more reliable methods for solving the problem.

41

HABIT 3
Ignore What Doesn't Change

This is an important idea, but one that is almost impossible to teach in a formal way.

In any reaction of an organic molecule, there will be a change to one functional group (or sometimes more), while the rest of the molecule is untouched. The key skill is to learn to recognize what has changed and what has not.

This will allow you to streamline your working, so that you spend your time looking at the right parts of the structure.

I'll show you an example. Here is a single step from the synthesis of an antibiotic compound, erythronolide B. The synthesis was reported in 1978.[19] The starting material is moderately complex, with 35 carbon atoms and 9 stereogenic centres. We will worry about stereogenic centres in Section 3.

This is a reaction type that we aren't going to cover in this book. Don't worry about that! In fact, the only thing that is happening is that two benzoate esters (Bz is a standard abbreviation for benzoyl, PhCO—yes, I explained one abbreviation by using another abbreviation) are hydrolysed.

Here it is again with the relevant functional groups shown in purple.

---

[19] You may think of this as ancient history. I remember it well!

When you look at more complex reactions, you need to recognize the need to be able to determine which parts of the molecule you need to worry about, and which parts you do not. For example, you might draw the reaction as follows.

There are advantages to doing this. For a start, it is quicker to draw. But there are definitely disadvantages. You never quite know when one of the functional groups in one of the 'R groups' will be needed in a subsequent reaction. If you are planning a synthesis, you might find that you ignore the possibility that the reaction will not be as selective as you like. We will explore the concept of selectivity later.

For now, I simply want to highlight the nature of the problem. What can you do about it? The honest answer is 'not much'!

Initially, you need to be mindful of the problem, and you should take every opportunity to look at reactions of more complex molecules, rather than avoiding them in favour of simple examples.

And then gradually, you will get quicker at spotting the functional groups that have changed. It will just happen over time and with experience.

BASICS 5
Electronegativity, Bond Polarization, and Inductive Effects

When we look at a structure, we need to consider how the electrons are distributed within the structure, so that we can predict how the structure will react with different reagents under various different reaction conditions.

The simplest parameter that affects electron distribution is electronegativity.

ELECTRONEGATIVITY

Electronegativity is a fundamental concept in chemistry, and it has implications throughout. To give a very brief summary, electronegativity is a measure of how much a particular atom will pull electron density towards itself. Once we exclude the unreactive noble gases, the most electronegative elements are at the top right of the periodic table, and the most electropositive (least electronegative) elements are at the bottom left of the periodic table.

Let's have a look at a few values, and consider the implications of them.

Element	Electronegativity
C	2.55
H	2.20
N	3.04
O	3.44
Cl	3.16
F	3.98
Li	0.98
Mg	1.31

BOND POLARIZATION

First of all, carbon and hydrogen have very similar electronegativity values. Hydrocarbons have very little polarization of bonds.

It gets more interesting when we consider elements such as O, N, or Cl. If we consider the structure of CH_3–Cl, we can see that the Cl is considerably more electronegative than the C. This means that the Cl will withdraw electron density towards itself, so there will be a polarization in the bond. We can represent this as follows:

$$\overset{\delta+}{H_3C} - \overset{\delta-}{Cl}$$

The Greek letter delta is used to indicate a partial charge on the atom. The carbon atom has a partial positive charge, and is susceptible to attack by a **nucleophile**, something that has an excess of electron density. The term **nucleophile**, and the corresponding term **electrophile**, will be fully defined in Basics 9.

If we now consider a compound that you will probably not have encountered, methyl-lithium (CH_3–Li), the relative electronegativity of the elements tells us that the bond polarization will be as follows:

$$\overset{\delta-}{H_3C} - \overset{\delta+}{Li}$$

In this case the carbon has a partial negative charge, and it will tend to react with something that is deficient in electron density.

You need to be comfortable with trends in electronegativity, and to understand how they affect the electron distribution in organic compounds.

INDUCTIVE EFFECTS

This bond polarization is referred to as an **inductive** effect. This is a ground state polarization of the σ bonded framework of the molecule. It is a relatively short-range effect, and it diminishes within a couple of bonds.

We will encounter a corresponding electronic effect, the **mesomeric** (or resonance) effect in Basics 10.

It is important that you recognize and can distinguish these two types of electronic effect, and know how to represent them.

45

4

PRACTICE 4
Bond Polarization and Electronegativity

THE PURPOSE OF THE EXERCISE

You need to be thinking about the distribution of charge in organic structures. You need to be able to do this no matter how complex the structure. In order to do this, you will need to 'see' functional groups even when they are not explicitly drawn out. Strangely enough, it's all about practice.

THE EXERCISE

Here are the natural product structures we saw in Practice 3. Redraw these structures and indicate with a δ+ or δ− the partial charge on 'key' atoms.

You don't need to mess around 'balancing' the delta charges. It's perfectly okay to have one δ− and three δ+ if needed.

I want you to think about what I mean by 'key' atoms. Does it make sense to put a δ on every atom, or are some more important than others? We are going to come back to this in Practice 6.

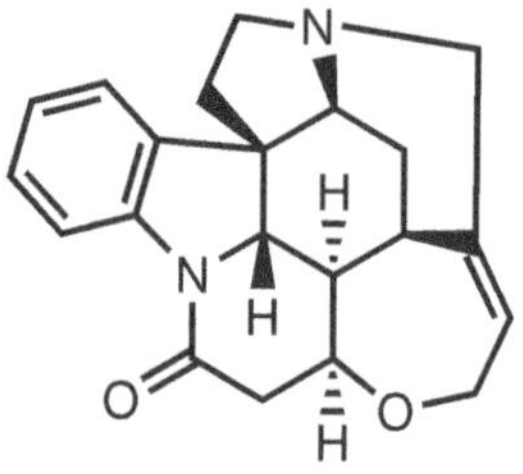

Strychnine

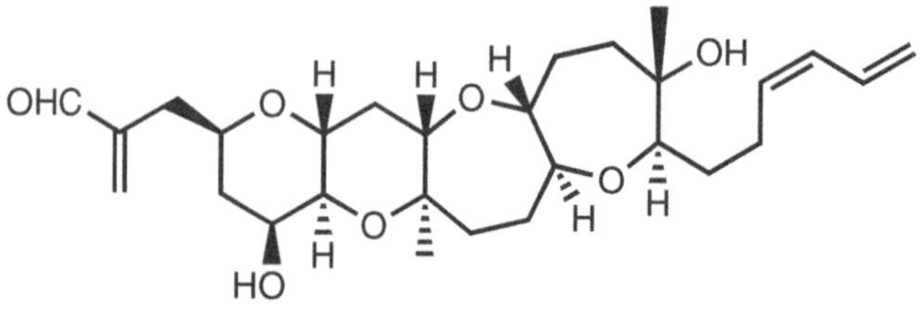

Hemibrevetoxin B

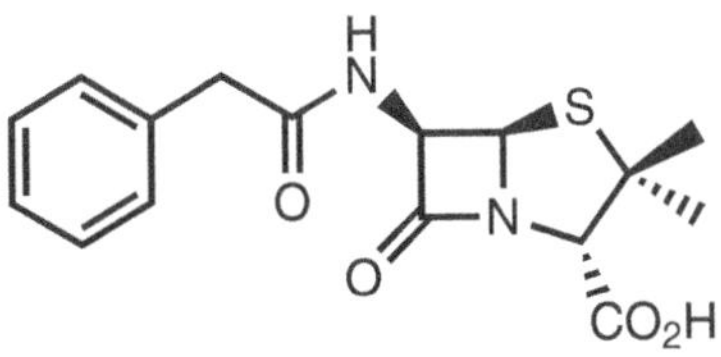

Penicillin G

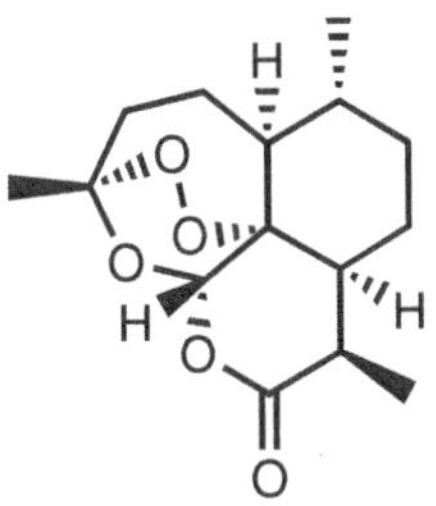

Artemisinin

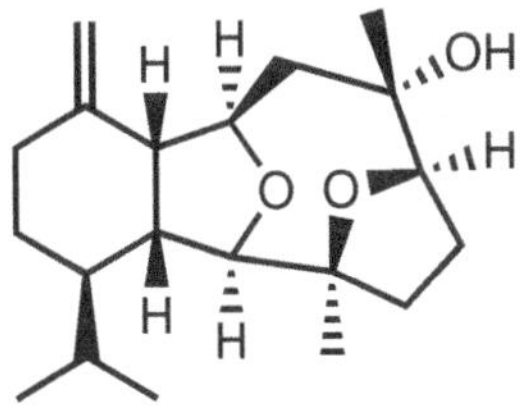

Vigulariol

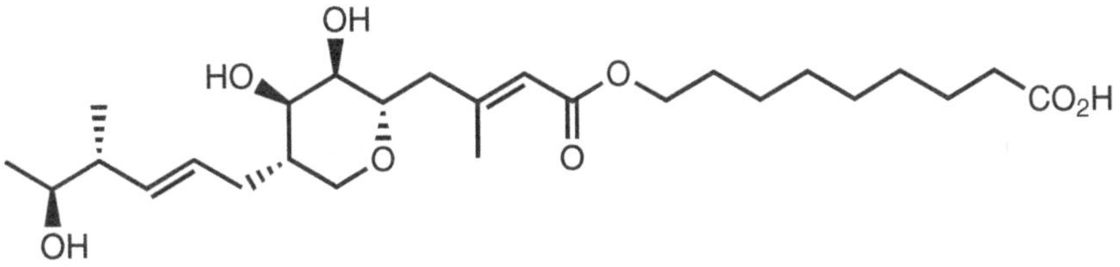

Pseudomonic acid C

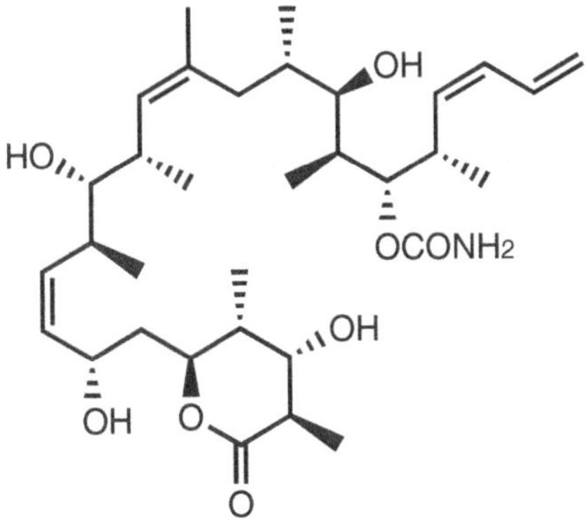

Discodermolide

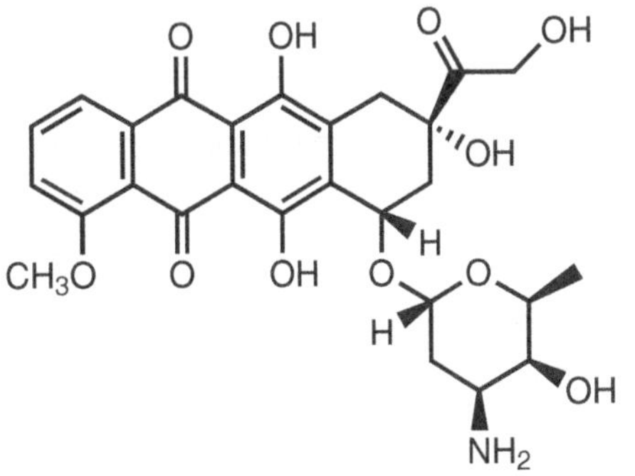

Doxorubicin

You will see that some of the functional groups on some of the structures are drawn differently to the same compounds in Practice 3. This gives you a chance to check 1.9 them!

47

49

BASICS 6
Bonding in Organic Compounds

We need to cover the basics of bonding and molecular orbitals here, to give you the tools you need to describe bonds and their reactivity.

We are going to discuss the overlap of orbitals to form σ (sigma) and π (pi) bonds. Before we do that, we need to look at some trends in bond dissociation energy. The bond dissociation energy is the energy needed to break a bond. Actually, this is a bit of a simplification, but it will do for now. We will sort out the detail in Basics 12.

BOND STRENGTHS, LENGTHS, AND ANGLES

Here are the structures of ethane (**1**), ethene (**2**), and acetylene (**3**), with the bond dissociation energies in kJ mol^{-1} and the bond lengths in angstrom units (Å). So, most chemical bonds are between 1 and 2 Å long. Single bonds in organic compounds have a bond strength of about 400 kJ mol^{-1}. Make sure you are familiar with these numbers. It is so easy to be a factor of 10 or even 100 out, and that would make a massive difference in any calculations you do.

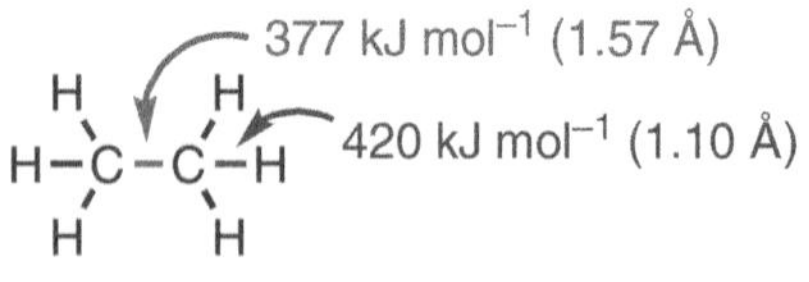

377 kJ mol^{-1} (1.57 Å)
420 kJ mol^{-1} (1.10 Å)

1

728 kJ mol^{-1} (1.35 Å)
458 kJ mol^{-1} (1.07 Å)

2

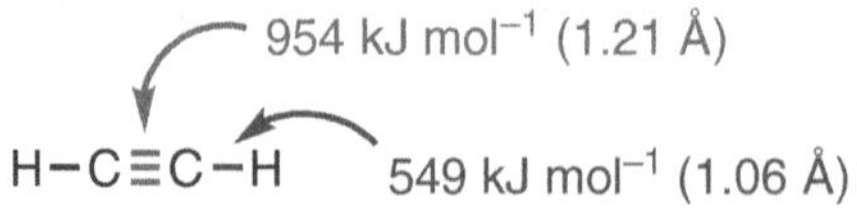

954 kJ mol^{-1} (1.21 Å)

H−C≡C−H   549 kJ mol^{-1} (1.06 Å)

3

Before we move on, I will state that these are homolytic bond dissociation energies. We aren't going to worry about this subtlety now, but we will come back to it in Basics 12. Furthermore, we will encounter 'average' bond dissociation energies in Basics 13, and find that the numbers are somewhat different to those above. We will explore the reasons for this in Perspective 1.

There are quite a few important points here. Let's try to summarize them, but also look a little deeper. Remember, we want to understand, not just to learn.

A C–C double bond is stronger/shorter than a C–C single bond. A C–C triple bond is even stronger/shorter.

This is all pretty obvious really. A C–C double bond is about 1.9 times as strong as a C–C single bond. A C–C triple bond is about 2.6 times as strong as a C–C single bond. So, on the face of it, each additional bond contributes less. This is true, but these numbers do not give the whole story.

A C–H bond in an alkene is stronger than a C–H bond in an alkane. A C–H bond in an alkyne is even stronger. Now this is a little surprising. Remember that we are only talking about the bonds to H directly from the alkene/alkyne carbon atoms.

And here are the bond angles, as we saw in Habit 1. In ethane, all bond angles are essentially tetrahedral, 109.5°. For ethene, the atoms are all in a plane, with approximate angles of 120°. For acetylene (ethyne) the bonds are linear, 180°.

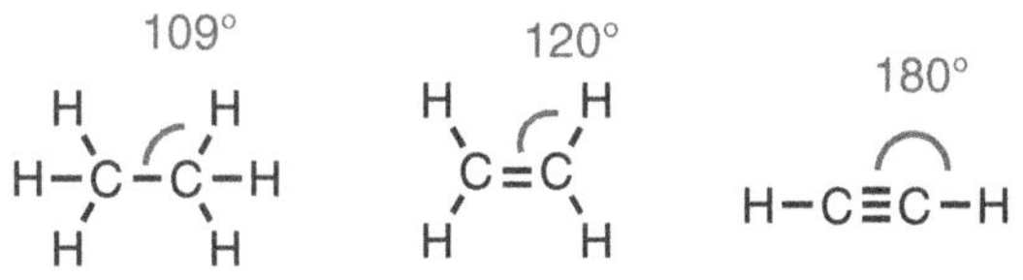

TYPES OF BOND IN ORGANIC COMPOUNDS

Before we go any further with theories of bonding, let's just make sure we know about the different types of bonds that we will encounter in organic compounds. There are plenty of textbooks that talk about atomic orbitals, so we are not going to explain them here (but we will use them).

It is easy to get bogged down with detail. Let's focus on one thing at a time. A C–H bond is a σ bond. A C–H bond in an alkene is stronger than a C–H bond in an alkane. A C–H bond in an alkyne is even stronger. We need an explanation for the bonding that leads naturally to this conclusion.

Let's look at the structures of ethane, ethene, and ethyne again, and focus on the conventions.

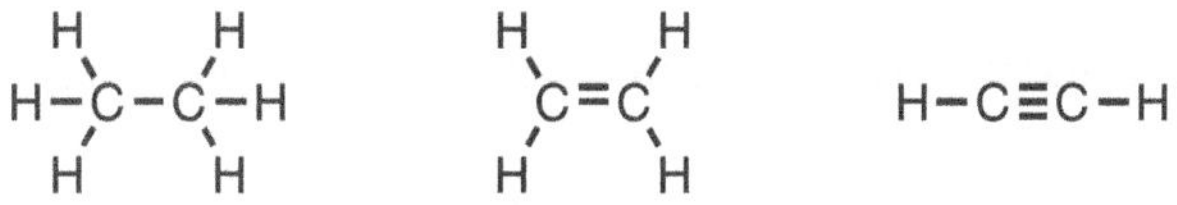

If we have a single line between two carbon atoms, it is a σ bond. If we have two lines between two carbon atoms, it represents a σ bond and a π bond. If we have three lines between two carbon atoms, it represents a σ bond and two π bonds.

49

*Whatever else is happening, we **always** have a σ bond.*

Since hydrogen only ever forms one bond, it must be a σ bond.

ATOMIC ORBITALS

All of the bonds we will encounter in organic compounds are **covalent** bonds formed by the **sharing** of electrons. Electrons in atoms and molecules reside in **orbitals**. An orbital is quite complicated, but we can represent the different types of orbitals as a shape. The shape shows us where there is a high probability of finding an electron at any given time.

We have to talk about probability, since electrons in orbitals are governed by the laws of quantum mechanics, which tell us that we cannot tell exactly where an electron is at any given time. We can only tell where it is more likely to be.

As organic chemists, most of the time we only need to consider **s orbitals** and **p orbitals**.

An s orbital is spherically symmetrical. The probability of finding an electron is the same in any direction from the nucleus. We can show it as follows, where the nucleus (not shown) is at the centre of the sphere.

s orbital

A p orbital looks like a dumb-bell. The nucleus is situated between the two coloured **lobes**.

p orbital

A p orbital has a **node**.

A node is a plane (in this case, although it can be a different shape) in which the probability of finding an electron is zero.

There is a change of **symmetry** at the node, and we represent this by the two coloured lobes.

The symmetry is a rather abstract idea, but one that you need to get used to. Eventually, you will encounter reactions for which the orbital symmetry determines the outcome in various ways. In depicting p orbitals, I have coloured one lobe of the orbital in green and the other one in yellow. Other textbooks may use different colours, or even + and − signs for the same purpose.

50

*I don't like the use of + and − signs for this purpose. Let's be absolutely clear. Orbitals are filled with electrons, and electrons **always** have a negative charge. When we talk about a change of symmetry, we are **not** talking about a change of charge.*

In fact, atoms such as carbon (p block!) have three p orbitals, each of which is directed along a different Cartesian axis. The plane of the node is defined by the other two axes.

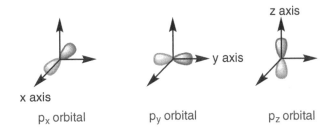

PI (π) BONDS

Let's start with π bonds. They are easier, and this will allow us to focus on the difference between the σ bonds in the various compounds. A π bond is formed from the overlap of p orbitals on two adjacent atoms. For now, we will assume that they are both carbon, although this is not always the case. We can show this in two different ways as follows:

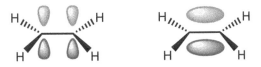

On the left, we are clear that this bonding is coming from the p orbitals on carbon. On the right we are showing that this overlap has occurred. Most organic chemists would use these representations interchangeably. We will look at π bonding in more detail later (Basics 10).

A π bond has a **node** in the plane of the atoms, and a change of **symmetry** which is a direct consequence of the symmetry of the p orbitals from which it is made.

Now let's make a statement that is obvious but profound. If we have a π bond that is formed by overlap of p orbitals on adjacent carbon atoms, each carbon atom has a p orbital that is not involved in σ bonding.

We will see what difference this makes!

51

SIGMA (σ) BONDS

A σ bond is spherically symmetrical around the axis of the bond. We will see in a moment that the representation below is not quite complete, but it will do for now.

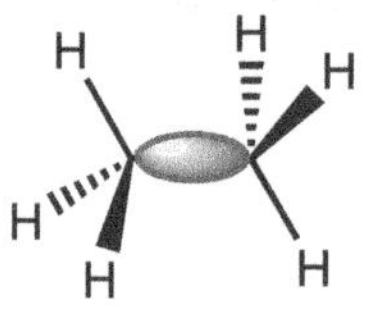

DESCRIPTIONS OF BONDING IN ORGANIC COMPOUNDS

There are many levels of theory that can be used to describe chemical bonds. Ultimately, everything derives from quantum mechanical equations that cannot be solved fully for any 'meaningful' molecule. Therefore, everything is a model or an approximation. Some are more approximate than others.

Most problems stem from the fact that we tend to look at the 'big world' and molecules and atoms are the 'small world'. Our own perspectives and language do not translate well to this world. It is very alien to us that we cannot know precisely where an electron is at any given time, and yet this is fundamental to quantum mechanics. We now have to think in terms of electron distributions and probability.

It helps to have a mental image of a chemical bond. **All** organic chemists use a theory called 'hybridization' to describe bonding. Even though this is only a model (and as such you can identify limitations with it, and eventually use other models of bonding where appropriate) it is very useful.

HYBRIDIZATION

Let's look at the fundamental problem. If we look at the structure of methane, we find that all C−H bonds are identical in length, and all H−C−H bond angles are identical (109.5°). However, we have four atomic orbitals on carbon, one spherically symmetrical s orbital and three p orbitals, which are at 90° to one another.[20] So how can we use these orbitals to make bonds at 109.5° to one another?

---

[20] We are ignoring the '1s' electrons, and focusing on the '2' shell which is what we will be using to form bonds.

52

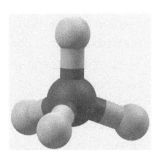

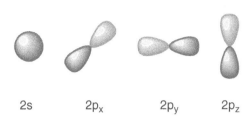

2s   2p$_x$ 2p$_y$ 2p$_z$

In order to get around this problem, we use a **model** known as hybridization. What we do, conceptually, is to make 'new' orbitals from the original atomic orbitals.

*This may seem like quite a random thing to do, but it turns out that the hybridized orbitals and the original orbitals are all valid solutions to the quantum mechanical equations, so there is no reason not to do this. Remember, this is **only a model** for the bonding.*

Hybridization is easier to explain using an example. We will continue to use the simplest possible hydrocarbon, methane, as our compound in the first instance, before looking at more complex molecules.

HYBRIDIZATION IN ALKANES—sp^3 HYBRIDIZATION

Carbon is in group IV of the second row of the periodic table. Therefore, a carbon atom has the electronic configuration

$$1s^2 2s^2 2p^2$$

We will ignore the 1s shell, as it does not get involved in bonding (at least in any way we need to worry about).

So, there are 4 'outer' electrons. As there are three 2p orbitals available, and they can each have 2 electrons, the '2' shell requires 4 more electrons to completely fill it with 8 electrons—an octet. It can accomplish this by gaining one electron from each of four hydrogen atoms (or indeed from other atoms). Carbon can therefore form four bonds.

Imagine that we promote an electron from the 2s orbital to a 2p orbital. We now have one electron in the 2s orbital and three electrons in the 2p orbitals—we will have one electron in each of p$_x$, p$_y$ and p$_z$.

So, we now think of carbon as having the electronic configuration

$$1s^2 2s^1 2p_x{}^1 2p_y{}^1 2p_z{}^1$$

Now the funky bit! We **hybridize** the orbitals. Conceptually, we can add up the 2s and the three 2p orbitals, divide by 4 and get 4 new orbitals. Since each of these is derived

53

from one s and three p orbitals, we will refer to them as sp³ hybrid orbitals. We can view these orbitals as 25 per cent s and 75 per cent p as far as the carbon atom is concerned (we only have the 1s orbital on H). In effect, we now have an electronic configuration of

$$1s^2 2(sp^3)^4$$

Here is a picture of methane with one of the sp³ hybrid orbitals drawn in. It does look a lot like a p orbital. However, there is more electron density between the C and the H. Rather than being symmetrical, like a p orbital, the lobes of the orbital are distorted, with only a small lobe behind the carbon.

As a result of hybridization, we have four new bonding orbitals that are disposed tetrahedrally, whereas the p orbitals that we started with were at 90° to one another and the s orbital was spherically symmetrical.

This is the bit that I didn't like! If we have three orbitals that are at 90° to one another, how do we end up with four orbitals at 109.5° to one another?

Remember that the hybridized molecular orbitals are a perfectly valid solution to the quantum mechanical equations for methane. They are as good as 'any other' orbitals. However, if you really don't like this, there is an alternative model presented in Perspective 3, in which the tetrahedral shape arises more naturally from the atomic orbitals.

I hope that when you read Perspective 3, you will relate it to hybridization, and you will become more comfortable with hybridization as a result.

HYBRIDIZATION IN ALKENES—sp² HYBRIDIZATION

What about the bonding in alkenes? The double bond in an alkene is made up of a σ bond and a π bond. The π bond is formed from the overlap of p orbitals on the two carbon atoms. We have already seen what these orbitals look like.

Now we should think about the σ bonded framework. Let's have another look at our electronic configuration after we promoted an electron. We have

$$1s^2 2s^1 2p_x^{\,1} 2p_y^{\,1} 2p_z^{\,1}$$

Let's 'reserve' the $2p_z$ orbitals on the carbon atoms to form the π bond. So, we have the 2s, $2p_x$, and $2p_y$ orbitals left. Let's now add them up and divide by three. This will give

54

us three new orbitals that we would describe as sp^2 hybrid orbitals (33 per cent s and 67 per cent p).

$$1s^2 2(sp^2)^3 2p_z^{\,1}$$

We can now do what we did before, and form bonds using the 1s orbitals of hydrogen and now the sp^2 orbitals on carbon. We also need to form a C–C σ bond using one sp^2 orbital on each carbon atom. We might draw the sp^2 hybrid orbitals in much the same way as we drew the sp^3 hybrid orbitals.

There isn't an easy way to show the difference between 25 per cent s and 33 per cent s.

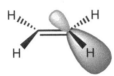

We would do the same thing for a C=O carbonyl double bond. We will come back to this.

HYBRIDIZATION IN ALKYNES–sp HYBRIDIZATION

The triple bond in an alkyne is made up of one σ bond and two π bonds. We will treat the π bonds as above, but this time we reserve two of the p orbitals, so that all we have left are the s orbital and one of the p orbitals. The π bonding in an alkyne can be shown schematically as follows.

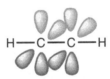

The σ bond is then made up by 'hybridizing' the s orbital and the remaining p orbital. We get two sp hybrid orbitals (50 per cent s, 50 per cent p) that are at an angle of 180° to one another. The alkyne is linear.

THE ADVANTAGE OF HYBRIDIZATION

Hybridization allows us to describe a single entity that we can call 'the bond'. We can think about two electrons that are associated with 'this' bond and with no other, and we can think about the shape of the bonding (and as we will see, the antibonding) orbital.

As we will see in due course, there are other ways to describe the bonding in organic compounds. **For now, accept hybridization as a fact.**

55

57

HYBRIDIZATION AS PART OF YOUR LANGUAGE

When you listen to organic chemists discussing a compound, or pointing out one specific carbon atom, you will hear expressions such as 'the sp³ hybridized carbon there'. In this situation, they are probably not trying to tell you anything about the bonding *per se*. They could equally say 'the tetrahedral carbon', or 'the saturated carbon'. This is really what they are meaning.

There is a consequence to this. As you become more confident talking about the hybridization state of a carbon atom, you will use the terminology fluently, but often without thinking about the actual bonding.

Personally, I think this is a good thing.

If you want to talk about specific bonding and antibonding hybridized orbitals, it doesn't really matter much whether they are sp, sp², or sp³ hybridized. The general shape and symmetry (which is what really matters) are the same.

HOW DOES HYBRIDIZATION AFFECT SIGMA BOND STRENGTH?

Now we can return to the bond lengths/strengths that we saw at the start of the chapter. The C–H bond in acetylene is considerably stronger than the C–H bond in methane. Methane is sp³ hybridized, so as far as the carbon atom is concerned, the C–H bond is 25 per cent s and 75 per cent p. Acetylene is sp hybridized, so as far as the carbon atom is concerned, the C–H bond is 50 per cent s and 50 per cent p. The s orbitals are lower in energy and more tightly held by the nucleus. Therefore, the more s character a bond has, the stronger the bond. This leads us to what we saw above. The C–H bond in an alkyne is stronger than that in an alkane.

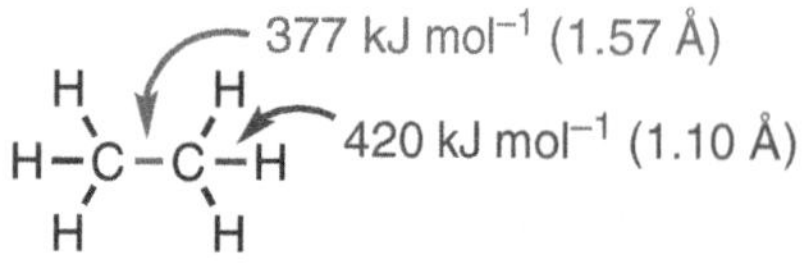

377 kJ mol⁻¹ (1.57 Å)
420 kJ mol⁻¹ (1.10 Å)

1

728 kJ mol⁻¹ (1.35 Å)
458 kJ mol⁻¹ (1.07 Å)

2

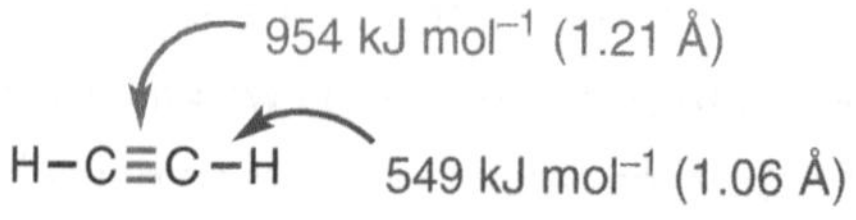

954 kJ mol⁻¹ (1.21 Å)
549 kJ mol⁻¹ (1.06 Å)

H–C≡C–H

3

56

A VERY SUBTLE POINT

We can draw a further conclusion from this. The C–C bond in ethane has a bond dissociation energy of 377 kJ mol^{-1}. That of the C=C bond in ethene is 728 kJ mol^{-1}. However, this doesn't mean that the strength of the π bond is 351 kJ mol^{-1} (the difference between these numbers) because the σ bond component of the double bond in ethene is stronger than the C–C σ bond in ethane. It might feel a bit odd at first thinking about the two separate components of the bond. However, you'll get used to it soon. The strength of the π bond in ethene is considerably less than 351 kJ mol^{-1}.

HYBRIDIZATION OF NITROGEN AND OXYGEN

Hybridization isn't just something carbon does! Let's consider an amine, for example trimethylamine.

$$H_3C-N\overset{CH_3}{\underset{CH_3}{\diagup}}$$

Nitrogen has the electronic configuration

$$1s^2 2s^2 2p^3$$

Again, we can ignore the 1s shell, and focus on the outer electrons. It isn't as easy in this case to think about promoting one electron from 2s to 2p. Each of the different (x, y, and z) 2p orbitals are already singly occupied. I would take a direct step to sp^3 hybridization and then consider how many electrons will be in the orbitals. We get

$$1s^2 2(sp^3)^2 (sp^3)^1 (sp^3)^1 (sp^3)^1$$

One of the sp^3 hybrid orbitals already has two electrons. It cannot participate in bonding. It is a lone pair! The other three sp^3 hybrid orbitals can form bonds to another element, carbon in the example we are using. This has clear implications for the shape of trimethylamine. It is tetrahedral.[21]

$$H_3C-\overset{\displaystyle N}{\underset{CH_3}{\diagdown}}{}^{\!\!\prime}CH_3$$

We could do the same for oxygen in water or dimethyl ether, and we come to a similar conclusion.

---

[21] There is a little bit more complexity to this, but I don't want to get bogged down right now. I haven't told you any lies here. I just haven't told you the whole truth!

 Have a go at working this out!

What about a double-bonded oxygen such as a carbonyl? The electronic configuration of oxygen is

$$1s^2 2s^2 2p^4$$

We could expand this as

$$1s^2 2s^2 2p_x^2 2p_y^1 2p_z^1$$

If we want to form a π bond, we need to reserve a single electron in a p orbital. This is highlighted in purple above. We then have to hybridize the $2s^2 2p_x^2 2p_y^1$ part to give three new orbitals. The new electronic configuration would be

$$1s^2 2(sp^2)^2 2(sp^2)^2 2(sp^2)^1 2p_z^1$$

We have five electrons in the sp^2 hybrid orbitals because we have five electrons in the orbitals that we are constructing these from. Because two of these orbitals are full, we can only form one bond. The oxygen in a carbonyl group is sp^2 hybridized and is trigonal planar. The hybridized lone pairs are in the same plane as the methyl groups.

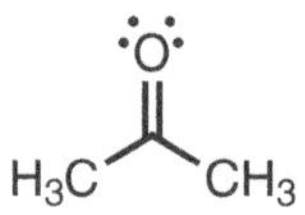

The key point here is that there is no difference between hybridization of carbon and hybridization of other elements.

There is another way to think about bonding. We can use molecular orbital theory. We will see that this leads us to broadly the same conclusions, but with minor differences, in Perspective 3.

Whether you like hybridization or not, you should still find yourself using it as part of the language of organic chemistry. When you compare the molecular orbital theory explanation, you will find that it does not rely on some of the more 'random' parts of hybridization, but it loses some of the key advantages.

 Can you remember what the advantages are? If you can't, re-read this chapter. They have been mentioned!

60

PRACTICE 5
Hybridization

THE PURPOSE OF THE EXERCISE

Most of the time, when an organic chemist states the hybridization of a carbon (or nitrogen or oxygen) atom in an organic compound, they are really describing the number of bonded atoms and the geometry. Stating the hybridization (it will always be 'sp^3', 'sp^2', or 'sp'!) needs to become automatic!

It's also a good point to take stock of where you are. By now, you should be drawing skeletal structures automatically. You will probably be drawing the structures faster, more confidently, and more accurately. You will still make mistakes (for example, the wrong length carbon chain) but you will notice and correct it without really thinking about it.

If you aren't 'there' yet, don't worry about it. Just keep going back to the earlier problems and keep doing them until it all clicks. It's important that you know how to measure your success.

THE EXERCISE

Here are some of the compound names we looked at in Practice 1. I have changed a couple of them. You will understand why once you look back at the original names, and you answer the questions below.

1. 4-Chlorohexanal
2. 4-Bromo-7-methylnonan-2-one
3. 5-Chloro-2-propylphenol
4. 4-Aminobenzonitrile
5. Methyl propanoate
6. Ethyl benzoate
7. 2-Hydroxyhept-5-yne
8. 2-Aminoheptan-4-ol
9. 4-Chloro-1-methylcyclohexan-2-ol
10. *N*-Ethyl-*N*-propyl-(2-hydroxypentylamine)

You might not need to draw them all out to answer the question. When you come back and repeat this (and repeat it!), you'll find that you instinctively go to the right answer.

Here are the questions.

1. Which compounds only contain **sp³** hybridized carbon atoms?

2. Which of these compounds contain two or more **sp²** hybridized carbon atoms?

1.10

3. Which of these compounds contain one or more **sp** hybridized carbon atoms?

Next up, we will make some connections. Here are the natural product structures that we looked at in **Practice 3**. For each structure, determine the number of **sp²** and **sp** hybridized carbon, nitrogen, and oxygen atoms by inspection.

Is it more difficult to spot them on the more complicated structures? It won't be if you practise.

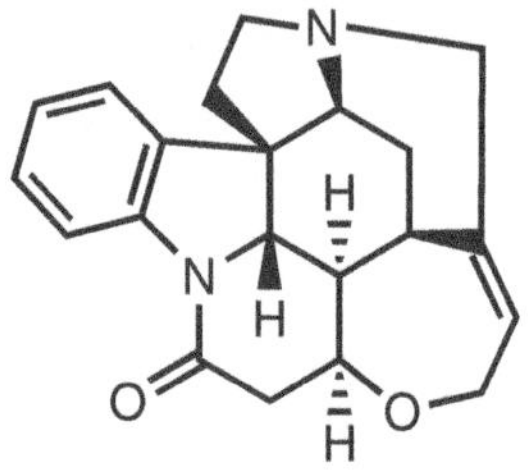

Strychnine

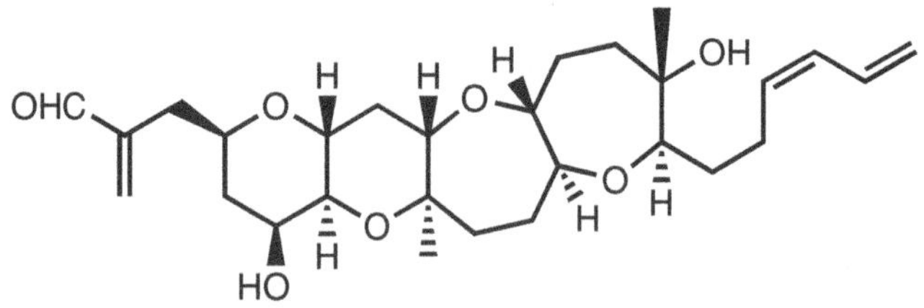

Hemibrevetoxin B

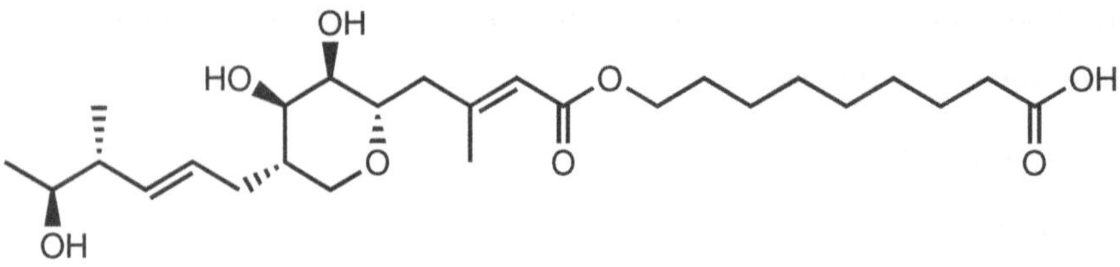

Pseudomonic acid C

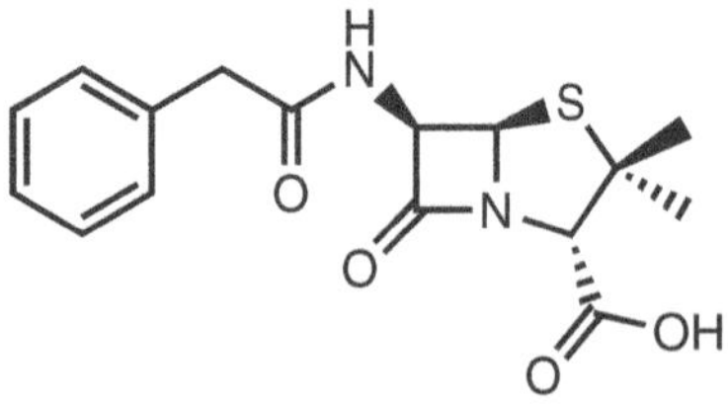

Penicillin G

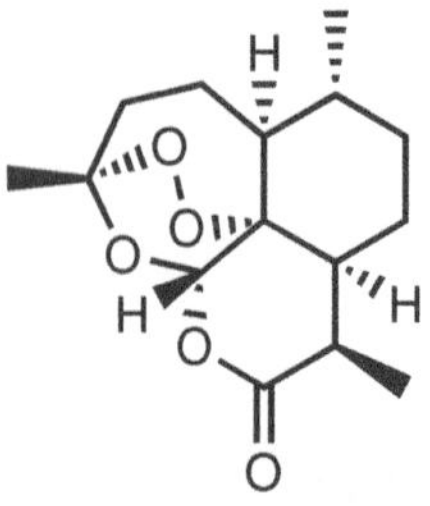

Artemisinin

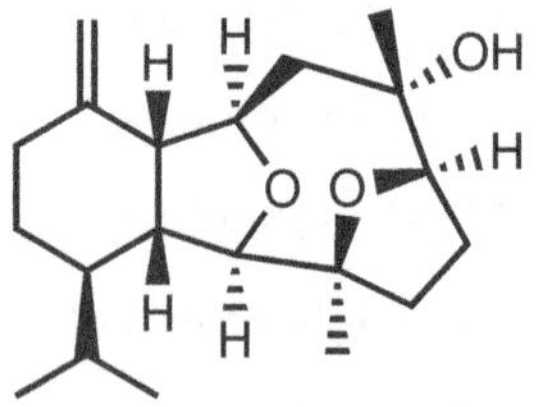

Vigulariol

60

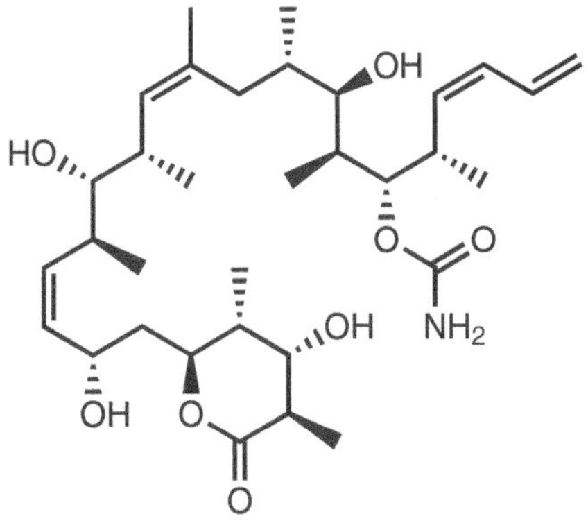

5

Discodermolide 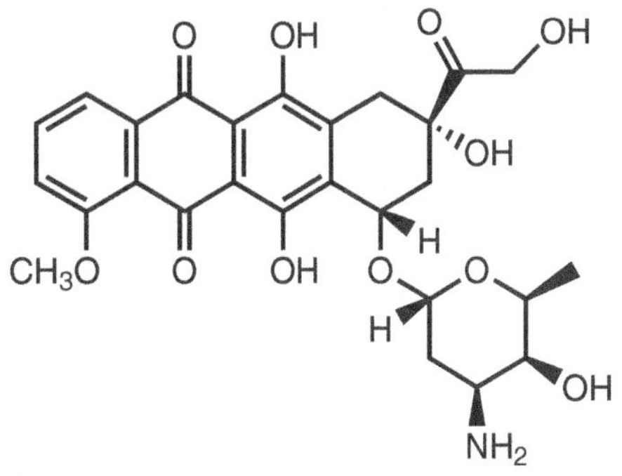   Doxorubicin

Now, use the number of sp^2 and sp hybridized atoms, along with the number of rings, to determine the number of double bond equivalents.[22]

1.11

This is quite an artificial problem. You don't really need to consider hybridization in order to work out how many double bonds you have. The real focus is getting you looking at more complex structures and doing 'something' with them.

---

[22] I haven't actually told you 'how' you should do this. That's okay. You can work it out! If you are struggling, talk to a friend or a lecturer.

BASICS 7

Bonding and Antibonding Orbitals

Electrons reside in orbitals. In stable molecules, these tend to be bonding orbitals—there is a high probability of finding electron density between the nuclei involved in the bond.

Whenever we form a bonding orbital by overlap of electrons from two atoms, we always form a corresponding antibonding orbital.

It turns out that the antibonding orbital is really important when we want to make the connection between curly arrow mechanisms (Basics 8) and molecular orbitals (Basics 9).

BONDING AND ANTIBONDING ORBITALS

Let's recap some simple molecular orbital theory. You know that hydrogen gas exists as a diatomic H_2 molecule, but helium is monatomic.

Here is a molecular orbital diagram showing the formation of H_2 from two hydrogen atoms, each with one electron.

We form a bonding (σ) orbital and an antibonding (σ^*) orbital from the combination of two 1s atomic orbitals (two hydrogen radicals). Only the σ orbital is filled, so that energetically this situation is favourable compared to two separate hydrogen radicals.

You will understand the reference to hydrogen radicals when we talk about bond dissociation energy (Basics 12).

The bonding and antibonding orbitals of H_2 look like this:

1.12

The antibonding orbital always has one more node, between the atoms, than the bonding orbital.

Let's compare this to the situation with a hypothetical diatomic helium (He_2) molecule.

Each helium atom has two electrons, so now we must fill the bonding and the antibonding orbital. Whatever energy we 'gain' by putting electrons into the bonding orbital is offset by also putting electrons into the antibonding orbital.

Therefore, there is no net bonding.

BONDING AND ANTIBONDING ORBITALS IN ORGANIC COMPOUNDS—SIGMA BONDS

We saw sp³ hybrid orbitals of methane in Basics 6, but we glossed over one important point. We didn't consider the formation of the bonding and antibonding orbitals from the sp³ hybrid orbital on C and a 1s orbital on H.

When we explicitly consider the overlap, there are two possible outcomes, related by symmetry. These are shown below.

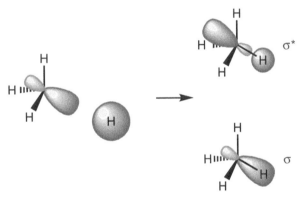

The antibonding orbital (σ*) has a significant lobe (orbital coefficient) behind the atom. There is also a node between the atoms.

We will see how this determines one key aspect of the outcome of S_N2 substitution reactions in due course.

MOLECULAR ORBITALS FOR ETHENE

We saw the π bonding molecular orbital for ethene in the previous chapter. Recall that the π bond is made by overlap of p orbitals on adjacent carbon atoms as shown below. If we consider the overlap of two p orbitals to make the π bond, there are two possibilities that differ in symmetry.

63

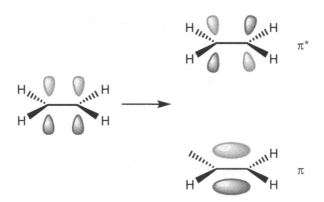

One (π) is bonding, and it has overlap of the p orbitals on the two carbon atoms. The other (π*) is antibonding, and it has a node between the two atoms in addition to the node in the plane of the double bond. Only the bonding orbital is occupied normally.

We will look at the implications of antibonding orbitals in reaction mechanisms at various points. They really are important!

BASICS 8
Introduction to Curly Arrows

There are few ideas more important in organic chemistry than curly arrows. We use them to represent 'electron flow' in reactions.

A good organic chemist will draw a curly arrow mechanism with an understanding of the meaning, but a great deal of what they do will be borne of habit rather than direct application of understanding.

Let's take this from the absolute basics. A σ or π bond contains two electrons.[23] In principle, when we form a bond between two compounds, we could have both electrons coming from one compound, or one electron from each. For the vast majority of reactions, we can consider that two electrons come from one of the reacting partners. We represent this with a curly arrow, showing the flow of electrons, and indicating where the bond is being formed.

Let's be absolutely clear about this. Most students who fail organic chemistry do so because they draw incorrect curly arrows. They will draw curly arrows going in the wrong direction, or draw curly arrows that would lead to impossible structures (e.g. those with five-valent carbon). It actually gets worse than that. Usually, if you draw one step in a mechanism incorrectly, you'll then need to draw another step incorrectly to get out of the hole you have dug yourself into. You absolutely need the correct drawing of curly arrows to become a habit.

For this reason, we are going to focus on the basic rules in this chapter, looking at only a small number of reaction types. Since there are only two fundamental reactions covered in this book, we won't have much opportunity to practise drawing curly arrows. As you progress in your studies, it will be important to keep thinking about what the curly arrows actually mean, rather than just memorizing them.

PROTONATION OF WATER

Let's start with a simple example, that isn't even an organic reaction (although the same process is found throughout organic chemistry). You know that if you add H⁺ (acid) to water, you will protonate it. Here's an equation for this reaction.

[23] Let's assume hybridization (Basics 6), so we can think of a single bond being two electrons in one orbital that contributes fully to the bond.

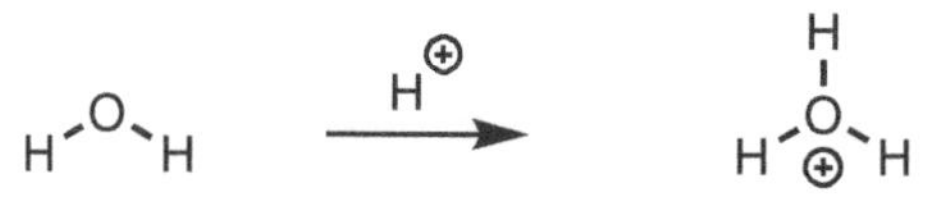

8

We are definitely making a new bond between O and H. We must be **sharing**[24] two electrons from the oxygen atom, because there is one thing that H+ doesn't have, and that is electrons. We must be using lone pair electrons, because we are still keeping the two existing O–H bonds. Here is the equation again, with the lone pairs added. Note that the product does still have a lone pair.

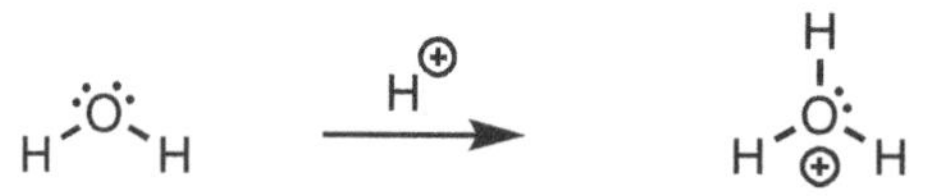

Now we have established where the electrons are coming from, we can add a curly arrow.

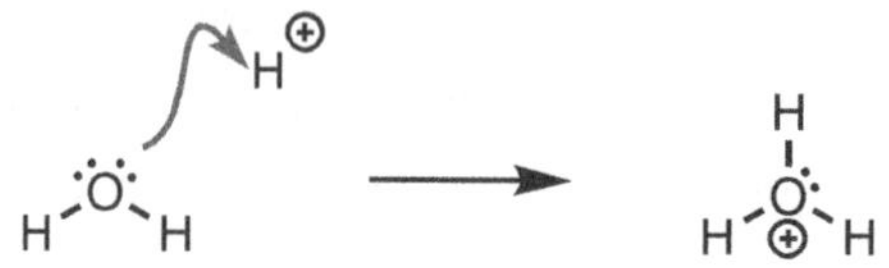

The curly arrow starts where the electrons are, and it finishes where the electrons are going. There isn't much more to it than this.

Curly arrows can start at non-bonded lone pairs of electrons.

DEPROTONATION OF THE HYDRONIUM ION

Now let's look at the reverse reaction.

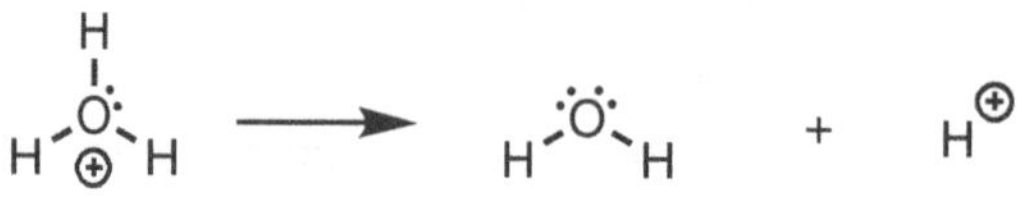

Here, we need to break an O–H bond, giving both electrons in this bond back to the oxygen atom. In saying this, we have established where the curly arrow needs to start (the bond) and where it needs to finish (the oxygen atom).

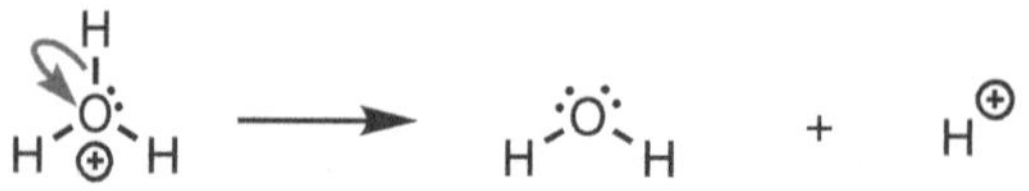

Curly arrows can start at bonds.

---

[24] I am being careful not to use words such as 'giving' or 'donating'. Since we are forming a bond between O and H, the two electrons are shared between the two atoms. The O retains a share!

We could add a little more complexity to this, by considering the possibility that the proton is removed by a base. We will use hydroxide as base. This will give us two water molecules as products.

Here we are forming a new bond from the hydroxide ion to the proton that is being 'lost'. The curly arrow to break the O–H bond is the same as before (purple). We just need to add one more curly arrow (blue) to form a new bond.

Curly arrows can start at non-bonded pairs of electrons in negative charges.

Whenever we have two curly arrows going to/from a single atom, we must count electrons. A hydrogen atom can have a share in two outer electrons (the 1s shell). The blue curly arrow above 'gives' the hydrogen atom a share in two more electrons. This cannot happen unless we take a share in two electrons away from it at the same time (the purple curly arrow).

We have now established the ground rules of curly arrow drawing, and we should consolidate these ideas with a couple more examples.

NUCLEOPHILIC SUBSTITUTION

This is one of the two reaction types that we are covering in this book, and you have almost certainly encountered it before. It takes place at saturated (sp³ hybridized) carbon.

Here's a simple example of this reaction, with the hydroxide anion reacting with iodomethane.

$$HO^{\ominus} \ + \ H_3C-I \ \longrightarrow \ HO-CH_3 \ + \ I^{\ominus}$$

We are forming a new C–O bond and breaking a C–I bond. Looking at the charges, there is no doubt that we need a curly arrow going to the iodine. After all, it has a negative charge on the product side of the equation. Similarly, the oxygen on the left has a negative charge, but is neutral on the right. It needs to lose electrons.[25] Here is the reaction with curly arrows added.

[25] Here's an important point. If we consider the O to be 'giving away' two electrons, then it would go from minus to plus. Because it is sharing two electrons, it is (in effect) giving away one electron – hence the change from minus to neutral.

67

$$HO^{\ominus} \quad + \quad H_3C-I \quad \longrightarrow \quad HO-CH_3 \quad + \quad I^{\ominus}$$

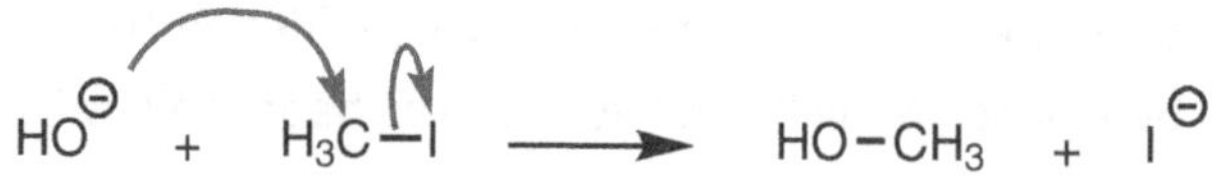

So far, what we have done is connect the given reaction outcome with the curly arrows necessary to achieve that outcome. This is important.

But there is something else to consider. The iodine atom in iodomethane is electronegative. It will have a $\delta-$ charge. The carbon will have a $\delta+$ charge. It is far more natural to draw a reaction in which a nucleophile attacks the more $\delta+$ centre, and 'gives' electrons to the more electronegative element.

We will formally define 'nucleophile' in Basics 9, but you will already have the idea by then! That's the beauty of seeing nucleophiles in action and not worrying too much!

It turns out that there is another way we can draw this reaction, which doesn't break and form bonds in the same step. We are going to deal with this one later (**Fundamental Reaction Type 1**), and we will build further on this once we have covered carbocation stability.

Now we are going to look at some curly arrows for reactions we won't be discussing in this book.

ADDITION REACTIONS OF ALKENES

If we consider the reaction of an alkene with HBr, the following outcome is typical.

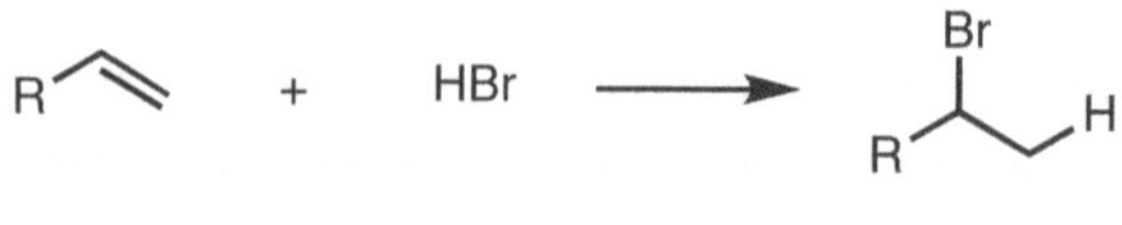

We should consider a number of factors. We are forming two bonds (C–Br and C–H) and breaking two bonds (H–Br and half of the C=C double bond). Do these events all happen at the same time, or in separate steps? If they happen in separate steps, which happens first, and which way do the electrons 'flow'?

Since we know that alkenes are electron-rich, it is natural to consider a curly arrow going from the C=C bond to somewhere. In fact, since we are breaking this bond, it is essential.

Similarly, we are breaking the H–Br bond. We know that Br is more electronegative than H, so that this bond is polarized as follows.

$$\overset{\delta+}{H}-\overset{\delta-}{Br}$$

68

If we are going to push a curly arrow from this bond, either to the H or the Br, it has to go to the Br, since this is the end of the bond that 'wants' the electrons (this is what electronegativity means!).

We are almost there now. We need to convert the C=C double bond into a single bond, so we need to start a curly arrow at this bond. We need to form a new C–H bond, so we will need to push this curly arrow to the H. We then need to break the H–Br bond. Putting all of this together looks like this:[26]

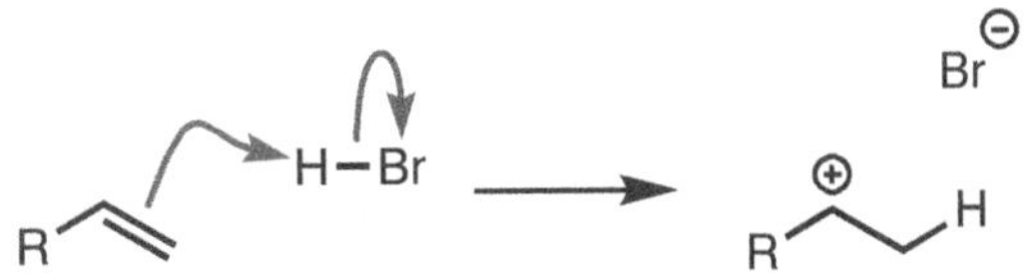

Finally, we need to form a C–Br bond. Because we got the first step right, there is only one way to do this.

Of course, if you got the first step wrong, and didn't realize it, you would probably get the second step wrong as well.

CARBONYL ADDITION REACTIONS

The carbonyl group is without doubt the most important functional group in organic chemistry. You can do so much with it. For now, we are just going to look at the curly arrows. Here is the reaction of the cyanide anion with an aldehyde.

Oxygen is considerably more electronegative than carbon, so the carbonyl bond is polarized. The carbon has a partial positive charge and the oxygen has a partial negative charge. Therefore, the carbon atom is electron-deficient.

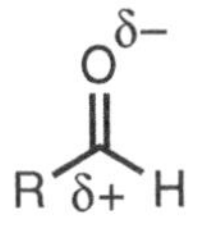

---

[26] We aren't going to worry about why the proton attaches to the carbon at the end, although perhaps you can propose a sensible reason for this once you have read Basics 16.

69

We have converted the C=O double bond into a single bond in this reaction, and the oxygen has gained a negative charge. There is little doubt that we need a curly arrow going from the carbonyl bond to the oxygen atom.

Of course we do! Oxygen is electronegative!

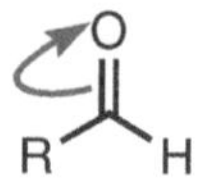

This would leave a carbon atom with only three bonds, and a positive charge. We know that we need to form a new bond from the cyanide carbon to this carbon atom, and this defines the next curly arrow. The complete curly arrow mechanism is shown below.

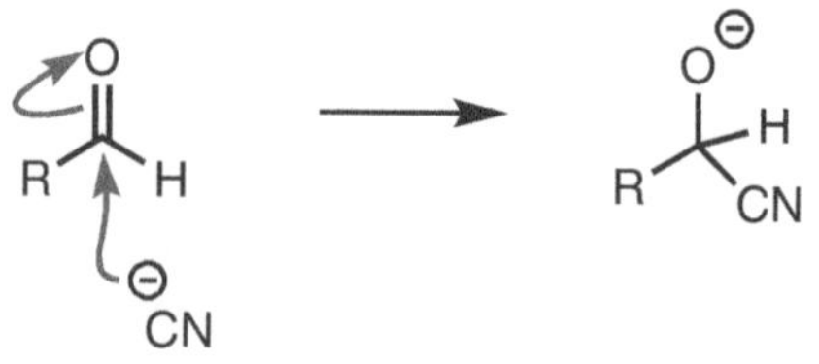

A DIFFERENT TYPE OF CURLY ARROW—FREE-RADICAL REACTIONS

Up to now, we have stated that a curly arrow is used to represent the movement of a **pair** of electrons. This is true for the types of reaction we have been looking at, which we could broadly class as ionic.

However, there are some reactions where we need to think about the movement of single electrons.

When we shine a light[27] on chlorine, Cl_2, the Cl–Cl bond can be broken. However, it doesn't give Cl^+ and Cl^-. Instead, it breaks in such a way that each chlorine atom retains one electron from the Cl–Cl bond.

We can show this as follows:

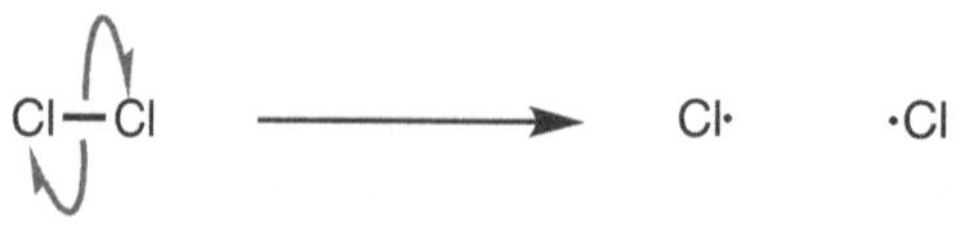

We get two Cl radicals, 'Cl dot'. Note that I haven't shown all the 'lone pair' electrons on Cl. We often focus on the key point rather than showing everything. You should be getting used to this by now.

---

[27] It needs to be the correct frequency (energy) of light!

Because the curly arrow we have been using so far represents the movement of a pair of electrons, we need a different type of curly arrow to represent the movement of a single electron. We use a 'fishhook arrow' for this purpose, as shown above.

When free-radical chemistry became prominent in the 1950s, books on radical chemistry were banned in some parts of the world! It sounds bonkers but it is true. Some politicians didn't like the idea of radicals of any sort, particularly free ones!

HOMOLYTIC BOND CLEAVAGE

The reaction we saw above (here it is again) is an example of homolytic bond cleavage. The bond is broken in an 'equal' way, so that each atom keeps one electron.

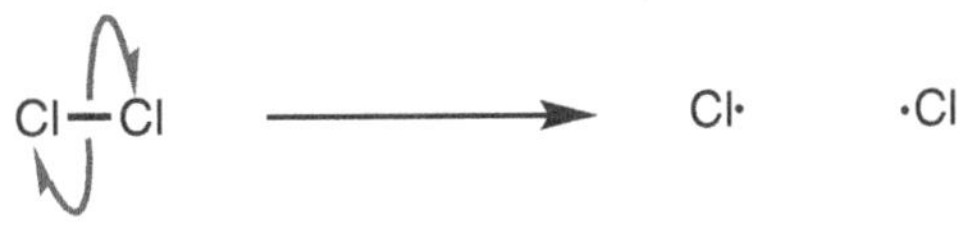

$$Cl{-}Cl \longrightarrow Cl\cdot \quad \cdot Cl$$

We won't encounter any reactions that feature homolytic cleavage of bonds in this book. However, when we talk about bond dissociation energies, we are referring to the enthalpy change for a homolytic bond cleavage of this type (Basics 12).

HETEROLYTIC BOND CLEAVAGE

This is when one atom in a bond cleavage gets both the electrons. For example,

$$\text{—Br} \longrightarrow \oplus \quad Br^{\ominus}$$

The Br atom ends up with a negative charge because it starts out with an equal share in two electrons (so effectively it has one of the electrons) and it gets both the electrons.

FUNDAMENTAL REACTION TYPE 1

Nucleophilic Substitution at Saturated Carbon

WHAT IS A SUBSTITUTION REACTION?

Quite simply, a substitution reaction is where one group on an atom is replaced with another group.

When we are talking about a substitution reaction at saturated (sp³) carbon, the following process is a substitution. We are replacing the iodine atom with a hydroxyl group to give an alcohol product.

$$HO^{\ominus} \;+\; H_3C-I \;\longrightarrow\; HO-CH_3 \;+\; I^{\ominus}$$

In this context, the hydroxide anion is acting as a **nucleophile** (we will define the term in Basics 9, but you are already getting the hang of what it means!). We call the iodide anion the **leaving group**. In this case, the 'organic' bit is simply methyl, but it could be a whole host of other alkyl groups with various other substituents.

We are going to work up to looking at the impact of each of these variables, as well as reaction conditions such as solvent, in Reaction Detail 1. For now, though, I want to guide you through a way of thinking about what possible mechanisms you might be able to draw. By considering all possibilities, you can then start to analyse what is good or bad about each.

ORDERS OF BOND FORMATION[28]

Let's define very clearly what is happening in this reaction. We are forming a new C−O bond and breaking a C−I bond.

---

[28] Do not confuse these conceptual 'orders of bond formation' with the idea of 'order of reaction'. In the latter case we are talking about the kinetics, and how the concentrations of the various reactants affect the overall rate of reaction. All we are doing here is looking at which bonds break/form first.

From a **purely conceptual** point of view, this could happen in three different ways.

1. The C–O bond forms and the C–I bond breaks at the same time.

2. The C–O bond forms first, and then the C–I bond breaks.

3. The C–I bond breaks, and then the C–O bond forms.

There are no other 'possibilities'. Now we need to determine whether all of these are actually possible in reality.

We can exclude option 2. If the C–O bond was to form first, this is what would have to happen.

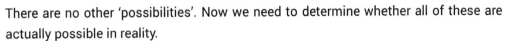

NOT POSSIBLE!

We have just drawn a structure with a carbon atom with a negative charge and five bonds. The negative charge in this case would not represent a non-bonded pair of electrons (as it does in a carbanion—more about these in Basics 16). The best way to count the electrons is to start with the iodomethane carbon having a share in eight outer-shell electrons. The curly arrow indicates that we are sharing a pair of electrons from oxygen with this carbon atom. Therefore, by having five bonds, the carbon has a share in ten outer-shell electrons. It cannot have more than eight outer-shell electrons. This mechanism is simply impossible!

We therefore only have two ways in which a nucleophilic substitution reaction can take place at a saturated carbon atom. Either the leaving group leaves first, followed by attack of the nucleophile, or the nucleophile can attack at the same time as the leaving group leaves. You will probably already have seen this before. These are the classic S_N1 and S_N2 reactions.

We will now look at each one in turn.

THE S_N1 MECHANISM

The basic reaction pathway is as follows:

1.14

We want to avoid equations as much as possible, but we do need to consider the rate equation for the basic reaction types.

An S_N1 reaction has rate = k[RX]

where k is the 'rate constant' and [RX] is the concentration of the substrate. The nucleophile does not enter into this equation because it isn't involved in the reaction until after the (slow) rate-determining step. To put this another way, since the second step is faster than the first, a 'better' nucleophile (we will see what this means later) won't make the overall reaction any faster.

The '1' in S_N1 refers to the fact that the rate of reaction is proportional to the concentration of only one of the components. It is a **first order** reaction.

You will probably have noticed that I switched from using a specific example at the start of this chapter, to a more general (R, X and Nu) one. There are a number of reasons for this, but let's just focus on one of them for now. If I had stuck with the same example, we would have had CH_3^+ being generated. In Basics 16 we will see that this is a very unstable carbocation. Therefore, you will never see an S_N1 reaction at a methyl group. It requires too much energy. We will consider the energetics of substitution reactions in detail in Reaction Detail 1.

THE S_N2 MECHANISM

In this mechanism, the attack of the nucleophile and the loss of leaving group occur at the same time. This reaction takes place by a 5-coordinate transition state, where the C−X bond is breaking and the C−Nu bond is forming at the same time. We have actually seen this a couple of times while introducing fundamental principles.

Now we can use the example of iodomethane again, because it turns out to be a perfectly good S_N2 reaction.

1.14

$$HO^\ominus \;+\; H_3C-I \;\longrightarrow\; \left[HO\text{-}\text{-}\underset{H\;H}{\overset{H}{C}}\text{-}\text{-}\text{-}I \right]^{\ddagger\ominus} \;\longrightarrow\; HO-CH_3 \;+\; I^\ominus$$

At a first glance, this might look a lot like the reaction we said could not happen. However, it is different in subtle but important ways. Because we are breaking the C−I bond **at the same time** as we form the C−O bond, we don't really get a species that has five bonds to carbon.

*The species in square brackets in the middle is a **transition state**, which is given the symbol '‡'. We will look at these in more detail in Basics 23.*

74

Instead of giving the carbon a share in two more electrons (from oxygen) *before* taking a share in two electrons away (to iodine), we are giving and taking away at the same time, so that the carbon always has a share in eight outer electrons.

This may seem like a subtle distinction, but it is an important one. When you are drawing curly arrow representations of reactions, you have to consider the precise sequence of breaking/forming bonds.

For now, we are just focusing on the very basic points that you **must** know. Normally we would not draw the transition state, so the reaction would look like this.

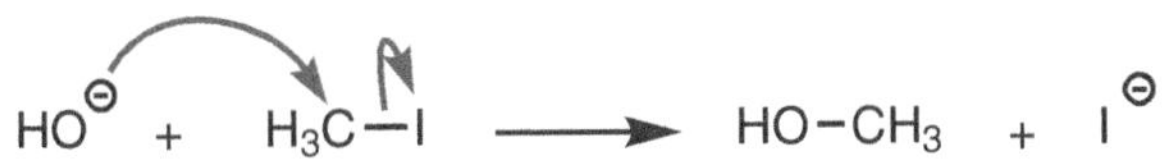

$$HO^{\ominus} \quad + \quad H_3C-I \quad \longrightarrow \quad HO-CH_3 \quad + \quad I^{\ominus}$$

This S_N2 reaction has rate = $k[CH_3I][HO^-]$

There is only one step, so it must be rate-determining. Since it involves both species, changing the concentration of either will alter the rate of the reaction. This is a '**second order**' reaction in terms of kinetics.

CAUTION

It is not uncommon for students to mix up S_N1 and S_N2 mechanisms, because the S_N1 mechanism has **two** steps, whereas the S_N2 mechanism has **one** step. You will probably make this mistake if you read this section once, feel happy that you understand it, and decide that you will have another look at it in the few days before the exam, in order to *refresh* your memory. At this point, you will be trying to remember which reaction is which. It is actually easier, and far more reliable, to keep reinforcing the basic points until you cannot forget or confuse them.

I would say the problem is a little more subtle. If you confuse S_N1 and S_N2, you will probably then try to explain, in your exam, why a given reaction is S_N1, even if it is not! At that point, you will be giving incorrect information about the stability of intermediates and transition states to try to justify something that is simply incorrect.

75

PRACTICE 6
Electronegativity in Context

THE PURPOSE OF THE EXERCISE

We have looked at substitution reactions, so you are now in a position to consider whether you could attack a particular carbon in a structure and have 'something' leave. That 'something' needs to be able to cope with a negative charge (it needs to be electronegative!). At this point, we are not yet ready to consider whether the reaction will actually take place.

By doing this, you will be drawing more structures. Make sure you keep them tidy. You might need to move parts of the molecule out of the way so that you can fit the new bits in. Make sure your curly arrows are clean and tidy. A nucleophile should always be attacking a carbon atom with a δ+ charge. Make sure the curly arrows reflect this.

THE EXERCISE

Here are those natural product structures again. Now consider which of the δ+ carbon atoms could be attacked by a nucleophile. Don't break carbon–carbon bonds! Which of the δ– atoms could be attacked by an electrophile (use a proton)? Draw sensible curly arrows!

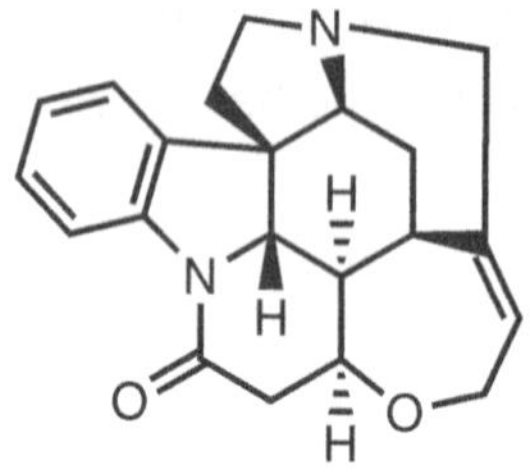

Strychnine

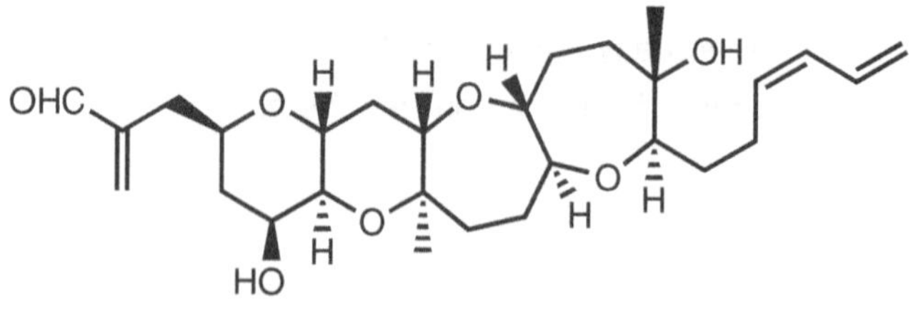

Hemibrevetoxin B

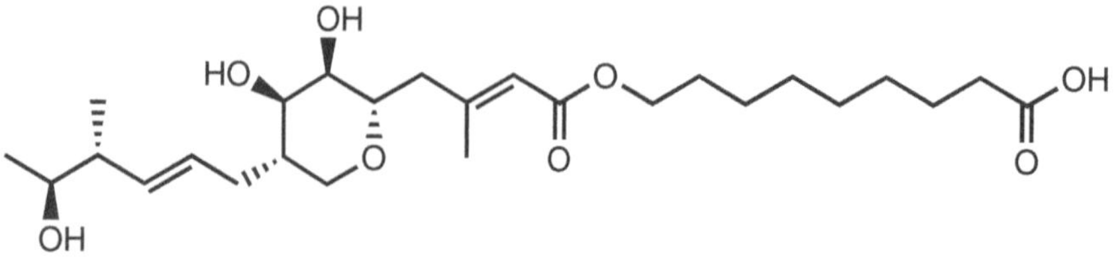

Pseudomonic acid C

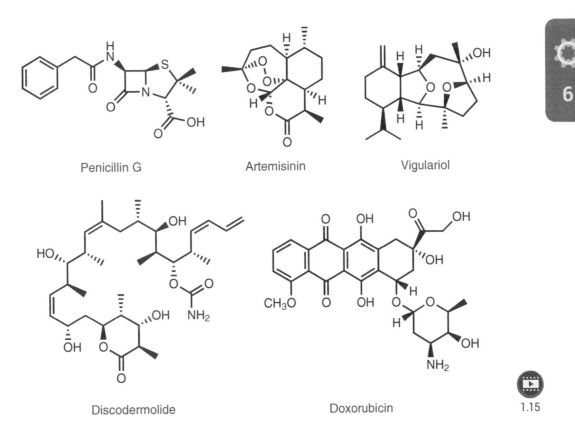

Penicillin G   Artemisinin   Vigulariol

Discodermolide   Doxorubicin   1.15

CAUTION

Come back to your answers from this when we have covered nucleophilic substitution in more detail, and consider whether any of the reactions you have drawn are particularly good or particularly bad.[29]

An experienced organic chemist will know exactly what we mean by this. A reaction that is particularly bad simply won't happen. A good reaction is more likely. It still doesn't mean it will happen quickly at room temperature.

[29] Have you looked up what these compounds do yet? Why would organic chemists want to spend years developing neat ways to make them, apart from the challenge?

77

FUNDAMENTAL REACTION TYPE 2
Elimination Reactions

INTRODUCTION

The very first thing we need to do is define what we mean by an elimination reaction. The second thing we need to do is get to grips with the electron flow and curly arrow drawing. We will look at these together.

First of all, when we use the word '**elimination**', we are generally talking about a '1,2-elimination' where two substituents are removed from adjacent carbon atoms. Let's call these groups X and Y for now. All of the elimination reactions which we will be talking about are polar in nature, so one of the groups (let's say X) must be lost as a positive species, X^+, and the other must be lost as a negative species, Y^-. Here is the overall reaction without curly arrows.

We have formed a new double bond between the two carbon atoms which X and Y were bonded to.

It is important that you are clear about this point. I often find that when I set an exam question stating that it is about elimination, but without drawing the reaction out, some students will draw a substitution reaction, probably because it is 'eliminating' the leaving group. You need to know the basic reaction types. That's why I am covering them with just the absolute basics for now.

Let's not beat about the bush. If you cannot confidently tell the difference between substitution and elimination, you're going to be in a mess. If you make this mistake in an exam, chances are it is because you didn't fully internalize the explanation. Sorry if this offends, but it's the truth!

Now let's think about the curly arrows we need. Since we are forming a new double bond between these carbon atoms, we definitely need a curly arrow which *ends* at the **middle** of this carbon–carbon bond. Where should this arrow start? Well, we are losing the X group as X⁺, so we know that the electrons from the C−X bond are staying with the carbon. These are the electrons which are going to form the new double bond, so the arrow needs to start at the **middle** of the C−X bond.

What about the other arrow? Well, we are losing Y as Y⁻, so we need to take the electrons from the C−Y bond and move them onto Y.

The curly arrows we need are shown below.

1.16

I know you are thinking by now that I am laying this on a bit thick, but I do see a lot of mistakes with these arrows, and your continued success in organic chemistry depends on you being able to draw curly arrows correctly without needing to think about it. Remember how you learned to write in the first place. You did it by endless repetition until you could form the letters and words clearly and correctly. You are now learning to write again, this time using bonds and arrows. Get practising!

WHAT ARE THE POSSIBLE MECHANISMS?

We will do the same as we did with substitution reactions in Fundamental Reaction Type 1, and consider the possible order of steps. We have two groups leaving from adjacent carbons. Let's simplify things a bit, and assume that the X group is a hydrogen atom—it usually is. You probably guessed that this was coming. After all, what is the easiest group to remove as a positively charged species? A proton, of course!

Conceptually, either X⁺ or Y⁻ could leave first, or they could both leave at the same time. This gives us three possible mechanisms.

This directly parallels the situation we encountered for substitution reactions. There is just one small but significant difference—this time, all three mechanisms are actually possible.

We will simply classify them according to their mechanisms. Here they are, complete with curly arrows:

79

THE E1 ELIMINATION MECHANISM

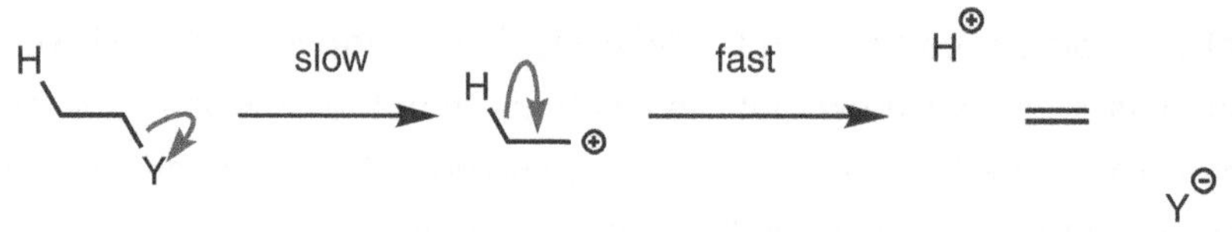

This one is referred to as the E1 mechanism. There is only one species involved in the rate-determining step—the substrate. For this mechanism to operate, the carbocation will need to be stabilized. We will start to see the factors that stabilize carbocations in Basics 16, and at that point you will immediately be able to speculate about the types of substrate which will undergo E1 elimination.

The primary carbocation shown above would not form. All we are doing for the moment is drawing curly arrow mechanisms.

THE E1cB ELIMINATION MECHANISM

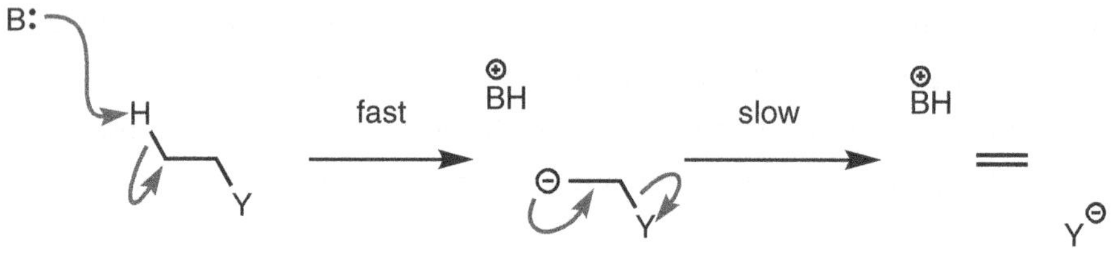

We now need a base to remove this proton, and this occurs rapidly. Then, the slow step is elimination of the Y group. Because the base is involved prior to the rate-determining step, the reaction has second order kinetics.

This is the least common of the three mechanisms for elimination, and will only be favoured in cases where the intermediate anion is stabilized (see Basics 18 for the sort of carbanions we would need). After all, why else would a substrate undergo fast deprotonation?

THE E2 ELIMINATION MECHANISM

These first two mechanisms can be considered extreme cases, where either the carbocation or carbanion intermediates are particularly stable. However, in most cases, we see a middle ground where neither is sufficiently stable to dominate the reaction pathway. In this case, loss of both groups occurs simultaneously, and the reaction only has one step, and proceeds through a transition state rather than an intermediate. We will cover the distinction between transition states and intermediates, along with their energy profiles, in Basics 14 and Basics 23.

You would normally need a base to remove the proton.

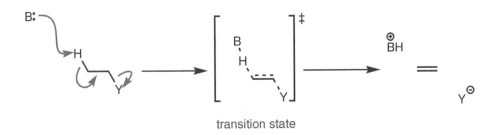

transition state

COMPETING REACTIONS

The E1 elimination shares a first step with the S_N1 substitution. If we add a nucleophile to the carbocation intermediate, we get a substitution reaction overall. If we lose a proton, we get elimination.

The E2 elimination is a one-step process, with a transition state—exactly like the S_N2 substitution. If a nucleophile attacks directly at carbon, we get a substitution reaction. If a base attacks a suitably disposed hydrogen, we get an elimination. We will look at the relationship between nucleophiles and bases in Basics 9.

There is often competition between substitution and elimination reactions. We will look at this in a series of worked problems in Worked Problem 2 and Worked Problem 3.

Substrates which undergo substitution by an S_N1 mechanism will undergo elimination by an E1 mechanism. This shouldn't surprise you too much, as the mechanisms only diverge after the rate-determining step. Substrates which undergo substitution by an S_N2 mechanism will undergo elimination by an E2 mechanism.

ONE MORE THING

When we look at the E1cB elimination mechanism in more detail, we can see that this corresponds to the 'impossible' mechanism in substitution reactions—therefore there is no direct parallel.

You should make sure you understand how to draw all three mechanisms before we add further complexity.

 Draw them all out a few times before you go any further!

Nucleophilic addition to the carbonyl group

6

Connections

Building on

- Functional groups, especially the C=O group ch2
- Identifying the functional groups in a molecule spectroscopically ch3
- How molecular orbitals explain molecular shapes and functional groups ch4
- How, and why, molecules react together and using curly arrows to describe reactions ch5

Arriving at

- How and why the C=O group reacts with nucleophiles
- Explaining the reactivity of the C=O group using molecular orbitals and curly arrows
- What sorts of molecules can be made by reactions of C=O groups
- How acid or base catalysts improve the reactivity of the C=O group

Looking forward to

- Additions of organometallic reagents ch9
- Substitution reactions of the C=O group's oxygen atom ch11
- How the C=O group in derivatives of carboxylic acids promotes substitution reactions ch10
- C=O groups with an adjacent double bond ch22

Molecular orbitals explain the reactivity of the carbonyl group

We are now going to leave to one side most of the reactions you met in the last chapter—we will come back to them all again later in the book. In this chapter we are going to concentrate on just one of them—probably the simplest of all organic reactions—the addition of a nucleophile to a carbonyl group. The carbonyl group, as found in aldehydes, ketones, and many other compounds, is without doubt the most important functional group in organic chemistry, and that is another reason why we have chosen it as our first topic for more detailed study.

You met nucleophilic addition to a carbonyl group on pp. 115 and 121, where we showed you how cyanide reacts with aldehydes to give an alcohol. As a reminder, here is the reaction again, this time with a ketone, with its mechanism.

■ We will frequently use a device like this, showing a reaction scheme with a mechanism for the same reaction looping round underneath. The reagents and conditions above and below the arrow across the top tell you how you might carry out the reaction, and the pathway shown underneath tells you how it actually works.

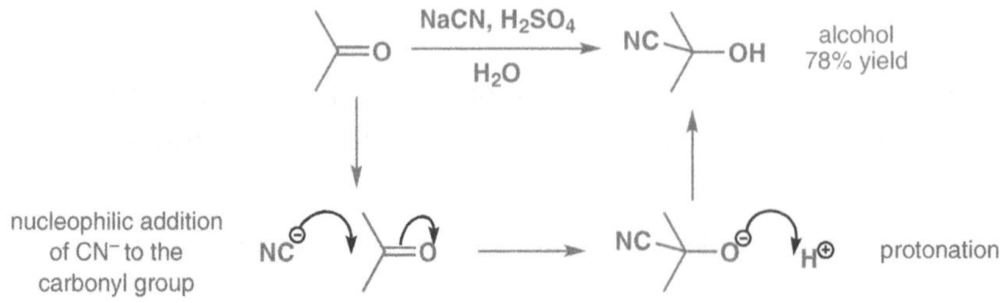

The reaction has two steps: nucleophilic addition of cyanide, followed by protonation of the anion. In fact, this is a general feature of all nucleophilic additions to carbonyl groups.

Online support. The icon in the margin indicates that accompanying interactive resources are provided online to help your understanding: just type **www.chemtube3d.com/clayden/123** into your browser, replacing **123** with the number of the page where you see the icon. For pages linking to more than one resource, type **123-1, 123-2** etc. (replacing **123** with the page number) for access to successive links.

> ● **Additions to carbonyl groups generally consist of two mechanistic steps:**
>   • **nucleophilic attack on the carbonyl group**
>   • **protonation of the anion that results.**

The addition step is more important, and it forms a new C–C σ bond at the expense of the C=O π bond. The protonation step makes the overall reaction addition of HCN across the C=O π bond.

Why does cyanide, in common with many other nucleophiles, attack the carbonyl group? And why does it attack the *carbon* atom of the carbonyl group? To answer these questions we need to look in detail at the structure of carbonyl compounds in general and the orbitals of the C=O group in particular.

The carbonyl double bond, like that found in alkenes (whose bonding we discussed in Chapter 4), consists of two parts: one σ bond and one π bond. The σ bond between the two sp² hybridized atoms—carbon and oxygen—is formed from two sp² orbitals. The other sp² orbitals on carbon form the two σ bonds to the substituents while those on oxygen are filled by the two lone pairs. The sp² hybridization means that the carbonyl group has to be planar, and the angle between the substituents is close to 120°. The diagram illustrates all this for the simplest carbonyl compound, formaldehyde (or methanal, CH₂O). The π bond then results from overlap of the remaining p orbitals—again, you can see this for formaldehyde in the diagram.

🔊 Interactive bonding orbitals in formaldehyde

➡ You were introduced to the polarization of orbitals in Chapter 4 and we discussed the case of the carbonyl group on p. 104.

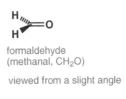

formaldehyde
(methanal, CH₂O)

viewed from a slight angle

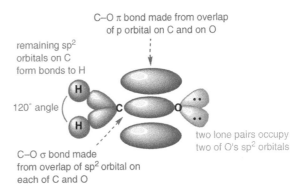

C–O π bond made from overlap of p orbital on C and on O

remaining sp² orbitals on C form bonds to H

120° angle

C–O σ bond made from overlap of sp² orbital on each of C and O

two lone pairs occupy two of O's sp² orbitals

When we introduced the bonding in the carbonyl group in Chapter 4 we explained how polarization in the π bond means it is skewed towards oxygen, because oxygen is more electronegative than carbon. Conversely, the unfilled π* antibonding orbital is skewed in the opposite direction, with a larger coefficient at the carbon atom. This is quite hard to represent with the π bond represented as a single unit, as shown above, but becomes easier to visualize if instead we represent the π and π* orbitals using individual p orbitals on C and O. The diagrams in the margin show the π and π* orbitals represented in this way.

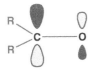

empty, antibonding π* orbital

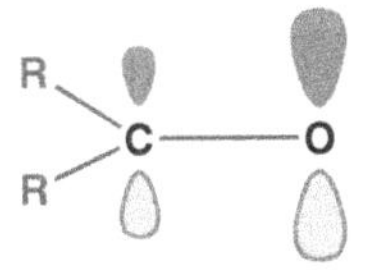

filled σ orbital

Electronegativities, bond lengths, and bond strengths					
Representative bond energy, kJ mol⁻¹		Representative bond length, Å		Electronegativity	
C–O	351	C–O	1.43	C	2.5
C=O	720	C=O	1.21	O	3.5

Because there are two types of bonding between C and O, the C=O double bond is rather shorter than a typical C–O single bond, and also over twice as strong—so why is it so reactive? Polarization is the key. The polarized C=O bond gives the carbon atom some degree of positive charge, and this charge attracts negatively charged nucleophiles (like cyanide) and encourages reaction. The polarization of the antibonding π* orbital towards carbon is also important because, when the carbonyl group reacts with a nucleophile, electrons move from the HOMO of the nucleophile (an sp orbital in the case of cyanide) into the LUMO of the electrophile—in other words the π* orbital of the C=O bond. The greater coefficient of the π* orbital at carbon means a better HOMO–LUMO interaction, so this is where the nucleophile attacks.

As our nucleophile—which we are representing here as 'Nu⁻'—approaches the carbon atom, the electron pair in its HOMO starts to interact with the LUMO (antibonding π*) to form a new σ bond. Filling antibonding orbitals breaks bonds and, as the electrons enter the antibonding π* of the carbonyl group, the π bond is broken, leaving only the C–O σ bond intact. But electrons can't just vanish, and those that were in the π bond move off on to the electronegative oxygen, which ends up with the negative charge that started on the nucleophile. You can see all this happening in the diagram below.

■ The HOMO of the nucleophile will depend on what the nucleophile is, and we will meet examples in which it is an sp or sp³ orbital containing a lone pair, or a B–H σ orbital or metal–carbon σ orbital. We shall shortly discuss cyanide as the nucleophile; cyanide's HOMO is an sp orbital on carbon.

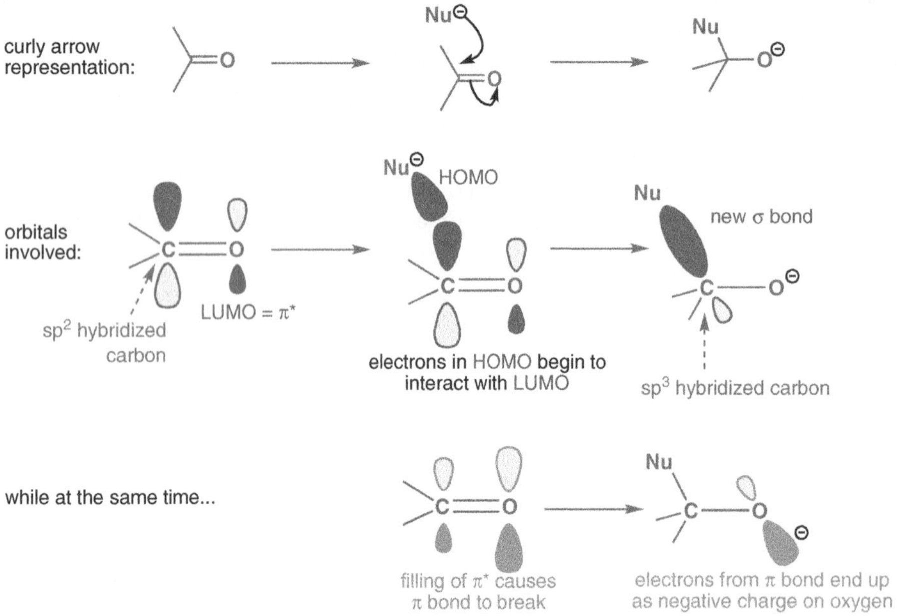

Notice how the trigonal, planar sp² hybridized carbon atom of the carbonyl group changes to a tetrahedral, sp³ hybridized state in the product. For each class of nucleophile you meet in this chapter, we will show you the HOMO–LUMO interaction involved in the addition reaction. These interactions also show you how the orbitals of the starting materials change into the orbitals of the product as they combine. Most importantly here, the lone pair of the nucleophile combines with the π* of the carbonyl group to form a new σ bond in the product.

Attack of cyanide on aldehydes and ketones

Now that we've looked at the theory of how a nucleophile attacks a carbonyl group, let's go back to the real reaction with which we started this chapter: cyanohydrin formation from a carbonyl compound and sodium cyanide. Cyanide contains sp hybridized C and N atoms, and its HOMO is an sp orbital on carbon. The reaction is a typical nucleophilic addition reaction to a carbonyl group: the electron pair from the HOMO of the CN⁻ (an sp orbital on carbon) moves into the C=O π* orbital; the electrons from the C=O π orbital move on to the oxygen atom. The reaction is usually carried out in the presence of acid, which protonates the resulting alkoxide to give the hydroxyl group of the composite functional group known as a cyanohydrin. The reaction works with both ketones and aldehydes, and the mechanism below shows the reaction of a general aldehyde. This reaction appeared first in Chapter 5.

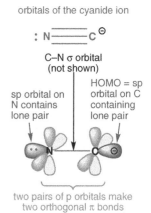

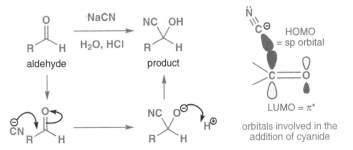

orbitals involved in the
addition of cyanide

Interactive mechanism for
cyanohydrin formation

Cyanohydrins in synthesis

Cyanohydrins are important synthetic intermediates, for example the cyanohydrin formed from this cyclic amino ketone is the first intermediate in a synthesis of some medicinal compounds known as 5HT$_3$ agonists, which were designed to reduce nausea in chemotherapy patients.

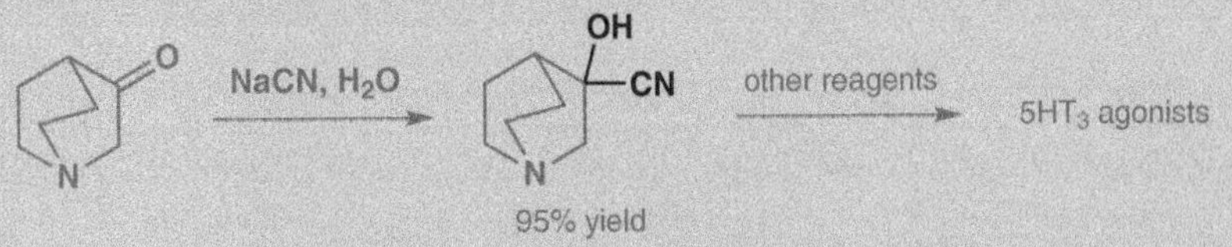

NaCN, H$_2$O

95% yield

other reagents

5HT$_3$ agonists

Cyanohydrins are also components of many natural and industrial products, such as the insecticide cypermethrin (marketed as 'Ripcord' and 'Barricade').

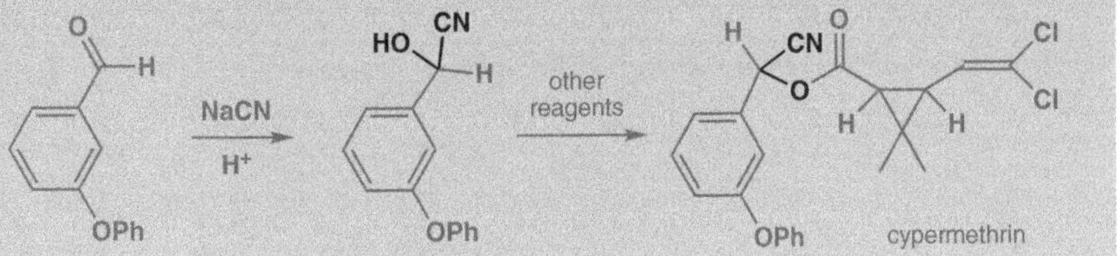

NaCN
H$^+$

other reagents

cypermethrin

Cyanohydrin formation is reversible: just dissolving a cyanohydrin in water can give back the aldehyde or ketone you started with, and aqueous base usually decomposes cyanohydrins completely. This is because cyanide is a good *leaving group*—we'll come back to this type of reaction in more detail in Chapter 10.

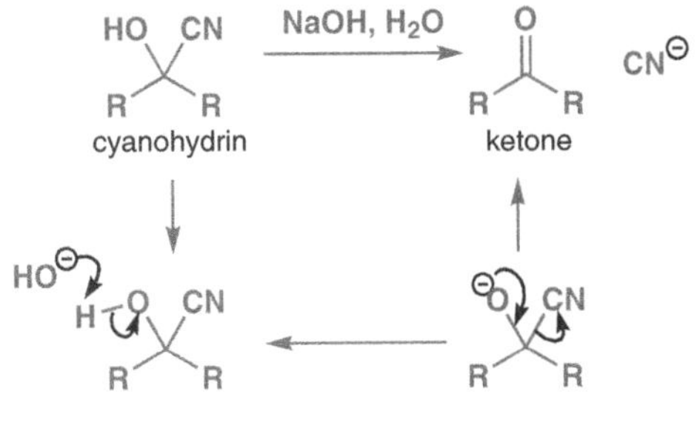

NaOH, H$_2$O

cyanohydrin   ketone

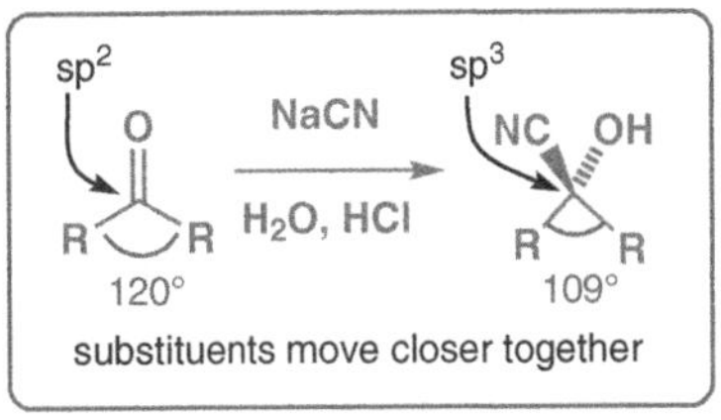

sp^2 NaCN   sp^3

H$_2$O, HCl

120°   109°

substituents move closer together

Some equilibrium constants

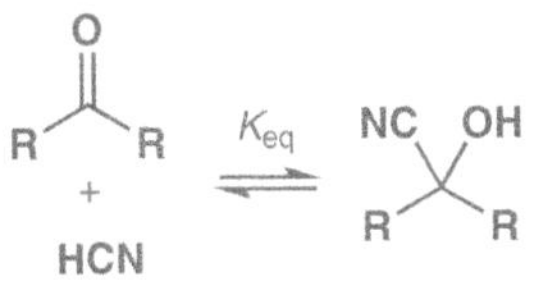

aldehyde or ketone	K_{eq}
PhCHO	212
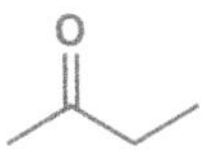	28

Cyanohydrin formation is therefore an equilibrium between starting materials and products, and we can get good yields only if the equilibrium favours the products. The equilibrium is more favourable for aldehyde cyanohydrins than for ketone cyanohydrins, and the reason is the size of the groups attached to the carbonyl carbon atom. As the carbonyl carbon atom changes from sp^2 to sp^3, its bond angles change from about 120° to about 109°—in other words, the substituents it carries move closer together. This reduction in bond angle is not a problem for aldehydes, because one of the substituents is just a (very small) hydrogen atom, but for ketones, especially ones that carry larger alkyl groups, this effect can disfavour the addition reaction. Effects that result from the size of substituents and the repulsion between them are called steric effects, and we call the repulsive force experienced by large substituents steric hindrance. Steric hindrance (not 'hinderance') is a consequence of repulsion between the electrons in all the filled orbitals of the alkyl substituents.

Steric hindrance

The size of substituents plays a role in very many organic reactions—it's the reason aldehydes (with an H next to the C=O group) are more reactive than ketones, for example. Steric hindrance affects reaction rates, but also makes molecules react by completely different mechanisms, as you will see in the substitution reactions in Chapter 15. You will need to get used to thinking about whether the presence of large substituents, with all their filled C—H and C—C bonds, is a factor in determining how well a reaction will go.

Cyanohydrins and cassava

The reversibility of cyanohydrin formation is of more than theoretical interest. In parts of Africa the staple food is cassava. This food contains substantial quantities of the glucoside of acetone cyanohydrin (a glucoside is an acetal derived from glucose). We shall discuss the structure of glucose later in this chapter, but for now, just accept that it stabilizes the cyanohydrin.

The glucoside is not poisonous in itself, but enzymes in the human gut break it down and release HCN. Eventually 50 mg HCN per 100 g of cassava can be released and this is enough to kill a human being after a meal of unfermented cassava. If the cassava is crushed with water and allowed to stand ('ferment'), enzymes in the cassava will do the same job and then the HCN can be washed out before the cassava is cooked and eaten.

The cassava is now safe to eat but it still contains some glucoside. Some diseases found in eastern Nigeria can be traced to long-term consumption of HCN. Similar glucosides are found in apple pips and the kernels inside the stones of fruit such as peaches and apricots. Some people like eating these, but it is unwise to eat too many at one sitting!

The angle of nucleophilic attack on aldehydes and ketones

Having introduced you to the sequence of events that makes up a nucleophilic attack at C=O (interaction of HOMO with LUMO, formation of new σ bond, breakage of π bond), we should now tell you a little more about the *direction* from which the nucleophile approaches the carbonyl group. Not only do nucleophiles always attack carbonyl groups at carbon, but they also always approach from a particular angle. You may at first be surprised by this angle, since nucleophiles attack not from a direction perpendicular to the plane of the carbonyl group but at about 107° to the C=O bond—close to the angle at which the new bond will form. This approach route is known as the Bürgi–Dunitz trajectory after the authors of the elegant crystallographic methods that revealed it. You can think of the angle of attack as the result of a compromise between maximum orbital overlap of the HOMO with π* and minimum repulsion of the HOMO by the electron density in the carbonyl π bond. But a better explanation is that π* does not have parallel atomic orbitals as there is a node halfway down the bond (Chapter 4) so the atomic orbitals are already at an angle. The nucleophile attacks along the axis of the larger orbital in the HOMO.

➡ We pointed this out in Chapter 4 on p. 104.

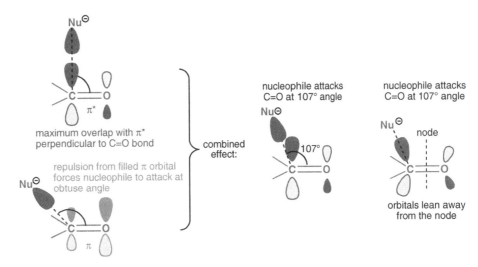

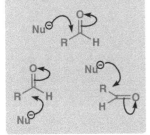

■ Although we now know precisely from which direction the nucleophile attacks the C=O group, this is not always easy to represent when we draw curly arrows. As long as you bear the Bürgi–Dunitz trajectory in mind, you are quite at liberty to write any of the variants shown here, among others.

Any other portions of the molecule that get in the way of (or, in other words, that cause *steric hindrance* to) the Bürgi–Dunitz trajectory will greatly reduce the rate of addition and this is another reason why aldehydes are more reactive than ketones. The importance of the Bürgi–Dunitz trajectory will become more evident later, particularly in Chapter 33.

Bürgi and Dunitz deduced this trajectory by examining crystal structures of compounds containing both a nucleophilic nitrogen atom and an electrophilic carbonyl group. They found that, when the two got close enough to interact, but were not free to undergo reaction, the nitrogen atom always lay on or near the 107° trajectory described here. Theoretical calculations later gave the same 107° value for the optimum angle of attack.

Nucleophilic attack by 'hydride' on aldehydes and ketones

Nucleophilic attack by the hydride ion, H⁻, is an almost unknown reaction. This species, which is present in the salt sodium hydride, NaH, has such a high charge density that it only ever reacts as a base. The reason is that its filled 1s orbital is of an ideal size to interact with the hydrogen atom's contribution to the σ* orbital of an H–X bond (X can be any atom), but much too small to interact easily with carbon's more diffuse 2p orbital contribution to the LUMO (π*) of the C=O group.

nucleophilic attack by H⁻ almost *never happens*   H⁻ almost *always* reacts as a base

Nevertheless, adding H⁻ to the carbon atom of a C=O group would be a very useful reaction, as the result would be the formation of an alcohol. This process would involve going down from the aldehyde or ketone oxidation level to the alcohol oxidation level (Chapter 2, p. 32) and would therefore be a reduction. It cannot be done with NaH, but it can be done with some other compounds containing nucleophilic hydrogen atoms.

reduction of a ketone to an alcohol

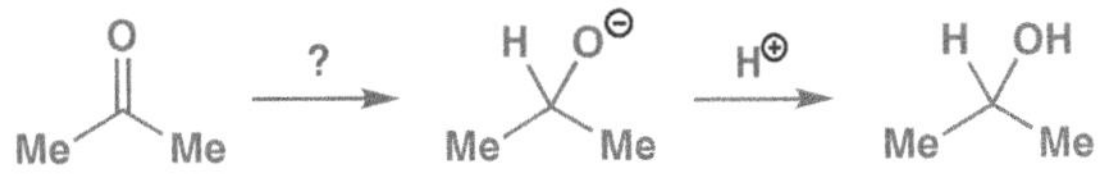

The most important of these compounds is sodium borohydride, NaBH₄. This is a water-soluble salt containing the tetrahedral BH₄⁻ anion, which is isoelectronic with methane but has a negative charge since boron has one less proton in the nucleus than does carbon.

In Chapter 4 we looked at isoelectronic borane BH_3 and the cation CH_3^+. Here we have effectively added a hydride ion to each of them.

But beware! Remember (p. 115) there is no lone pair on boron: you must not draw an arrow coming out of this negative charge to form another bond. If you did, you would get a pentacovalent B(V) compound, which would have ten electrons in its outer shell. Such a thing is impossible with a first-row element as there are only four available orbitals ($1 \times 2s$ and $3 \times 2p$). Instead, since all of the electrons (including those represented by the negative charge) are in B–H σ orbitals, it is from a B–H bond that we must start any arrow to indicate reaction of BH_4^- as a nucleophile. By transferring this pair of electrons we make the boron atom neutral—it is now trivalent with just six electrons.

borohydride anion   methane

arrow cannot start on negative charge: no lone pair on B

eight electrons in B–H bonds

impossible structure: ten electrons in bonds to B

electrons must be transferred from a bond

eight electrons in B–H bonds

six electrons in B–H bonds and one empty p orbital

■ Just as we have used Nu⁻ to indicate any (undefined) nucleophile, here E⁺ means any (undefined) electrophile.

What happens when we carry out this reaction using a carbonyl compound as the electrophile? The hydrogen atom, together with the pair of electrons from the B–H bond, will be transferred to the carbon atom of the C=O group. Although no hydride ion, H⁻, is actually involved in the reaction, the transfer of a hydrogen atom with an attached pair of electrons can be regarded as a 'hydride transfer'. You will often see it described this way in books. But be careful not to confuse BH_4^- with the hydride ion itself. To make it quite clear that it is the hydrogen atom that is forming the new bond to C, this reaction may also be helpfully represented with a curly arrow *passing through* the hydrogen atom.

Interactive mechanism for borohydride reduction

You met this reaction in Chapter 5 but there is more to say about it. The oxyanion produced in the first step can help stabilize the electron-deficient BH_3 molecule by adding to its empty p orbital. Now we have a tetravalent boron anion again, which could transfer a second hydrogen atom (with its pair of electrons) to another molecule of aldehyde.

This process can continue so that, in principle, all four hydrogen atoms could be transferred to molecules of aldehyde. In practice the reaction is rarely as efficient as that, but aldehydes and ketones are usually reduced in good yield to the corresponding alcohol by sodium borohydride in water or alcoholic solution. The water or alcohol solvent provides the proton needed to form the alcohol from the alkoxide.

examples of reductions with sodium borohydride

benzaldehyde   acetophenone

Aluminium is more electropositive (more metallic) than boron
and is therefore more ready to
give up a hydrogen atom (and
the associated negative charge),
whether to a carbonyl group or
to water. Lithium aluminium
hydride reacts violently and dangerously with water in an exothermic reaction that produces
highly flammable hydrogen.

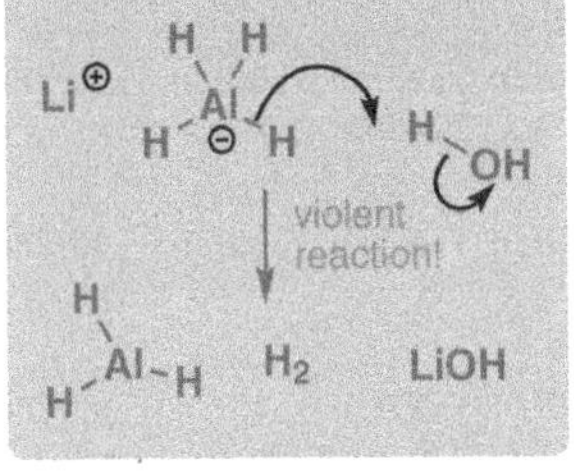

violent
reaction!

H_2 LiOH

Sodium borohydride is one of the weaker hydride donors. The fact that it can be used in water is evidence of this: more powerful hydride donors such as lithium aluminium hydride, $LiAlH_4$, react violently with water. Sodium borohydride reacts with both aldehydes and ketones, although the reaction with ketones is slower: for example, benzaldehyde is reduced about 400 times faster than acetophenone in isopropanol. This is because of steric hindrance (see above).

Sodium borohydride does not react at all with less reactive carbonyl compounds such as esters or amides: if a molecule contains both an aldehyde and an ester, only the aldehyde will be reduced.

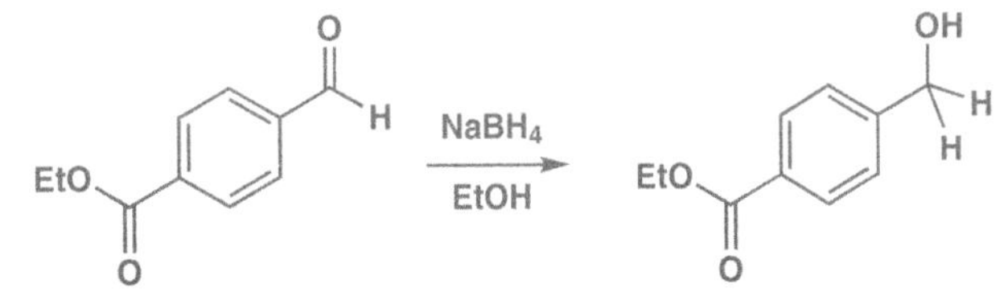

The next two examples illustrate the reduction of aldehydes and ketones in the presence of other reactive functional groups. No reaction occurs at the nitro group in the first case or at the alkyl halide in the second.

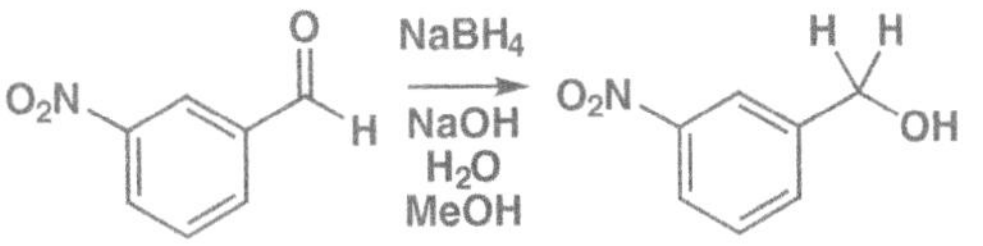

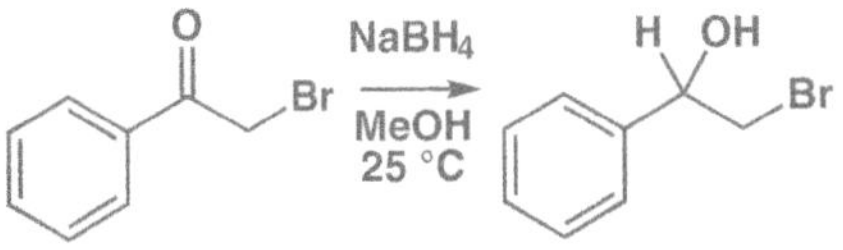

Addition of organometallic reagents to aldehydes and ketones

Organometallic compounds have a carbon–metal bond. Lithium and magnesium are very electropositive metals, and the Li–C or Mg–C bonds in organolithium or organomagnesium reagents are highly polarized towards carbon. They are therefore very powerful nucleophiles, and attack the carbonyl group to give alcohols, forming a new C–C bond. For our first example, we shall take one of the simplest of organolithiums, methyllithium, which is commercially available as a solution in Et_2O, shown here reacting with an aldehyde. The orbital diagram of the addition step shows how the polarization of the C–Li bond means that it is the carbon atom of the nucleophile that attacks the carbon atom of the electrophile and we get a new C–C bond. We explained on p. 113 the polarization of bonds between carbon and more electropositive elements. The relevant electronegativities are C 2.5, Li 1.0, and Mg 1.2 so both metals are much more electropositive than carbon. The orbitals of MeLi are discussed in Chapter 4.

HOMO = Li–C σ
polarized towards C

LUMO = π*

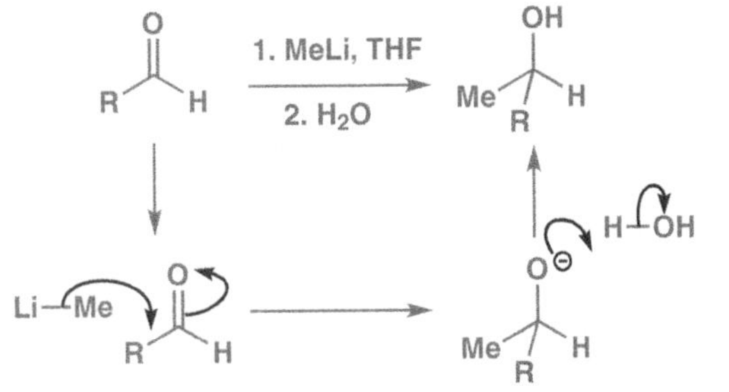

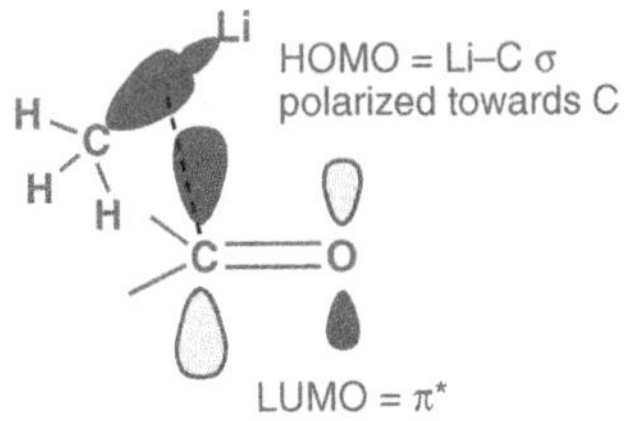

orbitals involved in the addition
of methyllithium

🖱 Interactive mechanism for
methyllithium addition

The course of the reaction is much the same as you have seen before, but we need to high-light a few points where this reaction scheme differs from those you have met earlier in the chapter. First of all, notice the legend '1. MeLi, THF; 2. H_2O'. This means that, first, MeLi is added to the aldehyde in a THF solvent. Reaction occurs: MeLi adds to the aldehyde to give an alkoxide. Then (and only then) water is added to protonate the alkoxide. The '2. H_2O' means that water is added in a separate step only when all the MeLi has reacted: it is not present at the start of the reaction as it was in the cyanide reaction and some of the borohydride addition reactions. In fact, water *must not be* present during the addition of MeLi (or of any other orga-nometallic reagent) to a carbonyl group because water destroys organometallics very rapidly by protonating them to give alkanes (organolithiums and organomagnesiums are strong bases as well as powerful nucleophiles). The addition of water, or sometimes dilute acid or ammonium chloride, at the end of the reaction is known as the *work-up*.

Because they are so reactive, organolithiums are usually used at low temperature, often $-78\ ^{\circ}C$ (the sublimation temperature of solid CO_2), in aprotic solvents such as Et_2O or THF. Protic solvents such as water or alcohols have acidic protons but aprotic solvents such as ether do not. Organolithiums also react with oxygen, so they have to be handled under a dry, inert atmosphere of nitrogen or argon. Other common, and commercially available, organolithium reagents include *n*-butyllithium and phenyllithium, and they react with both aldehydes and ketones. Note that addition to an aldehyde gives a secondary alcohol while addition to a ketone gives a tertiary alcohol.

> **Low-temperature baths**
>
> Cooling reaction mixtures is generally the job of a cooling bath of ice and water for around 0 °C, or baths of solid CO_2 in organic solvents such as acetone or ethanol down to about –78 °C. Small pieces of solid CO_2 are added slowly to the solvent until vigorous bubbling ceases. Few chemists then measure the temperature of the bath, which may be anywhere from –50 to –80 °C. The temperature given in publications is often –78 °C, about the lower limit. Lower temperatures require liquid nitrogen. Practical handbooks give details.

Organomagnesium reagents known as Grignard reagents (RMgX) react in a similar way. Some simple Grignard reagents, such as methyl magnesium chloride, MeMgCl, and phenyl magnesium bromide, PhMgBr, are commercially available, and the scheme shows PhMgBr reacting with an aldehyde. The reactions of these two classes of organometallic reagent— organolithiums and Grignard reagents—with carbonyl compounds are among the most important ways of making carbon–carbon bonds, and we will consider them in more detail in Chapter 9.

Addition of water to aldehydes and ketones

Nucleophiles don't have to be highly polarized or negatively charged to react with aldehydes and ketones: neutral ones will as well. How do we know? This ^{13}C NMR spectrum was obtained by dissolving formaldehyde, $H_2C{=}O$, in water. You will remember from Chapter 3 that the carbon atoms of carbonyl groups give ^{13}C signals typically in the region of 150–200 ppm. So where is formaldehyde's carbonyl peak? Instead we have a signal at 83 ppm—where we would expect tetrahedral carbon atoms singly bonded to oxygen to appear.

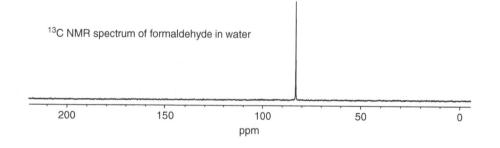

^{13}C NMR spectrum of formaldehyde in water

ppm

organometallics are destroyed by water

fast and exothermic

methane

Grignard reagents were discovered by Victor Grignard (1871– 1935) at the University of Lyon, who got the Nobel prize for his discovery in 1912. They are made by reacting alkyl or aryl halides with magnesium 'turnings'.

🐭 Interactive mechanism for Grignard addition

What has happened is that water has added to the carbonyl group to give a compound known as a hydrate or 1,1-diol.

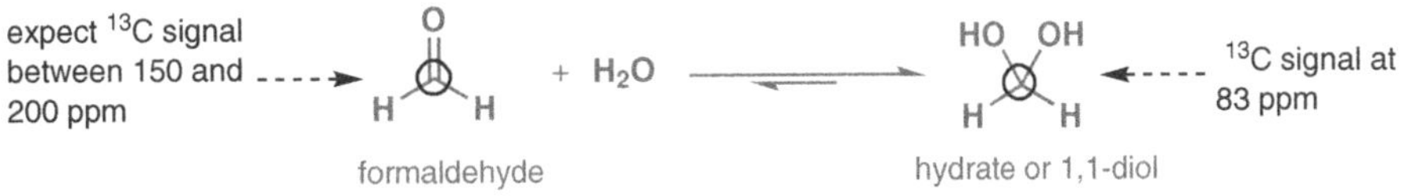

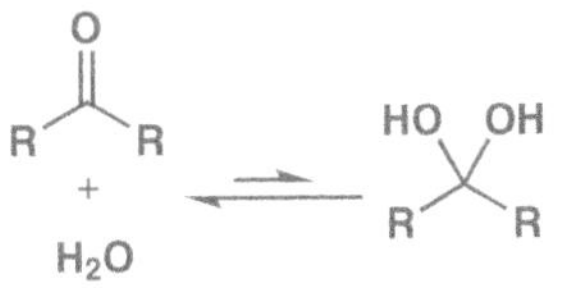

expect ^{13}C signal between 150 and 200 ppm

formaldehyde

+ H_2O

hydrate or 1,1-diol

^{13}C signal at 83 ppm

This reaction, like the addition of cyanide we discussed at the beginning of the chapter, is an equilibrium, and is quite general for aldehydes and ketones. But, as with the cyanohydrins, the position of the equilibrium depends on the structure of the carbonyl compound. Generally, the same steric factors (p. 129) mean that simple aldehydes are hydrated to some extent while simple ketones are not. However, special factors can shift the equilibrium towards the hydrated form even for ketones, particularly if the carbonyl compound is reactive or unstable.

Formaldehyde is an extremely reactive aldehyde as it has no substituents to hinder attack—it is so reactive that it is rather prone to polymerization. And it is quite happy to move from sp^2 to sp^3 hybridization because there is very little increased steric hindrance between the two hydrogen atoms as the bond angle changes from 120° to 109° (p. 129). This is why our aqueous solution of formaldehyde contains essentially no CH_2O—it is completely hydrated. A mechanism for the hydration reaction is shown below. Notice how a proton has to be transferred from one oxygen atom to the other, mediated by water molecules.

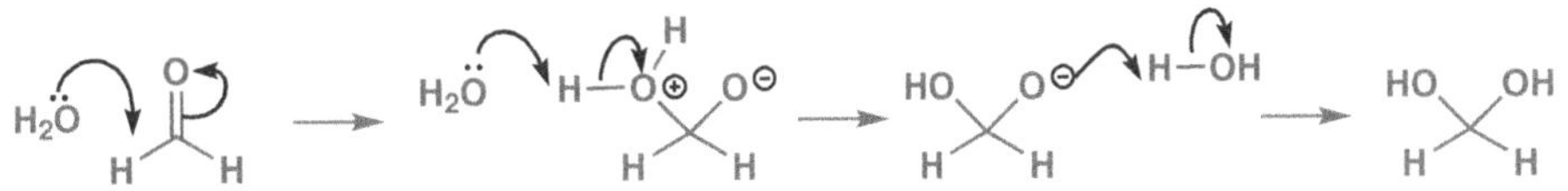

Formaldehyde reacts with water so readily because its substituents are very small: a steric effect. Electronic effects can also favour reaction with nucleophiles—electronegative atoms such as halogens attached to the carbon atoms next to the carbonyl group can increase the extent of hydration by the inductive effect according to the number of halogen substituents and their electron-withdrawing power. They increase the polarization of the carbonyl group, which already has a positively polarized carbonyl carbon, and make it even more prone to attack by water. Trichloroacetaldehyde (chloral, Cl_3CCHO) is hydrated completely in water, and the product 'chloral hydrate' can be isolated as crystals and is an anaesthetic. You can see this quite clearly in the two IR spectra below. The first one is a spectrum of chloral hydrate from a bottle—notice there is no strong absorption between 1700 and 1800 cm^{-1} (where we would expect C=O to appear) and instead we have the tell-tale broad O–H peak at 3400 cm^{-1}. Heating drives off the water, and the second IR spectrum is of the resulting dry chloral: the C=O peak has reappeared at 1770 cm^{-1} and the O–H peak has gone.

significant concentrations of hydrate are generally formed only from aldehydes

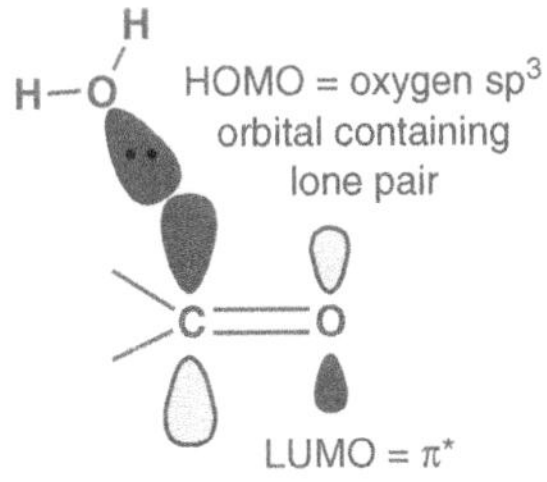

HOMO = oxygen sp^3 orbital containing lone pair

LUMO = π^*

orbitals involved in the addition of water

Interactive mechanism for hydrate formation

Monomeric formaldehyde

The hydrated nature of formaldehyde poses a problem for chemistry that requires anhydrous conditions such as the organometallic additions we have just been talking about. Fortunately, cracking (heating to decomposition) the polymeric 'paraformaldehyde' can provide monomeric formaldehyde in anhydrous solution.

polymeric 'paraformaldehyde'

CH_2O

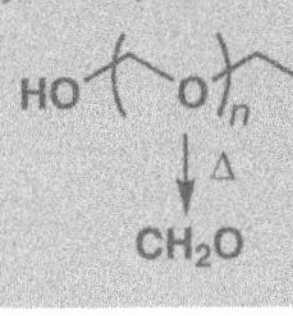

Chloral hydrate is the infamous 'knockout drops' of Agatha Christie or the 'Mickey Finn' of prohibition gangsters.

● **Steric and electronic effects**

• **Steric effects** are concerned with the size and shape of groups within molecules.

• **Electronic effects** result from the way that electronegativity differences between atoms affect the way electrons are distributed in molecules. They can be divided into *inductive effects*, which are the consequence of the way that electronegativity differences lead to polarization of σ bonds, and *conjugation* (sometimes called *mesomeric effects*) which affects the distribution of electrons in π bonds and is discussed in the next chapter.

Steric and electronic effects are two of the main factors dominating the reactivity of nucleophiles and electrophiles.

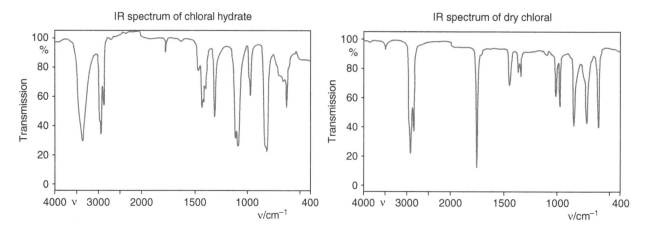

The chart shows the extent of hydration (in water) of a small selection of carbonyl compounds: hexafluoroacetone is probably the most hydrated carbonyl compound possible! The larger the equilibrium constant, the more the equilibrium is to the right.

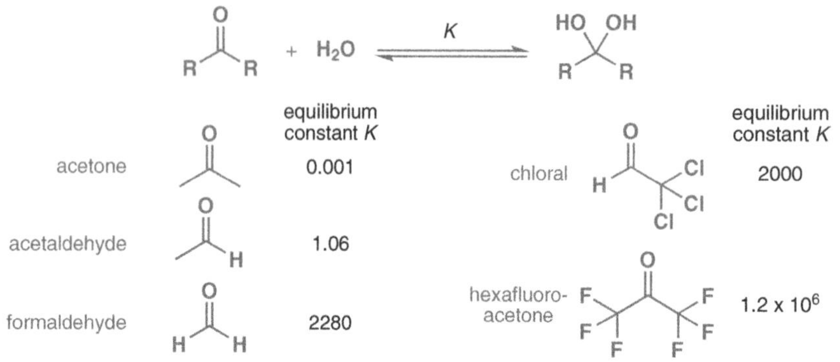

Interactive structures of carbonyl compounds and hydrates

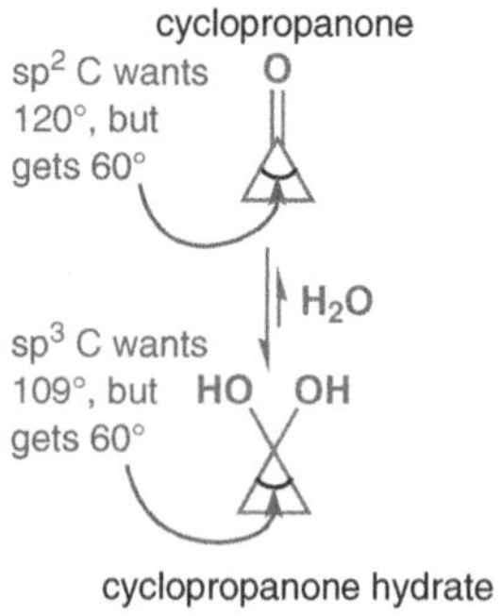

Cyclopropanones—three-membered ring ketones—are also hydrated to a significant extent, but for a different reason. You saw earlier how *acyclic* ketones suffer increased steric hindrance when the bond angle changes from 120° to 109° on moving from sp^2 to sp^3 hybridization. Cyclopropanones (and other small-ring ketones) conversely prefer the small bond angle because their substituents are already confined within a ring. Look at it this way: a three-membered ring is really very strained, with bond angles forced to be 60°. For the sp^2 hybridized ketone this means bending the bonds 60° away from their 'natural' 120°. But for the sp^3 hybridized hydrate the bonds have to be distorted by only 49° (= 109° – 60°). So addition to the C=O group allows some of the strain inherent in the small ring to be released—hydration is favoured, and indeed cyclopropanone and cyclobutanone are very reactive electrophiles.

cyclopropanone
sp^2 C wants 120°, but gets 60°

sp^3 C wants 109°, but gets 60°

cyclopropanone hydrate

● The same structural features that favour or disfavour hydrate formation are important in determining the reactivity of carbonyl compounds with other nucleophiles, whether the reactions are reversible or not. Steric hindrance and more alkyl substituents make carbonyl compounds less reactive towards any nucleophile; electron-withdrawing groups and small rings make them more reactive.

Hemiacetals from reaction of alcohols with aldehydes and ketones

Since water adds to (at least some) carbonyl compounds, it should come as no surprise that alcohols do too. The product of the reaction is known as a hemiacetal, because it is halfway to

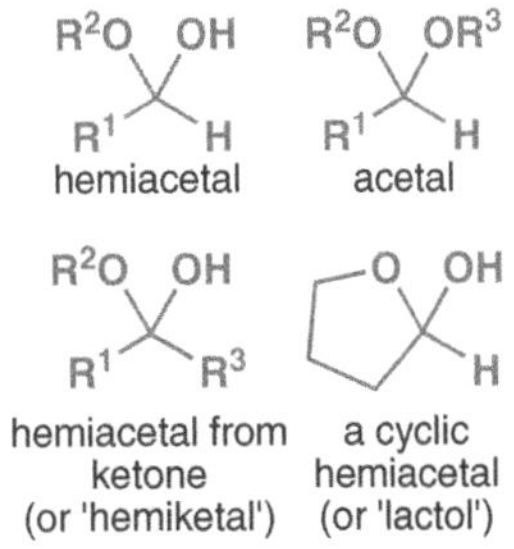

hemiacetal · acetal

hemiacetal from ketone (or 'hemiketal') · a cyclic hemiacetal (or 'lactol')

names for functional groups

Interactive mechanism for hemiacetal formation

an acetal, a functional group that you met in Chapter 2 (p. 32) and that will be discussed in detail in Chapter 11. The mechanism follows in the footsteps of hydrate formation: just use ROH instead of HOH.

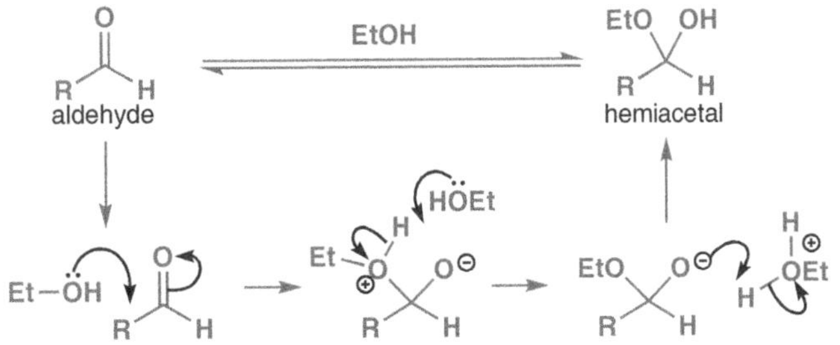

In the mechanism above, as in the mechanism of hydrate formation on p. 134, a proton has to be transferred between one oxygen atom and the other. We have shown a molecule of ethanol (or water) doing this, but it is impossible to define exactly the path taken by any one proton as it transfers between the oxygen atoms. It might not even be the same proton: another possible mechanism is shown below on the left, where a molecule of ethanol simultaneously gives away one proton and takes another. In the simplest case, the proton just hops from one oxygen to another, as shown in the right, and there is no shame in writing this mechanism: it is no more or less correct than the others.

two more (and equally correct) mechanisms for proton transfer between the oxygen atoms:

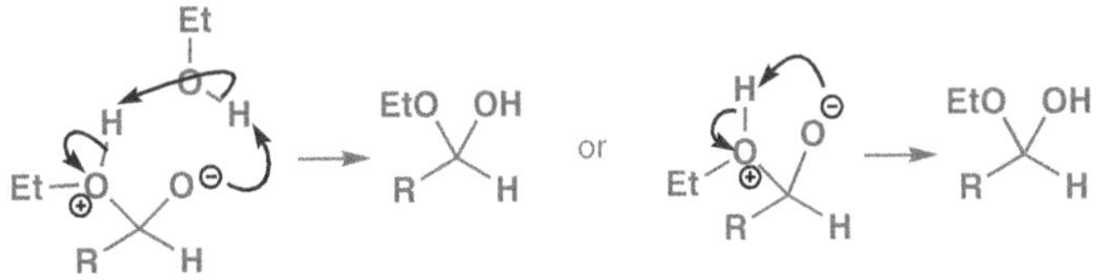

What is certain is that proton transfers between oxygen atoms are very fast and are reversible, and for that reason we don't need to be concerned with the details—the proton can always get to where it needs to be for the next step of the mechanism. As with all these carbonyl group reactions, what is really important is the addition step, not what happens to the protons.

Hemiacetal formation is reversible, and hemiacetals are stabilized by the same special structural features as those of hydrates. However, hemiacetals can also gain stability by being cyclic—when the carbonyl group and the attacking hydroxyl group are part of the same molecule. The reaction is now an intramolecular (within the same molecule) addition, as opposed to the intermolecular (between two molecules) ones we have considered so far.

Intermolecular reactions occur between two molecules.
Intramolecular reactions occur within the same molecule. We shall discuss the reasons why intramolecular reactions are more favourable and why cyclic hemiacetals and acetals are more stable in Chapters 11 and 12.

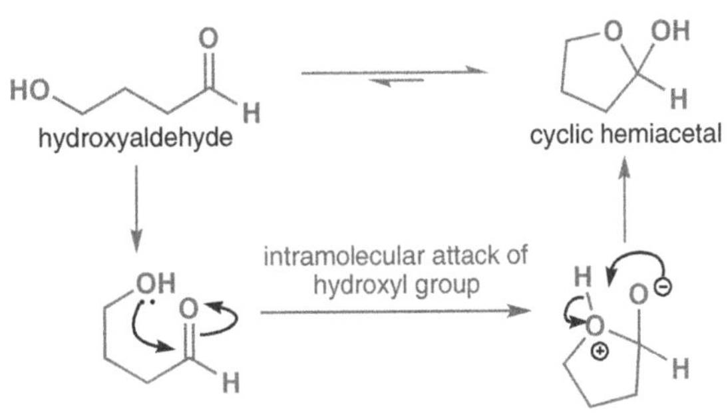

Although the cyclic hemiacetal (also called lactol) product is more stable, it is still in equilibrium with some of the open-chain hydroxyaldehyde form. Its stability, and how easily it forms, depends on the size of the ring: five- and six-membered rings are free from strain (their bonds are free to adopt 109° or 120° angles—compare the three-membered rings on p. 135), and five- or six-membered hemiacetals are common. Among the most important examples are many sugars. Glucose, for example, is a hydroxyaldehyde that exists mainly as a six-membered cyclic hemiacetal (>99% of glucose is cyclic in solution), while ribose exists as a five-membered cyclic hemiacetal.

can be drawn as

hydroxyaldehyde   hydroxyaldehyde   cyclic glucose: >99% in this form

can be drawn as

hydroxyaldehyde   hydroxyaldehyde   cyclic ribose

■ The way we have represented some of these molecules may be unfamiliar to you, although we first mentioned it in Chapter 2: we have shown **stereochemistry** (whether bonds come out of the paper or into it—the wiggly lines indicate a mixture of both) and, for the cyclic glucose, **conformation** (the actual shape the molecules adopt). These are very important in the sugars: we devote Chapter 14 to stereochemistry and Chapter 16 to conformation.

Ketones also form hemiacetals

Hydroxyketones can also form hemiacetals but, as you should expect, they usually do so less readily than hydroxyaldehydes. But we know that this hydroxyketone must exist as the cyclic hemiacetal as it has no C=O stretch in its IR spectrum. The reason? The hydroxyketone is already cyclic, with the OH group poised to attack the ketone—it can't get away so cyclization is highly favoured.

hydroxyketone   hemiacetal

Acid and base catalysis of hemiacetal and hydrate formation

In Chapter 8 we shall look in detail at acids and bases, but at this point we need to tell you about one of their important roles in chemistry: they act as catalysts for a number of carbonyl addition reactions, among them hemiacetal and hydrate formation. To see why, we need to look back at the mechanisms of hemiacetal formation on p. 138 and hydrate formation on p. 134. Both involve proton-transfer steps, which we can choose to draw like this:

ethanol acting as a base   ethanol acting as an acid

In the first proton-transfer step, ethanol acts as a **base**, removing a proton; in the second it acts as an **acid**, donating a proton. You saw in Chapter 5 how water can also act as an acid or a base. Strong acids or strong bases (for example HCl or NaOH) increase the rate of hemiacetal or hydrate formation because they allow these proton-transfer steps to occur *before* the addition to the carbonyl group.

In acid (dilute HCl, say), the mechanism is different in detail. The first step is now protonation of the carbonyl group's lone pair: the positive charge makes it much more electrophilic so the addition reaction is faster. Notice how the proton added at the beginning is lost again at the end—it is really a catalyst.

■ In acid it is also possible for the hemiacetal to react further with the alcohol to form an acetal, but this is dealt with in Chapter 11 and need not concern you at present.

hemiacetal formation in acid

protonation makes carbonyl group more electrophilic

in acid solution, the acetal may form (reactions discussed in Chapter 11)

hemiacetal

acetal

proton regenerated

The mechanism in basic solution is slightly different again. The first step is now deprotonation of the ethanol by hydroxide, which makes the addition reaction faster by making the ethanol more nucleophilic. Again, base (hydroxide) is regenerated in the last step, making the overall reaction catalytic in base.

■ As you will see in Chapter 11, the reaction in base always stops with the hemiacetal—acetals never form in base.

hemiacetal formation in base

deprotonation makes ethanol more nucleophilic (as ethoxide)

base regenerated

acetals are never formed in base

The final step could equally well involve deprotonation of ethanol to give alkoxide—and alkoxide could equally well do the job of catalysing the reaction. In fact, you will often come across mechanisms with the base represented just as 'B⁻' because it doesn't matter what the base is.

● For nucleophilic additions to carbonyl groups:

  • acid catalysts work by making the carbonyl group more electrophilic
  • base catalysts work by making the nucleophile more nucleophilic
  • both types of catalysts are regenerated at the end of the reaction.

Bisulfite addition compounds

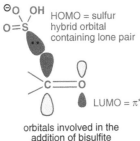

HOMO = sulfur hybrid orbital containing lone pair

LUMO = π^*

orbitals involved in the addition of bisulfite

The last nucleophile of this chapter, sodium bisulfite ($NaHSO_3$) adds to aldehydes and some ketones to give what is usually known as a **bisulfite addition compound**. The reaction occurs by nucleophilic attack of a lone pair on the carbonyl group, just like the attack of cyanide. This leaves a positively charged sulfur atom but a simple proton transfer leads to the product.

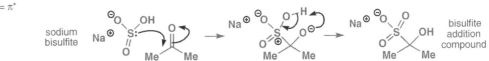

sodium bisulfite

bisulfite addition compound

The products are useful for two reasons. They are usually crystalline and so can be used to purify liquid aldehydes by recrystallization. This is of value only because this reaction, like several you have met in this chapter, is reversible. The bisulfite compounds are made by mixing the aldehyde or ketone with saturated aqueous sodium bisulfite in an ice bath, shaking, and crystallizing. After purification the bisulfite addition compound can be hydrolysed back to the aldehyde in dilute aqueous acid or base.

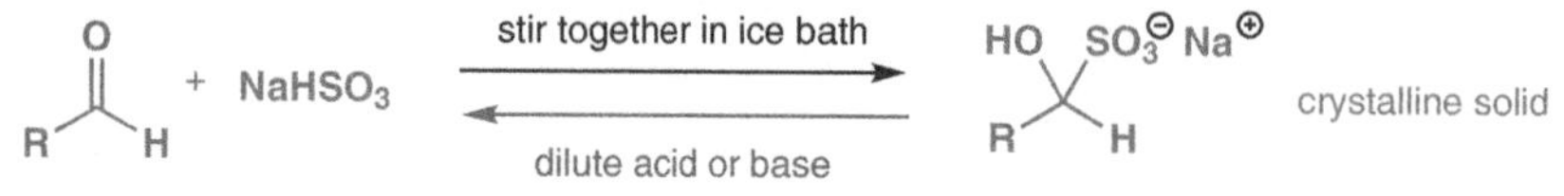

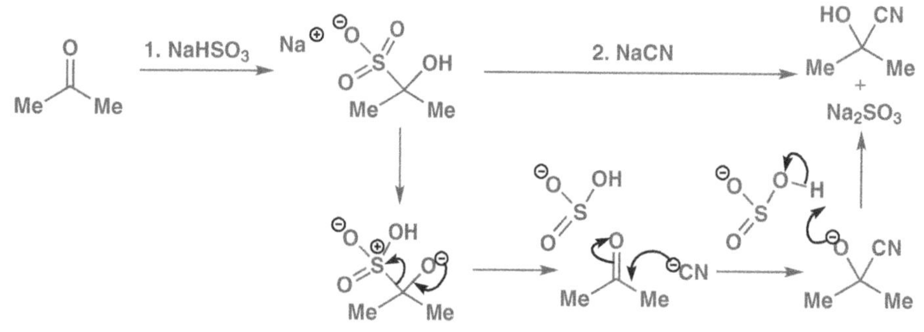

■ The structure of NaHSO₃, sodium bisulfite, is rather curious. It is an oxyanion of a sulfur(IV) compound with a lone pair of electrons—the HOMO—on the sulfur atom, but the charge is formally on the more electronegative oxygen. As a 'second-row' element (second row of the periodic table, that is) sulfur can have more than just eight electrons—it's all right to have four, five, or six bonds to S or P, unlike, say, B or O. Second-row elements have d orbitals as well as s and p so they can accommodate more electrons.

The reversibility of the reaction makes bisulfite compounds useful intermediates in the synthesis of other adducts from aldehydes and ketones. For example, one practical method for making cyanohydrins involves bisulfite compounds. The famous practical book 'Vogel' suggests reacting acetone first with sodium bisulfite and then with sodium cyanide to give a good yield (70%) of the cyanohydrin.

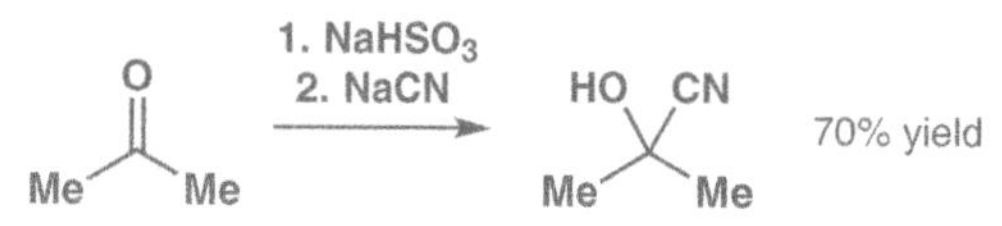

What is happening here? The bisulfite compound forms first, but only as an intermediate on the route to the cyanohydrin. When the cyanide is added, reversing the formation of the bisulfite compound provides the single proton necessary to give back the hydroxyl group at the end of the reaction. No dangerous HCN is released (always a hazard when cyanide ions and acid are present together).

Other compounds from cyanohydrins

Cyanohydrins can be converted by simple reactions into hydroxyacids or amino alcohols. Here is one example of each, but you will have to wait until Chapter 10 for the details and the mechanisms of the reactions. Note that one cyanohydrin was made by the simplest method—simply NaCN and acid—while the other came from the bisulfite route we have just discussed.

hydroxyacids by hydrolysis of CN in cyanohydrin

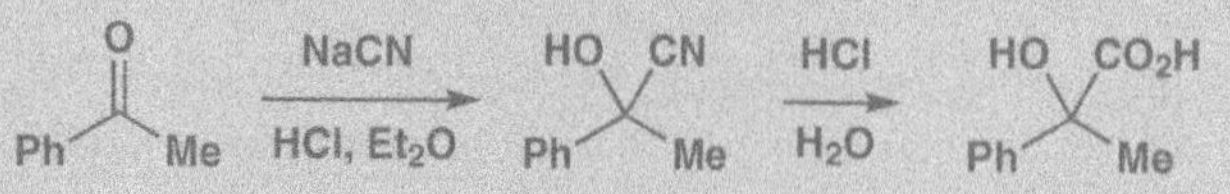

amino alcohols by reduction of CN in cyanohydrin

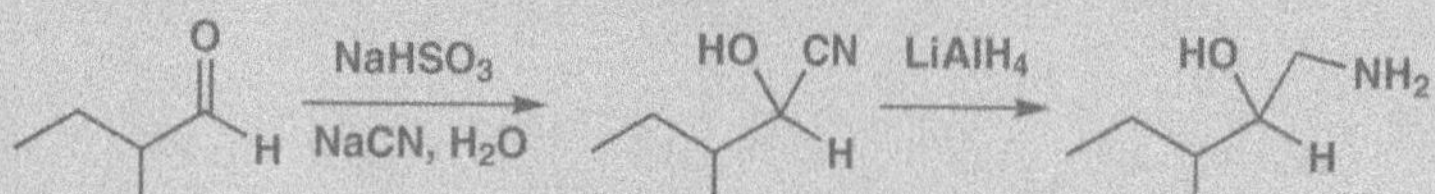

The second reason that bisulfite compounds are useful is that they are soluble in water. Some small (that is, low molecular weight) aldehydes and ketones are water-soluble—acetone is an example. But most larger (more than four or so carbon atoms) aldehydes and ketones are not.

This does not usually matter to most chemists as we often want to carry out reactions in organic solvents rather than water. But it can matter to medicinal chemists, who make compounds that need to be compatible with biological systems. And in one case, the solubility of bisulfite adduct in water is literally vital.

Dapsone is an antileprosy drug. It is a very effective one too, especially when used in combination with two other drugs in a 'cocktail' that can be simply drunk as an aqueous solution by patients in tropical countries without any special facilities, even in the open air. But there is a problem! Dapsone is insoluble in water. The solution is to make a bisulfite compound from it. You may ask how this is possible since dapsone has no aldehyde or ketone—just two amino groups and a sulfone. The trick is to use the formaldehyde bisulfite compound and exchange the OH group for one of the amino groups in dapsone.

dapsone: antileprosy drug; insoluble in water   water-soluble pro-drug

Now the compound will dissolve in water and release dapsone inside the patient. The details of this sort of chemistry will come in Chapter 11, when you will meet imines as intermediates. But at this stage we just want you to appreciate that even the relatively simple chemistry in this chapter is useful in synthesis, in commerce, and in medicine.

Further reading

Section 1, 'Nucleophilic addition to the carbonyl group' in S. Warren, *Chemistry of the Carbonyl Group*, Wiley, Chichester, 1974, and P. Sykes, *A Guidebook to Mechanism in Organic Chemistry*, 6th edn, Longman, Harlow, 1986, pp. 203–219. For a more theoretical approach, we suggest J. Keeler and P. Wothers, *Why Chemical Reactions Happen*, OUP, Oxford, 2003, especially pp. 102–106.

For further, more advanced, details of the cassava–HCN problem: D. Siritunga, D. Arias-Garzon, W. White, and R. T. Sayre, *Plant Biotechnology Journal*, 2004, **2**, 37. For details of cyanohydrin formation using sodium bisulfite: B. S. Furniss, A. J. Hannaford, P. W. G. Smith, and A. T. Tatchell, *Vogel's Textbook of Practical Organic Chemistry*, 5th edn, Longman, Harlow, 1989, pp. 729–730.

Check your understanding

To check that you have mastered the concepts presented in this chapter, attempt the problems that are available in the book's Online Resource Centre at **http://www.oxfordtextbooks.co.uk/orc/clayden2e/**

Suggested solutions for Chapter 6

PROBLEM 1

Draw mechanisms for these reactions:

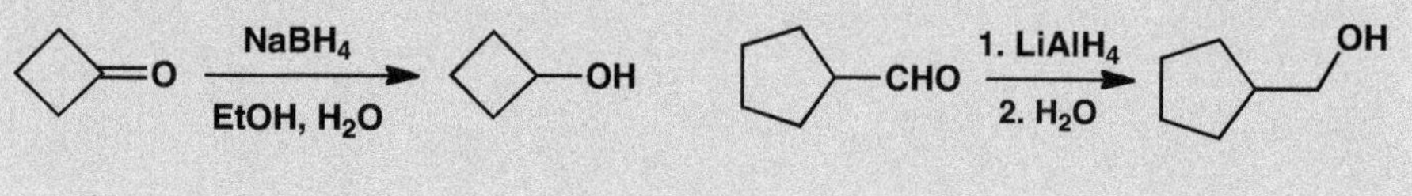

Purpose of the problem

Rehearsal of a simple but important mechanism that works for all aldehydes and ketones.

Suggested solution

Draw out the BH_4 and AlH_4 anions, with the carbonyl compound positioned so that one of the hydrogens can be transferred to the carbonyl group, and then transfer the hydrogen from B or Al to C. A proton transfer is needed to make the alcohol: from the solvent in the first case and during the work-up with water in the second.

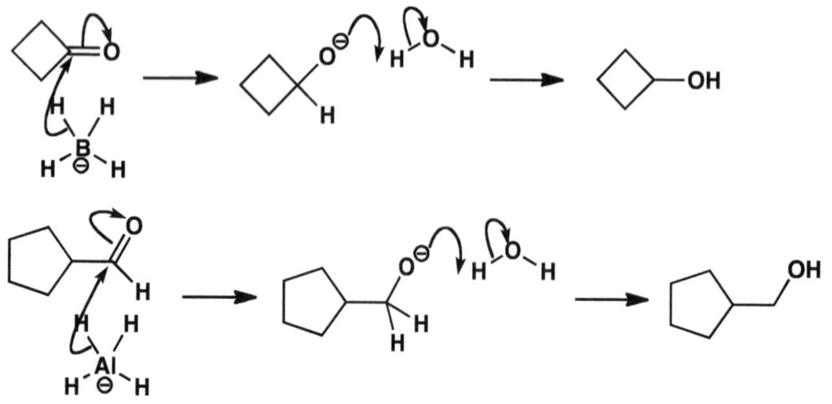

■ This reaction shows that you *can* reduce aldehydes with lithium aluminium hydride, even if you would usually prefer the more practical sodium borohydride.

PROBLEM 2

Cyclopropanone exists as the hydrate in water but 2-hydroxyethanal does not exist as the hemiacetal. Explain.

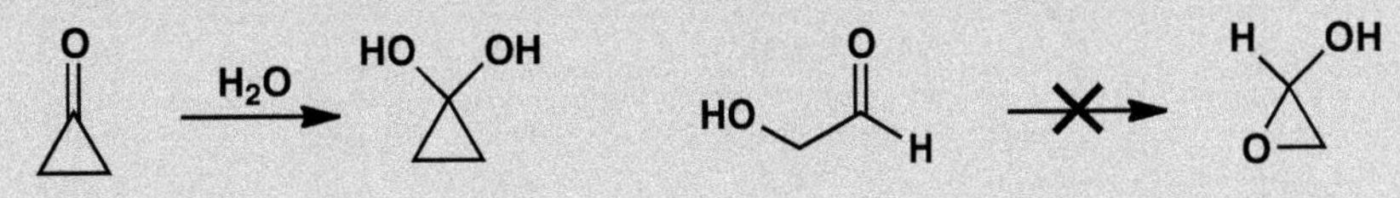

Purpose of the problem

To get you thinking about equilibria and hence the stability of compounds.

Suggested solution

Hydration is an equilibrium reaction so the mechanism is not strictly relevant to the question, though there is no shame in including mechanisms whenever you can. To answer the question we must consider the effect of the three-membered ring on the relative stability of starting material and product. All three-membered rings are very strained because the bond angles are 60° instead of 109° or 120°. Cyclopropanone is particularly strained because the sp^2 carbonyl carbon would like a bond angle of 120°—there is '60° of strain.' In the hydrate that carbon atom is sp^3 hybridized and so there is only about '49° of strain.' Not much gain, but the hydrate is more stable than the ketone.

The second case is totally different. The hydroxy-aldehyde is not strained at all but the hemiacetal has '49° of strain' at each atom. Even without strain, hydrates and hemiacetals are usually less stable than their aldehydes or ketones because one C=O bond is worth more than two C–O bonds. In this case the hemiacetal is even less stable and, unlike the cyclopropanone, can escape strain by breaking a C–O ring bond.

PROBLEM 3

One way to make cyanohydrins is illustrated here. Suggest a detailed mechanism for the process.

Purpose of the problem

To help you get used to mechanisms involving silicon and revise an important way to promote additions to the carbonyl group.

Suggested solution

The silyl cyanide is an electrophile while the cyanide ion in the catalyst is the nucleophile. Cyanide adds to the carbonyl group and the oxyanion product is captured by silicon, liberating another cyanide ion for the next cycle.

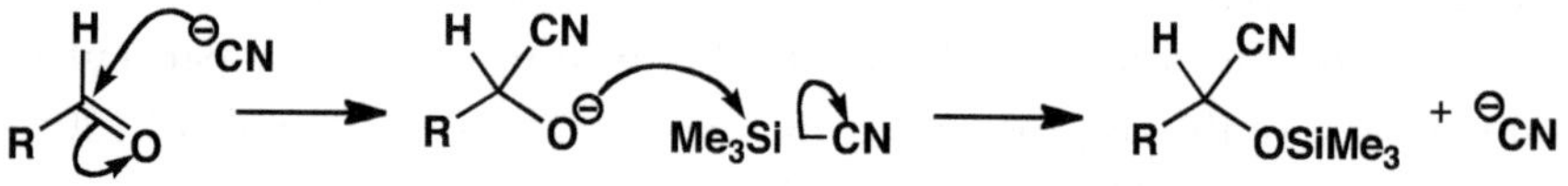

PROBLEM 4

There are three possible products from the reduction of this compound with sodium borohydride. What are their structures? How would you distinguish them spectroscopically, assuming you can isolate pure compounds?

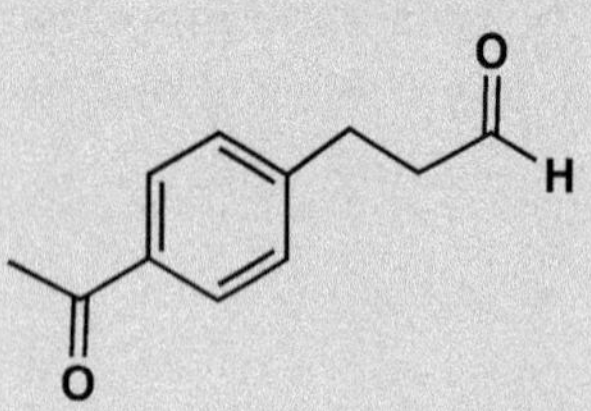

Purpose of the problem

To let you think practically about reactions that may give more than one product.

Suggested solution

The three compounds are easily drawn: one or other carbonyl group, or both, may be reduced.

■ Calculations from D. H. Williams and Ian Fleming (2007), *Spectroscopic methods in organic chemistry* (6th edn) McGraw Hill, London, 2007 suggest about 80 ppm for the C–OH carbon in the ketone and about 60 ppm for the aldehyde.

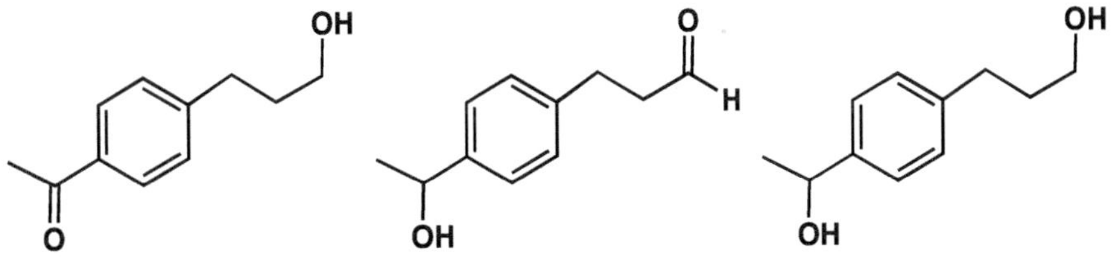

The third compound, the diol, has no carbonyl group in the ^{13}C NMR spectrum or the infrared and has a molecular ion two mass units higher than the other two products. Distinguishing those is more tricky, and needs techniques you will meet in detail in chapter 18. The hydroxyketone has a conjugated carbonyl group (C=O stretch at about 1680 cm^{-1} in the infrared spectrum) while the hydroxyaldehyde is not conjugated (C=O stretch at about 1730 cm^{-1} in the infrared). The chemical shift of the C–OH carbons will also be different because the benzene ring is joined to this carbon in the aldehyde but not in the ketone.

PROBLEM 5

The triketone shown here is called 'ninhydrin' and is used for the detection of amino acids. It exists in aqueous solution as a hydrate. Which ketone is hydrated and why?

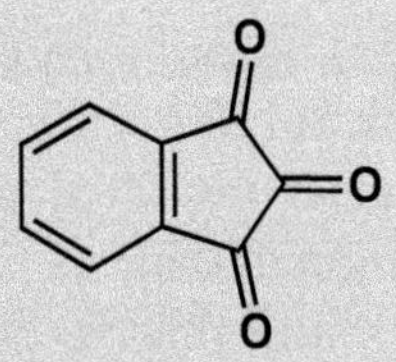

Purpose of the problem

To let you think practically about reactions that may give more than one product.

Suggested solution

The two ketones next to the benzene ring are stabilized by conjugation with it but also destabilized by the central ketone—two electron-withdrawing groups next to each other is a bad thing. The central carbonyl group is not stabilized by conjugation and is destabilized by *two* other ketones so it forms the hydrate. Did you remember that hydrate formation is thermodynamically controlled?

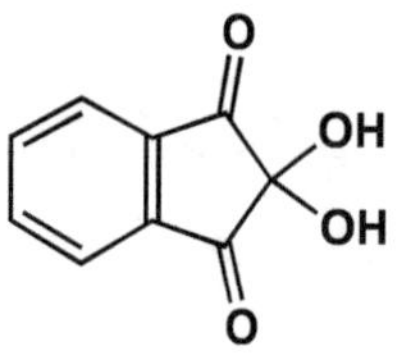

PROBLEM 6

This hydroxyketone shows no peaks in its infrared spectrum between 1600 and 1800 cm^{-1}, but it does show a broad absorption at 3000–3400 cm^{-1}. In the ^{13}C NMR spectrum there are no peaks above 150 ppm but there is a peak at 110 ppm. Suggest an explanation.

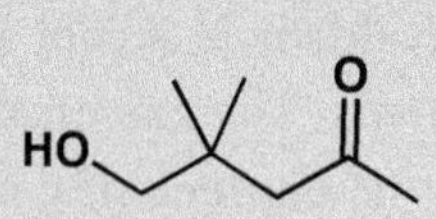

Purpose of the problem

Structure determination to solve a conundrum.

Suggested solution

The evidence shows that there is no carbonyl group in the molecule but that there is an OH group. The peak at 110 ppm looks at first sight like an alkene, but it could also be an unusual saturated carbon atom bonded to two oxygens. You might have argued that an alcohol and a ketone could combine to give a hemiacetal, and that is, of course, just what it is. The compound exists as a stable hemiacetal because it has a favourable five-membered ring.

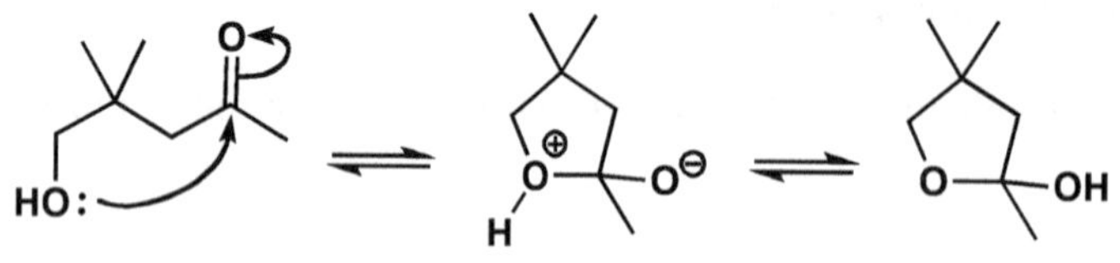

■ P. 136 of the textbook explains why cyclic hemiacetals are stable.

PROBLEM 7

Each of these compounds is a hemiacetal and therefore formed from an alcohol and a carbonyl compound. In each case give the structures of the original materials.

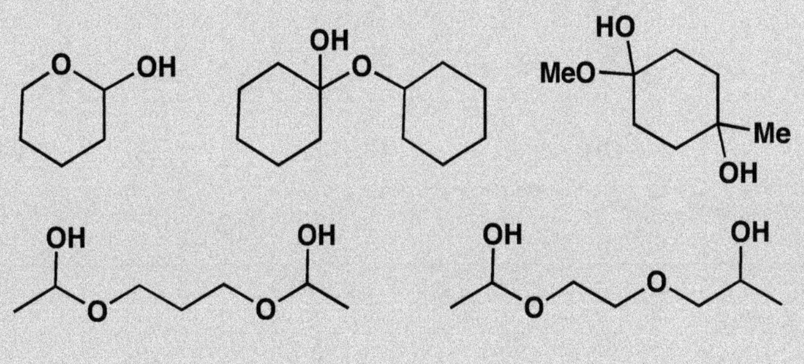

Purpose of the problem

To give you practice in seeing the underlying structure of a hemiacetal.

Suggested solution

Each OH group represents a carbonyl group in disguise (marked with a grey circle). Just break the bond between this carbon and the other oxygen atom and you will see what the hemiacetal was made from. The first example shows how it is done.

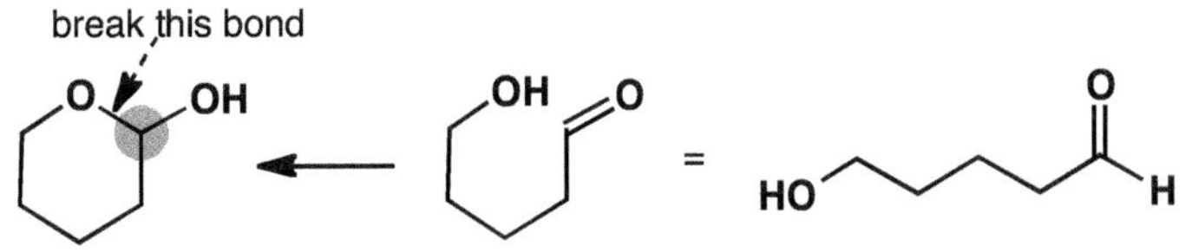

The next is similar but the alcohol is a different molecule.

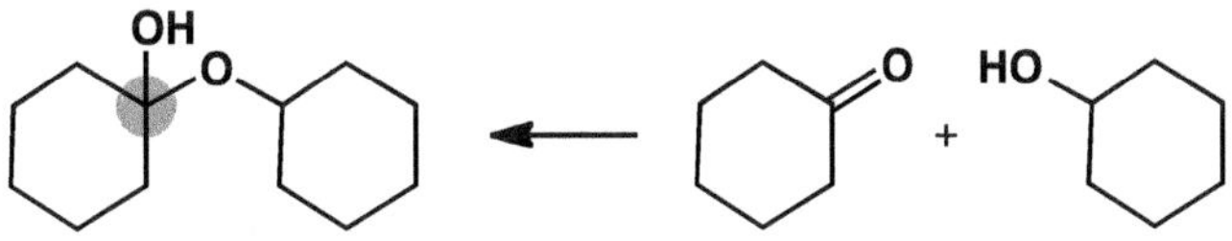

Do not be deceived by the third example. There is one hemiacetal (two oxygens joined to the same carbon atom) but the other OH is just a tertiary alcohol.

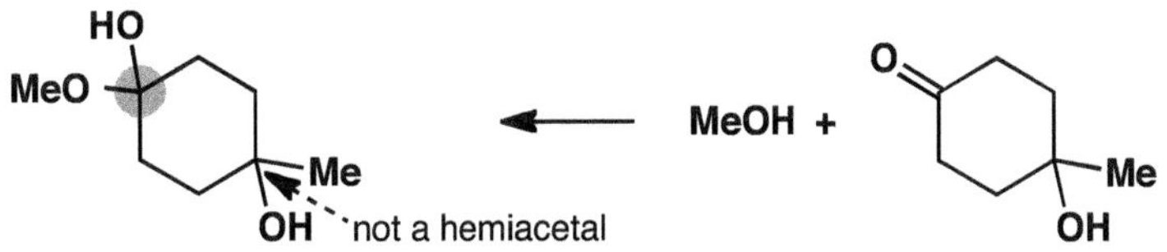

The last two examples are not quite the same. The first is indeed symmetrical but the second has one oxygen atom in a different position so that there is only one hemiacetal. Note that these hemiacetals may not be stable.

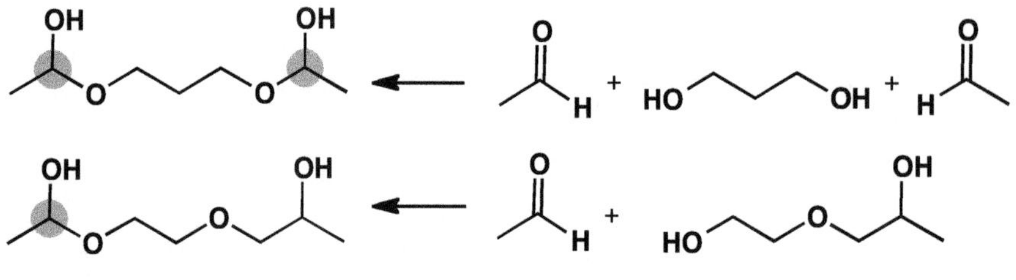

PROBLEM 8

Trichloroethanol my be prepared by the direct reduction of chloral hydrate in water with sodium borohydride. Suggest a mechanism for this reaction. Take note that sodium borohydride does not displace hydroxide from carbon atoms!

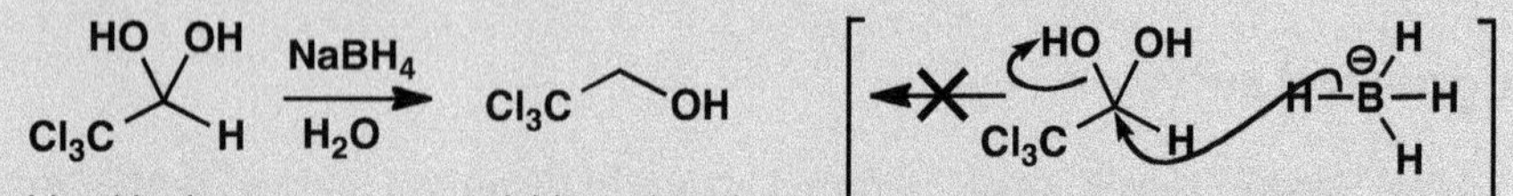

chloral hydrate   trichloroethanol   *this is not the mechanism*

Purpose of the problem

To help you detect bad mechanisms and find concealed good ones.

Suggested solution

If sodium borohydride doesn't displace hydroxide from carbon atoms, then what does it do? We know it attacks carbonyl groups to give alcohols and to get trichloroethanol we should have to reduce chloral. Hemiacetals are in equilibrium with their carbonyl equivalents, so...

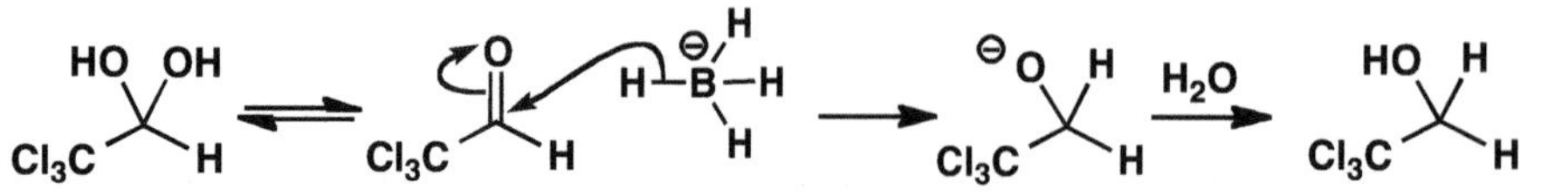

PROBLEM 9

It has not been possible to prepare the adducts from simple aldehydes and HCl. What would be the structure of such compounds, if they could be made, and what would be the mechanism of their formation? Why can't these compounds be made?

Purpose of the problem

More revision of equilibria to help you develop a judgement on stability.

Suggested solution

This time we need a mechanism so that we can work out what would be formed. Protonation of the carbonyl group and then nucleophilic addition of chloride ion would give the supposed products.

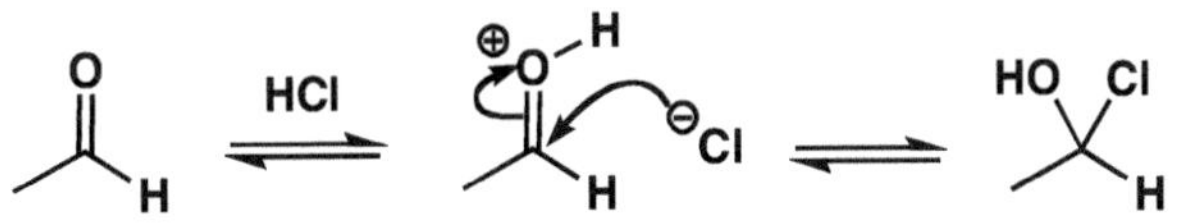

There's nothing wrong with the mechanism, it's just that the reaction is an equilibrium that will run backwards. Hemiacetals are unstable because they decompose back to carbonyl compounds. Chloride ion is very stable and decomposition will be faster than it is for hemiacetals.

PROBLEM 10

What would be the products of these reactions? In each case give a mechanism to justify your prediction.

Purpose of the problem

Giving you practice in the art of predicting products—more difficult than simply justifying a known answer.

Suggested solution

The Grignard reagent will add to the carbonyl group and the work-up will give a tertiary alcohol as the final product.

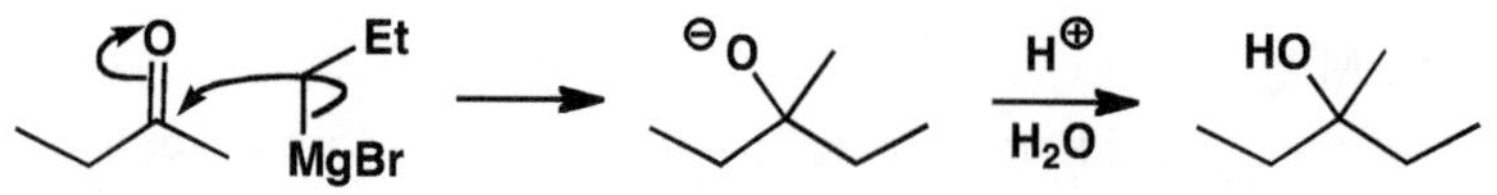

The second reaction should give you brief pause for thought as you need to recall that borohydride reduces ketones but not esters.

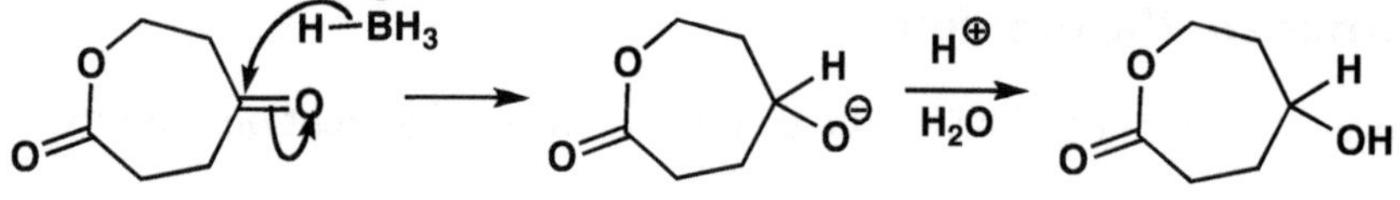

Delocalization and conjugation

7

Connections

➡ **Building on**

- Orbitals and bonding ch4
- Representing mechanisms by curly arrows ch5
- Ascertaining molecular structure spectroscopically ch3

Arriving at

- Interaction between orbitals over many bonds
- Stabilization by the sharing of electrons over more than two atoms
- Where colour comes from
- Molecular shape and structure determine reactivity
- Representing one aspect of structure by curly arrows
- Structure of aromatic compounds

➡ **Looking forward to**

- Acidity and basicity ch8
- How conjugation affects reactivity ch10, ch11, & ch15
- Conjugate addition and substitution ch22
- Chemistry of aromatic compounds ch21 & ch22
- Enols and enolates ch20, ch24–ch27
- Chemistry of heterocycles ch29 & ch30
- Chemistry of dienes and polyenes ch34 & ch35
- Chemistry of life ch42

Introduction

As you look around you, you will be aware of many different colours—from the greens and browns outside to the bright blues and reds of the clothes you are wearing. All these colours result from the interaction of light with the pigments in these different things—some frequencies of light are absorbed, others scattered. Inside our eyes, chemical reactions detect these different frequencies and convert them into electrical nerve impulses sent to the brain. All these pigments have one thing in common—lots of double bonds. For example, the pigment responsible for the red colour in tomatoes, lycopene, is a long-chain polyalkene.

lycopene, the red pigment in tomatoes, rose hips, and other berries

Lycopene contains only carbon and hydrogen; many pigments contain other elements. But nearly all contain double bonds—and many of them. This chapter is about the properties, including colour, of molecules that have several double bonds. These properties depend on the way the double bonds join up, or *conjugate*, and the resulting *delocalization* of the electrons within them.

In earlier chapters we talked about carbon skeletons made up of σ bonds. In this chapter we shall see how, in some cases, we can also have a large π framework spread over many atoms and how this dominates the chemistry of such compounds. We shall see how this π framework is responsible for the otherwise unexpected *stability* of certain cyclic polyunsaturated

Online support. The icon ✥ in the margin indicates that accompanying interactive resources are provided online to help your understanding: just type **www.chemtube3d.com/clayden/123** into your browser, replacing **123** with the number of the page where you see the icon. For pages linking to more than one resource, type **123-1, 123-2** etc. (replacing **123** with the page number) for access to successive links.

109

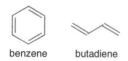

benzene   butadiene

compounds, including benzene, but also *reactivity* in others, such as butadiene. We shall also see how this π framework gives rise to colour. To understand such molecules properly, we need to start with the simplest of all unsaturated compounds, ethene.

The structure of ethene (ethylene, CH₂=CH₂)

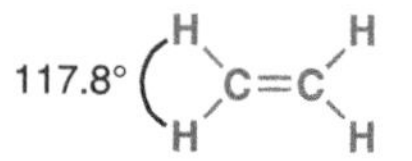

The structure of ethene (ethylene) is well known. It has been determined by electron diffraction and is **planar** *(all atoms are in the same plane),* with the bond lengths and angles shown on the left. The carbon atoms are roughly trigonal and the C=C bond distance is shorter than that of a typical C–C single bond. The electronic structure of ethene, you will recall from Chapter 4, can be considered in terms of two sp^2 hybridized C atoms with a σ bond between them and four σ bonds linking them each to two H atoms. The π bond is formed by overlap of a p orbital on each carbon atom.

117.8°

C–H bond length 108 pm
C=C bond length 133 pm

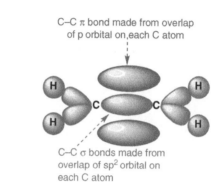

C–C π bond made from overlap of p orbital on each C atom

C–C σ bonds made from overlap of sp^2 orbital on each C atom

🖱 Interactive bonding orbitals in ethene

Ethene is chemically more interesting than *ethane* because of the π system. As you saw in Chapter 5, alkenes can be nucleophiles because the electrons in the π bond are available for donation to an electrophile. But remember that when we combine *two* atomic orbitals we get *two* molecular orbitals, from combining the p orbitals either in phase or out of phase. The in-phase combination accounts for the bonding molecular orbital (π), whilst the out-of-phase combination accounts for the antibonding molecular orbital (π*). The shapes of the orbitals as they were introduced in Chapter 4 are shown below, but in this chapter we will also represent them in the form shown in the brown boxes—as the constituent p orbitals.

➡ We described the structure of ethene in Chapter 4 (p. 101)

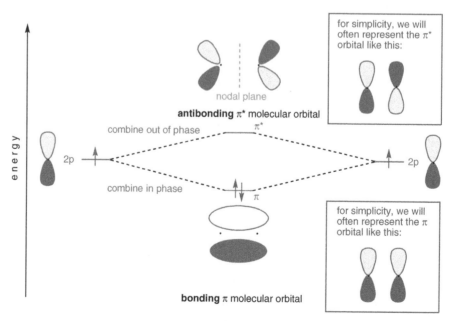

nodal plane

antibonding π* molecular orbital

for simplicity, we will often represent the π* orbital like this:

combine out of phase

π*

2p

energy

combine in phase

π

for simplicity, we will often represent the π orbital like this:

bonding π molecular orbital

Molecules with more than one C=C double bond

Benzene has three strongly interacting double bonds

The rest of this chapter concerns molecules with more than one C=C double bond and what happens to the π orbitals when they interact. To start, we shall take a bit of a jump and look at the structure of benzene. Benzene has been the subject of considerable controversy since its discovery in 1825. It was soon worked out that the formula was C_6H_6, but how were these atoms arranged? Some strange structures were suggested until Kekulé proposed the correct structure in 1865.

Shown below are the molecular orbitals for Kekulé's structure. As in simple alkenes, each of the carbon atoms is sp[2] hybridized, leaving the remaining p orbital free.

> The two early proposals for the structure of benzene were wrong, but nonetheless are stable isomers of benzene (they are both C_6H_6) which have since been synthesized. For more on the Kekulé structure, see p. 24.
>
> prismane
> synthesized
> 1973
>
> Dewar
> benzene
> synthesized
> 1963

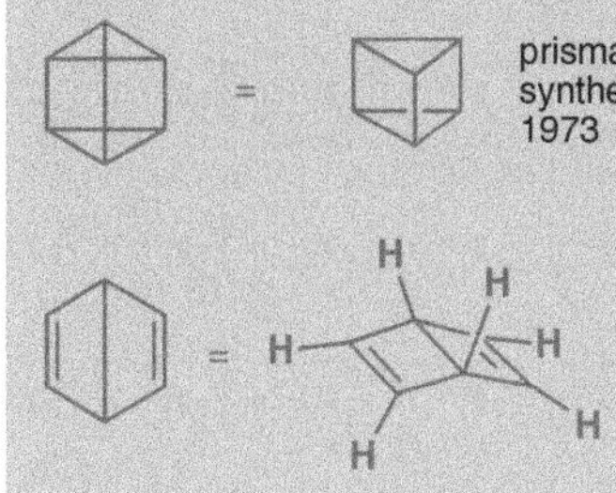

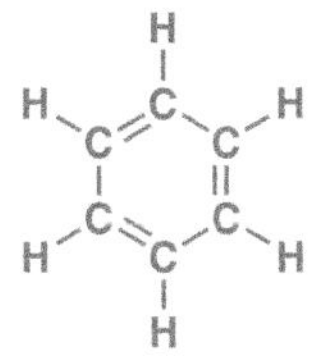

Kekulé's structure for benzene

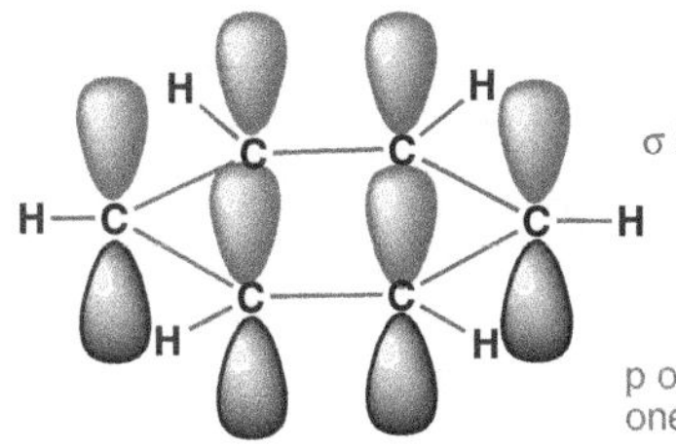

σ bonds shown in green

p orbitals shown with
one phase red, **one phase black**

The σ framework of the benzene ring is like the framework of an alkene, and for simplicity we have just represented the σ bonds as green lines. The difficulty comes with the p orbitals—which pairs do we combine to form the π bonds? There seem to be two possibilities.

combining different pairs of p orbitals
puts the double bonds in different positions

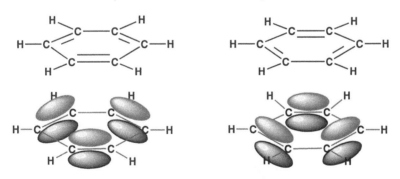

With benzene itself, these two forms are identical but, if we had a 1,2- or a 1,3-disubstituted benzene compound, these two forms would be different. A synthesis was designed for the two compounds in the box on the right but it was found that both compounds were identical. This posed a problem to Kekulé—his structure didn't seem to work after all. His solution—which we now know to be incorrect—was that benzene rapidly equilibrates, or 'resonates', between the two forms to give an averaged structure in between the two.

The molecular orbital answer to this problem is that all six p orbitals can combine to form (six) new molecular orbitals, and the electrons in these orbitals form a ring of electron density above and below the plane of the molecule. Benzene *does not resonate* between the two Kekulé structures—the electrons are in molecular orbitals spread equally over all the carbon atoms. However, the term 'resonance' is still sometimes used (but not in this book) to describe the averaging effect of this mixing of molecular orbitals. We shall describe the π electrons in benzene as **delocalized**, that is, no longer localized in specific double bonds between two particular carbon atoms but spread out, or delocalized, over all six atoms in the ring.

> ■ For example, if the double bonds were localized then these two compounds would be chemically different. (The double bonds are drawn shorter than the single bonds to emphasize the difference.)
>
> in reality these are
> the same compound
>
> CO_2H CO_2H
>
> 2-bromo   '6'-bromo
> benzoic acid   benzoic acid

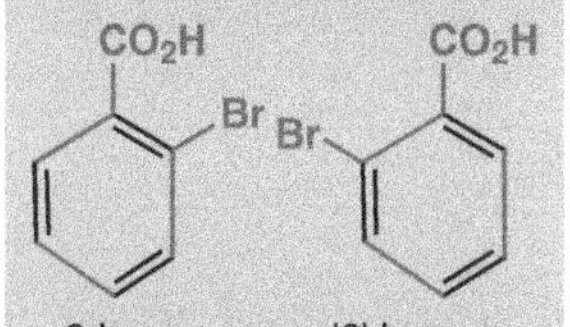

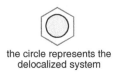

the circle represents the
delocalized system

The alternative drawing on the left shows the π system as a ring and does not put in the double bonds: you may feel that this is a more accurate representation, but it does present a problem when it comes to writing mechanisms. As you saw in Chapter 5, the curly arrows we use represent two electrons. The circle here represents *six* electrons, so in order to write reasonable mechanisms we still need to draw benzene *as though* the double bonds were localized. However, when you do so, you must keep in mind that the electrons are delocalized, and it does not matter which of the two arrangements of double bonds you draw.

If we want to represent delocalization using these 'localized' structures, we can do so using curly arrows. Here, for example, are the two 'localized' structures corresponding to 2-bromo-carboxylic acid. The double bonds are not localized, and the relationship between the two structures can be represented with curly arrows which indicate how one set of bonds map onto the other.

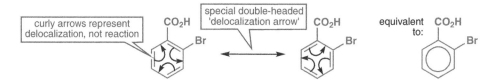

These curly arrows are similar to the ones we introduced in Chapter 5, but there is a crucial difference: here, there is no reaction taking place. In a real reaction, electrons move. Here, they do not: the only things that 'move' are the double bonds in the structures. The curly arrows just show the link between alternative representations of exactly the same molecule. You must not think of them as showing 'movement round the ring'. To emphasize this difference we also use a different type of arrow connecting them—a delocalization arrow made up of a single line with an arrow at each end. Delocalization arrows remind us that our simple fixed-bond structures do not tell the whole truth and that the real structure is a mixture of both.

The fact that *the π electrons are not localized* in alternating double bonds but are *spread out over the whole system* in a ring is supported by theoretical calculations and confirmed by experimental observations. Electron diffraction studies show benzene to be a regular planar hexagon with all the carbon–carbon bond lengths identical (139.5 pm). This bond length is in between that of a carbon–carbon single bond (154.1 pm) and a full carbon–carbon double bond (133.7 pm). A further strong piece of evidence for this ring of electrons is revealed by proton NMR and discussed in Chapter 13.

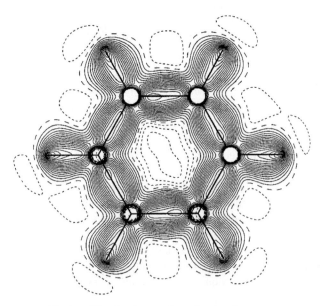

Electron diffraction image of a molecule of benzene

How to describe delocalization?

What words should be used to describe delocalization is a vexed question. Terms such as *resonance, mesomerism, conjugation,* and *delocalization* are only a few of the ones you will find in books. You will already have noticed that we're avoiding *resonance* because it carries a suggestion that the structure is somehow oscillating between localized structures. We shall use the words *conjugation* and *delocalization*: conjugation focuses on the way that double bonds link together into a single π system, while delocalization focuses on the electrons themselves. Adjacent double bonds, as you will see, are *conjugated*; the electrons in them are *delocalized*.

Multiple double bonds not in a ring

Are electrons still delocalized even when there is no ring? To consider this, we'll look at hexatriene—three double bonds and six carbons, like benzene, but without the ring. There are two isomers of hexatriene, with different chemical and physical properties, because the central double bond can adopt a *cis* or a *trans* geometry. The structures of both *cis-* and *trans*-hexatriene have been determined by electron diffraction and two important features emerge:

cis-hexatriene

trans-hexatriene

- Both structures are essentially planar.
- Unlike benzene, the double and single bonds have different lengths, but the central double bond in each case is slightly longer than the end double bonds and the single bonds are slightly shorter than a 'standard' single bond.

■ The terminal double bonds can't have two forms because they have only one substituent.

Here's the most stable structure of *trans*-hexatriene, with benzene shown for comparison.

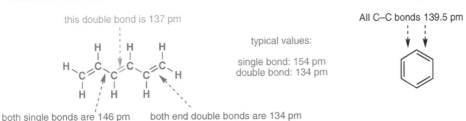

this double bond is 137 pm

typical values:

single bond: 154 pm
double bond: 134 pm

All C–C bonds 139.5 pm

both single bonds are 146 pm   both end double bonds are 134 pm

The reason for the deviation of the bond lengths from typical values and the preference for planar structures is again due to the molecular orbitals which arise from the combination of the six p orbitals. Just as in benzene, these orbitals can combine to give one molecular orbital stretching over the whole molecule. The p orbitals can overlap and combine only if the molecule is planar.

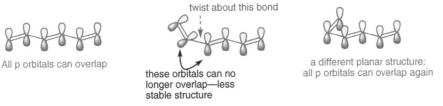

All p orbitals can overlap

twist about this bond

these orbitals can no longer overlap—less stable structure

a different planar structure: all p orbitals can overlap again

If the molecule is twisted about one of the single bonds, then some overlap is lost, making it harder to twist about the single bonds in this structure than in a simple alkene. Other planar arrangements are stable, however, and *trans*-hexatriene can adopt any of the planar conformations shown in the margin.

conformations of *trans*-hexatriene

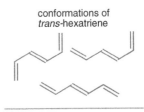

➡ Conformation is the topic of Chapter 16.

Conjugation

In benzene and hexatriene every carbon atom is sp² hybridized with one p orbital available to overlap with its neighbours. The uninterrupted chain of p orbitals is a consequence of having alternate double and single bonds. When two double bonds are separated by just one single bond, the two double bonds are said to be *conjugated*. Conjugated double bonds have different properties from isolated double bonds, both physically (they are often longer, as you have just seen) and chemically (see Chapters 22).

You have already met several conjugated systems: lycopene at the start of this chapter and β-carotene in Chapter 3, for example. Each of the 11 double bonds in β-carotene is separated

Conjugation

In the dictionary 'conjugated' is defined, among other ways, as 'joined together, especially in pairs' and 'acting or operating as if joined'. This does indeed fit very well with the behaviour of such conjugated double bonds, since the properties of a conjugated system are different from those of the component parts.

from its neighbour by only one single bond. We again have a long chain in which all the p orbitals can overlap to form molecular orbitals.

propenal (acrolein):
C=C and C=O are
conjugated

β-carotene—all eleven double bonds are conjugated

It is not necessary to have two C=O double bonds in order to have a conjugated system—the C=C and C=O double bonds of propenal (acrolein) are also conjugated. What is important is that the double bonds are separated by *one and only one* single bond. Here's a counter-example: arachidonic acid is one of the fabled 'polyunsaturated' fatty acids. None of the four double bonds in this structure are conjugated since in between any two double bonds there is an sp^3 carbon. This means that there is no p orbital available to overlap with the ones from the double bonds. The saturated carbon atoms 'insulate' the double bonds from each other and prevent conjugation.

> ■ The chemistry of such conjugated carbonyl compounds is significantly different from the chemistry of the component parts. The alkene in propenal, for example, is electrophilic and not nucleophilic. This will be explained in Chapter 22.

these four double bonds are not conjugated—
they are all separated by *two* single bonds

arachidonic acid

these tetrahedral (sp^3) carbon atoms prevent
overlap of the p orbitals in the double bonds

If an atom has two double bonds directly attached to it, that is, there are no single bonds separating them, again no conjugation is possible. The simplest compound with such an arrangement is allene. The arrangement of the p orbitals in allene means that no delocalization is possible because the two π bonds are perpendicular to each other.

$H_2C=C=CH_2$

allene

end carbons are
sp^2 hybridized

central carbon is sp hybridized

the π bonds formed as a result
of the overlap of the individual p orbitals (shown
here) must be at right angles to each other

not only are the two π bonds shown
in this diagram perpendicular,
but the two CH_2 groups are too

> Interactive bonding orbitals in allene

> ● **Requirements for conjugation**
> * **Conjugation requires double bonds separated by one single bond.**
> * **Double bonds separated by two single bonds or no single bonds are not conjugated.**

The conjugation of two π bonds

To understand the effects of conjugation on molecules, we need now to look at their molecular orbitals. We'll concentrated only on the electrons in π orbitals—you can take it that all the C–C and C–H σ bonds are essentially the same as those of all the other molecules you met in Chapter 4. We'll start with the simplest compound that can have two conjugated π bonds: butadiene. As you would expect, butadiene prefers to be planar to maximize overlap between its p orbitals. But exactly how does that overlap happen, and how does it give rise to bonding?

The molecular orbitals of butadiene

Butadiene has two π bonds, each made up of two p orbitals: a total of four atomic orbitals. We'd therefore expect four molecular orbitals, housing four electrons. Just like

> **Isomers of butadiene**
> Butadiene normally refers to 1,3-butadiene. It is also possible to have 1,2-butadiene, which is another example of an allene.
>
> 1,2-butadiene
> an allene
>
> 1,3-butadiene
> a conjugated diene

hexatriene above, these orbitals extend over the whole molecule, but we can easily work out what these molecular orbitals look like simply by taking the orbitals of two alkenes and interacting them side by side. We have two π orbitals and two π* orbitals, and we can interact them in phase or out of phase. Here are the first two, made by interacting the two π orbitals:

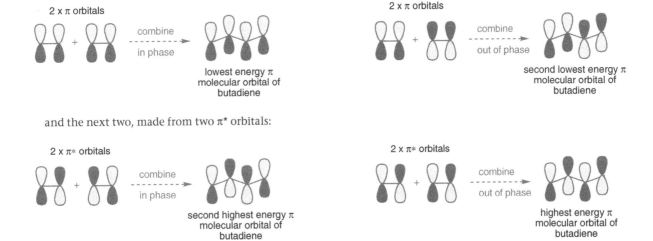

and the next two, made from two π* orbitals:

We can represent all four molecular orbitals like this, stacked up in order of their energy in a molecular orbital energy level diagram. With four orbitals, we can't just use '*' to represent antibonding orbitals, so conventionally they are numbered ψ_1–ψ_4 (ψ is the Greek letter psi).

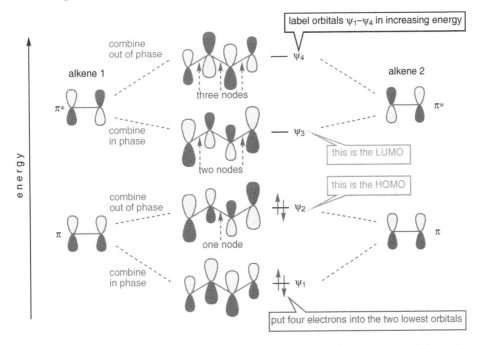

It's worth noticing a couple of other things about the way we have represented these four molecular orbitals before we move on. Firstly, the number of nodes (changes in phase as you move from one orbital to the next) increases from zero in ψ_1 to three in ψ_4. Secondly, notice that the p orbitals making up the π system are not all shown as the same size—their *coefficients* vary according to the orbital they are in. This is a mathematical consequence of the way the orbitals sum together, and you need not be concerned with the details, just the general principle that ψ_1 and ψ_4 have the largest coefficients in the middle; ψ_2 and ψ_3 the largest coefficients at the ends.

Now for the electrons: each orbital holds two electrons, so the four electrons in the π system go into orbitals ψ_1 and ψ_2.

Interactive bonding orbitals in butadiene

➡ The idea that higher energy orbitals have more nodes is familiar to you from Chapter 4—see p. 88.

The term 'coefficient' describes the contribution of an individual atomic orbital to a molecular orbital. It is represented by the size of the lobes on each atom.

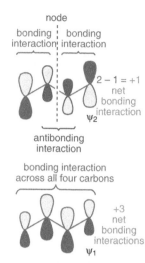

node

bonding interaction | bonding interaction

$2 - 1 = +1$ net bonding interaction

ψ_2

antibonding interaction

bonding interaction across all four carbons

$+3$ net bonding interactions

ψ_1

A closer look at these filled orbitals shows that in ψ_1, the lowest energy bonding orbital, the electrons are spread out over all four carbon atoms (above and below the plane) in one continuous orbital. There is bonding between all four C atoms—three net bonding interactions. ψ_2 has bonding interactions between carbon atoms 1 and 2, and also between 3 and 4 but an *antibonding* interaction between carbons 2 and 3—in other words, $2 - 1 = 1$ net bonding interaction. For the unoccupied orbitals there is a net -1 antibonding interaction in ψ_3 and a net -3 antibonding interaction in ψ_4.

Overall, in both the occupied π orbitals there are electrons between carbons 1 and 2 and between 3 and 4, but the antibonding interaction between carbons 2 and 3 in ψ_2 partially cancels out the bonding interaction in ψ_1. Only 'partially', because the coefficients of the antibonding pair of orbitals in ψ_2 are smaller than the coefficients of the bonding pair in ψ_1. This explains why all the bonds in butadiene are not the same, and also why the middle bond is like a single bond but with a little bit of double-bond character. Its double-bond character extends to its preference for planarity, the fact that it takes more energy to rotate about this bond than about a typical single bond, and the fact that it is slightly shorter (1.45 Å) than a typical C–C single bond (around 1.54 Å).

■ In our glimpse of hexatriene earlier in this chapter we saw similar effects: a tendency to be planar and restriction to rotation about the slightly shortened single bonds.

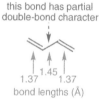

this bond has partial double-bond character

1.37 1.45 1.37
bond lengths (Å)

it takes 30 kJ mol^{-1} to rotate about this bond

it takes 3 kJ mol^{-1} to rotate about this single bond

rotation about a full double bond needs more than 260 kJ mol^{-1}: it's essentially impossible

The molecular orbital diagram also helps us explain some aspects of the reactivity of butadiene. Notice that we have marked on for you the HOMO (ψ_2) and the LUMO (ψ_3). On either side you can see the equivalent HOMO (π orbital) and LUMO (π* orbital) for the isolated alkene (i.e. ethene). Some relevant features to note:

- The overall energy of the two bonding butadiene molecular orbitals is lower than that of the two molecular orbitals for ethene. This means that conjugated butadiene is more thermodynamically stable than just two isolated double bonds.

- The HOMO of butadiene is *higher* in energy than the HOMO for ethene. This is consistent with the fact that butadiene is *more* reactive than ethene towards electrophiles.

- The LUMO for butadiene is *lower* in energy than the LUMO for ethene. This is consistent with the fact that butadiene is *more* reactive than ethene towards nucleophiles.

So conjugation makes butadiene more stable, but it also makes it more reactive to both nucleophiles and electrophiles! This superficially surprising result is revisited in detail in Chapter 19.

UV and visible spectra

■ To understand this section well you will need to remember the formulae linking energy and wavelength, $E = h\nu$, and energy and frequency, $E = hc/\lambda$. See p. 53 for more on these.

In Chapter 2 you saw how, if given the right amount of energy, electrons can be promoted from a low-energy atomic orbital to a higher energy one and how this gives rise to an atomic absorption spectrum. Exactly the same process can occur with molecular orbitals: energy of the right wavelength can promote an electron from a filled orbital (for example the HOMO) to an unfilled one (for example the LUMO), and plotting the absorption of energy against wavelength gives rise to a new type of spectrum called, for obvious reasons which you will see in a moment, a UV–visible spectrum.

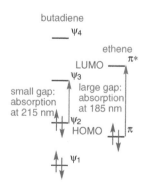

butadiene

ψ_4

ethene

LUMO — π*

ψ_3

small gap: absorption at 215 nm

large gap: absorption at 185 nm

ψ_2 HOMO — π

ψ_1

You have just seen that the energy difference between the HOMO and LUMO for butadiene is less than that for ethene. We would therefore expect butadiene to absorb light of longer wavelength than ethene (the longer the wavelength the lower the energy). This is indeed the case: butadiene absorbs at 215 nm compared to 185 nm for ethene. The conjugation in butadiene means it absorbs light of a longer wavelength than ethene. One of the consequences of conjugation is to lessen the gaps between filled and empty orbitals, and so allow absorption of light of a longer wavelength.

🖰 UV absorption in ethene and butadiene

> ● **The more conjugated a compound is, the smaller the energy transition between its HOMO and LUMO, and hence the longer the wavelength of light it can absorb. UV–visible spectroscopy can tell us about the conjugation present in a molecule.**

Both ethene and butadiene absorb in the UV region of the electromagnetic spectrum. If we extend the conjugation further, the gap between HOMO and LUMO will eventually be small enough to allow the compound to absorb visible light and hence have a colour. Lycopene, the pigment in tomatoes, which we introduced at the start of the chapter, has 11 conjugated double bonds (plus two unconjugated ones). It absorbs blue–green light at about 470 nm: consequently tomatoes are red. Chlorophyll, in the margin, has a cyclic conjugated system: it absorbs at long wavelengths and is green.

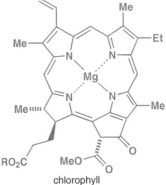

chlorophyll

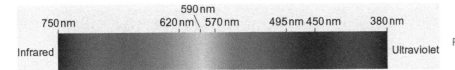

The colour of pigments depends on conjugation

It is no coincidence that these and many other highly conjugated compounds are coloured. All dyes and pigments based on organic compounds are highly conjugated.

The table below shows the approximate wavelengths of light absorbed by a polyene conjugated system containing various numbers n of double bonds. Note that the colour absorbed is complementary to the colour transmitted—a red compound must absorb blue and green light to appear red.

In colour chemistry, *dye* is a soluble colorant while a *pigment* is made of insoluble coloured particles. In biology the word pigment is used for any coloured compound. Dyeing pigments are often inorganic compounds, which are coloured for reasons other than conjugation, but nonetheless to do with the gaps between orbitals.

Approximate wavelengths for different colours

Absorbed frequency, nm	Colour absorbed	Colour transmitted	$R(CH{=}CH)_nR$, $n =$
200–400	ultraviolet	—	< 8
400	violet	yellow–green	8
425	indigo–blue	yellow	9
450	blue	orange	10
490	blue–green	red	11
510	green	purple	
530	yellow–green	violet	
550	yellow	indigo–blue	
590	orange	blue	
640	red	blue–green	
730	purple	green	

Fewer than about eight conjugated double bonds, and the compound absorbs only in the UV. With more than eight conjugated double bonds, the absorption creeps into the visible and, by the time it reaches 11, the compound is red. Blue or green polyenes are rare, and dyes of these colours rely on more elaborate conjugated systems.

🖰 UV absorption in linear conjugated polyenes

Blue jeans

Transitions from bonding to antibonding π orbitals are called π→π* transitions. If electrons are instead promoted from a non-bonding lone pair (*n* orbital) to a π* orbital (an *n*→π* transition) smaller energy gaps may be available, and many dyes make use of *n*→π* transitions to produce colours throughout the whole spectrum. For example, the colour of blue jeans comes from the pigment indigo. The two nitrogen atoms provide the lone pairs that can be excited into the π* orbitals of the rest of the molecule. These are low in energy because of the two carbonyl groups. Yellow light is absorbed by this pigment and indigo–blue light is transmitted.

Jeans are dyed by immersion in a vat of reduced indigo, which is colourless since the conjugation is interrupted by the central single bond. When the cloth is hung up to dry, the oxygen in the air oxidizes the pigment to indigo and the jeans turn blue.

colourless precursor to indigo

indigo: the pigment of blue jeans

The allyl system

The allyl anion

In butadiene, four atomic p orbitals interact to make four molecular orbitals; in hexatriene (and you will soon see benzene too) six atomic orbitals interact to make six molecular orbitals. We are now going to consider some common conjugated systems made up of *three* interacting p orbitals. We'll start with the structure we get from treating propene with a very strong base—one strong enough to remove one of the protons from its methyl group. H⁺ is removed, so the product must have a negative charge, which formally resides on the carbon of what was the methyl group. That carbon atom started off sp³ hybridized (i.e. tetrahedral: it had four substituents), but after it has been deprotonated it must become trigonal (sp²), with only three substituents plus a p orbital to house the negative charge.

> ➡ You will meet such super-strength bases in the next chapter.

> ■ Of course the anion doesn't really exist 'free' like this; it will most likely have a metal cation to which it is coordinated in some way. The arguments we are going to apply about its structure are still valid whether or not there is a metal associated with it.

this carbon becomes sp² hybridized

two filled p orbitals make up the alkene

propene   proton removed   the allyl anion   addtional p orbital   p orbitals of the allyl anion

We could work out the orbitals of the allyl anion by combining this p orbital with a ready-made π bond, but instead this time we will start with the three separate p atomic orbitals and combine them to get three molecular orbitals. At first we are not concerned about where the electrons are—we are just building up the molecular orbitals.

The lowest energy orbital (ψ_1) will have them all combining in phase. This is a bonding orbital since all the interactions are bonding. The next orbital (ψ_2) requires one node, and the only way to include a node and maintain the symmetry of the system is to put the node through the central atom. This means that when this orbital is occupied there will be no electron density on this central atom. Since there are no interactions between adjacent atomic orbitals (either bonding or antibonding), this is a non-bonding orbital. The final molecular orbital (ψ_3) must have two nodal planes. All the interactions of the atomic orbitals are out of phase so the resulting molecular orbital is an antibonding orbital.

the bonding molecular orbital of the allyl system, ψ_1 (net bonding interactions = +2)

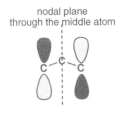

nodal plane through the middle atom

non-bonding ψ_2 (net bonding interactions = 0)

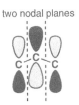

two nodal planes

antibonding ψ_3 (net bonding interactions = –2)

We can summarize all this information in a molecular orbital energy level diagram, and at the same time put the electrons into the orbitals. We need four electrons—two from the alkene π bond and two more for the anion (these were the two in the C–H bond, and they are still there because only a proton, H⁺, was removed). The four electrons go into the lowest two orbitals, ψ_1 and ψ_2, leaving ψ_3 vacant. Notice too that the energy of two of the electrons is lower than it would have been if they had remained in unconjugated p orbitals: conjugation lowers the energy of filled orbitals and makes compounds more stable.

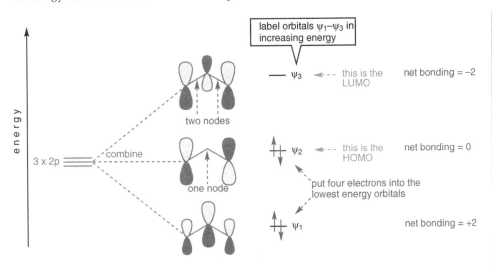

label orbitals ψ_1–ψ_3 in increasing energy

two nodes

ψ_3 · · · this is the LUMO   net bonding = –2

combine

one node

ψ_2 · · · this is the HOMO   net bonding = 0

put four electrons into the lowest energy orbitals

$3 \times 2p$

ψ_1 net bonding = +2

■ This diagram shows only the π orbitals of the allyl system. We have ignored all the molecular orbitals from the σ framework because the bonding σ orbitals are considerably lower in energy than the molecular orbitals of the π system and the vacant antibonding σ^* orbitals are much higher in energy than the π antibonding molecular orbital.

🖱 Interactive bonding orbitals in the allyl anion

Where is the electron density in the allyl anion π system? We have two filled π molecular orbitals and the electron density comes from a sum of both orbitals. This means there is electron density on all three carbon atoms. However, the coefficients of the end carbons are of a significant size in both orbitals, but in ψ_2 the middle carbon has no electron density at all—it lies on a node. So overall, even though the negative charge is spread over the whole molecule, the end carbons carry more of the electron density than the middle one. We can represent this in two ways—the first structure below emphasizes the delocalization of the charge over the whole molecule, but fails to get across the important point that the negative charge resides principally at the ends. Curly arrows do this much better: we can use them to show that the negative charge is not localized, but principally divided between the two end carbons.

■ A reminder: this is not an equilibrium—the arrows do not represent the movement of charge. The two structures are alternative, imperfect representations of an 'averaged' structure, and they are linked by a double-headed delocalization arrow.

these structures emphasize the equivalence of the bonds and delocalization of charge

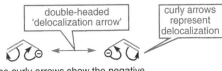

double-headed 'delocalization arrow'

curly arrows represent delocalization

the curly arrows show the negative charge concentrated on the end carbon atoms

$(-)\diagup\!\!\diagdown\!\!\diagup(-)$

The problem with these structures carrying curly arrows is that they seem to imply that the negative charge (and the double bond for that matter) is jumping from one end of the molecule to the other. This, as we have seen, is just not so. Another and perhaps better picture uses dotted lines and partial charges. But the structure with the dotted bonds, as with the representation of benzene with a circle in the middle, is no good for writing mechanisms. Each of the representations has its strong and weak points: we shall use each as the occasion demands.

Using NMR to study delocalization

Delocalization of the allyl anion, and the localization of the negative charge mainly on the end carbons, is clear from its ^{13}C NMR spectrum as well. In Chapter 3 we explained that ^{13}C NMR gives us a good measure of the amount of electron density around a C atom—the extent to which it is deshielded and therefore exposed to the applied magnetic field. If you need reminding about the terminology, theory, and practice of NMR, turn back now to Chapter 3, pp. 52–63.

It is possible to record a ^{13}C NMR spectrum of an allyl anion with a lithium counterion. The spectrum shows only two signals: the middle carbon at 147 ppm and the two end carbons both at 51 ppm. This confirms two things: (i) both end carbons are the same and the structure is delocalized, and (ii) most of the negative charge is on the end carbons—they are more shielded (have a smaller chemical shift) as a result of the greater electron density. In fact, the central carbon's shift of 147 ppm is not far from that of a normal double-bond carbon (compare the signals in propene). The end carbons' shift is in between that of a double bond and a saturated carbon directly bonded to a metal (e.g. methyllithium, whose negative chemical shift results from the highly polarized Li–C bond).

allyl lithium	propene	methyllithium
147 p.p.m.	134 ppm	–15 ppm

both end carbons resonate at 51 ppm

116 ppm   19.5 ppm

The allyl cation

What if, instead of taking just a proton, we had also taken away two electrons from propene? In reality we can get such a structure quite straightforwardly from allyl bromide (prop-2-enyl bromide or 1-bromoprop-2-ene). Carbon 1 in this compound has four atoms attached to it (a carbon, two hydrogens, and a bromine atom) so it is tetrahedral (or sp^3 hybridized).

allyl bromide

Bromine is more electronegative than carbon and so the C–Br bond is polarized towards the bromine. It is quite easy to break this bond completely, with the bromine keeping both electrons from the C–Br bond to become bromide ion, Br^-, leaving behind an *allyl cation*. The positively charged carbon now has only three substituents so it becomes trigonal (sp^2 hybridized). It must therefore have a vacant p orbital.

this carbon carries a vacant p orbital

bromide leaves

two filled p orbitals make up the alkene

vacant p orbital

allyl bromide   the allyl cation   p orbitals of the allyl cation

Like the allyl anion, the orbitals in the allyl cation are a combination of three atomic p orbitals, one from each carbon. So we can use the same molecular orbital energy level diagram as we did for the anion, simply by adjusting the number of electrons we put into the orbitals. This time, there are only two electrons, from the alkene, as those which were in the C–Br bond have left with anionic bromide.

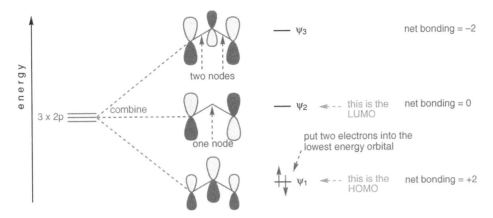

The two electrons in the filled orbital are in a lower energy orbital than they would have been if they had stayed in an unconjugated p orbital: as with the anion, *conjugation* leads to *stabilization*.

The two electrons are spread over three carbon atoms. Overall, the allyl cation has a positive charge. But where is the positive charge concentrated? What we need to do is look to see where there is a *deficit* of charge. The only orbital with any electrons in it is the bonding molecular orbital ψ_1. From the relative sizes of the coefficients on each atom we can see that the middle carbon has more electron density on it than the end ones, so the end carbons must be more positive than the middle one.

We expect both end carbons to be identical, and ^{13}C NMR tells us that this is so (see below). Again we need a way of showing this delocalization, either on a single structure or as a pair of localized structures linked by a delocalization arrow.

these structures emphasize the
equivalence of the bonds and
delocalization of charge

curly arrows show the positive charge is
shared over both the end atoms

Notice how we draw the curly arrows here: we want to show the positive charge 'moving', and it is tempting to draw a curly arrow starting from the positive charge. But curly arrows must always start on something representing a pair of electrons. So we must move the positive charge as a consequence of the movement of the electrons in the double bond: as we pull them away from one end, they leave behind a positive charge.

➡ The guidelines for drawing curly arrows are given on p. 123.

The NMR spectrum of a delocalized cation

The reaction below shows the formation of a cation close in structure to the allyl cation. A very strong acid (called 'super-acid'—see Chapter 15) protonates the OH group of 3-cyclohexenol, which can then leave as water. The resulting cation is, not surprisingly, unstable and would normally react rapidly with a nucleophile. However, at low temperatures and if there are no nucleophiles present, the cation is relatively stable and it is even possible to record a ^{13}C NMR spectrum (at –80 °C).

OH FSO$_3$H-SbF$_5$
liquid SO$_2$, –80 °C

–H$_2$O

delocalized cation

^{13}C NMR shifts in ppm—
notice the plane of
symmetry down the middle

141.9
224.4   224.4
37.1   37.1
17.5

The ^{13}C NMR spectrum of this allylic cation reveals a plane of symmetry, which confirms that the positive charge is spread over two carbons. The large shift of 224 ppm for these carbons indicates very strong deshielding (that is, lack of electrons) but is nowhere near as large as that of a localized cation (which would resonate at about 330 ppm). The middle carbon's shift of 142 ppm is almost typical of a normal double bond, indicating that it is neither significantly more nor less electron-rich than normal. In fact, it is interesting to note that the middle carbon of this cation and the allyl anion we described above have almost exactly the same chemical shift—proof that the charge lies mainly on the ends of the allyl system.

Delocalization over three atoms is a common structural feature

The carboxylate anion

You may already be familiar with one anion very much like the allyl anion—the carboxylate ion, which forms when a carboxylic acid reacts with a base. In this structure we again have a negatively charged atom separated from a double bond by a single bond adjacent to a single bond: it's analogous to an allyl anion with oxygen atoms replacing two of the carbon atoms.

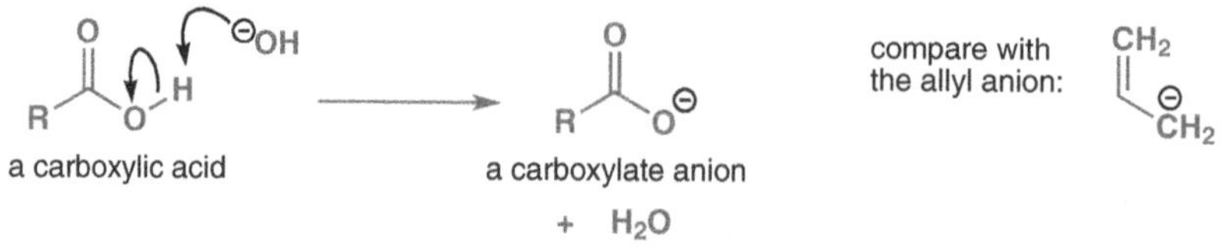

a carboxylic acid   a carboxylate anion

+ H₂O

compare with the allyl anion:

X-ray crystallography shows both carbon–oxygen bond lengths in this anion to be the same (136 pm), in between that of a normal carbon–oxygen double bond (123 pm) and single bond (143 pm). The negative charge is spread out equally over the two oxygen atoms, and we can represent this in two ways—as before, the one on the left shows the equivalence of the two C–O bonds, but you would use the one on the right for writing mechanisms. The delocalization arrow tells us that both localized forms contribute to the real structure.

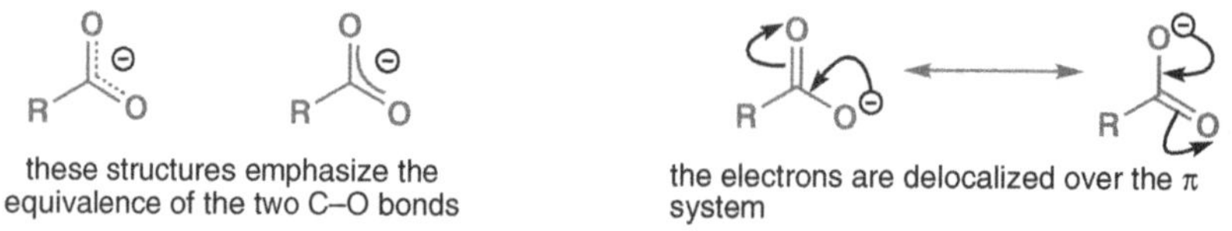

these structures emphasize the equivalence of the two C–O bonds

the electrons are delocalized over the π system

The nitro group

The nitro group consists of a nitrogen bonded to two oxygen atoms and a carbon (for example an alkyl group). There are two ways of representing the structure: one using formal charges, the other (which we suggest you avoid) using a dative bond. Notice in each case that one oxygen is depicted as being doubly bonded, the other singly bonded. Drawing both oxygen atoms doubly bonded is incorrect—*nitrogen cannot participate in five bonds*: this would require ten bonding electrons around the N atom, and there are not enough s and p orbitals to put them in.

two ways of representing the nitro group

incorrect drawing of the nitro group nitrogen cannot have five bonds

The problem even with the 'correct' structure on the left is that the equivalence of the two N–O bonds is not made clear. The nitro group has exactly the same number of electrons as a carboxylate anion (although it's neutral of course because nitrogen already has one more electron than carbon) and the delocalized structure can be shown with curly arrows in the same way.

We have not shown molecular orbital energy level diagrams for the carboxylate and nitro groups, since they are similar to that of the allyl anion. Only the absolute energies of the molecular orbitals are different since different elements with different electronegativities are involved in each case.

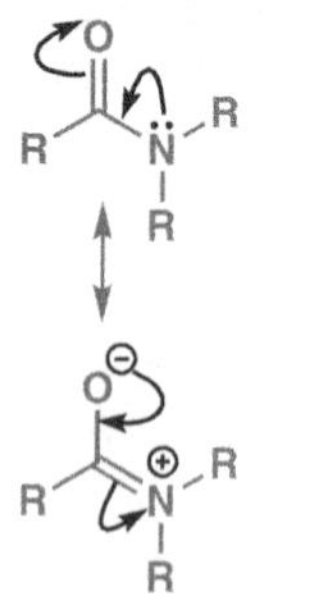

The amide group

Life is built of amides, because the amide group is the link through which amino acids join together to form the proteins which make up much of the structural features of living systems.

> ➡ This important principle was explained on p. 30.

> We call structures such as a nitro and a carboxylate group *isoelectronic*: the atoms may be different but the number of and arrangement of the bonding electrons are the same.

delocalization in the amide group

Nylon is a synthetic polyamide, and shares with many proteins the property of durability. The structure of this deceptively simple functional group has an unexpected feature which is responsible for much of the stability it confers.

The allyl anion, carboxylate, and nitro groups have four electrons in a π system spread out over three atoms. The nitrogen in the amide group also has a pair of electrons that can conjugate with the π bond of the carbonyl group. For effective overlap with the π bond, this lone pair of electrons must be in a p orbital. This in turn means that the nitrogen must be sp² hybridized.

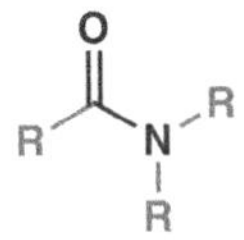

an amide

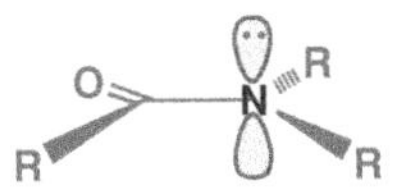

nitrogen is trigonal (sp²) with its lone pair in a p orbital

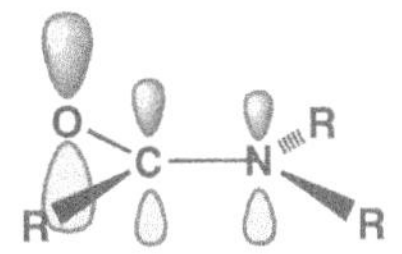

the lowest enegy π orbital of the amide

■ Contrast this with the lone pair of a typical amine, which lies in an sp³ orbital (see p. 103): an amine N is pyramidal (sp³) while an amide N is trigonal planar (sp²).

In the carboxylate ion, a negative charge is shared (equally) between two oxygen atoms. In an amide there is no charge as such—the lone pair on nitrogen is shared between the nitrogen and the oxygen. The delocalization can be shown as usual by using curly arrows, as shown in the margin.

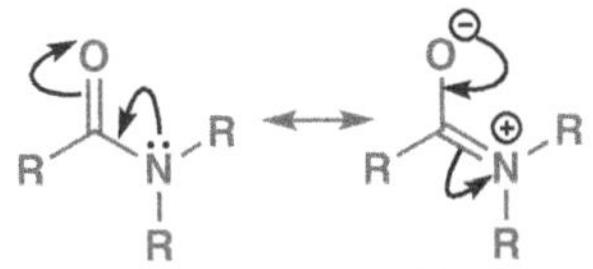

This representation suffers from the usual problems. Curly arrows usually show electron movement, but here they do not: they simply show how to get from one of the alternative representations to the other. The molecular orbital picture of the amide tells us that the electrons are unevenly distributed over the three atoms in the π system with a greater electron density on the oxygen: you can see this in the delocalized structure on the right, which has a full negative charge on O and a positive charge on N. (We also indicated this in the diagram of the lowest energy π orbital above, which has a greatest coefficient, and therefore greatest electron density, on O.) Another aspect of the structure of the amide group that this pair of structures indicates correctly is that there is partial double bond character between the C atom and the N atom. We will come back to this shortly.

delocalization in the amide group

The real structure of the amide group lies in between the two extreme structures linked by the delocalization arrow: a better representation might be the structure on the right. The charges in brackets indicate substantial, although not complete, charges, maybe about a half plus or minus charge. However, we cannot draw mechanisms using this structure.

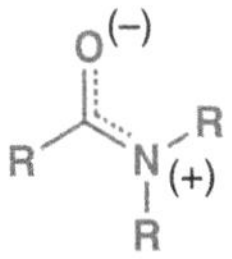

We can summarize several points about the structure of the amide group, and we will then return to each in a little more detail

- The amide group is planar—this includes the first carbon atoms of the R groups attached to the carbonyl group and to the nitrogen atom.
- The lone pair of electrons on nitrogen is delocalized into the carbonyl group.
- The C–N bond is strengthened by this interaction—it takes on partial double bond character. This also means that we no longer have free rotation about the C–N bond, which we would expect if it were only a single bond.
- The oxygen is more electron-rich than the nitrogen. Hence we might expect the oxygen rather than the nitrogen to be the site of electrophilic attack.
- The amide group as a whole is made more stable as a result of the delocalization.

How do we know the amide group is planar? X-ray crystal structures are the simplest answer. Other techniques such as electron diffraction also show that simple (non-crystalline) amides have planar structures. *N,N*-Dimethylformamide (DMF) is an example.

The N–CO bond length in DMF (135 pm) is closer to that of a standard C–N double bond (127 pm) than to that of a single bond (149 pm). This partial double bond character, which the delocalized structures led us to expect, is responsible for restricted rotation about this C–N bond. We must supply 88 kJ mol⁻¹ if we want to rotate the C–N bond in DMF (remember a single bond only takes about 3 kJ mol⁻¹, while a full C–C double bond takes about 260 kJ mol⁻¹). The amount of energy available at room temperature is only enough to allow this bond

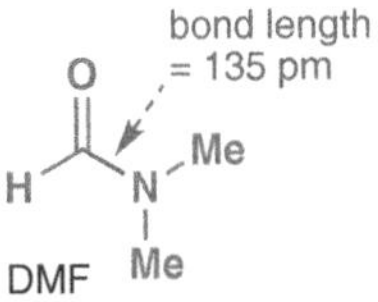

bond length = 135 pm

DMF
dimethylformamide

to rotate slowly, and the result is quite clear in the ^{13}C NMR spectrum of DMF. There are three carbon atoms altogether and three signals appear—the two methyl groups on the nitrogen are different. If free rotation were possible about the C–N bond, we would expect to see only two signals, since the two methyl groups would become identical.

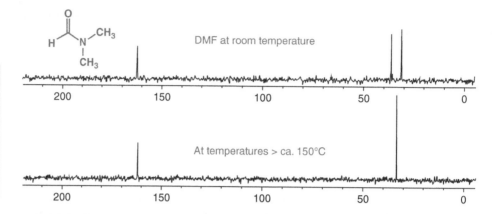

■ In fact, if we record the spectrum at higher temperatures, we do indeed only see two signals since now there is sufficient energy available to overcome the rotational barrier and allow the two methyl groups to interchange.

Amides in proteins

Proteins are composed of many amino acids joined together with amide bonds. The amino group of one can combine with the carboxylic acid group of another to give an amide known as a peptide—two amino acids join to form a dipeptide; many join to give a polypeptide.

The peptide unit so formed is a planar, rigid structure because of restricted rotation about the C–N bond. This rigidity confers organizational stability on protein structures.

Conjugation and reactivity: looking forward to Chapter 10

Just as delocalization stabilizes the allyl cation and anion (at least some of the electrons in conjugated systems end up in lower energy orbitals than they would have done without conjugation) so too is the amide group stabilized by the conjugation of the nitrogen's lone pair with the carbonyl group. This makes an amide C=O one of the least reactive carbonyl groups (we shall discuss this in Chapter 10). Furthermore, the nitrogen atom of an amide group is very different from that of a typical amine. Most amines are easily protonated. However, since the lone pair on the amide's nitrogen is conjugated into the π system, it is less available for protonation or, indeed, reaction with any electrophile. As a result, when an amide is protonated (and it is not protonated easily, as you will see in the next chapter) it is protonated on oxygen rather than nitrogen. The consequences of conjugation for reactivity extend far and wide, and will be a running theme through many chapters in this book.

Aromaticity

It's now time to go back to the structure of benzene. Benzene is unusually stable for an alkene and is not normally described as an alkene at all. For example, whereas normal alkenes (whether conjugated or not) readily react with bromine to give dibromoalkane *addition* products, benzene reacts with bromine only with difficulty—it needs a catalyst (iron will do) and then the product is a *monosubstituted* benzene and not an addition compound.

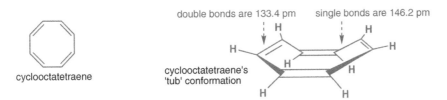

Bromine reacts with benzene in a substitution reaction (a bromine atom replaces a hydrogen atom), *keeping the benzene structure intact*. This ability to retain its conjugated structure through all sorts of chemical reactions is one of the important differences between benzene and other alkenes.

What makes benzene special?

You might assume benzene's special feature is its ring structure. To see whether this is the case, we'll look at another cyclic polyene, cyclooctatetraene, with four double bonds in a ring. Given what we have explained about the way that π systems gain stability by allowing overlap between their p orbitals, you may be surprised to find that cyclooctatetraene, unlike benzene, is *not* planar. There is no conjugation between any of the double bonds—there are indeed alternate double and single bonds in the structure, but conjugation is possible only if the p orbitals of the double bonds can overlap and here they do not. The fact that there is no conjugation is shown by the alternating C–C bond lengths in cyclooctatetraene—146.2 and 133.4 pm—which are typical for single and double C–C bonds. If possible, make a model of cyclooctatetraene for yourself—you will find the compound naturally adopts the shape on the right below. This shape is often called a 'tub'.

double bonds are 133.4 pm   single bonds are 146.2 pm

cyclooctatetraene

cyclooctatetraene's
'tub' conformation

Interactive structures of cyclooctatetraene, the dianion and dication

Chemically, cyclooctatetraene behaves like an alkene, not like benzene. With bromine, for example, it forms an addition product and not a substitution product. So benzene is not special just because it is cyclic—cyclooctatetraene is cyclic too but does not behave like benzene.

Heats of hydrogenation of benzene and cyclooctatetraene

C=C double bonds can be reduced using hydrogen gas and a metal catalyst (usually nickel or palladium) to produce fully saturated alkanes. This process is called hydrogenation and it is exothermic (that is, energy is released) since a thermodynamically more stable product, an alkane, is produced.

When *cis*-cyclooctene is hydrogenated to cyclooctane, 96 kJ mol⁻¹ of energy is released. Cyclooctatetraene releases 410 kJ mol⁻¹ on hydrogenation. This value is approximately four times one double bond's worth, as we might expect. However, whereas the heat of hydrogenation for cyclohexene is 120 kJ mol⁻¹, on hydrogenating benzene only 208 kJ mol⁻¹ is given out, which is much less than the 360 kJ mol⁻¹ that we would have predicted by multiplying the figure for cyclohexene by 3. Benzene has something to make it stable which cyclooctatetraene does not have.

More on this in Chapter 23.

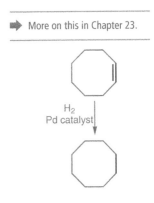

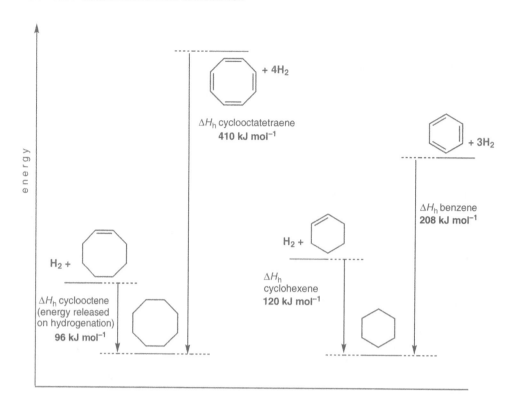

Varying the number of electrons

The mystery deepens when we look at what happens when we treat cyclooctatetraene with powerful oxidizing or reducing agents. If 1,3,5,7-tetramethylcyclooctatetraene is treated at low temperature (–78 °C) with SbF$_5$/SO$_2$ClF (strongly oxidizing conditions) a dication is formed. This cation, unlike the neutral compound, is *planar* and all the C–C bond lengths are the same.

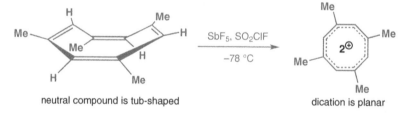

Interactive structures of cyclooctatetraene, the dianion and dication

neutral compound is tub-shaped   dication is planar

This dication still has the same number of atoms as the neutral species, only fewer electrons. The electrons have come from the π system, which is now two electrons short. We could draw a structure showing two local-ized positive charges, but the charge is in fact spread over the whole ring.

It is also possible to *add* electrons to cyclooctatetraene by treating it with alkali metals and a *dianion* results. X-ray structures reveal this dianion to be planar, again with all C–C bond lengths the same (140.7 pm). The difference between the anion and cation of cyclooctatetraene on the one hand and cyclooctatetraene on the other is the number of electrons in the π system. Neutral, non-planar, cyclooctatetraene has eight π electrons, the planar dication has six π electrons (as does benzene), and the planar anion has ten.

Can you see a pattern forming? The important point is not the number of conjugated atoms but the *number of electrons in the π system*.

● When they have four or eight π electrons, both six- and eight-membered rings adopt non-planar structures; when they have six or ten π electrons, a planar structure is preferred.

If you made a model of cyclooctatetraene, you might have tried to force it to be flat. If you managed this you probably found that it didn't stay like this for long and that it popped back into the tub shape. The strain in planar cyclooctatetraene can be overcome by the molecule adopting the tub conformation. The strain is due to the numbers of atoms and double bonds in the ring—it has nothing to do with the number of electrons. The planar dication and dianion of cyclooctatetraene still have this strain. The fact that these ions do adopt planar structures must mean there is some other form of stabilization that outweighs the strain of being planar. This extra stabilization is called *aromaticity*.

Benzene has six π molecular orbitals

The difference between the amount of energy we expect to get from benzene on hydrogenation (360 kJ mol⁻¹) and what is observed (208 kJ mol⁻¹) is about 150 kJ mol⁻¹. This represents a crude measure of just how extra stable benzene really is relative to what it would be like with three localized double bonds. In order to understand the origin of this stabilization, we must look at the molecular orbitals. We can think of the π molecular orbitals of benzene as resulting from the combination of the six p orbitals in a ring and, as with butadiene, each successively higher energy orbital contains one more node. This is what we get for benzene:

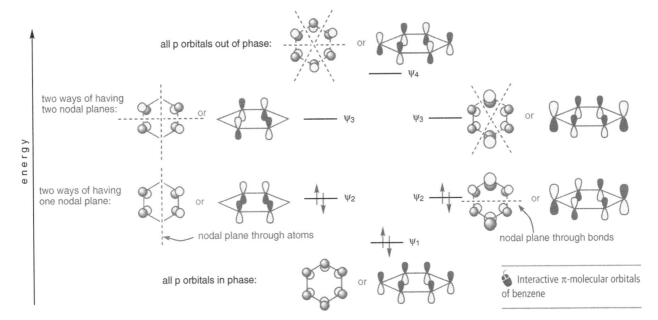

The molecular orbital lowest in energy, ψ_1, has no nodes, with all the orbitals combining in phase. The next lowest molecular orbital will have one nodal plane, which can be arranged in two ways depending on whether or not the nodal plane passes through a bond or an atom. It turns out that these two different molecular orbitals both have exactly the same energy, that is, they are degenerate, and we call them both ψ_2. There are likewise two ways of arranging two nodal planes and again there are two degenerate molecular orbitals ψ_3. The final molecular orbital ψ_4 will have three nodal planes, which must mean all the p orbitals combining out of phase. Six electrons slot neatly into the three lowest energy bonding orbitals.

The π molecular orbitals of other conjugated cyclic hydrocarbons

Notice that the layout of the energy levels in benzene is a regular hexagon with its apex pointing downwards. It turns out that the energy level diagram for the molecular orbitals resulting from the combination of *any* regular cyclic arrangement of p orbitals can be deduced from the appropriately sided polygon with an apex pointing downwards. The horizontal diameter (the red line) represents the energy of a carbon p orbital and any energy levels on this line represent non-bonding molecular orbitals. All molecular orbitals with energies below this line are bonding; all those above are antibonding.

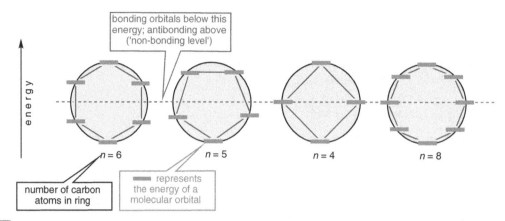

Aromaticity of cyclic polyenes

It's worth noting a few points about these energy level diagrams:

• The method predicts the energy levels for the molecular orbitals of planar, cyclic arrangements of identical atoms (usually all C) only.

• There is always one single molecular orbital lower in energy than all the others. This is because there is always one molecular orbital where all the p orbitals combine in phase.

• If there is an even number of atoms, there is also a single molecular orbital highest in energy; otherwise there will be a pair of degenerate molecular orbitals highest in energy.

• All the molecular orbitals come in degenerate pairs except the one lowest in energy and (for even-numbered systems) the one highest in energy.

Molecular orbitals and aromaticity

Now we can begin to put all the pieces together and make sense of what we know so far. We'll compare the way that the electrons fit into the energy level diagrams for benzene and planar cyclooctatetraene. We are not concerned with the actual shapes of the molecular orbitals involved, just their energies.

Benzene has six π electrons, which means that all its three bonding molecular orbitals are fully occupied, giving what we can call a 'closed shell' structure. Cyclooctatetraene's eight electrons, on the other hand, do not fit so neatly into its orbitals. Six of these fill up the bonding molecular orbitals but there are two electrons left. These must go into the degenerate pair of non-bonding orbitals. Hund's rule (Chapter 4) would suggest one in each. Planar cyclooctatetraene would not have the closed shell structure that benzene has—to get one it must either lose or gain two electrons. This is exactly what we have already seen—both the dianion and dication from cyclooctatetraene are planar, allowing delocalization all over the ring, whereas neutral cyclooctatetraene avoids the unfavourable arrangement of electrons shown below by adopting a tub shape with localized bonds.

■ You can draw an analogy here with the stability of 'closed shell' electronic arrangements in atoms.

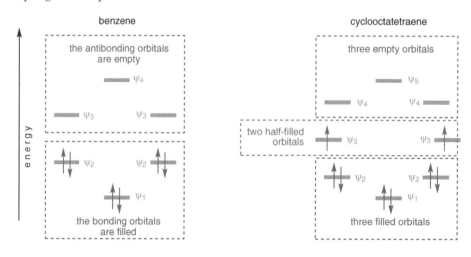

Hückel's rule tells us if compounds are aromatic

As we pointed out on the previous page, all the cyclic conjugated hydrocarbons have a single lowest energy molecular orbital, and then a stack of degenerate pairs of orbitals of increasing energy. Since the single low energy orbital holds two electrons, and then the successive degenerate pairs four each, a 'closed shell' arrangement in which all the orbitals below a certain level are filled will always contain $(4n + 2)$ electrons (where n is an integer—0, 1, 2, etc.—corresponding to the number of degenerate orbital pairs). This is the basis of Hückel's rule.

> ● **Hückel's rule**
>
> Planar, fully conjugated, monocyclic systems with **($4n + 2$) π electrons** have a closed shell of electrons all in bonding orbitals and are exceptionally stable. Such systems are said to be **aromatic**.
>
> Analogous systems with **$4n$ π electrons** are described as **anti-aromatic**.

■ This is not a strict definition of aromaticity: it is actually very difficult to define aromaticity precisely, but all aromatic systems obey Hückel's ($4n + 2$) rule.

Annulenes are compounds with alternating double and single bonds. The number in brackets tells us how many carbon atoms there are in the ring. Using this nomenclature, you could call benzene [6]annulene and cyclooctatetraene [8]annulene–but don't.

The next $(4n + 2)$ number after six is ten so we might expect this cyclic alkene, [10]annulene, to be aromatic. But if a compound with five *cis* double bonds were planar, each internal angle would be 144°. Since a normal double bond has bond angles of 120°, this would be far from ideal. This compound can be made but it does *not* adopt a planar conformation and therefore is not aromatic even though it has ten π electrons.

all-*cis*-[10]annulene

obeys the ($4n + 2$) rule but is not aromatic because it is too strained when planar

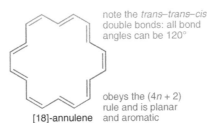

note the *trans–trans–cis* double bonds: all bond angles can be 120°

obeys the ($4n + 2$) rule and is planar and aromatic
[18]-annulene

[20]-annulene

has $4n$ electrons: is not planar and not aromatic

By contrast, [18]annulene, which is also a $(4n + 2)$ π electron system ($n = 4$), does adopt a planar conformation and *is* aromatic. The *trans–trans–cis* double bond arrangement allows all bond angles to be 120°. [20]Annulene presumably could become planar (it isn't quite) but since it is a $4n$ π electron system rather than a $4n + 2$ system, it is not aromatic and the structure shows localized single and double bonds.

When the conjugated systems are not monocyclic, the situation becomes a little less clear. Naphthalene, for example, has ten electrons but you can also think of it as two fused benzene rings. From its chemistry, it is very clear that naphthalene has aromatic character (it does substitution reactions) but is less aromatic than benzene itself. For example, naphthalene can easily be reduced to tetralin (1,2,3,4-tetrahydronaphthalene), which still contains a benzene ring. Also, in contrast to benzene, all the bond lengths in naphthalene are not the same. 1,6-Methano[10]annulene is rather like naphthalene but with the middle bond replaced by a methylene bridging group. This compound is almost flat and shows aromatic character.

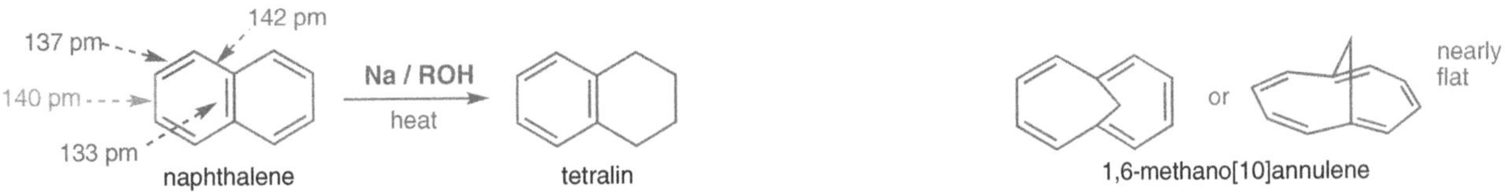

137 pm
142 pm
140 pm
133 pm
naphthalene

Na / ROH
heat

tetralin

or

nearly flat

1,6-methano[10]annulene

Hückel's rule helps us predict and understand the aromatic stability of numerous other systems. Cyclopentadiene, for example, has two conjugated double bonds but the conjugated system is not cyclic since there is an sp³ carbon in the ring. However, this compound is relatively easy to deprotonate to give a very stable anion in which all the bond lengths are the same.

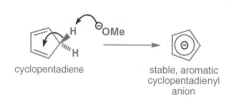

cyclopentadiene

stable, aromatic cyclopentadienyl anion

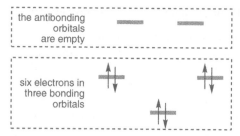

the antibonding orbitals are empty

six electrons in three bonding orbitals

Each of the double bonds contributes two electrons and the negative charge (which must be in a p orbital to complete the conjugation) contributes a further two, making six altogether. The energy level diagram shows that six π electrons completely fill the bonding molecular orbitals, thereby giving a stable aromatic structure.

Heterocyclic aromatic compounds

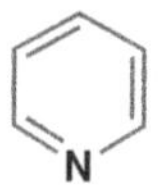

So far all the aromatic compounds you have seen have been hydrocarbons. However, most aromatic systems are heterocyclic—that is, they contain atoms other than just carbon and hydrogen. (In fact the majority of *all organic compounds* are aromatic heterocycles!) A simple example is pyridine, in which a nitrogen replaces one of the CH groups of benzene. The ring still has three double bonds and thus six π electrons.

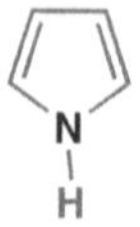

Consider the structure shown on the left, pyrrole. This is also aromatic but it's not enough just to use the electrons in the double bonds: in pyrrole the nitrogen's lone pair contributes to the six π electrons needed for the system to be aromatic. Aromatic chemistry makes several more appearances in this book: in Chapter 21 we shall look at the chemistry of benzene and in Chapters 30 and 31 we shall discuss heterocyclic aromatic compounds in much more detail.

■ Not only are most aromatic systems heterocyclic, but more than 50% of *all organic compounds* contain an aromatic heterocycle.

pyridine

pyrrole

Further reading

Molecular Orbitals and Organic Chemical Reactions: Student Edition by Ian Fleming, Wiley, Chichester, 2009, gives an excellent account of delocalization.

Check your understanding

To check that you have mastered the concepts presented in this chapter, attempt the problems that are available in the book's Online Resource Centre at http://www.oxfordtextbooks.co.uk/orc/clayden2e/

Suggested solutions for Chapter 7

PROBLEM 1

Are these molecules conjugated? Explain your answer in any reasonable way.

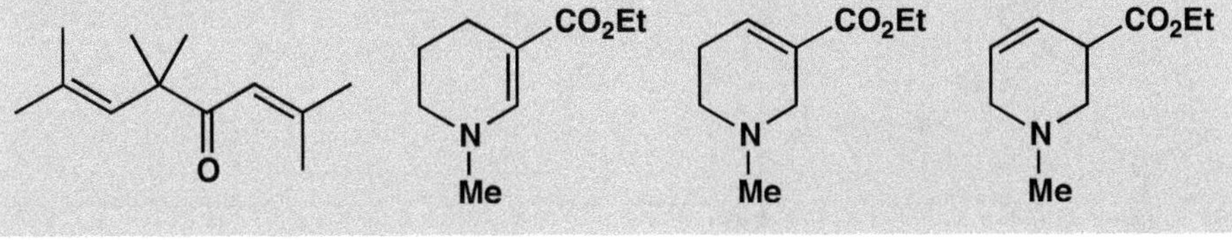

Purpose of the problem

Revision of the basic kinds of conjugation and how to show conjugation with curly arrows.

Suggested solution

The first compound is straightforward with one conjugated system (an enone) and a non-conjugated alkene. You could draw curly arrows to show the conjugation, like this, and/or give a diagram to show the distribution of the electrons.

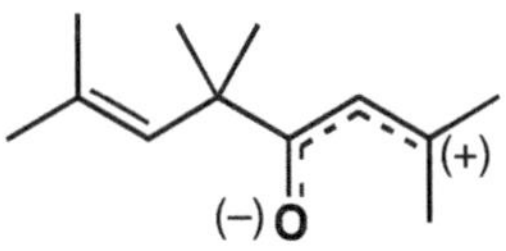

The last three compounds obviously form a related group with the same skeleton and only the alkene moved round. There is of course ester conjugation in all three and this is the only conjugation in the last molecule. The first has extended conjugation between the nitrogen lone pair and the carbonyl group and the second has simple conjugation between the alkene and the ester.

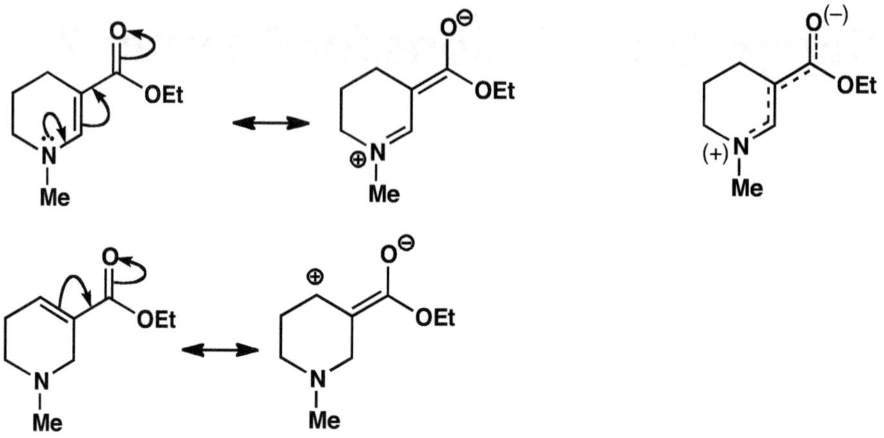

The only conjugation in the last compound is the delocalization of the ester oxygen lone pair. This is of course there in all the other compounds too.

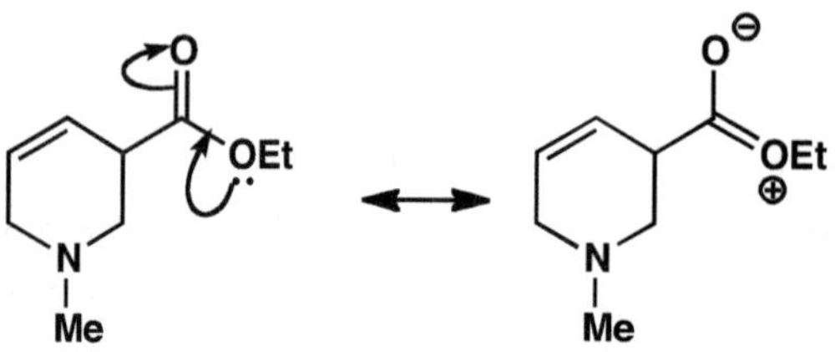

PROBLEM 2

How extensive is the conjugated system(s) in these compounds?

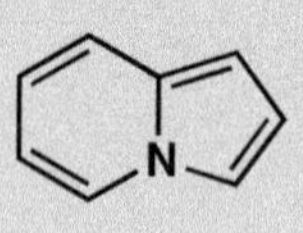

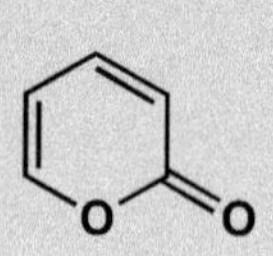

Purpose of the problem

To explore more extensive conjugated systems.

Suggested solution

Both compounds are completely conjugated: even including the nitrogen atom in the first and the carbonyl group in the second. You can draw the arrows going either way round the ring to give different ways of writing the same structure. The arrows on the second compound should end on the carbonyl oxygen.

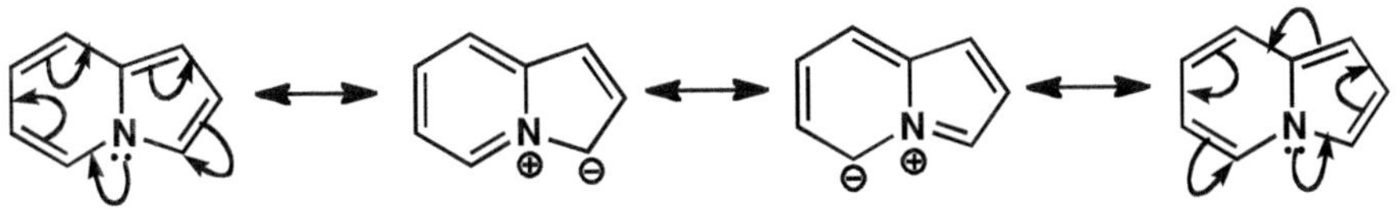

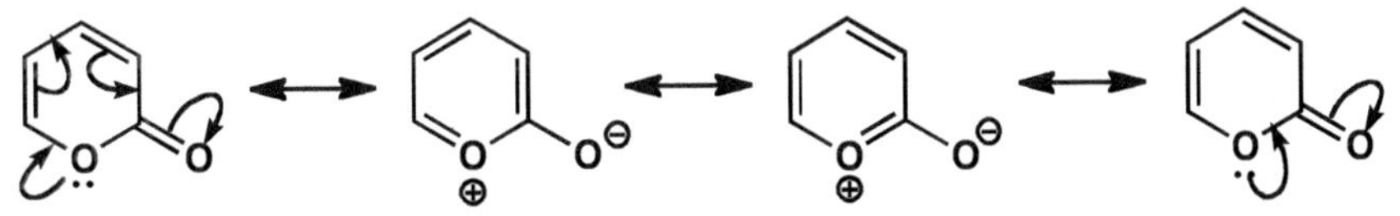

PROBLEM 3

Draw diagrams to represent the conjugation in these molecules. Draw two types of diagram:

(a) Show curly arrows linking at least two different ways of representing the molecule

(b) Indicate with dotted lines and partial charges (where necessary) the partial double bond (and charge) distribution.

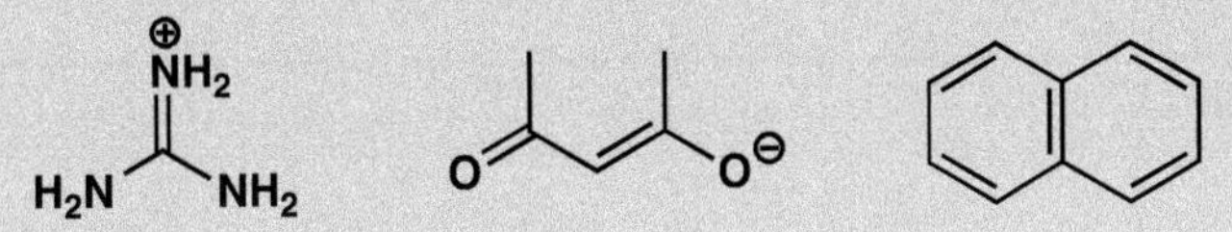

Purpose of the problem

A more exacting exploration of the precise details of conjugation.

Suggested solution

Treating each compound separately in the styles demanded by the question, the first (the guanidinium ion) is a very stable cation because of conjugation. The charge is delocalized onto all three nitrogen atoms as the first three structures show. Each nitrogen has an equal positive charge so our fourth diagram shows one third + on each.

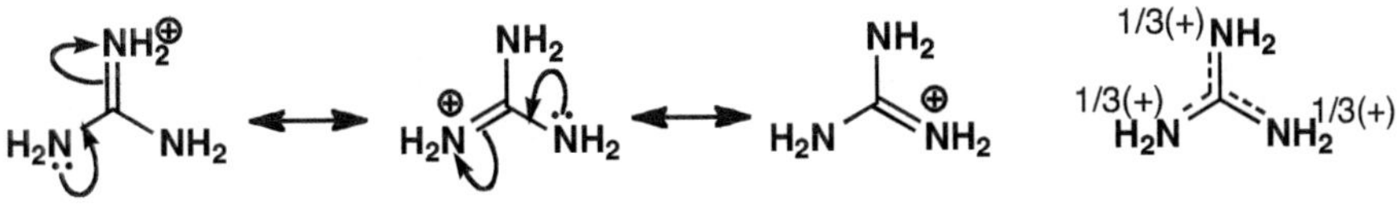

The second compound is what you will learn to call an enolate anion. The negative charge is delocalized throughout the molecule, mostly on the oxygens but some on carbon. It is difficult to represent this with partial charges but the charges on the oxygens will be nearly a half each.

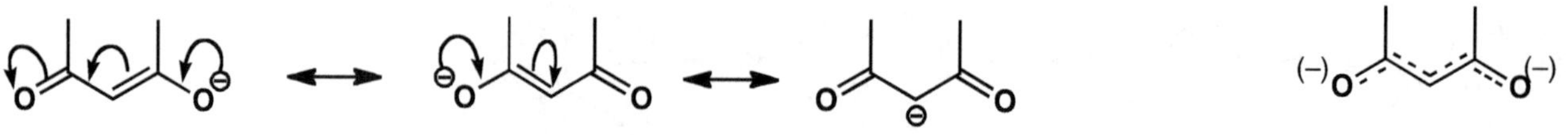

The third compound is naphthalene. The structure drawn in the question is the best as both rings are benzene rings. The results of curly arrow diagrams show how naphthalene is delocalized all round the outer ring. In fact these diagrams show the ten electrons in the outer ring – this is a $4n + 2$ number and all three diagrams show that naphthalene is aromatic.

PROBLEM 4

Draw curly arrows linking alternative structures to show the delocalization in
(a) diazomethane CH_2N_2
(b) nitrous oxide, N_2O
(c) dinitrogen tetroxide, N_2O_4

Purpose of the problem

Delocalization in some neutral molcules we nonetheless have to draw using charges.

Suggested solution

You saw on page 154 of the textbook that the nitro group, although it is neutral, can be represented as a pair of delocalized structures containing charges. The same is true for the explosive gas diazomethane. It has a linear structure, and we can draw two alternative structures, both with charges, even though it is a neutral compound. They're linked with the double headed arrow used for alternative representations for the same compound. We hope you remembered to avoid the trap of giving nitrogen five bonds!

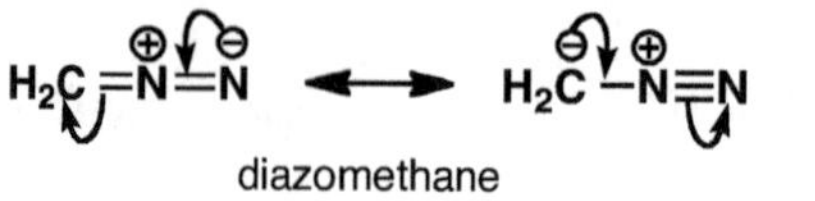

diazomethane    nitric oxide

■ The idea of isoelectronic structures is introduced on p. 102 of the textbook. These compounds are also isoelectronic with carbon dioxide and azide (N_3^-).

Nitric oxide is very similar—in fact it is isoelectronic with diazomethane. You can think of is as a nitrogen molecule in which an oxygen atom has captured one of the lone pairs.

Dinitrogen tetroxide is a gas which decomposes to the more familiar brown air pollutant nitrogen dioxide (NO_2) at higher temperatures. The only way we can draw it seems most unsatisfactory: both nitrogens with positive charges! Even though these are not full positive charges, and this molecule does bring into focus the inadequacy of some valence bond representations, perhaps our discomfort with the structure is an indication of why this N–N bond is so weak...

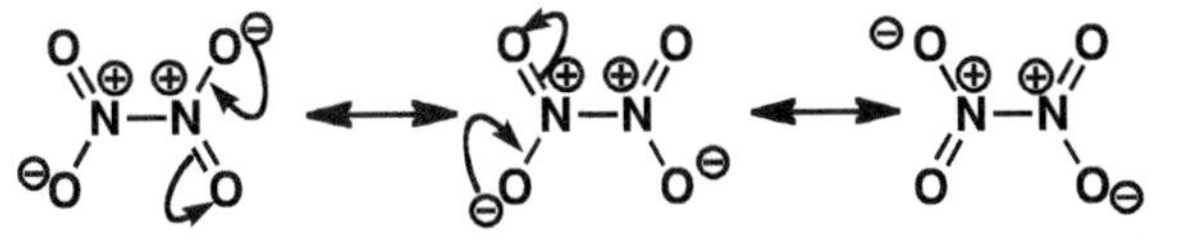

PROBLEM 5

Which (parts) of these compounds are aromatic? Justify your answer with some electron counting. You may treat rings separately or together as you wish. You may notice that two of them are compounds we met in problem 2 of this chapter.

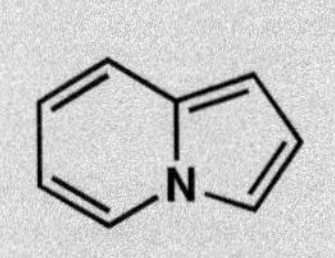

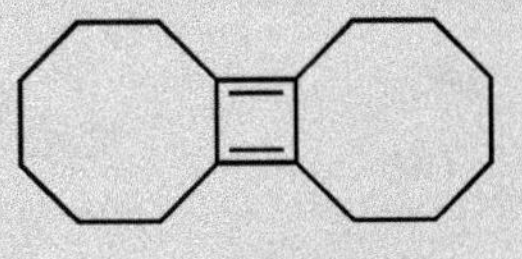

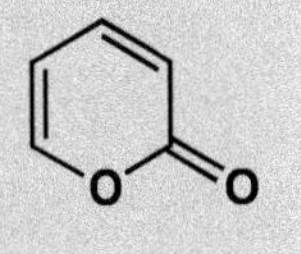

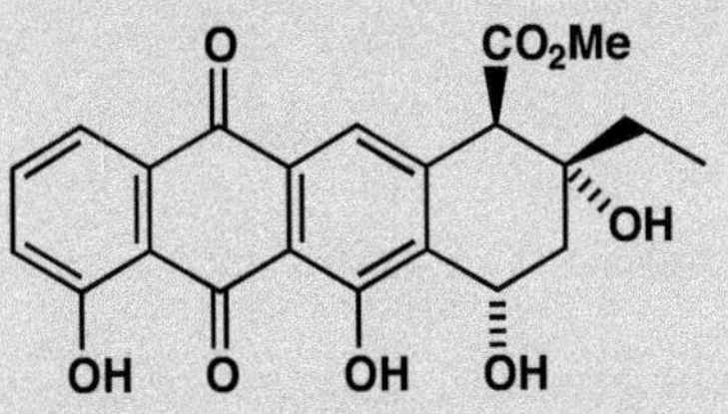

aklavinone: a tetracycline antibiotic

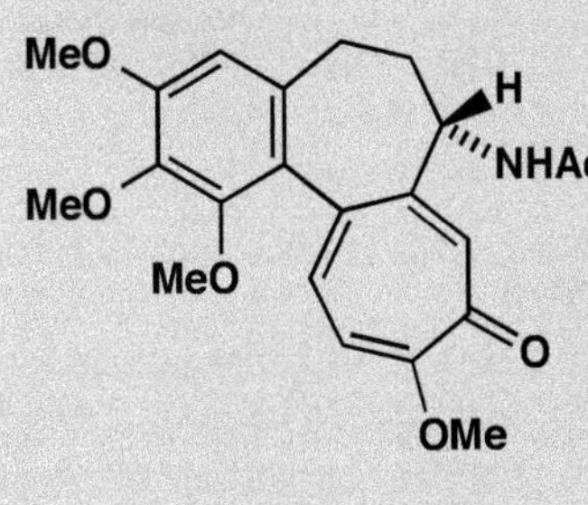

colchicine: a compound from the autumn crocus used to treat gout

Purpose of the problem

A simple exploration of the idea of aromaticity: can you count up to six? Remember: count only those π electrons in the ring and on no account put one electron on both atoms at either end of the bond: put the electrons where they are—in the bond.

Suggested solution

The numbers show how many π electrons there are in each bond or at each atom. The first compound has a lone pair on nitrogen in a p-orbital shared between both rings. Each ring has six electrons and the periphery of the whole molecule has ten electrons. Both rings and the entire molecule are aromatic. The second has four π electrons only so there is no aromaticity anywhere. The third has six π electrons in the ring including the lone pair on oxygen but not including the carbonyl group which is outside the ring. The compound is aromatic.

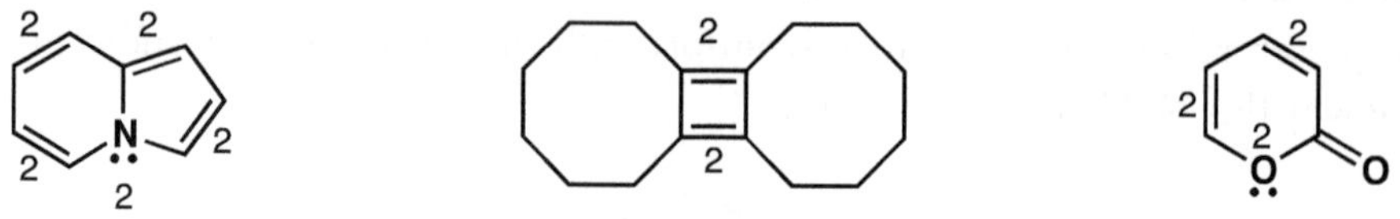

For the rest we have put the number of π electrons inside each ring and there are two aromatic rings in each compound. Again we don't count carbonyl group electrons as they are outside the ring. So one ring in aklavinone has only four electrons and is not aromatic while one of the seven-membered rings in colchicine is aromatic. Each compound has one saturated ring that cannot be aromatic.

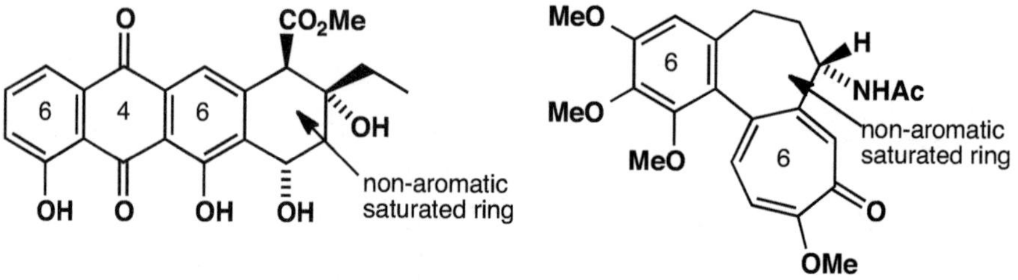

PROBLEM 6

The following compounds are considered to be aromatic. Account for this by identifying the appropriate number of delocalized electrons.

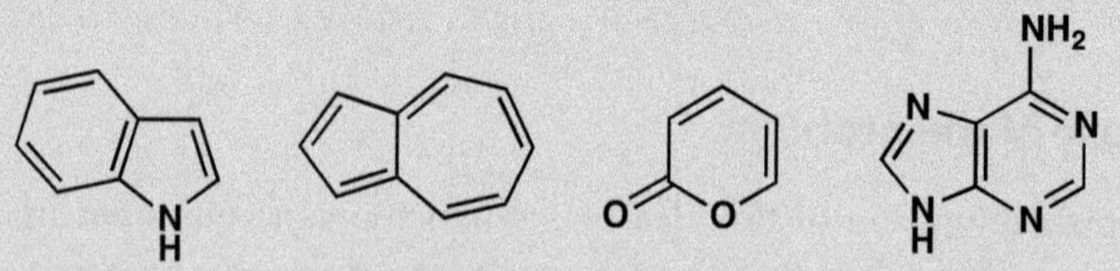

| indole | azulene | α-pyrone | adenine |

Purpose of the problem

Accounting for aromaticity in less familiar circumstances.

Suggested solution

Indole, as drawn here, has eight double bonds, which give eight delocalized electrons. To be aromatic, it needs 2n+2, so two more electrons must come from the lone pair of the nitrogen atom.

Azulene is isomeric with naphthalene, and it's quite easy to find the ten electrons – four in one ring and six in the other.

Pyridone has six electrons—two from the double bonds in the ring and two from nitrogen. That means that the carbonyl group, whose double bond is not part of the ring, does not contribute to the aromatic sextet. This is generally true for double bonds which stick out of the ring—see problem 2.

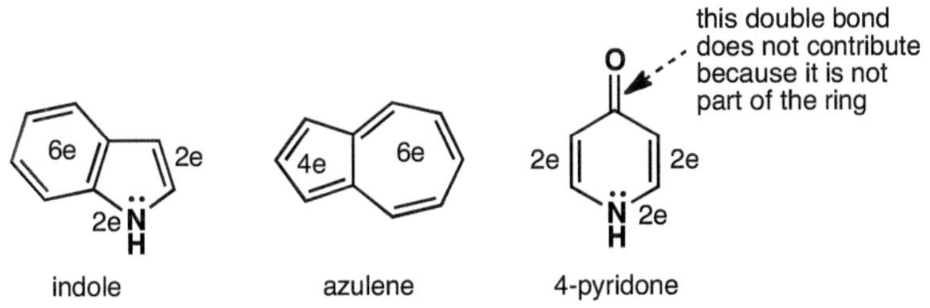

indole   azulene   4-pyridone

Adenine is one of the four bases which carry the genetic code in DNA. Its ten electrons arise as shown: eight from the double bonds and two from one of the nitrogen atoms in the five-membered ring. The other three nitrogens don't contribute their lone pairs, because they are not delocalized—like the lone pair in pyridine.

■ The details of bonding in pyridine will be explained in chapter 29, on page 724 of the textbook.

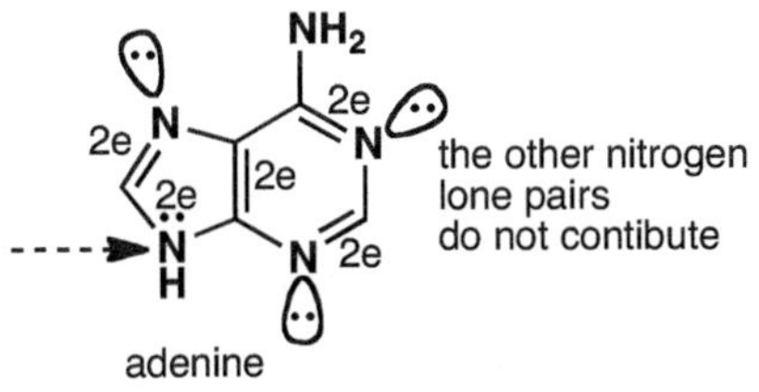

adenine

PROBLEM 7

Cyclooctatetraene (see p. 158 of the textbook) reacts readily with potassium metal to form a salt, K_2[cyclooctatetraene]. What shape do you expect the ring to have in this compound? A similar reaction of hexa(trimethylsilyl)benzene with lithium also gives a salt. What shape do you expect this ring to have?

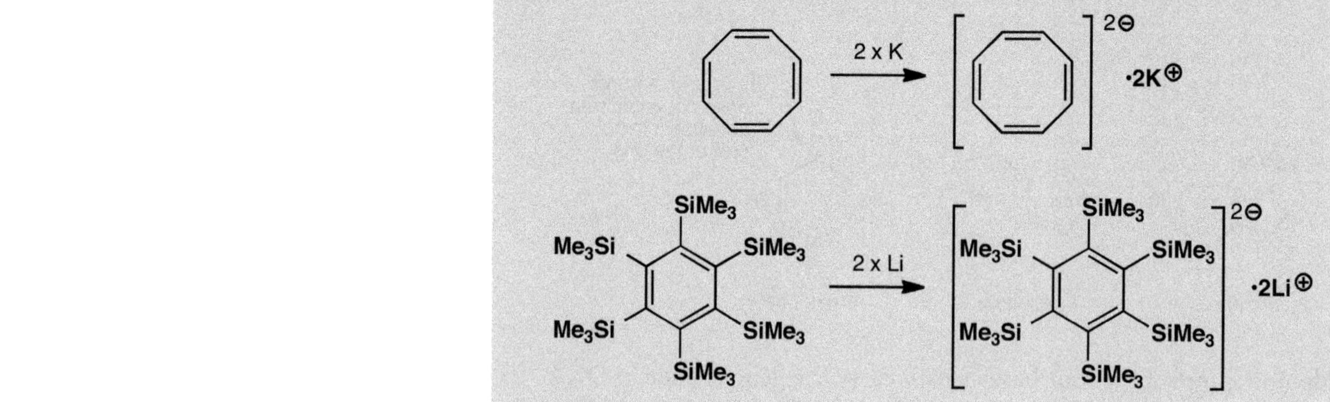

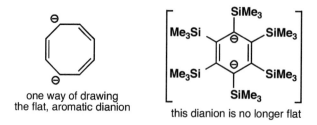

Purpose of the problem

The consequences of aromaticity and 'antiaromaticity'.

Suggested solution

Cyclooctatetraene, as explained on p. 158 of the textbook, is 'tub-shaped' and not planar, because its eight π electrons do not form a 2n+2 number. However, two atoms of potassium can reduce the cyclooctatetraene to a dianion by giving it two electrons, so now it has ten electrons, is aromatic, and is planar. Just one of the many possible delocalized structures of the product is shown here.

one way of drawing
the flat, aromatic dianion

this dianion is no longer flat

When lithium reduces hexa(trimethylsilyl)benzene, the aromatic sextet is increased to a total of eight delocalized electrons, so the compound is no longer aromatic. The six membered ring in the salt is no longer flat.

PROBLEM 8

How would you expect the hydrocarbon below to react with bromine, Br₂?

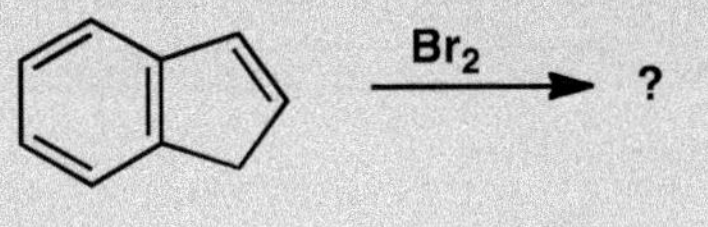

Purpose of the problem

The consequences of aromaticity for reactivity.

Suggested solution

Aromatic rings typically react by substitution, so that they can retain the aromatic sextet. By contrast, alkenes react by electrophilic addition—the classic test for an alkene is that they decolourize bromine water. So, how will our hydrocarbon (known as indene) react? It contains an aromatic ring, but the five-membered ring is not aromatic—it contains a saturated carbon atom. So there is a choice of substitution on the six-membered ring or addition to the alkene in the five-membered ring. Alkenes are more reactive than benzene, so the alkene reacts first:

■ This contrasting behaviour was one of the pieces of evidence for aromaticity, and is on page 157 of the textbook.

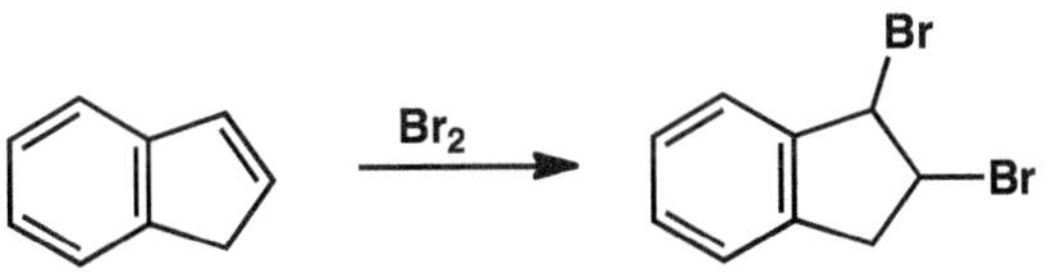

product of addition to the alkene

PROBLEM 9

In aqueous solution, acetaldehyde (ethanal) is about 50% hydrated. Draw the structure of the hydrate of acetaldehyde. Under the same conditions, the hydrate of *N,N*-dimethylformamide is undetectable. Why the difference?

acetaldehyde   *N,N*-dimethylformamide

Purpose of the problem

The consequences of delocalization for reactivity.

Suggested solution

As you saw in chapter 6, aldehydes are readily hydrated. For amides, however, there is a price to pay: the delocalization that contributes to the stability of the amide would be lost on hydration, so dimethylformamide is not hydrated in aqueous solution.

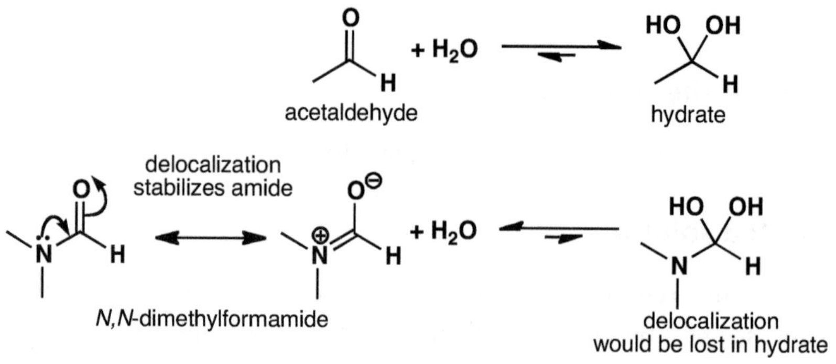

Acidity, basicity, and pK_a

8

Connections

⇒ Building on

- Conjugation and molecular stability ch7
- Curly arrows represent delocalization and mechanisms ch5
- How orbitals overlap to form conjugated systems ch4

Arriving at

- Why some molecules are acidic and others basic
- Why some acids are strong and others weak
- Why some bases are strong and others weak
- Estimating acidity and basicity using pH and pK_a
- Structure and equilibria in proton transfer reactions
- Which protons in more complex molecules are more acidic
- Which lone pairs in more complex molecules are more basic
- Quantitative acid/base ideas affecting reactions and solubility
- Effects of quantitative acid/base ideas on medicine design

⇒ Looking forward to

- Acid and base catalysis in carbonyl reactions ch10 & ch11
- The role of catalysts in organic mechanisms ch12
- Making reactions selective using acids and bases ch23
- More details on acid and base catalysis ch39

Organic compounds are more soluble in water as ions

Most organic compounds are insoluble in water. But sometimes it's necessary to make them dissolve, perhaps by converting them to anions or cations. Water can solvate both cations and anions, unlike some of the solvents you will meet later. A good way of dissolving an organic acid is to put it in basic solution: the base deprotonates the acid to give an anion. A simple example is aspirin: whilst the acid itself is not very soluble in water, the sodium salt is much more soluble. The sodium salt forms with the weak base, sodium hydrogencarbonate.

Water is special for many reasons, and it falls into a class of solvents we call *polar protic* solvents. We will discuss other solvents in this class, as well as *polar aprotic* solvents (such as acetone and DMF) and *nonpolar* solvents (such as toluene and hexane) in Chapter 12.

aspirin:
not very soluble
in water

NaHCO$_3$

the sodium salt
of aspirin is
more soluble
in water

The sodium or calcium salt of 'normal' aspirin is sold as 'soluble aspirin'. But when the pH of a solution of aspirin's sodium salt is lowered, the amount of the 'normal' acidic form present increases and the solubility decreases. In the acidic environment of the stomach (around pH 1–2), soluble aspirin will be converted back to the normal acidic form and precipitate out of solution.

Online support. The icon ⌨ in the margin indicates that accompanying interactive resources are provided online to help your understanding: just type **www.chemtube3d.com/clayden/123** into your browser, replacing **123** with the number of the page where you see the icon. For pages linking to more than one resource, type **123-1, 123-2** etc. (replacing **123** with the page number) for access to successive links.

In the same way, organic bases such as amines can be dissolved by *lowering* the pH. Codeine (7,8-didehydro-4,5-epoxy-3-methoxy-17-methylmorphinan-6-ol) is a commonly used pain-killer. Codeine itself is not very soluble in water but it does contain a basic nitrogen atom that can be protonated to give a more soluble salt. It is usually encountered as a phosphate salt. The structure is complex, but that doesn't matter.

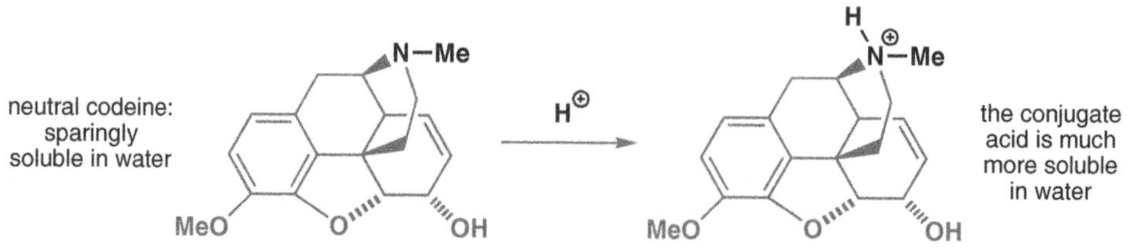

neutral codeine: sparingly soluble in water

the conjugate acid is much more soluble in water

Charged compounds can be separated by acid–base extraction

Adjusting the pH of a solution often provides an easy way to separate compounds. Separating a mixture of benzoic acid ($PhCO_2H$) and toluene (PhMe) is easy: dissolve the mixture in CH_2Cl_2, add aqueous NaOH, shake the mixture of solutions, and separate the layers. The CH_2Cl_2 layer contains all the toluene. The aqueous layer contains the sodium salt of benzoic acid. Addition of HCl to the aqueous layer precipitates the insoluble benzoic acid.

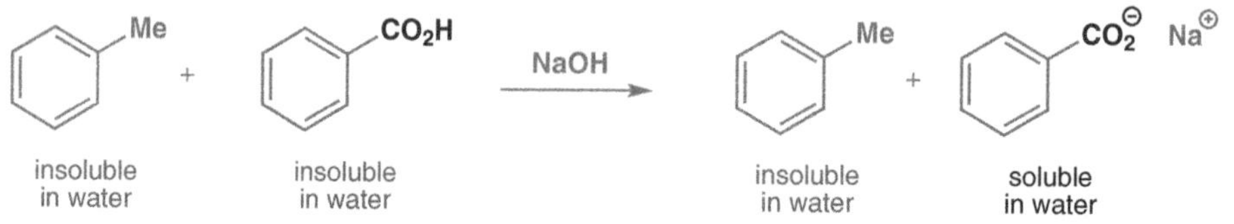

| insoluble in water | insoluble in water | | insoluble in water | soluble in water |

A more realistic separation is given in a modern practical book after a Cannizzaro reaction. You will meet this reaction in Chapters 26 and 39 but all you need to know now is that there are two products, formed in roughly equal quantities. Separation of these from starting material and solvent, as well as from each other, makes this a useful reaction.

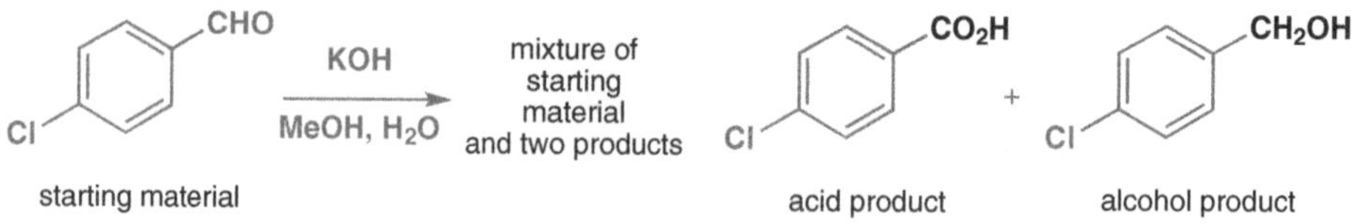

starting material

acid product

alcohol product

The products under the basic reaction conditions are the *salt* of the acid (soluble in water) and the alcohol (not soluble in water). Extraction with dichloromethane removes the alcohol and leaves the salt in the aqueous layer along with solvent methanol and residual KOH. Rotary evaporation of the CH_2Cl_2 layer gives crystalline alcohol and acidification of the aqueous layer precipitates the neutral acid.

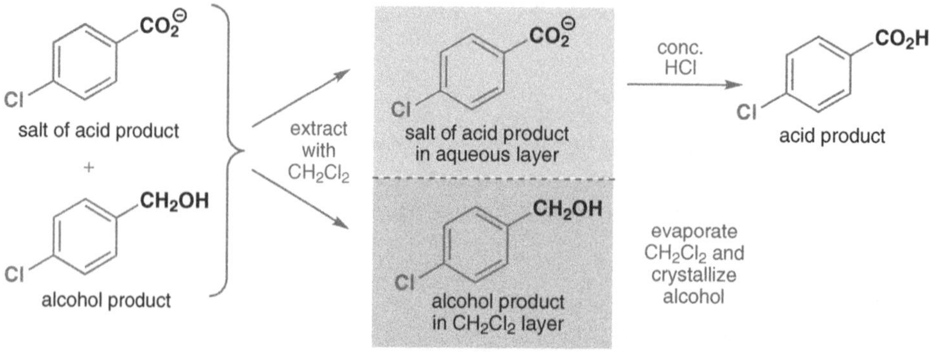

salt of acid product

+

alcohol product

extract with CH_2Cl_2

salt of acid product in aqueous layer

alcohol product in CH_2Cl_2 layer

conc. HCl

acid product

evaporate CH_2Cl_2 and crystallize alcohol

In the same way, any basic compounds dissolved in an organic layer can be extracted by washing the layer with dilute aqueous acid and recovered by raising the pH, which will precipitate out the less soluble neutral compound. A general way to make amines is by 'reductive amination.' Ignore the details of this reaction for now (we come back to them in Chapter 11) but consider how the amine might be separated from starting material, by-products, and solvent.

aldehyde: not basic — imine weakly basic — amine: basic

As the reaction mixture is weakly acidic, the amine will be protonated and will be soluble in water. The starting material and intermediate (of which very little is present anyway) are soluble in organic solvents. Extracting the aqueous layer and neutralizing with NaOH gives the amine.

Whenever you do any extractions or washes in practical experiments, just stop and ask yourself: 'What is happening here? In which layer is my compound and why?' You will then be less likely to throw away the wrong layer (and your precious compound)!

Acids, bases, and pK_a

If we are going to make use of the acid–base properties of compounds as we have just described, we are going to need a way of measuring *how acidic* or *how basic* they are. Raising the pH leads to deprotonation of aspirin and lowering the pH leads to protonation of codeine, but *how far* do we have to raise or lower the pH to do this? The measure of acidity or basicity we need is called pK_a. The value of pK_a tells us how acidic (or not) a given hydrogen atom in a compound is. Knowing about pK_a tells us, for example, that the amine product from the reaction just above will be protonated at weakly acidic pH 5, or that only a weak base (sodium hydrogen carbonate) is needed to deprotonate a carboxylic acid such as aspirin. It is also useful because many reactions proceed through protonation or deprotonation of one of the reactants (you met some examples in Chapter 6), and it is obviously useful to know what strength acid or base is needed. It would be futile to use too weak a base to deprotonate a compound but, equally, using a very strong base where a weak one would do risks the result of cracking open a walnut with a sledge hammer.

The aim of this chapter is to help you to understand *why* a given compound has the pK_a that it does. Once you understand the trends involved, you should have a good feel for the pK_a values of commonly encountered compounds and also be able to predict roughly the values for unfamiliar compounds.

Benzoic acid preserves soft drinks

Benzoic acid is used as a preservative in foods and soft drinks (E210). Like acetic acid, it is only the acid form that is effective as a bactericide. Consequently, benzoic acid can be used as a preservative only in foodstuffs with a relatively low pH, ideally less than its pK_a of 4.2. This isn't usually a problem: soft drinks, for example, typically have a pH of 2–3. Benzoic acid is often added as the sodium salt (E211), perhaps because this can be added to the recipe as a concentrated solution in water. At the low pH in the final drink, most of the salt will be protonated to give benzoic acid proper, which presumably remains in solution because it is so dilute.

Acidity

Let's start with two simple, and probably familiar, definitions:

- An acid is a species having a tendency to lose a proton.
- A base is a species having a tendency to accept a proton.

'The proton is a unique chemical species, being a bare nucleus. As a consequence it has no independent existence in the condensed state and is invariably found bound by a pair of electrons to another atom.' Ross Stewart, *The Proton: Applications to Organic Chemistry*, Academic Press, Orlando, 1985, p. 1.

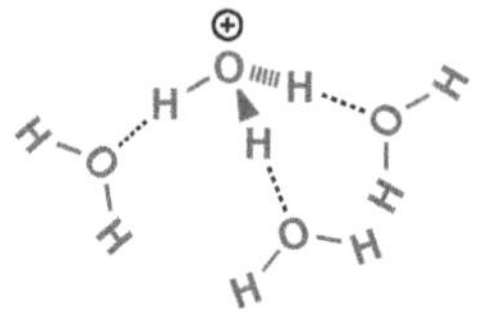

a structure for a solvated hydronium ion in water: the dashed bonds represent hydrogen bonds

An isolated proton is extremely reactive—formation of H_3O^+ in water

Gaseous HCl is not an acid at all—it shows no tendency to dissociate into H^+ and Cl^- as the H–Cl bond is strong. But hydrochloric acid—that is, a solution of HCl in water—*is* a strong acid. The difference is that an isolated proton H^+ is too unstable to be encountered under normal conditions, but in water the hydrogen of HCl is transferred to a water molecule and not released as a free species.

The chloride anion is the same in both cases: the only difference is that a very unstable naked proton would have to be the other product in the gas phase but a much more stable H_3O^+ cation would be formed in water. In fact it's even better than that, as other molecules of water cluster round ('solvate') the H_3O^+ cation, stabilizing it with a network of hydrogen bonds.

That is why HCl is an acid in water. But how strong an acid is it? This is where chloride plays a role: hydrochloric acid is a strong acid because chloride ion is a stable anion. The sea is full of it! Water is needed to reveal the acidic quality of HCl, and acidity is determined in water as the standard solvent. If we measure acidity in water, what we are really measuring is how much our acid transfers a proton to a water molecule.

HCl transfers its proton almost completely to water, and is a strong acid. But the transfer of protons to water from carboxylic acids is only partial. That is why carboxylic acids are weak acids. Unlike the reaction of HCl with water, the reaction below is an equilibrium.

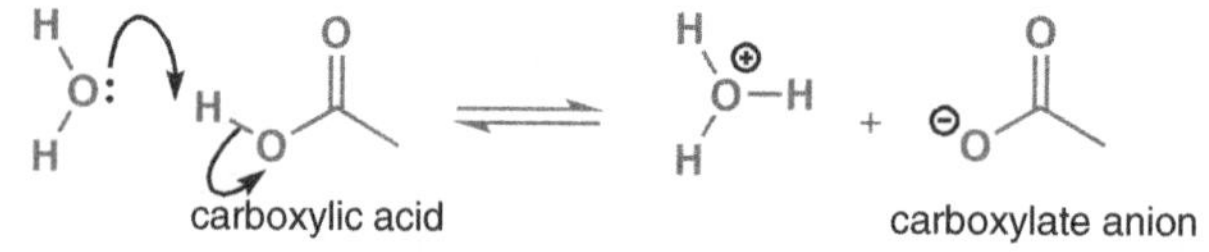

The pH scale and pK_a

The amount of H_3O^+ in any solution in water is described using the pH scale. pH is simply a measure of the concentration of H_3O^+ on a logarithmic scale, and it is characteristic of any aqueous acid—it depends not only on what the acid is (hydrochloric, acetic, etc.) but also on how concentrated the acid is.

- **pH is the negative logarithm of the H_3O^+ concentration.**

$pH = -\log[H_3O^+]$

You will already know that neutrality is pH 7 and that below pH 7 water is increasingly acidic while above pH 7 it is increasingly basic. At higher pH, there is little H_3O^+ in the solution and more hydroxide ion, but at lower pH there is more H_3O^+ and little hydroxide.

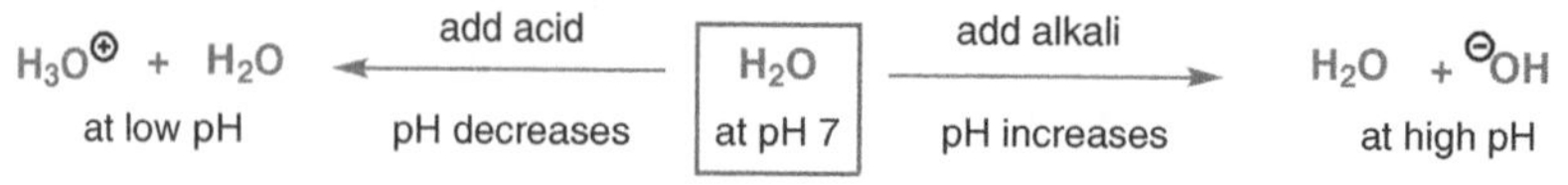

The reason that higher pH means less H_3O^+ is because the arbitrary definition of pH is the *negative* logarithm (to the base 10) of the H_3O^+ concentration. To summarize in a diagram:

■ We will explain later why this scale seems to stop at pH 0 and 14—in fact these numbers are approximate, but easy to remember.

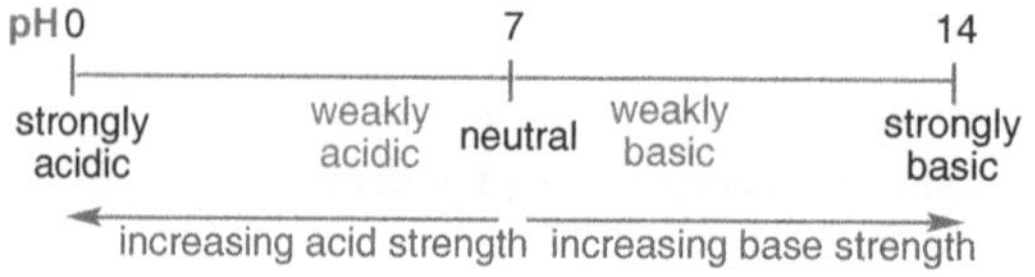

pH is used to measure the acidity of aqueous solutions, but what about the inherent tendency of an acidic compound to give up H^+ to water and form these acid solutions? A good way of measuring this tendency is to find the pH at which a solution contains exactly the same amount of the protonated, acidic form and its deprotonated, basic form. This number, which is characteristic of any acid, is known as the pK_a. In the example just above, this would be the pH where the amount of the carboxylic acid is matched by the amount of its carboxylate salt—which happens to be at about pH 5: the pK_a of acetic acid is 4.76.

We'll come back to a more formal definition of pK_a later, but first we need to look more closely at this pair of species—the protonated acid and its deprotonated, basic partner.

Every acid has a conjugate base

Looking back at the equilibrium set up when acetic acid dissolves in water, but drawing the mechanism of the back reaction, we see acetate ion acting as a base and H_3O^+ acting as an acid. In all equilibria involving just proton transfer a species acting as a base on one side acts as an acid on the other. We describe H_3O^+ as the *conjugate acid* of water and water as the *conjugate base* of H_3O^+. In the same way, acetic acid is the conjugate acid of acetate ion and acetate ion is the conjugate base of acetic acid.

acetate as base

> ● **For any acid and any base:**
>
> $$B: + \ HA \ \rightleftharpoons \ BH^{\oplus} + \ A^{\ominus}$$
>
> **AH is an acid and A⁻ is its conjugate base and B is a base and BH⁺ is its conjugate acid. That is, every acid has a conjugate base associated with it and every base has a conjugate acid associated with it.**

Water doesn't have to be one of the participants—if we replace water in the reaction we have been discussing with ammonia, we now have ammonia as the conjugate base of NH_4^+ (the ammonium cation) and the ammonium cation as the conjugate acid of ammonia. What is different is the position of equilibrium: ammonia is more basic than water and now the equilibrium will be well over to the right. As you will see, pK_a will help us assess where equilibria like these lie.

The amino acids you met in Chapter 2 have carboxylic acid and amine functional groups within the same molecule. When dissolved in water, they transfer a proton from the CO_2H group to the NH_2 group and form a *zwitterion*. This German term describes a double ion having positive and negative charges in the same molecule.

Water can behave as an acid or as a base

So far we have seen water acting as a (very weak) base to form H_3O^+. If we added a strong base, such as sodium hydride, to water, the base would deprotonate the water to give hydroxide ion,

HO$^-$, and here the water would be acting as an acid. It's amusing to notice that hydrogen gas is the conjugate acid of hydride ion, but more important to note that hydroxide ion is the conjugate base of water.

$$NaH \rightleftharpoons Na^{\oplus} + H^{\ominus} \quad H\overset{\,\,}{O}-H \longrightarrow H_2 + {}^{\ominus}OH$$

Water is a weak acid and a weak base so we need a strong acid like HCl to give much H_3O^+, and a strong base, like hydride ion, to give much hydroxide ion.

The ionization of water

The concentration of H_3O^+ ions in water is very low indeed at 10^{-7} mol dm^{-3}. Pure water at 25 °C therefore has a pH of 7.00. Hydronium ions in pure water can arise only from water protonating (and deprotonating) itself. One molecule of water acts as a base, deprotonating another that acts as an acid. For every H_3O^+ ion formed, a hydroxide ion must also be formed, so that in pure water at pH 7 the concentrations of H_3O^+ and hydroxide ions must be equal: $[H_3O^+] = [HO^-] = 10^{-7}$ mol dm^{-3}.

$$H_2O: \quad H\overset{\,\,}{O}-H \rightleftharpoons H_3O^{\oplus} + {}^{\ominus}OH$$

■ Water is still safe to drink because the concentrations of hydronium and hydroxide ions are *very* small (10^{-7} mol dm^{-3} corresponds to about 2 parts per billion). This very low concentration means that there are not enough free hydronium or hydroxide ions in water to do any harm when you drink it, but neither are there enough to provide acid or base catalysts for reactions which need them.

The product of these two concentrations is known as the *ionization constant* (or as the *ionic product*) of water, K_W, with a value of 10^{-14} mol^2 dm^{-6} (at 25 °C). This is a constant in aqueous solutions, so if we know the hydronium ion concentration (which we can get by measuring the pH), we also know the hydroxide concentration since the product of the two concentrations always equals 10^{-14}.

So, roughly at what pH does water become mostly H_3O^+ ions and at what pH mostly hydroxide ions? We can now add two additional pieces of information to the approximate chart we gave you before. At pH 7, water is almost entirely H_2O. At about pH 0, the concentrations of water and H_3O^+ ions are about the same and at about pH 14, the concentrations of hydroxide ions and water are about the same.

■ The figures 0 and 14 are approximate—there is a simple reason why this is so, which we will explain shortly. But you see now why we end the scale at these points—below 0 and above 14 there is little scope for varying the concentration of H_3O^+.

$$H_3O^{\oplus} \rightleftharpoons H_2O \qquad H_2O \qquad H_2O \rightleftharpoons {}^{\ominus}OH$$

at pH ~ 0   pH 0   7   14   at pH ~ 14
$[H_2O] = [H_3O^+]$ $[H_2O] = [HO^-]$

strongly acidic   weakly acidic   neutral   weakly basic   strongly basic

increasing acid strength   increasing base strength

Acids as preservatives

Acetic acid is used as a preservative in many foods, for example pickles, mayonnaise, bread, and fish products, because it prevents bacteria and fungi growing. However, its fungicidal nature is not due to any lowering of the pH of the foodstuff. In fact, it is the undissociated acid that acts as a bactericide and a fungicide in concentrations as low as 0.1–0.3%. Besides, such a low concentration has little effect on the pH of the foodstuff anyway.

Although acetic acid can be added directly to a foodstuff (disguised as E260), it is more common to add vinegar, which contains between 10 and 15% acetic acid. This makes the product more 'natural' since it avoids the nasty 'E numbers'. Actually, vinegar has also replaced other acids used as preservatives, such as propionic (propanoic) acid (E280) and its salts (E281, E282, and E283).

The definition of pK_a

When we introduced you to pK_a on p. 167, we said it is the pH at which an acid and its conjugate base are present in equal concentrations. We can now be more precise about the definition of pK_a. pK_a is the log (to the base ten) of the equilibrium constant for the dissociation of the acid. For an acid HA this is:

$$HA + H_2O \overset{K_a}{\rightleftharpoons} H_3O^{\oplus} + A^{\ominus} \qquad pK_a = -\log K_a \qquad K_a = \frac{[H_3O^+][A^-]}{[AH]}$$

The concentration of water is ignored in the definition because it is also constant (at 25 °C).

Because of the minus sign in the definition (it's there too in the definition of pH) the lower the pK_a the larger the equilibrium constant and the stronger the acid. You may find the way we introduced pK_a more helpful as a concept for visualizing pK_a: any acid is half dissociated in a solution whose pH matches the acid's pK_a. At a pH above the pK_a the acid exists largely as its conjugate base (A⁻) but at a pH below the pK_a the acid largely exists as HA.

With pK_a we can put figures to the relative strengths of hydrochloric and acetic acid we introduced earlier. HCl is a much stronger acid than acetic acid: the pK_a of HCl is around –7 compared to 4.76 for acetic acid. This tells us that in solution K_a for hydrogen chloride is 10^7 mol dm⁻³. This is an enormous number: only one molecule in 10,000,000 is not dissociated, so it is essentially fully dissociated. But K_a for acetic acid is only $10^{-4.76} = 1.74 \times 10^{-5}$ mol dm⁻³ so it is hardly dissociated at all: only a few molecules in every million of acetic acid are present as the acetate ion.

$$HCl + H_2O \longrightarrow H_3O^{\oplus} + Cl^{\ominus} \qquad K_a = 10^7$$

$$H_2O + HO\!-\!\!\overset{O}{\overset{\|}{\diagup}} \rightleftharpoons H_3O^{\oplus} + {}^{\ominus}O\!-\!\!\overset{O}{\overset{\|}{\diagup}} \qquad K_a = 1.74 \times 10^{-5}$$

What about the pK_a of water? You know the figures already: K_a for water is $[H_3O^+] \times [HO^-]/[H_2O] = 10^{-14}/55.5$. So p$K_a = -\log[10^{-14}/55.5] = 15.7$. Now you see why water isn't really quite half dissociated at pH 14—the concentration of water in the equation means that the two ends of the scale on p. 168 are not at 0 and 14, but at –1.7 and 15.7.

A graphical description of the pK_a of acids and bases

For both cases, adjusting the pH alters the proportions of the acid form and of the conjugate base. The graph plots the concentration of the free acid AH (green curve) and the ionized conjugate base A⁻ (red curve) as percentages of the total concentration as the pH is varied. At low pH the compound exists entirely as AH and at high pH entirely as A⁻. At the pK_a the concentration of each species, AH and A⁻, is the same. At pHs near the pK_a the compound exists as a mixture of the two forms.

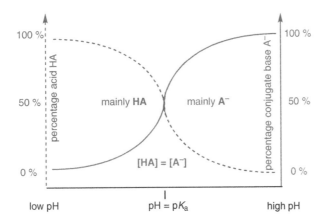

Now we have established why you need to understand acids and bases, we must move on to consider why some acids are stronger than other acids and some bases stronger than other bases. To do this we must be able to estimate the pK_a of common classes of organic compounds.

How concentrated is water? One mole of pure water has a mass of 18 g and occupies 18 cm³. So, in 1 dm³, there are 1000/18 = 55.56 mol. Water is a 55.56 mol dm⁻³ solution of water...in water.

You do not need to learn exact figures for pK_a values, but you will certainly need to develop a feel for approximate values—we will guide you towards which figures are worth learning and which you can leave to be looked up when you need them.

An acid's pK_a depends on the stability of its conjugate base

The stronger the acid, the easier it is to ionize, which means that it must have a stable conjugate base. Conversely, a weak acid is reluctant to ionize because it has an *unstable* conjugate base. The other side of this coin is that unstable anions A⁻ make strong bases and their conjugate acids AH are weak acids.

> ● **Acid and conjugate base strength**
>
> **The stronger the acid HA, the weaker its conjugate base A⁻.**
>
> **The stronger the base A⁻, the weaker its conjugate acid AH.**

For example, hydrogen iodide has a very low pK_a, about –10. This means that HI is a strong enough acid to protonate almost anything. Its conjugate base, iodide ion, is therefore not basic at all—it will not deprotonate anything. A very powerful base is methyllithium, MeLi. Although it is actually a covalent compound, as we discuss in Chapter 9, for the purpose of the discussion here you can think of MeLi as $CH_3^- Li^+$. CH_3^- can accept a proton to become neutral methane, CH_4. Methane is therefore the conjugate acid. Clearly, methane isn't at all acidic—its pK_a is estimated to be 48. The table below gives a few inorganic compounds and their approximate pK_a values.

The approximate pK_a values of some inorganic compounds

Acid	pK_a	Conjugate base	Acid	pK_a	Conjugate base	Acid	pK_a	Conjugate base
H_2SO_4	–3	HSO_4^-	H_3O^+	–1.7	H_2O	NH_4^+	9.2	NH_3
HCl	–7	Cl⁻	H_2O	15.7	HO^-	NH_3	33	NH_2^-
HI	–10	I⁻	H_2S	7.0	HS^-			

Notice that the lower down the periodic table we go, the stronger the acid. Notice also that oxygen acids are stronger than nitrogen acids. We have also put down more exact pK_a values for water but you need remember only the approximate values of 0 and 14. Over the next few pages we shall be considering the reasons for these differences in acid strength but we are first going to consider the simple consequences of mixing acids or bases of different strengths. Notice the vast range covered by pK_a values: from around –10 for HI to nearly 50 for methane. This corresponds to a difference of 10^{60} in the equilibrium constant.

The choice of solvent limits the pK_a range we can use

In water, we can measure the pK_a of an acid only if the acid does not completely protonate water to give H_3O^+ or completely deprotonate it to give HO^-. We are restricted roughly to pH –1.7 to 15.7, beyond which water is more than 50% protonated or deprotonated. The strength of acids or bases we can use in any solvent is limited by the acidity and basicity of the solvent itself. Think of it this way: say you want to remove the proton from a compound with a high pK_a, say 25–30. It would be impossible to do this in water since the strongest base we can use is hydroxide. If you add a base stronger than hydroxide, it won't deprotonate your compound, it will just deprotonate water and make hydroxide anyway. Likewise, acids stronger than H_3O^+ can't exist in water: they just protonate water completely to make H_3O^+. If you do need a stronger base than OH^- (or a stronger acid than H_3O^+, but this is rarer) you must use a different solvent.

Let's take acetylene as an example. Acetylene (ethyne) has pK_a 25. This is remarkably low for a hydrocarbon (see below for why) but, even so, hydroxide (the strongest base we could have in aqueous solution, pK_a 15.7) would establish an equilibrium where only 1 in $10^{9.3}$ ($10^{15.7}/10^{25}$), or about 1 in 2 billion, ethyne molecules are deprotonated. We can't use a stronger base than hydroxide, since, no matter what strong base we dissolve in water, we will only at best get hydroxide ions. So, in order to deprotonate ethyne to any appreciable extent, we must use a different solvent—one that does not have a pK_a less than 25.

Conditions often used to do this reaction are sodium amide (NaNH$_2$) in liquid ammonia. Using the pK_a values of NH$_3$ (ca. 33) and ethyne (25) we would estimate an equilibrium constant for this reaction of 10^8 ($10^{-25}/10^{-33}$)—well over to the right. Amide ions can be used to deprotonate alkynes.

Since we have an upper and a lower limit on the strength of an acid or base that we can use in water, this poses a bit of a problem: how do we know that the pK_a for HCl is more negative than that of H$_2$SO$_4$ if both completely protonate water? How do we know that the pK_a of methane is greater than that of ethyne since both the conjugate bases fully deprotonate water? The answer is that we can't simply measure the equilibrium for the reaction in water—we can do this only for pK_a values that fall between the pK_a values of water itself. Outside this range, pK_a values are determined in other solvents and the results are extrapolated to give a value for what the pK_a in water might be.

■ Because the pK_a values for very strong acids and bases are so hard to determine, you will find that they often differ in different texts—sometimes the values are no better than good guesses! However, while the absolute values may differ, the relative values (which is the important thing because we need only a rough guide) are usually consistent.

Constructing a pK_a scale

We now want to look at ways to rationalize, and estimate, the different pK_a values for different compounds—we wouldn't want to have to memorize all the values. You will need to get a feel for the pK_a values of different compounds and if you know what factors affect them it will make it much easier to predict an approximate pK_a value, or at least understand why a given compound has the pK_a value that it does.

$$\text{AH (solvent)} \rightleftharpoons \text{A}^- \text{ (solvent)} + \text{H}^+ \text{ (solvent)}$$

A number of factors affect the strength of an acid AH. These include:

1. The intrinsic stability of the conjugate base, anion A$^-$. Stability can arise by having the negative charge on an electronegative atom or by spreading the charge over several atoms (delocalization) groups. Either way, the more stable the conjugate base, the stronger the acid HA.

2. Bond strength A–H. Clearly, the easier it is to break this bond, the stronger the acid.

3. The solvent. The better the solvent is at stabilizing the ions formed, the easier it is for the reaction to occur.

● **Acid strength**

The most important factor in the strength of an acid is the stability of the conjugate base—the more stable the conjugate base, the stronger the acid.

An important factor in the stability of the conjugate base is which element the negative charge is on—the more electronegative the element, the more stable the conjugate base.

The negative charge on an electronegative element stabilizes the conjugate base

The pK_a values for the 'hydrides' of the first row elements CH$_4$, NH$_3$, H$_2$O, and HF are about 48, 33, 16, and 3, respectively. This trend is due to the increasing electronegativities across the period: F$^-$ is much more stable than CH$_3^-$, because fluorine is much more electronegative than carbon.

Acid	Conjugate base	pK_a
methane CH$_4$	CH$_3^-$	~48
ammonia NH$_3$	amide ion NH$_2^-$	~33
water H$_2$O	hydroxide ion HO$^-$	~16
HF	fluoride ion F$^-$	3

Weak A–H bonds make stronger acids

However, on descending group VII (group 17), the pK_a values for HF, HCl, HBr, and HI decrease: 3, –7, –9, and –10. Since the electronegativities decrease on descending the group we might expect an increase in pK_a. The decrease is due to the weakening bond strengths on descending the group and to some extent the way in which the charge can be spread over the increasingly large anions.

Acid	Conjugate base	pK_a
HF	fluoride ion F$^-$	3
HCl	chloride ion Cl$^-$	–7
HBr	bromide ion Br$^-$	–9
HI	iodide ion I$^-$	–10

Delocalization of the negative charge stabilizes the conjugate base

The acids HClO, HClO$_2$, HClO$_3$, and HClO$_4$ have pK_a values 7.5, 2, –1, and about –10, respectively. In each case the acidic proton is on an oxygen attached to chlorine, that is, *we are removing a proton from the same environment in each case*. Why then is perchloric acid, HClO$_4$, some 17 orders of magnitude stronger in acidity than hypochlorous acid, HClO? Once the proton is removed, we end up with a negative charge on oxygen. For hypochlorous acid, this is localized on the one oxygen. With each successive oxygen, the charge can be more delocalized, and this makes the anion more stable. For example, with perchloric acid, the negative charge can be delocalized over all four oxygen atoms.

Acid	Conjugate base	pK_a
hypochlorous acid HO–Cl	ClO$^-$	7.5
chlorous acid HO–ClO	ClO$_2^-$	2
chloric acid HO–ClO$_2$	ClO$_3^-$	–1
perchloric acid HO–ClO$_3$	ClO$_4^-$	–10

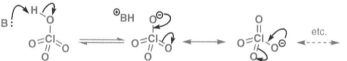

the negative charge on the perchlorate anion is delocalized over all four oxygens

That the charge is spread out over all the oxygen atoms equally is shown by electron diffraction studies: whereas perchloric acid has two types of Cl–O bond, one 163.5 pm and the other three 140.8 pm long, in the perchlorate anion all Cl–O bond lengths are the same, 144 pm, and all O–Cl–O bond angles are 109.5°. *Just to remind you:* these delocalization arrows do not indicate that the charge is actually moving from atom to atom. We discussed this in Chapter 7. These structures simply show that the charge is spread out in the molecular orbitals and mainly concentrated on the oxygen atoms.

Looking at some organic acids, we might expect alcohols to have a pK_a not far from that of water, and for ethanol that is correct (pK_a 15.9). If we allow the charge in the conjugate base to be delocalized over two oxygen atoms, as in acetate, acetic acid is indeed a much stronger acid (pK_a 4.8). The difference is huge: the conjugation makes acetic acid about 10^{10} times stronger.

ethoxide

charge localized on one oxygen

acetate

charge delocalized over two oxygens

It is even possible to have a negative charge of an organic acid delocalized over *three* atoms—as in the anions of the sulfonic acids. Methanesulfonic acid has a pK_a of –1.9.

charge delocalized over three oxygens

Even delocalization into a hydrocarbon part of the molecule increases acid strength. In phenol, PhOH, the OH group is directly attached to a benzene ring. On deprotonation, the negative charge can again be delocalized, not onto other oxygen atoms but into the aromatic ring itself. The effect of this is to stabilize the phenoxide anion relative to the conjugate base of cyclohexanol, where no delocalization is possible, and this is reflected in the pK_a values of the two compounds: 10 for phenol but 16 for cyclohexanol.

these lone pairs in sp^2 orbitals do not overlap with the π system of the ring

| cyclohexanol | localized | phenol | phenoxide | lone pair in p orbital overlaps with the π system of the ring | delocalization stabilizes the negative charge |
| pK_a 16 | anion | pK_a 10 | | | |

So now we can expand our chart of acid and base strengths to include the important classes of alcohols, phenols, and carboxylic acids. They conveniently, and memorably, have pK_a values of about 0 for the protonation of alcohols, about 5 for the deprotonation of carboxylic acids, about 10 for the deprotonation of phenols, and about 15 for the deprotonation of alcohols. The equilibria above each pK_a shows that at approximately that pH, the two species each form 50% of the mixture. You can see that carboxylic acids are weak acids, alkoxide ions (RO$^-$) are strong bases, and that it will need a strong acid to protonate an alcohol.

> ■ equilibrium arrow: ⇌
> delocalization arrow ↔
> Reminder: the equilibrium arrows mean two interconverting compounds. The double-headed arrow means two ways of drawing a conjugated structure.

> ■ It is worthwhile learning these approximate values.

If we need to make the anion of a phenol, a base such as NaOH will be good enough, but if we want to make an anion from an alcohol, we need a stronger base. Vogel (p. 986) suggests potassium carbonate (K$_2$CO$_3$) is strong enough to make an ether from phenol. The base strength of carbonate anion is about the same as that of phenoxide ion (PhO$^-$) so the two will be in equilibrium but enough phenoxide ion will be present for the reaction.

| phenol | phenoxide ion | phenyl allyl ether |

On the other hand, if we want to make the OH group into a good leaving group, we need to protonate it and a very strong acid will be needed. Sulfuric acid is used to make ethers from alcohols. Protonation of the OH groups leads to loss of water and formation of a cation. This reacts with more alcohol to give the ether. There is another example of this reaction in Chapter 5.

> ■ As you will discover in Chapters 10 and 15, a *leaving group* is simply a functional group that will leave the molecule, taking with it the pair of electrons that formed the bond. Leaving groups may be anions, such as bromide Br$^-$, or protonated groups such as the protonated alcohol in this example, which leaves as water.

151

The reaction scheme at the top shows:

Ph₂CH—OH (alcohol) →(H₂SO₄) Ph₂CH—OH₂⁺ →(– H₂O) Ph₂CH⁺ (cation) →(Ph₂CHOH) Ph₂CH—O—CHPh₂ (ether)

Nitrogen compounds as acids and bases

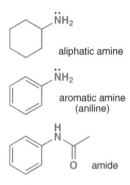

aliphatic amine

aromatic amine (aniline)

amide

The most important organic nitrogen compounds are amines and amides. Amine nitrogens can be joined to alkyl or aryl groups (in which case the amines are called anilines). They all have lone pairs on nitrogen and may have hydrogen atoms on nitrogen too. As nitrogen is less electronegative than oxygen, you should expect amines to be less acidic and more basic than alcohols. And they are. The pK_a values for the protonated amines are about 10 (this value is about 0 for water and alcohol) and the pK_a values for amines acting as acids are very high, something like 35 (compared with about 15 for an alcohol). So ammonium salts are about as acidic as phenols and amines will be protonated at pH 7 in water. This is why amino acids (p. 167) exist as zwitterions in water.

$$RNH_3^\oplus \underset{\text{at pH 7}}{\overset{pK_a = \sim 10}{\rightleftharpoons}} RNH_2 \underset{\text{at pH >10}}{\overset{pK_a = \sim 35}{\rightleftharpoons}} RNH^\ominus$$
at pH >35

$H_3N^\oplus \!\!-\!\! CO_2^\ominus$ amino acid zwitterion

diisopropyl-amine →(BuLi / THF) LDA: lithium diisopropylamide

Removing a proton from an amine is very difficult as the anion (unfortunately called an 'amide' anion) is very unstable and very basic. The only way to succeed is to use a very strong base, usually an alkyllithium. The 'anion' then has a N–Li bond and is soluble in organic solvents. This example, known as LDA, is commonly used as a strong base in organic chemistry.

The basicity of amines as neutral compounds is measured by the pK_a of their conjugate acids—so, for example, the pK_a associated with the protonation of triethylamine, a commonly used tertiary amine, is 11.0.

The 'pK_as' of bases

Chemists often say things like 'the pK_a of triethylamine is about 10.' (It's actually 11.0 but 10 is a good number to remember for typical amines). This may surprise you as triethylamine has no acidic hydrogens. What they mean is of course this: 'the pK_a of the conjugate acid of triethylamine is about 10.' Another way to put this is to write 'the pK_{aH} of triethylamine is about 10.' The subscript 'aH' refers to the conjugate acid.

the triethylammonium ion: triethylamine's conjugate acid   triethylamine

Et₃NH⁺   **pK_a = 11**   Et₃N
+ H₂O   + H₃O⁺

It's OK to say 'the pK_a of triethylamine is about 10' as long as you understand that what is really meant is 'the pK_a of the triethylammonium ion is about 10', which can also be expressed thus: 'the pK_{aH} of triethylamine is about 10'

When a molecule is both acidic and basic, as for example aniline, it is important to work out which pK_a is meant as again chemists will loosely refer to 'the pK_a of aniline is 4.6' when they mean 'the pK_a of the conjugate acid of aniline is 4.6.' Aniline is much less basic than ammonia or triethylamine because the lone pair on nitrogen is conjugated into the ring and less available for protonation.

conjugate acid of aniline   **pK_a = 4.6**   aniline   delocalized lone pair is less basic

But for the same reason, aniline is also more acidic than ammonia (pK_a 33) and has a genuine pK_a in which one of the protons on nitrogen is lost. So we can say correctly that 'the pK_a of aniline is about 28.' Just be careful to check which pK_a is meant in such compounds. The full picture is:

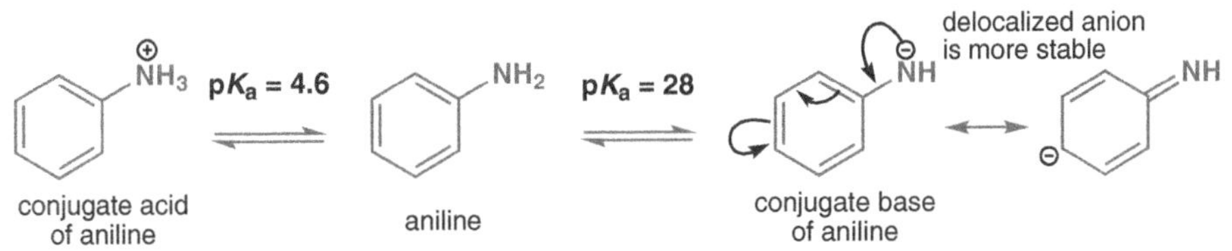

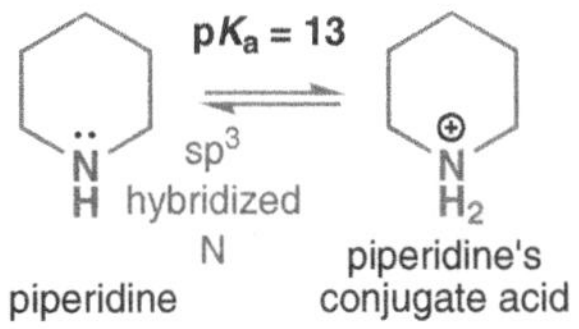

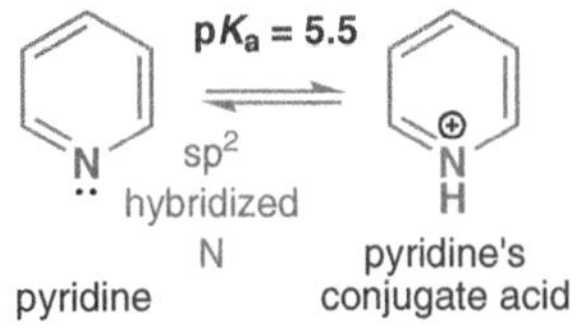

The pK_a associated for protonation of piperidine, a typical secondary amine, is about 13. The equivalent pK_a for protonation of pyridine—a compound with a similar heterocyclic structure, but with its lone pair in an sp^2 rather than an sp^3 orbital, is only 5.5: pyridine is a weaker base than piperidine (its conjugate acid is a stronger acid). Nitriles, whose lone pair is sp hybridized, are not basic at all. Lone pairs with more p character (sp^3 orbitals are 3/4 p, while sp orbitals are 1/2 p) are higher in energy—they spend more time further from the nucleus—and are therefore more basic.

Amides are very different because of the delocalization of the lone pair into the carbonyl group. This makes amides more acidic but less basic and protonation occurs on oxygen rather than nitrogen. Amides have pK_a values of around 15 when they act as acids, making them some 10^{10} times more acidic than amines. The pK_a of protonated amides is around 0, making them some 10^{10} times weaker as bases.

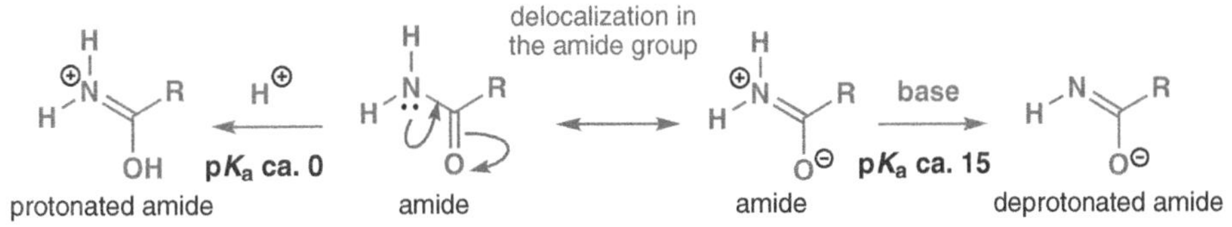

⮞ Delocalization in amides was discussed on p. 155.

If we replace the carbonyl oxygen atom in an amide by nitrogen we get an amidine. Amidines are conjugated, like amides, but unlike amides they are *stronger* bases than amines, by about 2–3 pK_a units, because the two nitrogens work together to donate electron density onto each other. The bicyclic amidine DBU is often used as a strong organic base (see Chapter 17).

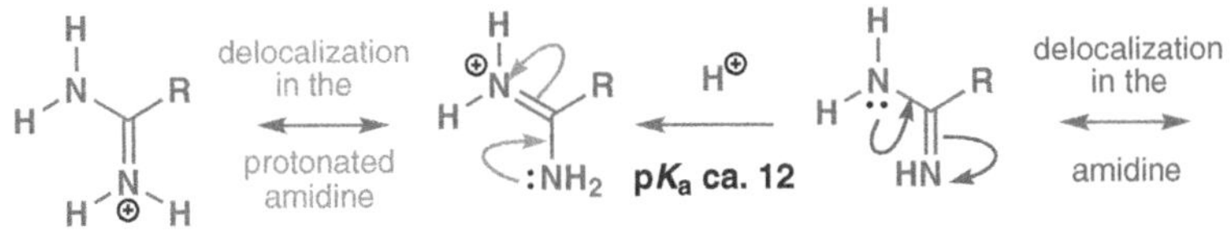

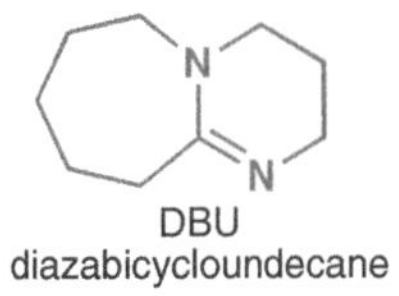

But the champions are the guanidines, with three nitrogens all donating lone pair electrons at once. A guanidine group (shown in green) makes arginine the most basic of the amino acids.

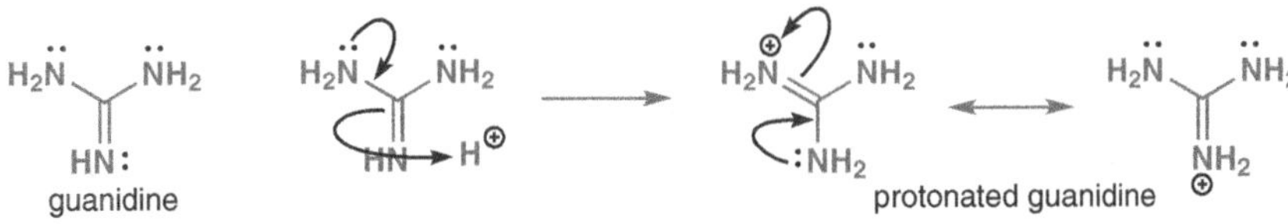

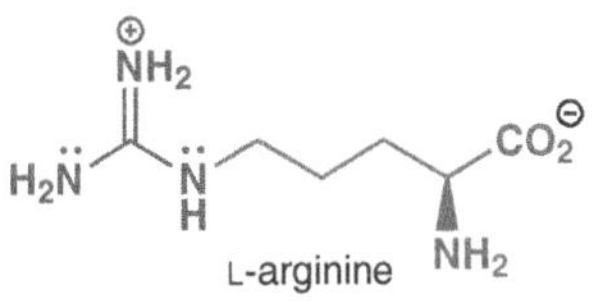

Substituents affect the pK_a

Substituents that are conjugated with the site of proton gain or loss, and even substituents that are electronegative but not conjugating, can have significant effects on pK_a values. Phenol has pK_a 10 but phenols with anions stabilized by extra conjugation can have much lower pK_as.

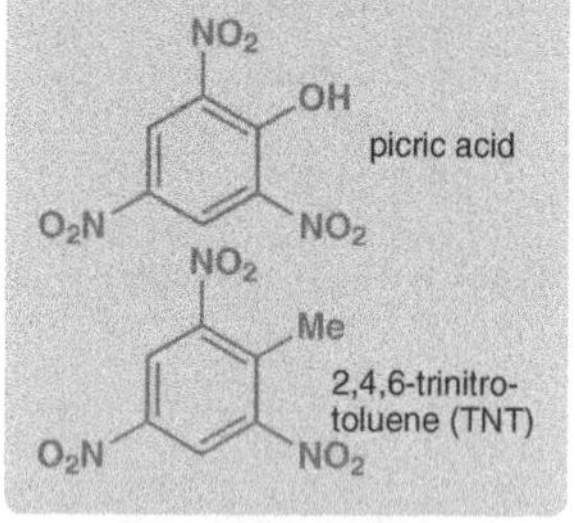

One nitro group, as in *p*-nitrophenol, lowers the pK_a to 7.14, nearly a thousand-fold increase in acidity. This is because the negative charge on oxygen is delocalized into the very electron-withdrawing nitro group. By contrast 4-chlorophenol, with only inductive withdrawal in the C–Cl bond, has pK_a 9.38, hardly different from phenol itself.

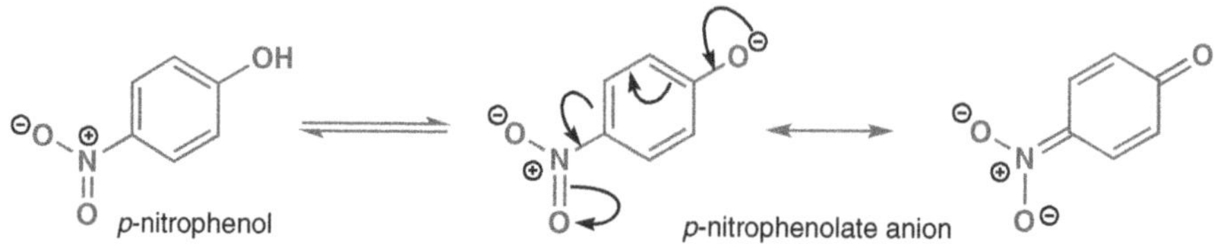

Inductive effects of nearby electronegative atoms can also have marked effects on the pK_a of acids. Adding fluorines to acetic acid reduces the pK_a from about 5 by smallish steps. Trifluoroacetic acid (TFA) is a very strong acid indeed, and is commonly used as a convenient strong acid in organic reactions. Inductive effects occur by polarization of σ bonds when the atom at one end is more electronegative than at the other. Fluorine is much more electronegative than carbon (indeed, F is the most electronegative element of all) so each σ bond is very polarized, making the carbon atom more electropositive and stabilizing the carboxylate anion.

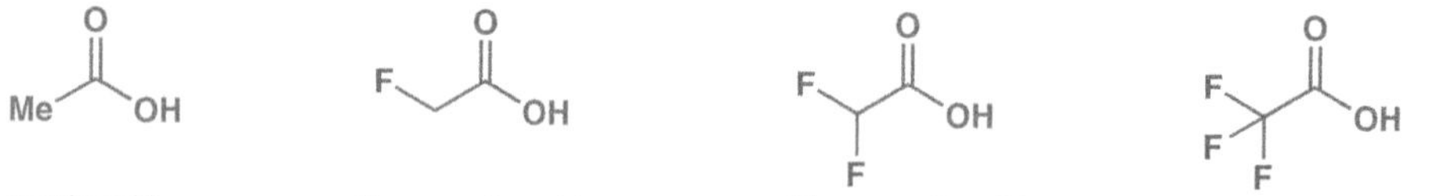

| acetic acid | fluoroacetic acid | difluoroacetic acid | trifluoroacetic acid |
| pK_a 4.76 | pK_a 2.59 | pK_a 1.34 | pK_a ~ −1 |

Carbon acids

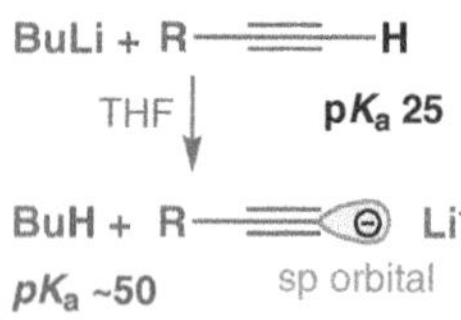

Hydrocarbons are not acidic. We have already established that methane has a pK_a of about 48 (p. 170 above)—it's essentially impossible to deprotonate. Alkyllithiums are for this reason among the strongest bases available. But some hydrocarbons *can* be deprotonated, the most important example being alkynes—you saw on p. 171 that acetylene has a pK_a of 25 and can be deprotonated by NH_2^- (as well as other strong bases such as BuLi). The difference is one of hybridization—an idea we introduced with the nitrogen bases above. Making the acetylide anion, whose negative charge resides in an sp orbital, is much easier than making a methyl anion, with a negative charge in an sp³ orbital, because electrons in sp orbitals spend a lot of their time closer to the nucleus than electrons in sp³ orbitals.

C–H bonds can be even more acidic than those of acetylene if stabilization of the resulting anion is possible by *conjugation*. Conjugation with a carbonyl group has a striking effect. One carbonyl group brings the pK_a down to 13.5 for acetaldehyde so that even hydroxide ion can produce the anion. You will discover in Chapter 20 that we call this the 'enolate anion' and that the charge is mostly on oxygen, although the anion can be drawn as a carbanion.

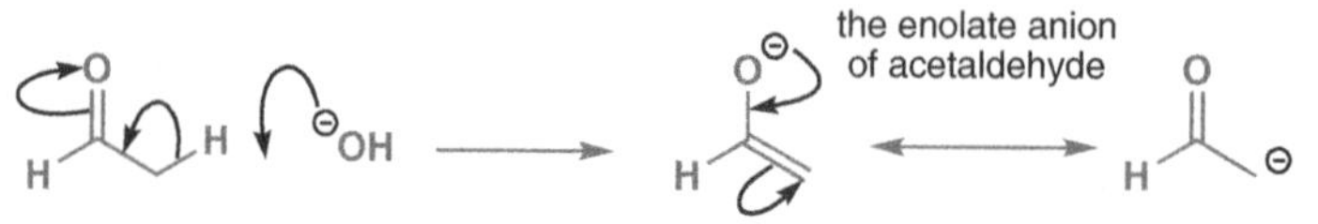

It is interesting to compare the strengths of the carbon, nitrogen, and oxygen acids of similar structure below. The ketone (acetone) is of course least acidic, the amide is more acidic, and the carboxylic acid most acidic. The oxyanion conjugate bases are all delocalized but delocalization onto a second very electronegative oxygen atom is much (~10 pH units) more effective than delocalization onto nitrogen, which is 4 pH units more effective than delocalization onto carbon.

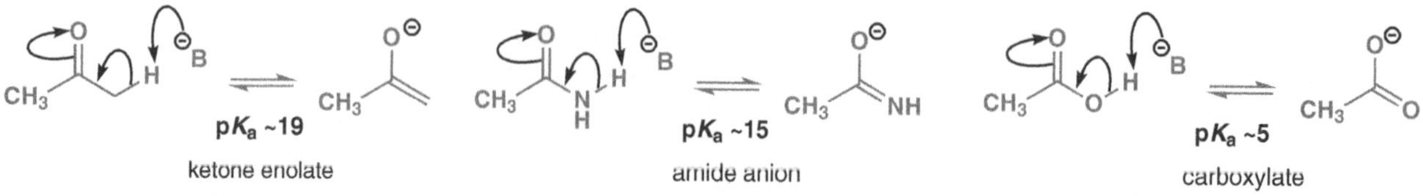

ketone enolate pK_a ~19   amide anion pK_a ~15   carboxylate pK_a ~5

Nevertheless, the effect of conjugation on the carbon acid compared with methane is enormous (~30 pH units) and brings proton removal from carbon within the range of accessible bases

The nitro group is even more effective: nitromethane, with a pK_a of 10, dissolves in aqueous NaOH. The proton is removed from carbon, but the negative charge in the conjugate base is on oxygen. The big difference is that the nitrogen atom has a positive charge throughout. If the anion is protonated in water by some acid (HA) the 'enol' form of nitromethane is the initial product and this slowly turns into nitromethane itself. Whereas proton transfers between electronegative atoms (O, N, etc.) are fast, proton transfers to or from carbon can be slow.

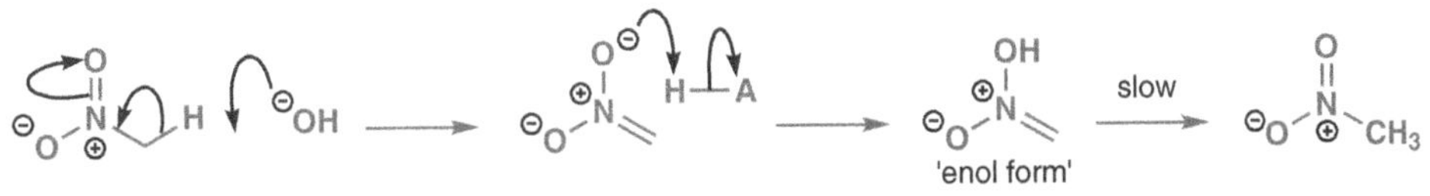

Carbon acids are very important in organic chemistry as they allow us to make carbon–carbon bonds and you will meet many more of them in later chapters of this book.

Why do we need to compare acid strengths of O and N acids?

The rates of nucleophilic addition to carbonyl groups that you met in Chapter 6 depend on the basicity of nucleophiles. As nitrogen bases are much stronger than oxygen bases (or, if you prefer, ammonium ions are much weaker acids than H_3O^+), amines are also much better nucleophiles than water or alcohols. This is dramatically illustrated in an amide synthesis from aniline and acetic anhydride in aqueous solution.

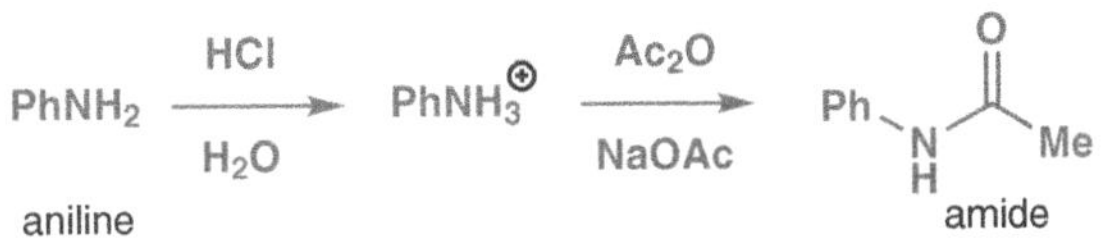

Aniline is not very soluble in water but addition of HCl converts it into the soluble cation by protonation at nitrogen. The solution is now warmed and equal amounts of acetic anhydride and aqueous sodium acetate are added. The pK_a of acetic acid is about 5, as is the pK_a of $PhNH_3^+$, so an equilibrium is set up and the solution now contains these species:

The only electrophile is acetic anhydride, with its two electrophilic carbonyl groups. The nucleophiles available are water, aniline, and acetate. Water is there in great abundance and does react with acetic anhydride but can't compete with the other two as they are more basic (by about 10^5). If acetate attacks the anhydride, it simply regenerates acetate. But if aniline attacks, the amide is formed as acetate is released.

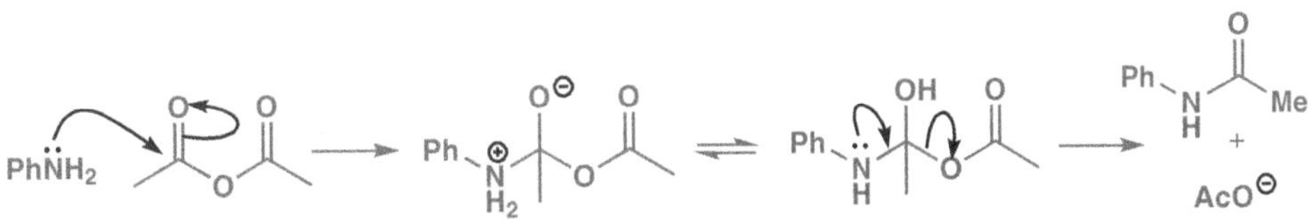

The isolation of the product is easy as the amide is insoluble in water and can be filtered off. Environmental considerations suggest that we should not use organic solvents so much and should use water when possible. If we have some idea about pK_as we can estimate whether water will interfere in a reaction we are planning and decide whether it is a suitable solvent or not. It is even possible to acylate amines with the more reactive acid chlorides in aqueous solution, and we will return in detail to acylation reactions such as these in Chapter 10.

pK_a in action—the development of the drug cimetidine

Histamine is an agonist in the production of gastric acid. It binds to specific sites (receptor sites) in the stomach cells and triggers the production of gastric acid (mainly HCl).

An antagonist works by binding to the receptor but not stimulating acid secretion. It therefore inhibits acid secretion by blocking the receptor sites.

The development of the anti-peptic ulcer drug cimetidine gives a fascinating insight into the important role of pK_a in chemistry. Peptic ulcers are a localized erosion of the mucous membrane, resulting from overproduction of gastric acid in the stomach. One of the compounds that controls the production of the acid is histamine. (Histamine is also responsible for the symptoms of hay fever and allergies.)

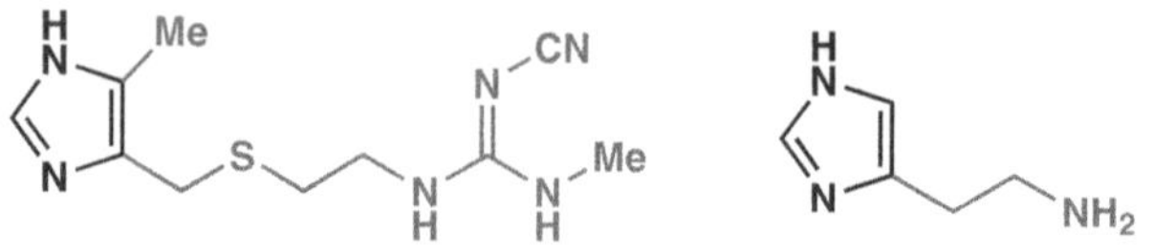

cimetidine   histamine

Histamine works by binding into a receptor in the stomach lining and stimulating the production of acid. What the developers of cimetidine at Smith, Kline and French wanted was a drug that would bind to these receptors without activating them and thereby prevent histamine from binding but not stimulate acid secretion itself. Unfortunately, the antihistamine drugs successfully used in the treatment of hay fever did not work—a different histamine receptor was involved.

■ When the drug was invented, the company was called Smith, Kline and French (SKF) but after a merger with Beechams the company became SmithKline Beecham (SB). SB and GlaxoWelcome later merged to form GlaxoSmithKline (GSK). Things may have changed further by the time you read this book.

Notice that cimetidine and histamine both have the same nitrogen-containing ring (shown in black) as part of their structures. This ring is known as an imidazole—imidazole itself is quite a strong base whose protonated form is delocalized as shown below. This is not coincidence—cimetidine's design was centred around the structure of histamine.

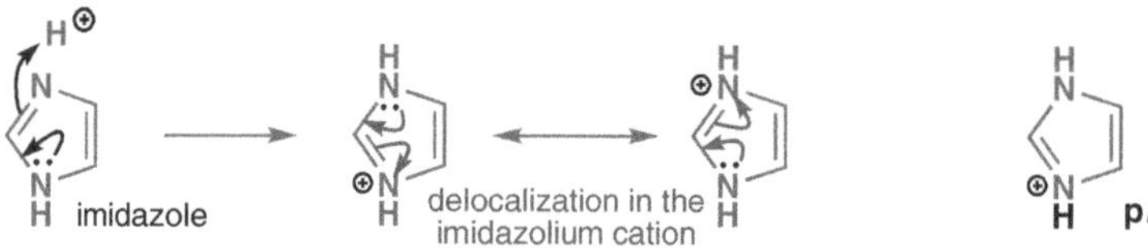

imidazole   delocalization in the
imidazolium cation   pK_a 6.8

➡ Guanidine was introduced to you on p. 175.

In the body, most histamine exists as a salt, being protonated on the primary amine and the early compounds modelled this. The guanidine analogue was synthesized and tested to see if it had any antagonistic effect (that is, if it could bind in the histamine receptors and prevent histamine binding). It did bind but unfortunately it acted as an *agonist* rather than an *antagonist* and stimulated acid secretion rather than blocking it. Since the guanidine analogue has a pK_a even greater than histamine (about 14.5 compared to about 10), it is effectively all protonated at physiological pH.

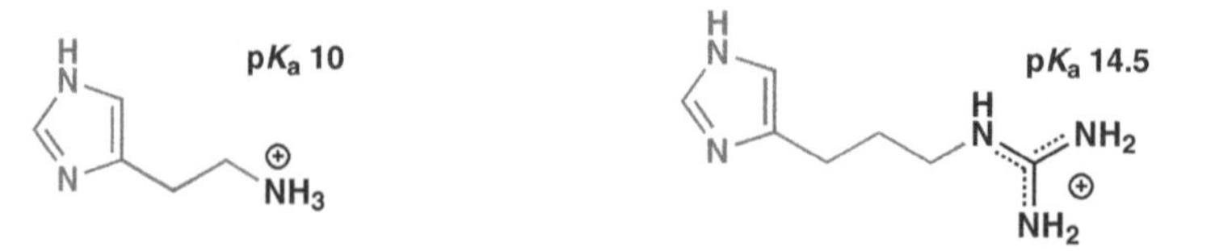

pK_a 10   pK_a 14.5

the major form of histamine   the guanidine analogue: the extra carbon in
at physiological pH (7.4)   the chain increased the efficacy of the drug

■ Remember that amidines and guanidines, p. 175, are basic but that amides aren't. The thiourea, and indeed a urea, is more like an amide.

The agonistic behaviour of the drug clearly had to be suppressed. The thought occurred to the chemists that perhaps the positive charge made the compound agonistic, and so a polar but much less basic compound was sought. Eventually, they came up with burimamide. The most important change is the replacement of the C=NH in the guanidine compound by C=S. Now instead of a guanidine we have a thiourea, which is much less basic. Other adjustments were to increase the chain length, insert a second sulfur atom on the chain, and add methyl groups to the thiourea and the imidazole ring, to give metiamide with increased efficacy.

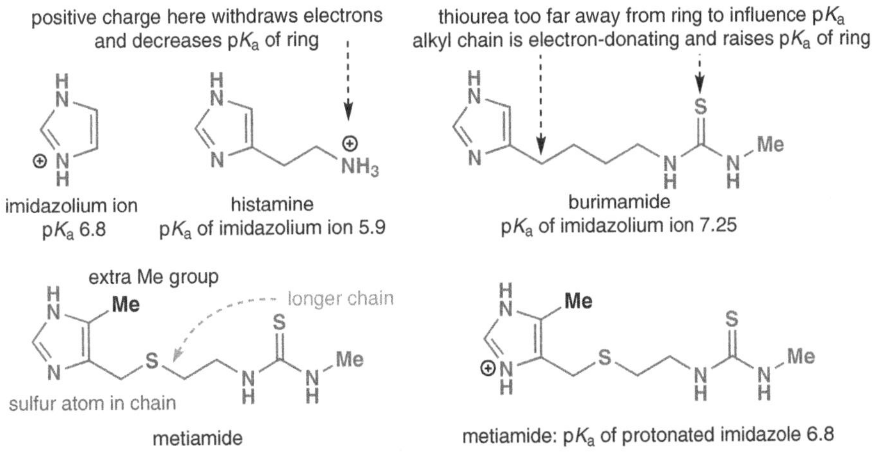

positive charge here withdraws electrons and decreases pK_a of ring

thiourea too far away from ring to influence pK_a
alkyl chain is electron-donating and raises pK_a of ring

imidazolium ion
pK_a 6.8

histamine
pK_a of imidazolium ion 5.9

burimamide
pK_a of imidazolium ion 7.25

extra Me group

longer chain

sulfur atom in chain

metiamide

metiamide: pK_a of protonated imidazole 6.8

The new drug, metiamide, was ten times more effective than burimamide when tested in humans. However, there was an unfortunate side-effect: in some patients: the drug caused a decrease in the number of white blood cells, leaving the patient open to infection. This was eventually traced back to the thiourea group. The sulfur had again to be replaced by oxygen, to give a normal urea and, just to see what would happen, by nitrogen to give another guanidine.

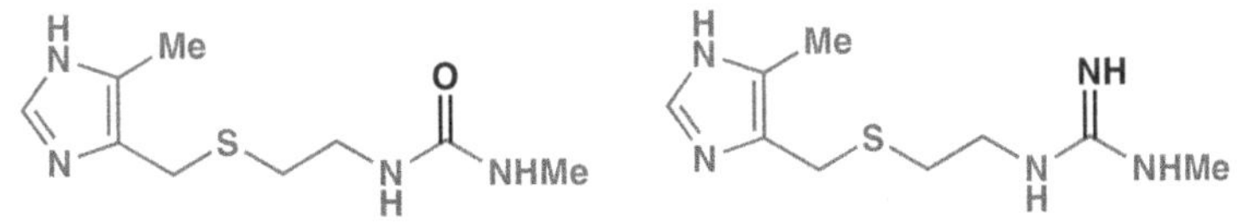

urea analogue of metiamide

guanidine analogue of metiamide

Neither was as effective as metiamide but the important discovery was that the guanidine analogue no longer showed the agonistic effects of the earlier guanidine. Of course, the guanidine would also be protonated so we had the same problem we had earlier—how to decrease the pK_a of the guanidinium ion. A section of this chapter considered the effect of electron-withdrawing groups on pK_a and showed that they make a base less basic. This was the approach now adopted—the introduction of electron-withdrawing groups on to the guanidine to lower its pK_a. The table below shows the pK_as of various substituted guanidinium ions.

substituted guanidinium ion   substituted guanidine

pK_as of substituted guanidinium ions

R	H	Ph	CH$_3$CO	NH$_2$CO	MeO	CN	NO$_2$
pK_a	14.5	10.8	8.33	7.9	7.5	−0.4	−0.9

Clearly, the cyano and nitro-substituted guanidines would not be protonated at all. These were synthesized and found to be just as effective as metiamide but without the side-effects. Of the two, the cyanoguanidine compound was slightly more effective and this was developed and named 'cimetidine'.

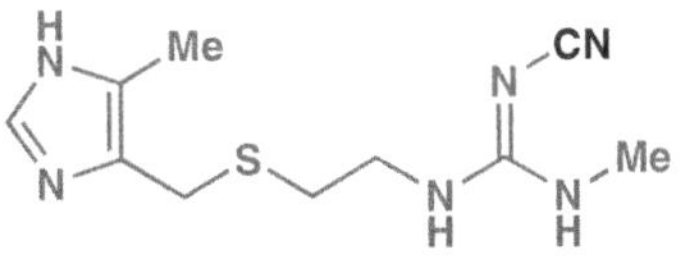

the end result
cimetidine
(Tagamet)

The development of cimetidine by Smith, Kline and French from the very start of the project up to its launch on the market took 13 years. This enormous effort was well rewarded—Tagamet (the trade name of the drug cimetidine) became the best-selling drug in the world and the first to gross more than one billion dollars per annum. Thousands of ulcer patients worldwide no longer had to suffer pain, surgery, or even death. The development of cimetidine followed a rational approach based on physiological and chemical principles and it was for this that one of the scientists involved, Sir James Black, received a share of the 1988 Nobel Prize for Physiology or Medicine. None of this would have been possible without an understanding of pK_as.

Lewis acids and bases

> Johannes Nicolaus Brønsted (1879–1947) was a Danish physical chemist who, simultaneous with Thomas Lowry, introduced the protic theory of acid–base reactions in 1923.

All the acids and bases we have been discussing so far have been protic, or *Brønsted*, acids and bases. In fact, the definition of an acid and a base we gave you on p. 165 is a definition of a Brønsted acid and a Brønsted base. When a carboxylic acid gives a proton to an amine, it is acting as a Brønsted acid while the amine is a Brønsted base. The ammonium ion produced is a Brønsted acid while the carboxylate anion is a Brønsted base.

- **Brønsted acids donate protons.**
- **Brønsted bases accept protons.**

> The American chemist Gilbert Lewis (1875–1946) introduced his electronic theory of acid–base interactions in 1924.

But there is another important type of acid: the Lewis acid. These acids don't donate protons—indeed they usually have no protons to donate. Instead they accept electrons. It is indeed a more general definition of acids to say that they accept electrons and of bases that they donate electrons. Lewis acids are usually halides of the higher oxidation states of metals, such as BF_3, $AlCl_3$, $ZnCl_2$, SbF_5, and $TiCl_4$. By removing electrons from organic compounds, Lewis acids act as important catalysts in important reactions such as the Friedel–Crafts alkylation and acylation of benzene (Chapter 21), the S_N1 substitution reaction (Chapter 15), and the Diels–Alder reaction (Chapter 34).

- **Lewis acids accept electrons.**
- **Lewis bases donate electrons.**

A simple Lewis acid is BF_3. As you saw in Chapter 5, monomeric boron compounds have three bonds to other atoms and an empty p orbital, making six electrons only in the outer shell. They are therefore not stable and BF_3 is normally used as its 'etherate': a complex with Et_2O. Ether donates a pair of electrons into the empty p orbital of BF_3 and this complex has tetrahedral boron with eight electrons. In this reaction the ether donates electrons (it can be described as a Lewis base) and BF_3 accepts electrons: it is a Lewis acid. No protons are exchanged. The complex is a stable liquid and is the form usually available from suppliers.

Lewis acids often form strong interactions with electronegative atoms such as halides or oxygen. In the Friedel–Crafts acylation, which you will meet in Chapter 21, for example, $AlCl_3$ removes the chloride ion from an acyl chloride to give a species, the acylium ion, which is reactive enough to combine with benzene.

Lewis acid–base interactions are very common in chemistry and are often rather subtle. You are about to meet, in the next chapter, an important way of making C–C bonds by adding organometallics to carbonyl compounds, and in many of these reactions there is an interaction at some point between a Lewis acidic metal cation and a Lewis basic carbonyl group.

Further reading

The quote at the start of the chapter comes from Ross Stewart, *The Proton: Applications to Organic Chemistry*, Academic Press, Orlando, 1985, p 1.

More detailed information about acid/base extraction can be found in any organic practical book. The details of the Cannizzaro reaction are from J. C. Gilbert and S. F. Martin, *Experimental Organic Chemistry*, Harcourt, Fort Worth, 2002. The reduction of amides to amines comes from B. S. Furniss, A. J. Hannaford, P. W. G. Smith,

and A. R. Tatchell, *Vogel's Textbook of Practical Organic Chemistry*, 5th edn, Longman, Harlow, 1989.

Details about the acylation of amines with anhydrides and acid chlorides are in L. M. Harwood, C. J. Moody, and J. M. Percy, *Experimental Organic Chemistry*, 2nd edn, Blackwell, Oxford, 1999, p 279.

There is more about the discovery of cimetidine in W. Sneader, *Drug Discovery: a History*, Wiley, Chichester, 2005.

Check your understanding

To check that you have mastered the concepts presented in this chapter, attempt the problems that are available in the book's Online Resource Centre at http://www.oxfordtextbooks.co.uk/orc/clayden2e/

Suggested solutions for Chapter 8

8

PROBLEM 1

How would you separate a mixture of these three compounds?

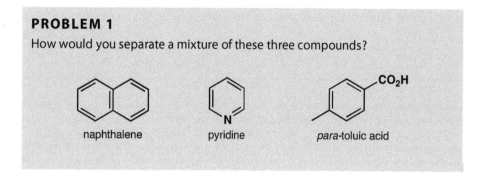

naphthalene   pyridine   *para*-toluic acid

Purpose of the problem

Revision of simple acidity and basicity in a practical situation.

Suggested solution

Pyridine is a weak base (pK_a of the pyridinium ion is about 5.5) and can be dissolved in aqueous acid. Naphthalene is neither an acid nor a base and is not soluble in water at any pH. *p*-Toluic acid is a weak acid (pK_a about 4.5) and can be dissolved in aqueous base. So dissolve the mixture in an organic solvent immiscible with water (say ether Et_2O or dichloromethane CH_2Cl_2) and extract with aqueous acid. This will dissolve the pyridine as its cation. Then extract the remaining organic layer with aqueous base such as $NaHCO_3$ which will remove the toluic acid as its water-soluble anion. You now have three solutions. Evaporate the organic solution to give crystalline naphthalene. Acidify the basic solution of *p*-toluic acid and the free acid will precipitate out and can be recrystallized. Add base to the pyridine solution, extract the pyridine with an organic solvent, and distil the pyridine. It doesn't matter if you extract the original solution with base first and acid second.

PROBLEM 2

In the separation of benzoic acid from toluene on p. 164 of the textbook we suggested using KOH solution. How concentrated a solution would be necessary to ensure that the pH was above the pK_a of benzoic acid (4.2)? How would you estimate how much KOH solution to use?

Purpose of the problem

To ensure you understand the relationship between pH and concentration.

Suggested solution

Even a very weak solution of KOH has a pH>4.2. If we want a pH of 5 (just above the pK_a of benzoic acid) we must ensure that we have $[H_3O^+] = 10^{-5}$ mol dm^{-3}. The ionic product of water is $[H_3O^+]$ x $[HO^-] = 10^{-14}$ and so we need 10^{-9} mol dm^{-3} of KOH. This is very dilute! The trouble would be that you need one hydroxide ion for each molecule of benzoic acid and so if you had, say, 1.22 g PhCO$_2$H (= 0.01 equiv.) you would need 1000 litres (dm^3) of KOH solution. It makes more sense to use a much more concentrated solution, say 0.1M. This would give an unnecessarily high pH (13) but you would need only 100 ml (0.1 dm^3) to extract your benzoic acid.

PROBLEM 3

What species would be present in a solution of this hydroxy-acid in (a) water at pH 7, (b) aqueous alkali at pH 12, and (c) in concentrated mineral acid?

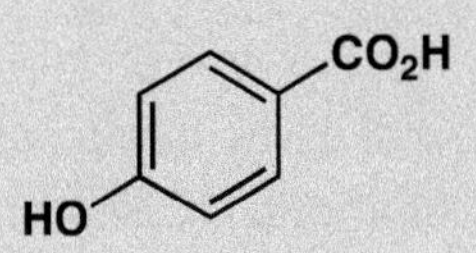

Purpose of the problem

To get you thinking about what really is present in solution using rough pK_a as a guide in a practical situation.

Suggested solution

The CO$_2$H group will have a pK_a of about 4–5 and the phenolic OH a pK_a of about 10. So the carboxylic acid but not the phenol will be ionized at pH 7, they will both be ionized at pH 12, and there will be a mixture of free acid and protonated acid at very low pH. The proton will be on the carbonyl oxygen atom as this gives a delocalized cation.

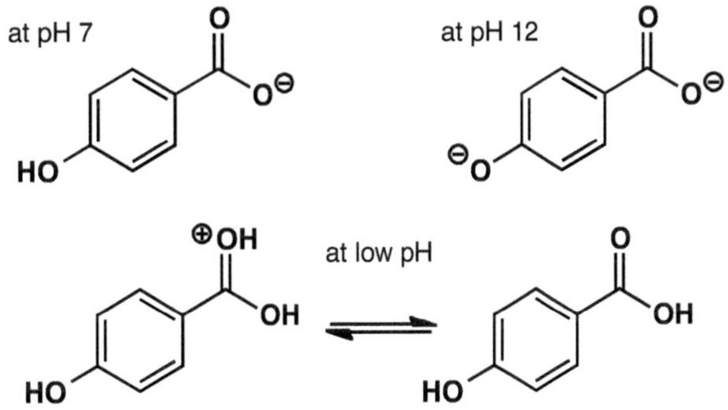

■ See page 173 of the textbook for the pK_a of phenol.

PROBLEM 4

What would you expect to be the site of (a) protonation and (b) deprotonation if these compounds were treated with the appropriate acid or base? In each case suggest a suitable acid or base and give the structure of the products.

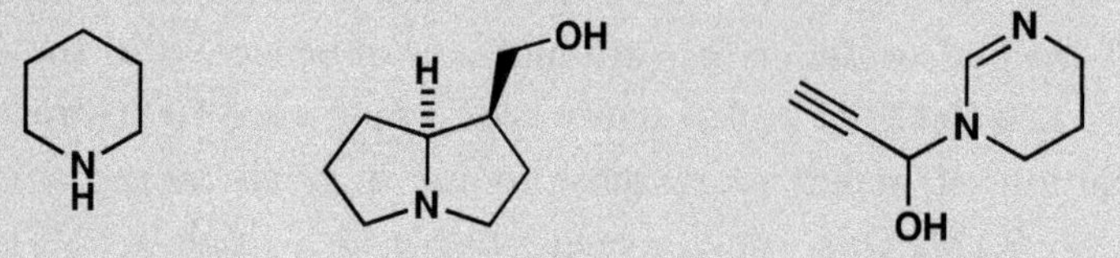

Purpose of the problem

Progressing to more taxing judgements on more interesting molecules.

Suggested solution

The simple amine piperidine will easily be protonated by even weak acids as the conjugate base has a pK_a of about 11. Any mineral acid such as HCl will do the job as would weaker acids such as RCO_2H. Deprotonation will remove the NH proton as nitrogen is more electronegative than carbon but a very strong base such as BuLi will be needed as the pK_a will be about 30–35. You could represent the product with an N–Li bond or as an anion.

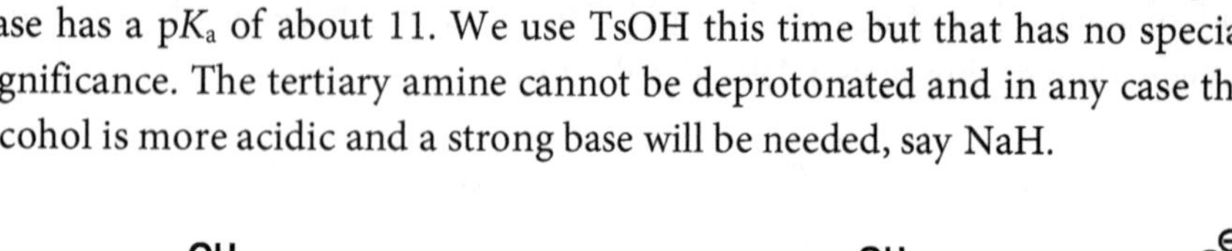

The second example is more complicated but contains a normal tertiary amine so protonation will occur there with most acids as as the conjugate base has a pK_a of about 11. We use TsOH this time but that has no special significance. The tertiary amine cannot be deprotonated and in any case the alcohol is more acidic and a strong base will be needed, say NaH.

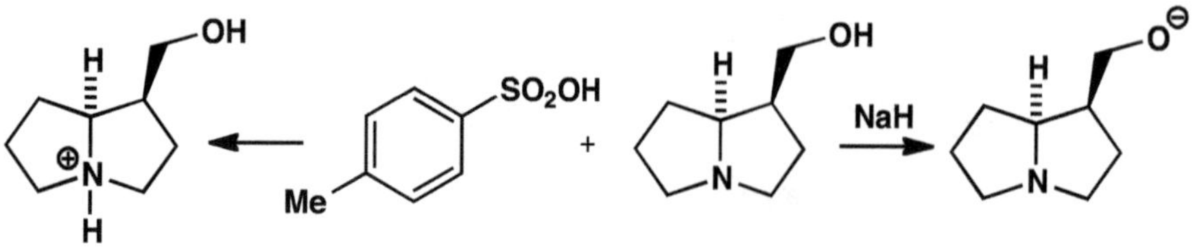

The third example is more complicated still. There is a normal OH group (pK_a of about 16) and a slightly acidic alkyne (pK_a of about 32). The basic group is not a simple amine but a delocalized amidine. Protonation occurs on the top (imine) nitrogen as the positive charge is then delocalized over both nitrogens. Protonation on the other nitrogen does not occur. The pK_a of the conjugate base is about 12.

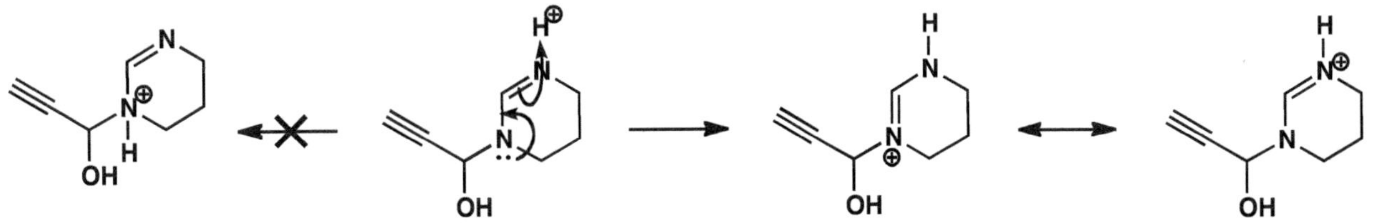

The first proton to be removed by base will be from the alcohol and this will need a reasonably strong base such as NaH. Removal of the alkyne proton requires a much stronger base such as BuLi. You might represent the product as an alkyne anion or a covalently bonded alkyllithium.

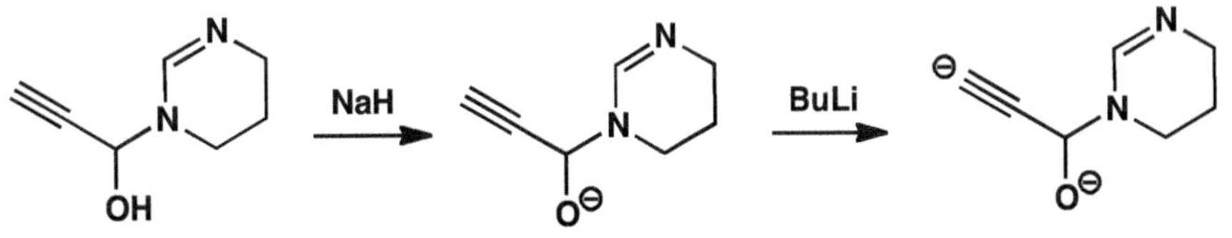

PROBLEM 5

Suggest what species would be formed by each of these combinations of reagents. You are advised to use estimated pK_a values to help you and to beware of those cases where nothing happens.

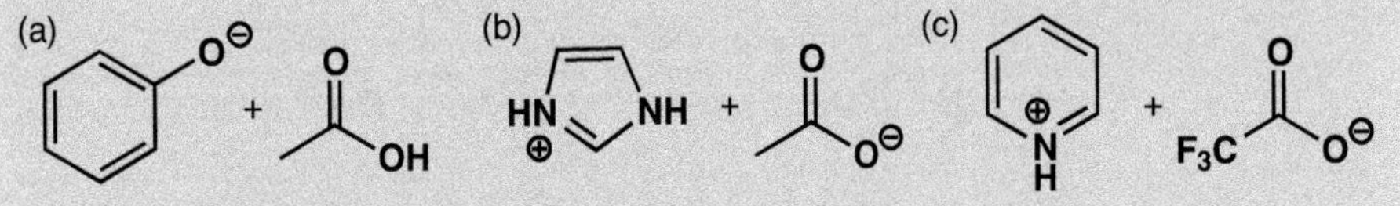

Purpose of the problem

Learning to compare species of similar acidity or basicity.

Suggested solution

In each case one of the reagents might take a proton from the other. In example (a), would the phenolate anion remove a proton from acetic acid? The answer is *yes* because acetic acid is a much stronger acid than phenol The difference is five pH units so the equilibrium constant would be about 10^5 and the equilibrium would lie far across to the right.

(a)

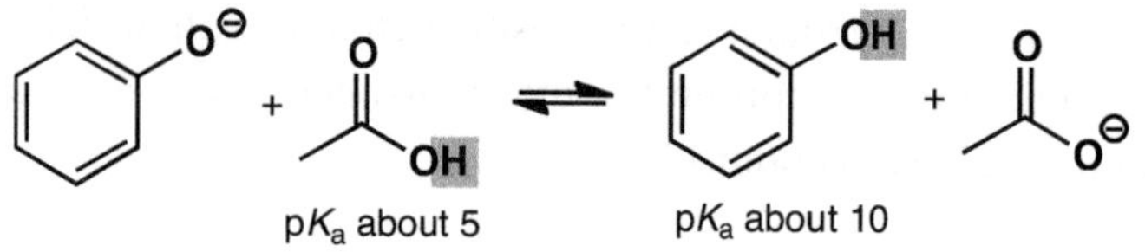

pK_a about 5   pK_a about 10

Example (b) has a similar possible reaction but this time the pK_a difference is much smaller and the other way so the equilibrium constant is 100 and favours the starting materials.

(b)

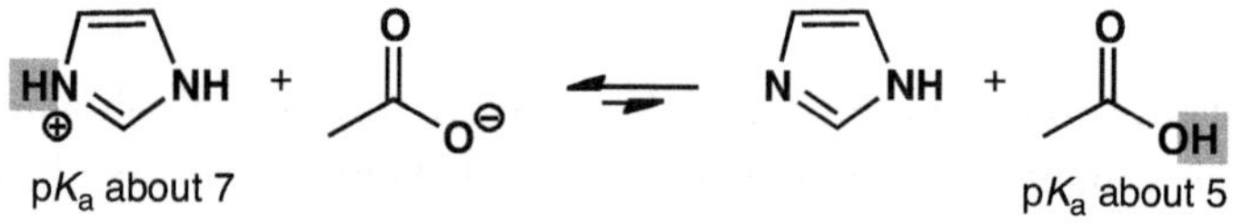

pK_a about 7   pK_a about 5

Example (c) is rather different. We do have another carboxylic acid but this is a much stronger acid because of the three fluorine atoms and the equilibrium is far over to the left.

■ See page 178 of the textbook for the effect of fluorine on pK_a.

(c)

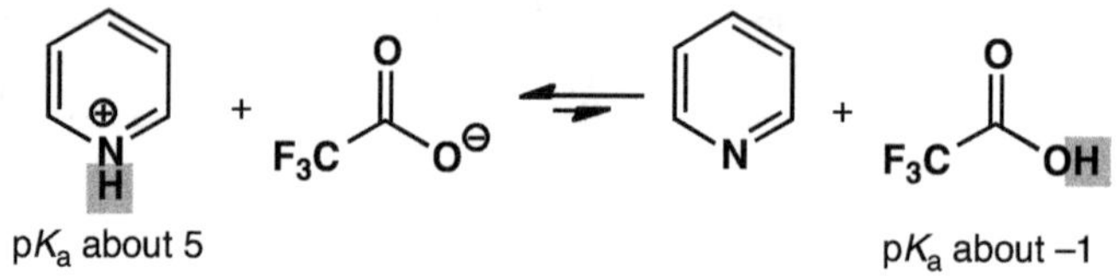

pK_a about 5   pK_a about −1

PROBLEM 6

What is the relationship between these two molecules? Discuss the structure of the anion that would be formed by the deprotonation of each compound.

Purpose of the problem

To help you recognize that conjugation may be closely related to tautomerism.

Suggested solution

They are tautomers: they differ only in the position of one hydrogen atom. It is on nitrogen in the first structure and on oxygen in the second. As it happens the first structure is more stable. They are both aromatic (check that you see why) but the first has a strong carbonyl group while the second has a weaker imine. Deprotonation may appear to give two different anions but they are actually the same because of delocalization. Note the different 'reaction' arrows: equilibrium sign for deprotonation and double headed arrow for delocalization.

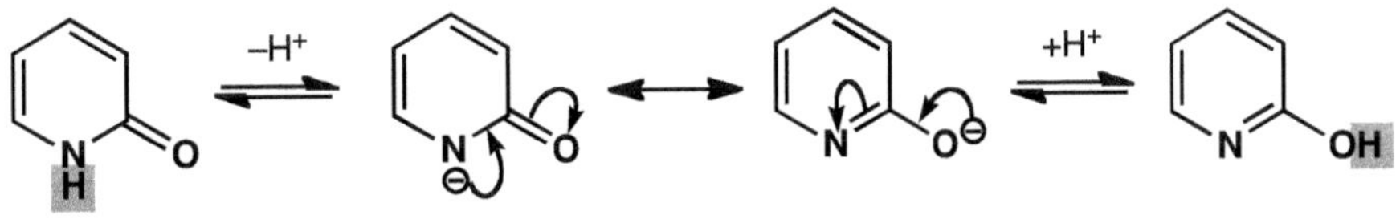

PROBLEM 7

The carbon NMR spectrum of these compounds can be run in D_2O under the conditions shown. Why are these conditions necessary? What spectrum would you expect to observe?

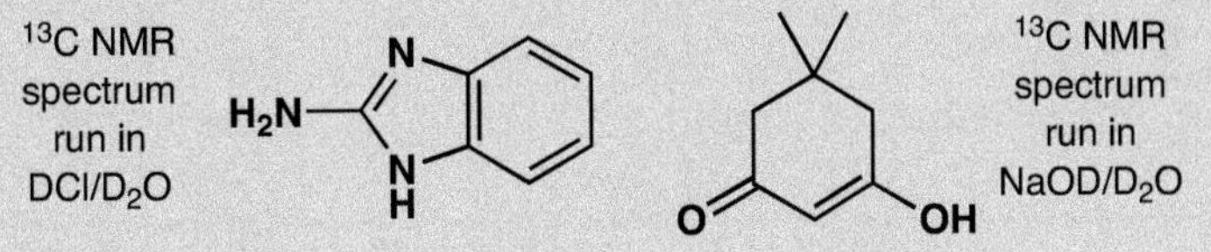

Purpose of the problem

NMR revision and practice at judging the states of compounds at different pHs. Observation of hidden symmetry from conjugation.

Suggested solution

Both compounds are quite polar and not very soluble in the usual NMR solvents. In addition they have NH or OH protons that exchange in solution and broaden the spectrum. With acid or base catalysis the NH and OH protons are exchanged with deuterium and sharp signals appear. But in the strong acid or base used here, ions are formed. The first compound, a strongly basic guanidine (see p. 167 of the textbook) forms a cation in DCl. The cation is symmetrical, unlike the original guanidine, and a very simple spectrum results: just three types of carbon in the benzene ring and one very low field carbon (at large δ) for the carbon in the middle of the cation.

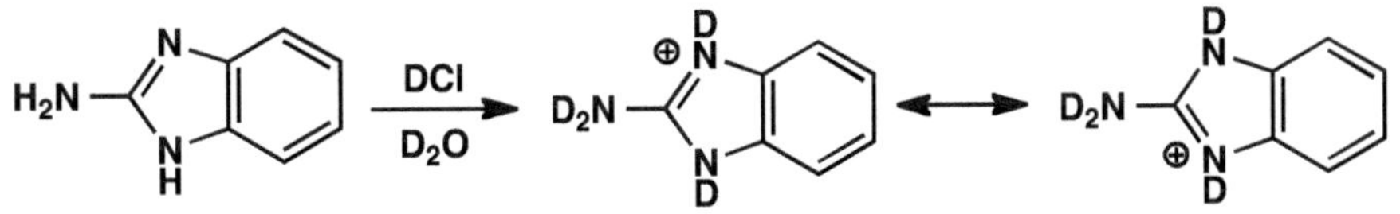

The second compound loses a proton from the OH group to give a delocalized symmetrical anion. There will be five signals in the NMR: the two methyl groups are the same (at small δ) as are the two CH$_2$ groups in the ring (at slightly larger δ). There is one unique carbon joined to the two methyl groups (at small δ) and another in the middle of the anion (at large δ). Finally both carbonyl groups are the same (at even larger δ).

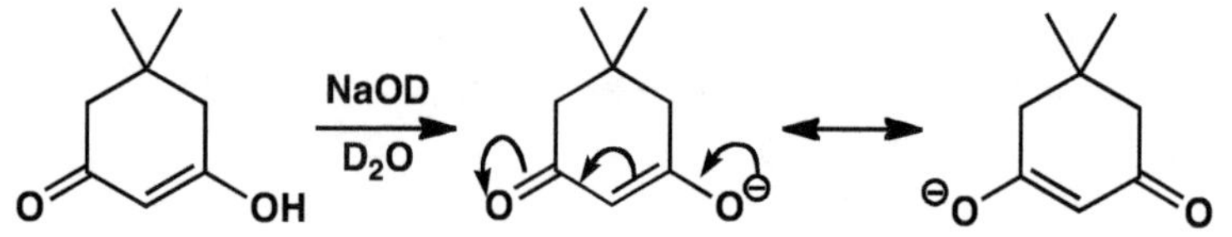

PROBLEM 8

These phenols have approximate pK_a values of 4, 7, 9, 10, and 11. Suggest with explanations which pK_a value belongs to which phenol.

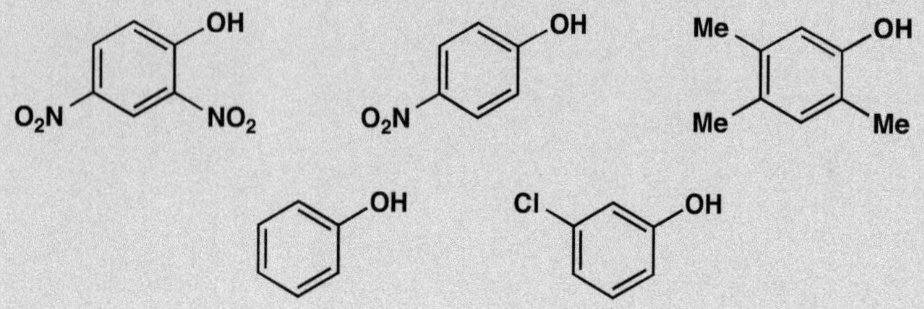

Purpose of the problem

Detailed examination of electronic effects to estimate pK_a values.

Suggested solution

Electron-withdrawing effects make phenols more acidic and electron-donating effects make them less acidic. Phenol itself (the fourth example) has a pK_a of 10. The only compound less acidic than phenol must be the third with three weakly electron-donating methyl groups. One chlorine atom has an inductive electron-withdrawing effect so the last compound has pK_a 9. The remaining two have the powerful electron-withdrawing nitro group. So the first compound, with two nitro groups, must have pK_a of 4 (making it as strong an acid as acetic acid) and the second, with one nitro group, must have pK_a of 7.

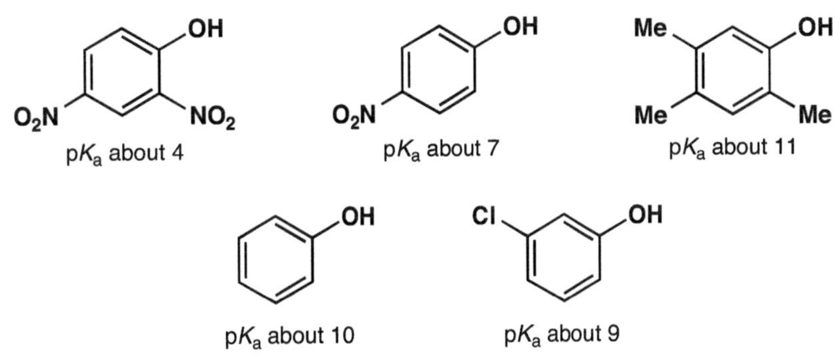

pK$_a$ about 4

pK$_a$ about 7

pK$_a$ about 11

pK$_a$ about 10

pK$_a$ about 9

PROBLEM 9

The pK$_a$ values of these two amino acids are as follows:

(a) cysteine: 1.8, 8.3, and 10.8

(b) arginine: 2.2, 9.0, and 13.2.

Assign these pK$_a$ values to the functional groups in each amino acid and draw the most abundant structure that each molecule will have at pH 1, 7, 10, and 14.

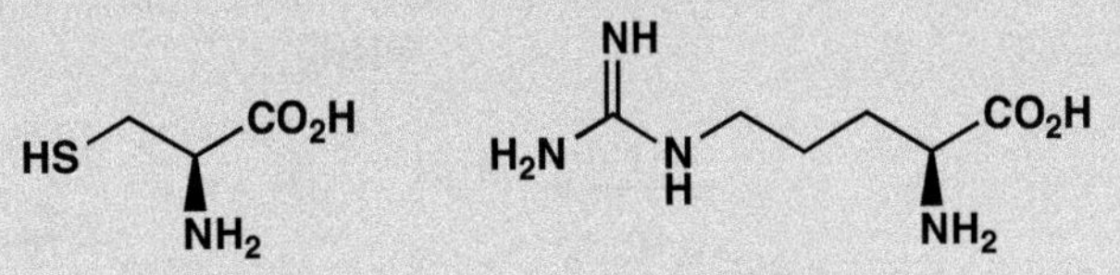

cysteine

arginine

Purpose of the problem

Further revision in thinking about acidity and basicity of functional groups, and reinforcement of expected pK$_a$ values for functional groups. Amino acids are particularly important.

Suggested solution

At high pH cysteine exists as a dianion as both the thiol and the carboxylic acid are anions. If we now add acid, at about pH 10 (actually 10.8) the amine will get protonated, then the thiol will be protonated at about pH 8 (actually 8.3) and finally the carboxylic acid will be protonated at low pH, rather lower than say MeCO$_2$H, as the electron-withdrawing ammonium group increases its acidity (actually at 1.8).

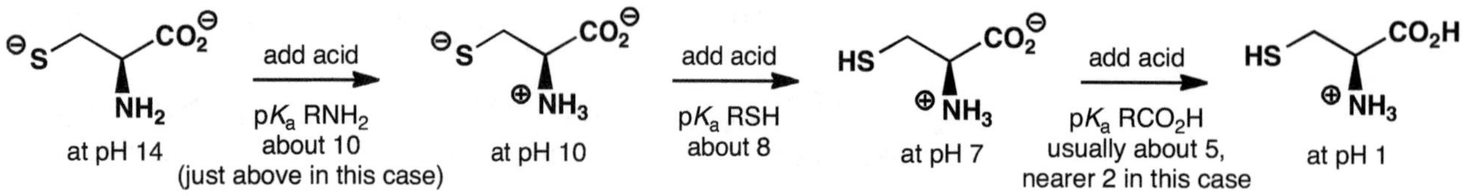

At high pH arginine exists as a monoanion—even the very basic guanidine group cannot be protonated at pH 14. If we now add acid, at about pH 13 the guanidine will get protonated, then the amino group will be protonated at about pH 10 (actually 9.0) and finally the carboxylic acid will be protonated at low pH. This carboxylic acid is rather more acidic than you might expect, but not surprisingly it is harder to protonate an anion in a molecule which already has an overall positive charge.

■ The basicity of arginine, and of the guanidine functional group it contains, was discussed on p. 175 of the textbook.

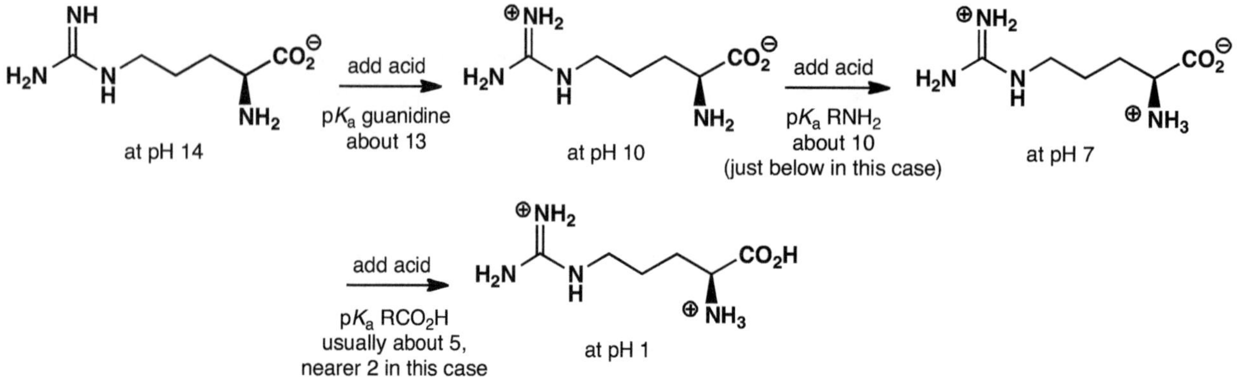

PROBLEM 10

Neither of these two methods for making pentan-1,4-diol will work. What will happen instead?

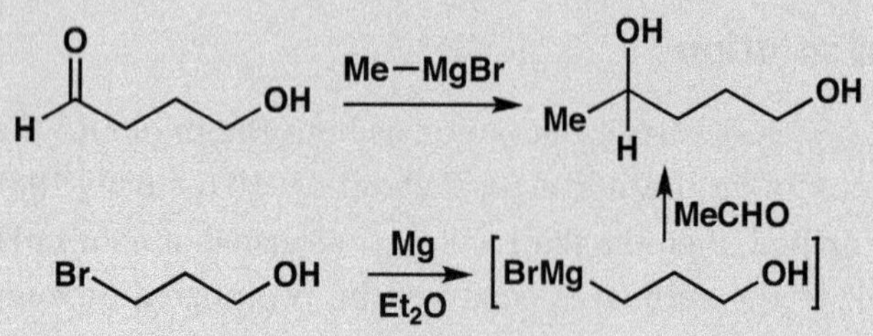

Purpose of the problem

To help you appreciate the disastrous effects that innocent-looking groups may have because of their weak acidity.

Suggested solution

The OH group is the Wicked Witch of the West in this problem. Whoever planned these syntheses expected it to lie quietly and do nothing. All chemists have to learn that things don't go our way just because we want them to do so. Here, although the OH group is only a weak acid (pK$_a$ about 16) it will give up its proton to the very basic Grignard reagents. In the first case, one molecule of Grignard is destroyed but the reaction might succeed if two equivalents were used.

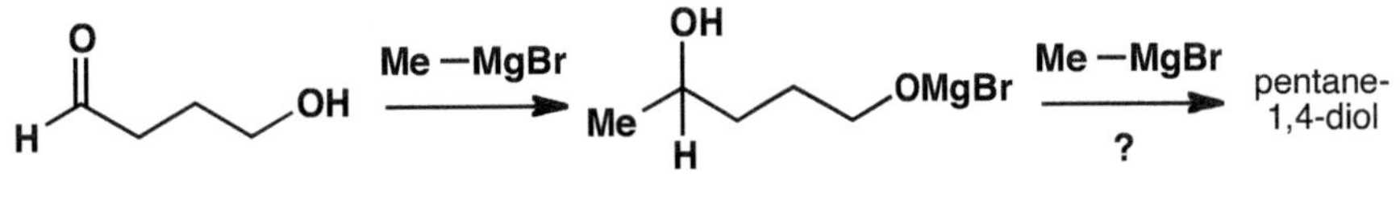

■ 'Protecting groups' are discussed in chapter 23.

The second case is hopeless as the Grignard reagent destroys itself by intramolecular deprotonation. This synthesis could be rescued by putting a protecting group on the OH.

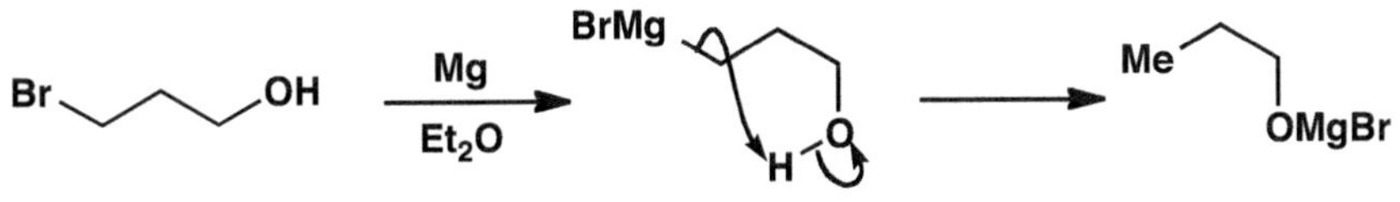

PROBLEM 11

Which of these bases would you choose to deprotonate the following molecules? Very strong bases are more challenging to handle so you should aim to use a base which is just strong enough for the job, but not unnecessarily strong.

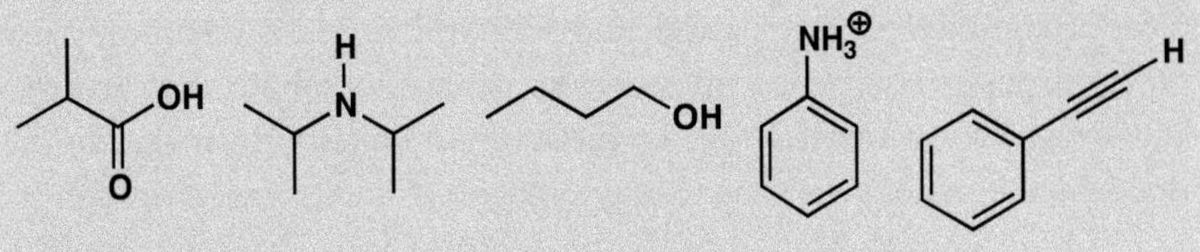

Choice of bases:

   KOH   NaH   BuLi   NaHCO$_3$

Purpose of the problem

To help you match basicity to acidity—an important part of choosing reagents for many reactions.

Suggested solution

We can start by estimating the pK_a of the most acidic proton in each of the substrates to be deprotonated, and likewise estimating the pK_a of the conjugate acids of the proposed bases. Most of these values were discussed in chapter 8, and you were encouraged to commit some of them to memory.

■ Bicarbonate may be new to you, but you might reasonably guess that it has basicity similar to a carboxylate anion. The base sodium hydride, and the pK_a of its conjugate acid, appears on p. 237 of the textbook.

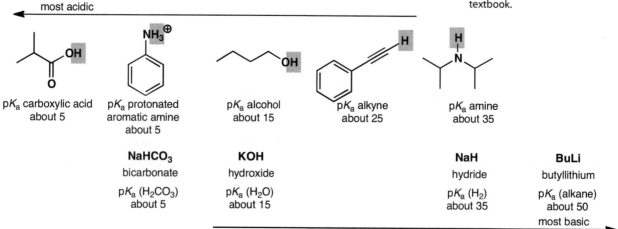

Any of the bases will deprotonate compounds above and to the left of them. So, to deprotonate the least acidic of these, the amine, you would choose to use butyllithium. To deprotonate the alkyne (a reaction which is commonly used to make C–C bonds) you could use BuLi, or alternatively sodium hydride (NaH). BuLi has to be handled under an inert atmosphere, while NaH, although it reacts with water, can be spooned out safely as a suspension in oil.

■ The lithium derivative that results from deprotonating this amine with BuLi is known as 'LDA' and features frequently in later chapters of the textbook.

The alcohol has a pK_a close to that of water, so hydroxide is not a good choice for complete deprotonation, and sodium hydride is commonly used for this purpose. Hydroxide will however deprotonate both the carboxylic acid, to make a carboxylate salt, or the ammonium ion, to make the free amine. Bicarbonate is also commonly used for this purpose: although it is only just basic enough to do the job, the deprotonation reaches completion because it is not an equilibrium: protonated bicarbonate forms carbonic acid which decomposes irreversibly to water and carbon dioxide.

9 Using organometallic reagents to make C–C bonds

Connections

Building on

- Electronegativity and the polarization of bonds ch4
- Grignard reagents and organolithiums attack carbonyl groups ch6
- C–H deprotonated by very strong bases ch8

Arriving at

- Organometallics: nucleophilic and often strongly basic
- Making organometallics from halo-compounds
- Making organometallics by deprotonating carbon atoms
- Using organometallics to make new C–C bonds from C=O groups

Looking forward to

- More about organometallics ch24 & ch40
- More ways to make C–C bonds from C=O groups ch25, ch26, & ch27
- Synthesis of molecules ch28

Introduction

In Chapters 2–8 we covered basic chemical concepts concerning *structure* (Chapters 2–4 and 7) and *reactivity* (Chapters 5, 6, and 8). These concepts are the bare bones supporting all of organic chemistry, and now we shall start to put flesh on these bare bones. In Chapters 9–22 we shall tell you about the most important classes of organic reaction in more detail.

One of the things organic chemists do, for all sorts of reasons, is to make molecules, and making organic molecules means making C–C bonds. In this chapter we are going to look at one of the most important ways of making C–C bonds: using organometallics, such as organolithiums and Grignard reagents, in combination with carbonyl compounds. We will consider reactions such as these:

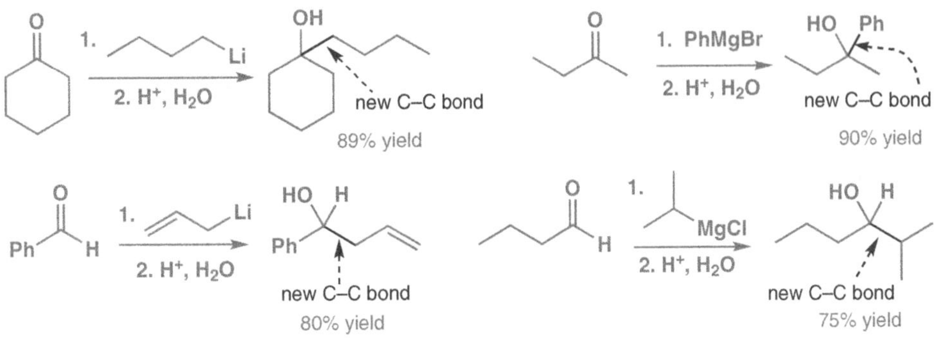

You met these types of reactions in Chapter 6: in this chapter we will be adding more detail with regard to the nature of the organometallic reagents and what sort of molecules can be made using the reactions. The organometallic reagents act as nucleophiles towards the

Online support. The icon in the margin indicates that accompanying interactive resources are provided online to help your understanding: just type **www.chemtube3d.com/clayden/123** into your browser, replacing **123** with the number of the page where you see the icon. For pages linking to more than one resource, type **123-1, 123-2** etc. (replacing **123** with the page number) for access to successive links.

electrophilic carbonyl group, and this is the first thing we need to discuss: why are organo-metallics nucleophilic? We then move on to, firstly, how to make organometallics, then to the sorts of electrophiles they will react with, and finally to the sort of molecules we can make with them.

Organometallic compounds contain a carbon–metal bond

The polarity of a covalent bond between two different elements is determined by electronega-tivity. The more electronegative an element is, the more it attracts the electron density in the bond. So the greater the *difference* between the electronegativities, the greater the difference between the attraction for the bonding electrons, and the more polarized the bond becomes. In the extreme case of complete polarization, the covalent bond ceases to exist and is replaced by electrostatic attraction between ions of opposite charge. We discussed this in Chapter 4 (p. 96), where we considered the extreme cases of bonding in NaCl.

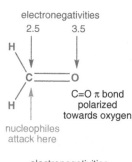

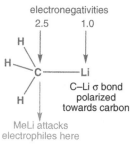

How important are organometallics for making C–C bonds?

As an example, let's take a molecule known as 'juvenile hormone'. It is a compound that prevents several species of insects from maturing and can be used as a means of controlling insect pests. Only very small amounts of the naturally occurring compound can be isolated from the insects, but it can instead be made in the laboratory from simple starting materials. At this stage you need not worry about how, but we can tell you that, in one synthesis, of the 16 C–C bonds in the final product, seven were made by reactions of organometallic reagents, many of them the sort of reactions we will describe in this chapter. This is not an isolated example. As further proof, take an important enzyme inhibitor, closely related to arachidonic acid which you met in Chapter 7. It has been made by a succession of C–C bond-forming reactions using organometallic reagents: eight of the 20 C–C bonds in the product were formed using organometallic reactions.

Cecropia juvenile hormone

black bonds made by organometallic reactions

an enzyme inhibitor

When we discussed (in Chapter 6) the electrophilic nature of carbonyl groups we saw that their reactivity is a direct consequence of the polarization of the carbon–oxygen bond towards the more electronegative oxygen, making the carbon a site for nucleophilic attack. In Chapter 6 you also met the two most important organometallic compounds—organolithiums and organomagnesium halides (known as Grignard reagents). In these organometallic rea-gents the key bond is polarized in the opposite direction—*towards* carbon—making carbon a nucleophilic centre. This is true for most organometallics because, as you can see from this edited version of the periodic table, metals (such as Li, Mg, Na, and Al) all have lower electro-negativity than carbon.

Pauling electronegativities of selected elements

Li 1.0			B 2.0	C 2.5	N 3.0	O 3.5	F 4.0
Na 0.9	Mg 1.3		Al 1.6	Si 1.9	P 2.2	S 2.6	Cl 3.2

Interactive display of polarity of organometallics

The molecular orbital energy level diagram—the kind you met in Chapter 4—represents the C–Li bond in methyllithium in terms of the sum of the atomic orbitals of carbon and lithium. The more electronegative an atom is, the lower in energy are its atomic orbitals (p. 96). The filled C–Li σ orbital is closer in energy to the carbon's sp^3 orbital than to the lithium's 2s orbital, so we can say that the carbon's sp^3 orbital makes a greater contribution to the C–Li σ bond and that the C–Li bond has a larger coefficient on carbon. Reactions involving the filled

We explained this reasoning on p. 104.

σ orbital will therefore take place at C rather than Li. The same arguments hold for the C–Mg bond of organo-magnesium or Grignard reagents, named after their inventor Victor Grignard.

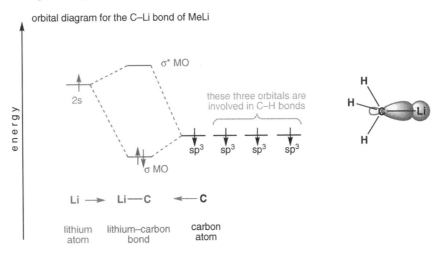

We can also say that, because the carbon's sp³ orbital makes a greater contribution to the C–Li σ bond, the σ bond is close in structure to a filled C sp³ orbital—a lone pair on carbon. This useful idea can be carried too far: methyl lithium is not an ionic compound Me⁻Li⁺—although you may sometimes see MeLi or MeMgCl represented in mechanisms as Me⁻.

■ Carbon atoms that carry a negative charge are known as **carbanions**.
You have already met cyanide (p. 121), a carbanion that really does have a lone pair on carbon. Cyanide's lone pair is stabilized by being in a lower-energy sp orbital (rather than sp3) and by having the electronegative nitrogen atom triply bonded to the carbon.

The true structure of organolithiums and Grignard reagents is rather more complicated! Even though these organometallic compounds are extremely reactive towards water and oxygen, and have to be handled under an atmosphere of nitrogen or argon, some have been studied by X-ray crystallography in the solid state and by NMR in solution. It turns out that they generally form complex aggregates with two, four, six, or more molecules bonded together, often with solvent molecules, one reason why apparently polar compounds such as BuLi dissolve in hydrocarbons. In this book we shall not be concerned with these details, and we shall represent organometallic compounds as simple monomeric structures.

Making organometallics

How to make Grignard reagents

R can be alkyl, allyl, or aryl   **R—X**   X can be I, Br, or Cl

↓ **Mg, Et₂O**

R–Mg–X

alkylmagnesium halide (Grignard reagent)

Grignard reagents are made by reacting magnesium turnings with alkyl halides in ether solvents to form solutions of alkylmagnesium halide. Iodides, bromides, and chlorides can be used, as can both aryl and alkyl halides. Our examples include methyl, primary, secondary, and tertiary alkyl halides, aryl and allyl halides. They cannot contain any functional groups that would react with the Grignard reagent once it is formed. The final example has an acetal functional group as an example of one that does not react with the Grignard reagent. (See Chapter 23 for further discussion.)

🖱 Interactive mechanism for Grignard addition

The solvents in these examples are all ethers, either diethyl ether Et₂O or THF. Other solvents that are sometimes used include the diethers dioxane and dimethoxyethane (DME).

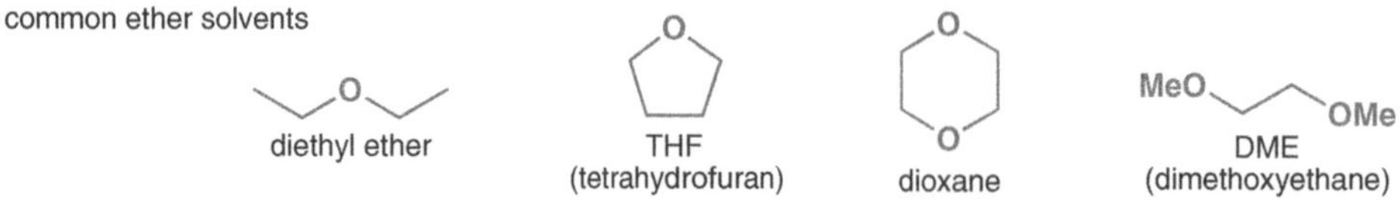

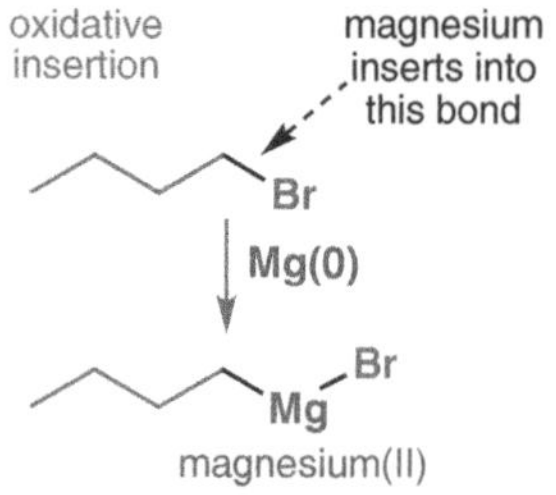

common ether solvents

diethyl ether | THF (tetrahydrofuran) | dioxane | DME (dimethoxyethane)

The reaction scheme is easy enough to draw, but what is the mechanism? Overall it involves an *insertion* of magnesium into the carbon–halogen bond. There is also a change in oxidation state of the magnesium, from Mg(0) to Mg(II). The reaction is therefore known as an oxidative insertion or oxidative addition, and is a general process for many metals such as Mg, Li (which we meet shortly), Cu, and Zn. Mg(II) is much more stable than Mg(0) and this drives the reaction.

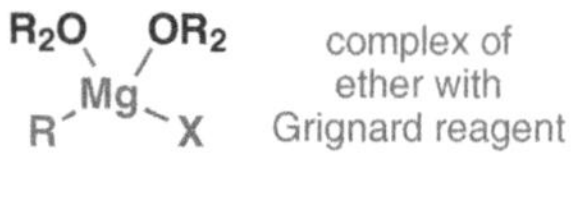

The mechanism of the reaction is not completely understood, and probably involves radical intermediates. But what is sure is that by the end of the reaction the magnesium has surrendered its lone pair of electrons and gained two σ bonds. The true product is a complex between the Grignard reagent and, probably, two molecules of the ether solvent, as Mg(II) prefers a tetrahedral structure.

oxidative insertion — magnesium inserts into this bond

magnesium(II)

complex of ether with Grignard reagent

More on making Grignard reagents

The reaction takes place not in solution but on the surface of the metal, and how easy it is to make a Grignard reagent can depend on the state of the surface—how finely divided the metal is, for example. Magnesium is usually covered by a thin coating of magnesium oxide, and Grignard formation generally requires 'initiation' to allow the metal to come into direct contact with the alkyl halide. Initiation usually means adding a small amount of iodine or 1,2-diiodoethane, or using ultrasound to dislodge the oxide layer. Once the Grignard starts to form, it catalyses further reactions of Mg(0), perhaps by this mechanism:

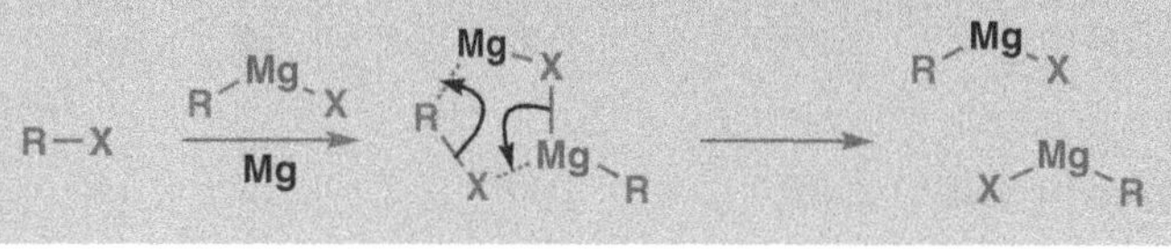

How to make organolithium reagents

Organolithium compounds may be made by a similar oxidative insertion reaction from lithium metal and alkyl halides. Each inserting reaction requires two atoms of lithium and generates one equivalent of lithium halide salt. As with Grignard formation, there is really very little limit on the types of organolithium that can be made this way.

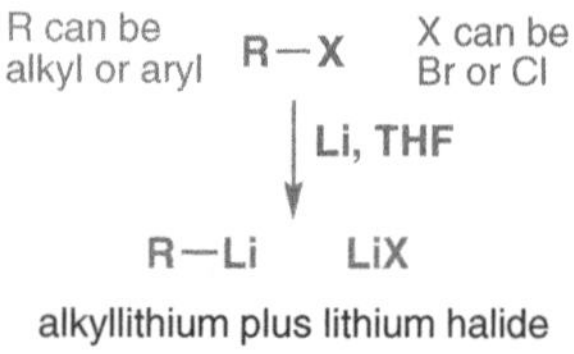

R can be alkyl or aryl · R—X · X can be Br or Cl · Li, THF · R—Li LiX

alkyllithium plus lithium halide

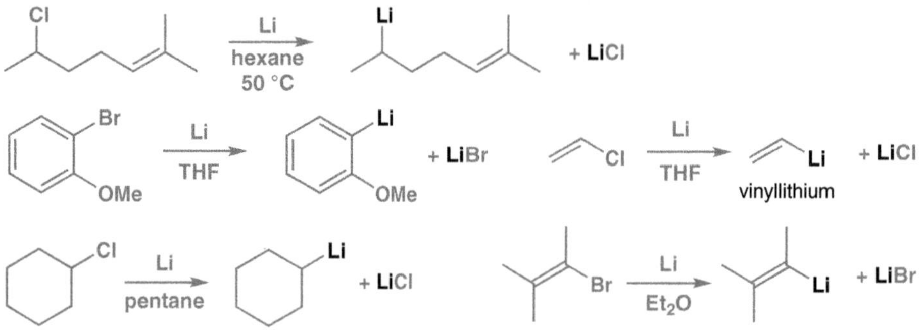

vinyllithium

Interactive mechanism for organolithium addition

You will notice secondary alkyllithiums, an aryllithium, and two vinyllithiums. The only other functional groups are alkenes and an ether. So far, that is quite like the formation of Grignard reagents. However, there are differences. Lithium goes from Li(0) to Li(I) during the reaction and there is no halide attached to the Li. Instead a second Li atom has to be used to make the Li halide. Again, Li(I) is very much more stable than Li(0) so the reaction is irreversible. Although ether solvents are often used, there is less need for extra coordination and hydrocarbon solvents such as pentane or hexane are also good.

Commercially available organometallics

Some Grignard and organolithium reagents are commercially available. Most chemists (unless they were working on a very large scale) would not usually make the simpler organolithiums or Grignard reagents by these methods, but would buy them in bottles from chemical companies (who, of course, do use these methods). The table lists some of the most important commercially available organolithiums and Grignard reagents.

methyllithium (MeLi) in Et_2O or DME

methylmagnesium chloride, bromide, and iodide (MeMgX) in Et_2O, or THF

n-butyllithium (*n*-BuLi or just BuLi)

in cyclohexane or hexanes

ethylmagnesium bromide (EtMgBr)

sec-butyllithium (*sec*-BuLi or *s*-BuLi) in pentane or cyclohexane

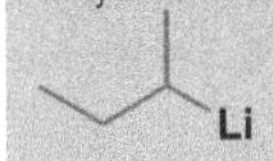

butylmagnesium chloride (BuMgCl) in Et_2O or THF

tert-butyllithium (*tert*-BuLi or *t*-BuLi) in pentane

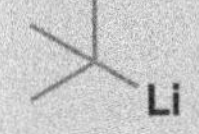

allylmagnesium chloride and bromide

in Et_2O

phenyllithium (PhLi) in (*n*-Bu)$_2$O

phenylmagnesium chloride and bromide (PhMgCl or PhMgBr) in Et_2O or THF

Organometallics as bases

Organometallics need to be kept absolutely free of moisture—even moisture in the air will destroy them. The reason is that they react very rapidly and highly exothermically with water to produce alkanes. Anything that can protonate them will do the same thing. The organometallic reagent is a strong base, and is protonated to form its conjugate acid—methane or benzene in these cases. The pK_a of methane (Chapter 8) is somewhere around 50: it isn't an acid at all and essentially nothing will remove a proton from methane.

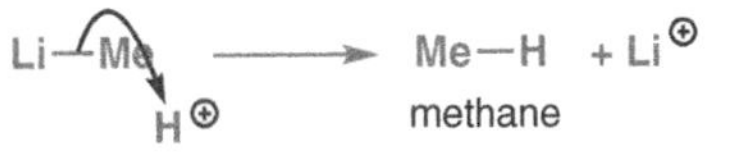

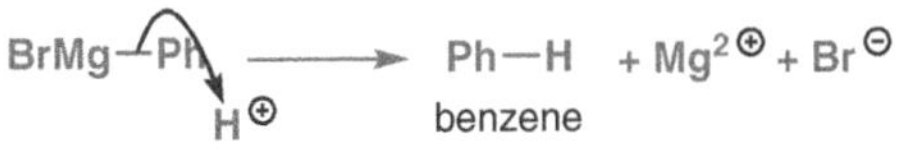

The equilibria lie vastly to the right: methane and Li$^+$ are much more stable than MeLi while benzene and Mg^{2+} are much more stable than PhMgBr. Some of the most important uses of organolithiums—butyllithium, in particular—are as bases and, because they are so strong, they will deprotonate almost anything. That makes them very useful as reagents for making *other* organolithiums.

Making organometallics by deprotonating alkynes

In Chapter 8 (p. 175) we talked about how hybridization affects acidity. Alkynes, with their C–H bonds formed from sp orbitals, are the most acidic of hydrocarbons, with pK_as of about 25.

They can be deprotonated by more basic organometallics such as butyllithium or ethylmagnesium bromide. Alkynes are sufficiently acidic to be deprotonated even by nitrogen bases and you saw on p. 171 that a common way of deprotonating alkynes is to use NaNH$_2$ (sodium amide), obtained by reacting sodium with liquid ammonia. An example of each is shown here. Propyne and acetylene are gases, and can be bubbled through a solution of the base.

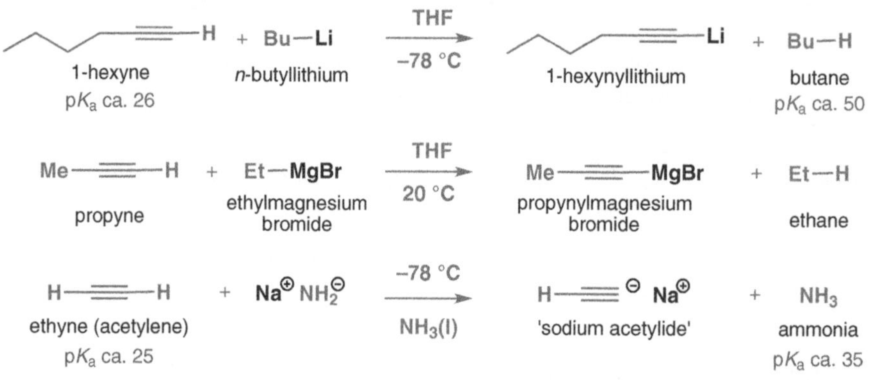

> ■ We have chosen to represent the alkynyl lithium and alkynyl magnesium halides as organometallics and the alkynyl sodium as an ionic salt. Both probably have some covalent character but lithium is less electropositive than sodium so alkynyl lithiums are more covalent and usually used in non-polar solvents while the sodium derivatives are more ionic and usually used in polar solvents.

The metal derivatives of alkynes can be added to carbonyl electrophiles, as in the following examples. The first (we have reminded you of the mechanism for this) is the initial step of an important synthesis of the antibiotic erythronolide A, and the second is the penultimate step of a synthesis of the widespread natural product farnesol.

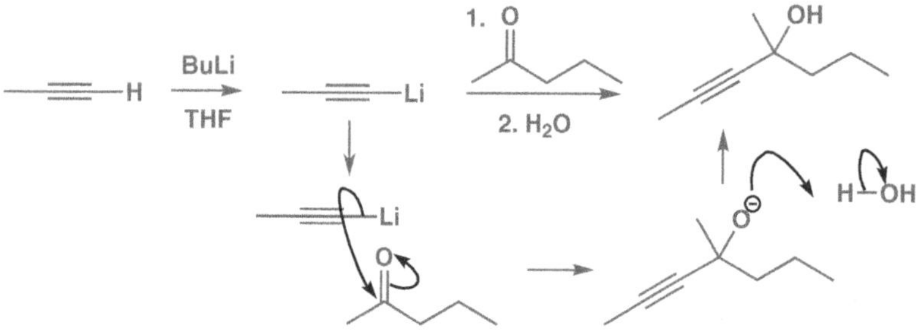

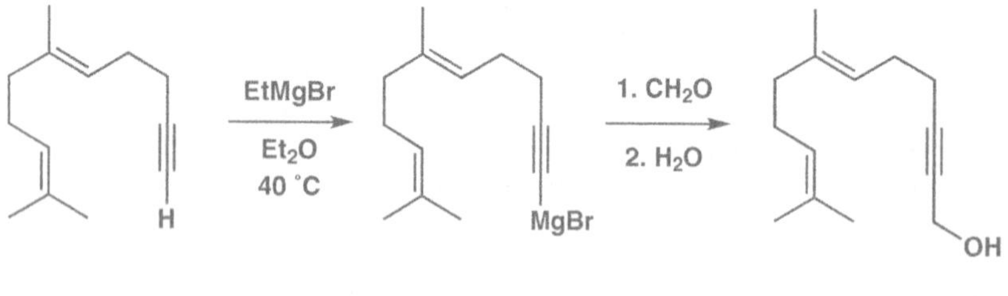

Ethynyloestradiol

The ovulation-inhibiting component of almost all oral contraceptive pills is a compound known as ethynyloestradiol, and this compound too is made by an alkynyllithium addition to the female sex hormone oestrone. A range of similar synthetic analogues of hormones containing an ethynyl unit are used in contraceptives and in treatments for disorders of the hormonal system.

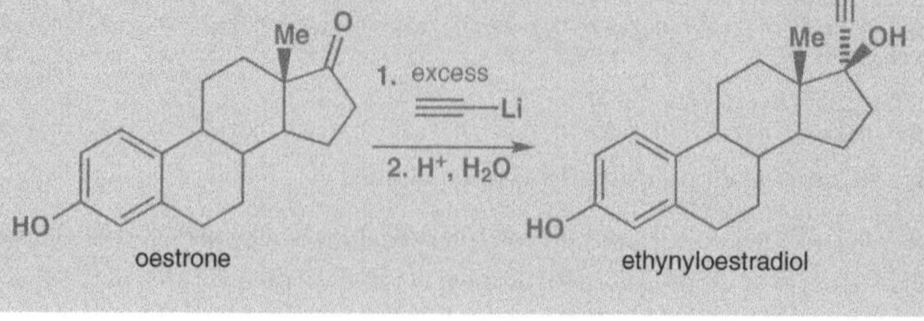

Triple bonds: stability and acidity

You have now met all the more important compounds with triple bonds. They all have electrons in low-energy sp hybrid orbitals (shown in green on the diagrams below), a feature which gives them stability or even unreactivity. Remember, an sp orbital has 50% s character, so electrons in this orbital are on average closer to the nucleus, and therefore more stable, than electrons in an sp^2 or sp^3 orbital.

Nitrogen, N_2, has sp orbitals at both ends and is almost inert. It is neither basic nor nucleophilic and a major achievement of life is the 'fixing' (trapping in reductive chemical reactions) of nitrogen by bacteria such as those in the roots of leguminous plants (peas and beans). HCN has an sp orbital on nitrogen and a C–H σ bond at the other end. The nitrogen's sp lone pair is not at all basic, but HCN is quite acidic with a pK_a of 10 because the negative charge in the conjugate base (CN^-) is in an sp orbital. Nitriles have similar bonds and they are non-nucleophilic and non-basic. Finally, we have just met alkynes, which are among the most acidic of hydrocarbons, again because of the stability of an anion with its charge in an sp orbital.

Halogen–metal exchange

Deprotonation is not the only way to use one simple organometallic reagent to generate another more useful one. Organolithiums can also remove halogen atoms from alkyl and aryl halides in a reaction known as halogen–metal exchange.

The bromine and the lithium simply swap places. As with many of these organometallic processes, the mechanism is not altogether clear, but can be represented as a nucleophilic attack on bromine by the butyllithium. But why does the reaction work? The product of our 'mechanism' is not PhLi and BuBr but a phenyl anion and a lithium cation. These could obviously combine to give PhLi and BuBr. But is this a reasonable interpretation and why does the reaction go that way and not the other? The key, again, is pK_a. We can think of the organolithiums as a complex between Li^+ and a carbanion.

■ The reason for this is again that the anion lies in an sp^2 orbital rather than an sp^3 orbital. See Chapter 8, p. 175.

The lithium cation is the same in all cases: only the carbanion varies. So the stability of the complex depends on the stability of the carbanion ligand. Benzene, (pK_a about 43) is more acidic than butane (pK_a about 50) so the phenyl complex is more stable than the butyl complex and the reaction is a way to make PhLi from available BuLi. Vinyllithiums (the lithium must be bonded directly to the alkene) can also be made this way and a R_2N– substituent is acceptable. Bromides or iodides react faster than chlorides.

Halogen–metal exchange tolls the knell of one appealing way to make carbon–carbon bonds. It may already have occurred to you that we might make a Grignard or organolithium reagent and combine it with another alkyl halide to make a new carbon–carbon σ bond.

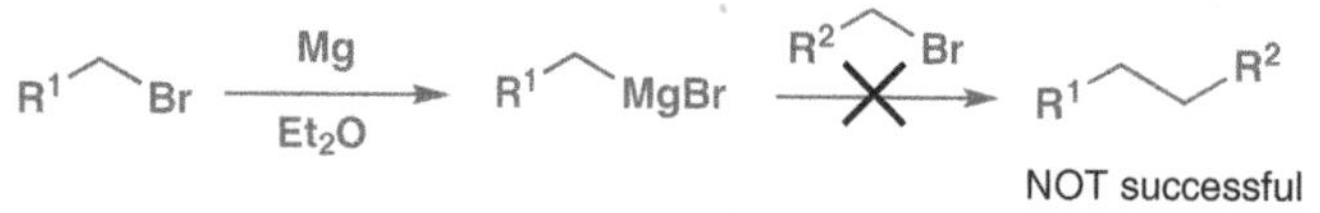

This reaction does not work because of transmetallation. The two alkyl bromides and their Grignard reagents will be in equilibrium with each other so that, even if the coupling were successful, three coupled products will be formed.

You will see later that transition metals are needed for this sort of reaction. The only successful reactions of this kind are couplings between metal derivatives of alkynes and alkyl halides. These do not exchange the metal as the alkynyl metal is much more stable than the alkyl metal.

A good example is the synthesis of a substituted alkyne starting from acetylene (ethyne) itself. One alkylation uses $NaNH_2$ as the base to make sodium acetylide and the other uses BuLi to make a lithium acetylide.

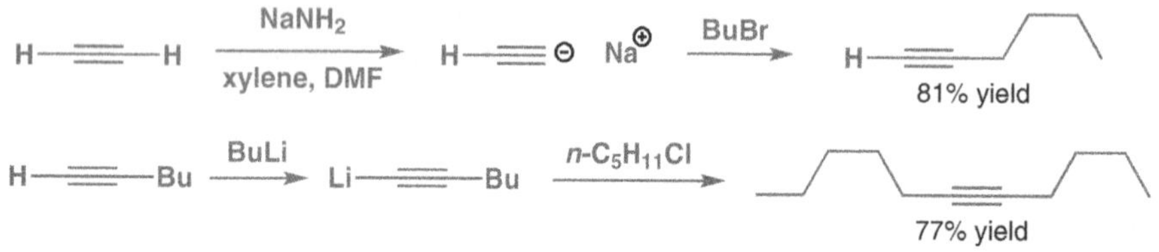

Transmetallation

Organolithiums can be converted to other types of organometallic reagents by transmetallation—simply treating with the salt of a less electropositive metal. The more electropositive Mg or Li goes into solution as an ionic salt, while the less electropositive metal such as Zn takes over the alkyl group.

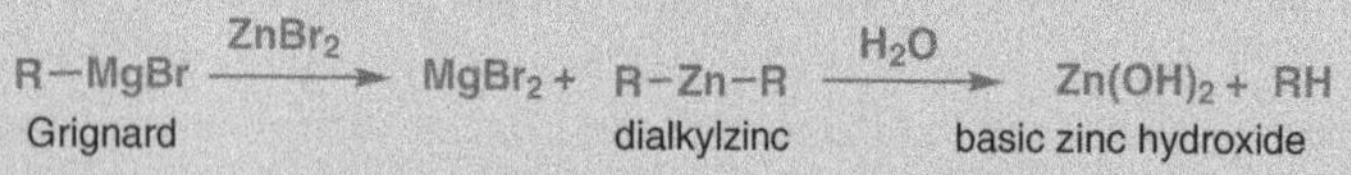

But why bother? Well, the high reactivity—and in particular the basicity—of Grignard reagents and organolithiums sometimes causes unwanted side reactions. Their combination with very strong electrophiles like acid chlorides usually results in a violent uncontrolled reaction. If a much less reactive organozinc compound is used instead, the reaction is more under control. These organozinc compounds can be made from either Grignard reagents or organolithium compounds. E. Negishi, a pioneer of organozinc chemistry, got the Nobel Prize for Chemistry in 2010 with R. F. Heck and A. Suzuki for their work on organometallic compounds.

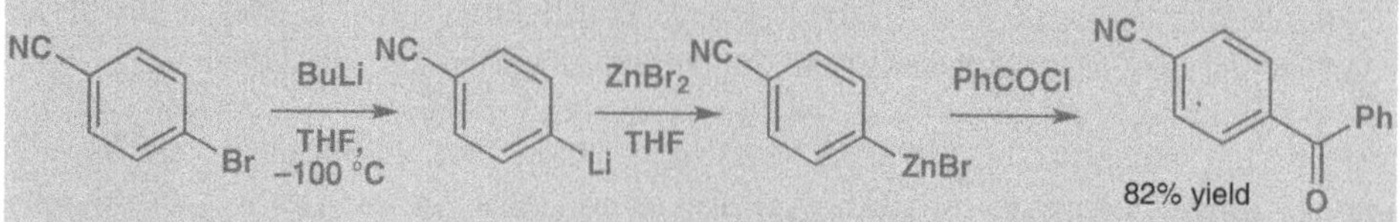

Using organometallics to make organic molecules

Now that you have met all of the most important ways of making organometallics (summarized here as a reminder), we shall move on to consider how to use them to make molecules:

what sorts of electrophiles do they react with and what sorts of products can we expect to get from their reactions? Having told you how you can make other organometallics, we shall really be concerned for the rest of this chapter only with Grignard reagents and organo-lithiums. In nearly all of the cases we shall talk about, the two classes of organometallics can be used interchangeably.

- **Ways of making organometallics**
  - **Oxidative insertion of Mg into alkyl halides**

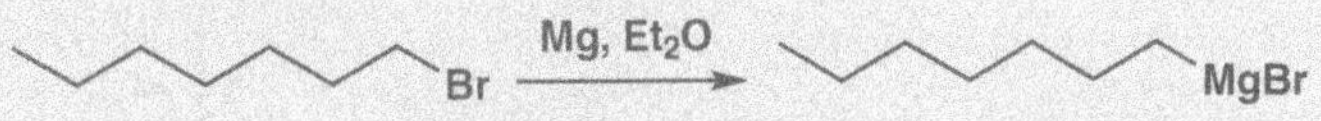

  - **Oxidative insertion of Li into alkyl halides**

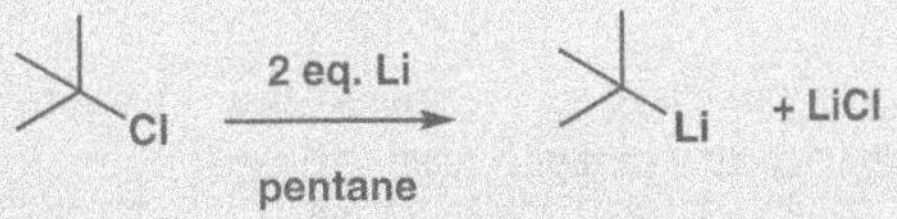

  - **Deprotonation of alkynes**

  - **Halogen–metal exchange**

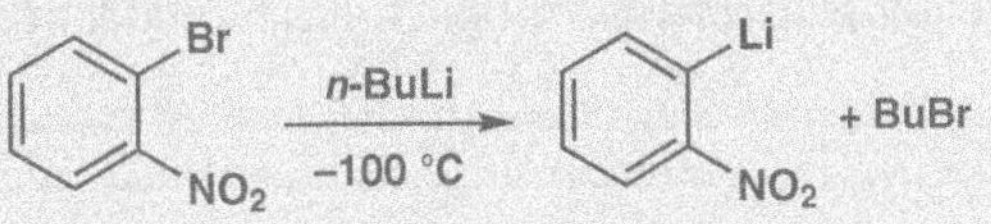

  - **Transmetallation**

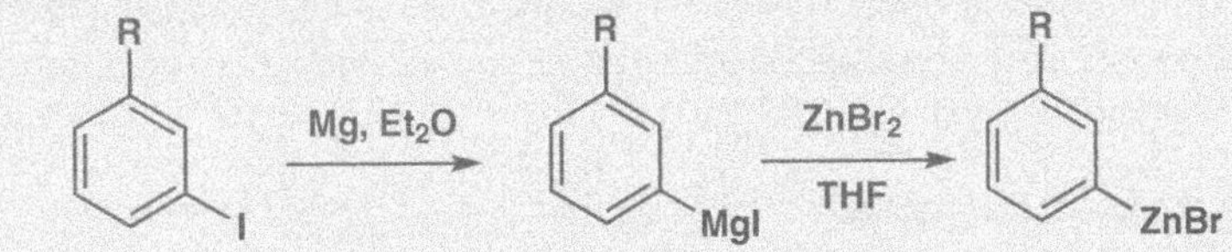

Making carboxylic acids from organometallics and carbon dioxide

Carbon dioxide reacts with organolithiums and Grignard reagents to give carboxylate salts. Protonating the salt with acid gives a carboxylic acid with one more carbon atom than the starting organometallic. The reaction is usually done by adding solid CO_2 to a solution of the organolithium in THF or ether, but it can also be done using a stream of dry CO_2 gas.

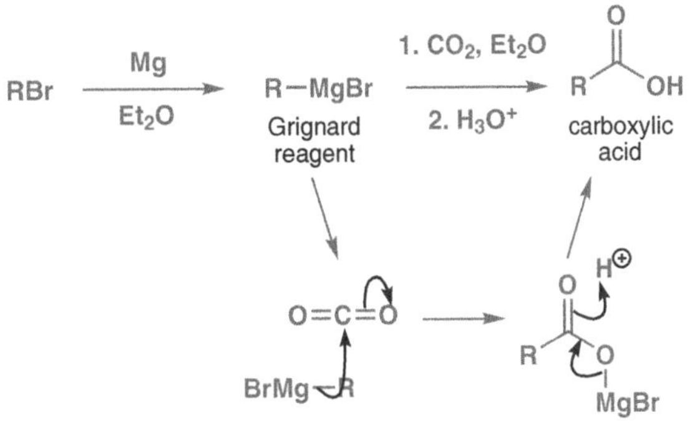

The example belows shows the three stages of the reaction: (1) forming the organometallic, (2) reaction with the electrophile (CO_2), and (3) the acidic work-up or quench, which protonates the product and destroys any unreacted organometallic. The three stages of the reaction have to be monitored carefully to make sure that each is finished before the next is begun. In particular it is absolutely essential that there is no water present during either of the first two stages—water must be added only at the end of the reaction, *after* the organometallic has all been consumed by reaction with the electrophile. You may occasionally see schemes written out without the quenching step included, but it is nonetheless always needed.

carboxylic acids from organometallics

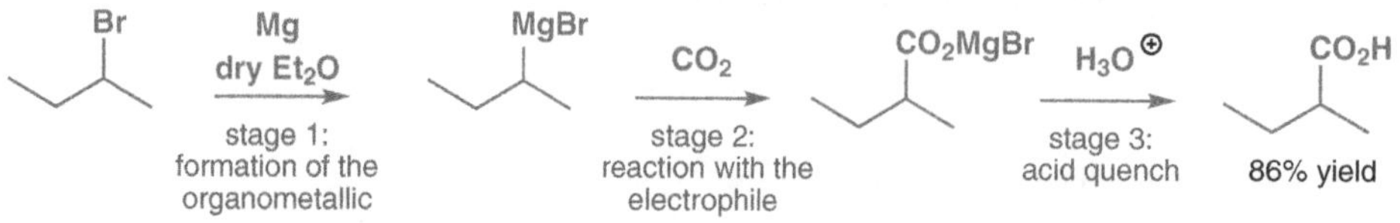

This next example shows that even very hindered chlorides can be used successfully. The significance of this will be clearer when you reach Chapter 15.

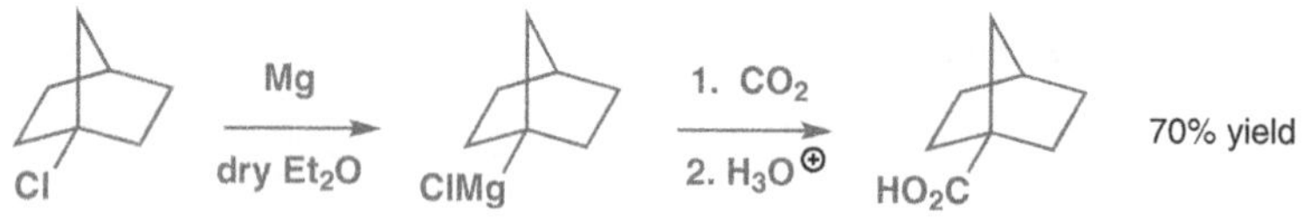

Making primary alcohols from organometallics and formaldehyde

You met formaldehyde, the simplest aldehyde, in Chapter 6, where we discussed the difficulties of using it in anhydrous reactions: it is either hydrated or a polymer paraformaldehyde, $(CH_2O)_n$, and in order to get pure, dry formaldehyde it is necessary to heat ('crack') the polymer to decompose it. But formaldehyde is a remarkably useful reagent for making primary alcohols, in other words alcohols that have just one carbon substituent on the hydroxy-bearing C atom. Just as carbon dioxide adds one carbon and makes an acid, formaldehyde adds one carbon and makes an alcohol.

a primary alcohol from formaldehyde

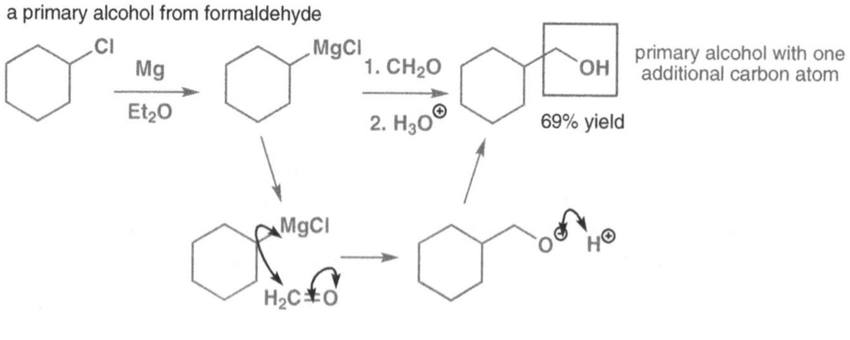

In the next two examples, formaldehyde makes a primary alcohol from two deprotonated alkynes. The second reaction here (for which we have shown organolithium formation, reaction, and quench simply as a series of three consecutive reagents) forms one of the last steps of the synthesis of *Cecropia* juvenile hormone, whose structure you met right at the beginning of the chapter.

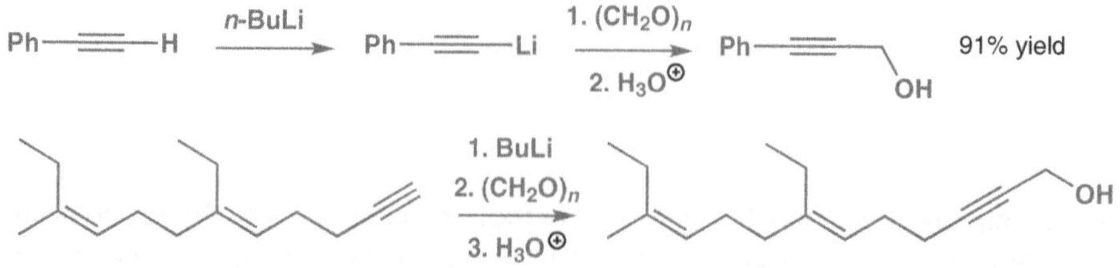

● Something to bear in mind with all organometallic additions to carbonyl compounds is that the addition takes the oxidation level down one (oxidation levels were described in Chapter 2, p. 33). In other words, if you start with an aldehyde, you end up with an alcohol. More specifically,

• additions to CO_2 give carboxylic acids

• additions to formaldehyde (CH_2O) give primary alcohols

• additions to other aldehydes (RCHO) give secondary alcohols

• additions to ketones give tertiary alcohols

Secondary and tertiary alcohols: which organometallic, which aldehyde, which ketone?

Aldehydes and ketones react with organometallic reagents to form secondary and tertiary alcohols, respectively, and some examples are shown with the general schemes here.

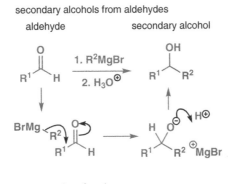

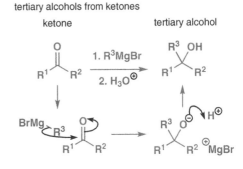

Fenarimol

Fenarimol is a fungicide that works by inhibiting the fungus's biosynthesis of important steroid molecules. It is made by reaction of a diarylketone with an organolithium derived by halogen–metal exchange.

To make any secondary alcohol, however, there may be a choice of two possible routes, depending on which part of the molecule you choose to make the organometallic and which part you choose to make the aldehyde. For example, the first example here shows the synthesis of a secondary alcohol from isopropylmagnesium chloride and acetaldehyde. But it is equally possible to make this same secondary alcohol from isobutyraldehyde and methyllithium or a methylmagnesium halide.

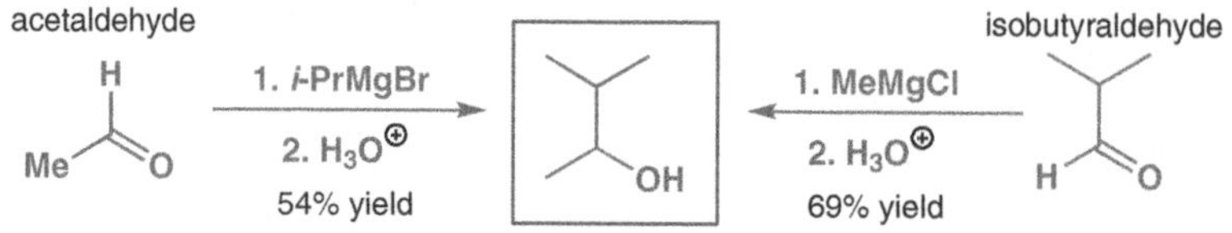

Indeed, back in 1912, when this alcohol was first described in detail, the chemists who made it chose to start with acetaldehyde, while in 1983, when it was needed as a starting material for a synthesis, it was made from isobutyraldehyde. Which way is better? The 1983 chemists probably chose the isobutyraldehyde route because it gave a better yield. But, if you were making a secondary alcohol for the first time, you might just have to try both in the laboratory and see which one gave a better yield.

Or you might be more concerned about which uses the cheaper, or more readily available, starting materials—this was probably also a factor in the choice of methylmagnesium chloride and the unsaturated aldehyde in the second example. Both can be bought commercially, while the alternative route to this secondary alcohol would require a vinyllithium or vinylmagnesium bromide reagent that would have to be made from a vinyl halide, which is itself not commercially available, along with difficult-to-dry acetaldehyde.

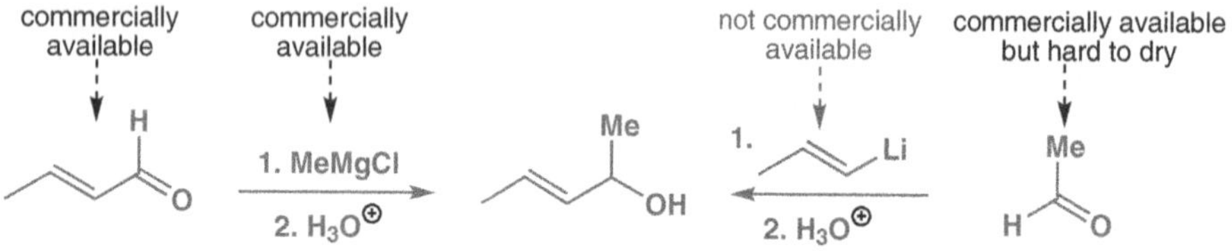

There is another choice for secondary alcohols: the reduction of a ketone. The ketone reacts with sodium borohydride to give a secondary alcohol. An obvious case where this would be a good route is the synthesis of a cyclic alcohol. This bicyclic ketone gives the secondary alcohol in good yield, and in the second example a diketone has both its carbonyl groups reduced.

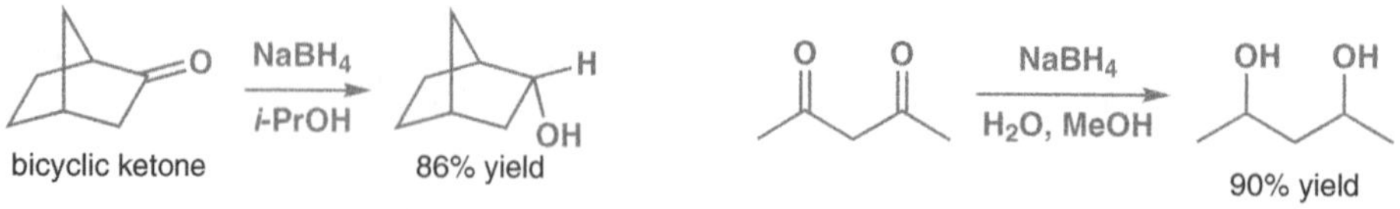

Flexibility in the synthesis of alcohols

As an illustration of the flexibility available in making secondary alcohols, one synthesis of bongkrekic acid, a highly toxic compound that inhibits transport across certain membranes in the cell, requires both of these (very similar) alcohols. The chemists making the compound at Harvard University chose to make each alcohol from quite different starting materials: an unsaturated aldehyde and an alkyne-containing organolithium in the first instance, and an alkyne-containing aldehyde and vinyl magnesium bromide in the second.

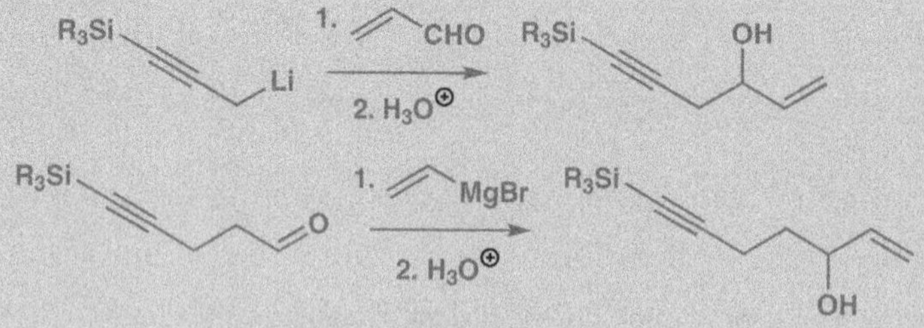

With tertiary alcohols, there is even more choice. The example below is a step in a synthesis of the natural product, nerolidol. But the chemists in Paris who made this tertiary alcohol could in principle have chosen any of three routes. Note that we have dropped the aqueous quench step from these schemes to avoid cluttering them.

three routes to a tertiary alcohol

Only the reagents in orange are commercially available, but, as it happens, the green Grignard reagent can be made from an alkyl bromide, which is itself commercially available, making route 1 on the left the most reasonable.

Now, do not be dismayed! We are not expecting you to remember a chemical catalogue and to know which compounds you can buy and which you can't. All we want you to appreciate at this stage is that there are usually two or three ways of making any given secondary or tertiary alcohol, and you should be able to suggest alternative combinations of aldehyde or ketone and Grignard or organolithium reagent that will give the same product. You are not expected to be able to assess the relative merits of the different possible routes to a compound. That is a topic we leave for a much later chapter on retrosynthetic analysis, Chapter 28.

Oxidation of alcohols

So far the metals we have used have had one oxidation state other than zero: Li(I), Mg(II), and Zn(II). If we want to oxidize organic compounds we need metals that have at least two higher oxidation states and that means transition metals. The most important by far is chromium, with Cr(III) and Cr(VI) as the useful oxidation states. Orange Cr(VI) compounds are good oxidizing agents: they remove hydrogen from organic compounds and are themselves reduced to green Cr(III). There are many Cr(VI) reagents used in organic chemistry, some of the more important ones are related to the polymeric oxide CrO_3. This is the anhydride of chromic acid and water breaks up the polymer to give a solution of chromic acid. Pyridine also breaks up the polymer to give a complex. This (Collins' reagent) was used to oxidize organic compounds but it is rather unstable and pyridinium dichromate (PDC) and pyridinium chlorochromate (PCC) are usually now preferred, especially as they are soluble in organic solvents such as CH_2Cl_2.

chromium(VI) oxide   chromic acid   pyridine–CrO$_3$ complex (Collins reagent)   PCC pyridinium chlorochromate   PDC pyridinium dichromate

Oxidation by these reagents of the various primary and secondary alcohols we have been making in this chapter takes us to a higher oxidation level. Oxidation of primary alcohols gives aldehydes and then carboxylic acids, while oxidation of secondary alcohols gives ketones. Note that you can't oxidize tertiary alcohols (without breaking a C–C bond).

■ The symbol [O] means an unspecified oxidizing agent.

primary alcohol   aldehyde   carboxylic acid   secondary alcohol   ketone

You will notice that the oxidation steps involve the removal of two hydrogen atoms and/or the addition of one oxygen atom. In Chapter 6 you saw that reduction meant the addition of hydrogen (and can also mean the removal of oxygen). Hiding behind these observations is the more fundamental idea that reduction requires the addition of electrons while oxidation requires the removal of electrons. If we used basic reagents, we could remove the OH proton from a primary alcohol, but to get the aldehyde we should have to remove a C–H proton as well *with a pair of electrons*. We should have to expel a hydride ion H^- and this doesn't happen. So we need some reagent that can remove a hydrogen atom *and a pair of electrons*. That defines an oxidizing agent.

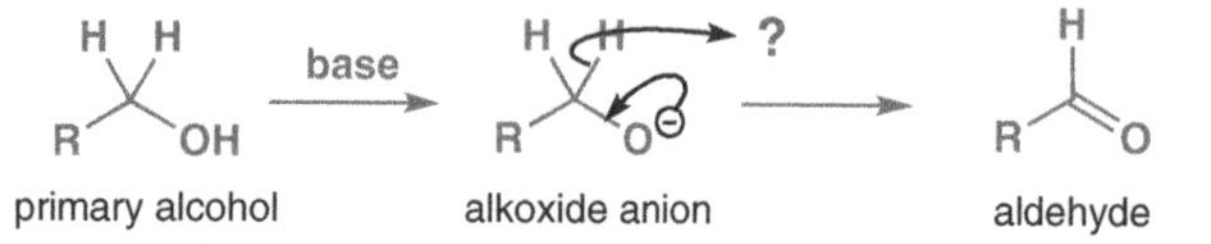

Here Cr(VI) can remove electrons to make Cr(III). It does so by a cyclic mechanism on a Cr(VI) ester. One hydrogen atom is removed (from the OH group) to make the ester and the second is removed (from carbon) in the cyclic mechanism. Notice how the arrows stop on the Cr atom and start again on the Cr=O bond, so two electrons are added to the chromium. This actually makes Cr(IV), an unstable oxidation state, but this gives green Cr(III) by further reactions.

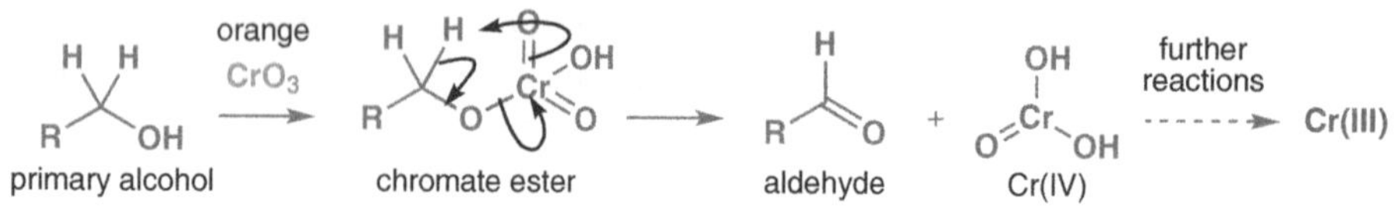

Interactive mechanism for chromium (VI) oxidation of alcohols

Two examples of the use of PCC in these oxidations come from Vogel. Hexanol is oxidized to hexanal in dichloromethane solution and commercial carveol (an impure natural product) to pure carvone with PCC supported on alumina in hexane solution. In both cases the pure aldehyde or ketone was isolated by distillation.

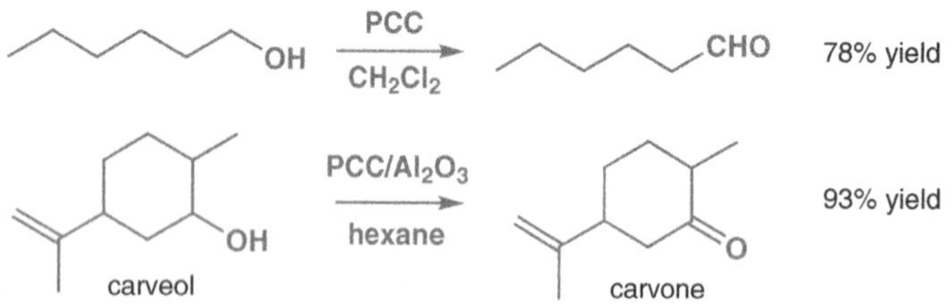

But a word of warning: stronger oxidizing agents like calcium hypochlorite or sodium hypochlorite (bleach) may oxidize primary alcohols all the way to carboxylic acids, especially in water. This is the case with *p*-chloro benzyl alcohol and the solid acid is easily isolated by the type of acid/base extraction we met in the previous chapter.

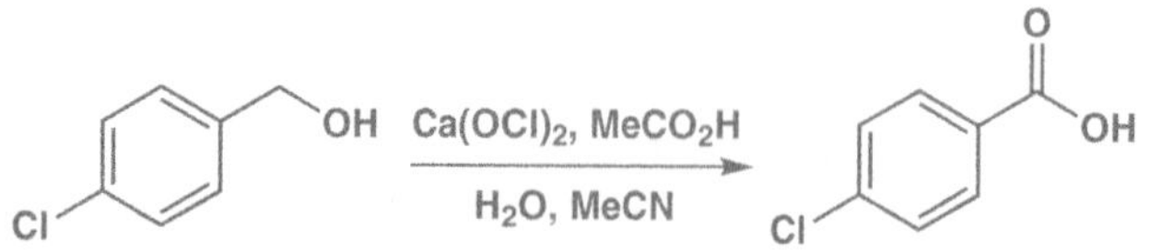

You will find further discussion of oxidizing agents in later chapters of the book. We have introduced them here so that you can see how primary and secondary alcohols, made by addition of organometallic reagents, can be oxidized to aldehydes or ketones so that the process can be repeated. A secondary alcohol, which could be made in two ways, can be oxidized with the pyridine–CrO_3 complex to the ketone and reacted with any Grignard or organolithium compound to give a range of tertiary alcohols.

Looking forward

In this chapter we have covered interconversions between ketones, aldehydes, and alcohols by forming C–C bonds using organometallics. We looked at oxidation and reduction as ways of complementing these methods—you should now be able to suggest at least one way of making any primary, secondary, or tertiary alcohol from simple precursors. In the next two chapters we will broaden our horizons beyond aldehydes and ketones to look at the reactivity of other carbonyl compounds—carboxylic acids and their derivatives such as esters and amides—and other nucleophiles. But the idea that we study organic reactions not only for their own sake but also so we can use them to make things should stay with you. We will come back to how to design ways of making molecules in Chapter 28. Many of these methods will employ the organometallics you have just met. We will then devote Chapter 40 to a broader range of more complex organometallic methods.

Further reading

For more on the detailed structures of Grignard reagents, see P. G. Williard in *Comprehensive Organic Synthesis*, vol. 1, 1999, p. 1. The alkylation of alkynes is described by P. J. Garratt in *Comprehensive Organic Synthesis*, vol. 1, 3rd edn, 1999, p. 271. The examples come from T. F. Rutledge, *J. Org. Chem.*, 1959, **24**, 840, D. N. Brattesoni and C. H. Heathcock, *Synth. Commun.* 1973, **3**, 245, R. Giovannini and P. Knochel, *J. Am. Chem. Soc.*, 1998, **120**, 11186, C. E. Tucker, T. N. Majid, and P. Knochel, *J. Am. Chem. Soc.*, 1992, **114**, 3983. For a rather advanced review of organozinc compounds, see P. Knochel, J. J. Almena Perea, and P. Jones, *Tetrahedron*, 1998, **54**, 8275.

Discovery of pyridinium chlorochromate (PCC): G. Piancatelli, A. Scettri, and M. D'Auria, *Synthesis*, 1982, 245; H. S. Kasmai, S. G. Mischke, and T. J. Blake, *J. Org. Chem.*, 1995, **60**, 2267 and PDC: E. J. Corey and J. W. Suggs, *Tetrahedron Lett.*, 1975, 2647. Details of oxidation experiments: B. S. Furniss, A. J Hannaford, P. W. G. Smith, and A. R. Tatchell, *Vogel's Textbook of Practical Organic Chemistry*, 5th edn, Longman, Harlow, 1989, pp. 590 and 610; J. C. Gilbert and S. F. Martin, *Experimental Organic Chemistry*, Harcourt, Fort Worth, 2002, p. 507.

Check your understanding

To check that you have mastered the concepts presented in this chapter, attempt the problems that are available in the book's Online Resource Centre at http://www.oxfordtextbooks.co.uk/orc/clayden2e/

Suggested solutions for Chapter 9

9

PROBLEM 1

Propose mechanisms for the first four reactions in the chapter.

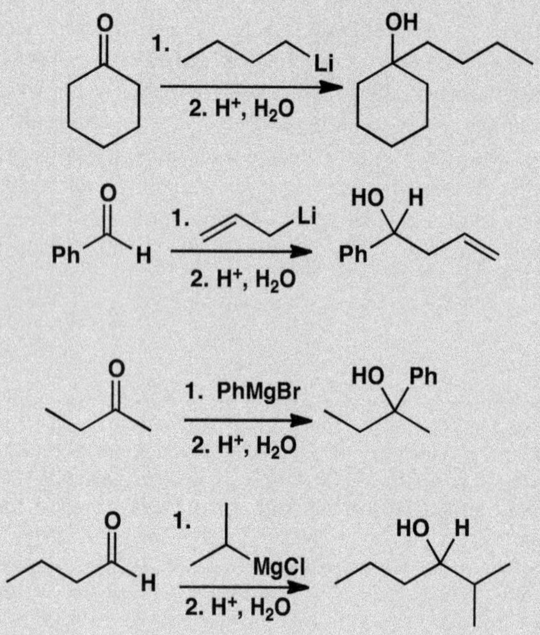

Purpose of the problem

Rehearsal of the basic mechanisms from chapter 9.

Suggested solution

Each reaction involves nucleophilic attack of the organometallic reagent on the aldehyde or ketone followed by protonation. You may draw the intermediate as an anion or with an O–metal bond as you please. Note the atom-specific arrows to show which atom is the nucleophile. In the third reaction the allyl-Li might attack through either end.

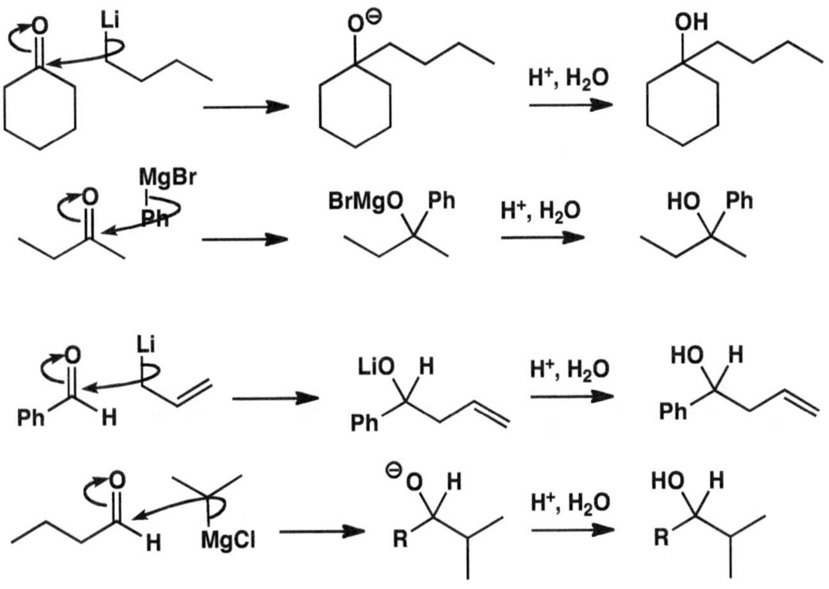

PROBLEM 2

What products would be formed in these reactions?

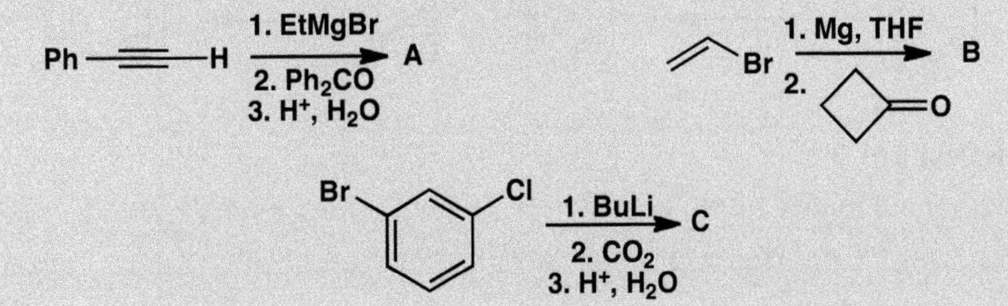

Purpose of the problem

The toughest test: predicting the product. The sooner you get practice the better.

Suggested solution

Though prediction is harder than explanation, you should get these right the first time as only the last one has a hint of difficulty. In the first example, the ethyl Grignard reagent acts as a base to remove a proton from the alkyne. Whether you draw the intermediate as an alkyne anion or a Grignard reagent is up to you. Notice that sometimes we give the protonation step at the end and sometimes not. This is the general practice among organic chemists and you may or you may not bother to draw the mechanism of this step.

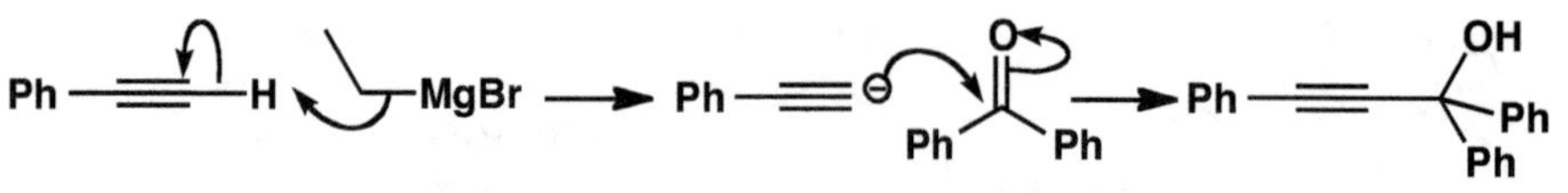

For the second example, just make the organometallic reagent and add it to the carbonyl group. Cyclobutanone is more electrophilic than many other ketones because of the strain of a carbonyl group in a four-membered ring. This time the protonation is shown.

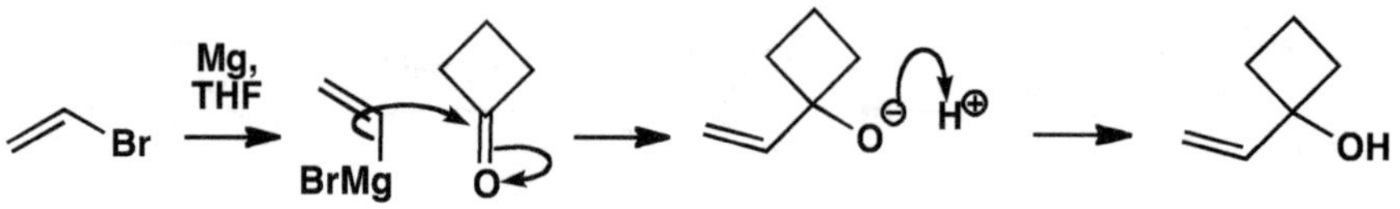

The third example raises the question of which halogen is replaced. In fact bromine is more easily replaced than chlorine. Iodine is more easily replaced than either and fluorine usually does not react. Don't be disappointed if you failed to see this.

■ This synthesis of carboxylic acids is on pp. 190-191 of the textbook.

PROBLEM 3

Suggest alternative routes to fenarimol different from the one in the textbook on p. 192. Remind yourself of the answer to problem 2 above.

Purpose of the problem

Practice in choosing alternative routes.

Suggested solution

Three aromatic rings are joined to a tertiary alcohol in fenarimol, so the alternatives are to make organometallic reagents from different aromatic compounds. Two aromatic compounds must be joined to form a ketone and the third added as an organometallic reagent. You will meet ways to make ketones like these in chapter 21. You'll need the insight from problem 2 above to help you choose Br as the functional group to be lithiated or converted into a Grignard reagent. Here are two possible methods:

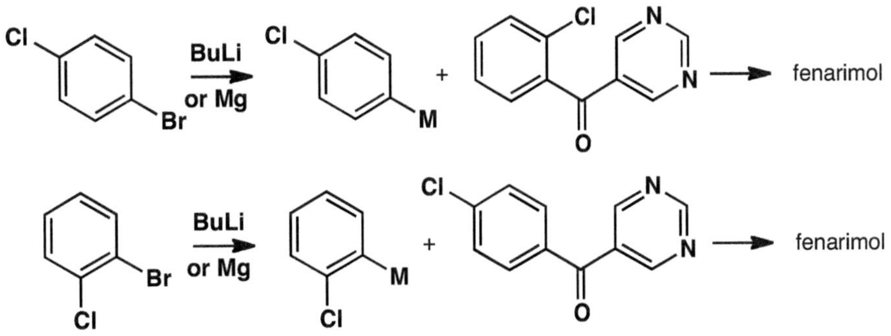

PROBLEM 4

Suggest two syntheses of the bee pheromone heptan-2-one.

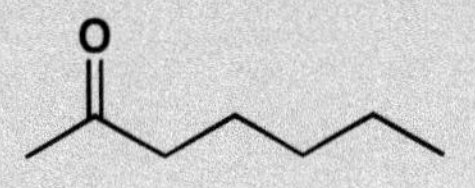

Purpose of the problem

Further exploration of the use of organometallic compounds. This time you'll probably need the oxidation of alcohols from p. 194 of the textbook.

Suggested solution

There are of course many different solutions but the most obvious are to make the corresponding secondary alcohol and oxidize it. Two alternatives are shown here.

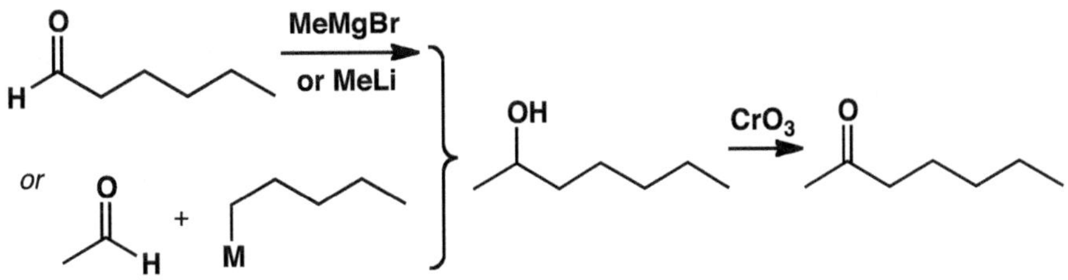

PROBLEM 5

The antispasmodic drug biperidin is made by the Grignard addition reaction shown here. What is the structure of the drug? Do not be put off by the apparent complexity of the structure: just use the chemistry of Chapter 9.

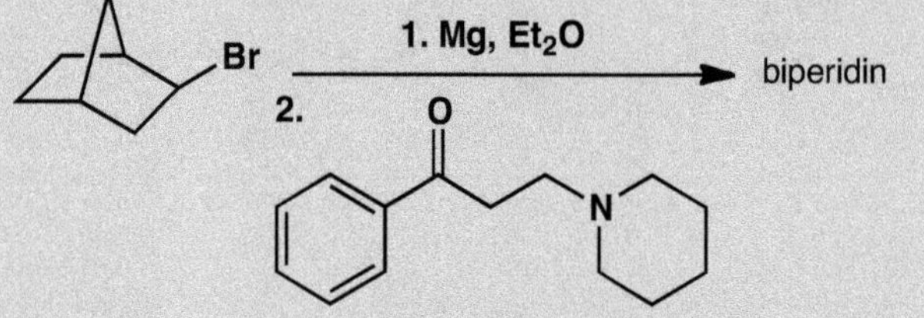

How would you suggest that the drug procyclidine should be made?

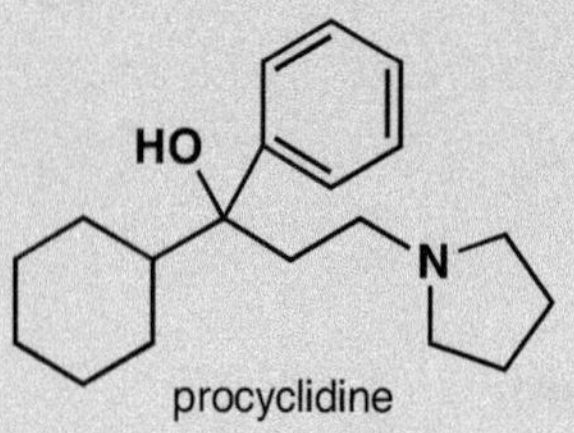

procyclidine

Purpose of the problem

Exercise in product prediction in a more complicated case and a logical extension to something new.

Suggested solution

A Grignard reagent must be formed from the alkyl bromide and this must add to the ketone. Aqueous acidic work up (not mentioned in the problem as is often the case) must give a tertiary alcohol and that is biperidin.

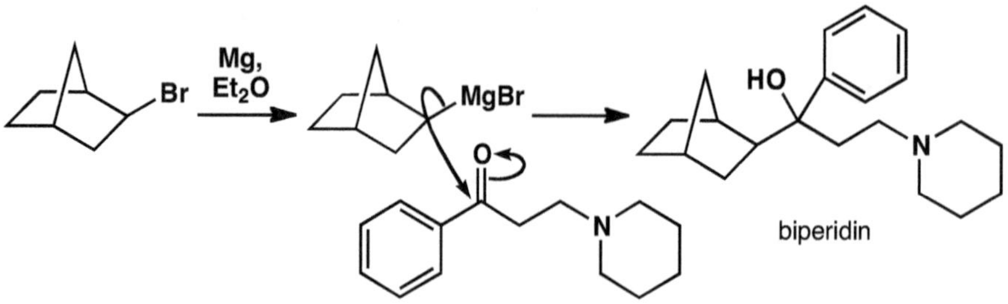

To get procyclidine, we must change both the alkyl halide and the ketone but the reaction is very similar.

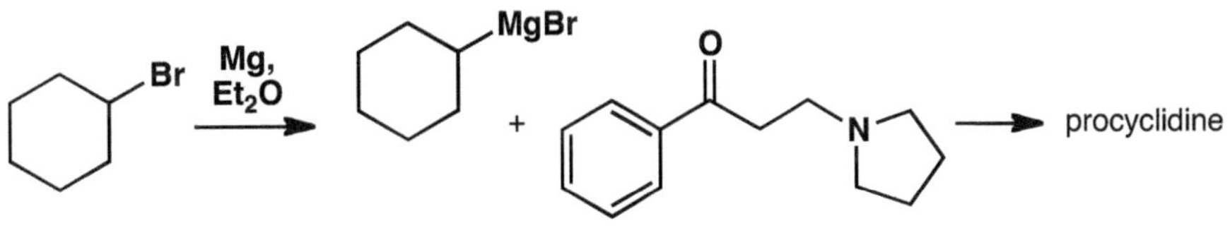

PROBLEM 6

The synthesis of the gastric antisecretory drug rioprostil requires this alcohol.

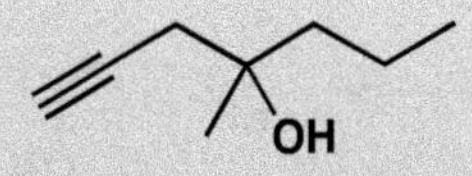

(a) Suggest possible syntheses starting from ketones and organometallics.
(b) Suggest possible syntheses of the ketones in part (a) from aldehydes and organometallics (don't forget about CrO₃ oxidation).

Purpose of the problem

Your first introduction to sequences of reactions where more complex molecules are created.

Suggested solution

There are three one-step syntheses from ketones and organometallic compounds. We have used 'M' to indicate the metal—it might be Li or MgX (in other words, the organometallic could be an organolithium or a Grignard reagent).

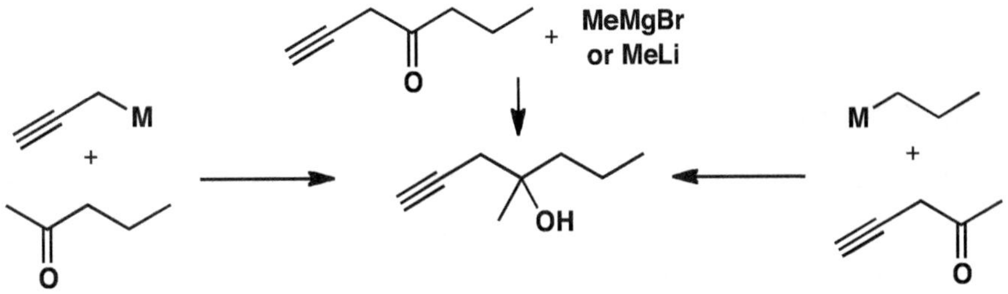

Each of these ketones can be made by oxidation of an alcohol that can in turn be made from an organometallic compound and an aldehyde.

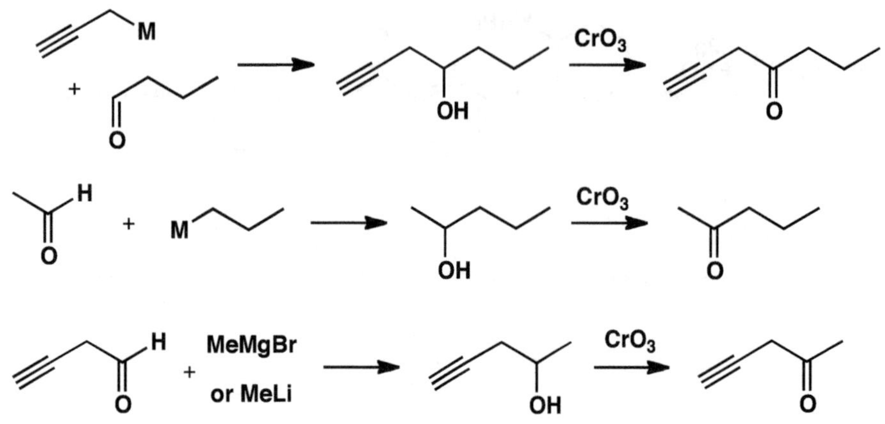

PROBLEM 7

Why is it possible to make the lithium derivative A by Br/Li exchange, but not the lithium derivative B?

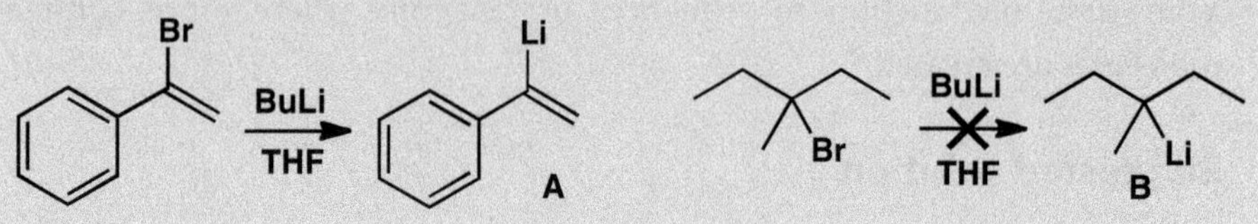

Purpose of the problem

Revision of the stability of carbanions and its relevance to lithium/bromine exchange.

Suggested solution

The first example is a vinyl bromide and vinyl (sp^2) carbanions are more stable than saturated (sp^3) carbanions because of the greater s-character in the C–Li bond. The second example is saturated, like BuLi, but it is a *tertiary* alkyl bromide. The *t*-alkyl carbanion would be less stable than the primary one and its lithium derivative less stable than BuLi, so it is not formed.

PROBLEM 8

How could you use these four commercially available starting materials

PhCHO   EtI CO_2 —Br

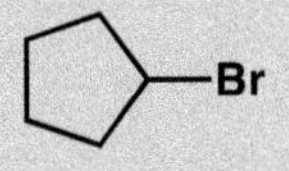

to make the following three compounds?

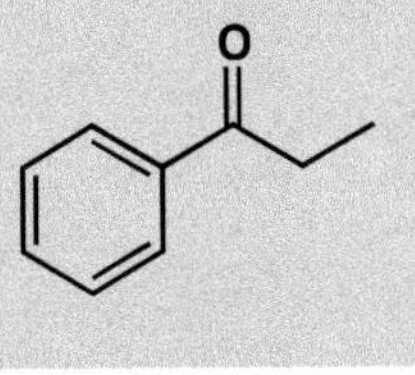

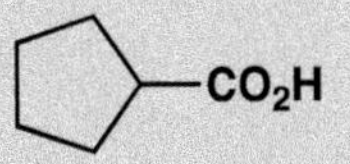

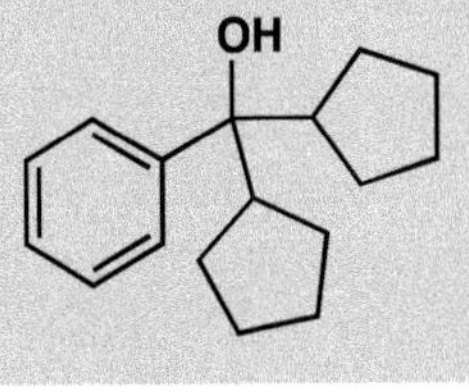

Purpose of the problem

Thinking about synthesis: how to put molecules together

Suggested solution

The first compound contains a phenyl and an ethyl group, so you could convert the ethyl iodide to a Grignard reagent and add it to the aldehyde. The product is an alcohol, so you need to use CrO_3 to oxidize it to the ketone

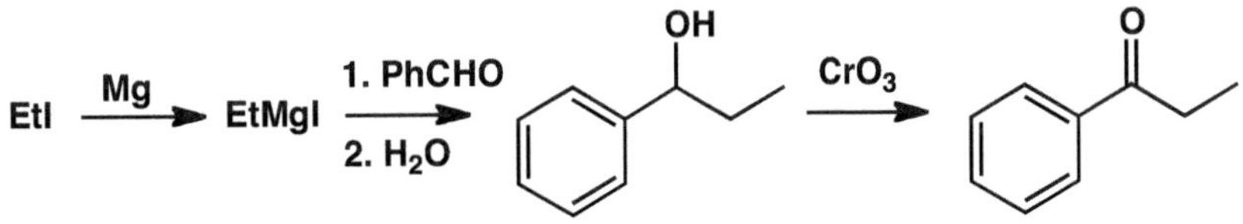

The second compound is a carboxylic acid, which can come from addition of the Grignard reagent derived from cyclopentyl bromide to carbon dioxide.

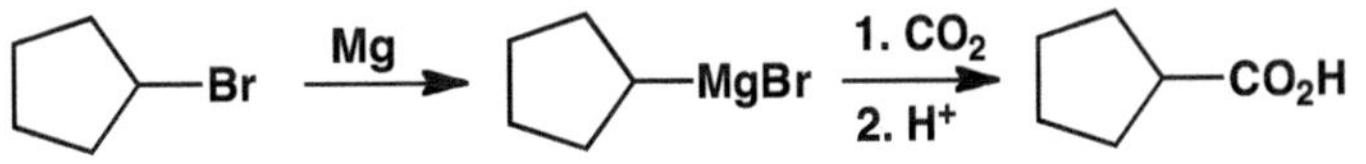

The third compound is a tertiary alcohol, which you could make by addition of the same cyclopentyl Grignard reagent to a ketone. The ketone will also need to be made by oxidation of an alcohol, itself derived from benzaldehyde and the cyclopentyl Grignard reagent.

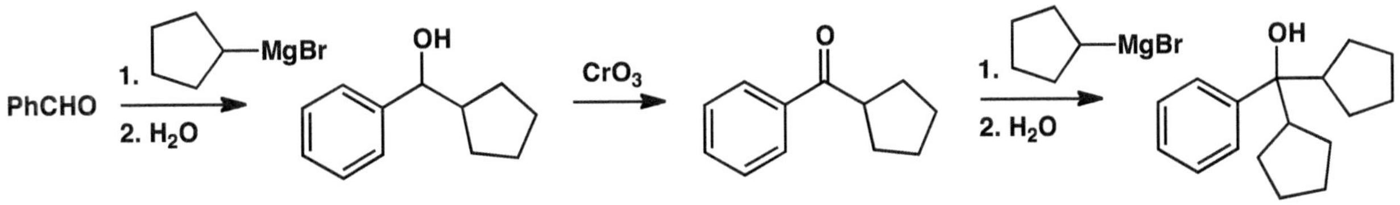

Nucleophilic substitution at the carbonyl group

10

Connections

➡ **Building on**

Drawing mechanisms **ch5**

Nucleophilic attack on carbonyl groups **ch6 & ch9**

- Acidity and pK_a **ch8**

Grignard and RLi addition to C=O groups **ch9**

Arriving at

- Nucleophilic attack followed by loss of leaving group
- What makes a good nucleophile
- What makes a good leaving group
- There is always a tetrahedral intermediate
- How to make acid derivatives
- Reactivity of acid derivatives
- How to make ketones from acids
- How to reduce acids to alcohols

➡ **Looking forward to**

- Loss of carbonyl oxygen **ch11**
- Kinetics and mechanism **ch12**
- Reactions of enols **ch20, ch25, & ch26**
- Chemoselectivity **ch23**

You are already familiar with reactions of compounds containing carbonyl groups. Aldehydes and ketones react with nucleophiles at the carbon atom of their carbonyl group to give products containing hydroxyl groups. Because the carbonyl group is such a good electrophile, it reacts with a wide range of different nucleophiles: you have met reactions of aldehydes and ketones with (in Chapter 6) cyanide, water, and alcohols, and (in Chapter 9) organometallic reagents (organolithiums and organomagnesiums, or Grignard reagents).

In this chapter and Chapter 11 we shall look at some more reactions of the carbonyl group—and revisit some of the ones we touched on in Chapter 6. It is a tribute to the importance of this functional group for organic chemistry that we have devoted four chapters of this book to its reactions. Just like the reactions in Chapters 6 and 9, the reactions in Chapters 10 and 11 all involve attack of a nucleophile on a carbonyl group. The difference is that this step is followed by other mechanistic steps, which means that the overall reactions are not just *additions* but also *substitutions*.

The product of nucleophilic addition to a carbonyl group is not always a stable compound

Addition of a Grignard reagent to an aldehyde or ketone gives a stable alkoxide, which can be protonated with acid to produce an alcohol (you met this reaction in Chapter 9). The same is not true for addition of an alcohol to a carbonyl group in the presence of base—in Chapter 6 we drew a reversible, equilibrium arrow for this transformation and said that the product, a hemiacetal, is formed to a significant extent only if it is cyclic.

The reason for this instability is that RO⁻ is easily expelled from the molecule. We call groups that can be expelled from molecules, usually taking with them a negative charge, **leaving groups**. We'll look at leaving groups in more detail later in this chapter and again in Chapter 15.

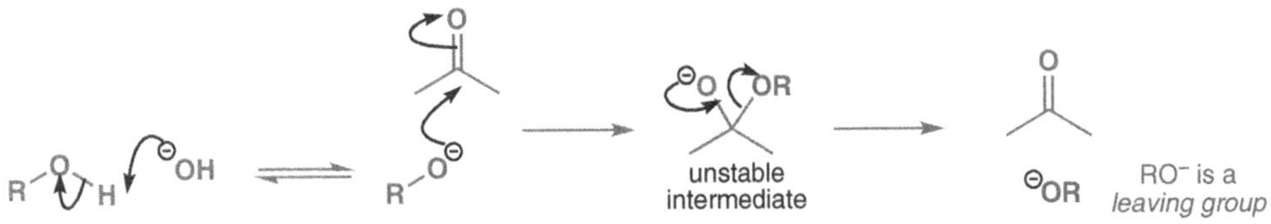

• **Leaving groups**

Leaving groups are anions such as Cl⁻, RO⁻, and RCO₂⁻ that can be expelled from molecules taking their negative charge with them.

So, if the nucleophile is also a leaving group, there is a chance that it will be lost again and that the carbonyl group will reform—in other words, the reaction will be reversible. The energy released in forming the C=O bond (bond strength 720 kJ mol⁻¹) makes up for the loss of two C–O single bonds (about 350 kJ mol⁻¹ each), one of the reasons for the instability of the hemiacetal product in this case.

The same thing can happen if the starting carbonyl compound contains a potential leaving group. The unstable negatively charged intermediate in the red box below is formed when a Grignard reagent is added to an ester.

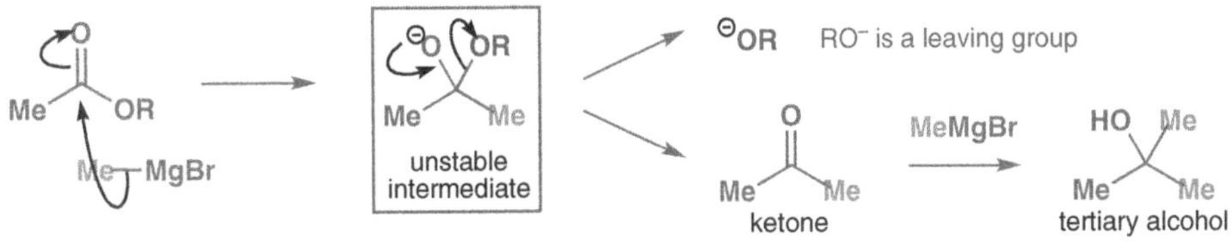

Again, it collapses with loss of RO⁻ as a leaving group. This time, though, we have not gone back to starting materials: instead we have made a new compound (a ketone) by a **substitution reaction**—the OR group of the starting material has been substituted by the Me group of the product. In fact the ketone product can react with the Grignard reagent a second time to give a tertiary alcohol. Later in this chapter we'll discuss why the reaction doesn't stop at the ketone.

Carboxylic acid derivatives

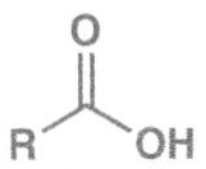

Most of the starting materials for, and products of, these substitutions will be carboxylic acid derivatives, with the general formula RCOX. You met the most important members of this class in Chapter 2: here they are again as a reminder.

Carboxylic acid derivatives

Carboxylic acid	Derivative of RCO₂H			
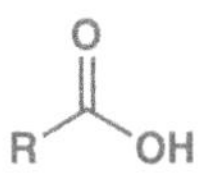	acid chloride or acyl chloride* 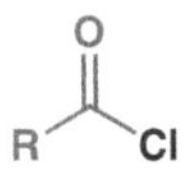	ester 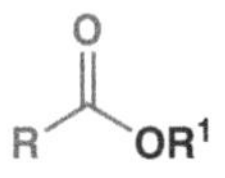	acid anhydride 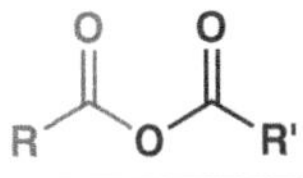	amide 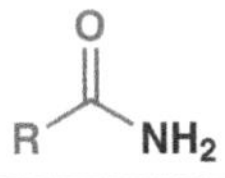

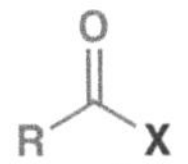

*We shall use these two terms interchangeably.

Acid chlorides and acid anhydrides react with alcohols to make esters

Acetyl chloride will react with an alcohol in the presence of a base to give an acetate ester and we get the same product if we use acetic anhydride.

■ The reactions of alcohols with acid chlorides and with acid anhydrides are the most important ways of making esters, but not the only ways. We shall see later how carboxylic acids can be made to react directly with alcohols.

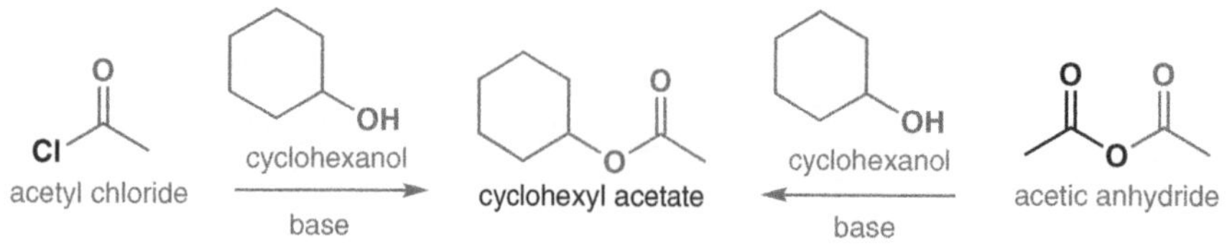

■ Remember the symbol for acetyl? $Ac=CH_3CO$. You can represent the acetate of an alcohol ROH as ROAc but not as RAc as this would be a ketone.

■ You will notice that the terms 'acid chloride' and 'acyl chloride' are used interchangeably.

In each case, a substitution (of the black part of the molecule, Cl^- or AcO^-, by cyclohexanol) has taken place—but how? It is important that you learn not only the *fact* that acyl chlorides and acid anhydrides react with alcohols but also the *mechanism* of the reaction. In this chapter you will meet a lot of reactions, but relatively few mechanisms—once you understand one, you should find that the rest follow on quite logically.

The first step of the reaction is, as you might expect, addition of the nucleophilic alcohol to the electrophilic carbonyl group—we'll take the acyl chloride first. The base is important because it removes the proton from the alcohol once it attacks the carbonyl group. A base commonly used for this is pyridine. If the electrophile had been an aldehyde or a ketone, we would have got an unstable hemiacetal, which would collapse back to starting materials by eliminating the alcohol. With an acyl chloride, the alkoxide intermediate we get is also unstable. It collapses again by an elimination reaction, this time losing chloride ion, to form the ester. Chloride is the *leaving group* here—it leaves with its negative charge.

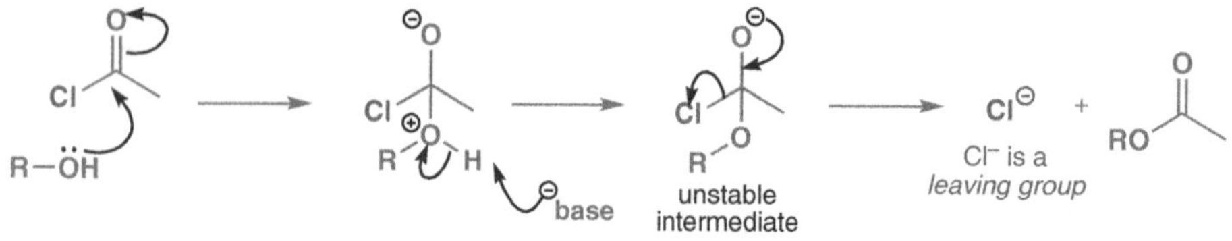

With this reaction as a model, you should be able to work out the mechanism of ester formation from acetic anhydride and an alcohol. Try to write it down without looking at the acyl chloride mechanism above, and certainly not at the answer below. Here it is, with pyridine as the base. Again, addition of the nucleophile gives an unstable intermediate, which undergoes an elimination reaction, this time losing a carboxylate anion to give an ester.

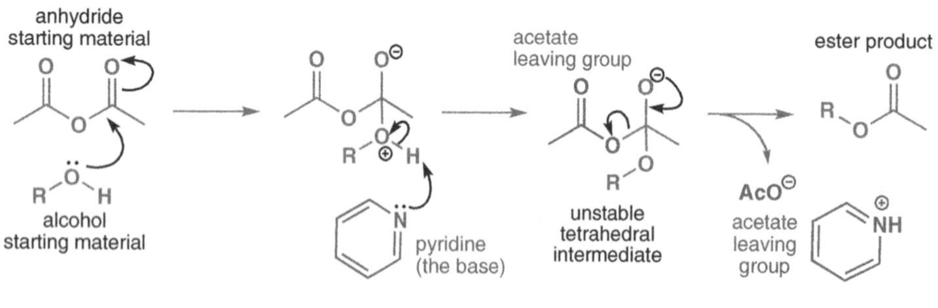

We call the unstable intermediate formed in these reactions the **tetrahedral intermediate** because the trigonal (sp^2) carbon atom of the carbonyl group has become a tetrahedral (sp^3) carbon atom.

● Tetrahedral intermediates

Substitutions at trigonal carbonyl groups go through a tetrahedral intermediate and then on to a trigonal product.

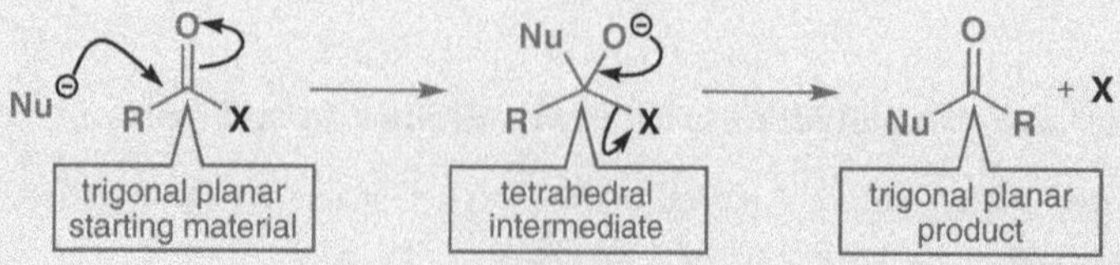

More details of this reaction

Acylation with acyl chlorides in the presence of pyridine has more subtleties than first meet the eye. If you are reading this chapter for the first time, you might skip this box, as it is not essential to the general flow of what we are saying. There are three more points to notice.

Pyridine is consumed during both of these reactions, since it ends up protonated. One whole equivalent of pyridine is therefore necessary and, in fact, the reactions are often carried out with pyridine as solvent.

The observant among you may also have noticed that the (weak—pyridine) base catalyst in this reaction works very slightly differently from the (strong—hydroxide) base catalyst in the hemiacetal-forming reaction on p. 197: pyridine removes the proton *after* the nucleophile has added; hydroxide removes the proton *before* the nucleophile has added. This is deliberate, and will be discussed further in Chapters 12 and 40. The basicities of pyridine (pK_a for protonation 5.5) and hydroxide (pK_a of water 15.7) were discussed in Chapter 8.

Pyridine is, in fact, more nucleophilic than the alcohol, and it attacks the acyl chloride rapidly, forming a highly electrophilic (because of the positive charge) intermediate. It is then this intermediate that subsequently reacts with the alcohol to give the ester. Because pyridine is acting as a nucleophile to speed up the reaction, yet is unchanged by the reaction, it is called a **nucleophilic catalyst**.

nucleophilic catalysis in ester formation

Interactive mechanism for pyridine nucleophilic catalysis

Why are the tetrahedral intermediates unstable?

The alkoxide formed by addition of a Grignard reagent to an aldehyde or ketone is stable, lasting long enough to be protonated on work-up in acid to give an alcohol as product.

Tetrahedral intermediates are similarly formed by addition of a nucleophile, say ethanol in base, to the carbonyl group of acetyl chloride, but these tetrahedral intermediates are unstable. Why are they *unstable*? The answer is to do with leaving group ability. Once the nucleophile has added to the carbonyl compound, the stability of the product (or tetrahedral intermediate) depends on how good the groups attached to the new tetrahedral carbon atom are at leaving with the negative charge. In order for the tetrahedral intermediate to collapse (and therefore be just an intermediate and not the final product) one of the groups has to be able to leave and carry off the negative charge from the alkoxide anion formed in the addition.

The most stable anion will be the best leaving group. There were three choices for the leaving group: Cl⁻, EtO⁻, or Me⁻. We can make MeLi but not Me⁻ because it is very unstable so Me⁻ must be a very bad leaving group. EtO⁻ is not so bad—alkoxide salts are stable, but they are still strong, reactive bases. But Cl⁻ is the best leaving group: Cl⁻ ions are perfectly stable and quite unreactive, and happily carry off the negative charge from the oxygen atom.

You probably eat several grams of Cl⁻ every day but you would be unwise to eat EtO⁻ or MeLi. So neither of these reactions occurs:

How do we know that the tetrahedral intermediate exists?

We don't expect you to be satisfied with the bland statement that tetrahedral intermediates are formed in these reactions: of course, you wonder how we know that this is true. The first evidence for tetrahedral intermediates in the substitution reactions of carboxylic acid derivatives was provided by Bender in 1951. He made carboxylic acid derivatives RCOX that had been 'labelled' with an isotope of oxygen, ^{18}O. This is a non-radioactive isotope that is detected by mass spectrometry. He then reacted these derivatives with water to make labelled carboxylic acids. By any reasonable mechanism, the products would have one ^{18}O atom from the labelled starting material. Because the proton on a carboxylic acid migrates rapidly from one oxygen to another, both oxygens are labelled equally.

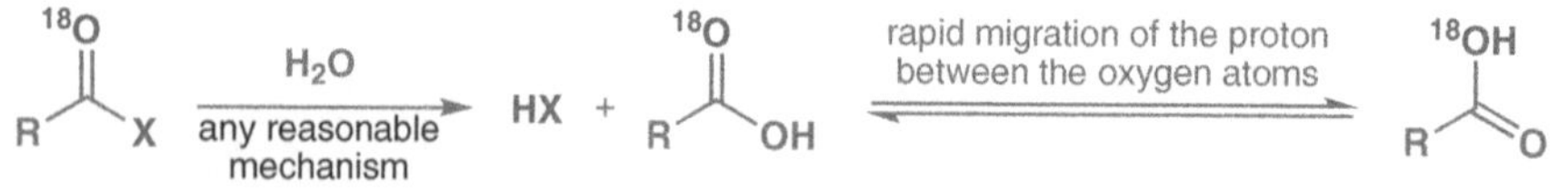

In Bender's original work, X was an alkoxy group (i.e. RCOX was an ester).

He then reacted these derivatives with insufficient water for complete consumption of the starting material. At the end of the reaction, he found that the proportion of labelled molecules in the *remaining starting material* had decreased significantly: in other words, it was no longer completely labelled with ^{18}O; some contained 'normal' ^{16}O. The formation of the tetrahedral intermediate would be as before but rapid proton transfer would also mean that the two oxygen atoms would be the same. Now you may see the next step in the argument.

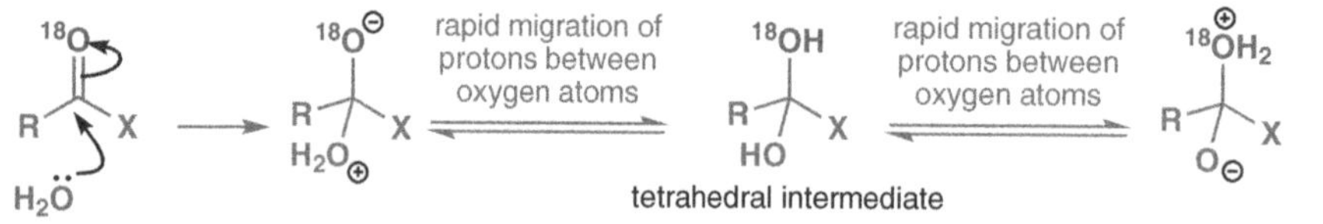

This result cannot be explained by direct substitution of X by H_2O, but is consistent with the existence of an intermediate in which the unlabelled ^{16}O and labelled ^{18}O can 'change places'. This intermediate is the *tetrahedral intermediate* for this reaction. Either isomer can lose X and in each case labelled carboxylic acid is formed.

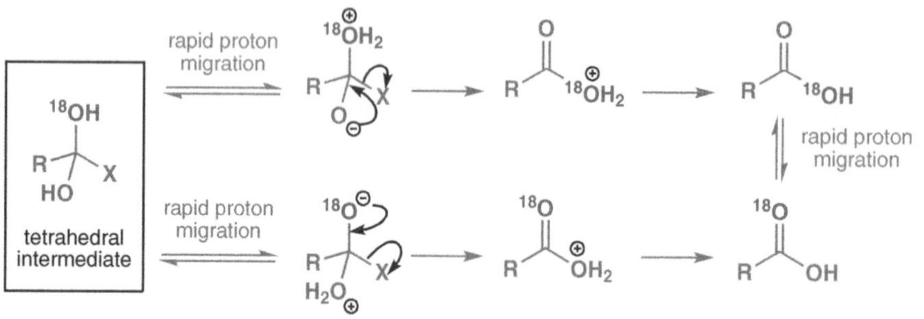

But either tetrahedral intermediate could lose water instead. In one case (top line below) the original starting material is regenerated complete with label. But in the second case, labelled water is lost and *unlabelled starting material is formed*. This result would be difficult to explain without a tetrahedral intermediate with a lifetime long enough to allow for proton exchange. This 'addition–elimination' mechanism is now universally accepted.

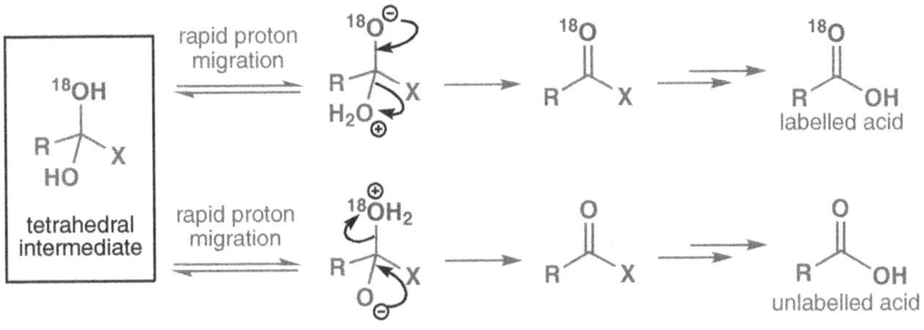

pKₐ is a useful guide to leaving group ability

It's useful to be able to compare leaving group ability quantitatively. This is impossible to do exactly, but a good guide is the pK_a of the conjugate acid (Chapter 8). If X^- is the leaving group, the lower the pK_a of HX, the better X^- is as a leaving group. If we go back to the example of ester formation from acyl chloride plus alcohol, there's a choice of Me^-, EtO^-, and Cl^-. HCl is a stronger acid than EtOH, which is a much stronger acid than methane. So Cl^- is the best leaving group and EtO^- the next best. These observations apply only to reactions at the carbonyl group.

> ● **Leaving group ability**
> The lower the pK_a of HX, the better the leaving group of X^- in carbonyl substitution reactions.

The most important substituents in carbonyl reactions are alkyl or aryl groups (R), amino groups in amides (NH_2), alkoxy groups in esters (RO^-), carboxylate groups (RCO_2^-) in anhydrides, and chloride (Cl^-) in acyl chlorides. The order of leaving group ability is then:

carboxylic acid derivative	leaving group, X⁻	conjugate acid, HX	pKa of HX	leaving group?
acyl chloride	Cl⁻	HCl	<0	excellent
anhydride	RCOO⁻	RCO₂H	about 5	good
ester	RO⁻	ROH	about 15	poor
amide	NH₂⁻	NH₃	about 25	very poor
alkyl or aryl derivative	R⁻	RH	>40	not a leaving group

We can use pK_a to predict what happens if we react an acyl chloride with a carboxylate salt. We expect the carboxylate salt (here, sodium formate or sodium methanoate, HCO_2Na) to act as the nucleophile to form a tetrahedral intermediate, which could collapse in any one of three ways. We can straightaway rule out loss of Me^- and we might guess that Cl^- is a better leaving group than HCO_2^- as HCl is a much stronger acid than a carboxylic acid, and we'd be right. Sodium formate reacts with acetyl chloride to give a mixed anhydride.

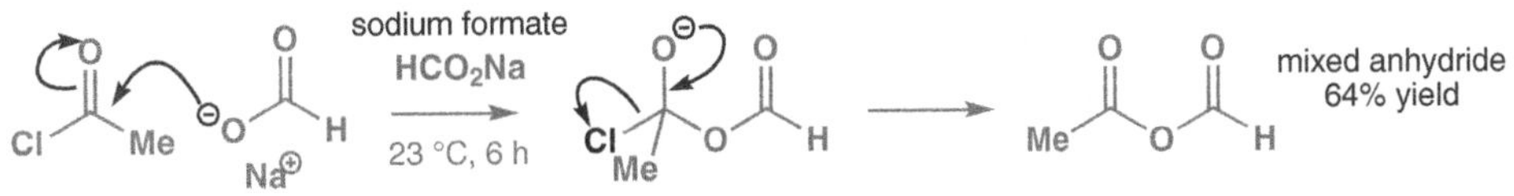

Amines react with acyl chlorides to give amides

Using the principles we've outlined above, you should be able to see how these compounds can be interconverted by substitution reactions with appropriate nucleophiles. We've seen that acid chlorides react with carboxylic acids to give acid anhydrides, and with alcohols to give esters. They also react with amines (such as ammonia) to give amides.

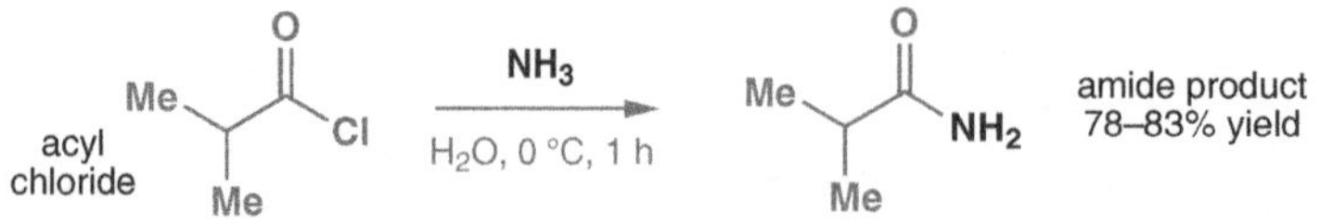

The mechanism is very similar to the mechanism of ester formation. Notice the second molecule of ammonia, which removes a proton, and the loss of chloride ion—the leaving group—to form the amide. Ammonium chloride is formed as a by-product in the reaction.

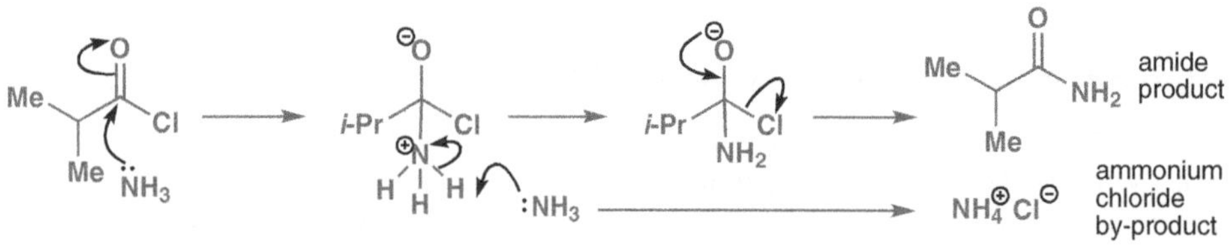

Interactive mechanism for amide formation

Here is another example, using a secondary amine, dimethylamine. Try writing down the mechanism now without looking at the one above. Again, two equivalents of dimethylamine are necessary, although the chemists who published this reaction added three for good measure.

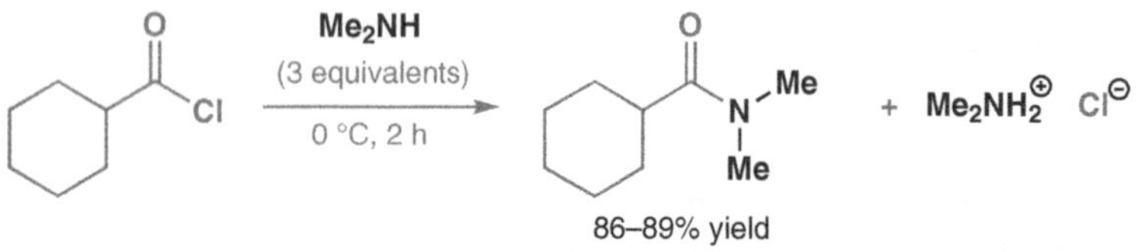

Schotten–Baumann synthesis of an amide

As these mechanisms show, the formation of amides from acid chlorides and amines is accompanied by production of one equivalent of HCl, which needs to be neutralized by a second equivalent of amine. An alternative method for making amides is to carry out the reaction in the presence of another base, such as NaOH, which then does the job of neutralizing the HCl. The trouble is, OH⁻ also attacks acyl chlorides to give carboxylic acids. Schotten and Baumann, in the late nineteenth century, published a way round this problem by carrying out these reactions in two-phase systems of immiscible water and dichloromethane. The organic amine (not necessarily ammonia) and the acyl chloride remain in the (lower) dichloromethane layer, while the base (NaOH) remains in the (upper) aqueous layer. Dichloromethane and chloroform are two common organic solvents that are heavier (more dense) than water. The acyl chloride reacts only with the amine, but the HCl produced can dissolve in, and be neutralized by, the aqueous solution of NaOH.

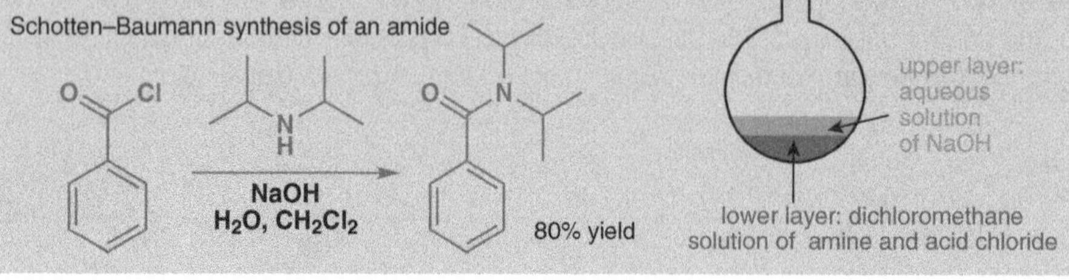

Schotten–Baumann synthesis of an amide

Using base strength to predict the outcome of substitution reactions of carboxylic acid derivatives

You saw that acid anhydrides react with alcohols to give esters: they will also react with amines to give amides. But would you expect esters to react with amines to give amides, or amides to react with alcohols to give esters? Both appear reasonable.

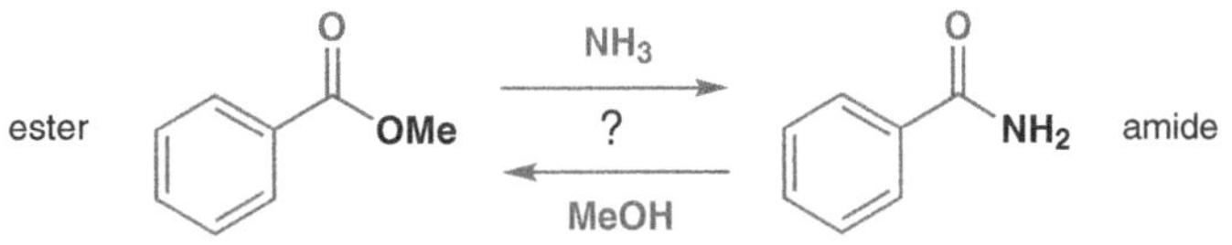

In fact only the top reaction works: amides can be formed from esters but esters cannot be formed from amides. The key question is: which group will leave from the common tetrahedral intermediate? The answer is MeO⁻ and not NH₂⁻. You should have worked this out from the stability of the anions. Alkoxides are reasonably strong bases (pK_a of ROH about 15) so they are not good leaving groups. But NH₂⁻ is a very unstable anion (pK_a of NH₃ about 25) and is a very bad leaving group.

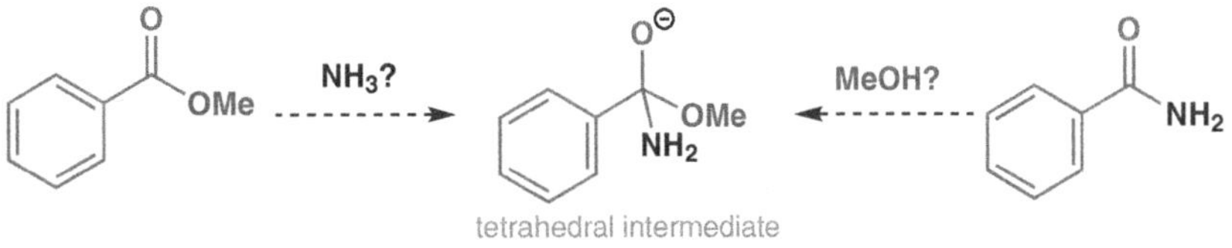

So MeO⁻ leaves and the amide is formed. The base used to deprotonate the first formed intermediate may be either the MeO⁻ produced in the reaction or, to start with, another molecule of NH₃.

> You will meet many more mechanisms like this, in which an unspecified base removes a proton from an intermediate. As long as you can satisfy yourself that there is a base available to perform the task, it is quite acceptable to write any of these shorthand mechanisms.

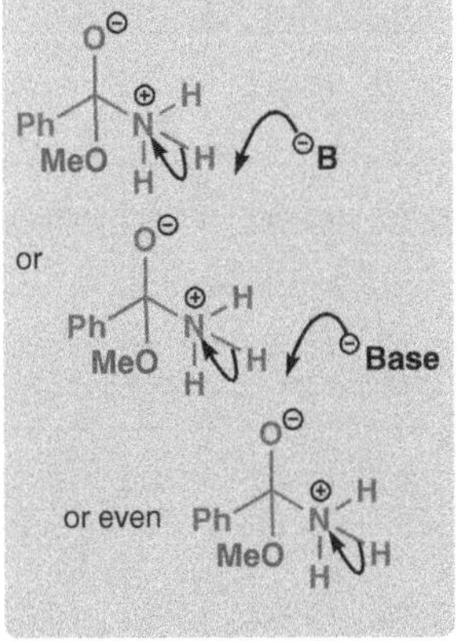

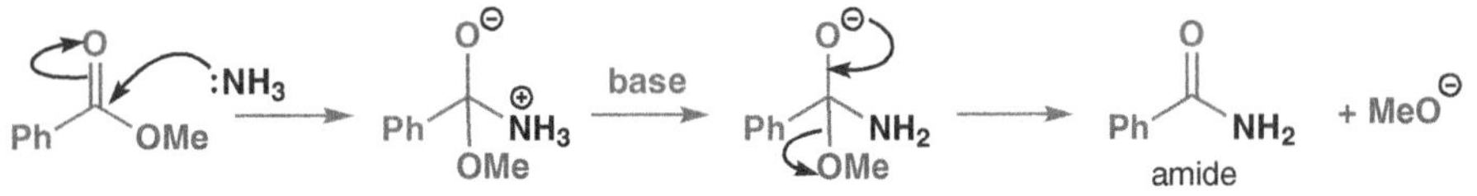

Here is a slightly unusual example in that there is a ketone present in the molecule as well. Later in the book we shall consider how to work out whether another functional group might interfere with the reaction we want to do.

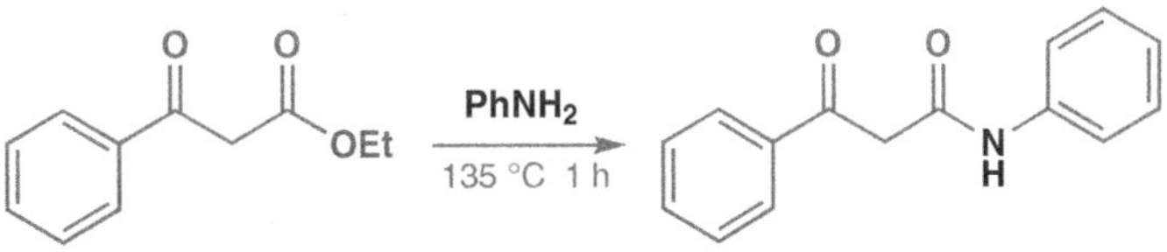

Factors other than leaving group ability can be important

In fact, the tetrahedral intermediate would simply never form from an amide and an alcohol; the amide is too bad an electrophile and the alcohol not a good enough nucleophile. We've looked at leaving group ability: next we'll consider the strength of the nucleophile Y and then the strength of the electrophile RCOX.

● **Conditions for reaction**

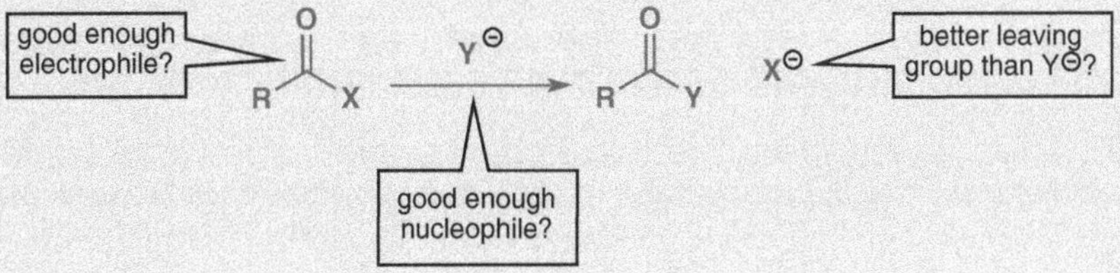

If this reaction is to go:

1 X⁻ must be a better leaving group than Y⁻ (otherwise the reverse reaction would take place).
2 Y⁻ must be a strong enough nucleophile to attack RCOX.
3 RCOX must be a good enough electrophile to react with Y⁻.

Strength of nucleophile and leaving group ability are related and pK_a is a guide to both

We have seen how pK_a gives us a guide to leaving group ability: it is also a good guide to how strong a nucleophile will be. These two properties are the reverse of each other: good nucleophiles are bad leaving groups. A stable anion is a good leaving group but a poor nucleophile. Anions of weak acids (HA has high pK_a) are bad leaving groups but good nucleophiles towards the carbonyl group.

⇒ We will come back to this concept again in Chapter 15, where you will see that this principle does not apply to substitution at saturated carbon atoms.

● **Guide to nucleophilicity**

In general, the higher the pK_a of AH the better A⁻ is as a nucleophile.

But just a moment—we've overlooked an important point. We have sometimes used anions as nucleophiles (for example when we made acid anhydrides from acid chlorides plus carboxylate salts, we used an anionic nucleophile RCO$_2^-$) but on other occasions we have used neutral nucleophiles (for example when we made amides from acid chlorides plus amines, we used a neutral nucleophile NH$_3$). Anions are better nucleophiles for carbonyl groups than are neutral compounds so we can choose our nucleophilic reagent accordingly.

For proper comparisons, we should use the pK_a of NH$_4^+$ (about 10) if we are using neutral ammonia, but the pK_a of RCO$_2$H (about 5) if we're using the carboxylate anion. Ammonia is a good nucleophile and we don't usually need its anion but carboxylic acids are very weak nucleophiles and we often use their anions. You will see later in this chapter that we can alter this with acid catalysts. So this reaction works badly in either direction. We don't make or hydrolyse esters this way.

While amines react with acetic anhydride quite rapidly at room temperature (reaction complete in a few hours), alcohols react extremely slowly in the absence of a base. On the other hand, an alkoxide anion reacts with acetic anhydride extremely rapidly—the reactions are often complete within seconds at 0 °C. We don't have to deprotonate an alcohol completely to increase its reactivity: just a catalytic quantity of a weak base can do this job. All the pK_as you need are in Chapter 8.

Not all carboxylic acid derivatives are equally reactive

We can list the common carboxylic acid derivatives in a 'hierarchy' of reactivity, with the most reactive at the top and the least reactive at the bottom. The nucleophile is the same in each case (water), as is the product, the carboxylic acid, but the electrophiles vary from very reactive to unreactive. The conditions needed for successful reaction show just how large is the variation on reactivity. Acid chlorides react violently with water. Amides need refluxing with 10% NaOH or concentrated HCl in a sealed tube at 100 °C overnight. We've seen that this hierarchy is partly due to how good the leaving group is (the ones at the top are best). But it also depends on the reactivity of the acid derivatives. Why is there such a large difference?

202

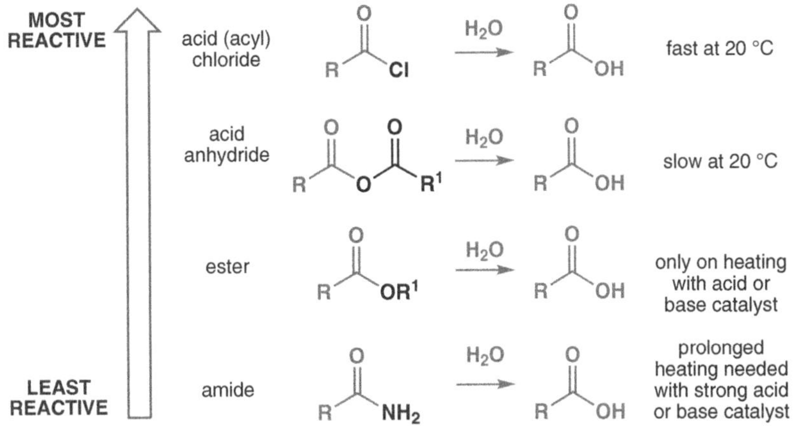

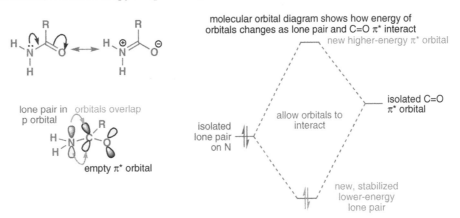

MOST REACTIVE — acid (acyl) chloride — $\xrightarrow{H_2O}$ — fast at 20 °C

acid anhydride — $\xrightarrow{H_2O}$ — slow at 20 °C

ester — $\xrightarrow{H_2O}$ — only on heating with acid or base catalyst

LEAST REACTIVE — amide — $\xrightarrow{H_2O}$ — prolonged heating needed with strong acid or base catalyst

Delocalization and the electrophilicity of carbonyl compounds

Amides are the least reactive towards nucleophiles because they exhibit the greatest degree of delocalization. You met this concept in Chapter 7 and we shall return to it many times more. In an amide, the lone pair on the nitrogen atom can be stabilized by overlap with the π^* orbital of the carbonyl group—this overlap is best when the lone pair occupies a p orbital (in an amine, it would occupy an sp^3 orbital).

molecular orbital diagram shows how energy of orbitals changes as lone pair and C=O π^* interact

new higher-energy π^* orbital

lone pair in p orbital   orbitals overlap

isolated C=O π^* orbital

isolated lone pair on N

allow orbitals to interact

empty π^* orbital

new, stabilized lower-energy lone pair

The molecular orbital diagram shows how this interaction both lowers the energy of the bonding orbital (the delocalized nitrogen lone pair), making it neither basic nor nucleophilic, and raises the energy of the π^* orbital, making it less ready to react with nucleophiles. Esters are similar, but because the oxygen lone pairs are lower in energy, the effect is less pronounced. The degree of delocalization depends on the electron-donating power of the substituent and increases along the series of compounds below from almost no delocalization from Cl to complete delocalization in the carboxylate anion, where the negative charge is equally shared between the two oxygen atoms.

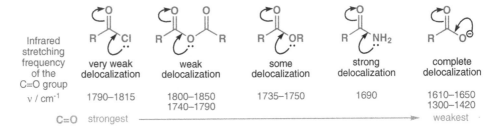

Infrared stretching frequency of the C=O group	very weak delocalization	weak delocalization	some delocalization	strong delocalization	complete delocalization
ν / cm^{-1}	1790–1815	1800–1850 1740–1790	1735–1750	1690	1610–1650 1300–1420

C=O   strongest ⟶ weakest

The greater the degree of delocalization, the weaker the C=O bond becomes. This is most clearly evident in the stretching frequency of the carbonyl group in the IR spectra of carboxylic acid derivatives—remember that the stretching frequency depends on the force constant of the bond, itself a measure of the bond's strength. The carboxylate anion is included because it represents the limit of the series, with complete delocalization of the negative charge over the two oxygen atoms. There are two frequencies for the anhydride and the carboxylate anion because of symmetric and antisymmetric stretching of identical bonds.

Amides react as electrophiles only with powerful nucleophiles such as HO^-. Acid chlorides, on the other hand, react with even quite weak nucleophiles: neutral ROH, for example. They are more reactive because the electron-withdrawing effect of the chlorine atom increases the electrophilicity of the carbonyl carbon atom.

➡ Infrared spectroscopy was introduced in Chapter 3.

Bond strengths and reactivity

You may think that a weaker C=O bond should be more reactive. This is not so because the partial positive charge on carbon is also lessened by delocalization and because the molecule as a whole is stabilized by the delocalization. Bond strength is not always a good guide to reactivity!

For example, in acetic acid the bond strengths are surprising. The strongest bond is the O–H bond and the weakest is the C–C bond. Yet very few reactions of acetic acid involve breaking the C–C bond, and its characteristic reactivity, as an acid, involves breaking O–H, the strongest bond of them all!

The reason is that polarization of bonds and solvation of ions play an enormously important role in determining the reactivity of molecules. In Chapter 37 you will see that radicals are relatively unaffected by solvation and that their reactions follow bond strengths much more closely.

Carboxylic acids do not undergo substitution reactions under basic conditions

Substitution reactions of RCO_2H require a leaving group OH^-. The pK_a of water is about 15, so acids should be about as electrophilic as esters. Esters react well with ammonia to give amides. However, if we try to react carboxylic acids with amines to give amides no substitution occurs: an ammonium salt is formed because the amines themselves are basic and remove the acidic proton from the acid.

Once the carboxylic acid is deprotonated, substitutions are prevented because (almost) no nucleophile will attack the carboxylate anion. Under neutral conditions, alcohols are just not reactive enough to add to the carboxylic acid but, with *acid* catalysis, esters can be formed from alcohols and carboxylic acids.

■ In fact, amides *can* be made from carboxylic acids plus amines, but only if the ammonium salt is heated strongly to dehydrate it. This is not usually a good way of making amides!

Acid catalysts increase the reactivity of a carbonyl group

We saw in Chapter 6 that the lone pairs of a carbonyl group may be protonated by acid. Only strong acids are powerful enough to protonate carbonyl groups: the pK_a of protonated acetone is –7 so, for example, even 1M HCl (pH 0) would protonate only 1 in 10^7 molecules of acetone. However, even proportions as low as this are sufficient to increase the rate of substitution reactions at carbonyl groups enormously because those carbonyl groups that are protonated become extremely powerful electrophiles.

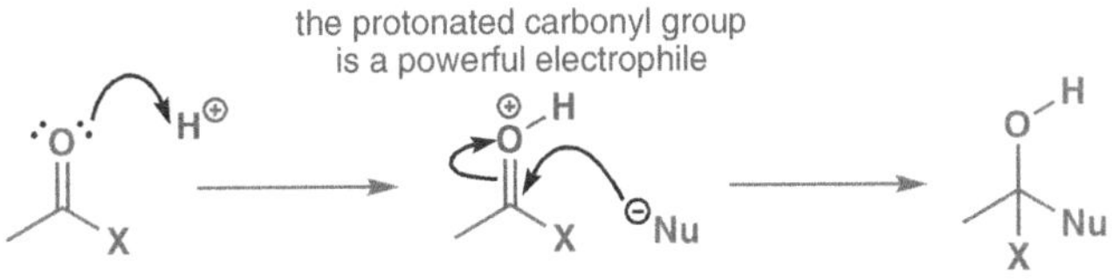

the protonated carbonyl group
is a powerful electrophile

It is for this reason that alcohols will react with carboxylic acids under acid catalysis. The acid (usually HCl or H_2SO_4) reversibly protonates a small percentage of the carboxylic acid molecules, and the protonated carboxylic acids are extremely susceptible to attack by even a weak nucleophile such as an alcohol. This is the first half of the reaction:

acid-catalysed ester formation: forming the tetrahedral intermediate

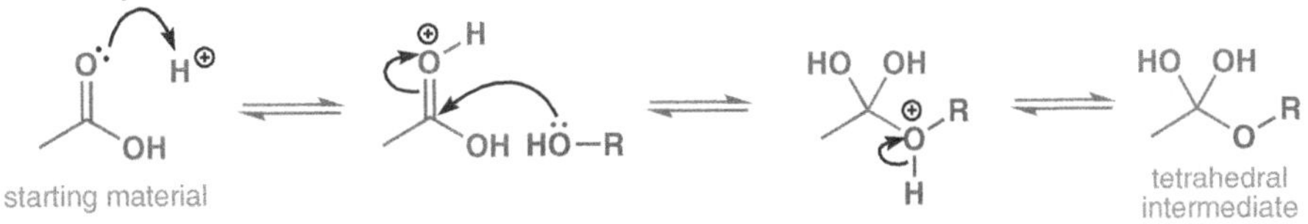

Acid catalysts can make bad leaving groups into good ones

■ Average bond strength C=O
720 kJ mol⁻¹.
  Average bond strength C—O
350 kJ mol⁻¹.

This tetrahedral intermediate is unstable because the energy to be gained by re-forming a C=O bond is greater than that used in breaking two C—O bonds. As it stands, none of the leaving groups (R⁻, HO⁻, or RO⁻) is very good. However, help is again at hand in the acid catalyst. It can protonate any of the oxygen atoms reversibly. Again, only a very small proportion of molecules are protonated at any one time but, once the oxygen atom of, say, one of the OH groups is protonated, it becomes a much better leaving group (water instead of HO⁻). Loss of ROH from the tetrahedral intermediate is also possible: this leads back to starting materials—hence the equilibrium arrow in the scheme above. Loss of H_2O is more fruitful, and takes the reaction forwards to the ester product.

acid-catalysed ester formation: decomposition of the tetrahedral intermediate

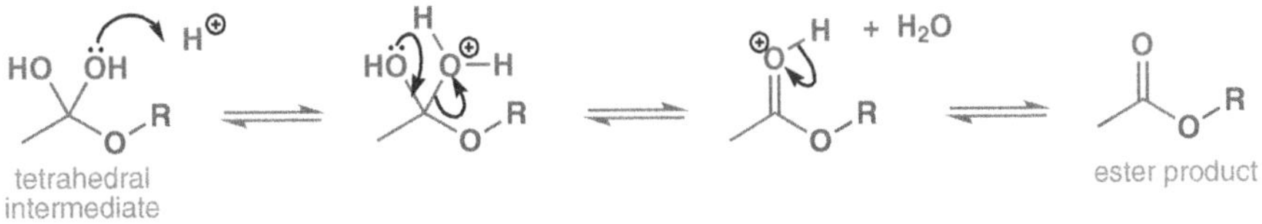

- Acid catalysts catalyse substitution reactions of carboxylic acids.
  - They make the carbonyl group more electrophilic by protonation at *carbonyl* oxygen.
  - They make the leaving group better by protonation there too.

Ester formation is reversible: how to control an equilibrium

Loss of water from the tetrahedral intermediate is reversible too: just as ROH will attack a protonated carboxylic acid, H_2O will attack a protonated ester. In fact, every step in the sequence from carboxylic acid to ester is an equilibrium, and the overall equilibrium constant is about 1. In order for this reaction to be useful, it is therefore necessary to ensure that the equilibrium is pushed towards the ester side by using an excess of alcohol or carboxylic acid (usually the reactions are done in a solution of the alcohol or the carboxylic acid). In this reaction, for example, no water is added and an excess of alcohol is used. Using less than three equivalents of ethanol gave lower yields of ester.

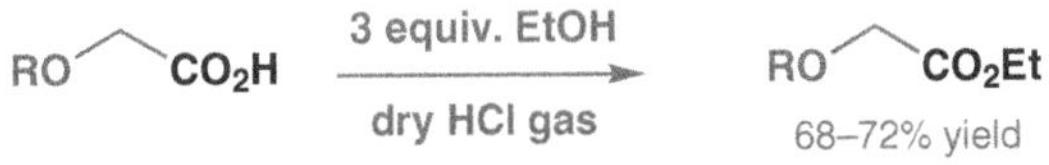

Alternatively, the reaction can be done in the presence of a dehydrating agent (concentrated H_2SO_4, for example, or silica gel) or the water can be distilled out of the mixture as it forms.

lactic acid
cat. H_2SO_4
benzene (solvent) remove water by distillation
89–91% yield

AcOH
cat. H_2SO_4 silica gel (drying agent)
57% yield

■ Lactic acid must be handled in solution in water. Can you see why, bearing in mind what we have said about the reversibility of ester formation?

● **Making esters from alcohols**
You have now met three ways of making esters from alcohols:

- **with acyl chlorides**
- **with acid anhydrides**
- **with carboxylic acids.**

Try to appreciate that different methods will be appropriate at different times. If you want to make a few milligrams of a complex ester, you are much more likely to work with a reactive acyl chloride or anhydride, using pyridine as a weakly basic catalyst, than to try to distil out a minute quantity of water from a reaction mixture containing a strong acid that may destroy the starting material. On the other hand, if you are a chemist making simple esters (such as those in Chapter 2, p. 31) for the flavouring industry on a scale of many tons, you might prefer the cheaper option of carboxylic acid and a strong acid (e.g. H_2SO_4) in alcohol solution.

Acid-catalysed ester hydrolysis and transesterification

By starting with an ester, an excess of water, and an acid catalyst we can persuade the reverse reaction to occur: formation of the carboxylic acid plus alcohol with consumption of water. Such a reaction is known as a hydrolysis reaction because water is used to break up the ester into carboxylic acid plus alcohol (*lysis*=breaking).

excess water gives ester hydrolysis

acid-catalysed ester hydrolysis

acid-catalysed ester formation

excess alcohol or removal of water gives ester formation

Acid-catalysed ester formation and hydrolysis are the exact reverse of one another: the only way we can control the reaction is by altering concentrations of reagents to drive the reaction the way we want it to go. The same principles can be used to convert an ester of one alcohol into an ester of another, a process known as transesterification. It is possible, for example, to force this equilibrium to the right by distilling methanol (which has a lower boiling point than the other components of the reaction) out of the mixture.

🖱 Interactive mechanism for acid-catalysed ester formation

OMe + OH
catalytic HCl
O + MeOH

The mechanism for this transesterification simply consists of adding one alcohol (here BuOH) and eliminating the other (here MeOH), both processes being acid-catalysed. Notice how easy it is now to confirm that the reaction is *catalytic* in H^+.

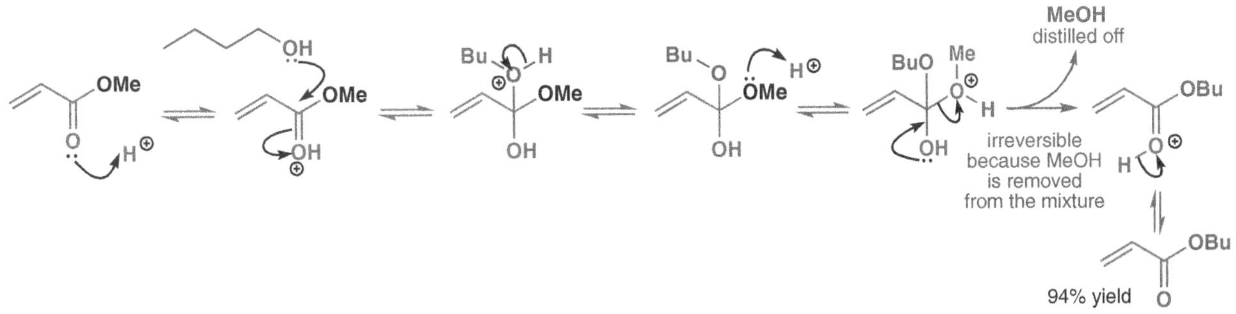

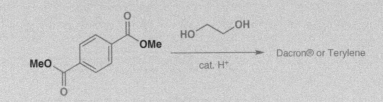

MeOH
distilled off

irreversible
because MeOH
is removed
from the mixture

94% yield

Polyester fibre manufacture

A transesterification reaction is used to make the polyester fibres that are used for textile production. Terylene, or Dacron, for example, is a polyester of the dicarboxylic acid terephthalic acid and the diol ethylene glycol.

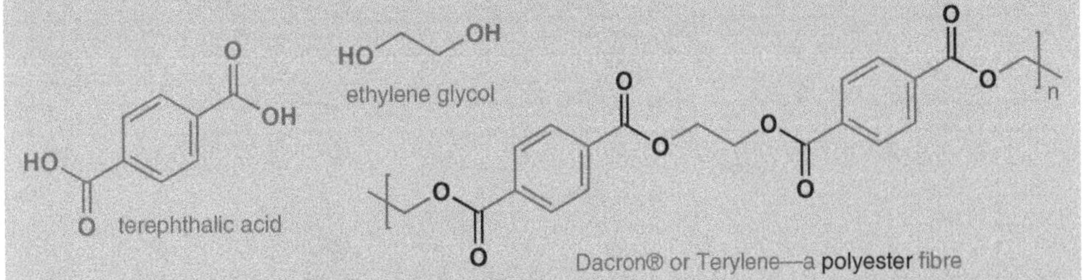

ethylene glycol

terephthalic acid

Dacron® or Terylene—a **polyester** fibre

Terylene is actually made by ester exchange: dimethyl terephthalate is heated with ethylene glycol and an acid catalyst, distilling off the methanol as it is formed.

Dacron® or Terylene

cat. H⁺

🖱 Interactive structure of polyester fibres

Base-catalysed hydrolysis of esters is irreversible

You can't make esters from carboxylic acids and alcohols under basic conditions because the base deprotonates the carboxylic acid (see p. 207). However, you can reverse that reaction and hydrolyse an ester to a carboxylic acid (more accurately, a carboxylate salt) and an alcohol.

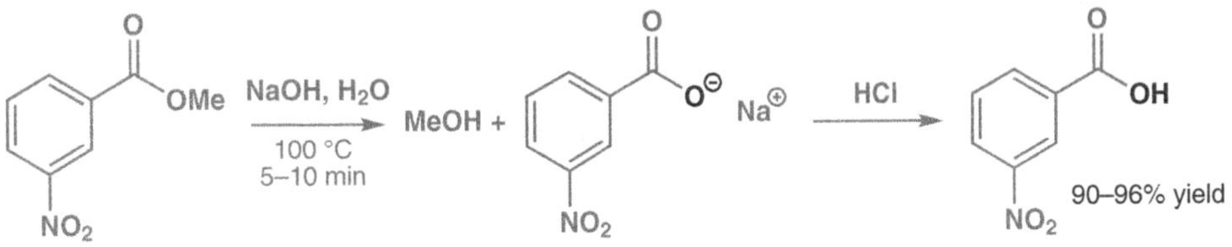

NaOH, H₂O
100 °C
5–10 min

MeOH +

HCl

90–96% yield

This time the ester is, of course, not protonated first as it would be in acid, but the unprotonated ester is a good enough electrophile because OH⁻, and not water, is the nucleophile. The tetrahedral intermediate can collapse either way, giving back ester or going forward to acid plus alcohol.

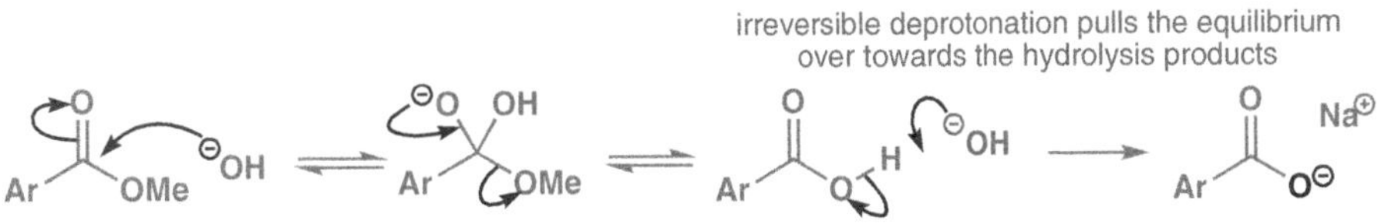

irreversible deprotonation pulls the equilibrium
over towards the hydrolysis products

The backward reaction is impossible because the basic conditions straightaway deprotonate the acid to make a carboxylate salt (which, incidentally, consumes the base, making at least one equivalent of base necessary in the reaction). Carboxylate salts do not usually react with nucleophiles, even those a good deal stronger than alcohols.

How do we know this is the mechanism?

Ester hydrolysis is such an important reaction that chemists have spent a lot of time and effort finding out exactly how it works. Many of the experiments that tell us about the mechanism involve oxygen-18 labelling. The starting material is an ester enriched in the heavy oxygen isotope ^{18}O. By knowing where the heavy oxygen atoms start off, and following (by mass spectrometry—Chapter 3) where they end up, the mechanism can be established.

1. An ^{18}O label in the 'ether' oxygen of the ester ends up in the alcohol product.

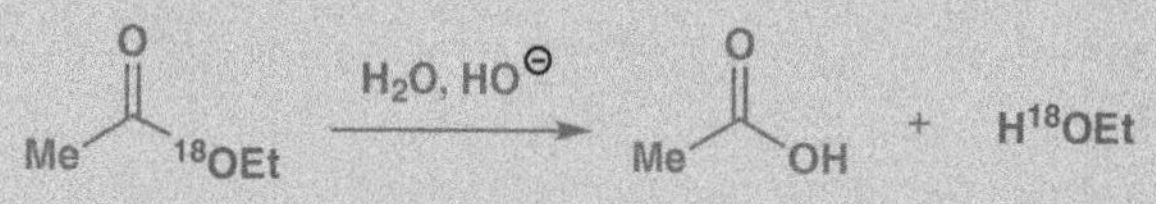

2. Hydrolysis with $^{18}OH_2$ gives ^{18}O-labelled carboxylic acid, but no ^{18}O-labelled alcohol.

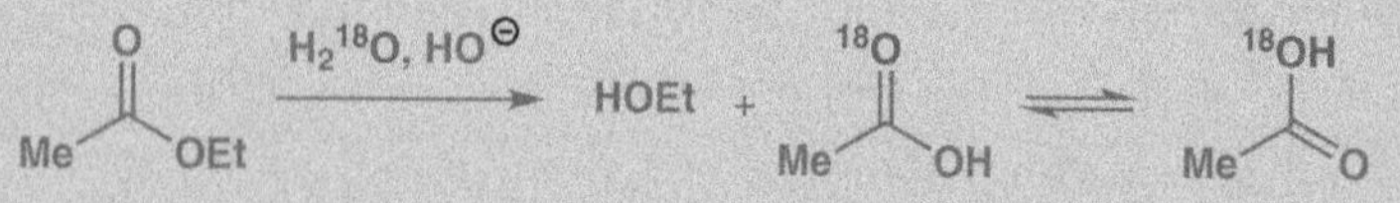

These experiments tell us that a displacement (substitution) has occurred at the carbonyl carbon atom, and rule out the alternative displacement at saturated carbon.

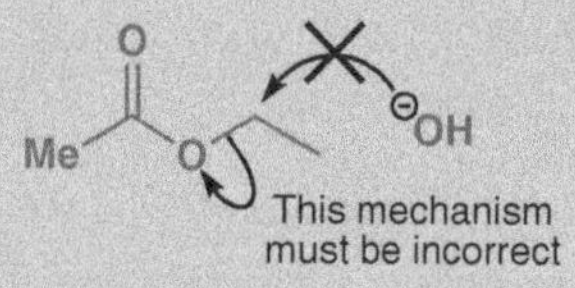

This mechanism must be incorrect

One further labelling experiment showed that a tetrahedral intermediate must be formed: an ester labelled with ^{18}O in its carbonyl oxygen atom passes some of its ^{18}O label to the water. We discussed this on p. 201.
There is more on the mechanism of ester hydrolysis in Chapter 12.

The saturated fatty acid tetradecanoic acid (also known as myristic acid) is manufactured commercially from coconut oil by hydrolysis in base. You may be surprised to learn that coconut oil contains more saturated fat than butter, lard, or beef dripping: much of it is the trimyristate ester of glycerol. Hydrolysis with aqueous sodium hydroxide, followed by reprotonation of the sodium carboxylate salt with acid, gives myristic acid. Notice how much longer it takes to hydrolyse this branched ester than it did to hydrolyse a methyl ester (p. 210).

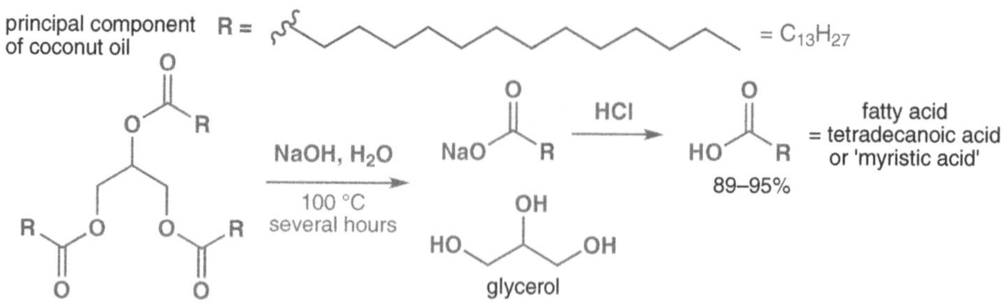

Saponification

The alkaline hydrolysis of esters to give carboxylate salts is known as saponification because it is the process used to make soap. Traditionally, beef tallow (the tristearate ester of glycerol—stearic acid is octadecanoic acid, $C_{17}H_{35}CO_2H$) was hydrolysed with sodium hydroxide to give sodium stearate, $C_{17}H_{35}CO_2Na$, the principal component of soap. Finer soaps are made from palm oil and contain a higher proportion of sodium palmitate, $C_{15}H_{31}CO_2Na$. Hydrolysis with KOH gives potassium carboxylates, which are used in liquid soaps. Soaps like these owe their detergent properties to the combination of polar (carboxylate group) and non-polar (long alkyl chain) properties.

14   tetradecanoic acid = myristic acid $\quad CO_2H$

16   hexadecanoic acid = palmitic acid $\quad CO_2H$

18   octadecanoic acid = stearic acid $\quad CO_2H$

Amides can be hydrolysed under acidic or basic conditions too

In order to hydrolyse amides, the least reactive of the carboxylic acid derivatives, we have a choice: we can persuade the amine leaving group to leave by protonating it, or we can use brute force and forcibly eject it with concentrated hydroxide solution.

Amides are very unreactive as electrophiles, but they are also rather more basic than most carboxylic acid derivatives: a typical protonated amide has a pK_a of –1; most other carbonyl compounds are much less basic. You might therefore imagine that the protonation of an amide would take place on nitrogen—after all, *amine* nitrogen atoms are readily protonated. And, indeed, the reason for the basicity of amides is the nitrogen atom's delocalized lone pair, making the carbonyl group unusually electron rich. But amides are always protonated on the oxygen atom of the carbonyl group, never the nitrogen, because protonation at nitrogen would disrupt the delocalized system that makes amides so stable. Protonation at oxygen gives a delocalized cation (Chapter 8).

protonation at O

delocalization of charge over N and O

protonation at N (does not happen)

no delocalization possible

Protonation of the carbonyl group by acid makes the carbonyl group electrophilic enough for attack by water, giving a neutral tetrahedral intermediate. The amine nitrogen atom in the tetrahedral intermediate is much more basic than the oxygen atoms, so now *it* gets protonated, and the RNH_2 group becomes really quite a good leaving group. Once it has left, it will immediately be protonated again, and therefore become completely non-nucleophilic. The conditions are very vigorous—70% sulfuric acid for 3 hours at 100 °C.

■ Notice that this means that one equivalent of acid is used up in this reaction—the acid is not solely a catalyst.

amide hydrolysis in acid

3 hours at 100 °C with 70% H_2SO_4 in water gives 70% yield of the acid

protonation of the amine prevents reverse reaction $PhNH_3^{\oplus}$

Hydrolysis of amides in base requires similarly vigorous conditions. Hot solutions of hydroxide are sufficiently powerful nucleophiles to attack an amide carbonyl group, although even when the tetrahedral intermediate has formed NH_2^- (pK_a of the ammonium ion 35) has only a slight chance of leaving when HO^- (pK_a of water 15) is an alternative. Nonetheless, at high temperatures amides are slowly hydrolysed by concentrated base since one product is the carboxylate salt and this does not react with nucleophiles. The 'base' for the irreversible step might be hydroxide or NH_2^-.

amide hydrolysis in base

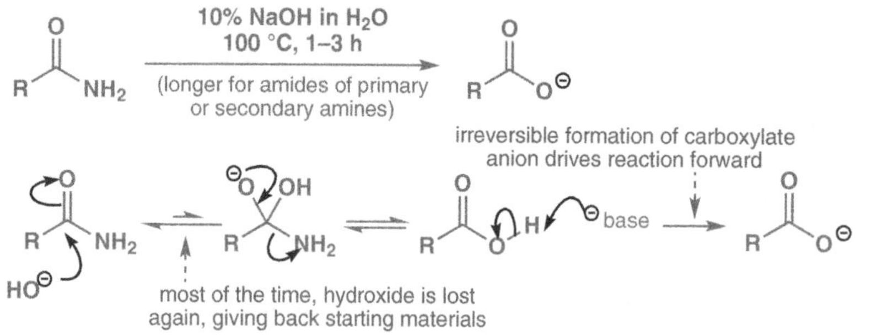

Secondary and tertiary amides hydrolyse much more slowly under these conditions. With all these amides a second mechanism kicks in if the hydroxide concentration is large enough. More hydroxide deprotonates the tetrahedral anion to give a dianion that must lose NH_2^- as the only alternative is O^{2-}. This leaving group deprotonates water so the second molecule of hydroxide ion is simply a catalyst.

■ You've not seen the option of O^{2-} as a leaving group before but this is what you would need if you want to break the bond to O^-. Asking O^{2-} to be a leaving group is like asking HO^- to be an acid.

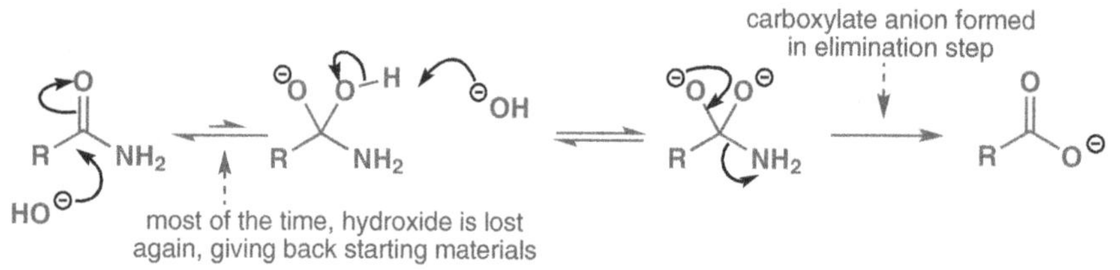

A similar mechanism is successful with only a little water and plenty of strong base, Then even tertiary amides can be hydrolysed at room temperature. Potassium *tert*-butoxide is a strong enough base (pK_a of *t*-BuOH about 18) to deprotonate the tetrahedral intermediate.

hydrolysis of amides using *t*-BuOK

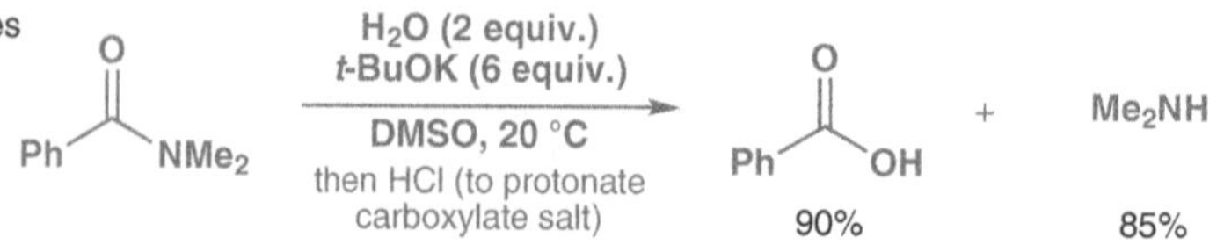

Hydrolysing nitriles: how to make the almond extract, mandelic acid

Closely related to the amides are nitriles. You can view them as primary amides that have lost one molecule of water and, indeed, they can be made by dehydrating primary amides.

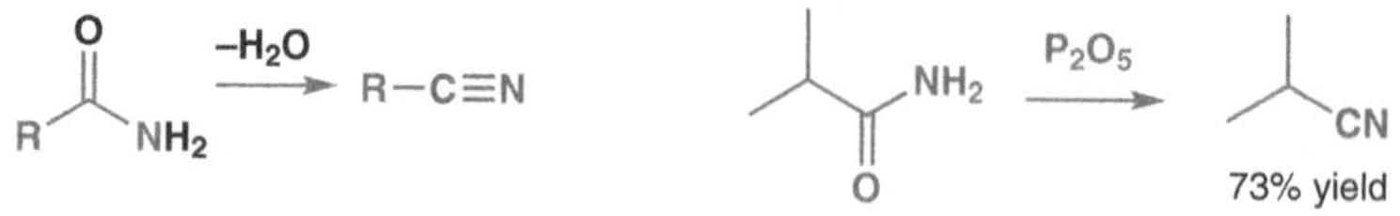

They can be hydrolysed just like amides too. Addition of water to the protonated nitrile gives a primary amide, and hydrolysis of this amide gives carboxylic acid plus ammonia.

■ Don't be put off by the number of steps in this mechanism—look carefully and you will see that most of them are simple proton transfers. The only step that isn't a proton transfer is the addition of water.

You met a way of making nitriles—from HCN (or NaCN + HCl) plus aldehydes—in Chapter 6: the hydroxynitrile products are known as **cyanohydrins**. With this in mind, you should be able to suggest a way of making mandelic acid, an extract of almonds, from benzaldehyde.

■ You have just designed your first total synthesis of a natural product. We return to designing syntheses much later in this book, in Chapter 28.

This is how some chemists did it.

Acid chlorides can be made from carboxylic acids using SOCl$_2$ or PCl$_5$

We have looked at a whole series of interconversions between carboxylic acid derivatives and, after this next section, we shall summarize what you need to understand. We said that it is always easy to move down the series of acid derivatives we listed early in the chapter, and so far that is all we have done. But some reactions of carboxylic acids also enable us to move upwards in the series. What we need is a reagent that changes the bad leaving group HO$^-$ into a good leaving group. Strong acid does this by protonating the OH$^-$, allowing it to leave as H$_2$O. In this section we look at two more reagents, SOCl$_2$ and PCl$_5$, which convert the OH group of a carboxylic acid and also turn it into a good leaving group. Thionyl chloride, SOCl$_2$, reacts with carboxylic acids to make acyl chlorides.

This volatile liquid with a choking smell is electrophilic at the sulfur atom (as you might expect with two chlorine atoms and an oxygen atom attached) and is attacked by carboxylic acids to give an unstable, and highly electrophilic, intermediate.

■ Note that it is the more nucleophilic carbonyl oxygen which actually attacks S. If you follow the fate of the two oxygens right through the mechanism you will see which fact it is the oxygen that starts off in the C=O group which is replaced by Cl. You may also be surprised to see the way we substituted at S=O without forming a 'tetrahedral intermediate'. Well, this trivalent sulfur atom is already tetrahedral (it still has one lone pair), and substitution can go by a direct substitution at sulfur.

Reprotonation of the unstable intermediate (by the HCl just produced, i.e. reversal of the last step above) gives an electrophile powerful enough to react even with the weak nucleophile Cl$^-$ (HCl is a strong acid, so Cl$^-$ is a poor nucleophile). The tetrahedral intermediate collapses to the acyl chloride, sulfur dioxide, and hydrogen chloride. This step is irreversible because SO$_2$ and HCl are gases that are lost from the reaction mixture.

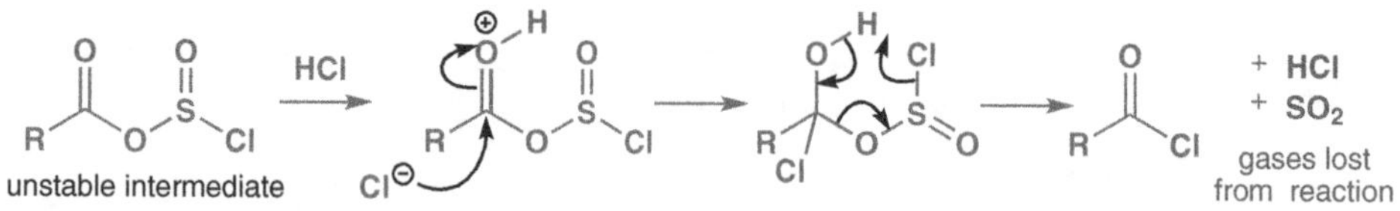

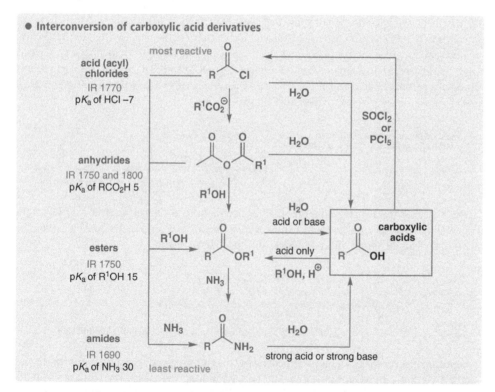

Although HCl is involved in this reaction, it cannot be used as the sole reagent for making acid chlorides. It is necessary to have a sulfur or phosphorus compound to remove the oxygen. An alternative reagent for converting RCO_2H into $RCOCl$ is phosphorus pentachloride, PCl_5. The mechanism is similar—try writing it out before looking at the scheme below.

acid chlorides can be made from carboxylic acids with phosphorus pentachloride

PCl₅ → 90–96% yield

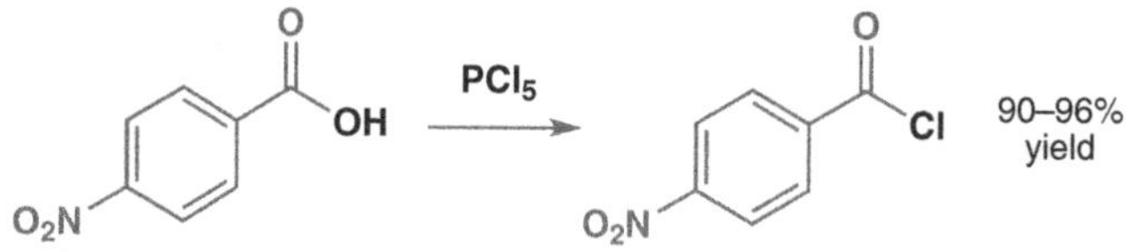

The mechanism is closely related to the previous one, except that the formation of a very stable P=O bond is the vital factor rather than the loss of two gaseous reagents.

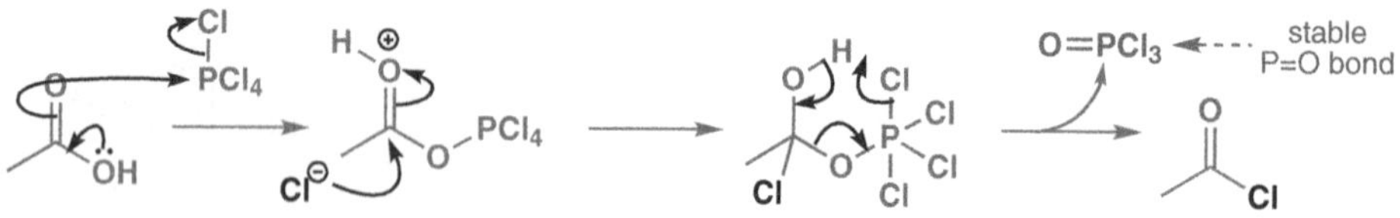

These conversions of acids into acid chlorides complete all the methods we need to convert acids into any acid derivatives. You can convert acids directly to esters and now to acid chlorides, the most reactive of acid derivatives, and can make any other derivative from them. The chart below adds reaction conditions, relevant pK_as, and infrared stretching frequencies to the reactivity order we met earlier.

● **Interconversion of carboxylic acid derivatives**

acid (acyl) chlorides
IR 1770
pK_a of HCl −7

anhydrides
IR 1750 and 1800
pK_a of RCO_2H 5

esters
IR 1750
pK_a of R^1OH 15

amides
IR 1690
pK_a of NH₃ 30

most reactive

$R^1CO_2^⊖$

H₂O

H₂O

SOCl₂ or PCl₅

R^1OH

R^1OH

H₂O
acid or base

acid only

$R^1OH, H^⊕$

NH₃

NH₃

carboxylic acids

H₂O

strong acid or strong base

least reactive

Interactive mechanism for acid chloride formation with SOCl₂

Interactive mechanism for acid chloride formation with PCl₅

■ We will explore the link between infrared stretching frequency and reactivity in Chapter 18.

All these acid derivatives can, of course, be hydrolysed to the acid itself with water alone or with various levels of acid or base catalysis depending on the reactivity of the derivative.

To climb the reactivity order therefore, the simplest method is to hydrolyse to the acid and convert the acid into the acid chloride. You are now at the top of the reactivity order and can go down to whatever level you require.

Making other compounds by substitution reactions of acid derivatives

■ Five 'oxidation levels'—(1) hydrocarbon, (2) alcohol, (3) aldehyde and ketone, (4) carboxylic acid, and (5) CO_2—were defined in Chapter 2.

We've talked at length about the interconversions of acid derivatives, explaining the mechanism of attack of nucleophiles such as ROH, H_2O, and NH_3 on acyl chlorides, acid anhydrides, esters, acids, and amines, with or without acid or base present. We shall now go on to talk about substitution reactions of acid derivatives that take us out of this closed company of compounds and allow us to make compounds containing functional groups at other oxidation levels, such as ketones and alcohols.

Making ketones from esters: the problem

Substitution of the OR group of an ester by an R group would give us a ketone. You might therefore think that reaction of an ester with an organolithium or Grignard reagent would be a good way of making ketones. However, if we try the reaction, something else happens, as you saw at the start of this chapter.

Two molecules of Grignard have been incorporated and we get an alcohol! If we look at the mechanism we can understand why this should be so. First, as you would expect, the nucleophilic Grignard reagent attacks the carbonyl group to give a tetrahedral intermediate. The only reasonable leaving group is RO⁻, so it leaves to give us the ketone we set out to make.

Now, the next molecule of Grignard reagent has a choice. It can react with either the ester starting material or the newly formed ketone. Ketones are more electrophilic than esters so the Grignard reagent prefers to react with the ketone in the manner you saw in Chapter 9. A stable alkoxide anion is formed, which gives the tertiary alcohol on acid work-up.

Making alcohols instead of ketones

In other words, the problem here lies in the fact that the ketone product is more reactive than the ester starting material. We shall meet more examples of this general problem later (in Chapter 23, for example): in the next section we shall look at ways of overcoming it. Meanwhile, why not see it as a useful reaction? This compound, for example, was needed by some chemists in the course of research into explosives.

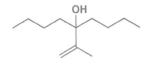

It is a tertiary alcohol with the hydroxyl group flanked by two identical R (= butyl) groups. The chemists who wanted to make the compound knew that an ester would react twice with the same organolithium reagent, so they made it from this unsaturated ester (known as methyl methacrylate) and butyllithium.

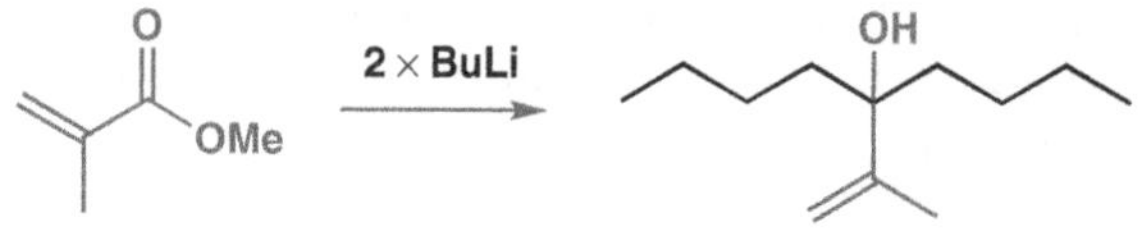

> ● **Tertiary alcohol synthesis**
> Tertiary alcohols with two identical R² groups can be made from ester R¹CO₂R plus two
> equivalents of organolithium R²Li or Grignard reagent R²MgBr.

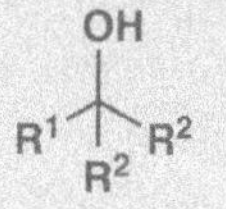

$$
\begin{array}{c}
\text{OH} \\
R^1 \overset{|}{\underset{R^2}{\text{—C—}}} R^2
\end{array}
$$

This reaction works in reduction too if we use lithium aluminium hydride, LiAlH₄. This is a powerful reducing agent that readily attacks the carbonyl group of an ester. Again, collapse of the tetrahedral intermediate gives a compound, this time an aldehyde, which is more reactive than the ester starting material, so a second reaction takes place and the ester is converted (reduced) into an alcohol. Sodium borohydride, often used for the reduction of ketones, does not usually reduce esters.

reduction of esters by LiAlH₄

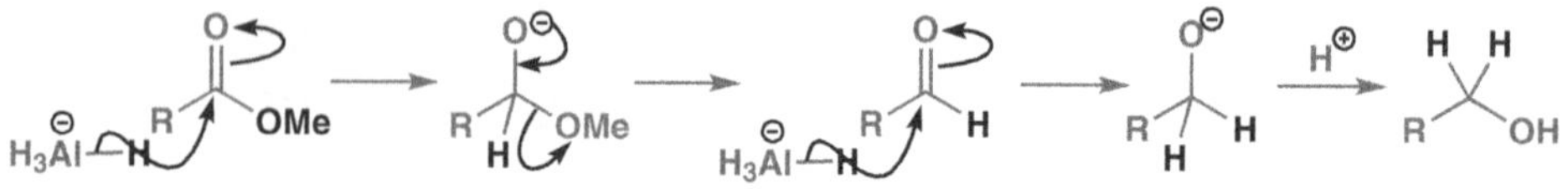

This is an extremely important reaction, and one of the best ways of making alcohols from esters. Stopping the reaction at the aldehyde stage is more difficult: we shall discuss this in Chapter 23.

A bit of shorthand

Before we go any further, we should introduce to you a little bit of chemical shorthand that makes writing many mechanisms easier. As you now appreciate, all substitution reactions at a carbonyl group go via a tetrahedral intermediate.

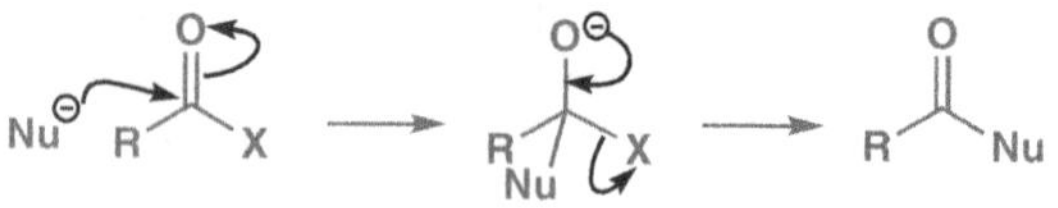

A convenient way to save writing a step is to show the formation and collapse of the tetrahedral intermediate in the same structure, by using a double-headed arrow, as in the diagrams below. Now, this is a useful shorthand, but it is not a substitute for understanding the true mechanism. Certainly, you must never ever write the reaction as a single step not involving the carbonyl group.

Here's the 'shorthand' at work in the LiAlH₄ reduction you have just met.

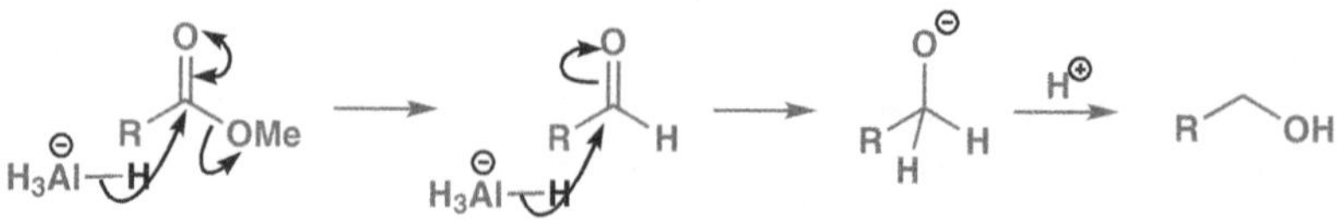

Making ketones from esters: the solution

We diagnosed the problem with our intended reaction as one of reactivity: the product ketone is more reactive than the starting ester. To get round this problem we need to do one of two things:

1. make the starting material more reactive *or*
2. make the product less reactive.

Making the starting materials more reactive

A more reactive starting material would be an acyl chloride: how about reacting one of these with a Grignard reagent? This approach can work—for example this reaction is successful.

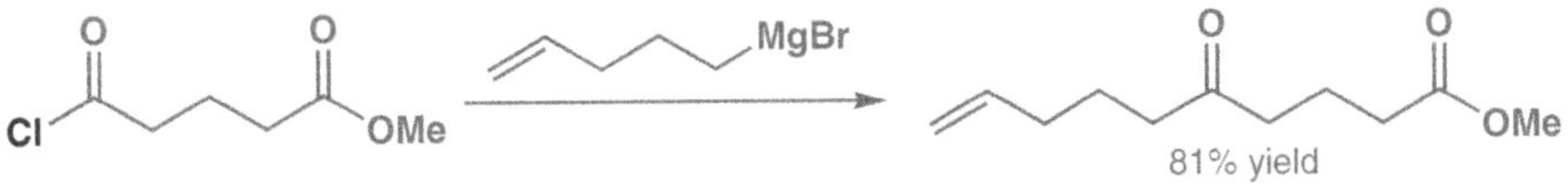

Often, better results are obtained by transmetallating (see Chapter 9) the Grignard reagent, or the organolithium, with copper salts. Organocopper reagents are too unreactive to add to the product ketones, but they react well with the acyl chloride. Consider this reaction, for example: the product was needed for a synthesis of the antibiotic septamycin.

■ Notice how this reaction illustrates the difference in reactivity between an acyl chloride functional group and an ester functional group.

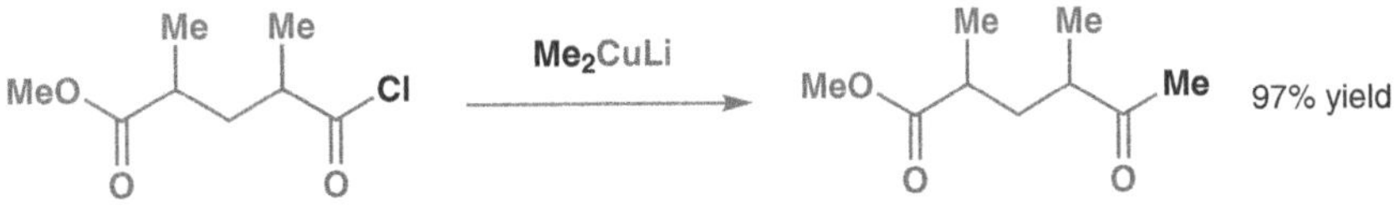

Making the products less reactive

This alternative solution is often better. With the right starting material, the tetrahedral intermediate can become stable enough not to collapse to a ketone during the reaction; it therefore remains completely unreactive towards nucleophiles. The ketone is formed only when the reaction is finally quenched with acid but the nucleophile is also destroyed by the acid and none is left for further addition.

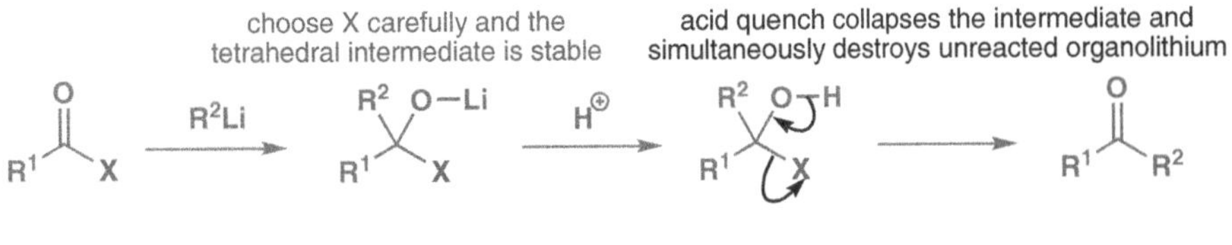

We can illustrate this concept with a reaction of an unlikely looking electrophile, a lithium carboxylate. Towards the beginning of the chapter we said that carboxylic acids were bad electrophiles and that carboxylate salts were even worse. Well, that is true, but with a sufficiently powerful nucleophile (an organolithium) it is just possible to get addition to the carbonyl group of a lithium carboxylate.

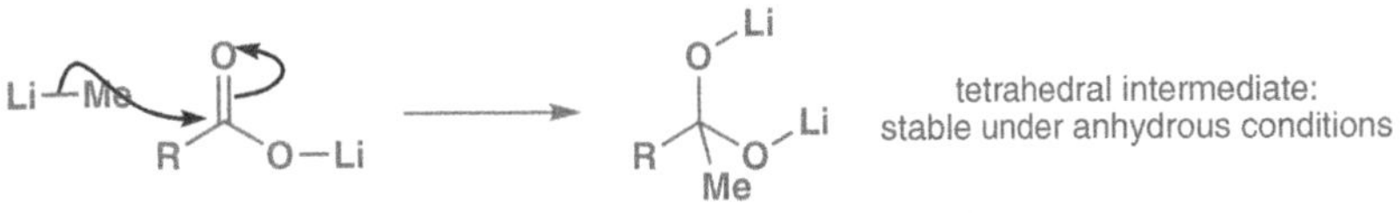

We could say that the affinity of lithium for oxygen means that the Li–O bond has considerable covalent character, making the CO_2Li less of a true anion. And the intermediate after addition of MeLi is probably best represented as a covalent compound too. Anyway, the product of this addition is a dianion of the sort that we met during one of the mechanisms of base-catalysed amide hydrolysis. But in this case there is no possible leaving group, so there the dianion sits. Only at the end of the reaction, when water is added, are the oxygen atoms protonated to give a hydrated ketone, which collapses immediately (remember Chapter 6) to give the ketone that we wanted. The water quench also destroys any remaining organolithium, so the ketone is safe from further attack.

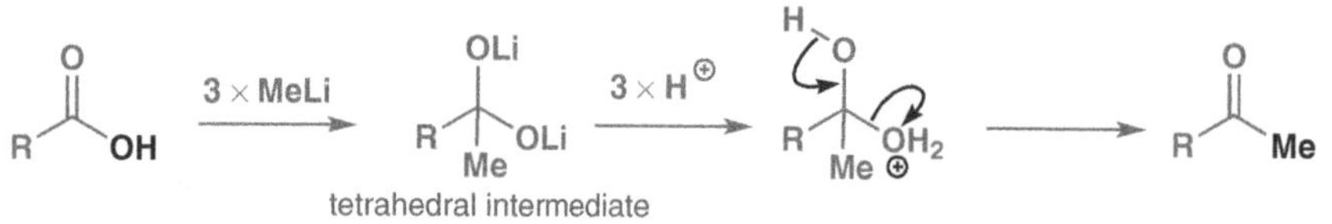

This method has been used to make some ketones that are important starting materials for making cyclic natural products known as macrolides.

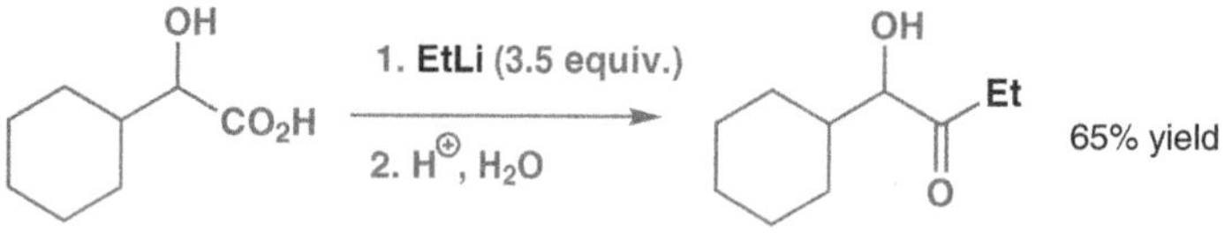

> ■ Notice that three equivalents of organolithium are needed in this reaction: one to deprotonate the acid, one to deprotonate the hydroxyl group, and one to react with the lithium carboxylate. These chemists added a further 0.5 for good measure.

Another good set of starting materials that lead to non-collapsible tetrahedral intermediates is known as the **Weinreb amides**, after their inventor, S. M. Weinreb. Addition of organolithium or organomagnesium reagents to *N*-methoxy-*N*-methyl amides gives the tetrahedral intermediate shown, stabilized by *chelation* of the magnesium atom by the two oxygen atoms. Chelation means the coordination of more than one electron-donating atom in a molecule to a single metal atom.

> ■ The word chelation derives from *chele*, the Greek for 'claw'.

a Weinreb amide (an *N*-methoxy-*N*-methyl amide)

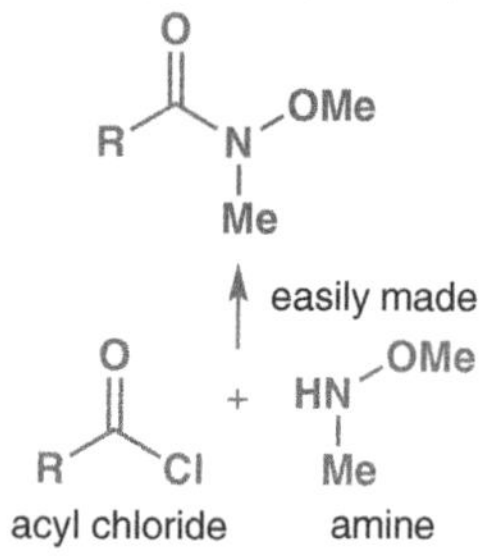

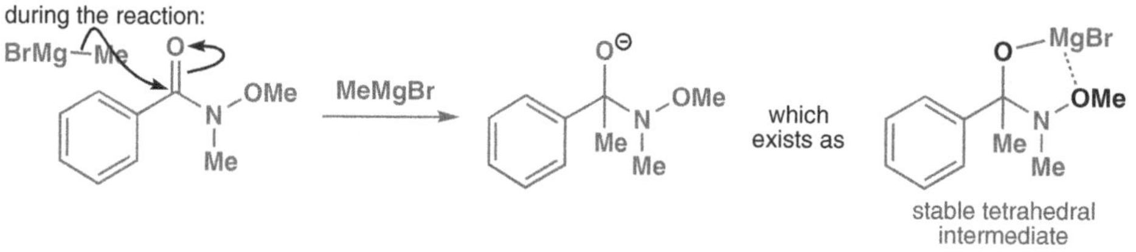

This intermediate collapses to give a ketone only when acid is added at the end of the reaction.

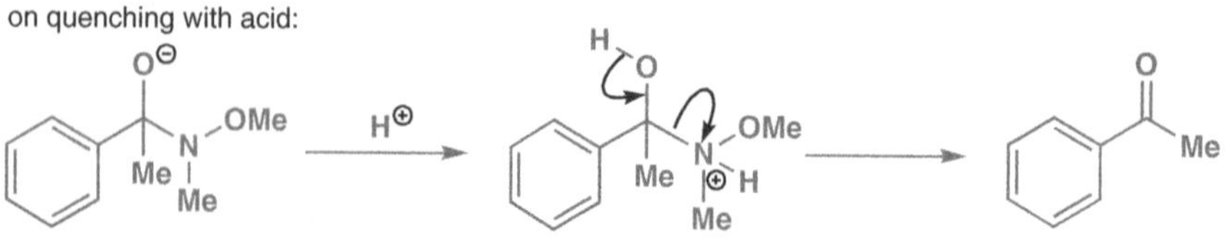

The mechanism looks complicated but the reaction is easy to do:

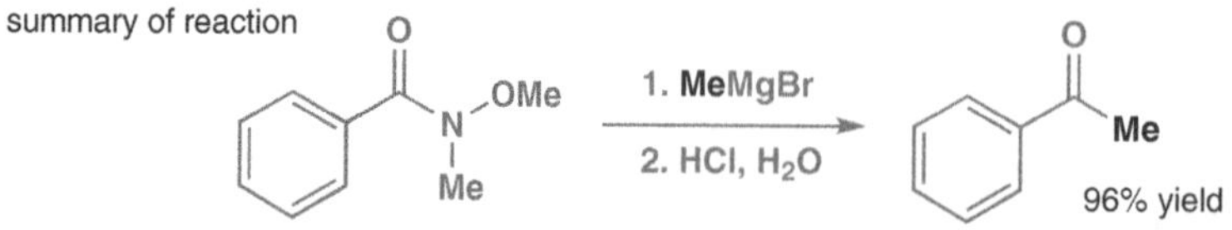

This strategy even works for making aldehydes, if the starting material is dimethylformamide (DMF, Me_2NCHO). This is an extremely useful way of adding electrophilic CHO groups to organometallic nucleophiles. Once again, the tetrahedral intermediate is stable until acid is added at the end of the reaction and the protonated tetrahedral intermediate collapses.

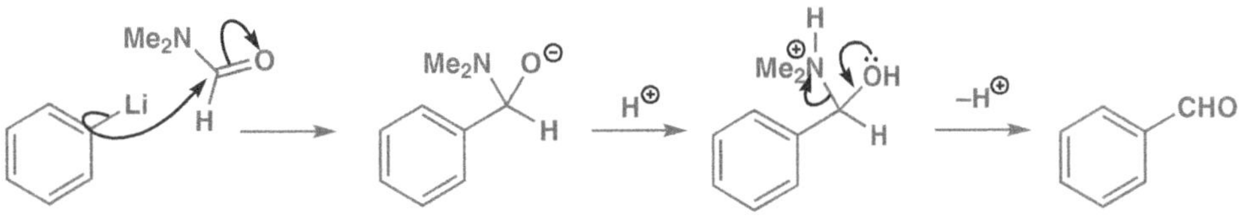

A final alternative is to use a nitrile instead of an ester. The intermediate is the anion of an imine (see Chapter 12 for more about imines), which is not electrophilic at all—in fact, it's quite nucleophilic, but there are no electrophiles for it to react with until the reaction is quenched with acid. It gets protonated and hydrolyses (we'll discuss this in the next chapter) to the ketone.

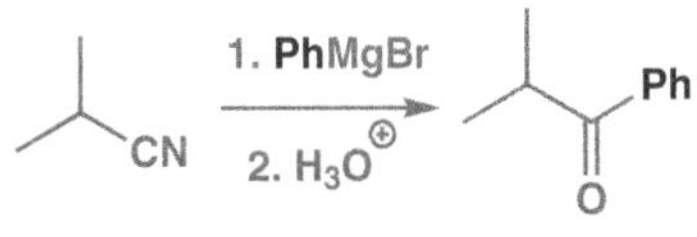

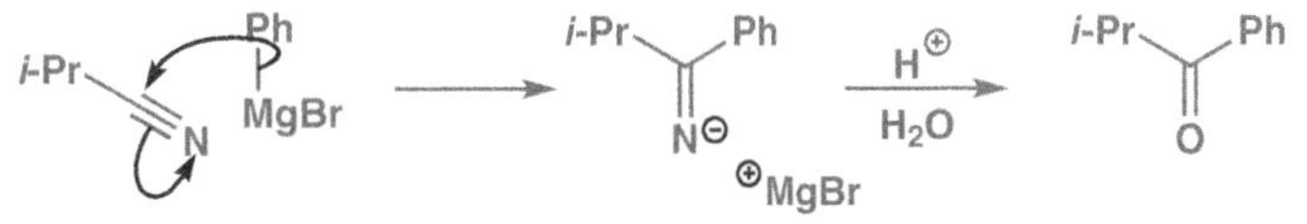

To summarize...

To finish, we should just remind you of what to think about when you consider a nucleophilic substitution at a carbonyl group.

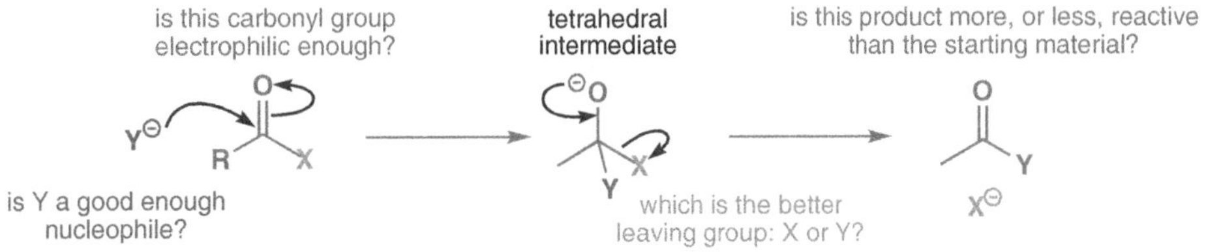

And to conclude...

In this chapter you have been introduced to some important reactions—you can consider them to be a series of facts if you wish, but it is better to see them as the logical outcome of a few simple mechanistic steps. Relate what you have seen to what you gathered from Chapters 6 and 9, when we first started looking at carbonyl groups. All we did in this chapter was to build some subsequent transformations on to the simplest organic reaction, addition to a carbonyl group. You should have noticed that the reactions of all acid derivatives are related and are very easily explained by writing out proper mechanisms, taking into account the presence of acid or base. In the next two chapters we shall see more of these acid- and base-catalysed reactions of carbonyl groups. Try to view them as closely related to the ones in this chapter—the same principles apply to their mechanisms.

Further reading

Section 2, 'Nucleophilic substitution to the carbonyl group' in S. Warren, *Chemistry of the Carbonyl Group*, Wiley, Chichester, 1974.

The dehydration of amides to give nitriles is described in *Vogel*, p. 716.

Check your understanding

 To check that you have mastered the concepts presented in this chapter, attempt the problems that are available in the book's Online Resource Centre at http://www.oxfordtextbooks.co.uk/orc/clayden2e/

Suggested solutions for Chapter 10

10

PROBLEM 1

Suggest reagents to make the drug phenaglycodol by the route below.

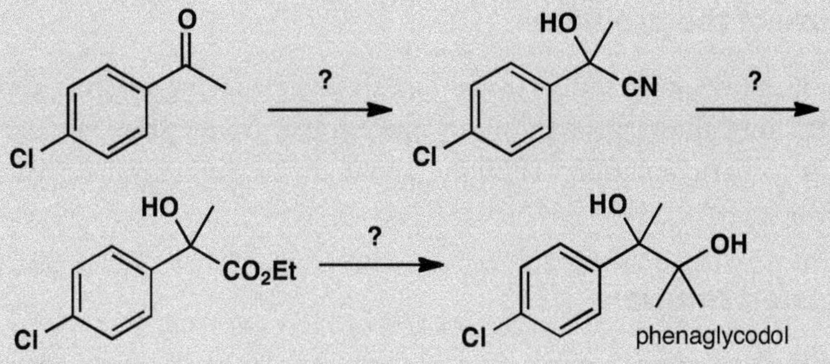

Purpose of the problem

Simple revision of addition to carbonyl groups from chapter 6.

Suggested solution

The first step is a simple addition of cyanide to a ketone (p. 127 of the textbook) usually carried out with NaCN and an acid, such as acetic acid. The second step is an acid-catalysed addition of an alcohol to a nitrile (p.-213 of the textbook). Finally there is a double addition of an organometallic reagent to an ester (p. 216 of the textbook). One way of doing all this is shown below.

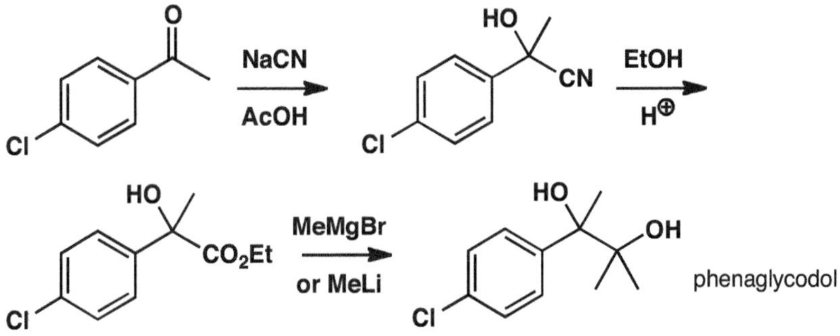

Direct ester formation from carboxylic acids (R^1CO_2H) and alcohols (R^2OH) works in acid solution but not in basic solution. Why not? By contrast, ester formation from alcohols (R^2OH) and acid anhydrides [$(R^1CO)_2O$)] or chlorides (R^1COCl) is commonly carried out in basic solution in the presence of bases such as pyridine. Why does this work?

Purpose of the problem

These questions may sound trivial but students starting organic chemistry often fall into the trap of trying to make esters from carboxylic acids and alcohols in basic solution. Thinking about the reasons my help you avoid this error.

Suggested solution

■ This mechanism is described in detail on p. 208 of the textbook.

The direct reaction works in acid solution as the carboxylic acid is protonated (at the carbonyl group, note) and becomes a good electrophile. Later the tetrahedral intermediate is protonated and can lose a molecule of water.

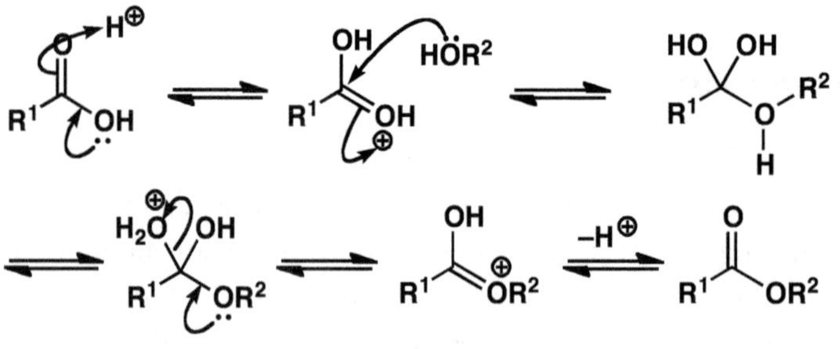

In basic solution, the first thing that happens is the removal of the proton from the carboxylic acid to form a stable delocalized anion. Nucleophiles cannot attack this anion and no further reaction occurs.

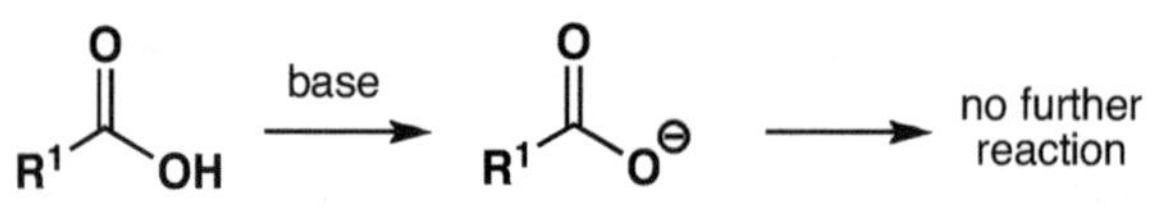

Acid anhydrides and acid chlorides do not have this acidic hydrogen so the alcohol attacks them readily and the base is helpful in removing the acidic proton from the intermediate. The weak base pyridine (pK_a of the conjugate acid 5.5) is ideal. The product from the uncatalysed reaction would be HCl from the acid chloride and the base also removes that.

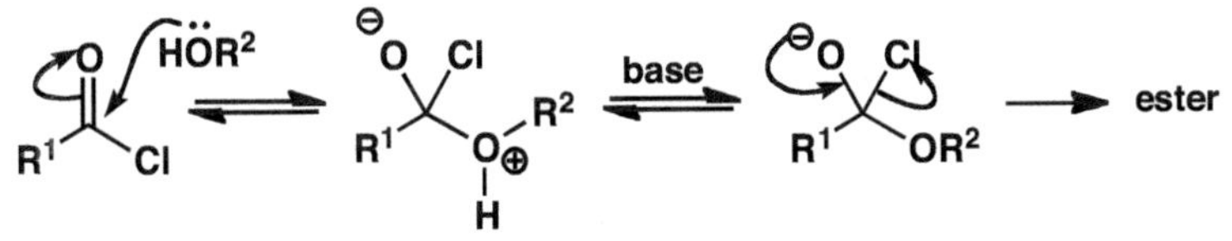

PROBLEM 3

Predict the success or failure of these attempted substitutions at the carbonyl group. You should use estimated pK_a values in your answer and, of course, draw mechanisms.

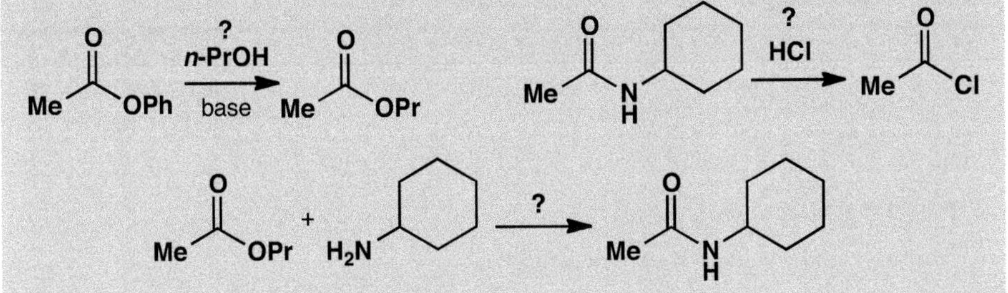

Purpose of the problem

A chance to try out the correlation between leaving group ability and pK_a explained in the textbook (p. 205).

Suggested solution

You need to draw mechanisms for the formation of the tetrahedral intermediate and check that it is in the right protonated form. Then you need to check which potential leaving group is the best, using appropriate estimated pK_a values. The first and the last proposals will succeed but the second will not as chloride ion is a better leaving group than even a protonated amine and this reaction would go backwards.

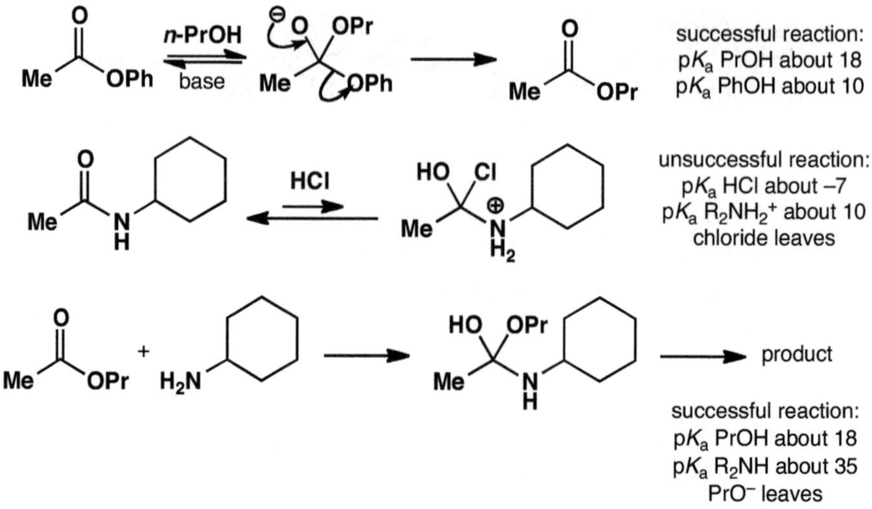
n-PrOH
base
successful reaction:
pKa PrOH about 18
pKa PhOH about 10
HCl
unsuccessful reaction:
pKa HCl about –7
pKa R2NH2+ about 10
chloride leaves
product
successful reaction:
pKa PrOH about 18
pKa R2NH about 35
PrO− leaves

PROBLEM 4

Suggest mechanisms for these reactions.

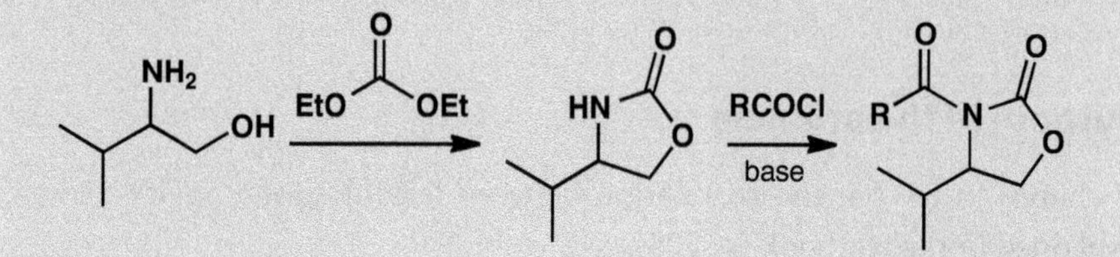
RCOCl
base

Purpose of the problem

Drawing mechanisms for nucleophilic substitution on important compounds including cyclic and dicarbonyl compounds.

Suggested solution

In the first reaction there are two nucleophilic substitutions and you must decide which nucleophile attacks first. The amine is a better nucleophile than the alcohol. The cyclization occurs because, in the intermediate for the second substitution, there are two alcohols as potential leaving groups. Either can leave but when the ring opens again, the alcohol is still part of the molecule and will re-cyclize, but if the ethoxide leaves it is lost into solution and does not come back.

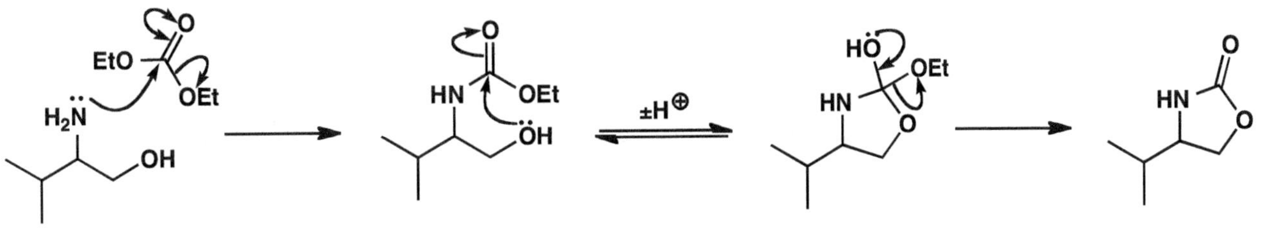

The second reaction is more straightforward. The amide proton is quite acidic and will be removed by the base making a better nucleophile. Notice that in these suggested solutions we are using the shorthand of the double-headed arrow on the carbonyl group.

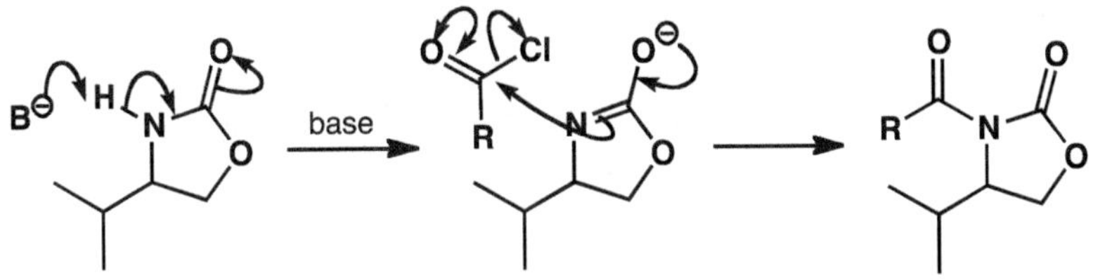

■ The use of the 'double-headed' arrow shorthand is explained on p. 217 of the textbook.

PROBLEM 5

In making esters of the naturally occurring amino acids (general structure below) it is important to keep them as their hydrochloride salts. What would happen to these compounds if they were neutralized?

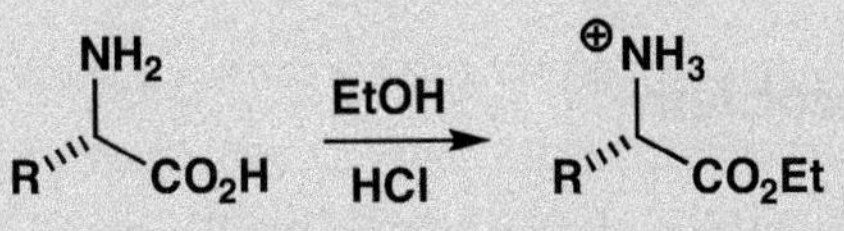

Purpose of the problem

Exploration of a simple reaction that can go seriously wrong if we do not think about what we are doing.

Suggested solution

The amino acids do not usually react with themselves as they exist mostly as the zwitterion. But after the acid is esterified it is much more electrophilic and the amino group is now nucleophilic.

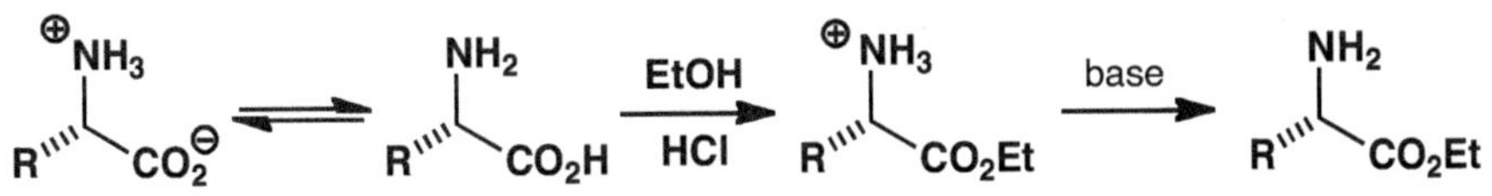

The amine of one compound attacks the ester group of another to form a dimer (a peptide) which may cyclize to form a double amide, known as a diketopiperazine. The cyclization is usually faster than the dimerization as it is an intramolecular reaction forming a stable six-membered ring.

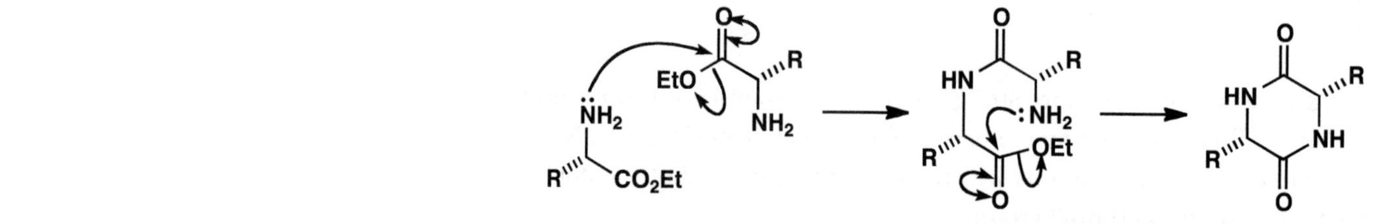

PROBLEM 6

It is possible to make either the diester or the monoester of butanedioic acid (succinic acid) from the cyclic anhydride as shown. Why does one method give the diester and one the monoester?

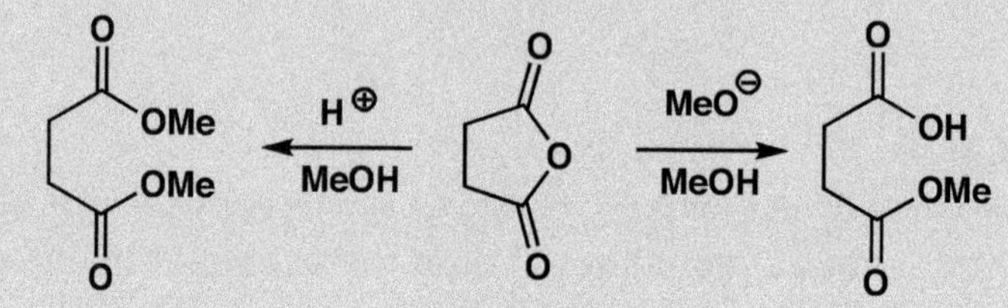

Purpose of the problem

An exploration of selectivity in carbonyl substitutions. Mechanistic thinking allows you to say confidently whether a reaction will happen or not. This problem builds on problem 2.

Suggested solution

In basic solution the nucleophile is methoxide ion. This strong nucleophile attacks the carbonyl group to give a tetrahedral intermediate having two possible leaving groups. The ester anion is preferred (pK_a of RCO_2H about 5) to the alkoxide ion (pK_a of ROH about 15). This carboxylate anion cannot be protonated in basic solution and is not attacked by methoxide ion.

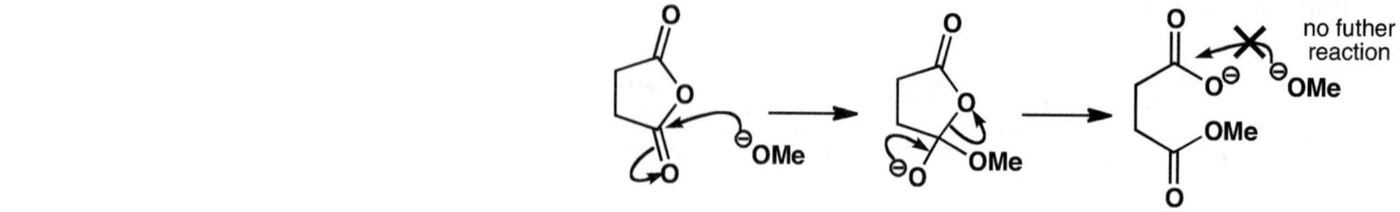

In acid solution the first reaction is similar, though the tetrahedral intermediate is neutral, and the carboxyl is still the better leaving group. The second esterification is now all right because methanol can attack the protonated carboxylic acid and water can be driven out after a second protonation. The second step is an equilibrium, with water and methanol about equal as leaving groups, but methanol is present in large excess as the solvent and drives the equilibrium across. We have omitted proton transfer steps.

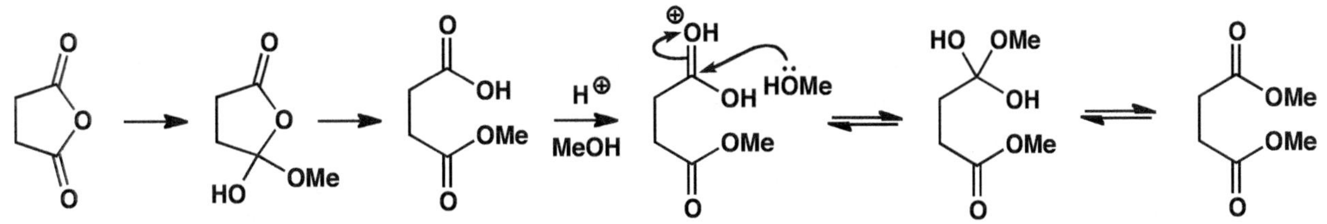

PROBLEM 7

Suggest mechanisms for these reactions, explaining why these particular products are formed.

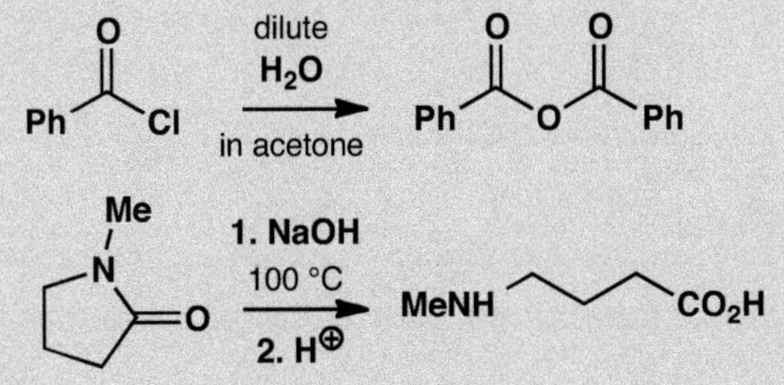

Purpose of the problem

A contrast between very reactive (acid chloride), less reactive (anhydride) and unreactive (amide) carbonyl compounds.

Suggested solution

The acid chloride reacts rapidly with the water and the carboxylic acid produced reacts rapidly with a second molecule of acid chloride. The anhydride reacts much more slowly (pK_a of HCl is –7 but the pK_a of RCO_2H is about 5) with water so there is a good chance of stopping the reaction there, especially when we use a low concentration of water in acetone solution. In this instance the chance is made a certainty because the anhydride precipitates from the solution and is no longer in equilibrium

■ The reactivity sequence of carboxylic acid derivatives is explained on pp. 202 and 206 of the textbook.

with the other reagents. It is usually possible to descend the reactivity sequence of acid derivatives.

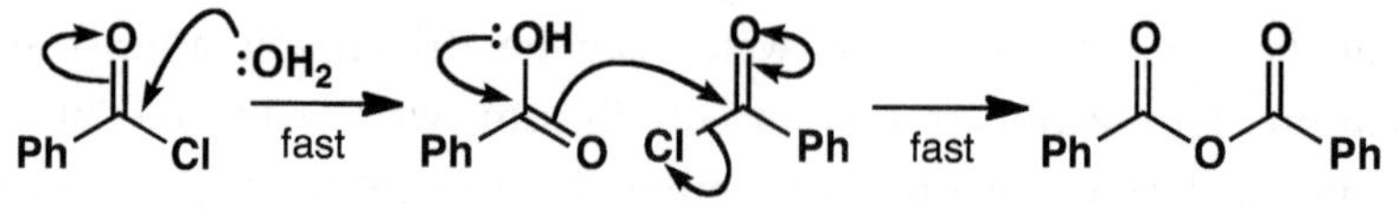

■ The alkaline hydrolysis of amides is on p. 213 of the textbook.

The second reaction is an example of the alkaline hydrolysis of amides. Though the nitrogen atom is never a good leaving group, it will leave from the dianion and, once gone, it is quickly protonated and does not come back. This example also benefits from the release of the slight strain in the five-membered ring.

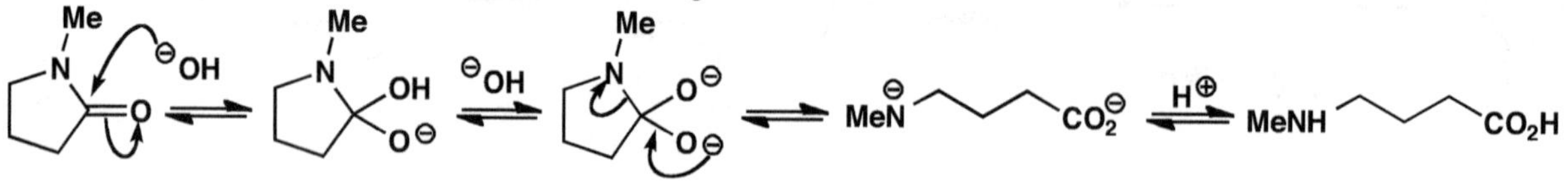

PROBLEM 8

Give mechanisms for these reactions, explaining the selectivity (or lack of it!) in each case.

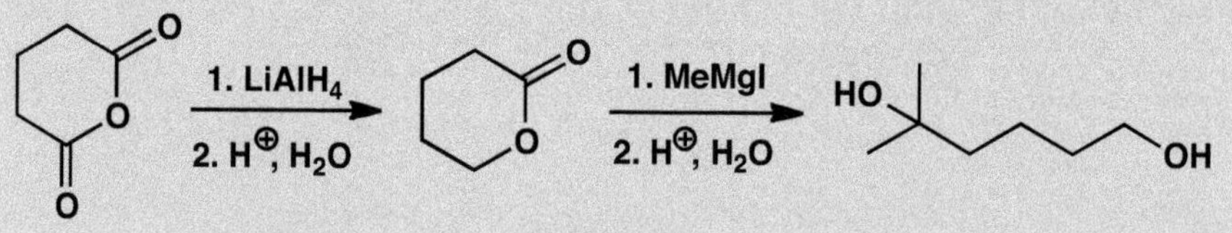

Purpose of the problem

Analysis of a sequence of reactions where the first stops at the halfway stage but the second does not.

Suggested solution

One of the carbonyl groups of the anhydride must be attacked by LiAlH$_4$ and we need to follow that reaction through to see what happens next. The first addition of AlH$_4^-$ produces a tetrahedral intermediate that decomposes with the loss of the only possible leaving group, the carboxylate ion, to give an aldehyde. That too is quickly reduced by AlH$_4^-$ to give the hydroxy-acid as its anion, which is resistant to further reduction. In the acidic aqueous work-up, excess LiAlH$_4$ is instantly destroyed and the hydroxy-acid cyclizes to the lactone. The fact that the lactone is not formed under the reaction conditions is important: if it were, then it too would be reduced by the LiAlH$_4$.

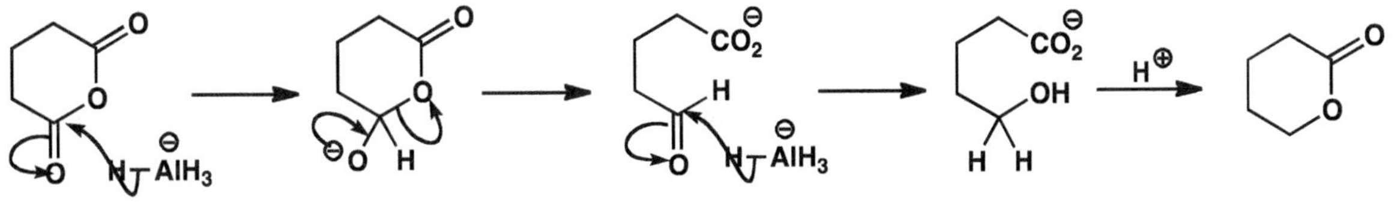

The second reaction starts similarly with the Grignard reagent adding to the ester carbonyl group and the tetrahedral intermediate losing the only possible leaving group. Again, a reactive carbonyl compound is produced: a ketone that is more electrophilic than the ester, so it adds the Grignard reagent even faster. Work-up in aqueous acid gives the diol.

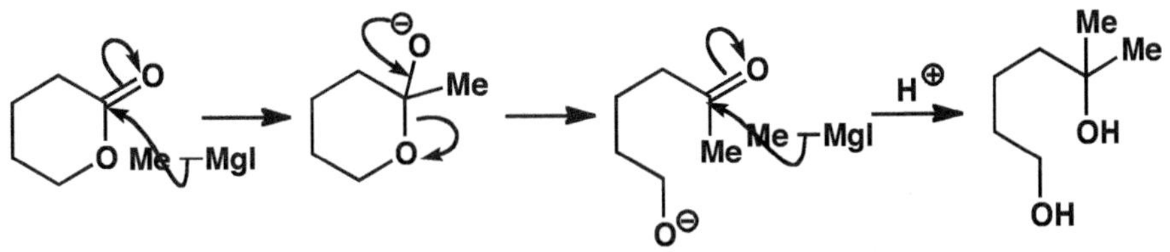

PROBLEM 9

This reaction goes in one direction in acid solution and in the other direction in basic solution. Draw mechanisms for the reactions and explain why the product depends on the conditions.

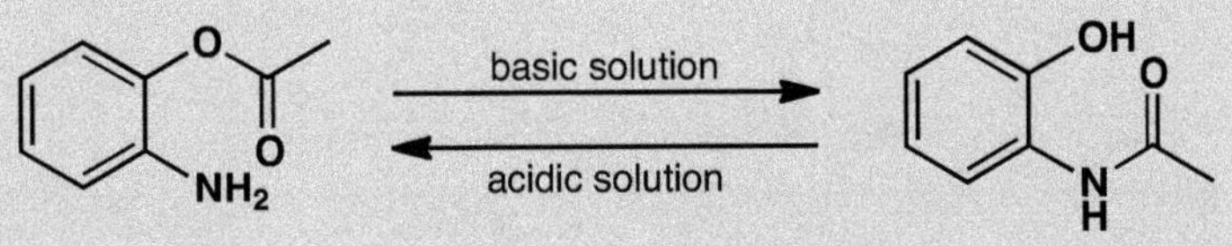

Purpose of the problem

A reminder that carbonyl substitutions are equilibria and that removal of a product from an equilibrium may decide which way the reaction goes. Practice at drawing mechanisms for intramolecular reactions.

Suggested solution

The equilibrium we are concerned with is that between the two products and we can draw what would happen in neutral solution.

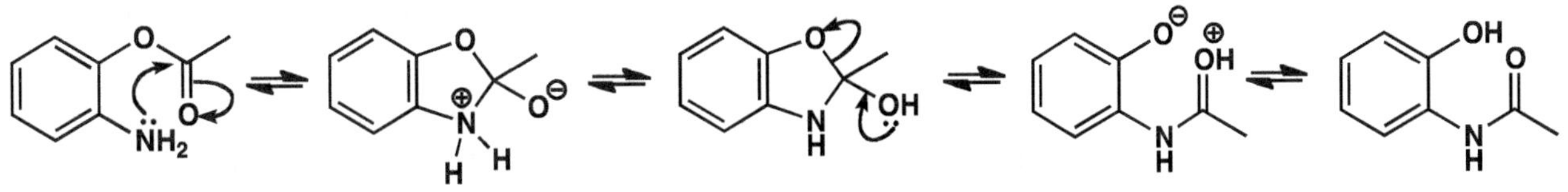

The amine attacks the ester in the usual way to give the tetrahedral intermediate which decomposes with the loss of the better leaving group: phenols are reasonably acidic (pK_a PhOH = 10) so the phenoxy anion is a much better leaving group than ArNH⁻. In strongly basic solution, the phenol product is fully deprotonated, so again, the equilibrium lies to the right. In acidic solution the starting amine is fully protonated, pulling the equilibrium back over to the left.

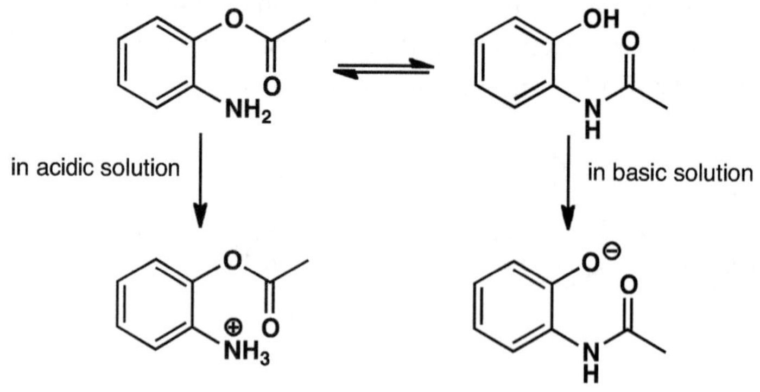

in acidic solution   in basic solution

PROBLEM 10

Amelfolide is a drug used to treat cardiac arrhythmia. Suggest how it could be made from 4-nitrobenzoic acid and 2,5-dimethylaniline.

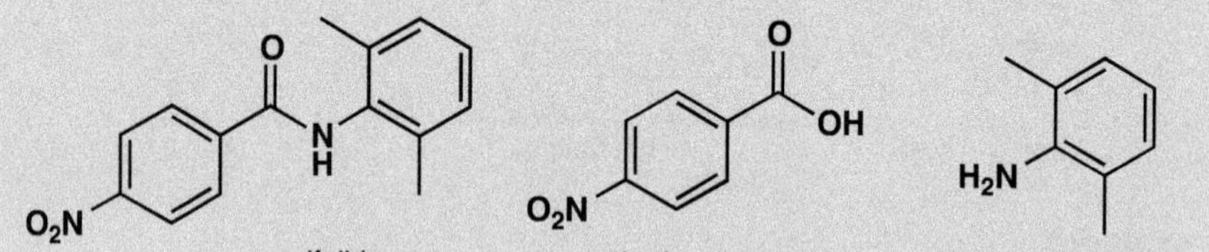

amelfolide   4-nitrobenzoic acid   2,6-dimethylaniline

Purpose of the problem

A reminder to avoid a common error in proposed reactions of carboxylic acids.

Suggested solution

It is tempting to try and react the amine directly with the acid, but unfortunately the only product this would give is the ammonium carboxylate salt: the amine deprotonates the acid, and the carboxylate anion that results is no longer electrophilic. With alcohols, esters can be formed from carboxylic acids under acid catalysis, but with amines the acid catalyst just protonates the amine, and it is no longer nucleophilic! The simplest solution is to convert the carboxylic acid to an acid chloride and allow that to react with the amine. Additional base will neutralize the HCl by-product.

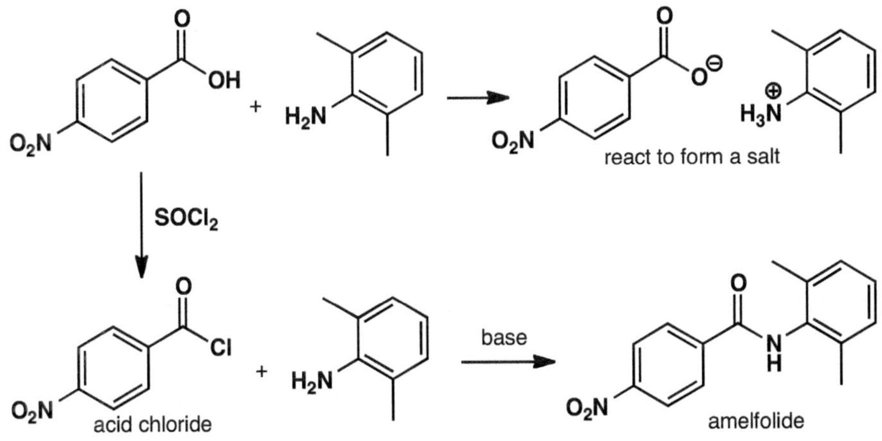

react to form a salt

PROBLEM 11

Given that the pK_a of tribromomethane, $CHBr_3$ (also known as bromoform) is 13.7, suggest what will happen when this ketone is treated with sodium hydroxide.

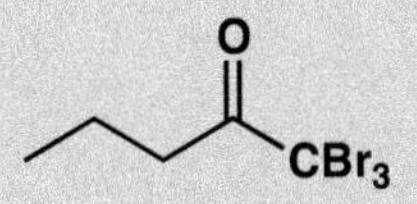

Purpose of the problem

Predicting reactivity with an unusual leaving group.

Suggested solution

The best approach to new reactions is to start drawing curly arrows for steps you know are reasonable, and to see where they take you. Here, we are treating a carbonyl compound, an electrophile, with hydroxide, a nucleophile, so the first step is likely to be addition of hydroxide to the C=O group. You have seen many, many reactions that start this way.

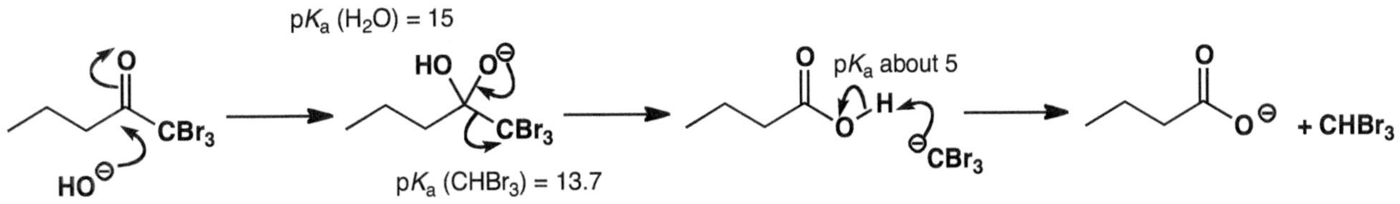

The result looks like a tetrahedral intermediate: the only possible leaving groups are the hydroxide (which takes us back to starting materials) or the anion Br_3C^-. You learnt in chapter 10 to use pK_a to estimate leaving group ability, so the relatively low value of 13.7 should encourage you to eject it, giving a carboxylic acid as well. Neutralization of the acid by the tribromomethyl anion gives the products—the carboxylate anion and tribromomethane.

■ This reaction is known as the 'bromoform' reaction and is described on pp. 462-3 of the textbook.

PROBLEM 12

This sequence of reactions is used to make a precursor to the anti-asthma drug montelukast (Singulair). Suggest structures for compounds **A** and **B**.

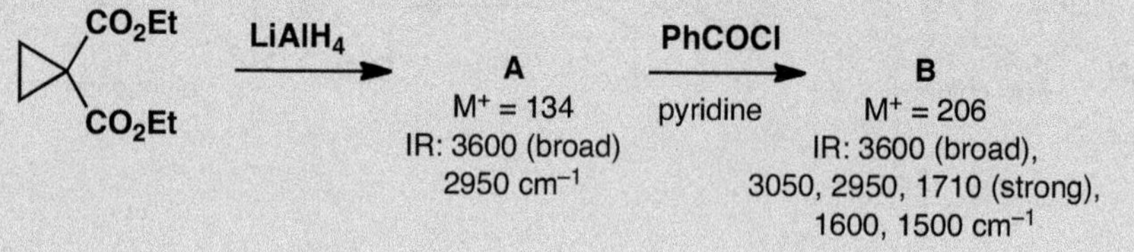

Purpose of the problem

Deducing the presence of functional groups from mass and infra-red spectra.

Suggested solution

Lithium aluminium hydride reduces esters to alcohols, so the only question here is whether it reduces one, or both esters. The IR tells us that there is an alcohol (3600 cm^{-1}) and no carbonyl group (which you would expect around 1700 cm^{-1}) so we can assume that both esters have been reduced. The diol structure below is consistent with the mass of the molecular ion.

Alcohols react with acid chlorides to form esters, so again we have the choice between a single or double ester formation. The IR tells us that one of the alcohols is still present, along with a carbonyl at 1710 cm^{-1}, and the mass of the product is consistent with the structure below.

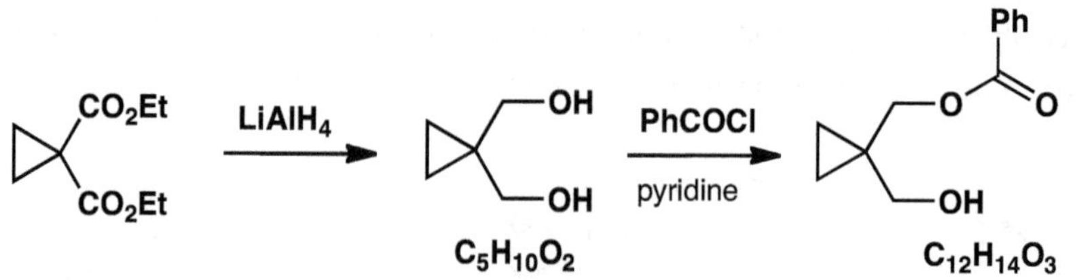

¹H NMR: Proton nuclear magnetic resonance

13

Connections

➡ Building on

- X-ray crystallography, mass spectrometry, NMR, and infrared spectroscopy ch3

Arriving at

- Proton (or ¹H) NMR spectra and their regions
- How ¹H NMR compares with ¹³C NMR: integration
- How 'coupling' in ¹H NMR provides most of the information needed to find the structure of an unknown molecule

➡ Looking forward to

- Using ¹H NMR with other spectroscopic methods to solve structures rapidly ch18
- Using ¹H NMR to investigate the detailed shape (stereochemistry) of molecules ch31
- ¹H NMR spectroscopy is referred to in most chapters of the book as it is the most important tool for determining structure; you must understand this chapter before reading further

The differences between carbon and proton NMR

We introduced nuclear magnetic resonance (NMR) in Chapter 3 as part of a three-pronged attack on the problem of determining molecular structure. We showed that mass spectrometry weighs the molecules, infrared spectroscopy tells us about functional groups, and ¹³C and ¹H NMR tell us about the hydrocarbon skeleton. We concentrated on ¹³C NMR because it's simpler, and we were forced to admit that we were leaving the details of the most important technique of all—proton (¹H) NMR—until a later chapter because it is more complicated than ¹³C NMR. This is that chapter and we must now tackle those complications. We hope you will see ¹H NMR for the beautiful and powerful technique that it surely is. The difficulties are worth mastering for this is the chemist's primary weapon in the battle to solve structures.

● We will make use of ¹H and ¹³C NMR evidence for structure throughout this book, and it is essential that you are familiar with the explanations in this chapter before you read further.

Proton NMR differs from ¹³C NMR in a number of ways.

- ¹H is the major isotope of hydrogen (99.985% natural abundance), while ¹³C is only a minor isotope (1.1%).
- ¹H NMR is quantitative: the area under the peak tells us the number of hydrogen nuclei, while ¹³C NMR may give strong or weak peaks from the same number of ¹³C nuclei.
- Protons interact magnetically ('couple') to reveal the connectivity of the structure, while ¹³C is too rare for coupling between ¹³C nuclei to be seen.

■ '¹H NMR' and 'proton NMR' are interchangeable terms. All nuclei contain protons of course, but chemists often use 'proton' specifically for the nucleus of a hydrogen atom, either as part of a molecule or in its 'free' form as H⁺. This is how it will be used in this chapter.

Online support. The icon 🖱 in the margin indicates that accompanying interactive resources are provided online to help your understanding: just type **www.chemtube3d.com/clayden/123** into your browser, replacing **123** with the number of the page where you see the icon. For pages linking to more than one resource, type **123-1, 123-2** etc. (replacing **123** with the page number) for access to successive links.

- ¹H NMR shifts give a more reliable indication of the local chemistry than that given by ¹³C spectra.

We shall examine each of these points in detail and build up a full understanding of proton NMR spectra.

Proton NMR spectra are recorded in the same way as ¹³C NMR spectra: radio waves are used to study the energy level differences of nuclei in a magnetic field, but this time they are ¹H and not ¹³C nuclei. Hydrogen nuclei in a magnetic field have two energy levels: they can be aligned either with or against the applied magnetic field.

➡ In Chapter 3 we illustrated the alignment of nuclei using the analogy of a compass needle in a magnetic field.

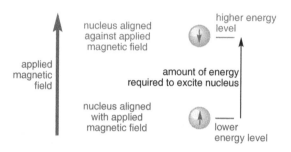

■ All nuclei are characterized by their 'nuclear spin', a value known as I. The number of energy levels available to a nucleus of spin I is $2I + 1$. ¹H and ¹³C both have $I = 1/2$.

■ This 10 ppm scale is not the same as any part of the ¹³C NMR spectrum. It is at a different frequency altogether.

¹H and ¹³C spectra have many similarities: the scale runs from right to left and the zero point is given by the same reference compound, although it is the proton resonance of Me₄Si rather than the carbon resonance that defines the zero point. You will notice at once that the scale is much smaller, ranging over only about 10 ppm instead of the 200 ppm needed for carbon. This is because the variation in the chemical shift is a measure of the shielding of the nucleus by the electrons around it. There is inevitably less change possible in the distribution of two electrons around a hydrogen nucleus than in that of the eight valence electrons around a carbon nucleus. Here is the ¹H NMR spectrum of acetic acid, which you first saw in Chapter 3.

■ A reminder from Chapter 3: ignore the peak at 7.25 shown in brown. This is from the solvent, as explained on p.272.

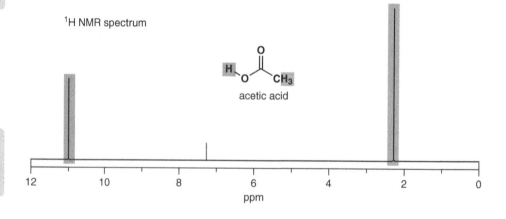

¹H NMR spectrum

acetic acid

Integration tells us the number of hydrogen atoms in each peak

You know from Chapter 3 that the position of a signal in an NMR spectrum tells us about its environment. In acetic acid the methyl group is next to the electron-withdrawing carbonyl group and so is slightly deshielded at about δ 2.0 ppm and the acidic proton itself, attached to O, is very deshielded at δ 11.2 ppm. The same factor that makes this proton acidic—the O–H bond is polarized towards oxygen—also makes it resonate at low field. So far things are much the same as in ¹³C NMR. Now for a difference. In ¹H NMR the *size* of the peaks is also important: the area under the peaks is exactly proportional to the number of protons. Proton spectra are normally integrated, that is, the area under the peaks is computed and recorded as a line with steps corresponding to the area, like this.

■ It is not enough simply to measure the relative heights of the peaks because, as here, some peaks might be broader than others. Hence the area under the peak is measured.

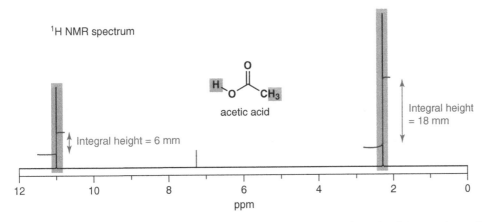

Simply measuring the height of the steps with a ruler gives you the *ratio* of the numbers of protons represented by each peak. In many spectra this will be measured for you and reported as a number at the bottom of the spectrum. Knowing the atomic composition (from the mass spectrum) we also know the distribution of protons of various kinds. Here the heights are 6 mm and 18 mm, a ratio of about 1:3. The compound is $C_2H_4O_2$ so, since there are four H atoms altogether, the peaks must contain $1 \times H$ and $3 \times H$, respectively.

In the spectrum of 1,4-dimethoxybenzene there are just two signals in the ratio of 3:2. This time the compound is $C_8H_{10}O_2$ so the true ratio must be 6:4. The positions of the two signals are exactly where you would expect them to be from our discussion of the regions of the NMR spectrum in Chapter 3: the 4H aromatic signal is in the left-hand half of the spectrum, between 5 and 10 ppm, where we expect to see protons attached to sp^2 C atoms, while the 6H signal is in the right-hand half of the spectrum, where we expect to see protons attached to sp^3 C atoms.

➡ We will come back to the regions of the ^{1}H NMR spectrum in more detail in just a moment, but we introduced them in Chapter 3 on p. 60.

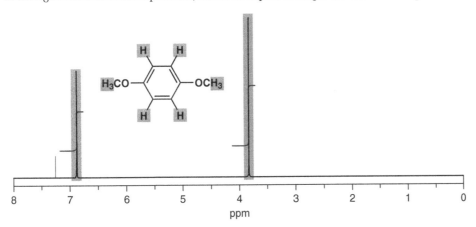

In this next example it is easy to assign the spectrum simply by measuring the steps in the integral. There are two identical methyl groups (CMe$_2$) with six Hs, one methyl group by itself with three Hs, the OH proton (1 H), the CH$_2$ group next to the OH (two Hs), and finally the CH$_2$CH$_2$ group between the oxygen atoms in the ring (four Hs).

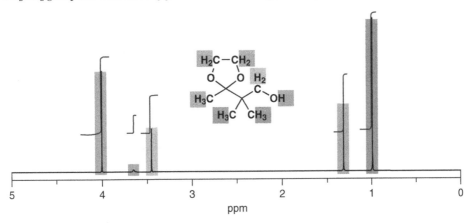

Before we go on, a note about the solvent peaks shown in brown in these spectra. Proton NMR spectra are generally recorded in solution in deuterochloroform ($CDCl_3$)—that is, chloroform ($CHCl_3$) with the ¹H replaced by ²H (deuterium). The proportionality of the size of the peak to the number of protons tells you why: if you ran a spectrum in $CHCl_3$, you would see a vast peak for all the solvent Hs because there would be much more solvent than the compound you wanted to look at. Using $CDCl_3$ cuts out all extraneous protons. ²H atoms have different nuclear properties and so don't show up in the ¹H spectrum. Nonetheless, $CDCl_3$ is always unavoidably contaminated with a small amount of $CHCl_3$, giving rise to the small peak at 7.25 ppm. Spectra may equally well be recorded in other deuterated solvents such as water (D_2O), methanol (CD_3OD), or benzene (C_6D_6).

Regions of the proton NMR spectrum

All the H atoms in the last example were attached to sp³ carbons, so you will expect them to fall between 0 and 5 ppm. However, you can clearly see that H atoms that are nearer to oxygen are shifted downfield within the 0–5 ppm region, to larger δ values (here as far as 3.3 and 3.9 ppm). We can use this fact to build some more detail into our picture of the regions of the ¹H NMR spectrum.

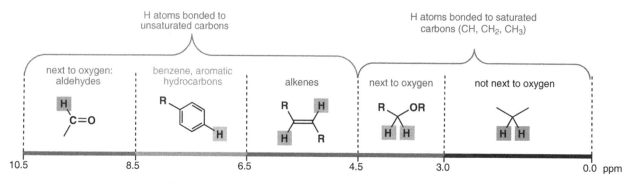

These regions hold for protons attached to C: protons attached to O or N can come almost anywhere on the spectrum. Even for C–H signals the regions are approximate and overlap quite a lot. You should use the chart as a basic guide, and you should aim to learn these regions. But you will also need to build up a more detailed understanding of the factors affecting proton chemical shift. To help you achieve this understanding, we now need to examine the classes of proton in more detail and examine the reasons for their particular shifts. It is important that you grasp these reasons.

In this chapter you will see a lot of numbers—chemical shifts and differences in chemical shifts. We need these to show that the ideas behind ¹H NMR are securely based in fact. *You do not need to learn these numbers.* Comprehensive tables can be found at the end of Chapter 18, which we hope you will find useful for reference while you are solving problems.

➡ These tables can be found on pp. 422–426.

Protons on saturated carbon atoms

Chemical shifts are related to the electronegativity of substituents

We shall start with protons on saturated carbon atoms. The top half of the diagram below shows how the protons in a methyl group are shifted more and more as the atom attached to them gets more electronegative.

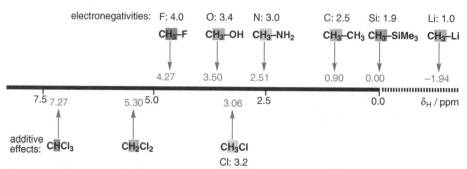

When we are dealing with single atoms as substituents, these effects are straightforward and more or less additive. If we go on adding electronegative chlorine atoms to a carbon atom, electron density is progressively removed from it and the carbon nucleus and the hydrogen atoms attached to it are progressively deshielded. You can see this in the bottom half of the diagram above. Dichloromethane, CH_2Cl_2, and chloroform, $CHCl_3$, are commonly used as solvents and their shifts will become familiar to you if you look at a lot of spectra.

Proton chemical shifts tell us about chemistry

The truth is that shifts and electronegativity are not perfectly correlated. The key property is indeed electron withdrawal but it is the electron-withdrawing power of the whole substituent in comparison with the carbon and hydrogen atoms in the CH skeleton that matters. Methyl groups joined to the same element—nitrogen, say—may have very different shifts if the substituent is an amino group (CH_3–NH_2 has δ_H for the CH_3 group = 2.41 ppm) or a nitro group (CH_3–NO_2 has δ_H 4.33 ppm). A nitro group is much more electron-withdrawing than an amino group.

What we need is a quick guide rather than some detailed correlations, and the simplest is this: all functional groups except very electron-withdrawing ones shift methyl groups from 1 ppm (where you find them if they are not attached to a functional group) downfield to about 2 ppm. Very electron-withdrawing groups shift methyl groups to about 3 ppm. This is the sort of thing it *is* worth learning.

> ■ You have seen δ used as a symbol for chemical shift. Now that we have two sorts of chemical shift—in the ^{13}C NMR spectrum and in the ^1H NMR spectrum—we need to be able to distinguish them. δ_H means chemical shift in the ^1H NMR spectrum, and δ_C is chemical shift in the ^{13}C NMR spectrum.

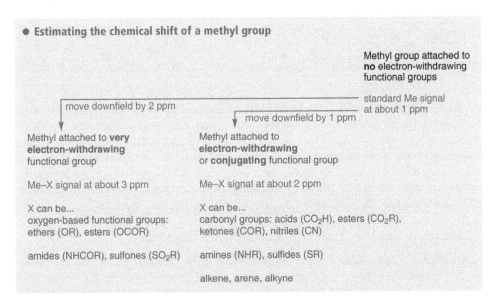

● **Estimating the chemical shift of a methyl group**

Methyl group attached to **no** electron-withdrawing functional groups

standard Me signal at about 1 ppm

move downfield by 2 ppm

move downfield by 1 ppm

Methyl attached to **very electron-withdrawing** functional group

Me–X signal at about 3 ppm

X can be...
oxygen-based functional groups:
ethers (OR), esters (OCOR)

amides (NHCOR), sulfones (SO_2R)

Methyl attached to **electron-withdrawing** or **conjugating** functional group

Me–X signal at about 2 ppm

X can be...
carbonyl groups: acids (CO_2H), esters (CO_2R),
ketones (COR), nitriles (CN)

amines (NHR), sulfides (SR)

alkene, arene, alkyne

Rather than trying to fit these data to some atomic property, even such a useful one as electronegativity, we should rather see these shifts as a useful measure of the electron-withdrawing power of the group in question. The NMR spectra are telling us about the chemistry. The largest shift you are likely to see for a methyl group is that caused by the nitro group, 3.43 ppm, at least twice the size of the shift for a carbonyl group. This gives us our first hint of some important chemistry: one nitro group is worth two carbonyl groups when it come to electron-withdrawing power. You have already seen that electron withdrawal and acidity are related (Chapter 8) and in later chapters you will see that we can correlate the anion-stabilizing power of groups like carbonyl, nitro, and sulfone with proton NMR.

Methyl groups give us information about the structure of molecules

It sounds rather unlikely that the humble methyl group could tell us much that is important about molecular structure—but just you wait. We shall look at four simple compounds and their NMR spectra—just the methyl groups, that is.

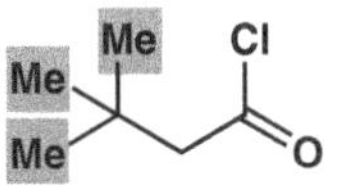

The first compound, the acid chloride in the margin, shows just one methyl signal containing nine Hs at δ_H 1.10. This tells us two things. All the protons in each methyl group are identical,

δ_H 1.10

■ Rotation about single bonds is generally very fast (you are about to see an exception); rotation about double bonds is generally very, very slow (it just doesn't happen). We talked about rotation rates in Chapter 12.

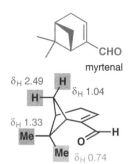

and all three methyl groups in the tertiary butyl (*t*-butyl, or Me₃C–) group are identical. This is because rotation about C–C single bonds, both about the CH₃–C bond and about the (CH₃)₃C–C bond, is fast. Although at any one instant the hydrogen atoms in one methyl group, or the methyl groups in the *t*-butyl group, may differ, on average they are the same. The time-averaging process is fast rotation about a σ bond.

The second compound shows two 3H signals, one at 1.99 and one at 2.17 ppm. Unlike a C–C bond, the C=C double bond does not rotate at all and so the two methyl groups are different. One is on the same side of the alkene as (or '*cis* to') the –COCl group while the other is on the opposite side (or '*trans*').

The next pair of compounds contain the CHO group. One is a simple aldehyde, the other an amide of formic acid: it is DMF, dimethylformamide. The first has two sorts of methyl group: a 3H signal at δ_H 1.81 for the SMe group and a 6H signal at δ_H 1.35 for the CMe₂ group. The two methyl groups in the 6H signal are the same, again because of fast rotation about a C–C σ bond. The second compound also has two methyl signals, at 2.89 and 2.98 ppm, each 3H, and these are the two methyl groups on nitrogen. Restricted rotation about the N–CO bond must be making the two Me groups different. You will remember from Chapter 7 (p. 155) that the N–CO amide bond has considerable double-bond character because of conjugation: the lone pair electrons on nitrogen are delocalized into the carbonyl group.

Like double bonds, cage structures prevent bond rotation and can make the two protons of a CH₂ group appear different. There are many flavouring compounds (terpenoids) from herbs that have structures like this. In the example here—myrtenal, from the myrtle bush—there is a four-membered ring bridged across a six-membered ring. The methyl groups on the other bridge are different because one is over the alkene while one is over the CH₂. No rotation of any bonds within the cage is possible, so these methyl groups resonate at different frequencies (0.74 and 1.33 ppm). The same is true for the two H atoms of the CH₂ group.

CH and CH₂ groups have higher chemical shift than CH₃ groups

Electronegative substituents have a similar effect on the protons of CH₂ groups and CH groups, but with the added complication that CH₂ groups have *two* other substituents and CH groups *three*. A simple CH₂ (methylene) group resonates at 1.3 ppm, about 0.4 ppm further downfield than a comparable CH₃ group (0.9 ppm), and a simple CH group resonates at 1.7 ppm, another 0.4 ppm downfield. Replacing each hydrogen atom in the CH₃ group by a carbon atom causes a small downfield shift as carbon is slightly more electronegative (C 2.5; H 2.2) than hydrogen and therefore shields less effectively.

● **Chemical shifts of protons in CH, CH₂, and CH₃ groups with no nearby electron-withdrawing groups.**

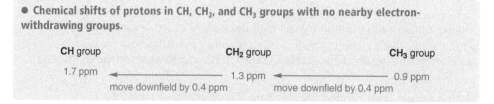

The benzyl group (PhCH₂–) is very important in organic chemistry. It occurs naturally in the amino acid phenylalanine, which you met in Chapter 2. Phenylalanine has its CH₂ signal at 3.0 ppm and is moved downfield from 1.3 ppm mostly by the benzene ring.

Amino acids are often 'protected' as the Cbz (carboxybenzyl) derivatives by reaction with an acid chloride (we'll discuss this more in Chapter 23). Here is a simple example together with the NMR spectrum of the product. Now the CH_2 group has gone further downfield to 5.1 ppm as it is next to both oxygen and phenyl.

amino acid + "Cbz chloride" (benzyl chloroformate) → "Cbz-protected" amino acid

> ⇒ You met this sort of amide-forming reaction in Chapter 10—here the amide is actually a carbamate as the C=O group is flanked by both O and N.

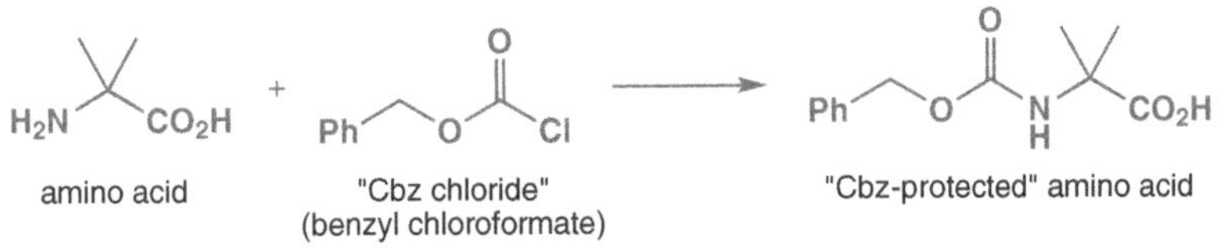

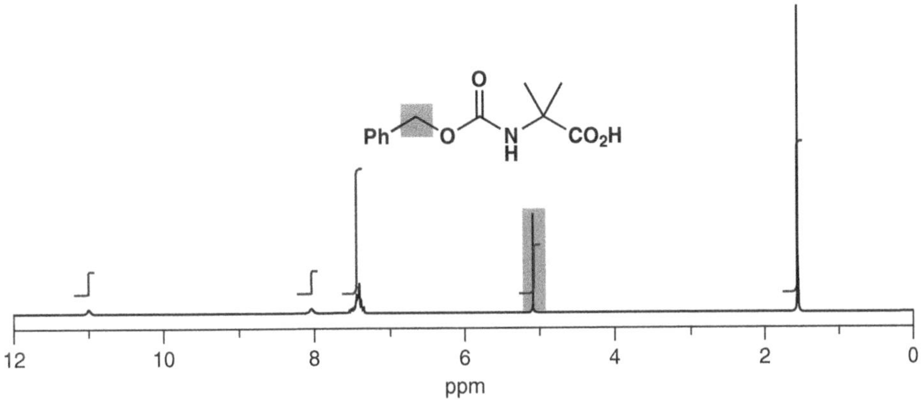

Chemical shifts of CH groups

A CH group in the middle of a carbon skeleton resonates at about 1.7 ppm—another 0.4 ppm downfield from a CH_2 group. It can have up to three substituents and these will cause further downfield shifts of about the same amount as we have already seen for CH_3 and CH_2 groups. Three examples from nature are nicotine, the methyl ester of lactic acid, and vitamin C. Nicotine, the compound in tobacco that causes the craving (although not the death, which is doled out instead by the carbon monoxide and tars in the smoke), has one hydrogen atom trapped between a simple tertiary amine and an aromatic ring at 3.24 ppm. The ester of lactic acid has a CH proton at 4.3 ppm. You could estimate this with reasonable accuracy using the guidelines in the two summary boxes on pp. 273 and 274. Take 1.7 (for the CH) and add 1.0 (for C=O) plus 2.0 (for OH) = 4.7 ppm—not far out. Vitamin C (ascorbic acid) has two CHs. One at 4.05 ppm is next to an OH group (estimate 1.7 + 2.0 for OH = 3.7 ppm) and one is next to a double bond and an oxygen atom at 4.52 ppm (estimate 1.7 + 1 for double bond + 2 for OH = 4.7 ppm). Again, not too bad for a rough estimate.

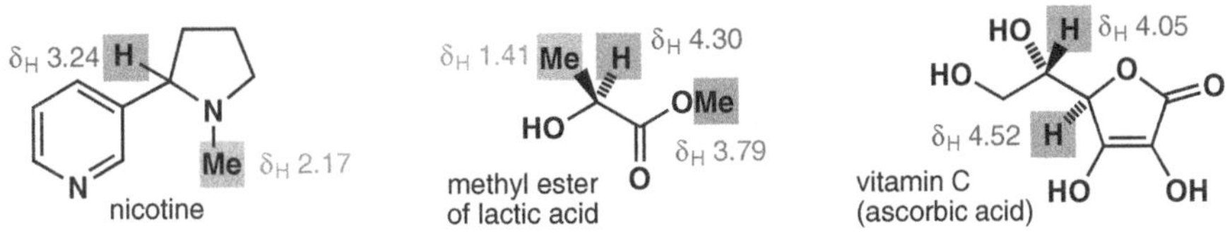

An interesting case is the amino acid phenylalanine whose CH_2 group we looked at a moment ago. It also has a CH group between the amino and the carboxylic acid groups. If we record the ^1H NMR spectrum in D_2O, in either basic (NaOD) or acidic (DCl) solutions, we see a large shift of that CH group. In basic solution the CH resonates at 3.60 ppm and in acidic solution it resonates at 4.35 ppm. There is a double effect here: CO_2H and NH_3^+ are both more electron-withdrawing than CO_2^- and NH_2 so both move the CH group downfield.

> ■ D_2O, NaOD, and DCl have to be used in place of their ^1H equivalents to avoid swamping the spectrum with H_2O protons. All acidic protons are replaced by deuterium in the process—more on this later.

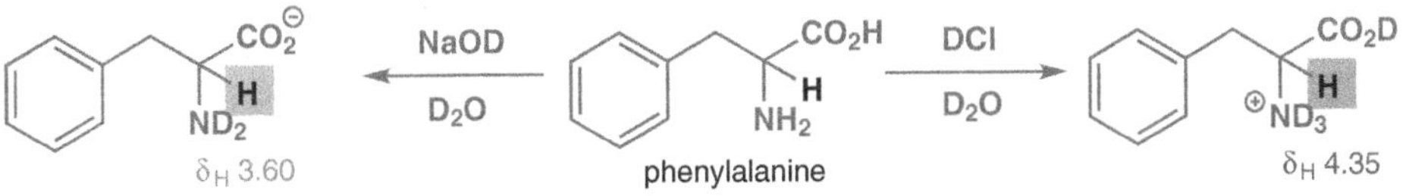

● A simple guide to estimating chemical shifts

We suggest you start with a very simple (and therefore necessarily oversimplified) picture, which should be the basis for any further refinements. Start methyl groups at 0.9, methylenes (CH₂) at 1.3, and methines (CH) at 1.7 ppm. Any functional group is worth a 1 ppm downfield shift except oxygen and halogen which are worth 2 ppm. This diagram summarizes this approach.

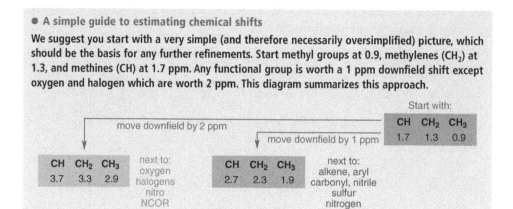

The guide above is very rough and ready, but is easily remembered and you should aim to learn it. However, if you want to, you can make it slightly more accurate by adding further subdivisions and separating out the very electron-withdrawing groups (nitro, ester OCOR, fluoride), which shift by 3 ppm. This gives us the summary chart on this page, which we suggest you use as a reference. If you want even more detailed information, you can refer to the tables in Chapter 18 or better still the more comprehensive tables in any specialized text (see the Further reading section).

Summary chart of proton NMR shifts

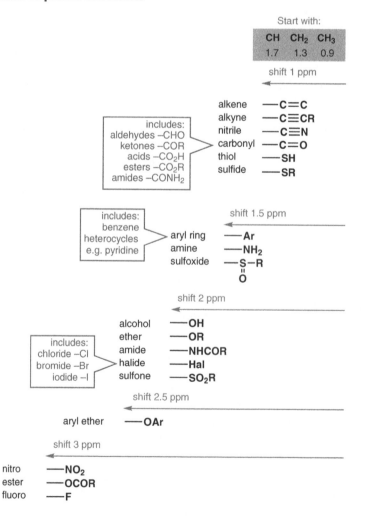

Answers deduced from this chart won't be perfect but will give a good guide. Remember—these shifts are additive. Take a simple example, the ketoester below. There are just three signals and the integration alone distinguishes the two methyl groups from the CH_2 group. One methyl has been shifted from 0.9 ppm by about 1 ppm, the other by more than 2 ppm. The first must be next to C=O and the second next to oxygen. More precisely, 2.14 ppm is a shift of 1.24 ppm from our standard value (0.9 ppm) for a methyl group, about what we expect for a methyl ketone, while 3.61 ppm is a shift of 2.71 ppm, close to the expected 3.0 ppm for an ester joined through the oxygen atom. The CH_2 group is next to an ester and a ketone carbonyl group and so we expect it at $1.3 + 1.0 + 1.0 = 3.3$ ppm, an accurate estimate, as it happens. We shall return to these estimates when we look at the spectra of unknown compounds.

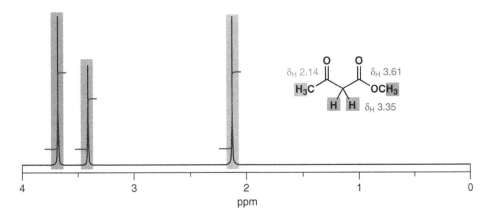

The alkene region and the benzene region

In ^{13}C NMR, alkene and benzene carbons came in the same region of the spectrum, but in the ^1H NMR spectrum the H atoms attached to arene C and alkene C atoms sort themselves into two groups. To illustrate this point, look at the ^{13}C and ^1H chemical shifts of cyclohexene and benzene, shown in the margin. The two carbon signals are almost the same (1.3 ppm difference, < 1% of the total 200 ppm scale) but the proton signals are very different (1.6 ppm difference = 16% of the 10 ppm scale). There must be a fundamental reason for this.

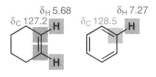

The benzene ring current causes large shifts for aromatic protons

A simple alkene has an area of low electron density in the plane of the molecule because the π orbital has a node there, and the carbons and hydrogen nuclei lying in the plane gain no shielding from the π electrons.

The benzene ring looks similar at first sight, and the plane of the molecule is indeed a node for all the π orbitals. However, as we discussed in Chapter 7, benzene is aromatic—it has extra stability because the six π electrons fit into three very stable orbitals and are delocalized round the whole ring. The applied field sets up a ring current in these delocalized electrons that produces a local field rather like the field produced by the electrons around a nucleus. Inside the benzene ring the induced field opposes the applied field, but outside the ring it reinforces the applied field. The carbon atoms are in the ring itself and experience neither effect, but the hydrogens are outside the ring, feel a stronger applied field, and appear less shielded (i.e. more deshielded; larger chemical shift).

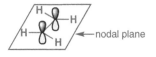

■ Magnetic fields produced by circulating electrons are all around you: electromagnets and solenoids are exactly this.

benzene has *six* delocalized π electrons:

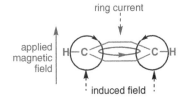

Cyclophanes and annulenes

You may think that it is rather pointless imagining what goes on inside an aromatic ring as we cannot have hydrogen atoms literally inside a benzene ring. However, we can get close. Compounds called cyclophanes have loops of saturated carbon atoms attached at both ends to the same benzene rings. You see here a structure for [7]-*para*-cyclophane, which has a string of seven CH_2 groups attached to the *para* positions of the same benzene ring. The four H atoms on the benzene ring itself appear as one signal at 7.07 ppm—a typical ring-current deshielded value for a benzene ring. The two CH_2 groups joined to the benzene ring (C1) are also deshielded by the ring current at 2.64 ppm. The next two sets of CH_2 groups on C2 and C3 are neither shielded nor deshielded at 1.0 ppm. But the middle CH_2 group in the chain (C4) must be pointing towards the ring in the middle of the π system and is heavily shielded by the ring current at −0.6 ppm).

[7]-*para*-cyclophane

[18]-annulene

H outside the ring δ_H +9.28

H inside the ring δ_H −2.9

δ_H 1.0

δ_H −0.6

δ_H 1.0

δ_H 2.84

With a larger aromatic ring it is possible actually to have hydrogen atoms inside the ring. Compounds are aromatic if they have $4n + 2$ delocalized electrons and this ring with nine double bonds, that is, 18 π electrons, is an example. The hydrogens outside the ring resonate in the aromatic region at rather low field (9.28 ppm) but the hydrogen atoms inside the ring resonate at an amazing −2.9 ppm, showing the strong shielding by the ring current. Such extended aromatic rings are called *annulenes*: you met them in Chapter 7.

🖱 Interactive structures of cyclophane and annulene

Uneven electron distribution in aromatic rings

δ_H 2.89

δ_H 6.38

δ_H 2.28

■ The greater electron density around the ring more than compensates for any change in the ring current.

The ¹H NMR spectrum of this simple aromatic amine has three peaks in the ratio 1:2:2, which must correspond to 3H:6H:6H. The 6.38 ppm signal clearly belongs to the protons round the benzene ring, but why are they at 6.38 and not at around 7.2 ppm? We must also distinguish the two methyl groups at 2.28 ppm from those at 2.89 ppm. The chart on p. 276 suggests that these should both be at about 2.4 ppm, close enough to 2.28 ppm but not to 2.89 ppm. The solution to both these puzzles is the distribution of electrons in the aromatic ring. Nitrogen feeds electrons into the π system, making it electron rich: the ring protons are more shielded and the nitrogen atom becomes positively charged and its methyl groups more deshielded. The peak at 2.89 ppm must belong to the NMe_2 group.

nitrogen's lone pair in p orbital

delocalization of lone pair

N becomes more electron deficient (more deshielding)

ring becomes more electron rich (more shielding)

■ Why should you usually expect to see *three* types of protons for a monosubstituted phenyl ring?

Other groups, such as simple alkyl groups, hardly perturb the aromatic system at all and it is quite common for all five protons in an alkyl benzene to appear as one signal instead of the three we might expect. Here is an example with some non-aromatic protons too: there is another on p. 275—the Cbz-protected amino acid.

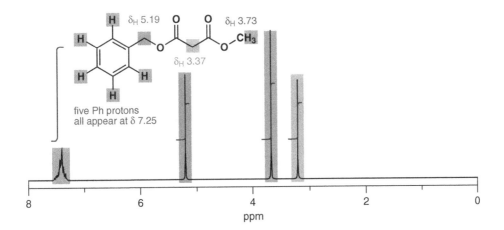

The five protons on the aromatic ring all have the same chemical shift. Check that you can assign the rest. The OCH_3 group (green) is typical of a methyl ester (the chart on p. 276 suggests 3.9 ppm). One CH_2 group (yellow) is between two carbonyl groups (compare 3.35 ppm for the similar CH_2 group on p. 277). The other (red) is next to an ester and a benzene ring: we calculate $1.3 + 1.5 + 3.0 = 5.8$ ppm for that—reasonably close to the observed 5.19 ppm. Notice how the Ph and the O together act to shift the Hs attached to this sp^3 C downfield into what we usually expect to be the alkene region. Don't interpret the regions on p. 272 too rigidly!

How electron donation and withdrawal change chemical shifts

We can get an idea of the effect of electron distribution by looking at a series of benzene rings with the same substituent in the 1 and 4 positions. This pattern makes all four hydrogens on the ring identical. Here are a few compounds listed in order of chemical shift: largest shift (lowest field; most deshielded) first. Conjugation is shown by the usual curly arrows, and inductive effects by a straight arrow by the side of the group. Only one hydrogen atom and one set of arrows are shown.

> Conjugation, as discussed in Chapter 7, is felt through π bonds, while inductive effects are the result of electron withdrawal or donation felt simply by polarization of the σ bonds of the molecule. See p. 135.

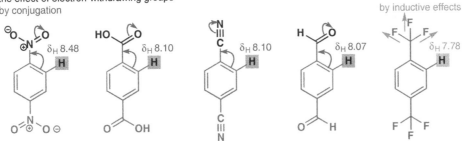

The largest shifts come from groups that withdraw electrons by conjugation. Nitro is the most powerful—this should not surprise you as we saw the same in non-aromatic compounds in both ^{13}C and ^1H NMR spectra. Then come the carbonyl and nitrile group followed by groups showing simple inductive withdrawal. CF_3 is an important example of this kind of group— three fluorine atoms combine to exert a powerful effect.

➡ This all has very important consequences for the reactivity of differently substituted benzene rings: their reactions will be discussed in Chapter 21.

In the middle of our sequence, around the position of benzene itself at 7.27 ppm, come the halogens, whose inductive electron withdrawal and lone pair donation are nearly balanced.

balance between withdrawal by inductive effect and donation of lone pairs by conjugation

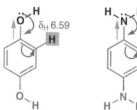

Alkyl groups are weak inductive donators, but the groups which give the most shielding—perhaps surprisingly—are those containing the electronegative atoms O and N. Despite being inductively electron withdrawing (the C–O and C–N σ bonds are polarized with δ + C), on balance conjugation of their lone pairs with the ring (as you saw on p. 278) makes them net electron donors. They *increase* the shielding at the ring hydrogens. Amino groups are the best. Note that one nitrogen-based functional group (NO_2) is the best electron withdrawer while another (NH_2) is the best electron donor.

the effect of electron-donating groups

by inductive effect

balance between withdrawal by inductive effect and donation of lone pairs by conjugation—electron donation wins

As far as the donors with lone pairs are concerned (the halogens plus O and N), two factors are important—the size of the lone pairs and the electronegativity of the element. If we look at the four halides at the top of this page the lone pairs are in 2p (F), 3p (Cl), 4p (Br), and 5p (I) orbitals. In all cases the orbitals on the benzene ring are 2p so the fluorine orbital is of the right size to interact well and the others too large. Even though fluorine is the most electronegative, it is still the best donor. The others don't pull so much electron density away, but they can't give so much back either.

If we compare the first row of the p block elements—F, OH, and NH_2—all have lone pairs in 2p orbitals so now electronegativity is the only variable. As you would expect, the most electronegative element, F, is now the weakest donor.

Electron-rich and electron-deficient alkenes

The same sort of thing happens with alkenes. We'll concentrate on cyclohexene so as to make a good comparison with benzene. The six identical protons of benzene resonate at 7.27 ppm; the two identical alkene protons of cyclohexene resonate at 5.68 ppm. A conjugating and electron-withdrawing group such as a ketone removes electrons from the double bond as expected—but unequally. The proton nearer the C=O group is only slightly downfield from cyclohexene but the more distant one is over 1 ppm downfield. The curly arrows show the electron distribution, *which we can deduce* from the NMR spectrum.

Oxygen as a conjugating electron donor is even more dramatic. It shifts the proton next to it downfield by the inductive effect but pushes the more distant proton upfield a whole 1 ppm by donating electrons. The separation between the two protons is nearly 2 ppm.

For both types of substituent, the effects are more marked on the more distant (β) proton. If these shifts reflect the true electron distribution, we should be able to deduce something about the chemistry of the following three compounds. You might expect that nucleophiles will attack the electron-deficient site in the nitroalkene, while electrophiles will be attacked by the electron-rich sites in silyl enol ethers and enamines. These are all important reagents and do indeed react as we predict, as you will see in later chapters. Look at the difference—there are nearly 3 ppm between the shifts of the same proton on the nitro compound and the enamine!

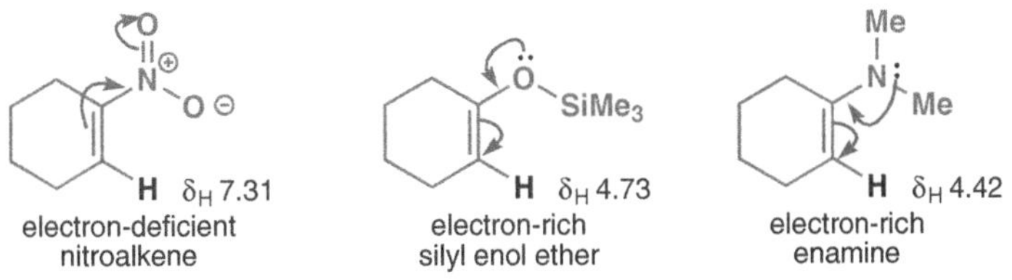

Structural information from the alkene region

Alkene protons on different carbon atoms can obviously be different if the carbon atoms themselves are different and we have just seen examples of that. Alkene protons can also be different if they are on the same carbon atom. All that is necessary is that the substituents at the other end of the double bond should themselves be different. The silyl enol ether and the unsaturated ester below both fit into this category. The protons on the double bond must be different, because each is *cis* to a different group. We may not be able to assign which is which, but the difference alone tells us something. The third compound is an interesting case: the different shifts of the two protons on the ring prove that the N–Cl bond is at an angle to the C=N bond. If it were in line, the two hydrogens would be identical. The other side of the C=N bond is occupied by a lone pair and the nitrogen atom is trigonal (sp^2 hybridized).

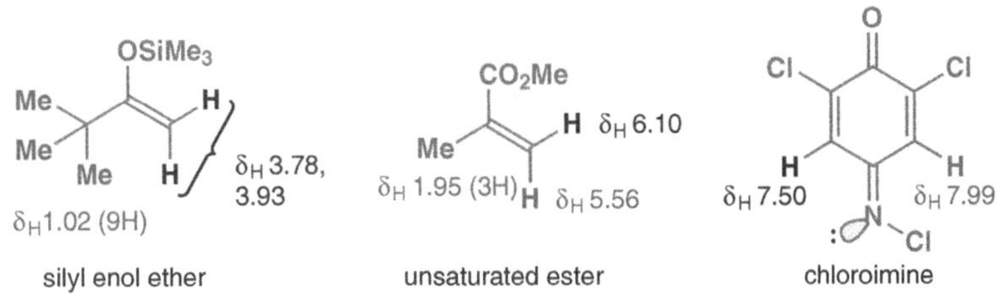

The aldehyde region: unsaturated carbon bonded to oxygen

The aldehyde proton is unique. It is directly attached to a carbonyl group—one of the most electron-withdrawing groups that exists—and is very deshielded, resonating with the largest shifts of any CH protons, in the 9–10 ppm region. The examples below are all compounds that we have met before. Two are just simple aldehydes—aromatic and aliphatic. The third is the solvent DMF. Its CHO proton is less deshielded than most—the amide delocalization that feeds electrons into the carbonyl group provides some extra shielding.

■ *Aliphatic* is a catch-all term for compounds that are not aromatic.

243

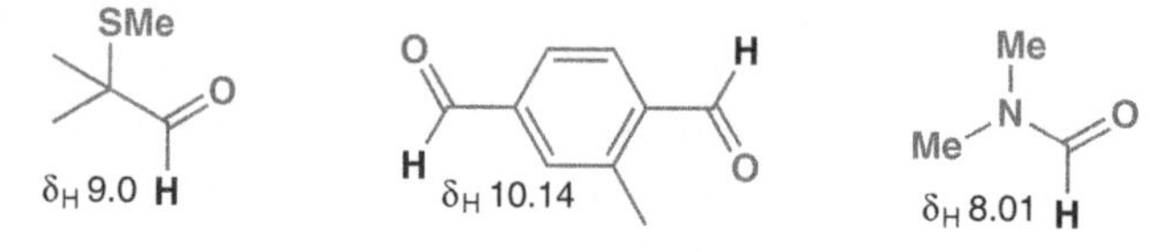

δ_H 9.0 **H**

an alphatic aldehyde

δ_H 10.14

an aromatic aldehyde

δ_H 8.01 **H**

DMF

Conjugation with an oxygen lone pair has much the same effect—formate esters resonate at about 8 ppm—but conjugation with π bonds does not. The aromatic aldehyde above, simple conjugated aldehyde below, and myrtenal all have CHO protons in the normal region (9–10 ppm).

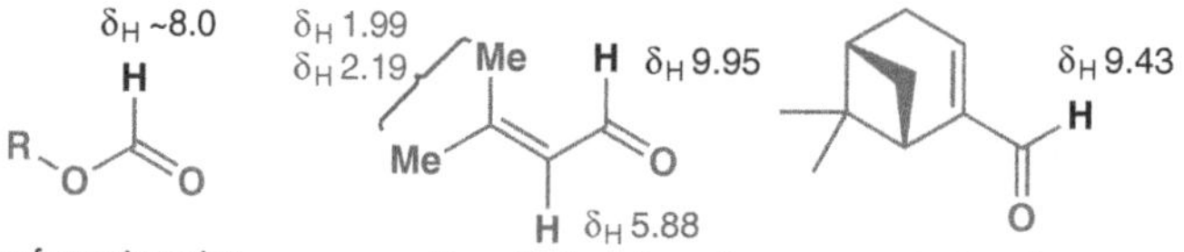

δ_H ~8.0

a formate ester

δ_H 1.99
δ_H 2.19
δ_H 9.95
H δ_H 5.88

3-methylbut-2-enal

δ_H 9.43

myrtenal

Non-aldehyde protons in the aldehyde region: pyridines

Two other types of protons resonate in the region around 9–10 ppm: some aromatic protons and some protons attached to heteroatoms like OH and NH. We will deal with NH and OH protons in the next section, but first we must look at some electron-deficient aromatic rings with distinctively large shifts.

Protons on double bonds, even very electron-deficient double bonds like those of nitro-alkenes, hardly get into the aldehyde region. However, some benzene rings with very electron-withdrawing groups do manage it because of the extra downfield shift of the ring current, so look out for nitrobenzenes as they may have signals in the 8–9 ppm region.

More important molecules with signals in this region are the aromatic heterocycles such as pyridine, which you saw functioning as a base in Chapters 8 and 10. The NMR shifts clearly show that pyridine is aromatic: one proton is at 7.1 ppm, essentially the same as benzene, but the others are more downfield and one, at C2, is in the aldehyde region. This is not because pyridine is 'more aromatic' than benzene but because nitrogen is more electronegative than carbon. Position C2 is like an aldehyde—a proton attached to sp² C bearing a heteroatom—while C4 is electron deficient due to conjugation (the electronegative nitrogen is electron withdrawing). Isoquinoline is a pyridine and a benzene ring fused together and has a proton even further downfield at 9.1 ppm—this is an imine proton that experiences the ring current of the benzene ring.

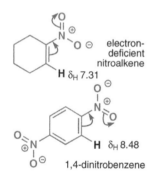

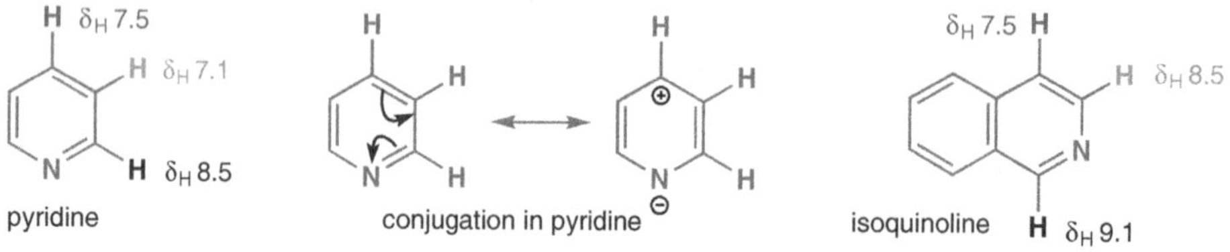

electron-deficient nitroalkene

H δ_H 7.31

H δ_H 8.48

1,4-dinitrobenzene

➡ There is more on the electron-withdrawing nature of the nitro group on p. 176.

■ Note that the alternative 'conjugation' shown in the structure below is wrong. The structure with two adjacent double bonds in a six-membered ring is impossible and, in any case, as you saw in Chapter 8, the lone pair electrons on nitrogen are in an sp² orbital orthogonal to the p orbitals in the ring. There is no interaction between orthogonal orbitals.

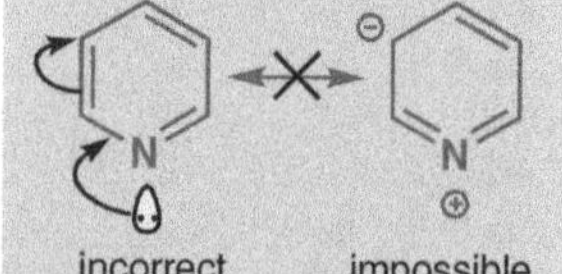

incorrect delocalization   impossible structure

H δ_H 7.5
H δ_H 7.1
H δ_H 8.5

pyridine

conjugation in pyridine

δ_H 7.5 **H**
H δ_H 8.5

isoquinoline   **H** δ_H 9.1

Protons on heteroatoms have more variable shifts than protons on carbon

Protons directly attached to O, N, or S (or any other heteroatom, but these are the most important) also have signals in the NMR spectrum. We have avoided them so far because the positions of these signals are less reliable and because they are affected by exchange.

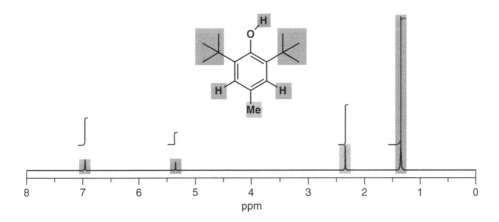

In Chapter 2 you met the antioxidant BHT. Its proton NMR is very simple, consisting of just four lines with integrals 2, 1, 3, and 18. The chemical shifts of the *tert*-butyl group (brown), the methyl group on the benzene ring (orange), and the two identical aromatic protons (green) should cause you no surprise. What is left, the 1 H signal at 5.0 ppm (pink), must be the OH. Earlier on in this chapter we saw the spectrum of acetic acid, CH_3CO_2H, which showed an OH resonance at 11.2 ppm. Simple alcohols such as *tert*-butanol have OH signals in $CDCl_3$ (the usual NMR solvent) at around 2 ppm. Why such big differences?

This is a matter of acidity. The more acidic a proton is—that is, the more easily it can escape as H^+ (this is the definition of acidity from Chapter 8)—the more the OH bond is polarized towards oxygen. The more the RO–H bond is polarized, the closer we are to free H^+, which would have no shielding electrons at all, and so the further the proton goes downfield. The OH chemical shifts and the acidity of the OH group are—to a rough extent at least—related.

Thiols (RSH) behave in a similar way to alcohols but are not so deshielded, as you would expect from the smaller electronegativity of sulfur (phenols are all about 5.0 ppm, PhSH is at 3.41 ppm). Alkane thiols appear at about 2 ppm and arylthiols at about 4 ppm. Amines and amides show a big variation, as you would expect for the variety of functional groups involved, and are summarized below. Amides are slightly acidic, as you saw in Chapter 8, and amide protons resonate at quite low fields. Pyrroles are special—the aromaticity of the ring makes the NH proton unusually acidic—and they appear at about 10 ppm.

	ROH[a]	ArOH[b]	RCO_2H[c]
pK_a	16	10	5
δ_H (OH), ppm	2.0	5.0	>10

[a]alcohol [b]phenol [c]carboxylic acid

chemical shifts of NH protons

Alkyl—NH₂	Aryl—NH₂			
δ_{NH} ~ 3	δ_{NH} ~6	δ_{NH} ~5	δ_{NH} ~7	δ_{NH} ~10	δ_{NH} ~10
amines		amides		pyrrole

Exchange of acidic protons is revealed in proton NMR spectra

Compounds with very polar groups often dissolve best in water. NMR spectra are usually run in $CDCl_3$, but heavy water, D_2O, is an excellent NMR solvent. Here are some results in that medium.

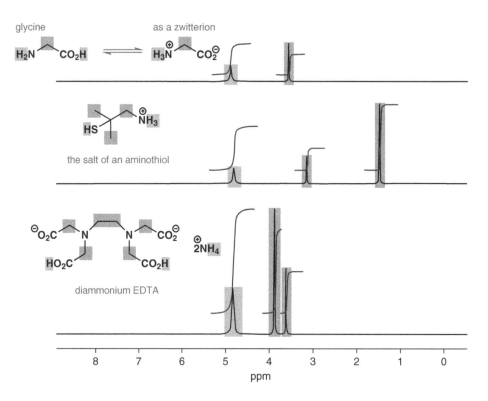

Glycine is expected to exist as a zwitterion (Chapter 8, p. 167). It has a 2H signal (green) for the CH_2 between the two functional groups, which would do for either form. The 3H signal at 4.90 ppm (orange) might suggest the NH_3^+ group, but wait a moment before making up your mind.

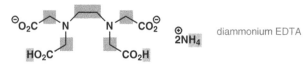

The aminothiol salt has the CMe_2 and CH_2 groups about where we would expect them (brown and green), but the SH and NH_3^+ protons appear as one 4H signal.

■ EDTA is ethylenediamine tetraacetic acid, an important complexing agent for metals. This is the salt formed with just two equivalents of ammonia.

The double salt of EDTA has several curious features. The two (green) CH_2 groups in the middle are fine, but the other four CH_2 (brown) groups all appear identical, as do all the protons on both the CO_2H and NH_3^+ groups.

The best clue to why this is so comes from the strange coincidence of the chemical shifts of the OH, NH, and SH protons in these molecules. They are all the same within experimental error: 4.90 ppm for glycine, 4.80 ppm for the aminothiol, and 4.84 ppm for EDTA. In fact all correspond to the same species: HOD, or monodeuterated water. Exchange between XH (where X=O, N, or S) protons is extremely fast, and the solvent, D_2O, supplies a vast excess of exchangeable *deuteriums*. These immediately replace all the OH, NH, and SH protons in the molecules with D, forming HOD in the process. Recall that we do not see signals for deuterium atoms (that's why deuterated solvents are used). They have their own spectra at a different frequency.

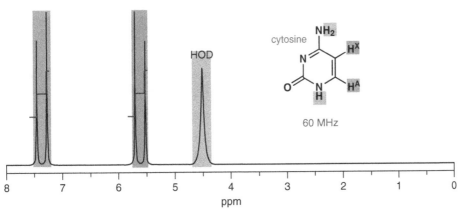

The same sort of exchange between OH or NH protons with each other or with traces of water in the sample means that the OH and NH peaks in most spectra in CDCl$_3$ are rather broader than the peaks for CH protons.

Two questions remain. First, can we tell whether glycine is a zwitterion in water or not? Not really: the spectra fit either or an equilibrium between both—other evidence leads us to expect the zwitterion in water. Second, why are all four CH$_2$CO groups in EDTA the same? This we can answer. As well as the equilibrium exchanging the CO$_2$H protons with the solvent, there will be an equally fast equilibrium exchanging protons between CO$_2$D and CO$_2^-$. This makes all four 'arms' of EDTA the same.

You should leave this section with an important chemical principle firmly established in your mind.

> ● **Proton exchange between heteroatoms is fast**
> Proton exchange between heteroatoms, particularly O, N, and S, is a very fast process in comparison with other chemical reactions, and often leads to averaged peaks in the ¹H NMR spectrum.

➡ We mentioned this fact before in the context of the mechanism of addition to a C=O group (p. 136), and we will continue to explore its mechanistic consequences throughout this book.

Coupling in the proton NMR spectrum

Nearby hydrogen nuclei interact and give multiple peaks

So far proton NMR has been not unlike carbon NMR on a smaller scale. However, we have yet to discuss the real strength of proton NMR, something more important than chemical shifts and something that allows us to look not just at individual atoms but also at the way the C–H skeleton is joined together. This is the result of the interaction between nearby protons, known as *coupling*.

An example we could have chosen in the last section is the nucleic acid component cytosine, which has exchanging NH$_2$ and NH protons giving a peak for HOD at 4.5 ppm. We didn't choose this example because the other two peaks would have puzzled you. Instead of giving just one line for each proton, they give two lines each—doublets as you will learn to call them—and it is time to discuss the origin of this 'coupling'.

■ Cytosine is one of the four bases that, in combination with deoxyribose and phosphate, make up DNA. It is a member of the class of heterocycles called pyrimidines. We come back to the chemistry of DNA towards the end of this book, in Chapter 42.

You might have expected a spectrum like that of the heterocycle below, which like cytosine is also a pyrimidine. It too has exchanging NH_2 protons and two protons on the heterocyclic ring. But these two protons give the expected two lines instead of the four lines in the cytosine spectrum. It is easy to assign the spectrum: the green proton labelled H^A is attached to an aldehyde-like C=N and so comes at lowest field. The red proton labelled H^X is *ortho* to two electron-donating NH_2 groups and so comes at high field for an aromatic proton (p. 272). These protons do not couple with each other because they are too far apart. They are separated by five bonds whereas the ring protons in cytosine are separated by just three bonds.

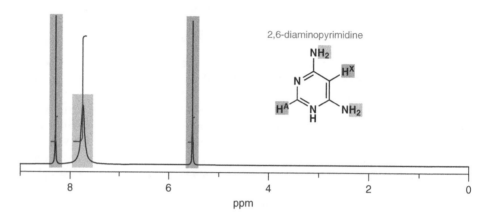

Understanding this phenomenon is so important that we are going to explain it in three different ways—you choose which appeals to you most. Each method offers a different insight.

The diaminopyrimidine spectrum you have just seen has two single lines (*singlets* we shall call them from now on) because each proton, H^A or H^X, can be aligned either with or against the applied magnetic field. The cytosine spectrum is different because each proton, say H^A, is near enough to experience the small magnetic field of the other proton H^X as well as the field of the magnet itself. The diagram shows the result.

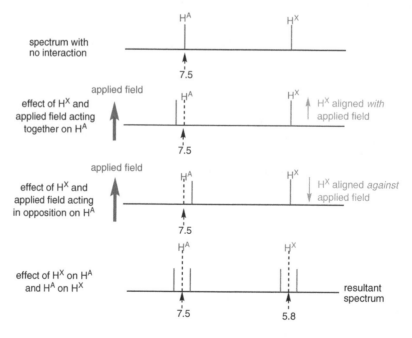

If each proton interacted only with the applied field we would get two singlets. But proton H^A actually experiences two slightly different fields: the applied field *plus* the field of

H^X or the applied field *minus* the field of H^X. H^X acts either to increase or decrease the field experienced by H^A. The position of a resonance depends on the field experienced by the proton so these two situations give rise to two slightly different peaks—a *doublet* as we shall call it. And whatever happens to H^A happens to H^X as well, so the spectrum has two doublets, one for each proton. Each couples with the other. The field of a proton is a very small indeed in comparison with the field of the magnet and the separation between the lines of a doublet is very small. We shall discuss the size of the coupling later (pp. 294–300).

The second explanation takes into account the energy levels of the nucleus. In Chapter 4, when we discussed chemical bonds, we imagined electronic energy levels on neighbouring atoms interacting with each other and splitting to produce new molecular energy levels, some higher in energy and some lower in energy than the original atomic energy levels. When hydrogen *nuclei* are near each other in a molecule, the nuclear energy levels also interact and split to produce new energy levels. If a single hydrogen nucleus interacts with a magnetic field, we have the picture on p. 270 of this chapter: there are *two* energy levels as the nucleus can be aligned with or against the applied magnetic field, there is one energy jump possible, and there is a resonance at one frequency. This you have now seen many times and it can be summarized as shown below.

energy levels for one isolated nucleus

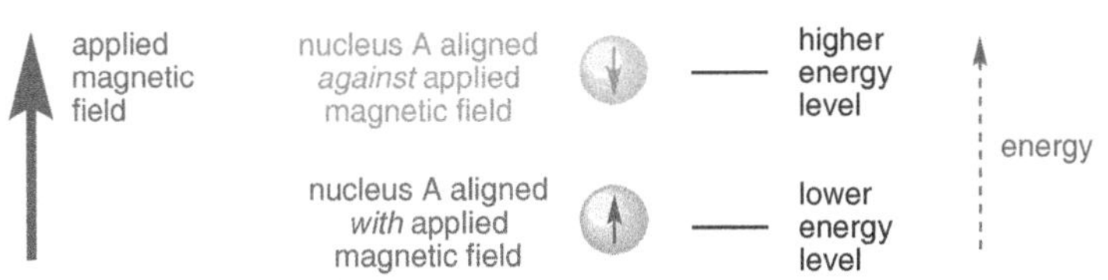

The spectrum of the pyrimidine on p. 286 shows exactly this situation: two protons well separated in the molecules and each behaving independently. Each has two energy levels, each gives a singlet, and there are two lines in the spectrum. But in cytosine, whose spectrum is shown on p. 285, the situation is different: each hydrogen atom has another hydrogen nucleus nearby and there are now *four* energy levels. Each nucleus H^A and H^X can be aligned with or against the applied field. There is one (lower) energy level where they are both aligned with the field and one (higher) level where they are both aligned against. In between there are two different energy levels in which one nucleus is aligned with the field and one against. Exciting H from alignment with to alignment against the applied field can be done in two slightly different ways, shown as A_1 and A_2 on the diagram. The result is two resonances very close together in the spectrum.

energy levels for two interacting nuclei

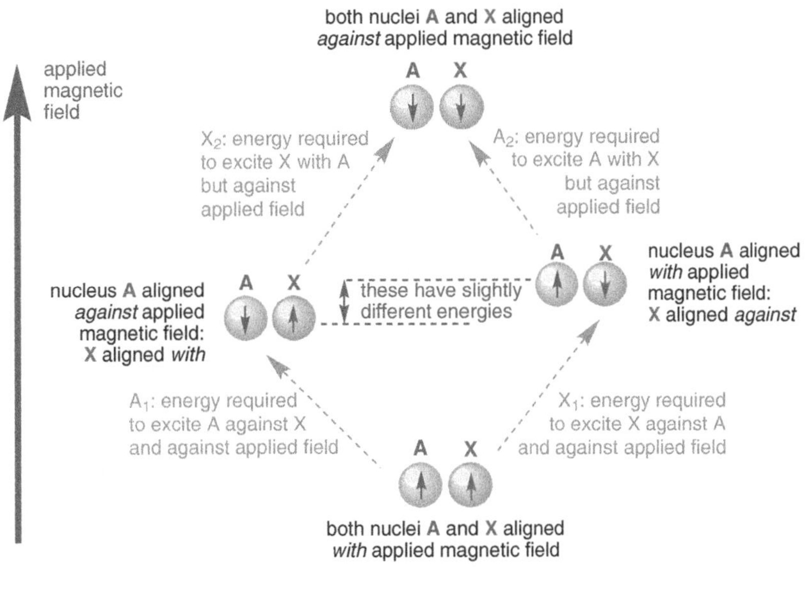

Please notice carefully that we cannot have this discussion about H^A without discussing H^X in the same way. If there are two slightly different energy jumps to excite H^A, there must also be two slightly different energy jumps to excite H^X. A_1, A_2, X_1, and X_2 are all different, but the *difference* between A_1 and A_2 is exactly the same as the *difference* between X_1 and X_2. Each proton now gives two lines (a doublet) in the NMR spectrum and the splitting of the two doublets is *exactly the same*. We describe this situation as coupling. We say 'A and X are coupled' or 'X is coupled to A' (and vice versa, of course). We shall be using this language from now on and so must you.

Now look back at the spectrum of cytosine at the beginning of this section. You can see the two doublets, one for each of the protons on the aromatic ring. Each is split by the same amount (this is easy to check with a ruler). The separation of the lines is the **coupling constant** and is called *J*. In this case *J* = 4 Hz. Why do we measure *J* in hertz and not in ppm? We pointed out on p. 55 (Chapter 3) that we measure chemical shifts in ppm because we get the same number regardless of the rating of the NMR machine in MHz. We measure *J* in Hz because we also get the same number regardless of the machine.

The spectra below show ¹H NMR spectra of the same compound run on two different NMR machines—one a 90 MHz spectrometer and one a 300 MHz spectrometer (these are at the lower and upper ends of the range of field strengths in common use). Notice that the peaks stay in the same place on the chemical shift scale (ppm) but the size of the coupling appears to change because 1 ppm is worth 90 Hz in the top spectrum but 300 Hz in the bottom.

■ Spectrometers in common use typically have field strengths of 200–500 MHz.

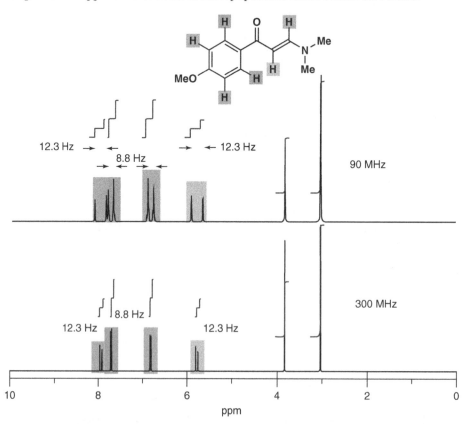

Measuring coupling constants in hertz

To measure a coupling constant it is essential to know the rating of the NMR machine in MHz (megahertz). This is why you are told that each illustrated spectrum is, say, a '400 MHz ¹H NMR spectrum'. Couplings may be marked on the spectrum, electronically, but if not then to measure the coupling, measure the distance between the lines by ruler or dividers and use the horizontal scale to find out the separation in ppm. The conversion is then easy—to turn parts per million of megahertz into hertz you just leave out the million! So 1 ppm on a 300 MHz machine is 300 Hz. On a 500 MHz machine, a 10 Hz coupling is a splitting of 0.02 ppm.

When you change from one machine to another, say, from a 200 MHz to a 500 MHz NMR machine, chemical shifts (δ) stay the same in ppm and coupling constants (J) stay the same in Hz.

Now for the third way to describe coupling. If you look again at what the spectrum would be like without interaction between H^A and H^X you will see the pattern on the right, with the chemical shift of each proton clearly obvious.

But you don't see this because each proton couples with the other and splits its signal by an equal amount either side of the true chemical shift. The true spectrum has a pair of doublets each split by an identical amount. Note that no line appears at the true chemical shift, but it is easy to measure the chemical shift by taking the midpoint of the doublet.

So this spectrum would be described as $δ_H$ 7.5 (1H, d, J 4 Hz, H^A) and 5.8 (1H, d, J 4 Hz, H^X). The main number gives the chemical shift in ppm and then, in brackets, comes the integration as the number of Hs, the shape of the signal (here 'd' for doublet), the size of coupling constants in Hz, and the assignment, usually related to a diagram. The integration refers to the combined area under both peaks in the doublet. If the doublet is exactly symmetrical, each peak integrates to half a proton. The combined signal, however complicated, integrates to the right number of protons.

We have described these protons as A and X with a purpose in mind. A spectrum of two equal doublets is called an AX spectrum. A is always the proton you are discussing and X is another proton with a different chemical shift. The alphabet is used as a ruler: nearby protons (on the chemical shift scale—not necessarily nearby in the structure!) are called B, C, etc. and distant ones are called X, Y, etc. You will see the reason for this soon.

If there are more protons involved, the splitting process continues. Here is the NMR spectrum of a famous perfumery compound supposed to have the smell of 'green leaf lilac'. The compound is an acetal with five nearly identical aromatic protons at the normal benzene position (7.2–7.3 ppm) and six protons on two identical OMe groups.

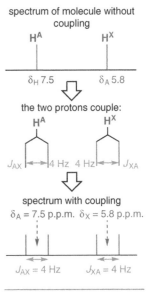

spectrum of molecule without coupling

$δ_H$ 7.5 $δ_A$ 5.8

the two protons couple:

H^A H^X

J_{AX} ◄──►4 Hz 4 Hz◄──► J_{XA}

spectrum with coupling

$δ_A$ = 7.5 p.p.m. $δ_X$ = 5.8 p.p.m.

J_{AX} = 4 Hz J_{XA} = 4 Hz

➡ We introduced integrals in ^1H NMR spectra on p. 270.

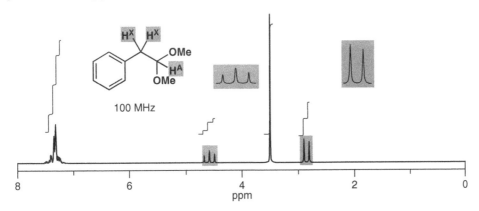

100 MHz

It is the remaining three protons that interest us. They appear as a 2H doublet at 2.9 ppm and a 1H *triplet* at 4.6 ppm. In NMR talk, triplet means three equally spaced lines in the ratio 1:2:1. The triplet arises from the three possible states of the two identical protons in the CH_2 group.

If one proton H^A interacts with two protons H^X, it can experience protons H^X in three different possible states. Both protons H^X can be aligned with the magnet or both against. These states will increase or decrease the applied field just as before. But if one proton H^X is aligned with and one against the applied field, there is no net change to the field experienced by H^A. There are two arrangements for this (see diagram overleaf). We'll therefore see a signal of double intensity for H^A at the correct chemical shift, one signal at higher field and one at lower field. In other words, a 1:2:1 triplet.

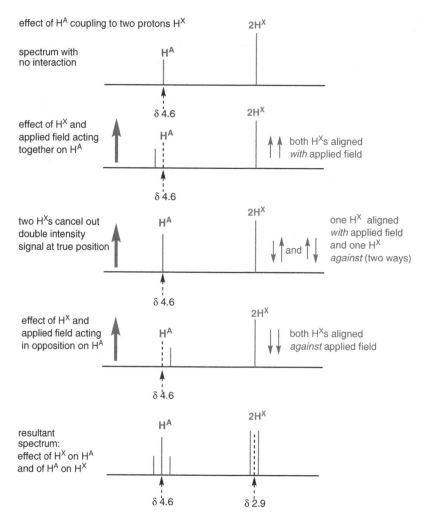

We could look at this result by our other methods too. There is one way in which both nuclei can be aligned with and one way in which both can be aligned against the applied field, but two ways in which they can be aligned one with and one against. Proton H^A interacts with each of these states. The result is a 1:2:1 triplet.

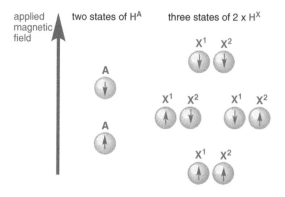

Using our third way of seeing coupling to see how the triplet arises, we can just make the peaks split in successive stages:

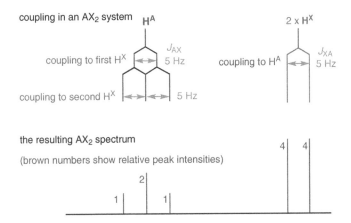

coupling in an AX_2 system

H^A

coupling to first H^X — J_{AX} 5 Hz

coupling to second H^X — 5 Hz

$2 \times H^X$

coupling to H^A — J_{XA} 5 Hz

the resulting AX_2 spectrum

(brown numbers show relative peak intensities)

If there are more protons involved, we continue to get more complex systems, but the intensities can all be deduced simply from Pascal's triangle, which gives the coefficients in a binomial expansion. If you are unfamiliar with this simple device, here it is.

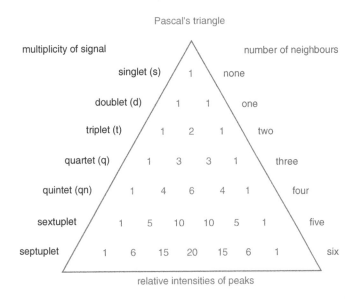

Pascal's triangle

multiplicity of signal

number of neighbours

singlet (s)						1							none
doublet (d)					1		1						one
triplet (t)				1		2		1					two
quartet (q)			1		3		3		1				three
quintet (qn)		1		4		6		4		1			four
sextuplet	1		5		10		10		5		1		five
septuplet	1	6		15		20		15		6		1	six

relative intensities of peaks

■ **Constructing Pascal's triangle**

Put '1' at the top and then add an extra number in each line by adding together the numbers on either side of the new number in the line above. If there is no number on one side, that counts as a zero, so the lines always begin and end with '1'.

You can read off from the triangle what pattern you may expect when a proton is coupled to *n* equivalent neighbours. There are always $n + 1$ peaks with the intensities shown by the triangle. So far, you've seen 1:1 doublets (line 2 of the triangle) from coupling to 1 proton, and 1:2:1 triplets (line 3) from coupling to 2. You will often meet ethyl groups (CH_3CH_2X), where the CH_2 group couples to three identical protons and appears as a 1:3:3:1 quartet and the methyl group as a 1:2:1 triplet. In isopropyl groups, $(CH_3)_2CHX$, the methyl groups appear as a 6H doublet and the CH group as a septuplet.

Here is a simple example: the four-membered cyclic ether oxetane. Its NMR spectrum has a 4H triplet for the two identical CH_2 groups next to oxygen and a 2H quintet for the CH_2 in the middle. Each proton H^X 'sees' four identical neighbours (H^A) and is split equally by them all to give a 1:4:6:4:1 quintet. Each proton H^A 'sees' two identical neighbours H^X and is split into a 1:2:1 triplet. The combined integral of all the lines in the quintet together is 2 and of all the lines in the triplet is 4.

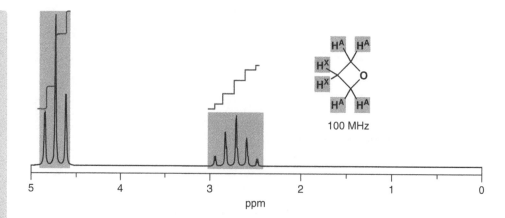

A slightly more complicated example is the diethyl acetal below. It has a simple AX pair of doublets for the two protons on the 'backbone' (red and green) and a typical ethyl group (2H quartet and 3H triplet). An ethyl group is attached to only one substituent through its CH_2 group, so the chemical shift of that CH_2 group tells us what it is joined to. Here the peak at 3.76 ppm can only be an OEt group. There are, of course, two identical CH_2 groups in this molecule.

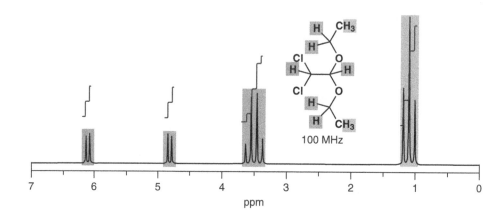

In all of these molecules, a proton may have had several neighbours, but all those neighbours have been the same. And therefore all the *coupling constants* have been the same. What happens when coupling constants differ? Chrysanthemic acid, the structural core of the insecticides produced by pyrethrum flowers, gives an example of the simplest situation—where a proton has two different neighbours.

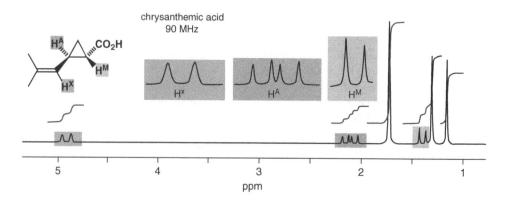

Chrysanthemic acid has a carboxylic acid, an alkene, and two methyl groups on the three-membered ring. Proton H^A has two neighbours, H^X and H^M. The coupling constant to H^X is 8 Hz, and that to H^M is 5.5 Hz. We can construct the splitting pattern as shown on the right.

The result is four lines of equal intensity called a **double doublet** (or sometimes a **doublet of doublets**), abbreviation dd. The smaller coupling constant can be read off from the separation between lines 1 and 2 or between lines 3 and 4, while the larger coupling constant is between lines 1 and 3 or between lines 2 and 4. The separation between the middle two lines is not a coupling constant. You could view a double doublet as an imperfect triplet where the second coupling is too small to bring the central lines together: alternatively, look at a triplet as a special case of a double doublet where the two couplings are identical and the two middle lines coincide.

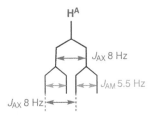

Coupling is a through-bond effect

Do neighbouring nuclei interact through space or through the electrons in the bonds? We know that coupling is in fact a 'through-bond effect' because of the way coupling constants vary with the shape of the molecule. The most important case occurs when the protons are at either end of a double bond. If the two hydrogens are *cis*, the coupling constant J is typically about 10 Hz, but if they are *trans*, J is much larger, usually 15–18 Hz. These two chloro acids are good examples.

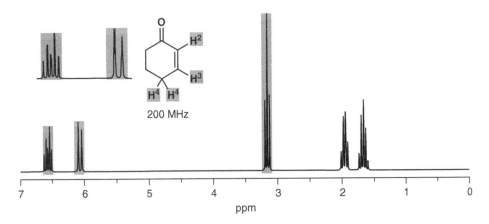

hydrogens are *trans* H atoms distant hydrogens are *cis* H atoms close
$J = 15$ Hz but bonds parallel $J = 9$ Hz but bonds not parallel

If coupling were through space, the nearer *cis* hydrogens would have the larger J. In fact, coupling occurs *through the bonds* and the more perfect parallel alignment of the bonds in the *trans* compound provides better communication and a larger J.

Coupling is at least as helpful as chemical shift in assigning spectra. When we said (p. 280) that the protons on cyclohexenone had the chemical shifts shown, how did we know? It was coupling that told us the answer. The proton next to the carbonyl group (H^2 in the diagram) has one neighbour (H^3) and appears as a doublet with $J = 11$ Hz, just right for a proton on a double bond with a *cis* neighbour. The proton H^3 itself appears as a double triplet. Inside each triplet the separation of the lines is 4 Hz and the two triplets are 11 Hz apart.

200 MHz

The coupling of H^3 is as complex as you have seen yet, but it can be represented diagrammatically by the same approach we have taken before.

Abbreviations used for style of signal

Abbre-viation	Meaning	Comments
s	singlet	
d	doublet	equal in height
t	triplet	should be 1:2:1
q	quartet	should be 1:3:3:1
dt	double triplet	other combinations too, such as dd, dq, tt
m	multiplet	a signal too complicated to resolve*

* Either because it contains a complex coupling pattern or because the signals from different protons overlap.

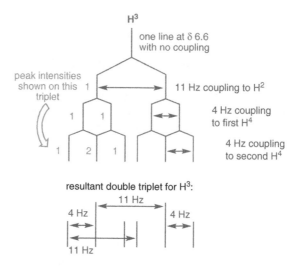

As coupling gets more and more complicated it can be hard to interpret the results, but *if you know what you are looking for* things do become easier. Here is the example of heptan-2-one. The green protons next to the carbonyl group are a 2H triplet (coupled to the two red protons) with *J* 7 Hz. The red protons themselves are next to four protons, and although these four protons are not identical the coupling constants are about the same: the red protons therefore appear as a 2H quintet, with a coupling constant also of 7 Hz. The brown signal is more complicated: we might call it a '4H multiplet' but in fact we know what it must be: the signals for the four brown protons on carbons 5 and 6 overlap, and must be made up of a 2H quintet (protons on C5) and a 2H sextet (protons on C6). We can see the coupling of the protons on C6 with the terminal methyl group because the methyl group (orange) is a 3H triplet (also with a 7 Hz coupling constant).

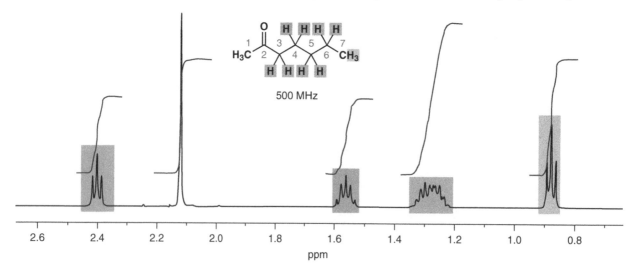

Coupling constants depend on three factors

The coupling constants in cyclohexenone were different, but all the coupling constants in heptanone are about the same—around 7 Hz. Why?

- **Factors affecting coupling constants**
  - **Through-bond distance between the protons.**
  - **Angle between the two C–H bonds.**
  - **Electronegative substituents.**

The coupling constants we have seen so far have all been between hydrogen atoms on neighbouring carbon atoms—in other words, the coupling is through three bonds (H–C–C–H) and is designated $^3J_{HH}$. These coupling constants $^3J_{HH}$ are usually about 7 Hz in an open-chain, freely rotating system such as we have in heptanone. The C–H bonds vary little in length but in cyclohexenone the C–C bond is a double bond, significantly shorter than a single bond. Couplings ($^3J_{HH}$) across double bonds are usually larger than 7 Hz (11 Hz in cyclohexenone). $^3J_{HH}$ couplings are called *vicinal couplings* because the protons concerned are on neighbouring carbon atoms.

Something else is different too: in an open-chain system we have a time average of all rotational conformations (we will look at this in the next chapter). But across a double bond there is no rotation and the angle between the two C–H bonds is fixed: they are always in the same plane. In the plane of the alkene, the C–H bonds are either at 60° (*cis*) or 180° (*trans*) to each other. Coupling constants in benzene rings are slightly less than those across *cis* alkenes because the bond is longer (bond order 1.5 rather than 2).

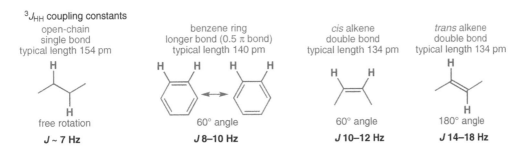

$^3J_{HH}$ coupling constants

| open-chain single bond typical length 154 pm | benzene ring longer bond (0.5 π bond) typical length 140 pm | *cis* alkene double bond typical length 134 pm | *trans* alkene double bond typical length 134 pm |

free rotation — **J ~ 7 Hz**   60° angle — **J 8–10 Hz**   60° angle — **J 10–12 Hz**   180° angle — **J 14–18 Hz**

In naphthalenes, there are unequal bond lengths around the two rings. The bond between the two rings is the shortest, and the lengths of the others are shown. Coupling across the shorter bond (8 Hz) is significantly stronger than coupling across the longer bond (6.5 Hz).

The effect of the third factor, electronegativity, is easily seen in the comparison between ordinary alkenes and alkenes with alkoxy substituents, known as enol ethers. We are going to compare two pairs of compounds with a *cis* or a *trans* double bond. One pair has a phenyl group at one end of the alkene and the other has an OPh group. For either pair, the *trans* coupling is larger than the *cis*, as you would now expect. But if you compare the two pairs, the enol ethers have much smaller coupling constants. The *trans* coupling for the enol ethers is only just larger than the *cis* coupling for the alkenes. The electronegative oxygen atom is withdrawing electrons from the C–H bond in the enol ethers and weakening communication through the bonds.

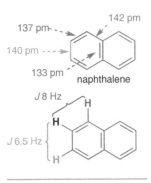

137 pm
142 pm
140 pm
133 pm
naphthalene

J 8 Hz
J 6.5 Hz

➡ Conjugation in naphthalene was discussed in Chapter 7, p. 161.

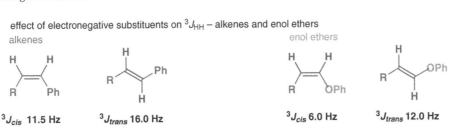

effect of electronegative substituents on $^3J_{HH}$ – alkenes and enol ethers

alkenes   enol ethers

$^3J_{cis}$ **11.5 Hz** $^3J_{trans}$ **16.0 Hz** $^3J_{cis}$ **6.0 Hz** $^3J_{trans}$ **12.0 Hz**

Long-range coupling

When the through-bond distance gets longer than three bonds, coupling is not usually seen. To put it another way, four-bond coupling $^4J_{HH}$ is usually zero. However, it is seen in some special cases, the most important being *meta* coupling in aromatic rings and allylic coupling in alkenes. In both, the orbitals between the two hydrogen atoms can line up in a zig-zag

meta coupling

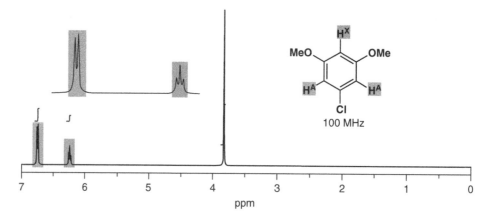

allylic coupling

$0 < {}^4J_{HH} < 3\ Hz$

fashion to maximize interaction. This arrangement looks rather like a letter 'W' and this sort of coupling is called W-coupling. Even with this advantage, values of ${}^4J_{HH}$ are usually small, about 1–3 Hz.

Meta coupling is very common when there is *ortho* coupling as well, but here is an example where there is no *ortho* coupling because none of the aromatic protons have immediate neighbours—the only coupling is *meta* coupling. There are two identical H^As, which have one *meta* neighbour and appear as a 2H doublet. Proton H^X between the two MeO groups has two identical *meta* neighbours and so appears as a 1H triplet. The coupling is small ($J \sim 2.5$ Hz).

We have already seen a molecule with allylic coupling. We discussed in some detail why cyclohexenone has a double triplet for H³. But it also has a less obvious double triplet for H². The triplet coupling is less obvious since *J* is small (about 2 Hz) because it is ${}^4J_{HH}$—allylic coupling to the CH₂ group at C4. Here is a diagram of the coupling, which you would be able to spot in an expansion of the cyclohexenone spectrum on p. 293.

Coupling between similar protons

Identical protons do not couple with each other. The three protons in a methyl group may couple to some other protons, but *never* couple with each other. They are an A₃ system. Identical neighbours do not couple either. Turn back to p. 271 and you'll see that even though each of the four protons on the *para*-disubstituted benzenes has one neighbour, they appear as one singlet because every proton is identical to its neighbour.

We have also seen how two different protons forming an AX system give two separate doublets. Now we need to see what happens to protons in between these two extremes. What happens to two similar neighbours? As two protons get closer and closer together, do the two doublets you see in the AX system suddenly collapse to the singlet of the A₂ system? You have probably guessed that they do not. The transition is gradual. Suppose we have two different neighbours on an aromatic ring. The spectra below show what we see. These are all 1,4-disubstituted benzene rings with different groups at the 1 and 4 positions.

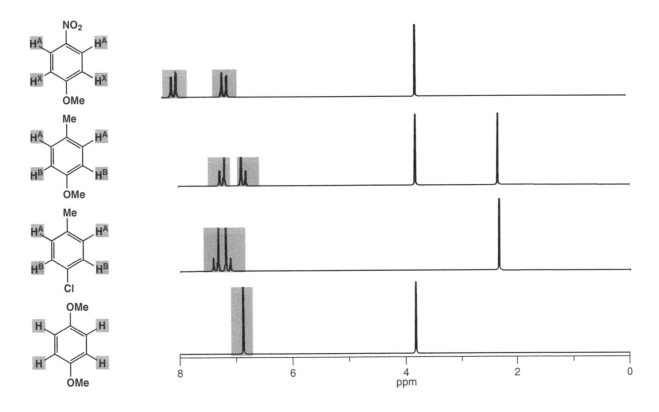

You'll notice that when the two doublets are far apart, as in the first spectrum, they look like normal doublets. But as they get closer together the doublets get more and more distorted, until finally they are identical and collapse to a 4H singlet.

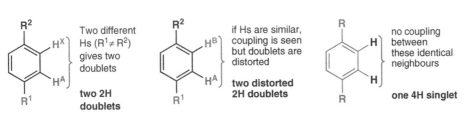

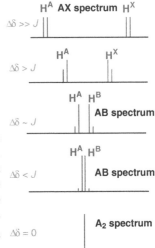

The critical factor in the shape of the peak is how the difference between the chemical shifts of the two protons ($\Delta\delta$) compares with the size of the coupling constant (J) for the machine in question. If $\Delta\delta$ is much larger than J there is no distortion: if, say, $\Delta\delta$ is 2 ppm at 500 MHz (= 1000 Hz) and the coupling constant is a normal 7 Hz, then this condition is fulfilled and we have an AX spectrum of two 1:1 doublets. As $\Delta\delta$ approaches J in size, so the inner lines of the two doublets increase and the outer lines decrease until, when $\Delta\delta$ is zero, the outer lines vanish away altogether and we are left with the two superimposed inner lines—a singlet or an A_2 spectrum. You can see this progression in the diagram on the right.

We call the last stages, where the distortion is great but the protons are still different, an AB spectrum because you cannot really talk about H^A without also talking about H^B. The two inner lines may be closer than the gap between the doublets or the four lines may all be equally spaced. Two versions of an AB spectrum are shown in the diagram—there are many more variations.

It is a generally useful tip that a distorted doublet 'points' towards the protons with which it is coupled.

Or, to put it another way, the AB system is 'roofed' with the usual arrangement of low walls and a high middle to the roof. Look out for doublets (or any other coupled signals) of this kind.

We shall end this section with a final example illustrating *para*-disubstituted benzenes and roofing as well as an ABX system and an isopropyl group. The aromatic ring protons form a pair of distorted doublets (2H each), showing that the compound is a *para*-disubstituted benzene. Then the alkene protons form the AB part of an ABX spectrum. They are coupled to each other with a large (*trans*) *J* = 16 Hz and one is also coupled to another distant proton. The large doublets are distorted (AB) but the small doublets within the right-hand half of the AB system are equal in height. The distant proton X is part of an *i*-Pr group and is coupled to H^B and the six identical methyl protons. Both *J*s are nearly the same so it is split by seven protons and is an octuplet. It looks like a sextuplet because the intensity ratios of the lines in an octuplet would be 1:7:21:35:35:21:7:1 (from Pascal's triangle) and it is hardly surprising that the outside lines disappear.

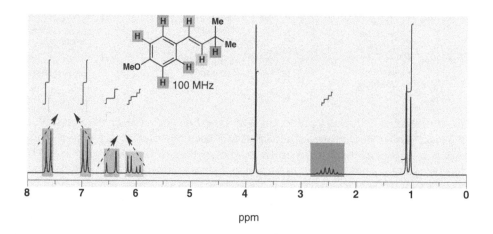

Coupling can occur between protons on the same carbon atom

We have seen cases where protons on the same carbon atom are different: compounds with an alkene unsubstituted at one end. If these protons are different (and they are certainly near to each other), then they should couple. They do, but in this case the coupling constant is usually very small. Here is the spectrum of an example you met on p. 281.

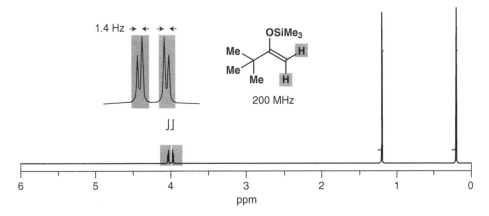

The small 1.4 Hz coupling is a $^2J_{HH}$ coupling between two protons on the same carbon that are different because there is no rotation about the double bond. $^2J_{HH}$ coupling is called *geminal coupling*.

This means that a monosubstituted alkene (a vinyl group) will have characteristic signals for each of the three protons on the double bond. Here is the example of ethyl acrylate (ethyl propenoate, a monomer for the formation of acrylic polymers). The spectrum looks rather complex at first, but it is easy to sort out using the coupling constants.

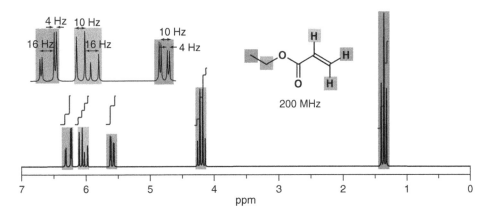

The largest J (16 Hz) is obviously between the orange and green protons (*trans* coupling), the medium J (10 Hz) is between the orange and red (*cis* coupling), and the small J (4 Hz) must be between the red and green (geminal). This assigns all the protons: red, 5.60 ppm; green, 6.40 ppm; orange, 6.11 ppm Assignments based on coupling are more reliable than those based on chemical shift alone.

● **Coupling constants in a vinyl group**

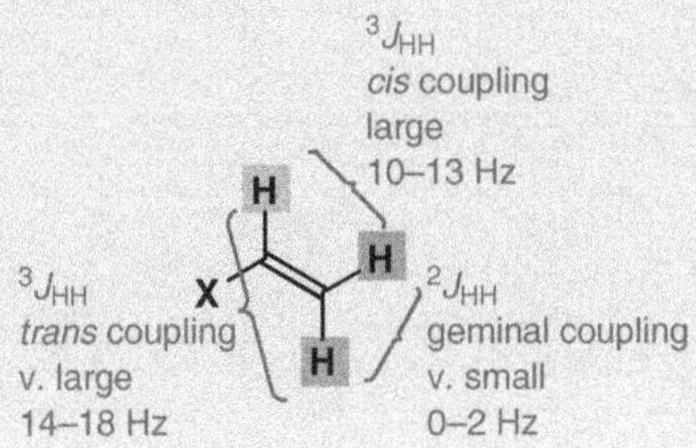

$^3J_{HH}$
cis coupling
large
10–13 Hz

$^3J_{HH}$
trans coupling
v. large
14–18 Hz

$^2J_{HH}$
geminal coupling
v. small
0–2 Hz

Ethyl vinyl ether is a reagent used for the protection of alcohols. All its coupling constants are smaller than is usual for an alkene because of the electronegativity of the oxygen atom, which is now joined directly to the double bond. It is still a simple matter to assign the protons of the vinyl group because couplings of 13, 7, and 2 Hz must be *trans*, *cis*, and geminal, respectively. In addition, the orange H is on a carbon atom next to oxygen and so goes downfield while the red and green protons have extra shielding from the conjugation of the oxygen lone pairs (see p. 281).

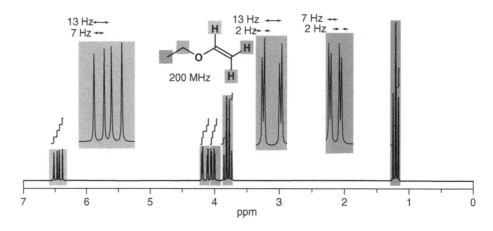

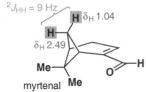

myrtenal

Geminal coupling on saturated carbons can be seen only if the hydrogens of a CH$_2$ group are different. The bridging CH$_2$ group of myrtenal (p. 274) provides an example. The coupling constant for the protons on the bridge, J_{AB}, is 9 Hz. Geminal coupling constants in a saturated system can be much larger (typically 10–16 Hz) than in an unsaturated one.

● **Typical coupling constants**

  ● **geminal** $^2J_{HH}$

  saturated   10–16 Hz

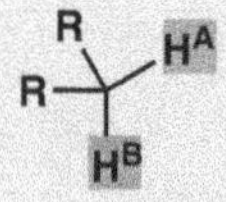

  unsaturated   0–3 Hz

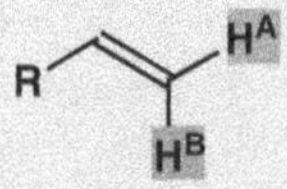

  ● **vicinal** $^3J_{HH}$

  saturated   6–8 Hz

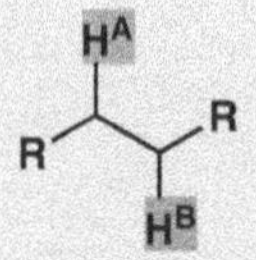

  unsaturated *trans*   14–18 Hz

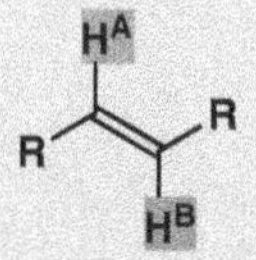

  unsaturated *cis*   10–12 Hz

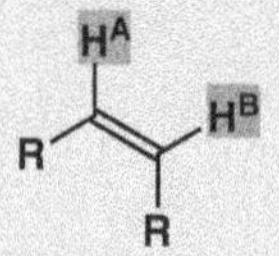

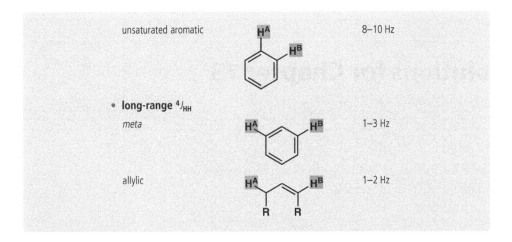

unsaturated aromatic		8–10 Hz
long-range $^4J_{HH}$		
meta		1–3 Hz
allylic		1–2 Hz

To conclude

You have now met, in Chapter 3 and this chapter, all of the most important spectroscopic techniques available for working out the structure of organic molecules. We hope you can now appreciate why proton NMR is by far the most powerful of these techniques, and we hope you will be referring back to this chapter as you read the rest of the book. We shall talk about proton NMR a lot, and specifically we will come back to it in detail in Chapter 18, where we will look at using all of the spectroscopic techniques in combination, and in Chapter 31, when we look at what NMR can tell us about the shape of molecules.

Further reading

A reminder: you will find it an advantage to have one of the short books on spectroscopic analysis to hand as they give explanations, comprehensive tables of data, and problems. We recommend *Spectroscopic Methods in Organic Chemistry* by D. H. Williams and Ian Fleming, McGraw-Hill, London, 6th edn, 2007.

A simple introduction is the Oxford Primer *Introduction to Organic Spectroscopy*, L. M. Harwood and T. D. W. Claridge, OUP, Oxford, 1996. A more advanced source of practical uses of stereochemistry is the Oxford Primer *Stereoselectivity in Organic Synthesis*, Garry Procter, OUP, Oxford, 1998.

Check your understanding

To check that you have mastered the concepts presented in this chapter, attempt the problems that are available in the book's Online Resource Centre at http://www.oxfordtextbooks.co.uk/orc/clayden2e/

Suggested solutions for Chapter 13

PROBLEM 1

How many signals will there be in the ^{1}H NMR spectrum of each of these compounds? Estimate the chemical shifts of the signals.

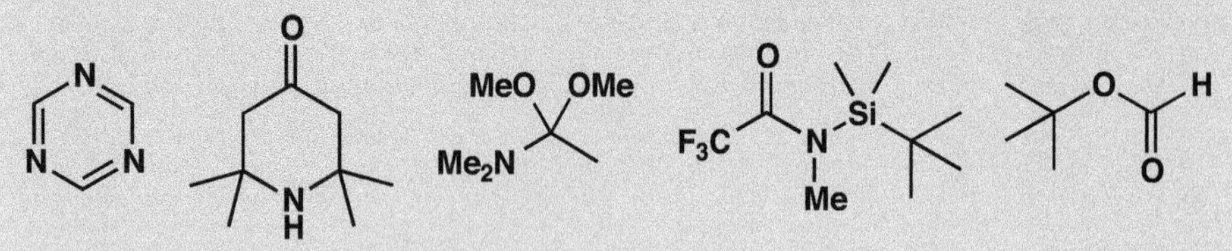

Purpose of the problem

Considering the effects of symmetry on proton (rather than carbon) NMR, and practice in estimating chemical shift.

Suggested solution

Considerations of symmetry apply equally to ^{1}H and to ^{13}C NMR. This is the answer, with different types of proton marked with different letters.

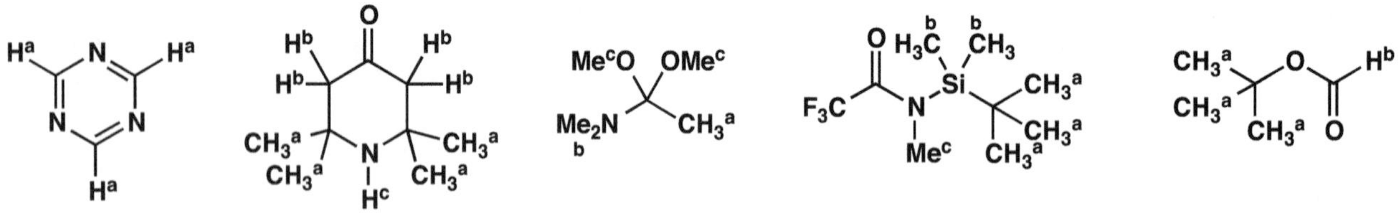

Estimating the chemical shift in ^{1}H NMR requires you to modify your experience of ^{13}C NMR to the narrower range of proton shifts and to consider that aromatic protons are in a distinct region from alkene protons. In each case we give a reasonable estimate and then the actual values. If your values agree with our estimates, you have done well. If you get something near the actual values, be very proud of yourself. The first compound has hydrogens on sp^2 carbon atoms bonded to two nitrogen atoms—hence the very large shift. The fourth molecule has two methyl groups directly bonded to electropositive silicon—hence the very small shift. The rest are more easily explained.

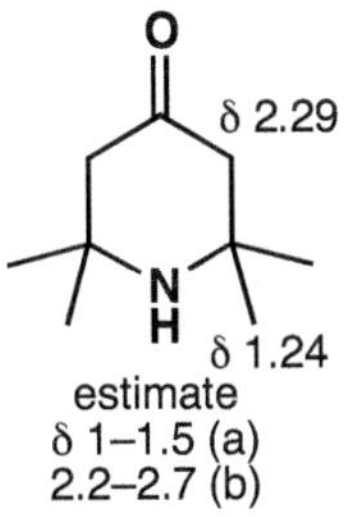

δ 3.20
MeO　OMe

Me₂N　Me
δ 2.27　δ 1.20
estimate
δ 1.0–1.5 (a)
2.2–2.7 (b)
3–3.5 (c)

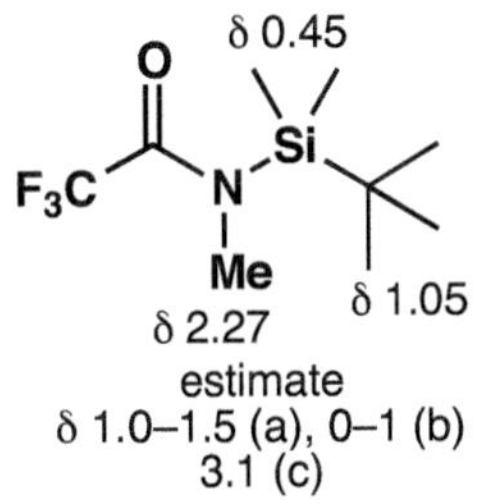

δ 0.45

F₃C

Me
δ 2.27

δ 1.05

estimate
δ 1.0–1.5 (a), 0–1 (b)
3.1 (c)

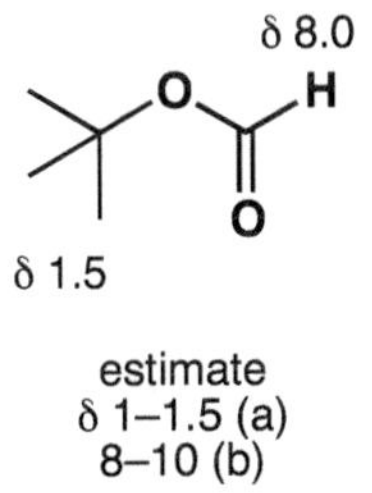

δ 8.0

δ 1.5

estimate
δ 1–1.5 (a)
8–10 (b)

δ 2.29

δ 1.24
estimate
δ 1–1.5 (a)
2.2–2.7 (b)

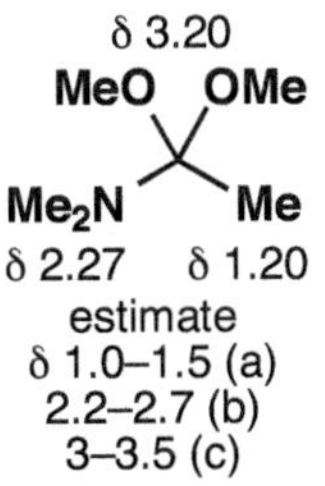

PROBLEM 2

The following products might possibly be formed from the reaction of MeMgBr with the cyclic anhydride shown. How would you tell the difference between these compounds using IR and ^{13}C NMR? With ^{1}H NMR available as well, how would your task be easier?

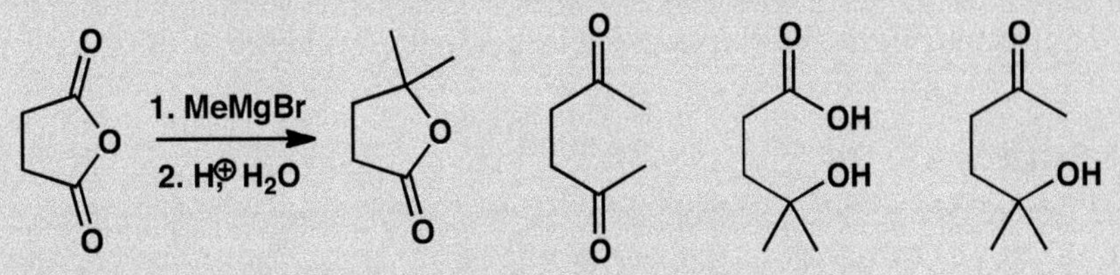

Purpose of the problem

Further thinking the other way round—from structure to data. Contrasting the limitations of ^{13}C NMR with data from ^{1}H NMR spectra.

Suggested solution

The molecular formula of the compounds varies so a mass spectrum would be useful. The compounds with an OH group would show a broad U-shaped band at above 3000 cm^{-1}. The cyclic ester would have a C=O stretch at about 1775 cm^{-1}, the ketones at about 1715 cm^{-1}, and the CO$_2$H group a band at about 1715 cm^{-1} as well as a very broad band from 2500 to 3500 cm^{-1}. In the ^{13}C NMR the acid and ester would have a carbonyl peak at about 170–180 ppm, but the ketones would have one at about 200 ppm. The number and position of the other signals would also vary.

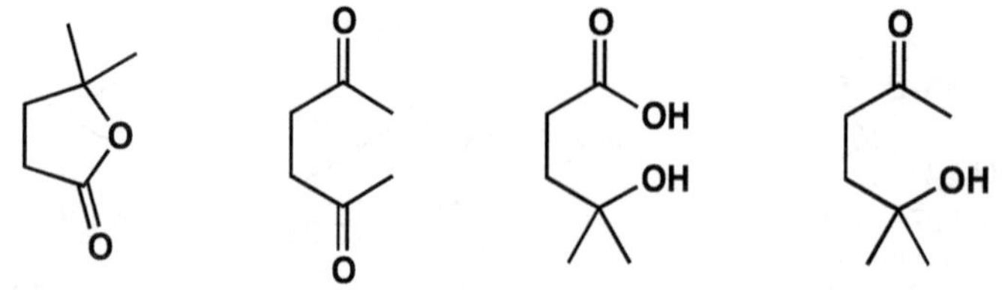

^{13}C NMR (ppm):	5 signals: 1 at about 170 1 at 50–100 3 at 0–50	3 signals: 1 at about 200 2 at 0–50	5 signals: 1 at about 170 1 at 50–100 3 at 0–50	6 signals: 1 at about 200 1 at about 170 1 at 50–100 3 at 0–50

In the proton NMR, all compounds would show two linked CH_2 groups as a pair of triplets except in the second compound as there the symmetry makes the two the same and would give a singlet. All except the second have a 6H singlet for the CMe_2 group. The second compound has two singlets because of the symmetry. The last has an isolated Me group. The OH and CO_2H protons might show up as broad signals at any chemical shift.

PROBLEM 3

One isomer of dimethoxybenzoic acid has the ^{1}H NMR spectrum δ_H (ppm) 3.85 (6H, s), 6.63 (1H, t, *J* 2 Hz), and 7.17 (2H, d, *J* 2 Hz). One isomer of coumalic acid has the ^{1}H NMR spectrum δ_H (ppm) 6.41 (1H, d, *J* 10 Hz), 7.82 (1H, dd, *J* 2, 10 Hz), and 8.51 (1H, d, *J* 2Hz). In each case, which isomer is it? The bonds sticking into the centre of the ring can be to any carbon atom.

dimethoxy benzoic acid

coumalic acid

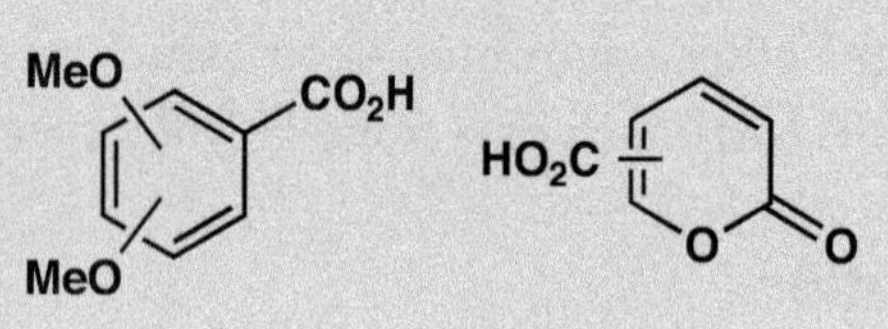

Purpose of the problem

First steps in using coupling to decide structure.

Suggested solution

The coupling constants in the first spectrum are all too small to be between hydrogens on neighbouring carbon atoms, and there must be symmetry in the molecule. There is only one structure that answers these criteria: 3,5-dimethoxybenzoic acid.

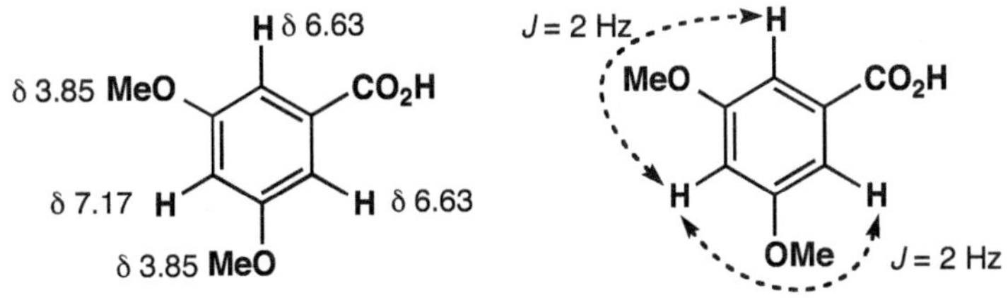

The second compound has one coupling of 10 Hz, and this must be between protons on neighbouring carbon atoms. The other coupling is 2 Hz and this is too small to be anything but *meta* coupling. There are two structures that might be right. In fact the first one is correct and you might have worked this out from the very large chemical shift—almost in the aldehyde region—of the isolated proton with only a 2 Hz coupling. This proton is on an alkene carbon bonded to oxygen in the first structure, but on a simple alkene carbon in the second.

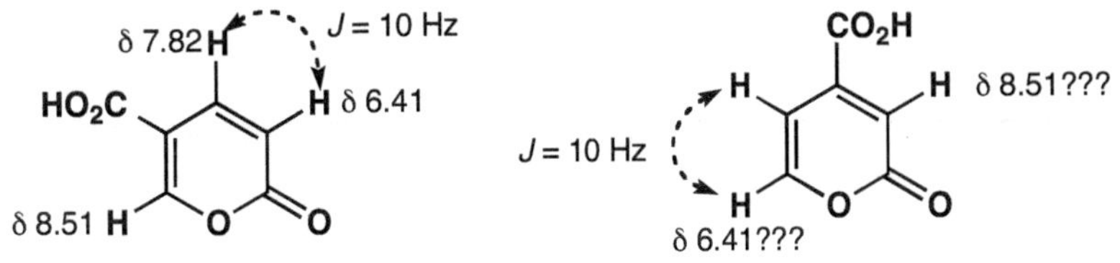

PROBLEM 4

Assign the NMR spectra of this compound and justify your assignments. 'Assign' means 'say which signal belongs to which atom'.

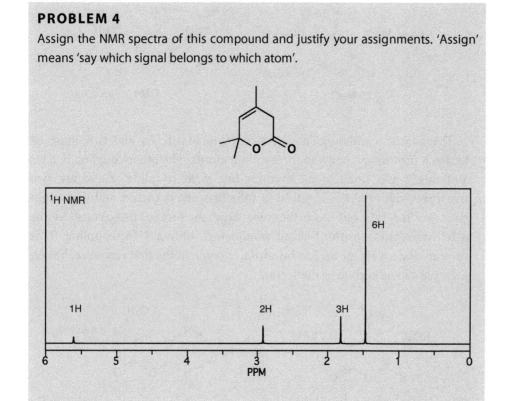

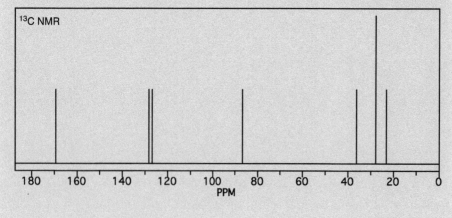

Purpose of the problem

Practice in the interpretation of real NMR spectra – this is harder than if the spectra have already been analysed and reported as a list of peaks.

Suggested solution

There is no coupling in this proton NMR spectrum which makes it much easier, but you should measure the chemical shifts and estimate the number of protons in each signal from the integration. Here we have: δ_H (ppm) 1.4

(6H), 1.8 (3H), 2.9 (2H)and 5.6 (1H). The peak at 7.5 is CHCl₃ impurity in the CDCl₃ solvent. This is enough to assign the spectrum but we should check that the chemical shifts are right and they are.

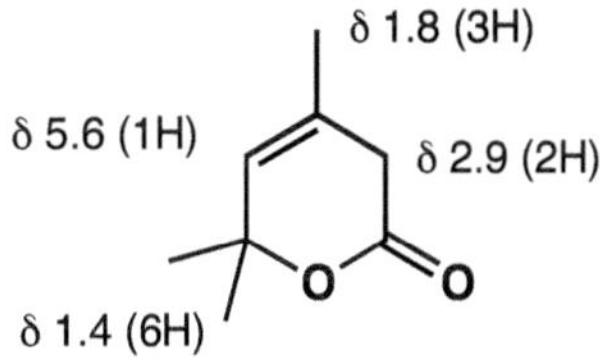

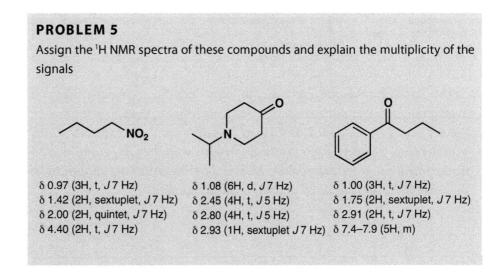

The carbon spectrum is more familiar to you from chapter 3 and you will remember that integration means little here. There are three peaks in the 0–50 ppm region corresponding to the methyl group on the alkene, the CH₂ group and the pair of methyls on the same carbon atom. The 1:1:1 triplet at 77 ppm is the solvent CDCl₃. The other signal in the 50–100 ppm region must be the carbon next to oxygen in the Me₂C group. The two signals in the 100–150 ppm region are the two carbons of the alkene and the very small peak at 150 ppm is the carbonyl group. No further assignment is necessary.

PROBLEM 5

Assign the ¹H NMR spectra of these compounds and explain the multiplicity of the signals

δ 0.97 (3H, t, J 7 Hz)
δ 1.42 (2H, sextuplet, J 7 Hz)
δ 2.00 (2H, quintet, J 7 Hz)
δ 4.40 (2H, t, J 7 Hz)

δ 1.08 (6H, d, J 7 Hz)
δ 2.45 (4H, t, J 5 Hz)
δ 2.80 (4H, t, J 5 Hz)
δ 2.93 (1H, sextuplet J 7 Hz)

δ 1.00 (3H, t, J 7 Hz)
δ 1.75 (2H, sextuplet, J 7 Hz)
δ 2.91 (2H, t, J 7 Hz)
δ 7.4–7.9 (5H, m)

Purpose of the problem

First serious practice in correlating splitting patterns and chemical shift.

Suggested solution

Redrawing the molecules with all the hydrogens showing probably helps at this stage, though you will not do this for long. The spectrum of 1-nitrobutane can be assigned by integration and splitting pattern without even looking at the chemical shifts! Just counting the number of neighbours and adding one gives the multiplicity and leads to the assignment. Alternatively you could inspect the chemical shifts which get smaller the further the protons are from the nitro group. Everything fits.

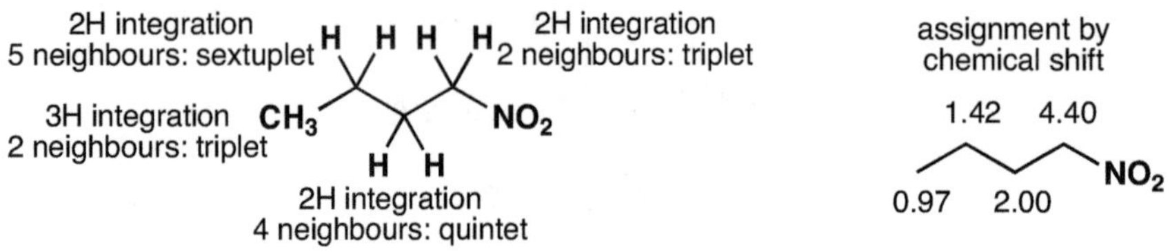

The next compound has an isopropyl group, typically a 6H doublet at about δ 1 ppm and a 1H septuplet with a larger chemical shift. Assigning the two triplets for the two CH_2 groups in the ring is not so easy as they are very similar. It doesn't really matter which is which as this uncertainty does not affect our identification of the compound.

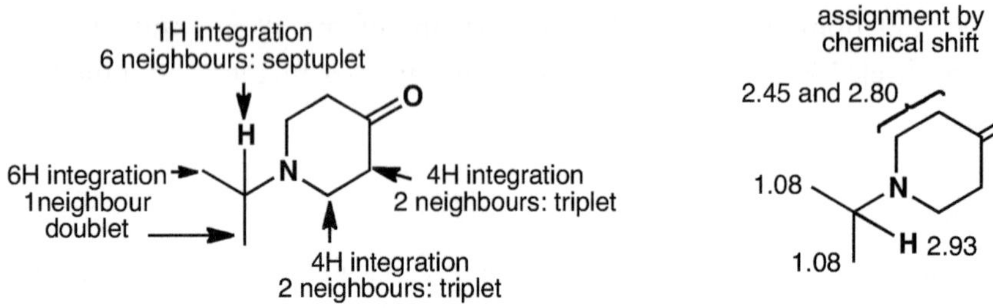

The aromatic ketone happens to have all five aromatic protons overlapping so they cannot be sorted out. This is not unusual and a signal in the 6.5–8 region described as '5H, m' usually means a monosubstituted benzene ring. The side chain is straightforward with the CH_2 group next to the ketone having the largest shift. All the coupling constants happen to be the same (7 Hz) as is usual in an open-chain compound.

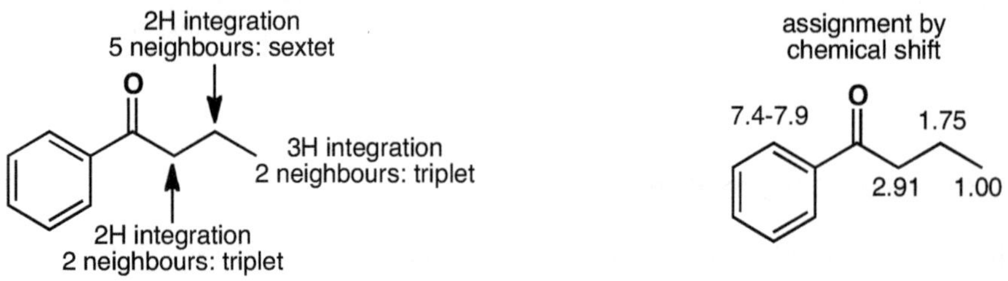

PROBLEM 6

The reaction below was expected to give the product **A** and did indeed give a compound with the correct molecular formula by its mass spectrum. However the NMR spectrum of this product was:

δ_H (ppm) 1.27 (6H, s), 1.70 (4H, m), 2.88 (2H, m), 5.4–6.1 (2H, broad s, exchanges with D_2O) and 7.0–7.5 (3H, m).

Though the detail is missing from this spectrum, how can you already tell that this is not the expected product?

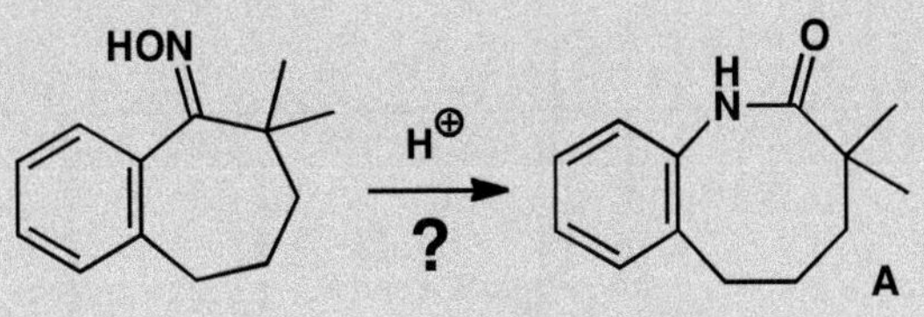

Purpose of the problem

To show that it is helpful to predict the NMR spectrum of an expected product provided that the structure is rejected if the NMR is 'wrong'.

Suggested solution

The spectrum is all wrong. There are only three aromatic Hs instead of the four expected. There are two exchanging hydrogens, presumably in NH_2 and not the one expected. The only thing that is expected is the chain of three CH_2 groups. If you managed to work out the product that was actually formed, you should be very pleased.

■ This surprising result was reported by B. Amit and A. Hassner, *Synthesis*, 1978, 932. The expected reaction was a Beckmann rearrangement (pp. 958–960) but what actually happened was a Beckmann fragmentation (pp. 959–960) followed by intramolecular Friedel-Crafts alkylation.

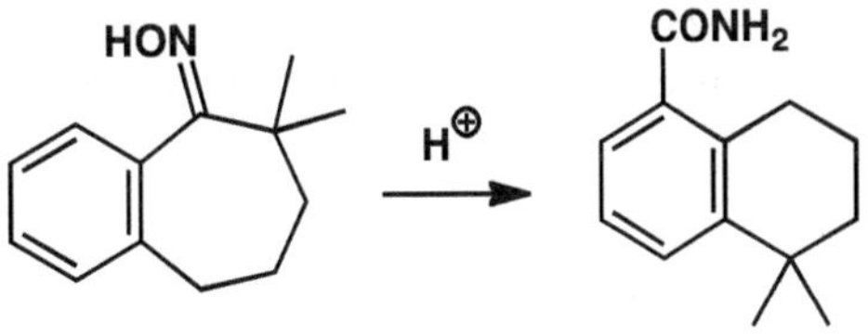

Now you know the structure of the product, you should be able to assign the spectrum and confirm the result.

PROBLEM 7

Assign the 400 MHz ¹H NMR spectrum of this enynone as far as possible, justifying both chemical shifts and coupling patterns.

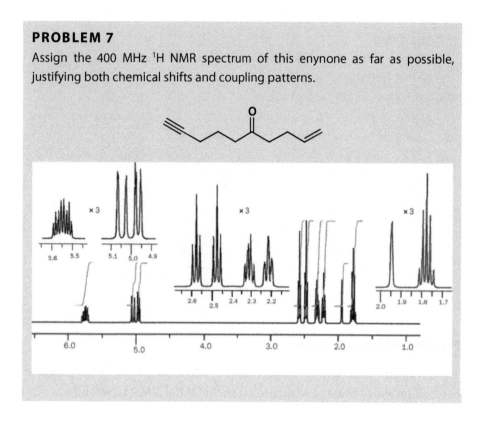

Purpose of the problem

Practice at interpretation of more complicated ¹H NMR spectra.

Suggested solution

First measure the spectrum and list the data. The expansions make it easier to see the coupling but even so we are going to have to call the signal at 5.6 ppm a multiplet. For the rest of the signals you should have measured the J values. Coupling is measured in Hz and at 400 MHz each chemical shift unit of 1 ppm is 400 Hz, so each subunit of 0.1 ppm is 40 Hz.

δ /ppm	integration	multiplicity	coupling, J / Hz	comments
5.6	1H	m	?	alkene region
5.05	1H	d with fine splitting	16.3	alkene region
4.97	1H	d with fine splitting	10.4	alkene region
2.58	2H	t with fine splitting	6.5	next to C=O or C=C
2.47	2H	t with fine splitting	6.5	next to C=O or C=C
2.32	2H	q with fine splitting	6.5	next to C=O or C=C
2.21	2H	t with fine splitting	6.5	next to C=O or C=C
1.95	1H	broad s	-	alkyne?
1.77	2H	q	6.5	not next to anything

That gives us three protons in the alkene region, five CH₂ groups and one solitary proton which must be on the alkyne. In the alkene region, the multiplet must be H² which couples to the the CH₂ at C3 and the other two alkene Hs. On C1, H¹ᵃ has a large *trans* coupling (16 Hz) to H² while H¹ᵇ has a smaller *cis* coupling (10 Hz). The coupling between H¹ᵃ and H¹ᵇ is very small.

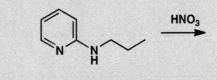

Of the five CH₂ groups, the quintet at small chemical shift must be C7. Those at C4, C6, and C8 have two neighbours and are basically triplets, but that at C3 couples to three protons and must be the quartet at 2.32 ppm.

PROBLEM 8

A nitration product (C₈H₁₁N₃O₂) of this pyridine has been isolated which has a nitro group somewhere in the molecule. From the spectrum deduce where the nitro group is and give a full analysis of the spectrum.

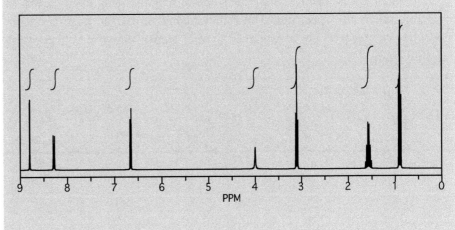

Purpose of the problem

Practice at working out the structure of a reaction product.

Suggested solution

The nitro group might go on the pyridine ring or on the aliphatic side chain or even, perhaps, on the nitrogen atom. Checking the integral shows that it must have gone on the pyridine: the propyl side chain is still there (CH_3 triplet, CH_2 quintet, and a CH_2 triplet with a large chemical shift). The NH proton is till there at 4.0 ppm. But there are now only three protons on the pyridine ring (at 6.7, 8.3, and 8.8 ppm).

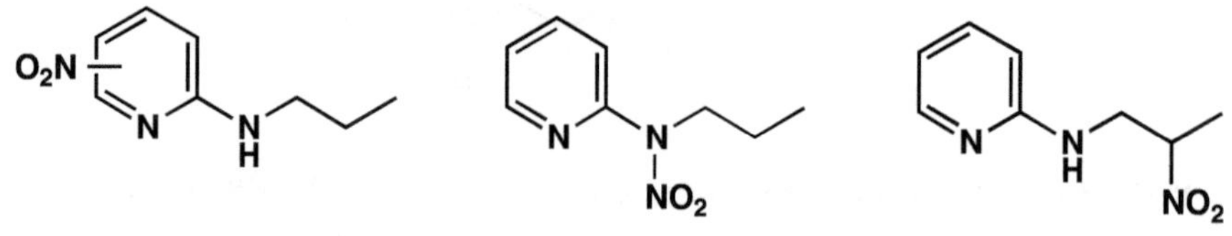

There are four possible structures. The most significant feature of the aromatic ring is the proton at very large chemical shift (8.8) with only very small coupling. Protons next to nitrogen in pyridine rings have very large chemical shifts so this rules out all the structures except the second.

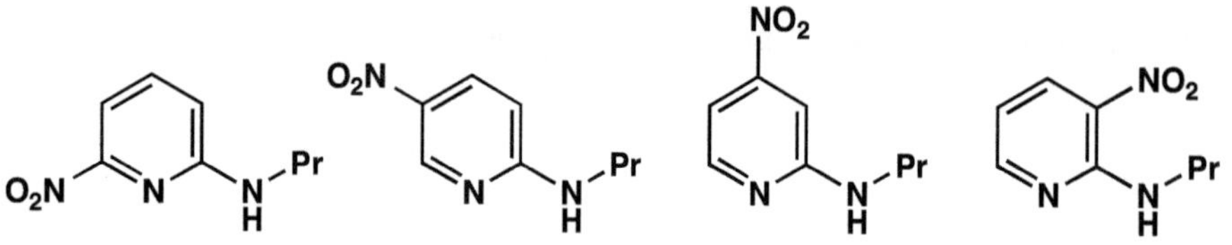

The nitro group also increases the shifts of neighbouring protons and so we can assign the spectrum. The rather high field of the proton on the pyridine ring at 6.6 ppm is explained by the electron-donating effect of the amino group.

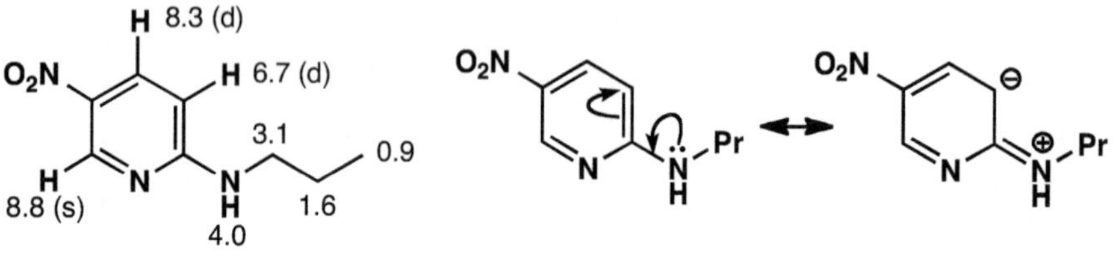

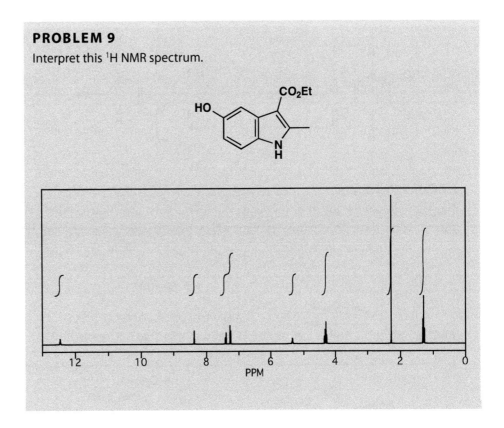

PROBLEM 9

Interpret this ¹H NMR spectrum.

Purpose of the problem

Further correlation of chemical shift and coupling with interpretation of longer-range coupling.

Suggested solution

The ethyl group is easy to find—a typical 3H triplet at 1.2 ppm and a 2H quartet at 4.3 ppm. The large shift of the CH_2 group tells us it is next to O. The methyl group is also easy—a 3H singlet at 2.3 ppm, typical of a methyl group on an alkene. At the other end of the spectrum, the broad singlet at 12.5 ppm can be only the OH or the NH; the other is at 5.4 ppm. That leaves the three signals in the aromatic region. You may not be able to see clearly the couplings, but they are: δ_H (ppm) 7.2 (1H, dd, *J* 9, 2 Hz), 7.5 (1H, d, *J* 9 Hz), and 8.4 (1H, d, *J* 2 Hz). The larger coupling is typical *ortho* and the small coupling typically *meta* so we can assign the whole spectrum.

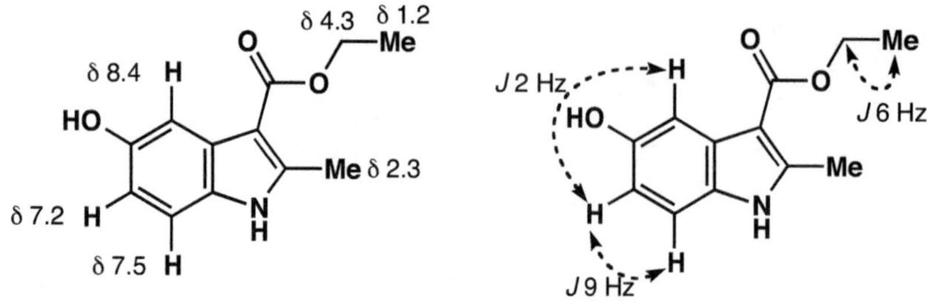

δ 4.3 δ 1.2

δ 8.4 H

HO

δ 7.2 H

δ 7.5 H

—Me δ 2.3

J 2 Hz

J 6 Hz

—Me

J 9 Hz

PROBLEM 10

Suggest structures for the products of these reactions, interpreting the spectroscopic data. Most of the reactions will be new to you, and you should aim to solve the structures from the data, not by guessing what might happen.

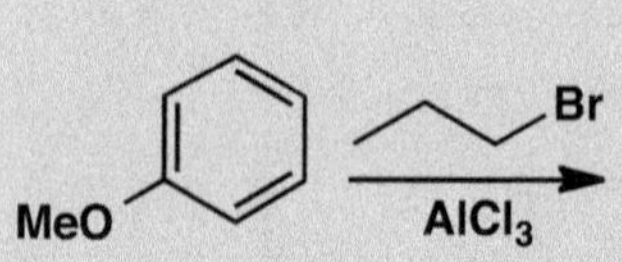

A, $C_{10}H_{14}O$

ν_{max} (cm⁻¹) C–H and fingerprint only

δ_C (ppm) 153, 141, 127, 115, 59, 33, 24

δ_H (ppm) 1.21 (6H, d, J 7 Hz), 2.83 (1H, septuplet, J 7 Hz), 3.72 (3H, s), 6.74 (2H, d, J 9 Hz) and 7.18 (2H, d, J 9 Hz)

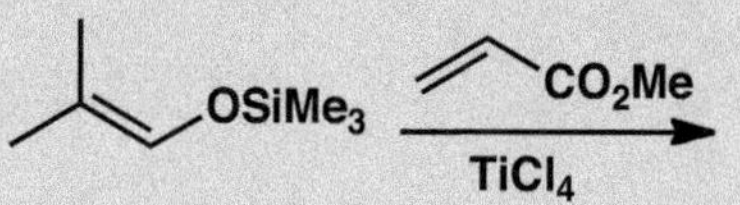

B, $C_8H_{14}O_3$

ν_{max} (cm⁻¹) 1745, 1730

δ_C (ppm) 202, 176, 62, 48, 34, 22, 15

δ_H (ppm) 1.21 (6H, s), 1.8 (2H, t, J 7 Hz), 2.24 (2H, t, J 7 Hz), 4.3 (3H, s) and 10.01 (1H, s)

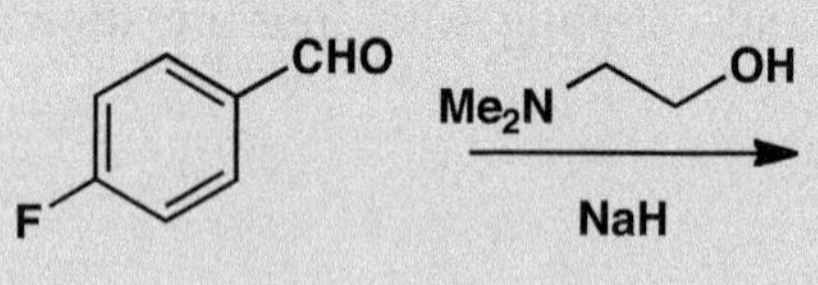

C, $C_{14}H_{15}NO_2$

ν_{max} (cm⁻¹) 1730

δ_C (ppm) 191, 164, 132, 130, 115, 64, 41, 29

δ_H (ppm) 2.32 (6H, s), 3.05 (2H, t, J 6 Hz) 4.20 (2H, t, J 6 Hz), 6.97 (2H, d, J 7 Hz), 7.82 (2H, d, J 7 Hz) and 9.97 (1H, s)

Purpose of the problem

Practice at determining the structures of reaction products of moderate complexity. This is a very common pastime of real chemists!

Suggested solution

Compound **A** contains the two reagents combined with the loss of HBr. The four Hs at 6.4 and 7.18 suggest the other reagent is attached to the benzene ring. The OMe group is still there (3H singlet at 3.72 ppm) and the new signals are a coupled 6H doublet and 1H septuplet—an isopropyl group. The compound is one of three isomers.

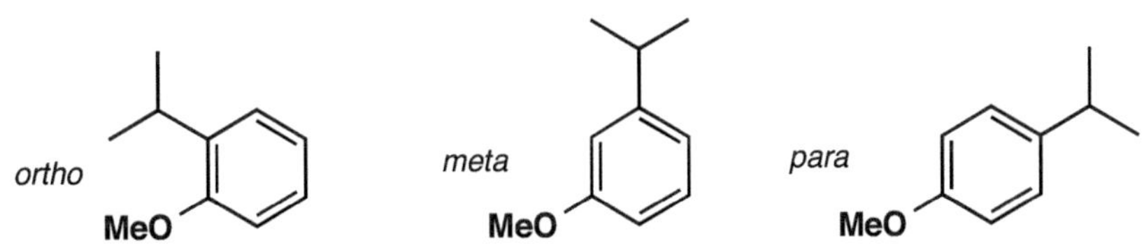

ortho   MeO   meta   MeO   para   MeO

The two 2H doublets coupled with *J* 9 Hz show that the product has symmetry and only the *para* isomer will fit, as both the *ortho* and *meta* compounds have four different protons. We have used the proton NMR alone but we could have mentioned that there are no functional groups other than the ether and that the four aromatic signals in the ^{13}C NMR reflect the symmetry of the product.

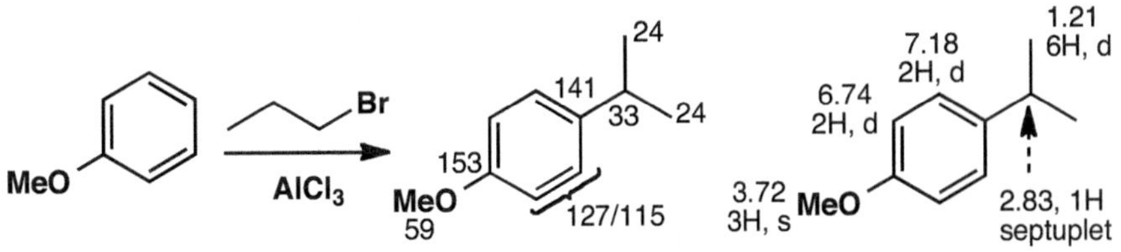

Compound **B** combines the two reagents with the loss of Me$_3$Si and the gain of H. Both the IR and the ^{13}C NMR show the appearance of a second carbonyl group—the ester (1745 cm^{-1} and 176 ppm) has been joined by an aldehyde or ketone (1730 cm^{-1} and 202 ppm). The proton NMR shows it is an aldehyde (10.01, 1H, s). There is also a CMe$_2$ group but it is no longer part of an alkene (proton and carbon NMR show that the alkene has gone). The OMe of the ester has survived. Finally, and very helpfully, there are two open chain CH$_2$ groups linked together (the two triplets with *J* 7 Hz). One of them (2.24 ppm) has to be next to something and that can only be a carbonyl group as there is nothing else. So we have:

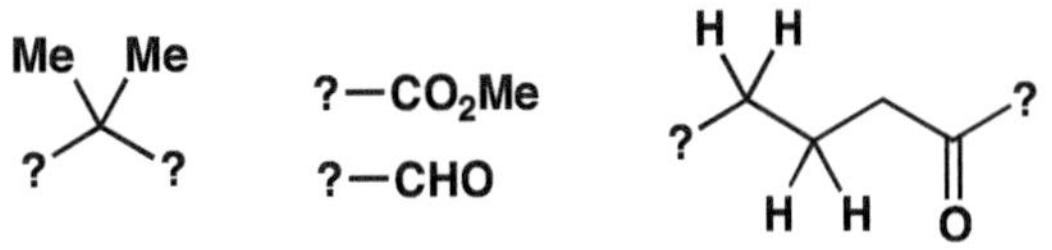

Though saying what 'ought to happen' is not always helpful, it obviously makes much more sense to consider first a solution in which the ester group stays where it is, on a chain of two carbon atoms, than one in which it moves mysteriously to the other end of the molecule. We prefer the first of these two possibilities:

Real evidence comes from the lack of coupling of the aldehyde proton which would surely be a triplet in the second structure. The first structure is indeed correct.

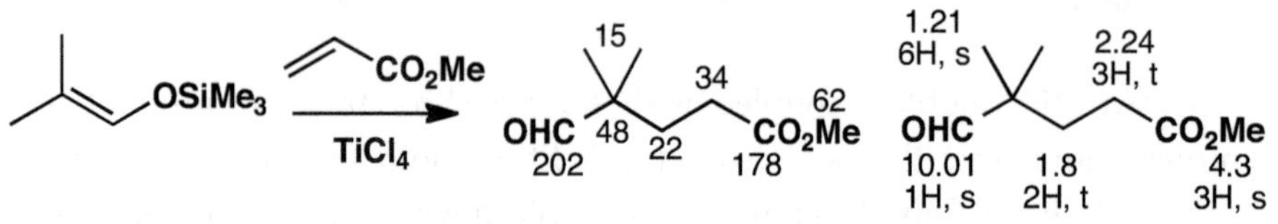

Adding up the atoms for compound **C** reveals that the two reagents have joined together with the loss of HF. The 1,4-disubstituted benzene ring is still there (same pattern as compound **A**) as is the aldehyde (1730 cm^{-1}, 191 ppm and 9.97 ppm). The NMe$_2$ group and the CH$_2$–CH$_2$ chain from the other reagent have also survived. It looks as though the fluoride has been displaced by the oxygen of the alcohol and this is indeed what has happened.

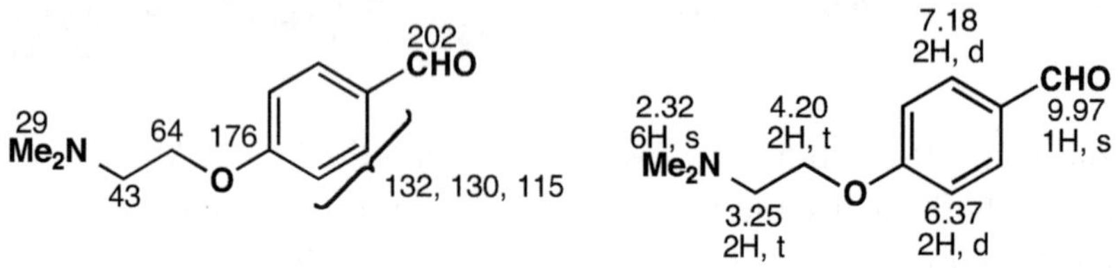

14 Stereochemistry

Connections

⇒ Building on

- Drawing organic molecules ch2
- Organic structures ch4
- Nucleophilic addition to the carbonyl group ch6
- Nucleophilic substitution at carbonyl groups ch10 & ch11

Arriving at

- Three-dimensional shape of molecules
- Molecules with mirror images
- Molecules with symmetry
- How to separate mirror-image molecules
- Diastereoisomers
- Shape and biological activity
- How to draw stereochemistry

⇒ Looking forward to

- Nucleophilic substitution at saturated C ch15
- Conformation ch16
- Elimination ch18
- Controlling alkene geometry ch27
- Controlling stereochemistry with cyclic compounds ch32
- Diastereoselectivity ch33
- Asymmetric synthesis ch41
- Chemistry of life ch42

Some compounds can exist as a pair of mirror-image forms

One of the very first reactions you met, back in Chapter 6, was between an aldehyde and cyanide. The product was a compound containing a nitrile group and a hydroxyl group.

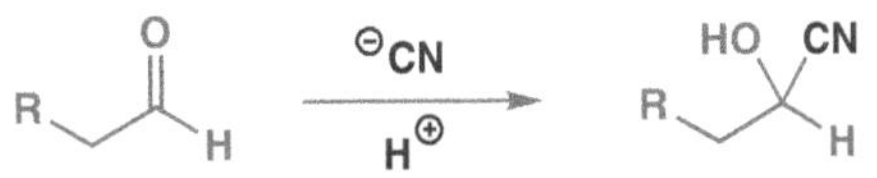

How many products are formed in this reaction? Well, the straightforward answer is one—there's only one aldehyde, only one cyanide ion, and only one reasonable way in which they can react. But this analysis is not *quite* correct. One point that we ignored when we first talked about this reaction, because it was irrelevant at that time, is that the carbonyl group of the aldehyde has two faces. The cyanide ion could attack either from the front face or the back face, giving, in each case, a distinct product.

🖱 Interactive results of cyanide addition to carbonyls

■ The bold wedges represent bonds coming towards you, out of the paper, and the cross-hatched bonds represent bonds going away from you, into the paper.

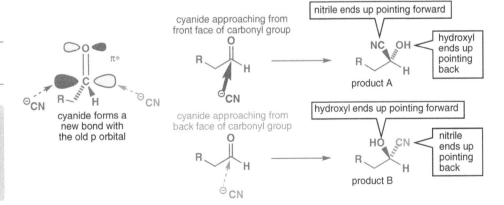

Online support. The icon 🖱 in the margin indicates that accompanying interactive resources are provided online to help your understanding: just type www.chemtube3d.com/clayden/123 into your browser, replacing **123** with the number of the page where you see the icon. For pages linking to more than one resource, type **123-1, 123-2** etc. (replacing **123** with the page number) for access to successive links.

As we explained in Chapter 6 (pp. 125–7), the cyanide attacks the π* orbital of the aldehyde more or less at right angles to the plane of the molecule as it forms a new bond with the old p orbital on C. This translates into 'front' and 'back' on a diagram on paper. Compare the diagram on the left with the others to make sure this is clear.

Are these two products different? If we lay them side by side and try to arrange them so that they look identical, we find that we can't—you can verify this by making models of the two structures. The structures are non-superimposable—so they are not identical. In fact, they are mirror images of each other: if we reflected one of the structures, **A**, in a mirror, we would get a structure that *is* identical with **B**.

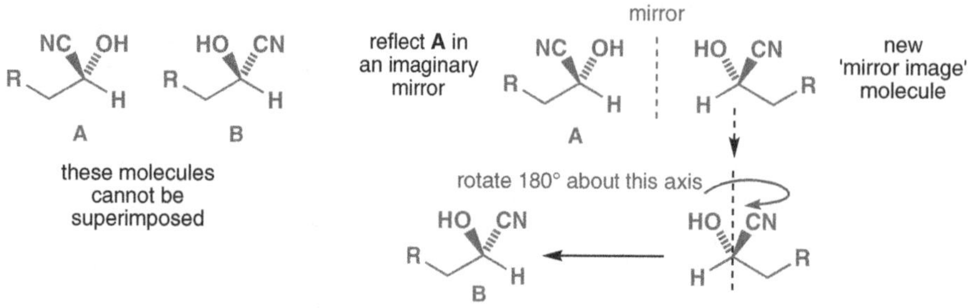

We call two structures that are not identical but are mirror images of each other (like these two) **enantiomers**. Structures that are not superimposable on their mirror image, and can therefore exist as two enantiomers, are called **chiral**. In this reaction, the cyanide ions are just as likely to attack the 'front' face of the aldehyde as they are the 'back' face, so we get a 50:50 mixture of the two enantiomers.

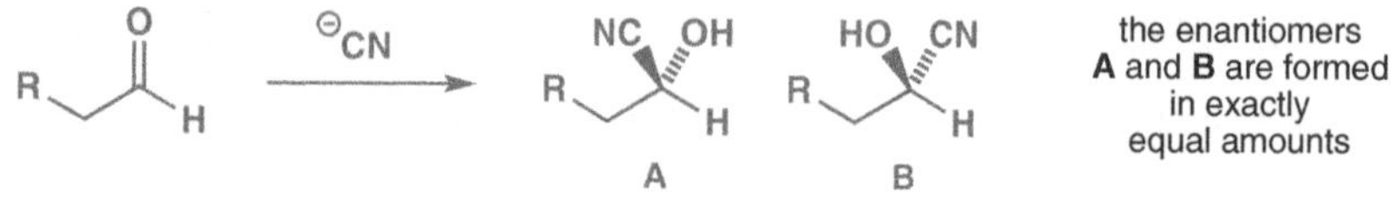

> In reading this chapter, you will have to do a lot of mental manipulation of three-dimensional shapes. Because we can represent these shapes in the book only in two dimensions, we suggest that you make models, using a molecular model kit, of the molecules we talk about. With some practice, you will be able to imagine the molecules you see on the page in three dimensions.

Interactive aldehyde cyanohydrin structures—chiral

● **Enantiomers and chirality**

- Enantiomers are structures that are not identical, but are *mirror images* of each other.
- Structures are *chiral* if they cannot be superimposed on their mirror image.

Now consider another similar reaction—the addition of cyanide to acetone.

Again an adduct (a cyanohydrin) is formed. You might imagine that attacking the front or the back face of the acetone molecule could again give two structures, C and D.

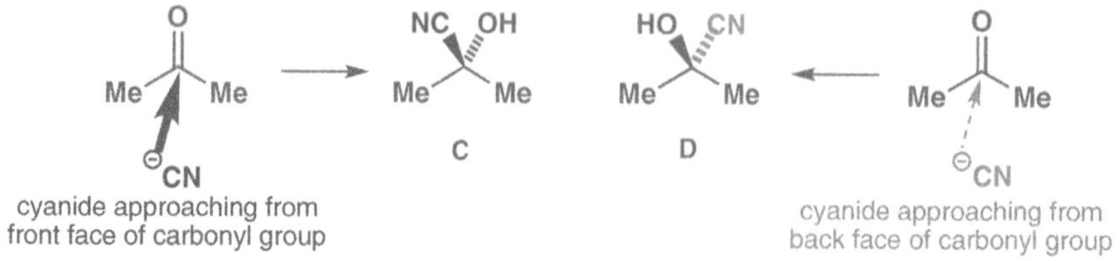

However, this time rotating one to match the other shows that they are superimposable and therefore identical.

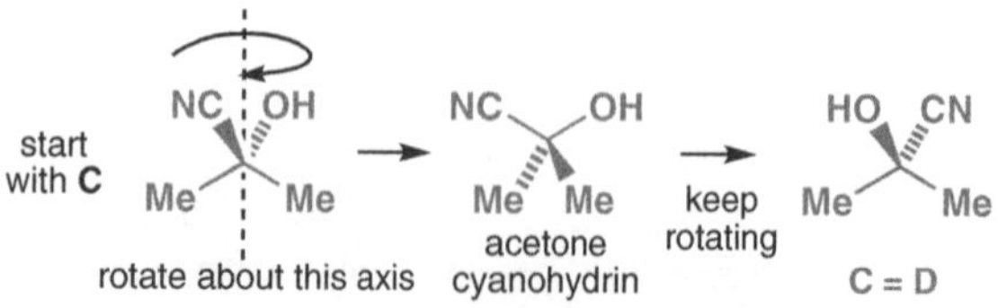

Interactive acetone cyanohydrin structure—achiral

Make sure that you are clear about this: **C** and **D** are identical molecules, while **A** and **B** are mirror images of each other. Reflection in a mirror makes no difference to **C** or **D**; they are superimposable on their own mirror images and therefore cannot exist as two enantiomers. Structures that are superimposable on their mirror images are called *achiral*.

> ● *Achiral* structures are superimposable on their mirror images.

Chiral molecules have no plane of symmetry

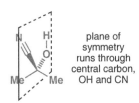

plane of symmetry runs through central carbon, OH and CN

What is the essential difference between these two compounds that means one is super-imposable on its mirror image and one is not? The answer is *symmetry*. Acetone cyanohydrin has a plane of symmetry running through the molecule. This plane cuts the central carbon and the OH and CN groups in half, and has one methyl group on each side. All planar molecules (such as our simple aldehyde) cannot be chiral as the plane of the molecule must be a plane of symmetry. Cyclic molecules may have a plane of symmetry passing through two atoms of the ring, as in the cyclohexanone below. The plane passes through both atoms of the carbonyl group and bisects the methyl group as well as the hydrogen atom (not shown) on the same carbon atom. The bicyclic acetal looks more complicated but a plane of symmetry passes between the two oxygen atoms and the two ring-junction carbon atoms while bisecting the two methyl groups. None of these molecules is chiral.

molecules with planes of symmetry

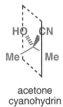

acetone cyanohydrin

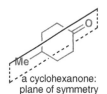

any planar molecule: the plane of the paper is a plane of symmetry

a cyclohexanone: plane of symmetry is orthogonal to the paper

a bicyclic acetal: plane of symmetry is orthogonal to the paper

On the other hand, the aldehyde cyanohydrin has no plane of symmetry: the plane of the paper has OH on one side and CN on the other while the plane at right angles to the paper has H on one side and RCH_2 on the other. This compound has no plane of symmetry (has *asymmetry*) and has two enantiomers.

aldehyde cyanohydrin

plane of paper not a plane of symmetry

plane through OH and CN not a plane of symmetry

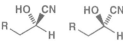

so the molecule is chiral with two enantiomers

■ Later in this chapter we shall meet a much less important type of symmetry that also means molecules are not chiral if they possess it. This is a centre of symmetry.

> ● **Planes of symmetry and chirality**
>
> • **Any structure that has no plane of symmetry is chiral and can exist as two mirror-image forms (*enantiomers*).**
> • **Any structure with a plane of symmetry is not chiral and cannot exist as two enantiomers.**

By 'structure', we don't just mean chemical structure: the same rules apply to everyday objects. Some examples from among more familiar objects in the world around us should help make these ideas clear. Look around you and find a chiral object—a pair of scissors, a screw (but not the screwdriver), a car, and anything with writing on it, like this page. Look again for achiral objects with planes of symmetry—a plain mug, saucepan, chair, most simple manufactured objects without writing on them. The most significant chiral object near you is the hand you write with.

Gloves, hands, and socks

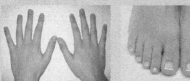  

Most gloves exist in pairs of non-identical mirror-image forms: only a left glove fits a left hand and only a right glove fits a right hand. This property of gloves and of the hands inside them gives us the word 'chiral'—*cheir* is Greek for 'hand'. Hands and gloves are chiral; they have no plane of symmetry, and a left glove is not superimposable on its mirror image (a right glove). Feet are chiral too, as are shoes. But socks (usually!) are not. Although we all sometimes have problems finding two socks of a matching colour, once you've found them, you never have to worry about which sock goes on which foot because socks are achiral. A pair of socks is manufactured as two identical objects, each of which has a mirror plane.

The ancient Egyptians had less care for the chirality of hands and their paintings often show people, even Pharaohs, with two left hands or two right hands—they just didn't seem to notice.

Tennis racquets and golf clubs

If you are left-handed and want to play golf, you either have to play in a right-handed manner or get hold of a set of left-handed golf clubs. Golf clubs are clearly therefore chiral; they can exist as either of two enantiomers. You can tell this just by looking at a golf club. It has no plane of symmetry, so it must be chiral. But left-handed tennis players have no problem using the same racquets as right-handed tennis players and tennis players of either chirality sometimes swap the racquet from hand to hand. Look at a tennis racquet: it has a plane of symmetry (indeed, usually two), so it's achiral. It can't exist as two mirror-image forms.

■ These statements are slightly incomplete but will serve you well in almost all situations: we will come to centres of symmetry shortly (p. 321).

● **To summarize**

- A structure *with* a plane of symmetry is *achiral* and *superimposable* on its mirror image and *cannot* exist as two enantiomers.
- A structure *without* a plane of symmetry is *chiral* and *not superimposable* on its mirror image and *can* exist as two enantiomers.

Stereogenic centres

Back to chemistry, and the product from the reaction of an aldehyde with cyanide. We explained above that this compound, being chiral, can exist as two enantiomers. Enantiomers are clearly isomers; they consist of the same parts joined together in a different way. In particular, enantiomers are a type of isomer called **stereoisomers** because the isomers differ not in the connectivity of the atoms, but only in the overall shape of the molecule.

● **Stereoisomers and constitutional isomers**

Isomers are compounds that contain the same atoms bonded together in different ways. If the connectivity of the atoms in the two isomers is different, they are constitutional isomers. If the connectivity of the atoms in the two isomers is the same, they are stereoisomers. Enantiomers are stereoisomers, and so are *E* and *Z* double bonds. We shall meet other types of stereoisomers shortly.

constitutional isomers: the way the atoms are connected up (their *connectivity*) differs

enantiomers
E/Z isomers (double bond isomers)
stereoisomers: the atoms have the same connectivity, but are arranged differently

We should also introduce you briefly to another pair of concepts here, which you will meet again in more detail in Chapter 16: *configuration* and *conformation*. Two stereoisomers really are different molecules: they cannot be interconverted without breaking a bond somewhere. We therefore say that they have different configurations. But any molecule can exist in a number of conformations: two conformations differ only in the temporary way the molecule happens to arrange itself, and can easily be interconverted just by rotating around bonds. Humans all have the same *configuration*: two arms joined to the shoulders. We may have different *conformations*: arms folded, arms raised, pointing, waving, etc.

● **Configuration and conformation**

- Changing the *configuration* of a molecule always means that bonds are broken.
- A different configuration is a different molecule.
- Changing the *conformation* of a molecule means rotating about bonds, but not breaking them.
- Conformations of a molecule are readily interconvertible, and are all the same molecule.

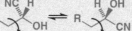

two configurations: going from one enantiomer to the other requires a bond to be broken

three conformations of the same enantiomer: getting from one to the other just requires rotation about a bond: all three are the same molecule

🖱 Interactive cyanohydrin conformations

The aldehyde cyanohydrin is chiral because it does not have a plane of symmetry. In fact, it *cannot* have a plane of symmetry because it contains a tetrahedral carbon atom carrying four different groups: OH, CN, RCH$_2$, and H. Such a carbon atom is known as a **stereogenic** or **chiral centre**. The product of cyanide and acetone is not chiral; it has a plane of symmetry and no chiral centre because two of the groups on the central carbon atom are the same.

HO CN   1   2
R   HO CN stereogenic centre
     H   3 R   or chiral centre
aldehyde cyanohydrin H 4
           four different groups

           1   2 only three
           HO CN different
                   groups
           3 Me   Me 3
           acetone
           cyanohydrin

> • If a molecule contains one carbon atom carrying four different groups it will not have a plane of symmetry and must therefore be chiral. A carbon atom carrying four different groups is a stereogenic or chiral centre.

■ You will see shortly that compounds with *more than one* chiral centre are not always chiral.

We saw how the two enantiomers of the aldehyde cyanohydrin arose by attack of cyanide on the two faces of the carbonyl group of the aldehyde. We said that there was nothing to favour one face over the other, so the enantiomers must be formed in equal quantities. A mixture of equal quantities of a pair of enantiomers is called a **racemic mixture**.

the enantiomers are formed in exactly equal amounts: the product is a racemic mixture

> • A **racemic mixture** is a mixture of two enantiomers in equal proportions. This principle is very important. If all the starting materials and reagents in a reaction are achiral and the products are chiral they will be formed as a racemic mixture of two enantiomers.

Here are some more reactions you have come across that make chiral products from achiral starting materials. In each case, the principle must hold—equal amounts of the two enantiomers (racemic mixtures) are formed.

1. MeMgCl
2. H_3O^+

containing exactly 50% of this: HO H

and 50% of this: H OH

containing exactly 50% of this: HO H

and 50% of this: H OH

■ When we don't show bold and dashed bonds to indicate the three-dimensional structure of the molecule, we mean that we are talking about both enantiomers of the molecule. Another useful way of representing this is with wiggly bonds. Wiggly bonds are in fact slightly ambiguous: here the wiggly bond means both stereoisomers. Elsewhere a wiggly bond might mean just one stereoisomer, but with unknown stereochemistry.

HO CN
R   H

Many chiral molecules are present in nature as single enantiomers

Let's turn to some simple, but chiral, molecules—the natural amino acids. All amino acids have a carbon carrying an amino group, a carboxyl group, a hydrogen atom, and the R group, which varies from amino acid to amino acid. So unless R=H (this is the case for glycine), amino acids always contain a chiral centre and lack a plane of symmetry.

amino acids are chiral

4 H NH₂ 1
3 R   CO₂H 2

3 H NH₂ 1
3 H   CO₂H 2

except glycine— plane of paper is a plane of symmetry through C, N, and CO_2H

It is possible to make amino acids quite straightforwardly in the laboratory. The scheme below shows a synthesis of alanine, for example. It is a version of the Strecker synthesis you met in Chapter 11.

laboratory synthesis of racemic alanine from acetaldehyde

O
Me   H   NH₄Cl
           KCN

NH
Me   H
unstable imine

H NH₂
Me   CN
           H₂O, H⊕

H NH₂
Me   CO₂H
amino acid

acetaldehyde

alanine extracted from plants
is only this enantiomer

Alanine made in this way must be racemic because the starting material and all reagents are achiral. However, if we isolate alanine from a natural source—by hydrolysing vegetable protein, for example—we find that this is not the case. Natural alanine is solely one enantiomer, the one drawn in the margin. Samples of chiral compounds that contain only one enantiomer are called **enantiomerically pure**. We know that 'natural' alanine contains only this enantiomer from X-ray crystal structures.

> **Enantiomeric alanine**
>
> In fact, nature does sometimes (but very rarely) use the other enantiomer of alanine, for example in the construction of bacterial cell walls. Some antibiotics (such as vancomycin) owe their selectivity to the way they can recognize these 'unnatural' alanine components and destroy the cell wall that contains them.

Chiral and enantiomerically pure

Before we go further, we should just mention one common point of confusion. Any compound whose molecules do not have a plane of symmetry is chiral. Any sample of a chiral compound that contains molecules all of the same enantiomer is enantiomerically pure. *All* alanine is chiral (the structure has no plane of symmetry) but *laboratory-produced* alanine is racemic (a 50:50 mixture of enantiomers) whereas *naturally isolated* alanine is enantiomerically pure.

> ● **'Chiral' does not mean 'enantiomerically pure'.**

■ Remember—we use the word *configuration* to describe the arrangement of bonds around an atom. Configurations cannot be changed without breaking bonds.

Most of the molecules we find in nature are chiral—a complicated molecule is much more likely not to have a plane of symmetry than to have one. Nearly all of these chiral molecules in living systems are found not as racemic mixtures, but as single enantiomers. This fact has profound implications, for example in the chemistry of drug design, and we will come back to it later.

R and *S* can be used to describe the configuration of a chiral centre

Before going on to talk about single enantiomers of chiral molecules in more detail, we need to explain how chemists describe which enantiomer they're talking about. We can, of course, just draw a diagram, showing which groups go into the plane of the paper and which groups come out of the plane of the paper. This is best for complicated molecules. Alternatively, we can use the following set of rules to assign a letter, *R* or *S*, to describe the configuration of groups at a chiral centre in the molecule.

Here again is the enantiomer of alanine you get if you extract alanine from living things.

natural alanine

🖱 Interactive configuration assignment

1. Assign a priority number (1–4) to each substituent at the chiral centre. Atoms with higher atomic numbers get higher priority.

   Alanine's chiral centre carries one N atom (atomic number 7), two C atoms (atomic number 6), and one H atom (atomic number 1). So, we assign priority 1 to the NH_2 group, because N has the highest atomic number. Priorities 2 and 3 will be assigned to the CO_2H and the CH_3 groups, and priority 4 to the hydrogen atom; but we need a way of deciding which of CO_2H and CH_3 takes priority over the other. If two (or more) of the atoms attached to the chiral centre are identical, then we assign priorities to these two by assessing the atoms attached to those atoms. In this case, one of the carbon atoms carries oxygen atoms (atomic number 8) and one carries only hydrogen atoms (atomic number 1). So CO_2H is higher priority that CH_3; in other words, CO_2H gets priority 2 and CH_3 priority 3.

2. Arrange the molecule so that the lowest priority substituent is pointing away from you. In our example, naturally extracted alanine, H is priority 4, so we need to look at the molecule with the H atom pointing into the paper, like this.

■ These priority rules are also used to assign *E* and *Z* to alkenes, and are sometimes called the Cahn–Ingold–Prelog (CIP) rules, after their devisors. You can alternatively use atomic weights—for isotopes you have to (D has a higher priority than H)—apart from in the vanishingly rare case of a chiral centre bearing Te and I (look at the data in the periodic table at the front of this book to see why).

3. Mentally move from substituent priority 1 to 2 to 3. If you are moving in a clockwise manner, assign the label *R* to the chiral centre; if you are moving in an anticlockwise manner, assign the label *S* to the chiral centre.

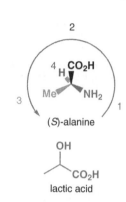

(*S*)-alanine

lactic acid

A good way of visualizing this is to imagine turning a steering wheel in the direction of the numbering. If you are turning your car to the right, you have *R*; if you are turning to the left you have *S*. For our molecule of natural alanine, if we move from NH_2 (1) to CO_2H (2) to CH_3 (3) we're going anticlockwise (turning to the left), so we call this enantiomer (*S*)-alanine.

You can try working the other way, from the configurational label to the structure. Take lactic acid as an example. Lactic acid is produced by bacterial action on milk; it's also produced in your muscles when they have to work with an insufficient supply of oxygen, such as during bursts of vigorous exercise. Lactic acid produced by fermentation is often racemic, although certain species of bacteria produce solely (*R*)-lactic acid. On the other hand, lactic acid produced by anaerobic respiration in muscles has the *S* configuration.

As a brief exercise, try drawing the three-dimensional structure of (*R*)-lactic acid. You may find this easier if you draw both enantiomers first and then assign a label to each.

You should have drawn:

H OH   or   OH

Me CO_2H Me CO_2H

(*R*)-lactic acid   (*R*)-lactic acid

Remember that, if we had made lactic acid in the laboratory from simple achiral starting materials, we would have got a racemic mixture of (*R*)- and (*S*)-lactic acid. Reactions in living systems can produce enantiomerically pure compounds because they make use of enzymes, themselves enantiomerically pure compounds of (*S*)-amino acids.

Is there a chemical difference between two enantiomers?

The short answer is *no*.* Take (*S*)-alanine (in other words, alanine extracted from plants) and (*R*)-alanine (the enantiomer found in bacterial cell walls) as examples. They have identical NMR spectra, identical IR spectra, and identical physical properties with a single important exception. If you shine plane-polarized light through a solution of (*S*)-alanine, you will find that the light is rotated to the right. A solution of (*R*)-alanine rotates plane-polarized light to the left and by the same amount. Racemic alanine doesn't rotate such light at all.

The rotation of plane-polarized light is known as optical activity

Observation of the rotation of plane-polarized light is known as polarimetry; it is a straightforward way of finding out if a sample is racemic or if it contains more of one enantiomer than the other. Polarimetric measurements are carried out in a polarimeter, which has a single-wavelength (monochromatic) light source with a plane-polarizing filter, a sample holder, where a cell containing a solution of the substance under examination can be placed, and a detector with a read-out that indicates by how much the light is rotated. Rotation to the right is given a positive value, rotation to the left a negative one.

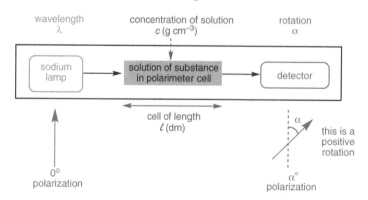

* The longer answer is more involved, and we go into it in more detail in Chapter 41.

> ■ Remember how, in Chapter 2 (p. 21) we showed you that hydrogen atoms at stereogenic centres (we didn't call them that then) could be missed out—we just assume that they take up the fourth vertex of the imagined tetrahedron at the stereogenic centre.
>
> This also brings us to another point about drawing stereogenic centres: always try to have the carbon skeleton lying in the plane of the paper: in other words,
>
> H OH
>
> Me CO_2H
>
> (*R*)-lactic acid
> rather
> than
> say:
>
> Me CO_2H
>
> H OH
>
> (*R*)-lactic acid
>
> Both are correct but the first will make things a lot easier when we are talking about molecules with several chiral centres!

> ■ Plane-polarized light can be considered as a beam of light in which all of the light waves have their direction of vibration aligned parallel. It is produced by shining light through a polarizing filter.

The angle through which a sample of a compound (usually a solution) rotates plane-polarized light depends on a number of factors, the most important ones being the path length (how far the light has to pass through the solution), concentration, temperature, solvent, and wavelength. Typically, optical rotations are measured at 20 °C in a solvent such as ethanol or chloroform, and the light used is from a sodium lamp, with a wavelength of 589 nm.

The observed angle through which the light is rotated is given the symbol α. By dividing this value by the path length ℓ (in dm) and the concentration c (in g cm^{-3}) we get a value, $[\alpha]$, which is specific to the compound in question. Indeed, $[\alpha]$ is known as the compound's **specific rotation**. The choice of units is eccentric and arbitrary but is universal so we must live with it.

$$[\alpha] = \frac{\alpha}{c\ell}$$

Most $[\alpha]$ values are quoted as $[\alpha]_D$ (where the D indicates the wavelength of 589 nm, the 'D line' of a sodium lamp) or $[\alpha]_D^{20}$, the 20 indicating 20 °C. These define the remaining variables.

Here is an example. A simple acid, known as mandelic acid, can be obtained from almonds in an enantiomerically pure form. When 28 mg was dissolved in 1 cm^3 of ethanol and the solution placed in a 10-cm-long polarimeter cell, an optical rotation α of –4.35° was measured (that is, 4.35° to the left) at 20 °C with light of wavelength 589 nm. What is the specific rotation of the acid?

First, we need to convert the concentration to grammes per cubic centimetre: 28 mg in 1 cm^3 is the same as 0.028 g cm^{-3}. The path length of 10 cm is 1 dm, so

$$[\alpha]_D^{20} = \frac{\alpha}{c\ell} = \frac{-4.35}{0.028 \times 1} = -155.4$$

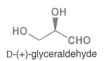
(R)-mandelic acid

■ Note that the units of a measured optical rotation α are degrees, but, by convention, specific rotation $[\alpha]$ is quoted without units.

■ $[\alpha]_D$ values can be used as a guide to the enantiomeric purity of a sample, in other words, to how much of each enantiomer it contains. We will come back to this in Chapter 41.

Enantiomers can be described as (+) or (–)

We can use the fact that two enantiomers rotate plane-polarized light in opposite directions to assign each a label that doesn't depend on knowing its configuration. We call the enantiomer that rotates plane-polarized light to the right (gives a positive rotation) the (+)-enantiomer (or the *dextrorotatory* enantiomer) and the enantiomer that rotates plane-polarized light to the left (gives a negative rotation) the (–)-enantiomer (or the *laevorotatory* enantiomer). The direction in which light is rotated is not dependent on whether a stereogenic centre is R or S. An (R) compound is equally as likely to be (+) as (–)—of course, if it is (+) then its (S) enantiomer must be (–). The enantiomer of mandelic acid we have just discussed, for example, is (R)-(–)-mandelic acid because its specific rotation is negative, and (S)-alanine happens to be (S)-(+)-alanine. The labels (+) and (–) were more useful before the days of X-ray crystallography, when chemists did not know the actual configuration of the molecules they studied, and could distinguish two enantiomers only by the signs of their specific rotations.

Enantiomers can be described as D or L

Long before the appearance of X-ray crystallography as an analytical tool, chemists had to discover the detailed structure and stereochemistry of molecules by a complex series of degradations. A molecule was gradually broken down into its constituents, and from the products that were formed the overall structure of the starting molecule was deduced. As far as stereochemistry was concerned, it was possible to measure the specific rotation of a compound, but not to determine its configuration. However, by using series of degradations it was possible to tell whether certain compounds had the same or opposite configurations.

Glyceraldehyde is one of the simplest chiral compounds in nature. Because of this, chemists took it as the standard against which the configurations of other compounds could be compared. The two enantiomers of glyceraldehyde were given the labels D (for dextro—because it was the (+)-enantiomer) and L (for laevo—because it was the (–)-enantiomer). Any enantiomerically pure compound that could be related, by a series of chemical degradations and transformations, to D-(+)-glyceraldehyde was labelled D, and any compound that could be

D-(+)-glyceraldehyde

related to L-(–)-glyceraldehyde was labelled L. The processes concerned were slow and laborious (the scheme below shows how (–)-lactic acid was shown to be D-(–)-lactic acid) and are never used today. D and L are now used only for certain well-known natural molecules, where their use is established by tradition, for example, the L-amino acids or the D-sugars. These labels, D and L, are in *small capital* letters.

> ● Remember that the *R/S*, +/–, and D/L nomenclatures all arise from different observations and the fact that a molecule has, say, the *R* configuration gives no clue as to whether it will have + or – optical activity or be labelled D or L. Never try and label a molecule as D/L or +/– simply by working it out from the structure. Likewise, never try and predict whether a molecule will have a + or – specific rotation by looking at the structure.

> ### The correlation between D-(–)-lactic acid and D-(+)-glyceraldehyde
>
> Here, for example, is the way that (–)-lactic acid was shown to have the same configuration as D-(+)-glyceraldehyde. We do not expect you to have come across the reactions used here.
>
>
> Notice from this scheme that the three intermediates all have the 'same' stereochemistry and yet one is (*R*) and two are (*S*). This is merely the consequence of the priority of the elements. (*R*) can be D or L and (+) or (–).

Diastereoisomers are stereoisomers that are not enantiomers

Two enantiomers are chemically identical because they are mirror images of one another. Other types of stereoisomers may be chemically (and physically) quite different. These two alkenes, for example, are geometrical isomers (or *cis–trans* isomers). Their physical chemical properties are different, as you would expect, since they are quite different in shape. Neither is chiral of course as they are planar molecules.

butenedioic acids

trans-butenedioic acid (fumaric acid)
m.p. 299–300 °C

cis-butenedioic acid (maleic acid)
m.p. 140–142 °C

A similar type of stereoisomerism can exist in cyclic compounds. In one of these, 4-*t*-butyl-cyclohexanols, the two substituents are on the same side of the ring; in the other, they are on opposite sides of the ring. Again, the two compounds have chemical and physical properties that are quite different.

4-*t*-butylcyclohexanol

cis 4-*t*-butyl-
cyclohexanol
mp 82–83 °C
^{1}H NMR: δ_H of
green proton 4.02

cis isomer

trans 4-*t*-butyl-
cyclohexanol
mp 80–81 °C
^{1}H NMR: δ_H of
green proton 3.50

trans isomer

> ■ Notice that we do not always write in all the hydrogen atoms. If the *t*-butyl group is forward, as in these diagrams, the hydrogen atom must be back.

Stereoisomers that are not mirror images of one another are called *diastereoisomers*. Both of these pairs of isomers fall into this category. Notice how the physical and chemical properties of a pair of diastereoisomers differ.

> ● The physical and chemical properties of enantiomers are identical; the physical and chemical properties of diastereoisomers differ. 'Diastereoisomer' is sometimes shortened to 'diastereomer'.

Diastereoisomers can be chiral or achiral

This pair of epoxides was produced by chemists in Pennsylvania in the course of research on drugs intended to alleviate the symptoms of asthma. Clearly, they are again diastereoisomers, and again they have different properties. Although the reaction they were using to make these compounds gave some of each diastereoisomer, the chemists working on these compounds only wanted to use the first (*trans*) epoxide. They were able to separate it from its *cis* diastereoisomer by chromatography because the diastereoisomers differ in polarity.

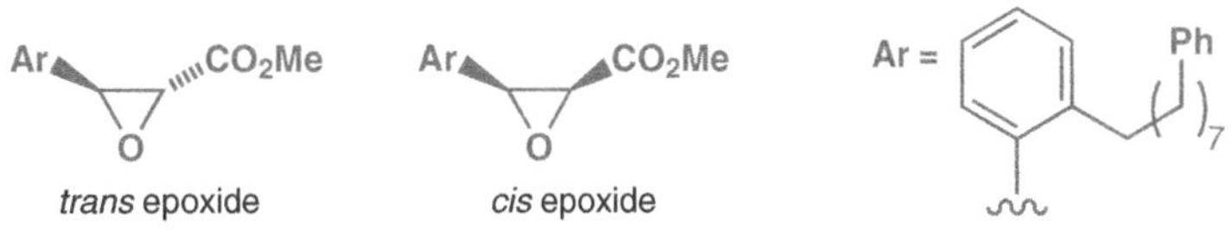

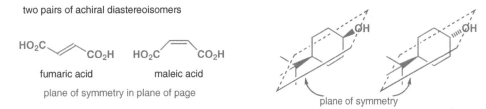

This time, the diastereoisomers are a little more complex than the examples above. The first two pairs of diastereoisomers we looked at were achiral—they each had a plane of symmetry through the molecule.

two pairs of achiral diastereoisomers

fumaric acid   maleic acid

plane of symmetry in plane of page

plane of symmetry

The epoxide diastereoisomers, on the other hand, are chiral. We know this because they do not have a plane of symmetry and we can check that by drawing the mirror image of each one: it is not superimposable on the first structure.

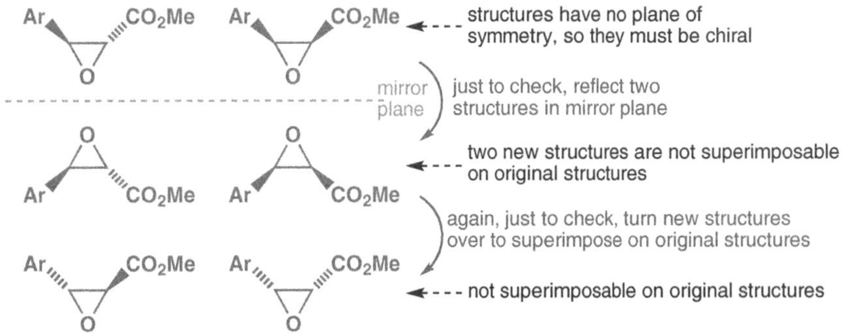

structures have no plane of symmetry, so they must be chiral

mirror plane } just to check, reflect two structures in mirror plane

two new structures are not superimposable on original structures

again, just to check, turn new structures over to superimpose on original structures

not superimposable on original structures

If a compound is chiral, it can exist as two enantiomers. We've just drawn the two enantiomers of each of the diastereoisomers of our epoxide. This set of four structures contains two diastereoisomers (stereoisomers that are not mirror images). These are the two different chemical compounds, the *cis* and *trans* epoxides, that have different properties. Each can exist as two enantiomers (stereoisomers that are mirror images) indistinguishable except for rotation. We have two pairs of diastereoisomers, each being a pair of enantiomers. When you are considering the stereochemistry of a compound, always distinguish the diastereoisomers first and then split these into enantiomers if they are chiral.

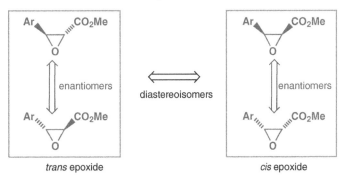

trans epoxide   *cis* epoxide

In fact, the chemists working on these compounds wanted only one enantiomer of the *trans* epoxide—the top left stereoisomer. They were able to separate the *trans* epoxide from the *cis* epoxide by chromatography because they are diastereoisomers. However, because they had made both diastereoisomers in the laboratory from achiral starting materials, both diastereoisomers were racemic mixtures of the two enantiomers. Separating the top enantiomer of the *trans* epoxide from the bottom one was much harder because enantiomers have identical physical and chemical properties. To get just the enantiomer they wanted the chemists had to develop some completely different chemistry, using enantiomerically pure compounds derived from nature.

➡ We shall discuss how chemists make enantiomerically pure compounds later in this chapter, and in more detail in Chapter 41.

Absolute and relative stereochemistry

When we talk about two chiral diastereoisomers, we have no choice but to draw the structure of one enantiomer of each diastereoisomer because we need to include the stereochemical information to distinguish them, even if we're talking about a racemic mixture of the two enantiomers. To avoid confusion, it's best to write something definite under the structure, such as ' ± ' (meaning racemic) under a structure if it means 'this diastereoisomer' but not 'this enantiomer of this diastereoisomer'. So in this case we should say that the chemists were able to separate these two diastereoisomers, but wanted only one enantiomer of the *trans* diastereoisomer and that this could not be separated by physical means.

cis and trans diastereisomers easily separated

this enantiomer of the
trans epoxide is wanted

When the stereochemistry drawn on a molecule means 'this diastereoisomer', we say that we are representing **relative stereochemistry**; when it means 'this enantiomer of this diastereoisomer' we say we are representing its **absolute stereochemistry**. Relative stereochemistry tells us only how the stereogenic centres *within a molecule* relate to each other.

● **Enantiomers and diastereoisomers**

- **Enantiomers** are stereoisomers that are mirror images. A pair of enantiomers are mirror-image forms of the same compound and have opposite **absolute stereochemistry**.
- **Diastereoisomers** are stereoisomers that are not mirror images. Two diastereoisomers are different compounds, and have different **relative stereochemistry**.
- Diastereoisomers may be achiral (have a plane of symmetry) or they may be chiral (have no plane of symmetry).

these diastereoisomers are *achiral*

these diastereoisomers are *chiral*

Diastereoisomers can arise when structures have more than one stereogenic centre

Let's analyse our set of four stereoisomers a little more closely. You may have already noticed that these structures all contain stereogenic centres—two in each case. Go back to the diagram of the four structures at the bottom of p. 312 and, without looking at the structures overleaf, assign an *R* or *S* label to each of the stereogenic centres.

You should have made assignments of *R* and *S* like this.

➡ You need to know, and be able to use, the rules for assigning *R* and *S*; they were explained on p. 308. If you get any of the assignments wrong, make sure you understand why.

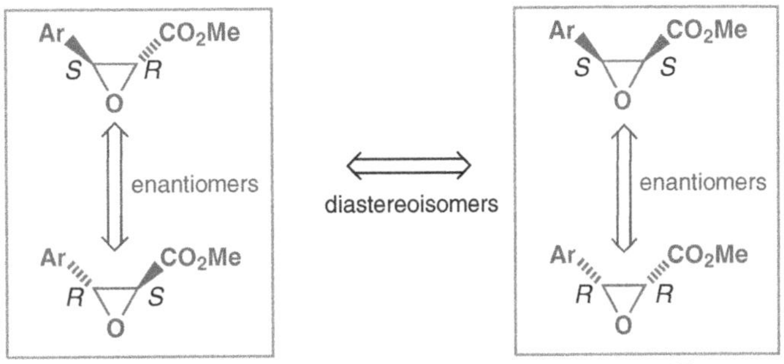

trans epoxide   cis epoxide

● Converting enantiomers and diastereoisomers

- To go from one *enantiomer* to another, *both* stereogenic centres are inverted.
- To go from one *diastereoisomer* to another, only *one* of the two is inverted.

■ If you are asked to explain some stereochemical point in an examination, choose a cyclic example—it makes it much easier.

All the compounds that we have talked about so far have been cyclic because the diastereoisomers are easy to visualize: two diastereoisomers can be identified because the substituents are either on the same side or on opposite sides of the ring (*cis* or *trans*). But acyclic compounds can exist as diastereoisomers too. Take these two, for example. Both ephedrine and pseudoephedrine are members of the amphetamine class of stimulants, which act by imitating the action of the hormone adrenaline.

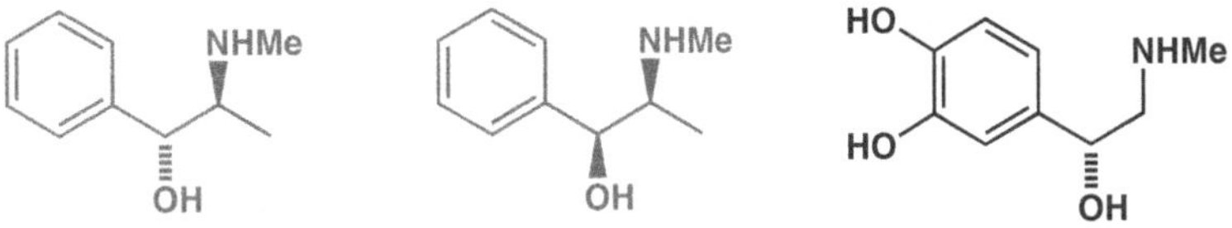

ephedrine   pseudoephedrine   adrenaline

Ephedrine and pseudoephedrine are stereoisomers that are clearly not mirror images of each other—only one of the two stereogenic centres in ephedrine is inverted in pseudoephedrine—so they must be diastereoisomers. Thinking in terms of stereogenic centres is useful because, just as this compound has two stereogenic centres and can exist as two diastereoisomers, any compound with more than one stereogenic centre can exist in more than one diastereoisomeric form.

Both ephedrine and pseudoephedrine are produced in enantiomerically pure form by plants, so, unlike the anti-asthma intermediates above, in this case we are talking about single enantiomers of single diastereoisomers. Adrenaline (also known as epinephrine) is also chiral. In nature it is a single enantiomer but it cannot exist as other diastereoisomers as it has only one stereogenic centre.

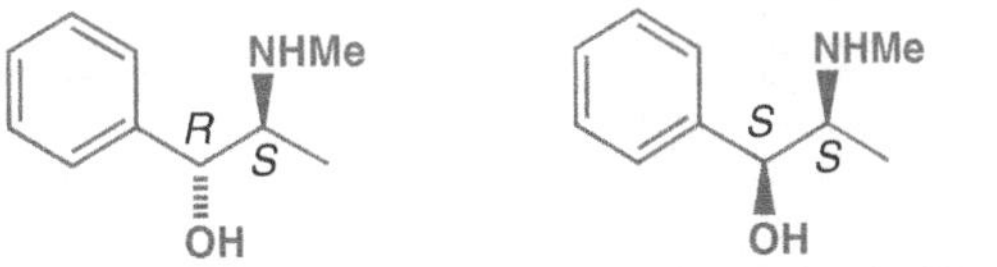

(1R,2S)-(–)-ephedrine   (1S,2S)-(+)-pseudoephedrine

Ephedrine and pseudoephedrine

Ephedrine is a component of the traditional Chinese remedy 'Ma Huang', extracted from *Ephedra* species. It is also used in nasal sprays as a decongestant. Pseudoephedrine is the active component of the decongestant Sudafed.

The 'natural' enantiomers of the two diastereomers are (–)-ephedrine and (+)-pseudoephedrine, which does not tell you which is which, or (1R,2S)-(–)-ephedrine and (1S,2S)-(+)-

pseudoephedrine, which does. From that you should be able to deduce the corresponding structures.

Here are some data on (1*R*,2*S*)-(–)-ephedrine and (1*S*,2*S*)-(+)-pseudoephedrine and their 'unnatural' enantiomers (which have to be made in the laboratory), (1*S*,2*R*)-(+)-ephedrine and (1*R*,2*R*)-(–)-pseudoephedrine.

■ Remember that (+) and (–) refer to the sign of the specific rotation, while *R* and *S* are derived simply by looking at the structure of the compounds. There is no simple connection between the two!

	(1*R*,2*S*)-(–)-ephedrine	(1*S*,2*R*)-(+)-ephedrine	(1*S*,2*S*)-(+)-pseudoephedrine	(1*R*,2*R*)-(–)-pseudoephedrine
mp	40–40.5 °C	40–40.5 °C	117–118 °C	117–118 °C
$[\alpha]_D^{20}$	–6.3	+6.3	+52	–52

● **The two diastereoisomers are different compounds with different names and different properties, while the pairs of enantiomers are the same compound with the same properties, differing only in the direction in which they rotate polarized light.**

We can illustrate the combination of two stereogenic centres in a compound by considering what happens when you shake hands with someone. Hand-shaking is successful only if you each use the same hand! By convention, this is your right hand, but it's equally possible to shake left hands. The overall pattern of interaction between two right hands and two left hands is the same: a right-handshake and a left-handshake are enantiomers of one another; they differ only in being mirror images. If, however, you misguidedly try to shake your right hand with someone else's left hand you end up holding hands. Held hands consist of one left and one right hand; a pair of held hands have totally different interactions from a pair of shaking hands; we can say that holding hands is a diastereoisomer of shaking hands. We can summarize the situation when we have two hands, or two chiral centres, each one *R* or *S*.

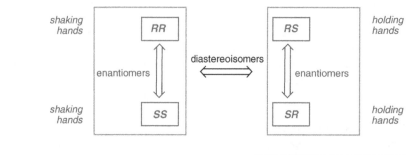

What about compounds with more than two stereogenic centres? The family of sugars provides lots of examples. Ribose is a five-carbon sugar that contains three stereogenic centres. The enantiomer shown here is the one used in the metabolism of all living things and, by convention, is known as D-ribose. The three stereogenic centres of D-ribose have the *R* configuration. For convenience, we will consider ribose in its open-chain form, but more usually it would be cyclic, as shown underneath.

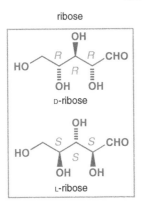

D-ribose,
open-chain form

D-ribose,
cyclic hemiacetal form

In theory we can work out how many 'stereoisomers' there are of a compound with three stereogenic centres simply by noting that there are 8 ($= 2^3$) ways of arranging Rs and Ss.

RRR	*RRS*	*RSR*	*RSS*
SSS	*SSR*	*SRS*	*SRR*

But this method blurs the all-important distinction between diastereoisomers and enantiomers. In each case, the combination in the top row and the combination directly below it are enantiomers (all three centres are inverted); the four columns are diastereoisomers. Three stereogenic centres therefore give four diastereoisomers, each a pair of two enantiomers. Going back to the example of the C_5 aldoses, each of these diastereoisomers is a different sugar. In these diagrams each diastereoisomer is in a frame but the top line shows one enantiomer (D) and the bottom line the other (L).

ribose	arabinose	xylose	lyxose
D-ribose	D-arabinose	D-xylose	D-lyxose
L-ribose	L-arabinose	L-xylose	L-lyxose

■ You do not need to remember the names of these sugars.

Structure of sugars

A sugar has the empirical formula $C_nH_{2n}O_n$, and consists of a chain of carbon atoms, one being a carbonyl group and the rest carrying OH groups. If the carbonyl group is at the end of the chain (in other words, it is an aldehyde), the sugar is an aldose. If the carbonyl group is not at the end of the chain, the sugar is a ketose. We come back to all this in detail in Chapter 42. The number of carbon atoms, n, can be 3–8: aldoses have n − 2 stereogenic centres and ketoses n − 3 stereogenic centres. In fact, most sugars exist as an equilibrium mixture of this open-chain structure and a cyclic hemiacetal isomer (Chapter 6).

an aldose   a ketose

■ This is an oversimplification to be used cautiously because it works only if all diastereoisomers are chiral and fails with the sort of symmetrical molecules we are about to describe.

You've probably recognized that there's a simple mathematical relationship between the number of stereogenic centres and the number of stereoisomers a structure can have. **Usually**, a structure with n stereogenic centres can exist as 2^n stereoisomers. These stereoisomers consist of 2^{n-1} diastereoisomers, each of which has a pair of enantiomers.

Fischer projections

The stereochemistry of sugars used to be represented by Fischer projections. The carbon backbone was laid out in a vertical line and twisted in such a way that all the substituents pointed towards the viewer. Fischer projections are so unlike real molecules that you should never use them. However, you may see them in older books, and you should have an idea about how to interpret them. Just remember that all the branches down the side of the central trunk are effectively bold wedges (coming towards the viewer), while the central trunk lies in the plane of the paper. By mentally twisting the backbone into a realistic zig-zag shape you should end up with a reasonable representation of the sugar molecule.

D-ribose   used to be drawn as: means

D-ribose
(Fischer projection)

Why only *usually?*—achiral compounds with more than one stereogenic centre

Sometimes, symmetry in a molecule can cause some stereoisomers to be degenerate, or 'cancel out'—there aren't as many stereoisomers as you'd *expect*. Take tartaric acid, for example. This stereoisomer of tartaric acid is found in grapes, and its salt, potassium hydrogen tartrate, can precipitate out as crystals at the bottom of bottles of wine. It has two stereogenic centres, so you'd expect $2^2 = 4$ stereoisomers; two diastereoisomers, each a pair of enantiomers.

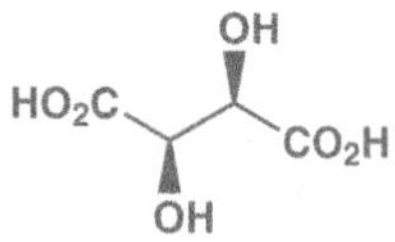

(+)-tartaric acid

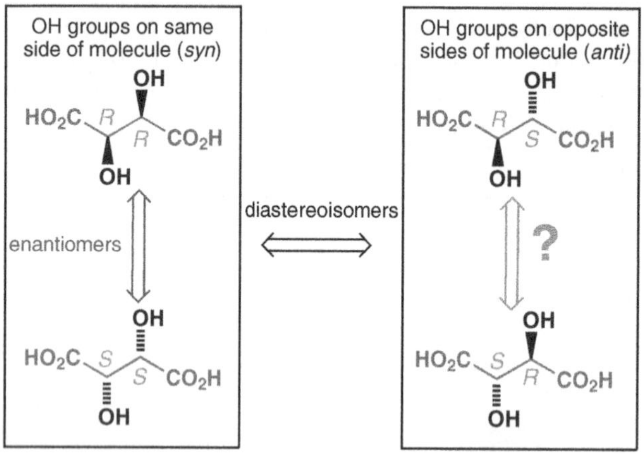

Interactive stereoisomers of tartaric acid

While the pair of structures on the left are certainly enantiomers, if you look carefully at the pair of structures on the right, you'll see that they are, in fact, not enantiomers but identical structures. To prove it, just rotate the top one through 180° in the plane of the paper.

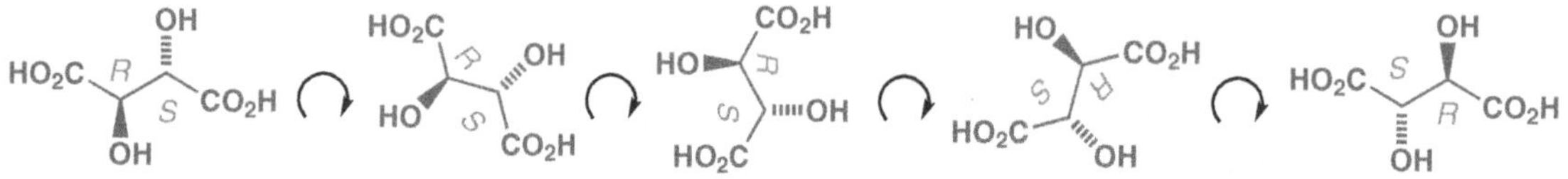

$(1R,2S)$-Tartaric acid and $(1S,2R)$-tartaric acid are not enantiomers, but they are identical because, even though they contain stereogenic centres, they are achiral. By drawing $(1R,2S)$-tartaric acid after a 180° rotation about the central bond, you can easily see that it has a mirror plane, and so must be achiral. Since the molecule has a plane of symmetry, and R is the mirror image of S, the R,S diastereoisomer cannot be chiral.

Interactive display of meso form of tartaric acid

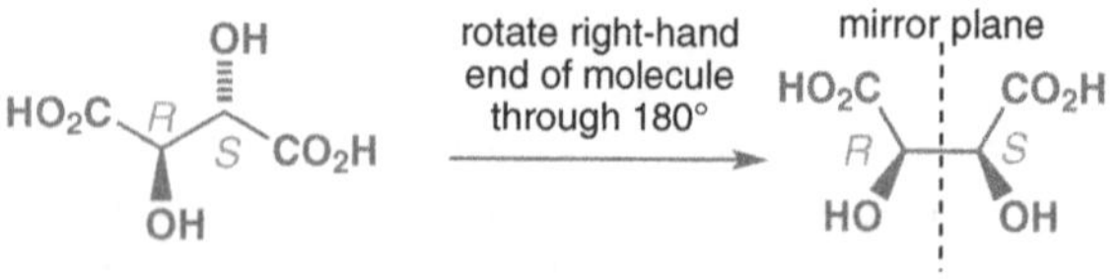

■ These two structures are the same molecule drawn in two different conformations—to get from one to the other just rotate half of the molecule about the central bond.

● **Compounds that contain stereogenic centres but are themselves achiral are called *meso* compounds. This means that there is a plane of symmetry with *R* stereochemistry on one side and *S* stereochemistry on the other.**

So tartaric acid can exist as two diastereoisomers, one with two enantiomers and the other achiral (a *meso* compound). It's worth noting that the formula stating that a compound with n stereogenic centres has 2^{n-1} diastereoisomers has worked but not the formula that states there are 2^n 'stereoisomers'. In general, it's safer not to count up total 'stereoisomers' but to work out first how many diastereoisomers there are, and then to decide whether or not each one is chiral, and therefore whether or not it has a pair of enantiomers.

Meso hand-shaking

We can extend our analogy between hand-shaking and diastereoisomers to *meso* compounds as well. Imagine a pair of identical twins shaking hands. They could be shaking their left hands or their right hands and there would be a way to tell the two handshakes apart because they are enantiomers. But if the twins hold hands, you will not be able to distinguish left holds right from right holds left, because the twins themselves are indistinguishable—this is the *meso* hand-hold!

	Chiral diastereoisomer		Achiral diastereoisomer
	(+)-tartaric acid	(−)-tartaric acid	meso-tartaric acid
$[\alpha]_D^{20}$	+12	−12	0
m.p.	168–170 °C	168–170 °C	146–148 °C

inositol

pentane-2,3,4-triol

Meso diastereoisomers of inositol

Look out for *meso* diastereoisomers in compounds that have a degree of symmetry in their overall structure. Inositol, one of whose diastereoisomers is an important growth factor, has six stereogenic centres. It's a challenge to work out how many diastereoisomers it has—in fact all but one of them are *meso*.

■ *syn* and *anti*: These refer to substituents on the same side (*syn*) or on opposite (*anti*) sides of a chain or ring. They must be used only in reference to a diagram.

Investigating the stereochemistry of a compound

When you want to describe the stereochemistry of a compound our advice is to identify the diastereoisomers and then think about whether they are chiral or not. Don't just count up 'stereoisomers'—to say that a compound with two stereogenic centres has four 'stereoisomers' is rather like saying that 'four hands are getting married'. Two people are getting married, each with two hands.

Let's work through how you might think about the stereochemistry of a simple example, the linear triol 2,3,4-trihydroxypentane or pentane-2,3,4-triol.

This is what you should do.

1

1. Draw the compound with the carbon skeleton in the usual zig-zag fashion running across the page, **1**.

2

2. Identify the chiral centres, **2**.

all up or *syn,syn*   outside one down, others up or *anti,syn*   inside one down, others up or *anti,anti*

3. Decide how many diastereoisomers there are by putting the substituents at those centres up or down. It often helps to give each diastereoisomer a 'tag' name. In this case there are three diastereoisomers. The three OH groups can be all on the same side or else one of the end OHs or the middle one can be on the opposite side to the rest. We can call the first *syn,syn* because the two pairs of chiral centres (1 & 2, and 2 & 3) groups are both arranged with the OHs on the same side of the molecule (*syn*).

plane of symmetry   chiral   plane of symmetry
achiral (*meso*)   achiral (*meso*)

4. By checking on possible planes of symmetry, see which diastereoisomers are chiral. In this case only the plane down the centre can be a plane of symmetry.

the two enantiomers of the *anti,syn* diastereoisomer

5. Draw the enantiomers of any chiral diastereoisomer by inverting *all* the stereogenic centres. This can easily be achieved by reflecting the molecule in the plane of the paper, as if it were a mirror. Everything that was 'up' is now 'down' and vice versa.

6. Announce the conclusion. You could have said that there are four 'stereoisomers' but the following statement is much more helpful. There are three diastereoisomers, the *syn,syn*, the *syn,anti*, and the *anti,anti*. The *syn,syn* and the *anti,anti* are achiral (*meso*) compounds but the *syn,anti* is chiral and has two enantiomers.

The mystery of Feist's acid

It is hard nowadays to realize how difficult structure solving was before the days of spectroscopy. A celebrated case was that of 'Feist's acid', discovered by Feist in 1893 from a deceptively simple reaction. Early work without spectra led to two suggestions for its structure, both based on a three-membered ring, which gave the compound some fame because unsaturated three-membered rings were rare. The favoured structure was the cyclopropene.

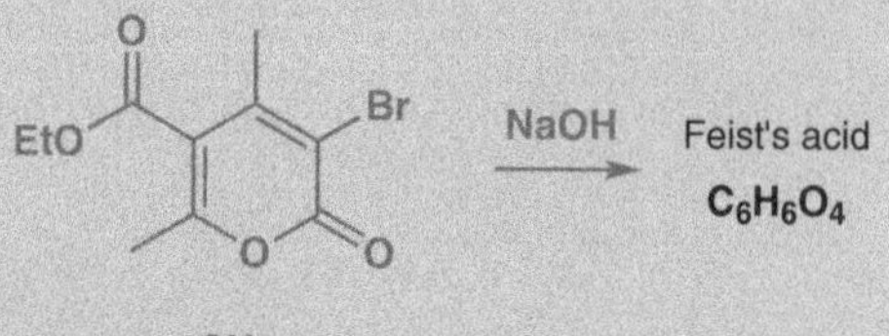

The argument was still going on in the 1950s when the first NMR spectrometers appeared. Although infrared appeared to support the cyclopropene structure, one of the first problems resolved by the primitive 40 MHz instruments available was that of Feist's acid, which had no methyl group signal but did have two protons on a double bond and so had to be the exomethylene isomer after all.

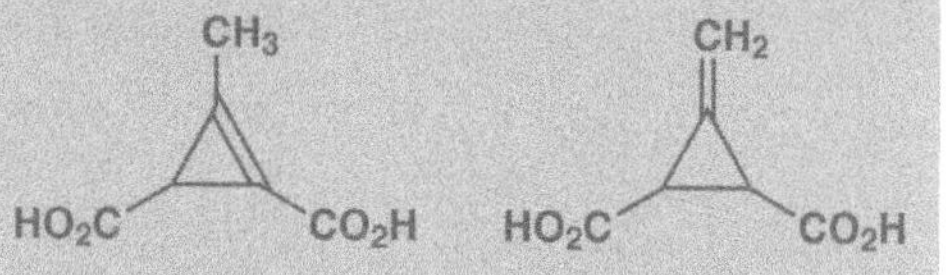

This structure has two chiral centres, so how will we know which diastereoisomer we have? The answer was simple: the stereochemistry has to be *trans* because Feist's acid is chiral: it can be resolved (see later in this chapter) into two enantiomers. Now, the *cis* diacid would have a plane of symmetry, and so would be achiral—it would be a *meso* compound. The *trans* acid on the other hand is chiral. If you do not see this, try superimposing it on its mirror image—you will find that you cannot. In fact, Feist's acid has an *axis* of symmetry, and you will see shortly that axes of symmetry *are* compatible with chirality.

Modern NMR spectra make the structure easy to deduce. There are only two proton signals as the CO_2H protons exchange in the DMSO solvent needed. The two protons on the double bond are identical (5.60 ppm) and so are the two protons on the three-membered ring, which come at the expected high field (2.67 ppm). There are four carbon signals: the C=O at 170 ppm, two alkene signals between 100 and 150 ppm, and the two identical carbons in the three-membered ring at 25.45 ppm.

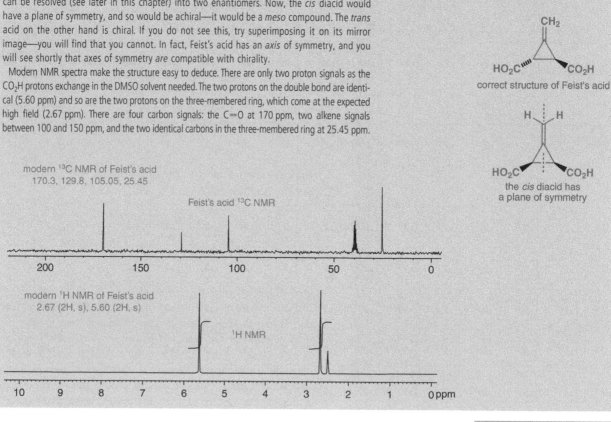

modern ^{13}C NMR of Feist's acid
170.3, 129.8, 105.05, 25.45

Feist's acid ^{13}C NMR

modern ^1H NMR of Feist's acid
2.67 (2H, s), 5.60 (2H, s)

^1H NMR

cyclopropene isomer

exomethylene isomer

correct structure of Feist's acid

the *cis* diacid has a plane of symmetry

🖱 Interactive possible structures for Feist's acid

Chiral compounds with no stereogenic centres

A few compounds are chiral, yet have no stereogenic centres. Try making a model of the allene in the margin. It has no stereogenic centre, but these mirror images are not superimposable and so the allene is chiral: the structures shown are enantiomers. Similarly, some biaryl compounds, such as the important bisphosphine below, known as BINAP, exist as two separate enantiomers because rotation about the green bond is restricted. If you were to look at this molecule straight down along the green bond, you would see that the two flat rings are at right angles to each other and so the molecule has a twist in it rather like the 90° twist in the allene. Compounds that are chiral because of restricted rotation about a single bond are called *atropisomers* (from the Greek for 'won't turn').

a chiral allene

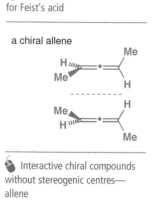

🖱 Interactive chiral compounds without stereogenic centres—allene

➡ We come back to BINAP in Chapter 41.

🖱 Interactive chiral compounds without stereogenic centres— BINAP

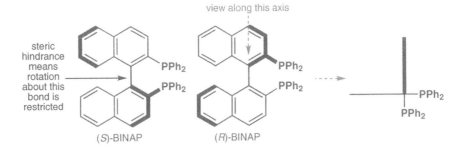

(S)-BINAP   (R)-BINAP

These two examples rely on the rigidity of π systems but this simple saturated system is also chiral. These two rings have to be orthogonal because of the tetrahedral nature of the central carbon atom. There is no chiral centre, but there is no plane of symmetry. Cyclic compounds like this with rings joined at a single C atom are called spiro compounds. Spiro compounds are often chiral even when at a first glance they look quite symmetrical, and you should look particularly carefully for planes of symmetry when you think about their stereochemistry.

🖱 Interactive chiral compounds without stereogenic centres—spiro amide

non-superimposable enantiomers

Axes and centres of symmetry

rotate 180° about this axis and the molecule is the same: it has C_2 symmetry

(R)-BINAP

🖱 Interactive BINAP showing C_2 axis of symmetry

■ The subscript 2 means twofold axis of symmetry. Other orders of axial symmetry are possible in chemistry but are much rarer in simple organic compounds.

The fact that the three compounds we have just introduced (along with Feist's acid in the box on p. 319) were chiral might have surprised you, because at first glance they do look quite 'symmetrical'. In fact, they do all have an element of symmetry, and it is only one which is compatible with chirality: an *axis* of symmetry. If a molecule can be rotated through 180° about an axis to give exactly the same structure then it has twofold axial symmetry, or C_2 symmetry. Compounds with an axis of symmetry will still be chiral, provided they lack either a plane or a centre of symmetry.

C_2 symmetry is common in many more everyday molecules than the ones in the last section. Below is an example of a compound with two diastereoisomers. One (we call it the *syn* diastereoisomer here—the two phenyl rings are on the same side) has a plane of symmetry—it must be achiral (and as it nonetheless has chiral centres we can also call it the *meso* diastereoisomer). The other has some degree of symmetry, but it has axial symmetry and can therefore be chiral. The C_2 axis of symmetry is shown in orange. Rotating 180° gives back the same structure, but reflecting in a mirror plane (brown) gives a non-superimposable mirror image.

🖱 Interactive epoxide diastereoisomers showing plane of symmetry

this diastereoisomer is achiral

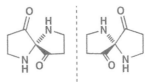

the plane of symmetry in the *syn* diastereoisomer

this diastereoisomer is chiral

rotate 180°

the axis of symmetry in the *anti* diastereoisomer

non-superimposable mirror image

■ We warned you that these statements (pp. 304 and 312) were incomplete: we are now about to complete them.

So far we have used a *plane of symmetry* as the defining characteristic of an achiral molecule: we have said several times that a molecule is chiral if it lacks a plane of symmetry. We are now going to introduce a second type of symmetry that is not compatible with chirality. If a molecule has a *centre of symmetry* it is not chiral. We will now explain how to spot a centre of symmetry.

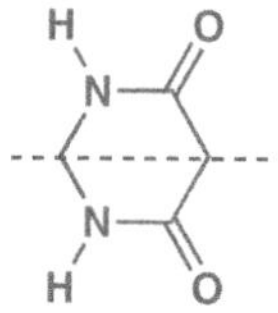

The diamide skeleton in the margin has a plane of symmetry in the plane of the page and also a plane of symmetry at right angles to that plane passing through the two saturated carbon atoms (represented by the green dotted line). If we add substituents R to this structure we can have two diastereoisomers with the two R groups on the same side (*syn*) of the flat ring or on opposite (*anti*) sides. Although the plane of the paper is no longer a plane of symmetry, neither isomer is chiral as the other plane bisects the substituents and is still a plane of symmetry. So far nothing new.

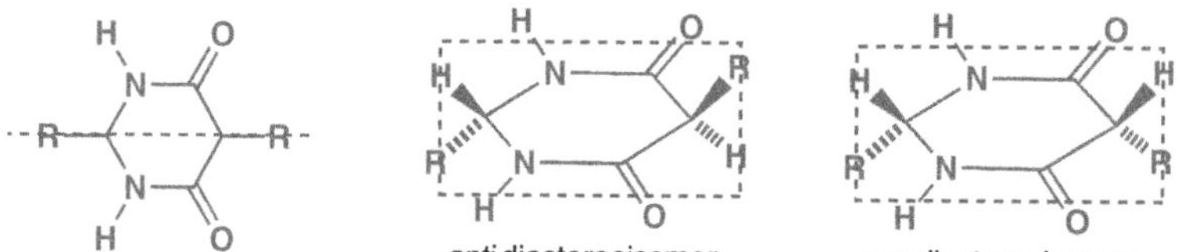

anti diastereoisomer *syn* diastereoisomer
both have a plane of symmetry: both are achiral

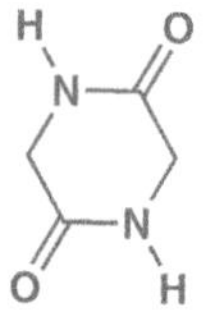

Now consider the related double amide below. The plane of the page is again a plane of symmetry but there is now no plane of symmetry at right angles. This heterocycle is called a 'diketopiperazine' and can be made by dimerizing an amino acid: the compound in the margin is the dimer of glycine. With substituted amino acids, such as those below where R ≠ H, there are again two diastereoisomers, *syn* and *anti*. But their symmetry properties are different. The *syn* isomer is chiral but the *anti* isomer is not.

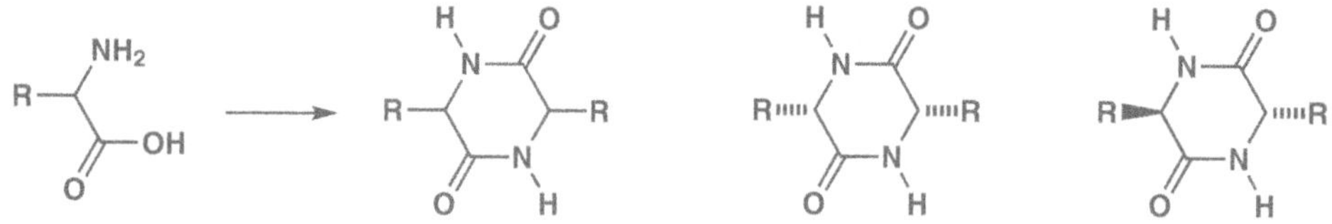

amino acid amino acid dimer *syn* diastereoisomer *anti* diastereoisomer

The *syn* diastereoisomer has no plane of symmetry but you should be able to spot a C_2 axis of symmetry running straight through the middle of the ring. The axis is compatible with chirality of course. In this compound both chiral centres are *S* and it has an enantiomer where both are *R*.

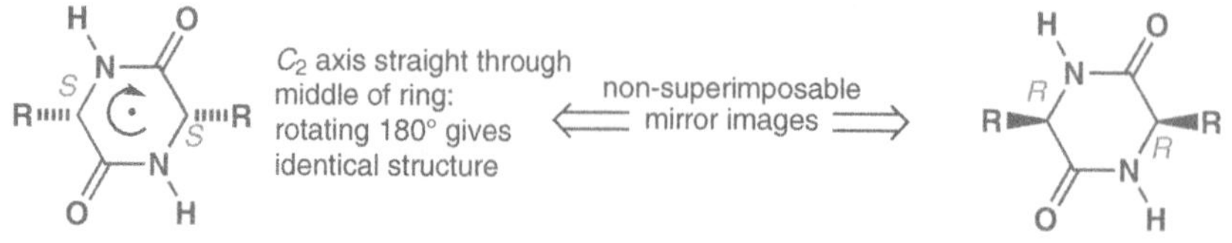

C_2 axis straight through middle of ring: rotating 180° gives identical structure

non-superimposable mirror images

syn diastereoisomer

the other enantiomer of the *syn* diastereoisomer

Interactive diamides showing centre of symmetry

The *anti* diastereoisomer has no plane of symmetry, nor does it have an axis. Instead it has a *centre* of symmetry. This is marked with a black dot in the middle of the molecule and means that if you go in any direction from this centre and meet, say, an R group, you will meet the same thing if you go in the opposite direction (green arrows). The same thing applies to the brown arrows and, of course, to the ring itself. There is no centre of symmetry in the *syn* isomer as the green or brown arrows would point to R on one side and H on the other. The *anti* isomer is superimposable on its mirror image and is achiral.

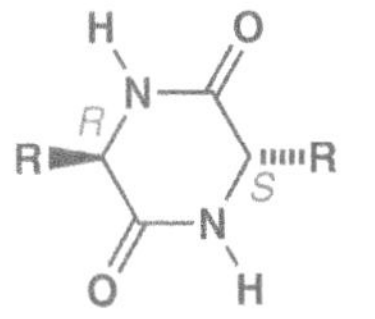

anti diastereoisomer

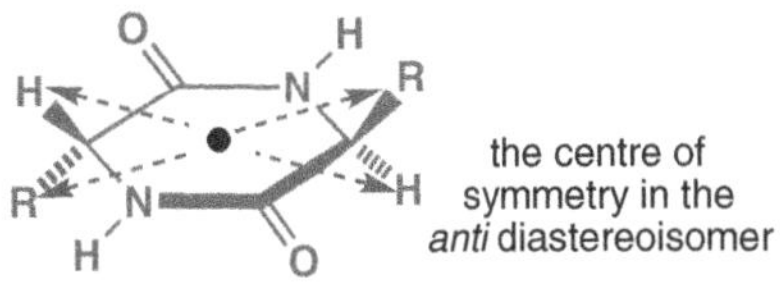

the centre of symmetry in the *anti* diastereoisomer

- **Chirality in terms of planes, centres, and axes of symmetry**

  - Any molecule which has a *plane* of symmetry or a *centre* of symmetry is *achiral*.
  - Any molecule which has an axis of symmetry is *chiral*, provided it does not also have a plane or a centre of symmetry. An axis of symmetry is the only symmetry element compatible with chirality.

Separating enantiomers is called resolution

Early in this chapter we said that most of the molecules in Nature are chiral, and that Nature usually produces these molecules as single enantiomers. We've talked about the amino acids, the sugars, ephedrine, pseudoephedrine, and tartaric acid—all compounds that can be isolated from natural sources as single enantiomers. On the other hand, in the laboratory, if we make chiral compounds from achiral starting materials we are doomed to get racemic mixtures. So how do chemists ever isolate compounds as single enantiomers, other than by extracting them from natural sources? We'll consider this question in much more detail in Chapter 41, but here we will look at the simplest way: using Nature's enantiomerically pure compounds to help us separate the components of a racemic mixture into its two enantiomers. This process is called resolution. Imagine the reaction between a chiral, but racemic, alcohol and a chiral, but racemic, carboxylic acid, to give an ester in an ordinary acid-catalysed esterification (Chapter 10).

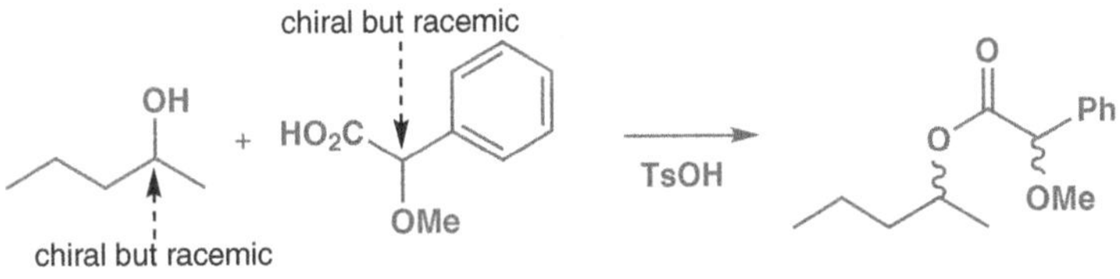

The product contains two chiral centres, so we expect to get two diastereoisomers, each a racemic mixture of two enantiomers. Diastereoisomers have different physical properties, so they should be easy to separate, for example by chromatography.

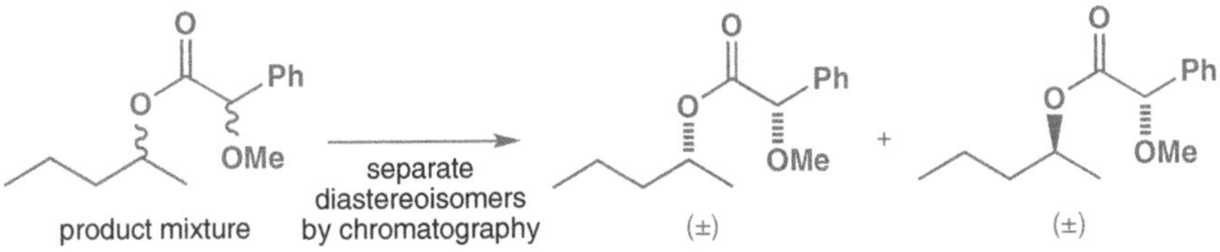

product mixture   separate diastereoisomers by chromatography   (±)   +   (±)

We could then reverse the esterification step and hydrolyse either of these diastereoisomers, to regenerate racemic alcohol and racemic acid.

■ Remember that (±) means the compounds are racemic: we're showing only relative, not absolute, stereochemistry.

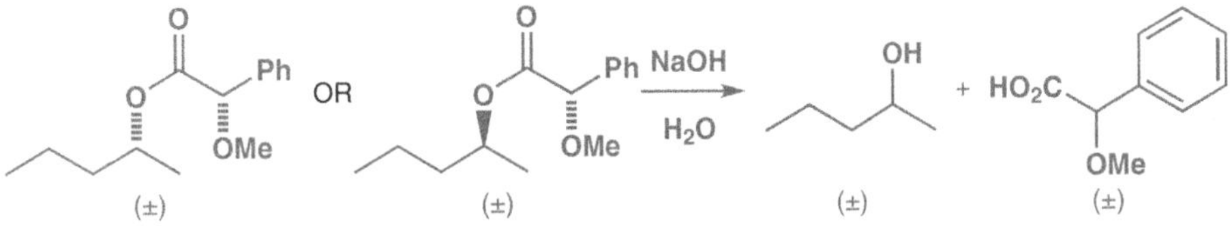

If we repeat this reaction, this time using an enantiomerically pure sample of the acid, available from (R)-mandelic acid, the almond extract you met on p. 310, we will again get two diastereoisomeric products, but this time each one will be enantiomerically pure. Note that the stereochemistry shown here *is* absolute stereochemistry.

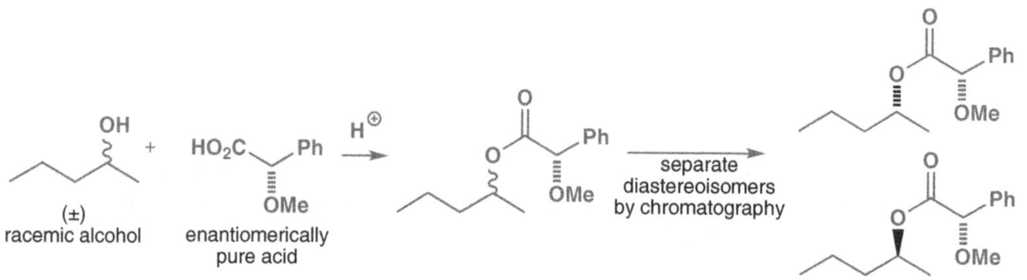

If we now hydrolyse each diastereoisomer separately, we have done something rather remarkable: we have managed to separate two enantiomers of the starting alcohol.

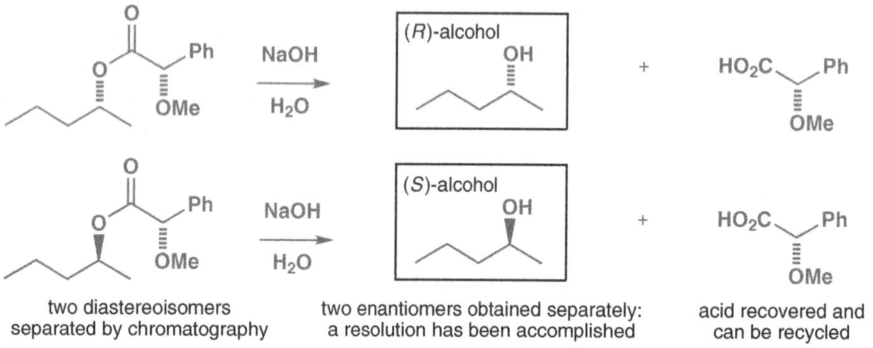

two diastereoisomers separated by chromatography — two enantiomers obtained separately: a resolution has been accomplished — acid recovered and can be recycled

A separation of two enantiomers is called a **resolution**. Resolutions can be carried out only if we make use of a component that is already enantiomerically pure: it is very useful that Nature provides us with such compounds; resolutions nearly always make use of compounds derived from nature.

Natural chirality

Why Nature uses only one enantiomer of most important biochemicals is an easier question to answer than how this asymmetry came about in the first place, or why L-amino acids and D-sugars were the favoured enantiomers, since, for example, proteins made out of racemic samples of amino acids would be complicated by the possibility of enormous numbers of diastereomers. Some have suggested that life arose on the surface of single chiral quartz crystals, which provided the asymmetric environment needed to make life's molecules enantiomerically pure. Or perhaps the asymmetry present in the spin of electrons released as gamma rays acted as a source of molecular asymmetry. Given that enantiomerically pure living systems should be simpler than racemic ones, maybe it was just chance that the L-amino acids and the D-sugars won out.

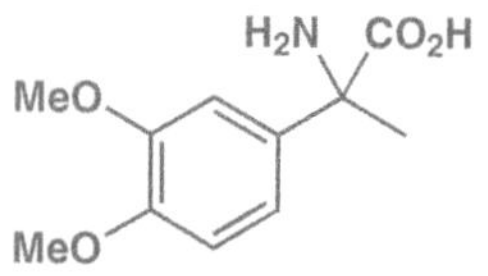

Now for a real example. Chemists studying the role of amino acids in brain function needed to obtain each of the two enantiomers of the amino acid in the margin. They made a racemic sample using the Strecker synthesis of amino acids that you met in Chapter 11. The racemic amino acid was treated with acetic anhydride to make the mixed anhydride and then with the sodium salt of naturally derived, enantiomerically pure alcohol menthol to give two diastereoisomers of the ester.

One of the diastereoisomers turned out to be more crystalline (that is, to have a higher melting point) than the other and, by allowing the mixture to crystallize, the chemists were able to isolate a pure sample of this diastereoisomer. Evaporating the diastereoisomer left in solution (the 'mother liquors') gave them the less crystalline diastereoisomer.

Next the esters were hydrolysed by boiling them in aqueous KOH. The acids obtained were enantiomers, as shown by their (nearly) opposite optical rotations and similar melting points. Finally, a more vigorous hydrolysis of the amides (boiling for 40 hours with 20% NaOH) gave them the amino acids they required for their biological studies (see bottom of p. 322).

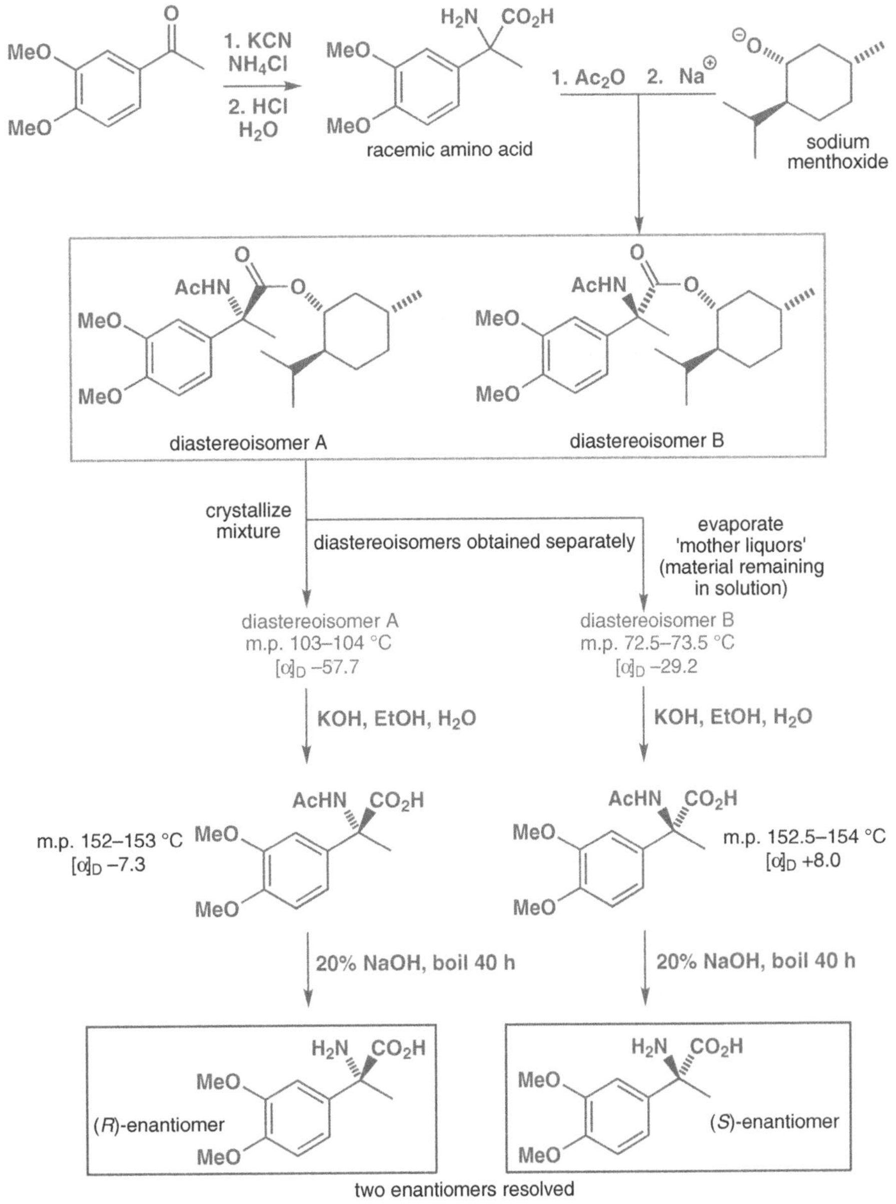

■ Note that the rotations of the pure diastereoisomers were not equal and opposite. These are single enantiomers of different compounds and there is no reason for them to have the same rotation.

diastereoisomer A
m.p. 103–104 °C
$[\alpha]_D$ −57.7

diastereoisomer B
m.p. 72.5–73.5 °C
$[\alpha]_D$ −29.2

m.p. 152–153 °C
$[\alpha]_D$ −7.3

m.p. 152.5–154 °C
$[\alpha]_D$ +8.0

(R)-enantiomer

(S)-enantiomer

two enantiomers resolved

Resolutions using diastereoisomeric salts

The key point about resolution is that we must bring together two stereogenic centres in such a way that there is a degree of interaction between them: separable diastereoisomers are created from inseparable enantiomers. In the last two examples, the stereogenic centres were brought together in covalent compounds, esters. Ionic compounds will do just as well—in fact, they are often better because it is easier to recover the compound after the resolution.

An important example is the resolution of the enantiomers of naproxen. Naproxen is a member of a family of compounds known as non-steroidal anti-inflammatory drugs (NSAIDs) which are 2-aryl propionic acids. This class also includes ibuprofen, the painkiller developed by Boots and marketed as Nurofen.

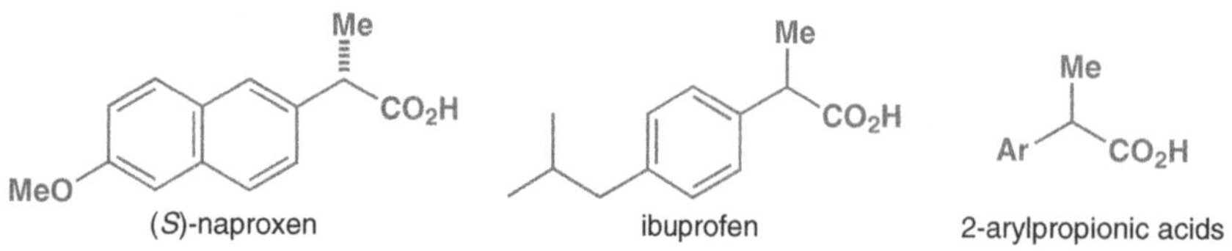

(S)-naproxen   ibuprofen   2-arylpropionic acids

Both naproxen and ibuprofen are chiral but, while both enantiomers of ibuprofen are effective painkillers, and the drug is sold as a racemic mixture (and anyway racemizes in the body) only the (S) enantiomer of naproxen has anti-inflammatory activity. When the American pharmaceutical company Syntex first marketed the drug they needed a way of resolving the racemic naproxen they synthesized in the laboratory.

Since naproxen is a carboxylic acid, they chose to make the carboxylate salt of an enantiomerically pure amine, and found that the most effective was a glucose derivative. Crystals were formed, which consisted of the salt of the amine and (S)-naproxen, the salt of the amine with (R)-naproxen (the diastereoisomer of the crystalline salt) being more soluble and so remaining in solution. These crystals were filtered off and treated with base, releasing the amine (which can later be recovered and reused) and allowing the (S)-naproxen to crystallize as its sodium salt. This is an unusual resolving agent as a simpler amine might usually be preferred. However, it makes the point that many resolving agents may have to be tried before one is found that works.

resolution of naproxen via an amine salt

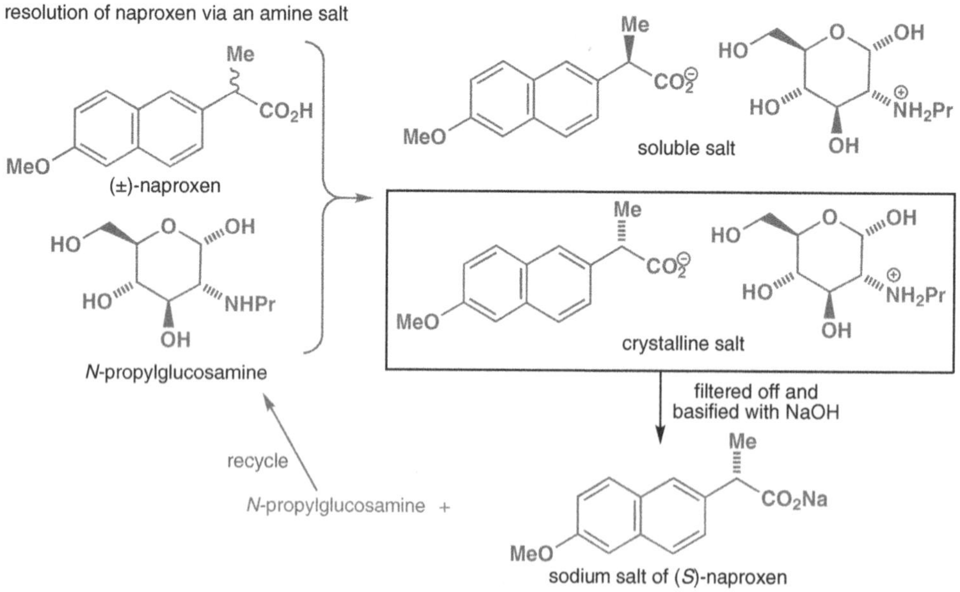

Chiral drugs

You may consider it strange that it was necessary to market naproxen as a single enantiomer, in view of what we have said about enantiomers having identical properties. The two enantiomers of naproxen do indeed have identical properties in the laboratory, but once they are inside a living system they, and any other chiral molecules, are differentiated by interactions with the enantiomerically pure molecules they find there. An analogy is that of a pair of gloves—the gloves weigh the same, are made of the same material, and have the same colour—in these respects they are identical. But interact them with a chiral environment, such as a hand, and they become differentiable because only one fits.

The way in which drugs interact with receptors mirrors this hand-and-glove analogy quite closely. Drug receptors, into which drug molecules fit like hands in gloves, are nearly always protein molecules, which are enantiomerically pure because they are made up of just L-amino acids. One enantiomer of a drug is likely to interact much better than the other, or perhaps in a different way altogether, so the two enantiomers of chiral drugs often have quite different pharmacological effects. In the case of naproxen, the (S)-enantiomer is 28 times as effective as the (R). Ibuprofen, on the other hand, is still marketed as a racemate because the compound racemizes in the bloodstream.

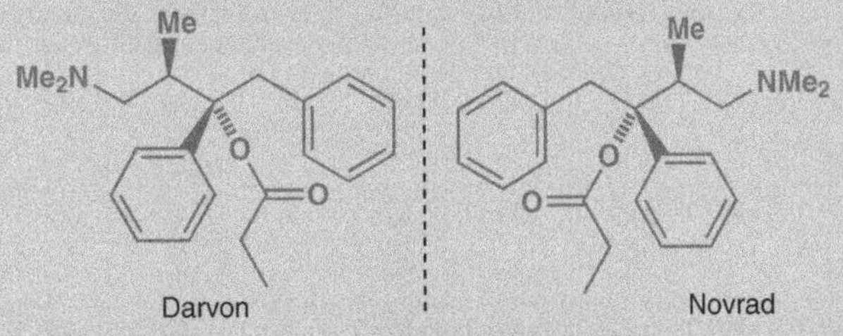

Darvon   Novrad

Sometimes, the enantiomers of a drug may have completely different therapeutic properties. One example is Darvon, which is a painkiller. Its enantiomer, known as Novrad, is an anticough agent. Notice how the enantiomeric relationship between these two drugs extends beyond their chemical structures! In Chapter 41 we will talk about other cases where two enantiomers have quite different biological effects.

Resolutions can be carried out by chromatography on chiral materials

■ Silica, SiO_2, is a macromolecular array of silicon and oxygen atoms. Its surface is covered with free OH groups, which can be used as an anchor for chiral derivatizing agents.

Interactions even weaker than ionic bonds can be used to separate enantiomers. Chromatographic separation relies on a difference in affinity between a stationary phase (often silica) and a mobile phase (the solvent travelling through the stationary phase, known as the eluent) mediated by, for example, hydrogen bonds or van der Waals interactions. If the stationary phase is made chiral by bonding it with an enantiomerically pure compound (often a derivative of an amino acid), chromatography can be used to separate enantiomers.

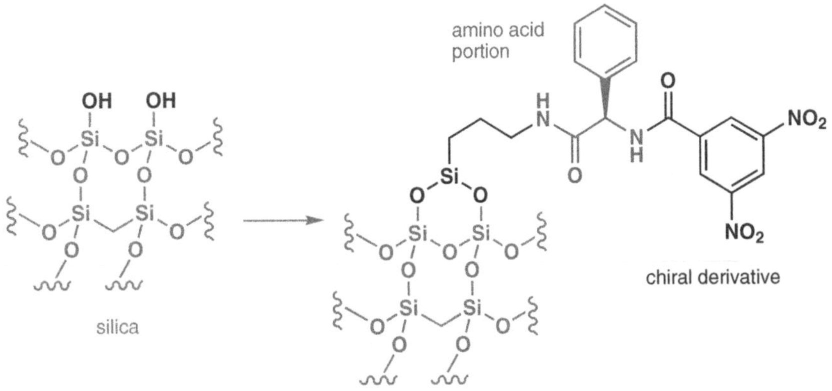

Chromatography on a chiral stationary phase is especially important when the compounds being resolved have no functional groups suitable for making the derivatives (usually esters or salts) needed for the more classical resolutions described above. For example, the two enantiomers of an analogue of the tranquillizer Valium were found to have quite different biological activities.

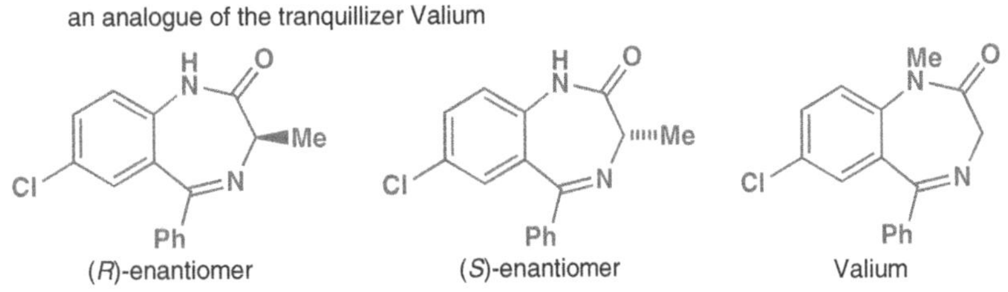

an analogue of the tranquillizer Valium

(R)-enantiomer   (S)-enantiomer   Valium

In order to study these compounds further, it was necessary to obtain them in enantiomerically pure form. This was done by passing a solution of the racemic compound through a column of silica bonded to an amino-acid-derived chiral stationary phase. The (R)-(–)-enantiomer showed a lower affinity for the stationary phase and therefore was eluted from the column first, followed by the (S)-(+)-enantiomer.

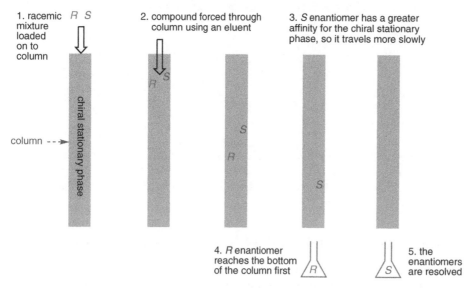

1. racemic mixture loaded on to column

column

chiral stationary phase

2. compound forced through column using an eluent

3. S enantiomer has a greater affinity for the chiral stationary phase, so it travels more slowly

4. R enantiomer reaches the bottom of the column first

5. the enantiomers are resolved

Two enantiomers of one molecule may be the same compound, but they are clearly different, although only in a limited number of situations. They can interact with biological systems differently, for example, and can form salts or compounds with different properties when reacted with a single enantiomer of another compound. In essence, enantiomers behave identically *except* when they are placed in a chiral environment. In Chapter 41 we will see how to use this fact to make single enantiomers of chiral compounds, but next we move on to three classes of reactions in which stereochemistry plays a key role: substitutions, eliminations, and additions.

■ You can think about chiral chromatography like this. Put yourself in this familiar situation: you want to help out a pensioner friend of yours who sadly lost his left leg in the war. A local shoe shop donates to you all their spare odd shoes, left and right, in his size (which happens to be the same as yours). You set about sorting the lefts from the rights, but are plunged into darkness by a power cut. What should you do? Well, you try every shoe on your right foot. If it fits you keep it; if not it's a left shoe and you throw it out.

Now this is just what chromatography on a chiral stationary phase is about. The stationary phase has lots of 'right feet' (one enantiomer of an adsorbed chiral molecule) sticking out of it and, as the mixture of enantiomers of 'shoes' flows past, 'right shoes' fit, and stick but 'left shoes' do not and flow on down the column, reaching the bottom first.

Further reading

There are very many books on stereochemistry. The most comprehensive is probably E. L. Eliel and S. H. Wilen, *Stereochemistry of Organic Compounds*, Wiley Interscience, Chichester, 1994. But you may find this too comprehensive at this stage. A more accessible introduction is the Oxford Primer *Organic Stereochemistry*, M. J. T. Robinson, OUP, Oxford, 2001.

The first announcement of the correct structure of Feist's acid was by M. G. Ettinger, *J. Am. Chem. Soc.*, 1952, **74**, 5805 and an interesting follow-up article gives the NMR spectrum: W. E. von Doering and H. D. Roth, *Tetrahedron*, 1970, **26**, 2825.

Check your understanding

 To check that you have mastered the concepts presented in this chapter, attempt the problems that are available in the book's Online Resource Centre at **http://www.oxfordtextbooks.co.uk/orc/clayden2e/**

Suggested solutions for Chapter 14

PROBLEM 1

Are these molecules chiral? Draw diagrams to justify your answer.

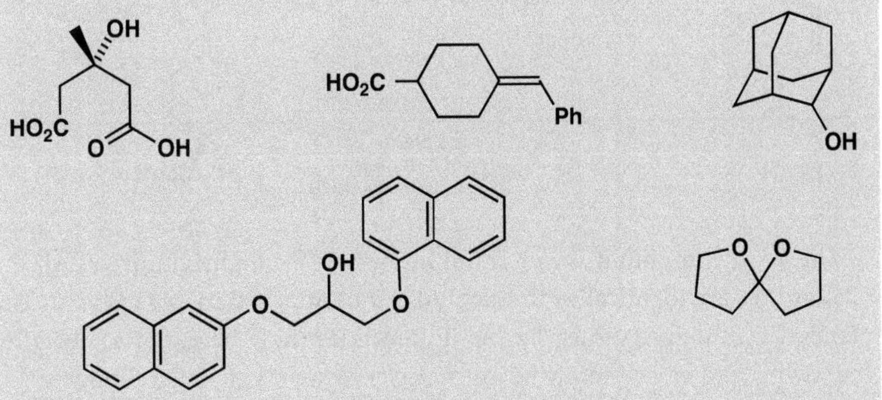

Purpose of the problem

Reinforcement of the very important criterion for chirality. Make sure you understand the answer.

Suggested solution

Only one thing matters: does the molecule have a plane of symmetry? We need to redraw some of them to see if they do. On no account look for chiral centres or carbon atoms with four different groups or anything else. *Just look for a plane of symmetry.* If the molecule has one, it isn't chiral. The first compound has been drawn with carboxylic acids represented in two different ways. The two CO_2H groups are in fact the same and the molecule has a plane of symmetry (shown by the dashed lines). It isn't chiral.

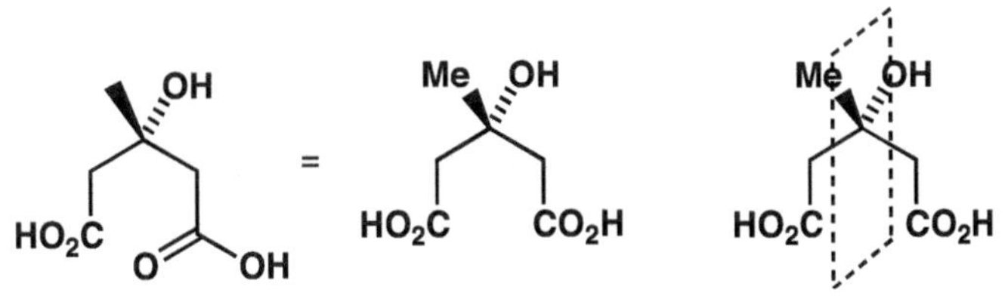

The second compound is chiral but if you got this wrong don't be dismayed. Making a model would help but there are only two plausible candidate planes of symmetry: the ring itself, in the plane of the page, and a plane at right angles to the ring. The molecule redrawn below with the tetrahedral centre displayed shows that the plane of the page isn't a plane of symmetry as the CO_2H is on one side and the H on the other, and neither is the plane perpendicular to the ring, as Ph is on one side and H on the other. No plane of symmetry: molecule is chiral.

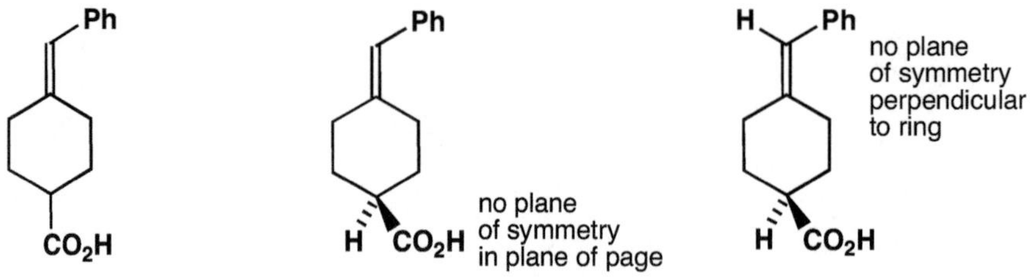

The third compound is not chiral because of its high symmetry. All the CH_2 groups are identical so the alcohol can be attached to any of them. The plane of symmetry (shown by the dotted lines) may be easier to see after redrawing, and will certainly be much easier to see if you make a model.

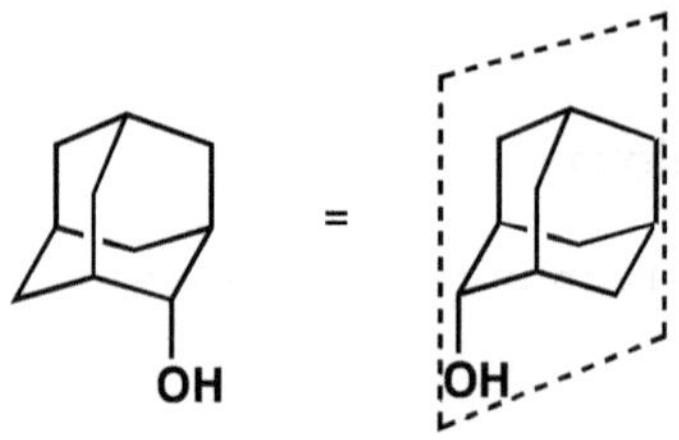

The fourth compound needs only the slightest redrawing to make it very clear that it is not chiral. The dashed line shows the plane of symmetry at right angles to the paper.

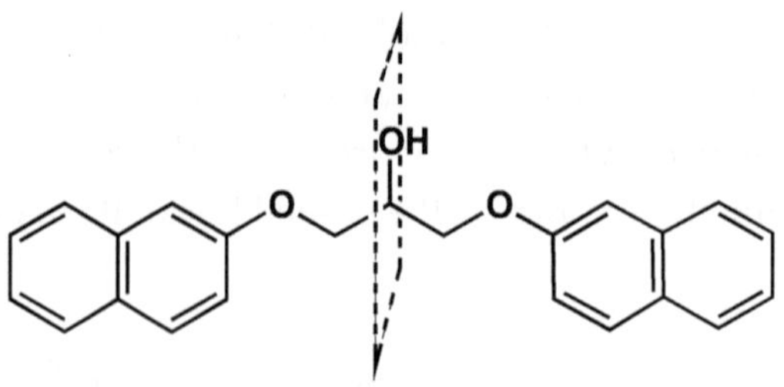

■ Spiro compounds, which contain two rings joined at a single atom, are discussed on p. 653 of the textbook.

The final acetal (which is a spiro compound) is drawn flat but the central carbon atom must in fact be tetrahedral so that the two rings are orthogonal. By drawing first one and then the other ring in the plane of the page it is easy to see that neither ring is a plane of symmetry for the other because of the oxygen atoms.

PROBLEM 2

If a solution of a compound has an optical rotation of +12, how could you tell if this was actually +12 or really –348 or +372?

Purpose of the problem

Revision of the meaning of optical rotation and what it depends on.

Suggested solution

Check the equation (p. 310 of the textbook) that states that rotation depends on three things: the rotating power of the molecule, the length of the cell used in the polarimeter, and the concentration of the solution. We can't change the first, we may be able to change the second, but the third is easiest to change. If we halve the concentration, the rotation will change to +6, – 174, or +186. That is not quite good enough as the last two figures are the same, but any other change of concentration will distinguish them.

PROBLEM 3

Cinderella's glass slipper was undoubtedly a chiral object. But would it have rotated the plane of polarized light?

Purpose of the problem

Revision of cause of rotation and optical activity.

Suggested solution

No. The macroscopic shape of an object is irrelevant. Only the molecular structure matters as light interacts with electrons in the molecules. Glass is not chiral (it is usually made up of inorganic borosilicates). Only if the slipper had been made of single enantiomers of a transparent substance would it have rotated the plane of polarized light. The molecules of Cinderella's left foot are the same as those in her right foot, despite both feet being macroscopically enantiomeric.

PROBLEM 4

Discuss the stereochemistry of these compounds. *Hint:* this means saying how many diastereoisomers there are, drawing clear diagrams of each, and stating whether they are chiral or not.

Purpose of the problem

Making sure you can handle this important approach to the stereochemistry of molecules.

Suggested solution

Just follow the hint in the question! Diastereoisomers are different compounds so they must be distinguished first. Then it is easy to say if each diastereoisomer is chiral or not. The first two are simple:

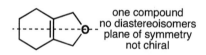 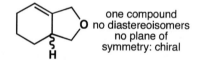

The third structure could exist as two diastereoisomers. The one with the *cis* ring junction has a plane of symmetry and is not chiral. The one with the *trans* ring junction has no plane of symmetry and is chiral (it has C_2 symmetry). Only one enantiomer is shown here.

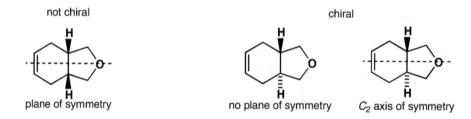

The last compound is most complicated as it has no symmetry at all. We can have two diastereoisomers and neither has a plane of symmetry. Both the *cis* compound and the *trans* compound can exist as two enantiomers.

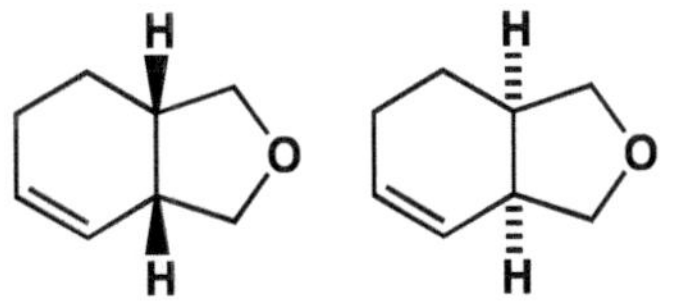

enantiomers of the *cis* compound 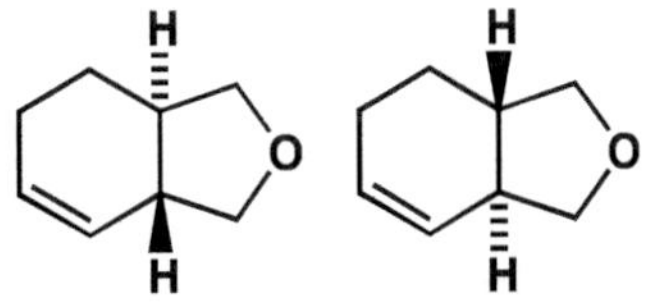   enantiomers of the *trans* compound

PROBLEM 5

In each case state, with explanations, whether the products of these reactions are chiral and/or enantiomericaly pure.

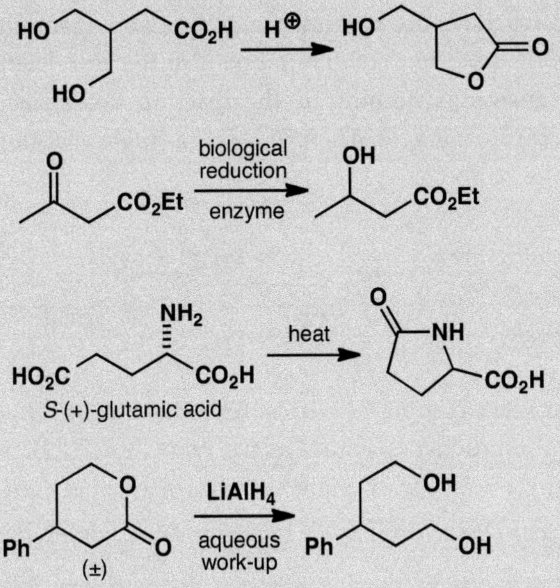

Purpose of the problem

Combining mechanism and stereochemical analysis for the first time.

Suggested solution

We need a mechanism for each reaction, a stereochemical description for each starting material (achiral, chiral? enantiomerically enriched?) and an analysis of what happens to the stereochemistry in each reaction. Don't forget: you can't get single enantiomers out of nothing—if everything that goes into a reaction is racemic or achiral, so is the product.

In the first reaction the starting material is achiral as the two CH₂OH side chains are identical. The product is chiral as it has no plane of symmetry but it cannot be one enantiomer as that would require one of the CH₂OH side chains to cyclize rather than the other. It must be racemic.

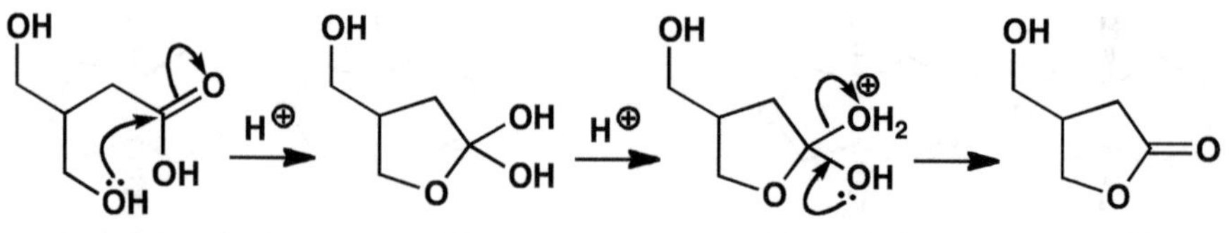

The starting material for the second reaction is planar and achiral. If the reagent had been sodium borohydride, the product would be chiral but racemic. But an enzyme, because it is made up of enantiomerically pure components (amino acids), can deliver hydride to one side of the ketone only. We expect the product to be enantiomerically enriched.

In the third reaction, the starting material is one enantiomer of a chiral compound. So we need to ask what happens to the chiral centre during the reaction. The answer is nothing as the reaction takes place between the amine and the carboxylic acid. The product is a single enantiomer too.

■ Amides don't usually form well from amines and carboxylic acids (see p. 207 of the textbook). But in this case the reaction is intramolecular, and with a fair degree of heating (the product is known trivially as 'pyroglutamic acid') the amide-forming reaction is all right.

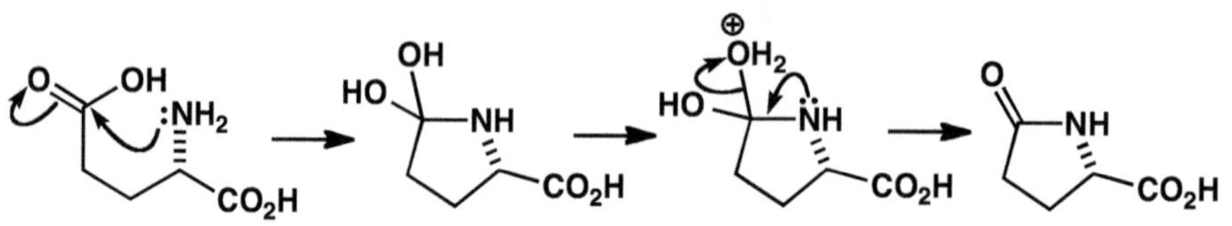

The final problem is a bit of a trick. The starting material is chiral, but racemic while the product is achiral as the two CH_2CH_2OH side chains are identical so there can be a plane of symmetry between them. The mechanism doesn't really matter but we might as well draw it.

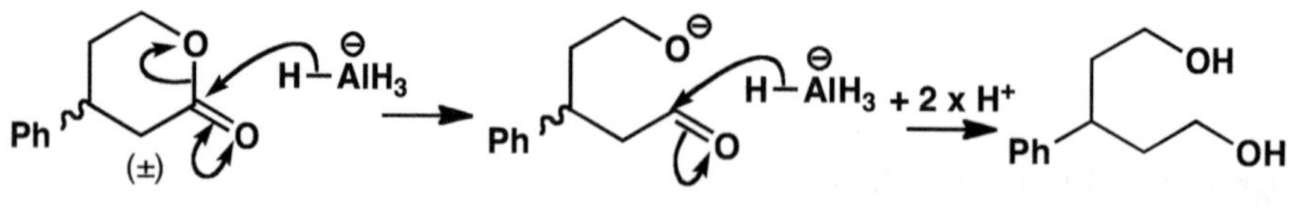

PROBLEM 6

This compound racemizes in base. Why is that?

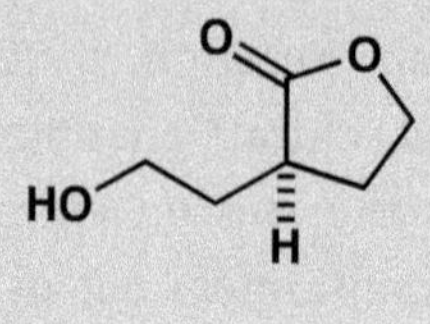

Purpose of the problem

To draw your attention to the dangers in working with nearly symmetrical molecules and revision of ester exchange (textbook p. 209).

Suggested solution

Ester exchange in base goes in this case through a symmetrical (achiral) tetrahedral intermediate with a plane of symmetry. Loss of the right hand leaving group gives one enantiomer of the ester and loss of the left hand leaving group gives the other.

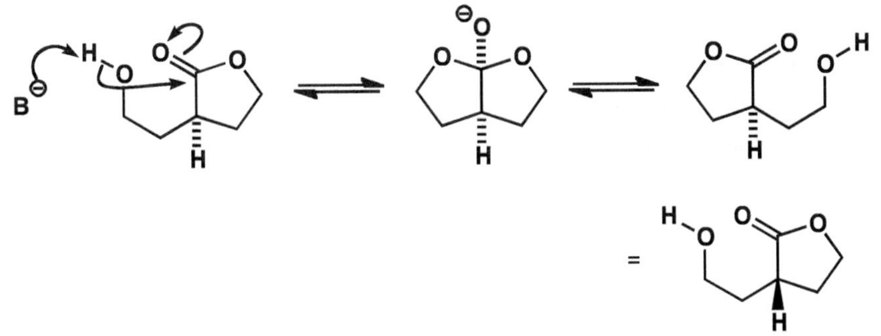

PROBLEM 7

Assign a configuration (*R* or *S*) to each of these compounds.

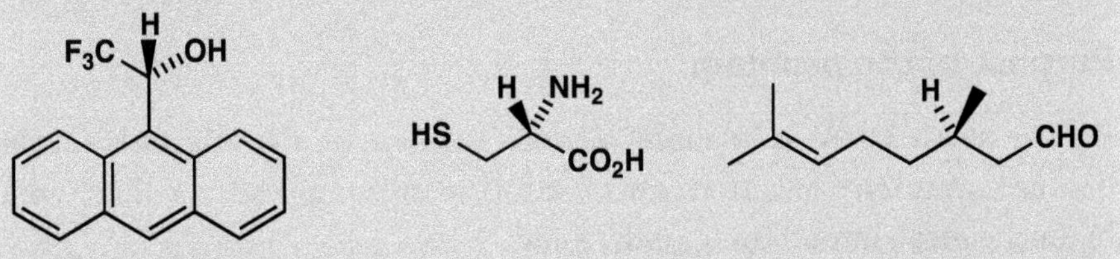

Purpose of the problem

Nomenclature may be the least important of the organic chemist's necessary skills, but giving *R* or *S* designation to simple compounds is an essential skill. These three examples check your basic knowledge of the rules.

Suggested solution

Carrying out the procedure given in the chapter (pp. 308–9 of the textbook) we prioritize the substituents 1–4 and deduce the configuration. In all these cases '4' is H and goes at the back when we work out the configuration. The first compound is Pirkle's chiral solvating agent, used to check the purity of enantiomerically enriched samples. The next is the amino acid cysteine and, despite being the natural enantiomer, is *R* because S ranks higher than O (all other natural amino acids are *S*). The third is natural citronellol having three carbon atoms on the chiral centre. They are easily ranked by the next atom along the chain or the atom beyond that if necessary.

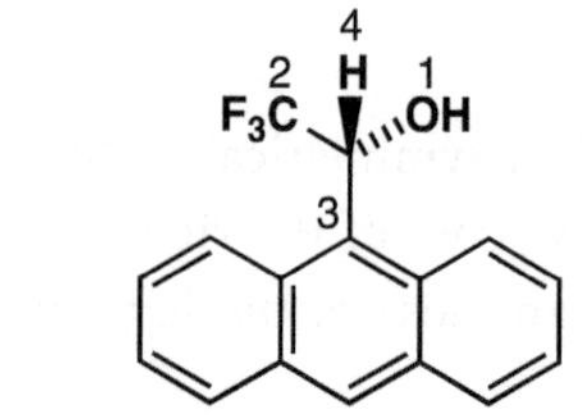

(R)-2,2,2-trifluoro-1-(9-anthryl)ethanol

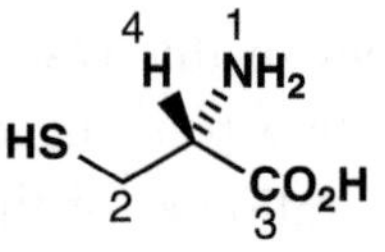

(R)-cysteine

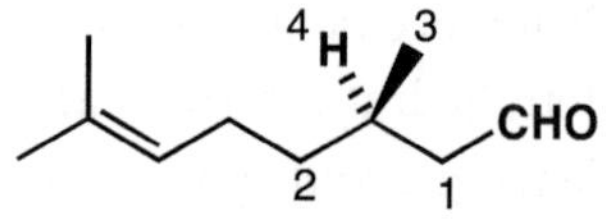

(R)-citronellol

PROBLEM 8

Just for fun, you might try and work out just how many diastereoisomers there are of inositol and how many of them are chiral.

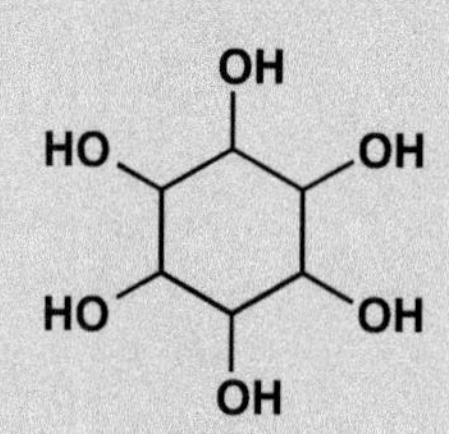

inositol

Purpose of the problem

Fun, it says! There is a more serious purpose in that the relationship between symmetry and stereochemistry is interesting, and, for this human brain chemical, important to understand.

Suggested solution

If we start with all the OH groups on one side and gradually move them over, we should get the answer. If you got too many diastereoisomers, check that some are not the same as others. There are eight diastereoisomers altogether and, remarkably, only one is chiral. All the others have at least one plane of symmetry (shown as dotted lines).

all OHs up

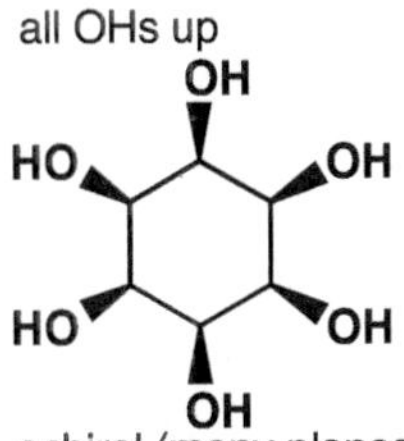

achiral (many planes)

four OHs up

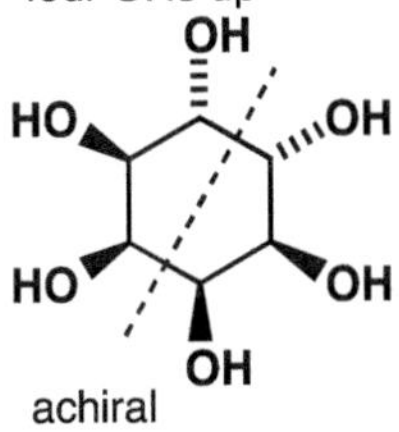

achiral

four OHs up

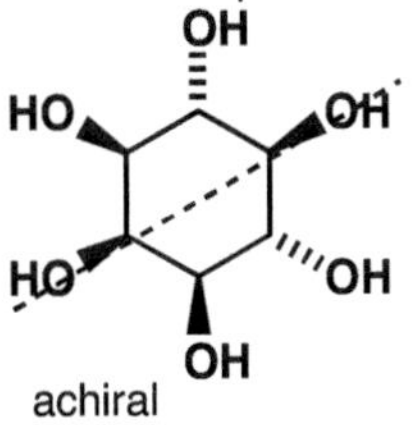

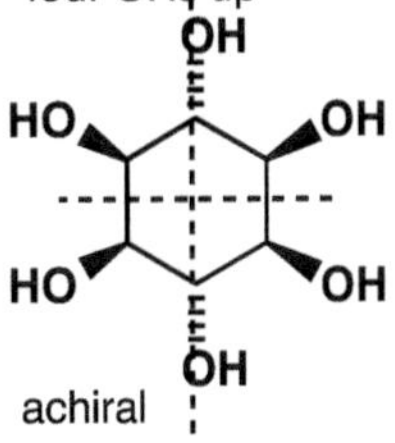

achiral

four OHs up

achiral

five OHs up

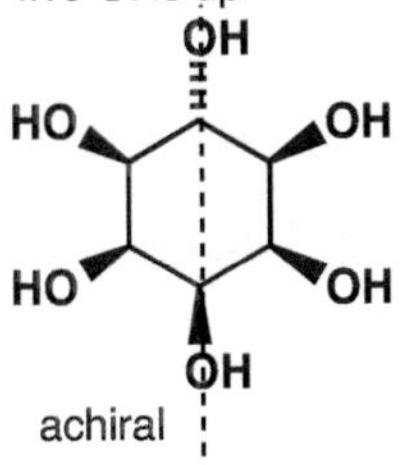

achiral

three OHs up

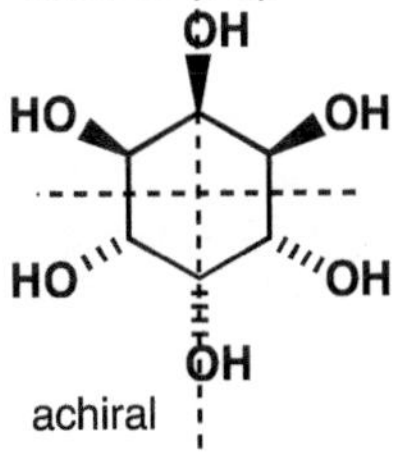

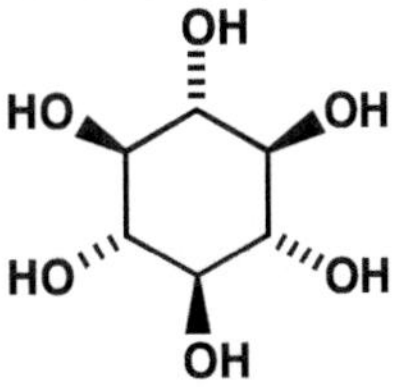

achiral

three OHs up

achiral (many planes)

three OHs up

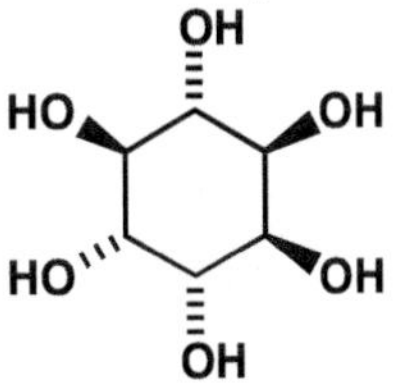

chiral

15 Nucleophilic substitution at saturated carbon

Connections

➡ Building on

- Attack of nucleophiles on carbonyl groups ch6 & ch9
- Substitution at carbonyl groups ch10
- Substitution of the oxygen atom of carbonyl groups ch11
- Reaction mechanisms ch12
- ¹H NMR ch13
- Stereochemistry ch14

Arriving at

- Nucleophilic attack on saturated carbon atoms, leading to substitution reactions
- How substitution at a saturated carbon atom differs from substitution at C=O
- Two mechanisms of nucleophilic substitution
- Intermediates and transition states in substitution reactions
- How substitution reactions affect stereochemistry
- What sort of nucleophiles can substitute, and what sort of leaving groups can be substituted
- The sorts of molecules that can be made by substitution, and what they can be made from

➡ Looking forward to

- Elimination reactions ch17
- Substitution reactions with aromatic compounds as nucleophiles ch21
- Substitution reactions with enolates as nucleophiles ch25
- Retrosynthetic analysis ch28
- Participation, rearrangement, and fragmentation reactions ch36

Mechanisms for nucleophilic substitution

Chapter 10...

This chapter...

Substitution is the replacement of one group by another. You met such reactions in Chapter 10, and an example is shown in the margin. This reaction is a substitution because the Cl group is replaced by the NH_2 group. You learnt to call the molecule of ammonia (NH_3) the **nucleophile** and the chloride you called the **leaving group**. In Chapter 10, the substitution reactions always took place at the *trigonal* (sp²) carbon atom of a carbonyl group.

In this chapter we shall be looking at reactions such as the second reaction in the margin. These are substitution reactions, because the Cl group is replaced by the PhS group. But the CH_2 group at which the reaction takes place is a *tetrahedral* (sp³), or *saturated*, carbon atom, rather than a C=O group. This reaction and the one above may look superficially the same but they are quite different in mechanism. The requirements of good reagents are also different in substitutions at carbonyl groups and at saturated carbon—that's why we changed the nucleophile from NH_3 to PhS⁻: ammonia would not give a good yield of $PhCH_2NH_2$ in the second reaction.

Let's have a look at why the mechanisms of the two substitutions must be different. Here's a summary of the mechanism of the first reaction.

Online support. The icon 🖱 in the margin indicates that accompanying interactive resources are provided online to help your understanding: just type www.chemtube3d.com/clayden/123 into your browser, replacing **123** with the number of the page where you see the icon. For pages linking to more than one resource, type **123-1, 123-2** etc. (replacing **123** with the page number) for access to successive links.

mechanism of nucleophilic substitution at the carbonyl group

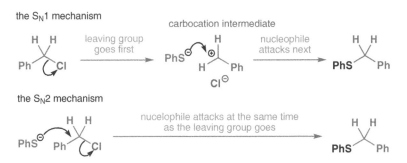

In the first step the nucleophile attacks the C=O π bond. It's immediately obvious that the first step is no longer possible at a saturated carbon atom. The electrons cannot be added to a π bond as the CH₂ group is fully saturated. In fact there is no way for the nucleophile to add before the leaving group departs (as it did in the reaction above) because this would give an impossible five-valent carbon atom.

Instead, two new and different mechanisms become possible. Either the leaving group goes first and the nucleophile comes in later, or the two events happen at the same time. The first of these possibilities you will learn to call the **S_N1 mechanism**. The second mechanism, which shows how the neutral carbon atom can accept electrons provided it loses some at the same time, you will learn to call the **S_N2 mechanism**. You will see later that both mechanisms are possible with this molecule, benzyl chloride.

Interactive mechanism for amide formation

the S_N1 mechanism

carbocation intermediate

the S_N2 mechanism

Interactive mechanisms for S_N1 and S_N2

Why is it important to know about the two mechanisms for substitution?

If we know which mechanism a compound reacts by, we know what sort of conditions to use to get good yields in substitutions. For example, if you look at a commonly used nucleophilic substitution, the replacement of OH by Br, you'll find that two quite different reaction conditions are used depending on the structure of the alcohol. *Tertiary* alcohols react rapidly with HBr to give tertiary alkyl bromides. *Primary* alcohols, on the other hand, react only very slowly with HBr and are usually converted to primary alkyl bromides with PBr₃. The reason is that the first example is an S_N1 reaction while the second is an S_N2 reaction: by the end of this chapter you will have a clear picture of how to predict which mechanism will apply and how to choose appropriate reaction conditions.

substitution of a tertiary alcohol   substitution of a primary alcohol

tert-butanol
(2-methylpropan-2-ol)

tert-butyl bromide
(2-bromo-2-methylpropane)

n-BuOH
(butan-1-ol)

n-BuBr
(1-bromobutane)

Kinetic evidence for the S_N1 and S_N2 mechanisms

Before we go any further we are going to look in a bit more detail at these two mechanisms because they allow us to explain and predict many aspects of substitution reactions. The evidence that convinced chemists that there are two different mechanisms for substitution at saturated carbon is kinetic: it relates to the rate of reactions such as the displacement of bromide by hydroxide, as shown in the margin.

It was discovered, chiefly by Hughes and Ingold in the 1930s, that some nucleophilic substitutions are first order (that is, the rate depends only on the concentration of the alkyl halide

For

the reaction is **second order** (its rate depends on both [R–Br] and [OH⁻])

For

the reaction is **first order** (its rate depends only on [R–Br] and not on [OH⁻])

Edward David Hughes (1906–63) and Sir Christopher Ingold (1893–1970) worked at University College, London in the 1930s. They first thought of many of the mechanistic ideas that chemists now take for granted.

■ There is more about the relationship between reaction rates and mechanisms in Chapter 12. Quantities in square brackets represent concentrations and the proportionality constant k is called the rate constant.

■ Please note how this symbol is written. The S and the N are both capitals and the N is a subscript.

and does not depend on the concentration of the nucleophile), while others are second order (the rate depends on the concentrations of both the alkyl halide and the nucleophile). How can we explain this result? In what we called the 'S$_N$2 mechanism' on p. 329 there is just one step. Here's the one-step S$_N$2 mechanism for substitution of n-butyl bromide by hydroxide:

With only one step, that step must be the **rate-determining step**. The rate of the overall reaction depends only on the rate of this step, and kinetic theory tells us that the rate of a reaction is proportional to the concentrations of the reacting species:

$$\text{rate of reaction} = k[\text{n-BuBr}][\text{HO}^-]$$

If this mechanism is right, then the rate of the reaction will be simply and linearly proportional to both [n-BuBr] and [HO$^-$]. And it is. Ingold measured the rates of reactions like these and found that they were proportional to the concentration of each reactant—in other words they were second order. He called this mechanism Substitution, Nucleophilic, 2nd order; S$_N$2 for short. The rate equation is usually given like this, with k_2 representing the second-order rate constant.

$$\text{rate} = k_2[\text{n-BuBr}][\text{HO}^-]$$

Significance of the S$_N$2 rate equation

This equation is useful for two reasons. Firstly, it gives us a test for the S$_N$2 mechanism. Let's illustrate this with another example: the reaction between NaSMe (an ionic solid—the nucleophile will be the anion MeS$^-$) and MeI to give Me$_2$S, dimethyl sulfide.

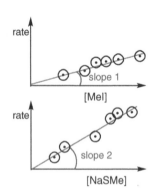

To study the rate equation, first, we keep the concentration of NaSMe constant and in a series of experiments vary that of MeI and see what happens to the rate. Then in another set of experiments we keep the concentration of MeI constant and vary that of MeSNa and see what happened to the rate. If the reaction is indeed S$_N$2 we should get a linear relationship in both cases: the graphs in the margin show a typical set of results.

The first graph tells us that the rate is proportional to [MeI], that is, rate = k_a[MeI] and the second graph that it is proportional to [MeSNa], that is, rate = k_b[MeSNa]. But why are the slopes different? If you look at the rate equation for the reaction, you will see that we have incorporated a constant concentration of one of the reagents into what appears to be the rate constant for the reaction. The true rate equation is

$$\text{rate} = k_2[\text{MeSNa}][\text{MeI}]$$

If [MeSNa] is constant, the equation becomes

$$\text{rate} = k_a[\text{MeI}], \text{where } k_a = k_2[\text{MeSNa}]$$

If [MeI] is constant, the equation becomes

$$\text{rate} = k_b[\text{MeSNa}], \text{where } k_b = k_2[\text{MeI}]$$

If you examine the graphs you will see that the slopes are different because

$$\text{slope 1} = k_a = k_2[\text{MeSNa}], \text{but slope 2} = k_b = k_2[\text{MeI}]$$

We can easily measure the true rate constant k_2 from these slopes because we know the constant values for [MeSNa] in the first experiment and for [MeI] in the second. The value of k_2

from both experiments should be the same. The mechanism for this reaction is indeed S_N2: the nucleophile MeS^- attacks as the leaving group I^- leaves.

The second reason that the S_N2 rate equation is useful is that it confirms that the performance of an S_N2 reaction depends **both on the nucleophile and on the carbon electrophile**. We can therefore make a reaction go better (speed it up or improve its yield) by changing either. For example, if we want to displace I^- from MeI using an oxygen nucleophile we might consider using any of those in the table below.

Oxygen nucleophiles in the S_N2 reaction

Oxygen nucleophile	pK_a of conjugate acid	Rate in S_N2 reaction
HO^-	15.7 (H_2O)	fast
RCO_2^-	about 5 (RCO_2H)	moderate
H_2O	−1.7 (H_3O^+)	slow
RSO_2^-	0 (RSO_2OH)	slow

> ➡ See Chapter 8 for discussion of pK_a values.

The same reasons that made hydroxide ion basic (chiefly that it is unstable as an anion and therefore reactive) make it a good nucleophile. Basicity can be viewed as nucleophilicity towards a proton, and nucleophilicity towards carbon must be related. So if we want a fast reaction, we should use NaOH rather than, say, Na_2SO_4 to provide the nucleophile. Even at the same concentration, the rate constant k_2 with HO^- as the nucleophile is much greater than the k_2 with SO_4^- as the nucleophile.

But that is not our only option. The reactivity and hence the structure of the carbon electrophile matter too. If we want reaction at a methyl group we can't change the carbon skeleton, but we can change the leaving group. The table below shows what happens if we use the various methyl halides in reaction with NaOH. The best choice for a fast reaction (greatest value of k_2) will be to use MeI and NaOH to give methanol.

> You saw in Chapter 10 that nucleophilicity towards the carbonyl group is closely related to basicity. The same is not quite so true for nucleophilic attack on the saturated carbon atom, as we shall see, but there is a relationship nonetheless.

> ■ We shall discuss nucleophilicity and leaving group ability in more detail later.

$$HO^{\ominus} \curvearrowright Me-I \xrightarrow{S_N2} HO-Me + I^{\ominus} \qquad rate = k_2\,[NaOH]\,[MeI]$$

Halide leaving groups in the S_N2 reaction

Halide X in MeX	pK_a of conjugate acid HX	Rate of reaction with NaOH
F	+3	very slow indeed
Cl	−7	moderate
Br	−9	fast
I	−10	very fast

● **The rate of an S_N2 reaction depends on:**

- **the nucleophile**
- **the carbon skeleton**
- **the leaving group**

along with the usual factors of temperature and solvent.

Significance of the S_N1 rate equation

If we replace the substitution of *n*-butyl bromide with a substitution of *t*-butyl bromide, we get the reaction shown in the margin. It turns out that, kinetically, this reaction is first order: its rate depends only on the concentration of *tert*-BuBr—it doesn't matter how much hydroxide you add: the rate equation is simply

$$\text{rate} = k_1[t\text{-BuBr}]$$

The reason for this is that the reaction happens in two steps: first the bromide leaves, to generate a carbocation, and only then does the hydroxide ion move in to attack, forming the alcohol.

the S_N1 mechanism: reaction of t-BuBr with hydroxide ion

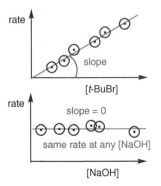

stage 1: formation of the carbocation   stage 2: reaction of the carbocation

In the S_N1 mechanism, the formation of the cation is the rate-determining step. This makes good sense: a carbocation is an unstable species and so it will be formed slowly from a stable neutral organic molecule. But once formed, being very reactive, all its reactions will be fast, regardless of the nucleophile. The rate of disappearance of t-BuBr is therefore simply the rate of the slow first step: the hydroxide nucleophile is not involved in this step and therefore does not appear in the rate equation and hence cannot affect the rate. If this is not clear to you, think of a crowd of people trying to leave a railway station or a football match through some turnstiles. It doesn't matter how fast they walk, run, or are driven away in taxis afterwards, it is only the rate of struggling through the turnstiles that determines how fast the station or stadium empties.

Once again, this rate equation is useful because we can determine whether a reaction is S_N1 or S_N2. We can plot the same graphs as we plotted before. If the reaction is S_N2, the graphs look like those we have just seen. But if it is S_N1, the graphs in the margin show what happens when we vary [t-BuBr] at constant [NaOH] and then vary [NaOH] at constant [t-BuBr].

The slope of the first graph is simply the first-order rate constant because rate = k_1[t-BuBr]. But the slope of the second graph is zero. The rate-determining step does not involve NaOH so adding more of it does not speed up the reaction. The reaction shows first-order kinetics (the rate is proportional to one concentration only) and the mechanism is called S_N1, that is, Substitution, Nucleophilic, 1st order.

This observation is very significant. The fact that the nucleophile does not appear in the rate equation means that not only does its *concentration* not matter—its *reactivity* doesn't matter either! We are wasting our time opening a tub of NaOH to add to this reaction—water will do just as well. All the oxygen nucleophiles in the table above react at the *same* rate with t-BuBr although they react at very different rates with MeI. Indeed, S_N1 substitution reactions are generally best done with weaker, non-basic nucleophiles to avoid the competing elimination reactions discussed in Chapter 17.

- The rate of an S_N1 reaction depends on:

  - the carbon skeleton
  - the leaving group

  along with the usual factors of temperature and solvent.
  But NOT the nucleophile.

How can we decide which mechanism (S_N1 or S_N2) will apply to a given organic compound?

So, substitution reactions at saturated C go via one of two alternative mechanisms, each with a very different dependence on the nature of the nucleophile. It's important to be able to predict which mechanism is likely to apply to any reaction, and rather than doing the kinetic experiments to find out, we can give you a few simple pointers to predict which will operate in which case. The factors that affect the mechanism of the reaction also help to explain why that mechanism operates.

The most important factor is **the structure of the carbon skeleton**. A helpful generalization is that compounds that can form relatively stable carbocations generally do so and react by the S$_N$1 mechanism, while the others have no choice but to react by the S$_N$2 mechanism. As you will see in a moment, the most stable carbocations are the ones that have the most substituents, so the more carbon substituents at the reaction centre, the more likely the compound is to react by the S$_N$1 mechanism.

As it happens, the structural factors that make cations stable usually also lead to slower S$_N$2 reactions. Heavily substituted compounds are good in S$_N$1 reactions, but bad in an S$_N$2 reaction because the nucleophile would have to squeeze its way into the reaction centre past the substituents. It is better for an S$_N$2 reaction if there are only hydrogen atoms at the reaction centre—methyl groups react fastest by the S$_N$2 mechanism. The effects of the simplest structural variations are summarized in the table below (where R is a simple alkyl group like methyl or ethyl).

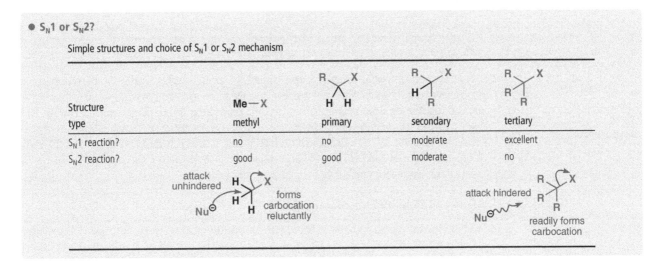

● S$_N$1 or S$_N$2?

Simple structures and choice of S$_N$1 or S$_N$2 mechanism

Structure type	Me—X methyl	primary	secondary	tertiary
S$_N$1 reaction?	no	no	moderate	excellent
S$_N$2 reaction?	good	good	moderate	no

The only doubtful case is the secondary alkyl derivative, which can react by either mechanism, although it is not very good at either. The first question you should ask when faced with a new nucleophilic substitution is this 'Is the carbon electrophile methyl, primary, secondary, or tertiary?' This will start you off on the right foot, which is why we introduced these important structural terms in Chapter 2.

Later in this chapter we will look in more detail at the differences between the two mechanisms and the structures that favour each, but all of what we say will build on the table above.

A closer look at the S$_N$1 reaction

In our discussion of the S$_N$1 reaction above, we proposed the *t*-butyl carbocation as a reasonable intermediate formed by loss of bromide from *t*-butyl bromide. We now need to explain the evidence we have that carbocations can indeed exist, and the reasons why the *t*-butyl carbocation is much more stable than, for example, the *n*-butyl cation.

In Chapter 12 we introduced the idea of using a reaction energy profile diagram to follow the progress of a reaction from starting materials to products, via transition states and any intermediates. The energy profile diagram for the S$_N$1 reaction between *t*-butyl bromide and water looks something like this:

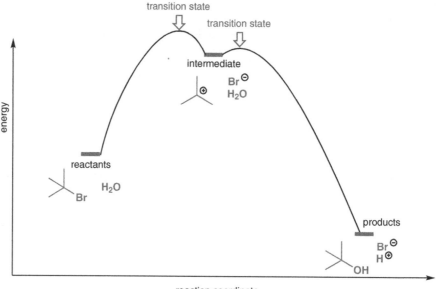

The carbocation is shown as an intermediate—a species with a finite (if short) lifetime for reasons we shall describe shortly. And because we know that the first step, the formation of the carbocation, is slow, that must be the step with the higher energy transition state. The energy of that transition state, which determines the overall rate of the reaction, is closely linked to the stability of the carbocation intermediate, and it is for this reason that the most important factor in determining the efficiency of an S_N1 reaction is the stability or otherwise of any carbocation that might be formed as an intermediate.

Shape and stability of carbocations

We discussed the planar shape of the methyl cation in Chapter 4 (p. 103), and the *tert*-butyl cation is similar in structure: the electron-deficient central carbon atom has only six electrons, which it uses to form three σ bonds, and therefore also carries an empty p orbital. Any carbocation will have a planar carbon atom with an empty p orbital. Think of it this way: only filled orbitals contribute to the energy of a molecule, so if you have to have an unfilled orbital (which a carbocation always does) it is best to make that unfilled orbital as high in energy as possible to keep the filled orbitals low in energy. p orbitals are higher in energy than s orbitals (or hybrid sp, sp^2, or sp^3 orbitals for that matter) so the carbocation always keeps the p orbital empty.

Carbocation stability

The *t*-butyl carbocation is *relatively* stable as far as carbocations go, but you would not be able to keep it in a bottle on the shelf! The concept of more and less stable carbocations is important in understanding the S_N1 reaction, but it is important to realize that these terms are all relative: even 'stable' carbocations are highly reactive electron-deficient species.

non-nucleophilic anions

We know that the *t*-butyl cation is stable enough to observe because of the work of George Olah, who won the Nobel Prize for Chemistry in 1994. The challenge is that carbocations are very reactive electrophiles, so Olah's idea was to have a solution containing no nucleophiles. Any cation must have an anion to balance the charge, so the important advance was to find anions, consisting of a negatively charged atom surrounded by tightly held halogen atoms, which are just too stable to be nucleophilic. Examples include BF_4^-, PF_6^-, and Sb_6^-. The first is small and tetrahedral and the others are larger and octahedral.

In these anions, the negative charge does not correspond to a lone pair of electrons (they are like BH_4^- in this respect) and there is no orbital high enough in energy to act as a nucleophile. By using a non-nucleophilic solvent, liquid SO_2, at low temperature, Olah was able to turn alcohols into carbocations with these counterions. This is what happens when *tert*-butanol is treated with SbF_5 and HF in liquid SO_2. The acid protonates the hydroxyl group, allowing it to leave as water, while the SbF$_5$ grabs the fluoride ion, preventing it from acting as a nucleophile. The cation is left high and dry.

Olah's preparation of the *t*-butyl cation in liquid SO$_2$

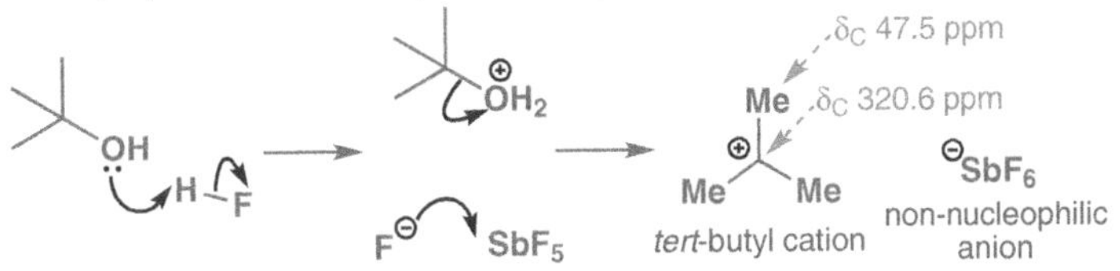

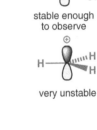

The proton NMR of this cation showed just one signal for the three methyl groups at 4.15 ppm, quite far downfield for C–Me groups. The ^{13}C spectrum also showed downfield Me groups at 47.5 ppm, but the key evidence that the cation was formed was the shift of the central carbon atom, which came at an amazing 320.6 ppm, way downfield from anything you have met before. This carbon is very deshielded—it is positively charged and extremely electron deficient.

From Olah's work we know what the *t*-butyl cation looks like by NMR, so can we use NMR to try to detect it as an intermediate in substitution reactions? If we mix *t*-BuBr and NaOH in an NMR tube and let the reaction run inside the NMR machine, we see no signals belonging to the cation. But this proves nothing. We would not expect a reactive intermediate to be present in any significant concentration. There is a simple reason for this. If the cation is unstable, it will react very quickly with any nucleophile around and there will never be any appreciable amount of cation in solution. Its rate of formation will be much slower than its rate of reaction.

Alkyl substituents stabilize a carbocation

Olah found that he could measure the spectrum of the *tert*-butyl cation, but he was never able to observe the methyl cation in solution. Why do those extra substituents stabilize the cationic centre?

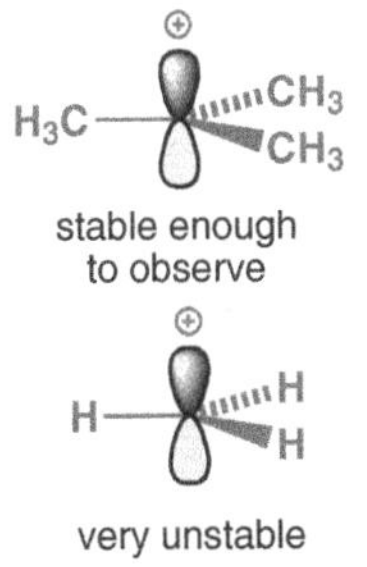

Any charged organic intermediate is inherently unstable because of the charge. A carbocation can be formed only if it has some extra stabilization, and extra stabilization can come to the planar carbocation structure from weak donation of σ bond electrons into the empty p orbital of the cation. In the *t*-butyl cation, three of these donations occur at any one time: it doesn't matter if the C–H bonds point up or down; one C–H bond of each methyl group must be parallel to one lobe of the empty p orbital at any one time. The first diagram shows one overlap in orbital terms and the second and third diagrams, three as dotted lines.

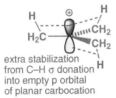

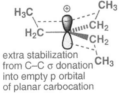

empty p orbital / filled C–H σ orbital

extra stabilization from C–H σ donation into empty p orbital of planar carbocation

extra stabilization from C–C σ donation into empty p orbital of planar carbocation

C–H s bonds and empty p orbital are perpendicular: no donation possible

There is nothing special about the C–H bond donating electrons into an empty orbital: a C–C bond is just as good and some bonds are better (C–Si, for example). But there must be a bond of some sort—a hydrogen atom by itself has no lone pairs and no σ bonds so it cannot stabilize a cation.

Planarity is so important to the structure of a carbocation that if a tertiary cation cannot become planar, it is not formed. A classic case is the structure in the margin, which does not react with nucleophiles either by S$_N$1 or by S$_N$2. It does not react by S$_N$1 because the cation cannot become planar, nor by S$_N$2 because the nucleophile cannot approach the carbon atom from the right direction.

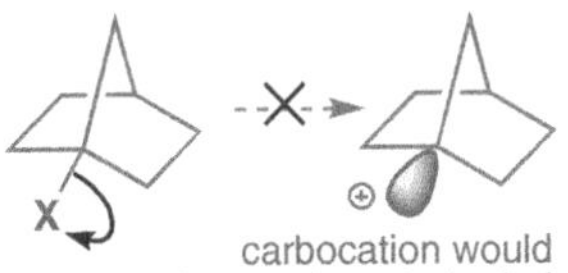

In general, though, simple tertiary structures undergo efficient S$_N$1 substitution reactions. With good leaving groups such as halides, substitutions can be done under neutral conditions; with less good leaving groups such as alcohols or ethers, acid catalysis is required. The following group of reactions give an idea of the types of S$_N$1 reactions that work well.

stable enough to observe

very unstable

Interactive display of stability and structure of carbocations

carbocation would have to be tetrahedral

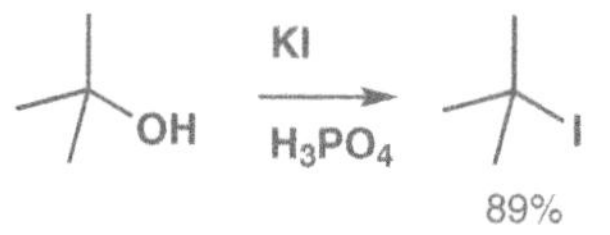

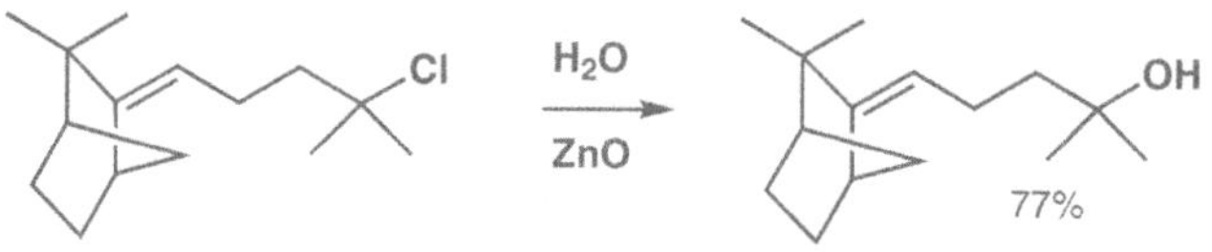

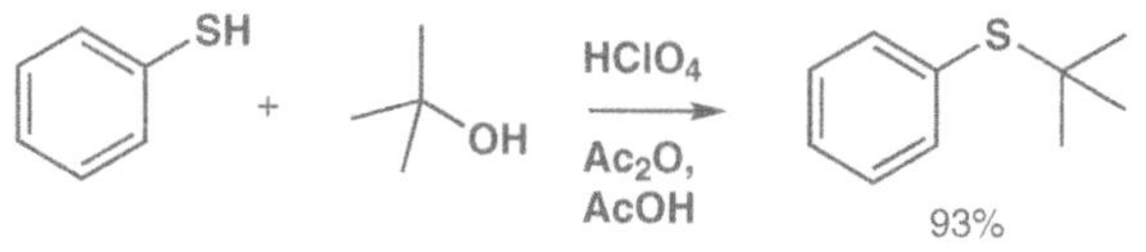

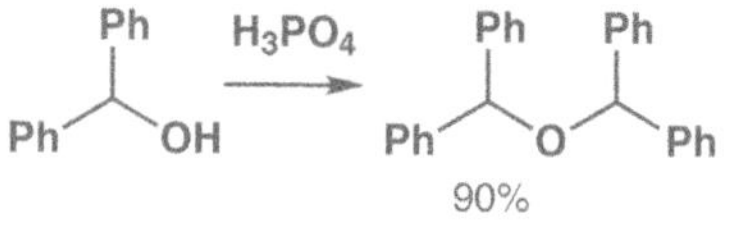

An adjacent C=C π system stabilizes a carbocation: allylic and benzylic carbocations

Tertiary carbocations are more stable than primary ones, but powerful stabilization is also provided when there is genuine conjugation between the empty p orbital and adjacent π or lone pair electrons. The allyl cation has a filled (bonding) orbital containing two electrons delocalized over all three atoms and an important empty orbital with coefficients on the end atoms only. It's this orbital that is attacked by nucleophiles. The curly arrow picture tells us the same thing.

We discussed conjugation in allyl cations in Chapter 7.

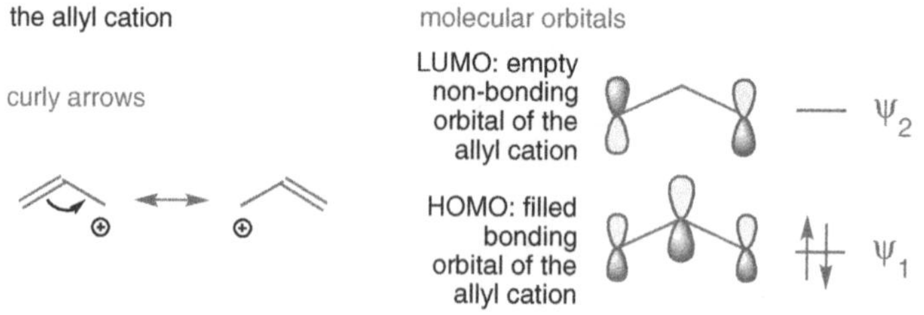

Allylic electrophiles react well by the S$_N$1 mechanism because the allyl cation is relatively stable. Here's an example of a reaction working in the opposite direction from most of those you have seen so far—we start with the alcohol and form the bromide. Treatment of cyclohexenol with HBr gives the corresponding allylic bromide.

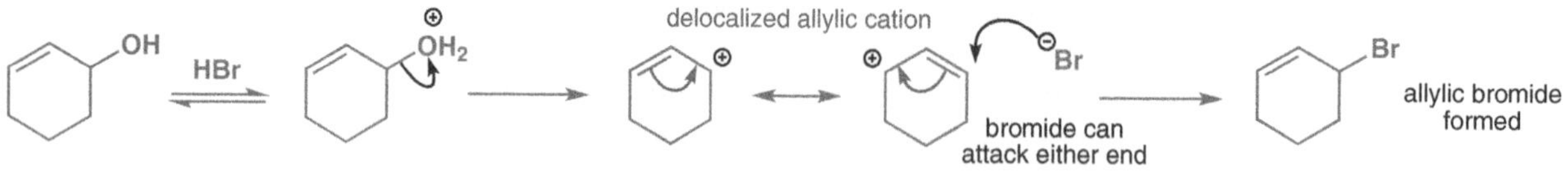

In this case, only one compound is formed because attack at either end of the allylic cation gives the same product. But when the allylic cation is unsymmetrical this can be a nuisance as a mixture of products may be formed. It doesn't matter which of these two butenols you treat with HBr, you get the same delocalized allylic cation.

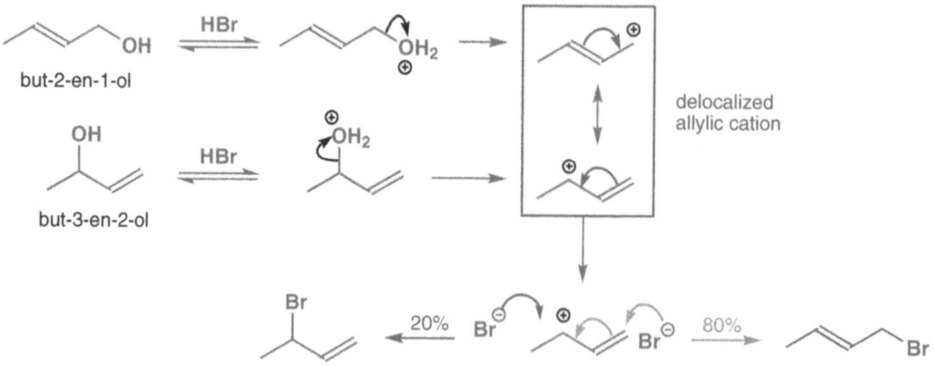

When this cation reacts with Br⁻, about 80% goes to one end and 20% to the other, giving a mixture of butenyl bromides. This **regioselectivity** (where the nucleophile attacks) is determined by steric hindrance: attack is faster at the less hindered end of the allylic system.

➡ The concept of regioselectivity is developed in more detail in Chapter 24.

Sometimes this ambiguity is useful. The tertiary allylic alcohol 2-methylbut-3-en-2-ol is easy to prepare and reacts well by the S$_N$1 mechanism because it can form a stable carbocation that is both tertiary and allylic. The allylic carbocation intermediate is unsymmetrical and reacts only at the less substituted end to give 'prenyl bromide'.

2-methylbut-3-en-2-ol

'prenyl bromide'
1-bromo-3-methylbut-2-ene

The benzyl cation is about as stable as the allyl cation but lacks its ambiguity of reaction. Although the positive charge is delocalized around the benzene ring, to three positions in particular, the benzyl cation always reacts on the side chain so that aromaticity is preserved.

delocalization in the benzyl cation

nucleophile attacks only at CH$_2$ group

An exceptionally stable cation is formed when three benzene rings can help to stabilize the same positive charge. The result is the triphenylmethyl cation or, for short, the trityl cation. Trityl chloride is used to form an ether with a primary alcohol group by an S$_N$1 reaction. You will notice that pyridine is used as solvent for the reaction. Pyridine (a weak base: the pK_a of its conjugate acid is 5.5—see Chapter 8) is not strong enough to remove the proton from the primary alcohol (pK_a about 15), and there would be no point in using a base strong enough to make RCH$_2$O⁻ as the nucleophile makes no difference to an S$_N$1 reaction. Instead the TrCl ionizes first to trityl cation, which now captures the primary alcohol and finally pyridine is able to remove the proton from the oxonium ion. Pyridine does not catalyse the reaction; it just stops it becoming too acidic by removing the HCl formed. Pyridine is also a convenient polar organic solvent for ionic reactions.

RCH$_2$OH + trityl chloride
primary alcohol

pyridine

■ The symbol Tr refers to the group Ph$_3$C.

trityl cation

rate-determining step

fast

fast

trityl ether

The table below shows the rates of solvolysis (i.e. a reaction in which the solvent acts as the nucleophile) in 50% aqueous ethanol for substituted allylic chlorides compared with benzylic chlorides and simple alkyl chlorides. The values give you an idea of the relative reactivity towards substitution of the different classes of compound. These rates are mostly S$_N$1, but there will be some S$_N$2 reactivity with the primary compounds.

Rates of solvolysis of alkyl chlorides in 50% aqueous ethanol at 44.6 °C		
Compound	Relative rate	Comments
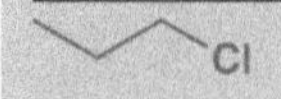	0.07	primary chloride: probably all S_N2
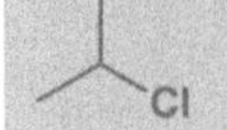	0.12	secondary chloride: can do S_N1 but not very well
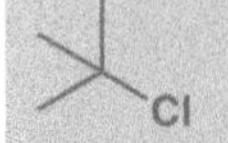	2100	tertiary chloride: very good at S_N1
	1.0	primary but allylic: S_N1 all right
	91	allylic cation is secondary at one end
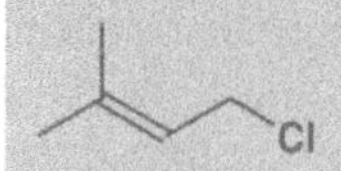	130000	allylic cation is tertiary at one end: compare with 2100 for simple tertiary
	7700	primary but allylic and benzylic

Carbocations are stabilized by an adjacent lone pair

The alkyl chloride known as methyl chloromethyl ether, $MeOCH_2Cl$, reacts very well with alcohols to form ethers. Being a primary alkyl chloride, you might think that its reactions would follow an S_N2 mechanism, but in fact it has characteristic S_N1 reactivity. As usual, the reason for its preference for the S_N1 mechanism is its ability to form a stabilized carbocation. Loss of the chloride ion is assisted by the adjacent lone pair, and we can draw the resulting cation either as an oxonium ion or as a carbocation.

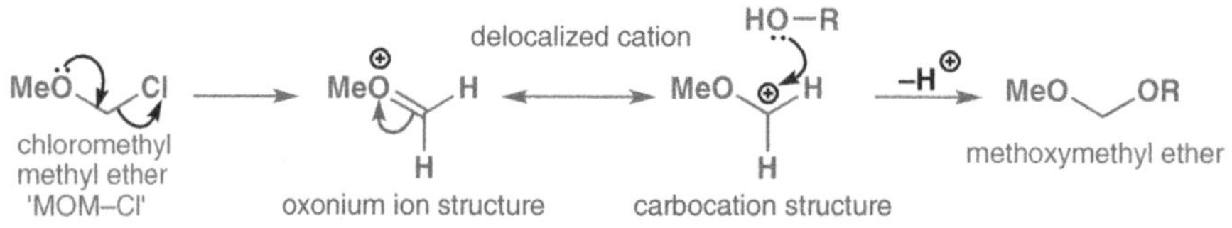

The methoxymethyl cation

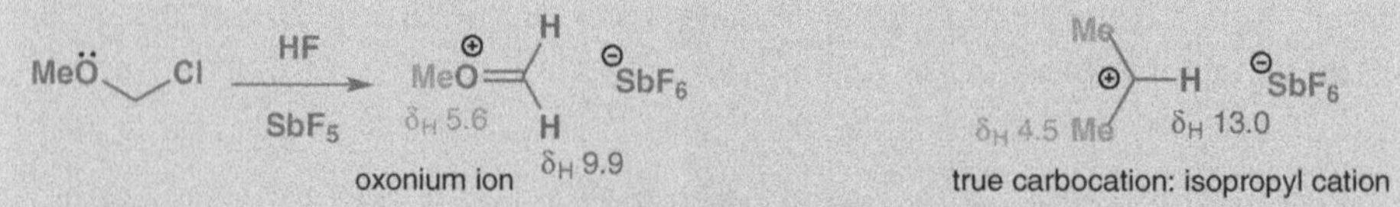

Olah has used the methods described above to make the methoxymethyl cation in solution. Although this cation can be drawn either as an oxonium ion or as a primary carbocation, the oxonium ion structure is the more realistic. The proton NMR spectrum of the cation compared with that of the isopropyl cation (this is the best comparison we can make) shows that the protons on the CH_2 group resonate at 9.9 ppm instead of at the 13.0 ppm of the true carbocation.

If you think back to Chapter 11, you will recall that the first step in the hydrolysis of an acetal is a similar reaction, with one alkoxy group replaced by water to give a hemiacetal. We considered the mechanism for this reaction in Chapter 11 but did not then concern ourselves with a label for the first step. It is in effect an S_N1 substitution reaction: the decomposition of the protonated acetal to give an oxonium ion. If you compare this step with the reaction of the chloroether we have just described you will see that they are very similar in mechanism.

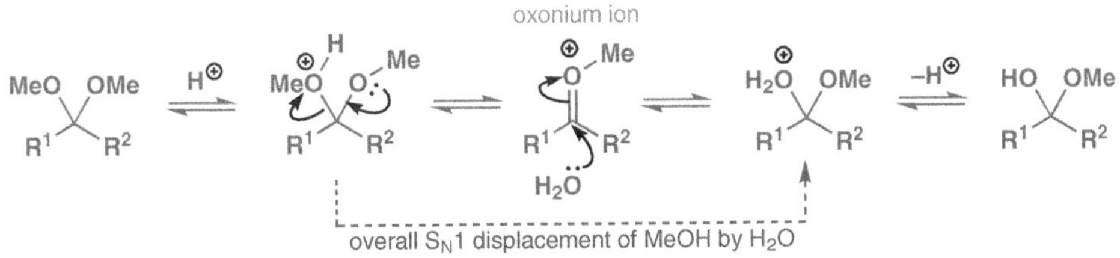

oxonium ion

overall S$_N$1 displacement of MeOH by H$_2$O

Interactive mechanism for
acetal hydrolysis

A common mistake

Don't be tempted to shortcut this mechanism by drawing the displacement of the first molecule of methanol by water as an S$_N$2 reaction.

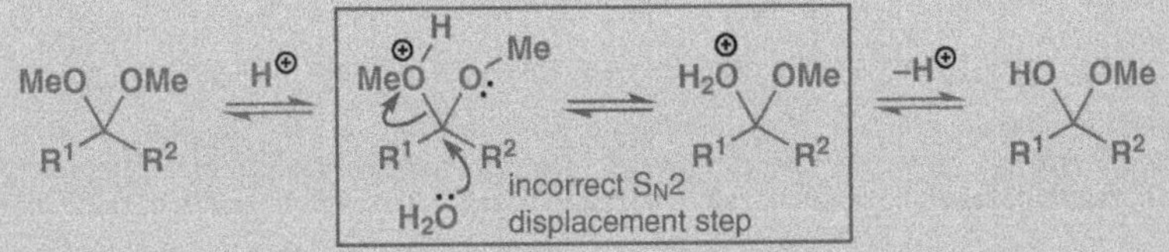

incorrect S$_N$2
displacement step

An S$_N$2 mechanism is unlikely at such a crowded carbon atom. However, the main reason why the S$_N$2 mechanism is wrong is that the S$_N$1 mechanism is so very efficient, with a neighbouring MeO group whose lone pair can stabilize the carbocation intermediate. The S$_N$2 mechanism doesn't get a chance.

This mechanism for the S$_N$1 replacement of one electronegative group at a carbon atom by a nucleophile where there is another electronegative group at the same carbon atom is very general. You should look for it whenever there are two atoms such as O, N, S, Cl, or Br joined to the same carbon atom. The better leaving groups (such as the halogens) need no acid catalyst but the less good ones (N, O, S) usually need acid.

$X = OR, SR, NR_2$

$Y = Cl, Br, \overset{\oplus}{O}H_2, \overset{\oplus}{O}HR$

➡ Think back to the formation
and reactions of iminium ions in
Chapter 11 for further examples.

We now have in the box below a complete list of the sorts of structures that normally react by the S$_N$1 mechanism rather than by the S$_N$2 mechanism.

Stable carbocations as intermediates in S$_N$1 reactions

Type of cations	Example 1	Example 2
simple alkyl	tertiary (good)	secondary (not so good)
	t-butyl cation	*i*-propyl cation
	$Me_3C^{\oplus}$ =	$Me_2CH^{\oplus}$ =
conjugated	allylic 	benzylic 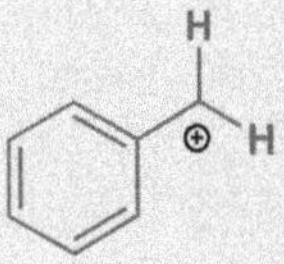
heteroatom-stabilized	oxygen-stabilized (oxonium ions) 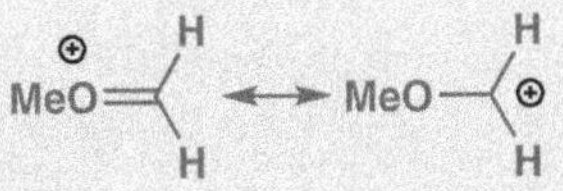	nitrogen-stabilized (iminium ions) 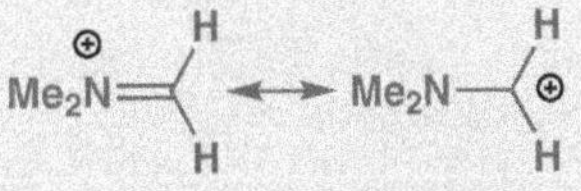

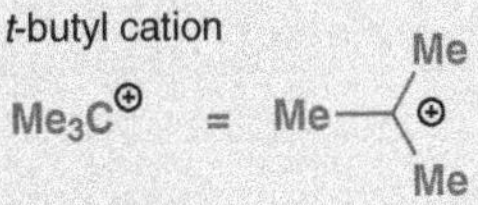

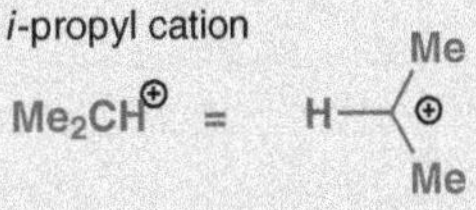

325

A closer look at the S$_N$2 reaction

Among simple alkyl groups, methyl and primary alkyl groups always react by the S$_N$2 mechanism and never by S$_N$1. This is partly because the cations are unstable and partly because the nucleophile can push its way in easily past the hydrogen atoms.

A common way to make ethers is to treat an alkoxide anion with an alkyl halide. If the alkyl halide is a methyl compound, we can be sure that the reaction will go by the S$_N$2 mechanism. A strong base, here NaH, will be needed to form the alkoxide ion, since alcohols are weak acids (pK_a about 16). Methyl iodide is a suitable electrophile.

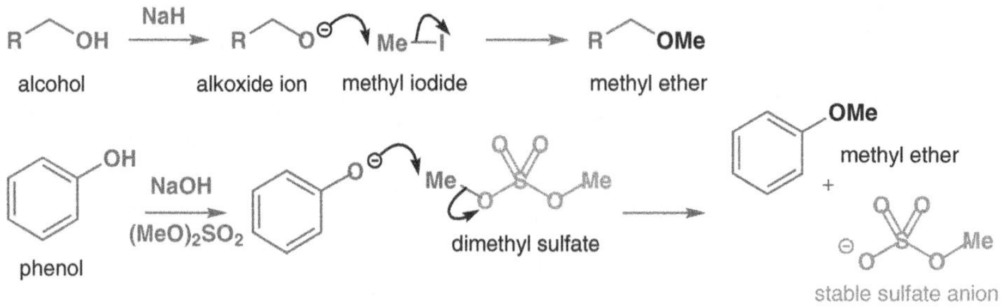

> ■ Notice that we said *simple* alkyl groups: of course, primary allylic, benzylic, and RO or R$_2$N substituted primary derivatives may react by S$_N$1!

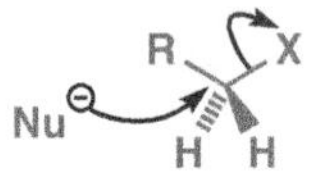

uncluttered approach of nucleophile in S$_N$2 reactions of methyl compounds (R=H) and primary alkyl compounds (R=alkyl)

With the more acidic phenols (pK_a about 10), NaOH is a strong enough base and dimethyl sulfate, the dimethyl ester of sulfuric acid, is often used as the electrophile. It is worth using a strong base to make the alcohol into a better nucleophile because as we discussed on p. 331 the rate equation for an S$_N$2 reaction tells us that the strength and concentration of the nucleophile affects the rate of the reaction.

The transition state for an S$_N$2 reaction

> ➡ We introduced the terms *transition state* and *intermediate* in Chapter 12.

Another way to put this would be to say that the nucleophile, the methyl group, and the leaving group are all present in the transition state for the reaction. The transition state is the highest energy point on the reaction pathway. In the case of an S$_N$2 reaction it will be the point where the new bond from the nucleophile is partly formed while the old bond to the leaving group is not yet completely broken. It will look something like this:

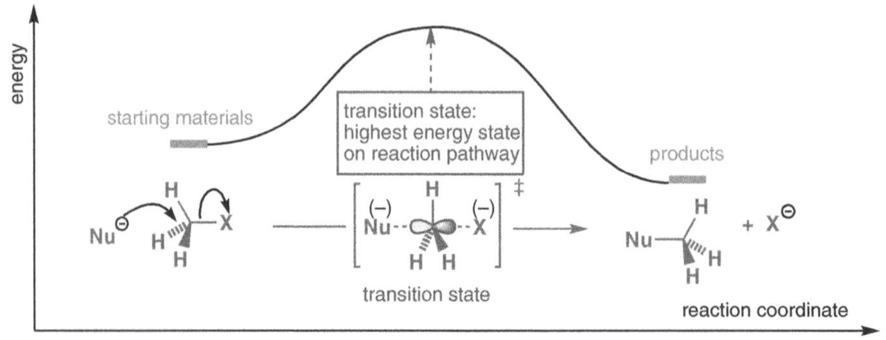

The dashed bonds in the transition state indicate partial bonds (the C–Nu bond is partly formed and the C–X bond partly broken) and the charges in brackets indicate substantial partial charges (about half a minus charge each in this case). Transition states are often shown in square brackets and marked with the symbol ‡.

> 🖱 Interactive mechanism for simple S$_N$2

Another way to look at this situation is to consider the orbitals. The nucleophile must have lone-pair electrons, which will interact with the σ* orbital of the C–X bond.

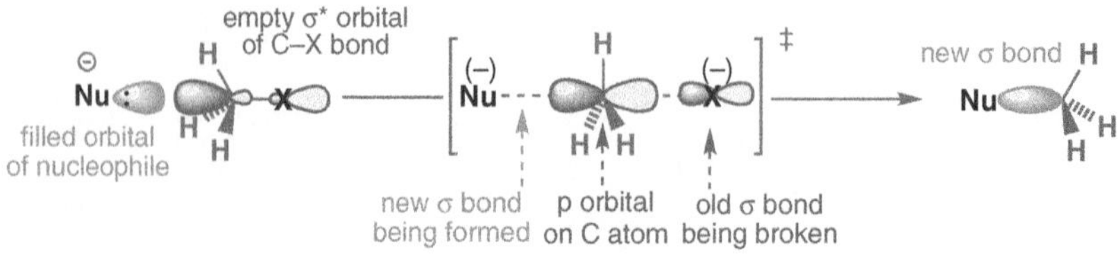

In the transition state the carbon atom in the middle has a p orbital that shares one pair of electrons between the old and the new bonds. Both these pictures suggest that the transition state for an S$_N$2 reaction has a more or less planar carbon atom at the centre with the nucleophile and the leaving group arranged at 180° to each other. This picture can help us explain two important observations concerning the S$_N$2 reaction—firstly the types of structures that react efficiently, and secondly the stereochemistry of the reaction.

Adjacent C=C or C=O π systems increase the rate of S$_N$2 reactions

We have already established that methyl and primary alkyl compounds react well by the S$_N$2 mechanism, while secondary alkyl compounds undergo S$_N$2 reactions only reluctantly. But there are other important structural features that also encourage the S$_N$2 mechanism. Two of these, allyl and benzyl groups, also encourage the S$_N$1 mechanism.

Allyl bromide reacts well with alkoxides to make ethers, and shown below is the typical S$_N$2 mechanism for this reaction. Also shown is the transition state for this reaction. Allyl compounds react rapidly by the S$_N$2 mechanism because the π system of the adjacent double bond can stabilize the transition state by conjugation. The p orbital at the reaction centre (shown in brown, and corresponding to the brown orbital in the diagram on p. 340) has to make two partial bonds with only two electrons—it is electron deficient, and so any additional electron density it can gather from an adjacent π system will stabilize the transition state and increase the rate of the reaction.

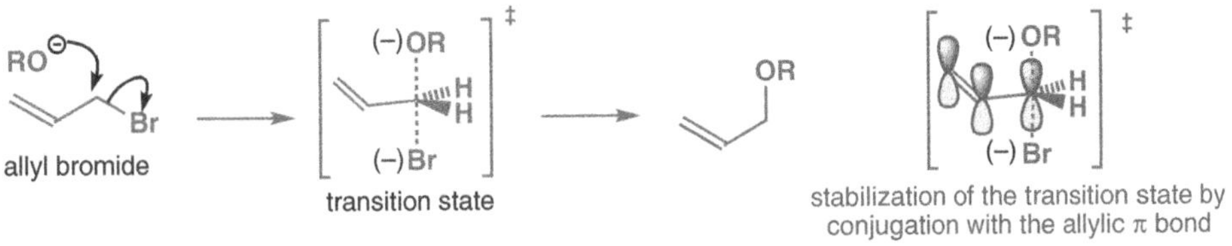

allyl bromide

transition state

stabilization of the transition state by conjugation with the allylic π bond

Interactive S$_N$2 mechanism at allylic and benzylic centres

The benzyl group acts in much the same way using the π system of the benzene ring for conjugation with the p orbital in the transition state. Benzyl bromide reacts very well with alkoxides to make benzyl ethers.

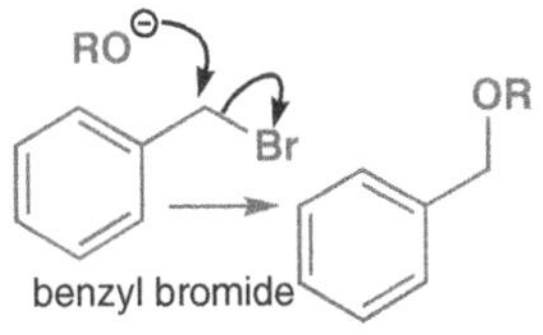

Among the fastest of all S$_N$2 reactions are those where the leaving group is adjacent to a carbonyl group. With α-bromo carbonyl compounds, two neighbouring carbon atoms are both powerfully electrophilic sites. Each has a low-energy empty orbital—π* from C=O and σ* from C–Br (this is what makes them electrophilic)—and these can combine to form a new LUMO (π* + σ*) lower in energy than either. Nucleophilic attack will occur easily where this new orbital has its largest coefficient, shown in orange on the diagram.

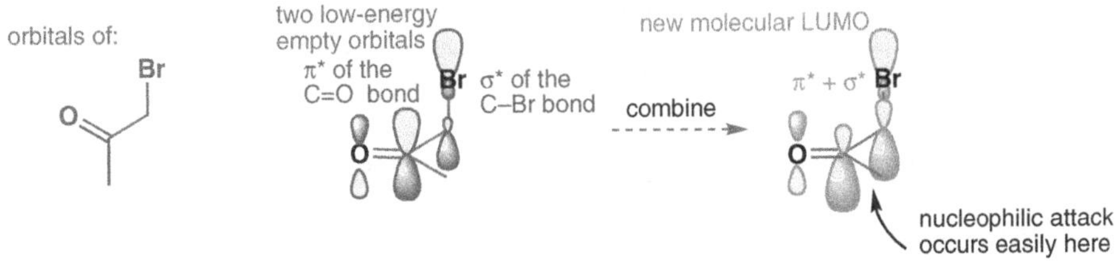

benzyl bromide

orbitals of:

two low-energy empty orbitals
π* of the C=O bond   σ* of the C–Br bond   combine

new molecular LUMO
π* + σ*

nucleophilic attack occurs easily here

The effect of this interaction between antibonding orbitals is that each group becomes more electrophilic because of the presence of the other—the C=O group makes the C–Br bond more reactive and the Br makes the C=O group more reactive. In fact, it may well be that the nucleophile will attack the carbonyl group, but this will be reversible whereas displacement of bromide is irreversible.

There are many examples of this type of reaction. Reactions with amines go well and the aminoketone products are widely used in the synthesis of drugs.

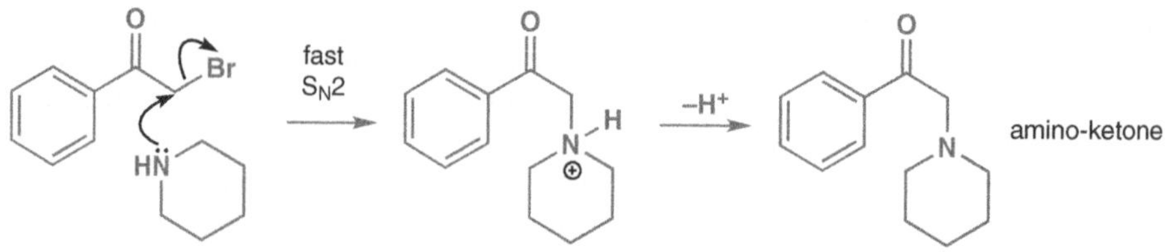

amino-ketone

Quantifying structural effects on S_N2 reactions

Some actual data may help at this point. The rates of reaction of the following alkyl chlorides with KI in acetone at 50 °C broadly illustrate the patterns of S_N2 reactivity we have just analysed. These are relative rates with respect to *n*-BuCl as a 'typical primary halide'. You should not take too much notice of precise figures but rather observe the trends and notice that the variations are quite large—the full range from 0.02 to 100,000 is eight powers of ten.

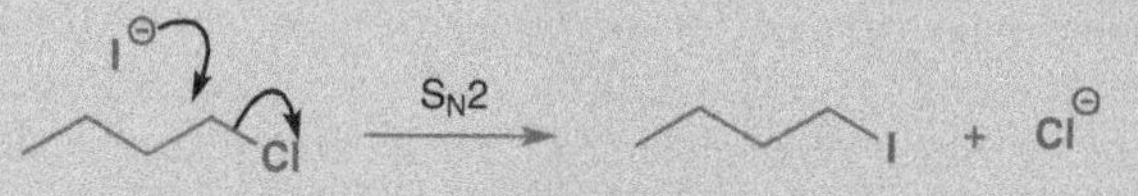

Relative rates of substitution reactions of alkyl chlorides with the iodide ion

Alkyl chloride	Relative rate	Comments
Me—Cl	200	least hindered alkyl chloride
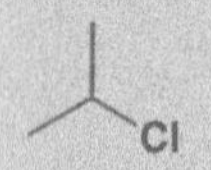 (isopropyl chloride)	0.02	secondary alkyl chloride; slow because of steric hindrance
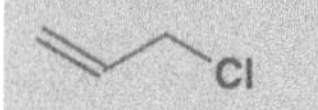 (allyl chloride)	79	allyl chloride accelerated by π conjugation in transition state
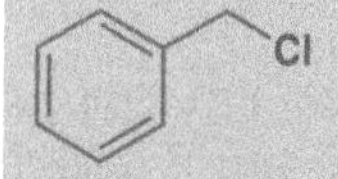 (benzyl chloride)	200	benzyl chloride a bit more reactive than allyl: benzene ring slightly better at π conjugation than isolated double bond
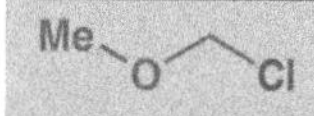 Me—O—CH₂Cl	920	conjugation with oxygen lone pair accelerates reaction (this is an S_N1 reaction)
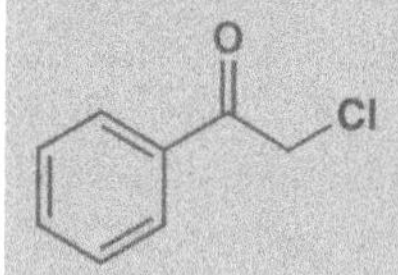 (phenacyl chloride)	100,000	conjugation with carbonyl group much more effective than with simple alkene or benzene ring; these α-halo carbonyl compounds are the most reactive of all

Contrasts between S_N1 and S_N2

You have now met the key features of both important mechanisms for substitution. You should at this stage in the chapter have a grasp of the kinetics, the nature of the intermediates and transition states, and the simple steric and electronic factors that control reactivity in S_N1 and S_N2 reaction pathways.

We are now going to look in more detail at some other aspects where there are significant contrasts between the mechanisms, either because they lead to different outcomes or because they lead to a change in reactivity towards one or the other of the two pathways.

A closer look at steric effects

We have already pointed out that having more alkyl substituents at the reaction centre makes a compound more likely to react by S_N1 than by S_N2 for two reasons: firstly they make a carbocation more stable, so favouring S_N1, and secondly they make it hard for a nucleophile to get close to the reaction centre in the rate-determining step, disfavouring S_N2. Let's look in more detail at the transition state for the slow steps of the two reactions and see how steric hindrance affects both.

In the approach to the S_N2 transition state, the carbon atom under attack gathers in another substituent and becomes (transiently) five-coordinate. The angles between the substituents decrease from tetrahedral to about 90°.

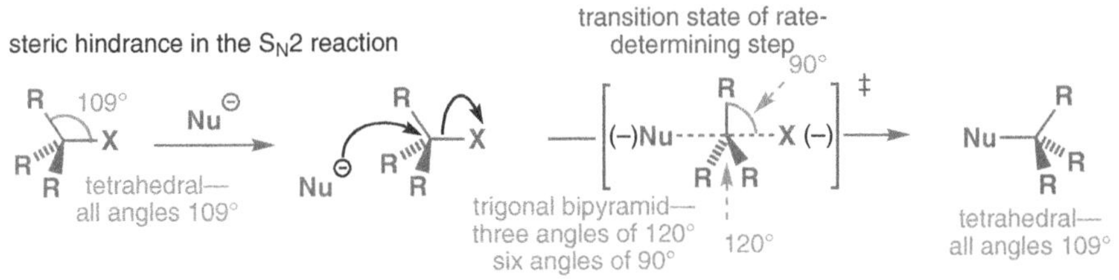

In the starting material there are four angles of about 109°. In the transition state (enclosed in square brackets and marked ‡ as usual) there are three angles of 120° and six angles of 90°, a significant increase in crowding. The larger the substituents R, the more serious this is, and the greater the increase in energy of the transition state. We can easily see the effects of steric hindrance if we compare these three structural types:

- methyl: CH_3–X: very fast S_N2 reaction
- primary alkyl: RCH_2–X: fast S_N2 reaction
- secondary alkyl: R_2CH–X: slow S_N2 reaction.

The opposite is true of the S_N1 reaction. The rate-determining step is simply the loss of the leaving group, and the transition state for this step will look something like the structure shown below—with a longer, weaker, and more polarized C–X bond than the starting material. The starting material is again tetrahedral (four angles of about 109°) and in the intermediate cation there are just three angles of 120°—fewer and less serious interactions. The transition state will be on the way towards the cation, and because the R groups are further apart in the transition state than in the starting material, large R groups will actually *decrease* the energy of the transition state relative to the starting material. S_N1 reactions are therefore accelerated by alkyl substituents both for this reason and because they stabilize the cation.

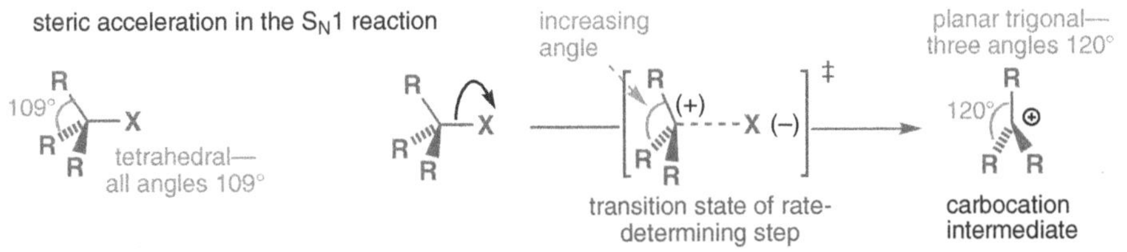

Stereochemistry and substitution

Look back at the scheme we showed you for the S_N2 reaction on p. 340. It shows the nucleophile attacking the carbon atom on the opposite side from the leaving group. Look carefully at the carbon atom it is attacking and you see that its substituents end up turning inside out as the reaction goes along, just like an umbrella in a high wind. If the carbon atom under attack is a stereogenic centre (Chapter 14), the result will be inversion of configuration. Something very different happens in the S_N1 reaction, and we will now illustrate the difference with a simple sequence of reactions.

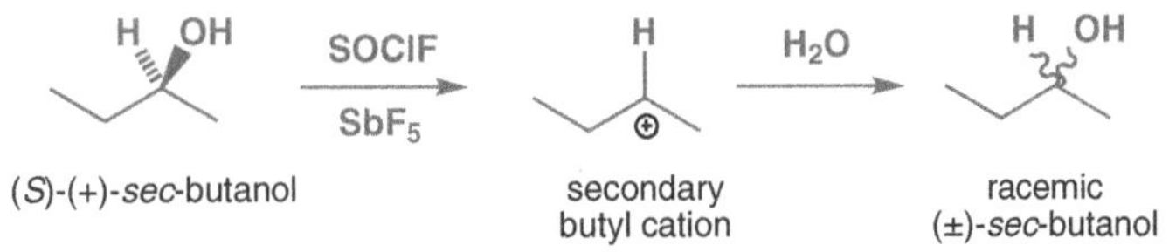

Starting with the optically active secondary alcohol *sec*-butanol (or butan-2-ol, but we want to emphasize that it is *secondary*), the secondary cation can be made by the method described on p. 338. Quenching this cation with water regenerates the alcohol but without any optical

You may hear it said that *tert*-alkyl compounds do not react by the S_N2 mechanism because the steric hindrance would be too great. In fact *t*-alkyl compounds react so fast by the S_N1 mechanism that the S_N2 mechanism wouldn't get a chance *even if it went as fast as it goes with methyl compounds*. You can see the figures that show this in the box on p. 338. Some tertiary alkyl compounds will react by slow S_N2 if, for example, S_N1 is prevented by an adjacent electron-withdrawing carbonyl group.

Likewise, primary alkyl halides that experience steric hindrance for other reasons react neither by S_N1 (because they are primary) nor S_N2 (because they are hindered)—a notorious example is the unreactivity of 'neopentyl halides'.

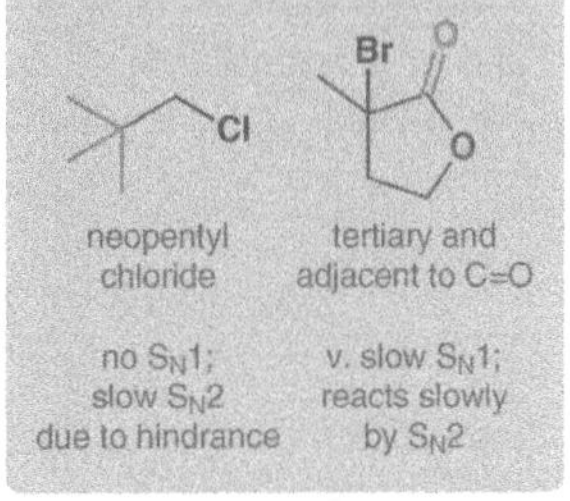

Notice how the transition state for the S_N1 reaction is really very close in structure to the carbocation, and you can see that they also lie very close to one another in the energy profile diagram on p. 334. When we say that the rate of an S_N1 reaction is increased by stabilization of the carbocation, what we really mean of course is that the rate is increased by stabilization of *the transition state leading to* the carbocation. However, they are so similar in structure that you can assume that steric and electronic effects on the carbocation will be very similar to those on the carbocation.

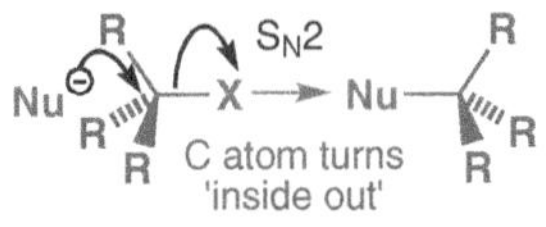

activity. Water must attack the two faces of the planar cation with exactly equal probability: the product is an exactly 50:50 mixture of (S)-butanol and (R)-butanol. It is *racemic*.

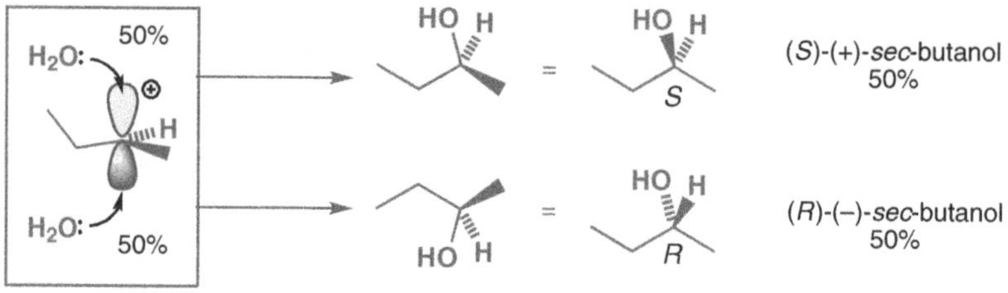

Alternatively, we can first make the hydroxyl group into a good enough leaving group to take part in an S_N2 reaction. The leaving group we shall use, a sulfonate ester, will be introduced to you in a few pages' time, but for now you just need to accept that nucleophilic attack of the OH group on a sulfonyl chloride in pyridine solution gives the sulfonate ester shown below in orange: no bonds have been formed or broken at the chiral carbon atom, which still has (S) stereochemistry.

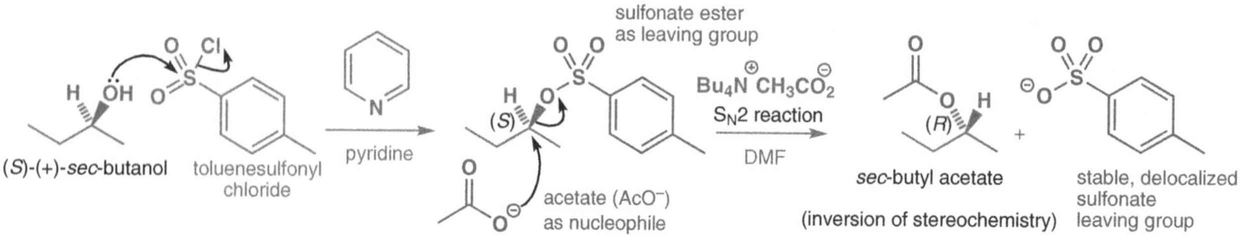

Optical rotation is described on p. 309.

Now we can carry out an S_N2 reaction on the sulfonate with an acetate anion. A tetra-alkyl ammonium salt is used in the solvent DMF to avoid solvating the acetate, making it as powerful a nucleophile as possible and getting a clean S_N2 reaction. This is the key step and we don't want any doubt about the outcome. The sulfonate is an excellent leaving group—the charge is delocalized across all three oxygen atoms.

The product *sec*-butyl acetate is optically active and we can measure its optical rotation. But this tells us nothing. Unless we know the true rotation for pure *sec*-butyl acetate, we don't yet know whether it is optically pure nor even whether it really is inverted. We expect it to have (R) stereochemistry, but we can easily find out for sure. All we have to do is to hydrolyse the ester and get the original alcohol back again. We know the true rotation of the alcohol—it was our starting material—and we know that ester hydrolysis (Chapter 10) proceeds by attack at the carbonyl carbon—it can't affect the stereochemistry of the chiral centre.

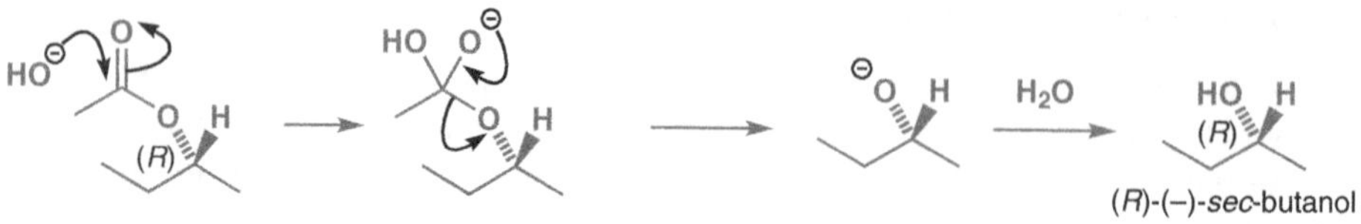

Now we really know where we are. This new sample of *sec*-butanol has the same rotation as the original sample, *but with the opposite sign*. It is (−)-(R)-*sec*-butanol. It is optically pure and inverted. Somewhere in this sequence there has been an inversion, and we know it wasn't in the formation of the sulfonate or the hydrolysis of the acetate as no bonds are formed or broken at the stereogenic centre in these steps. It must have been in the S_N2 reaction itself.

● An S_N2 reaction goes with inversion of configuration at the carbon atom under attack but an S_N1 reaction generally goes with racemization.

The effect of solvent

The different types of solvents were discussed in Chapter 12.

Why was the S_N2 reaction we have just shown you carried out in DMF? You will generally find S_N2 reactions are carried out in aprotic, and often less polar, solvents. S_N1 reactions are typically carried out in polar, protic solvents. A common solvent for an S$_N$2 reaction is acetone—just polar enough to dissolve the ionic reagents, but not as polar as, say, acetic acid, a common solvent for the S$_N$1 reaction.

It is fairly obvious why the S$_N$1 reaction needs a polar solvent: the rate-determining step involves the formation of ions (usually a negatively charged leaving group and a positively charged carbocation) and the rate of this process will be increased by a polar solvent that can solvate these ions. More precisely, the transition state is more polar than the starting materials (note the charges in brackets in the scheme above) and so is stabilized by the polar solvent. Hence solvents like water or carboxylic acids (RCO$_2$H) are ideal.

It is less obvious why a less polar solvent is better for the S$_N$2 reaction. The most common S$_N$2 reactions use an anion as the nucleophile. The transition state is then less polar than the localized anion as the charge is spread between two atoms. Here's an example: the formation of an alkyl iodide from an alkyl bromide. Acetone fails to solvate the iodide well, making it more reactive; the transition state is less in need of solvation, so overall the reaction is faster.

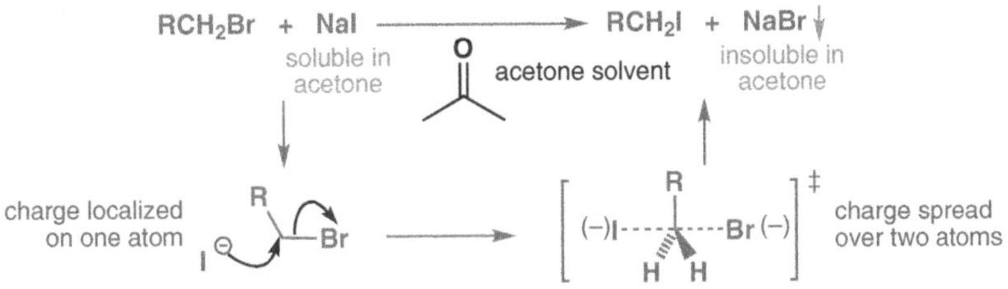

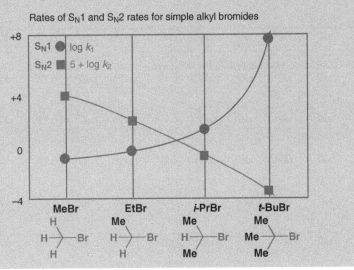

■ Acetone also assists this reaction because it dissolves sodium iodide but not the sodium bromide product, which precipitates from solution and prevents bromide acting as a competing nucleophile.

DMF and DMSO, the polar aprotic solvents we discussed in Chapter 12 (p. 255) are also good solvents for S$_N$2 reactions because they dissolve ionic compounds well but fail to solvate anions well, making them more reactive. The choice of Bu$_4$N$^+$—a large, non-coordinating cation—as the counterion for the reaction on p. 344 was also made with this in mind.

Quantifying the rates of S$_N$1 and S$_N$2 reactions

The data below illustrate the effect of structure on the rates of S$_N$1 and S$_N$2 reactions. The green curve on the graph shows the rates (k_1) of an S$_N$1 reaction: the conversion of alkyl bromides to alkyl formate esters in formic acid at 100 °C. Formic acid is a polar solvent and a weak nucleophile: perfect for an S$_N$1 reaction. The red curve shows the rates of displacement of Br$^-$ by radioactive ^{82}Br$^-$ in acetone at 25 °C. Acetone solvent and the good nucleophile Br$^-$ favour S$_N$2. The rates (k_2) are multiplied by 10^5 to bring both curves onto the same graph.

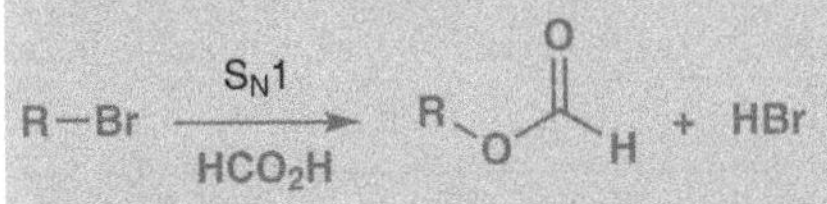

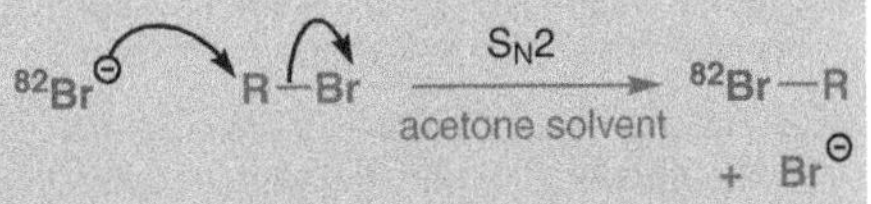

R—Br $\xrightarrow[\text{HCO}_2\text{H}]{\text{S}_N1}$ R$\overset{\text{O}}{\underset{}{\diagup}}O\diagdown$H + HBr ^{82}Br$^{\ominus}$ $\curvearrowright$ R—Br $\xrightarrow[\text{acetone solvent}]{\text{S}_N2}$ ^{82}Br—R + Br$^{\ominus}$

Both curves are plotted on a log scale, the log$_{10}$ of the actual rate being used on the y-axis. The x-axis has no real significance; it just shows the four points corresponding to the four basic structures: MeBr, MeCH$_2$Br, Me$_2$CHBr, and Me$_3$CBr.

Rates of S$_N$1 and S$_N$2 rates for simple alkyl bromides

S$_N$1 ● log k_1
S$_N$2 ■ 5 + log k_2

MeBr	EtBr	i-PrBr	t-BuBr

The values are also summarized in the table below, which gives the relative rates compared with that of the secondary halide, *i*-PrBr, set at 1.0 for each reaction.

Rates of S_N1 and S_N2 reactions of simple alkyl bromides

alkyl bromide type	CH₃Br methyl	CH₃CH₂Br primary	(CH₃)₂CHBr secondary	(CH₃)₃CBr tertiary
k_1 (s⁻¹)	0.6	1.0	26	10^8
$10^5 k_2$ (M⁻¹ dm⁻³ s⁻¹)	13,000	170	6	0.0003
relative k_1	2×10^{-2}	4×10^{-2}	1	4×10^6
relative k_2	6×10^3	30	1	5×10^{-5}

Although the reactions were chosen to give as much S_N1 reaction as possible in one case and as much S_N2 reaction as possible in the other, of course you will understand that we cannot prevent the molecules doing the 'wrong' reaction! The values for the 'S_N1' reaction of MeBr and MeCH₂ Br are actually the low rates of S_N2 displacement of the bromide ion by the weak nucleophile HCO₂H, while the 'S_N2' rate for *t*-BuBr may be the very small rate of ionization of *t*-BuBr in acetone.

A closer look at electronic effects

We mentioned above that adjacent π systems increase the rate of the S_N2 reaction by stabilizing the transition state, and likewise increase the rate of S_N1 reactions by stabilizing the carbocation. The effect on the S_N2 reaction applies to both C=C (electron-rich) and C=O (electron-deficient) π systems, but only C=C π systems increase the rate of S_N1 reactions. Adjacent C=O groups in fact significantly decrease the reactivity of alkyl halides towards S_N1 reactions because the electron-withdrawing effect of the carbonyl group greatly destabilizes the carbocation.

Electron-withdrawing or -donating groups can also tip finely balanced cases from one mechanism to another. For example, benzylic compounds react well by either S_N1 or S_N2, and a change of solvent, as just discussed, might switch them from one mechanism to another. Alternatively, a benzylic compound that has a well-placed electron-donating group able to stabilize the cation will also favour the S_N1 mechanism. Thus 4-methoxybenzyl chloride reacts by S_N1 for this reason: here we show the methoxy group stabilizing the cation intermediate by assisting departure of the chloride.

electron donation favours the S_N1 mechanism

On the other hand, an electron-withdrawing group, such as a nitro group, within the benzylic compound will decrease the rate of the S_N1 reaction and allow the S_N2 mechanism to take over.

Rate measurements for benzylic chlorides illustrate the importance of this effect. We can force them all to react by S_N1 by using methanol as the solvent (methanol is a poor nucleophile and a polar solvent: both disfavour S_N2). Comparing with the rate of substitution of benzyl chloride itself, PhCH₂Cl, 4-methoxybenzyl chloride reacts with methanol about 2500 times faster and the 4-nitrobenzyl chloride about 3000 times more slowly.

electron withdrawal disfavours the S_N1 mechanism

electron-withdrawing nitro group would destabilize cationic intermediate

the same benzylic chloride instead reacts by the S_N2 mechanism

transition state stabilized by adjacent electron-deficient system

● **Summary of structural variations and nucleophilic substitution**

We are now in a position to summarize the structural effects on both mechanisms we have been discussing over the last few pages. The table lists the structural types and rates each reaction qualitatively.

Electrophile	Me–X	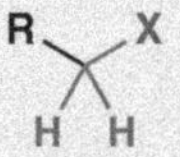 R–CX(H)(H)	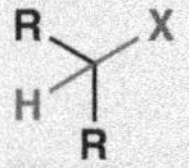 R–CX(H)(R)	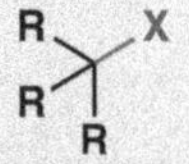 R–CX(R)(R)	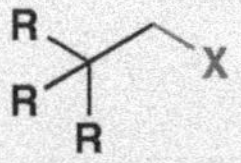 R–CX(R)(R)–X
	methyl	primary	secondary	tertiary	'neopentyl'
S$_N$1 mechanism?	bad	bad	poor	excellent	bad
S$_N$2 mechanism?	excellent	good	poor	bad	bad

Electrophile	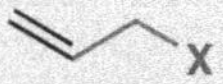 allyl–X	 Ar–X	 RO–X	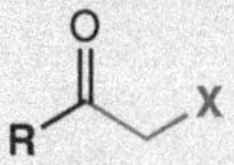 R(C=O)–X	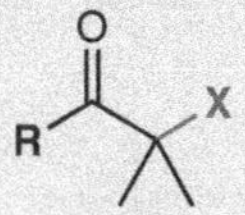 R(C=O)C(R)(R)–X
	allylic	benzylic	α-alkoxy (adj. lone pair)	α-carbonyl	α-carbonyl and tertiary
S$_N$1 mechanism?	good	good	good	bad	bad
S$_N$2 mechanism?	good	good	okay but S$_N$1 better	excellent	possible

We have considered the important effects of the basic carbon skeleton and of solvent on the course of S$_N$1 and S$_N$2 reactions and we shall now look at two final structural factors: the nucleophile and the leaving group. We shall tackle the leaving group first because it plays an important role in both S$_N$1 and S$_N$2 reactions.

The leaving group in S$_N$1 and S$_N$2 reactions

The leaving group is important in both S$_N$1 and S$_N$2 reactions because departure of the leaving group is involved in the rate-determining step of both mechanisms.

The leaving group in the S$_N$1 reaction

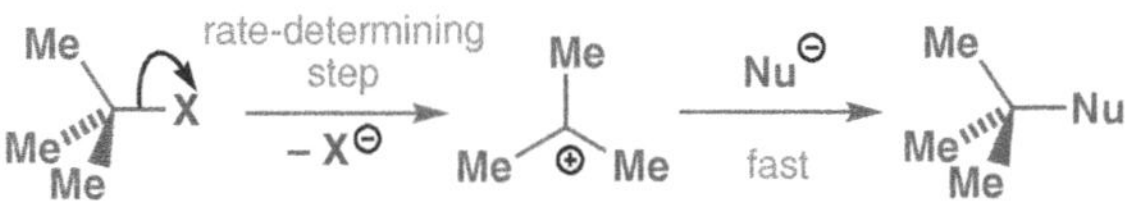

The leaving group in the S$_N$2 reaction

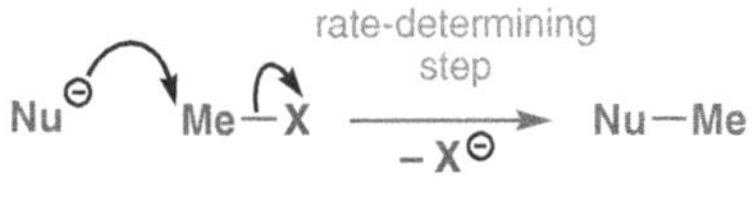

So far you have mostly seen halides and water (from protonated alcohols) as leaving groups. Leaving groups involving halides or oxygen atoms are by far the most important, and now we need to establish the principles that make for good and bad leaving groups. As a chemist, we want leaving groups to have some staying power, so that our compounds are not too unstable, but we also don't want them to outstay their welcome—they must have just the right level of reactivity.

Halides as leaving groups

With halide leaving groups two main factors are at work: the strength of the C–halide bond and the stability of the halide ion. The strengths of the C–X bonds can be measured easily, but how can we measure anion stability? One way, which you met in Chapter 8, was to use the pK_a values of the acids HX. pK_a quantifies the stability of an anion relative to its conjugate acid. We want to know about the stability of an anion relative to that anion bonded to C, not H, but pK_a will do as a guide.

Halide leaving groups in the S_N1 and S_N2 reactions

Halide (X)	Strength of C–X bond, kJ mol^{-1}	pK_a of HX
fluorine	118	+3
chlorine	81	−7
bromine	67	−9
iodine	54	−10

The table in the margin shows both bond strengths and pK_a. It is clearly easiest to break a C–I bond and most difficult to break a C–F bond. Iodide sounds like the best leaving group. We get the same message from the pK_a values: HI is the strongest acid, so it must ionize easily to H$^+$ and I$^-$. This result is quite correct—iodide is an excellent leaving group and fluoride a very bad one, with the other halogens in between.

Nucleophilic substitutions on alcohols: how to get an OH group to leave

Now what about leaving groups joined to the carbon atom by a C–O bond? There are many of these but the most important are OH itself, the carboxylic esters, and the sulfonate esters. First we must make one thing clear: alcohols themselves do *not* react with nucleophiles. In other words, OH$^-$ is never a leaving group. Why not? For a start hydroxide ion is very basic, and if the nucleophile were strong enough to displace hydroxide ion it would be more than strong enough to remove the proton from the alcohol.

■ You were given the same message in Chapter 10 in relation to substitutions at C=O: hydroxide is never a leaving group. There is one exception to this rule, in the E1cb reaction, which you will meet in Chapter 17, but it's rare enough to ignore at this stage.

S_N2 displacement of hydroxide never happens...

If the nucleophile reacts, it attacks the *proton* instead

But we do want to use alcohols in nucleophilic substitution reactions because they are easily made (by the reactions in Chapter 9, for example). The simplest solution is to protonate the OH group with strong acid. This will work only if the nucleophile is compatible with strong acid, but many are. The preparation of t-BuCl from t-BuOH simply by shaking it with concentrated HCl is a good example. This is obviously an S_N1 reaction with the t-butyl cation as intermediate.

t-butyl chloride from t-butanol

the mechanism

Similar methods can be used to make secondary alkyl bromides with HBr alone and primary alkyl bromides using a mixture of HBr and H_2SO_4.

substituting a secondary alcohol in acid

substituting a primary alcohol in acid

The second of these two reactions must be S_N2, with substitution of the protonated hydroxyl group by bromide.

Another way to approach the substitution of OH is to make it a better leaving group by combination with an element that forms very strong bonds to oxygen. The most popular choices are phosphorus and sulfur. Making primary alkyl bromides with PBr$_3$ usually works well.

The phosphorus reagent is first attacked by the OH group (an S_N2 reaction at phosphorus) and the displacement of an oxyanion bonded to phosphorus is now a good reaction because of the anion stabilization by phosphorus.

91% yield

Sulfonate esters—tosylates and mesylates—from alcohols

The most widely used way of making a hydroxyl group into a good leaving group is to make it into a sulfonate ester. Primary and secondary alcohols are easily converted to sulfonate esters by treating with sulfonyl chlorides and base. The sulfonate esters are often crystalline, and are so widely used that they have been given trivial names—*tosylates* for *p*-toluenesulfonates and *mesylates* for methanesulfonates—and the functional groups have been allocated the 'organic element' symbols Ts and Ms.

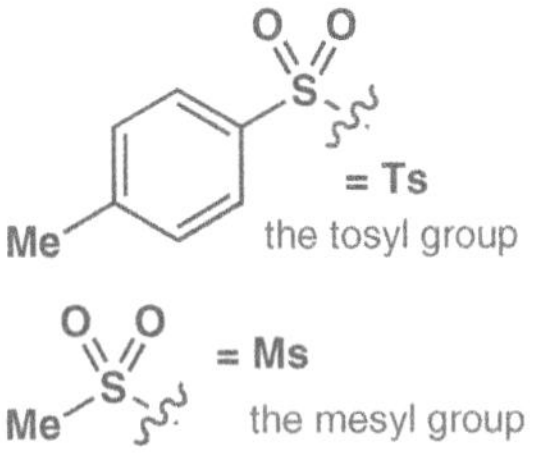

Tosylates (*p*-toluenesulfonates) are made by treating alcohols with *p*-toluenesulfonyl chloride (or tosyl chloride) in the presence of pyridine. A similar reaction (but with a different mechanism, which we will discuss in Chapter 17) with methanesulfonyl chloride (mesyl chloride) gives a mesylate (methanesulfonate).

➡ The mechanism by which sulfonate esters are formed is discussed in more detail in Chapter 17.

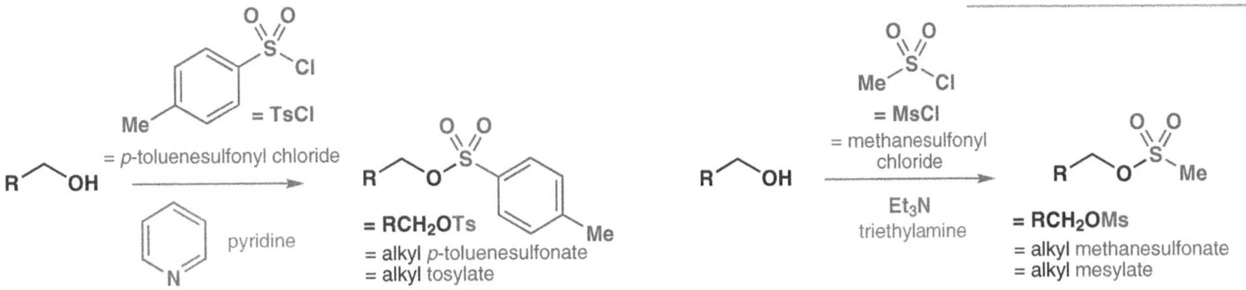

Sulfonic acids RSO$_3$H are strong acids (pK_a around 0) and so any sulfonate RSO$_3^-$ is a good leaving group: tosylates and mesylates can be displaced by almost anything. As you saw in Chapter 8, the lithium derivative of an alkyne can be prepared by deprotonation with the very strong base butyllithium. In the example below, the tosyl derivative of a primary alcohol reacts with this lithium derivative in an S$_N$2 reaction. Note that the tosylate leaving group is represented as TsO$^-$ (not Ts$^-$!).

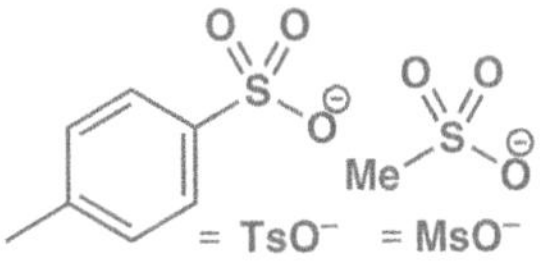

tosylate and mesylate: excellent leaving groups

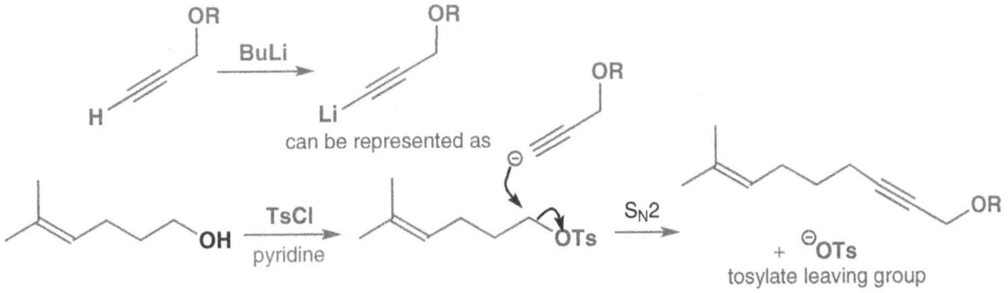

On p. 344 you saw a tosylate (we just called it a sulfonate ester then) being displaced by acetate in an S$_N$2 reaction. Acetate is not a very good nucleophile, and it is a testament to the power of the sulfonate esters that they are willing to act as leaving groups even with acetate, which is usually too weak to react by S$_N$2.

Substituting alcohols with the Mitsunobu reaction

Rather than use two steps to convert the OH group first to a sulfonate ester, and then displace it, it is possible to use a method that allows us to put an alcohol straight into a reaction mixture and get an S$_N$2 product in one operation. This is the Mitsunobu reaction. In this reaction, the alcohol becomes the electrophile, the nucleophile is usually relatively weak (the conjugate base of a carboxylic acid, for example), and there are two other reagents.

Oyo Mitsunobu (1934–2003) worked at the Aoyama Gakuin University in Tokyo. Western chemists often misspell his name: make sure you don't!

a Mitsunobu reaction

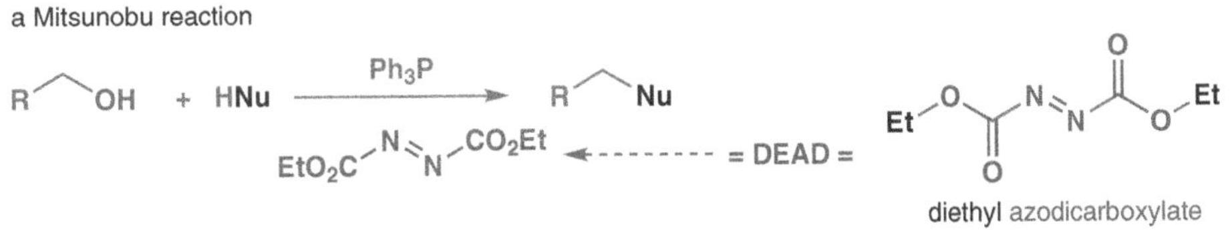

diethyl azodicarboxylate

One of these reagents, Ph_3P, triphenylphosphine, is the simple phosphine you met in Chapter 11. Phosphines are nucleophilic, but not basic like amines. The other reagent deserves more comment. Its full name is diethyl azodicarboxylate, or DEAD.

So how does the Mitsunobu reaction work? It's a long mechanism, but don't be discouraged: there is a logic to each step and we will guide you through it gently. The first stage involves neither the alcohol nor the added nucleophile. The phosphine adds to the weak $N=N$ π bond to give an anion stabilized by one of the ester groups.

stage 1 of the Mitsunobu reaction   stabilization of the nitrogen anion by the ester group

You will note that the nucleophile has been added as its conjugate acid 'HNu'—often this might be a carboxylic acid, for example benzoic acid. The anion produced by this first stage is basic enough to remove a proton from this acid, generating Nu^- ready for reaction.

stage 2 of the Mitsunobu reaction

$+ Nu^{\ominus}$
nucleophile revealed

Oxygen and phosphorus have a strong affinity, as we saw in the conversion of alcohols to bromides with PBr_3 (p. 348) and in the Wittig reaction (Chapter 11, pp. 237–8), and the positively charged phosphorus is now attacked by the alcohol, displacing a second nitrogen anion in an S_N2 reaction at phosphorus. The nitrogen anion generated in this step is stabilized by conjugation with the ester, but rapidly removes the proton from the alcohol to give an electrophilic $R–O–PPh_3^+$ species and a by-product, the reduced form of DEAD.

stage 3 of the Mitsunobu reaction

by-product

Finally, the anion of the nucleophile can now attack this phosphorus derivative of the alcohol in a normal S_N2 reaction at carbon with the phosphine oxide as the leaving group. We have arrived at the products.

stage 4 of the Mitsunobu reaction

$Nu^{\ominus}$

$Nu \frown R$ + $O=PPh_3$
S_N2 product   phosphine oxide

🖱 Interactive mechanism for the Mitsunobu reaction

The whole process takes place in one operation. The four reagents are all added to one flask and the products are the phosphine oxide, the reduced azo diester with two NH bonds replacing the $N=N$ double bond, and the product of an S_N2 reaction on the alcohol. Another way to look at this reaction is that a molecule of water must formally be lost: OH must be removed from the alcohol and H from the nucleophile. These atoms end up in very stable molecules—the P=O and N–H bonds are strong where the $N=N$ bond was weak, compensating for the sacrifice of the strong C–O bond in the starting alcohol.

If this is all correct, then the vital S_N2 step should lead to inversion as it always does in S_N2 reactions. This turns out to be one of the great strengths of the Mitsunobu reaction—it is a reliable way to replace OH by a nucleophile with inversion of configuration. The most dramatic example is probably the formation of esters from secondary alcohols with inversion. Normal ester formation leads to retention as the C–O bond of the alcohol is not broken: compare these two reactions and note the destination of the coloured oxygen (and hydrogen) atoms.

ester formation from a secondary alcohol with inversion by the Mitsunobu reaction

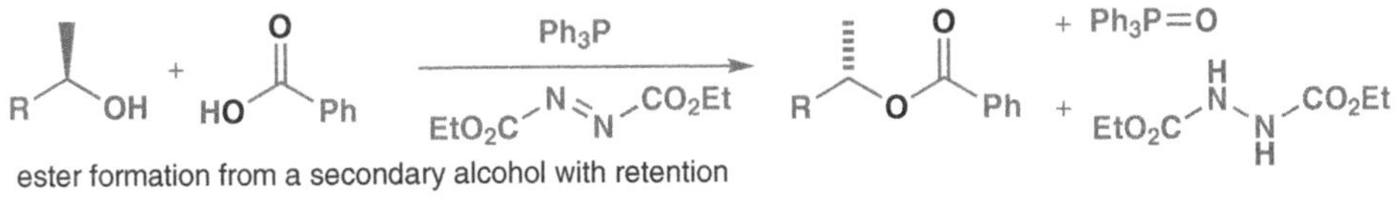

ester formation from a secondary alcohol with retention

Ethers as electrophiles

Ethers are stable molecules that do not react with nucleophiles: THF and Et$_2$O are widely used as solvents for this reason. To make them react, we need to make the oxygen positively charged so that it can accept electrons more readily, and we also need to use a very good nucleophile. A good way of doing both is to treat with HBr or HI, which protonate the oxygen. Iodide and bromide are excellent nucleophiles in S$_N$2 reactions (see below), and attack will occur preferentially at the carbon atom more susceptible to S$_N$2 reactions (usually the less hindered one). Aryl alkyl ethers cleave only on the alkyl side—you cannot get attack through the benzene ring.

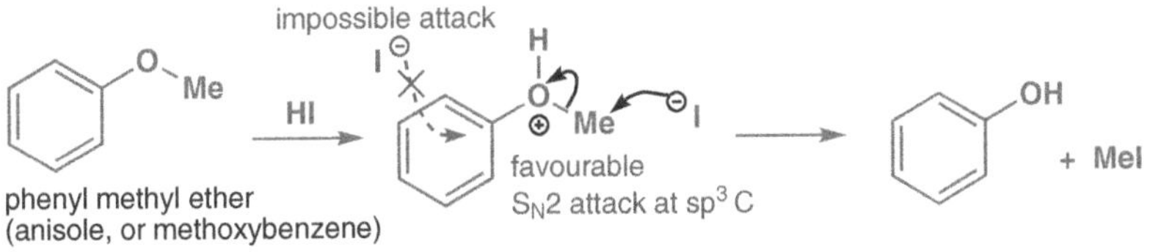

phenyl methyl ether
(anisole, or methoxybenzene)

So far we have used only protic acids to help oxygen atoms to leave. But Lewis acids—species other than H$^+$ that also have an empty orbital capable of accepting a lone pair—work well too, and the cleavage of aryl alkyl ethers with BBr$_3$ is a good example. Trivalent boron compounds have an empty p orbital so they are very electrophilic and prefer to attack oxygen. The resulting oxonium ion can be attacked by Br$^-$ in an S$_N$2 reaction.

⮕ Lewis acids were introduced in Chapter 8, p. 180.

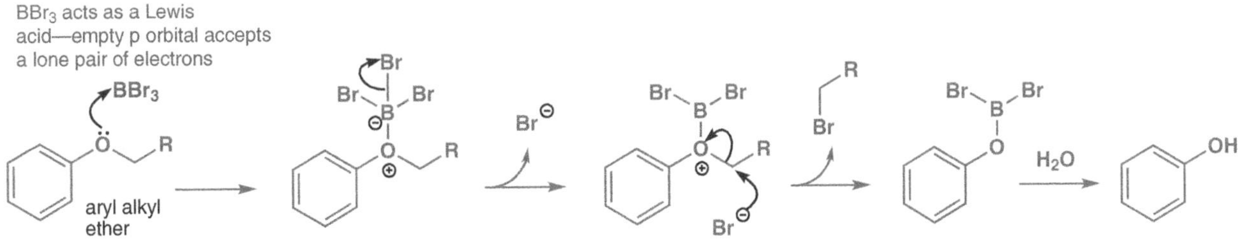

BBr$_3$ acts as a Lewis acid—empty p orbital accepts a lone pair of electrons

aryl alkyl ether

Epoxides as electrophiles

One family of ethers reacts in nucleophilic substitution even without protic or Lewis acids. They are the three-membered cyclic ethers called epoxides (or oxiranes). The leaving group is genuinely an alkoxide anion RO$^-$, so obviously some special feature must be present in these ethers making them unstable. This feature is ring strain, which comes from the angle between the bonds in the three-membered ring that has to be 60° instead of the ideal tetrahedral angle of 109°. You could subtract these numbers and say that there is '49° of strain' at each carbon atom, making about 150° of strain in the molecule. This is a lot: the molecule would be much

⮕ You will see how to make epoxides from alkenes in Chapter 19.

337

more stable if the strain were released by opening up to restore the ideal tetrahedral angle at all atoms. This can be done by one nucleophilic attack.

➡ Ring strain is discussed further in Chapter 16 on p. 368.

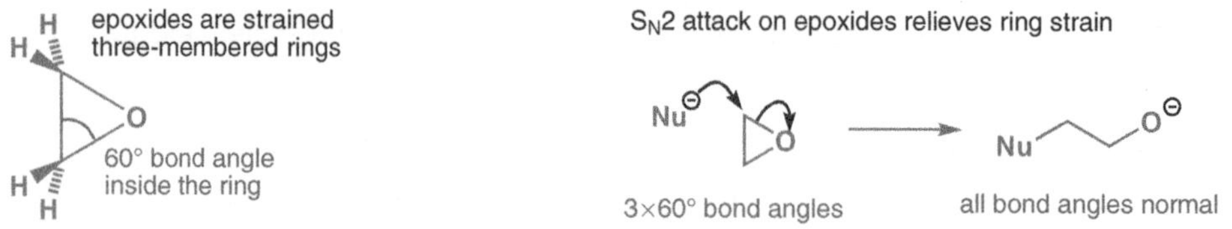

Epoxides react cleanly with amines to give amino alcohols. We have not so far featured amines as nucleophiles because their reactions with alkyl halides are often bedevilled by overreaction (see the next section), but with epoxides they give good results.

■ When epoxides are substituted differently at either end, the nucleophile has a choice of which end to attack. The factors that control this will be discussed in Chapter 24.

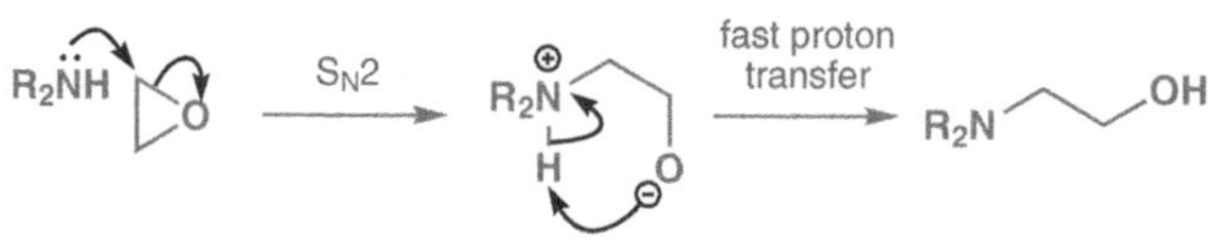

It is easy to see that inversion occurs in these S_N2 reactions when the epoxide is attached to (or 'fused with') another ring. With this five-membered ring nucleophilic attack with inversion gives the *trans* product. As the epoxide in the starting material is *up*, attack has to come from underneath. The new C–N bond is *down* and inversion has occurred.

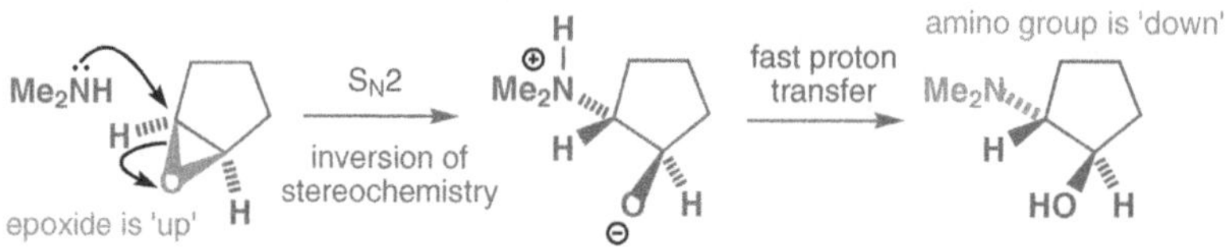

The nucleophile in S_N1 reactions

We established earlier that in an S_N1 reaction the nucleophile is not important with regard to *rate*. The rate-determining step of the reaction is loss of the leaving group, so good and bad nucleophiles all give products. We don't need to deprotonate the nucleophile to make it more reactive (water and hydroxide work just as well as each other) and this means that S_N1 reactions are often carried out under acidic conditions, to assist departure of a leaving group.

Compare, for example, these typical conditions used to make a methyl ether and a *tert*-butyl ether. The methyl ether is made, as you saw on p. 340, using methyl iodide in an S_N2 reaction. It needs a good nucleophile, so the alcohol is deprotonated to make an alkoxide with sodium hydride in DMF, which, as you saw on p. 345, is a good solvent for S_N2 reactions. The *tert*-butyl ether on the other hand is made simply by stirring the alcohol with *tert*-butanol and a little acid. No base is needed, and the reaction proceeds rapidly to give the *tert*-butyl ether.

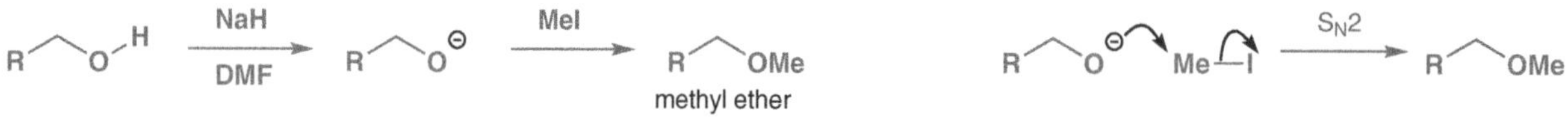

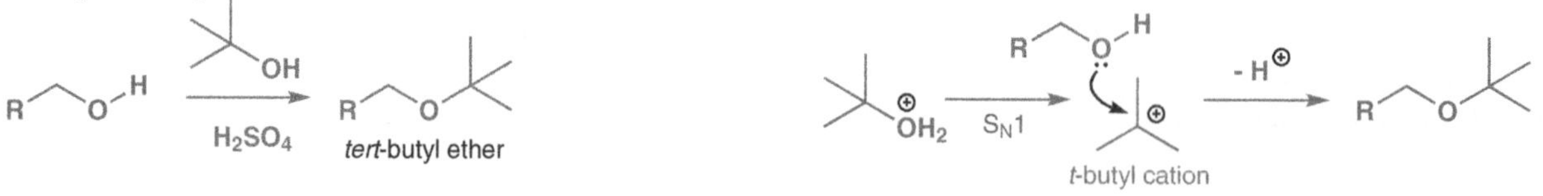

A very bad nucleophile in a good S$_N$1 reaction: the Ritter reaction

An interesting result of the unimportance of the nucleophile to the rate (and therefore the usefulness) of an S$_N$1 reaction is that very poor nucleophiles indeed may react in the absence of anything better. Nitriles, for example, are very poorly basic and nucleophilic because the lone pair of electrons on the nitrogen atom is in a low-energy sp orbital. However, if *t*-butanol is dissolved in a nitrile as solvent and strong acid is added, a reaction does take place. The acid does not protonate the nitrile, but does protonate the alcohol to produce the *t*-butyl cation in the usual first step of an S$_N$1 reaction. This cation is reactive enough to combine with even such a weak nucleophile as the nitrile.

nitriles are very weak bases and poor nucleophiles

$$R—\equiv N$$

lone pair in sp orbital

The resulting cation is captured by the water molecule released in the first step and an exchange of protons leads to a secondary amide. The overall process is called the Ritter reaction and it is one of the few reliable ways to make a C–N bond to a tertiary centre.

new C–N bond

The nucleophile in the S$_N$2 reaction

In an S$_N$2 reaction, a good nucleophile is essential. We finish this chapter with a survey of effective choices for forming new bonds to sp^3 by S$_N$2 reactions, and a description of the factors that determine how good a nucleophile will be.

Nitrogen nucleophiles: a problem and a solution

Amines are good nucleophiles, but reactions between ammonia and alkyl halides rarely lead cleanly to single products. The problem is that the product of the substitution is at least as nucleophilic as the starting material, so it competes for reaction with the alkyl halide.

primary amine formed
in reaction mixture

secondary amine formed
in reaction mixture

primary amine reacts again with alkyl halide

Even this is not all! The alkylation steps keep going, forming the secondary and tertiary amines, and stopping only when the non-nucleophilic tetra-alkylammonium ion R$_4$N$^+$ is formed. The problem is that the extra alkyl groups push more and more electron density onto N, making each product more reactive than the previous. The quaternary ammonium salt could probably be made cleanly if a large excess of alkyl halide RX is used, but other more controlled methods are needed for the synthesis of primary, secondary, and tertiary amines.

One solution for primary amines is to replace ammonia with azide ion N$_3^-$. This linear tri-atomic species, nucleophilic at both ends, is a slender rod of electrons able to insert itself into almost any electrophilic site. It is available as the water-soluble sodium salt NaN$_3$.

Sometimes these alkylations can work, but usually only if the alkylating agent or the amine is very hindered, or the alkylating agent contains an inductive electron-withdrawing group (such as the hydroxyl group generated when an epoxide is opened: epoxides *are* reliable alkylating agents for amines). With amine alkylations, you should nonetheless always expect the worst.

structure of azide ion N_3^-

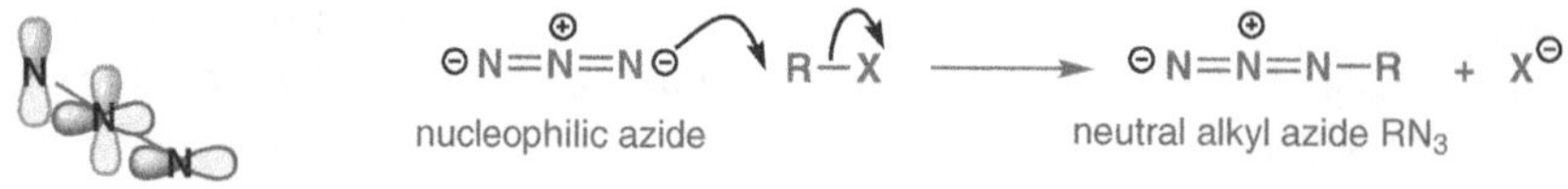

nucleophilic azide   neutral alkyl azide RN_3

■ Azide is isoelectronic with carbon dioxide, and has the same linear shape.

Azide reacts only once with alkyl halides because the product, an alkyl azide, is no longer nucleophilic. However, rarely is the azide product required: it is usually reduced to a primary amine by catalytic hydrogenation (H_2 over a Pd catalyst—see Chapter 23), $LiAlH_4$, or triphenylphosphine.

A warning about azides

Azides can be converted by heat—or even sometimes just by a sharp blow—suddenly into nitrogen gas. In other words they are potentially explosive, particularly inorganic (that is, ionic) azides and low molecular weight covalent organic azides.

$$RX + NaN_3 \longrightarrow RN_3 \xrightarrow[\text{or } H_2/Pd]{LiAlH_4} RNH_2$$

Azides react with epoxides too. This epoxide is one diastereoisomer (*trans*) but racemic and the symbol (±) under each structure reminds you of this (Chapter 14). Azide attacks at either end of the three-membered ring (the two ends are the same) to give the hydroxy-azide. The reaction is carried out in a mixture of water and an organic solvent with ammonium chloride as buffer to provide a proton for the intermediate. Triphenylphosphine in water is used for reduction to the primary amine.

➡ The mechanism of the reduction of azides by triphenylphosphine can be found on p. 1176.

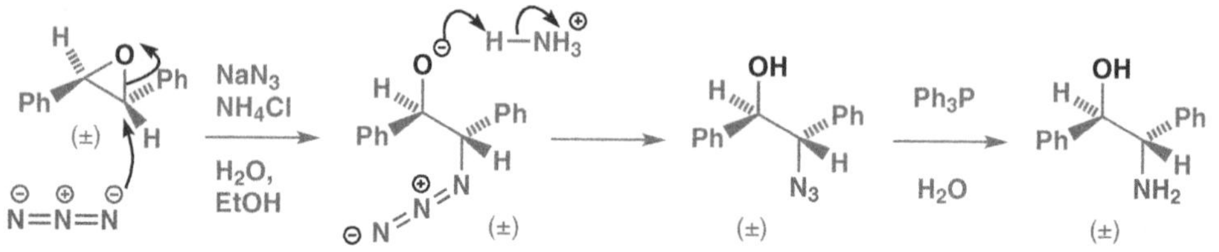

Sulfur nucleophiles are better than oxygen nucleophiles in S$_N$2 reactions

Thiolate anions RS$^-$ make excellent nucleophiles in S$_N$2 reactions on alkyl halides. It is enough to combine the thiol, sodium hydroxide, and the alkyl halide to get a good yield of the sulfide.

$$PhSH + NaOH + \textit{n-}BuBr \longrightarrow PhSBu + NaBr$$

thiol   sulfide

Thiols are more acidic than water (pK_a of RSH is typically 9–10, pK_a of PhSH is 6.4, pK_a of H_2O is 15.7) and rapid proton transfer from sulfur to oxygen gives the thiolate anion that acts as a nucleophile in the S$_N$2 reaction.

the S$_N$2 reaction with a thiolate anion as nucleophile

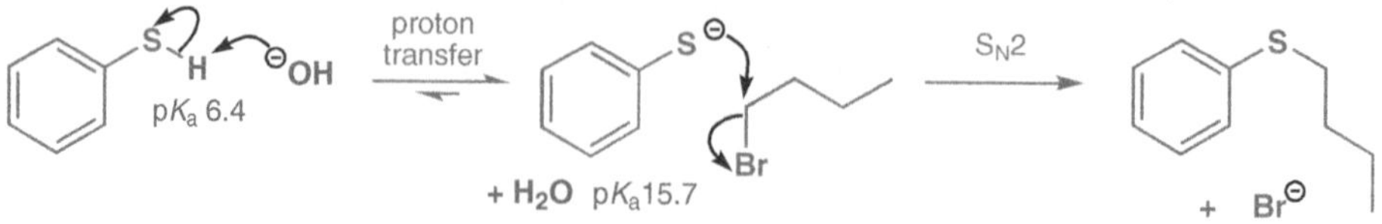

But how do you make a thiol in the first place? The obvious way to make aliphatic thiols would be by an S$_N$2 reaction using NaSH on the alkyl halide.

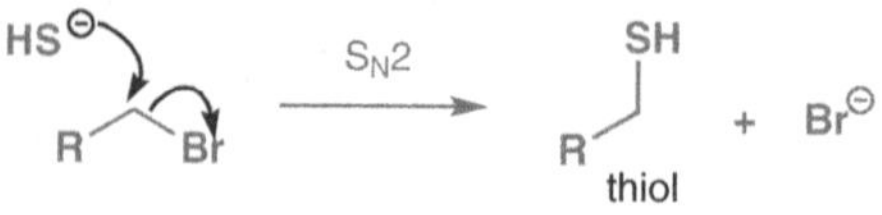

This works well but, unfortunately, the product easily exchanges a proton and the reaction normally produces the symmetrical sulfide—this should remind you of what happened with amines!

R~SH + HS$^\ominus$ ⇌ R~S$^\ominus$ $\xrightarrow{\text{S}_N2}$ R~S~R

thiol   sulfide

The solution is to use the anion of thioacetic acid, usually the potassium salt. This reacts cleanly through the more nucleophilic sulfur atom and the resulting ester can be hydrolysed in base to liberate the thiol.

thioacetate $\xrightarrow{\text{S}_N2}$ thioester $\xrightarrow[\text{H}_2\text{O}]{\text{NaOH}}$ acetate + HS~R thiol

Effectiveness of different nucleophiles in the S$_N$2 reaction

In Chapter 10 we pointed out that basicity is nucleophilicity towards protons. At that stage we said that nucleophilicity towards the carbonyl group parallels basicity almost exactly. We are able to use pK_a as a guide to the effectiveness of nucleophilic substitution reactions **at the carbonyl group**.

During this chapter you have had various hints that nucleophilicity towards saturated carbon is not so straightforward. Now we must look at this question seriously and try to give you helpful guidelines.

1. If the atom that is forming the new bond to carbon is the same over a range of nucleophiles—it might be oxygen, for example, and the nucleophiles might be HO$^-$, PhO$^-$, AcO$^-$, and TsO$^-$—then nucleophilicity does parallel basicity. The anions of the weakest acids are the best nucleophiles. The order for the nucleophiles we have just mentioned will be: HO$^-$ > PhO$^-$ > AcO$^-$ > TsO$^-$. The actual values for the rates of attack of the various nucleophiles on MeBr in EtOH relative to the rate of reaction with water (= 1) are given in the table below.

pK_a of HNu is a good guide to the rate of this sort of reaction

nucleophilic attack on C=O

but the story with this sort of reaction is more complicated

nucleophilic substitution at saturated C

Relative rates (water = 1) of reaction with MeBr in EtOH

Nucleophile X$^-$	pK_a of HX	Relative rate
HO$^-$	15.7	1.2×10^4
PhO$^-$	10.0	2.0×10^3
AcO$^-$	4.8	9×10^2
H$_2$O	−1.7	1.0
ClO$_4^-$	−10	0

2. If the atoms that are forming the new bond to carbon are *not* the same over the range of nucleophiles we are considering, then another factor is important. In the very last examples we have been discussing we have emphasized that RS$^-$ is an excellent nucleophile for saturated carbon. Let us put that another way: RS$^-$ is a better nucleophile for saturated carbon than RO$^-$, even though RO$^-$ is more basic than RS$^-$ (see table below).

Relative rates (water = 1) of reaction with MeBr in EtOH

Nucleophile X$^-$	pK_a of HX	Relative rate
PhS$^-$	6.4	5.0×10^7
PhO$^-$	10.0	2.0×10^3

Sulfur is plainly a better nucleophile than oxygen for saturated carbon. Why should this be? As we discussed back in Chapter 5, there are two main factors controlling bimolecular reactions: (1) electrostatic attraction (simple attraction of opposite charges or partial charges) and (2) bonding interactions between the HOMO of the nucleophile and the LUMO of the electrophile.

A proton is, of course, positively charged, so electrostatic attraction is the more important factor in nucleophilicity towards H⁺, or pK_a. The carbonyl group too has a substantial positive charge on the carbon atom, arising from the uneven distribution of electrons in the C=O π bond, and reactions of nucleophiles with carbonyl groups are also heavily influenced by electrostatic attraction, with HOMO–LUMO interactions playing a smaller role.

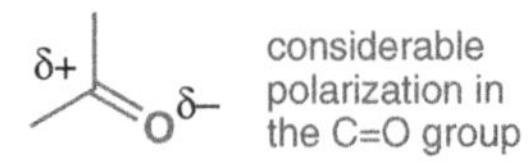

When it comes to saturated carbon atoms carrying leaving groups, polarization is typically much less important. There is, of course, some polarity in the bond between a saturated carbon atom and, say, a bromine atom, but the electronegativity difference between C and Br is less than half that between C and O. In alkyl iodides, one of the best classes of electrophiles in S_N2 reactions, there is in fact almost no dipole at all—the electronegativity of C is 2.55 and that of I is 2.66.

Electronegativities:

C: 2.55 I: 2.66 Br: 2.96 O: 3.44

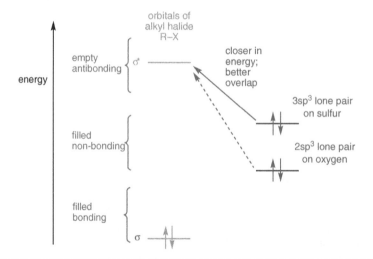

considerable polarization in the C=O group

much less polarization in the C–Br bond

HOMO of Nu⁻

LUMO of alkyl bromide

- **Electrostatic attraction is often unimportant in S_N2 reactions.**

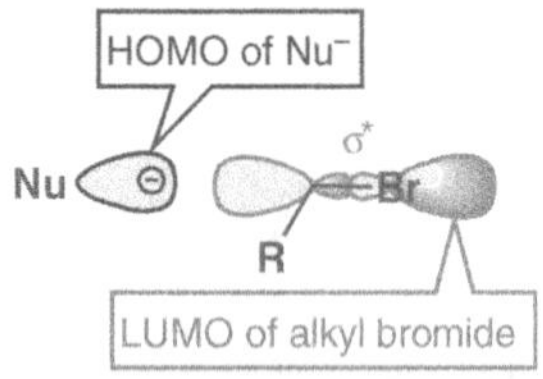

What does matter is the strength of the HOMO–LUMO interaction. In a nucleophilic attack on the carbonyl group, the nucleophile adds in to the low-energy π* orbital. In a nucleophilic attack on a saturated carbon atom, the nucleophile must donate its electrons to the σ* orbital of the C–X bond, as illustrated in the margin for an alkyl bromide reacting with the non-bonding lone pair of a nucleophile.

σ* antibonding orbitals are, of course, higher in energy than non-bonding lone pairs, but the higher the energy of the nucleophile's lone pair, the better the overlap. The $3sp^3$ lone-pair electrons of sulfur overlap better with the high-energy σ* orbital of the C–X bond than do the lower energy $2sp^3$ lone-pair electrons on oxygen because the higher energy of the sulfur electrons brings them closer in energy to the C–X σ* orbital. The conclusion is that nucleophiles from lower down the periodic table are more effective in S_N2 reactions than those from the top rows.

- **Typically, nucleophilic power towards saturated carbon goes like this:**

  I⁻ > Br⁻ > Cl⁻ > F⁻

  RSe⁻ > RS⁻ > RO⁻

  R_3P: > R_3N:

Nucleophiles in substitution reactions

Relative rates (water = 1) of reaction of nucleophiles with MeBr in EtOH

nucleophile	F⁻	H₂O	Cl⁻	Et₃N	Br⁻	PhO⁻	EtO⁻	I⁻	PhS⁻
relative rate	0.0	1.0	1100	1400	5000	2.0×10^3	6×10^4	1.2×10^5	5.0×10^7

Hard and soft nucleophiles

The fact that some nucleophiles, like R_3P: and RS^-, react very fast at saturated C atoms (they have high-energy lone pairs), but very poorly at C=O groups (they are either uncharged or have charge spread diffusely over large orbitals) gives them a different type of character from strongly basic nucleophiles like HO^- that attack C=O groups rapidly. We call nucleophiles that react well at saturated carbon **soft** nucleophiles; those that are more basic and react well with carbonyl groups are referred to as **hard** nucleophiles. These are useful and evocative terms because the soft nucleophiles are indeed rather large and flabby with diffuse high-energy electrons while the hard nucleophiles are small and spiky with closely held electrons and high charge density.

When we say 'hard' (nucleophile or electrophile) we refer to species whose reactions are dominated by electrostatic attraction and when we say 'soft' (nucleophile or electrophile) we refer to species whose reactions are dominated by HOMO–LUMO interactions.

■ Just to remind you: reactions dominated by electrostatic attraction also need to pass electrons from HOMO to LUMO, but reactions that are dominated by HOMO–LUMO interactions need have no contribution from electrostatic attraction.

● Summary of the characteristics of the two types of nucleophile.

Hard nucleophiles X	Soft nucleophiles Y
small	large
charged	neutral
basic (HX weak acid)	not basic (HY strong acid)
low-energy HOMO	high-energy HOMO
like to attack C=O	like to attack saturated carbon
such as RO^-, NH_2^-, MeLi	such as RS^-, I^-, R_3P

Nucleophiles and leaving groups compared

In Chapter 10 we explained that, in a nucleophilic attack on the carbonyl group, a good nucleophile is a bad leaving group and vice versa. We set you the challenge of predicting which way the following reaction would go.

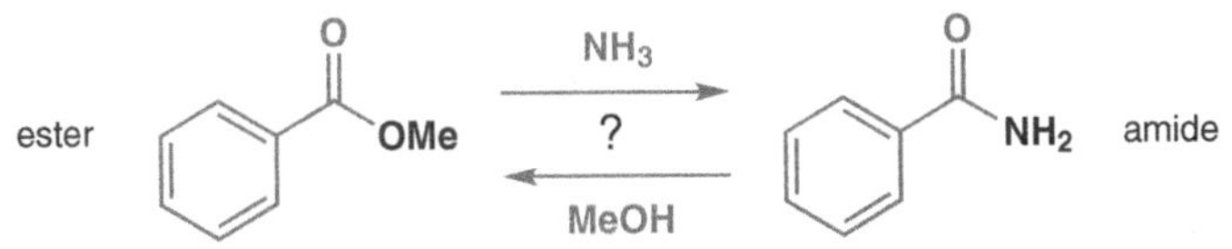

You should by now understand well that the reaction goes from ester to amide rather than the other way round, because NH_3 is a better nucleophile than MeOH and NH_2^- is a worse leaving group than MeO^-.

The S_N2 reaction is different: some of the best nucleophiles are also the best leaving groups. The most important examples of this are bromide and iodide. As the table on p. 356 showed, iodide ion is one of the best nucleophiles towards saturated carbon because it is at the bottom of its group in the periodic table and its lone-pair electrons are very high in energy. Alkyl iodides are readily formed from alkyl chlorides or tosylates. Here are two examples. The first is assisted by the solvent, acetone, which allows NaCl to precipitate and drives the reaction forward.

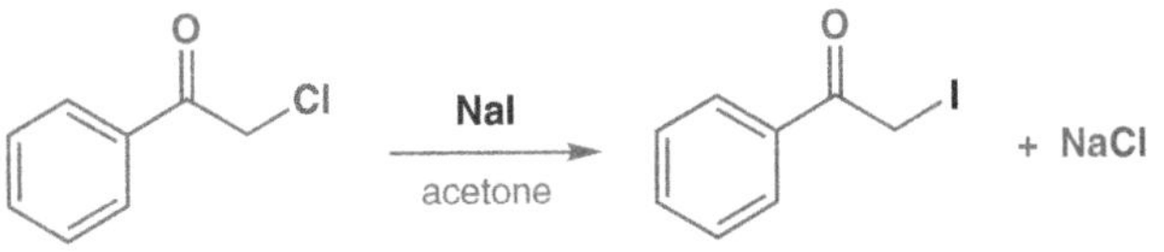

The second example is from the preparation of a phosphonium salt used in a synthesis of terpenes. An unsaturated primary alcohol was first made into its tosylate, the tosylate was converted into the iodide, and the iodide into the phosphonium salt.

■ We explained on p. 347 why I⁻ is such a good leaving group: the C–I bond is particularly weak. The poor overlap between the atomic orbitals on C and on I also mean that the σ* is low lying and easily accessible to the nucleophile's HOMO.

But why this roundabout route via the iodide? The answer is that as well as being an excellent nucleophile, iodide is such a good leaving group that alkyl iodides are often used as intermediates to encourage substitution with other nucleophiles. Yields are often higher if the alkyl iodide is prepared than if the eventual nucleophile is reacted directly with the alkyl tosylate or chloride.

However, iodine is expensive, and a way round that problem is to use a catalytic amount of iodide. The phosphonium salt below is formed slowly from benzyl bromide but the addition of a small amount of LiI speeds up the reaction considerably.

The solvent 'xylene' needs some explanation. Xylene is the trivial name for dimethyl benzene and there are three isomers. Mixed xylenes are isolated cheaply from oil and often used as a relatively high boiling solvent (b.p. about 140 °C) for reactions at high temperature. In this case, the starting materials are soluble in xylene but the product is a salt and conveniently precipitates out during the reaction. Non-polar xylene favours the S_N2 reaction (p. 345).

ortho-xylene
1,2-dimethyl benzene

meta-xylene
1,3-dimethyl benzene

para-xylene
1,4-dimethyl benzene

The iodide reacts both as a better nucleophile than Ph₃P and then as a better leaving group than Br⁻. Each iodide ion goes round the cycle many times as a **nucleophilic catalyst**.

A nucleophilic catalyst speeds up a reaction by acting as both a good nucleophile and a good leaving group. You saw pyridine performing a similar function in substitution reactions at the C=O group of acid anhydrides in Chapter 10.

Looking forward: elimination and rearrangement reactions

Simple nucleophilic substitutions at saturated carbon atoms are fundamental reactions found wherever organic chemistry is practised. They are used in industry on an enormous scale and in pharmaceutical laboratories to make important drugs. They are worth studying for their importance and relevance.

There is another side to this simple picture. These were among the first reactions whose mechanisms were thoroughly investigated by Ingold in the 1930s and since then they have probably been studied more than any other reactions. All our understanding of organic mechanisms begins with S_N1 and S_N2 reactions, and you need to understand these basic mechanisms properly.

The carbocations you met in this chapter are reactive intermediates not only in S_N1 substitutions but in other reactions too. One of the most convincing pieces of evidence for their formation is that they undergo reactions other than simple addition to nucleophiles. For example, the carbon skeleton of the cation may *rearrange*, as we will discuss in Chapter 36.

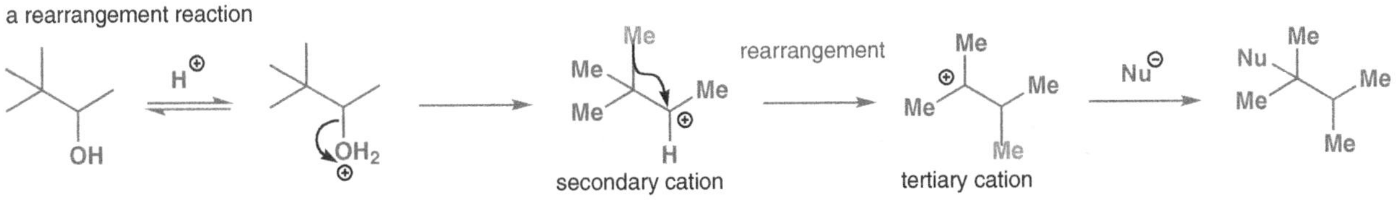

a rearrangement reaction

secondary cation   rearrangement   tertiary cation

Another common fate of cations, and something that may also happen instead of an intended S_N1 or S_N2 reaction, is *elimination*. Here an alkene is formed by the nucleophile acting as a base to remove HX instead of adding to the molecule.

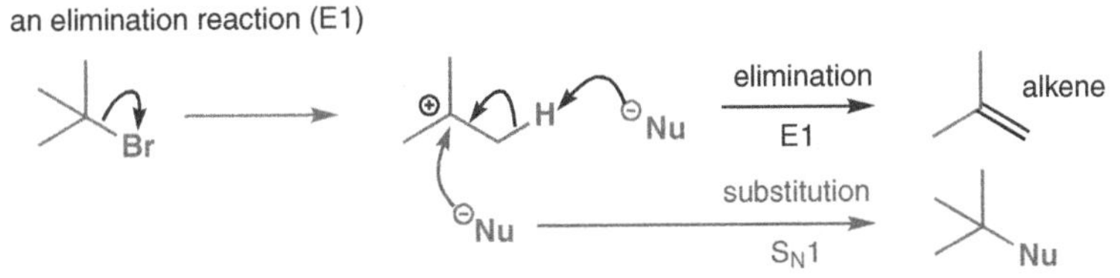

an elimination reaction (E1)

elimination
E1
alkene

substitution
S_N1

You will meet elimination reactions in the next chapter but one (17) after some further exploration of stereochemistry.

Further reading

This subject is treated in every organic chemistry textbook, often as the first reaction described. Good examples include: J. Keeler and P. Wothers, *Why Chemical Reactions Happen*, OUP, Oxford, 2003, chapter 11 and F. A. Carey and R. J. Sundberg, *Advanced Organic Chemistry A, Structure and Mechanisms*, 5th edn, Springer, 2007, chapter 4.

Check your understanding

 To check that you have mastered the concepts presented in this chapter, attempt the problems that are available in the book's Online Resource Centre at http://www.oxfordtextbooks.co.uk/orc/clayden2e/

345

Suggested solutions for Chapter 15

15

PROBLEM 1

Suggest mechanisms for the following reactions, commenting on your choice of S_N1 or S_N2.

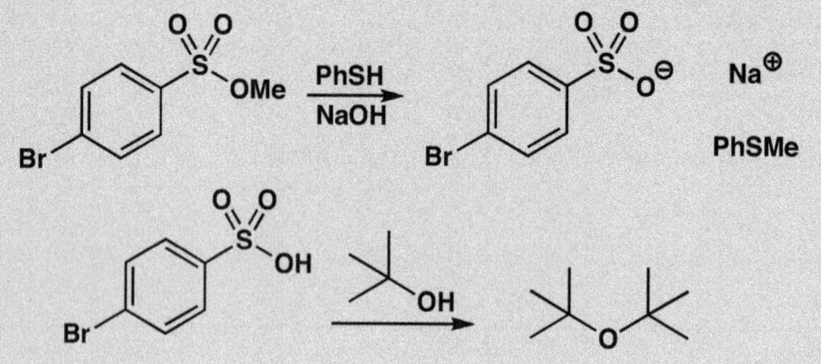

Purpose of the problem

Simple example of the two important mechanisms of chapter 15: S_N1 and S_N2.

Suggested solution

NaOH (pK_a of water about 16) removes the proton from PhSH (pK_a about 7) rapidly as this is a proton transfer between electronegative atoms. Clearly the methyl group must be transferred from O to S and this must be an S_N2 reaction.

The first reagent in the second reaction resembles the reagent in the first reaction but it is the free sulfonic acid and not the ester. The ether product must come from the displacement of OH from one molecule of t-BuOH by the OH group of the other and this can only be an S_N1 reaction. The OH group leaves as H_2O after being protonated by the sulfonic acid.

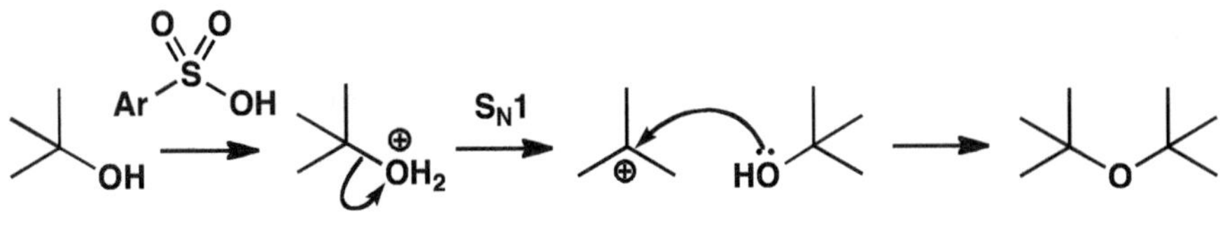

PROBLEM 2

Arrange the following in order of reactivity towards the nucleophile sodium azide. Give a brief comment for each compound to explain what factor influences its place in the reactivity scale.

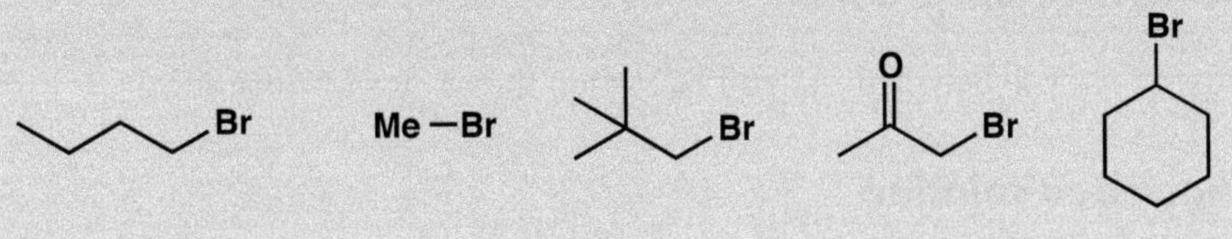

Purpose of the problem

Revision of the factors affecting reactivity in a series of S_N2-reactive molecules.

Suggested solution

None of these compounds has structural features necessary to promote S_N1 (not even the third: notice that the bromine is attached to a primary carbon, even though there is a *tert*-butyl group in the molecule), so we need to think about S_N2 reactivity only. In general, steric hindrance slows down S_N2 reactions, so we can start by saying that methyl bromide > *n*-butyl bromide > cyclohexyl bromide. But how do the other two fit into the scale? An adjacent carbonyl group accelerates S_N2 reactions enormously, so the ketone will react even faster than methyl bromide. On the other hand, a bulky *tert*-butyl group adjacent to a reaction centre leads to very slow substitution, so this compound ('neopentyl bromide') goes at the bottom of the scale.

■ The summary table on p. 347 of the textbook might be useful revision here.

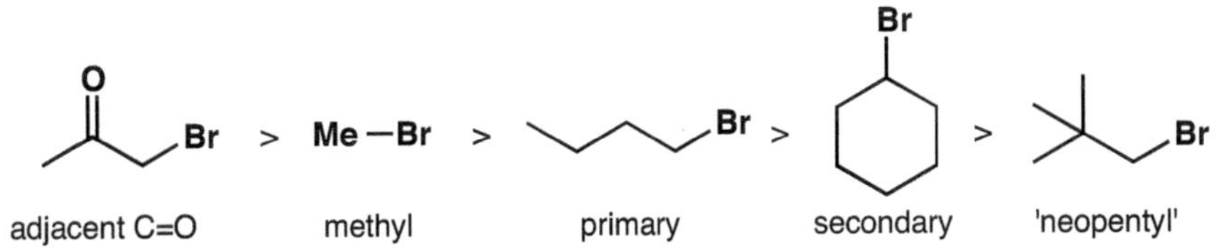

PROBLEM 3

Draw mechanisms for these reactions, explaining why these particular products are formed.

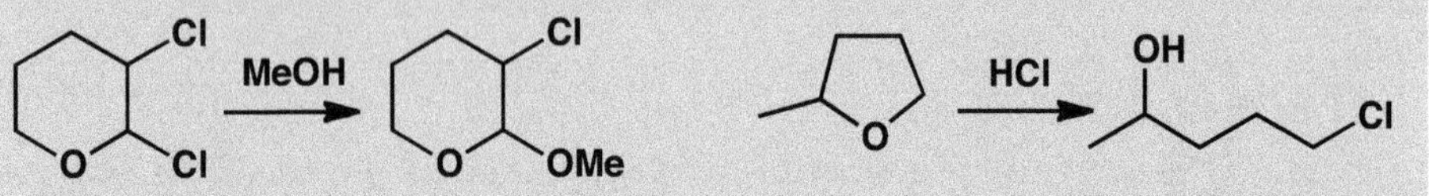

Purpose of the problem

How to choose between S_N1 and S_N2 when the choice is more subtle.

Suggested solution

The first compound has two leaving groups—both secondary chlorides. The one that leaves is next to oxygen so that suggests S_N1 (the oxygen lone pair can stabilize the cation) as does the reagent: surely MeO^- would be used for S_N2.

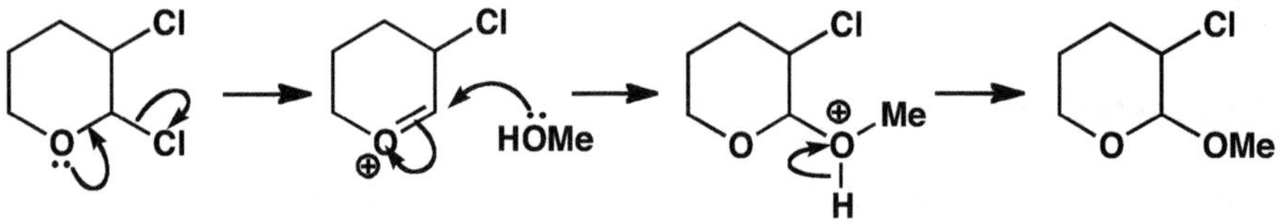

The second compound has only one leaving group and that must be protonated before it can leave. It has two possible sites for attack by the nucleophile (Cl^-), one primary and one secondary. As the primary is chosen, this must be S_N2.

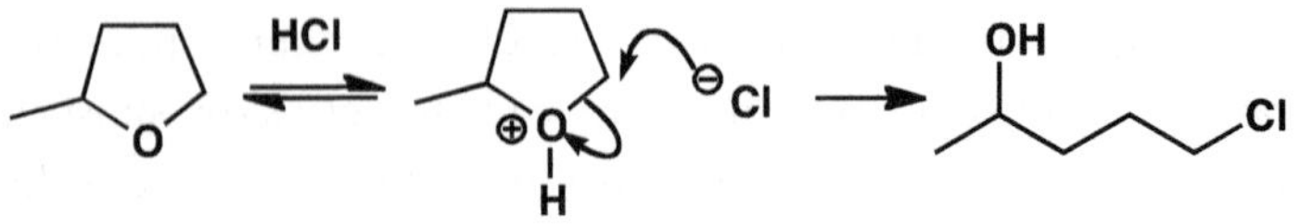

PROBLEM 4

Suggest how to carry out the following transformations.

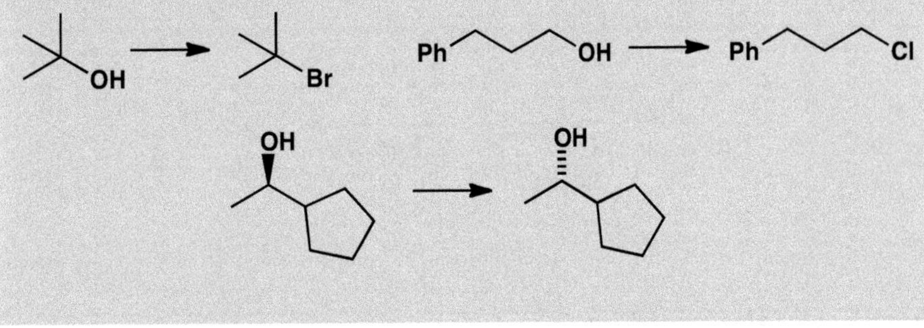

Purpose of the problem

Choosing reagents for substitution reactions of alcohols.

Suggested solution

The first compound can react by S_N1, so we have an opportunity to use an excellent reaction: just treating the alcohol with HBr will give the bromide. The second compound is primary, so we have to make it react by S_N2. But we can't simply try to get OH⁻ to leave (hydroxide is never a leaving group!). We have to convert the alcohol into a better leaving group, and we can do this just by treating with PCl₃ (or we could make a sulfonate leaving group and displace with chloride).

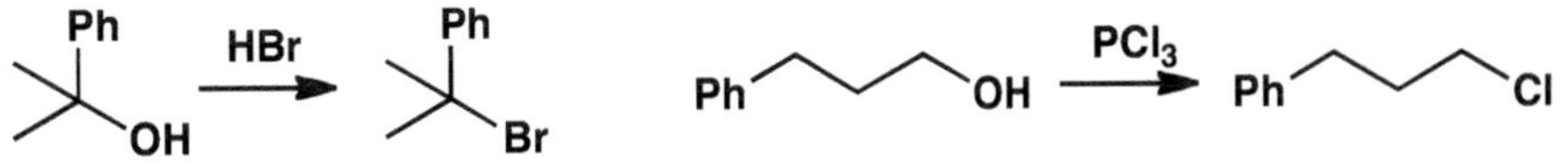

The final example must be an S_N2 reaction because it involves an inversion of configuration. Again, hydroxide cannot be a leaving group, so we have to make the alcohol into a tosylate first. We could then use hydroxide as a nucleophile, but to avoid competing elimination reactions the next step is usually done in two stages using acetate as the nucleophile and then hydrolysing to the alcohol.

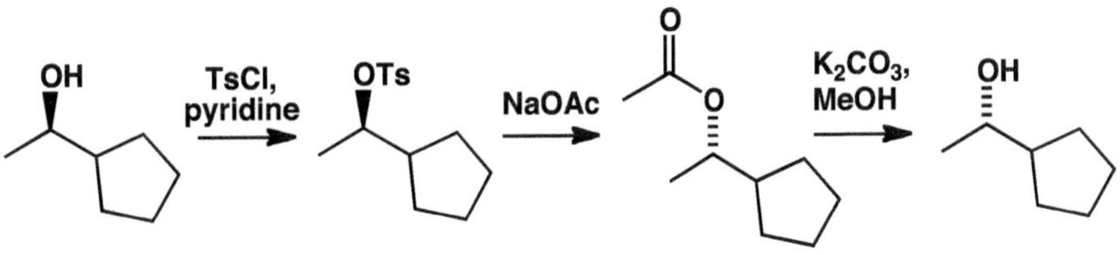

PROBLEM 5

Draw mechanisms for these reactions and give the stereochemistry of the product.

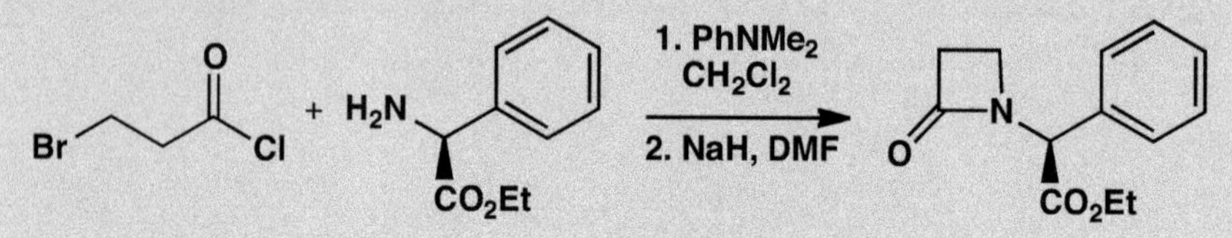

Purpose of the problem

Drawing mechanisms for two types of nucleophilic substitution in the same sequence to make a β-lactam antibiotic.

Suggested solution

We need an S_N2 reaction at a primary carbon and a nucleophilic substitution at the carbonyl group with the amino group as the nucleophile in both cases. The carbonyl group reaction probably happens first. Don't worry if you didn't deprotonate the amide before the S_N2 reaction. The stereochemistry is the same as that of the starting material (CO_2Et up as drawn) as no change has occurred at the chiral centre.

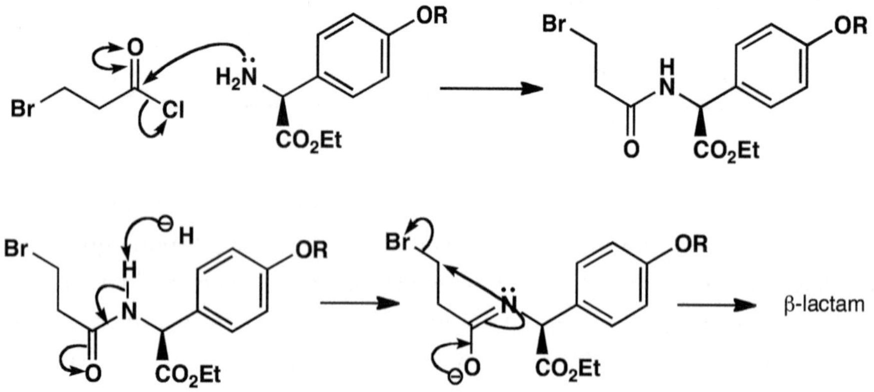

PROBLEM 6

Suggest a mechanism for this reaction. You will find it helpful first of all to draw good diagrams of reagents and products.

$$t\text{-BuNMe}_2 + (\text{MeCO})_2\text{O} \longrightarrow \text{Me}_2\text{NCOMe} + t\text{-BuO}_2\text{CMe}$$

Purpose of the problem

Revision of chapter 2 and practice at drawing mechanisms of unusual reactions.

Suggested solution

First draw good diagrams of the molecules as the question suggests.

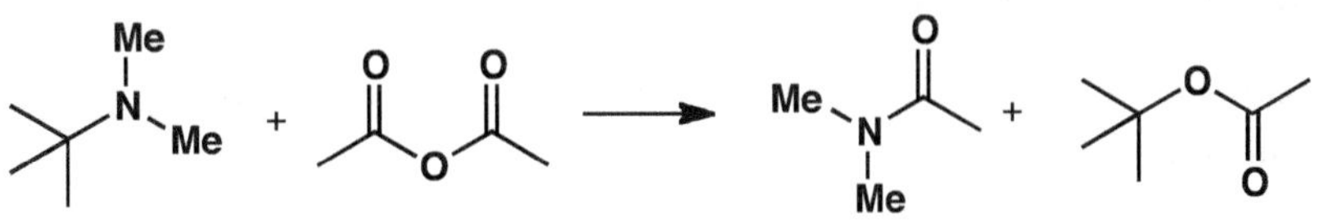

With an unfamiliar reaction, it is best to identify the nucleophile and the electrophile and see what happens when we unite them. The nitrogen atom is obviously the nucleophile and one of the carbonyl groups must be the electrophile.

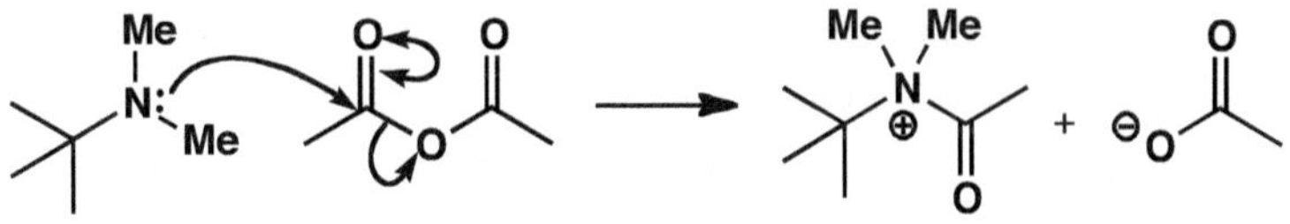

We must lose a *t*-butyl group from this intermediate to give one of the products and unite it with the acetate ion to give the other. This must be an S_N1 rather than an S_N2 at a *t*-butyl group.

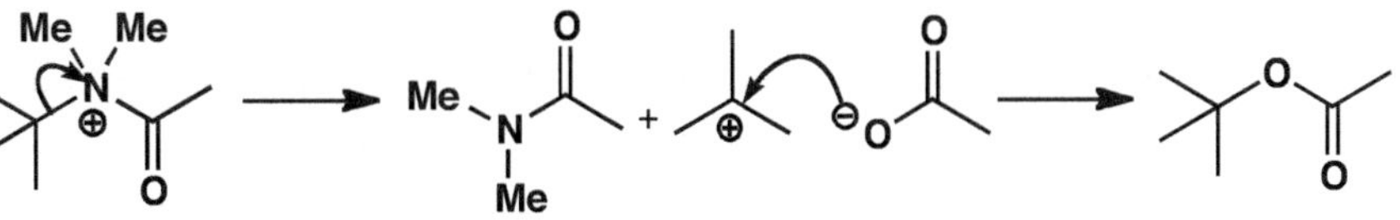

PROBLEM 7

Predict the stereochemistry of these products. Are they diastereoisomers, enantiomers, racemic or what?

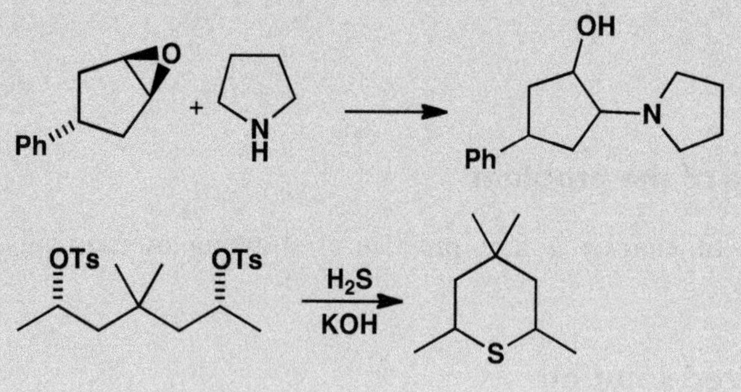

Purpose of the problem

Revision of stereochemistry from chapter 14 and practice at applying it to substitution reactions.

Suggested solution

The starting material in the first reaction has a plane of symmetry so it is achiral: the stereochemistry shows only which diastereoisomer we have. Attack by the amine nucleophile at either end of the epoxide (the two ends are the same) must take place from underneath for inversion to occur. The product is a single diastereoisomer but cannot, of course, be a single enantiomer so it doesn't matter which enantiomer you have drawn. The stereochemistry of the Ph group cannot change—it is just a spectator.

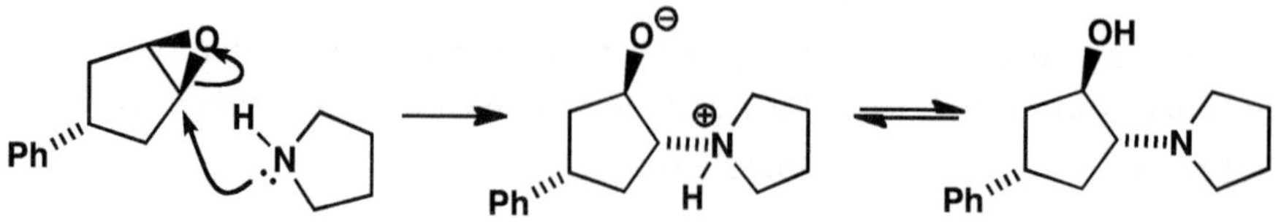

The starting material for the second reaction is also achiral as it too has a plane of symmetry. The stereochemistry merely shows that the two OTs groups are on the same side of the molecule as drawn. Displacement with sulfur will occur with inversion and it is wise to redraw the intermediate before the cyclization. This 'inverts' the chiral centre so that we can see that the stereochemistry of the product has the methyl groups *cis*. There are various ways to draw this.

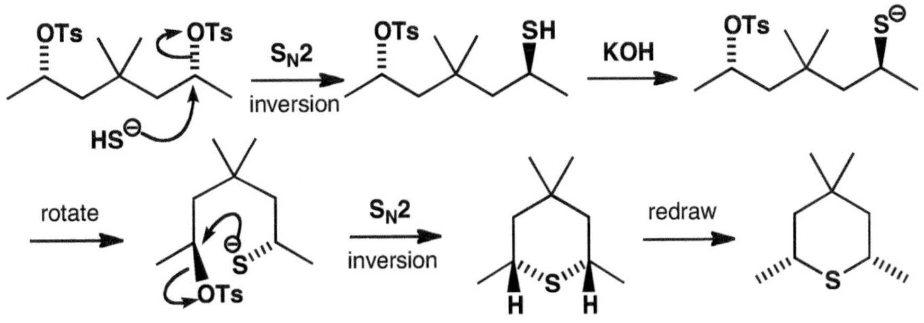

PROBLEM 8

What are the mechanisms of these reactions, and what is the role of the ZnCl$_2$ in the first step and the NaI in the second?

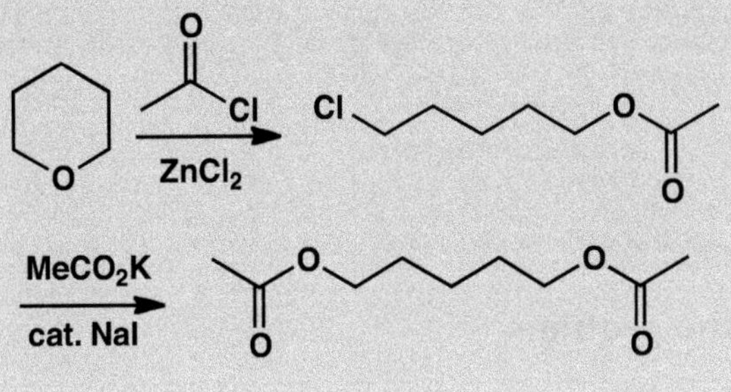

Purpose of the problem

Exploration of two different kinds of catalysis in substitution reactions.

Suggested solution

The ZnCl$_2$ acts as a Lewis acid and can be used either to remove chloride from MeCOCl or to complex with its carbonyl oxygen atom, in either case making it a better electrophile so that it can react with the unreactive oxygen atom of the cyclic ether. Ring cleavage by chloride follows.

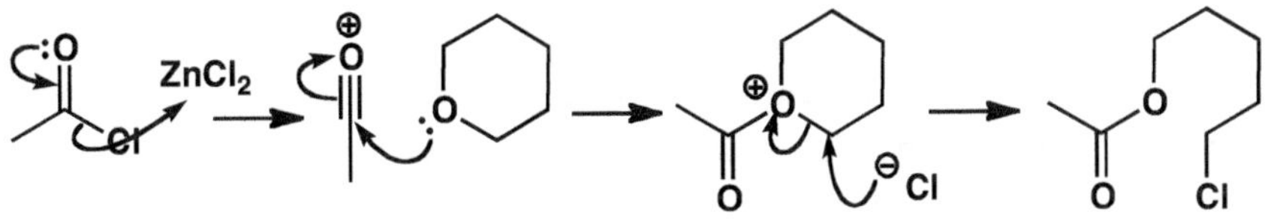

The second reaction is an S$_N$2 displacement of a reasonable leaving group (chloride) by a rather weak nucleophile (acetate). The reaction is very slow unless catalysed by iodide ion—a better nucleophile than acetate and a better leaving group than chloride.

■ Iodide behaves as a nucleophilic catalyst: see p. 358 of the textbook.

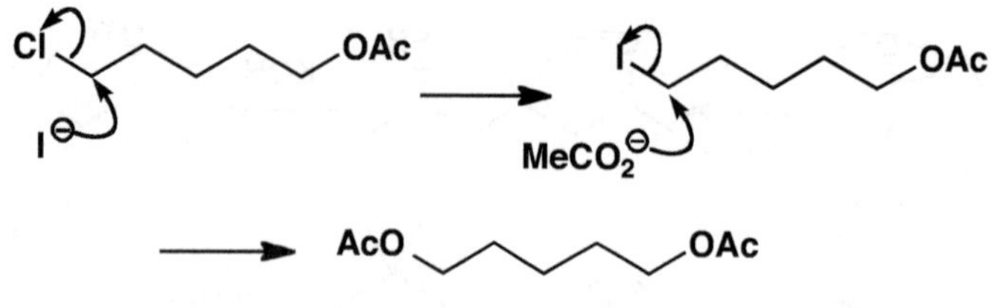

■ You can read more about this in B. S. Furniss *et al.*, *Vogel's textbook of organic chemistry* (5th edn), Longmans, Harlow, 1989, p. 492.

PROBLEM 9

Describe the stereochemistry of the products of these reactions.

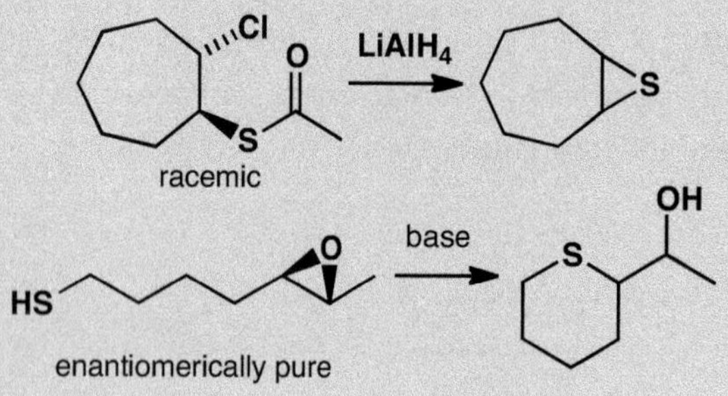

racemic

enantiomerically pure

Purpose of the problem

Nucleophilic substitution and stereochemistry, with a few extra twists.

Suggested solution

The ester in the first example is removed by reduction leaving an oxyanion that cyclizes by intramolecular S_N2 reaction with inversion giving one diastereoisomer (*cis*) of the product. The product is achiral.

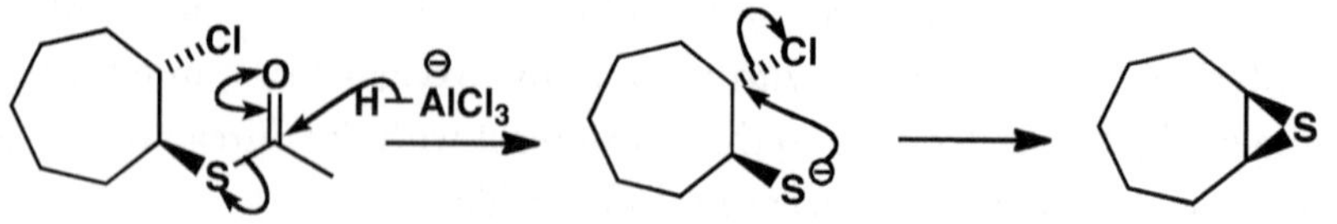

The second case involves an intramolecular S_N2 reaction on one end of the epoxide. The reaction occurs stereospecifically with inversion and so one enantiomer of one diastereoisomer of the product is formed. Some redrawing is needed and we have left the epoxide in its original position to avoid mistakes.

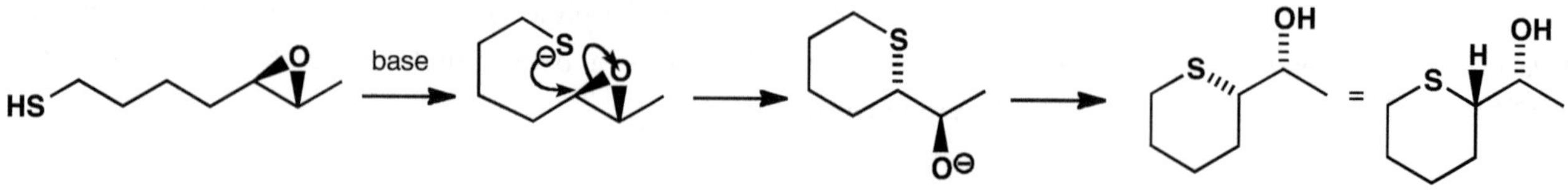

PROBLEM 10

State, with reasons, whether these reactions will be S_N1 or S_N2.

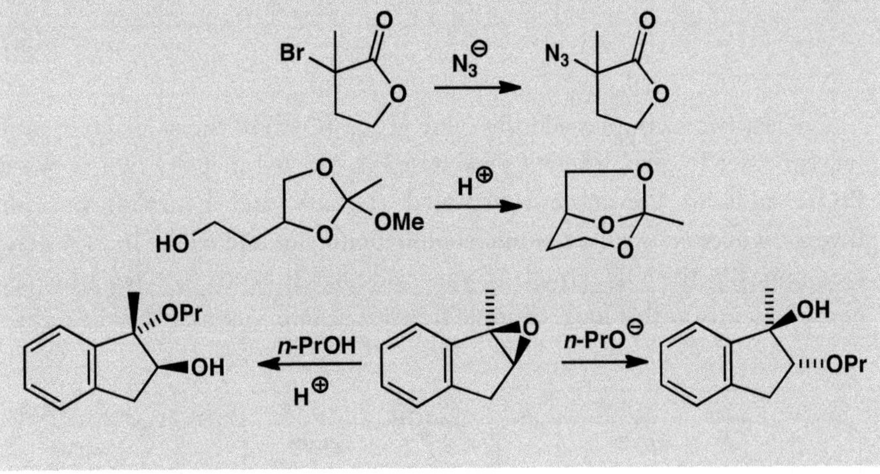

Purpose of the problem

Taxing examples of the choice between our two main mechanisms. The last two differ only in reaction conditions.

Suggested solution

The first reaction offers a choice between an S_N2 reaction at a tertiary carbon or an S_N1 reaction next to a carbonyl group. Neither looks very good but experiments have shown that these reactions go with inversion of configuration and they are about the only examples of S_N2 reactions at tertiary carbon. They work because the p orbital in the transition state is stabilized by conjugation with the carbonyl group: S_N2 reactions adjacent to C=O groups are usually fast.

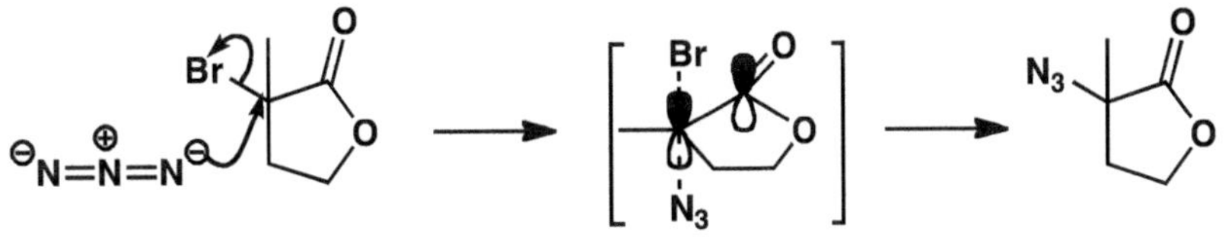

The moment that you see acetal-like compounds in the second example, you should suspect S_N1 with oxonium ion intermediates. In fact the compounds are orthoesters but this makes no difference to the mechanism. If you are not sure of this sort of chemistry, have a look at chapter 11. The OH group displaces the OMe group by an acid-catalysed S_N1 reaction.

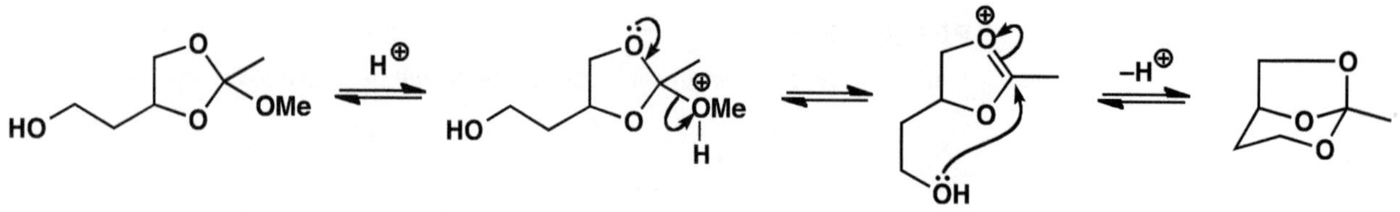

The last two examples add the same group (OPr) to the same compound (an epoxide) to give different products. We can tell that the first is S$_N$1 as PrOH adds to the more substituted (tertiary and benzylic) position. Inversion occurs because the nucleophile prefers to add to the less hindered face opposite the OH group. If you said that it is an S$_N$2 reaction at a benzylic centre with a loose cationic transition state, you may well be right.

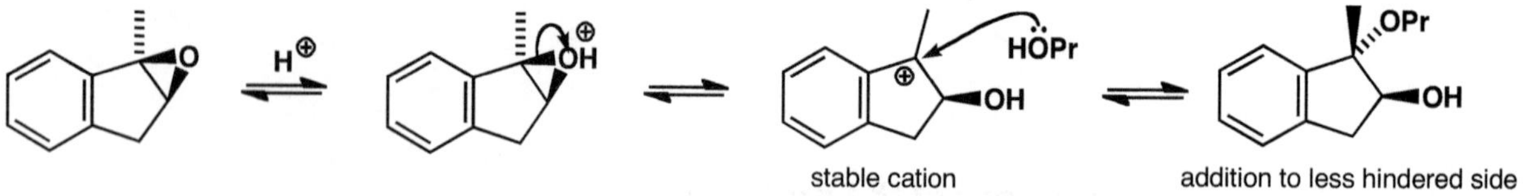

stable cation   addition to less hindered side

The second is easier as the more reactive anion adds to the less hindered centre with inversion and this must be S$_N$2.

PROBLEM 11

The pharmaceutical company Pfizer made the antidepressant reboxetine by the following sequence of reactions. Suggest a reagent for each step, commenting on aspects of stereochemistry or reactivity.

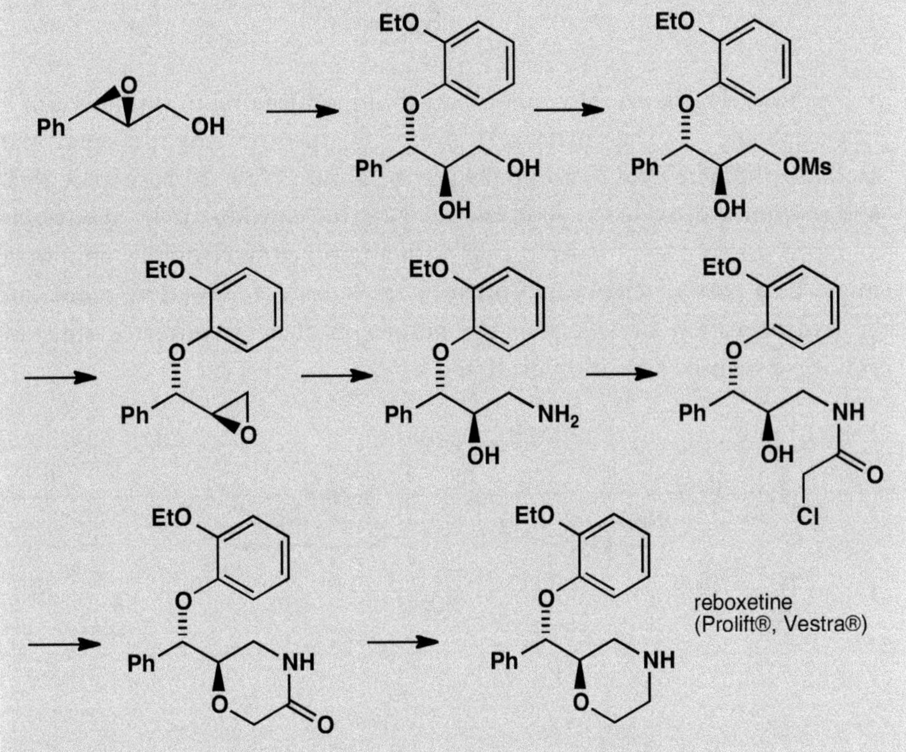

reboxetine
(Prolift®, Vestra®)

Purpose of the problem

Thinking about substitution reactions in a real synthesis. It might look challenging, but each step uses a reaction you have already met.

Suggested solution

The first step is the attack of a nucleophile on an epoxide. It's an S_N2 reaction, because it goes with inversion of configuration, and we need a phenol as the nucleophile. To make the phenol more reactive, we probably want to deprotonate it to make the phenoxide, and NaOH will do this. Why does this end of the epoxide react? Well, it is next to a phenyl ring, and benzylic S_N2 reactions are faster than reactions at 'normal' secondary carbons. Next the end hydroxyl group is made into a leaving group (a 'mesylate'), for which we need methanesulfonyl chloride (mesyl chloride) and triethylamine. The primary hydroxyl group must react faster than the secondary one because it is less hindered.

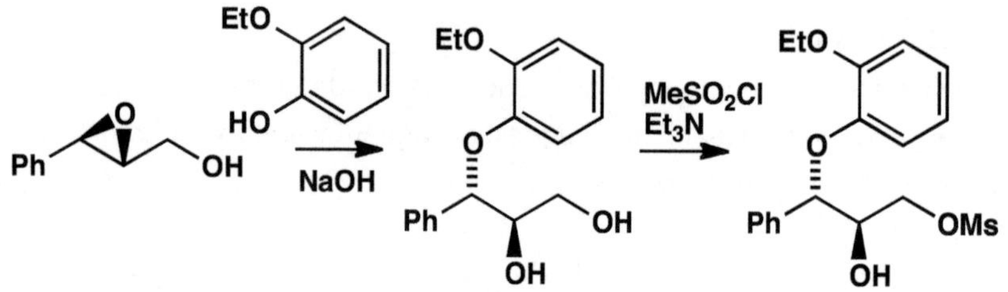

The next stage is an intramolecular subtitution leading to formation of a new epoxide.   The hydroxyl group is the nucleophile and the methanesulfonate (MsO⁻) group the leaving group. We need base to do this, and sodium hydroxide is a good choice. Now the epoxide can be opened (at its more reactive, less hindered end) with a nitrogen nucleophile: ammonia might be a possible choice, but often better is azide, followed by reduction by hydrogenation or LiAlH₄. The amine product is converted into an amide, so we need an acid chloride and base.

■ The use of azide as a nitogen nucleophile and as an alternative to ammonia is described on pp. 353-4 of the textbook.

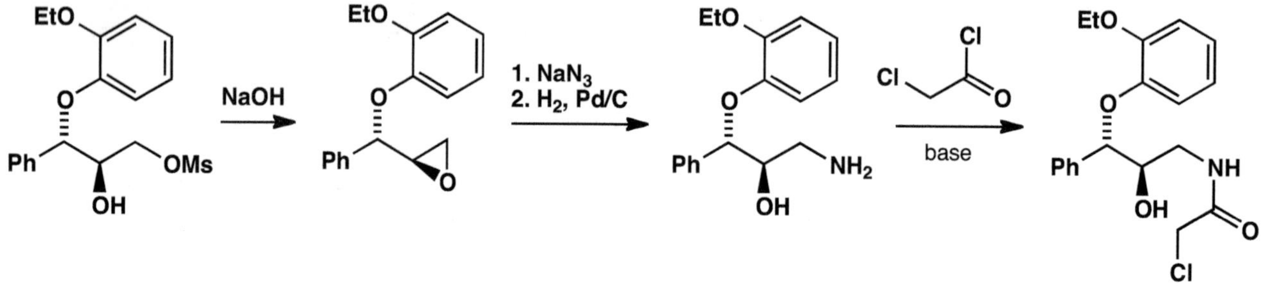

Another intramolecular substitution follows, this time with an alcohol nucleophile displacing a chloride leaving group to make a new ring. A strong base will make the alcohol nucleophilic by deprotonating it to form the epoxide, and KOt-Bu works here (though if you just suggested 'base' that is fine: only experimentation will show which works best). Finally, the amide is reduced to an amine, for which we need LiAlH₄.

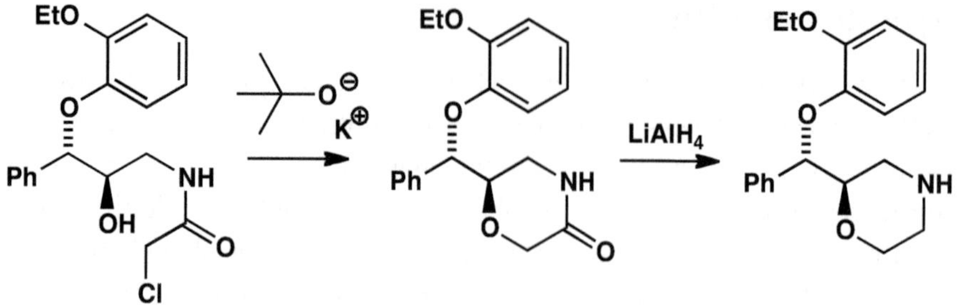

17

Elimination reactions

Connections

⇨ Building on	Arriving at	⇨ Looking forward to
• Stereochemistry ch14	• Elimination reactions	• Electrophilic additions to alkenes (the reverse of the reactions in this chapter) ch19
• Mechanisms of nucleophilic substitution at saturated carbon ch15	• What factors favour elimination over substitution	
• Conformation ch16	• The three important mechanisms of elimination reactions	• How to control double-bond geometry ch27
	• The importance of conformation in elimination reactions	
	• How to use eliminations to make alkenes (and alkynes)	

Substitution and elimination

⇨ Remember the turnstiles at the railway station (see p. 332).

Substitution reactions of *t*-butyl halides, you will recall from Chapter 15, invariably follow the S_N1 mechanism. In other words, the rate-determining step of their substitution reactions is unimolecular—it involves only the alkyl halide. This means that, no matter what the nucleophile is, the reaction goes at the same rate. You can't speed this S_N1 reaction up, for example, by using hydroxide instead of water, or even by increasing the concentration of hydroxide. You'd be wasting your time, we said (see p. 332).

nucleophilic substitution reactions of *t*-BuBr

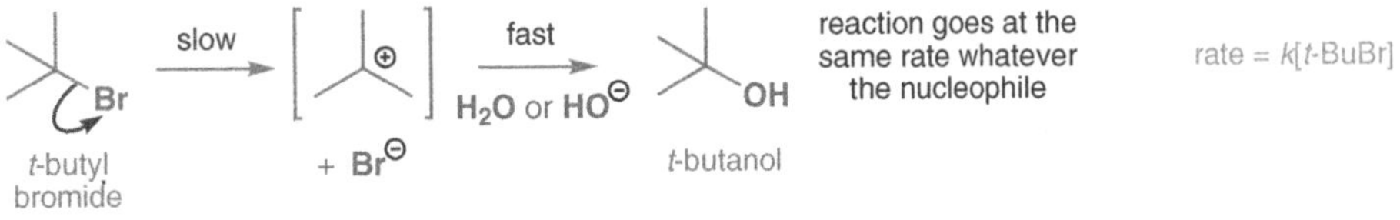

t-butyl bromide → [slow] + Br⁻ → fast, H_2O or $HO^{\ominus}$ → *t*-butanol

reaction goes at the same rate whatever the nucleophile   rate = $k[t\text{-BuBr}]$

You'd also be wasting your alkyl halide. This is what actually happens if you try the substitution reaction with a *concentrated* solution of sodium hydroxide.

reaction of *t*-BuBr with concentrated solution of NaOH

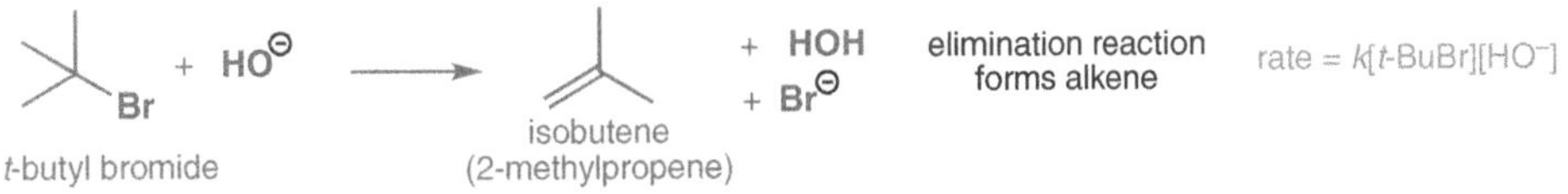

t-butyl bromide + $HO^{\ominus}$ → isobutene (2-methylpropene) + HOH + Br⁻

elimination reaction forms alkene   rate = $k[t\text{-BuBr}][HO^-]$

The reaction stops being a substitution and an alkene is formed instead. Overall, HBr has been lost from the alkyl halide, and the reaction is called an **elimination**.

In this chapter we will talk about the mechanisms of elimination reactions—as in the case of substitutions, there is more than one mechanism for eliminations. We will compare eliminations with substitutions—either reaction can happen from almost identical starting materials, and you will learn how to predict which is the more likely. Much of the mechanistic discussion relates very closely to Chapter 15, and we suggest that you make sure you understand all of the points in that chapter before tackling this one. This chapter will also tell you about uses for elimination reactions. Apart from a brief look at the Wittig reaction in Chapter 11, this is the first time you have met a way of making simple alkenes.

Elimination happens when the nucleophile attacks hydrogen instead of carbon

The elimination reaction of *t*-butyl bromide happens because the nucleophile is *basic*. You will recall from Chapter 10 that there is *some* correlation between basicity and nucleophilicity: strong bases are usually good nucleophiles. Being a good nucleophile doesn't get hydroxide anywhere in the substitution reaction because it doesn't appear in the first-order rate equation. But being a good base does get it somewhere in the elimination reaction because hydroxide is involved in the rate-determining step of the elimination, and so it appears in the rate equation. This is the mechanism.

■ The correlation between basicity and nucleophilicity is best for attack at C=O. In Chapter 15 you met examples of nucleophiles that are good at substitution at saturated carbon (such as I−, Br−, PhS−) but that are not strong bases.

The hydroxide is behaving as a base because it is attacking the hydrogen atom, instead of the carbon atom it would attack in a substitution reaction. The hydrogen atom is not acidic, but proton removal can occur because bromide is a good leaving group. As the hydroxide attacks, the bromide is forced to leave, taking with it the negative charge. Two molecules—*t*-butyl bromide and hydroxide—are involved in the rate-determining step of the reaction. This means that the concentrations of both appear in the rate equation, which is therefore second-order and this mechanism for elimination is termed E2, for *elimination, bimolecular*.

■ *Note*: No subscripts or superscripts, just plain old E2.

$$\text{rate} = k_2 \, [\text{t-BuBr}][\text{HO}^-]$$

Now let's look at another sort of elimination. We can approach it again by thinking about another S_N1 substitution reaction, the reverse of the one at the beginning of the chapter: an alcohol is converted into an alkyl halide.

Bromide, the nucleophile, is not involved in the rate-determining step, so we know that the rate of the reaction will be independent of the concentration of Br−. Indeed the first step, to form the cation, will happen just as fast even if there is *no bromide at all*. But what happens to the carbocation in such a case? To find out, we need to use an acid whose counterion is such a weak nucleophile that it won't even attack the positive carbon of the carbocation. Here is an example—*t*-butanol in sulfuric acid doesn't undergo substitution, but undergoes elimination instead.

E1 elimination of *t*-BuOH in H_2SO_4

Interactive E1 elimination
mechanism

t-butanol

isobutene
(2-methylpropene)

The HSO_4^- anion is not involved in the rate-determining formation of the carbocation, and is also a very bad nucleophile, so it does not attack the C atom of the carbocation. Neither is it basic, but you can see from the mechanism that it does behave as a base (that is, it removes a proton). It does this only because it is even more feeble as a nucleophile. The rate equation will not involve the concentration of HSO_4^-, and the rate-determining step is the same as that in the S_N1 reaction—unimolecular loss of water from the protonated *t*-BuOH. This elimination mechanism is therefore called E1.

We will shortly come back to these two mechanisms for elimination, plus a third, but it is worth noting at this stage that the choice between E1 and E2 is not based on the same grounds as the choice between S_N1 and S_N2: you have just seen both E1 and E2 elimination from a substrate that would only undergo S_N1. The difference between the two reactions was the strength of the base, so first we need to answer the question: when does a nucleophile start behaving as a base?

Elimination in carbonyl chemistry

We have left detailed discussion of the formation of alkenes until this chapter, but we used the term 'elimination' in Chapters 10 and 11 to describe the loss of a leaving group from a tetrahedral intermediate. For example, the final steps of acid-catalysed ester hydrolysis involve E1 elimination of ROH to leave a double bond: C=O rather than C=C.

E1 elimination of ROH during ester hydrolysis

tetrahedral
intermediate

new C=O
double bond

+ ROH

In Chapter 11 you even saw an E1 elimination giving an alkene. That alkene was an enamine—here is the reaction.

E1 elimination of H_2O during enamine formation

tetrahedral
intermediate

new C=C
double bond

+ H_2O

How the nucleophile affects elimination versus substitution

Basicity

attack here leads to
elimination

attack here leads to
substitution

You have just seen molecules bearing leaving groups being attacked at two distinct electrophilic sites: the carbon to which the leaving group is attached, and the hydrogen atoms on the carbon adjacent to the leaving group. Attack at carbon leads to substitution; attack at hydrogen leads to elimination. Since strong bases attack protons, it is generally true that, the more basic the nucleophile, the more likely that elimination is going to replace substitution as the main reaction of an alkyl halide.

Here is an example of this idea at work: a weak base (EtOH) leads to substitution while a strong base (ethoxide ion) leads to elimination.

weak base: substitution

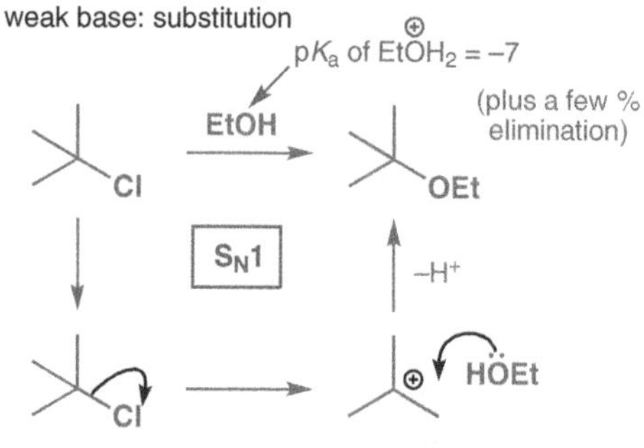

strong base: elimination

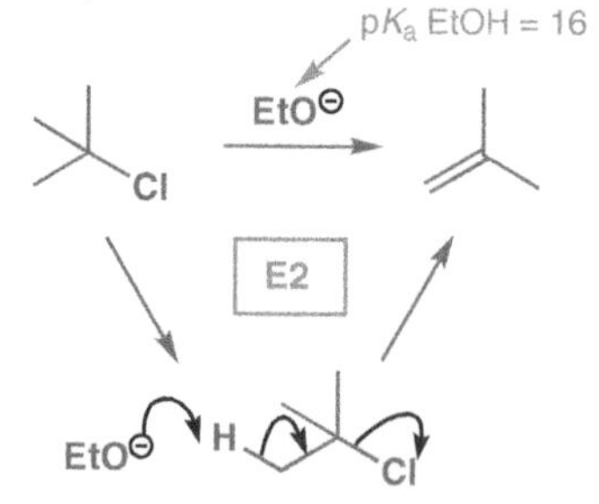

Elimination, substitution, and hardness

We can also rationalize selectivity for elimination versus substitution, or attack on H versus attack on C in terms of hard and soft electrophiles (p. 357). In an S_N2 substitution, the carbon centre is a soft electrophile—it is essentially uncharged, and with leaving groups such as halide the C–X σ^* is a relatively low-energy LUMO. Substitution is therefore favoured by nucleophiles whose HOMOs are best able to interact with this LUMO—in other words soft nucleophiles. In contrast, the C–H σ^* is higher in energy because the atoms are less electronegative. This, coupled with the hydrogen's small size, makes the C–H bond a hard electrophilic site, and as a result hard nucleophiles favour elimination.

Size

For a nucleophile, attacking a carbon atom means squeezing past its substituents—and even for unhindered primary alkyl halides there is still one alkyl group attached. This is one of the reasons that S_N2 is so slow on hindered alkyl halides—the nucleophile has difficulty getting to the reactive centre. Getting at a more exposed hydrogen atom in an elimination reaction is much easier, and this means that, as soon as we start using basic nucleophiles that are also bulky, elimination becomes preferred over substitution, even for primary alkyl halides. One of the best bases for promoting elimination and avoiding substitution is potassium *t*-butoxide. The large alkyl substituent makes it hard for the negatively charged oxygen to attack carbon in a substitution reaction, but it has no problem attacking hydrogen.

small nucleophile: substitution

large nucleophile: elimination

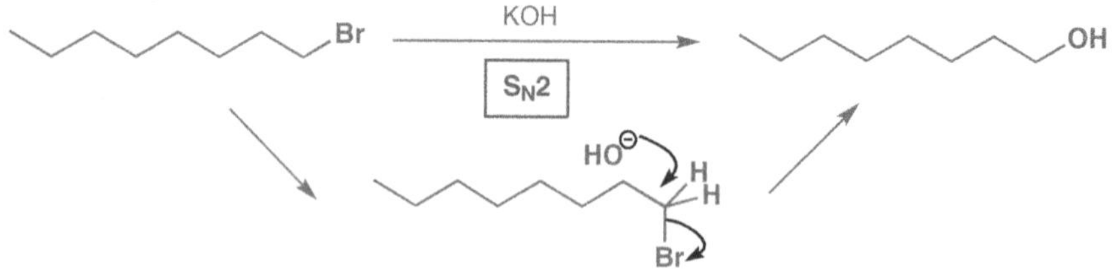

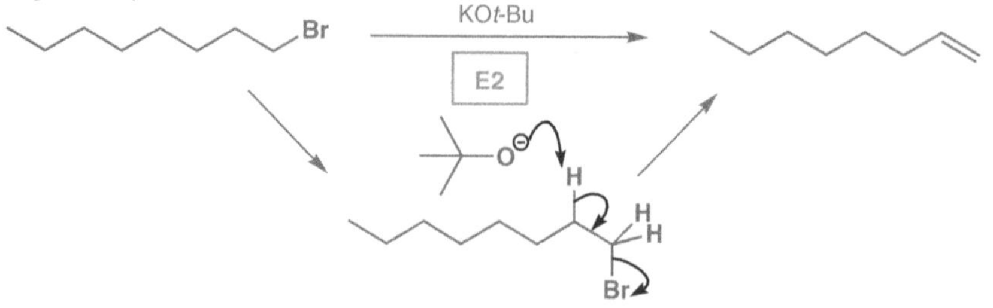

Temperature

Temperature has an important role to play in deciding whether a reaction is an elimination or a substitution. In an elimination, two molecules become three (count them). In a substitution, two molecules form two new molecules. The two reactions therefore differ in the change in entropy during the reaction: ΔS is greater for elimination than for substitution. In Chapter 12, we discussed the equation

■ This explanation is simplified because what matters is the rate of the reaction, not the stability of the products. A detailed discussion is beyond the scope of the book, but the general argument still holds.

$$\Delta G = \Delta H - T\Delta S$$

➡ For a related example see Chapter 12, p. 247

This equation says that a reaction in which ΔS is positive becomes more favourable (ΔG becomes more negative) at higher temperature. Eliminations should therefore be favoured at high temperature, and this is indeed the case: most eliminations you will see are conducted at room temperature or above.

● **Elimination versus substitution**

  • **Nucleophiles that are strong bases favour elimination over substitution.**
  • **Nucleophiles (or bases) that are bulky favour elimination over substitution.**
  • **High temperatures favour elimination over substitution.**

E1 and E2 mechanisms

Now that you have seen a few examples of elimination reactions, it is time to return to our discussion of the two mechanisms for elimination. To summarize what we have said so far:

  • E1 describes an elimination reaction (E) in which the rate-determining step is unimolecular (1) and does not involve the base. The leaving group leaves in this step, and the proton is removed in a separate second step.

general mechanism for E1 elimination

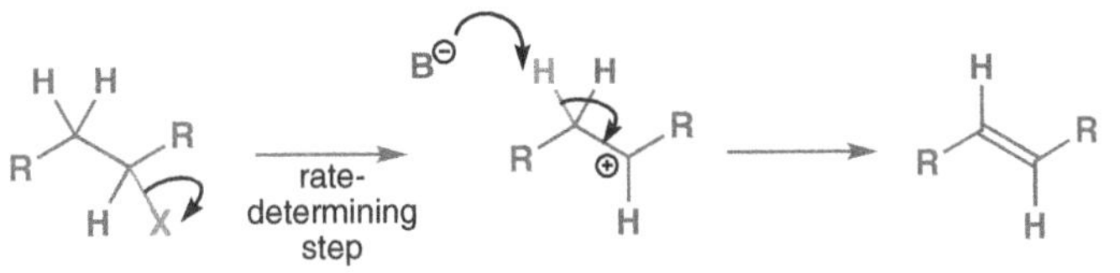

rate = k[alkyl halide]

■ In E2 eliminations the loss of the leaving group and removal of the proton are **concerted**.

  • E2 describes an elimination (E) that has a bimolecular (2) rate-determining step that must involve the base. Loss of the leaving group is simultaneous with removal of the proton by the base.

general mechanism for E2 elimination

rate = k[B⁻][alkyl halide]

There are a number of factors that affect whether an elimination goes by an E1 or E2 mechanism. One is immediately obvious from the rate equations: only the E2 is affected by the concentration of base, so at high base concentration E2 is favoured. The rate of an E1 reaction is not even affected by what base is present—so E1 is just as likely with weak as with strong bases, while E2 goes faster with strong bases than weak ones: strong bases at whatever concentration will favour E2 over E1. If you see that a strong base is required for an elimination, it is certainly an E2 reaction. Take the first elimination in this chapter as an example.

reaction of t-butyl bromide with concentrated hydroxide

With less hindered alkyl halides hydroxide would not be a good choice as a base for an elimination because it is rather small and still very good at S_N2 substitutions (and even with tertiary alkyl halides, substitution outpaces elimination at low concentrations of hydroxide). So what are good alternatives?

We have already mentioned the bulky *t*-butoxide—ideal for promoting E2 as it's both bulky and a strong base (pK_a of *t*-BuOH = 18). Here it is at work converting a dibromide to a diene with two successive E2 eliminations. Since dibromides can be made from alkenes (you will see how in the next chapter), this is a useful two-step conversion of an alkene to a diene.

synthesis of a diene by a double E2 elimination

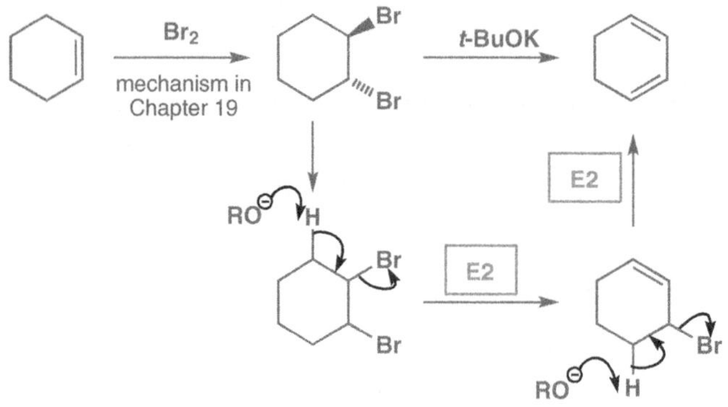

Interactive mechanism for double E2 to form diene

The product of the next reaction is a 'ketene acetal'. Unlike most acetals, this one can't be formed directly from ketene (ketene, CH_2=C=O, is too unstable), so instead the acetal is made by the usual method from bromoacetaldehyde and then HBr is eliminated using *t*-BuOK.

You will meet ketene, briefly, in the next chapter.

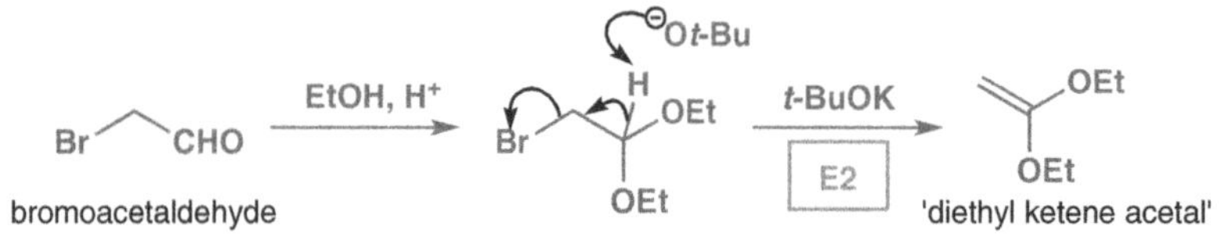

Among the most commonly used bases for converting alkyl halides to alkenes is one that you met in Chapter 8: DBU. This base is an amidine—delocalization of one nitrogen's lone pair onto the other, and the resulting stabilization of the protonated amidinium ion, makes it particularly basic, with a pK_a (of the protonated amidine) of about 12.5. There is not much chance of getting those voluminous fused rings into tight corners—so they pick off the easy-to-reach protons rather than attacking carbon atoms in substitution reactions.

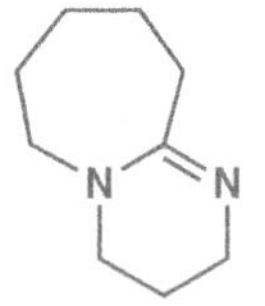

DBU
1,8-diazabicyclo-
[5.4.0]undecene-7

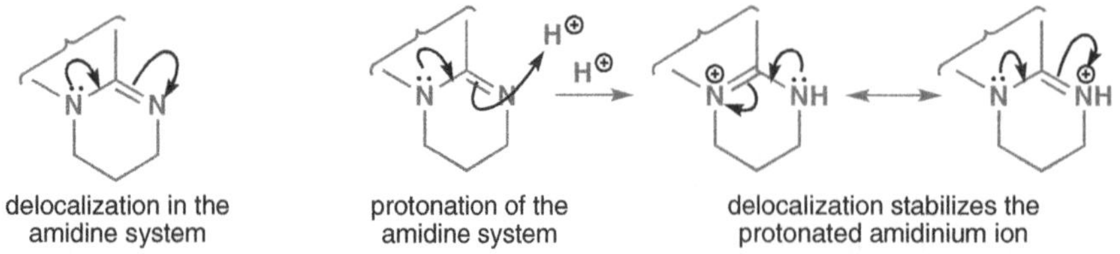

See p. 175 for more on DBU.

delocalization in the amidine system

protonation of the amidine system

delocalization stabilizes the protonated amidinium ion

DBU will generally eliminate HX from alkyl halides to give alkenes. In these two examples, the products were used as intermediates in the synthesis of natural products.

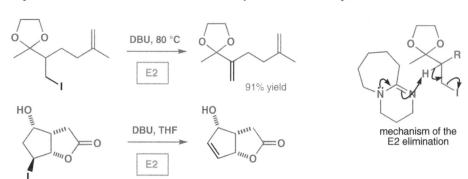

91% yield

■ Note the use of high temperature to drive the elimination.

mechanism of the E2 elimination

Substrate structure may allow E1

The first elimination of the chapter (*t*-BuBr plus hydroxide) illustrates something very important: the starting material is a tertiary alkyl halide and would therefore *substitute* only by S_N1, but it can *eliminate* by either E2 (with strong bases) or E1 (with weak bases). The steric factors that disfavour S_N2 at hindered centres don't exist for eliminations. Nonetheless, E1 can occur *only* with substrates that can ionize to give relatively stable carbocations—tertiary, allylic, or benzylic alkyl halides, for example. Secondary alkyl halides may eliminate by E1, while primary alkyl halides only ever eliminate by E2 because the primary carbocation required for E1 would be too unstable. The chart below summarizes the types of substrate that can undergo E1—but remember that any of these substrates, under the appropriate conditions (in the presence of strong bases, for example), may also undergo E2. For completeness, we have also included in this chart three alkyl halides that cannot eliminate by either mechanism simply because they do not have any hydrogens to lose from carbon atoms adjacent to the leaving group.

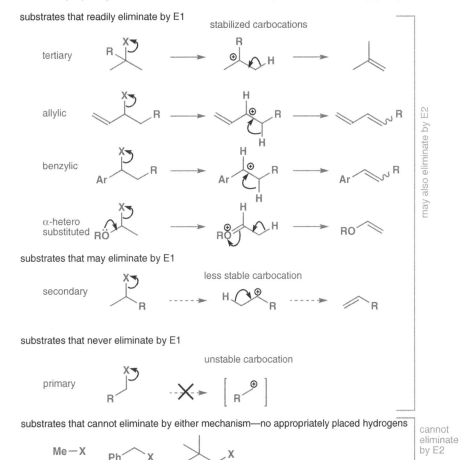

Can a proton just 'fall off' a cation?

In E1 mechanisms, once the leaving group has departed almost anything will serve as a base to remove a proton from the intermediate carbocation. Weakly basic solvent molecules (water or alcohols), for example, are quite sufficient, and you will often see the proton just 'falling off' in reaction mechanisms, with the assumption that there is a weak base somewhere to capture it. We showed the loss of a proton like this in the last example, and in the chart on this page.

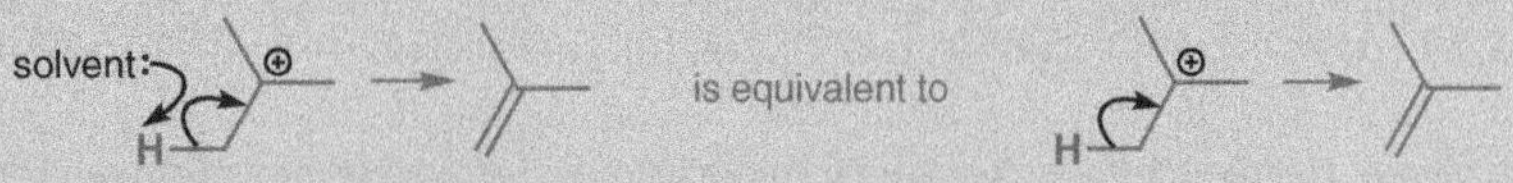

In very rare cases, such as the superacid solutions we described in Chapter 15 (p. 335), the cation is stable because counterions such as BF_4^- and SbF_6^- are not only non-nucleophilic but also so non-basic that they won't even accept a proton. This fact tells us that despite this common way of writing the E1 mechanism, *some* sort of weak base is necessary even for E1.

Polar solvents also favour E1 reactions because they stabilize the intermediate carbocation. E1 eliminations from alcohols in aqueous or alcohol solution are particularly common and very useful. An acid catalyst is used to promote loss of water, and in dilute H_2SO_4, H_3PO_4, or HCl the absence of good nucleophiles ensures that substitution does not compete. With phosphoric acid, for example, the secondary alcohol cyclohexanol gives cyclohexene.

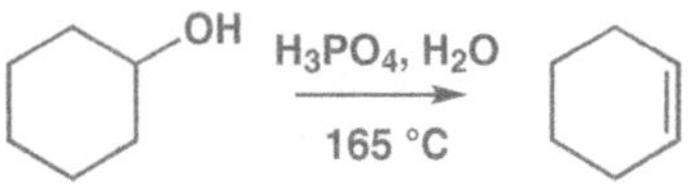

But the best E1 eliminations of all are with tertiary alcohols. The alcohols can be made using the methods of Chapter 9: nucleophilic attack by an organometallic on a carbonyl compound. Nucleophilic addition, followed by E1 elimination, is an excellent way of making this substituted cyclohexene, for example. Note that the proton required in the first step is recovered in the last—the reaction requires only catalytic amounts of acid.

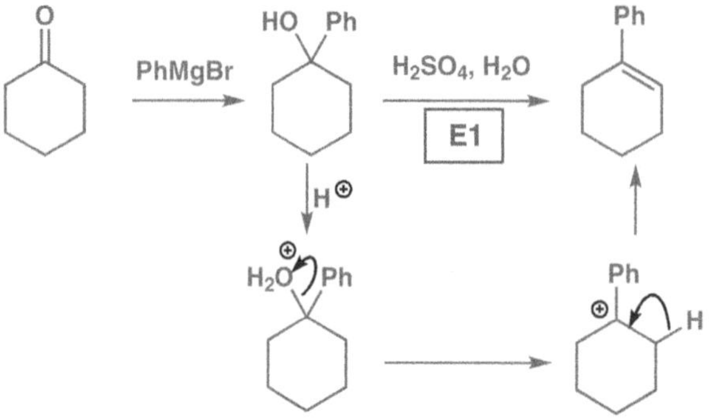

Cedrol is important in the perfumery industry—it has a cedar wood fragrance. Corey's synthesis includes this step—the acid (toluenesulfonic acid, see p. 227) catalyses both the E1 elimination and the hydrolysis of the acetal.

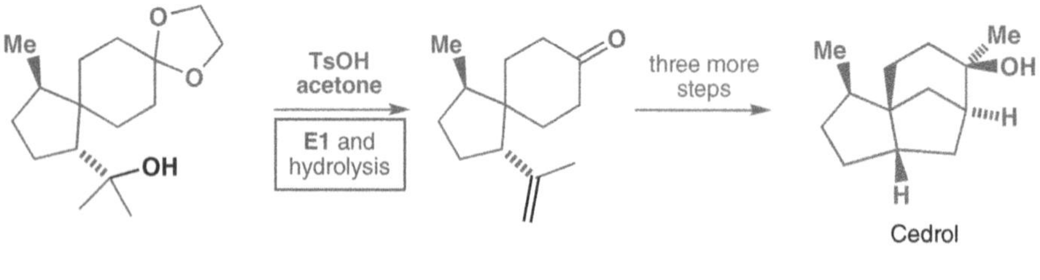

At the end of the last chapter you met some bicyclic structures. These sometimes pose problems for elimination reactions. For example, this compound will not undergo elimination by either an E1 or an E2 mechanism.

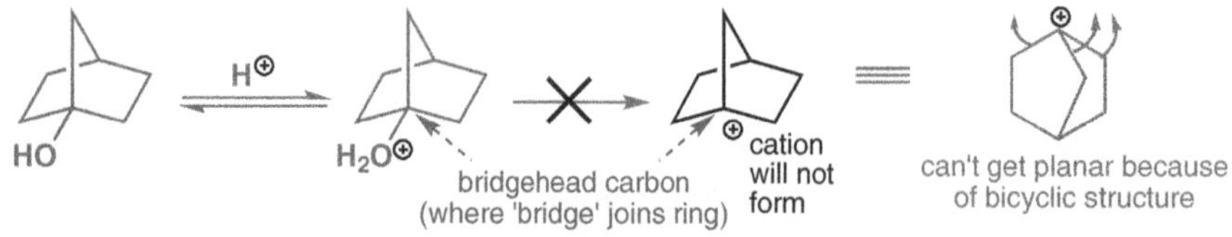

We shall see shortly what the problem with E2 is, but for E1 the hurdle to be overcome is the formation of a planar carbocation. The bicyclic structure prevents the bridgehead carbon becoming planar so, although the cation would be tertiary, it is very high in energy and does not form. You could say that the non-planar structure forces the cation to have an empty sp^3 orbital instead of an empty p orbital, and we saw in Chapter 4 that it is always best to leave the orbitals with the highest possible energy empty.

On p. 335 (Chapter 15) you saw a related example of an impossible S_N1 reaction with a non-planar cation.

> **Bredt's rule**
> The impossibility of planar bridgehead carbons means that double bonds can almost never be formed to bridgehead carbons in bicyclic systems. This principle is known as Bredt's rule, but, as with all rules, it is much more important to know the reason than to know the name, and Bredt's rule is simply a consequence of the strain induced by a planar bridgehead carbon.

The role of the leaving group

We haven't yet been very adventurous with our choice of leaving groups for eliminations: all you have seen so far are E2 from alkyl halides and E1 from protonated alcohols. This is deliberate: the vast majority of the two classes of eliminations use one of these two types of starting materials. But since the leaving group is involved in the rate-determining step of both E1 and E2, in general any good leaving group will lead to a fast elimination. You may, for example, see amines acting as leaving groups in eliminations of quaternary ammonium salts.

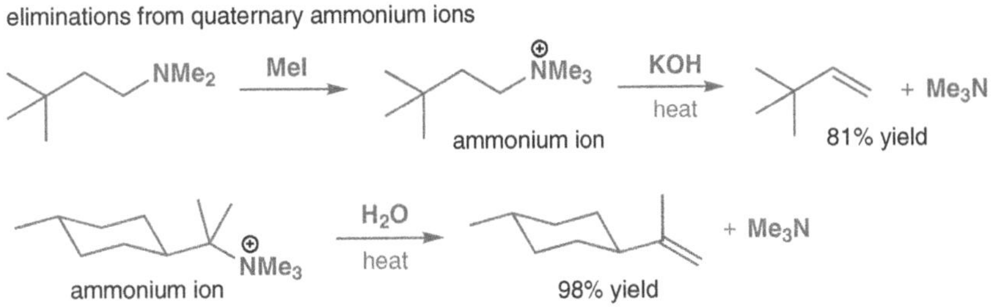

Both E1 and E2 are possible, and from what you have read so far you should be able to spot that there is one of each here: in the first example, a stabilized cation cannot be formed (so E1 is impossible), but a strong base is used, allowing E2. In the second, a stabilized tertiary cation could be formed (so *either* E1 or E2 might occur), but no strong base is present, so the mechanism must be E1.

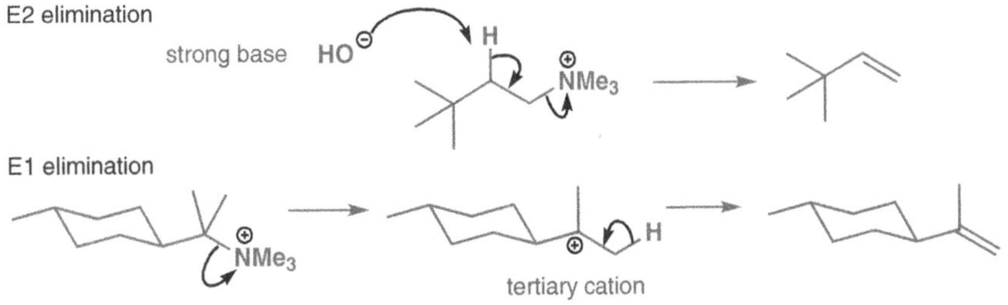

You have just seen that hydroxyl groups can be turned into good leaving groups in acid, but this is only useful for substrates that can react by E1 elimination. The hydroxyl group is *never* a leaving group in E2 eliminations, since they have to be done in base. A strong base would remove the proton from the OH group instead.

● **OH⁻ is never a leaving group in an E2 reaction.**

For primary and secondary alcohols, the hydroxyl is best made into a leaving group for elimination reactions by sulfonylation with *para*-toluenesulfonyl chloride (tosyl chloride, TsCl) or methanesulfonyl chloride (mesyl chloride, $MeSO_2Cl$ or MsCl).

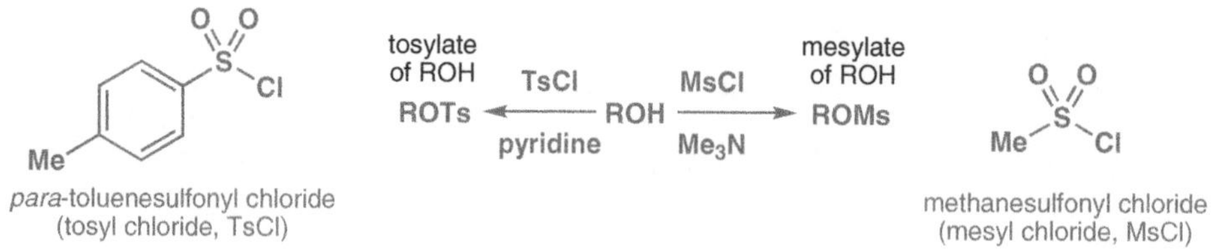

para-toluenesulfonyl chloride
(tosyl chloride, TsCl)

tosylate of ROH

ROTs ⟵ ROH ⟶ ROMs

TsCl / pyridine   MsCl / Me₃N

mesylate of ROH

methanesulfonyl chloride
(mesyl chloride, MsCl)

You met the sulfonate esters—toslylates and mesylates—in Chapter 15 (p. 344).

Toluenesulfonate esters (tosylates) can be made from alcohols (with TsCl, pyridine). We introduced tosylates in Chapter 15 because they are good electrophiles for substitution reactions with *non-basic* nucleophiles. With strong bases such as *t*-BuOK, NaOEt, or DBU they undergo very efficient elimination reactions. Here are two examples.

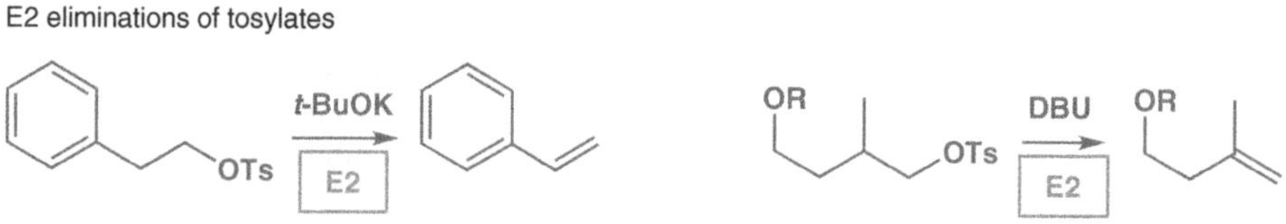

E2 eliminations of tosylates

OTs → [*t*-BuOK / E2] → (styrene)

OR ... OTs → [DBU / E2] → OR

Methanesulfonyl esters (or mesylates; Chapter 15) can be eliminated using DBU, but a good way of using MsCl to convert alcohols to alkenes is to do the mesylation and elimination steps in one go, using the same base (Et₃N) for both. Here are two examples making biologically important molecules. In the first, the mesylate is isolated and then eliminated with DBU to give a synthetic analogue of uracil, one of the nucleotide bases present in RNA. In the second, the mesylate is formed and eliminated in the same step using Et₃N, to give a precursor to a sugar analogue.

There is more about RNA bases and sugars in Chapter 42.

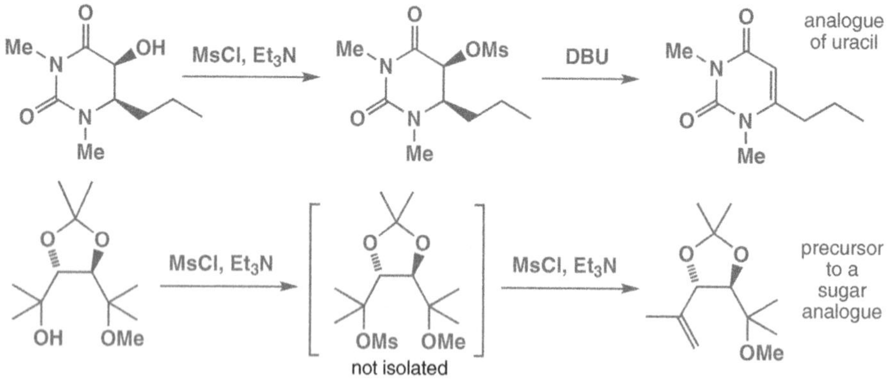

(scheme) ...OH → [MsCl, Et₃N] → ...OMs → [DBU] → analogue of uracil

(scheme) ...OH OMe → [MsCl, Et₃N] → [OMs OMe] not isolated → [MsCl, Et₃N] → OMe precursor to a sugar analogue

The second example here involves (overall) the elimination of a tertiary alcohol—so why couldn't an acid-catalysed E1 reaction have been used? The problem here, nicely solved by the use of the mesylate, is that the molecule contains an acid-sensitive acetal functional group. An acid-catalysed reaction would also have risked eliminating methanol from the other tertiary centre.

E1 reactions can be stereoselective

For some eliminations only one product is possible. For others, there may be a choice of two (or more) alkene products that differ either in the location or stereochemistry of the double bond. We shall now move on to discuss the factors that control the stereochemistry (geometry—*cis* or *trans*) and regiochemistry (that is, where the double bond is) of the alkenes, starting with E1 reactions.

only one alkene possible

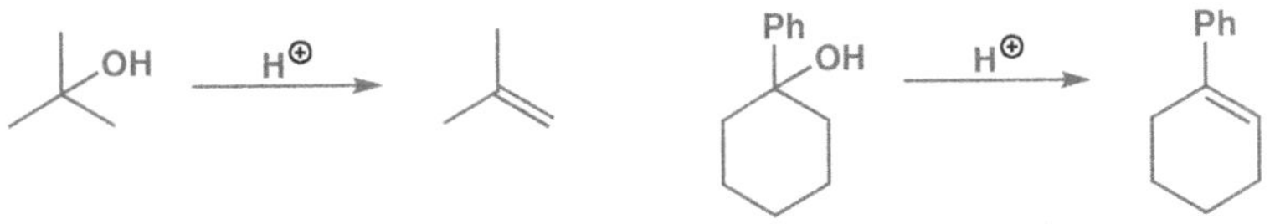

two regioisomeric alkenes possible

trisubstituted alkene   disubstituted alkene

and/or   regioisomers

two stereoisomeric alkenes possible

trans-alkene   and/or   cis-alkene   stereoisomers (geometrical isomers)

E and Z alkenes

You met the idea that alkenes can exist as geometrical *cis* and *trans* isomers in Chapters 3 and 7, and now that you have read Chapter 14 we can be more precise with our definitions. *cis* and *trans* are rather loosely defined terms (like *syn* and *anti*), although no less useful for it. But for formal assignment of geometry, we use the stereochemical descriptors *E* and *Z*. For disubstituted alkenes, *E* corresponds to *trans* and *Z* corresponds to *cis*. To assign *E* or *Z* to tri-or tetrasubstituted alkenes, the groups at either end of the alkene are given an order of priority according to the same rules as those outlined for *R* and *S* in Chapter 15. If the two higher priority groups are *cis*, the alkene is *Z*; if they are *trans* the alkene is *E*. Of course, molecules don't know these rules, and sometimes (as in the second example here) the *E* alkene is less stable than the *Z*.

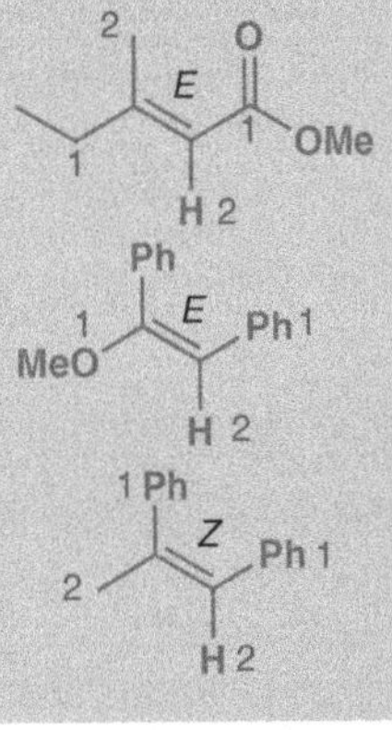

E alkenes (and transition states leading to *E* alkenes) are usually lower in energy than *Z* alkenes (and the transition states leading to them) for steric reasons: the substituents can get further apart from one another. A reaction that can choose which it forms is therefore likely to favour the formation of *E* alkenes. For alkenes formed by E1 elimination, this is exactly what happens: the less hindered *E* alkene is favoured. Here is an example.

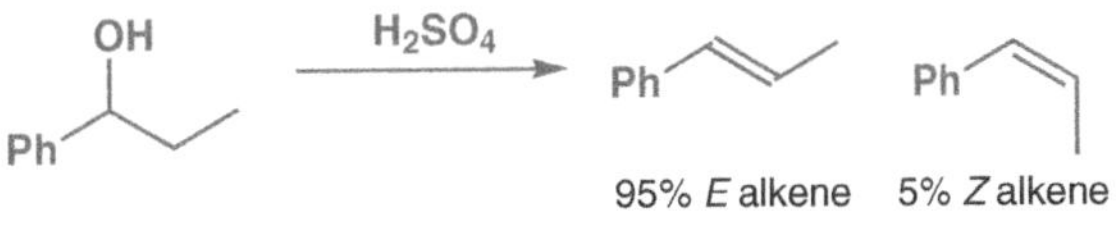

95% *E* alkene   5% *Z* alkene

The geometry of the product is determined at the moment that the proton is lost from the intermediate carbocation. The new π bond can only form if the vacant p orbital of the carbocation and the breaking C–H bond are aligned parallel. In the example shown there are two possible conformations of the carbocation with parallel orientations, but one is more stable than the other because it suffers less steric hindrance. The same is true of the transition states on the route to the alkenes—the one leading to the *E* alkene is lower in energy, and more *E* alkene than *Z* alkene is formed. The process is stereoselective because the reaction chooses to form predominantly one of two possible stereoisomeric products.

➡ In Chapter 39 we shall discuss why the transition states for the decomposition of high-energy intermediates like carbocations are very similar in structure to the carbocations themselves.

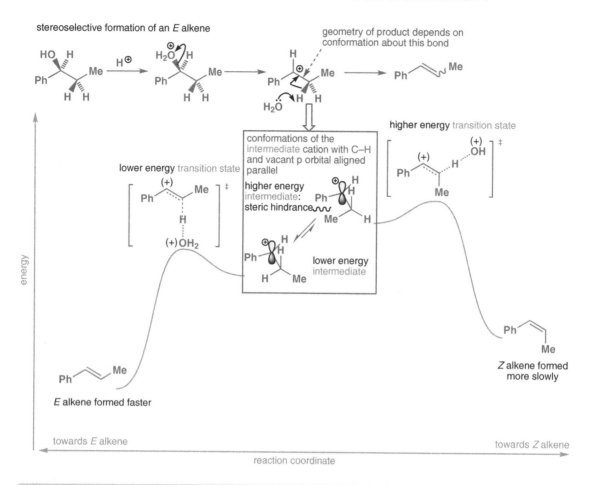

E alkene formed faster

Z alkene formed more slowly

towards *E* alkene

towards *Z* alkene

reaction coordinate

Tamoxifen

Tamoxifen is an important drug in the fight against breast cancer, one of the most common forms of cancer. It works by blocking the action of the female sex hormone oestrogen. The tetrasubstituted double bond can be introduced by an E1 elimination: there is no ambiguity about where the double bond goes, although the two stereoisomers form in about equal amounts. Tamoxifen is the *Z* isomer.

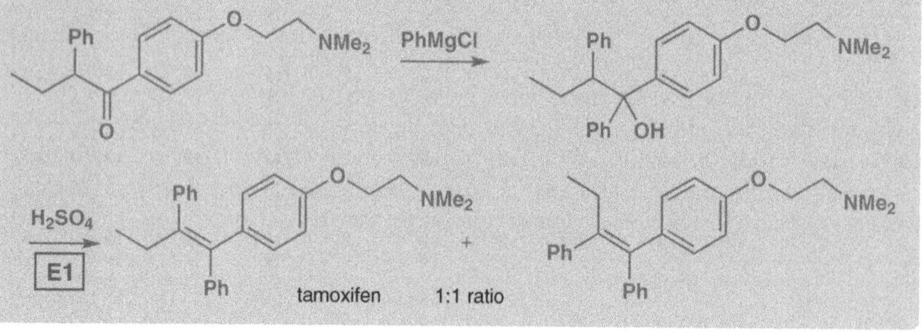

tamoxifen   1:1 ratio

➡ We will come back to the most useful ways of controlling the geometry of double bonds in Chapter 27.

E1 reactions can be regioselective

We can use the same ideas when we think about E1 eliminations that can give more than one regioisomeric alkene. Here is an example. The major product is the alkene that has the more substituents because this alkene is the more stable of the two possible products.

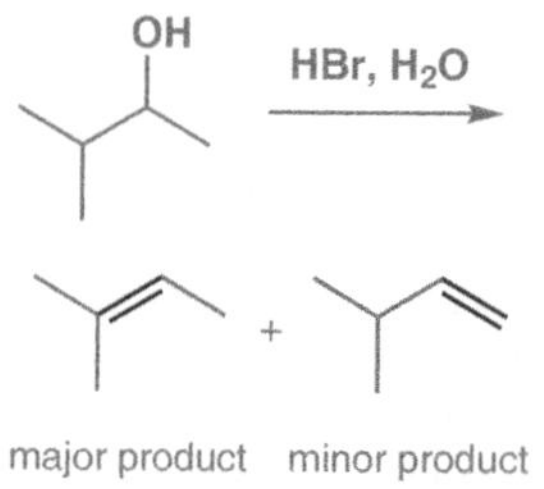

● **More substituted alkenes are more stable.**

HBr, H$_2$O

major product   minor product

This is quite a general principle. But why should it be true? The reason for this is related to the reason why more substituted carbocations are more stable. In Chapter 15 we said that the carbocation is stabilized when its empty p orbital can interact with the filled orbitals of parallel C–H and C–C bonds. The same is true of the π system of the double bond—it is stabilized when the empty π* antibonding orbital can interact with the filled orbitals of parallel C–H and C–C bonds. The more C–C or C–H bonds there are, the more stable the alkene.

■ This explanation of both stereo and regioselectivity in E1 reactions is based on *kinetic* arguments—which alkene forms faster. But it is also true that some E1 eliminations are reversible: the alkenes may be protonated in acid to re-form carbocations, as you will see in the next chapter. This reprotonation allows the more stable product to form preferentially under *thermodynamic* control. In any individual case, it may not be clear which is operating. However, with E2 reactions, which follow, only kinetic control applies: E2 reactions are never reversible.

The more substituted alkene is more stable, but this does not necessarily explain why it is the one that forms faster. To do that, we should look at the transition states leading to the two alkenes. Both form from the same carbocation, but which one we get depends on which proton is lost. Removal of the proton on the right (brown arrow) leads to a transition state in which there is a monosubstituted double bond partly formed. Removal of the proton on the left (orange arrow) leads to a partial double bond that is trisubstituted. This is more stable—the transition state is lower in energy, and the more substituted alkene forms faster.

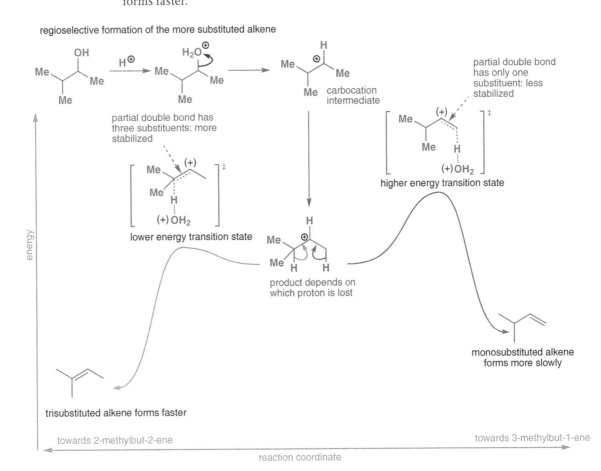

E2 eliminations have anti-periplanar transition states

Although E1 reactions show some stereo and regioselectivity, the level of selectivity in E2 reactions can be much higher because of the more stringent demands on the transition state for E2 elimination. In an E2 elimination, the new π bond is formed by overlap of the C–H σ bond with the C–X σ* antibonding orbital. The two orbitals have to lie in the same plane for best overlap, and now there are two conformations that allow this. One has H and X syn-periplanar, the other anti-periplanar. The anti-periplanar conformation is more stable because it is staggered (the syn-periplanar conformation is eclipsed) but, more importantly, only in the anti-periplanar conformation are the bonds (and therefore the orbitals) truly parallel.

➡ Look back to p. 365 if you need reminding of the shapes and names of the conformations of C–C single bonds.

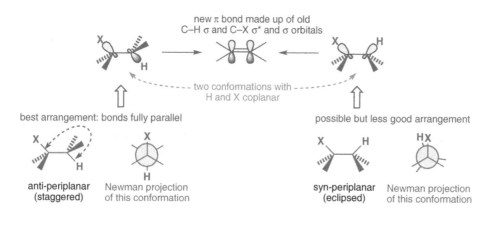

➡ Newman projections illustrate the conformation of molecules viewed along the length of a bond. See p. 364 if you need reminding of how to draw and interpret them.

E2 eliminations therefore take place preferentially from the anti-periplanar conformation. We shall see shortly how we know this to be the case, but first we consider an E2 elimination that gives mainly one of two possible stereoisomers. 2-Bromobutane has two conformations with H and Br anti-periplanar, but the one that is less hindered leads to more of the product, and the E alkene predominates.

There is a choice of protons to be eliminated—the stereochemistry of the product results from which proton is anti-periplanar to the leaving group when the reaction takes place, and the reaction is stereoselective as a result.

E2 eliminations can be stereospecific

In the next example there is only one proton that can take part in the elimination. Now there is no choice of anti-periplanar transition states. Whether the product is E or Z, the E2 reaction has only one course to follow. And the outcome depends on which diastereoisomer of the starting material is used. When the first diastereoisomer is drawn with the proton and

🖰 Interactive mechanism for stereoselective E2

bromine anti-periplanar, as required, and in the plane of the page, the two phenyl groups have to lie one in front and one behind the plane of the paper. As the hydroxide attacks the C–H bond and eliminates Br⁻, this arrangement is preserved and the two phenyl groups end up *trans* (the alkene is *E*). This is perhaps easier to see in the Newman projection of the same conformation.

The second diastereoisomer forms the *Z* alkene for the same reasons: the two phenyl groups are now on the same side of the H–C–C–Br plane in the reactive anti-periplanar conformation (again, this is clear in the Newman projection) and so they end up *cis* in the product. Each diastereoisomer gives a different alkene geometry, and they do so at different rates. The first reaction is about ten times as fast as the second because, although this anti-periplanar conformation is the only reactive one, it is not necessarily the most stable. The Newman projection for the second reaction shows clearly that the two phenyl groups have to lie synclinal (gauche) to one another: the steric interaction between these large groups will mean that, at any time, a relatively small proportion of molecules will adopt the right conformation for elimination, slowing the process down.

Reactions in which the stereochemistry of the product is determined by the stereochemistry of the starting material are called **stereospecific**.

■ A stereospecific reaction is not simply a reaction that is very stereoselective! The two terms have different mechanistic meanings, and are not just different degrees of the same thing.

● **Stereoselective or stereospecific?**

- **Stereoselective reactions** give one predominant product because the reaction pathway has a choice. Either the pathway of lower activation energy is preferred (kinetic control) or the more stable product is preferred (thermodynamic control).

- **Stereospecific reactions** lead to the production of a single isomer as a direct result of the mechanism of the reaction and the stereochemistry of the starting material. There is no choice. The reaction gives a different diastereoisomer of the product from each stereoisomer of the starting material.

E2 eliminations from cyclohexanes

The stereospecificity of the reactions you have just met is very good evidence that E2 reactions proceed through an anti-periplanar transition state. We know with which diastereoisomer we started, and we know which alkene we get, so there is no question over the course of the reaction.

➡ In the next chapter (p. 415) you will see how the fact that pairs of axial bonds have overlapping orbitals also gives rise to distinctively large ¹H NMR coupling constants.

More evidence comes from the reactions of substituted cyclohexanes. You saw in Chapter 16 that substituents on cyclohexanes can be parallel with one another only if they are both axial. An equatorial C–X bond is anti-periplanar only to C–C bonds and cannot take part in an elimination. For mono-substituted cyclohexyl halides treated with base, this is not a problem because, although the axial conformer is less stable, there is still a significant amount present (see the table on p. 375), and elimination can take place from this conformer.

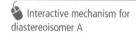

equatorial X is anti-periplanar only to C–C bonds and cannot be eliminated by an E2 mechanism

axial X is anti-periplanar to C–H bonds, so E2 elimination is possible

● **For E2 elimination in cyclohexanes, both C–H and C–X must be axial.**

These two diastereoisomeric cyclohexyl chlorides derived from menthol react very differently under the same conditions with sodium ethoxide as base. Both eliminate HCl but diastereoisomer A reacts rapidly to give a mixture of products, while diastereoisomer B (which differs only in the configuration of the carbon atom bearing chlorine) gives a single alkene product but very much more slowly. We can safely exclude E1 as a mechanism because the same cation would be formed from both diastereoisomers, and this would mean the ratio of products (although not necessarily the rate) would be the same for both.

elimination of diastereoisomer A

CH₃ — NaOEt → CH₃ + CH₃

ratio of 1:3

elimination of diastereoisomer B

CH₃ — NaOEt, 250 times slower → CH₃

The key to explaining reactions like this is to draw the conformation of the molecules. Both will adopt a chair conformation, and generally the chair having the largest substituent equatorial (or the largest *number* of substituents equatorial) is the more stable. In these examples the isopropyl group is most influential—it is branched and will have very severe 1,3-diaxial interactions if it occupies an axial position. In both diastereoisomers, an equatorial *i*-Pr also means an equatorial Me: the only difference is the orientation of the chlorine. For diastereoisomer A, the chlorine is forced axial in the major conformer: there is no choice because the relative configuration is fixed in the starting material. It's less stable than equatorial Cl, but is ideal for E2 elimination and there are two protons that are anti-periplanar available for removal by the base. The two alkenes are formed as a result of each of the possible protons with a 3:1 preference for the more substituted alkene.

For diastereoisomer B, the chlorine is equatorial in the lowest-energy conformation. Once again there is no choice. But equatorial leaving groups cannot be eliminated by E2: in this conformation there is no anti-periplanar proton. This accounts for the difference in rate between the two diastereoisomers. A has the chlorine axial virtually all the time ready for E2, while B has an axial leaving group only in the minute proportion of the molecules that happen not to be in the lowest-energy conformation, but that have all three substituents axial. The all-axial conformer is much higher in energy, but only in this confomer can Cl⁻ be eliminated. The concentration of reactive molecules is low, so the rate is also low. There is only one proton anti-periplanar and so elimination gives a single alkene.

■ This would be a good time to make sure you can reliably sketch a cyclohexane in the chair conformation. Our guidelines for helping you do this are on pp. 371–2.

🖱 Interactive mechanism for diastereoisomer A

🖱 Interactive mechanism for diastereoisomer B

conformation of diastereoisomer A

No C–H bonds anti-periplanar to the C–Cl bond: no elimination

two anti-periplanar C–H bonds: either can eliminate to give different products

ring inversion

disfavoured; axial *i*-Pr   favoured; equatorial *i*-Pr

conformation of diastereoisomer B

No C–H bonds anti-periplanar to the C–Cl bond: no elimination

one anti-periplanar C–H bond: single alkene formed

ring inversion

favoured; equatorial *i*-Pr   disfavoured: axial *i*-Pr

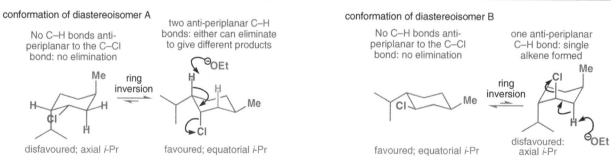

E2 elimination from vinyl halides: how to make alkynes

An anti-periplanar arrangement of C–Br and C–H is attainable with a vinylic bromide too, provided the Br and H are *trans* to one another. E2 elimination from the *Z* isomer of a vinyl bromide gives an alkyne rather faster than elimination from the *E* isomer because in the *E* isomer the C–H and C–Br bonds are syn-periplanar.

C–H and C–Br parallel (anti-periplanar): fast E2 elimination

C–H and C–Br syn-periplanar: slower E2 elimination

■ The base used here is LDA (lithium diisopropylamide) made by deprotonating *i*-Pr$_2$NH with BuLi (see p. 174). LDA is very basic (pK_a of *i*-Pr$_2$NH is about 35) but too hindered to be nucleophilic—ideal for promoting E2 elimination.

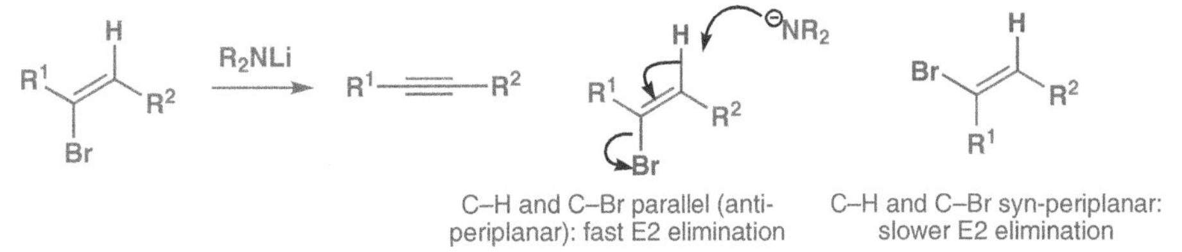

Vinyl bromides can themselves be made by elimination reactions of 1,2-dibromoalkanes. Watch what happens when 1,2-dibromopropane is treated with three equivalents of R$_2$NLi: first, elimination to the vinyl halide, then elimination of the vinyl halide to the alkyne. The terminal alkyne is amply acidic enough to be deprotonated by R$_2$NLi, and this is the role of the third equivalent. Overall, the reaction makes a lithiated alkyne (ready for further reactions) from a fully saturated starting material. This may well be the first reaction you have met that makes an alkyne from a starting material that doesn't already contain a triple bond.

making an alkyne from 1,2-dibromopropane

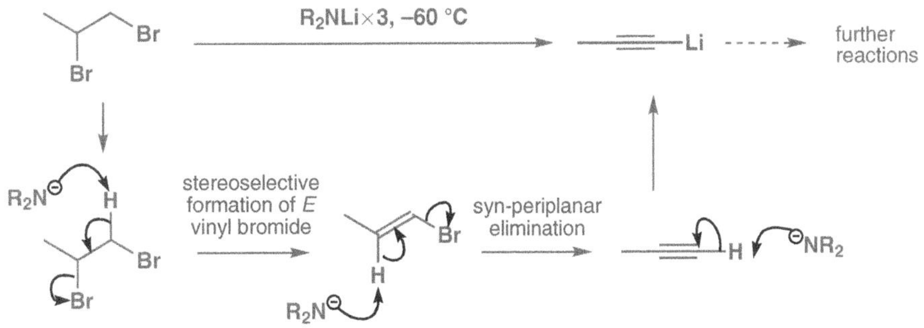

The regioselectivity of E2 eliminations

Here are two deceptively similar elimination reactions. The leaving group changes and the reaction conditions are very different but the overall process is elimination of HX to produce one of two alkenes.

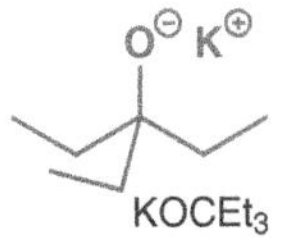

KOCEt$_3$

In the first example acid-catalysed elimination of water from a tertiary alcohol produces a trisubstitued alkene. Elimination of HCl from the corresponding tertiary alkyl chloride promoted by a very hindered alkoxide base (more hindered than *t*-BuOK because all the ethyl groups have to point away from one another) gives exclusively the less stable disubstituted alkene.

The reason for the two different regioselectivities is a change in mechanism. As we have already discussed, acid-catalysed elimination of water from tertiary alcohols is usually E1, and you already know the reason why the more substituted alkene forms faster in E1 reactions (p. 394). It should come as no surprise to you now that the second elimination, with a strong, hindered base, is an E2 reaction. But why does E2 give the less substituted product? This time, there is no problem getting C–H bonds anti-periplanar to the leaving group: in the conformation with the Cl axial there are two equivalent ring hydrogens available for elimination, and removal of either of these would lead to the trisubstituted alkene. Additionally, any of the three equivalent methyl hydrogens are in a position to undergo E2 elimination to form the disubstituted alkene whether the Cl is axial or equatorial—and yet it is these and only these that are removed by the hindered base. The diagram summarizes two of the possibilities.

two ring hydrogens anti-periplanar to Cl

ring hydrogens more hindered: no reaction

methyl hydrogen anti-periplanar to Cl

methyl hydrogens less hindered: this product formed

The base attacks the methyl hydrogens because they are less hindered—they are attached to a primary carbon atom, well away from the other axial hydrogens. E2 eliminations with hindered bases typically give the less substituted double bond because the fastest E2 reaction involves deprotonation at the least-substituted site. The hydrogens attached to a less substituted carbon atom are also more acidic. Think of the conjugate bases: a *t*-butyl anion is more basic (because the anion is destabilized by the three electron-donating alkyl groups) than a methyl anion, so the corresponding alkane must be less acidic. Steric factors are evident in the following E2 reactions, where changing the base from ethoxide to *t*-butoxide alters the major product from the more to the less substituted alkene.

28%   73%   *t*-BuOK ←   Br   → NaOEt   69%   31%

● **Elimination regioselectivity**

- E1 reactions give the more substituted alkene.
- E2 reactions may give the more substituted alkene, but become more regioselective for the less substituted alkene with more hindered bases.

Hofmann and Saytsev

Traditionally, these two opposite preferences—for the more or the less substituted alkenes—have been called Saytsev's rule and Hofmann's rule, respectively. You will see these names used (along with a number of alternative spellings—acceptable for Saytsev, whose name is transliterated from Russian, but not for Hofmann: this Hofmann had one f and two n's), but there is little point remembering which is which (or how to spell them)—it is far more important to understand the reasons that favour formation of each of the two alkenes.

Anion-stabilizing groups allow another mechanism—E1cB

To finish this chapter, we consider a reaction that at first sight seems to go against what we have told you so far. It's an elimination catalysed by a strong base (KOH), so it looks like E2. But the leaving group is hydroxide, which we categorically (and truthfully) stated cannot be a leaving group in E2 eliminations.

The key to what is going on is the carbonyl group. In Chapter 8 you met the idea that negative charges are stabilized by conjugation with carbonyl groups, and the list on p. 176 demonstrated how acidic a proton adjacent to a carbonyl group is. The proton that is removed in this elimination reaction is adjacent to the carbonyl group, and is therefore also rather acidic (pK_a about 20). This means that the base can remove it without the leaving group departing at the

■ This delocalized anion is called an **enolate**, and we will discuss enolates in more detail in Chapter 20 and beyond.

same time—the anion that results is stable enough to exist because it can be delocalized on to the carbonyl group.

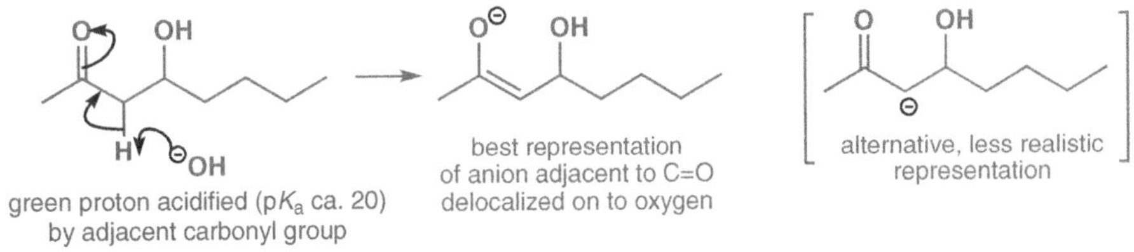

best representation
of anion adjacent to C=O
delocalized on to oxygen

alternative, less realistic
representation

green proton acidified (pK_a ca. 20)
by adjacent carbonyl group

Although the anion is stabilized by the carbonyl group, it still prefers to lose a leaving group and become an alkene. This is the next step.

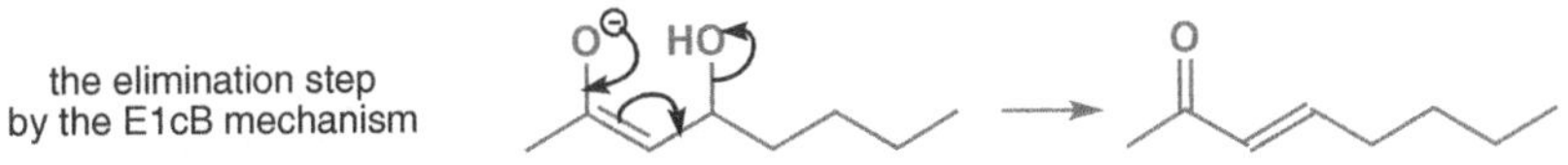

the elimination step
by the E1cB mechanism

This step is also the rate-determining step of the elimination—the elimination is unimolecular and so is some kind of E1 reaction. The leaving group is not lost from the starting molecule, but from the *conjugate base* of the starting molecule, so this sort of elimination, which starts with a deprotonation, is called E1cB (cB for conjugate base). Here is the full mechanism, generalized for other carbonyl compounds.

the E1cB mechanism

fast, reversible
deprotonation

rate-
determining
step

stabilized anionic
intermediate

Interactive mechanism for E1cB elimination

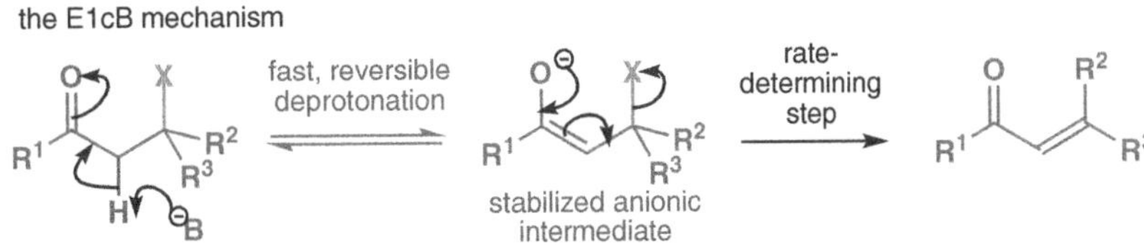

It's important to note that, while HO⁻ is never a leaving group in E2 reactions, it can be a leaving group in E1cB reactions. The anion it is lost from is already an alkoxide—the oxyanion does not need to be created as the HO⁻ is lost. The establishment of conjugation in the product also assists loss of HO⁻. As the scheme above implies, other leaving groups are possible too. Here are two examples with methanesulfonate leaving groups.

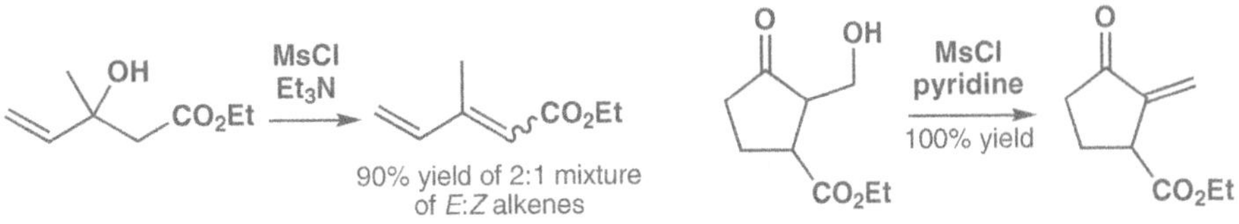

■ 'E1cB' is written with no super- or subscripts, a lower-case c, and an upper-case B.

MsCl
Et₃N

90% yield of 2:1 mixture
of E:Z alkenes

MsCl
pyridine

100% yield

The first looks E1 (stabilized cation), the second E2—but in fact both are E1cB reactions. The most reliable way to spot a likely E1cB elimination is to see whether the alkene in the product is conjugated with a carbonyl group. If it is, the mechanism is probably E1cB.

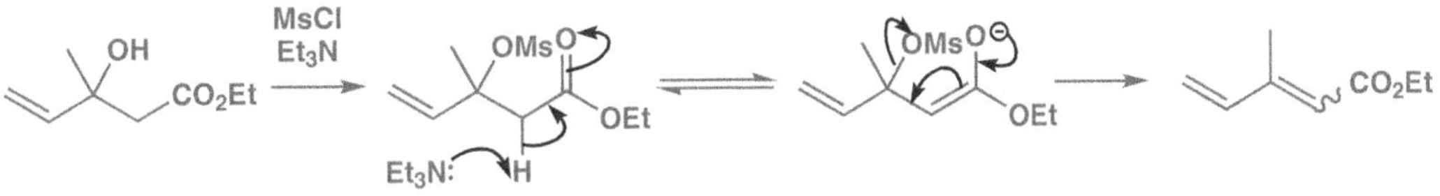

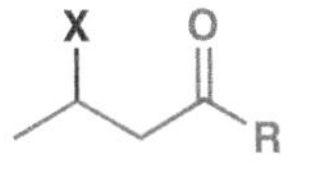

MsCl
Et₃N

Et₃N:

a β-halocarbonyl
compound

β-Halocarbonyl compounds can be rather unstable: the combination of a good leaving group and an acidic proton means that E1cB elimination is extremely easy. This mixture of diastereoisomers is first of all lactonized in acid (Chapter 10), and then undergoes E1cB elimination with triethylamine to give a product known as a butenolide. Butenolides are common structures in naturally occurring compounds.

You will have noticed that we have shown the deprotonation step in the last few mechanisms as an equilibrium. Both equilibria lie rather over to the left-hand side because neither triethylamine (pK_a of Et_3NH^+ about 10) nor hydroxide (pK_a of H_2O 15.7) is basic enough to remove completely a proton next to a carbonyl group ($pK_a > 20$). However, because the loss of the leaving group is essentially irreversible, only a small amount of deprotonated carbonyl compound is necessary to keep the reaction going. The important point about substrates that undergo E1cB is that there is some form of anion-stabilizing group next to the proton to be removed—it doesn't have to stabilize the anion very well but, as long as it makes the proton more acidic, an E1cB mechanism has a chance. Here is an important example with two phenyl rings helping to stabilize the anion, and a carbamate anion ($R_2N\!-\!CO_2^-$) as the leaving group.

six π electrons aromatic cyclopentadienyl anion

The proton to be removed has a pK_a of about 25 because its conjugate base is an aromatic cyclopentadienyl anion (we discussed this in Chapter 8). The E1cB elimination takes place with a secondary or tertiary amine as the base. Spontaneous loss of CO_2 from the eliminated product gives an amine, and you will meet this class of compounds again in Chapter 23, where we discuss the Fmoc protecting group.

The E1cB rate equation

The rate-determining elimination step in an E1cB reaction is unimolecular, so you might imagine it would have a first-order rate equation. In fact, the rate is also dependent on the concentration of base. This is because the unimolecular elimination involves a species—the anion—whose concentration is itself determined by the concentration of base by the equilibrium we have just been discussing. Using the following general E1cB reaction, the concentration of the anion can be expressed as shown.

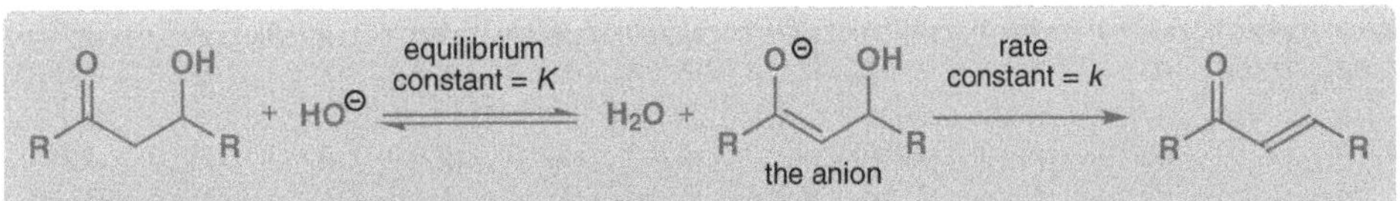

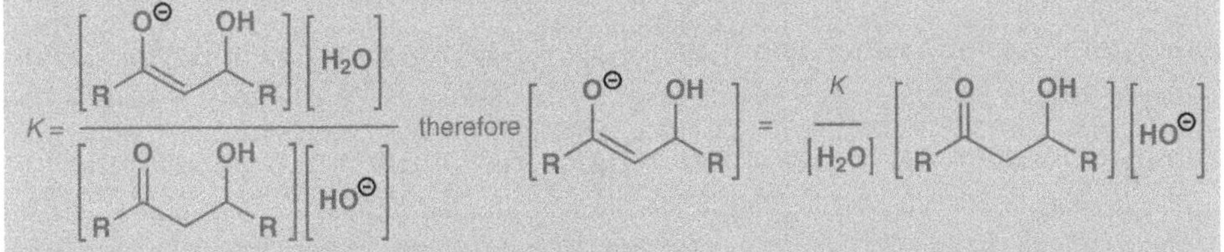

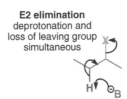

The rate is proportional to the concentration of the anion, and we now have an expression for that concentration. We can simplify it further because the concentration of water is effectively constant.

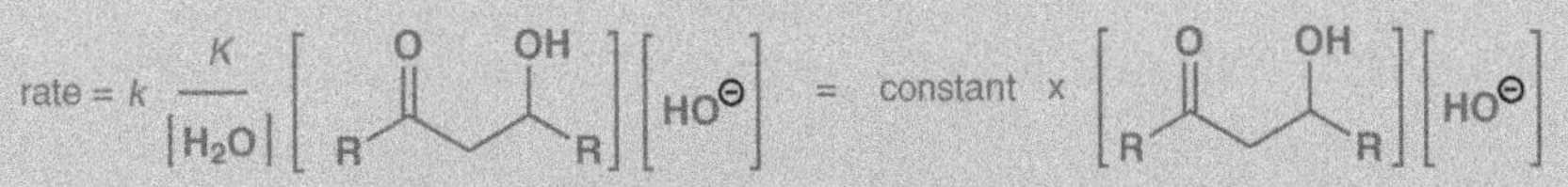

Just because the base (hydroxide) appears in this rate equation doesn't mean to say it is involved in the rate-determining step. Increasing the concentration of base makes the reaction go faster by increasing the amount of anion available to eliminate.

➡ You met this idea in Chapter 12 in the context of *third-order* rate equations.

E1cB eliminations in context

We can also compare the E1cB mechanism with the other elimination reactions you have met by thinking of the relative timing of proton removal and leaving group departure. E1 is at one end of the scale: the leaving group goes first and proton removal follows in a second step. In E2 reactions, the two events happen at the same time: the proton is removed as the leaving group leaves. In E1cB the proton removal moves in front of leaving group departure.

E1 elimination
leaving group first
deprotonation
follows

E2 elimination
deprotonation and
loss of leaving group
simultaneous

E1cB elimination
deprotonation first
leaving group
follows

We talked about regio- and stereoselectivity in connection with E1 and E2 reactions. With E1cB, the regioselectivity is straightforward: the location of the double bond is defined by the position of (a) the acidic proton and (b) the leaving group.

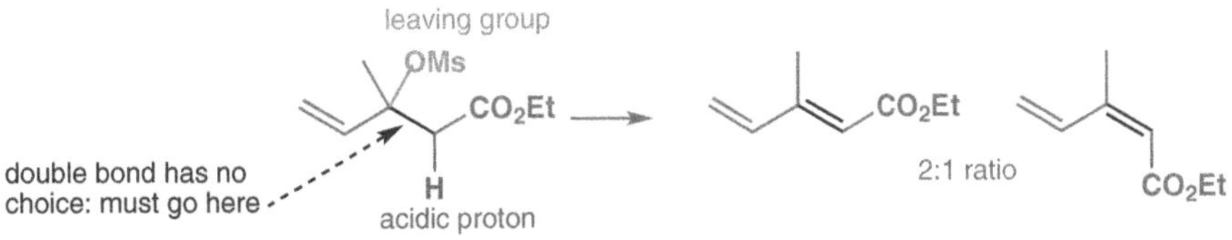

leaving group
OMs

double bond has no
choice: must go here
H
acidic proton

2:1 ratio

E1cB reactions may be stereoselective—the one above, for example, gives mainly the *E* alkene product (2:1 with *Z*). The intermediate anion is planar, so the stereochemistry of the starting materials is irrelevant, the less sterically hindered (usually *E*) product is preferred. This double E1cB elimination, for example, gives only the *E,E* product.

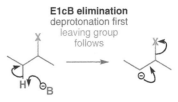

To finish this chapter we need to tell you about two E1cB eliminations that you may meet in unexpected places. We have saved them till now because they are unusual in that the leaving group is actually part of the anion-stabilizing group itself. First of all, try spotting the E1cB elimination in this step from the first total synthesis of penicillin V.

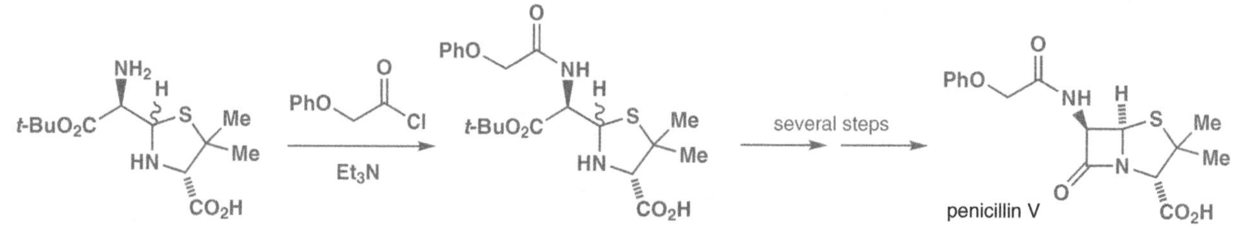

The reaction is deceptively simple—formation of an amide in the presence of base—and you would expect the mechanism to follow what we told you in Chapter 10. But the acyl chloride is, in fact, set up for an E1cB elimination—and you should expect this whenever you see an acyl chloride *with acidic protons next to the carbonyl group* used in the presence of a base such as triethylamine.

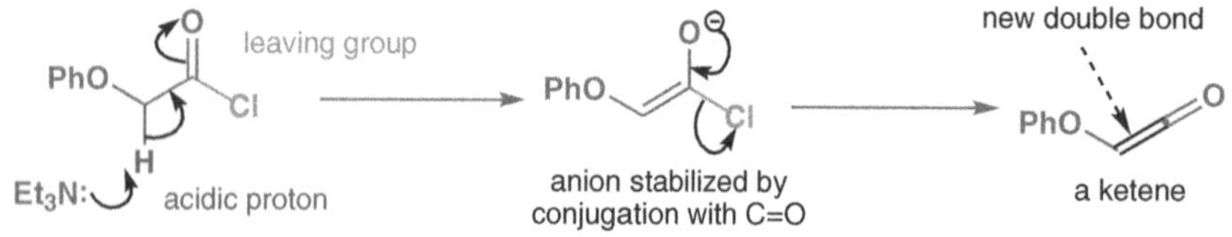

The product of the elimination is a substituted ketene—a highly reactive species whose parent structure is the molecule $CH_2=C=O$ that you will meet in the next chapter. It is the ketene that reacts with the amine to form the amide.

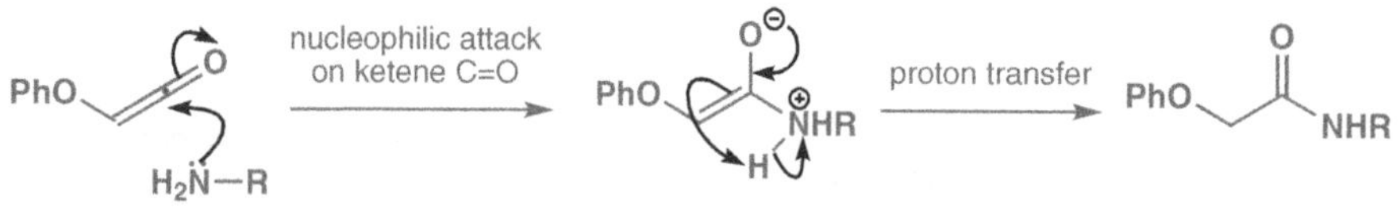

The second 'concealed' E1cB elimination is disguised in the mechanism of formation of methanesulfonates (mesylates). When we introduced sulfonate esters in Chapter 15, and revisited them on p. 391 of this chapter, we avoided (uncharacteristically, you may say) explaining the mechanism by which they are formed from sulfonyl chlorides. This was deliberate because, while TsCl reacts with alcohols by the mechanism you might predict, the reaction with MsCl involves an elimination step.

Here is the mechanism by which alkyl tosylates are formed from alcohols. The alcohol acts as a nucleophile towards the electrophilic sulfonyl chloride, and pyridine removes a proton to give the product.

formation of toluenesulfonates (tosylates): reagents ROH + TsCl + pyridine

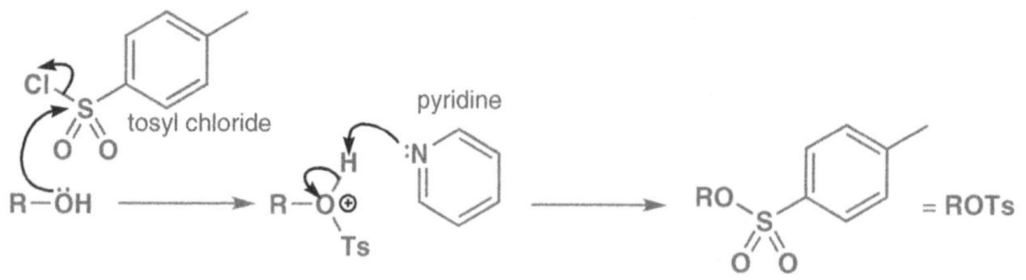

Methanesulfonyl chloride by contrast has a feature it shares with the acyl chlorides just above: a relatively acidic proton that can be removed by base. This deprotonation, followed by loss of chloride, is the first step in the formation of a mesylate ester. It is an E1cB elimination and the product is called a sulfene.

formation of methanesulfonates (mesylates): reagents ROH + MsCl + triethylamine

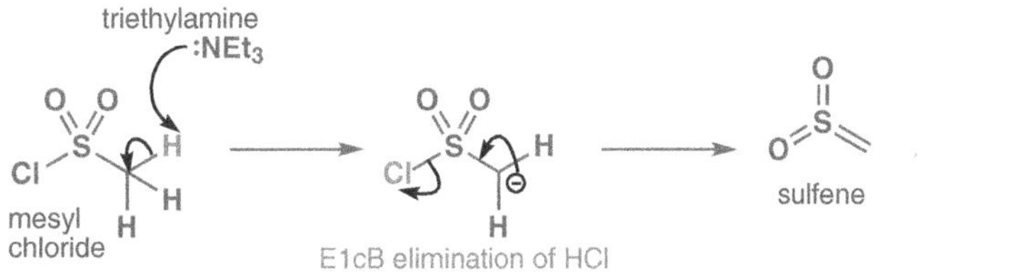

The sulfene is electrophilic in a slightly odd way: the alcohol acts as a nucleophile for sulfur and generates an anion of carbon which undergoes a proton transfer to give the mesylate. It is not uncommon for anions to form adjacent to sulfur, as you will see again in Chapter 27. Notice how similar the overall mechanism is to the acylation mechanism we showed you above.

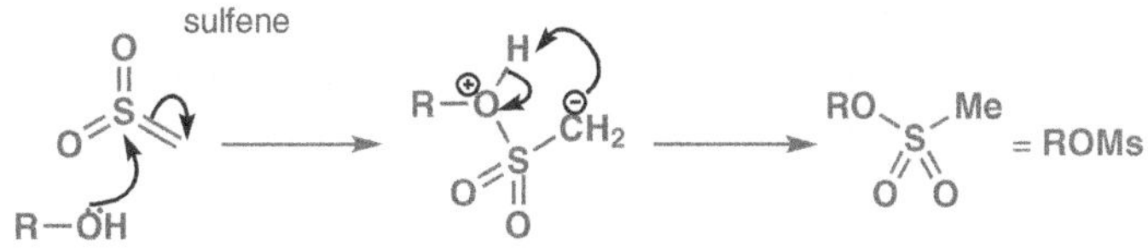

To conclude

We finish with brief summaries of three important discussions we have had in this chapter.

Elimination *versus* substitution

The table below summarizes the general pattern of reactivity expected from various structural classes of alkyl halides (or tosylates, mesylates) in reactions with a representative range of nucleophiles (which may behave as bases).

		Poor nucleophile (e.g. H_2O, ROH)	Weakly basic nucleophile (e.g. I^-, RS^-)	Strongly basic, unhindered nucleophile (e.g. RO^-)	Strongly basic, hindered nucleophile (e.g. DBU, t-BuO^-)
methyl	H_3C-X	no reaction	S_N2	S_N2	S_N2
primary (unhindered)		no reaction	S_N2	S_N2	E2
primary (hindered)	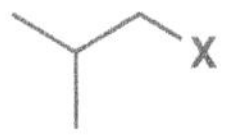	no reaction	S_N2	E2	E2
secondary		S_N1, E1 (slow)	S_N2	E2	E2
tertiary	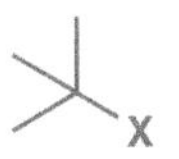	E1 or S_N1	S_N1, E1	E2	E2
β to anion-stabilizing group	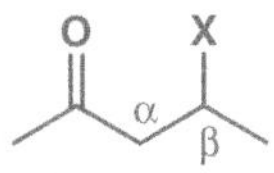	E1cB	E1cB	E1cB	E1cB

Some points about the table:

- Methyl halides cannot eliminate as there are no appropriately placed protons.
- Increasing branching favours elimination over substitution and strongly basic hindered nucleophiles always eliminate unless there is no option.
- Good nucleophiles undergo substitution by S_N2 unless the substrate is tertiary and then the intermediate cation can eliminate by E1 as well as substitute by S_N1.
- High temperatures favour elimination by gearing up the importance of entropy in the free energy of reaction ($\Delta G = \Delta H - T\Delta S$). This is a good way of ensuring E1 in ambiguous cases.

Summary of the stabilities of types of alkene

Alkenes are stabilized by:

- **conjugation**—anything that can conjugate with an alkene stabilizes it, including carbonyl groups, nitriles, benzene rings, RO or RNH groups, or another alkene. This is the strongest stabilization and usually dominates.
- **substitution**—alkyl groups stabilize alkenes weakly by σ-conjugation, so the more alkyl groups the better—but beware of the next point.
- **lack of steric hindrance**—as alkenes are planar, large, especially branched, substituents arranged *syn* on the alkene destabilize it, so tetra-substituted alkenes are usually less stable than tri-substituted ones. If the alkene is in a stable ring, this does not apply as the ring substituents have to be *syn* for the ring to exist.

Alkene stereochemistry: a summary of terminology

The official way of assigning alkene geometry is *E* and *Z*. *Z* comes from the German *zusammen* (together) and means that the two highest ranking substituents (by the same rules introduced in Chapter 14 for *R* and *S*) are on the same side of the alkene. The letter *Z* is a particularly unfortunate choice as it looks like a *trans* alkene! *E* comes from the German *entgegen* (opposed) and means that the two highest ranking substituents are on opposite sides (and, if anything also unfortunately looks like a *cis* alkene). The green numbers in the structures below show the relative rankings of the two substituents at each end of the alkene and the consequent assignment of geometry.

■ This terminology may be used only for alkenes and not for three-dimensional stereochemistry.

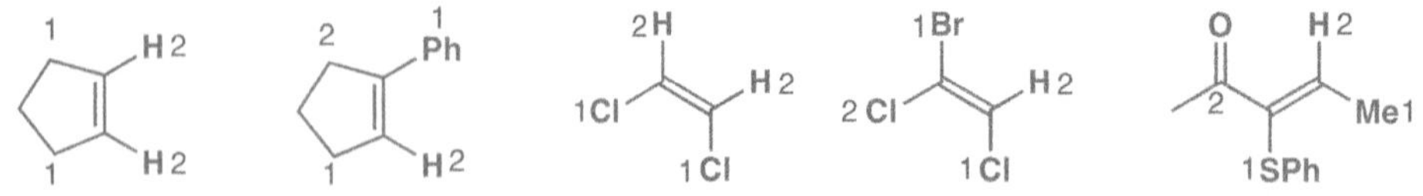

Z or *cis*   *E* or *cis* (*cis* within ring)   *Z* or *cis*   *E* or *cis* (as in '*cis* dichloride')   *Z* or *trans* (as in '*trans* enone')

But possibly the most common method of referring to alkene geometry is to use *cis* and *trans*. These require a diagram as they refer to two substituents being on the same (*cis*) or opposite (*trans*) sides of the alkene. There is no specified order of priority and the speaker chooses substituents that are significant to the structure or to the reaction under discussion, making this a more flexible and versatile way of talking about alkenes. We have assigned *cis* and *trans* to the alkenes above to indicate their most important features, but notice the ambiguities that may occur.

■ Much the same are the terms *syn* and *anti*, introduced on p. 317 and used for relative three-dimensional stereochemistry. There is no formal definition, and a diagram is needed for clarity.

Further reading

See J. Keeler and P. Wothers, *Why Chemical Reactions Happen*, OUP, Oxford, 2003, chapter 11 for a comparison between substitution and elimination and F. A. Carey and R. J. Sundberg, *Advanced Organic Chemistry A, Structure and Mechanisms*, 5th edn, Springer 2007, chapter 5.

DBU and other strong bases are described in T. Ishikawa, ed. *Superbases for organic synthesis: guanidines, amidines and phosphazenes and related organocatalysts*, Wiley, Chichester, 2009. The trityl protecting group, along with many others is described in P. J. Kocienski, *Protecting Groups*, 3rd edn, Thieme, 2003.

Check your understanding

 To check that you have mastered the concepts presented in this chapter, attempt the problems that are available in the book's Online Resource Centre at **http://www.oxfordtextbooks.co.uk/orc/clayden2e/**

Suggested solutions for Chapter 17

17

PROBLEM 1

Draw mechanisms for these elimination reactions.

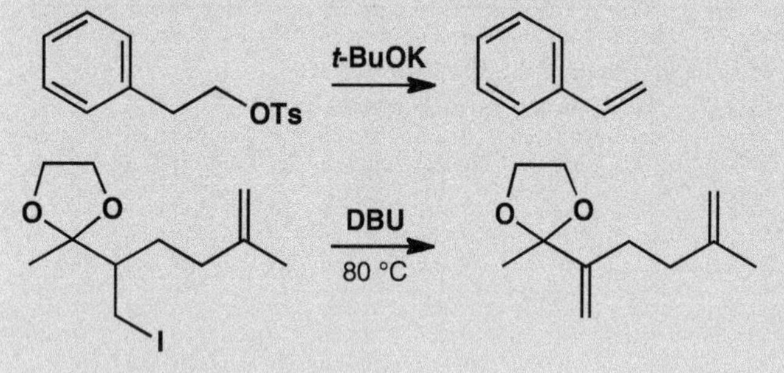

Purpose of the problem

Exercise in drawing simple eliminations.

Suggested solution

These are both E2 reactions as the leaving groups are on primary carbons. In fact both of these reaction are in the textbook (pp. 387 and 391 of the textbook).

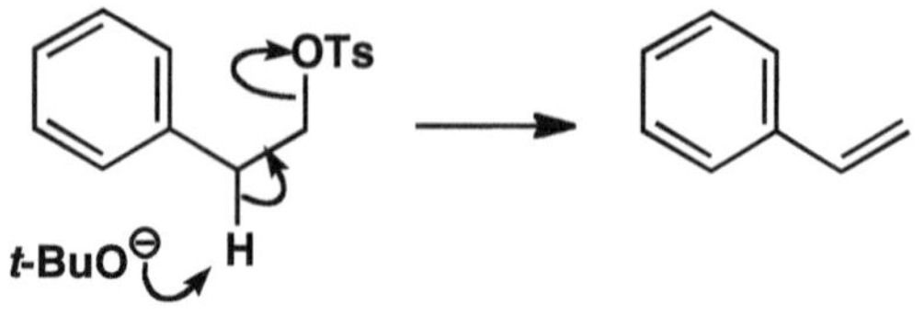

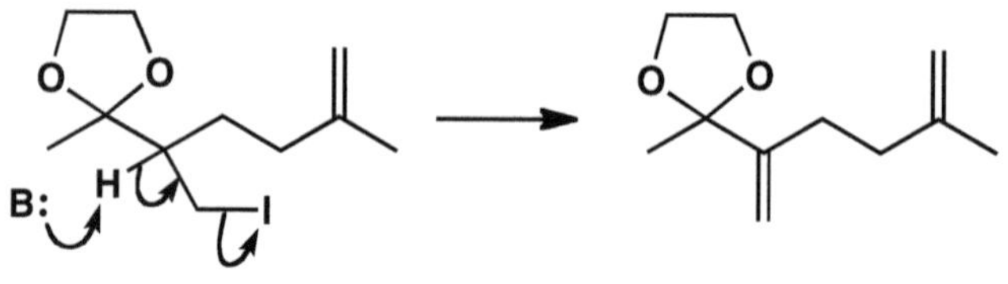

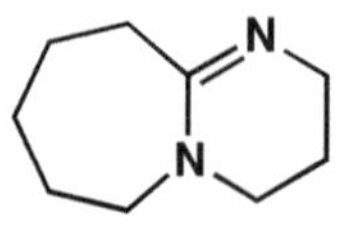

■ The structure of the amidine base, DBU, and why it is used in elimination reactions is discussed in the textbook on p. 387.

DBU

PROBLEM 2

Give a mechanism for the elimination reaction in the formation of tamoxifen, a breast cancer drug, and comment on the roughly 50:50 mixture of geometrical isomers (*cis*- and *trans*-alkenes)

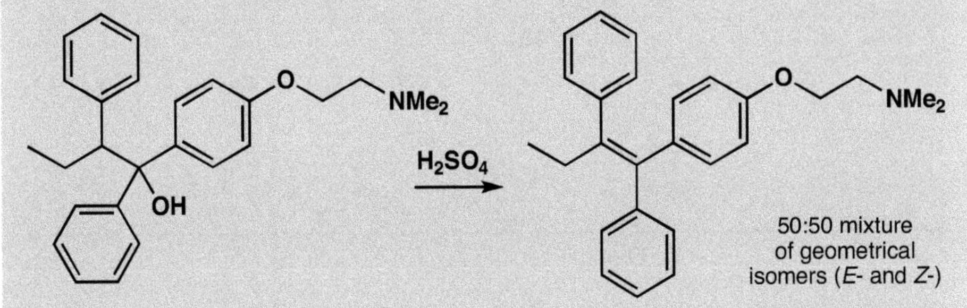

Purpose of the problem

Thinking about the stereochemical consequences of E1.

Suggested solution

■ The fact that equilibration of the products of E1 elimination gives the most stable possible alkene is discussed in the textbook on p. 394.

The tertiary alcohol leaving group, the acid catalyst, and the 50:50 mixture all suggest E1 rather than E2. There is only one proton that can be lost and, as there is very little difference between the isomeric alkenes, equilibration probably gives the 50:50 mixture.

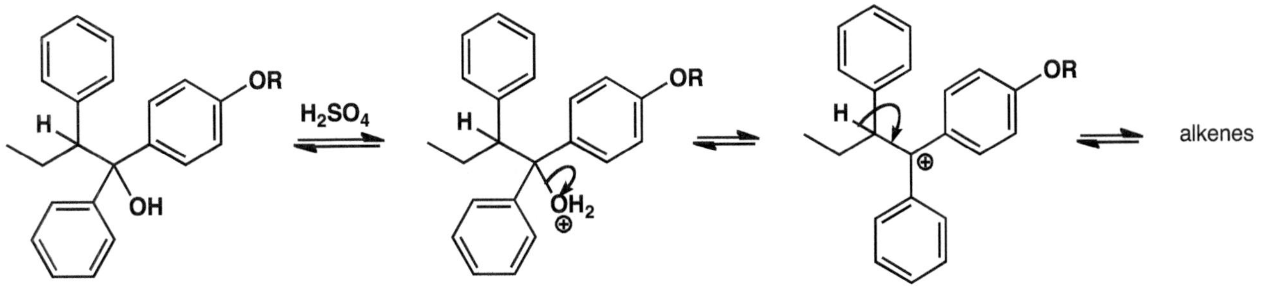

PROBLEM 3

Suggest mechanisms for these eliminations. Why does the first give a mixture and the second a single product?

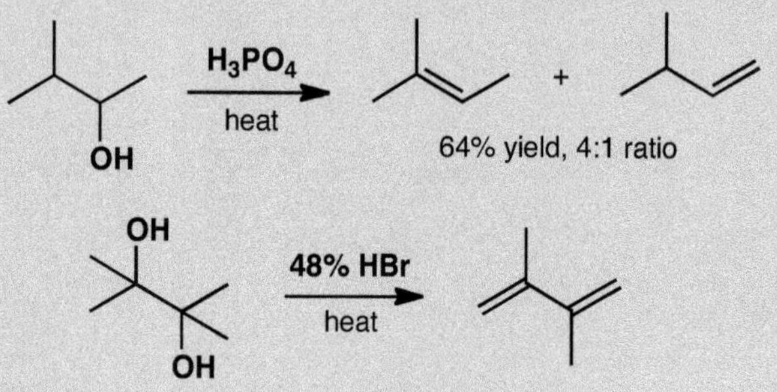

64% yield, 4:1 ratio

Purpose of the problem

Regioselectivity of eliminations.

Suggested solution

Whether the first reaction is E1 or E2, there are two sets of hydrogen atoms that could be lost in the elimination. The conditions suggest E1 and the major product may be so because of equilibration.

■ The fact that equilibration of the products of E1 elimination gives the most stable possible alkene is discussed in the textbook on p. 394.

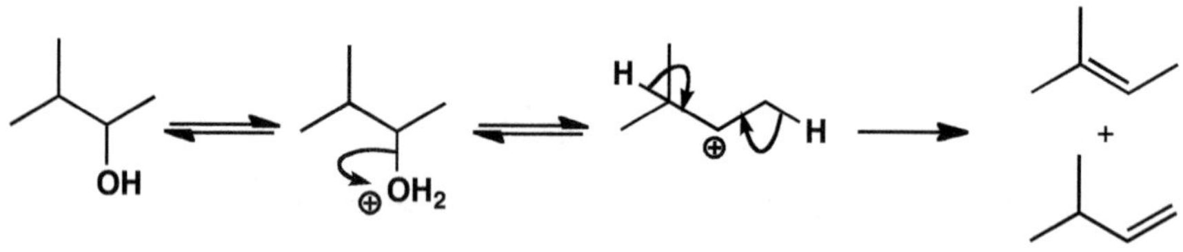

The second reaction produces a more stable tertiary cation from which any of six protons could be lost, but all give the same product. Repetition gives the diene.

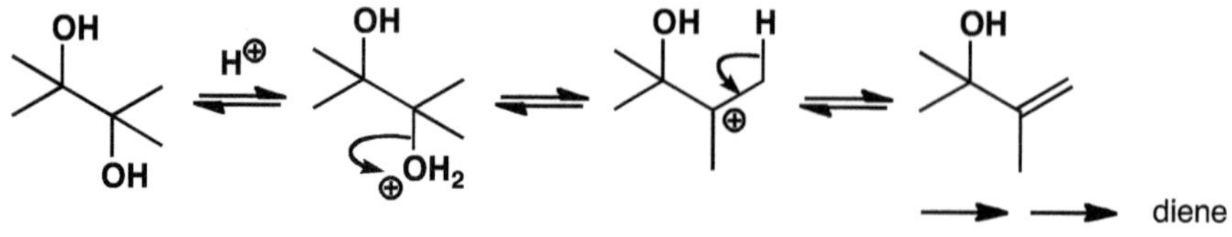

PROBLEM 4

Explain the position of the alkene in the products of these reactions. The starting materials are enantiomerically pure. Are the products also enantiomerically pure?

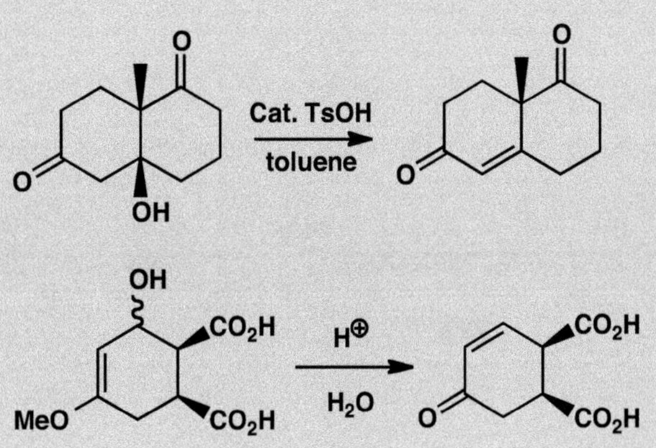

Purpose of the problem

Examples of E1cB in the context of absolute stereochemistry.

Suggested solution

■ E1cB reactions are on p. 399 in the textbook

The first reaction is an E1cB elimination of a β-hydroxy-ketone. The product is still chiral although it has lost one stereogenic centre. The other (quaternary) centre is not affected by the reaction so the product is enantiomerically pure.

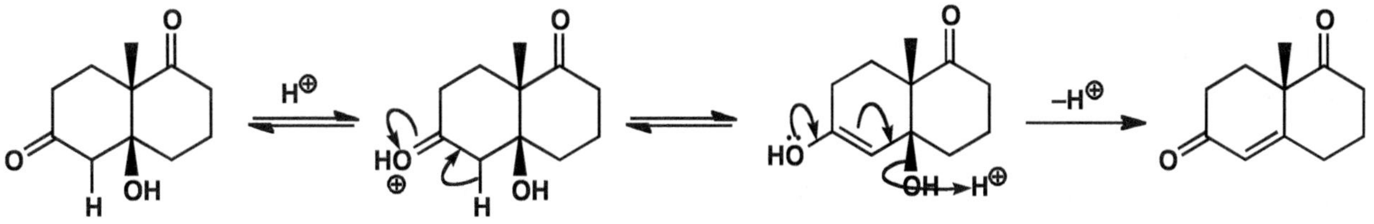

The second example already has an electron-rich alkene (an enol ether) present in the starting material so this is more of an E1 than an E1cB mechanism. The intermediate is a hemiacetal that hydrolyses to a ketone (p. 224 in the textbook). The product has two chiral centres unaffected by the reaction and is still chiral so it is also enantiomerically pure.

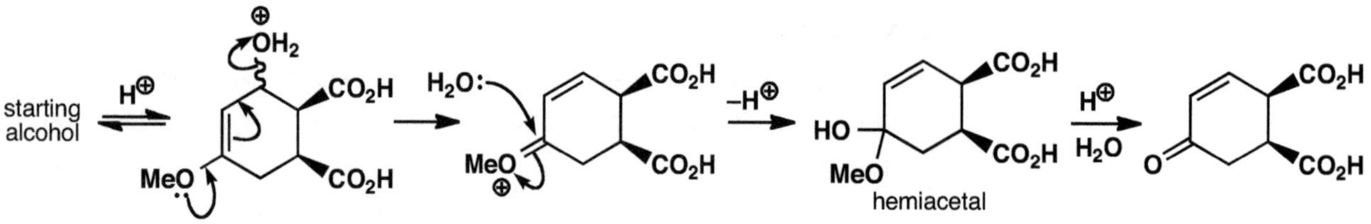

PROBLEM 5

Explain the stereochemistry of the alkenes in the products of these reactions.

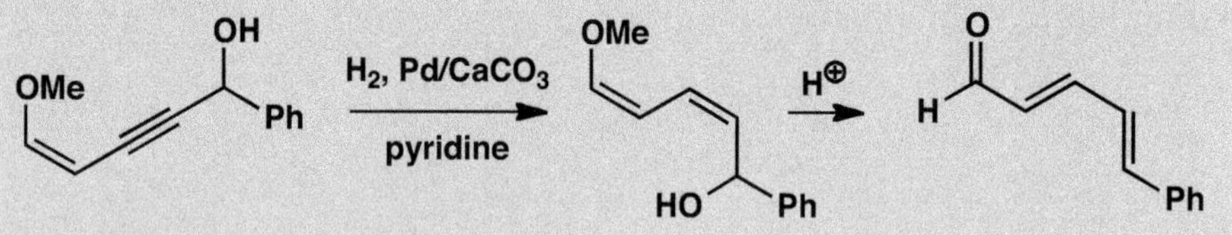

Purpose of the problem

Display your skill in a deceptive example of control of alkene geometry by elimination.

Suggested solution

The first reaction is stereospecific *cis* addition of hydrogen to an alkyne to give the *cis*-alkene. The intermediate is therefore a *cis,cis*-diene and it may seem remarkable that it should become a *trans,trans*-diene on elimination. However, when we draw the mechanism for the elimination, we see that there need be no relationship between the stereochemistry of the intermediate and the product as this is an E1 reaction and the cationic intermediate can rotate into the most stable shape before conversion to the aldehyde.

■ The hydrogenation of alkynes to give *cis* alkenes is described on p. 537 of the textbook.

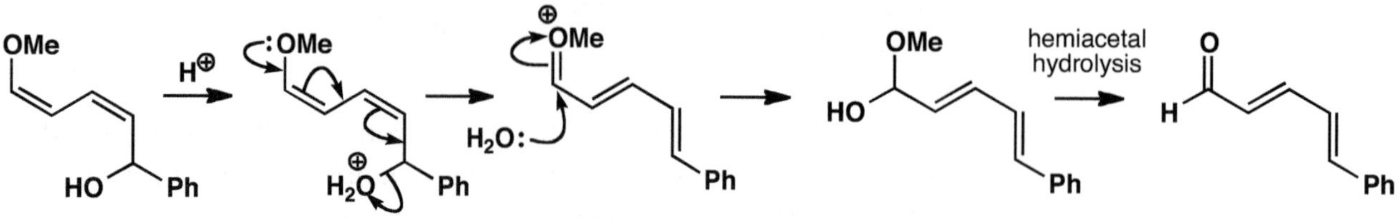

PROBLEM 6

Suggest a mechanism for this reaction and explain why the product is so stable.

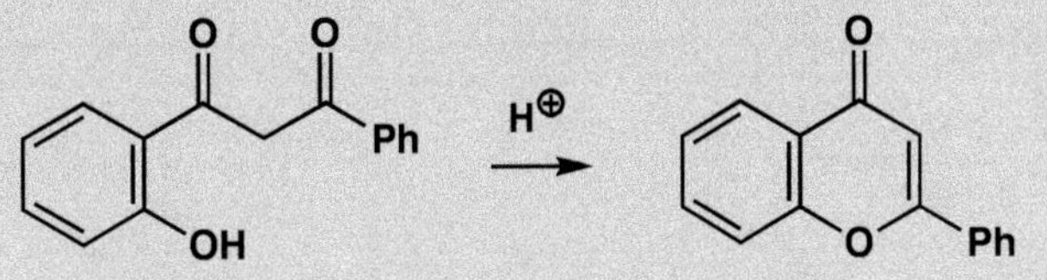

Purpose of the problem

Exploring what might happen on the way to an elimination and explaining special stability.

Suggested solution

The obvious place to start is cyclization of the phenol onto a ketone to form a six-membered ring. The product is a hemiacetal that will surely eliminate by a combination of hemiacetal hydrolysis and the E1cB mechanism.

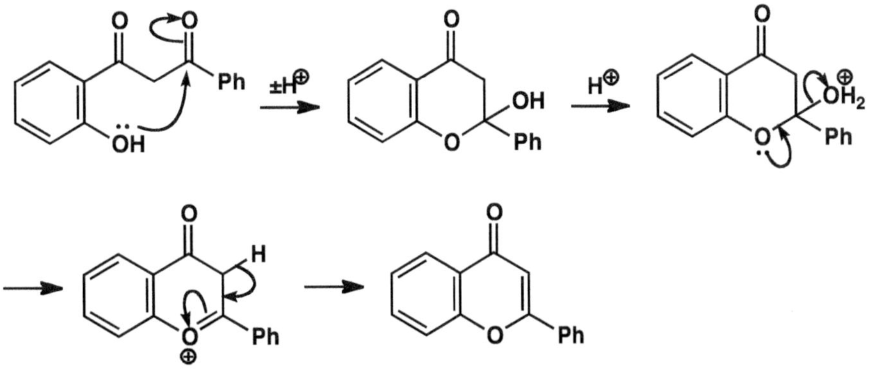

■ If you have already read chapter 20 you may have preferred to form the enol of the remaining ketone and eliminate directly.

The final product is particularly stable as the right hand ring is aromatic. It has two alkenes and a lone pair on oxygen, making six electrons in all. If you prefer you can show the delocalization to make the ring more benzene-like.

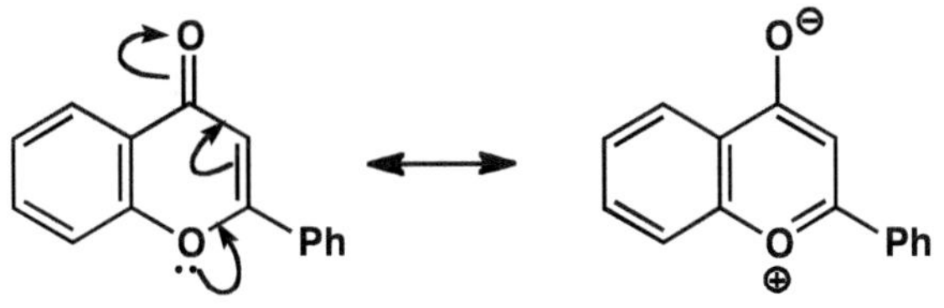

PROBLEM 7

Comment on the position taken by the alkene in these eliminations.

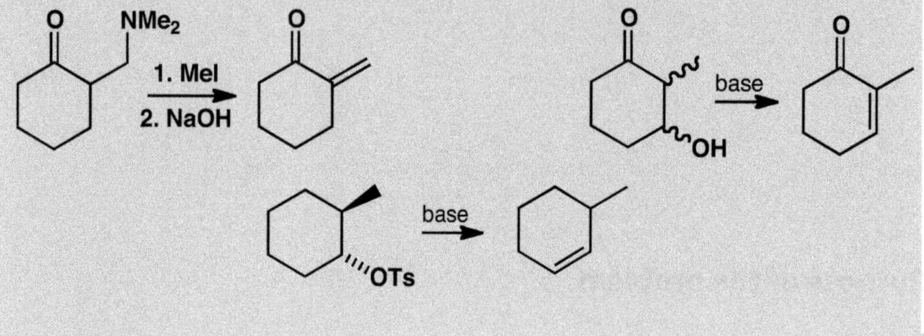

Purpose of the problem

Further exploration of the site occupied by the alkene after an elimination.

Suggested solution

The first is an E1cB reaction after methylation makes the amine into a leaving group. The alkene has to go where the amine was (and in conjugation with the ketone).

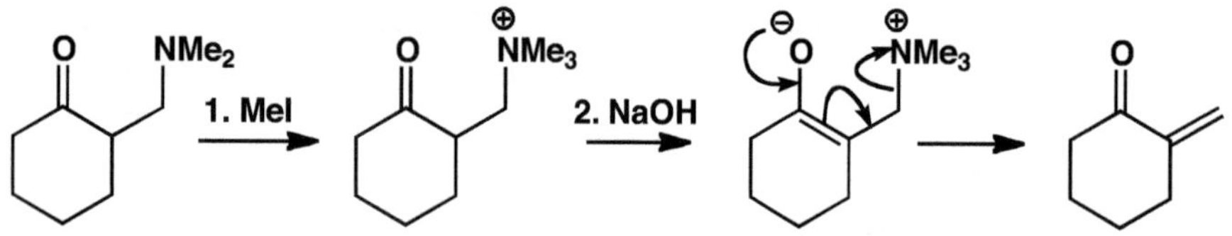

The second is also E1cB and so the alkene must end up conjugated with the ketone. But this time the leaving group is on the ring so that is where the alkene goes. The stereochemistry is irrelevant as the enolate has lost one chiral centre and there is no requirement in E1cB for H and OH to be antiperiplanar.

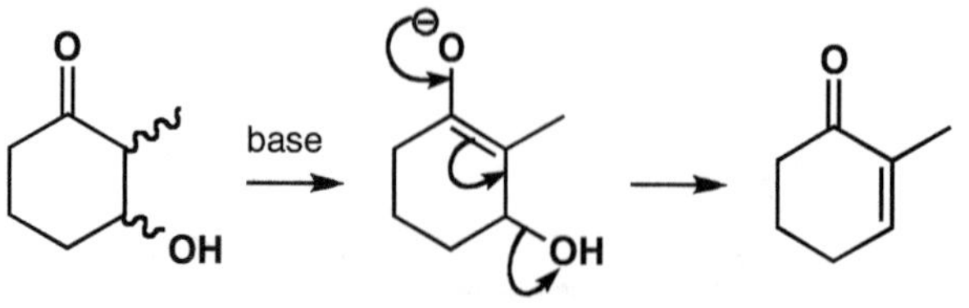

The third is an E2 reaction so there *is* now a requirement for H and Br to be *anti*-periplanar. This means that the Br must be axial and only one hydrogen is then in the right place.

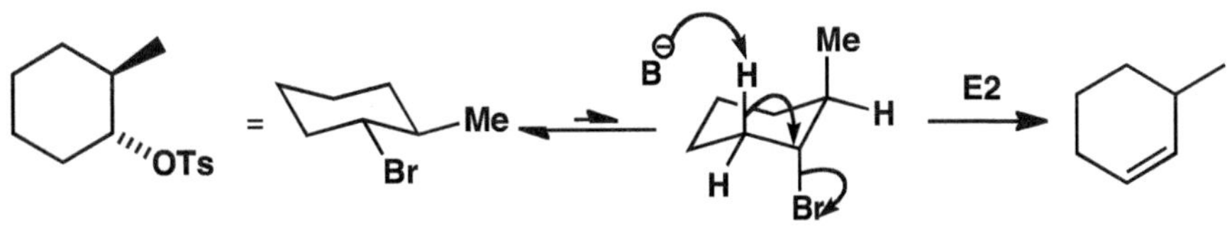

PROBLEM 8

Why is it difficult (though not impossible) for cyclohexyl bromide to undergo an E2 reaction? What conformational changes must occur during this reaction?

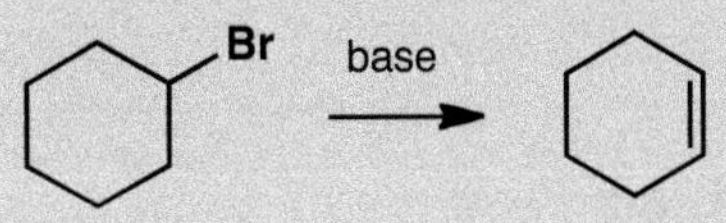

Purpose of the problem

Simple exploration of the relationship between conformation and mechanism.

Suggested solution

Cyclohexyl bromide prefers the chair conformation with the bromine equatorial. It cannot do an E2 reaction in this conformation as E2 requires the reacting C–H and C–Br bonds to be anti-periplanar. This can be achieved if the molecule first flips to put the C–Br bond in an unfavourable axial conformation.

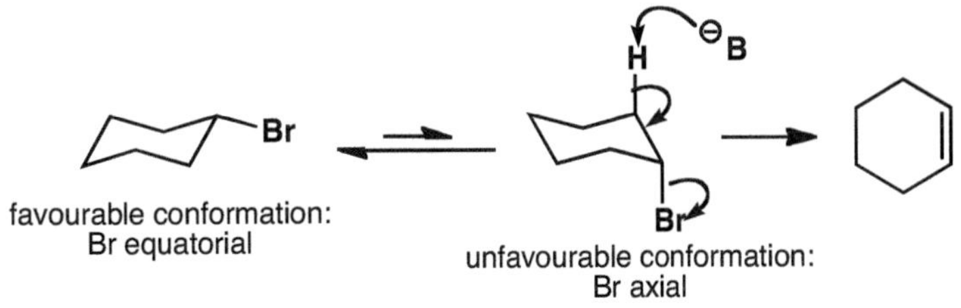

favourable conformation:
Br equatorial

unfavourable conformation:
Br axial

PROBLEM 9

Only one of these bromides eliminates to give alkene A. Why? Neither alkene eliminates to give alkene B. Why not?

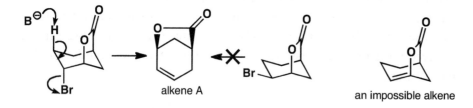

Purpose of the problem

Helping you to understand that cage molecules often have restricted opportunity for elimination.

Suggested solution

The first molecule has one H antiperiplanar to the Br atom so elimination can occur. The second has no hydrogens antiperiplanar to Br. Alkene B is a bridgehead alkene and cannot exist (see the textbook, pp. 389–390).

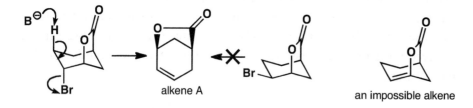

PROBLEM 10

Account for the constrasting results of these two reactions.

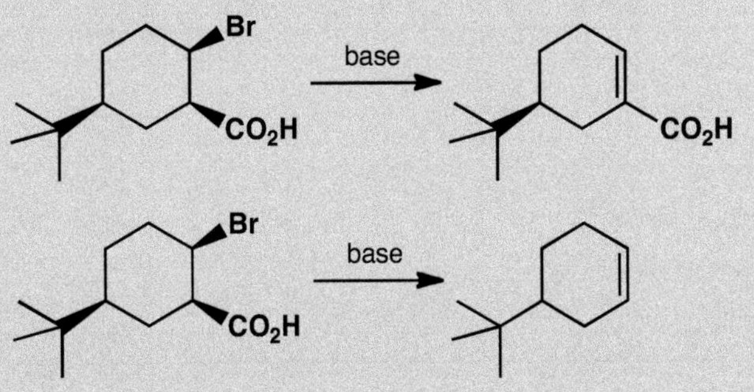

Purpose of the problem

How configuration controls mechanism, and an alternative type of elimination.

Suggested solution

The two compounds differ only in their configuration, and as they both have a *tert*-butyl group they have no choice about their conformation. The bromide must be the leaving group, and when you draw the molecules you find that it must also be axial. In the first case there is a proton antiperiplanar to it that can lead to a conjugated alkene. In the second case, the bond antiperiplanar to the bromine is a C–C bond, but that's OK on this occasion because decarboxylation can take place by the mechanism shown. There is an antiperiplanar C–H bond on the other side of course, but the decarboxylation must be faster than simple E2 elimination.

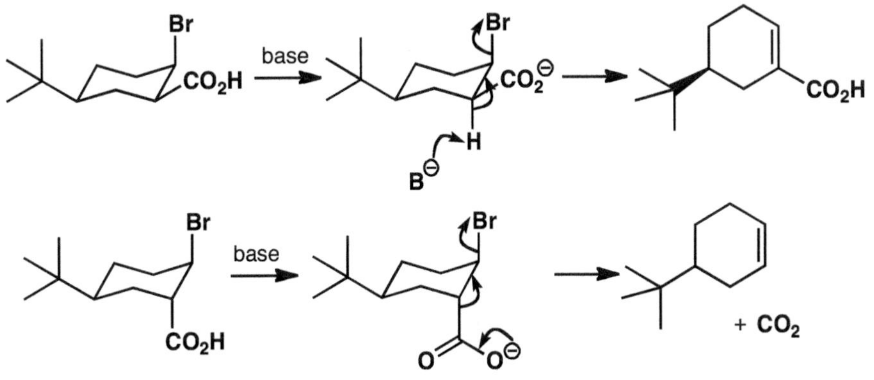

Review of spectroscopic methods

18

Connections

➡ Building on

- Mass spectrometry ch3
- Infrared spectroscopy ch3
- ^{13}C NMR ch3
- ^{1}H NMR ch13
- Stereochemistry ch14
- Conformation ch16
- Elimination ch17
- Carbonyl chemistry ch10 and ch12

Arriving at

- How spectroscopy explains the reactions of the C=O group
- How spectroscopy tells us about the reactivity of, and reaction products from, conjugated C=C and C=O bonds
- How spectroscopy tells us about the size of rings
- How spectroscopy solves the structure of unknown compounds
- Some guidelines for solving unknown structures

➡ Looking forward to

- A final review of spectroscopy, including what it tells us about the stereochemistry of molecules ch31
- Spectroscopy is an essential tool and will be referred to throughout the rest of the book

This is the first of two review chapters on spectroscopic methods taken as a whole. In Chapter 31 we shall tackle the complete identification of organic compounds, including the vital aspect of stereochemistry, introduced in Chapters 14 and 17. In this chapter we gather together some of the ideas introduced in previous chapters on spectroscopy and mechanism, and show how they are related. We shall explain the structure of the chapter as we go along.

There are three reasons for this chapter

1. To review the methods of structure determination we met in Chapters 3 and 13, to extend them a little further, and to consider the relationships between them.
2. To show how these methods may be combined to determine the structure of unknown molecules.
3. To provide useful tables of data for you to use when you are attempting to determine unknown structures.

The main tables of data appear at the end of the chapter (pp. 423–425) so that they are easy to refer to when you are working on problems. You may also wish to look at them, along with the tables in the text, as you work through this chapter.

We shall deal with points 1 and 2 together, looking first at the interplay between the chemistry of the carbonyl group (as discussed in Chapters 10 and 11) and spectroscopy, solving some structural problems, then moving on to discuss, for example, NMR of more

Online support. The icon 🔖 in the margin indicates that accompanying interactive resources are provided online to help your understanding: just type **www.chemtube3d.com/clayden/123** into your browser, replacing **123** with the number of the page where you see the icon. For pages linking to more than one resource, type **123-1, 123-2** etc. (replacing **123** with the page number) for access to successive links.

than one element in the same compound, doing some more problems, and so on. We hope that the lessons from each section will help in your overall understanding of structure solving. The first section deals with the assignment of carbonyl compounds to their various classes.

Spectroscopy and carbonyl chemistry

Chapters 10 and 11 completed our systematic survey of carbonyl chemistry, and we can now put together chemistry and spectroscopy on this most important of all functional groups.

We have divided carbonyl compounds into two main groups:

1. **aldehydes** (RCHO) and **ketones** (R^1COR^2)
2. **acids** (RCO_2H) and their derivatives (in order of reactivity):

    acid chlorides (RCOCl)

    anhydrides (RCO_2COR)

    esters ($R^1CO_2R^2$)

    amides ($RCONH_2$, R^1CONMe_2, etc.).

Which spectroscopic methods most reliably distinguish these two groups? Which help us to separate aldehydes from ketones? Which allow us to distinguish the various acid derivatives? Which offer the most reliable evidence on the chemistry of the carbonyl group? These are the questions we tackle in this section.

Distinguishing aldehydes and ketones from acid derivatives

The most consistently reliable method for doing this is ^{13}C NMR. It doesn't much matter whether the compounds are cyclic or unsaturated or have aromatic substituents, they all give carbonyl ^{13}C shifts in about the same regions. There is a selection of examples on the facing page which we now discuss. First, look at the shifts arrowed into the carbonyl group on each structure. All the aldehydes and ketones fall between 191 and 208 ppm regardless of structure, whereas all the acid derivatives (and these are very varied indeed!) fall between 164 and 180 ppm. These two sets do not overlap and the distinction is easily made. Assigning the spectrum of the ketoacid in the margin, for example, is easy.

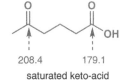

208.4   179.1

saturated keto-acid

● ^{13}C NMR distinguishes acid derivatives from aldehydes and ketones

The carbonyl carbons of all aldehydes and ketones resonate at about 200 ppm, while acid derivatives usually resonate at about 175 ppm.

^{13}C NMR shifts of carbonyl groups

Carbonyl group	δ_C, ppm
aldehydes	195–205
ketones	195–215
acids	170–185
acid chlorides	165–170
acid anhydrides	165–170
esters	165–175
amides	165–175

aldehydes

H 9.80

MeO

HO 191.0

aromatic aldehyde: vanillin

10.2 H

190.9

conjugated unsaturated aldehyde: all *trans* retinal

ketones 199.5

154.3

HO

cyclic conjugated ketone: (−)-carvone

208.8

saturated ketone: raspberry ketone

acids

180.1

saturated: lipoic (thioctic) acid

HO HO OH 160.0

conjugated: shikimic acid

172.2

aromatic: salicylic acid

181.4

non-conjugated: ibuprofen

acid chlorides

saturated: acetyl chloride 170.2

conjugated unsaturated 122.6

163.6 170.2

anhydrides

two signals 169.5 177.2

saturated cyclic

164.2

unsaturated conjugated cyclic: maleic anhydride

171.3

saturated cage tricyclic

esters

167.9

conjugated: methyl methacrylate

OMe

H$_2$N 165

ester of aromatic acid: benzocaine

OEt

amides

Me$_2$N H 167.9

simple amide: dimethyl formamide (DMF)

H$_2$N N N N OH

tetrapeptide: L-Ala-L-Ala-L-Ala-L-Ala
four C=O signals:168.9. 171.6. 171.8, 173.8

More on these structures

Aldehydes and ketones

The first aldehyde is vanillin, which comes from the vanilla pod and gives the characteristic vanilla flavour in, for example, ice cream. Vanilla is the seed pod of a South American orchid. 'Vanilla essence' is made with synthetic vanillin and tastes slightly different because the vanilla pod contains other flavour components in small quantities. The second aldehyde is retinal. As you look at this structure your eyes use the light reaching them to interconvert *cis* and *trans* retinal in your retina to create nervous impulses (see also Chapter 27).

The two ketones are all flavour compounds too. The first, (−)-carvone, is the chief component (70%) of spearmint oil. Carvone is an interesting compound: in Chapter 14 you met the mirror-image isomers known as enantiomers, and (−)-carvone's mirror image, (+)-carvone, is the chief component (35%) of dill oil. Our taste can tell the difference, although an NMR machine can't and both carvones have *identical NMR spectra*. See Chapter 14 for more detail! The second ketone is 'raspberry ketone', which is largely responsible for the flavour of raspberries. It is entirely responsible for the flavour of some 'raspberry' foods. The signal for the aromatic carbon joined to OH is at 154.3 ppm (in the 100–150 ppm region because it is an unsaturated carbon atom joined to oxygen) and cannot possibly be confused with the ketone signal at 208.8 ppm. Both ketones have C=O shifts at about 200 ppm, and both lack any signals in the proton NMR of $\delta > 8$.

Acid derivatives

Lipoic acid uses its S–S bond in redox reactions (Chapter 42), while shikimic acid is an intermediate in the formation of compounds with benzene rings, such as phenylalanine, in living things (Chapter 42). Salicylic acid's acetate ester is aspirin, which is, of course, like the last example ibuprofen, a painkiller.

The first acid chloride is a popular reagent for the synthesis of acetate esters and you have seen its reactions in Chapter 10. We have chosen three cyclic anhydrides as examples because they are all related to an important reaction (the Diels–Alder reaction), which you will meet in Chapter 34.

The first ester, methyl methacrylate, is a bulk chemical. It is the monomer whose polymerization gives Perspex, the rigid transparent plastic used in windows and roofs. The second ester is an important local anaesthetic used for minor operations.

One amide is the now-familiar DMF, but the other is a tetrapeptide and so contains one carboxylic acid group at the end and three amide groups. Although the four amino acids in this peptide are identical (alanine, Ala for short), the carbon NMR faithfully picks up four different C=O signals, all made different by being different distances from the end of the chain.

The distinction can be vital in structural problems. The symmetrical alkyne diol below cyclizes in acid with Hg(II) catalysis to a compound having, by proton NMR, the structural fragments shown. The product is unsymmetrical in that the two CMe$_2$ groups are still present, but they are now different. In addition, the chemical shift of the CH$_2$ group shows that it is next to C=O but not next to oxygen. This leaves us with two possible structures. One is an ester and one a ketone. The C=O shift is 218.8 ppm and so there is no doubt that the second structure is correct.

■ You need not, at this stage, worry about *how* the reaction works. It is more important that you realize how spectroscopy enables us to work out *what* has happened even before we have any idea *how*. Nonetheless, it is true that the second structure here also makes more sense chemically as the carbon skeleton is the same as in the starting material.

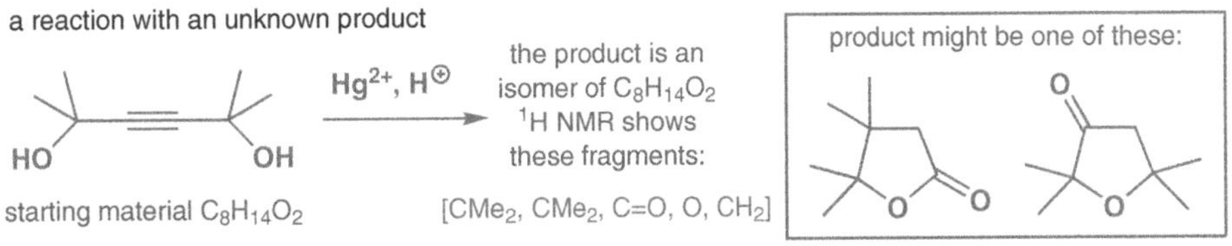

a reaction with an unknown product

starting material C$_8$H$_{14}$O$_2$

the product is an isomer of C$_8$H$_{14}$O$_2$
^{1}H NMR shows these fragments:

[CMe$_2$, CMe$_2$, C=O, O, CH$_2$]

product might be one of these:

Distinguishing aldehydes from ketones is simple by proton NMR

Now look at the first two groups, the aldehydes and ketones. The two aldehydes have smaller carbonyl shifts than the two ketones, but they are too similar for this distinction to be reliable. What distinguishes the aldehydes very clearly is the characteristic proton signal for CHO at 9–10 ppm. So you should identify aldehydes and ketones by C=O shifts in carbon NMR and then separate the two by proton NMR.

● **Aldehyde protons are characteristic**
A proton at 9–10 ppm indicates an aldehyde.

Distinguishing acid derivatives by carbon NMR is difficult

Now examine the other panels on p. 409. The four carboxylic acids are all important biologically or medicinally. Their C=O shifts are very different *from each other* as well as from those of the aldehydes or ketones.

The next five compounds (two acid chlorides and three anhydrides) are all reactive acid derivatives, and the five esters and amides below them are all unreactive acid derivatives and yet the C=O shifts of all ten compounds fall in the same range. The C=O chemical shift is obviously *not* a good way to check on chemical reactivity.

➡ The relative reactivity of carboxylic acid derivatives was discussed in Chapter 10.

What the carbon NMR fails to do is distinguish these types of acid derivative. There is more variation between the carboxylic acids on display than between the different classes of acid derivatives. This should be obvious if we show you some compounds containing two acid derivatives. Would you care to assign these signals?

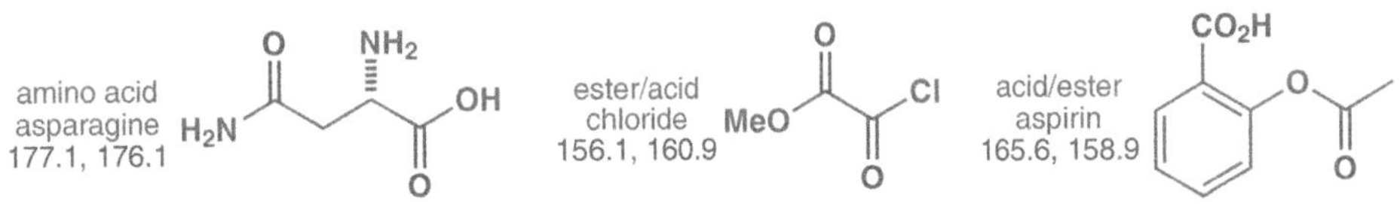

amino acid asparagine 177.1, 176.1

ester/acid chloride 156.1, 160.9

acid/ester aspirin 165.6, 158.9

No, neither would we. In each case the difference between the carbonyl signals is only a few ppm. Although acid chlorides are extremely reactive in comparison with esters or amides, the electron deficiency at the carbon nucleus as measured by deshielding in the NMR spectrum evidently does not reflect this. Carbon NMR reliably distinguishes acid derivatives as a group from aldehydes and ketones as another group but it fails to distinguish even very reactive (for example, acid chlorides) from very unreactive (for example, amides) acid derivatives. So how do we distinguish acid derivatives?

Acid derivatives are best distinguished by infrared

A much better measure is the difference in IR stretching frequency of the C=O group. We discussed this in Chapter 10 (p. 206), where we noted a competition between conjugation by lone-pair electron donation *into* the carbonyl from OCOR, OR, or NH$_2$ and inductive withdrawal *from* the C=O group because of the electronegativity of the substituent. Conjugation donates electrons into the π* orbital of the π bond and so lengthens and weakens it. The C=O bond becomes more like a single bond and its stretching frequency moves towards the single-bond region, that is, it goes *down*. The inductive effect removes electrons from the π orbital and so shortens and strengthens the π bond. It becomes more like a full double bond and moves *up* in frequency.

➡ For a reminder of the distinction between conjugation and inductive effects, see Chapter 8, p. 176.

These effects are balanced in different ways according to the substituent. Chlorine is poor at lone-pair electron donation (its lone pair is in a large 3p orbital and overlaps badly with the 2p orbital on carbon) but strongly electron-withdrawing so acid chlorides absorb at high frequency, almost in the triple-bond region. Anhydrides have an oxygen atom between two carbonyl groups. Inductive withdrawal is still strong but conjugation is weak because the lone pairs are pulled both ways. Esters have a well-balanced combination with the inductive effect slightly stronger (oxygen donates from a compatible 2p orbital but is very electronegative and so withdraws electrons strongly as well). Finally, amides are dominated by conjugation as nitrogen is a much stronger electron donor than oxygen because it is less electronegative.

Acid chlorides	Anhydrides	Esters	Amides
inductive effect dominates	tug-of-war for lone pair: inductive effect dominates	inductive effect slightly dominates	conjugation strongly dominates
1815 cm^{-1}	two peaks: ~1790, 1810 cm^{-1}	1745 cm^{-1}	~ 1650 cm^{-1}

■ The two peaks for anhydrides are the symmetrical and anti-symmetrical stretches for the two C=O groups; see Chapter 3, p. 70.

Conjugation with π electrons or lone pairs affects IR C=O stretches

We need to see how conjugation works when it is with a π bond rather than with a lone pair. This will make the concept more general as it will apply to aldehydes and ketones as well as to acid groups. How can we detect whether an unsaturated carbonyl compound is conjugated or not? Well, compare these two unsaturated aldehydes.

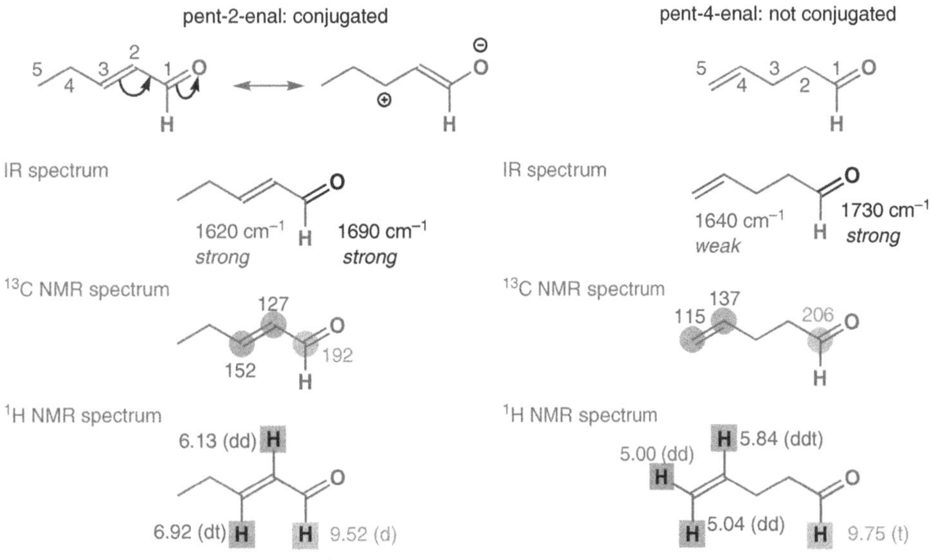

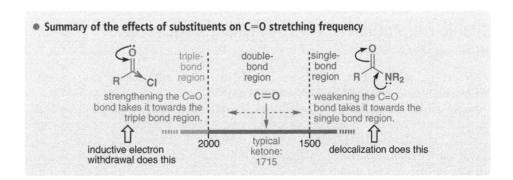

The key differences are the frequency of the C=O stretch (lowered by 40 cm⁻¹ by conjugation) and the strength (that is, the intensity) of the C=C stretch (increased by conjugation) in the IR. In the ^{13}C NMR, C3 in the conjugated enal is moved out of the alkene region just into the carbonyl region, showing how electron-deficient this carbon atom must be. In the proton NMR there are many effects but the downfield shift of the protons on the alkene, especially C3 (again!), is probably the most helpful.

● **Summary of the effects of substituents on C=O stretching frequency**

Because the infrared carbonyl frequencies follow such a predictable pattern, it is possible to make a simple list of correlations using just three factors. Two are the ones we have been discussing—conjugation (frequency-lowering) and the inductive effect (frequency-raising). The third is the effect of small rings and this we next need to consider in a broader context.

Small rings introduce strain inside the ring and higher s character outside it

■ The three-membered ring is, of course, flat. The others are not. Even the four-membered ring is slightly puckered, the five and especially the six-membered rings more so. This is all discussed, along with analysis of ring strain, in Chapter 16.

Cyclic ketones can achieve the perfect 120° angle at the carbonyl group only if the ring is at least six-membered. The smaller rings are 'strained' because the orbitals have to overlap at a less than ideal angle.

For a four-membered ring, the actual angle is 90°, so there is $120° − 90° = 30°$ of strain at the carbonyl group. The effects of this strain on five-, four-, and three-membered rings are shown here.

O
θ = 60°
60° of strain
1813 cm⁻¹

O
θ = 90°
30° of strain
1780 cm⁻¹

O
θ = 108°
12° of strain
1745 cm⁻¹

O
θ = 120°
0° of strain
1715 cm⁻¹

← - increasing strain;
increasing frequency

But why should strain raise the frequency of a carbonyl group? It is evidently shortening and strengthening the C═O bond as it moves it towards the triple-bond region (higher frequency), not towards the single-bond region (lower frequency). In a six-membered ring, the sp² orbitals forming the σ framework around the carbonyl group can overlap perfectly with the sp³ orbitals on neighbouring carbon atoms because the orbital angle and the bond angle are the same. In a four-membered ring the orbitals do not point towards those on the neighbouring carbon atoms, but point too far out, effectively forcing the bonds to be bent and lowering the degree of overlap.

Ideally, we should like the orbitals to have an angle of 90° as this would make the orbital angle the same as the bond angle. In theory it *would* be possible to have a bond angle of 90° if we used pure p orbitals instead of sp² hybrid orbitals. The diagram in the margin shows this hypothetical situation. If we did this, we should leave a pure s orbital for the σ bond to oxygen. This extreme is not possible, but a compromise is. *Some* more p character goes into the ring bonds—maybe they become s⁰·⁸p³·²—so that they can approach the 90° angle needed, and the same amount of extra s character goes into the σ bond to oxygen. The more s character there is in the orbital, the shorter it gets as s orbitals are much smaller than p orbitals.

Simple calculations of C═O stretching frequencies in IR spectra

The best way is to relate all our carbonyl frequencies to those for saturated ketones (1715 cm⁻¹). We can summarize what we have just learned in a table.

Notice in this simple table (for full details you should refer as usual to a specialist book) that the adjustment '30 cm⁻¹' appears quite a lot (−30 cm⁻¹ for both alkene and aryl, for example), that the increment for small rings is 35 cm⁻¹ each time (30 to 65 cm⁻¹ and then 65 to 100 cm⁻¹), and that the extreme effects of Cl and NH_2 are +85 and −85 cm⁻¹, respectively. These effects are additive. If you want to estimate the C═O frequency of a proposed structure, just add or subtract all the adjustments to 1715 cm⁻¹ and you will get a reasonable result.

Lactam C═O stretching frequencies

A further good example is the difference between C═O stretching frequencies in cyclic amides, or lactams. The penicillin class of antibiotics all contain a four-membered ring amide known as a β-lactam. The carbonyl stretching frequency in these compounds is way above the 1680 cm⁻¹ of the six-membered lactam, which is what you might expect for an unstrained amide.

β-lactam in penicillin
1715 cm⁻¹

unstrained lactam
1680 cm⁻¹

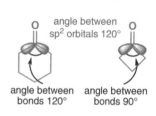

angle between sp² orbitals 120°

angle between bonds 120°

angle between bonds 90°

angle between p orbitals 90°

Incorrect but idealized arrangement

angle between bonds 90°

Effects of substituents on IR carbonyl frequencies

Effect	Group	C═O stretch, cm⁻¹	Frequency change[a], cm⁻¹
inductive effect	Cl	1800	+ 85
	OCOR	1765, 1815	+ 50, +100
	OR	1745	+ 30
	H	1730	+ 15
conjugation	C═C	1685	−30
	aryl	1685	−30
	NH_2	1630	−85
ring strain	five-membered ring	1745	+ 30
	four-membered ring	1780	+ 65
	three-membered ring	1815	+ 100

[a]Difference between stretching frequency of C═O and stretching frequency of a typical saturated ketone (1715 cm⁻¹).

Try this out with the five-membered unsaturated (and conjugated) lactone (cyclic ester) in the margin. We must add 30 cm⁻¹ for the ester, subtract 30 cm⁻¹ for the double bond, and add 30 cm⁻¹ for the five-membered ring. Two of those cancel out, leaving just $1715 + 30 = 1745$ cm⁻¹. These compounds absorb at 1740–1760 cm⁻¹. Not bad!

NMR spectra of alkynes and small rings

δ_H 0.22

δ_H 0.63

δ_H −0.44

This idea that small rings have more p character in the ring and more s character outside the ring also explains the effects of small rings on proton NMR shifts. These hydrogens, particularly on three-membered rings, resonate at unusually high fields, between 0 and 1 ppm in cyclopropanes instead of the 1.3 ppm expected for CH_2 groups, and may even appear at negative δ values. High p character in the framework of small rings also means high s character in C–H bonds outside the ring and this will mean shorter bonds, greater shielding, and small δ values.

Three-membered rings and alkynes

You have also seen the same argument used in Chapter 8 to justify the unusual acidity of C–H protons on triple bonds (such as alkynes and HCN), and alluded to in Chapter 3 to explain the stretching frequency of the same C–H bonds. Like alkynes, three-membered rings are also unusually easy to deprotonate in base.

Here is an example where deprotonation occurs at a different site in two compounds identical except for a C–C bond closing a three-membered ring. The first is an ortholithiation of the type discussed in Chapter 24.

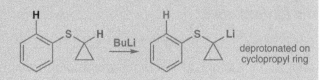

deprotonated on benzene ring (ortholithiated)

deprotonated on cyclopropyl ring

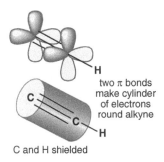

two π bonds make cylinder of electrons round alkyne

C and H shielded

Now what about the NMR spectra of alkynes? By the same argument, protons on alkynes ought to appear in the NMR at quite high field because the C atom is sp hybridized, so it makes its σ bonds with sp orbitals (i.e. 50% s character). Protons on a typical alkene have δ_H about 5.5 ppm, while the proton on an alkyne comes right in the middle of the protons on saturated carbons at about δ_H 2–2.5 ppm This is rather a large effect just for increased s character and some of it is probably due to better shielding by the triple bond, which surrounds the linear alkyne with π bonds without a nodal plane.

This means that the carbon atoms also appear at higher field than expected, not in the alkene region but from about δ_C 60–80 ppm. The s character argument is important, however, because shielding can't affect IR stretching frequencies, yet C≡C–H stretches are strong and at about 3300 cm⁻¹, just right for a strong C–H bond.

➡ In Chapter 13, p. 296, you saw that bonds aligned in a 'W' arrangement can give rise to a small $^4J_{HH}$ coupling.

A simple example is the ether 3-methoxyprop-1-yne. Integration alone allows us to assign the spectrum and the 1H signal at 2.42 ppm, the highest field signal, is clearly the alkyne proton. Notice also that it is a triplet and that the OCH_2 group is a doublet. This $^4J_{HH}$ is small (about 2 Hz) and, although there is nothing like a letter 'W' in the arrangement of the bonds, coupling of this kind is often found in alkynes.

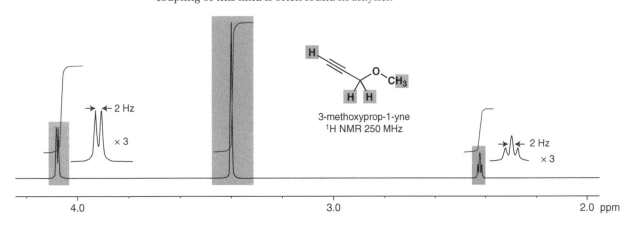

3-methoxyprop-1-yne
¹H NMR 250 MHz

2 Hz
× 3

2 Hz
× 3

4.0   3.0   2.0 ppm

A more interesting example comes from the base-catalysed addition of methanol to buta-1,3-diyne (diacetylene). The compound formed has one double and one triple bond and the ^{13}C NMR shows clearly the greater deshielding of the double bond.

You may have noticed that we have drawn the double bond with the *cis* (Z) configuration. We know that this is true because of the proton NMR, which shows a 6.5 Hz coupling between the two alkene protons (much too small for a *trans* coupling; see p. 295). There is also the longer-range coupling ($^4J = 2.5$ Hz) just described and even a small very long-range coupling ($^5J = 1$ Hz) between the alkyne proton and the terminal alkene proton.

Proton NMR distinguishes axial and equatorial protons in cyclohexanes

Coupling is a through-bond phenomenon, as we know from the couplings in *cis* and *trans* alkenes, where *trans* alkenes have much larger coupling constants as their orbitals are perfectly parallel. Another case of perfectly parallel orbitals occurs with *trans*-diaxial protons in cyclohexanes. Typical coupling constants are 10–12 Hz for *trans*-diaxial protons, but much smaller (2–5 Hz) for axial/equatorial and equatorial/equatorial protons.

➡ Coupling in alkenes is explained on p. 295.

This makes assignment of conformation easy. The simple ester below has a triple triplet for the black H, with two large coupling constants (8.8 Hz) that must be to axial protons (green) and two small coupling constants (3.8 Hz) that must be to equatorial Hs (brown). This is possible only if the black H is axial and the ester group must therefore be equatorial. The acetal ester on the right is very different: it is a simple triplet with two small coupling constants (3.2 Hz), which is too small for an axial/axial coupling. The only possibility therefore is that the black proton is equatorial, and one of the 3.2 Hz couplings is to its equatorial neighbour, and the other to its axial neighbour. The ester group must be axial in this compound.

➡ Proton–proton coupling in alkenes is discussed in Chapter 13 and the conformation of cyclohexanes is discussed in Chapter 16. The Karplus relationship, explaining precisely what affects the couplings in cyclohexanes, is discussed in Chapter 31.

➡ You will see in Chapter 31 why the ester group might prefer to be axial in this compound.

Interactions between different nuclei can give enormous coupling constants

We have looked at coupling between hydrogen atoms and you may have wondered why we have ignored coupling between other NMR active nuclei. Why does ^{13}C not cause similar couplings? In this section we are going to consider not only couplings between the same kind of nuclei, such as two protons, called **homonuclear coupling**, but also coupling between different nuclei, such as a proton and a fluorine atom or ^{13}C and ^{31}P, called **heteronuclear coupling**.

Two nuclei are particularly important, ^{19}F and ^{31}P, since many organic compounds contain these elements and both are at essentially 100% natural abundance and have spin $I = 1/2$. We shall start with organic compounds that have just one of these nuclei and see what happens to both the ^1H and the ^{13}C spectra. In fact, it is easy to find a ^{19}F or a ^{31}P atom in a molecule because these elements couple to all nearby carbon and hydrogen atoms. Since they can be directly bonded to either, ^1J coupling constants such as $^1J_{CF}$ or $^1J_{PH}$ become possible, as well as the more 'normal' couplings such as $^2J_{CF}$ or $^3J_{PH}$, and these ^1J coupling constants can be enormous.

We shall start with a simple phosphorus compound, the dimethyl ester of phosphorous acid (H_3PO_3). There is an uncertainty about the structure of both the acid and its esters. They could exist as P(III) compounds with a lone pair of electrons on phosphorus, or as P(V) compounds with a P=O double bond.

> ■ Note that these spectra with heteronuclear couplings provide the only cases where we can see *one* doublet in the proton NMR. Normally, if there is one doublet, there must be another signal with at least this complexity as all coupling appears twice (A couples to B and so B also couples to A!). If the coupling is to another element (here phosphorus) then the coupling appears once in each spectrum. The Wittig reagent has an A_3P ($CH_3–P$) system: proton A appears as a doublet, while the phosphorus atom appears as a quartet in the *phosphorus* spectrum at a completely different frequency, but with the same coupling constant measured in Hz.

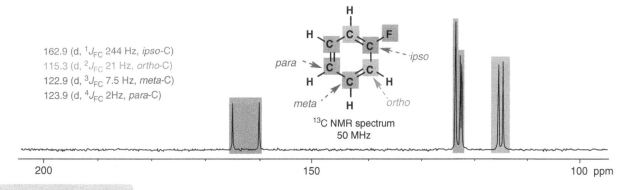

3.80 (6H, d, $^3J_{PH}$ 9 Hz)
6.77 (1H, d, $^1J_{PH}$ 693 Hz)

In fact, dimethyl phosphite has a 1H doublet with the amazing coupling constant of 693 Hz: on a 250 MHz machine the two lines are over 2 ppm apart and it is easy to miss that they are two halves of the same doublet. This can only be a $^1J_{PH}$ as it is so enormous. The compound has to have a P–H bond and the P(V) structure is correct. The coupling to the protons of the methyl group is much smaller but still large for a three-bond coupling ($^3J_{PC}$ of 18 Hz).

Next, consider the phosphonium salt you met at the end of Chapter 11 for use in the Wittig reaction, turning aldehydes and ketones to alkenes. It has a $^2J_{PH}$ of 18 Hz. There is no doubt about this structure—it is just an illustration of coupling to phosphorus. There is coupling to phosphorus in the carbon spectrum too: the methyl group appears at δ_C 10.6 ppm with a $^1J_{PC}$ of 57 Hz, somewhat smaller than typical $^1J_{PH}$. We haven't yet talked about couplings to ^{13}C: we shall now do so.

$Ph_3\overset{\oplus}{P}\!-\!CH_3 \;\; Br^{\ominus}$

methyltriphenylphosphonium
bromide
aromatic protons and
δ_H 3.25 (3H, d, $^2J_{PH}$ 18 Hz)

Coupling in carbon NMR spectra

We shall use coupling with fluorine to introduce this section. Fluorobenzenes are good examples because they have a number of different carbon atoms all coupled to the fluorine atom.

162.9 (d, $^1J_{FC}$ 244 Hz, *ipso*-C)
115.3 (d, $^2J_{FC}$ 21 Hz, *ortho*-C)
122.9 (d, $^3J_{FC}$ 7.5 Hz, *meta*-C)
123.9 (d, $^4J_{FC}$ 2Hz, *para*-C)

^{13}C NMR spectrum
50 MHz

> ■ *Ipso* can join the list (*ortho*, *meta*, *para*) of trivial names for positions on a substituted benzene ring. The *ipso* carbon is the one directly attached to a substituent.

The carbon directly joined to fluorine (the *ipso* carbon) has a very large $^1J_{CF}$ value of about 250 Hz. More distant coupling is evident too: all the carbons in the ring couple to the fluorine in PhF with steadily diminishing J values as the carbons become more distant.

Trifluoroacetic acid is an important strong organic acid (Chapter 8) and a good solvent for ^1H NMR. The carbon atom of the CF_3 group is coupled equally to all the three fluorines and so appears as a quartet with a large $^1J_{CF}$ of 283 Hz, about the same as in PhF. Even the carbonyl group is also a quartet, although the coupling constant is much smaller ($^2J_{CF}$ is 43 Hz). Notice too how far downfield the CF_3 carbon atom is!

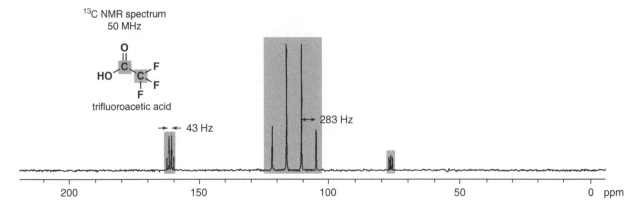

Coupling between protons and ^{13}C

In view of all this, you may ask why we don't apparently see couplings between ^{13}C and ^1H in either carbon or proton spectra. In proton spectra the answer is simple: we don't see coupling to ^{13}C because of the low abundance (1.1%) of ^{13}C. Most protons are bonded to ^{12}C: only 1.1% of protons are bonded to ^{13}C. If you look closely at proton spectra with very flat baselines, you may see small peaks either side of strong peaks at about 0.5% peak height. These are the ^{13}C 'satellites' for those protons that are bonded to ^{13}C atoms.

As an example, look again at the 500 MHz ^1H NMR spectrum of heptan-2-one that we saw on p. 294. When the baseline of this spectrum is vertically expanded, the ^{13}C satellites may be seen. The singlet due to the methyl protons is actually in the centre of a tiny doublet due to the 1% of protons coupling to ^{13}C. Similarly, each of the triplets in the spectrum is flanked by two tiny triplets. The two tiny triplets on either side make up a doublet of triplets with a large ^1J coupling constant to the ^{13}C (around 130 Hz) and smaller ^3J coupling to the two equivalent protons.

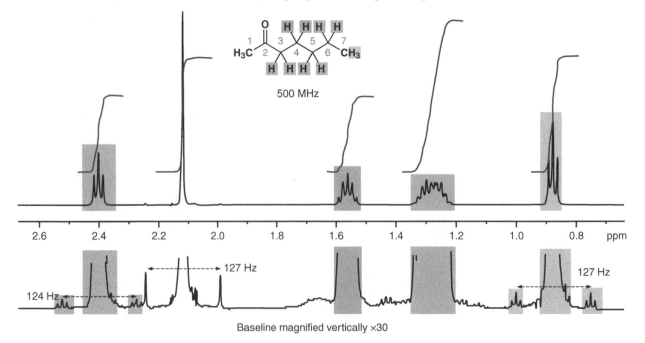

^{13}C satellites are usually lost in the background noise of the spectrum and need concern us no further. You do, however, see coupling in the ^1H NMR spectrum with compounds deliberately labelled with ^{13}C because the ^{13}C abundance can then approach 100%. The same Wittig reagent we saw a moment ago shows a 3H doublet of doublets with the typically enormous $^1J_{CH}$ of 135 Hz when labelled with pure ^{13}C in the methyl group.

$Ph_3\overset{\oplus}{P}$——$^{13}CH_3 \overset{\ominus}{Br}$

^{13}C-labelled phosphonium salt
δ_H 3.25 (3H, dd,
$^1J_{CH}$ 135, $^2J_{PH}$ 18 Hz)

But this begs the question—where is the 135 Hz coupling in the ^{13}C NMR? Surely we should see this coupling to the protons in the ^{13}C NMR spectrum too?

Why is there no coupling to protons in normal ^{13}C NMR spectra?

We get the singlets consistently seen in carbon spectra because of the way we record the spectra. The values of $^1J_{CH}$ are so large that, if we recorded ^{13}C spectra with all the coupling constants, we would get a mass of overlapping peaks. When run on the same spectrometer, the frequency at which ^{13}C nuclei resonate turns out to be about a quarter of that of the protons. Thus a '400 MHz machine' (remember that the magnet strength is usually described by the frequency at which the protons resonate) gives ^{13}C spectra at 100 MHz. Coupling constants ($^1J_{CH}$) of 100–250 Hz would cover 2–5 ppm and a CH$_3$ group with $^1J_{CH}$ of about 125 Hz would give a quartet covering nearly 8 ppm (see the example on the previous page).

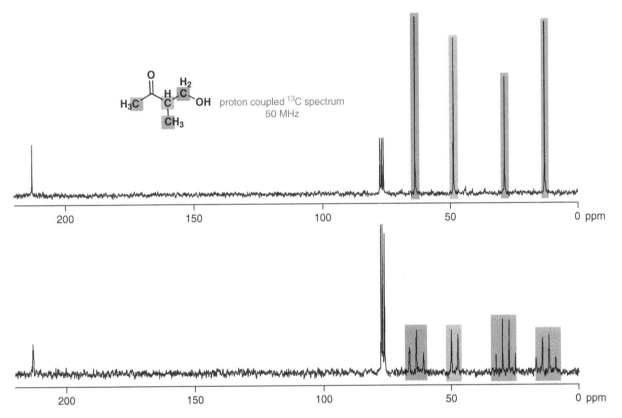

Since the proton-coupled ^{13}C spectrum can so easily help us to distinguish CH$_3$, CH$_2$, CH, and quaternary carbons, you might wonder why they are not used more. The above example was chosen very carefully to illustrate proton-coupled spectra at their best. Unfortunately, this is not a typical example. More usually, the confusion from overlapping peaks makes this just not worthwhile. So ^{13}C NMR spectra are recorded while the whole 10 ppm proton spectrum is being irradiated with a secondary radio frequency source. The proton energy levels are equalized by this process and all coupling disappears. Hence the singlets we are used to seeing.

For the rest of this chapter we shall not be introducing new theory or new concepts; we shall be applying what we have told you to a series of examples where spectroscopy enables chemists to identify compounds.

Identifying products spectroscopically

An ambiguous reaction product

This was the case of diazonamide A (p. 45).

In Chapter 3 we gave an example of a compound which was misidentified because an O atom and an N atom were mistaken for one another, even in the X-ray crystal structure.

Another famous case of ambiguity between structures containing O or N arises in the identification of the product of addition of hydroxylamine (NH_2OH) to a simple enone. This condensation reaction gives a compound with the formula $C_6H_{11}NO$. But what is its structure? We can first of all think about what we expect to happen: it is not always necessary to do this in order to identify a structure, but it can help. Nitrogen is more nucleophilic than oxygen so we might expect it to add first. But will it add directly to the carbonyl group or in the conjugate fashion we shall describe in Chapter 22? Either way, an intermediate will be formed that can cyclize.

conjugate addition by the nitrogen atom of hydroxylamine

direct addition by the nitrogen atom of hydroxylamine

■ Do not be concerned about the details of the mechanisms: note that we have used the '± H⁺' shorthand introduced in Chapter 11, and have abbreviated the mechanism where water is eliminated and the oxime formed—the full mechanism of imine (and oxime) formation can be found in Chapter 11, p. 229. In this chapter, we are much more concerned just with the structure of the products.

The two possible isomeric products were once the subject of a long-running controversy, but with IR and proton NMR spectra of the product, doubt vanished. The IR showed no NH stretch. The NMR showed no alkene proton but did have a CH_2 group at 2.63 ppm. Only the second structure is possible.

We need to look now at a selection of problems of different kinds to show how the various spectroscopic methods can cooperate in structure determination.

Reactive intermediates can be detected by spectroscopy

Some intermediates proposed in reaction mechanisms look so unlikely that it is comforting if they can be isolated and their structure determined. We feel more confident in proposing an intermediate if we are sure that it can really be made. Of course, this is not necessarily evidence that the intermediate is actually formed during reactions and it certainly does *not* follow that the failure to isolate a given intermediate disproves its involvement in a reaction. We shall use ketene as an example.

Ketene looks pretty unlikely! It is $CH_2=C=O$ with two π bonds (C=C and C=O) to the same carbon atom. The orbitals for these π bonds must be orthogonal because the central carbon atom is sp hybridized with two linear σ bonds and two p orbitals at right angles both to the σ bonds and to each other. Can such a molecule exist? When acetone vapour is heated to very high temperatures (700–750 °C) methane is given off and ketene is supposed to be the other product. What is isolated is a ketene dimer ($C_4H_4O_2$) and even the structure of this is in doubt as two reasonable structures can be written.

■ We used this logic in Chapter 15: carbocations were proposed as intermediates in S_N1 reactions long before they were observed spectroscopically, but it was reassuring to be able to see them by NMR once appropriate conditions were devised (see p. 335).

ketene

orthogonal π bonds of ketene

cyclic ester structure for diketene

cyclobuta-1,3-dione structure for diketene

ketene

The spectra fit the ester structure well, but not the more symmetrical diketone structure at all. There are *three* types of proton (cyclobuta-1,3-dione would have just *one*), with allylic coupling between one of the protons on the double bond and the CH_2 group in the ring. The carbonyl group has the shift (185 ppm) of an acid derivative (not that of a ketone, which would be about 200 ppm) and all four carbons are different.

■ The structure of *ketene* is analogous to that of *allene*, discussed in Chapter 7, p. 146. Ketene is isoelectronic (p. 354) with CO_2 and azide, N_3^-.

diketene

¹H NMR spectrum:
4.85 (1H, narrow t, $J \sim 1$)
4.51 (1H, s)
3.90 (2H, d, $J \sim 1$)

¹³C NMR spectrum:
185.1, 147.7, 67.0, 42.4

Ozonolysis or ozonation is the cleavage of an alkene by ozone (O$_3$). The reaction and its mechanism are discussed in Chapters 19 and 34: the only point to note now is that ozone is a powerful oxidant and cleaves the alkene to make two carbonyl compounds. Again, in this chapter we are concerned only with the structure of the products and how these can be determined.

Malonic anhydride cannot be made directly from malonic acid because attempted dehydration of the acid leads to the exotic molecule carbon suboxide C$_3$O$_2$.

malonic acid

$O=C=C=C=O$
carbon suboxide C$_3$O$_2$

Ozonolysis of ketene dimer gives a very unstable compound that can be observed only at low temperatures (–78 °C or below). It has two carbonyl bands in the IR and reacts with amines to give amides, so it looks like an anhydride (Chapter 10). Can it be the previously unknown cyclic anhydride of malonic acid?

The two carbonyl bands are of high frequency, as would be expected for a four-membered ring—using the table on p. 413 we estimate 1715 + 50 cm^{-1} (for the anhydride) + 65 cm^{-1} (for the four-membered ring) = 1830 cm^{-1}. Both the proton and the carbon NMR are very simple: just a 2H singlet at 4.12 ppm, shifted downfield by two carbonyls, a C=O group at 160 ppm, right for an acid derivative, and a saturated carbon shifted downfield but not as much as a CH$_2$O group.

All this is reasonably convincing, and is confirmed by allowing the anhydride to warm to –30 °C, at which temperature it loses CO$_2$ (detected by the ^{13}C peak at 124.5 ppm) and gives another unstable compound with the strange IR frequency of 2140 cm^{-1}. Could this be monomeric ketene? It's certainly not either of the possible ketene dimers as we know what their spectra are like, and this is quite different: just a 2H singlet at 2.24 ppm and ^{13}C peaks at 194.0 and, remarkably, 2.5 ppm. It is indeed monomeric ketene.

Squares and cubes: molecules with unusual structures

Some structures are interesting because we believe they can tell us something fundamental about the nature of bonding while others are a challenge because many people argue that they cannot be made. What do you think are the prospects of making cyclobutadiene, a conjugated four-membered ring, or the hydrocarbons tetrahedrane and cubane, which have, respectively, the shapes of the perfectly symmetrical Euclidean solids, the tetrahedron and the cube?

With four electrons, cyclobutadiene is anti-aromatic—it has 4n instead of 4n + 2 electrons. You saw in Chapter 7 that cyclic conjugated systems with 4n electrons (cyclooctatetraene, for example) avoid being conjugated by puckering into a tub shape. Cyclobutadiene cannot do this: it must be more or less planar, and so we expect it to be very unstable. Tetrahedrane has four fused three-membered rings. Although the molecule is tetrahedral in shape, each carbon atom is nowhere near a tetrahedron, with three bond angles of 60°. Cubane has six fused four-membered rings and is again highly strained.

In fact, cubane has been made, cyclobutadiene has a fleeting existence but can be isolated as an iron complex, and a few substituted versions of tetrahedrane have been made. The most convincing evidence that you have made any of these three compounds would be the extreme simplicity of the spectra. Each has only one kind of hydrogen and only one kind of carbon. They all belong to the family (CH)$_n$.

Cubane has a molecular ion in the mass spectrum at 104, correct for C$_8$H$_8$, only CH stretches in the IR at 3000 cm^{-1}, a singlet in the proton NMR at 4.0 ppm, and a single line in the carbon

cyclobutadiene

tetrahedrane

cubane

NMR at 47.3 ppm. It is a very symmetrical molecule and a stable one in spite of all those four-membered rings.

Stable compounds with a cyclobutadiene and a tetrahedrane core can be made if each hydrogen atom is replaced by a *t*-butyl group. The very large groups round the edge of the molecule repel each other and hold the inner core tightly together. Now another difficulty arises—it is rather hard to tell the compounds apart. They both have four identical carbon atoms in the core and four identical *t*-butyl groups round the edge. The starting material for a successful synthesis of both was the tricyclic ketone below identified by its strained C=O stretch and partly symmetrical NMR spectra. When this ketone was irradiated with UV light (indicated by '*hv*' in the scheme), carbon monoxide was evolved and a highly symmetrical compound (*t*-BuC)₄ was formed. But which compound was it?

IR 1762 (C=O) cm⁻¹
δ_H 1.37 (18H, s),
1.27 (18H, s)
δ_C 188.7 (C=O), 60.6,
33.2, 33.1, 31.0,
30.2, 29.3

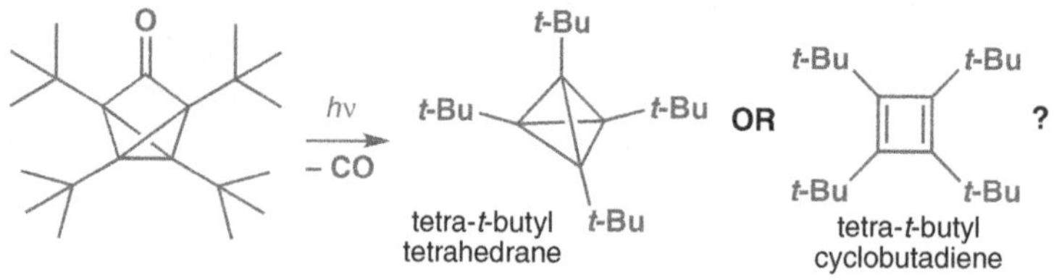

tetra-*t*-butyl tetrahedrane

tetra-*t*-butyl cyclobutadiene

➡ You can read more about the synthesis of cubane in Chapter 36, when we discuss the rearrangement reactions that were used to make it.

The story is made more complicated (but in the end easier!) by the discovery that this compound on heating turned into another very similar compound. There are only two possible structures for (*t*-BuC)₄, so clearly one compound must be the tetrahedrane and one the cyclobutadiene. The problem simplifies with this discovery because it is easier to distinguish two possibilities when you can make comparisons between two sets of spectra. Here both compounds gave a molecular ion in the mass spectrum, neither had any interesting absorptions in the IR, and the proton NMRs could belong to either compound as they simply showed four identical *t*-Bu groups. So did the carbon NMR, of course, but it showed the core too. The first product had only saturated carbon atoms, while the second had a signal at 152.7 ppm for the unsaturated carbons. The tetrahedrane is formed from the tricyclic ketone on irradiation but it isomerizes to the cyclobutadiene on heating.

Identifying compounds from nature

The next molecules we need to know how to identify are those discovered from nature—natural products. These often have biological activity and many useful medicines have been discovered this way. We shall look at a few examples from different fields. The first is the sex pheromone of the Trinidad butterfly *Lycorea ceres ceres*. The male butterflies start courtship by emitting a tiny quantity of a volatile compound. Identification of this type of compound is very difficult because of the minute amounts available but this compound was crystallized and gave enough for a mass spectrum and an IR. The highest peak in the mass spectrum was at 135. This is an odd number so we might have one nitrogen atom and a possible composition of C₈H₉ON. The IR showed a carbonyl peak at 1680 cm⁻¹. With only this meagre information, the first proposals were for a pyridine aldehyde.

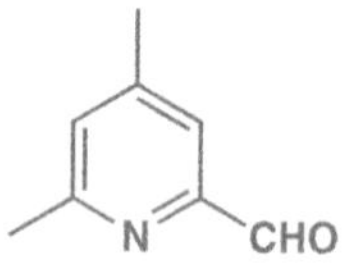

Eventually a little more compound (6 mg!) was available and a proton NMR spectrum was run. This showed at once that this structure was wrong. There was no aldehyde proton and only one methyl group. More positive information was the pair of triplets showing a –CH₂CH₂– unit between two electron-withdrawing groups (N and C=O?) and the pair of doublets for neighbouring protons on an aromatic ring, although the chemical shift and the coupling constant are both rather small for a benzene ring.

If we look at what we have got so far, we see that we have accounted for four carbon atoms in the methyl and carbonyl groups and the –CH₂CH₂– unit. This leaves only four carbon atoms for the aromatic ring. We must use nitrogen too as the only possibility is a pyrrole ring. Our fragments are now those shown below (the black dotted lines show joins to another fragment). These account for all the atoms in the molecule and suggest structures such as these.

possible structure for *Lycorea* sex pheromone

➡ Pyrrole was introduced in Chapter 7, p. 162.

put these fragments together to get structures such as these possible structures for *Lycorea* pheromone:

Now we need to use the known chemical shifts and coupling constants for these sorts of molecules. An N–Me group would normally have a larger chemical shift than 2.2 ppm so we prefer the methyl group on a carbon atom of the pyrrole ring. Typical shifts and coupling constants around pyrroles are shown below. Chemists do not, of course, remember these numbers; we look them up in tables. Our data, with chemical shifts of 6.09 and 6.69 ppm and a coupling constant of 2.5 Hz, clearly favour hydrogen atoms in the 2 and 3 positions, and suggest this structure for the sex pheromone, which was confirmed by synthesis and is now accepted as correct.

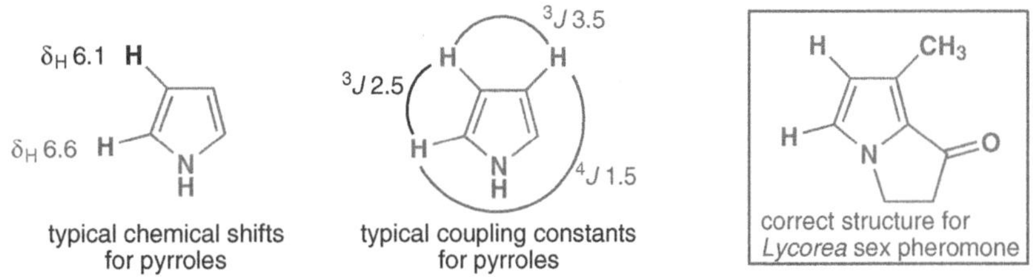

typical chemical shifts for pyrroles

typical coupling constants for pyrroles

correct structure for *Lycorea* sex pheromone

Tables

The final section of this chapter contains some tables of NMR data, which we hope you will find useful in solving problems. In Chapter 13 there were a few guides to chemical shift—summaries of patterns that you might reasonably be expected to remember. But we have left the main selections of hard numbers—tables that *you are not expected to remember*—until now. There are a few comments to explain the tables, but you will probably want to use this section as reference rather than bedtime reading. The first four tables give detailed values for various kinds of compounds and the final table gives a simple summary. We hope that you will find this last table particularly useful.

Effects of electronegativity

This table shows how the electronegativity of the atom attached directly to a methyl group affects the shifts of the CH_3 protons (δ_H) and the CH_3 carbon atom (δ_C) in their NMR spectra.

Chemical shifts of methyl groups attached to different atoms

Element	Electronegativity	Compound	δ_H, ppm	δ_C, ppm
Li	1.0	CH_3-Li	−1.94	−14.0
Si	1.9	CH_3-SiMe_3	0.0	0.0
I	2.7	CH_3-I	2.15	−23.2
S	2.6	CH_3-SMe	2.13	18.1
N	3.1	CH_3-NH_2	2.41	26.9
Cl	3.2	CH_3-Cl	3.06	24.9
O	3.4	CH_3-OH	3.50	50.3
F	4.0	CH_3-F	4.27	75.2

Effects of functional groups

Many substituents are more complicated than just a single atom and electronegativity is only part of the story. We need to look at all the common substituents and see what shifts they cause relative to the CH skeleton of the molecule. Our zero really ought to be at about 0.9 ppm for protons and at 8.4 ppm for carbon, that is, where ethane (CH_3–CH_3) resonates, and not at the arbitrary zero allocated to Me_4Si. In the table below we give such a list. The reason for this is that the shifts (from Me_4Si) themselves are not additive but the shift differences (from 0.9 or 8.4 ppm) are.

Chemical shifts of methyl groups bonded to functional groups

	Functional group	Compound	δ_H, ppm	δ_H – 0.9, ppm	δ_C, ppm	δ_C – 8.4, ppm
1	silane	Me_4Si	0.0	−0.9	0.0	−8.4
2	alkane	Me–Me	0.86	0.0	8.4	0.0
3	alkene	$Me_2C{=}CMe_2$	1.74	0.84	20.4	12.0
4	benzene	Me–Ph	2.32	1.32	21.4	13.0
5	alkyne	Me–C≡C–R[a]	1.86	0.96		
6	nitrile	Me–CN	2.04	1.14	1.8	−6.6
7	acid	Me–CO_2H	2.10	1.20	20.9	11.5
8	ester	Me–CO_2Me	2.08	1.18	20.6	11.2
9	amide	Me–CONHMe	2.00	1.10	22.3	13.9
10	ketone	$Me_2C{=}O$	2.20	1.30	30.8	21.4
11	aldehyde	Me–CHO	2.22	1.32	30.9	21.5
12	sulfide	Me_2S	2.13	1.23	18.1	9.7
13	sulfoxide	$Me_2S{=}O$	2.71	1.81	41.0	32.6
14	sulfone	Me_2SO_2	3.14	2.24	44.4	36.0
15	amine	Me–NH_2	2.41	1.51	26.9	18.5
16	amide	MeCONH–Me	2.79	1.89	26.3	17.9
17	nitro	Me–NO_2	4.33	3.43	62.5	53.1
18	ammonium salt	$Me_4N^+Cl^-$	3.20	2.10	58.0	49.6
19	alcohol	Me–OH	3.50	2.60	50.3	44.3
20	ether	Me–OBu	3.32	2.42	58.5	50.1
21	enol ether	Me–OPh	3.78	2.88	55.1	46.7
22	ester	Me–CO_2Me	3.78	2.88	51.5	47.1
23	phosphonium salt	Ph_3P^+–Me	3.22	2.32	11.0	2.2

[a]R=CH_2OH; compound is but-2-yn-1-ol.

The effects of groups based on carbon (the methyl group is joined directly to another carbon atom) appear in entries 2 to 11. All the electron-withdrawing groups based on carbonyl and cyanide have about the same effect (1.1–1.3 ppm downfield shift from 0.9 ppm). Groups based on nitrogen (Me–N bond) show a similar progression through amine, ammonium salt, amide, and nitro compound (entries 15–18). Finally, all the oxygen-based groups (Me–O bond) show large shifts (entries 19–22).

Effects of substituents on CH$_2$ groups

It is more difficult to give a definitive list for CH$_2$ groups as they have two substituents. In the table below we set one substituent as phenyl (Ph) just because so many compounds of this kind are available, and give the actual shifts relative to PhCH$_2$CH$_3$ for protons (2.64 ppm) and PhCH$_2$CH$_3$ for carbon (28.9 ppm), again comparing the substituent with the CH skeleton.

If you compare the shifts caused on a CH$_2$ group by each functional group in the table below with the shifts caused on a CH$_3$ group by the same functional group in the table on p. 423 you will see that they are broadly the same.

Chemical shifts of CH$_2$ groups bonded to phenyl and functional groups

	Functional group	Compound	δ_H, ppm	$\delta_H - 2.64$, ppm	δ_C, ppm	$\delta_C - 28.9$, ppm
1	silane	PhCH$_2$–SiMe$_3$	?	?	27.5	−1.4
2	hydrogen	PhCH$_2$–H	2.32	−0.32	21.4	−7.5
3	alkane	PhCH$_2$–CH$_3$	2.64	0.00	28.9	0.0
4	benzene	PhCH$_2$–Ph	3.95	1.31	41.9	13.0
5	alkene	PhCH$_2$–CH=CH$_2$	3.38	0.74	41.2	12.3
6	nitrile	PhCH$_2$–CN	3.70	1.06	23.5	−5.4
7	acid	PhCH$_2$–CO$_2$H	3.71	1.07	41.1	12.2
8	ester	PhCH$_2$–CO$_2$Me	3.73	1.09	41.1	12.2
9	amide	PhCH$_2$–CONEt$_2$	3.70	1.06	?	?
10	ketone	(PhCH$_2$)$_2$C=O	3.70	1.06	49.1	20.2
11	thiol	PhCH$_2$–SH	3.69	1.05	28.9	0.0
12	sulfide	(PhCH$_2$)$_2$S	3.58	0.94	35.5	6.6
13	sulfoxide	(PhCH$_2$)$_2$S=O	3.88	1.24	57.2	28.3
14	sulfone	(PhCH$_2$)$_2$SO$_2$	4.11	1.47	57.9	29.0
15	amine	PhCH$_2$–NH$_2$	3.82	1.18	46.5	17.6
16	amide	HCONH–CH$_2$Ph	4.40	1.76	42.0	13.1
17	nitro[a]	PhCH$_2$–NO$_2$	5.20	2.56	81.0	52.1
18	ammonium salt	PhCH$_2$–NMe$_3$$^+$	4.5/4.9		55.1	26.2
19	alcohol	PhCH$_2$–OH	4.54	1.80	65.3	36.4
20	ether	(PhCH$_2$)$_2$O	4.52	1.78	72.1	43.2
21	enol ether	PhCH$_2$–OAr[a]	5.02	2.38	69.9	41.0
22	ester	MeCO$_2$–CH$_2$Ph	5.10	2.46	68.2	39.3
23	phosphonium salt	Ph$_3$P$^+$–CH$_2$Ph	5.39	2.75	30.6	1.7
24	chloride	PhCH$_2$–Cl	4.53	1.79	46.2	17.3
25	bromide	PhCH$_2$–Br	4.45	1.81	33.5	4.6

[a] Compound is (4-chloromethylphenoxymethyl)benzene.

Shifts of a CH group

We can do the same with a CH group, and in the left-hand side of the table below we take a series of isopropyl compounds, comparing the measured shifts with those for the central proton (CHMe$_2$) or carbon (CHMe$_2$) of 2-methylpropane. We set two of the substituents as methyl groups and just vary the third. Yet again the shifts for the same substituent are broadly the same.

Effects of α and β substitution on ^{1}H and ^{13}C NMR shifts in Me$_2$CHX[a]

X	Effects on C$_\alpha$ (Me$_2$CH–X), ppm				Effects on C$_\beta$ (Me$_2$CH–X), ppm			
	δ_H	$\delta_H - 1.68$	δ_C	$\delta_C - 25.0$	δ_H	$\delta_H - 0.9$	δ_C	$\delta_C - 8.4$
Li			10.2	−14.8			23.7	17.3
H	1.33	−0.35	15.9	−9.1	0.91	0.0	16.3	7.9
Me	1.68	0.00	25.0	0.0	0.89	0.0	24.6	16.2
CH=CH$_2$	2.28	0.60	32.0	7.0	0.99	0.09	22.0	13.6
Ph	2.90	1.22	34.1	9.1	1.24	0.34	24.0	15.6
CHO	2.42	0.74	41.0	16.0	1.12	0.22	15.5	7.1
COMe	2.58	0.90	41.7	16.7	1.11	0.21	27.4	19.0
CO$_2$H	2.58	0.90	34.0	4.0	1.20	0.30	18.8	10.4
CO$_2$Me	2.55	0.87	33.9	8.9	1.18	0.28	19.1	10.7
CONH$_2$	2.40	0.72	34.0	9.0	1.08	0.18	19.5	11.1
CN	2.71	1.03	20.0	−5.0	1.33	0.43	19.8	11.4
NH$_2$	3.11	1.43	42.8	17.8	1.08	0.18	26.2	17.8
NO$_2$	4.68	3.00	78.7	53.7	1.56	0.66	20.8	12.4
SH	3.13	1.45	30.6	5.6	1.33	0.43	27.6	19.2
Si-Pr	3.00	1.32	33.5	8.5	1.27	0.37	23.7	15.3
OH	4.01	2.33	64.2	39.2	1.20	0.30	25.3	16.9
Oi-Pr	3.65	1.97	68.4	43.4	0.22	0.22	22.9	14.5
O$_2$CMe	5.00	3.32	67.6	42.6	1.22	0.32	21.4(8)	17.(0/4)
Cl	4.19	2.51	53.9	28.9	1.52	0.62	27.3	18.9
Br	4.29	2.61	45.4	20.4	1.71	0.81	28.5	20.1
I	4.32	2.36	31.2	6.2	1.90	1.00	21.4	13.0

[a] There is coupling between the CH and the Me$_2$ groups in the proton NMR.

Shifts in proton NMR are easier to calculate and more informative than those in carbon NMR

This final table, on p. 426, helps to explain something we have avoided so far. Correlations of shifts caused by substituents in proton NMR really work very well. Those in ^{13}C NMR work much less well and more complicated equations are needed. More strikingly, the proton shifts often seem to fit better with our understanding of the chemistry of the compounds. There are two main reasons for this.

First, the carbon atom is much closer to the substituent than the proton. In the compounds in the table on p. 423 the methyl carbon atom is directly bonded to the substituent, while the protons are separated from it by the carbon atom of the methyl group. If the functional group is based on a large electron-withdrawing atom like sulfur, the protons will experience a simple inductive electron withdrawal and have a proportional downfield shift. The carbon atom is close enough to the sulfur atom to be shielded as well by the lone-pair electrons in the large 3sp^3 orbitals. The proton shift caused by S in Me$_2$S is about the same (1.23 ppm) as that caused by a set of more or less equally strong electron-withdrawing groups like CN (1.14 ppm) or ester (1.18 ppm). The carbon shift (9.7 ppm) is less than that caused by an ester (11.2 ppm) but much *more* than that caused by CN, which actually shifts the carbon upfield (−6.6 ppm) relative to the effect of a methyl group.

Approximate additive functional group (X) shifts in ¹H NMR spectra

Entry	Functional group X	¹H NMR shift difference[a], ppm
1	alkene (–C=C)	1.0
2	alkyne(–C≡C)	1.0
3	phenyl (–Ph)	1.3
4	nitrile (–C≡N)	1.0
5	aldehyde (–CHO)	1.0
6	ketone (–COR)	1.0
7	acid (–CO_2H)	1.0
8	ester (–CO_2R)	1.0
9	amide (–$CONH_2$)	1.0
10	amine (–NH_2)	1.5
11	amide (–NHCOR)	2.0
12	nitro (–NO_2)	3.0
13	thiol (–SH)	1.0
14	sulfide (–SR)	1.0
15	sulfoxide (–SOR)	1.5
16	sulfone (–SO_2R)	2.0
17	alcohol (–OH)	2.0
18	ether (–OR)	2.0
19	aryl ether (–OAr)	2.5
20	ester (–O_2CR)	3.0
21	fluoride (–F)	3.0
22	chloride (–Cl)	2.0
23	bromide (–Br)	2.0
24	iodide (–I)	2.0

[a] To be added to 0.9 ppm for MeX, 1.3 ppm for CH_2X, or 1.7 ppm for CHX.

Second, the carbon shift is strongly affected not only by what is directly joined to that atom (α position), but also by what comes next (β position). The right-hand half of the table on p. 424 shows what happens to methyl shifts when substituents are placed on the next carbon atom. There is very little effect on the proton spectrum: all the values are much less than the shifts caused by the same substituent on a methyl group in the table on p. 423. Carbonyls give a down-field shift of about 1.2 ppm when directly joined to a methyl group, but only of about 0.2 ppm when one atom further away. By contrast, the shifts in the carbon spectrum are of the same order of magnitude in the two tables, and the β shift may even be greater than the α shift! The CN group shifts a directly bonded methyl group upfield (–6.6 ppm) when directly bonded, but downfield (14.4 ppm) when one atom further away. This is an exaggerated example, but the point is that these carbon shifts must *not* be used to suggest that the CN group is electron-donating in the α position and electron-withdrawing in the β position. The carbon shifts are erratic but the proton shifts give us useful information and are worth understanding as a guide to both structure determination and the chemistry of the compound.

When you use this table and are trying to interpret, say, a methyl group at 4.0 ppm then you have no problem. Only one group is attached to a methyl group so you need a single shift value—it might be a methyl ester, for example. But when you have a CH_2 group at 4.5 ppm and you are inter-preting a downfield shift of 3.2 ppm you must beware. There are *two* groups attached to each CH_2 group and you might need a single shift of about 3 ppm (say, an ester again) or two shifts of 1.5 ppm, and so on. The shifts are additive.

Further reading

A reminder: you will find it an advantage to have one of the short books on spectroscopic analysis to hand as they give explanations, comprehensive tables of data, and problems. We recommend *Spectroscopic Methods in Organic Chemistry* by D. H. Williams and Ian Fleming, McGraw-Hill, London, 6th edn, 2007.

Other books include R. M. Silverstein, F. X. Webster, and D. J. Kiemle, *Spectrometric Identification of Organic Compounds*, Wiley, 2005 and a book of problems: L. D. Field, S. Sternhell, and J. R. Kalman, *Organic Structures from Spectra*, 3rd edn, Wiley, 2003.

The ¹³C NMR of ketene was reported by J. Firl and W. Runger, *Angew. Chem. Int. Ed.*, 1973, **12**, 668, the tetrahedrane/cyclobutadiene story is expounded by G. Maier in *Angew. Chem. Int. Ed.*, 1988, **27**, 309, and the *Lycorea* sex pheromone story by G. Meinwald and team, *Science*, 1968, **164**, 1174.

Check your understanding

To check that you have mastered the concepts presented in this chapter, attempt the problems that are available in the book's Online Resource Centre at http://www.oxfordtextbooks.co.uk/orc/clayden2e/

Suggested solutions for Chapter 18

18

PROBLEM 1

A compound C_6H_5FO has a broad peak in the infrared at about 3100-3400 cm^{-1} and the following signals in its (proton decoupled) ^{13}C NMR spectrum. Suggest a structure for the compound and interpret the spectra.

δ_C (ppm) 157.4 (d, J 229 Hz), 151.2 (s), 116.3 (d J 7.5 Hz), and 116.0 (d, J 23.2 Hz).

Purpose of the problem

A reminder that coupling may occur in ^{13}C NMR spectra too and can be useful.

Suggested solution

All the signals are in the sp^2 region and two (at >150 ppm) are of carbons attached to electronegative elements. As the formula contains C_6, a benzene ring is strongly suggested. The IR spectrum tells us that we have an OH group, so the compound is one of these three:

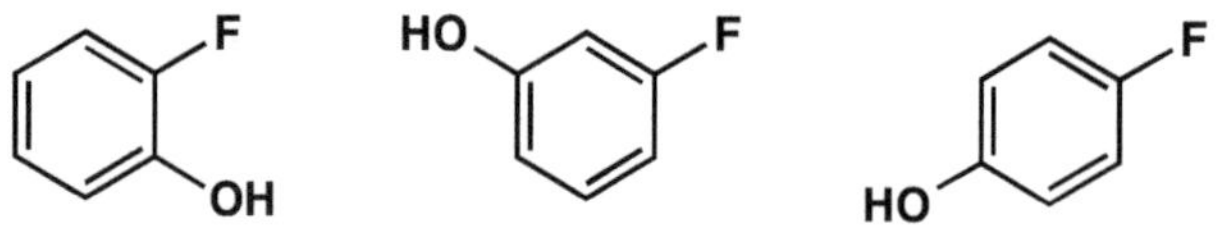

The symmetry of the spectrum suggests the *para* disubstituted compound as there are only four types of carbon atom. We can assign the spectrum by noting that the very large coupling (J 229) must be a $^2J_{CF}$ and the zero coupling must be the carbon furthest from F, i.e. the *para* carbon. The intermediate couplings are for the other two carbons and the CF coupling diminishes with distance.

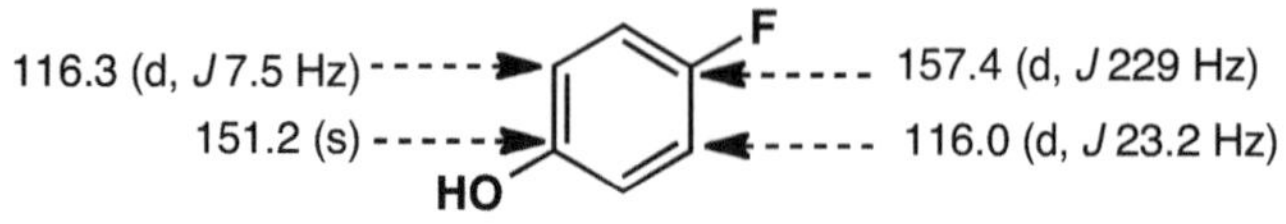

116.3 (d, J 7.5 Hz) ----▶ ◀---- 157.4 (d, J 229 Hz)
151.2 (s) ----▶ ◀---- 116.0 (d, J 23.2 Hz)

PROBLEM 2

The natural product bullatenone was isolated in the 1950s from a New Zealand myrtle and assigned the structure **A**. Then authentic compound **A** was synthesized and found not to be identical to natural bullatenone. Predict the expected ¹H NMR spectrum of **A**. Given the full spectroscopic data, not available in the 1950s, say why **A** is definitely wrong and suggest a better structure for bullatenone.

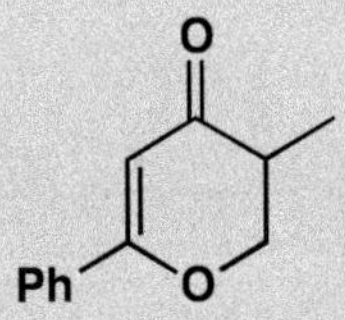

A: alleged bullatenone

Spectra of isolated bullatenone:

Mass spectrum: m/z 188 (10%) (high resolution confirms $C_{12}H_{12}O_2$), 105 (20%), 102 (100%), and 77 (20%)

Infrared: 1604 and 1705 cm^{-1}

¹H NMR: δ_H (ppm) 1.43 (6H, s), 5.82 (1H, s), 7.35 (3H, m), and 7.68 (2H, m).

Purpose of the problem

Detecting wrong structures teaches us to be alert to what the spectra are telling us rather than what we expect or want.

Suggested solution

The mass spectrum and IR are all right for **A** but the NMR shows at once that the structure is wrong. There is a monosubstituted benzene ring all right, but the aliphatic protons are a 6H singlet, presumably a CMe_2 group, and a 1H singlet in the alkene region at 5.82 ppm.

The fragments we have are Ph, carbonyl, a CMe_2 group, and an alkene with one proton on it. That adds up to $C_{12}H_{12}O$ leaving only one oxygen to fit in somewhere. There must still be a ring or there would not be enough hydrogen atoms and the ring must be five-membered (just try other possibilities yourself). There are three ring systems we can choose and each can have the Ph group at either end of the alkene, making six possibilities in all.

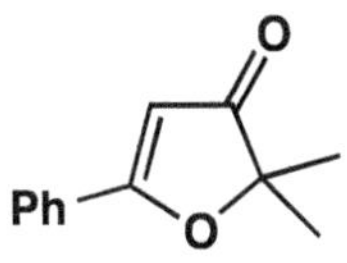

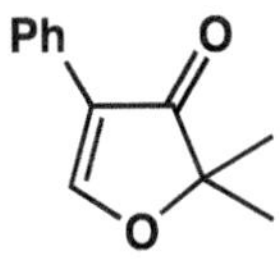

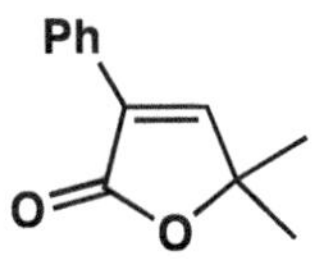

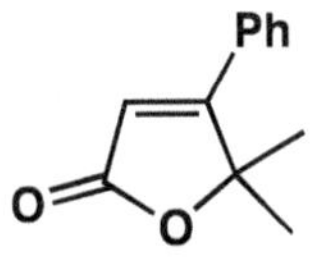

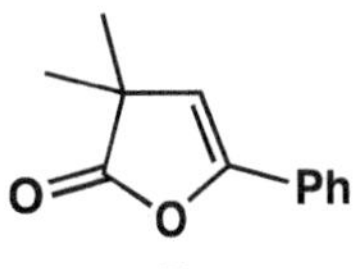

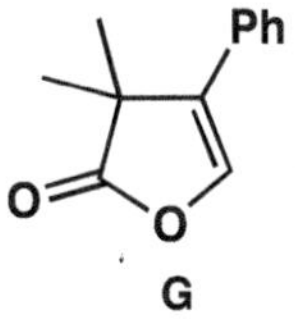

The last four are esters (cyclic esters or lactones) and they would have a C=O frequency at 1745–1780 cm^{-1} so **D–G** are all wrong. The hydrogen on the alkene cannot be next to oxygen as it would have a very large chemical shift indeed whereas it is close to the 'normal' alkene shift of 5.25 ruling out structure **C**. Structure **B** is correct and the spectrum can be assigned. Compound **B** has now been synthesized and proved identical to natural bullatenone.

■ You can read the full story in W. Parker *et al.*, *J. Chem. Soc.* 1958, 3871. See also T. Reffstrup and P. M. Boll, *Acta Chem. Scand.*, 1977, **31B**, 727; *Tetrahedron Lett.*, 1971, 4891, and R. F. W. Jackson and R. A. Raphael, *J. Chem. Soc., Perkin Trans. 1*, 1984, 535.

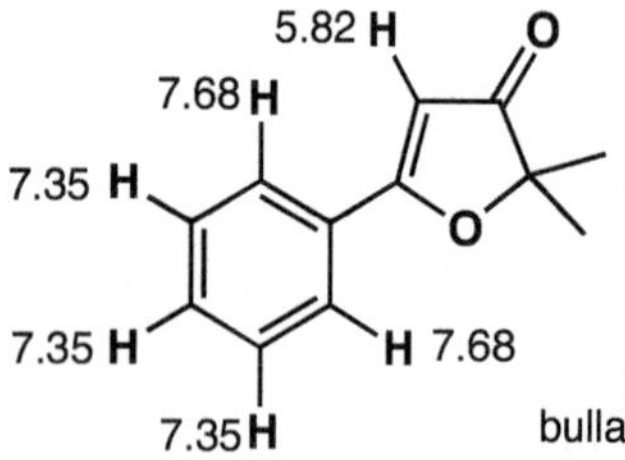

B:
bullatenone

PROBLEM 3

Suggest structures for each of these reaction products, interpreting the spectroscopic data. You are *not* expected to give mechanisms for the reactions and you must resist the temptation to say what 'ought to happen'. These are all unexpected products.

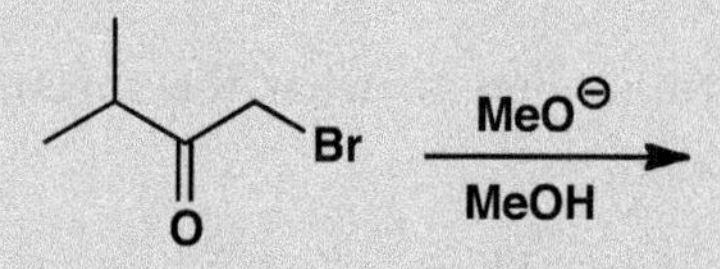

A, C$_6$H$_{12}$O$_2$
ν_{max} (cm^{-1}) 1745
δ_C (ppm) 179, 52, 39, 27
δ_H (ppm) 1.20 (9H, s) and 3.67 (3H, s)

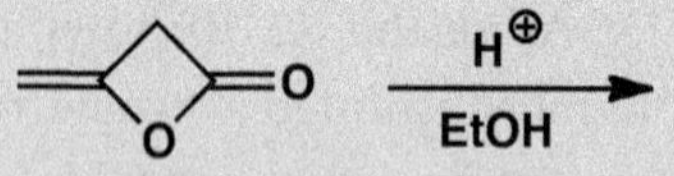

B, C$_6$H$_{10}$O$_3$
ν_{max} (cm^{-1}) 1745, 1710
δ_C (ppm) 203, 170, 62, 39, 22, 15
δ_H (ppm) 1.28 (3H, t, *J* 7 Hz), 2.21 (3H, s)
3.24 (2H, s) and 4.2 (2H, q, *J* 7 Hz)

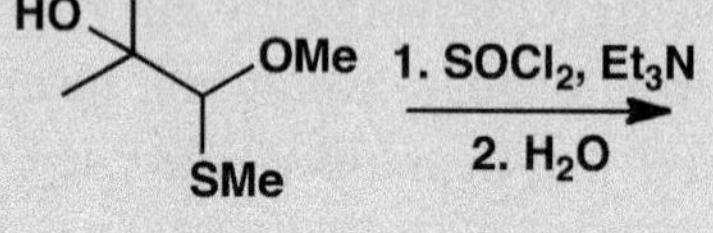

C, m/z 118
ν_{max} (cm^{-1}) 1730
δ_C (ppm) 202, 45, 22, 15
δ_H (ppm) 1.12 (6H, s), 2.28 (3H, s)
and 9.8 (1H, s)

Purpose of the problem

A common situation in real life—you carry out a reaction, isolate the product, and it's something quite different from what you were expecting. What is it?

Suggested solution

Compound **A** has a carbonyl group (IR) that is an acid derivative (179 ppm in the ^{13}C NMR). The 9H singlet in the proton NMR must be a *t*-Bu group and the 3H singlet at 3.67 ppm must be an OMe group. Putting these four fragments together we get a structure immediately. The IR is typical for an ester (1715 + 30 = 1745 cm^{-1}).

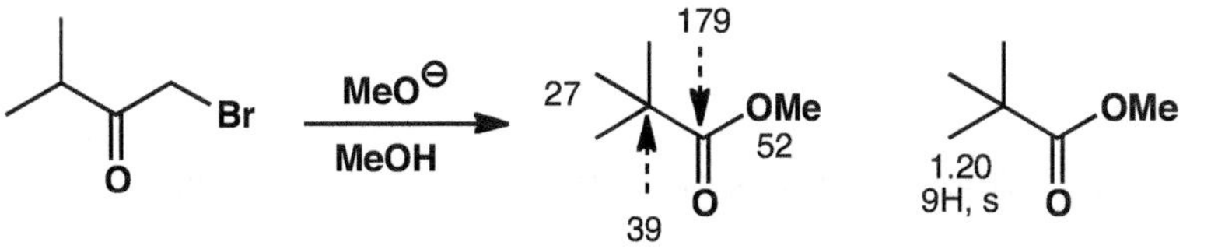

Compound **B** again has an ester (1745 cm^{-1} and 170 ppm) but it also has a ketone (1710 cm^{-1} and 203 ppm). The proton NMR shows an OEt group (3H triplet and 2H quartet) together with another methyl group next to something electron-withdrawing (that can only be a C=O as there isn't anything else), and a CH$_2$ group with no coupling at 3.24 ppm. This is 2 ppm away from a 'normal' CH$_2$ but it can't be next to O as we've used up all the O atoms already. It must be between two electron-withdrawing groups. These can only be carbonyls so this CH$_2$ is isolated between the two carbonyl groups and we have the structure.

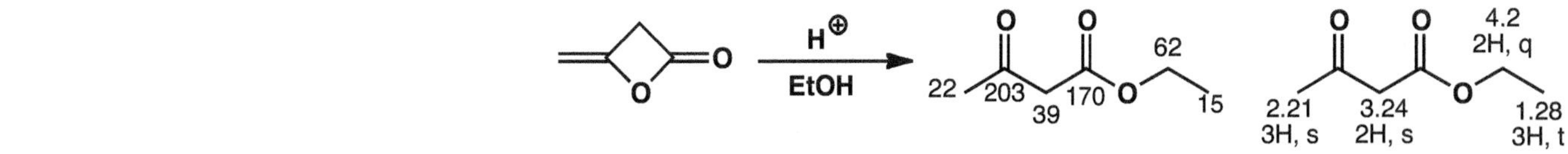

Compound **C** has no formula given, just a molecular ion in the mass spectrum. The most obvious formula is $C_5H_{10}O_3$ but S is 32 while O is 16 so it might be $C_5H_{10}OS$. We must look at the rest of the spectra for clarification. There is a carbonyl group (1730 cm^{-1}) that is an aldehyde or ketone (202 ppm). The proton NMR shows a CMe$_2$ group (6H, s), a methyl group at 2.8 ppm that doesn't look like an OMe (expected 3-3.5 ppm), but might be an SMe. The carbon spectrum also suggests SMe rather than OMe at 45 ppm, and one hydrogen atom at 9.8 ppm that looks like an aldehyde. We know we have these fragments:

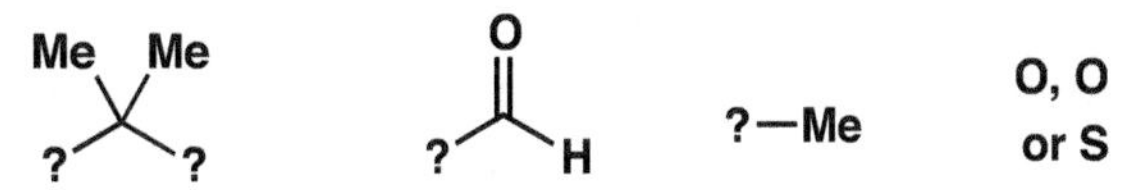

It is not possible to construct a molecule with two extra oxygen atoms but without an OMe, and those we could propose look rather unstable, such as:

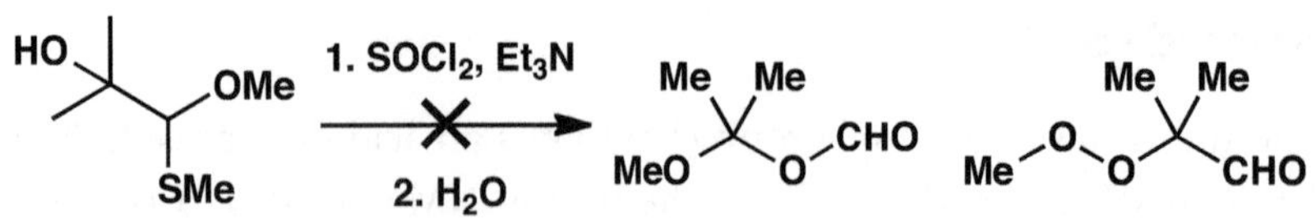

Only one compound is possible if we have an S atom—this fits the data very much better and indeed is the correct structure. It has a genuine aldehyde (not a formate ester) and SMe fits better than OMe the signal at δ_H 2.8 ppm and δ_C 45 ppm.

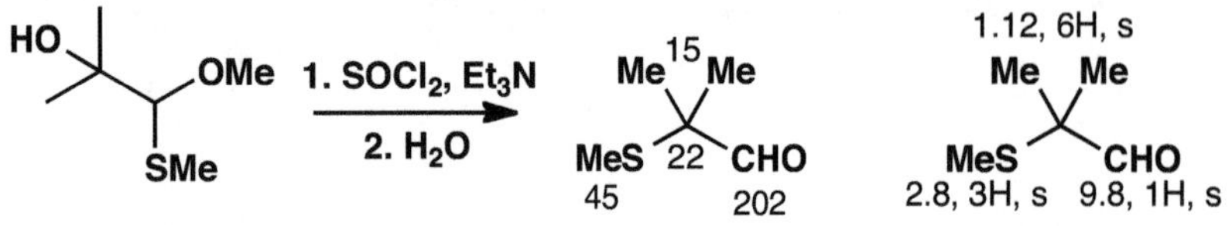

PROBLEM 4

Suggest structures for the products of these reactions.

Compound **A**: $C_7H_{12}O_2$; IR 1725 cm^{-1}; δ_H (ppm) 1.02 (6H, s), 1.66 (2H, t, J 7 Hz), 2.51 (2H, t, J 7 Hz), and 4.6 (2H, s).

Compound **B**: m/z 149/151 (M$^+$ ratio 1:3); IR 2250 cm^{-1}; δ_H (ppm) 2.0 (2H, quintet, J 7 Hz), 2.5 (2H, t, J 7 Hz), 2.9 (2H, t, J 7 Hz), and 4.6 (2H, s).

Purpose of the problem

More practice at the important skill of total structure determination.

Suggested solution

The starting material for **A** is $C_7H_{12}O_3$ and appears just to have lost an oxygen atom. As the reagent is NaBH$_4$, the chances are that two hydrogens have been added and the oxygen lost as a water molecule. The IR spectrum shows a carbonyl group and the frequency suggests an ester or a strained ketone. The NMR shows two joined CH$_2$ groups, one at 2.51 being next to a functional group, not O and so it must be C=O. There is also an unchanged CMe$_2$ group and an isolated CH$_2$ group next to oxygen at 3.9 ppm. There is only one reasonable structure.

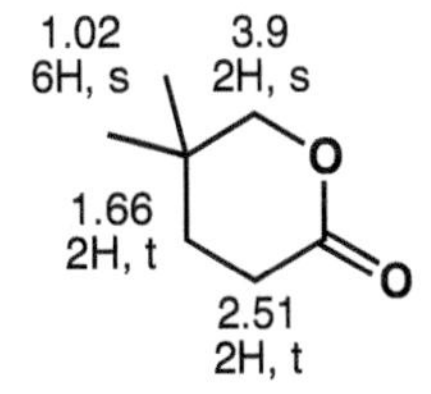

1.02
6H, s

3.9
2H, s

1.66
2H, t

2.51
2H, t

The mass spectrum of compound B shows that it has chlorine in it, the IR shows a CN group and the proton NMR shows eight Hs. If we assume that no carbons have been lost, the most reasonable formula is C_5H_8ClNS. The compound has lost a water molecule. The NMR shows three linked CH_2 groups with triplets at the ends and a quintet in the middle. The shifts of the terminal CH_2s show that they are next to functional groups but not Cl. This means we must have a unit $-SCH_2CH_2CH_2CN$. All that remains is the isolated CH_2 group with a large chemical shift evidently joined to both the S and Cl. The large shift comes from 1.5 + 1 (S) + 2 (Cl) = 4.5 ppm. Again only one structure emerges.

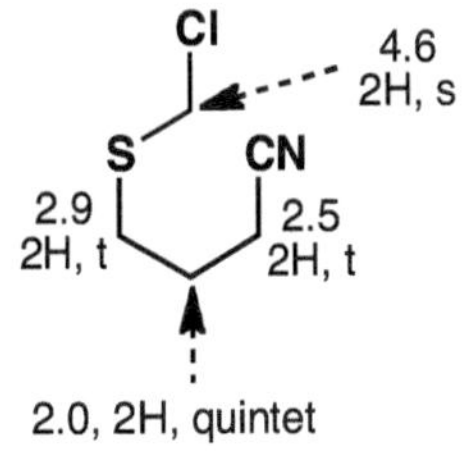

Cl
4.6
2H, s

S   CN

2.9
2H, t

2.5
2H, t

2.0, 2H, quintet

PROBLEM 5

Two alternative structures are shown for the products of these reactions. Explain in each case how you would decide which product is actually formed. Several pieces of evidence will be required and estimated values are better than general statements.

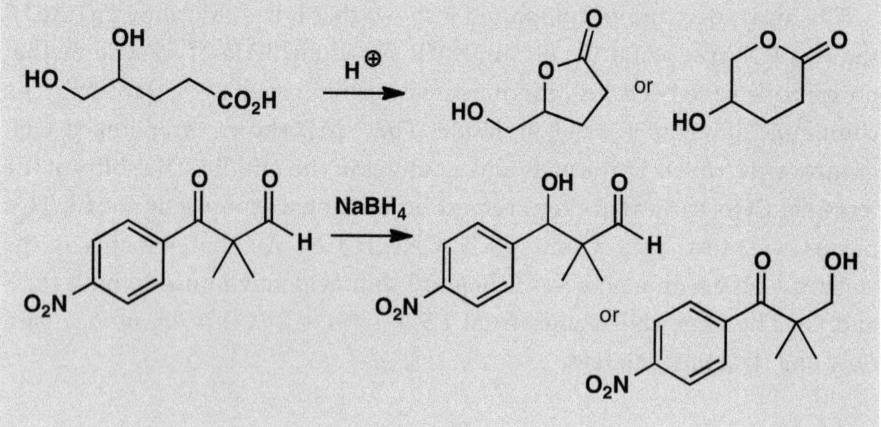

Purpose of the problem

To get you thinking the other way round: from structure to data. What are the important pieces of evidence?

Suggested solution

There are many acceptable ways in which you could answer this question ranging from choosing just one vital statistic for each pair to analysing all the data. We'll adopt a middle way and point out several important distinctions. In the first example, one main difference is the ring size, seen mainly in the IR. Both are esters (about 1745 cm^{-1}) but we should add 30 cm^{-1} for the five-membered ring. The functional group next to OCH$_2$ is also different—an OH in one case and an ester in another. There will be other differences too of course.

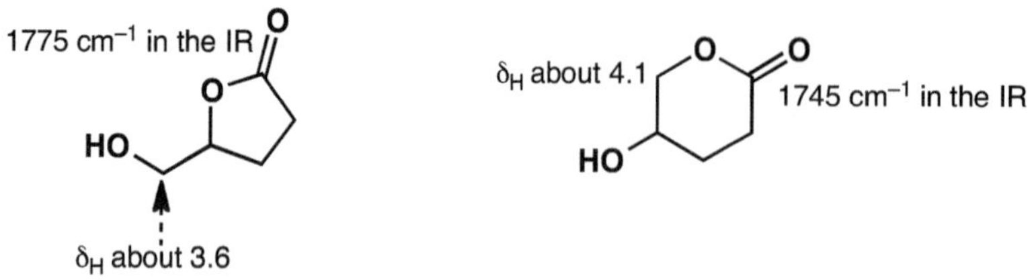

In the second case there are also differences in the IR C=O stretch between the aldehyde (about 1730 cm^{-1}) and the conjugated ketone (about

1680 cm⁻¹). The aldehyde proton and the number of protons next to oxygen make a clear distinction. There will also be differences in the ¹H and ¹³C NMR signals of the benzene rings as one is conjugated to a C=O group and the other is not. This reaction actually gave a mixture of both compounds.

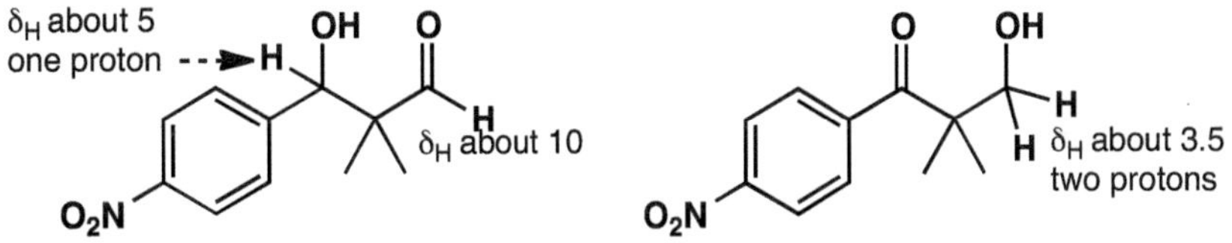

PROBLEM 6

The NMR spectra of sodium fluoropyruvate in D₂O are given below. Are these data compatible with the structure shown? If not, suggest how the compound might exist in this solution.

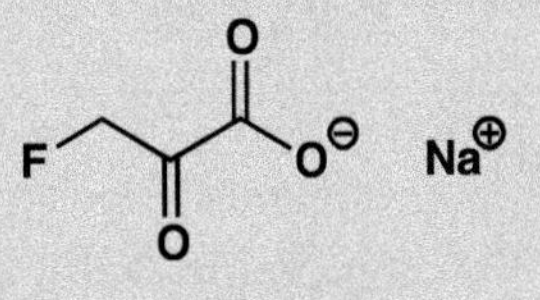

δ_H (ppm) 4.43 (2H, d, *J* 47 Hz);

δ_C (ppm) 83.5 (d, *J* 22 Hz), 86.1 (d, *J* 171 Hz), and 176.1 (d, *J* 2 Hz).

Purpose of the problem

To show how NMR spectra can reveal more than just the identity of a compound.

Suggested solution

The proton NMR spectrum is all right as we expect a large shift: from the chart on p. 276 of the textbook, we can predict 1.3 + 1 (C=O) + 2 (F) = 4.3 ppm and the coupling to fluorine is fine. The carbon NMR shows the carboxylate carbon at 176 ppm with a small coupling to F as it is so far away. The CH₂ carbon is at 86.1 ppm with a huge coupling as it is joined directly to F. So far, so good. But what about the C=O group itself? We should expect it at about 200 ppm but it is at 83.5 with the expected intermediate coupling. It cannot be a carbonyl group at all. So what could have happened in D₂O? The obvious answer is that a hydrate is formed from this very electrophilic carbonyl group.

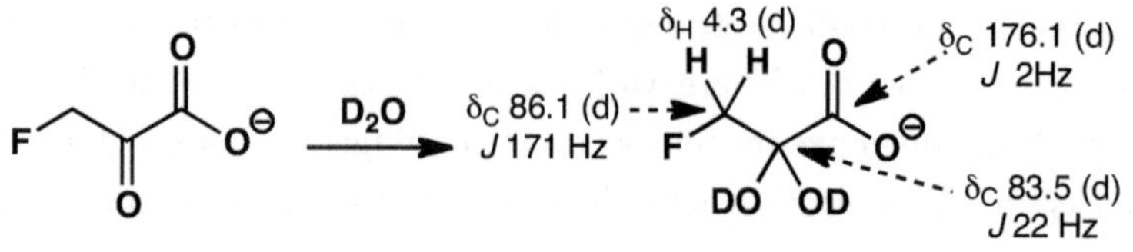

PROBLEM 7

An antibiotic isolated from a microorganism was crystallized from water and formed different crystalline salts in either acid or base. The spectroscopic data were:

Mass spectrum 182 (M^+, 9%), 109 (100%), and 74 (15%).

δ_H (ppm in D_2O at pH<1) 3.67 (2H, d, J 7), 4.57 (1H, t, J 7), 8.02 (2H, m), and 8.37 (1H, m).

δ_C (ppm in D_2O at pH<1) 33.5, 52.8, 130.1, 130.6, 130.9, 141.3, 155.9, and 170.2. Suggest a structure for the antibiotic.

Purpose of the problem

Structure determination of a compound with biological activity from a natural source.

Suggested solution

The solubility and salt formation suggest the presence of both acidic and basic groups, perhaps CO_2H and NH_2 as this is a natural compound. If so, the ^{13}C peak at 170.2 ppm is the CO_2H group. The five carbons in the sp^2 region and protons at 8.0 and 8.4 suggest an aromatic ring, probably a pyridine. The mass spectrum gives an even molecular ion (182) so there must be another nitrogen atom beyond the one in the pyridine. The two sets of aliphatic protons are coupled and the large shift of the ^1H signal at 4.57 ppm suggests a proton between CO_2H and NH_3^+ (pH <1). We have these fragments:

Presumably the aliphatic part must be X or Y, and that leaves just one oxygen atom for a formula of $C_8H_{10}N_2O_3 = 182$. Only six of the ten H atoms show up in the NMR because the OH, NH_3^+, and CO_2H protons all exchange rapidly at pH <1.

■ The details of the structure and spectra are in S. Inouye *et al.*, *Chem. Pharm. Bull.*, 1975, **23**, 2669; S. R. Schow *et al.*, *J. Org. Chem.*, 1994, **59**, 6850 and B. Ye and T. R. Burke, *J. Org. Chem.*, 1995, **60**, 2640.

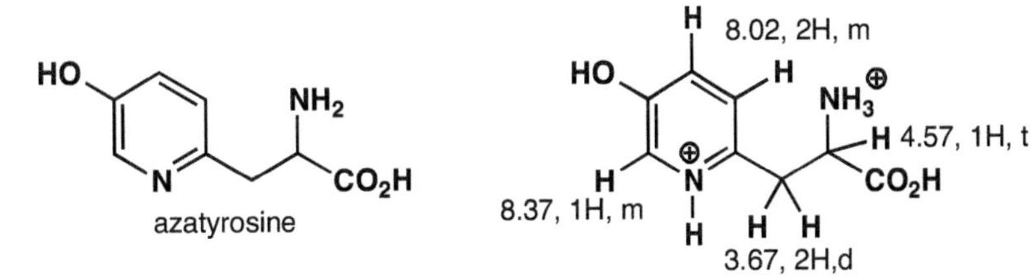

azatyrosine

PROBLEM 8

Suggest structures for the products of these two reactions.

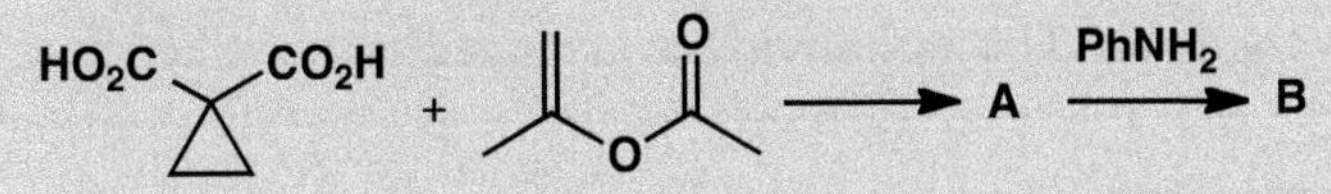

Compound A:

m/z 170 (M$^+$, 1%), 84 (77%), and 66 (100%);

IR 1773, 1754 cm^{-1};

δ_H (ppm, CDCl$_3$) 1.82 (6H, s) and 1.97 (4H, s);

δ_C (ppm, CDCl$_3$) 22, 23, 28, 105, and 169.

Compound B:

m/z 205 (M+, 40%), 161 (50%), 160 (35%), 105 (100%), and 77 (42%);

IR 1670, 1720 cm^{-1};

δ_H (ppm, CDCl$_3$) 2.55 (2H, m), 3.71 (1H, t, *J* 6 Hz), 3.92 (2H, m), 7.21 (2H, d, *J* 8 Hz), 7.35 (1H, t, *J* 8 Hz), and 7.62 (2H, d, *J* 8 Hz);

δ_C (ppm, CDCl$_3$) 21, 47, 48, 121, 127, 130, 138, 170, and 172.

Purpose of the problem

The other important kind of structure determination: compounds isolated from a chemical reaction.

Suggested solution

Compound **A** is much simpler so we start with that. The two reagents are C$_5$H$_6$O$_4$ and C$_5$H$_8$O$_2$ these add up to C$_{10}$H$_{14}$O$_6$ (230) so 60 has been lost. This looks like C$_2$H$_4$O$_2$ or, less likely (because it must be saturated—it has no double bond equivalents), C$_3$H$_8$O. If the first is right, **A** is C$_8$H$_{10}$O$_4$ which at least fits the proton NMR. The IR suggests two carbonyl groups. The ^{13}C NMR shows only one, but there must be some symmetry as there are only five signals for eight carbon atoms. The only unsaturation we have identified is the two carbonyl groups so the signal at 105 ppm is very strange. It must be next to two oxygen atoms to have such a large shift. Either 22 or 105 must be the C of CMe$_2$. C$_8$H$_{10}$O$_4$ would have four double bond equivalents, so the last two degrees of unsaturation must be rings. The cyclopropane provides one and the other must link the two oxygen atoms in the second part-structure. So we have:

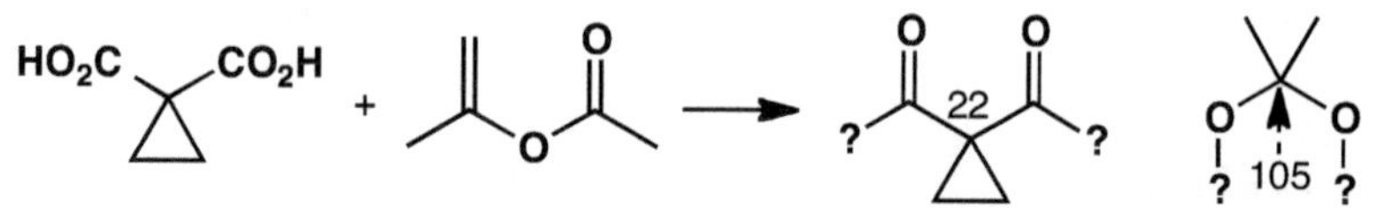

This accounts for all the atoms in **A** so all we need to do is join these two fragments together! The carbonyls are arranged rather like those in cyclic anhydrides and the two carbonyl peaks must be the symmetric and antisymmetric stretches.

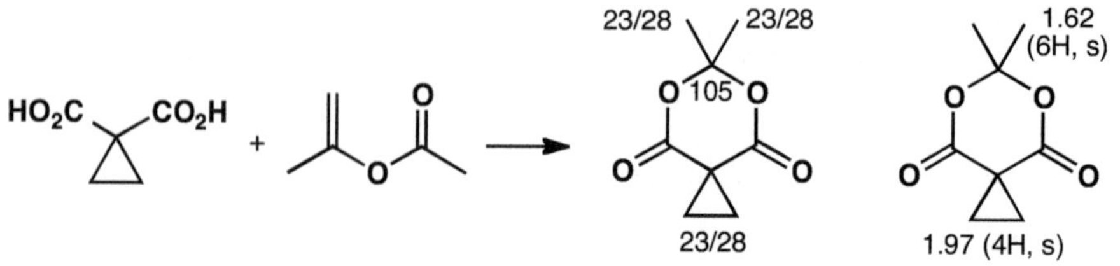

Compound **B** has nitrogen in it (it has an odd molecular weight) and clearly has a benzene ring from the NMR spectra, so we can put down PhN (= 91) as part of the structure. It also has two carbonyl groups (in the IR the one at 1670 cm^{-1} looks like an amide) and they are both acid derivatives (you can see that in the ^{13}C NMR). There are three aliphatic carbons, two CH$_2$s and one CH. Adding that together gives C$_{11}$H$_{10}$NO$_3$ = 188 so there is 17 missing that looks like OH. Since we need a second acid derivative and the OH is the only remaining heteroatom, it must be a carboxylic acid. Given that the CH is a triplet, it must be joined to one of the CH$_2$ groups and, as they are both multiplets, they must be joined to each other. There is one double bond equivalent to account for and that must be a ring. So we have:

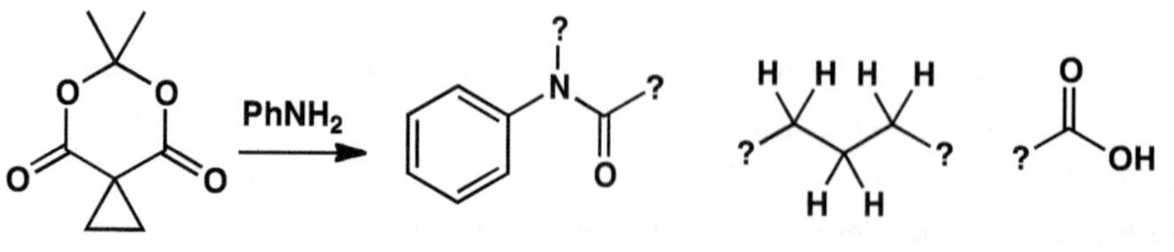

To assemble these three fragments into a molecule we must plug the amide into the C$_3$ fragment and put the CO$_2$H group in the last free position. We can do this in two ways. Proton NMR distinguishes them. The end CH$_2$ is attached either to the nitrogen atom (which would give an estimated shift

■ See S. Danishefsky and R. R. Singh, *J. Am. Chem. Soc.*, 1975, **97**, 3239.

of 3.2 ppm) or to the carbonyl group (estimated shift 2.2 ppm) of the amide. The observed value (3.92) fits the first better. A similar estimate for the CH gives the same answer and the first structure is indeed correct.

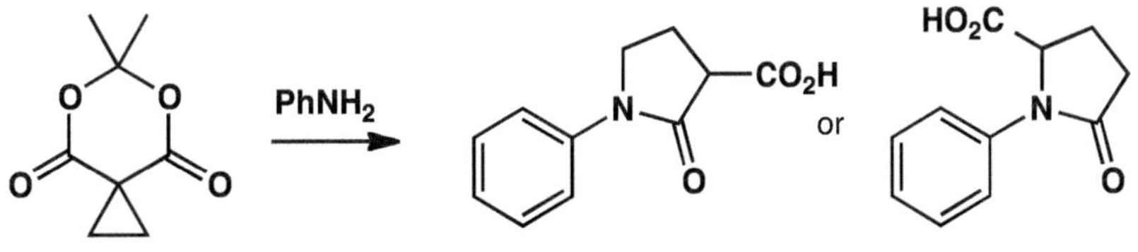

PROBLEM 9

Treatment of this epoxy-ketone with tosyl hydrazine gives a compound with the spectra shown below. What is its structure?

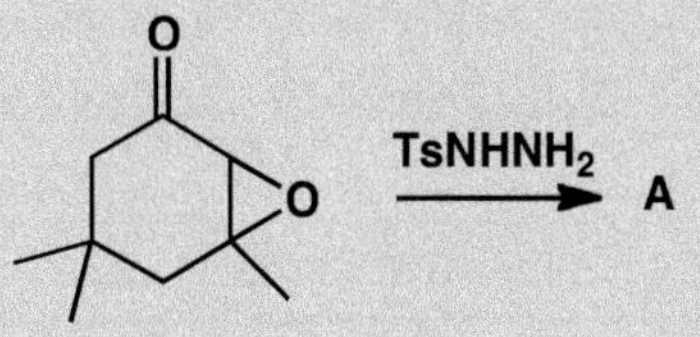

m/z 138 (M+, 12%), 109 (56%), 95 (100%), 81 (83%), 82 (64%), and 79 (74%);

IR 3290, 2115, 1710 cm^{-1};

δ_H (ppm in CDCl$_3$) 1.12 (6H, s), 2.02 (1H, t, *J* 3 Hz), 2.15 (3H, s), 2.28 (2H, d, *J* 3 Hz), and 2.50 (2H, s);

δ_C (ppm in CDCl$_3$) 26, 31, 32, 33, 52, 71, 82, 208.

Purpose of the problem

Further practice at structure determination, adding a curious chemical shift.

Suggested solution

■ See A. Eschenmoser *et al.*, *Helv. Chim. Acta*, 1971, **54**, 2896.

The compound is an alkyne formed by a reaction known as the Eschenmoser fragmentation. It is not possible to assign all the ^{13}C NMR signals but you can spot the alkyne carbons in the region 70–85 ppm and the alkyne CH at about 2 in the proton NMR. The triple bond signals in the IR at about 2150 cm^{-1} is a give-away too. Alkyne C–H bonds are strong and come well above 3000 in the IR. The lack of vicinal coupling in the ^{1}H NMR helps identify the rest of the skeleton of the molecule.

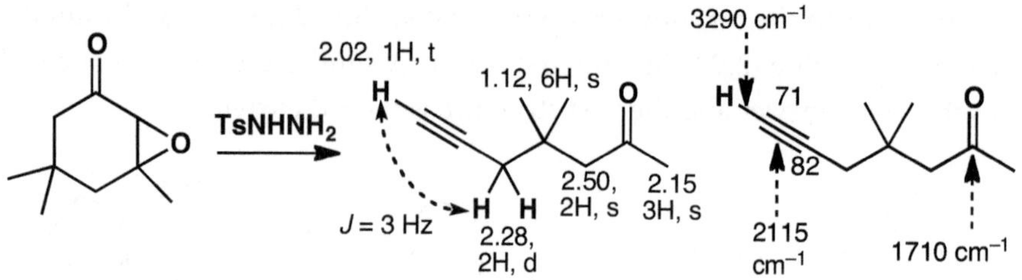

PROBLEM 10

Reaction of the epoxy-alcohol below with LiBr in toluene gave a 92% yield of compound **A**. Suggest a structure for this compound from the data:

mass spectrum gives $C_8H_{12}O$;

ν_{max} (cm^{-1}) 1685, 1618;

δ_H (ppm) 1.26 (6H, s), 1.83, 2H, t, J 7 Hz), 2.50 (2H, dt, J 2.6, 7 Hz), 6.78 (1H, t (J 2.6 Hz), and 9.82 (1H, s);

δ_C (ppm) 189.2, 153.4, 152.7, 43.6, 40.8, 30.3, and 25.9.

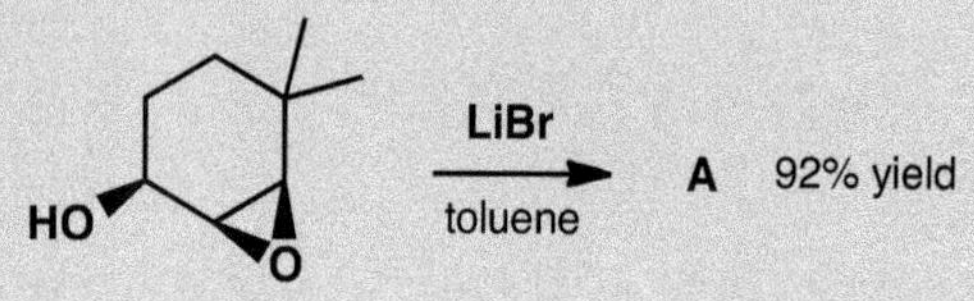

Purpose of the problem

Further practice at structure determination including a change in the carbon skeleton—a ring contraction.

Suggested solution

The compound **A** is a simple cyclopentenal. The ^{13}C NMR assignment is not at all certain.

■ G. Magnusson and S. Thorén, *J. Org. Chem.*, 1973, **38**, 1380.

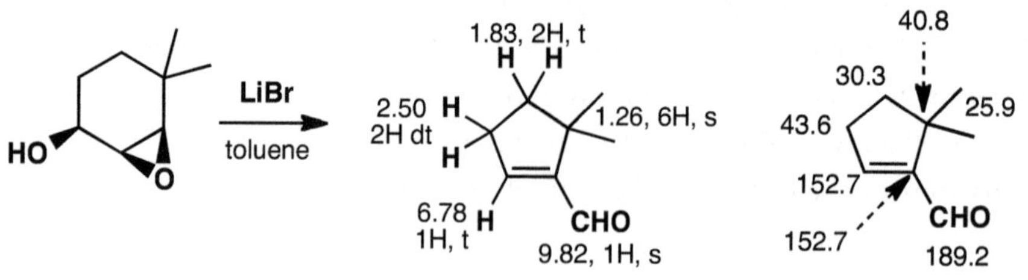

PROBLEM 11

Female boll weevils (a cotton pest) produce two isomeric compounds that aggregate the males for food and sex. A few mg of two isomeric active compounds, grandisol and *Z*-ochtodenol, were isolated from 4.5 million insects. Suggest structures for these compounds from the data below. Signals marked * exchange with D_2O.

Z-ochtodenol:

m/z 154 ($C_{10}H_{18}O$), 139, 136, 121, 107, 69 (100%);

ν_{max} (cm^{-1}) 3350, and 1660;

δ_H (ppm) 0.89 (6H, s), 1.35-1.70 (1H, broad m), 1.41* (1H, s), 1.96 (2H, s), 2.06 (2H, t, *J* 6 Hz), 4.11 (2H, d, *J* 7 Hz), and 5.48 (1H, t, *J* 7 Hz).

Grandisol:

m/z 154 ($C_{10}H_{18}O$), 139, 136, 121, 109, 68 (100%);

ν_{max} (cm^{-1}) 3630, 3520, 3550, and 1642;

δ_H (ppm) 1.15 (3H, s), 1.42 (1H, dddd, *J* 1.2, 6.2, 9.4, 13.4 Hz), 1.35-1.45 (1H, m), 1.55-1.67 (2H, m), 1.65 (3H, s), 1.70-1.81 (2H, m), 1.91-1.99 (1H, m), 2.52* (1H, broad t, *J* 9.0 Hz), 3.63 (1H, ddd, *J* 5.6, 9.4, 10.2 Hz), 3.66 (1H, ddd, *J* 6.2, 9.4, 10.2 Hz), 4.62 (1H, broad s), and 4.81 (1H, broad s);

δ_C (ppm) 19.1, 23.1, 28.3, 29.2, 38.8, 41.2, 52.4, 59.8, 109.6, and 145.1.

Purpose of the problem

Further practice at structure determination of natural products.

Suggested solution

These are the structures. If you have other answers, check that these structures fit the data better.

■ J. M. Tumlinson *et al.*, *Science*, 1969, **166**, 1010; but see K. Mori *et al.*, *Liebigs Annalen*, 1989, 969 for the spectra of ochtodenol and K. Narasaka, *et al.*, *Bull. Chem. Soc. Jap.*, 1991, **64**, 1471 for the spectra of grandisol.

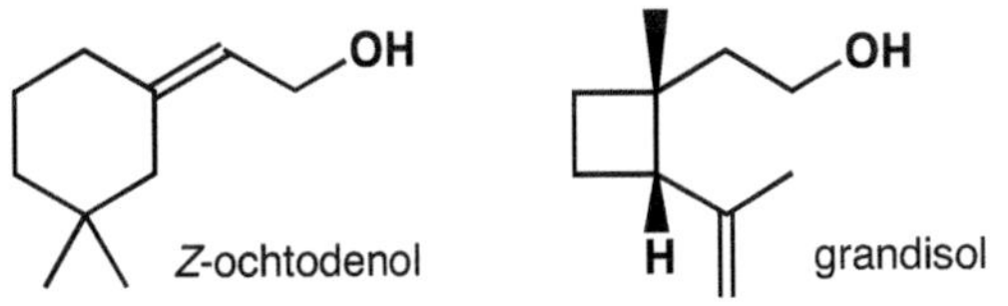

PROBLEM 12

Suggest structures for the products of these reactions.

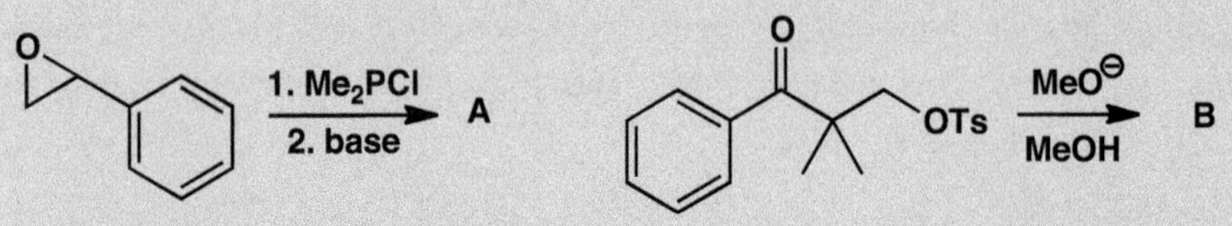

Compound **A**:

$C_{10}H_{13}OP$, IR (cm^{-1}) 1610, 1235;

δ_H (ppm) 6.5-7.5 (5H, m), 6.42 (1H, t, J 17 Hz), 7.47 (1H, dd, J 17, 23 Hz), and 2.43 (6H, d, J 25 Hz).

Compound **B**:

$C_{12}H_{17}O_2$, IR (cm^{-1}) C-H and fingerprint only;

δ_H (ppm) 7.25 (5H, s), 4.28 (1H, d, J 4.8 Hz), 3.91 (1H, d, J 4.8 Hz), 2.96 (3H, s), 1.26 (3H, s) and 0.76 (3H, s).

Purpose of the problem

Structure determination of reaction products with extra twists: a nucleus with spin (P) and protons on the same carbon atom that are different in the NMR.

Suggested solution

The coupling constants $^3J_{PH}$ across the alkene are very large. Typically *cis* $^3J_{PH}$ is about 20 and *trans* $^3J_{PH}$ about 40. Geminal ($^2J_{PH}$) are also large but more variable. In **B** there is a stereogenic centre, meaning that the hydrogen atoms and methyl groups in the ring are different: they are either on the same side as MeO or the same side as Ph. (The term we will introduce in chapter 31 to describe such groups is 'diastereotopic'). We cannot say which H gives which signal.

■ F. Nerdel *et al., Tetrahedron Lett.,* 1968, 5751.

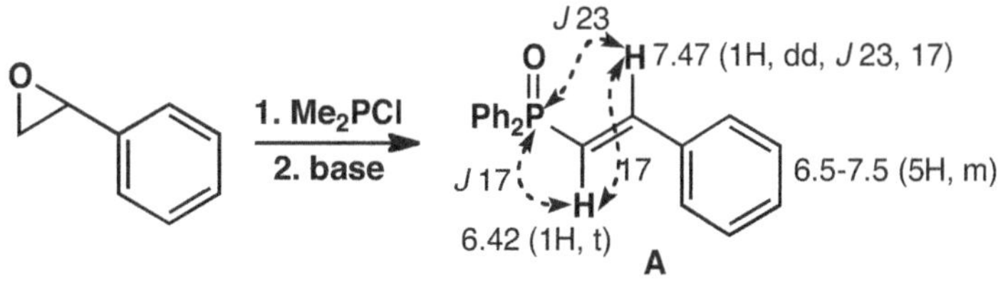
1. Me2PCl
2. base
J 23
H 7.47 (1H, dd, J 23, 17)
Ph2P
J 17
17
6.42 (1H, t)
6.5-7.5 (5H, m)
A

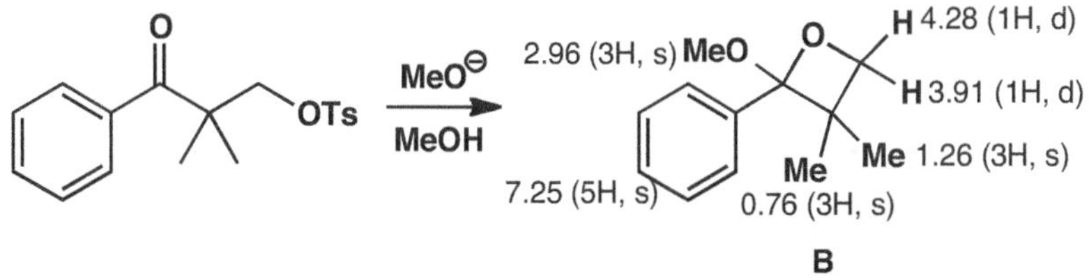
OTs
MeO⊖
MeOH
2.96 (3H, s) MeO
H 4.28 (1H, d)
H 3.91 (1H, d)
Me 1.26 (3H, s)
Me
0.76 (3H, s)
7.25 (5H, s)
B

PROBLEM 13

Identify the compounds produced in these reactions. Warning! Do not attempt to deduce the structures from the starting materials, but use the data. These molecules are so small that you can identify them from ^{1}H NMR alone.

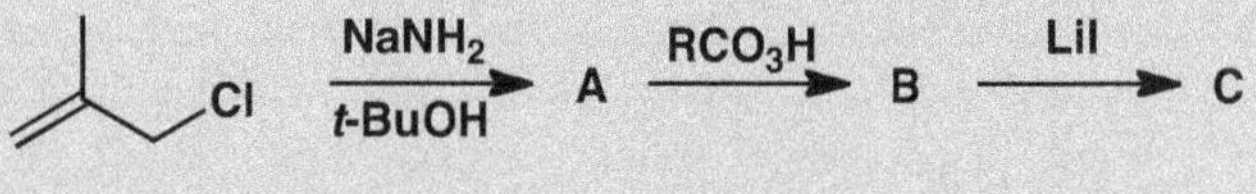
Cl
NaNH2
t-BuOH
A
RCO3H
B
LiI
C

Data for **A**: C_4H_6; δ_H (ppm) 5.35 (2H, s) and 1.00 (4H, s)

Data for **B**: C_4H_6O; δ_H (ppm) 3.00 (2H, s), 0.90 (2H, d, *J* 3 Hz) and 0.80 (2H, d, *J* 3 Hz)

Data for **C**: C_4H_6O; δ_H (ppm) 3.02 (4H, t, *J* 5 Hz) and 1.00 (2H, quintet, *J* 5 Hz).

Purpose of the problem

Structure determination of reaction products by ^{1}H NMR alone.

Suggested solution

The very small shifts of cyclopropane protons may have worried you but they often have shifts of less than 1 ppm. Compounds **A** and **C** are simple enough but **B** may have amazed you. It is unstable but can be isolated and the two three-membered rings sit at right angles to each other, so as in problem 12 the protons on each side of the cyclopropane ring are different.

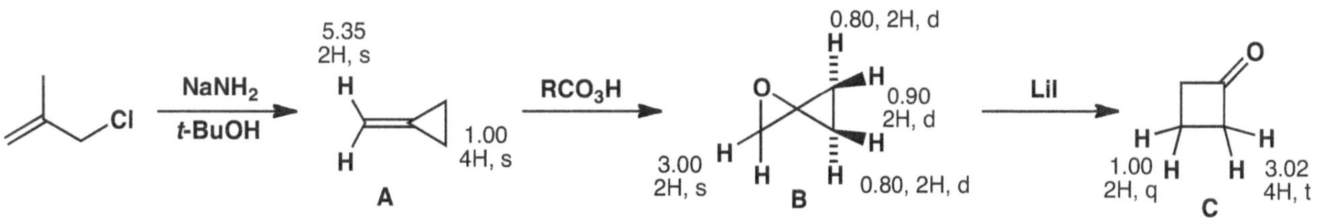
Cl
NaNH2
t-BuOH
5.35
2H, s
1.00
4H, s
A
RCO3H
0.80, 2H, d
0.90
2H, d
3.00
2H, s
0.80, 2H, d
B
LiI
1.00
2H, q
3.02
4H, t
C

PROBLEM 14

The yellow crystalline antibiotic frustulosin was isolated from a fungus in 1978 and it was suggested the structure was an equilibrium mixture of **A** and **B**. Apart from the difficulty that the NMR spectrum clearly shows one compound and not an equilibrium mixture of two compounds, what else makes you unsure of this assignment? Suggest a better structure. Signals marked * exchange with D_2O.

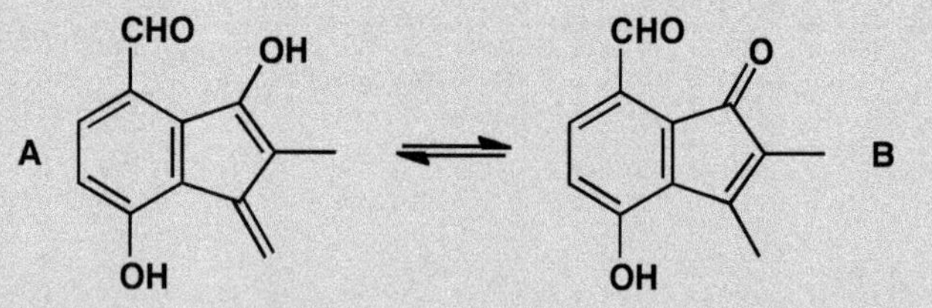

Frustulosin:

m/z 202 (100%), 174 (20%);

v_{max} (cm^{-1}) 3279, 1645, 1613, and 1522;

δ_H (ppm) 2.06 (3H, dd, J 1.0, 1.6 Hz), 5.44 (1H, dq, J 2.0, 1.6 Hz), 5.52 (1H, dq, J 2.0, 1.0 Hz), 4.5* (1H, broad s), 7.16 (1H, d, J 9.0 Hz), 6.88 (1H, dd, J 9.0, 0.4 Hz), 10.31 (1H, d, J 0.4 Hz), and 11.22* (1H, broad s);

δ_C (ppm) 22.8, 80.8, 100.6, 110.6, 118.4, 118.7, 112.6, 125.2, 129.1, 151.8, 154.5, and 196.6.

Warning! This is difficult—after all, the original authors got it wrong initially.
Hint: How might the DBEs be achieved without a second ring?

Purpose of the problem

A serious and difficult determination of a natural product as a final challenge.

Suggested solution

Structure **B** is definitely wrong because the NMR shows only one methyl group, not two, and only one carbonyl group, not two. Structure **A** looks unlikely because it appears to be unstable, but that is not evidence. The NMR shows two protons on the same end of a double bond (at 5.44 and 5.52 ppm) with the characteristic small coupling, but they are coupled to a methyl group, presumably by allylic coupling, and the methyl group is too far away in **B**. But what is the signal at 80.8 in the ^{13}C NMR? The 'hint' was meant to guide you towards suggesting an alkyne. That solves many of the problems even though the carbons of the alkene and the aromatic ring cannot be assigned with confidence. At least the revised structure is one compound and not two.

■ The true structure was later described with the help of NMR as you can read in R. C. Ronald *et al., J. Org. Chem.*, 1982, **47**, 2541 and M. S. Nair and M. Anchel, *Phytochemistry*, 1977, **16**, 390, revised from M. S. Nair and M. Anchel, *Tetrahedron Lett.*, 1975, 2641.

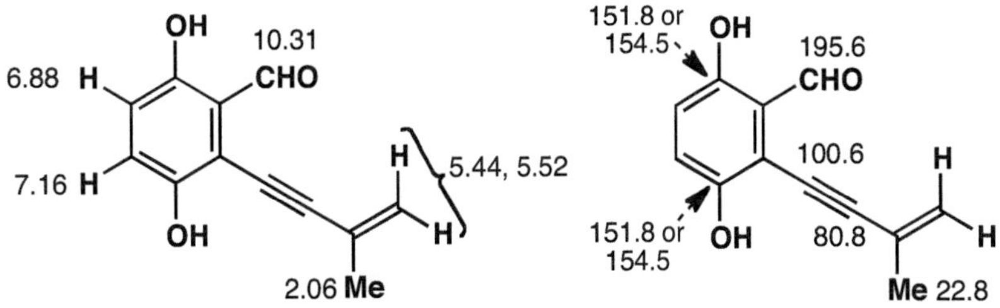

Electrophilic aromatic substitution

21

Connections

➡ Building on	➡ Arriving at	➡ Looking forward to
• Structure of molecules ch4	• Phenols as aromatic enols	• Nucleophilic aromatic substitution ch22
• Conjugation ch7	• Benzene and alkenes compared: what is special about aromatic compounds?	• Oxidation and reduction ch23
• Mechanisms and catalysis ch12		• Regioselectivity and ortholithiation ch24
• Electrophilic addition to alkenes ch19	• Electrophilic attack on benzene	• Retrosynthetic analysis ch28
• Enols and enolates ch20	• Activation and deactivation of the benzene ring	• Aromatic heterocycles ch29 & ch30
	• Position of substitution	• Rearrangements ch36
	• Elaborating aromatic structures: competition and cooperation	• Transition-metal catalysed couplings to aromatic compounds ch40
	• Problems with some aromatic substitution reactions and how to solve them	

Introduction: enols and phenols

In the last chapter you saw that many ketones have a nucleophilic 'alter ego' known as an enol tautomer. Formation of the enol tautomer is catalysed by acid or by base, and because the ketone and enol are in equilibrium, enolization in the presence of D_2O can lead to replacement of the protons in the α positions of ketones by deuterium atoms. This is what happens to pentan-3-one in acidic D_2O:

■ If you haven't just read Chapter 20, look back at p. 451 to remind yourself of how this works.

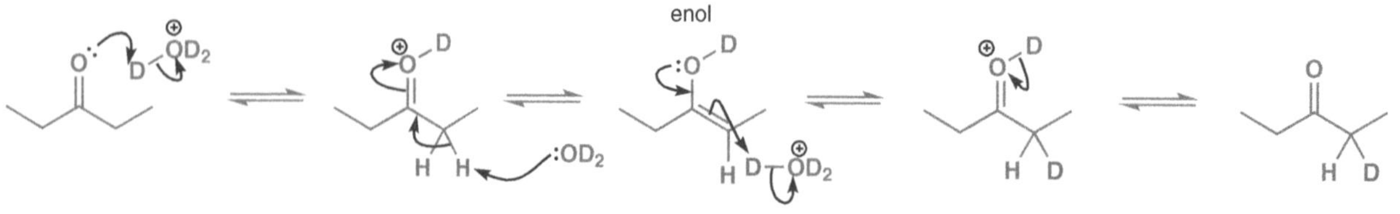

enol

Because the enolization and deuteration process can be repeated, eventually all of the α-protons are replaced by deuterium.

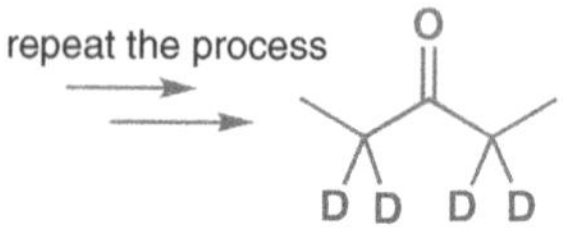

The way this ketone is deuterated provides evidence that its enol form exists, even though the keto/enol equilibrium greatly favours the ketone form at equilibrium. In this chapter we shall be discussing similar reactions of a compound that exists entirely in its enol form. That very stable enol is phenol and its stability is a consequence of the aromaticity of its benzene ring.

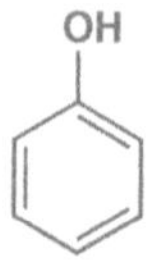

repeat the process

final product

phenol

The proton NMR spectrum for phenol is shown below. Before reading any further cover up the rest of this page and make sure you can assign the spectrum.

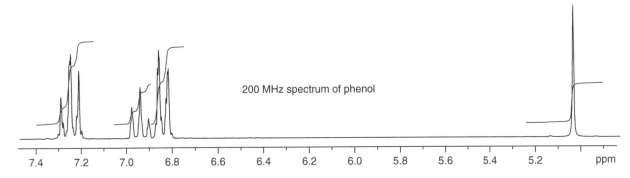

The next spectrum is the proton NMR after shaking phenol with acidic D_2O. Most of the peaks have almost disappeared because the H atoms have been replaced with D. Only one signal remains the same size, and even that is simplified because it has lost any coupling to adjacent protons it may have had previously.

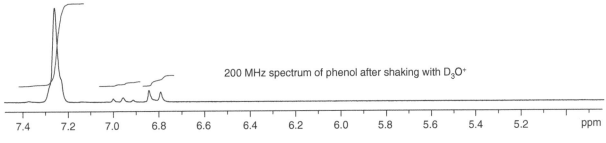

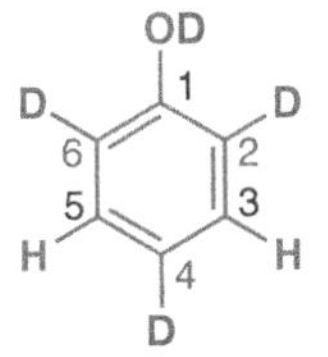

The signal that remains is the 2H signal for the protons in the 3 and 5 positions of the aromatic ring, so the product must be the one shown in the margin. We can explain why by using the same mechanism we used with the ketone on the previous page. Phenol is deuterated in the same way as other enols, except that the final product remains in the very stable, aromatic, enol form rather than reverting to the keto form. The first step (after initial replacement of the OH with OD) is addition of D_3O^+ to the enol.

➡ This equilibrium was discussed on p. 456.

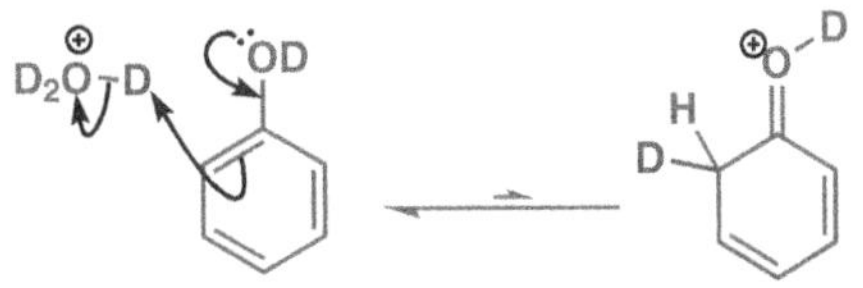

Now this cation could lose the D from oxygen to leave a ketone (brown arrow below), or it could lose the proton from carbon to leave the phenol (orange arrows below). Alternatively, it could just lose the D and go back to the starting material, which is why there is an equilibrium arrow in the scheme above.

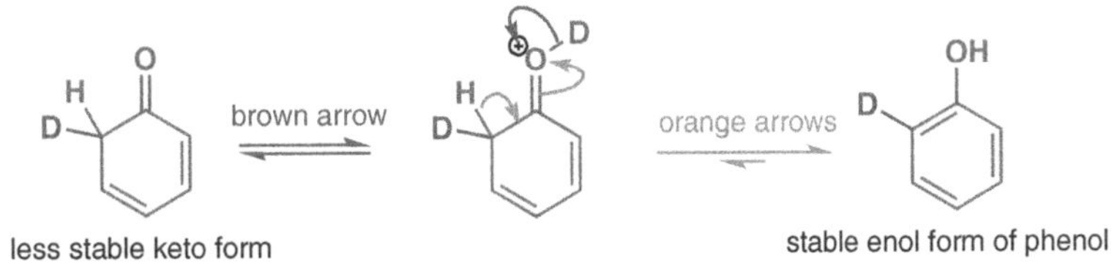

less stable keto form   stable enol form of phenol

Our spectrum tells us that *three* ring protons are replaced by D—the ones at the 2, 4, and 6 positions. It's not hard to see how the same process on the other side of the OH group replaces the proton at C-6. But how does the D at position 4 get there? The enol of phenol is conjugated, and we can push the curly arrows one stage further, like this:

The end product on treating phenol with D_3O^+ has the protons in the 2, 4, and 6 positions (that is, the *ortho* and *para* positions) substituted by deuterium. D_3O^+ is an electrophile, and the overall process is called *electrophilic substitution*. It is a reaction characteristic of not only phenol but of other aromatic compounds, and it forms the subject of this chapter.

> ● When aromatic compounds react with electrophiles they generally do so by **electrophilic aromatic substitution.**

Benzene and its reactions with electrophiles

We'll start with the most straightforward aromatic compound: benzene. Benzene is a planar symmetrical hexagon with six trigonal (sp^2) carbon atoms, each having one hydrogen atom in the plane of the ring. All the bond lengths are 1.39 Å (compare C–C 1.47 Å and C=C 1.33 Å). All the ^{13}C shifts are the same (δ_C 128.5).

two ways of drawing benzene   the π system   NMR data

The special stability of benzene (aromaticity) comes from the six π electrons in three molecular orbitals formed by the overlap of the six atomic p orbitals on the carbon atoms. The energy levels of these orbitals are arranged so that there is exceptional stability in the molecule (a notional 140 kJ mol^{-1} over a molecule with three conjugated double bonds), and the shift of the six identical hydrogen atoms in the NMR spectrum (δ_H 7.2) is evidence of a ring current in the delocalized π system.

Aromatic substituents

A reminder (see pp. 36 and 416) of the names we give to the positions around a benzene ring relative to any substituent:

Ortho, *meta*, and *para* are sometimes abbreviated to *o*, *m*, and *p*.

■ The concept of *aromaticity* is central to this chapter: we will elaborate considerably on the introduction to aromatic compounds we presented in Chapter 7.

➡ The orbitals of benzene were discussed in Chapter 7.

Drawing benzene rings

Benzene is symmetrical and the structure with a circle in the middle best represents this. However, it is impossible to draw curly arrow mechanisms using this representation so we shall usually make use of the Kekulé form with three double bonds. This does not mean that we think the double bonds are localized! It makes no difference which Kekulé structure you draw—any mechanism can be equally well drawn using either.

This circle structure best represents the six delocalized π electrons.

These Kekulé structures are best for drawing curly arrows. They are equivalent.

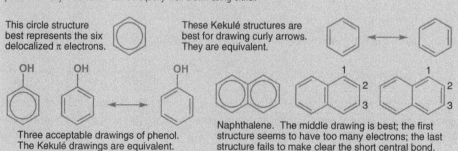

Three acceptable drawings of phenol. The Kekulé drawings are equivalent.

Naphthalene. The middle drawing is best; the first structure seems to have too many electrons; the last structure fails to make clear the short central bond.

In substituted aromatic molecules such as phenol, the C–C bond lengths in the ring are no longer exactly the same. However, it is still all right to use either representation, depending on the purpose of the drawing. With some aromatic compounds, such as naphthalene, it *does* matter which Kekulé structure you use as there is some alternation of bond lengths. Only the first Kekulé representation shows that the central bond is the strongest and shortest in the molecule and that the C1–C2 bond is shorter than the C2–C3 bond. And if a circle in a ring indicates six π electrons, then two circles suggests 12, even though naphthalene has only 10, making this representation less satisfactory too.

Electrophilic attack on benzene and on cyclohexene

Simple alkenes, including cyclohexene, react rapidly with electrophiles such as bromine or peroxy-acids (Chapter 19). Bromine gives the product of *trans* addition, peracids give epoxides by *cis* addition. Under the same conditions benzene reacts with neither reagent.

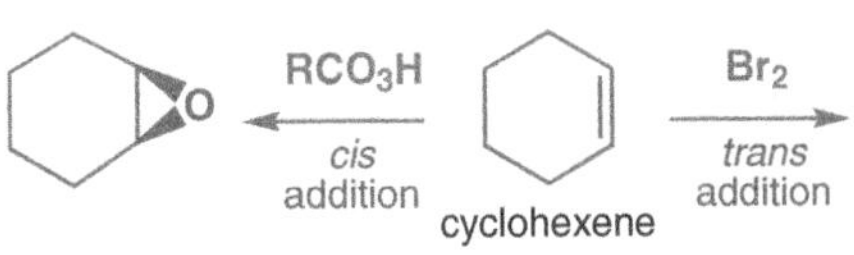

⮕ Lewis acids were described on p. 180.

Benzene can, however, be persuaded to react with bromine if a Lewis acid catalyst such as $AlCl_3$ is added. The product contains bromine but is not from either *cis* or *trans* addition.

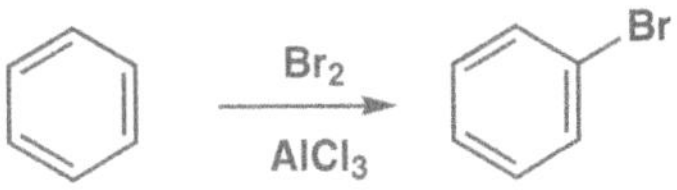

The bromine atom has replaced an atom of hydrogen, so this is a substitution reaction. The reagent (Br_2) is electrophilic and benzene is aromatic so the reaction is **electrophilic aromatic substitution**, the subject of this chapter.

We can compare the bromination of cyclohexene and of benzene directly.

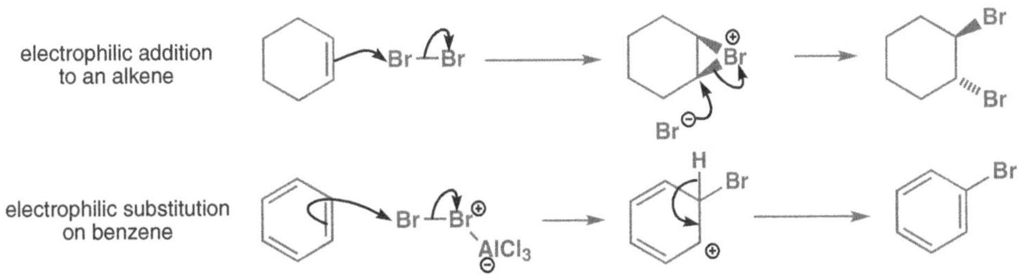

The intermediate in both reactions is a cation but the first (from cyclohexene) adds an anion while the second (from benzene) loses a proton so that the aromatic system can be restored. Notice also that neutral bromine reacts with the alkene but the cationic $AlCl_3$ complex is needed to get reaction with benzene. Bromine itself is a very reactive electrophile. It is indeed a dangerous compound and should be handled only with special precautions. Even so it does not react with benzene. It is difficult to get benzene to react with anything.

● **Benzene is very unreactive**

  • **It combines only with very reactive (usually cationic) electrophiles.**
  • **It gives substitution and not addition products.**

The intermediate in electrophilic aromatic substitution is a delocalized cation

We will return again and again to this mechanism of electrophilic aromatic substitution during this chapter. In its most general form the mechanism has two stages: attack by an electrophile to give an intermediate cation and loss of a proton from the cation to restore the aromaticity.

General mechanism for electrophilic aromatic substitution

The cationic intermediate is, of course, unstable compared with the starting materials or the product. But it is nonetheless stabilized by delocalization. The arrows below show how the positive charge can be delocalized to the two *ortho* positions and to the *para* position, or can be drawn as a single delocalized structure with partial (dotted) bonds and about one-third of a positive charge (+) at three atoms.

the brown H is drawn in to emphasize the non-aromaticity of this delocalized cation

It's very important to note that although it is delocalized, this cation is not aromatic: there is no cyclic array of p orbitals because the ring contains a single tetrahedral (sp³ hybridized) carbon atom. We have emphasized this tetrahedral atom by drawing in the hydrogen atom at the point of substitution—the one that will be lost when aromaticity is regained. We suggest that when you write mechanisms for electrophilic aromatic substitution you do the same. Given this loss of aromaticity, it is not surprising that formation of the cationic intermediate is the rate-determining step of an electrophilic aromatic substitution.

How do we know the cationic intermediate exists?

In strong acid, the electrophile is a proton and it is actually possible to observe this cationic intermediate. The trick is to pick a non-nucleophilic and non-basic counterion X⁻, such as SbF₆⁻. In this octahedral anion, the central antimony atom is surrounded by the fluorine atoms with the negative charge spread over all seven atoms. The protonation is carried out using FSO₃H and SbF₅ at −120 °C. A similar trick was described in Chapter 15 as a means to show the existence of simple carbocations as intermediates in the S_N1 mechanism.

Under these conditions it is possible to record the ¹H and ¹³C NMR spectra of the cation. The shifts show that the positive charge is spread over the ring but is greatest (i.e. the electron density is least) at the *ortho* and *para* positions. Using the data for the ¹H and ¹³C NMR shifts (δ_H and benzene δ_C, respectively), a charge distribution can be calculated that closely matches the predictions of the curly arrrows.

position	δ_H	δ_C
1	5.6	52.2
2,6	9.7	186.6
3,5	8.6	136.9
4	9.3	178.1
benzene (for comparison)	7.33	129.7

Nitration of benzene

Now we've introduced to you the general principles of electrophilic aromatic substitution we need to delve into the details a little more and show you some real reactions of benzene. In each case, a powerful cationic electrophile is needed to persuade the unreactive benzene to act as a nucleophile.

■ The delocalized structure of the nitro group was discussed in Chapter 2.

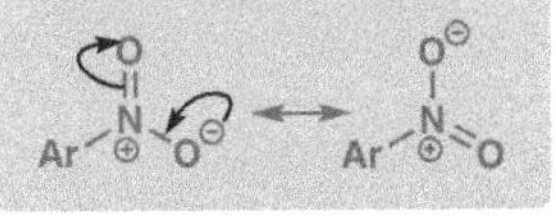

We'll start with nitration, the introduction of a nitro (NO_2) group. Nitration requires very powerful reagents, the most typical being a mixture of concentrated nitric and sulfuric acids.

nitration of benzene

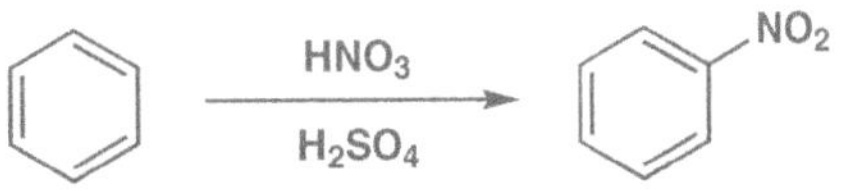

Sulfuric acid is the stronger acid and it produces the powerful electrophile NO_2^+ by protonating the nitric acid so that a molecule of water can leave.

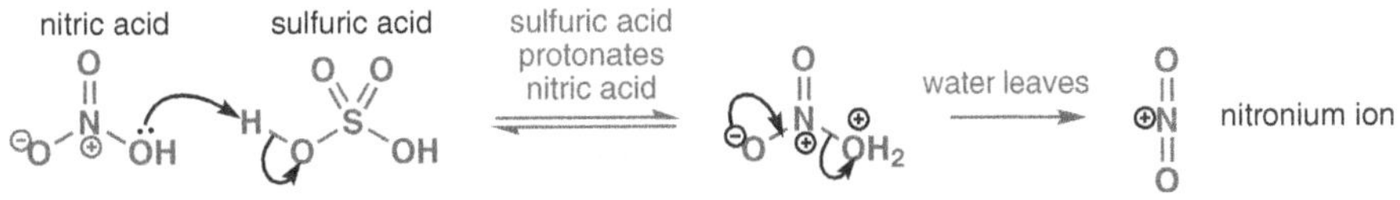

nitric acid   sulfuric acid   sulfuric acid protonates nitric acid   water leaves   nitronium ion

The nitronium ion (NO_2^+) is linear—it's isoelectronic with CO_2, with an sp-hybridized nitrogen atom at the centre. It's this nitrogen that is attacked by benzene, breaking one of the N=O bonds to avoid a five-valent nitrogen.

🖱 Interactive mechanism for nitration of benzene

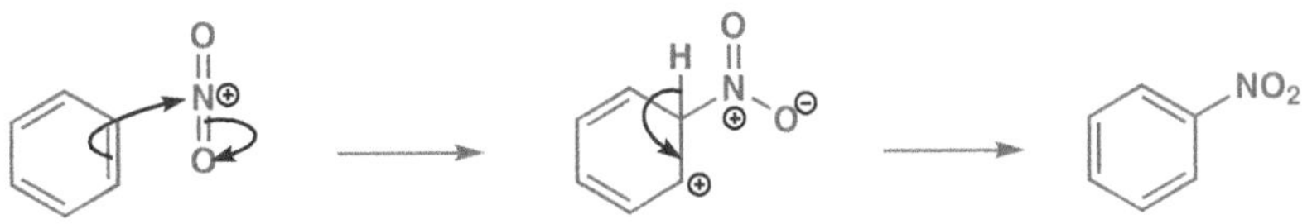

■ A reminder: electrophilic aromatic substitution mechanisms are easier to follow if you draw in the H at the point of substitution.

● **Nitration converts aromatic compounds (ArH) into nitrobenzenes (ArNO$_2$) using NO$_2^-$ from HNO$_3$ + H$_2$SO$_4$.**

Sulfonation of benzene

The cationic intermediate can also be formed by the protonation of sulfur trioxide, SO$_3$, and another way to do sulfonations is to use concentrated sulfuric acid with SO$_3$ added. These solutions have the industrial name **oleum**. It is possible that the sulfonating agent in all these reactions is not protonated SO$_3$ but SO$_3$ itself.

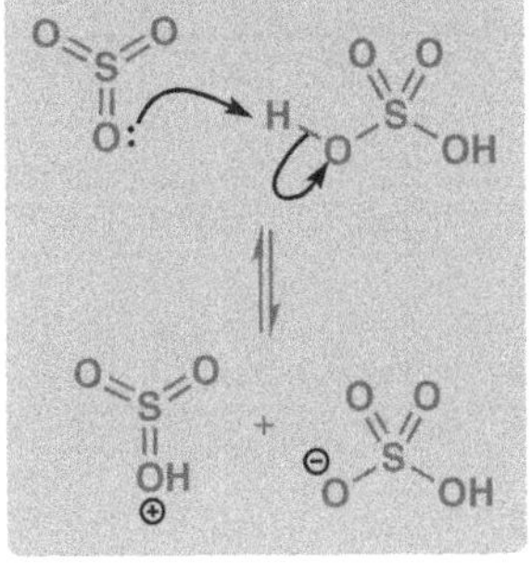

Benzene reacts slowly with sulfuric acid alone to give benzenesulfonic acid. One molecule of sulfuric acid protonates another and loses a molecule of water. Notice the similarity with the first step of the nitration above.

The cation produced is very reactive and attacks benzene by the same mechanism we have seen for bromination and nitration—slow addition to the π system followed by rapid loss of a proton to regenerate aromaticity.

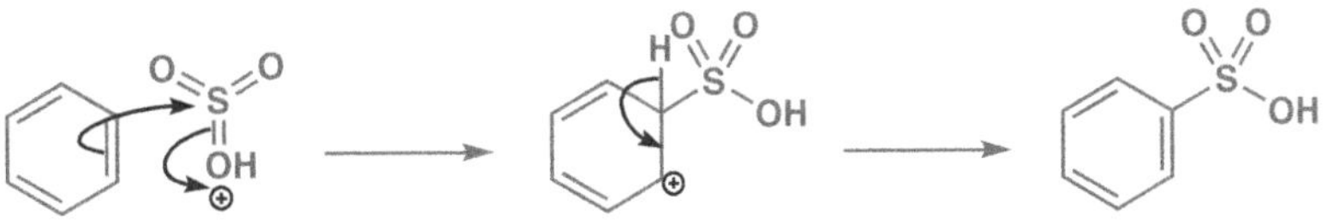

benzenesulfonic acid

🖱 Interactive mechanism for sulfonation of benzene

The product contains the sulfonic acid group $-SO_2OH$. Sulfonic acids are strong acids, about as strong as sulfuric acid itself. They are stronger than HCl, for example, and can be isolated from the reaction mixture as their crystalline sodium salts if an excess of NaCl is added. Not many compounds react with NaCl!

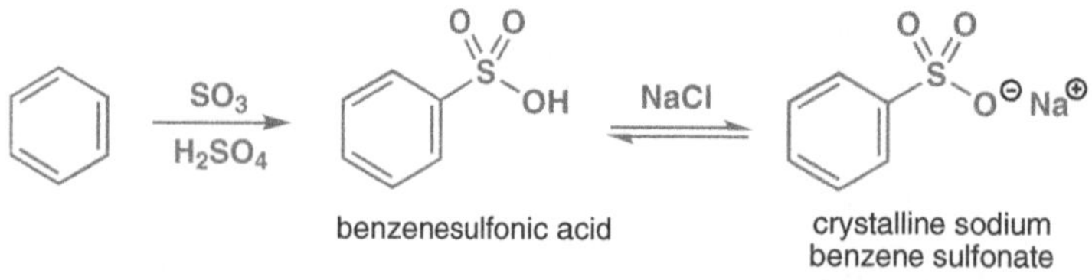

benzenesulfonic acid

crystalline sodium benzene sulfonate

> ➡ You met a related sulfonate anion in the guise of the excellent tosylate leaving group in Chapter 15.

> ● Sulfonation with H_2SO_4 or SO_3 in H_2SO_4 converts aromatic compounds (ArH) into aromatic sulfonic acids ($ArSO_2OH$). The electrophile is SO_3 or SO_3H^+.

Alkyl and acyl substituents can be added to a benzene ring by the Friedel–Crafts reaction

So far we have added heteroatoms only—bromine, nitrogen, or sulfur. Adding a carbon substituent to a reluctant aromatic nucleophile requires reactive carbon electrophiles and that means carbocations. In Chapter 15 you learned that any nucleophile, however weak, will react with a carbocation in the S_N1 reaction: benzene rings are no exception. The classic S_N1 electrophile is the *t*-butyl cation, which is generated from *tert*-butanol with acid.

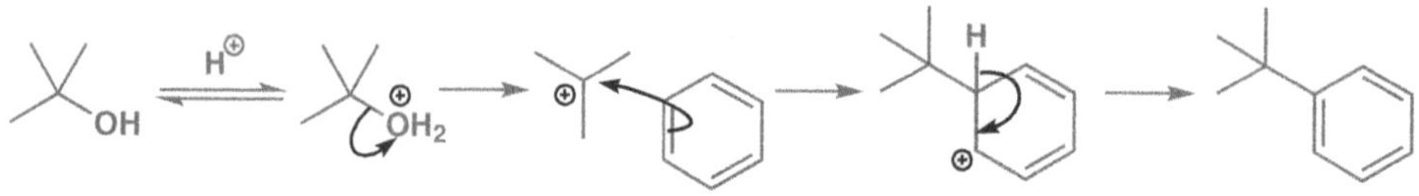

This is, in fact, an unusual way to carry out such reactions. The **Friedel–Crafts alkylation**, as this is known, usually involves treating benzene with a tertiary alkyl chloride and the Lewis acid $AlCl_3$. Rather in the manner of the reaction with bromine, $AlCl_3$ removes the chlorine atom from *t*-BuCl and releases the *t*-Bu cation for the alkylation reaction.

> Charles Friedel (1832–1899), a French chemist, and James Crafts (1839–1917), an American mining engineer, both studied with Wurtz and then worked together in Paris, where in 1877 they discovered the reaction which now carries their names.

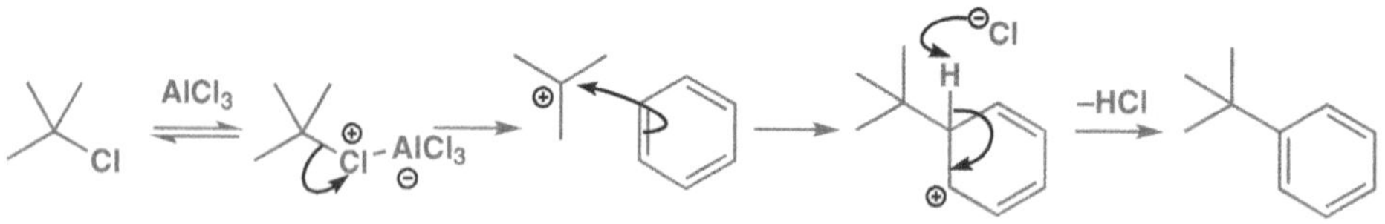

We have not usually bothered with the base that removes the proton from the intermediate. Here it is chloride ion as the by-product is HCl, so you can see that even a very weak base will do. Anything, such as water, chloride, or other counterions of strong acids, will do this job well enough and you need not in general be concerned with the exact agent.

A more important variation of this reaction is the **Friedel–Crafts acylation** with acid chlorides and $AlCl_3$. Aluminium chloride behaves with acyl chlorides much as it does with alkyl chlorides—it removes chloride to leave behind a cation. In this case the cation is a linear acylium ion, with the carbocation stabilized by the adjacent oxygen lone pair. When the acylium ion attacks the benzene ring it gives an aromatic ketone: the benzene ring has been acylated.

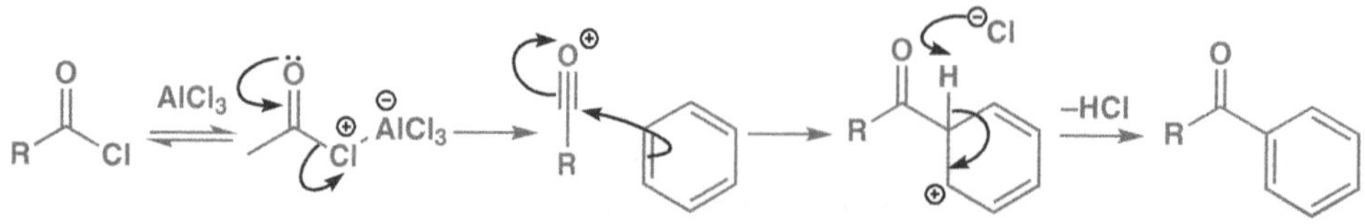

> 🖱 Interactive mechanism for Friedel–Crafts alkylation

> 🖱 Interactive mechanism for Friedel–Crafts acylation

➡ We'll come back (on p. 492) to why this is and what can be done about it.

The acylation is better than the alkylation because it does not require any particular structural feature in the acyl chloride—R can be almost anything. In the alkylation step it is essential that the alkyl group can form a cation, otherwise the reaction does not work very well. In addition, for reasons we are about to explore, the acylation stops cleanly after one reaction whereas the alkylation often gives mixtures of products.

● **Friedel–Crafts reactions**

Friedel–Crafts alkylation with *t*-alkyl chlorides and Lewis acids (usually $AlCl_3$) gives *t*-alkyl benzenes. The more reliable Friedel–Crafts acylation with acid chlorides and Lewis acids (usually $AlCl_3$) gives aryl ketones.

Summary of electrophilic substitution on benzene

This completes our preliminary survey of the most important reactions in aromatic electrophilic substitution. We shall switch our attention to the benzene ring itself now and see what effects various types of substituent have on these reactions. During this discussion we will return to each of the main reactions and discuss them in more detail. Meanwhile, we conclude this introduction with an energy profile diagram for a typical substitution.

Since the first step involves the temporary disruption of the aromatic π system, and is therefore rate determining, it must have the higher-energy transition state. The intermediate is unstable and has a much higher energy than either the starting material or the products, close to that of the transition states for its formation and breakdown. The two transition states will be similar in structure to the intermediate and we shall use the intermediate as a model for the important first transition state.

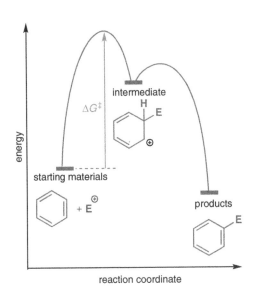

■ This argument is based on the **Hammond postulate**, which suggests that structures close in energy that transform directly into each other are also similar in structure. For more on this, see Chapter 39.

● **Summary of the main electrophilic substitutions on benzene**

Reaction	Reagents	Electrophile	Product
bromination	Br_2 and Lewis acid, e.g. $AlCl_3$, $FeBr_3$, Fe powder		Br
nitration	$HNO_3 + H_2SO_4$		NO_2
sulfonation	concentrated H_2SO_4 or $H_2SO_4 + SO_3$ (oleum)		SO_2OH
Friedel–Crafts alkylation	RX + Lewis acid usually $AlCl_3$		R
Friedel–Crafts acylation	RCOCl + Lewis acid usually $AlCl_3$		O ‖ R

Electrophilic substitution on phenols

We started this chapter by comparing phenols with enols and now we return to phenols and look at electrophilic substitution in full detail. You will find that the reaction is much easier than it was with benzene itself because phenols are like enols and the same reactions (bromination, nitration, sulfonation, and Friedel–Crafts reactions) occur more easily. There is a new question too: the positions round the phenol ring are no longer equivalent—so where does substitution take place?

Phenols react rapidly with bromine

Benzene does not react with bromine except with Lewis acid catalysis. Phenols react in a very different manner: no Lewis acid is needed, the reaction occurs very rapidly, and the product contains three atoms of bromine in specific positions. All that needs to be done is to add bromine dropwise to a solution of phenol in ethanol. Initially, the yellow colour of the bromine disappears but if, when the colour just remains, water is added, a white precipitate of 2,4,6-tribromophenol is formed.

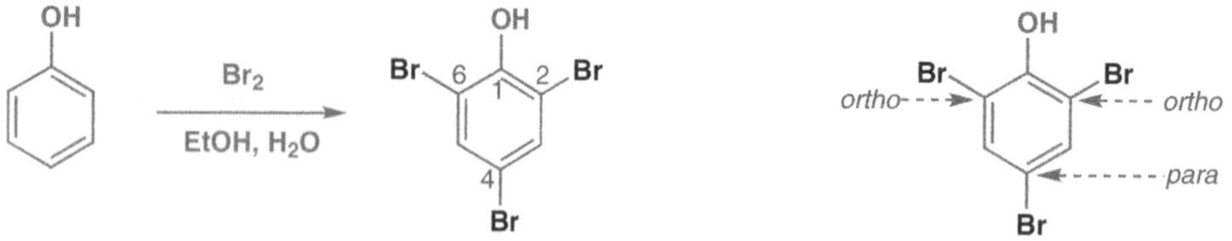

Why do we use numbers for some descriptions, such as 2,4-dibromophenol, but also use *ortho* and *para* in others? The numbers are best in naming compounds but we need *ortho* and *para* to describe the relationship between substituents. Phenol brominates in both *ortho* positions. In this molecule they happen to be positions 2 and 6. In other molecules, where the OH group is not at C1, they will have other numbers, but they will still be *ortho* to the OH group. Use whichever description suits the point you are making.

The product shows that bromination has occurred at the *para* position and at both *ortho* positions. What a contrast to benzene! Phenol reacts three times, without catalysis, at room temperature. Benzene reacts once, and needs a Lewis acid to make the reaction go at all. The difference is, of course, the enol nature of phenol. The non-bonding lone pair of electrons at oxygen contribute to a much higher-energy HOMO than the low-energy bonding electrons in a benzene ring. We should let our mechanism show this. Starting in the *para* position:

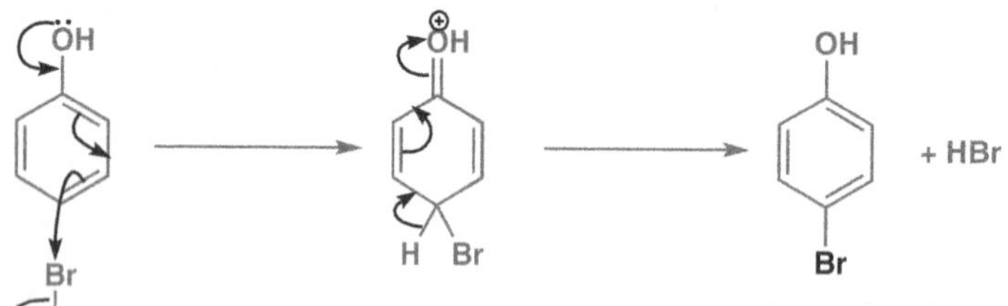

para-bromophenol
(4-bromophenol)

■ This mechanism should remind you of the bromination of enols in Chapter 20.

Notice that we start the chain of arrows with the lone pair electrons on the OH group and push them through the ring so that they emerge at the *para* position to attack the bromine molecule. The benzene ring is acting as a conductor, allowing electrons to flow from the OH group to the bromine molecule.

Now the reaction is repeated, but this time at one of the two equivalent *ortho* positions:

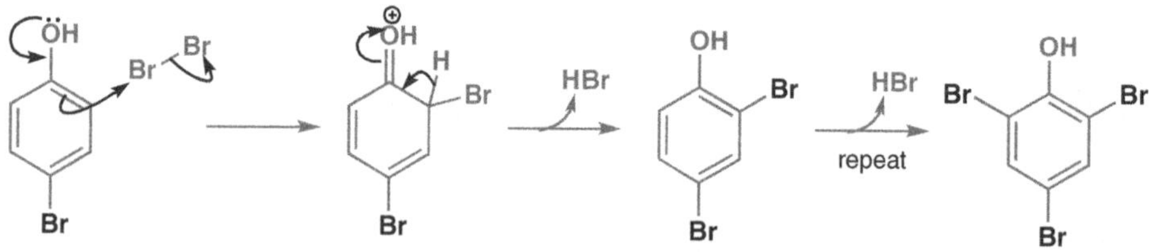

A similar reaction with chlorine is used to make the well-known antiseptic TCP (2,4,6-trichlorophenol). The characteristic smell of TCP is typical of the smell of many other phenols.

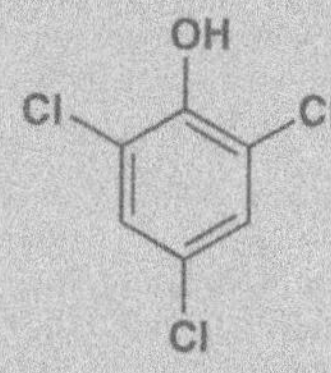

2,4,6-trichlorophenol (TCP)

Again the lone pair electrons on the OH group are fed through the benzene ring to emerge at the *ortho* position. A third bromination in the remaining *ortho* position—you could draw the mechanisms for this as practice—gives the final product 2,4,6-tribromophenol.

If you want to put just one bromine atom into a phenol, you must work at low temperature (<5 °C) and use just one equivalent of bromine. The best solvent is the rather dangerously inflammable carbon disulfide (CS_2), the sulfur analogue of CO_2. Under these conditions, *para*-bromophenol is formed in good yield as the main product (which is why we started the mechanism for bromination of phenol in the *para* position). The minor product is *ortho*-bromophenol.

4-bromophenol
85% yield

The OH group is said to be ***ortho*, *para*-directing** towards electrophiles. No substitution occurs in either *meta* position. We can understand this by looking at the curly arrow mechanisms or by looking at the molecular orbitals. In Chapter 20 (p. 453) we looked at the π system of an enolate and saw how the electron density is located mainly on the end atoms (the oxygen and the carbon). In phenol it is the *ortho* and *para* positions that are electron-rich (and, of course, the oxygen itself). We can show this using curly arrows.

The curly arrows actually give an indication of the electron distribution in the HOMO of the molecule. The reason is that the HOMO has large coefficients at *alternate atoms*, just as the allyl anion had large coefficients at its ends but not in the middle (Chapter 7).

NMR can give us some confirmation of the electron distribution

¹H NMR shifts in phenol and benzene

The ¹H NMR shifts of phenol give us an indication of the electron distribution in the π system. The more electron density that surrounds a nucleus, the more shielded it is and so the smaller the shift (see Chapter 13). All the chemical shifts for the ring protons in phenol are smaller than those for benzene (7.26 ppm), which means that overall there is greater electron density in the ring. There is little difference between the *ortho* and the *para* positions: these are where the electron density is greatest and hence these are the sites for electrophilic attack. The chemical shift at the *meta* positions is not significantly different from those in benzene—this is where the electron density is lowest.

● **Electrophilic attack on phenols**
OH groups on benzene rings are *ortho*, *para*-directing and activating.
You will get the right product if you start your arrows at a lone pair on the OH group.

Oxygen substituents activate a benzene ring

anisole

To brominate phenol, all we had to do was to mix bromine and phenol—if we do this with benzene itself, nothing happens. We therefore say that, relative to benzene, the OH group in phenol *activates* the ring towards electrophilic attack. The OH group is both activating and *ortho*, *para*-directing. Other groups that can donate electrons also activate and direct *ortho*, *para*. Anisole (methoxybenzene) is the 'enol ether' equivalent of phenol. It reacts faster than benzene with electrophiles.

2,4-D

The multiple chlorination of another oxygen-substituted compound, phenoxyacetic acid, leads to a useful product. Chlorination with two equivalents of chlorine provides 2,4-dichlorophenoxy acetic acid, which is the herbicide 2,4-D. The oxygen substituent again activates the ring and directs the chlorination to the *ortho* and *para* positions.

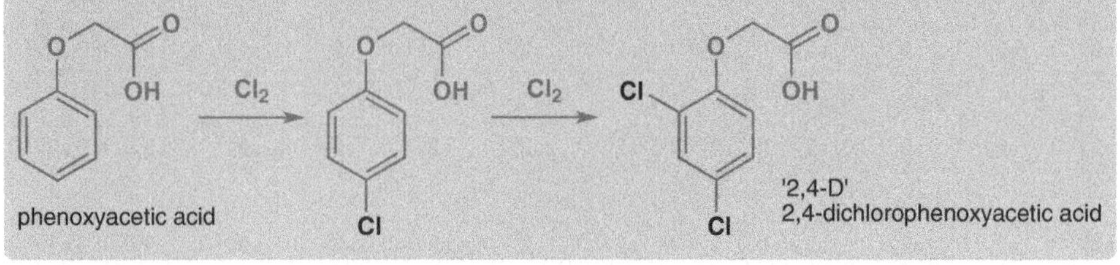

phenoxyacetic acid → Cl → '2,4-D' 2,4-dichlorophenoxyacetic acid

Nitration of phenol is also very fast and can be problematic under the usual nitration conditions (conc. HNO_3, conc. H_2SO_4) because concentrated nitric acid oxidizes phenols. The solution is to use dilute nitric acid. The concentration of NO_2^+ will be small but that does not matter with such a reactive benzene ring.

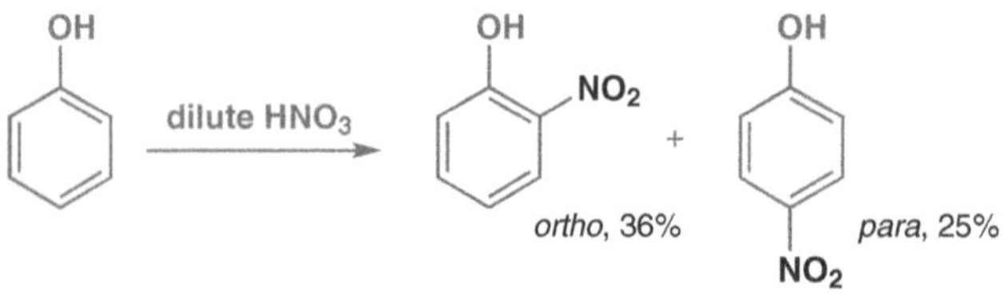

ortho, 36% + *para*, 25%

The product is a mixture of *ortho*- and *para*-nitrophenol from which the *ortho* compound can be separated by steam distillation. A strong intramolecular hydrogen bond reduces the availability of the OH group for intermolecular hydrogen bonding so the *ortho* compound has a lower boiling point.

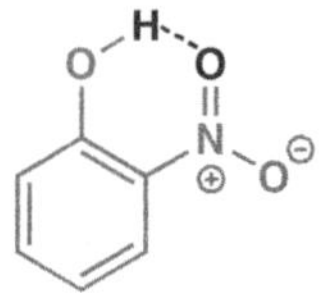

strong intramolecular H bond

Paracetamol from a phenol

The remaining *para*-nitrophenol is used in the manufacture of the painkiller paracetamol (also known as acetaminophen).

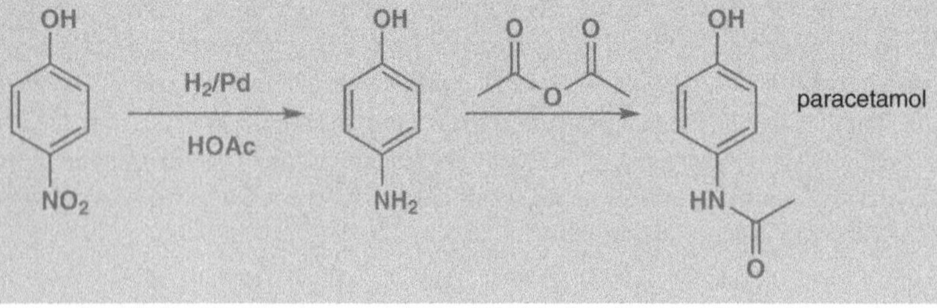

paracetamol

The phenoxide ion is even more reactive towards electrophilic attack than phenol. It manages to react with such weak electrophiles as carbon dioxide. This reaction, known as the **Kolbe–Schmitt process**, is used industrially to prepare salicylic acid (2-hydroxybenzoic acid), a precursor in making aspirin.

■ Salicylic acid is 2-hydroxybenzoic acid and is named after the willow trees (genus *Salix*) from which it was first isolated.

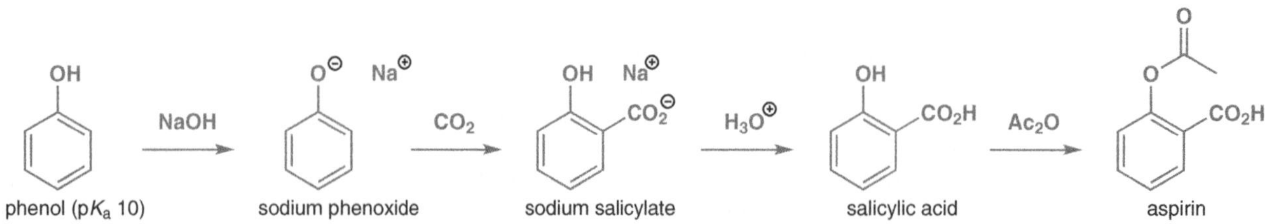

phenol (pK_a 10) → sodium phenoxide → sodium salicylate → salicylic acid → aspirin

The O⁻ substituent is *ortho, para*-directing but the electrophilic substitution step with CO_2 gives mostly the *ortho* product. There must be some coordination between the sodium ion and the oxygen atoms of both the phenoxide and CO_2 that delivers the electrophile to the *ortho* position.

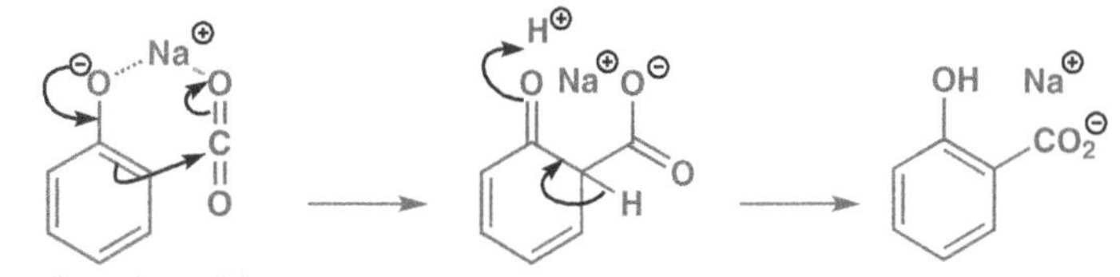

sodium phenoxide   sodium salicylate

A nitrogen lone pair activates even more strongly

Aniline (phenylamine) is even more reactive towards electrophiles than phenols, phenyl ethers, or phenoxide ions. Because nitrogen is less electronegative than oxygen, the lone pair is higher in energy and so even more available to interact with the π system than is the lone pair on oxygen. Reaction of aniline with bromine is very vigorous and rapidly gives 2,4,6-tribromoaniline. The mechanism is very similar to the bromination of phenol so we show only one *ortho* substitution to remind you of how it goes.

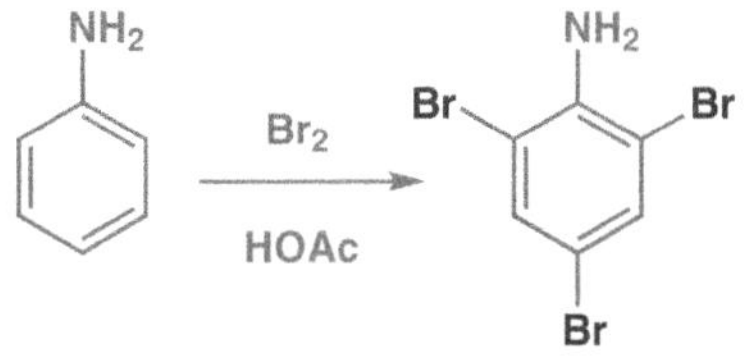

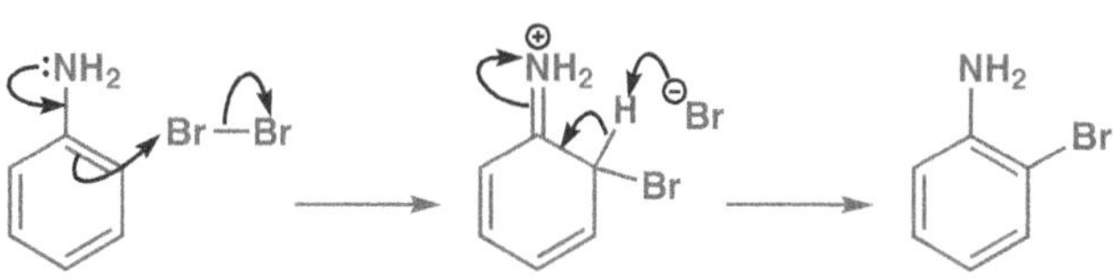

¹H NMR shifts in phenol and aniline

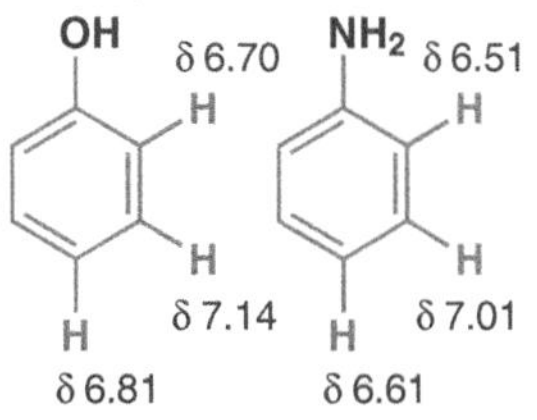

The ¹H NMR spectrum of aniline supports the increased electron density in the π system—the aromatic protons are even less deshielded than those of phenol as a result.

Just how good nitrogen is in donating electrons into the π system is shown by comparing the relative rates for the bromination of benzene, methoxybenzene (anisole), and *N,N*-dimethylaniline.

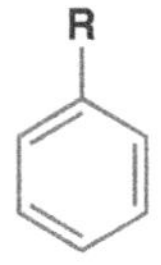

Compound	R	Relative rate of bromination
benzene	H	1
methoxybenzene (anisole)	OMe	10^9
N,N-dimethylaniline	NMe₂	10^{14}

Making aromatic amines less reactive

The high reactivity of aniline can actually be a problem. Suppose we wanted to put just one bromine atom onto the ring. With phenol, this is possible (p. 480)—if bromine is added slowly to a solution of phenol in carbon disulfide solution and the temperature is kept below 5 °C, the main product is *para*-bromophenol. Not so if aniline is used—the main product is the triply substituted product.

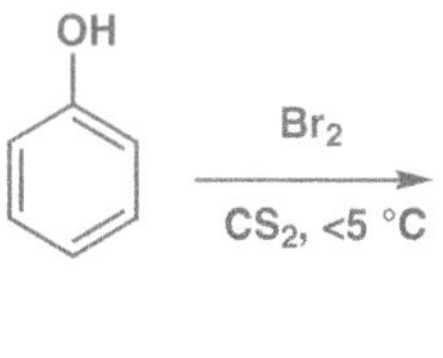

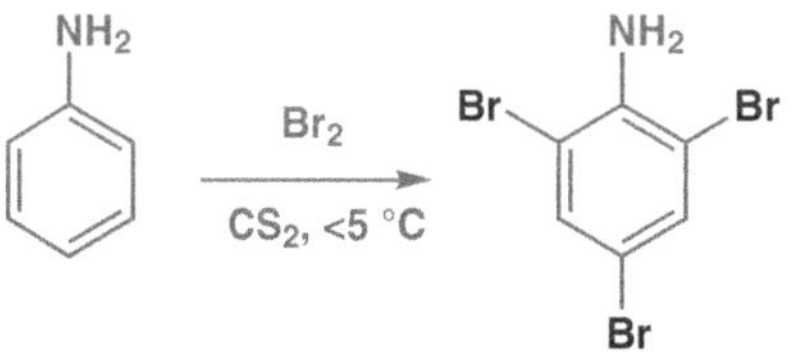

How then could we prevent oversubstitution from occurring? What we need is a way to make aniline less reactive by preventing the nitrogen lone pair from interacting so strongly with the π system of the ring. Fortunately, it is very simple to do this. In Chapter 8 (p. 175) we saw how the nitrogen atom in an amide is much less basic than a normal amine because it is conjugated with the carbonyl group. This is the strategy that we will use here—simply acylate the amine to form an amide. The lone pair electrons on the nitrogen atom of the amide are conjugated with the carbonyl group as usual but their delocalization into the benzene ring is weaker than in the amine. The amide nitrogen donates less electron density into the ring, so the electrophilic aromatic substitution is more controlled. The lone pair is still there, but its power is tamed. Reaction still occurs in the *ortho* and *para* positions (mainly *para*) but it occurs once only.

■ Amides formed by the acylation of anilines are sometimes called *anilides*. If they are acetyl derivatives they are called *acetanilides*. We shall not use these names but you may meet them elsewhere.

After the reaction, the amide can be hydrolysed (here, with aqueous acid) back to the amine.

● **Anilines react rapidly with electrophiles to give polysubstituted products. Their amide derivatives react in a more controlled manner to give *para*-substituted products.**

Selectivity between *ortho* and *para* positions

Phenols and anilines react in the *ortho* and/or *para* positions for electronic reasons. These are the most important effects in deciding where an electrophilic substitution will occur on a benzene ring. When it comes to choosing between *ortho* and *para* positions we need to consider steric effects as well. You will have noticed that we have seen one *ortho* selective reaction—the formation of salicylic acid from phenol—and several *para* selective reactions such as the bromination of an amide just discussed.

If the reactions occurred merely statistically, we should expect twice as much *ortho* as *para* product because there are two *ortho* positions. However, we should also expect more steric hindrance in *ortho* substitution since the new substituent must sit closely beside the one already there. With large substituents, such as the amide, steric hindrance will be significant and it is not surprising that we get more *para* product.

There is another effect that decreases the amount of *ortho* substitution, and that is the *inductive* electron-withdrawing effect of an electronegative substituent. As you've seen, oxygen and nitrogen, although they are electronegative, activate the ring towards attack by donating π electron density from their lone pairs. At the same time, the C–O or C–N σ bond is polarized back towards the O or N atom—in other words, they *donate* electron density to the π system but *withdraw* electron density from the σ framework. This is *inductive* electron withdrawal—it affects the atoms nearest the O or N atom the most, and has the effect of decreasing the likelihood that attack will happen in the *ortho* positions.

steric hindrance at *ortho* position

fast C–N rotation

no steric hindrance at *para* position

➡ Inductive effects were introduced on p. 135.

Alkyl benzenes also react at the *ortho* and *para* positions

This is what happens when toluene (methylbenzene) meets bromine:

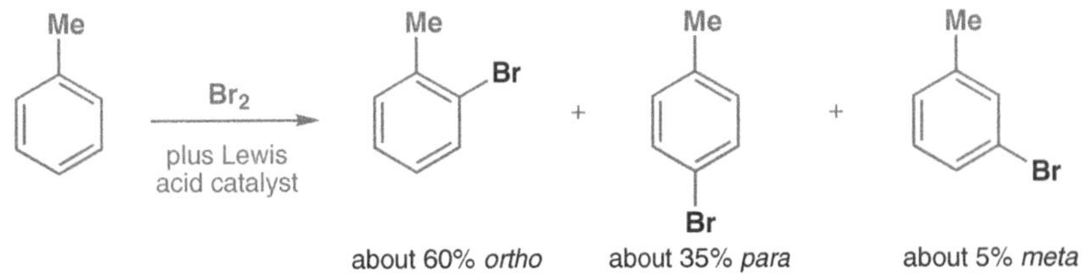

about 60% *ortho*   about 35% *para*   about 5% *meta*

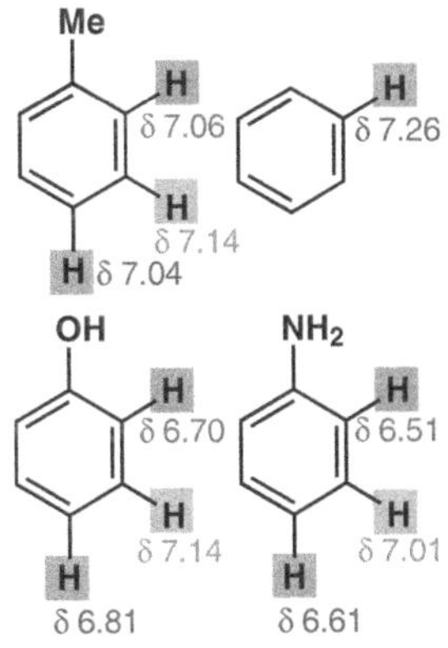

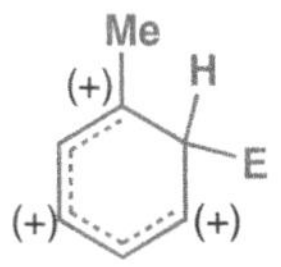

Toluene reacts 4000 times faster than benzene (this may sound like a lot, but the rate constant for *N,N*-dimethylaniline is 10^{14} times greater), and the electrophile attacks mostly the *ortho* and *para* positions. These two observations together suggest that the methyl groups may be increasing the electron density in the π system of the benzene ring, specifically in the *ortho* and *para* positions, rather like a weaker version of an OR group. The ^{1}H NMR chemical shifts for toluene (see margin) do suggest that there is slightly more electron density in the *para* position than in the *meta* positions. All the shifts are smaller than those of benzene (but not by much) and the shielding is much less than it is in phenols or anilines.

The methyl group donates electrons weakly by conjugation. In phenol, a lone pair on oxygen is conjugated with the π system. In toluene there is no lone pair but one of the C–H σ bonds can still interact with the π system in a similar way. This interaction is known as σ **conjugation**. Just as the conjugation of the oxygen lone pair increases the electron density at the *ortho* and *para* positions, so too does σ conjugation, but far less so.

σ conjugation also means toluene's π electrons—its HOMO—become slightly higher in energy than those of benzene. It is best to regard alkyl benzenes as rather reactive benzenes, and to draw mechanisms using their π electrons as the nucleophile, like this:

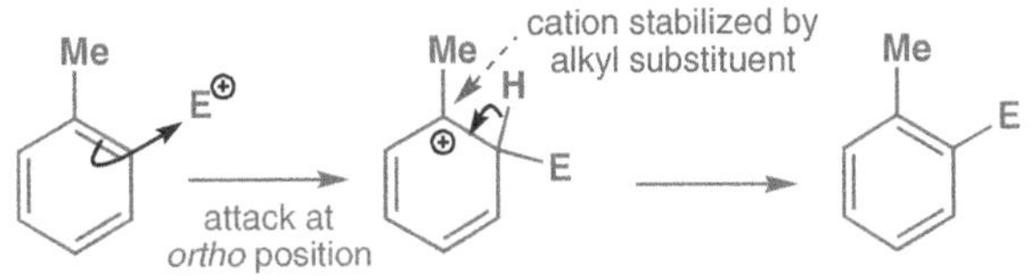

Electrophilic attack occurs on alkyl benzenes so that the positive charge ends up on the carbon bearing the alkyl group. This carbon is tertiary, making the cation there more stable. This condition is fulfilled if toluene is attacked at the *ortho* position, as shown above, but also at the *para* position, because in both cases the positive charge is delocalized onto the same three carbons atoms.

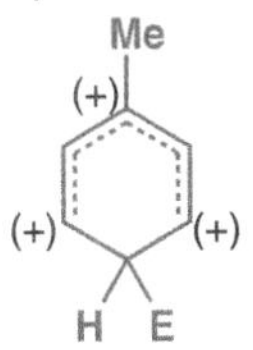

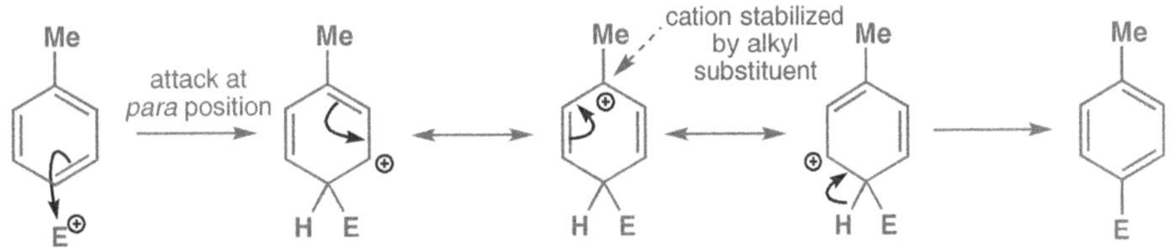

If, on the other hand, the electrophile were to attack at the *meta* position, the charge would end up delocalized over three carbon atoms, none of which are tertiary, so no stabilization by the alkyl group is possible. The situation is no worse than that of benzene, but given that toluene reacts some 10^3 times faster than benzene at the *ortho* and *para* positions these reactions win out. Nonetheless, unlike phenol, toluene does give trace amounts of *meta*-substituted products.

overlap of one C–H σ bond
with π system of the ring

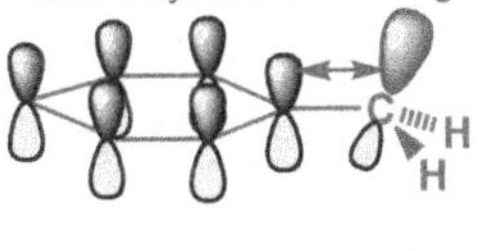

favourable intermediate
for *ortho* substitution

■ You are familiar with the idea that more substituted cations are more stable (Chapter 15, p. 335) and that more substituted alkenes are more stable (Chapter 17, p. 394). The effect we are discussing here is the same.

favourable intermediate
for *para* substitution

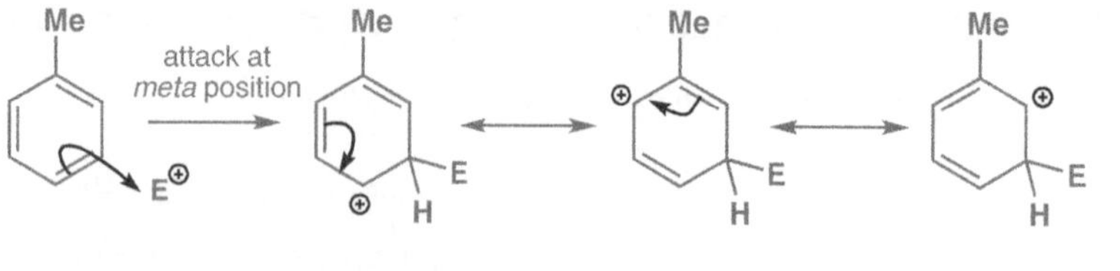

unfavourable intermediate
for *meta* substitution

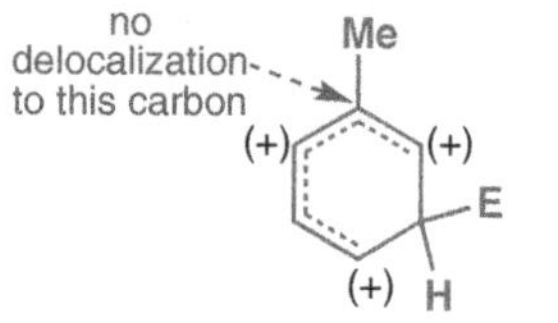

no
delocalization-
to this carbon

Protonating toluene with a superacid

On p. 475 we described how to observe the cationic intermediate in electrophilic substitution reactions of benzene by protonation in an NMR tube using a superacid. In benzene the cation which forms is symmetrical. Doing the same experiment with toluene leads to protonation in the *para* position.

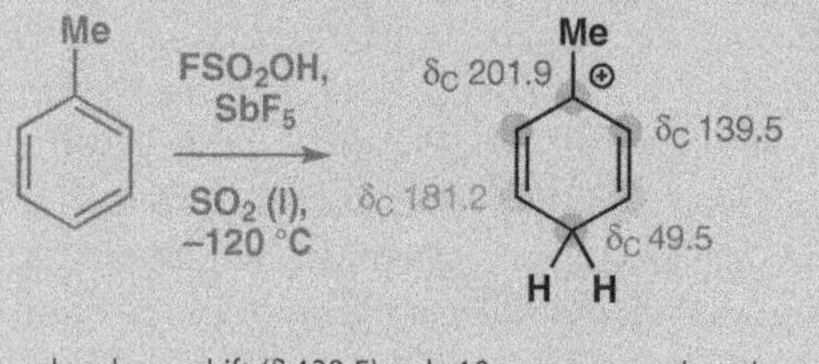

The *ortho* (to the Me group) carbon has a shift (δ 139.5) only 10 ppm greater than that of benzene (δ 129.7) but the *ipso* and *meta* carbons have the very large shifts that we associate with cations. The charge is mainly delocalized to these carbons but the greatest charge is at the *ipso* carbon.

The sulfonation of toluene

Direct sulfonation of toluene with concentrated sulfuric acid gives a mixture of *ortho* and *para* sulfonic acids from which about 40% of toluene *para* sulfonic acid can be isolated as the sodium salt.

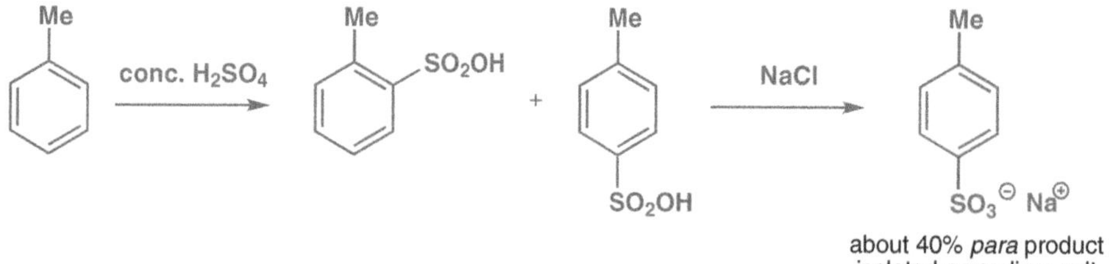

about 40% *para* product
isolated as sodium salt

We shall use SO_3 as the electrophile in this case and draw the intermediate with the charge at the *ipso* carbon to show the stabilization from the methyl group.

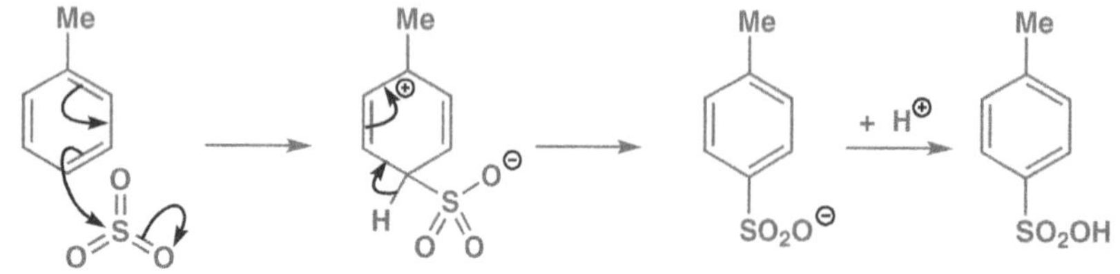

You met the *para*-toluenesulfonate group (tosylate, OTs) as an important leaving group if you want to carry out an S_N2 reaction on an alcohol (Chapter 15, p. 349) and the acid chloride (tosyl chloride, TsCl) needed to make tosylates can be made from the acid in the usual way (p. 215) with PCl_5. It can also be made directly from toluene by sulfonation with chlorosulfonic acid $ClSO_2OH$. This reaction favours the *ortho* sulfonyl chloride, which is isolated by distillation.

Toluenesulfonic acid

The product *para*-toluenesulfonic acid is important as a convenient solid acid, useful when a strong acid is needed to catalyse a reaction. Being much more easily handled than oily and corrosive sulfuric acid or syrupy phosphoric acid, it is useful for acetal formation (Chapter 11) and eliminations by the E1 mechanism on alcohols (Chapter 17). It also gets called tosic acid, TsOH, or PTSA, and its sulfonyl chloride derivative is tosyl chloride, TsCl (Chapter 15).

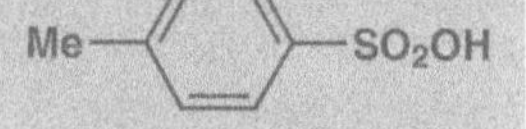

p-toluenesulfonic acid
= tosic acid = TsOH = PTSA

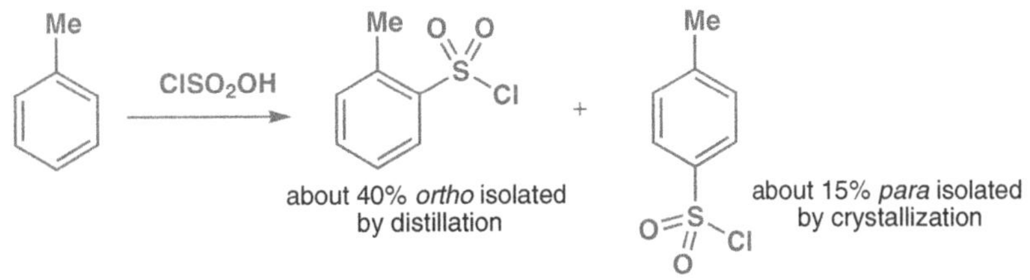

about 40% *ortho* isolated
by distillation

about 15% *para* isolated
by crystallization

No other acid is needed because chlorosulfonic acid is a very strong acid indeed and proto-nates itself to give the electrophile. This explains why OH is the leaving group rather than Cl and why chlorosulfonation rather than sulfonation is the result.

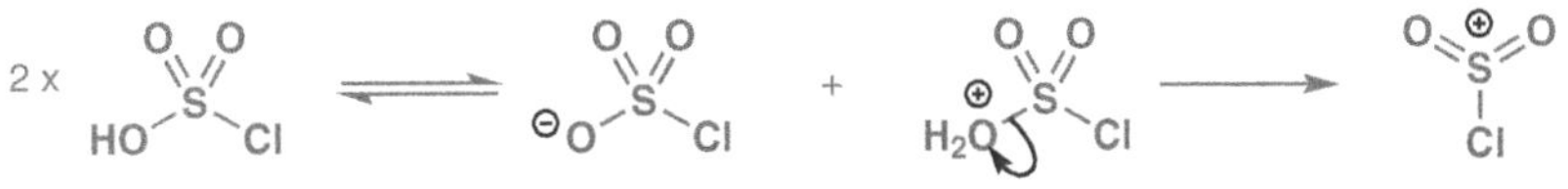

In drawing the mechanism we can again get the positive charge onto the tertiary *ipso* atom. No treatment with NaCl is needed in this reaction as the major product (the *ortho* acid chloride) is isolated by distillation.

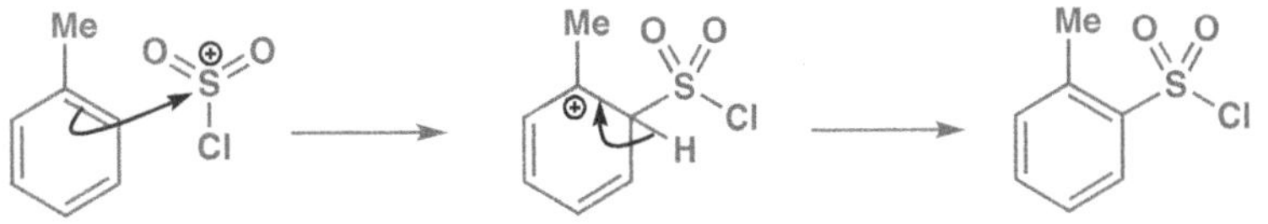

It is fortunate that the *ortho* acid chloride is the major product in the chlorosulfonation because it is needed in the synthesis of saccharin, the first of the non-fattening sweeteners. The formation of the sulfonamide is like that of an ordinary amide, but the oxidation of the methyl group with potassium permanganate is probably new to you. It's a rather vigorous reaction, but one which very usefully turns toluene derivatives into benzoic acid derivatives.

89% yield

not isolated

saccharin
58% yield

- The preference for *para* product in the sulfonation and *ortho* product in the chlorosulfonation is the first hint that sulfonation is reversible. This point is discussed, and exploited, in Chapter 24 (p. 566).

- As you will see in the next section, the substitution reactions of benzoic acid derivatives show different selectivity from the substitution reactions of toluene itself.

● Alkylbenzenes react with electrophiles faster than benzene and give mixtures of *ortho*- and *para*-substituted products.

Electron-withdrawing substituents give *meta* products

So far, all of the substituted benzene rings we have considered have carried substituents capa-ble of donating electron density to the ring: despite being electronegative atoms, oxygen and nitrogen have lone pairs which conjugate with the ring's π system; a similar but weaker effect results from σ conjugation from a methyl group. Two consequences arise from these substitu-ents: the ring becomes more reactive than benzene, and substitution takes place in the *ortho* and *para* positions.

So what happens with groups which pull electron density away from the ring? Such a group is the trimethylammonium substituent: the nitrogen is electronegative but unlike in aniline this electronegativity is not offset by donation of a lone pair—the nitrogen is tetrahedral and no longer has one to donate. Nitration of the phenyltrimethyl ammonium ion yields mainly the *meta* product. And it does so slowly too—this nitration proceeds about 10^7 times more slowly than that of benzene.

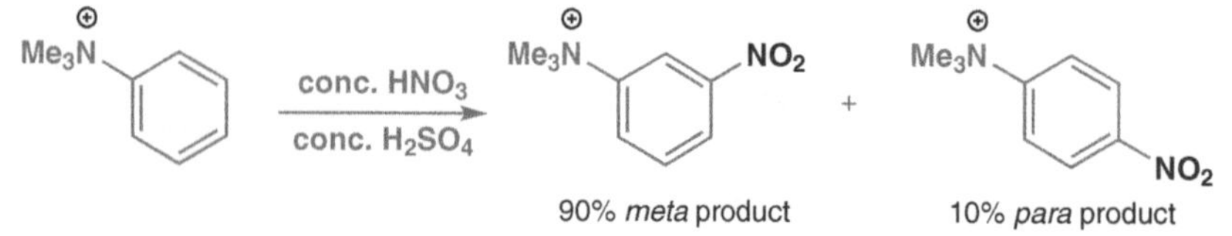

90% *meta* product   10% *para* product

The same thing happens with the CF_3 group. The three very electronegative fluorine atoms polarize the C–F bonds so much that the Ar–C bond is polarized too. Nitration of trifluoro-methylbenzene gives a nearly quantitative yield of *meta* nitro compound.

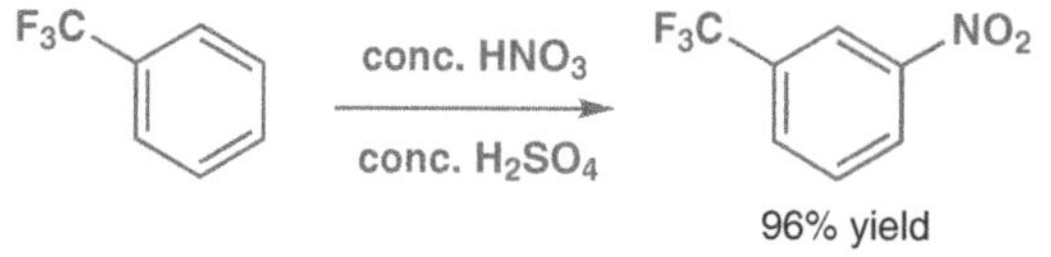

96% yield

Draw the mechanism for this reaction and you see the reason for the switch to *meta* selectivity.

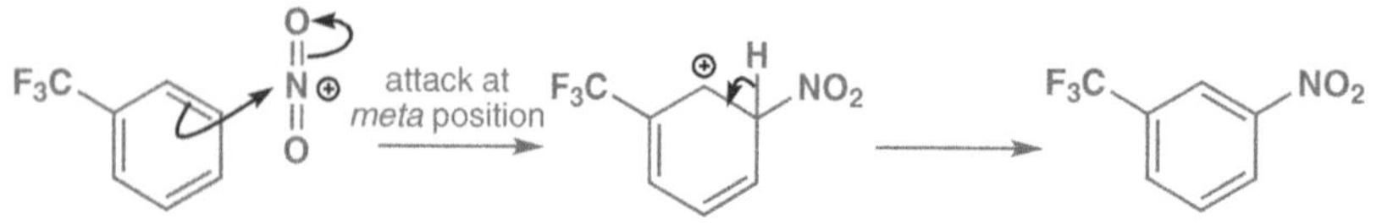

The intermediate cation is again delocalized over three carbons, but importantly none of these carbons is the one next to the CF_3 group.

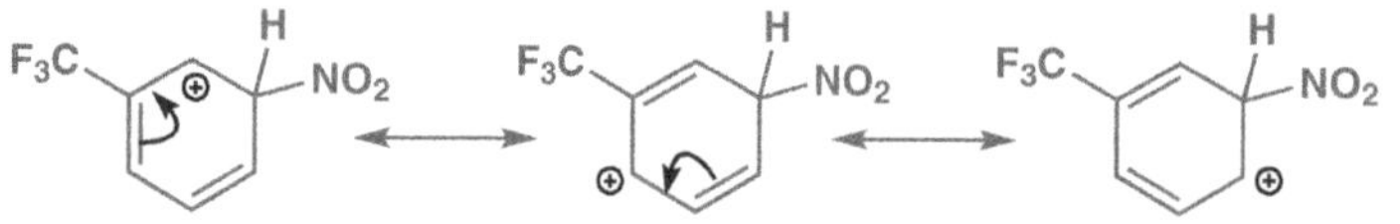

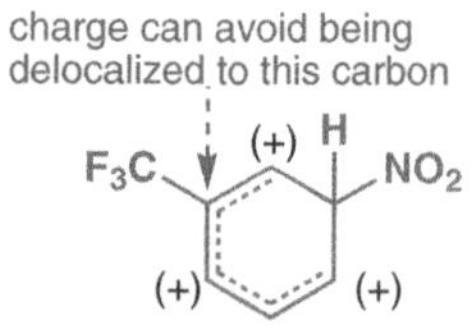

If, on the other hand, the electrophile were to attack the *ortho* or *para* position (the hypo-thetical reaction *para* to CF_3 is shown below) then the carbon next to CF_3 would *have* to carry a positive charge, which would be destabilized by the electron withdrawal, making this a high-energy intermediate.

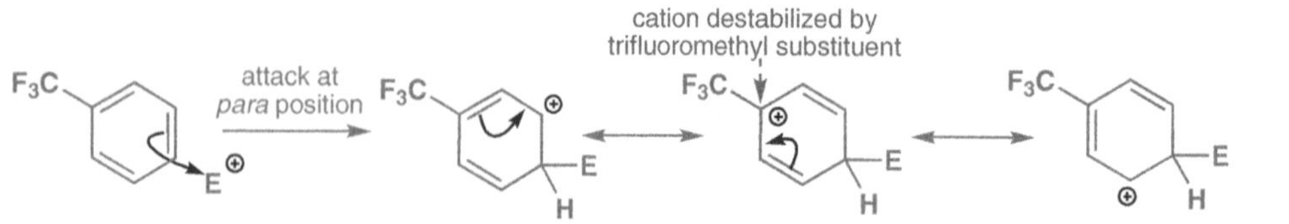

charge can avoid being delocalized to this carbon

favourable intermediate for *meta* substitution

cation destabilized by trifluoromethyl substituent

unfavourable intermediate for *para* substitution

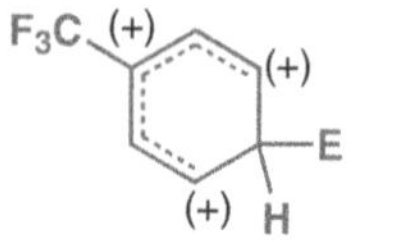

Think of it this way: the electron-deficient ring would really rather not react with an electro-phile (hence the slower rate) but if it has to (because the electrophile is so reactive) then it takes the least bad course of keeping the positive charge away from the electron-withdrawing groups—and that means *meta* substitution.

Some substituents withdraw electrons by conjugation

Aromatic nitration is important because it is a convenient way of adding a nitrogen sub-stituent to the ring and because it stops cleanly after one nitro group has been added. Double nitration of benzene is possible but stronger conditions must be used—fuming nitric acid instead of normal concentrated nitric acid—and the mixture must be refluxed at around 100 °C.

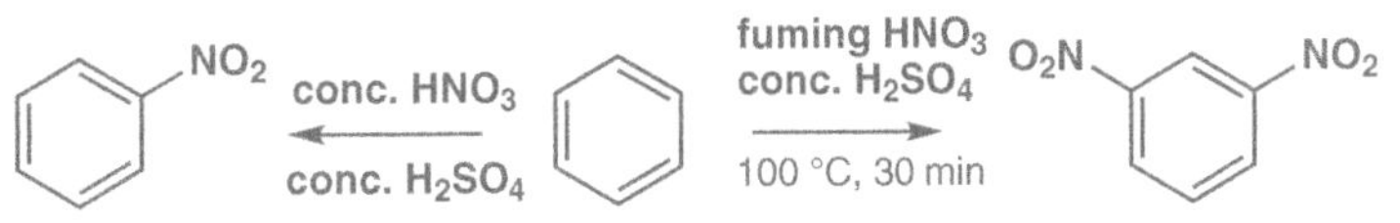

75% yield *meta*-dinitrobenzene

The second nitro group is introduced *meta* to the first: evidently the nitro group is deactivating and *meta*-directing.

The nitro group is conjugated with the π system of the benzene ring and is strongly electron *withdrawing*—and it withdraws electrons specifically from the *ortho* and *para* positions. We can use curly arrows to show this:

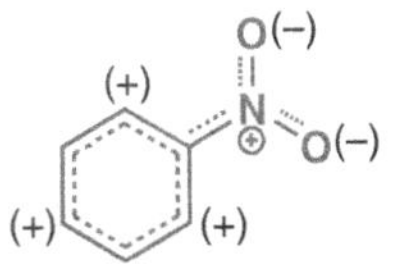

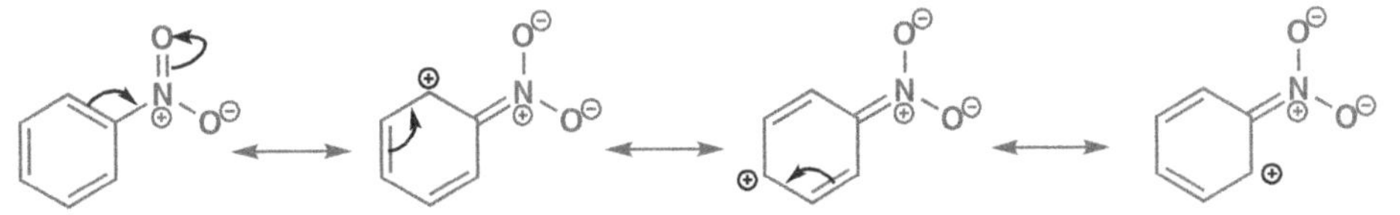

The nitro group withdraws electron density from the π system of the ring thereby making the ring less reactive towards an electrophile. Since more electron density is removed from the *ortho* and *para* positions, the least electron-deficient position is the *meta* position. Hence the nitro group is *meta* directing. In the nitration of benzene, it is much harder to nitrate a second time and, if we insist, the second nitro group goes in *meta* to the first.

Other reactions go the same way: bromination of nitrobenzene gives *meta*-bromonitrobenzene in good yield. The combination of bromine and iron powder provides the necessary Lewis acid catalyst ($FeBr_3$) while the high temperature needed for this unfavourable reaction is easily achieved as the boiling point of nitrobenzene is over 200 °C.

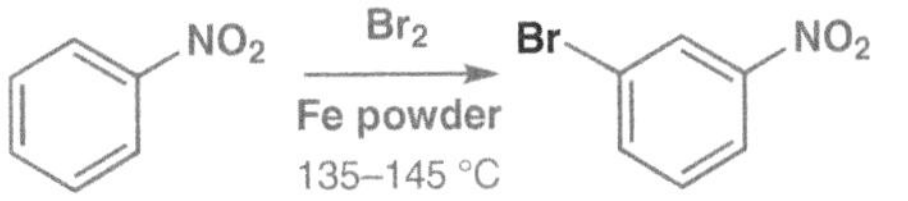

In drawing the mechanism it is best to draw the intermediate and to emphasize that the positive charge must not be delocalized to the carbon atom bearing the nitro group.

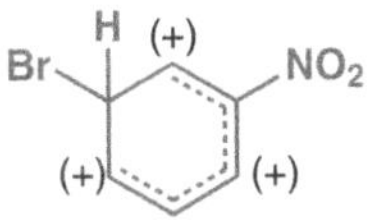

delocalization of
intermediate cation

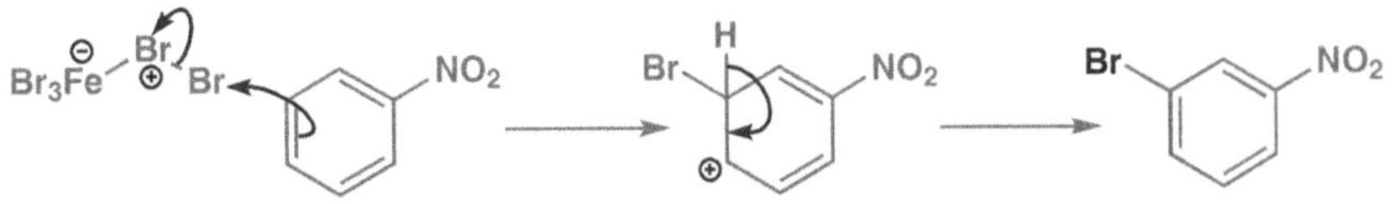

Nitro is just one of a number of groups that are also deactivating towards electrophiles and *meta* directing because of electron withdrawal by conjugation. Others include carbonyl groups (aldehydes, ketones, esters, etc.), nitriles, and sulfonates. The ¹H NMR shifts of rings carrying these substituents confirm that they remove electrons principally from the *ortho* and *para* positions.

¹H NMR chemical shifts

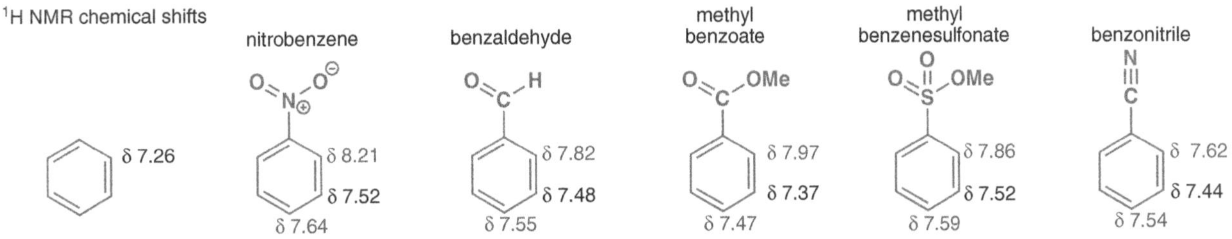

Points to note:

- Each of the compounds contains the unit Ph–X=Y, where Y is an electronegative element, usually oxygen.
- In each compound, *all* the protons have larger chemical shifts than benzene because the electron density at carbon is less.
- The protons in the *meta* position have the smallest shift and so the greatest electron density.

Nitro is the most electron-withdrawing of these groups and some of the other compounds are nearly as reactive (in the *meta* position, of course) as benzene itself. It is easy, for example, to nitrate methyl benzoate and the *m*-nitro ester can then be hydrolysed to *meta*-nitrobenzoic acid very easily.

84% yield
methyl *meta*-nitrobenzoate

96% yield
meta-nitrobenzoic acid

> ● Electron-withdrawing groups make aromatic rings more reluctant to undergo electrophilic substitution, but when they do react, they react in the *meta* position.

One group of substituents remains and they are slightly odd. They are *ortho, para*-directing but they are also *deactivating*. They are the halogens.

Halogens show evidence of both electron withdrawal and donation

So far we have steered clear of the reactions of halogenated derivatives of benzene. Before we explain their reactions, have a look at the table, which shows the rates of nitration of fluoro, chloro, bromo, and iodobenzene relative to benzene itself, and also gives an indication of the products formed in each case.

| Compound | Products formed (%) | | | Nitration rate (relative to benzene) |
	ortho	meta	para	
PhF	13	0.6	86	0.18
PhCl	35	0.9	64	0.064
PhBr	43	0.9	56	0.060
PhI	45	1.3	54	0.12

We'll come back to this table a few times in the next page or so, but the first thing to note is that **all the halobenzenes react more slowly than benzene itself**. Evidently, electron withdrawal by the electronegative halogen deactivates the ring towards attack. But the second thing that should strike you is that, unlike the deactivating groups we have just been discussing, **halogens are *ortho, para* directing**—very few *meta*-nitrated products are formed.

The only way this makes sense is if there are two opposing effects: electron donation by conjugation and electron withdrawal by induction. The halogen has three lone pairs, one of which may conjugate with the ring just like in phenol or aniline. Yet the conjugation is much less good than in phenol or aniline, for one of two reasons. When Cl, Br, or I is the substituent, the problem is size: the 2p orbitals from the carbon atoms overlap poorly with the bigger p orbitals from the halogen (3p for chlorine, 4p for bromine, and 5p for iodine). This size mismatch is clearly illustrated by comparing the reactivities of aniline and chlorobenzene:

chlorine and nitrogen have approximately the same electronegativity, but aniline is much more reactive than chlorobenzene because of the better overlap between the carbon and nitrogen 2p orbitals. Fluorine 2p orbitals are the right size to overlap well with the carbon 2p orbitals, but now there is another problem: the orbitals of fluorine are much lower in energy than the orbitals of carbon since fluorine is so electronegative.

So, all four halogens are less good at donating electrons to the ring than an OH or NH_2 group, but not only are the halobenzenes less reactive than phenol or aniline, they are even less reactive than benzene itself. Now, when we looked at aniline and phenol, we didn't worry about any electron withdrawal by induction, even though both oxygen and nitrogen are of course rather electronegative. Electron donation from their N and O lone pairs is evidently much more important. But with the conjugation in the halobenzenes already weak, inductive electron withdrawal takes over as the dominant factor in determining reactivity.

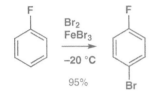

With all this in mind, how would you expect fluorobenzene to react? Most electron density is removed first from the *ortho* positions by induction, then from the *meta* positions, and then from the *para* position. Any conjugation of the lone pairs on fluorine with the π system would increase the electron density in the *ortho* and *para* positions. Both effects favour the *para* position and this is where most substitution occurs. But is the ring more or less reactive than benzene? This is hard to say and the honest answer is that sometimes fluorobenzene is more reactive in the *para* position than benzene (for example, in proton exchange and in acetylation—see later) and sometimes it is less reactive than benzene (for example, in nitration, as shown by the table above). In all cases, fluorobenzene is significantly more reactive than the other halobenzenes. We appreciate that this is a rather surprising conclusion, but the evidence supports it. For example, fluorobenzene reacts with bromine and an iron catalyst (it does need a catalyst: it is not as reactive as phenol) at only –20 °C to give the *para*-bromo derivative.

Let's now look back in bit more detail at the table above. We can now also explain two other features of the results:

➡ We mentioned inductive effects as a factor controlling *ortho* vs *para* reactivity on p. 483.

- The percentage of the *ortho* product increases from fluorobenzene to iodobenzene. We might have expected the amount to decrease as the size of the halide increases because of increased steric hindrance at the *ortho* position but this is clearly not the case. Instead the greater inductive effect of the more electronegative atoms (F, Cl) withdraws electron density mostly from the *ortho* positions, lessening their reactivity.

- The rates of the reactions fall into two pairs and follow a 'U-shaped' sequence: fluorobenzene nitrates most quickly, followed closely by iodobenzene; chloro-, and bromobenzene nitrate at around half these rates. Chlorine and bromine suffer because both are quite electronegative and neither has good lone pair overlap: in fluorine, overlap is good; in iodine, electronegativity is much less.

In practical terms, it is usually possible to get high yields of *para* products from electrophilic substitution reactions of halobenzenes. Both nitration and sulfonation of bromobenzene give enough material to make the synthesis worthwhile. Although mixtures of products are always bad in a synthesis, electrophilic aromatic substitution is usually simple to carry out on a large enough scale to make separation of the major product, ideally by crystallization, a workable method. A 68% yield of sodium *p*-bromobenzenesulfonate can be achieved by recrystallization of the sodium salt from water and a 70% yield of *p*-bromonitrobenzene by separation from the *ortho* isomer by recrystallization from EtOH.

● **Summary of directing and activating effects**

Now we can summarize the stage we have reached in terms of *activation* and *direction*.

Electronic effect	Example	Activation	Direction
donation by conjugation	$-NR_2$, $-OR$	very activating	*ortho*, *para* only
donation by inductive effect	alkyl	activating	mostly *ortho*, *para* but some *meta*
donation by conjugation *and* withdrawal by inductive effect	F, Cl, Br, I	deactivating	*ortho* and (mostly) *para*
withdrawal by inductive effect	$-CF_3$, $-NR_3^+$	deactivating	*meta* only
withdrawal by conjugation	$-NO_2$, $-CN$, $-COR$, $-SO_3R$	very deactivating	*meta* only

Two or more substituents may cooperate or compete

The directing effects of two or more substituents can work with or against one another. Bromoxynil and ioxynil are contact herbicides especially used in spring cereals to control weeds resistant to other weedkillers, and both are synthesized from *p*-hydroxybenzaldehyde by double halogenation. The aldehyde directs *meta* and the OH group directs *ortho*: both effects work together to promote bromination or iodination at the same two positions.

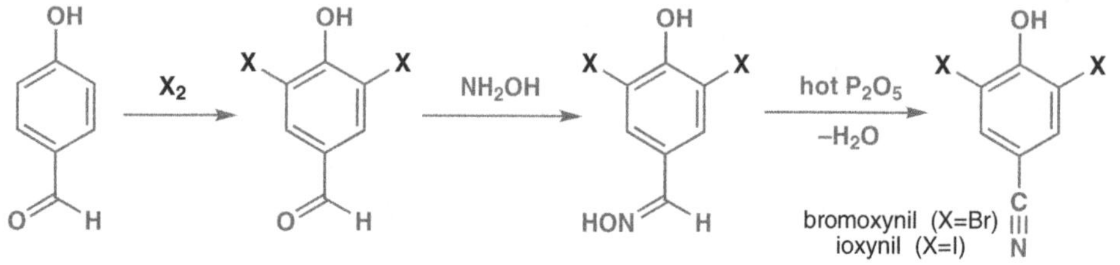

bromoxynil (X=Br)
ioxynil (X=I)

■ The reaction with NH_2OH is the formation of an oxime from the aldehyde and hydroxylamine and was dealt with in Chapter 11. The reaction with P_2O_5 is a dehydration—phosphorus is used to form the nitrile by removing water from the oxime.

In other cases substituents compete by directing to different positions. The antioxidant BHT (p. 58) is made from 4-methylphenol (known as *p*-cresol) by a Friedel–Crafts alkylation. Usually, both the methyl and OH groups are *ortho, para* directors. The *para* positions are obviously both blocked, but the positions *ortho* to each of the groups are different. Since the –OH group is much more powerfully directing than the methyl group it 'wins' and directs the electrophile (a *t*-butyl cation) *ortho* to itself.

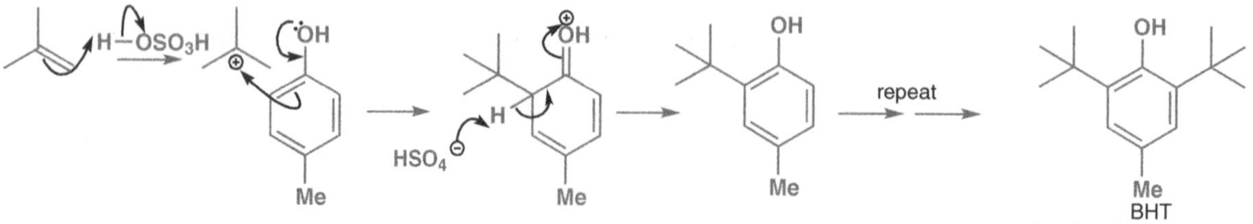

BHT
butylated hydroxytoluene

In this case the *t*-butyl cation is made from the alkene and protic acid; alternative reagents would be *t*-butanol with protic acid or *t*-butyl chloride with $AlCl_3$.

Even a 'watered-down' activating group like the amide –NHCOMe, which provides an extra pair of electrons, will 'win' over a deactivating group or an activating alkyl group. Bromination of this amide goes *ortho* to the –NHCOMe group but *meta* to the methyl group.

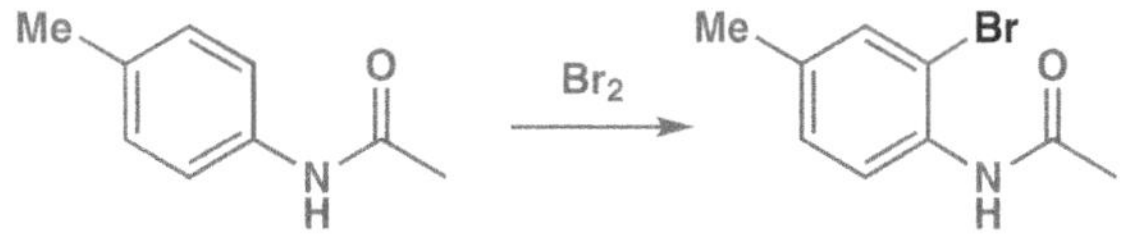

When looking at any compound where competition is an issue it is sensible to consider electronic effects first and then steric effects. For electronic effects, in general, any activating effects are more important than deactivating ones. For example, the aldehyde below has three groups—two methoxy groups that direct *ortho* and *para* and an aldehyde that directs *meta*.

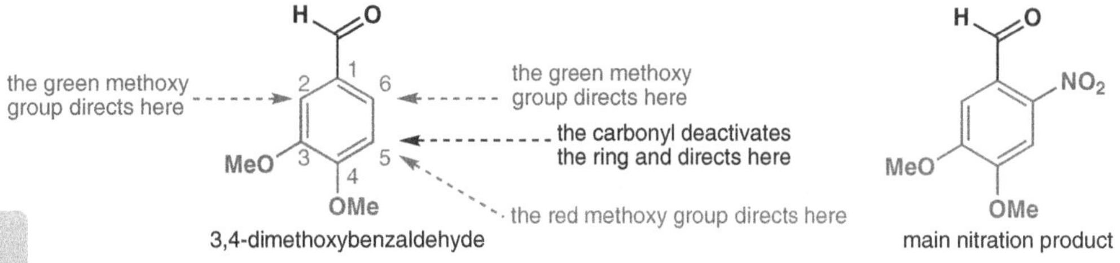

3,4-dimethoxybenzaldehyde   main nitration product

Despite the fact that the aldehyde group withdraws electron density from positions 2 and 6, C6 is still the position for nitration. The activating methoxy groups dominate electronically and the choice is really between C2, C5, and C6. Now consider steric factors: reaction at C2 or C5 would lead to three adjacent substituents. Substitution occurs at position 6.

Some problems and some opportunities

You've seen plenty of electrophilic aromatic substitution reactions in this chapter that are reliable and widely used—bromination and nitration, for example. But others pose problems:

- Friedel–Crafts alkylation works only when the intermediate cation is stable, so how do we add an *n*-alkyl chain to an aromatic ring?
- There is no good way of introducing an oxygen electrophile to an aromatic ring, so how do we make Ar–O bonds?
- Electron-donating groups always direct *ortho, para*, so how do we put in a group *meta* to, for example, an amino group?

We will consider some answers to these questions in this last section of this chapter.

A closer look at Friedel–Crafts chemistry

Reactions such as nitration and sulfonation add a very deactivating substituent. They usually stop cleanly after a single substitution unless there is also a strongly activating substituent. Even then it may be possible to stop after a single substitution. Weakly electron-withdrawing substituents like the halogens can be added once, but multiple substitution is common when the starting arene carries strongly activating substituents like OH and NH₂.

Two reasons to avoid a Friedel–Crafts alkylation

When electron-donating substituents are added, multiple substitution is always a threat. The principal reaction where multiple substitution is a genuine problem is the Friedel–Crafts alkylation reaction. Here's an example: preparation of diphenylmethane from benzene and benzyl chloride is a useful reaction but the product has two benzene rings, each more reactive than benzene itself. A 50% yield is the best we can do and that requires a large excess of benzene to ensure that it competes successfully for the reagent with the reactive, electron-rich product.

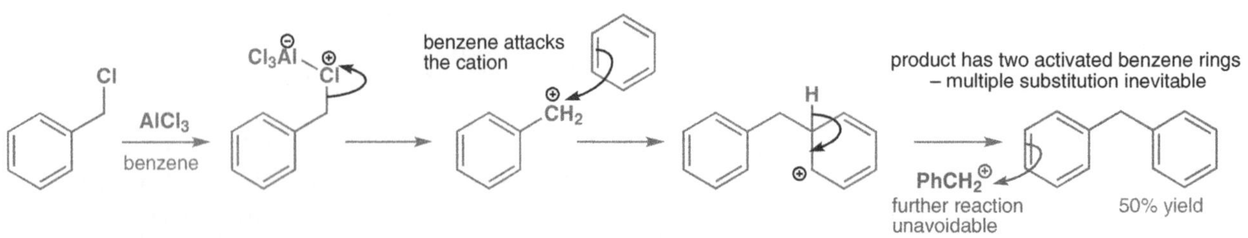

Multiple substitution is just one of the potential pitfalls of Friedel–Crafts alkylations. The other is important to be aware of too: **Friedel–Crafts alkylations work well only with stable cations.** This is what happens when we try a Friedel–Crafts reaction with *n*-propyl chloride.

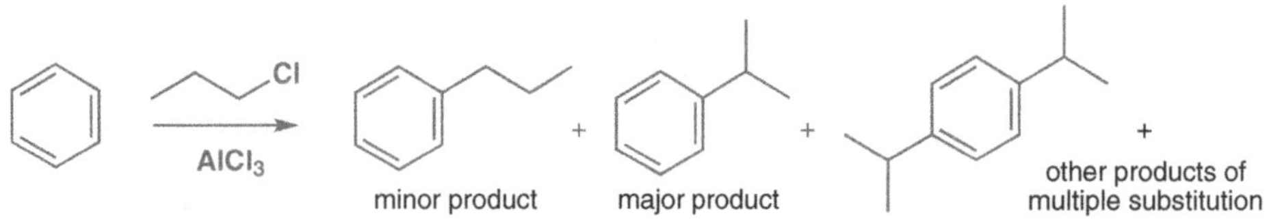

Recall from Chapter 15 that primary halides don't form cations easily, so the Friedel–Crafts reaction with *n*-propyl chloride has to go via an S_N2 mechanism.

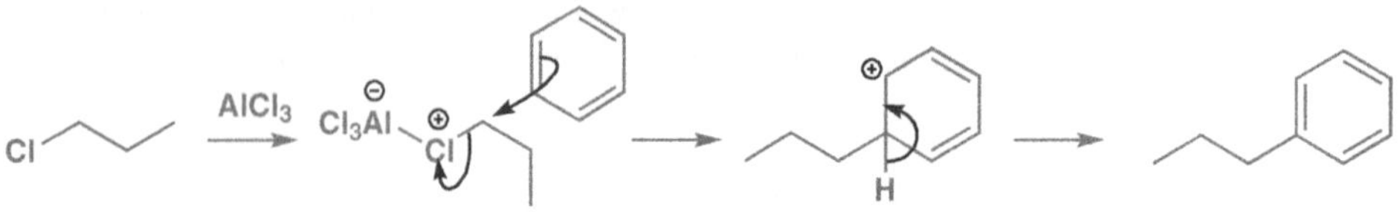

So where does the major product of the reaction come from? The three carbons are arranged not as an *n*-propyl group but as an *iso*-propyl group: a *rearrangement* has occurred. This is the mechanism:

We'll deal with rearrangements in much greater detail in Chapter 36.

rearrangement (migration of green H) leads to isopropyl benzene

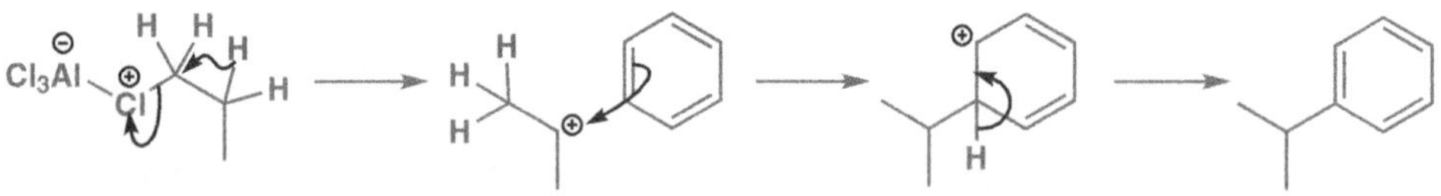

The green hydrogen migrates to allow a secondary rather than a primary alkyl cation to be formed, and *iso*-propylbenzene results. This leaves us with a problem: how can you add primary alkyl groups to benzene rings?

The solution: use Friedel–Crafts acylation instead

We can kill two birds with one stone here: both problems common to the Friedel–Crafts alkylation are solved when the acylation is used instead. Firstly, the product of the acylation is a ketone: the reaction introduces a deactivating, electron-withdrawing, conjugating carbonyl group to the ring, so the product is *less* reactive than the starting material. Reaction will stop cleanly after one acylation. Here's benzene reacting with propionyl chloride.

We introduced the Friedel–Crafts acylation on p. 477.

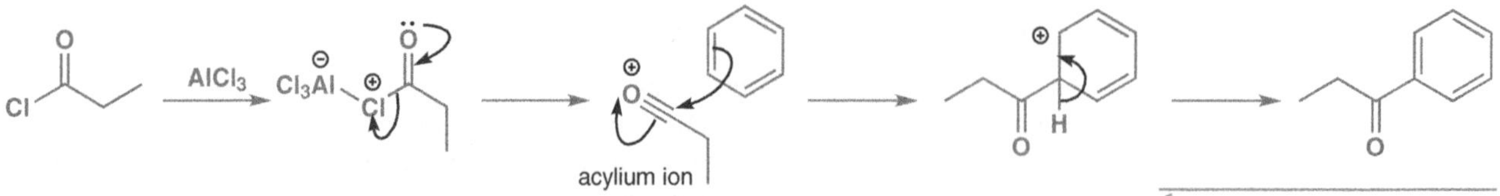

acylium ion

Interactive mechanism for Friedel–Crafts acylation

If we want the ketone then all well and good. But a simple reduction also allows us to get the alkylated product—this compound (trivially called propiophenone) is reduced to

You may also meet the trivial names acetophenone and benzophenone.

R = Me: acetophenone
R = Ph: benzophenone

More reductions like this—which get rid of the carbonyl group completely—are discussed in Chapter 23.

propylbenzene using any of a number of reduction methods, for example zinc amalgam in hydrochloric acid.

n-propylbenzene

The reduction of a Friedel–Crafts acylation product like this always gives an *n*-alkylbenzene, exactly the sort of compound that causes the problems in Friedel–Crafts alkylation. Friedel–Crafts acylations also work well when anhydrides are used in the place of acid chlorides. The acylium ion is formed in the same way:

anhydride   acylium ion

If a cyclic anhydride is used, the product is a keto-acid.

Notice how much AlCl$_3$ is needed: in Friedel–Crafts alkylations using an alkyl chloride, the Lewis acid is used in catalytic quantities. In an acylation, however, the Lewis acid can also complex to any oxygen atoms present, to the carbonyl in the product, for example. As a result, in acylation reactions more Lewis acid is required—just over one equivalent per carbonyl group.

succinic anhydride (the anhydride of butanedioic acid, or succinic acid)

AlCl$_3$
2.2 equivalents

3-benzoylpropanoic acid

Reduction of the ketone can give a simpler carboxylic acid, but we can go one step further and do another acylation—because the reaction is intramolecular, it goes even with just a strong acid (phosphoric acid): the strong acid makes the OH into a good leaving group (water) and the acylium ion is again an intermediate.

Make sure you can see how this reaction works.

Ph

Zn / Hg
HCl

Ph

4-phenylbutanoic acid

H$_3$PO$_4$

● **The advantages of acylation over alkylation**

Two problems in Friedel–Crafts alkylation do not arise with acylation.

• The acyl group in the product withdraws electrons from the π system, making multiple substitutions harder. Indeed, if the ring is too deactivated to start off with, Friedel–Crafts acylation may not be possible at all—nitrobenzene is inert to Friedel–Crafts acylation and is often used as a solvent for these reactions.

• Rearrangements are also no longer a problem because the electrophile, the acylium cation, is already relatively stable.

• The acyl groups of the products can be reduced to primary alkyl groups, which are impossible to introduce cleanly by Friedel–Crafts alkylation.

Exploiting the chemistry of the nitro group

The nitro group is remarkably useful in a number of ways:

• It is easy to introduce by nitration chemistry (p. 476).

• Unlike most N- or O-based functional groups, it is a *meta* director (p. 488).

• It can be reduced to an amino group.

• It can be replaced with other substituents using diazonium chemistry.

You have met the first two of these features, but the last two may be new to you. An aromatic nitro group is easy to turn into an amino group—a number of reagents will do this, but the most common are tin in dilute HCl or hydrogenation with a palladium catalyst supported on charcoal (written as Pd/C).

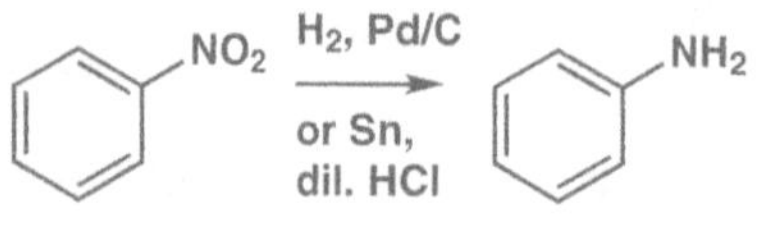

➡ There is more on these selective reducing agents in Chapter 23.

This simple transformation is extremely important because it turns the *meta*-directing nitro group into an *ortho, para*-directing amino group (although as you saw on p. 483, the amino group may need 'taming' to make its reactivity useful). The sequence of nitration–reduction allows us to introduce a useful NH_2^+ equivalent into an aromatic molecule, and can let us make otherwise difficult-to-form *meta*-substituted amino compounds.

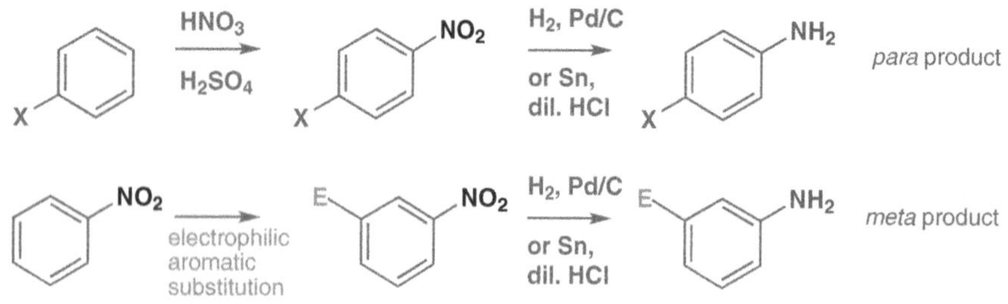

The reduction to an amino group also opens up the possibility of replacing the nitrogen substituent completely, by converting it first to a diazonium group. Treatment of an amine with nitrous acid converts it to an unstable diazonium salt, whose mechanism of formation and chemistry we will discuss in the next chapter. Not surprisingly, diazonium salts very readily lose nitrogen gas, and this substitution of N_2 by a nucleophile opens yet more opportunities to compounds derived from nitrobenzene derivatives. It also involves *nucleophilic* substitution at the aromatic ring, which forms the subject of the next chapter.

➡ Diazonium salts are discussed on p. 520. Chapter 40 introduces the idea of using transition metals in the formation of bonds to aromatic rings, while Chapter 24 revisits the methods available when control of regiochemistry (i.e. *ortho, meta*, or *para* selectivity) is needed.

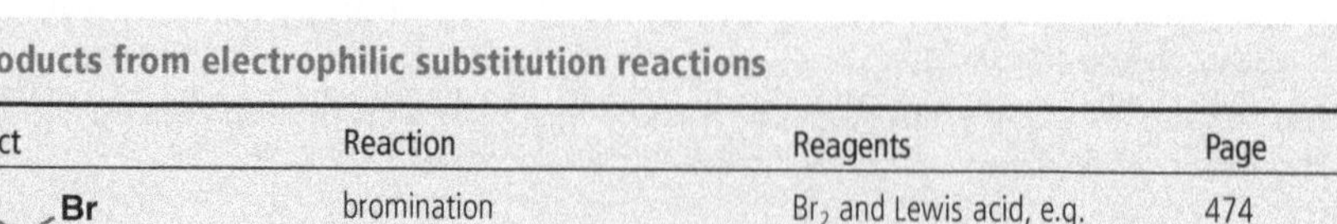

Summary

● Products from electrophilic substitution reactions

Product	Reaction	Reagents	Page
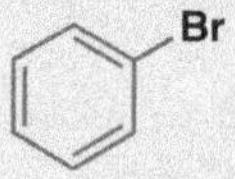 Br	bromination	Br_2 and Lewis acid, e.g. $AlCl_3$, $FeBr_3$, Fe powder	474
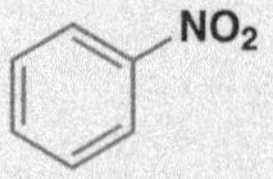 NO_2	nitration	$HNO_3 + H_2SO_4$	476
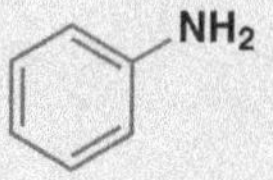 NH_2	reduction of nitro compounds	From $ArNO_2$: Sn, HCl or H_2, Pd/C	495

(continued) Products from electrophilic substitution reactions

Product	Reaction	Reagents	Page
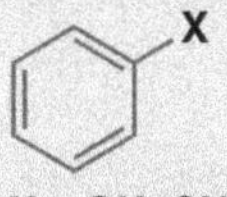 X **X = OH, CN, Br, I...**	substitution of diazonium salts	From ArNH$_2$: 1. NaNO$_2$, HCl; 2. X$^-$	See Chapter 22, p. 520
 SO$_3$H	sulfonation	concentrated H$_2$SO$_4$ or H$_2$SO$_4$ + SO$_3$ (oleum)	476
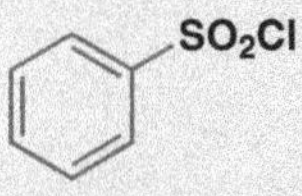 SO$_2$Cl	chlorosulfonation	ClSO$_3$H	486
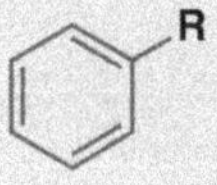 R	Friedel–Crafts alkylation	RX + Lewis acid, usually AlCl$_3$	477
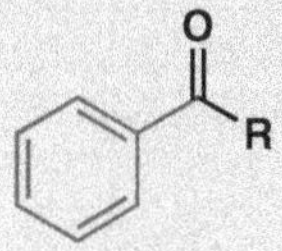 O R	Friedel–Crafts acylation	RCOCl + Lewis acid, usually AlCl$_3$	477
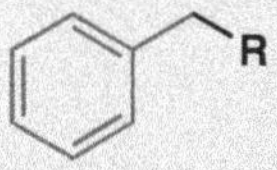 R	Friedel–Crafts acylation and reduction	From ArCOR: Zn/Hg, HCl	493

● **Reactions of aromatic compounds in this chapter**

Starting material	Example	Activating/deactivating	Directing effect	Page
benzene, PhH	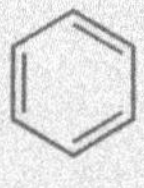	–	–	474
phenol, PhOH	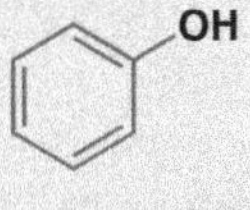 OH	activating	ortho, para	479
anisole, PhOMe	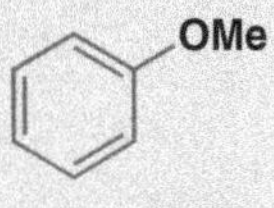 OMe	activating	ortho, para	480
aniline, PhNH$_2$	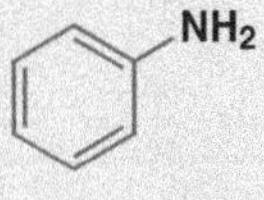 NH$_2$	activating	ortho, para	482
ArNHCOR (anilides)	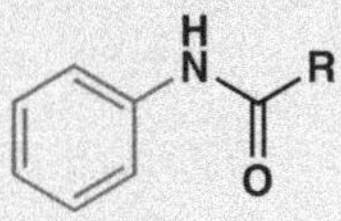 H N R O	activating	ortho, para	483
toluene and alkyl-benzenes, PhR	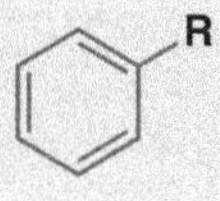 R	activating	ortho, para	484

(*continued*) Reactions of aromatic compounds in this chapter

Starting material	Example	Activating/deactivating	Directing effect	Page
nitrobenzene, PhNO$_2$	NO$_2$	deactivating	*meta*	488*
acylbenzenes, PhCOR (acetophenone, benzophenone)	O / R	deactivating	*meta*	489
benzonitrile, PhCN	C≡N	deactivating	*meta*	488
halobenzenes, PhX	Br	deactivating	*ortho, para*	489

*For methods of converting nitro substituents to other groups by reduction, diazotization and substitution, see pp. 520 and 567, and Chapters 22 and 24.

Further reading

Every big organic chemistry text has a chapter on this topic. One of the best is: F. A. Carey and R. J. Sundberg, *Advanced Organic Chemistry A, Structure and Mechanisms*, 5th edn, Springer, 2007, chapter 9 and B, *Reactions and Synthesis*, chapter 11. B. S. Furniss, A. J. Hannaford, P. W. G. Smith, and A. T. Tatchell, *Vogel's Textbook of Practical Organic Chemistry*, Longman, 5th edn, 1989, sections 6.1–6.4 and 6.10–6.13 gives many practical examples of the reactions in this chapter.

Check your understanding

To check that you have mastered the concepts presented in this chapter, attempt the problems that are available in the book's Online Resource Centre at http://www.oxfordtextbooks.co.uk/orc/clayden2e/

Suggested solutions for Chapter 21

21

PROBLEM 1

All you have to do is to spot the aromatic rings in these compounds. It may not be as easy as you think and you should give some reasons for questionable decisions.

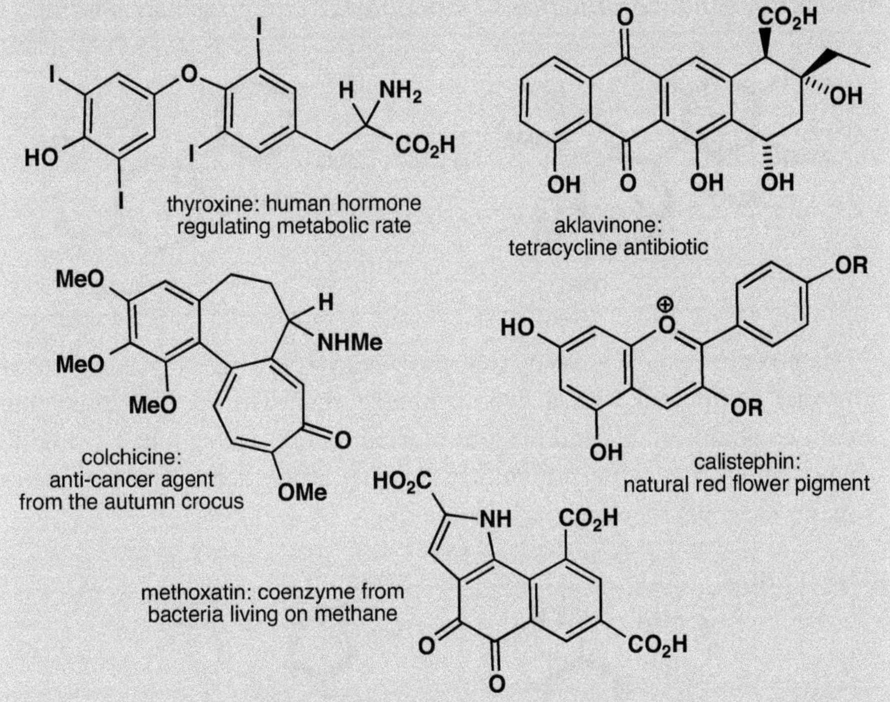

thyroxine: human hormone regulating metabolic rate

aklavinone: tetracycline antibiotic

colchicine: anti-cancer agent from the autumn crocus

calistephin: natural red flower pigment

methoxatin: coenzyme from bacteria living on methane

Purpose of the problem

Simple exercise in counting electrons with a few hidden tricks.

Suggested solution

Truly aromatic rings are marked with bold lines. Thyroxine has two benzene rings—obviously aromatic—and that's that. Aklavinone also has two aromatic benzene rings and we might argue about ring 2. It has four electrons as drawn, and you might think that you could push electrons round from the OH groups to give ring 2 six electrons as well. But if you try it, you'll find you can't.

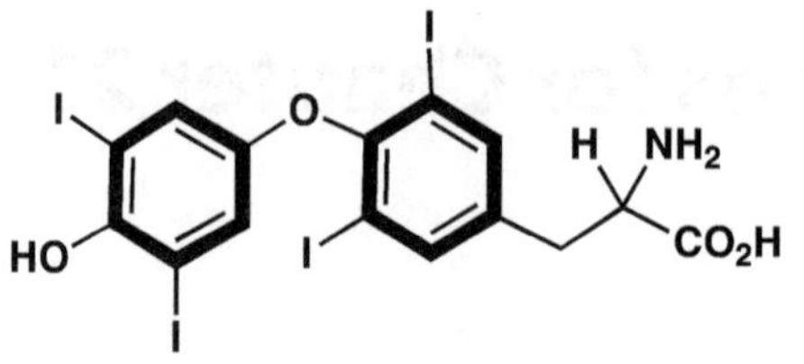

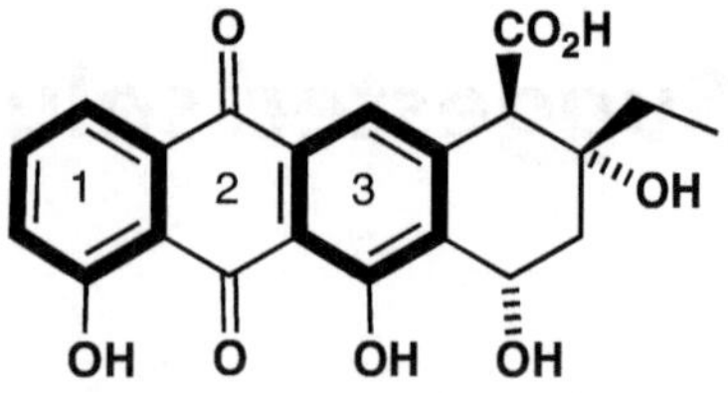

Colchicine has one benzene ring and a seven-membered conjugated ring with six electrons in double bonds (don't count the carbonyl electrons as they are out of the ring). It perhaps looks more aromatic if you delocalize the electrons and represent it as a zwitterion. Either representation is fine.

Methoxatin has one benzene ring and one pyrrole ring—an example of an aromatic compound with a five-membered ring. The six electrons come from two double bonds and the lone pair on the nitrogen atom. The middle ring is not aromatic—even if you try drawing other delocalized structures, you can never get six electrons into this ring.

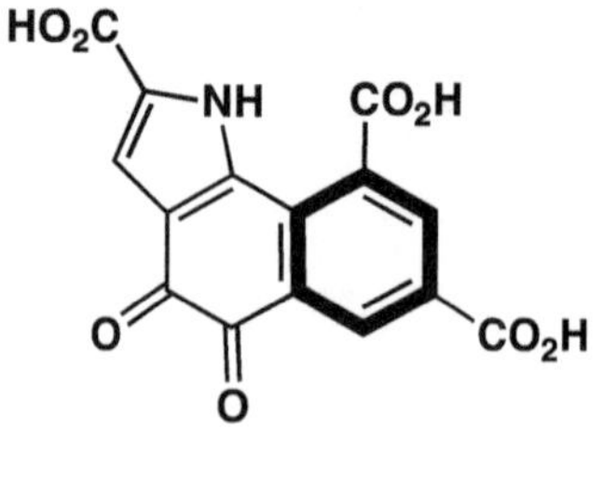

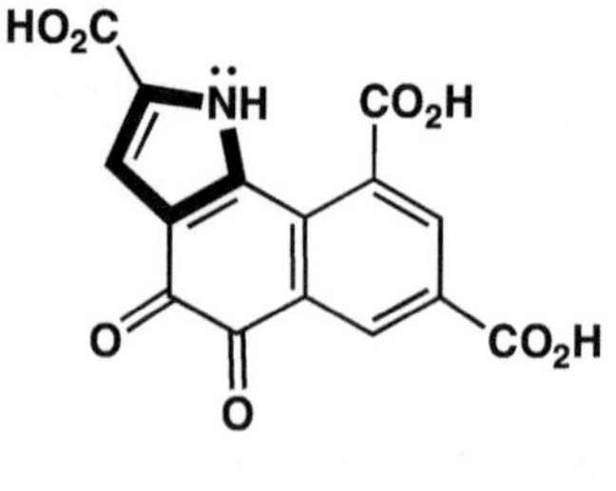

PROBLEM 2

First, as some revision, write out the detailed mechanism for these steps.

$$HNO_3 + H_2SO_4 \longrightarrow \oplus NO_2$$

In a standard nitration reaction with, say, HNO_3 and H_2SO_4, each of these compounds forms a single nitration product. What is its structure? Explain your answer with at least a partial mechanism.

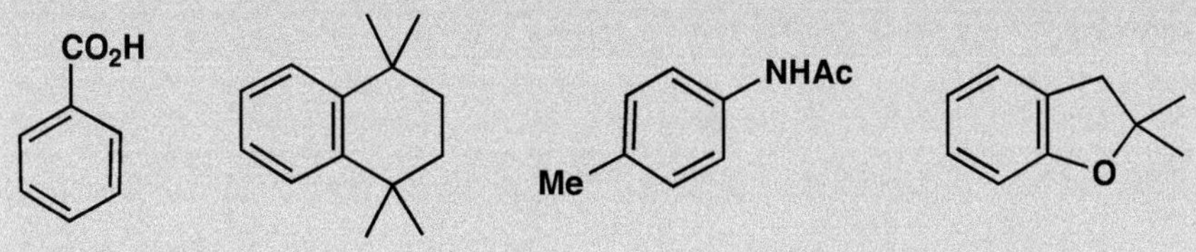

Purpose of the problem

Revision of the basic nitration mechanism and extension to compounds where selectivity is an issue.

Suggested solution

The basic mechanisms for the formation of NO_2^+ and its reaction with benzene appear on p. 476 of the textbook. Benzoic acid has an electron-withdrawing substituent so it reacts in the *meta* position. The second compound is activated in all positions by the weakly electron-donating alkyl groups (all positions are either *ortho* or *para* to one of these groups) but will react at one of the positions more remote from the alkyl groups because of steric hindrance.

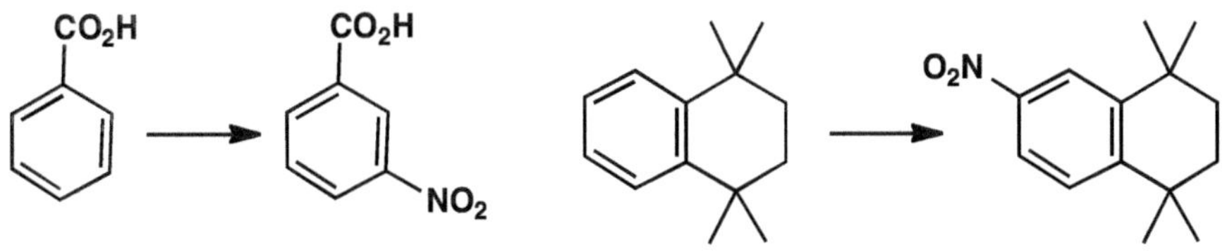

The remaining two compounds have competing *ortho,para*-directing substituents but in each case the one with the lone pair of electrons (N or O) is a more powerful director than the simple alkyl group. In the first case nitrogen directs *ortho* but in the second oxygen activates both *ortho* and *para* and steric hindrance makes the *para* position marginally more reactive.

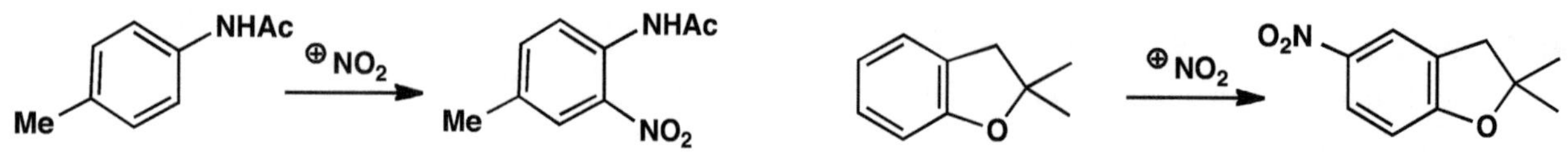

PROBLEM 3

How reactive are the different sites in toluene? Nitration of toluene produces the three possible products in the ratios shown. What would be the ratios if all the sites were equally reactive? What is the actual relative reactivity of the three sites? You could express this as x:y:1 or as a:b:c where a+b+c = 100. Comment on the ratio you deduce.

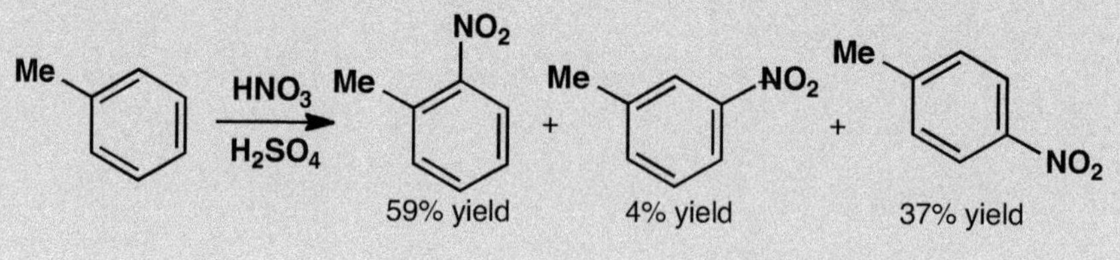

Purpose of the problem

A more quantitative assessment of relative reactivities.

Suggested solution

As there are two *ortho* and two *meta* sites, the ratio if all were equally reactive would be 2:2:1 *o:m:p*. The observed reactivity is 30:2:37 or 15:1:18 or 43:3:54 depending on how you expressed it. The *ortho* and *para* positions are roughly equally reactive because the methyl group is electron-donating. The *para* is slightly more reactive than the *ortho* because of steric hindrance. The *meta* position is an order of magnitude less reactive because the intermediate is not stabilized by electron-donation (σ-conjugation) from the methyl group.

reaction in the *ortho* position

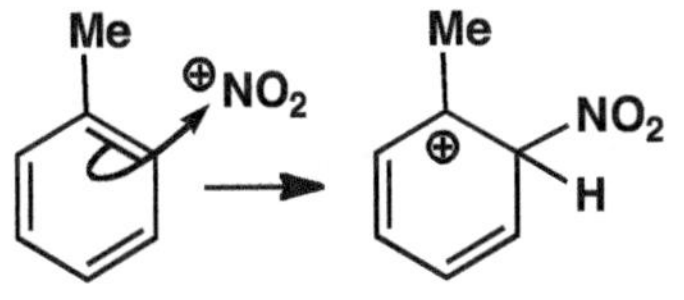

reaction in the *meta* position

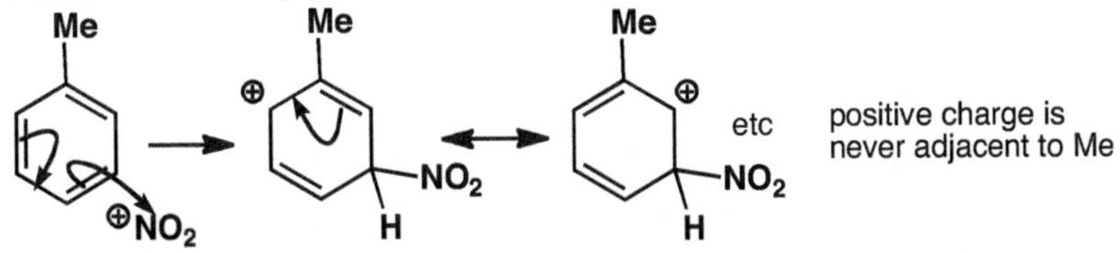

etc   positive charge is
       never adjacent to Me

PROBLEM 4

Draw mechanisms for these reactions and explain the positions of substitution.

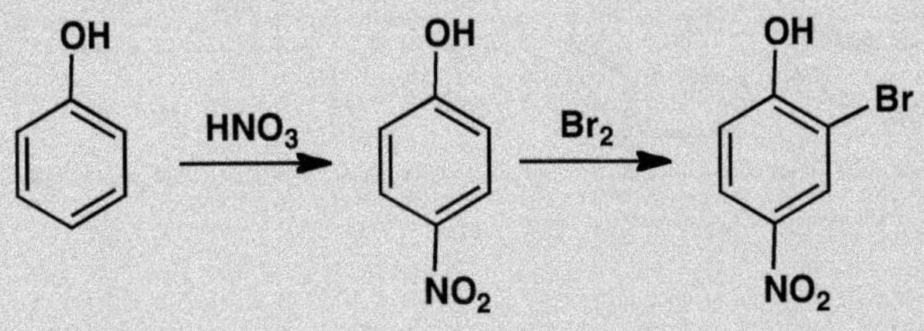

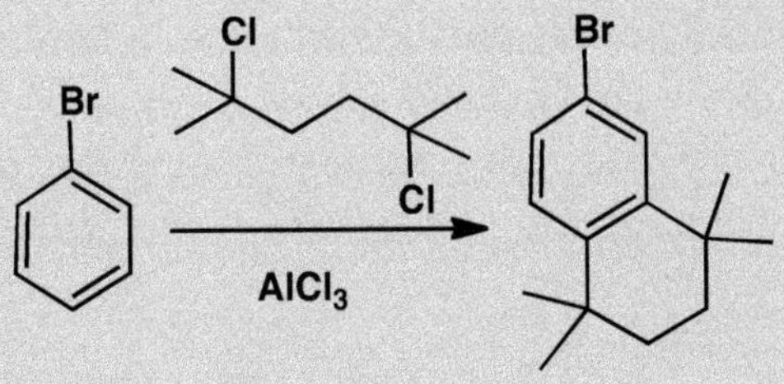

Purpose of the problem

More advanced questions of orientation with more powerful electron-donating groups.

Suggested solution

The OH group has a lone pair of electrons and dominates reactivity and selectivity. Steric hindrance favours the *para* product in the first reaction. The bromination has to occur *ortho* to the phenol as the *para* position is blocked.

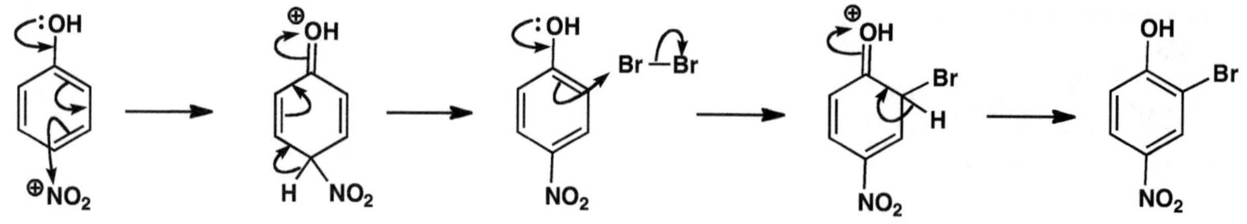

The second example has two Friedel-Crafts alkylations with *tertiary* alkyl halides. The first occurs *para* to bromine, a deactivating but *ortho,para-*directing group (see p. 489 in the textbook), preferring *para* because of steric hindrance. The second is a cyclization—the new ring cannot stretch any further than the next atom.

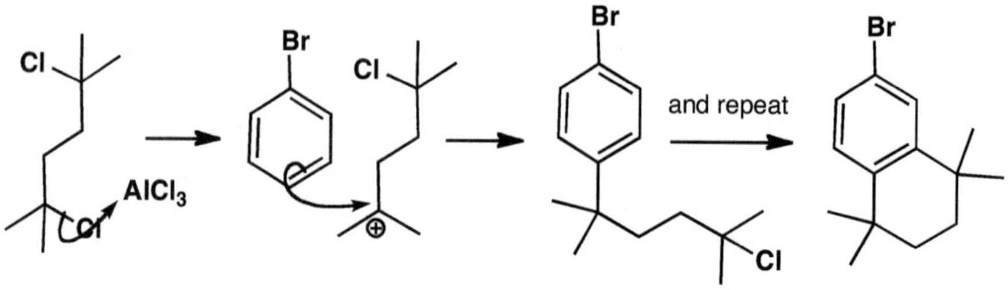

PROBLEM 5

Nitration of these compounds gives products with the ^1H NMR spectra shown. Deduce the structures of the products and explain the position of substitution. WARNING: do not decide the structure by saying where the nitro group 'ought to go'! Chemistry has many surprises and it is the evidence that counts.

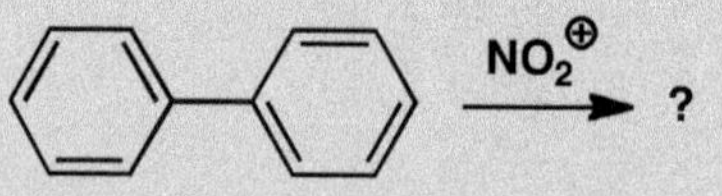

δ_H
7.77 (4H, d, J 10)
8.26 (4H, d, J 10)

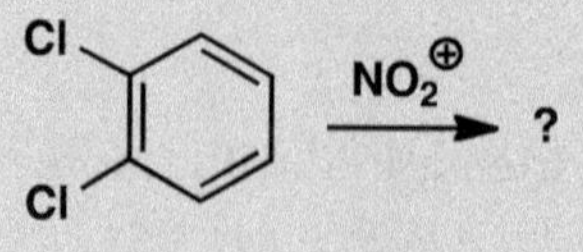

δ_H
7.6 (1H, d, J 10)
8.1 (1H, dd, J 10,2)
8.3 (1H, d, J 2)

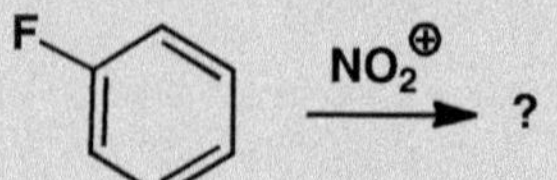

δ_H
7.15 (2H, dd, J 7,8)
8.19 (2H, dd, J 6,8)

Purpose of the problem

Revision of the relationship between NMR and substitution pattern.

Suggested solution

The first product has only eight hydrogens so two nitro groups must have been added. The molecule is clearly symmetrical and the coupling constant is right for neighbouring hydrogens so a substitution on each ring must have occurred in the *para* position. Note that the hydrogen next to the nitro group has the larger shift. We can deduce that each benzene ring is an *ortho,para*-directing group on the other because the intermediate cation is stabilized by conjugation.

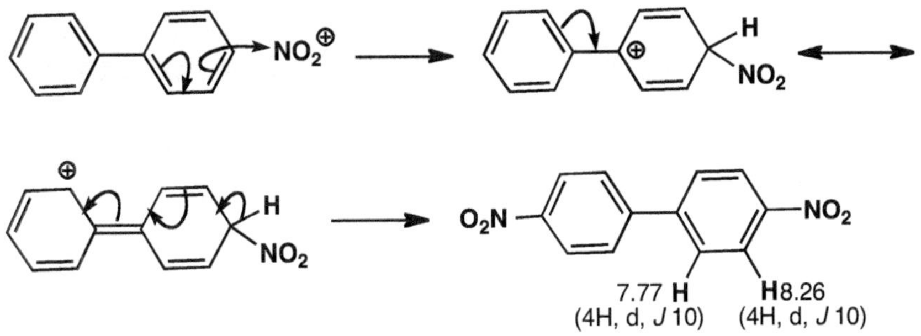

7.77 **H**   **H** 8.26
(4H, d, *J* 10)   (4H, d, *J* 10)

The hydrogen count reveals that the next two products are mono-nitro compounds. There are two hydrogens *ortho* to nitro in the second compound and one of them also has a typical *ortho* coupling to a neighbouring hydrogen while the other has only a small coupling (2 Hz) which must be a *meta* coupling. Substitution has occurred *para* to one of the chlorines and *ortho* to the other. The chlorines are *ortho,para*-directing thus activating all remaining positions so steric hindrance must explain the site of nitration.

■ Vicinal (*ortho*) coupling constants in benzene rings are typically 8–10 Hz; *meta* coupling constants are typically <2 Hz: see pp. 295-6 of the textbook.

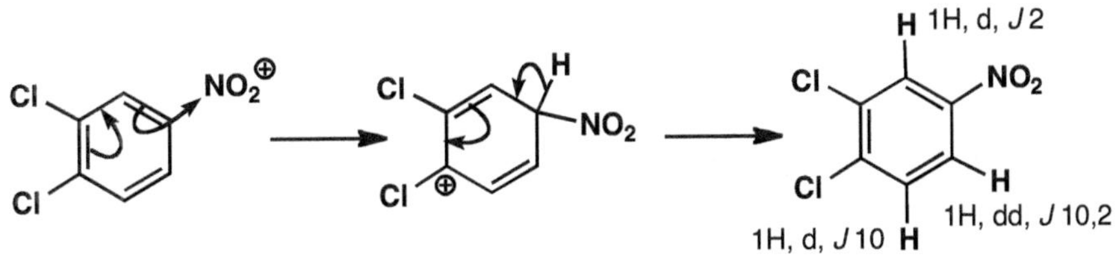

H 1H, d, *J* 2

1H, d, *J* 10 **H**   **H** 1H, dd, *J* 10,2

The third compound has the extra complication of couplings to fluorine. The coupling of 7 Hz shown by one hydrogen and 6 Hz shown by the other must be to fluorine as they occur once only. The symmetry of the compound and the typical *ortho* coupling between the hydrogens (8 Hz) shows that *para* substitution must have occurred.

■ The idea that heteronuclear couplings leave 'unpaired' coupling constants in the ^{1}H NMR spectrum is explained in the green box on p. 416 of the textbook.

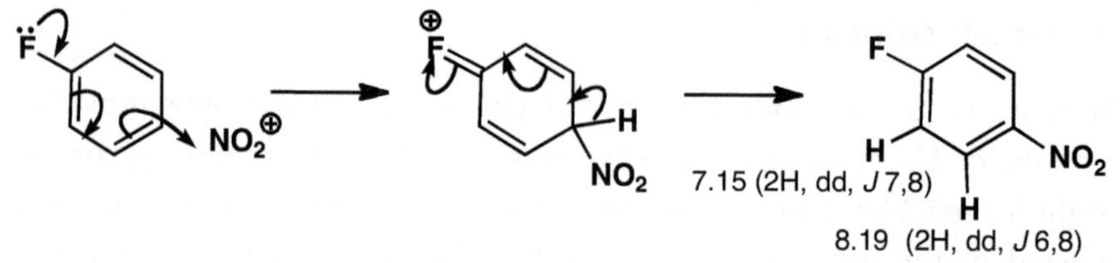

7.15 (2H, dd, *J* 7,8)

8.19 (2H, dd, *J* 6,8)

PROBLEM 6

Attempted Friedel-Crafts acylation of benzene with *t*-BuCOCl gives some of the expected ketone **A** as a minor product, as well as some *t*-butylbenzene **B**, but the major product is the substituted ketone **C**. Explain how these compounds are formed and suggest the order in which the two substituents are added to form compound **C**.

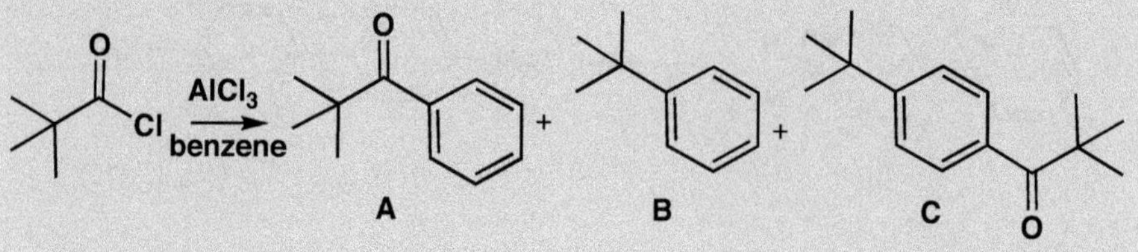

Purpose of the problem

Detailed analysis of a revealing example of the Friedel-Crafts reaction.

Suggested solution

■ Friedel-Crafts acylation is on p. 477 of the textbook.

The expected reaction to give **A** is a simple Friedel-Crafts acylation with the usual acylium ion intermediate.

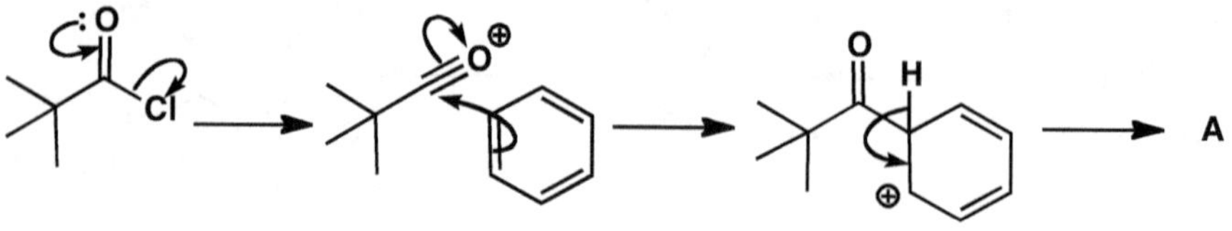

Product **B** must arise from a *t*-butyl cation and the only way that might be formed is by loss of carbon monoxide from the original acylium ion. Such a reaction happens only when the resulting carbocation is reasonably stable.

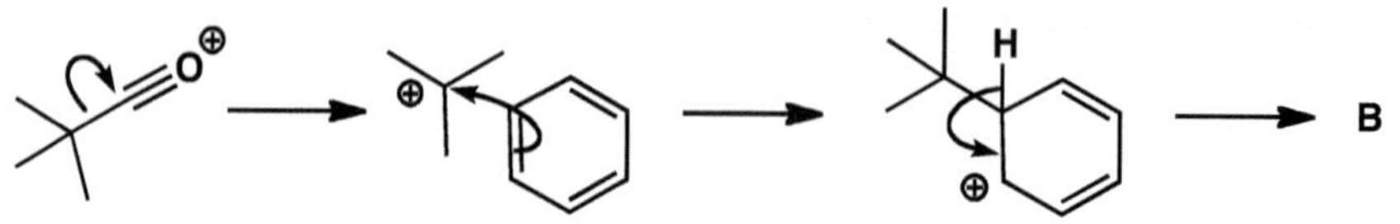

The main product **C** comes from the addition of both these electrophiles, but which adds first? The ketone in **A** is deactivating and *meta* directing but the *t*-butyl group in **B** is activating and *para*-directing so it must be added first.

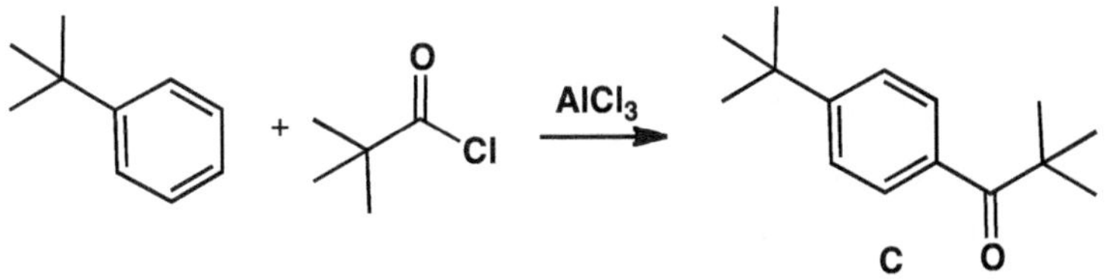

That answers the question but you might like to go further. Both **A** and **C** are formed by the alkylation of benzene as the first step. The decomposition of the acylium ion is evidently *faster* than the acylation of benzene. However, when **B** reacts further, it is mainly by acylation as only a small amount of di-*t*-butyl benzene is formed. Evidently the decomposition of the acylium ion is *slower* than the acylation of **B**! This is not unreasonable as the *t*-butyl group accelerates electrophilic attack—but it is a dramatic demonstration of that acceleration.

PROBLEM 7

Nitration of this heterocyclic compound with the usual HNO_3/H_2SO_4 mixture gives a single nitration product with the ^1H NMR spectrum shown below. Suggest which product is formed and why.

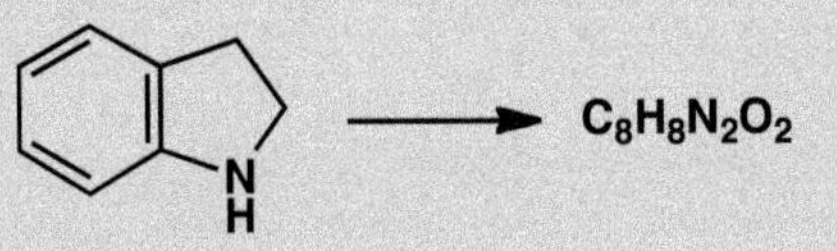

δ_H
3.04 (2H, t, *J* 7 Hz)
3.68 (2H, t, *J* 7 Hz)
6.45 (1H, d, *J* 8 Hz)
7.28 (1H, broad s)
7.81 (1H, d, *J* 1 Hz)
7.90 (1H, dd, *J* 8, 1 Hz)

Purpose of the problem

Revision of NMR and an attempt to convince you that the methods of chapter 21 can be applied to molecules you've not met before.

Suggested solution

The two 2H triplets and the broad NH signal show that the heterocyclic ring is intact. One nitro group has been added to the benzene ring. The proton at 7.81 with only one small (*meta*) coupling must be between the nitro group and the other ring and is marked on the two possible structures.

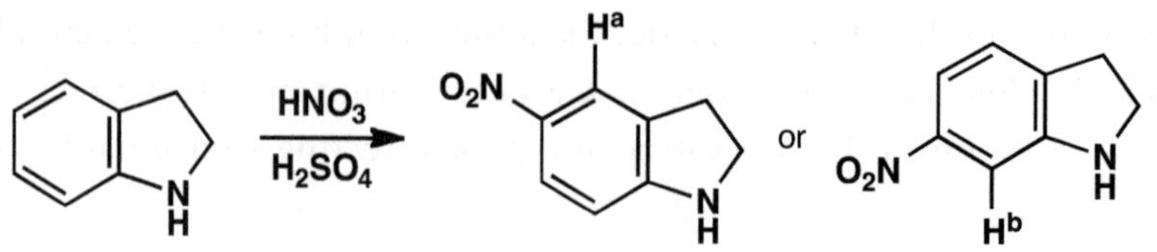

You could argue that NH is *ortho, para*-directing and so the second structure is more likely. But this is a risky argument as the reaction is carried out in strong acid solution where the nitrogen will mostly be protonated. It is safer to use the predicted δ_H from tables. Here we get:

Proton	*ortho*	*meta*	*para*	predicted δ_H
H^a	$NO_2 = +0.95$	$CH_2 = -0.14$	$NH = -0.25$	7.73
H^b	$NO_2 = +0.95$	$NH = -0.75$	$CH_2 = -0.06$	7.31

There's not much difference but H^a at 7.73 is closer to the observed 7.81, so it looks as though the small amount of unprotonated amine directs the reaction.

PROBLEM 8

What are the two possible isomeric products of this reaction? Which structure do you expect to predominate? What would be the bromination product from each?

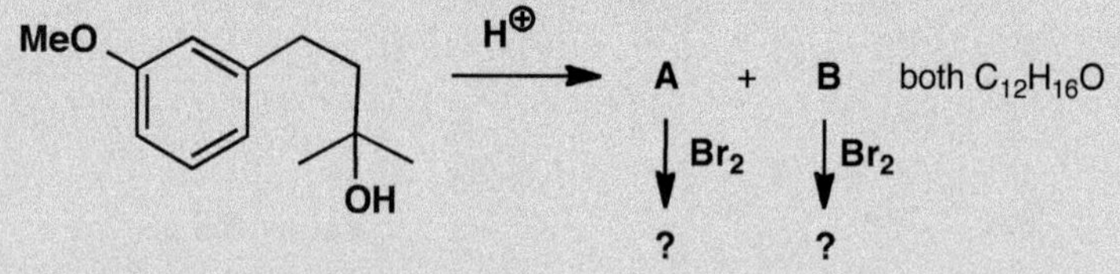

Purpose of the problem

Getting you to think about alternative products and possible reactions on compounds that haven't been made (yet).

Suggested solution

The reaction is a Friedel-Crafts cyclization, as you could have deduced by the simple loss of water. The resulting cation could cyclize in two ways, arbitrarily called **A** and **B**. Steric hindrance suggests that **A** would be the more likely product.

■ A Top Tip: when you have a formula for a product, but no structure, is to compare it with the formula for the starting material—in this case, $C_{12}H_{18}O_2$.

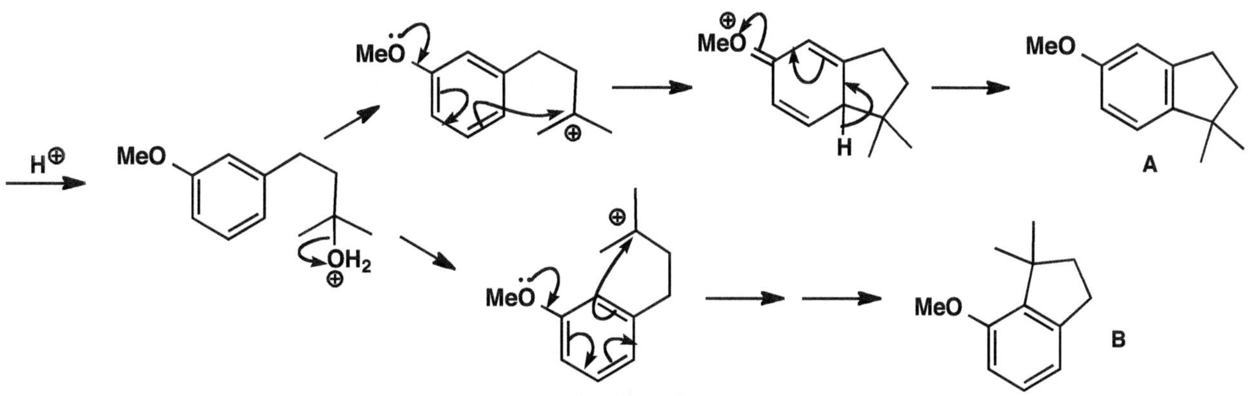

Bromination will go either *ortho* or *para* to the methoxy group: **A** has two different positions *ortho* to the OMe, but the *para* position is blocked. The least sterically hindered position gives a 1,2,4,5-tetrasubstituted ring. **B** might give a mixture of *ortho* and *para* substitution.

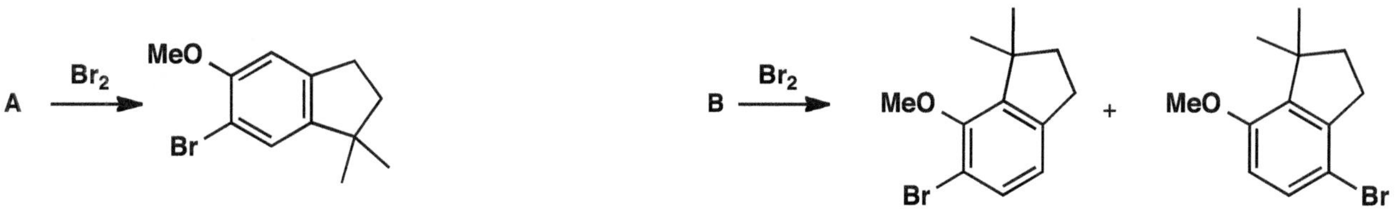

PROBLEM 9

On p. 479 of the textbook we explain the formation of 2,4,6-tribromophenol by bromination of phenol in water. It looks as though we can go no further as all the *ortho* and *para* positions are brominated. But we can if we treat the tribromo-compound with bromine in an organic solvent. Account for the formation of the tetrabromo-compound.

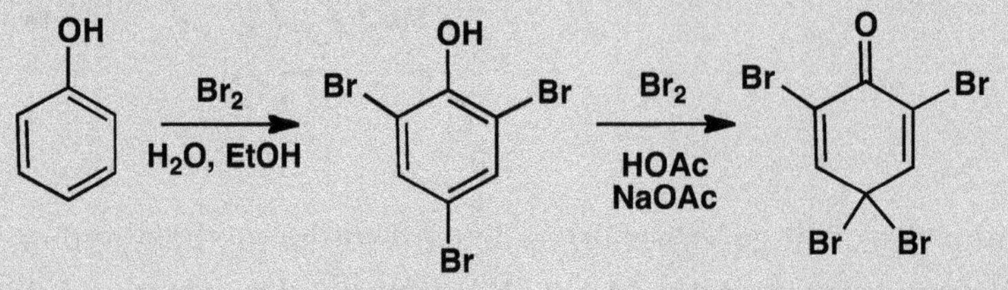

The product is useful in brominations as it avoids using unpleasant Br_2. Suggest a mechanism for the following bromination and account for the selectivity.

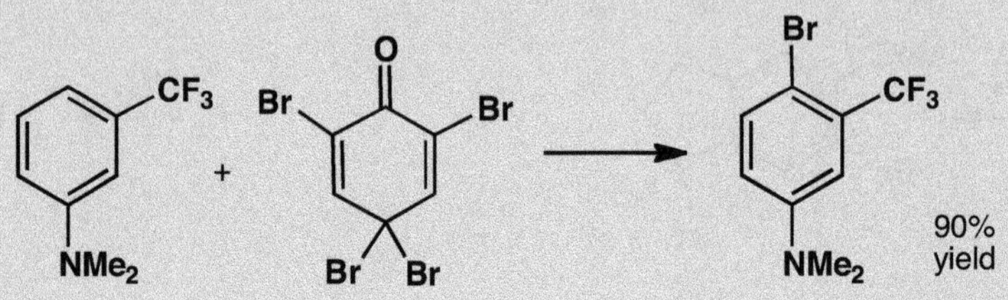

Purpose of the problem

Exploration of interesting chemistry associated with electrophilic substitution on benzene rings.

Suggested solution

Phenol is so reactive that the fourth bromine adds in the *para* position. Now the molecule has a problem as there is no hydrogen on that carbon to be lost. So the phenolic hydrogen is lost instead. It is surprising but revealing that this loss of aromaticity is preferred to the alternative bromination at the *meta* position.

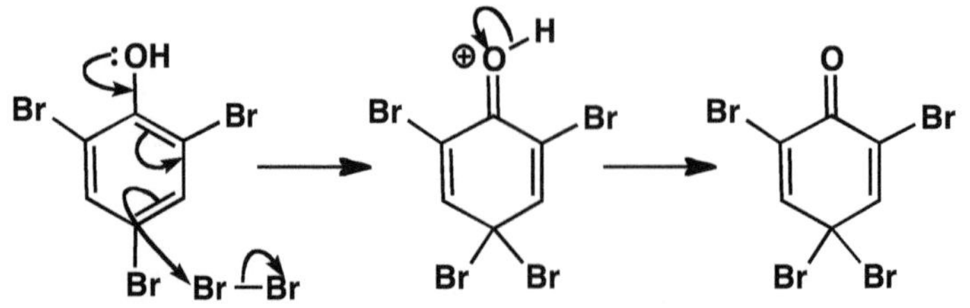

In the second reaction, one of the reactive bromines in the *para* position is transferred to the amine. It could have added *ortho* or *para* to the NMe$_2$ group but CF$_3$ is small and NMe$_2$ is large, because the two methyl groups lie in the plane of the ring, so steric hindrance rules. The other product is recovered tribromophenol.

■ Note that the *meta* directing effect of the deactivating CF$_3$ group is irrelevant (see p. 491 of the textbook).

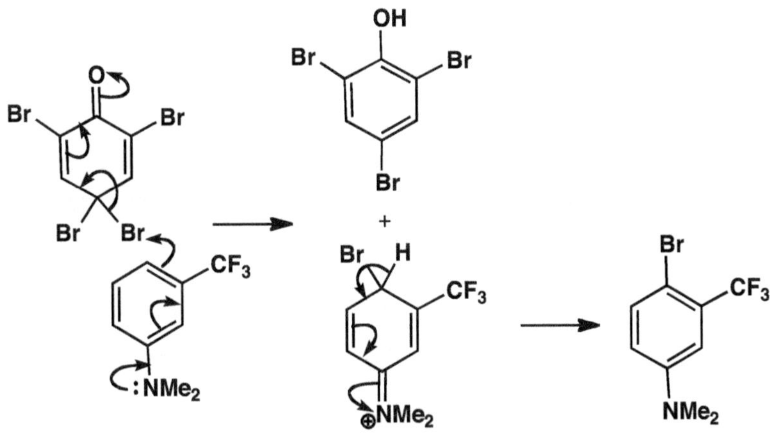

PROBLEM 10

How would you make each of the following compounds from benzene?

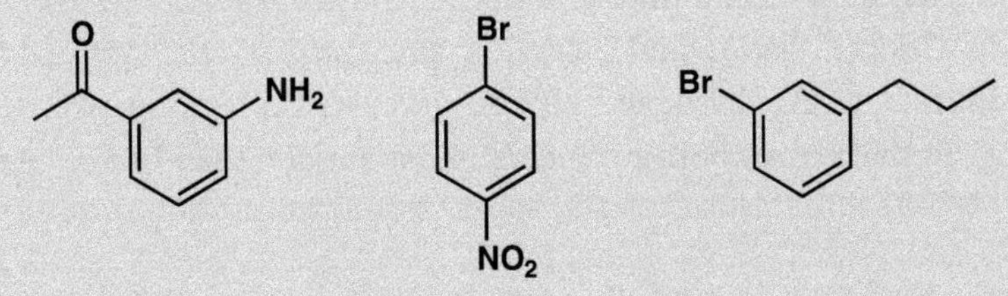

Purpose of the problem

Choosing a synthetic route, taking into account the directing effects of the substituents involved.

Suggested solution

The first compound has a ketone substituent, which is electron-withdrawing and therefore *meta*-directing, and an amino group, which is electron-donating and therefore *ortho,para*-directing. Aromatic amino groups are best made by reduction of nitro groups, which are also meta directing, so there are two possibilities. We can either start with a Friedel-Crafts acylation of benzene to give the ketone, which we can nitrate in the *meta* position and then reduce, or we can start by nitrating benzene, then do the acylation and then reduce. Either is a reasonable solution.

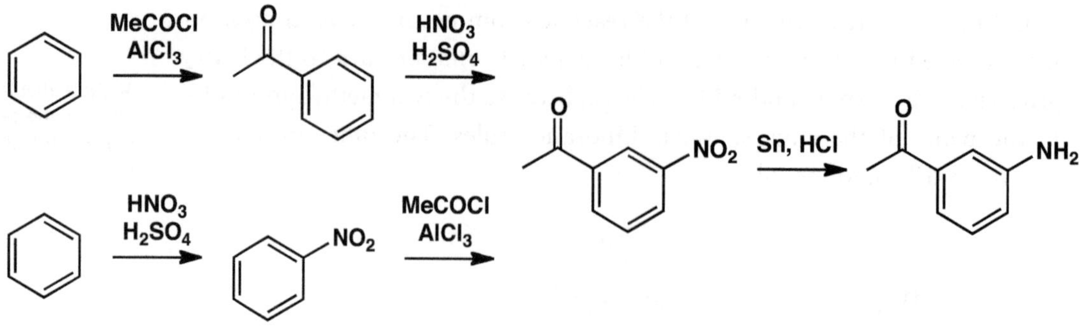

The second compound has a bromo substituent, which is *ortho,para*-directing, and a *meta*-directing nitro group. We need the *para* relationship, so we must put the bromine in first, then nitrate.

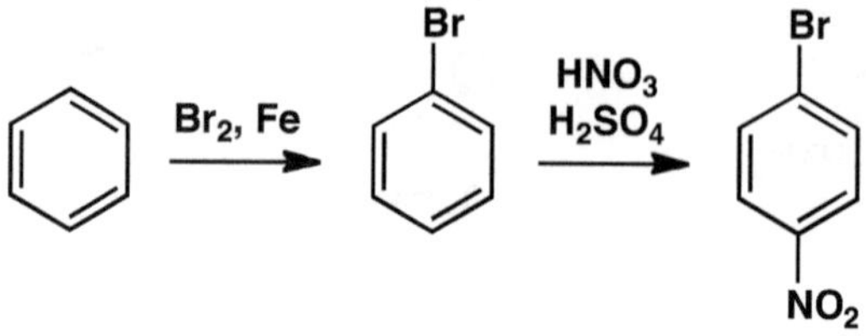

Finally, a compound with two *para*-directors arranged *meta* to one another. This may seem a problem, but we must introduce the alkyl group by Friedel-Crafts acylation and reduction, since primary alkyl groups cannot be introduced by Friedel-Crafts alkylation. The acyl group will be *meta* directing, so that solves both problems. First acylate, then brominate, then reduce.

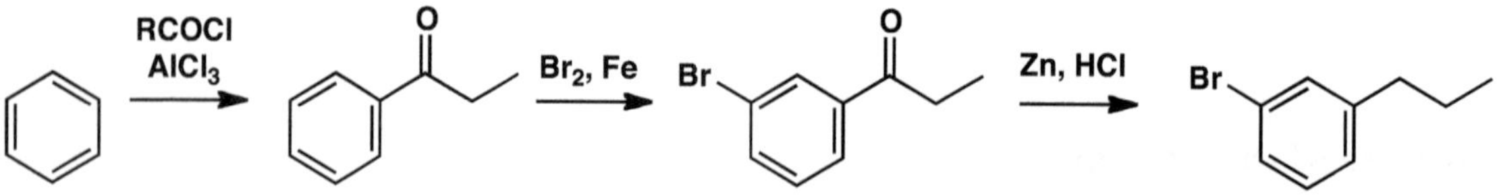

22 Conjugate addition and nucleophilic aromatic substitution

Connections

⮕ **Building on**

- Nucleophilic substitution at C=O ch10 and at saturated C ch15
- Electrophilic additions to alkenes ch19
- Electrophilic substitution on aromatic rings ch21

Arriving at

- Conjugate addition: conjugation of alkenes with electron-withdrawing groups makes them electrophilic and allows nucleophilic attack
- Conjugate substitution: electrophilic alkenes bearing leaving groups can promote substitution reactions at C=C related to those at C=O
- Nucleophilic aromatic substitution: electron-poor aromatic rings that allow substitution reactions with nucleophiles rather than the usual electrophiles
- Special leaving groups and nucleophiles that allow nucleophilic aromatic substitution on electron-rich rings

⮕ **Looking forward to**

- Regioselectivity ch24
- Conjugate addition of enolates ch26
- Reactions of heterocyclic aromatic compounds ch29 & ch30

This chapter is also the last chapter in the second cycle of chapters within this book, with which we complete our survey of the important elementary types of organic reactions. We follow it with two review chapters, where we bring together aspects of *selectivity*, before looking in more detail at enolate chemistry and how to make molecules.

Alkenes conjugated with carbonyl groups

To start this chapter, let us take you back to one of the first reactions we introduced: nucleophilic addition to carbonyl groups. Here are two examples, both of which give products which you should fully expect. We've included details of the IR spectra of the products to confirm firstly that the product has no carbonyl group and secondly that the alkene is still there.

■ If you need to review IR spectroscopy, turn back to Chapter 3. Any C=O peak would appear near 1700 cm⁻¹, but there isn't one. Instead there's an O–H peak at 3600 cm⁻¹. The 2250 cm⁻¹ peak is C≡N; C=C is at 1650 cm⁻¹.

IR: 3600 (broad), 2250, 1650 (weak)
no absorption near 1700

IR: 3600 (broad), 1640 (weak)
no absorption near 1700

Now let's tweak the conditions: we repeat the first reaction at a higher temperature, and we add to the second a small amount of a copper salt. Now the products are different:

Online support. The icon 🖱 in the margin indicates that accompanying interactive resources are provided online to help your understanding: just type www.chemtube3d.com/clayden/123 into your browser, replacing **123** with the number of the page where you see the icon. For pages linking to more than one resource, type **123-1, 123-2** etc. (replacing **123** with the page number) for access to successive links.

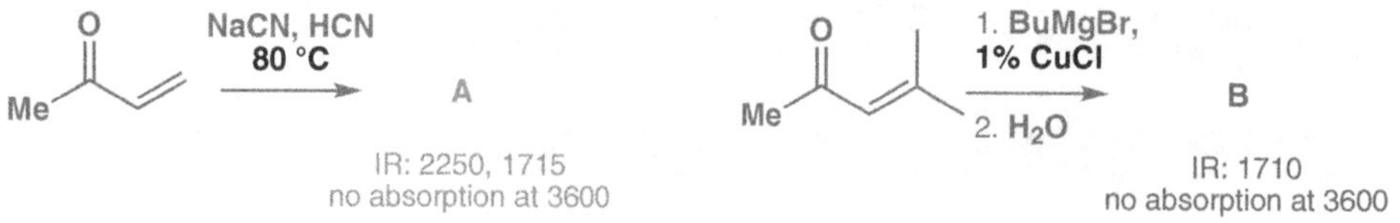

Both products A and B have kept their carbonyl group (IR peak at 1710–1715 cm^{-1}) but have lost the C=C. Yet A, at least, is unquestionably an addition product because it contains a C≡N peak at 2250 cm^{-1}.

Well, the identities of A and B are revealed here: they are the products of addition, not to the carbonyl group, but to the C=C bond. Here's a mechanism, for both reactions of cyanide: firstly the direct addition to C=O and secondly addition to the C=C bond.

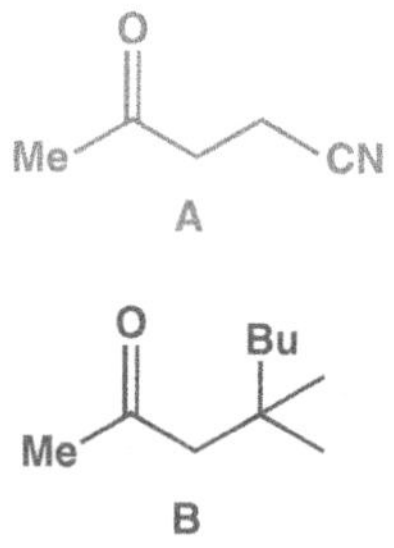

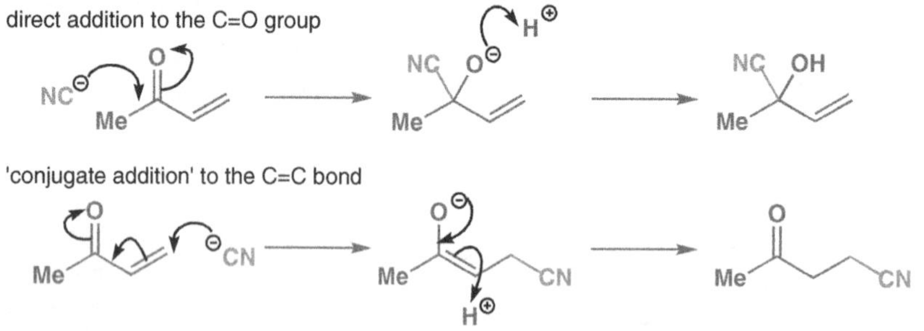

Interactive mechanism for conjugate addition of cyanide

This type of reaction, where a nucleophile adds to a C=C double bond, is called **conjugate addition**, and this chapter is about the sorts of alkenes (and arenes) that do this kind of thing. We will also explain how such small differences in reaction conditions (temperature, or the presence of CuCl) manage to change the outcome so dramatically.

But first we need to place these conjugate additions into context. As you found out in Chapter 19, **alkenes are nucleophilic**. Almost regardless of their substituents, they react with electrophiles such as bromine to form adducts in which the π bond of the alkene has been replaced by two σ bonds.

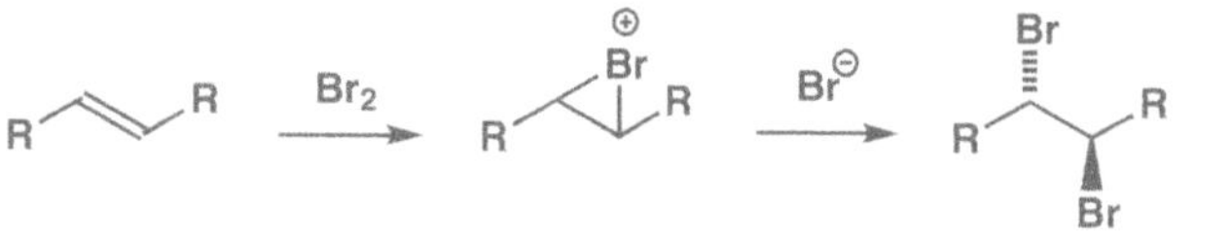

Reactions like this were discussed in Chapter 19.

Even when the alkene is conjugated with an electron-withdrawing group, as the alkenes on the last page were, bromine addition still occurs, although less readily. Never lose sight of this: alkenes are nucleophilic.

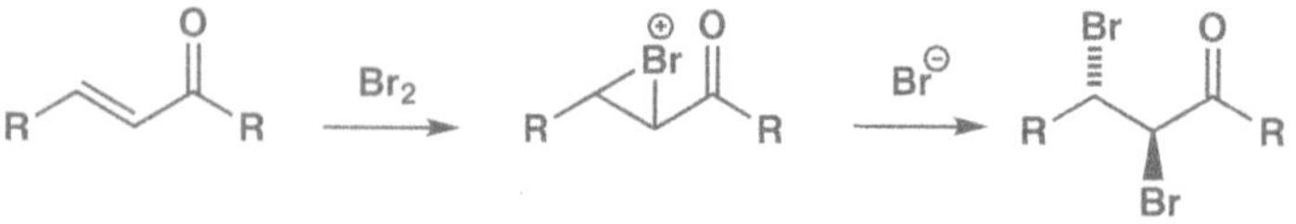

But as we have just seen, this last type of alkene also reacts with nucleophiles (such as cyanide, Grignard reagents, and, as you see below, more besides), and we now need to consider why.

Conjugated alkenes can be electrophilic

Conjugate additions occur only when the C=C double bond is immediately adjacent to the C=O group. They don't occur to C=C bonds that aren't conjugated (see the box on p. 501 for an illustration of this).

Compounds with double bonds adjacent to a C=O group are known as α,β-unsaturated carbonyl compounds. Many α,β-unsaturated carbonyl compounds have trivial names, and some are shown here. Some *classes* of α,β-unsaturated carbonyl compounds also have names such as 'enone', made up of 'ene' (for the double bond) + 'one' (for ketone).

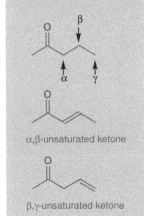

The α and β refer to the distance of the double bond from the C=O group: the α carbon is the one next to C=O (*not* the carbonyl carbon itself), the β carbon is one further down the chain, and so on.

α,β-unsaturated ketone

β,γ-unsaturated ketone

α,β-unsaturated aldehyde
(an enal)

propenal
(trivial name = acrolein)

α,β-unsaturated ketone
(an enone)

but-3-en-2-one
(trivial name =
methyl vinyl ketone)

α,β-unsaturated acid

propenoic acid
(trivial name =
acrylic acid)

α,β-unsaturated ester

ethyl propenoate
(trivial name =
ethyl acrylate)

Most types of nucleophiles can be made to undergo conjugate additions with α,β-unsaturated carbonyl compounds, and seven examples are shown below. Note that many of these nucleophiles would not add to a simple carbonyl group: we will explain why shortly. Conjugate addition is also known as Michael addition, and the reactive α,β-unsaturated carbonyl compounds shown here are often known as Michael acceptors.

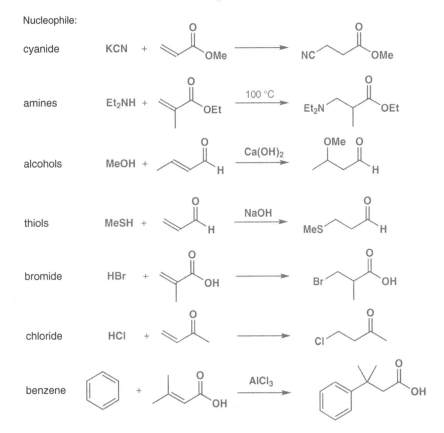

The reason that α,β-unsaturated carbonyl compounds react differently is conjugation, the phenomenon we discussed in Chapter 7. There we introduced you to the idea that bringing two π systems (two C=C bonds, for example, or a C=C bond and a C=O bond) close together leads to a stabilizing interaction. It also leads to modified reactivity, because the π bonds no longer react as independent functional groups but as a single, conjugated system.

● **Conjugation makes alkenes electrophilic**

● **C=C double bonds are nucleophilic**   ● **C=C double bonds conjugated with carbonyl groups can be electrophilic**

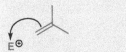

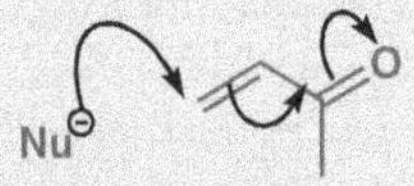

Termite self-defence and the reactivity of alkenes

Soldier termites of the species *Schedorhinotermes lamanianus* defend their nests with secretions of the enone shown below (compound 1), which is very effective at taking part in conjugate addition reactions with thiols (RSH). This makes it highly toxic, since many important biochemicals carry SH groups (one is described on p. 508). The worker termites of the same species—who build the nests—need to be able to avoid being caught in the crossfire, so they are equipped with an enzyme that allows them to reduce compound 1 to compound 2. This still has a double bond, but the double bond is completely unreactive towards nucleophiles because it is not conjugated with a carbonyl group. The workers escape unharmed.

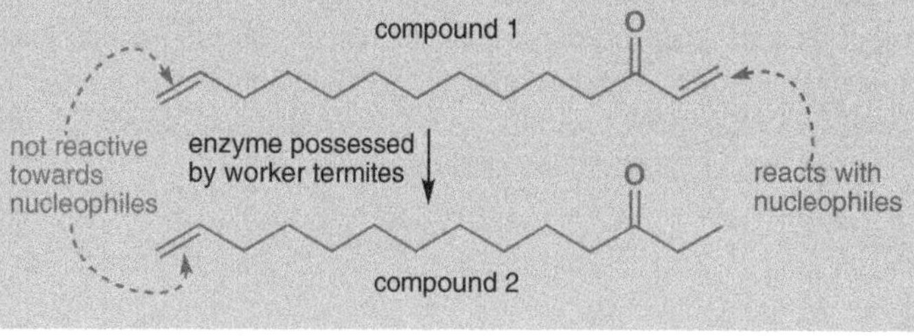

compound 1

not reactive towards nucleophiles

enzyme possessed by worker termites

reacts with nucleophiles

compound 2

Alkenes conjugated with carbonyl groups become polarized

To show why alkenes conjugated with carbonyl groups behave differently from unconjugated alkenes, we use curly arrows to indicate delocalization of the π electrons over the four atoms in the conjugated system. Both representations are extremes, and the true structure lies somewhere in between, but the polarized structure indicates why the conjugated C=C bond is electrophilic and why the β carbon is attacked by nucleophiles.

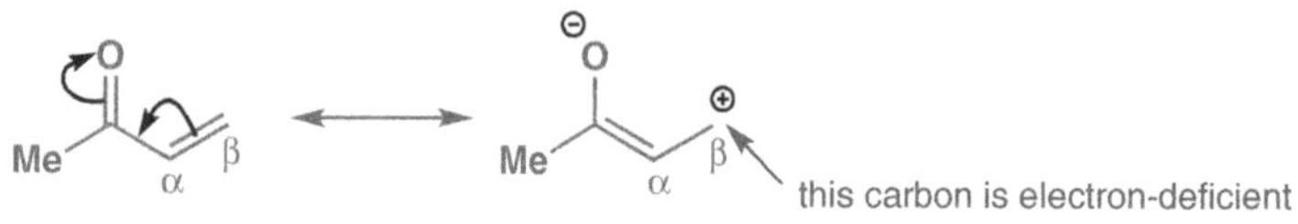

this carbon is electron-deficient

Polarization is detectable spectroscopically

IR spectroscopy provides us with evidence for polarization in C=C bonds conjugated to C=O bonds. An unconjugated ketone C=O absorbs at 1715 cm^{-1} while an unconjugated alkene C=C absorbs (usually rather weakly) at about 1650 cm^{-1}. Bringing these two groups into conjugation in an α,β-unsaturated carbonyl compound leads to two peaks at 1675 and 1615 cm^{-1}, respectively, both quite intense. The lowering of the frequency of both peaks is consistent with a weakening of both π bonds (notice that the polarized structure has only single bonds where the C=O and C=C double bonds were). The increase in the *intensity* of the C=C absorption is consistent with polarization brought about by conjugation with C=O: a conjugated C=C bond has a significantly larger dipole moment than its unconjugated cousins.

The polarization of the C=C bond is also evident in the ^{13}C NMR spectrum, with the signal for the sp^2 carbon atom furthest from the carbonyl group moving downfield relative to an unconjugated alkene to about 140 ppm, and the signal for the other double bond carbon atom staying at about 120 ppm.

Molecular orbitals control conjugate additions

We have spectroscopic evidence that a conjugated C=C bond is polarized, and we can explain this with curly arrows, but the actual bond-forming step must involve movement of electrons from the HOMO of the nucleophile to the LUMO of the unsaturated carbonyl compound. This example is an efficient (the reaction happens even at 0 °C) addition to acrolein (propenal) with methoxide as the nucleophile.

■ You may be asking yourself why we can't show the delocalization by moving the electrons the other way, like this.

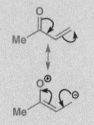

Think about electronegativities: O is much more electronegative than C, so it is quite happy to accept electrons, but here we have taken electrons away, leaving it with only six electrons. This structure therefore cannot represent the distribution of electrons in the conjugated system.

^{13}C NMR chemical shifts:

conjugated alkene $\quad$ non-conjugated alkene

δ_C 143 $\quad$ δ_C 132

δ_C 124 $\quad$ δ_C 119

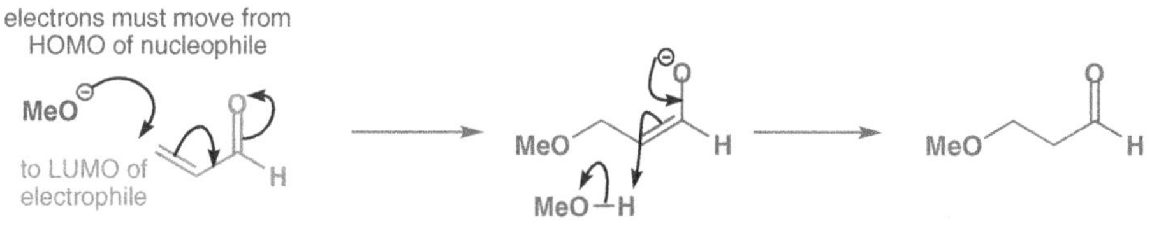

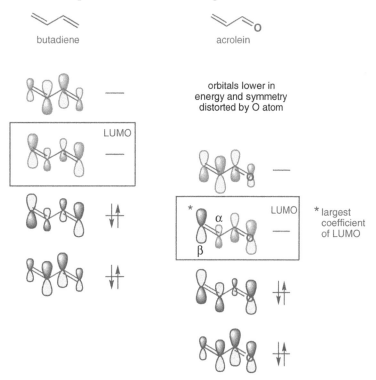

But what does this LUMO look like? It will certainly be more complicated than the π* LUMO of a simple carbonyl group. The nearest thing you have met so far (in Chapter 7) are the orbitals of butadiene (C=C conjugated with C=C), which we can compare with the α,β-unsaturated aldehyde acrolein (C=C conjugated with C=O). The orbitals in the π systems of butadiene and acrolein are shown here. They are different because acrolein's orbitals are perturbed (distorted) by the oxygen atom (Chapter 4). You need not be concerned with exactly how the sizes of the orbitals are worked out, but for the moment just concentrate on the shape of the LUMO, the orbital that will accept electrons when a nucleophile attacks.

butadiene   acrolein

orbitals lower in energy and symmetry distorted by O atom

LUMO

LUMO   * largest coefficient of LUMO

* largest coefficient of LUMO

α

β

Interactive molecular orbitals in acrolein

■ In acrolein, the HOMO is in fact not the highest filled π orbital you see here, but the lone pairs on oxygen. This is not important, however, because here we are considering acrolein as an electrophile, so we are interested only in its LUMO.

In the LUMO, the largest coefficient is on the β carbon of the α,β-unsaturated system, shown with an asterisk. And it is here, therefore, that nucleophiles attack. In the reaction you have just seen, the HOMO is the methoxide oxygen's lone pair, so this will be the key orbital interaction that gives rise to the new bond.

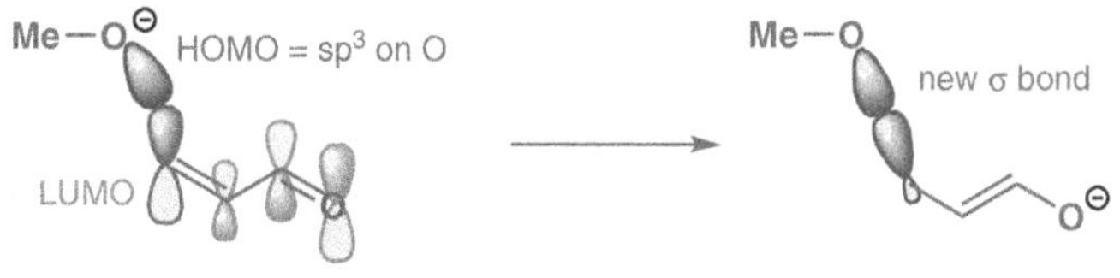

Me—O   HOMO = sp³ on O   Me—O   new σ bond

LUMO

The second largest coefficient is on the C=O carbon atom, so it's not surprising that some nucleophiles attack here as well—remember the example right at the beginning of the chapter where you saw cyanide attacking either the double bond or the carbonyl group depending on the conditions of the reaction. We shall next look at some conjugate additions with alcohols and amines as nucleophiles, before reconsidering the question of where the nucleophile attacks.

Conjugate additions have enolates or enols as intermediates

So much for the addition step of the reaction. But the product of this step is of course not the final product of the reaction—it is in fact an enolate. We hope you recognize these species from Chapter 20, where you saw them being made by treating carbonyl compounds with base. Conjugate addition is another way of generating an enolate, and as with all enolates, protonation gives back a carbonyl compound. The proton has to come from somewhere, so conjugate additions are usually done in protic solvents (such as alcohols or water). Here is another example with an alcohol as the nucleophile:

⮕ Enols and enolates were introduced in Chapter 20.

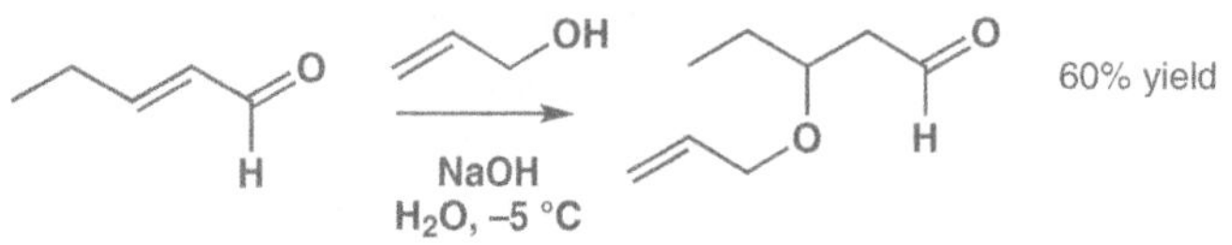

In alkaline solution, a small amount of alkoxide is produced (the pK_a of an alcohol is slightly higher than that of water), which attacks the C=C double bond in a conjugate addition. The product is an enolate, which is protonated by water to give the final aldehyde, and regenerates hydroxide as it does so: only a catalytic amount of base is required for this type of reaction.

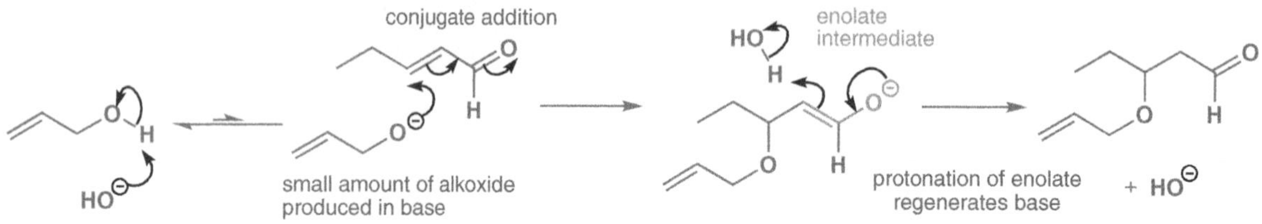

Amines are good nucleophiles for conjugate addition. In the reaction below, aqueous dimethylamine is used in a sealed system to stop the amine evaporating (dimethylamine is a gas even at room temperature).

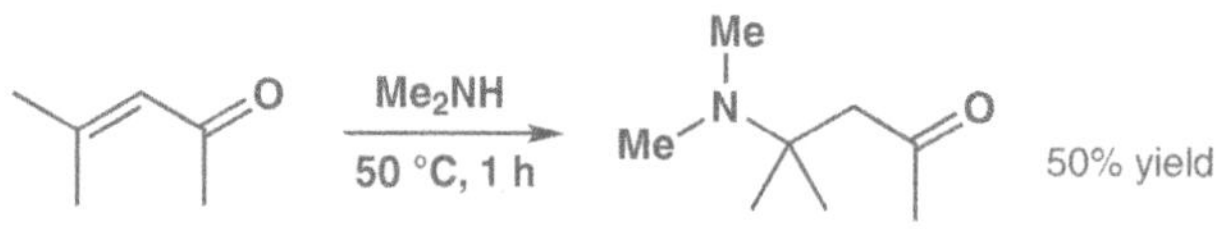

Amines are neutral nucleophiles, and the amine itself provides a proton for the enolate.

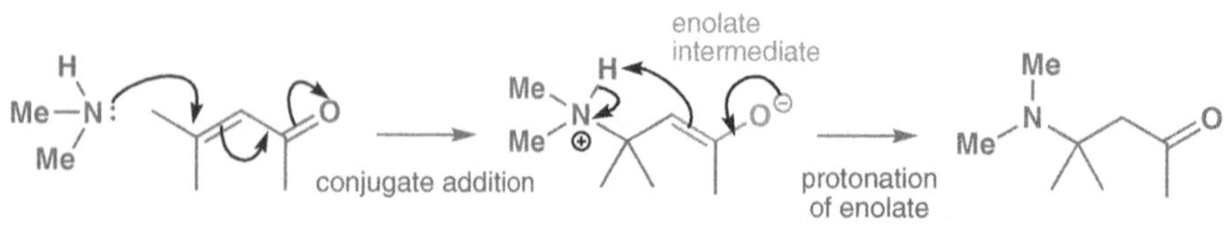

If you survey the initial overview of conjugate additions on p. 500 you will see that several take place under acidic conditions. Treatment of this α,β-unsaturated ketone with HCl, for example, gives a chloroketone. The first step must be protonation of the carbonyl group, which makes the enone even more electrophilic by giving it a positive charge. Chloride attacks the β carbon to give an *enol*.

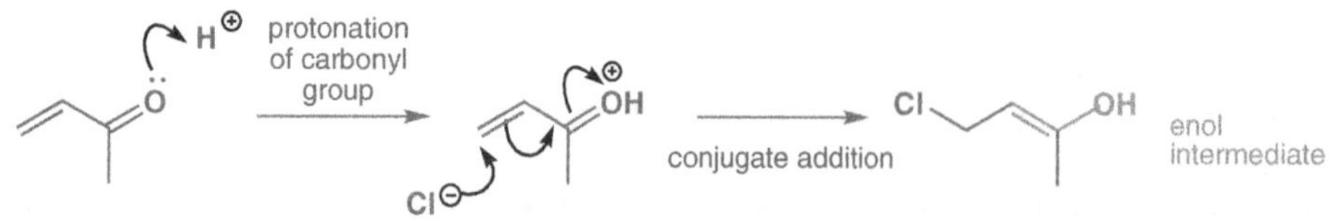

➡ *Tautomerism* is defined on p. 451.

All that remains to happen now is tautomerism of the enol to its keto form by proton transfer from O to C.

Conjugate addition or direct addition to the carbonyl group?

We have shown you several examples of conjugate additions using various nucleophiles and α,β-unsaturated carbonyl compounds, but we haven't yet addressed one important question. When do nucleophiles do conjugate addition (also called 1,4-addition) and when do they add directly to the carbonyl group (1,2-addition)? Several factors are involved—they are summarized here, and we will spend the next section of this chapter discussing them in turn.

- **Conjugate addition to C=C** (also called 1,4-addition) 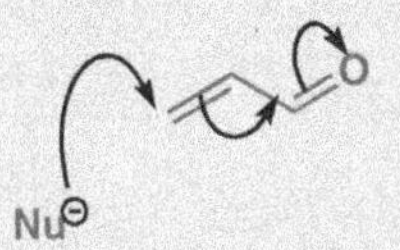
- **Direct addition to C=O** (also called 1,2-addition)

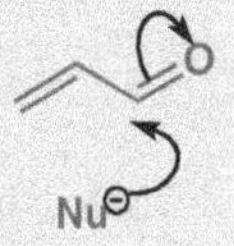

The way that nucleophiles react depends on:
- the conditions of the reaction
- the nature of the α,β-unsaturated carbonyl compound
- the type of nucleophile.

Reaction conditions

The very first conjugate addition reaction in this chapter depended on the conditions of the reaction. Treating an enone with cyanide and an acid catalyst at low temperature gives a cyanohydrin by direct attack at C=O, while heating the reaction mixture leads to conjugate addition. What is going on?

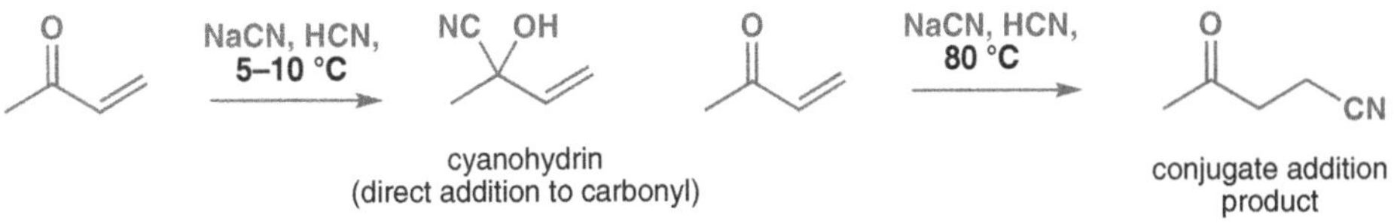

We'll consider the low-temperature reaction first. As you know from Chapter 6, it is quite normal for cyanide to react with a ketone under these conditions to form a cyanohydrin. You also know from Chapter 6 that cyanohydrin formation is reversible. Even if the equilibrium for cyanohydrin formation lies well over to the side of the products, there will always be a small amount of starting enone remaining. Most of the time, this enone will react to form more cyanohydrin and, as it does, some cyanohydrin will decompose back to enone plus cyanide—such is the nature of a dynamic equilibrium. But every now and then—at a much slower rate—the starting enone will undergo a *conjugate addition* with the cyanide.

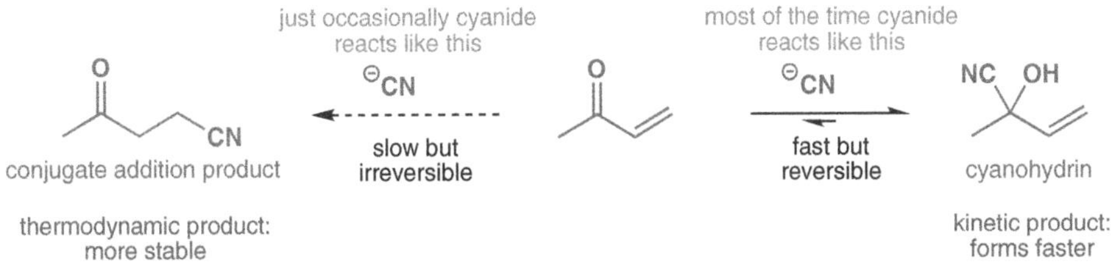

Now we have a different situation: conjugate addition is essentially an *irreversible* reaction, so once a molecule of enone has been converted to conjugate addition product, its fate is sealed: it cannot go back to enone again. Very slowly, therefore, the amount of conjugate addition product in the mixture will build up. In order for the enone–cyanohydrin equilibrium to be maintained, any enone that is converted to conjugate addition product will have to be replaced by reversion of cyanohydrin to enone plus cyanide. Even at room temperature, we can therefore expect the cyanohydrin to be converted bit by bit to conjugate addition product. This may take a very long time, but reaction rates are faster at higher temperatures, so at 80 °C this process does not take long at all and, after a few hours, the cyanohydrin has all been converted to conjugate addition product.

The contrast between the two products is this: the cyanohydrin is **formed faster** than the conjugate addition product, and is known as the product of kinetic control (or the kinetic product), but the conjugate addition product is the **more stable compound** and is the product of thermodynamic control (or the thermodynamic product). Typically, kinetic control involves lower temperatures and shorter reaction times, which ensures that only the fastest reaction has the chance to occur. And, typically, thermodynamic control involves higher temperatures and long reaction times to ensure that even the slower reactions have a chance to occur, and all the material is converted to the more stable compound.

➡ Kinetic and thermodynamic control were introduced in Chapter 12.

- ● **Kinetic and thermodynamic control**

  - • The product that forms faster is called the **kinetic product**.
  - • The product that is the more stable is called the **thermodynamic product**.

  Similarly,

  - • Conditions that give rise to the kinetic product are called **kinetic control**.
  - • Conditions that give rise to the thermodynamic product are called **thermodynamic control**.

Why is direct addition faster than conjugate addition? Well, although the carbon atom β to the C=O group carries some positive charge, the carbon atom of the carbonyl group carries more, and so electrostatic attraction for the charged nucleophiles will encourage it to attack the carbonyl group directly rather than undergo conjugate addition.

And why is the conjugate addition product the more stable? In the conjugate addition product, we gain a C–C σ bond, losing a C=C π bond, but keeping the C=O π bond. With direct addition, we still gain a C–C bond, but we lose the C=O π bond and keep the C=C π bond. C=O π bonds are stronger than C=C π bonds, so the conjugate addition product is more stable.

Practically, then, to get conjugate addition to occur you just have to give the reaction plenty of energy and maybe plenty of time to find its way to the most stable product. Here's an example: note the temperature!

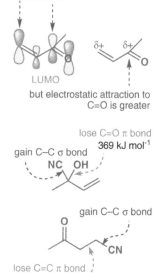

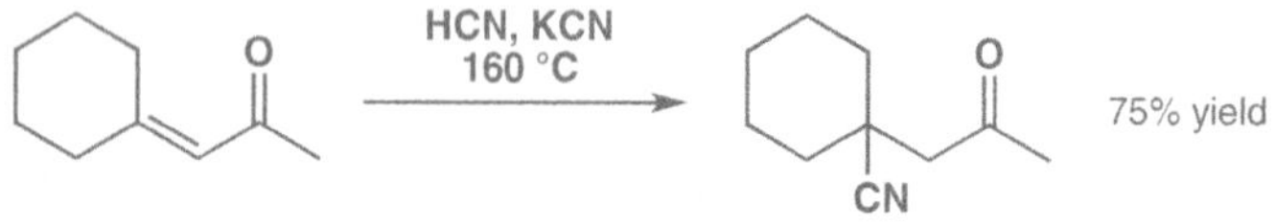

attack is possible at either site

LUMO

but electrostatic attraction to C=O is greater

lose C=O π bond
369 kJ mol⁻¹
gain C–C σ bond
NC OH

gain C–C σ bond
CN

lose C=C π bond
280 kJ mol⁻¹

HCN, KCN
160 °C
→
75% yield

CN

Structural factors

So far we have shown you conjugate additions mainly of α,β-unsaturated aldehydes and unsaturated α,β-ketones. You won't be at all surprised to learn, however, that unsaturated acids, esters, amides, and nitriles—in fact all carboxylic acid derivatives—can also take part in conjugate addition reactions. Two examples, an amide and an ester, are shown on the right below. But notice how the selectivity of these reactions depends on the structure of the unsaturated compound: compare the way butyllithium adds to this α,β-unsaturated aldehyde and α,β-unsaturated amide. Both additions are irreversible, and BuLi attacks the reactive carbonyl group of the aldehyde, but prefers conjugate addition to the less reactive amide. Similarly, ammonia reacts with this acyl chloride to give an amide product that derives from direct addition to the carbonyl group, while with the ester it undergoes conjugate addition to give an amine.

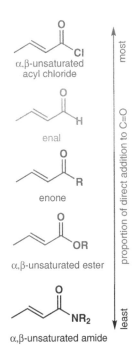

In both of these cases, the site of nucleophilic attack is determined simply by reactivity: the more reactive the carbonyl group, the more direct addition to C=O will result. The most reactive carbonyl groups, as you saw in Chapter 10, are those that are not conjugated with O or N (as they are in esters and amides), and particularly reactive are acyl chlorides and aldehydes. In general, the proportion of direct addition to the carbonyl group follows the reactivity sequence in the margin.

Sodium borohydride is a nucleophile that you have seen reducing simple aldehydes and ketones to alcohols, but it will also do conjugate addition reactions. Which of the alternatives actually takes place depends on the reactivity of the C=O group. NaBH$_4$ usually reacts with α,β-unsaturated aldehydes to give alcohols by direct addition to the carbonyl group.

Quite common with ketones, however, is the outcome below.

The borohydride has reduced not only the carbonyl group but the double bond as well. In fact, it's the double bond that's reduced first in a conjugate addition, followed by addition to the carbonyl group.

Luche reduction

It is possible to force NaBH$_4$ to attack only the C=O group by adding CeCl$_3$ to the reaction mixture. This modification is known as the Luche reduction, after its discoverer.

For esters and other less reactive carbonyl compounds conjugate addition is the only reaction that occurs because NaBH$_4$ doesn't reduce esters or amides.

The nature of the nucleophile: hard or soft

Among the best nucleophiles of all at doing conjugate addition are thiols, the sulfur analogues of alcohols. In this example, the nucleophile is thiophenol (phenol with the O replaced by S).

Remarkably, no acid or base catalyst is needed (as it was with the alcohol additions), and the product is obtained in 94% yield under quite mild reaction conditions.

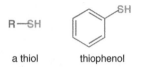

a thiol   thiophenol

■ We introduced the terms **hard** and **soft** in relation to nucleophiles in Chapter 15, p. 357,

So what's so special about a thiol? As you've seen already, attraction between nucleophiles and electrophiles is governed by two related interactions—electrostatic attraction between positive and negative charges and orbital overlap between the HOMO of the nucleophile and the LUMO of the electrophile. Successful reactions usually result from a combination of both, but sometimes reactivity can be dominated by one or the other. The dominant factor, be it electrostatic or orbital control, depends on the nucleophile and electrophile involved. Nucleophiles containing small, electronegative atoms (such as O or Cl), which we call 'hard', tend to react under predominantly electrostatic control, while 'soft' nuclophiles containing larger atoms (including the sulfur of thiols, but also P, I, and Se) are predominantly subject to control by orbital overlap.

The table below divides some nucleophiles into the two categories (plus some that lie in between)—but don't try to learn it! Rather, convince yourself that the properties of each one justify its location in the table. Most of these nucleophiles you have not yet seen in action, and the most important ones at this stage are indicated in **bold type**.

Hard and soft nucleophiles

Hard nucleophiles	Borderline	Soft nucleophiles
F⁻, OH⁻, RO⁻, SO_4^{2-}, Cl⁻	N₃⁻, **CN⁻**	I⁻, **RS⁻**, RSe⁻, S²⁻
H₂O, ROH, ROR', RCOR'	**RNH₂, R¹R²NH**	**RSH**, RSR', R₃P
NH₃, RMgBr, RLi	Br⁻	alkenes, aromatic rings

Not only can nucleophiles be classified as hard or soft, but electrophiles can too. For example, H^+ is a very hard electrophile because it is small and charged, while Br_2 is a soft electrophile: its orbitals are diffuse and it is uncharged. You saw Br_2 reacting with an alkene earlier in the chapter, and we explained in Chapter 5 that this reaction happens solely because of orbital interactions: no charges are involved.

● Hard/soft reactivity

- Reactions of hard species are dominated by charges and electrostatic effects.
- Reactions of soft species are dominated by orbital effects.
- Hard nucleophiles tend to react well with hard electrophiles.
- Soft nucleophiles tend to react well with soft electrophiles.

What has all this to do with the conjugate addition of thiols? Well, an α,β-unsaturated carbonyl compound is unusual in that it has two electrophilic sites, one of which is hard and one of which is soft. The carbonyl group has a high partial charge on the carbonyl carbon and will tend to react with hard nucleophiles, such as organolithium and Grignard reagents, that have a high partial charge on the nucleophilic carbon atom. Conversely, the β carbon of the α,β-unsaturated carbonyl system does not have a high partial positive charge but is the site of the largest coefficient in the LUMO. This makes the β carbon a soft electrophile and likely to react well with soft nucleophiles such as thiols.

● Hard/soft—direct/conjugate addition

- Hard nucleophiles tend to react at the carbonyl carbon (hard) of an enone.
- Soft nucleophiles tend to react at the β carbon (soft) of an enone and lead to conjugate addition.

Anticancer drugs that work by conjugate addition of thiols

Drugs to combat cancer act on a range of biochemical pathways, but most commonly on processes that cancerous cells need to use to proliferate rapidly. One class attacks DNA polymerase, an enzyme needed to make the copy of DNA that has to be provided for each new cell. Helenalin and vernolepin are two such compounds, and if you look closely at their structure you should be able to spot two α,β-unsaturated carbonyl groups in each. Biochemistry is just chemistry in very small flasks called cells, and the reaction between DNA polymerase and these drugs is simply a conjugate addition reaction between a thiol (the SH group of one of the enzyme's cysteine residues) and the unsaturated carbonyl groups. The reaction is irreversible and shuts down completely the function of the enzyme.

helenalin   vernolepin

For this reason any compound capable of conjugate addition is potentially dangerous to living things. Even simple compounds like ethyl acrylate are labelled 'cancer suspect agents'. They attack enzymes, particularly the DNA polymerase involved in cell division by conjugate addition to thiol and amino groups in the enzyme. Fortunately, we are offered some degree of protection by an important compound present in most tissues. The compound is glutathione, a tripeptide—a compound made from three amino acids. We shall discuss such compounds in more detail later in the book (Chapter 42) but notice for the moment that this compound can be divided into three at the two amide bonds.

glutathione

detoxification of carcinogens by glutathione

glutamic acid   cysteine   glycine

inactivated carcinogen cannot do conjugate addition

The business end of glutathione is the thiol (SH) group, which scavenges carcinogenic compounds by conjugate addition. If we use an 'exomethylene lactone'—a highly reactive Michael acceptor—as an example and represent glutathione as RCH₂SH, you can see the sort of thing that happens. If the normally abundant glutathione is removed by such processes as oxidation (Chapter 42) and cannot any longer scavenge toxins, then the organism is in danger. This is one reason why 'antioxidants' like vitamin C are so beneficial—they remove stray oxidizing agents and protect the supply of glutathione. Keep eating the fruit and vegetables!

Promoting conjugate addition with copper(I) salts

Grignard reagents add directly to the carbonyl group of α,β-unsaturated aldehydes and ketones to give allylic alcohols: you have seen several examples of this, and you can now explain it by saying that the hard Grignard reagent prefers to attack the harder C=O rather than the softer C=C electrophilic centre. Here is a further example—the addition of MeMgBr to a cyclic ketone to give an allylic alcohol, plus, as it happens, some of a diene that arises from this alcohol by loss of water (dehydration). Below this example is the same reaction to which a very small amount (just 0.01 equivalents, that is, 1%) of copper(I) chloride has been added. The effect of the copper is dramatic: it makes the Grignard reagent undergo conjugate addition, with only a trace of the diene.

Organocopper reagents undergo conjugate addition

The copper works by *transmetallating* the Grignard reagent to give an organocopper reagent—simply put, the magnesium is exchanged for copper. Organocoppers are softer than Grignard reagents, and add in a conjugate fashion to the softer C=C double bond. Once the organocopper has added, the copper salt is available to transmetallate some more Grignard, and only a catalytic amount is required.

The organocopper is shown here as 'Me–Cu' because its precise structure is not known. But there are other organocopper reagents that also undergo conjugate addition and are much better understood. The simplest result from the reaction of two equivalents of organolithium with one equivalent of a copper (I) salt such as CuBr in ether or THF solvent at low temperature. The lithium cuprates (R$_2$CuLi) that are formed are not stable and must be used immediately.

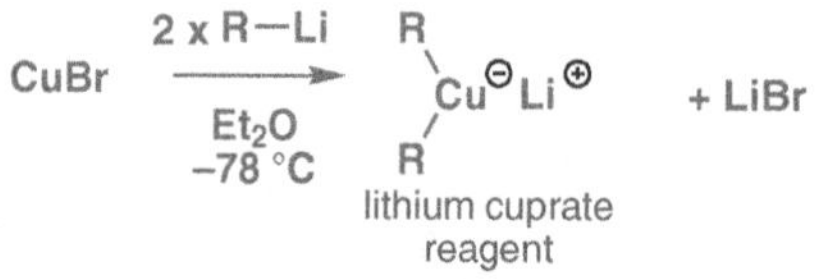

The addition of lithium cuprates to α,β-unsaturated ketones turns out to be much better if trimethylsilyl chloride is added to the reaction—we shall explain what this does shortly, but for the moment here are two examples of lithium cuprate additions.

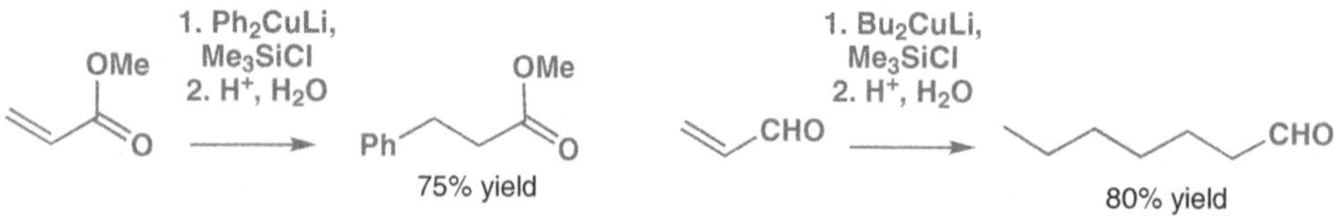

The silicon works by reacting with the negatively charged intermediate in the conjugate addition reaction to give a silyl enol ether—a type of molecule we met in Chapter 20. Here is a possible mechanism for a reaction between Bu$_2$CuLi and an α,β-unsaturated aldehyde in the presence of Me$_3$SiCl. The silyl enol ether simply hydrolyses to the ketone at the end of the reaction.

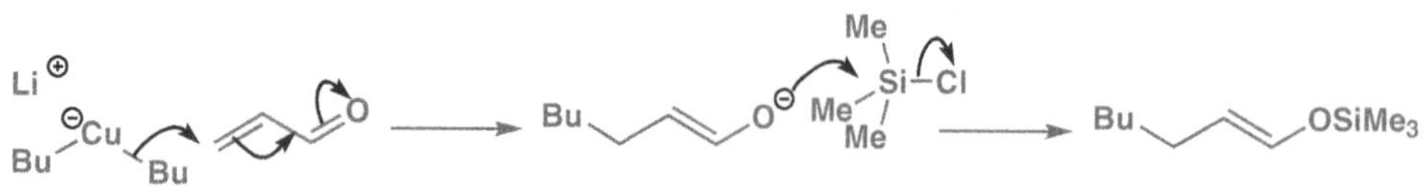

Summary: factors controlling conjugate addition

At this point in the chapter it is worthwhile talking stock of the factors controlling the two modes of addition to α,β-unsaturated carbonyl compounds.

> ■ Organocoppers are softer than Grignard reagents because copper is less electropositive than magnesium, so the C–Cu bond is less polarized than the C–Mg bond, giving the carbon atom less of a partial negative charge. Electronegativities: Mg, 1.3; Cu, 1.9.

> ■ As with many other organometallic compounds, the exact structure of these reagents is more complex than we imply here: they are probably tetramers (four molecules of R$_2$CuLi bound together), but for simplicity we will draw them as monomers. Organometallics (compounds with metal–carbon bonds) get a chapter to themselves (Chapter 40).

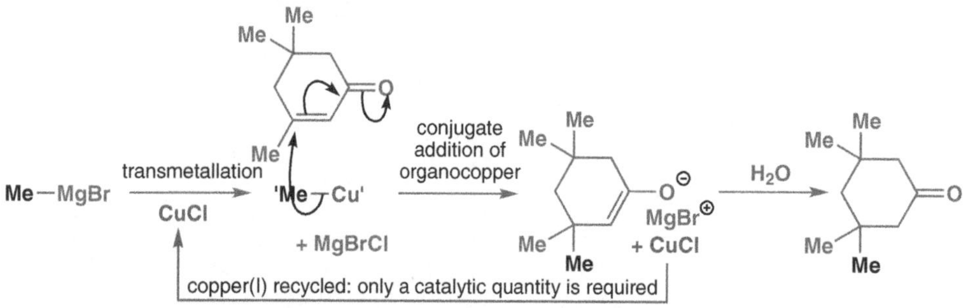

● Conjugate (1,4 or Michael) vs direct (1,2) addition

	Conjugate addition favoured by	Direct addition to C=O favoured by
Reaction conditions (for reversible additions):	thermodynamic control: high temperatures, long reaction times	kinetic control: low temperatures, short reaction times
Structure of α,β-unsaturated compound:	unreactive C=O group (amide, ester)	reactive C=O group (aldehyde, acyl chloride)
	unhindered β carbon	hindered β carbon
Type of nucleophile:	soft nucleophiles	hard nucleophiles
Organometallic:	organocoppers or catalytic Cu(I)	organolithiums, Grignard reagents

Extending the reaction to other electron-deficient alkenes

It's not only carbonyl-based groups that make alkenes react with nucleophiles rather than the more usual electrophiles. Other electron-withdrawing groups do just the same thing. Here are two examples: a nitrile and a nitro group. These compound classes appeared in Chapter 21 in the context of aromatic substitution reactions where we saw them pulling electron density away from the ring. The same thing happens here.

Unsaturated nitriles and nitro compounds

The simplest conjugated nitrile is acrylonitrile. This compound adds amines readily. No special conditions are needed to encourage attack at C=C rather than C≡N because the nitrile carbon is rather unreactive as an electrophilic centre.

➡ For reactions where strong nucleophiles do attack C≡N—partly because they have nothing else to attack—see pp. 220 and 231.

acrylonitrile   Et₂NH, 50 °C → Et₂N...CN   86% yield of an aminonitrile

🖱 Conjugate addition of amines to acrylonitrile

The amine first attacks the alkene in a typical conjugate addition to make an anion stabilized by being next to the nitrile. The anion can have its charge drawn on C or N: it is delocalized like an enolate. Do not be put off by the odd appearance of the 'enolate'. The dot between the two double bonds is a reminder that there is a linear sp carbon atom at this point.

◼ You will see a few mechanisms in this chapter where we have written an intramolecular deprotonation. This saves writing two steps—protonation of the enolate and deprotonation of N (in this case)—but quite possibly this is not the actual mechanism by which the proton transfer takes place. Any proton will do, as will any base—the protons are hopping around all the time, so as with any proton transfer it doesn't pay to take the arrows too literally. We discussed alternative but equivalent mechanisms for proton transfers in Chapter 12, p. 267.

stabilized, delocalized anion

Protonation at carbon restores the nitrile and gives the product—an amino-nitrile. The whole process adds a 2-cyano-ethyl group to the amine and is known industrially as cyanoethylation.

With a primary amine, the reaction need not stop at that stage as the product is still nucleophilic and a second addition can occur to replace the second hydrogen atom on nitrogen.

Other elements such as O, S, or P can add too. Phenyl phosphine can undergo a double addition just as in the last example, but alcohols can add only once. If there is competition between a first-row (for example N or O) and a second-row (for example S or P) element, the second-row element normally wins, for the reasons discussed above (p. 507).

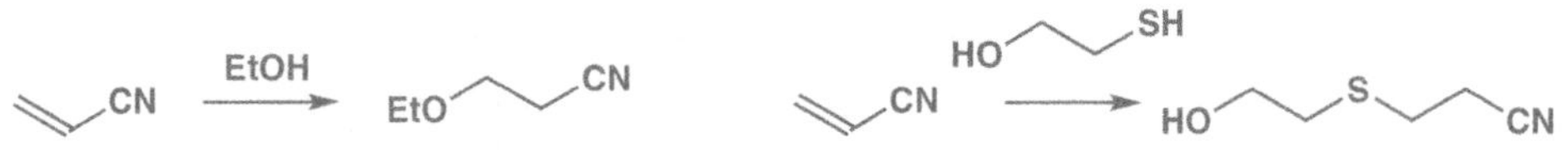

The nitro group (NO_2) is extremely electron-withdrawing—about twice as electron-withdrawing as a carbonyl group. It is also unreactive as an electrophilic centre, which makes conjugate addition to nitro-alkenes a very reliable reaction. In this example, sodium borohydride attacks the C=C bond in a conjugate manner to give an intermediate looking rather like an enolate anion, with a negatively charged oxygen atom conjugated to an (N=C) double bond. It reacts like an enolate too, picking up a proton on carbon to re-form the nitro group and give a stable product.

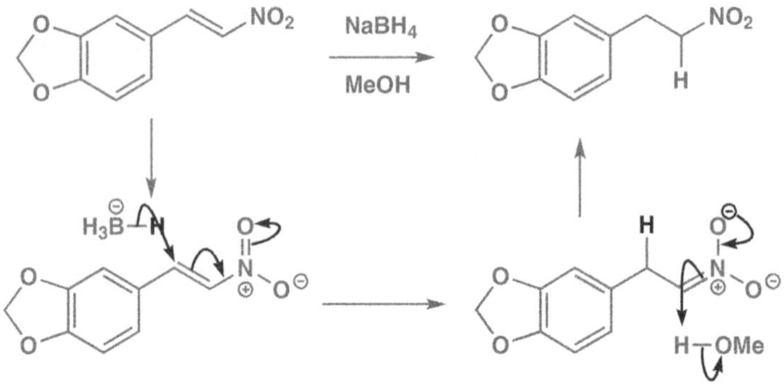

Conjugate substitution reactions

Just as direct addition to C=O (Chapter 6) becomes substitution at C=O (Chapter 10) when there is a leaving group at the carbonyl carbon, so conjugate *addition* becomes conjugate *substitution* if there is a leaving group, such as Cl, at the β carbon atom. Here is an example: substitution replaces Cl with OMe, just as it would have done in a reaction with an acyl chloride.

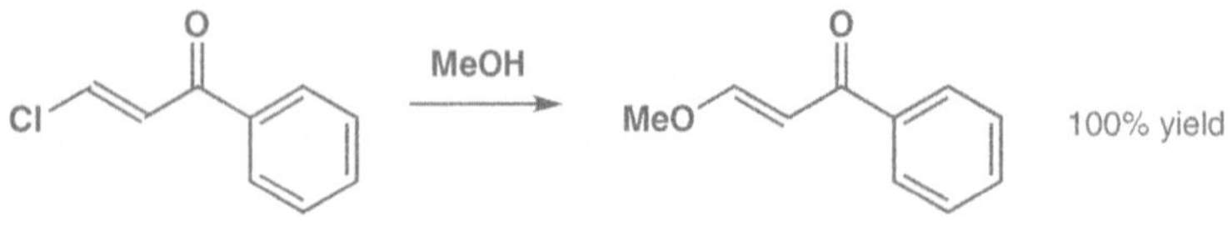

As with substitution at C=O, this apparently simple reaction does *not* involve a direct displacement of the leaving group in a single step. The mechanism starts in exactly the same way as for conjugate addition, giving an enol intermediate.

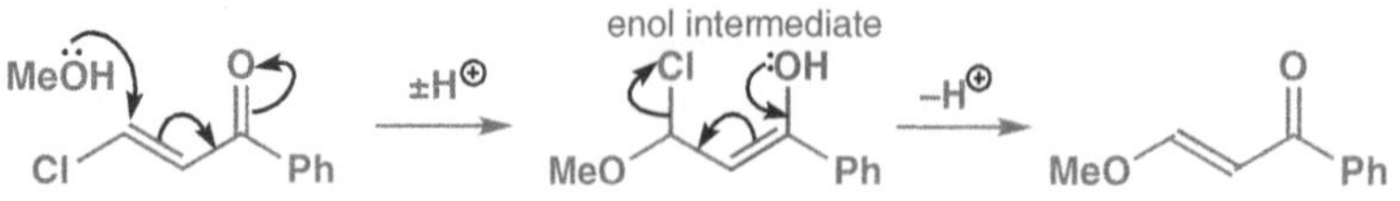

Now the leaving group can be expelled by the enol: the double bond moves back into its original position in an elimination reaction—the sequence is often called an addition–elimination reaction. The 'new' double bond has the more stable *E* configuration. In the next example, two consecutive conjugate substitution reactions give a 1,1-diamine.

98% yield

At first sight, the product looks rather unstable—sensitive to water, or traces of acid perhaps. But, in fact, it is remarkably resistant to reaction with both. The reason is conjugation: this isn't really an amine (or a diamine) at all because the lone pairs of the nitrogen atoms are delocalized through into the carbonyl group, very much as they are in an amide. This makes them less basic, and makes the carbonyl group less electrophilic.

delocalization of the nitrogen's lone pair

delocalization in an amide

Conjugate substitution and the synthesis of anti-ulcer drugs

Just as the cyanide (CN) and nitro (NO$_2$) groups can be used to bring about conjugate addition, so also can they initiate conjugate substitution. Examples of these reactions play important roles in the synthesis of two of the most significant drugs in the development of modern medicinal chemistry: the anti-ulcer compounds cimetidine (marketed as Tagamet) and ranitidine (Zantac). We looked at some aspects of the structure of these drugs in Chapter 8 (p. 178) and we are now going to see how conjugate addition is used in their synthesis.

histamine

cimetidine (Tagamet)

guanidine

ranitidine (Zantac)

The simple cyanoimine on the left, with two SMe groups as built-in leaving groups, is readily available and reacts with amines to give guanidines in two stages. Each of the reactions is a conjugate substitution. It will be clearer if we draw the reaction with a generalized primary amine RNH$_2$ first: conjugate addition, exactly as we saw with acrylonitrile, is followed by expulsion of the best leaving group. Thiols are acidic compounds, and MeS$^-$ is a better leaving group than RNH$^-$.

The reaction stops cleanly at this point and more vigorous conditions are required to displace the second MeS$^-$ group. This is because the first product is less reactive than the starting material: the new amino group is electron-donating and conjugation is established between it and the cyano group, deactivating the molecule towards a second conjugate substitution (as shown in the margin). But under more forcing conditions a second and different amine can be introduced and the second MeS$^-$ group displaced. In the synthesis of cimetidine the second amine is MeNH$_2$ and the molecule is complete.

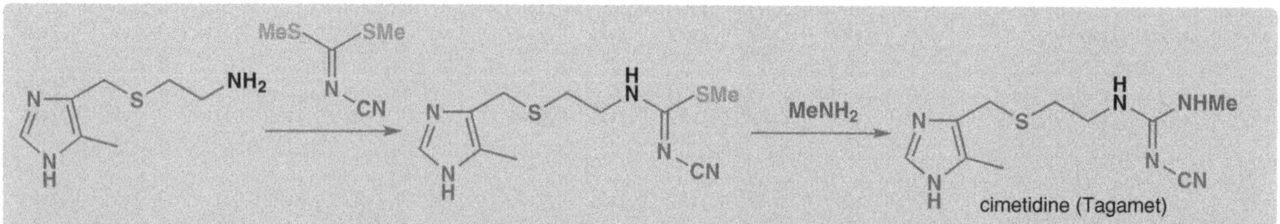

Ranitidine's right-hand portion is made in a similar way from an unsaturated nitro compound. This time the methylamine substitution is done first, followed by addition of the rest of the molecule.

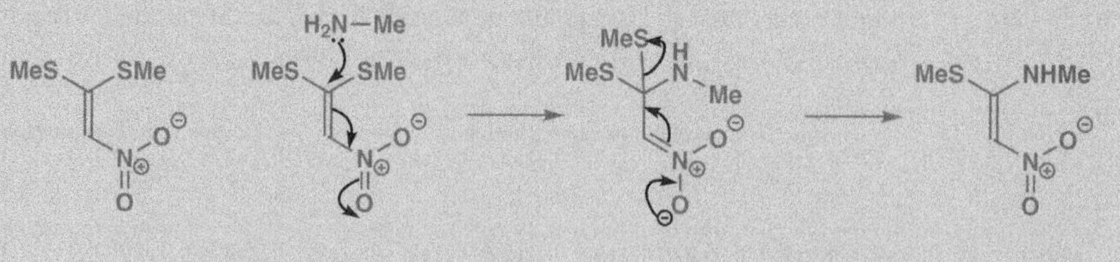

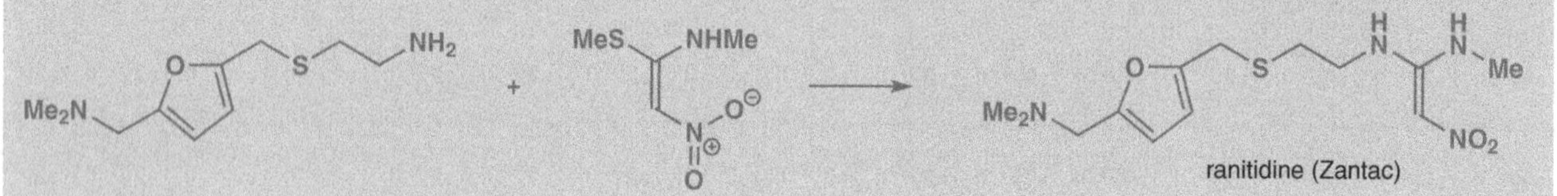

Nucleophilic epoxidation

The conjugate substitutions we have just been discussing rely on a starting material containing a leaving group. In this section we are going to look at what happens if the leaving group is not attached to the unsaturated carbonyl compound, but instead is attached to the nucleophile. We shall look at this class of compounds—nucleophiles with leaving groups attached—in more detail in Chapter 38), but for the moment the most important will be hydroperoxide, the anion of hydrogen peroxide.

Hydroperoxide is a good nucleophile because of the **alpha effect**: interaction of the two lone pairs on adjacent oxygen atoms raises the HOMO of the anion and makes it a better and softer nucleophile than hydroxide. Hydroperoxide is also less basic than hydroxide because of the inductive electron-withdrawing effect of the second oxygen atom. Basicity and nucleophilicity usually go hand in hand—not here though. This means that the hydroperoxide anion can be formed by treating hydrogen peroxide with aqueous sodium hydroxide.

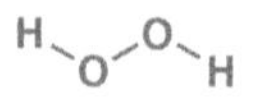

hydrogen peroxide
$pK_a = 11.6$

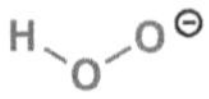

hydroperoxide
anion

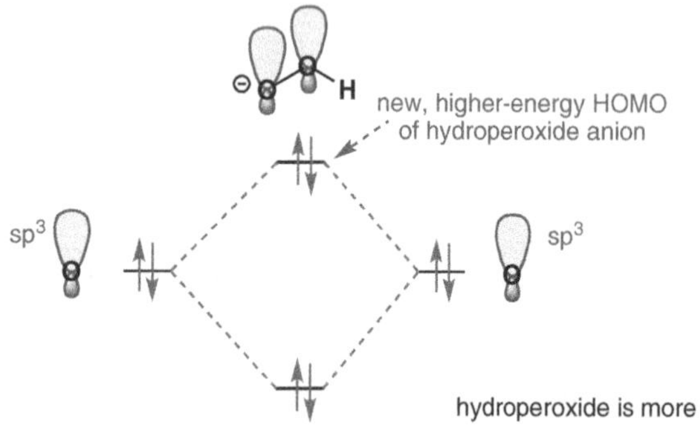

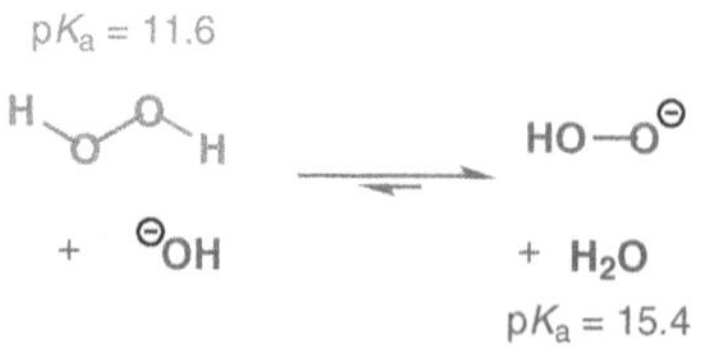

new, higher-energy HOMO
of hydroperoxide anion

$pK_a = 11.6$

$pK_a = 15.4$

hydroperoxide is more nucleophilic and less basic than hydroxide

■ The same effect explains why hydroxylamine and hydrazine are more nucleophilic than ammonia (see p. 232).

This is what happens when this mixture is added to an enone. First, there is the conjugate addition.

But the product is not stable because hydroxide can be lost from the oxygen atom that was the nucleophile. Hydroxide is fine as a leaving group here—after all, hydroxide is lost from enolates in E1cB eliminations, and here the bond breaking is a weak O–O bond. The product is an epoxide.

🖱 Interactive mechanism for nucleophilic epoxidation

The electrophilic epoxidizing agents such as *m*-CPBA, which you met in Chapter 19, work reliably only with nucleophilic alkenes, and for α,β-unsaturated carbonyl compounds and other electron-deficient alkenes, hydroperoxide—a nucleophilic epoxidizing agent—is often used instead.

There is another significant difference between hydrogen peroxide and *m*-CPBA, highlighted by the pair of reactions below.

cis-epoxide   ←*m*-CPBA—   *cis*-alkene   —H_2O_2, $OH^\ominus$→   *trans*-epoxide

➡ For a reminder of the meaning of the term *stereospecific*, see p. 396.

m-CPBA epoxidation is stereospecific because the reaction happens in one step. But nucleophilic epoxidation with hydroperoxide is a two-step reaction: there is free rotation about the bond marked in the anionic intermediate, and the more stable, *trans*-epoxide results, whatever the geometry of the starting alkene.

free rotation about this bond

Nucleophilic aromatic substitution

In this next section we are going to consider reactions related to conjugate substitutions but in which the double bond is part of an aromatic ring. We spent some considerable time in Chapter 21 explaining that aromatic rings are *nucleophilic*: electrophiles attack them, and typical aromatic reactivity is to undergo electrophilic substitution.

In general, nucleophilic substitutions of aromatic halides—such as the one proposed here in which hydroxide is attempting to displace bromide—**do not happen**. You might well ask, 'Why not?' The reaction looks all right and, if the ring were saturated, it *would be* all right.

This is an S_N2 reaction, and we know (Chapter 15) that attack must occur in line with the C–Br bond from the back, where the largest lobe of the σ* orbitals lies. That is perfectly all right for the aliphatic ring because the carbon atom is tetrahedral and the C–Br bond is not in the plane of the ring. Substitution of an equatorial bromine goes like this:

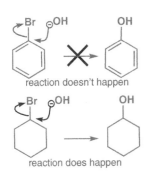

reaction doesn't happen

reaction does happen

line of attack is not in plane of ring

But in the aromatic compound, the C–Br bond *is* in the plane of the ring as the carbon atom is trigonal. To attack from the back, the nucleophile would have to appear inside the benzene ring and invert the carbon atom in an absurd way. This reaction is of course not possible.

This is another example of the general rule:

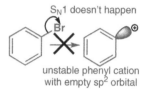

S_N2 can't happen

- **S_N2 at sp^2 C does not occur.**

If S_N2 is impossible, what about S_N1? This is possible but very unfavourable unless the leaving group is an exceptionally good one (see below for an example). It would involve the unaided loss of the leaving group and the formation of an aryl cation. All the cations we saw as intermediates in the S_N1 reaction (Chapter 15) were planar with an empty p orbital. This cation is planar but the p orbital is full—it is part of the aromatic ring—and the empty orbital is an sp^2 orbital outside the ring.

S_N1 doesn't happen

unstable phenyl cation with empty sp^2 orbital

Yet some aromatic compounds *do* undergo nucleophilic substitution. Just as normally nucleophilic alkenes can be made to undergo conjugate substitution if they carry electron-withdrawing substituents, so normally nucleophilic aromatic rings also become electrophilic if they have the right substituents. The mechanism by which they undergo nucleophilic substitution also closely parallels that of conjugate substitution which you have just seen.

The addition–elimination mechanism

Imagine a cyclic β-fluoro-enone reacting with a secondary amine in a conjugate substitution reaction. The normal addition to form the enolate followed by return of the negative charge to expel the fluoride ion gives the product.

Now imagine just the same reaction with two extra double bonds in the ring. These play no part in our mechanism; they just make what was an aliphatic ring into an aromatic one. Conjugate substitution has become **nucleophilic aromatic substitution**.

The mechanism involves *addition* of the nucleophile followed by *elimination* of the leaving group—the **addition–elimination mechanism**. It is not necessary to have a carbonyl group—any electron-withdrawing group will do—the only requirement is that the electrons must be able to get out of the ring into this anion-stabilizing group. Here is an example with a *para*-nitro group.

■ This mechanism is also abbreviated to S_NAr (for Substitution, Nucleophilic, Aromatic).

● Interactive mechanism for aromatic addition–elimination

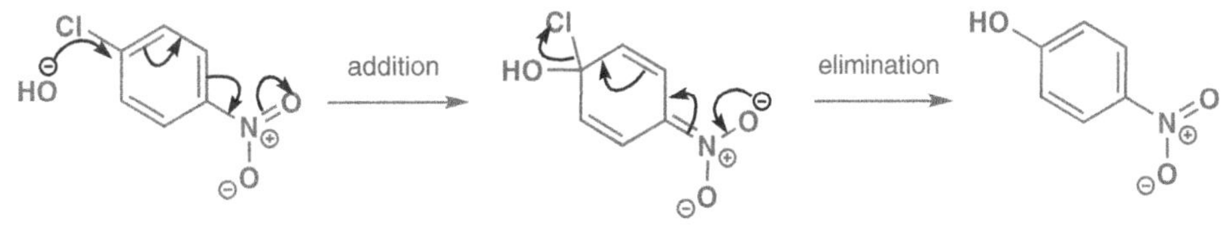

Everything is different about this example—the nucleophile (HO⁻), the leaving group (Cl⁻), the anion-stabilizing group (NO₂), and its position *(para)*—but the reaction still works. The nucleophile is a good one, the negative charge can be pushed through on to the oxygen atom(s) of the nitro group, and chloride is a better leaving group than OH.

> ● **A typical nucleophilic aromatic substitution has:**
>
> • an oxygen, nitrogen, or cyanide nucleophile
> • a halide for a leaving group
> • a carbonyl, nitro, or cyanide group *ortho* and/or *para* to the leaving group.

Since the *nitro* group is usually introduced by electrophilic aromatic substitution (Chapter 21) and halides direct *ortho/para* in nitration reactions, a common sequence is nitration followed by nucleophilic substitution.

■ If you try and do the same reaction with a *meta* anion-stabilizing group, it doesn't work. You can't draw the arrows to push the electrons through on to the oxygen atom. Try it yourself.

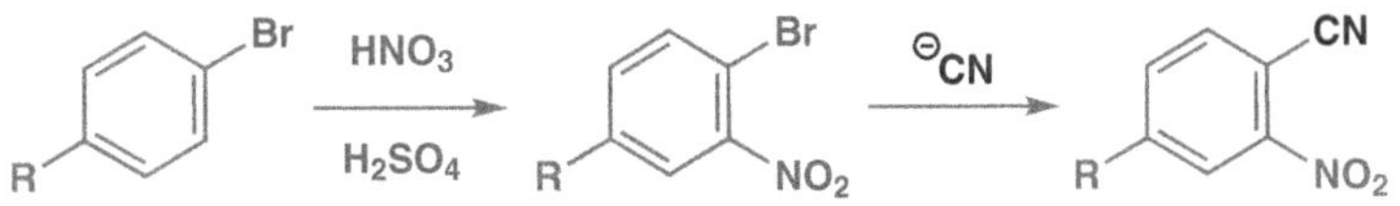

This sequence is useful because the nitro group could not be added directly to give the final product as nitration would go in the wrong position. The nitrile is *meta*-directing, while the alkyl group (R) is *ortho, para*-directing.

Two activating electron-withdrawing groups are better than one and dinitration of chlorobenzene makes a very electrophilic aryl halide. Reaction with hydrazine gives a useful reagent.

It also makes a very toxic one! This compound—2,4-dinitrophenylhydrazine—is carcinogenic. Nonetheless it forms coloured crystalline imines (hydrazones) with carbonyl compounds—before the days of spectroscopy these were used to characterize aldehydes and ketones (see p. 232).

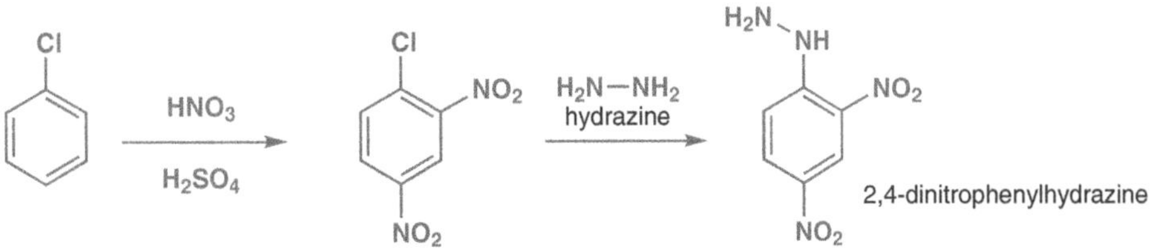

2,4-dinitrophenylhydrazine

The intermediate in the addition–elimination mechanism

What evidence is there for intermediates like the ones we have been using in this section? When reactions like this last example are carried out, a purple colour often appears in the reaction mixture and then fades away. In some cases the colour is persistent and thought to be due to the intermediate. Here is an example with RO⁻ attacking a nitrated aniline. This intermediate is persistent because neither potential leaving group (NR₂ or OR) is very good.

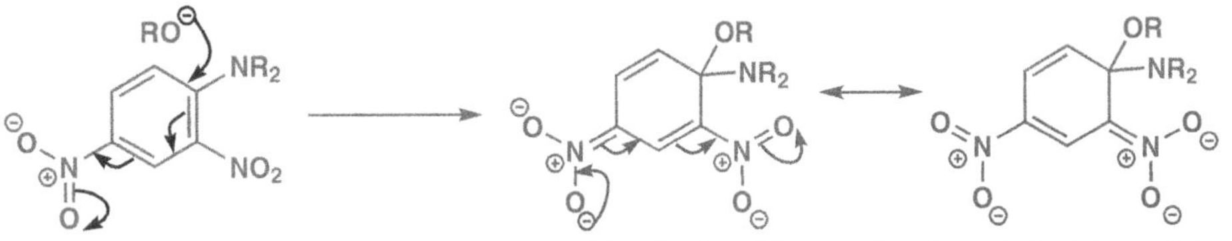

delocalization of the negative charge in the intermediate

What is the nature of this intermediate? Well, in essence it is an anion delocalized over five sp^2 hybridized carbons of a six-membered ring (the sixth, the point at which the nucleophile attacked, is sp^3 hybridized). It's possible to make a simple homologue of such a species by deprotonating cyclohexadiene. Delocalizing the anion generates the three structures below.

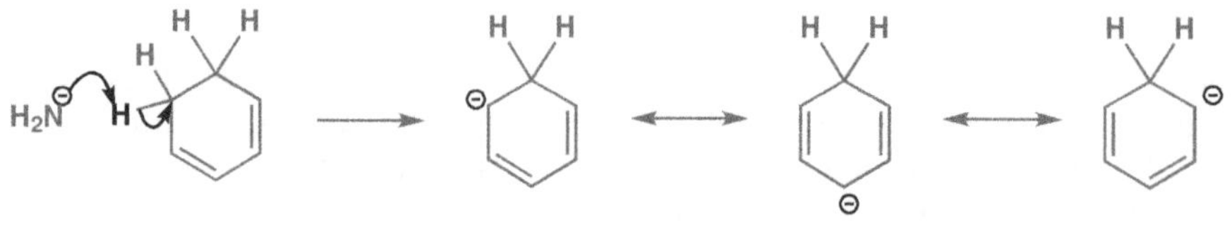

You've seen before that ^{13}C NMR spectra are revealing when it comes to distribution of charge, and the details of the ^{13}C NMR spectrum of this anion are shown below, along with those of benzene itself and also of the cation generated by protonating benzene (which, as you will remember from Chapter 21, corresponds to the intermediate generated in electrophilic aromatic substitution).

δ_C 3.0
δ_C 75.8 δ_C 75.8
δ_C 131.8 δ_C 131.8
δ_C 78.0

δ_C 128.5

δ_C 52.2
δ_C 186.6 δ_C 186.6
δ_C 136.9 δ_C 136.9
δ_C 178.1

■ *A reminder:* A larger shift means less electronic shielding and a smaller shift more electronic shielding.

These results are very striking. The shifts of the *meta* carbons in both ions are very slightly different from those of benzene itself (about 130 ppm). But the *ortho* and *para* carbons in the anion have gone upfield to much smaller shifts, indicating greater electron density. By contrast, *ortho* and *para* carbons in the cation have gone downfield to much larger shifts.

The differences are very great—about 100 ppm between the cation and the anion! It is very clear from these spectra that the ionic charge is delocalized almost exclusively to the *ortho* and *para* carbons in both cases. The alternative structures in the margin show this delocalization.

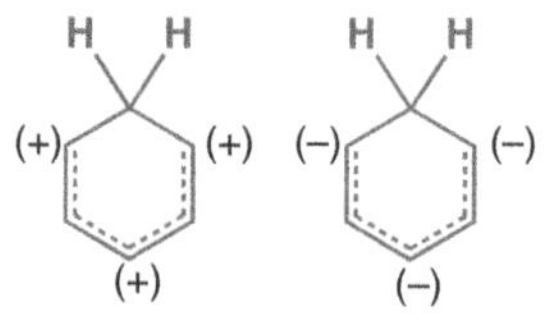

This means that stabilizing groups, such as nitro or carbonyl in the case of the anion, can only have an effect if they are on carbons *ortho* or *para* to the position being attacked by the nucleophile. A good illustration of this is the selective displacement of one chlorine atom out of these two. The chlorine *ortho* to the nitro group is lost; the one *meta* is retained.

(+)   (+)   (–)   (–)
(+)   (–)

■ Remember, charges in brackets show significant (here ca. 1/3) portions of a charge, in contrast with δ, which means a much smaller polarization.

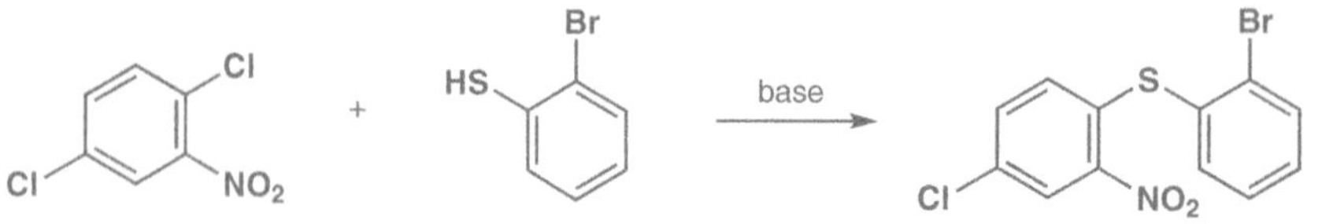

The mechanism works well if the nucleophile (the anion derived from the thiol) attacks the carbon bearing the chlorine *ortho* to the nitro group as the negative charge can then be pushed into the nitro group. Satisfy yourself that you cannot do this if you attack the other chlorine position. This is a very practical reaction and is used in the manufacture of a tranquillizing drug.

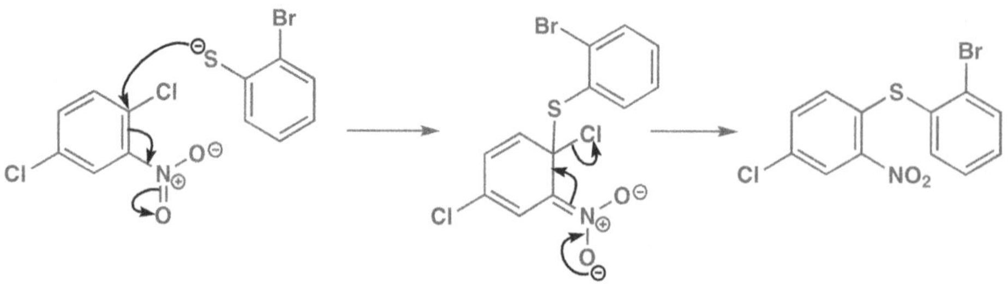

The leaving group and the mechanism

In the first nucleophilic aromatic substitution that we showed you, we used fluoride ion as a leaving group. Fluoride works very well in these reactions and even such a simple compound as 2-halo-1-nitrophenyl fluoride reacts efficiently with a variety of nucleophiles, as in these examples.

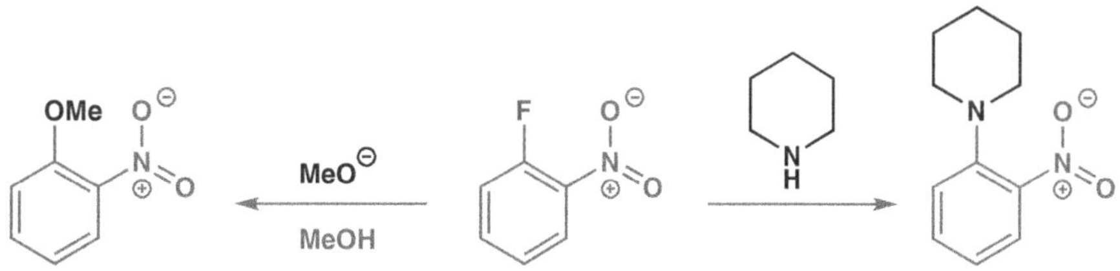

The same reactions happen with the other 2-halo-1-nitrobenzenes but less efficiently. The fluoro-compound reacts about 10^2–10^3 times faster than the chloro or bromo compounds and the iodo compound is even slower.

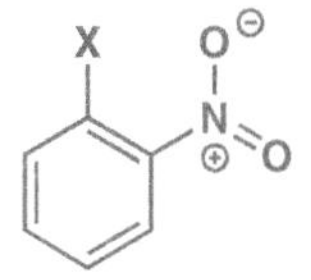

This ought to surprise you. When we were looking at other nucleophilic substitutions such as those at the carbonyl group or saturated carbon, we never used fluoride as a leaving group! The C–F bond is very strong—the strongest of all the single bonds to carbon—and it is difficult to break. As a consequence, these reactions are not a good prospect:

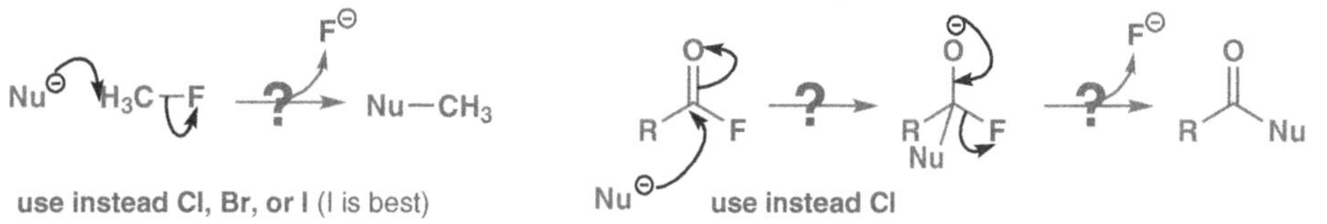

This reaction is never used:

use instead Cl, Br, or I (I is best)

This reaction is rarely used:

use instead Cl

So why is fluoride so good in nucleophilic aromatic substitution when the reverse is true with other reactions? You will notice that we have *not* said that fluoride is a better leaving group in nucleophilic aromatic substitution. It isn't! The explanation depends on a better understanding of the mechanism of the reaction. We shall use azide ion as our nucleophile because this has been well studied and because it is one of the best.

The mechanism is exactly the same as that we have been discussing all along—a two-stage addition–elimination sequence. In a two-step mechanism, one step is slower and rate determining; the other is unimportant to the rate. You may guess that, in the mechanism for nucleophilic aromatic substitution, it is the first step that is slower because it disturbs the aromaticity. The second step restores the aromaticity and is faster. The effect of fluoride, or any other leaving group, can only come from its effect on the first step. How good a leaving group it might be does not matter: the rate of the second step—the step where fluoride leaves—has no effect on the overall rate of the reaction.

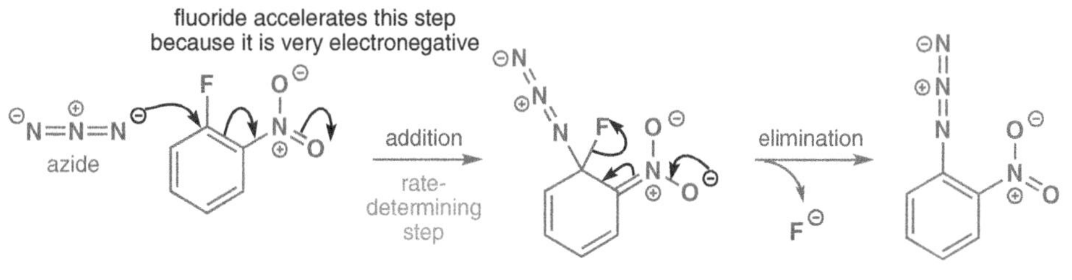

Fluoride accelerates the first step through its inductive effect. It is the most electronegative element of all and it stabilizes the anionic intermediate, assisting the acceptance of electrons by the benzene ring.

Sidebar (left margin)

reactivity of 2-halo-1-nitrobenzenes in nucleophilic aromatic substitution

F >> Cl ~ Br >> I

➡ You met the azide anion in Chapter 15, p. 354.

■ Note carefully that this is an *inductive* effect: there are no arrows to be drawn to show how fluorine withdraws electrons—it does it just by polarizing C–F bonds towards itself. Contrast the electron-withdrawing effect of the nitro group, which works (mainly) by conjugation.

The activating anion-stabilizing substituent

We have used nitro groups very extensively so far because they are the best at stabilizing the anionic intermediate. Others that work include carbonyl, cyanide, and sulfur-based groups such as sulfoxides and sulfones. A direct comparison of the different groups Z that can assist the displacement of bromide (by the secondary amine piperidine in this example) is shown in the margin.

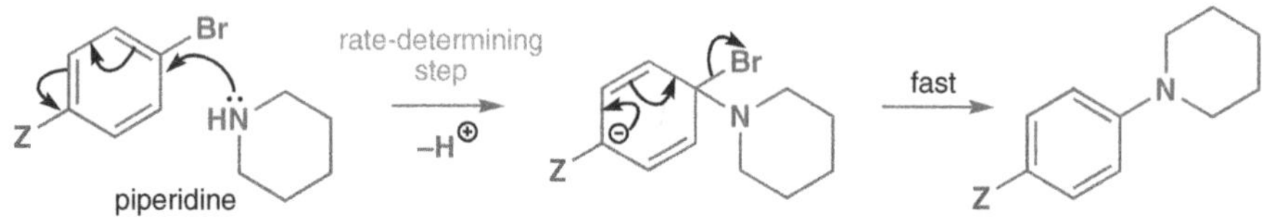

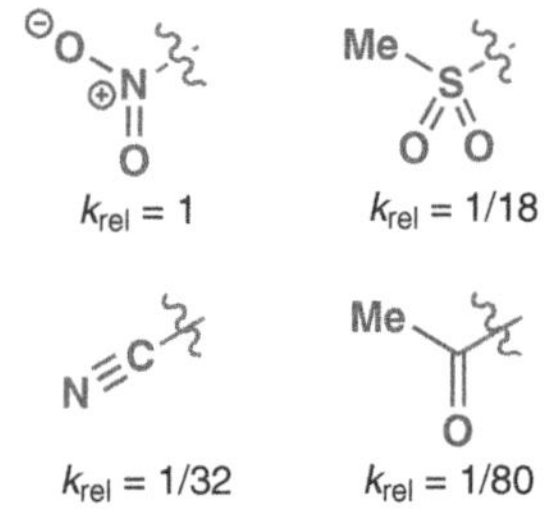

All the compounds react more slowly than the nitro compound. We have already mentioned (Chapters 8 and 21) the great electron-withdrawing power of the nitro group—here is a new measure of that power. The sulfone reacts 18 times slower, the nitrile 32 times slower, and the ketone 80 times slower.

Nitro is the best activating group, but the others will all perform well, especially when combined with a fluoride rather than a bromide as the leaving group. Here are two reactions that work well in a preparative sense with other anion-stabilizing groups. Note that the trifluoromethyl group works by using only its powerful inductive effect.

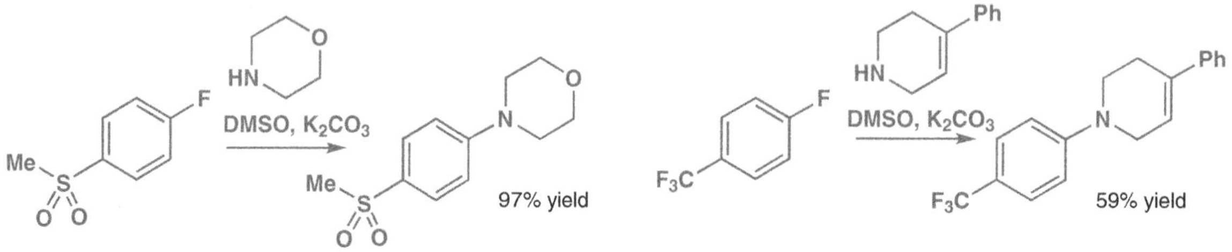

● **To summarize**

An anion-stabilizing (electron-withdrawing) group *ortho* or *para* to a potential leaving group can be used to facilitate nucleophilic aromatic substitution.

Conjugate and nucleophilic aromatic substitution reactions in action: the synthesis of an antibiotic

We want to convince you that this chemistry is useful and also that it works in more complicated molecules so we are going to describe in part the preparation of the antibiotic ofloxacin. The sequence starts with an aromatic compound having four fluorine atoms. Three are replaced sequentially by nucleophiles and the last is present in the antibiotic itself.

The first reaction is a conjugate substitution of the ethoxy group marked in orange. An amino alcohol is used as the nucleophile and it is the more nucleophilic amino group (rather than the hydroxyl group) that adds to the alkene.

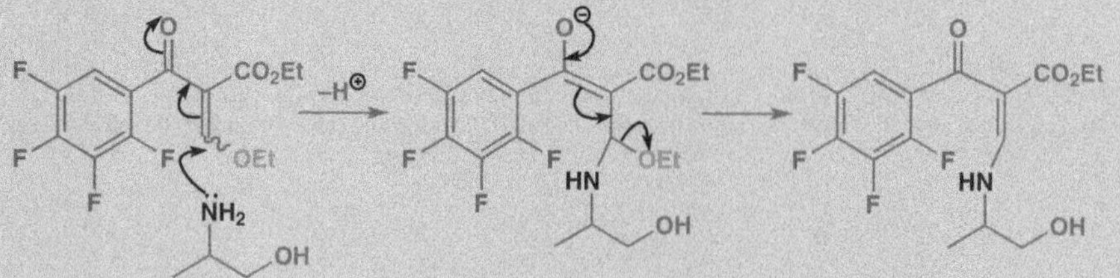

Now for the first nucleophilic aromatic substitution. The amino group attacks in the position *ortho* to the carbonyl group so that an enolate intermediate can be formed. The first fluoride is expelled in the elimination step.

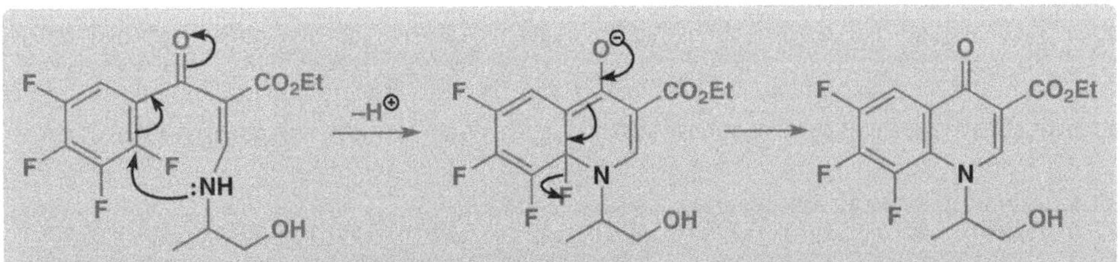

Treatment with base (NaH can be used) now converts the OH group into an alkoxide, which takes part in the next aromatic nucleophilic substitution. In this reaction we are attacking the position *meta* to the ketone so we cannot put the negative charge on the oxygen atom. The combined inductive effect of the remaining three fluorines is enough to stabilize the anion.

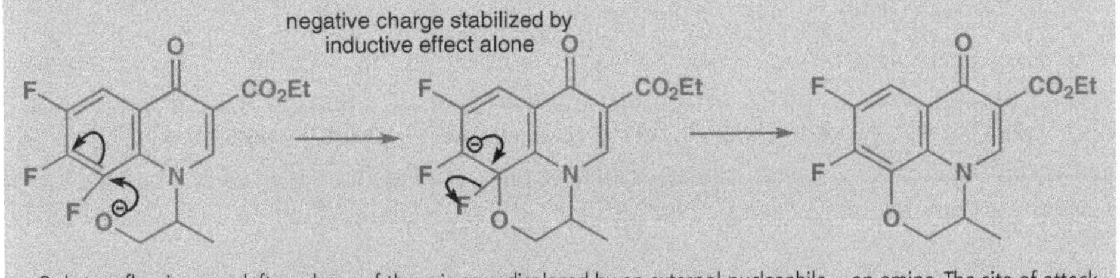

Only two fluorines are left, and one of these is now displaced by an external nucleophile—an amine. The site of attack of the amine is determined by the need to stabilize the charge in the intermediate, which is an enolate.

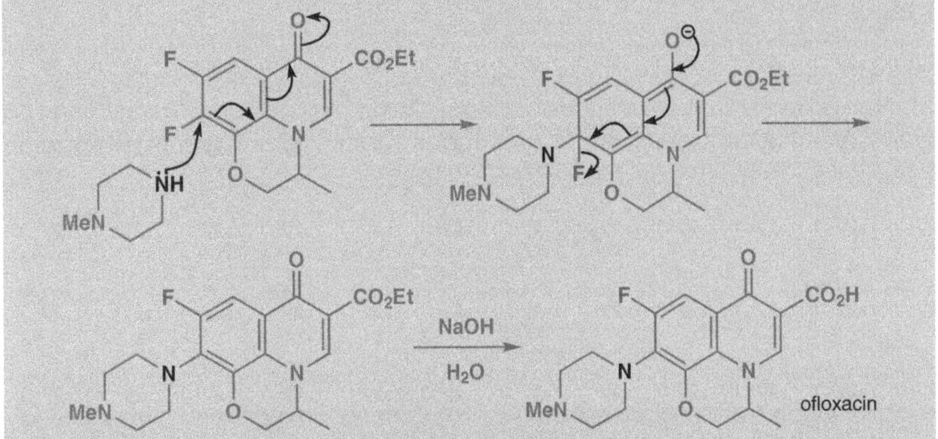

All that is left is to hydrolyse the ester to the free acid with aqueous base (Chapter 10). Every single reaction in this quite complicated sequence is one that you have met earlier in the book, and it illustrates the power of simple organic mechanisms to allow chemists to make important life-saving compounds.

Nucleophilic substitution on aromatic rings is possible by alternative mechanisms as well. We will now turn to these.

The S_N1 mechanism for nucleophilic aromatic substitution: diazonium compounds

If we really want to make aromatic compounds undergo nucleophilic substitution in a general way, the way to do it is to use absolutely the best leaving group of all—nitrogen gas. In fact, the diazonium compound below is so good at nucleophilic aromatic substitution that is does so even without activating groups. On warming, the nitrogen molecule just departs, leaving behind a cation, which is captured by a nucleophile, in this case water. Do you find this reminiscent of the S_N1 reaction? We hope so.

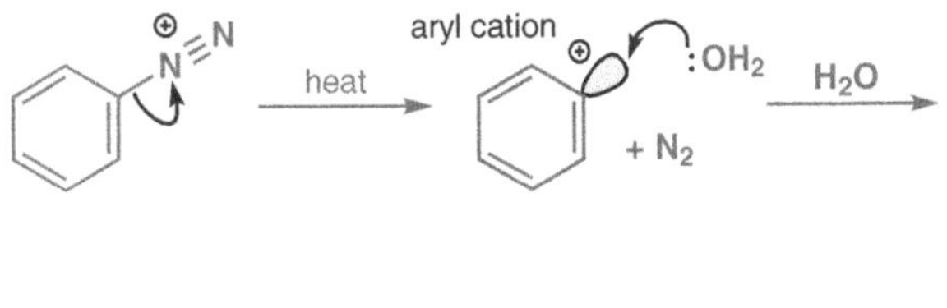

Before we talk about this group of aromatic S_N1 reactions in more detail, let's consider how to make the diazonium salt. The reagent we need is the reactive nitrogen electrophile NO^+. You met NO^+ in Chapter 20, but to remind you, it forms when the nitrite anion (usually sodium nitrite) is treated with acid at around 0 °C. Protonation of nitrite gives nitrous acid, HONO; protonation again gives a cation, which can lose water to form NO^+. Butyl nitrite (or other alkyl nitrites) can also be used as a source of NO^+.

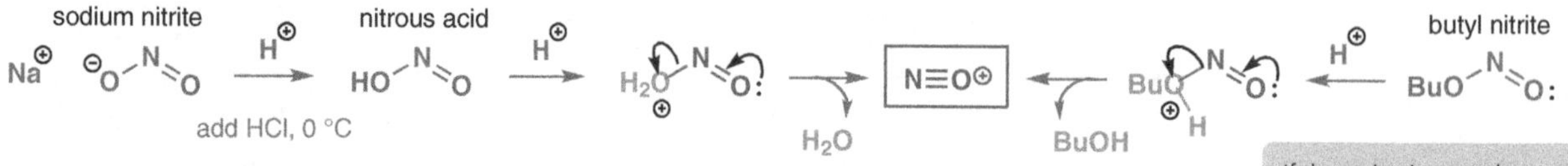

A diazonium salt is formed when NO^+ reacts with an amine. The lone pair of the amine attacks the NO^+ cation, and then water is lost. The mechanism is actually quite simple, but it does involve a lot of proton transfers. There is, of course, an anion associated with the nitrogen cation, and this will be the conjugate base (Cl⁻ usually) of the acid used to form NO^+. This reaction is known as *diazotization*.

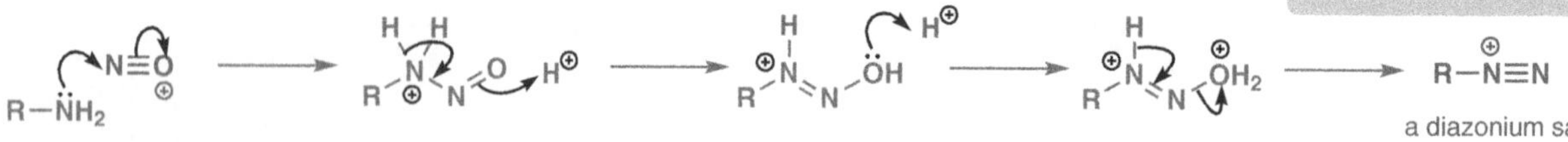

> If the amine is secondary, water can't be eliminated and a *nitros-amine* forms.
>
> a nitrosamine

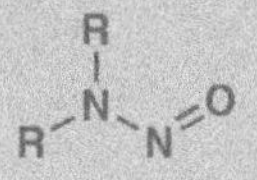

If the amine is an *alkyl* amine, this diazonium salt is very unstable and immediately loses nitrogen gas to give a planar carbocation, which normally reacts with a nucleophile in an S_N1 process (Chapter 15), loses a proton in an E1 process (Chapter 17), or rearranges (Chapter 36). It may, for example, react with water to give an alcohol:

> Interactive mechanism for formation of diazonium salt

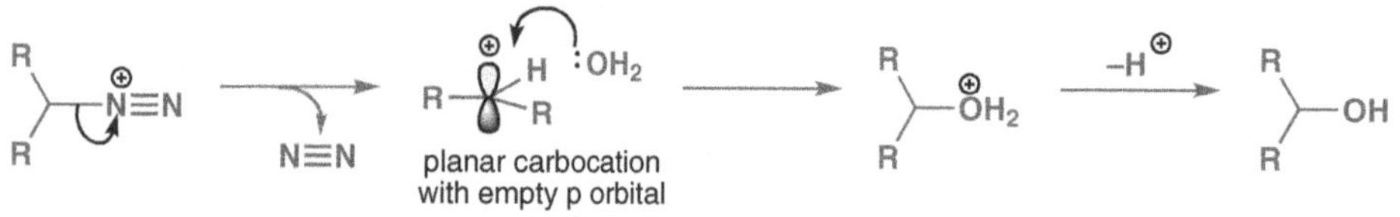

If the amine is an aryl amine, then the reaction you saw at the beginning of this section will take place and a phenol will form. This is in fact rather a useful reaction as it is difficult to add an oxygen atom to a benzene ring by normal electrophilic substitution: there is no good reagent for OH⁺. A nitrogen atom can be added easily by nitration, and reduction and diazotization provide a way of replacing the nitro group by a hydroxyl group.

> We alluded to this sequence at the end of Chapter 21.

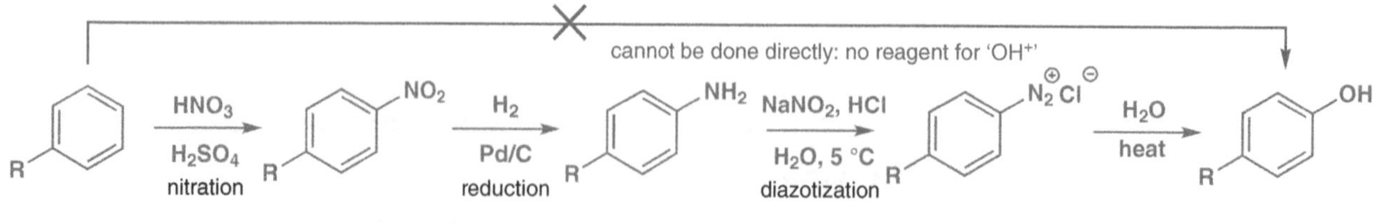

Substitution reactions in the synthesis of a drug

The synthesis of the drug thymoxamine (Moxysylyte) provides a practical example of how this reaction can be used.

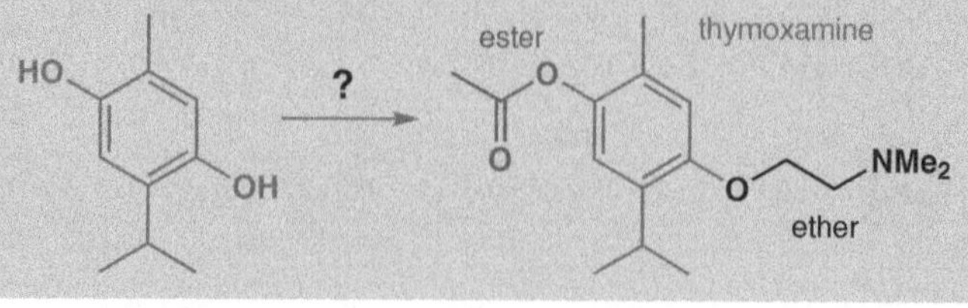

It seems obvious to make this compound by alkylation and acylation of a dihydroxybenzene, but how are we to make sure that the acylation and alkylation go on the right OH groups? French pharmaceutical chemists had an ingenious answer: start with a compound having only one OH group, alkylate that, and only then introduce the second using the diazonium salt method. They used a simple phenol and introduced nitrogen as a nitroso (NO) rather than a nitro (NO$_2$) group. This means using the same reagent as we have been using for diazotization. These were the first two steps.

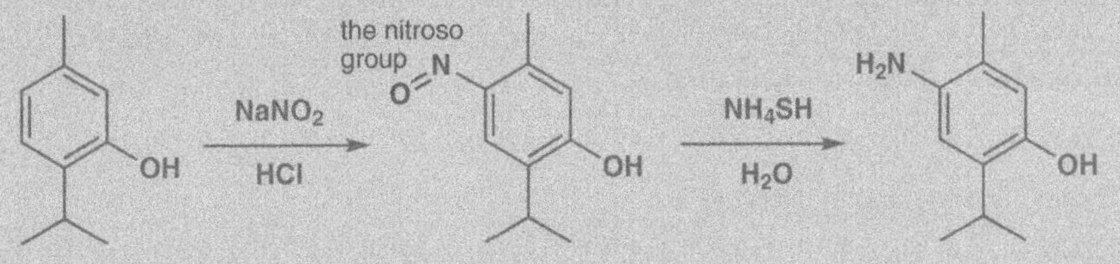

The reduction of NO is easier than that of NO$_2$, and H$_2$S is enough to do the job. The amine can now be converted to an amide to lessen its nucleophilicity so that alkylation of the phenol occurs cleanly—a form of protection (see Chapter 23).

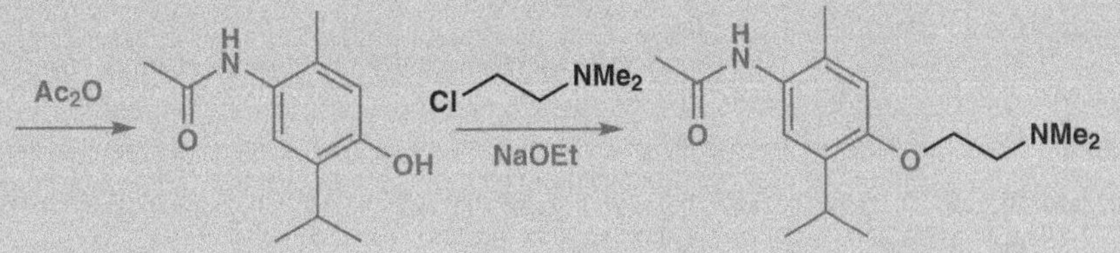

Finally, the amide must be hydrolysed, the amino converted into an OH group by diazotization and hydrolysis, and the new phenol acetylated.

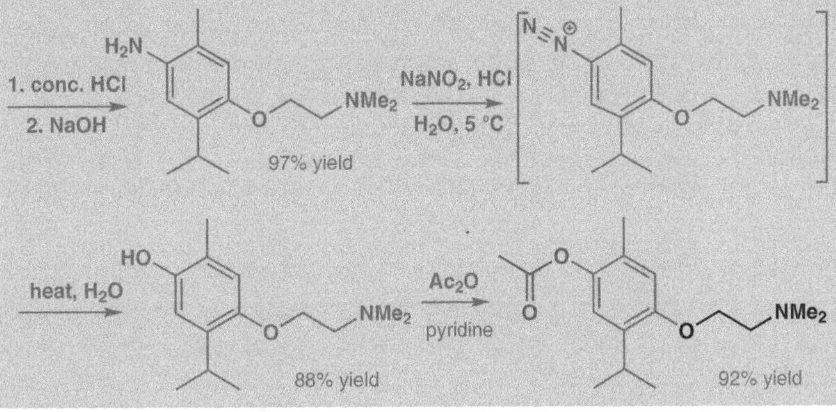

However, an aryl carbocation is much less stable than an alkyl carbocation because its empty orbital is an sp^2 rather than a p orbital. This makes the loss of nitrogen slower. If the diazotization is done at temperatures around 0 °C (classically at 5 °C), the diazonium salt is stable and can be reacted with various nucleophiles other than water.

Other nucleophiles

Aryl iodides are not as easy to make by electrophilic substitution as aryl chlorides or bromides because iodine is not reactive enough to attack benzene rings. But adding potassium iodide to the diazonium salt gives an aryl iodide by nucleophilic aromatic substitution.

■ Aryl iodides have wide utility in the coupling chemistry, catalysed by Pd and other transition metals, that you will meet in Chapter 40.

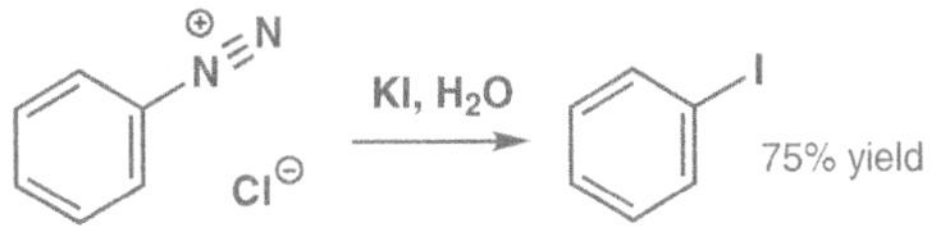

Other nucleophiles, such as chloride, bromide, and cyanide, are best added as copper(I) salts. Since aromatic amines are usually made by reduction of nitro compounds, a common sequence of reactions goes like this:

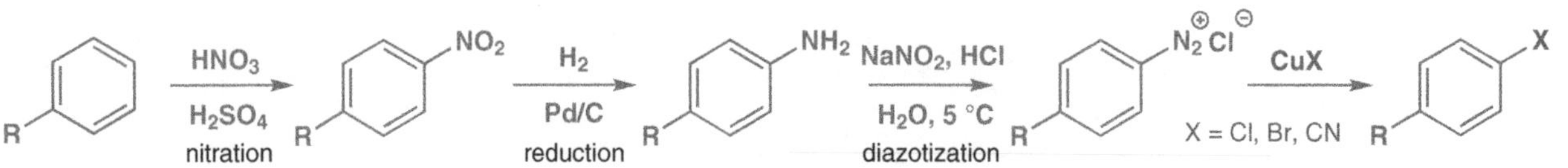

As often in aromatic chemistry, it's the versatility of the nitro group that makes this sequence work—easy introduction by electrophilic substitution, easy reduction, and easy nucleophilic substitution of its diazonium derivative.

The benzyne mechanism

We now need to introduce you to one last mechanism for aromatic nucleophilic substitution and you may well feel that this is the weirdest mechanism you have yet seen with the most unlikely intermediate ever! For our part, we hope to convince you that this mechanism is not only possible but also useful.

Earlier in this chapter we said that the displacement by nucleophiles of bromide from bromobenzene does not occur. In fact substitution reactions of bromobenzene *can* occur but only under the most vigorous conditions, such as when bromobenzene and NaOH are melted together (fused) at very high temperature. A similar reaction with the very powerful reagent $NaNH_2$ (which supplies NH_2^- ion) also happens, at a rather lower temperature.

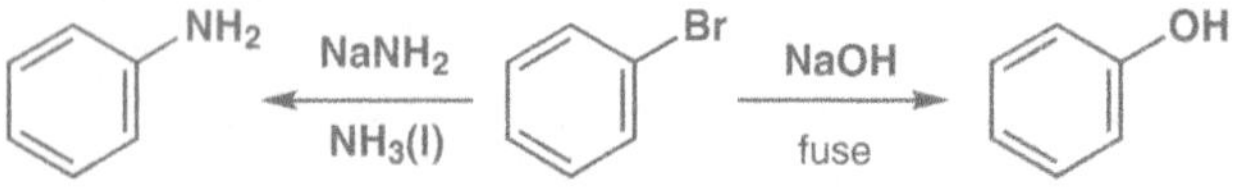

These reactions were known for a long time before anyone saw what was happening. They do not happen by an S_N2 mechanism, as we explained earlier, and they can't happen by the addition–elimination mechanism because there is nothing to stabilize the negative charge in the intermediate. The first clue to the true mechanism is that all the nucleophiles that react in this way are very basic. They start the reaction off by removing a proton *ortho* to the leaving group.

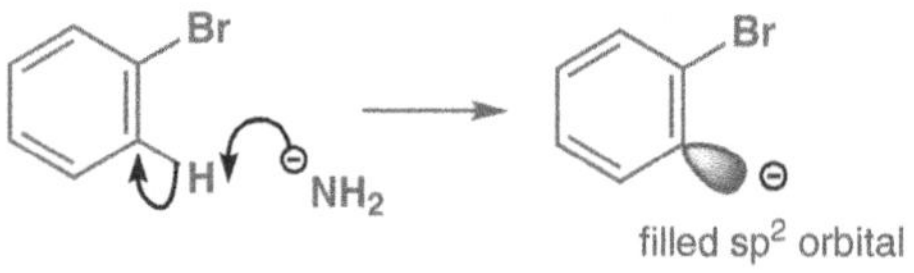

The carbanion is in an sp^2 orbital in the plane of the ring. Indeed, this intermediate is very similar to the aryl cation intermediate in the S_N1 mechanism from diazonium salts. That had no electrons in the sp^2 orbital; the carbanion has two. Why should this proton be removed rather than any other? The bromine atom is electronegative and the C–Br bond is in the plane of the sp^2 orbital and removes electrons from it. The stabilization is nonetheless weak and only exceptionally strong bases will do this reaction.

The next step is the loss of bromide ion in an elimination reaction. This is the step that is difficult to believe as the intermediate we are proposing looks impossible. The orbitals are bad for the elimination too—it is a *syn-* rather than an *anti*-periplanar elimination. But it happens.

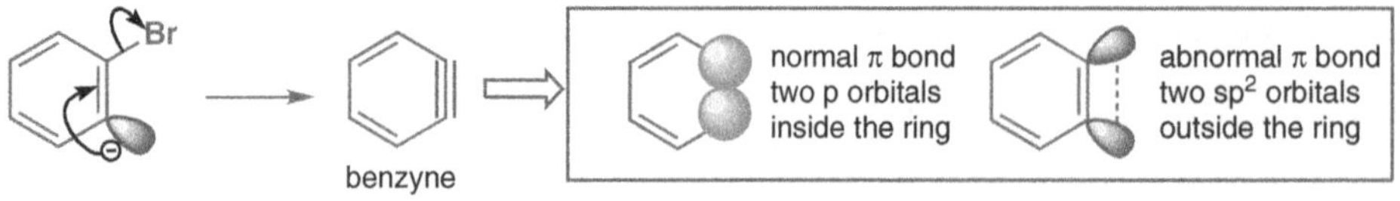

The intermediate is called benzyne as it is an alkyne with a triple bond in a benzene ring. But what does this triple bond mean? It certainly isn't a normal alkyne as these are linear. In fact one π bond is normal—it is just part of the aromatic system. One π bond—the new one—is abnormal and is formed by overlap of two sp^2 orbitals outside the ring. This external π bond is very weak and benzyne is a very unstable intermediate. Indeed, when the structure was proposed few chemists believed it and some pretty solid evidence was needed before they did. We shall come to that shortly, but let us first finish the mechanism. Unlike normal alkynes, benzyne is electrophilic as the weak third bond can be attacked by nucleophiles.

The whole mechanism from bromobenzene to aniline involves an elimination to give benzyne followed by an addition of the nucleophile to the triple bond of benzyne. In many ways, this mechanism is the reverse of the normal addition–elimination mechanism for nucleophilic aromatic substitution and it is sometimes called the **elimination–addition mechanism**.

Any nucleophile basic enough to remove the *ortho* proton can carry out this reaction. Known examples include oxyanions, amide anions (R_2N^-), and carbanions. The rather basic alkoxide *t*-butoxide will do the reaction on bromobenzene if the potassium salt is used in the dipolar aprotic solvent DMSO to maximize reactivity.

■ DMSO (see p. 255) solvates K^+ but not RO^-.

🖱 Interactive mechanism for benzyne formation and reaction

One rather special feature of the benzyne mechanism allows us to be certain that this proposed mechanism is correct, and this is the fact that the triple bond could in principle be attacked by nucleophiles at either end. This is of no consequence with bromobenzene as the products would be the same, but we can make the ends of the triple bond different and then we see something interesting. *ortho*-Chloro aryl ethers are easy to prepare by chlorination of the ether (Chapter 21). When these compounds are treated with $NaNH_2$ in liquid ammonia, a single amine is formed in good yield.

The new amino group finds itself in the *meta* position even though the chlorine was at the *ortho* position. It would be very difficult to explain this other than by the benzyne mechanism. Using the same elimination–addition sequence, this must be the mechanism:

■ Steric hindrance is not nearly as important in electrophilic substitution or in nucleophilic substitution by the addition–elimination mechanism. In both of these reactions, the reagent is attacking the p orbital *at right angles* to the ring and is some distance from an *ortho* substituent.

That shows *how* the *meta* product might be formed, but *why* should it be formed? Attack could also occur at the *ortho* position, so why is there no *ortho* product? There are two reasons: electronic and steric. Electronically, the anion next to the electronegative oxygen atom is preferred because oxygen is inductively electron-withdrawing. The same factor facilitates deprotonation next to Cl in the formation of the benzyne. Sterically, it is better for the amide anion to attack away from the OMe group rather than come in alongside it. Nucleophilic attack on a benzyne has to occur in the plane of the benzene ring because that is where the orbitals are. This reaction is therefore very sensitive to steric hindrance as the nucleophile must attack in the plane of the substituent as well.

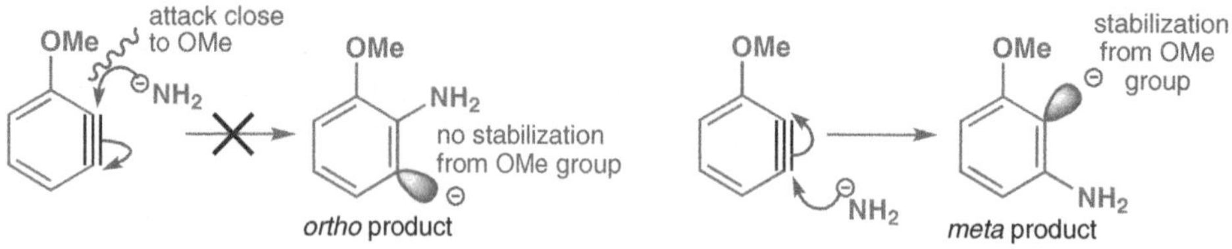

attack close
to OMe

no stabilization
from OMe group

ortho product

stabilization
from OMe
group

meta product

■ Oxygen is an electron-withdrawing group here because the anion is formed in the plane of the ring and has nothing to do with the benzene's π orbitals.

This is a useful way to make amino ethers with a *meta* relationship as both groups are *ortho, para*-directing and so the *meta* compounds cannot be made by electrophilic substitution.

para-Disubstituted halides can again give only one benzyne and most of them give mixtures of products. A simple alkyl substituent is too far from the triple bond to have much steric effect.

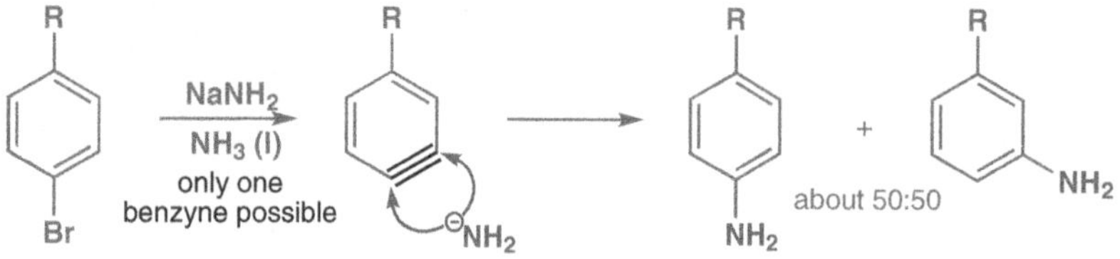

only one
benzyne possible

about 50:50

If the substituent is an electron-repelling anion, then the *meta* product is formed exclusively because this puts the product anion as far as possible from the anion already there. This again is useful as it creates a *meta* relationship between two *ortho, para*-directing groups.

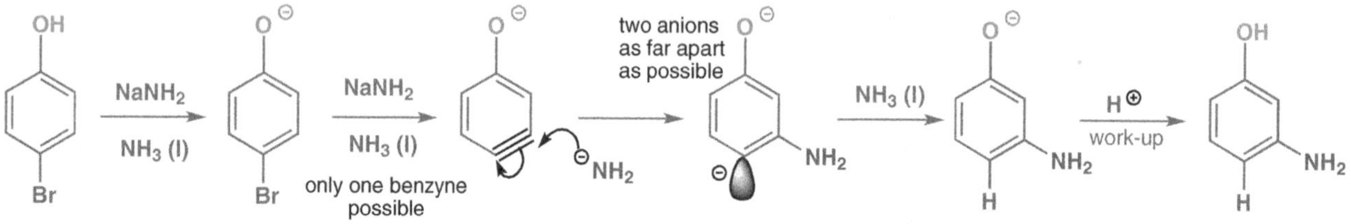

only one benzyne
possible

two anions
as far apart
as possible

work-up

Other evidence for benzyne as an intermediate

As you would expect, the formation of benzyne is the slow step in the reaction so there is no hope of isolating benzyne from the reaction mixture or even of detecting it spectroscopically. However, it can be made by other reactions where there are no nucleophiles to capture it, for example from this diazotization reaction.

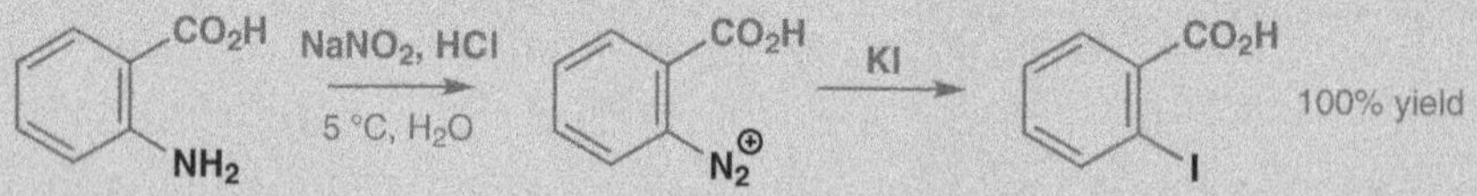

100% yield

This diazotization is particularly efficient as you can see by the quantitative yield of 2-iodobenzoic acid on capture of the diazonium salt with iodide ion. However, if the same diazonium salt is neutralized with NaOH, it gives a zwitterion with the negative charge on the carboxylate balancing the positive charge on the diazonium group. This diazotization is done with an alkyl nitrite in an organic solvent to avoid the chance that nucleophiles such as chloride or water might capture the product. When the zwitterion is heated it decomposes in an entropically favourable reaction to give carbon dioxide, nitrogen, and benzyne.

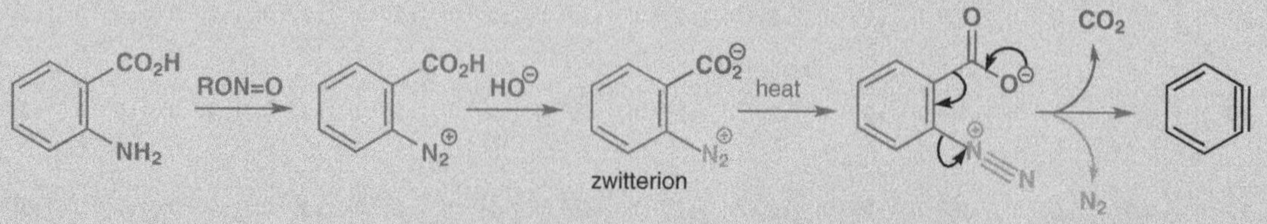

zwitterion

heat

You can't isolate the benzyne because it reacts with itself to give a benzyne dimer having a four-membered ring between two benzene rings. If the zwitterion is injected into a mass spectrometer, there is a peak at 152 for the dimer but also a strong peak at 76, which is benzyne itself. The lifetime of a particle in the mass spectrometer is about 2×10^{-8} s so benzyne can exist for at least that long in the gas phase.

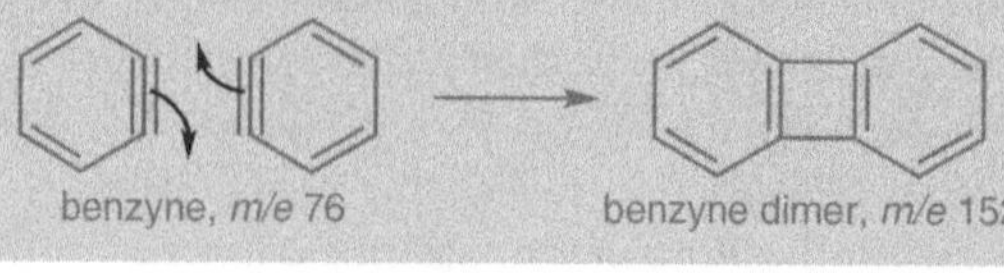

benzyne, *m/e* 76

benzyne dimer, *m/e* 152

To conclude...

Alkenes and arenes are usually nucleophiles. This chapter is about the occasions on which they are not, and instead react as electrophiles. Remember that, important though the reactions in this chapter are, the principal reactivity you can expect from these compound classes is nucleophilicity.

The table below summarizes these reactions and also other similar ones you will find elsewhere in the book.

Page	Type of alkene	Example	Reaction
500	unsaturated carbonyl compounds		conjugate addition
510	unsaturated nitriles and nitoalkenes	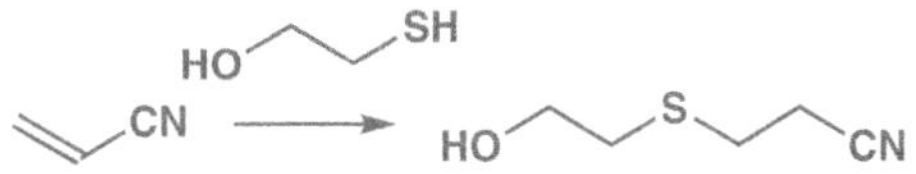	conjugate addition
511	enones, etc. with β leaving group	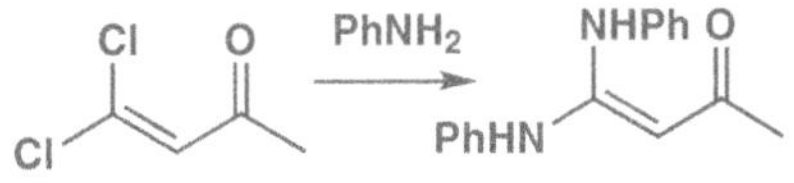	conjugate substitution
513	unsaturated carbonyl	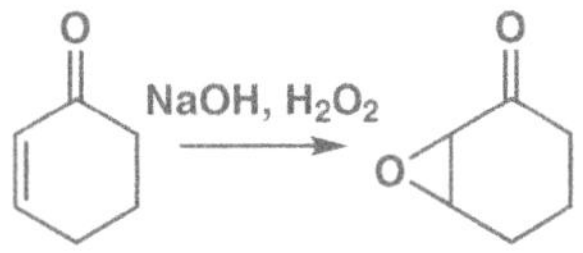	nucleophilic epoxidation
515	aryl chlorides/ fluorides/ethers with *ortho* or *para* electron-withdrawing groups	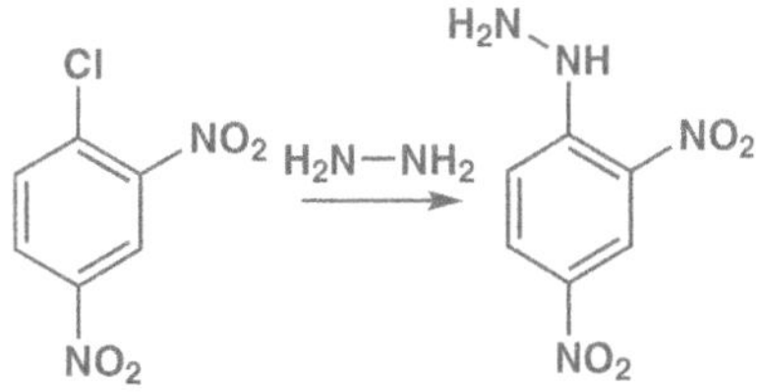	nucleophilic aromatic substitution: addition–elimination mechanism
520	aryl cations (from diazonium salts)	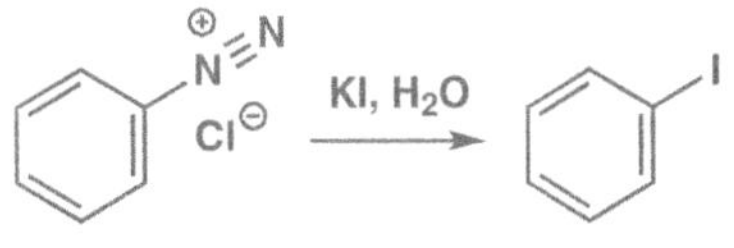	nucleophilic aromatic substitution: S_N1 mechanism
525	benzyne	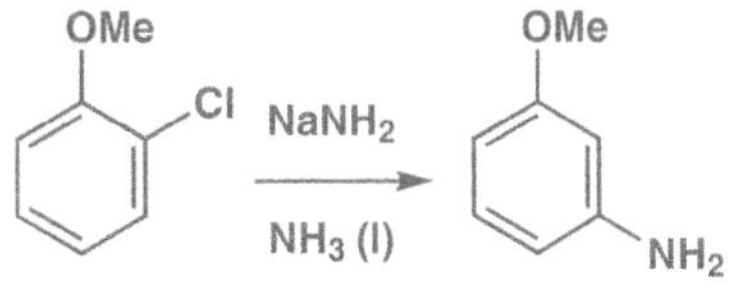	nucleophilic aromatic substitution: elimination–addition mechanism
ch. 26	enolates and enolate equivalents as nucleophiles		conjugate addition

Further reading

F. A. Carey and R. J. Sundberg, *Advanced Organic Chemistry A, Structure and Mechanisms*, 5th edn, Springer, 2007, chapter 9 and B, *Reactions and Synthesis*, chapter 11 also has a discussion of nucleophilic aromatic substitution. B. S. Furniss, A. J. Hannaford, P. W. G. Smith, and A. T. Tatchell, *Vogel's Textbook of Practical Organic Chemistry*, Longman, 5th edn, 1989, 6.6–6.7 gives many practical examples of the nucleophilic aromatic substitution. P. Wyatt and S. Warren, *Organic Synthesis: Strategy and Control*, Wiley, Chichester, 2007, chapter 9.

Check your understanding

To check that you have mastered the concepts presented in this chapter, attempt the problems that are available in the book's Online Resource Centre at http://www.oxfordtextbooks.co.uk/orc/clayden2e/

Suggested solutions for Chapter 22

PROBLEM 1

Draw a mechanism for this reaction. Why is base unnecessary?

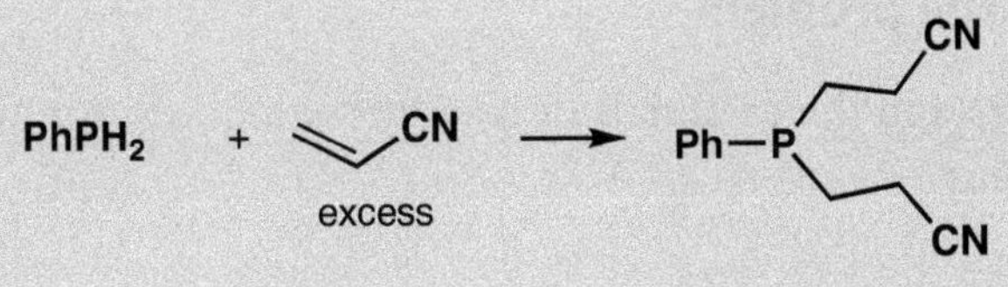

Purpose of the problem

Simple example of conjugate addition with a nucleophile from the second row of the periodic table.

Suggested solution

The phosphine is a good soft nucleophile with a high energy lone pair, well able to add in a conjugate fashion without help. In particular, the neutral phosphine does not need to be converted into its anion. The intermediate is a good base and removes a proton from itself, not necessarily intramolecularly.

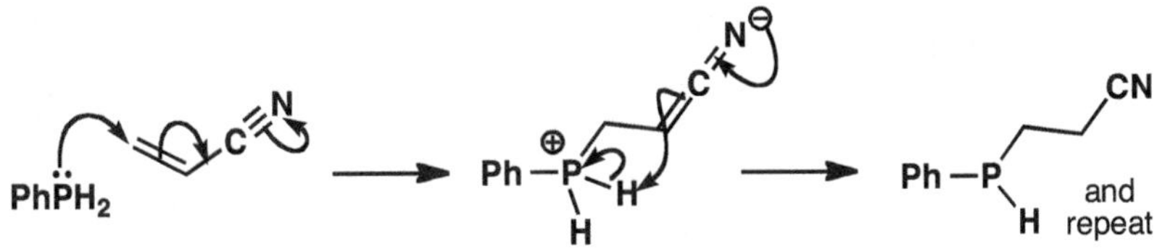

PROBLEM 2

Which of the two routes suggested here would actually lead to the product?

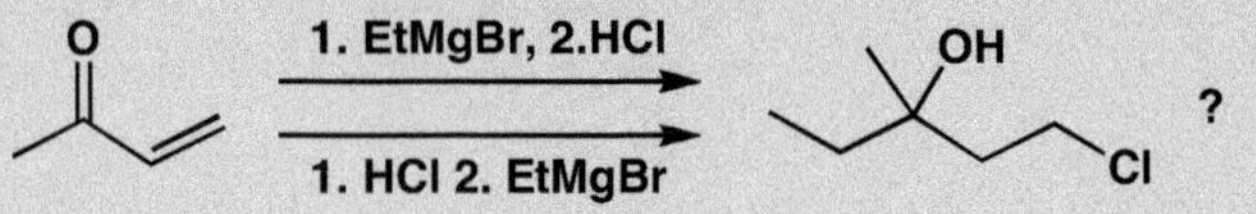

Purpose of the problem

Do you understand the essentials of conjugate addition? Can you say when it *won't* happen?

Suggested solution

To get the product, the chloride must add in a conjugate fashion and ethyl Grignard in a direct fashion that removes the carbonyl group. Conjugate addition can happen only if the carbonyl group is intact so HCl must be added first.

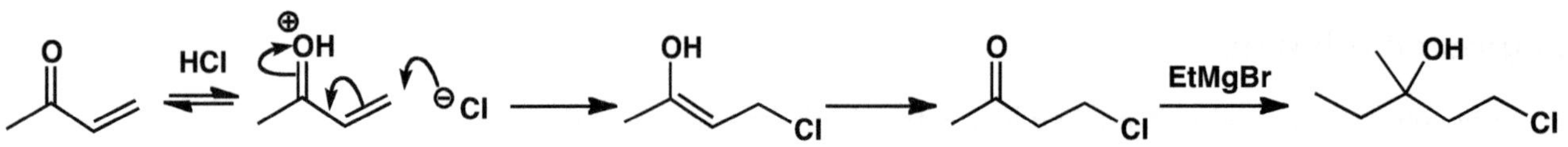

In the other sequence, EtMgBr is likely to add to the carbonyl group direct and further addition of HCl may either substitute on the allylic alcohol or add the 'wrong way round' to the alkene.

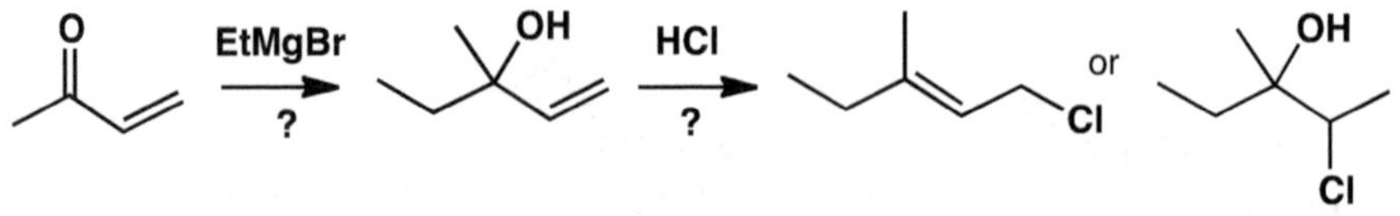

PROBLEM 3

Suggest reasons for the different outcome of each of these reactions. Your answer must of course be mechanistically based.

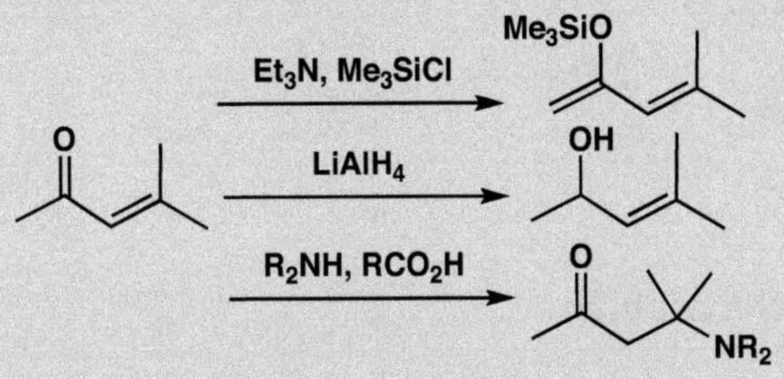

Purpose of the problem

A reminder of the reactions possible with enones.

Suggested solution

The three reactions are: enolization and trapping with silicon, direct addition with a hard irreversible nucleophile, and conjugate addition with a softer reversible nucleophile.

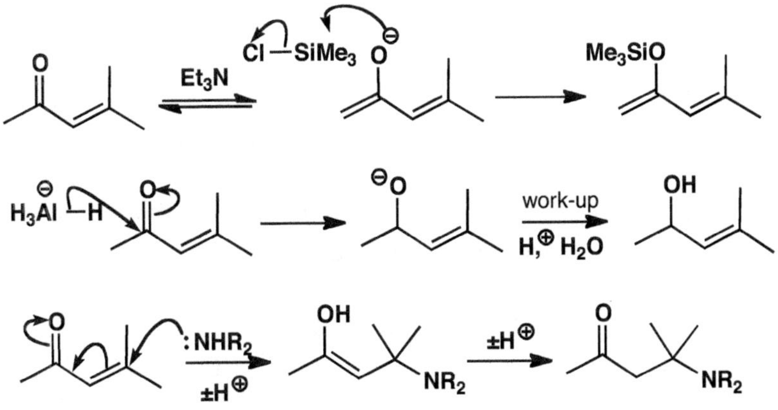

PROBLEM 4

Suggest a mechanism for this reaction.

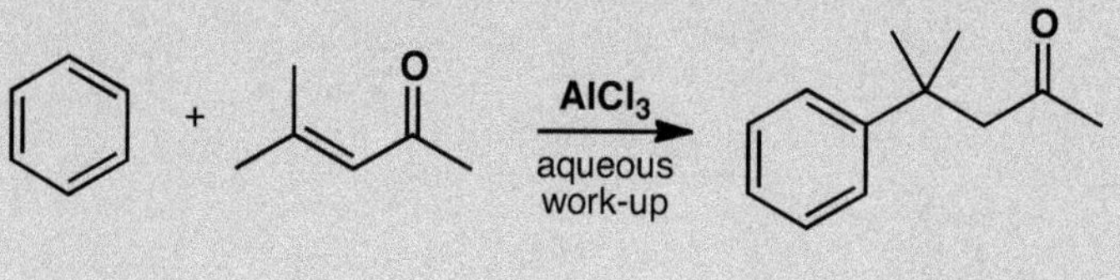

Purpose of the problem

Combination of conjugate addition and electrophilic aromatic substitution.

Suggested solution

The weakly nucleophilic benzene has evidently added in conjugate fashion to the enone in a kind of Friedel-Crafts reaction and we can use the Lewis acid to make the enone into the necessary cation.

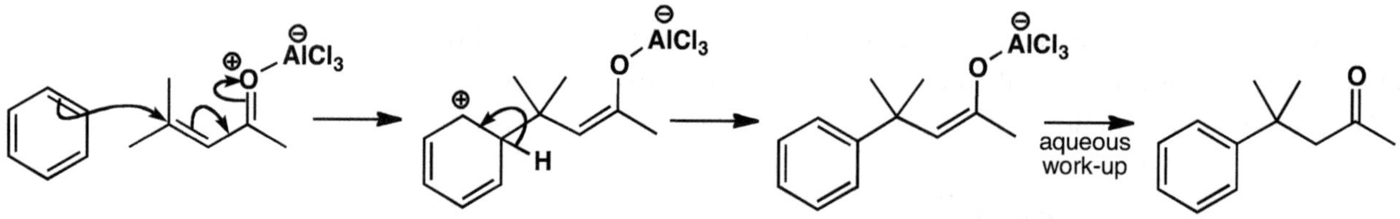

PROBLEM 5

What is the structure of the product of this reaction and how is it formed? It has δ_C 191, 164, 132, 130, 115, 64, 41, 29 and δ_H 2.32 (6H, s), 3.05 (2H, t, *J* 6 Hz), 4.20 (2H, t, *J* 6 Hz), 6.97 (2H, d, *J* 7 Hz), 7.82 (2H, d, *J* 7 Hz), 9.97 (1H, s). You should obviously interpret the spectra to get the structure.

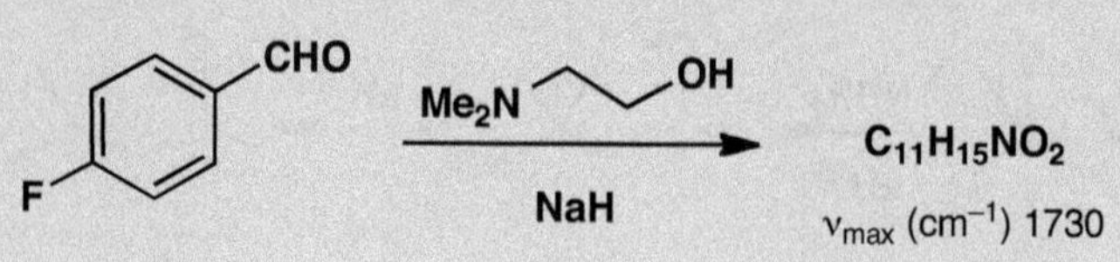

Purpose of the problem

Revision of NMR with an exercise in nucleophilic aromatic substitution.

Suggested solution

Summing the formulae of the two starting materials shows that this is a substitution of fluoride (the product is the sum of the starting materials less

HF). The aldehyde is still there (from the IR and the proton at 10 ppm) so the spectra are best interpreted by this structure:

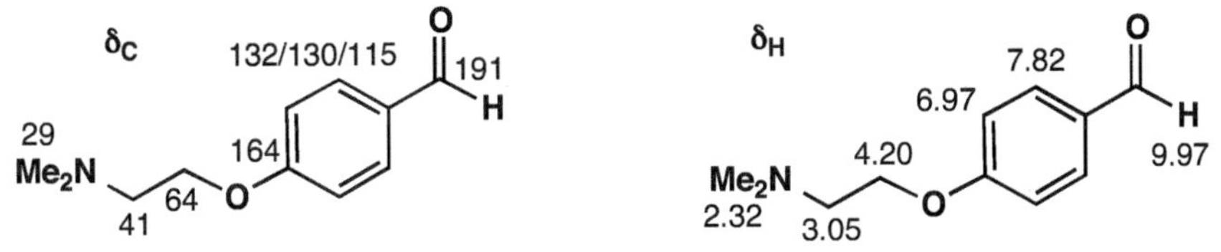

That suggests a simple nucleophilic aromatic substitution by the addition-elimination mechanism with both F and CHO assisting the first step.

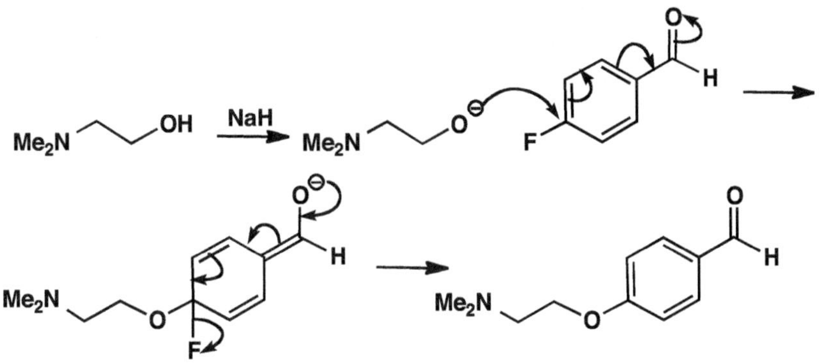

PROBLEM 6

Suggest a mechanism for this reaction, explaining the selectivity.

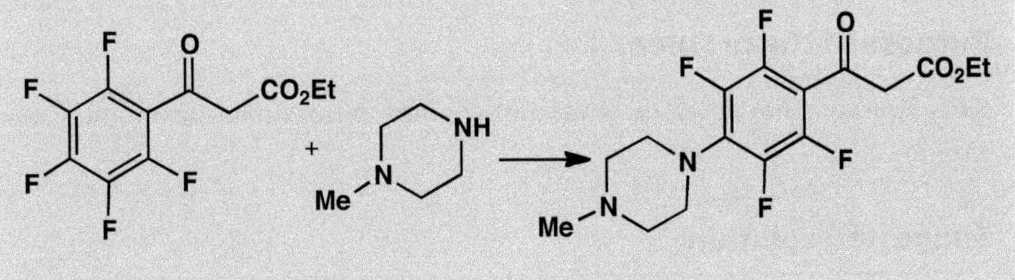

Purpose of the problem

Introduction to the mechanism and selectivity of nucleophilic aromatic substitution.

Suggested solution

Both *ortho* and *para* positions are activated by the ketone towards nucleophilic attack by the amine, but the *para* position is preferred because of steric hindrance between the large heterocyclic ring and the ketone. The substitution works because those five fluorine atoms make the ring very electron-deficient.

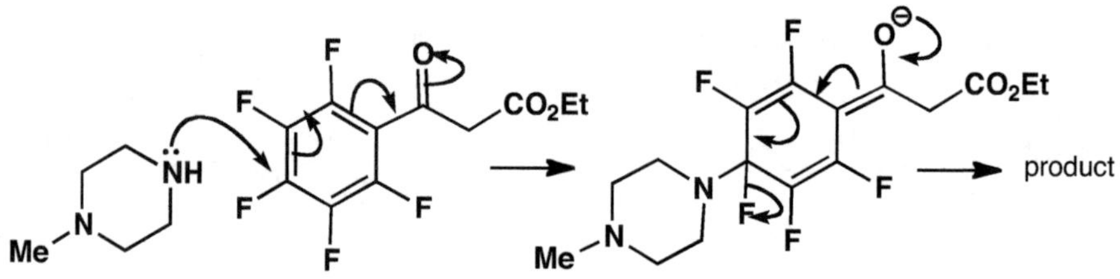

PROBLEM 7

Pyridine is a six-electron aromatic system like benzene. You have not yet been taught anything systematic about pyridine (that will come in chapter 29) but see if you can work out why 2- and 4-chloropyridines react with nucleophiles but 3-chloropyridine does not.

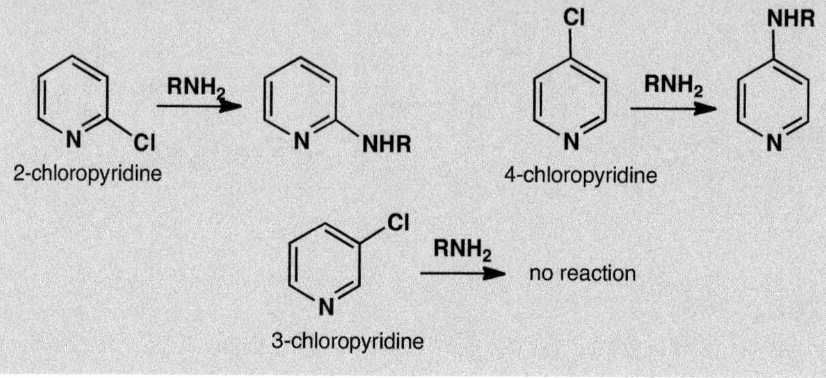

2-chloropyridine

4-chloropyridine

3-chloropyridine

RNH₂ → no reaction

Purpose of the problem

Extension of the ideas on nucleophilic aromatic substitution into new compounds.

Suggested solution

The problem is to find somewhere to park the negative charge in the intermediate and the only possible place is on the pyridine nitrogen atom. This is easy with 2- and 4-choropyridine but impossible with 3-chloropyridine. Using a general nucleophile:

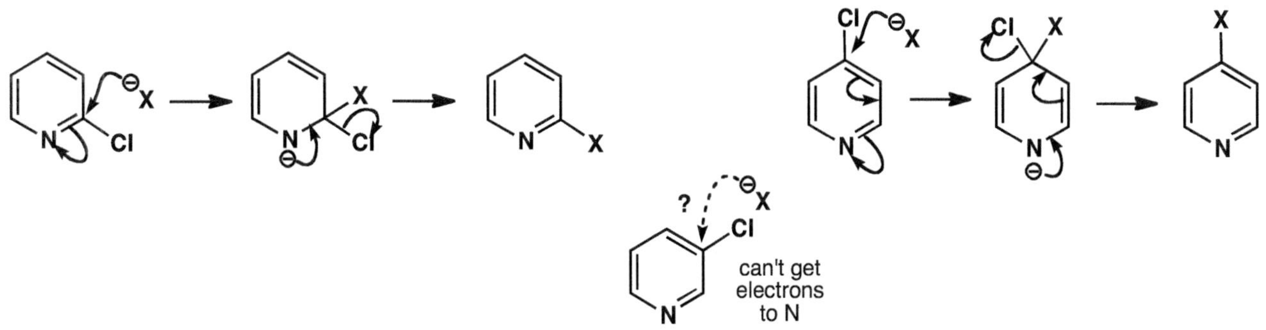

Amine formation by this reaction is particularly important as you will see in chapters 29 and 30. The mechanism is the same with a few proton transfers.

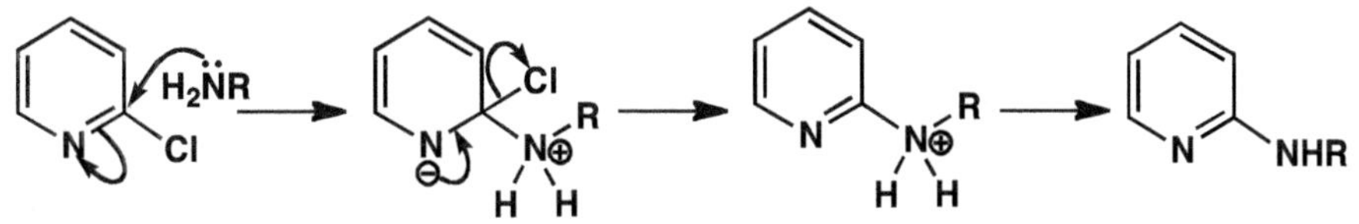

PROBLEM 8

How would you carry out these two conversions?

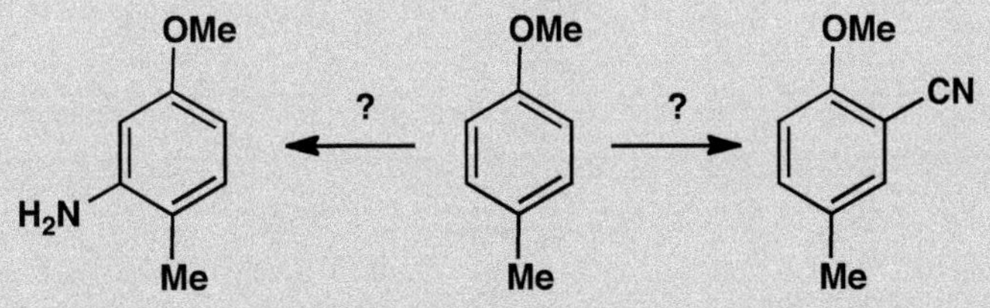

Purpose of the problem

Application of nucleophilic aromatic substitution in synthesis.

Suggested solution

Usually you would think of introducing NH_2 by nitration and reduction (chapter 21), but the regioselectivity is wrong for the first reaction: the methoxy group will direct nitration *ortho* to itself. An alternative is to introduce both NH_2 and CN as nucleophiles, but the ring is unactivated so we can't use the addition-elimination mechanism (there is nowhere for the negative charge to go). The successful alternatives are electrophilic aromatic substitution followed by diazonium salt formation and the benzyne method. Here are two possible routes. Nitration will insert the nitro group *ortho* to the more strongly electron-donating MeO group. Reduction, diazotization and substitution with copper cyanide by the S_N1 mechanism gives one product.

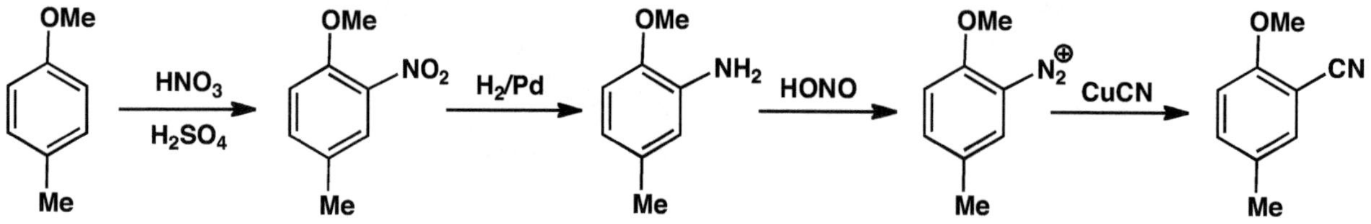

The other product could come from chlorination, elimination to give a benzyne, addition of amide anion to put the anion *ortho* to MeO (p. 524 in the textbook) and protonation.

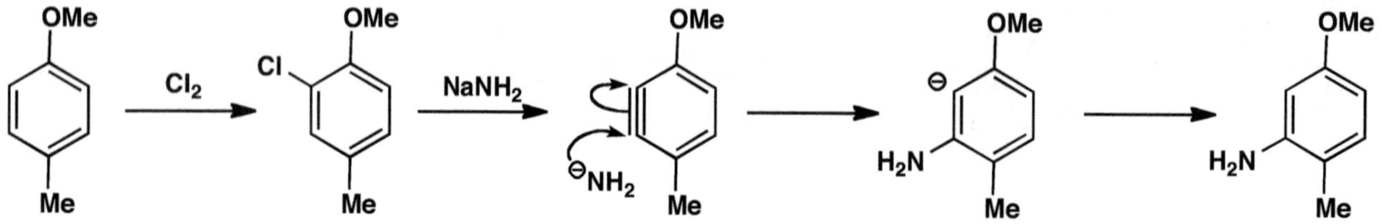

PROBLEM 9

Suggest mechanisms for these reactions, pointing out why you chose the pathways.

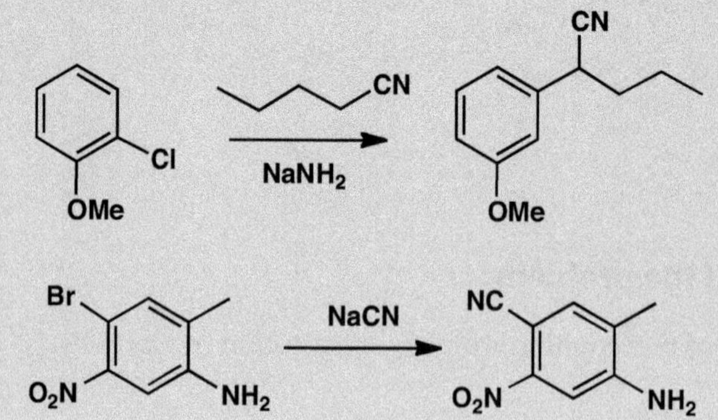

Purpose of the problem

Studies in selectivity and choosing the right mechanism.

Suggested solution

In the first reaction, the nucleophile adds in the 'wrong' position (i.e. where the leaving group isn't) so a benzyne mechanism is likely. Notice that the nucleophile and the benzyne are formed with the same strong base, that the anion is recycled and that the nucleophile adds to the benzyne to put the negative charge next to OMe (p. 524 in the textbook).

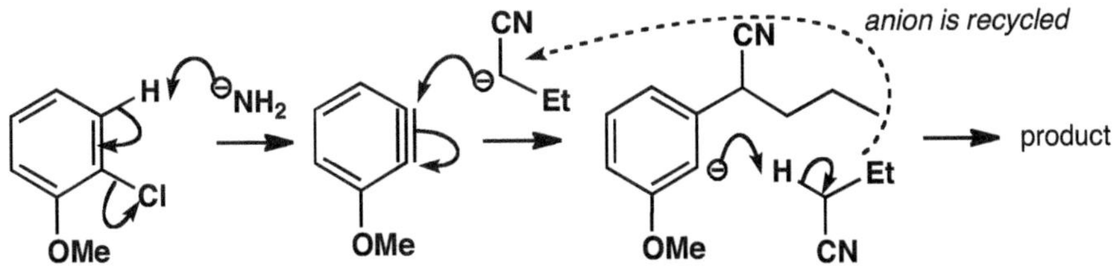

The second reaction is a straightforward substitution by the addition-elimination mechanism activated by the nitro group. The amino group is a spectator.

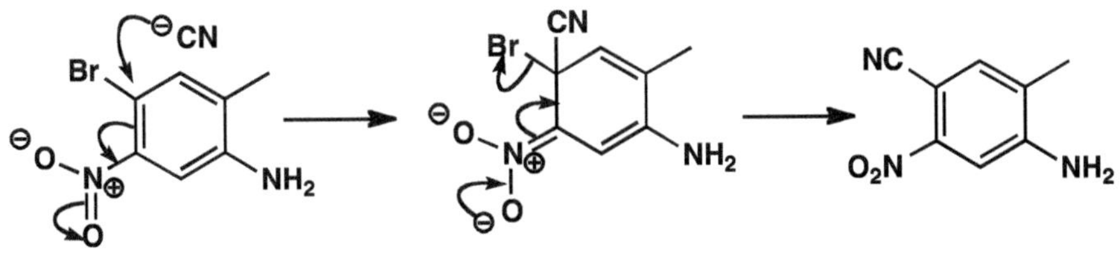

PROBLEM 10

When we discussed reduction of cyclopentenone to cyclopentanol, we suggested that conjugate addition of borohydride must occur before direct addition of borohydride: in other words, the scheme below must be followed. What is the alternative scheme? Why is the scheme shown definitely correct?

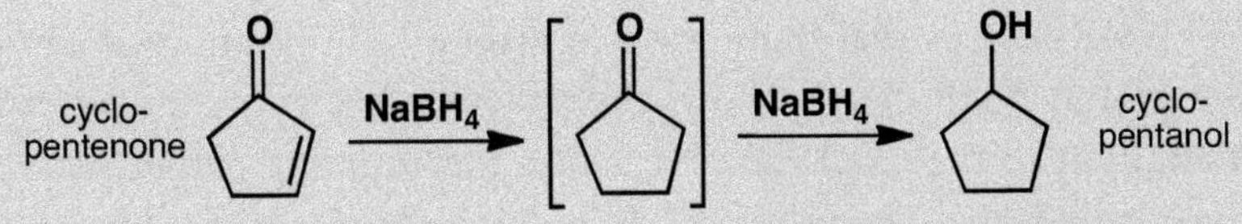

Purpose of the problem

Serious thinking about mechanisms is an advantage when reactions get more complex.

Suggested solution

The alternative scheme would be to reduce the ketone first and the alkene second. This order must be wrong though, because simple alkenes are nucleophilic and are not reduced by $NaBH_4$. $NaBH_4$ is a nucleophilic reducing agent and attacks alkenes only if they are conjugated with an electron-withdrawing group. The conjugate addition must always occur first so as to keep the carbonyl group intact for the second step.

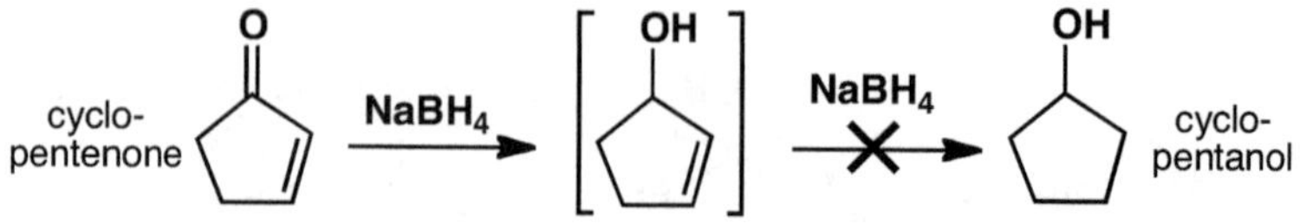

PROBLEM 11

Stirring thioacetic acid with acrolein (propenaldehyde) in acetone gives a compound with the NMR data shown below. What is the compound?

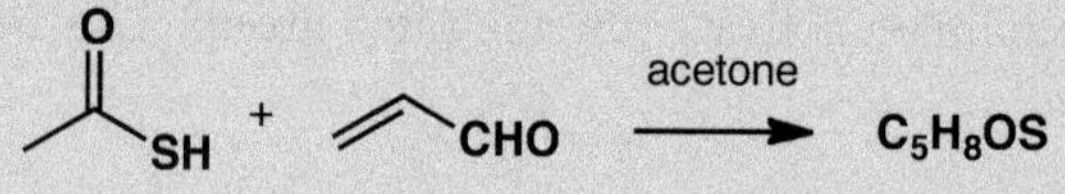

δ_H: 2.28 (3H, s), 3.58 (2H, d, *J* 8), 4.35 (1H, td, *J* 8, 6), 6.44 (1H, t, *J* 6, 7.67 (1H, d, *J* 6).

δ_C: 23.5, 31.0, 99.3, 144.2, 196.5.

Purpose of the problem

Using NMR to gain insight into a conjugate addition.

Suggested solution

The product formula is the sum of the reaction partners, and all 5 C and 8 H atoms are visible in the NMR spectra, so this looks like an addition reaction. The ^{13}C NMR tells us that there are two alkene carbons and one carbonyl, and the proton NMR clearly shows the aldehyde has gone. But it can't be direct addition to the C=O group, because the coupling pattern isn't right for a terminal alkene. The product is in fact the enol formed from conjugate addition of the sulfur, which is stable under these conditions. The low coupling constant across the alkene tells us it's formed unusually as the *Z*-isomer, probably because of an intramolecular proton transfer from the thioacid to the new OH group. The anhydrous conditions in dry acetone prevent the enol from tautomerizing back to the aldehyde.

■ This work is described by Lukas Hintermann in *J. Org. Chem.*, 2012, **77**, 11345.

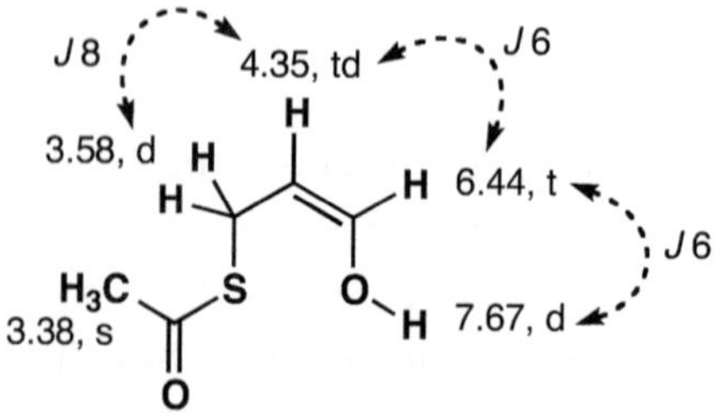

24 Regioselectivity

Connections

Building on

- Chemoselectivity ch23
- Electrophilic aromatic substitution ch21
- Addition to alkenes ch19
- Substitution at saturated C ch15
- Electrophilic alkenes and nucleophilic aromatic substitution ch22

Arriving at

- Selectivity of a new kind decided by mechanism
- Reagent and substrate are both important
- Controlling the arrangement of aromatic substituents
- How to get *ortho* selectivity: ortholithiation and sulfonation
- Radical as well as ionic reactions
- Reactions of allylic compounds
- Revisiting conjugate addition

Looking forward to

- Stereoselectivity ch32–ch33
- Enol and enolate reactions ch25 & ch26
- Reactions and synthesis of heterocycles ch29 & ch30
- Radical reactions ch37

Introduction

We met *chemo*selectivity—*which* group reacts—in the last chapter. Chemoselectivity means that there are two separate functional groups and that a reagent must choose between them. By contrast, *regio*selectivity implies that there is one functional group that can react in two different places and a reagent must choose where to react. Simple examples include addition of HX to an alkene (Chapter 19) and nucleophilic attack on the epoxide derived from that alkene (Chapter 15).

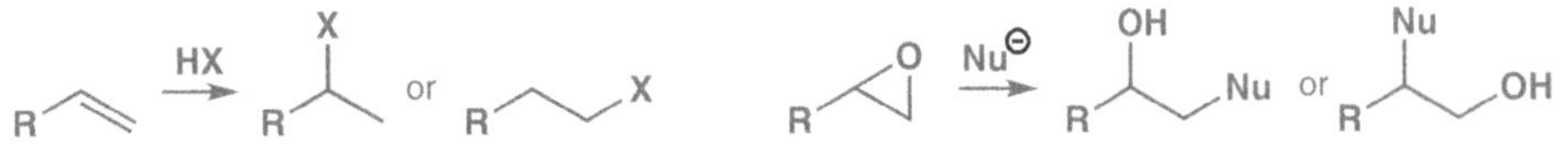

It might also mean that two functional groups are combined in a single conjugated system that can again react in two (or more) places. Examples include the addition of bromine to dienes (two conjugated alkenes) and addition of a nucleophile to a conjugated carbonyl compound (carbonyl group conjugated to an alkene).

Online support. The icon in the margin indicates that accompanying interactive resources are provided online to help your understanding: just type **www.chemtube3d.com/clayden/123** into your browser, replacing **123** with the number of the page where you see the icon. For pages linking to more than one resource, type **123-1, 123-2** etc. (replacing **123** with the page number) for access to successive links.

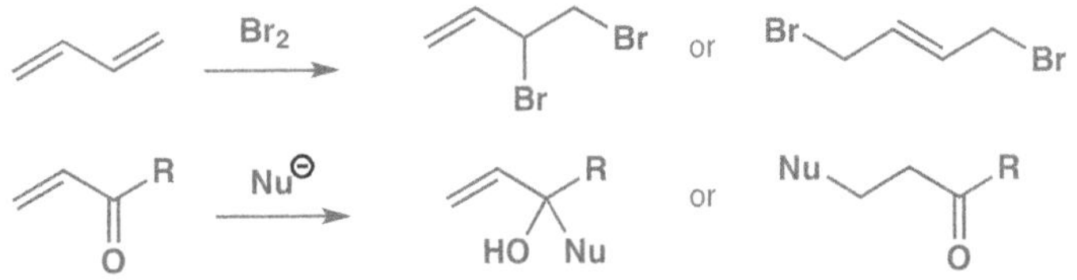

The choice between *ortho/para* and *meta* substitution when an electrophile attacks a benzene ring (Chapter 21) is also a matter of regioselectivity. We shall discuss all these examples in further detail in this chapter, and extend these ideas to new reactions as well.

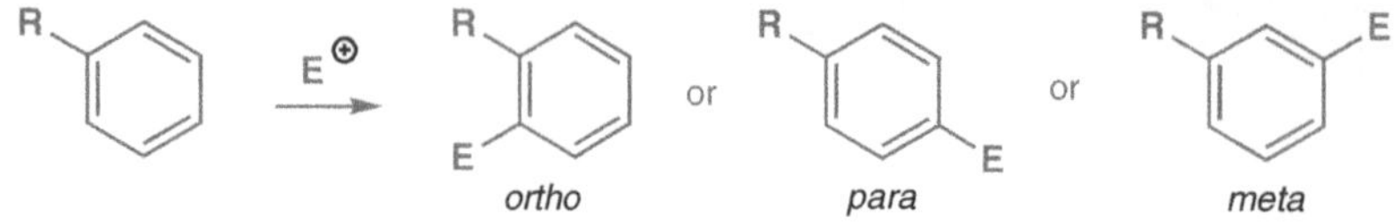

Regioselectivity in electrophilic aromatic substitution

We start with electrophilic aromatic substitution. It was established in Chapter 21 that an electron-donating substituent favours *ortho/para* and an electron-withdrawing substituent favours *meta* substitution. Although *meta* substitution is usually slower than *ortho/para* substitution (because electron-withdrawing groups deactivate the ring), it usually gives the *meta* product alone.

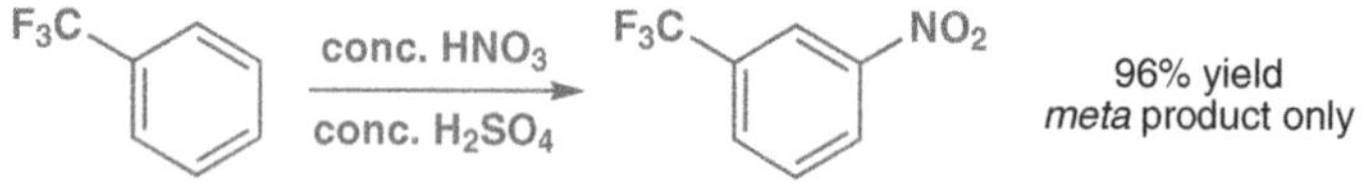

Most reactions of benzene rings with electron-donating substituents give *ortho/para* mixtures and, if the substituent is very electron-donating, may lead to both *ortho* and *para* substitution in the same molecule. Control in favour of the *para* product can usually be achieved by reducing the reactivity of the substituent and increasing its size.

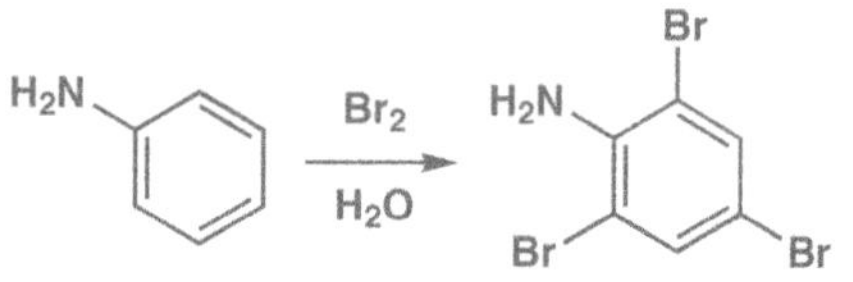

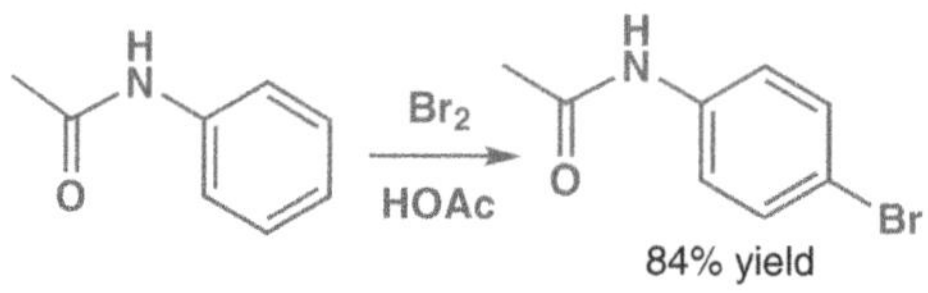

Of course, if the *para* position is blocked, *ortho* substitution is the only option, and we will come back to the idea of blocking substituents shortly. But there is a general way of directing electrophiles to the *ortho* position using activation by metallation.

➡ All these examples are drawn from Chapter 21.

Making organometallics by deprotonating aromatic rings: ortholithiation

Look at the reaction below: butyllithium deprotonates an sp^2 hybridized carbon atom to give an aryllithium. It works because the protons attached to sp^2 carbons are more acidic than protons attached to sp^3 carbons (although they are a lot less acidic than alkyne protons).

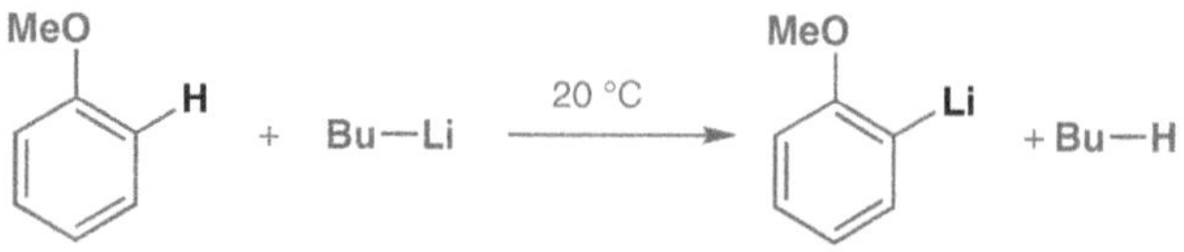

But there must be another factor involved to account for exclusive *ortho* substitution, which is after all the most hindered site. The functional group containing oxygen (sometimes nitrogen) is next to the proton to be removed. This functional group 'guides' the butyllithium, so that it attacks the adjacent protons. It does this by forming a complex with the Lewis acidic lithium atom, much as ether solvents dissolve Grignard reagents by complexing their Lewis-acidic metal ions. This mechanism means that it is only the protons *ortho* to the functional group that can be removed, and the reaction is known as an **ortholithiation**.

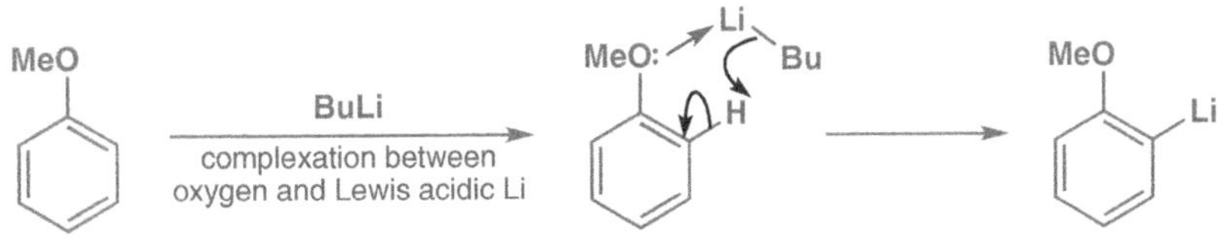

The example below shows ortholithiation, activated by the nitrogen atom of a tertiary amine, being used to make a new C–C bond. Here it is the nitrogen atom that directs attack of the butyllithium, again by complexation with the Li atom.

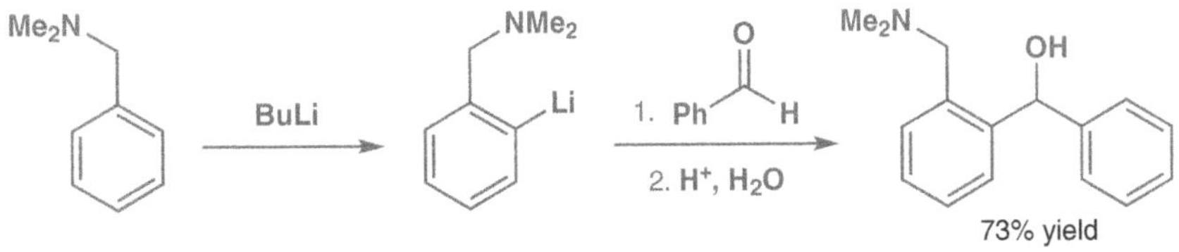

■ Compare the methods for making organolithiums or Grignard reagents that you met in Chapter 9—most of them rely on formation of an organometallic by reduction of an alkyl or aryl halide.

Ortholithiation is a useful way of making reactive organometallics because the starting material does not need to contain a halogen atom. But it is much less general than the other ways we have told you about for making organolithiums, as there are rather tight restrictions on what sorts of groups the aromatic ring must carry. The best ortholithiation-directing substituents have lone pairs to donate electrons to Li and are also electronegative so they withdraw electrons from the benzene ring and help stabilize the anion forming at the *ortho* position.

Fredericamycin

Fredericamycin is a curious aromatic compound extracted in 1981 from the soil bacterium *Streptomyces griseus*. It is a powerful antibiotic and antitumour agent, and its structure is shown below. The first time it was made in the laboratory, in 1988, the chemists in Boston started their synthesis with three consecutive lithiation reactions: two are ortholithiations, and the third is slightly different. You needn't be concerned about the reagents that react with the organolithiums; just look at the lithiation reactions themselves. In each one, one or more oxygen atoms (colour-coded green) directs a strongly basic reagent to remove a nearby proton (colour-coded black). As it happens, none of the steps uses *n*-BuLi itself, but instead its more reactive cousins, *sec*-BuLi and *tert*-BuLi (see the table on p. 186). The third lithiation step uses a different kind of base related to LDA, made by deprotonating an amine (pK_a about 35). The black proton removed in this third lithiation is more acidic because it is next to an aromatic ring.

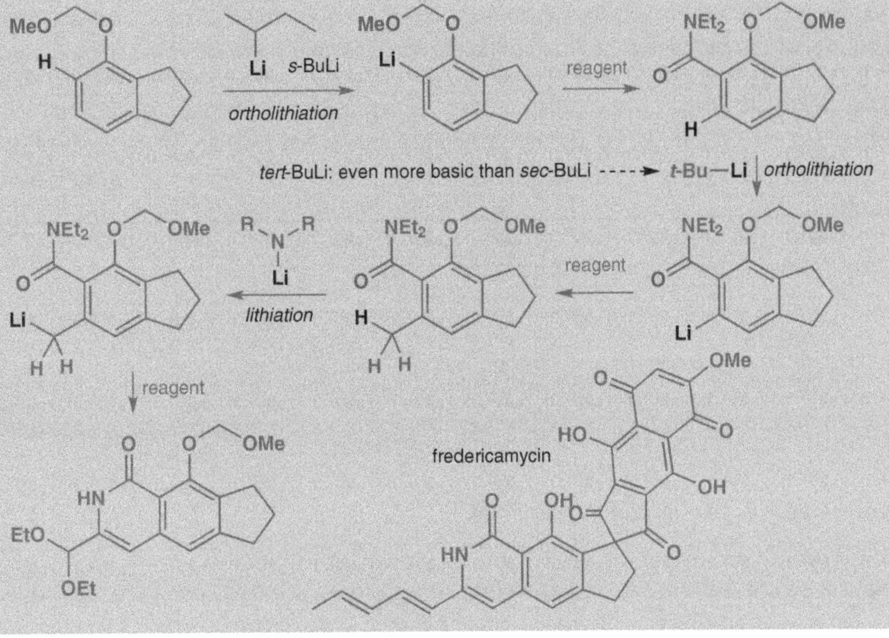

Sulfonation may lead to ortho selectivity without lithiation

We introduced sulfonation in Chapter 21 but have left detailed discussion until now because sulfonation has some features that make it more interesting than first meets the eye. One important difference between sulfonation and other examples of electrophilic substitution is that sulfonation is *reversible*. Heating an arenesulfonic acid causes it to decompose with loss of gaseous SO_3.

Here's an example of how we can exploit this to gain control of regioselectivity without resorting to lithiation. In stage 1 the phenol is sulfonated twice—the first sulfonic acid group (which adds *para* to the OH group) is electron-withdrawing and deactivates the ring, making the introduction of the second group (which goes *ortho* to the OH and *meta* to the first sulfonic acid) harder and that of the third group harder still, which is why we can isolate the disulfonated phenol.

In the second stage, the bromination, the OH directs to the *ortho* and *para* positions, but only one *ortho* position is vacant, so the bromine attacks there. Sodium hydroxide is needed to deprotonate the sulfonic acid groups to make them less deactivating. The sulfonation reaction is reversible, and in the third stage it is possible to drive the reaction over to the products by distilling out the relatively volatile 2-bromophenol at high temperature. The loss of SO_3 involves attack of H^+ on the aromatic ring.

Overall, we have succeeded in making 2-bromophenol where direct bromination of phenol itself would have given (at low temperatures) mainly *p*-bromophenol and at higher temperatures, 2,4,6-tribromophenol. The sulfonic acid groups are useful reversible blocking groups.

The same method can be used with anilines because *para*-sulfonation of aromatic amines is possible. This seems surprising because in sulfuric acid essentially all the amine will be protonated. You might expect the resulting ammonium ion to react in the *meta* position (because NH_3^+ is no longer electron-rich) but instead the *para*-sulfonic acid (sulfanilic acid) is formed. At the high temperature of the reaction, it is probable that any *meta*-substituted product reverts to the starting material, while the *para*-sulfonic acid accumulates because it is stabilized by delocalization and is less hindered.

Regioselective reactions of naphthalene

We introduced you to the 10-electron aromatic system of naphthalene in Chapter 7. As you would expect, it undergoes electrophilic aromatic substitution with the same reagents you met in Chapter 21, but the regioselectivity of its reactions is of a different type to the *ortho, meta,*

The sulfonating agent

The exact nature of the electrophile in sulfonation reactions seems to vary with the amount of water present. Certainly for oleum (fuming sulfuric acid, that is concentrated sulfuric acid with added sulfur trioxide) and solutions of sulfur trioxide in organic solvents, the electrophile is sulfur trioxide itself, SO_3. With more water around, $H_3SO_4^+$ and even $H_2S_2O_7$ have been suggested.

■ You might want to consider *why* sulfonation is reversible at high temperature in the light of our discussion of entropy and temperature on p. 248.

■ The reversibility of sulfonation with sulfuric acid may also account for the higher yield of *para* product in the sulfonation of toluene with H_2SO_4 as compared with $ClSO_2OH$ (p. 485).

Sulfa drugs

The product is important because the amides derived from it (sulfanilamides) were the first antibiotics, the sulfa drugs.

sulfanilamide

sulfapyridine, a sulfa drug

■ In Chapter 7 we pointed out that the middle bond is shorter than the rest, and for this reason we suggest you draw naphthalene with a double bond in this position—it makes mechanistic explanations more realistic.

naphthalene has two sorts of ring positions

➡ Unactivated rings such as benzene need a Lewis acid to react with bromine: see p. 474.

■ Another way of looking at the difference between these two delocalized cations is that the first can be shown delocalized into the double bond without disrupting the remaining aromatic ring; in the second, all other representations of delocalized structures must lose the aromatic ring.

para selectivity we have been talking about. Naphthalene has 10 carbons: two form the ring junction, and aren't available for substitution reactions, and the other eight are of just two types α (the 1-position, next to the ring junction) and β (the 2-position).

Electrophilic substitution on naphthalene normally occurs at a site next to the ring junction (α). This is because the HOMO has its largest coefficient at this atom, but you can rationalize the result by looking at the long, linear delocalization in the resulting cation, which can be represented by a single train of arrows. This extended conjugation makes naphthalene more nucleophilic than benzene. So, bromination occurs at the α-position in good yield even without a Lewis acid.

Reaction at the other position (β) is less favourable as the intermediate cation is cross-conjugated. The cation delocalizes into both rings, but no long linear chain of arrows is possible.

If the reaction is irreversible, the α-product is usually formed. But if the reaction is reversible, as is the case with sulfonation, the position of substitution may be determined by temperature. Sulfonation at low temperatures gives the α-product by kinetic control, while sulfonation at high temperatures gives the β-product by thermodynamic control. The β-product is formed more slowly but it is more stable as there is less steric hindrance between the large sulfonic acid group and the orange hydrogen on the other ring. Under conditions allowing reversible sulfonation, eventually all the product ends up β.

α-product   reversible at high temperature   β-product

Regiocontrol by choice of route

Choosing the right route to an aromatic product is essential if you want to get one particular isomer. We can illustrate this with the synthesis of the isomers of bromonitrobenzene. Because the bromo substituent is *ortho, para*-directing and the nitro group *meta*-directing, it's possible to make all three isomers, providing we exploit the regioselectivity of electrophilic substitution. Nitration of bromobenzene would give the *ortho* and *para* isomers while bromination of nitrobenzene would give the *meta* isomer. The selectivity of the first reaction is not good: bromine is small and not very electronegative, so steric hindrance is weak and the *ortho* positions are not deactivated. Furthermore there are two *ortho* positions but only one *para*: a typical result is about 37% *ortho*, 1% *meta*, and 62% *para*. Both compounds are industrial products, made by nitration and separated.

Bromination of nitrobenzene is remarkably good, considering the unreactivity of nitro-benzene in electrophilic aromatic substitution. One recipe uses iron powder and bromine at 140 °C and gives 74% of the *meta* product. We shall need these reactions in the next section.

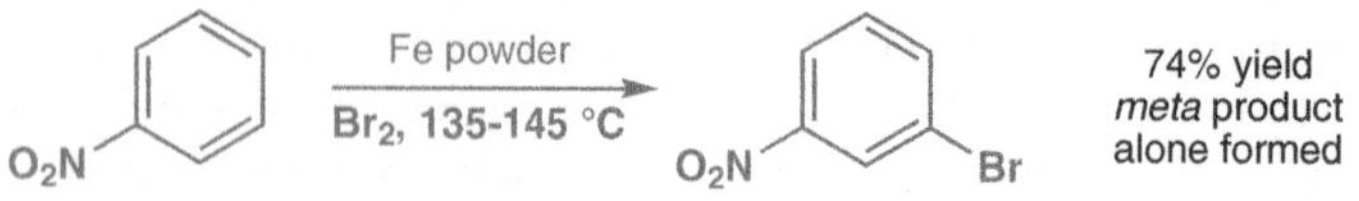

Before we move on, consider why this selectivity works: we can get all three isomers because we have one *ortho/para* director and one *meta* director. But what if we had two *ortho/para* directors—say, amino and bromo—and wanted the *meta* isomer?

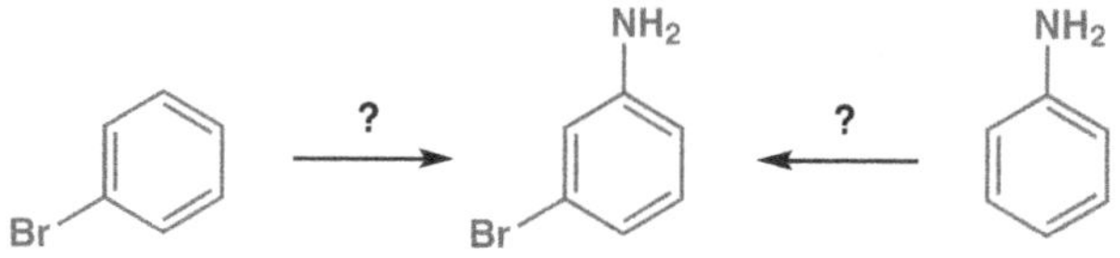

The solution in these cases is often to make use of the transformation of the nitro group (a *meta* director) into an amino group (a *para* director) by reduction.

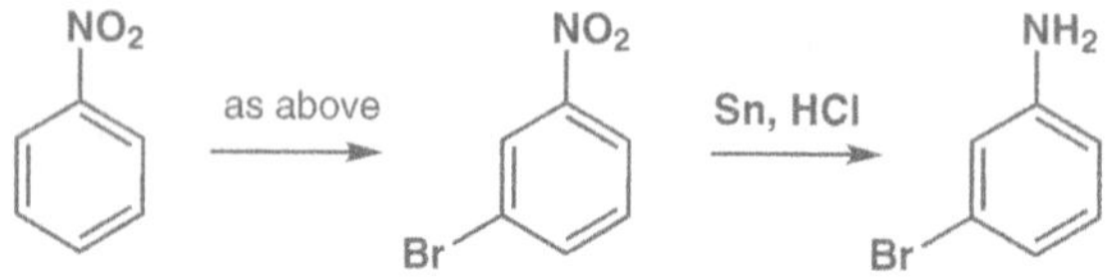

Since the amino group can be substituted by diazotization (p. 520), many problems of regio-selectivity can be solved by using nitro compounds as intermediates. You could, for example, use the product above to make the otherwise challenging 3-bromoiodobenzene:

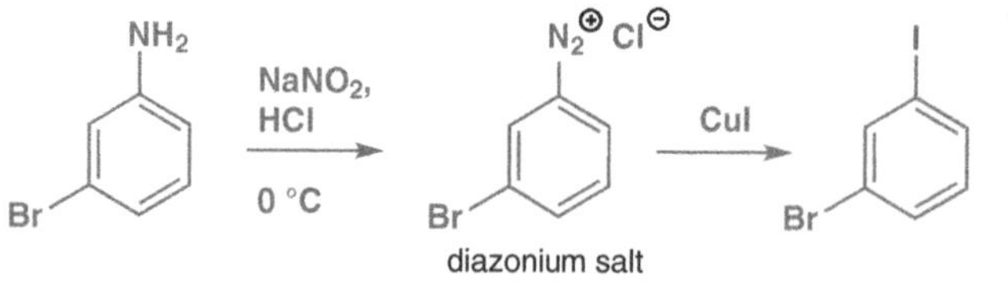

➡ Diazonium salts, and their use in the synthesis of aromatic compounds, were discussed on pp. 495 and 520.

Regioselectivity in nucleophilic aromatic substitution

As you saw in Chapters 21 and 22, diazonium salts need no activation to undergo nucleophilic aromatic substitution, but for other leaving groups a nitro group is commonly used as an activator. The three fluoronitrobenzenes are all commercial products but only the *ortho* and *para* isomers can do the nucleophilic substitution. This is because the nitro group must be able to stabilize the addition intermediate by accepting the negative charge.

➡ The various ways to carry out nucleophilic aromatic substitution are described on pp. 514–526.

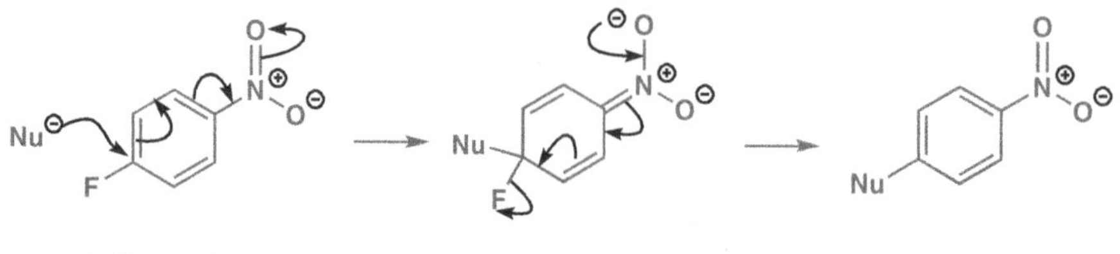

By carefully combining electrophilic and nucleophilic substitution it is possible to make aromatic compounds with substituents arranged in a precise and predictable fashion. So, if we nitrate *o*-dichlorobenzene, all positions are favourable but the nitro group goes in *para* to one Cl atom because of steric hindrance at the *ortho* positions. Although chlorine is small, two chlorines next to each other have a buttressing effect as each pushes the other away. It is difficult to get three adjacent substituents on a benzene ring. If we now do a nucleophilic aromatic substitution, only the Cl *para* to the nitro group is displaced. We can even reduce the nitro group to the corresponding amine.

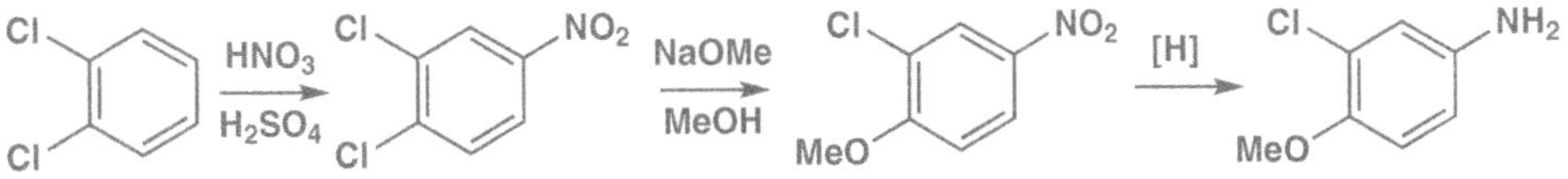

The last successful method for nucleophilic aromatic substitution uses a benzyne intermediate—on p. 524 you saw benzyne chemistry being used to make *meta*-aminoanisole, like this:

➡️ The regiochemistry of this reaction is explained in Chapter 22.

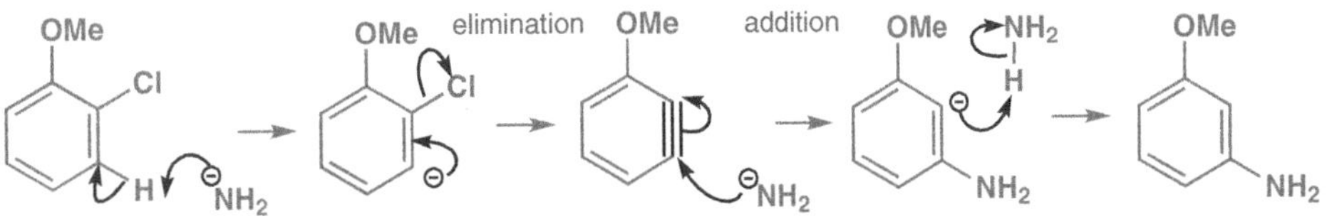

Now that the amino group is fixed; we can displace it via a diazonium salt using any chosen nucleophile—copper cyanide for example:

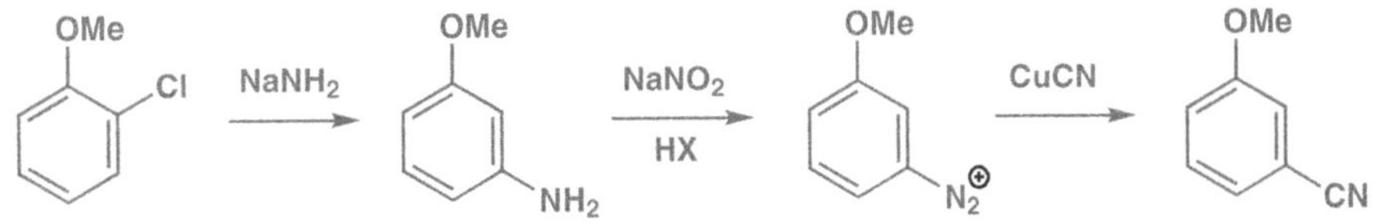

Regioselectivity of intramolecular reactions

A cunning way to get unusual regioselectivity is to make the reaction intramolecular. The synthesis from benzene of the cyclic ketone known as tetralone may look difficult as we must get an *ortho* relationship on the benzene ring. But if we make the final bond in the ring by a Friedel–Crafts acylation there is no problem. The alkyl group is *ortho,para*-directing and the acid cannot reach the *para* position.

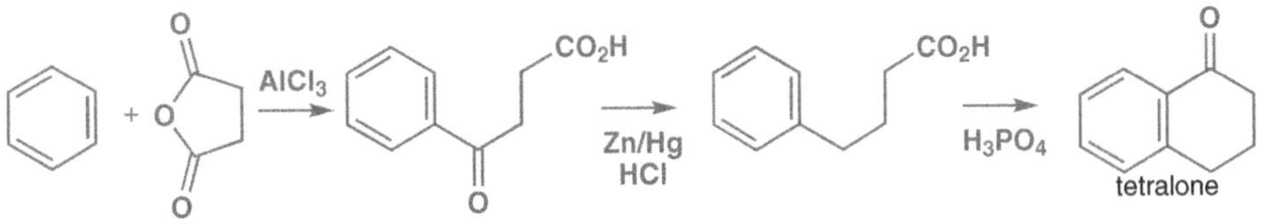

■ Usually a more powerful catalyst (AlCl₃) is needed, but intramolecular acylations are fast enough without this.

Notice the use of a cyclic anhydride in the first Friedel–Crafts acylation. It doesn't matter where the acylation occurs and the reaction stops there as the ring is deactivated by the ketone and the carboxylic acid released in the reaction is much less electrophilic than the anhydride. The ketone is then reduced to a CH_2 group by the Clemmensen method (see Chapter 23) and polyphosphoric acid is used to carry out the intramolecular acylation step.

A more subtle approach is to use a 'tether'—something that holds two reagents together and is afterwards cleaved. An example is halolactonization. The idea is simple. A halogen, say bromine, attacks an alkene and the bromonium ion intermediate is captured intramolecularly by the anion of a carboxylic acid. The reaction therefore uses bromine and $NaHCO_3$—a weak base, but one strong enough to deprotonate a carboxylic acid. The anion attacks the more highly substituted end of the bromonium ion, as explained in Chapter 19, and forms a five-membered ring.

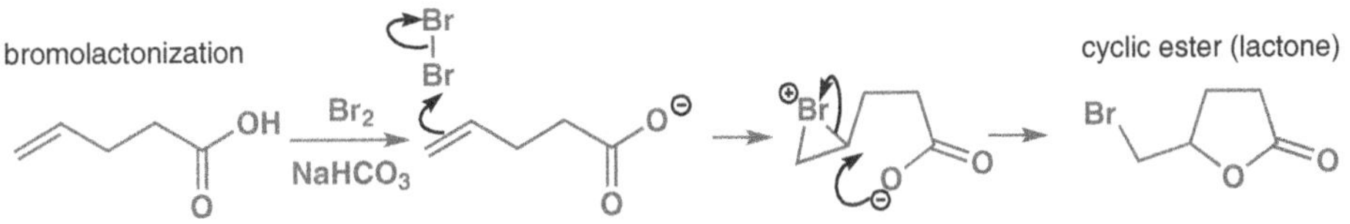

Although any halogen might be used in this reaction, iodine is the most versatile and the reaction is commonly called iodolactonization. The tether is the C–O bond of the lactone and this can be cleaved with an alkoxide.

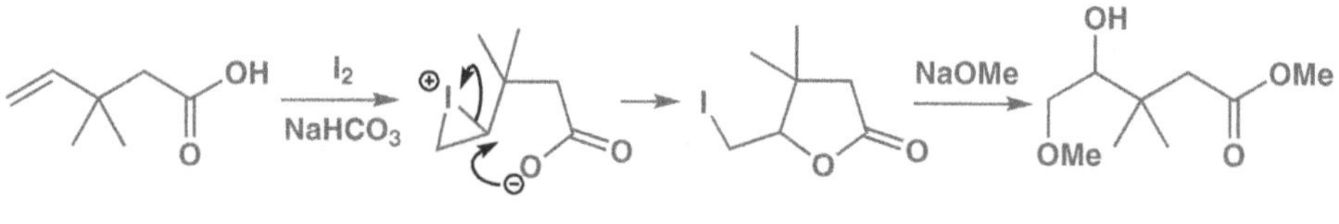

The reaction with methoxide needs some explanation. Attack on the carbonyl group cleaves the lactone, releasing an alkoxide that cyclizes to form an epoxide. A second molecule of methoxide now attacks the epoxide, opening it from the less hindered end as we should expect in an anionic reaction (Chapter 19).

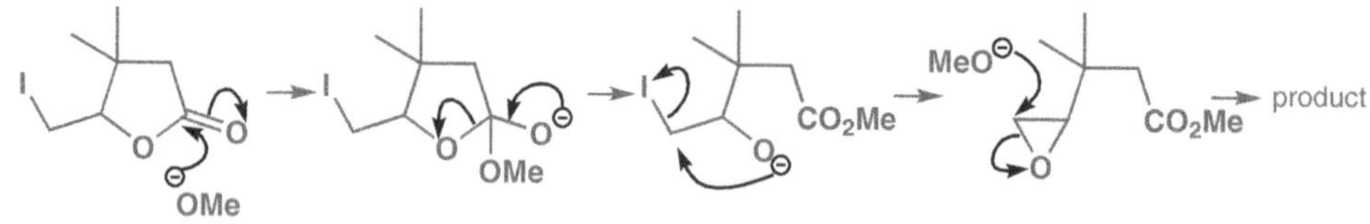

Another example shows that reaction may occur at the other end of the iodonium ion. Attack at the tertiary carbon would be difficult sterically and, in any case, would give an unstable four-membered ring. The lactone formed has the iodine β to the carbonyl group and so eliminates easily in base (pyridine works well) by the E1cB mechanism (Chapter 17) to give the unsaturated lactone. Although the relative stereochemistry of the iodolactone is controlled by the inversion in the opening of the iodonium ion, it is irrelevant as it disappears in the elimination step.

➡ We'll come back to the use of iodolactonization to control stereochemistry in Chapter 32.

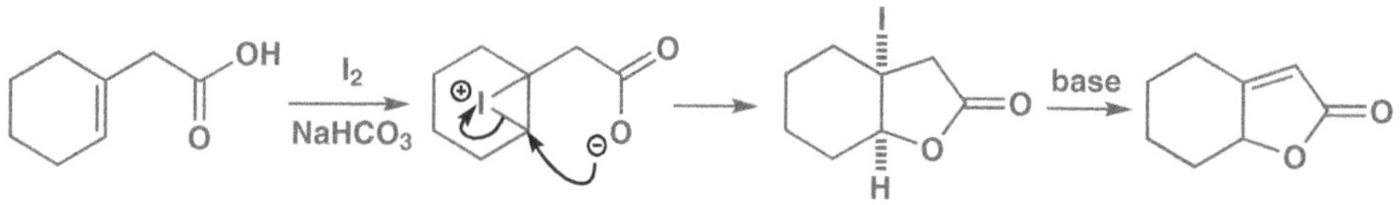

Regioselectivity in elimination reactions

This question was discussed in Chapter 17 but we can return to it here with more sophisticated examples. The regioselectivity in the last reaction of the sequence above dictates the position of the alkene in the product. Of all of the protons adjacent to the iodo group, only a black one is lost:

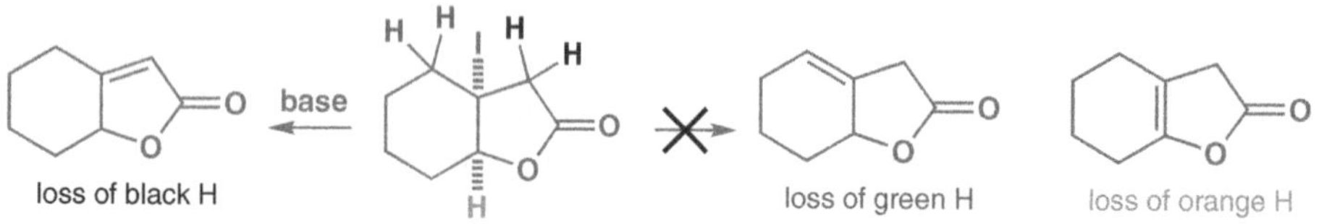

loss of black H   loss of green H   loss of orange H

The orange hydrogen cannot be lost by E2 as it is *cis* to the iodine and E2 reactions prefer a *trans* (anti-periplanar) arrangement. The green hydrogens are not lost because they are less acidic than the black hydrogens. In fact, this is not an E2 elimination at all. Because one of the black hydrogens can be lost in enolate formation, this is an E1cB elimination.

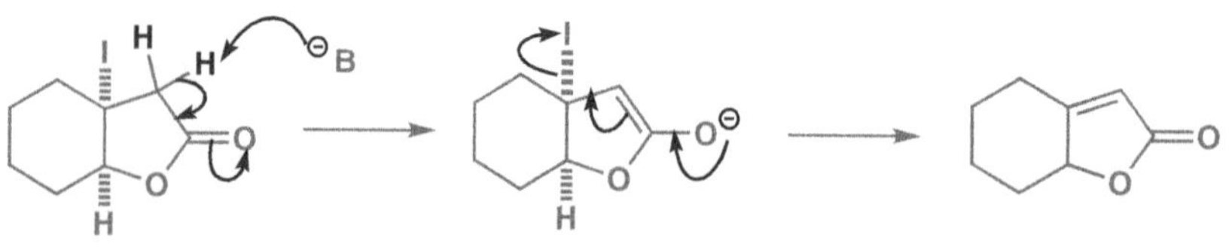

But now another regioselectivity question arises: if elimination occurs preferably towards the carbonyl group, how can we make the starting material for the iodolactonization sequence, which has the alkene not in conjugation with the carboxylic acid? It turns out that it is better to make the ester with the 'wrong' regioselectivity. This is easily done by a Horner–Wadsworth–Emmons reaction using a phosphonate ester. This Wittig-style reaction is explained in Chapter 27.

Now comes the remarkable regioselectivity. The ester is hydrolysed, as usual, in aqueous NaOH. On acidification to pH 3, the free acid is released and the double bond has moved into the ring.

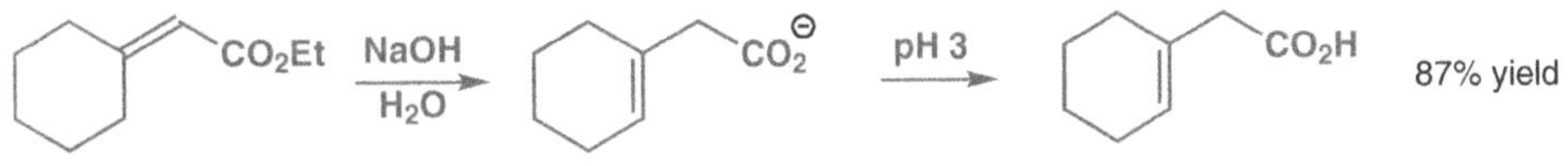

Alkenes like to be conjugated with carbonyl groups but they also prefer to be *inside* six-membered rings rather than outside—in this case presumably because the ester group otherwise has to eclipse a ring carbon. Conjugation with an ester group pulls the alkene out of the six-membered ring in the lactone we made above, but when the carbonyl group is a carboxylate anion, conjugation is very weak and the double bond moves into the ring.

Electrophilic attack on alkenes

You met electrophilic attack on alkenes in Chapter 19 and we shall just briefly revisit its regioselectivity. Unsymmetrical alkenes add HBr to give the more stable of the two possible cations. If R is alkyl or aryl, this means the more substituted cation.

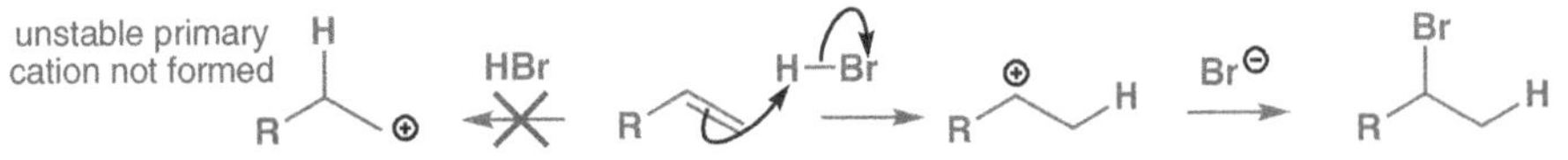

If you want to get the other regioisomer, with the heteroatom at the end, you can use hydroboration (Chapter 19) or the radical reactions described in the next section. Here is a brief reminder of hydroboration. Reaction between a borane having at least one B–H bond with an alkene gives an alkyl borane in which all the hydrogens are replaced by alkyl groups. Oxidation gives the terminal alcohol.

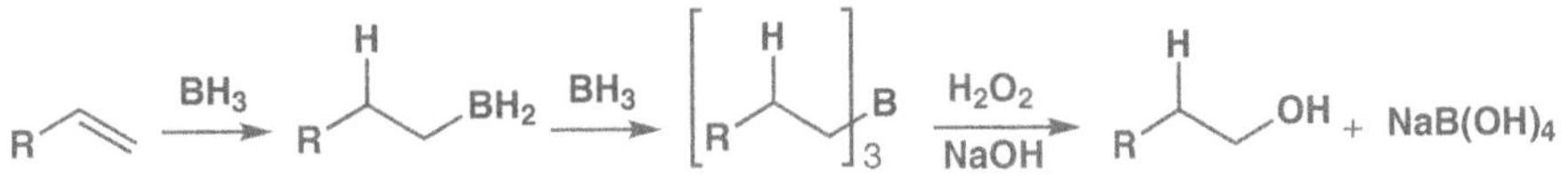

The regioselectivity comes from the first step. The boron's empty p orbital bonds to the more nucleophilic end of the alkene and hydride is transferred to give a borane. Reaction with alkaline H_2O_2 leads to migration of an alkyl group from boron to oxygen and eventually to the alcohol.

■ In these structures X can be R or H.

🖱 Interactive mechanism for hydroboration

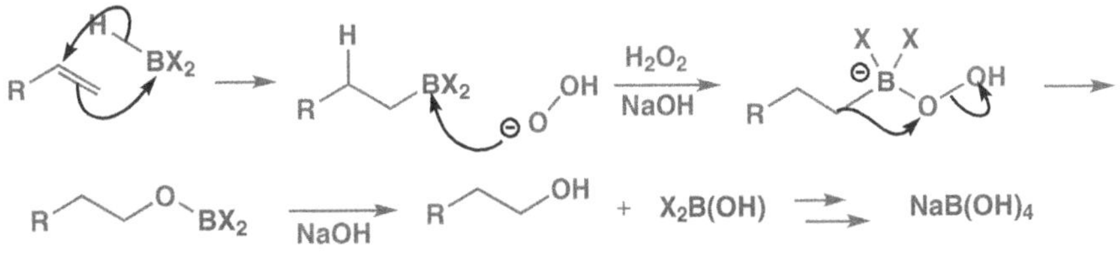

Borane is unstable but can easily be made from $NaBH_4$ and BF_3. In this synthesis of hexan-1-ol from hex-1-ene, a water molecule has been added to the alkene, but with the opposite regioselectivity to reactions with H_2O in acid or HBr.

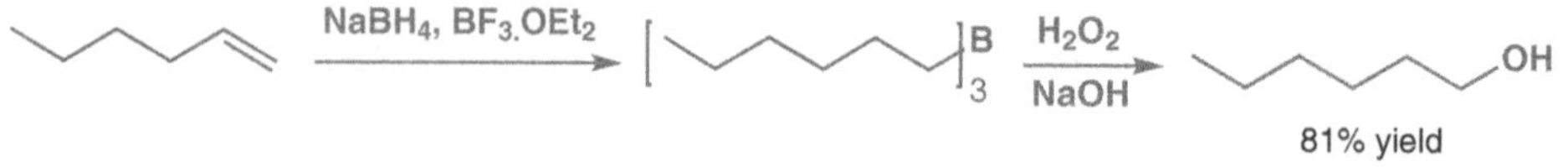

81% yield

Regioselectivity in radical reactions

Almost every reaction we have discussed so far has been ionic, but in this short section we need to give you a preview of another group of reactions we return to in Chapter 37—those of *radicals*. When HBr adds to an unsymmetrical alkene we use arrows that represent the movement of two electrons to give charged intermediates that combine in a second step to give a neutral product. The strong H–Br bond breaks to give a bromide ion and a stable alkyl cation. This bond breaks *heterolytically*—that is, unsymmetrically—as does the alkene bond. We can predict the regiochemistry of these reactions by making the most stable anions and cations as intermediates, in this case a tertiary alkyl cation and a bromide anion.

> ⮕ You'll meet radicals in much greater detail in Chapter 37.

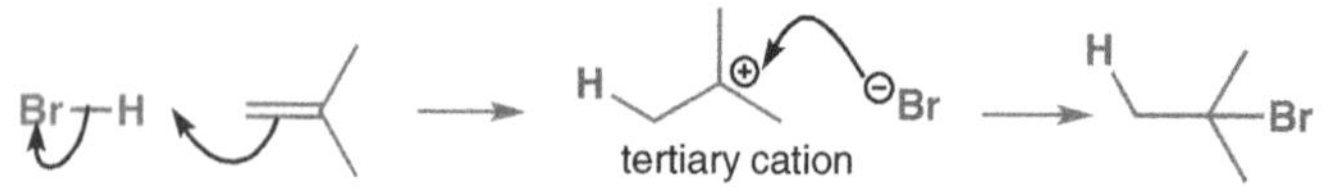

tertiary cation

Radical addition

The regioselectivity in the reaction below is opposite: a primary alkyl bromide is formed, by a different mechanism involving radicals.

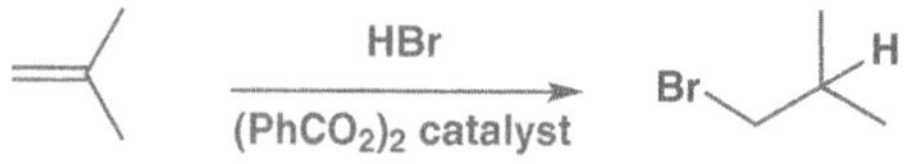

In radical reactions, bonds break *homolytically* with one electron going one way and one the other. The radicals that are formed have an odd number of electrons, one of which must be unpaired. This makes them very reactive and they are not usually isolated. Even strong bonds can break into ions provided they are polarized, but to make radicals we need weak symmetrical bonds such as O–O, Br–Br or I–I. Dibenzoyl peroxide, the $Ph(CO_2)_2$ catalyst in this reaction, readily undergoes homolysis like this—the one-electron movements are represented by 'fish-hook' arrows having one barb and odd electrons on atoms are represented by dots.

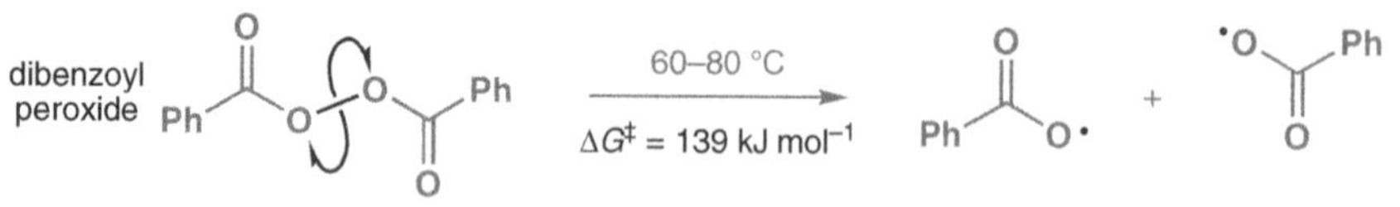

Now we can use the new radicals we have just made to cleave the strong HBr bond homolytically because a new and very strong OH bond will be formed. As we start with one radical intermediate that must have an unpaired electron, we must finish with another radical with an unpaired electron. In this case, it is a bromine radical.

If we do this reaction in the presence of the alkene we have just reacted with HBr, the bromine radical adds to the alkene in one of the two possible ways. Although radicals are neutral, they are electron-deficient (the C atom is one electron short) and, rather like cations, are more stable the more substituents they have. So the tertiary radical is formed rather than the primary radical, and the bromine ends up at the primary position.

We have still not reached the end of the reaction as our product is still a radical. How can it become a molecule with only paired electrons? The answer is simple. It reacts with another HBr molecule to produce more bromine radicals. Now you see something important to all radical reactions: only a small amount of the radical is needed as more radicals are produced every time the reaction gives product. The overall process is a *radical chain reaction*.

<table>
<tr><td>
Interactive mechanism for radical addition of HBr to alkenes
</td></tr>
</table>

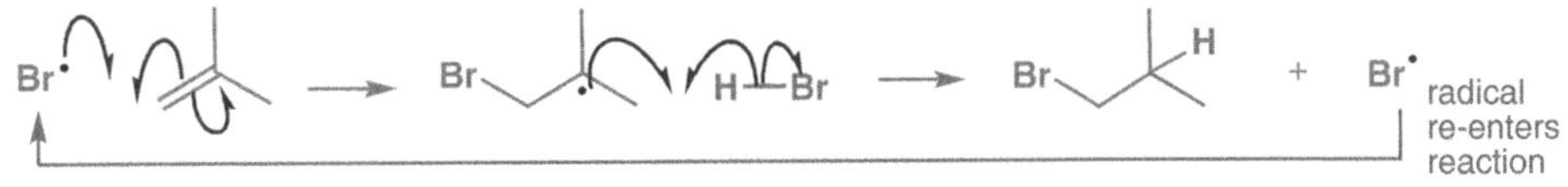

Because of this we also need only very small amounts of dibenzoyl peroxide, the radical initiator, which is just as well as it is potentially explosive, like many radical generators. Here is the reaction being used to make a bromoacid:

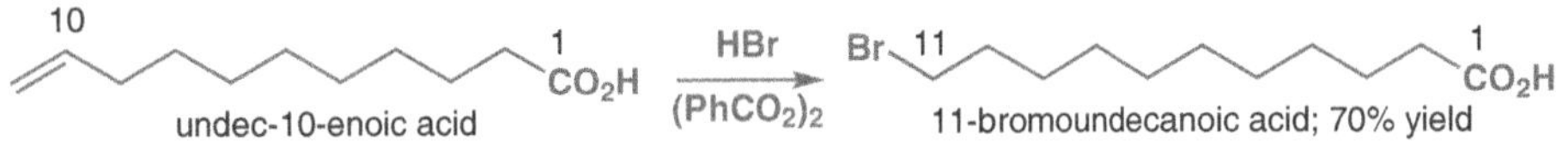

Radical abstraction

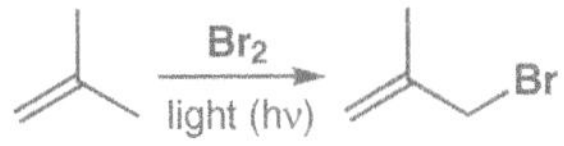

<table>
<tr><td>
</td></tr>
</table>

We sneaked a new reaction into that sequence. The removal of a hydrogen atom (note: not a proton) from HBr by the peroxide radical is an *abstraction* reaction. The bromine radical will also abstract hydrogen atoms and will do so from the same alkene we have just used but with yet another different outcome, as you see in the margin.

When light shines on bromine, the weak Br–Br bond breaks to give two bromine radicals. Heat will do the job too but light is cleaner and, as bromine is brown, it absorbs most wavelengths of visible light.

■ Note that the Br–Br bond is more stable than the O–O bond in the peroxide.

$$Br-Br \xrightarrow{\text{light } (h\nu)} 2 \times Br^\bullet \qquad \Delta G^\ddagger = 192 \text{ kJ mol}^{-1}$$

homolytic cleavage

Radicals are very unstable and reactive, and these bromine radicals may simply recombine or they may react with other compounds. You already know that bromide anions are good nucleophiles in S_N2 reactions, but bromine radicals do two quite different reactions: abstraction and addition. The Br radical may abstract a hydrogen atom from the alkene or it may add to the π bond. Notice that each reaction produces a new carbon-centred radical and, in the first case, a molecule of HBr. Whereas the Br–Br bond is weak, the H–Br bond is much stronger (366 kJ mol^{-1}) and, unlike ionic reactions, radical reactions are dominated by bond strength.

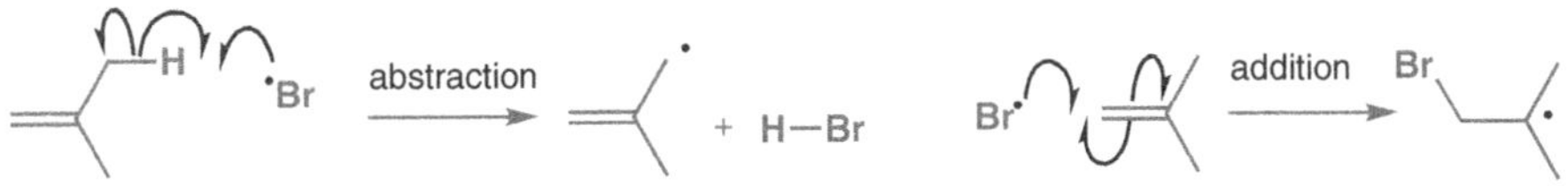

The first reaction introduces another important aspect of regioselectivity: why does the radical abstract that H atom, and not one from the alkene?

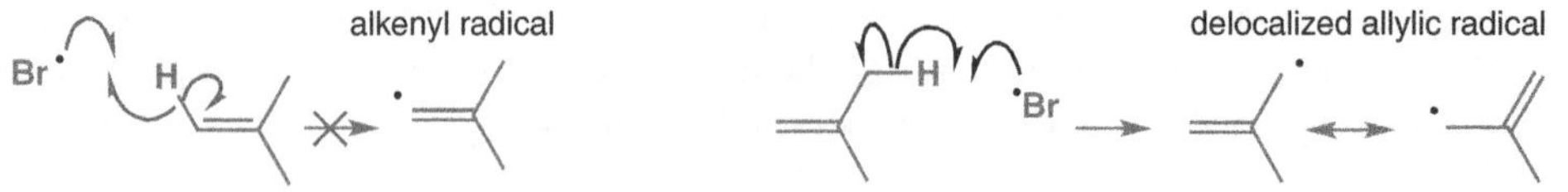

Removal of an alkene H gives a carbon-centred radical localized on the sp^2 atom but the removal of an H from a methyl group gives a much more stable delocalized allylic radical. In addition there are six such H atoms but only two alkene H atoms.

The reaction obviously cannot end there with the formation of another radical, however stable, and this allylic radical collects a bromine atom from a bromine molecule. Note that the allylic radical doesn't react with a bromine *radical* in this step: radicals are very unstable and the concentration of radicals at any one time is so low that it is rare for two of them to meet.

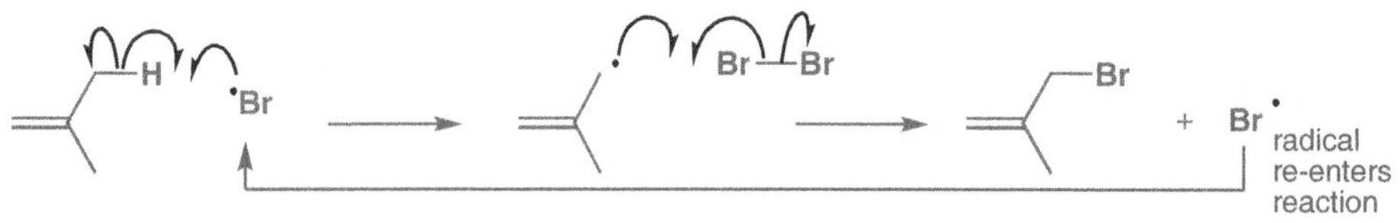

Interactive mechanism for allylic bromination

This step also produces a new bromine radical that can start a new series of reactions. Like the addition of HBr above, the reaction is a radical chain reaction, and only a small amount of Br_2 needs to break down to $Br^\bullet$ to get the reaction going. This is important as you already know what happens when bromine molecules react with alkenes: addition occurs by an ionic mechanism. Add too much Br_2 and the bromine molecules attack the alkene directly and do not abstract H atoms.

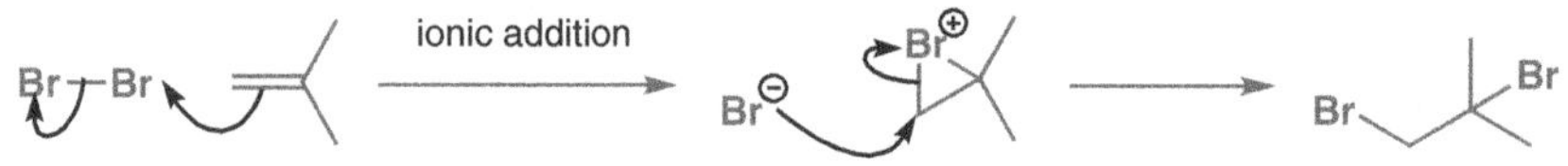

If we want to make the dibromide, we use plenty of bromine, but if we want to use a radical process to make the allylic bromide we must take advantage of the greater reactivity of the radical and keep the bromine concentration low. A good way to do this is to use the compound NBS (*N*-bromosuccinimide), which you met in Chapter 19. NBS acts as a sort of turnstile which only lets a molecule of Br_2 out when a molecule of HBr is formed (and of course HBr is the by-product in the radical bromination).

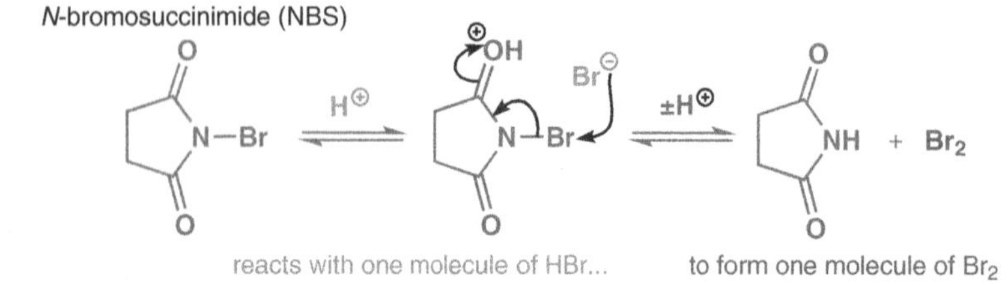

Br₂ is slowly released into the reaction as it proceeds, and the concentration never builds up enough to generate the dibromide. In this example, dibenzoyl peroxide is the initiator and allylic bromination gives the useful cyclohexenyl bromide.

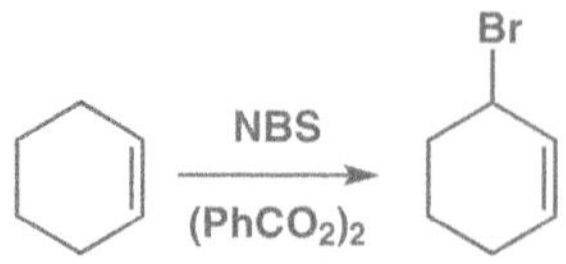

These radical reactions will be described in much greater detail in Chapter 37. For the moment you need only notice that they can have quite different regioselectivity from ionic reactions with the same reagents.

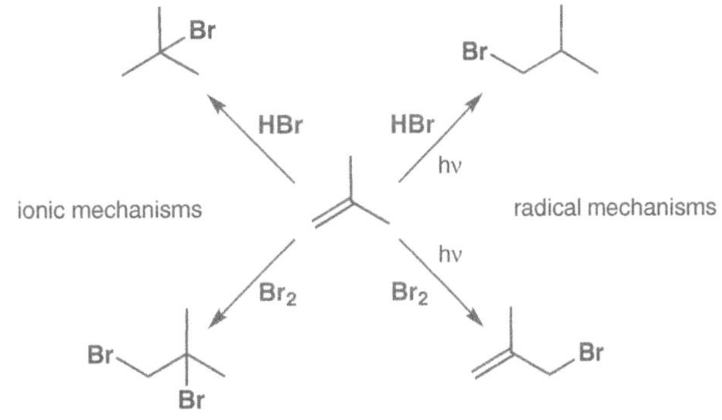

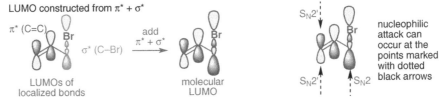

Nucleophilic attack on allylic compounds

The allylic bromides that can be made by these radical reactions display interesting regio-selectivity. We shall start with some substitution reactions with which you are familiar from Chapter 15. There we said that allyl bromide is about 100 times more reactive towards simple S_N2 reactions than is propyl bromide or other saturated alkyl halides.

The double bond stabilizes the S_N2 transition state by conjugation with the p orbital at the carbon atom under attack. This full p orbital (shown in orange in the diagram below) forms a partial bond with the nucleophile and with the leaving group in the transition state. Any stabilization of the transition state will, of course, accelerate the reaction by lowering the energy barrier.

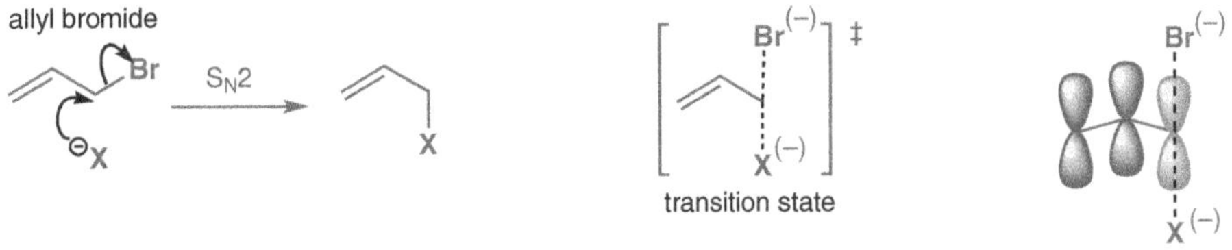

There is an alternative mechanism for this reaction that involves nucleophilic attack on the alkene instead of on the saturated carbon atom. This mechanism leads to the same product and is often called the S_N2' (pronounced 'S-N-two-prime') mechanism.

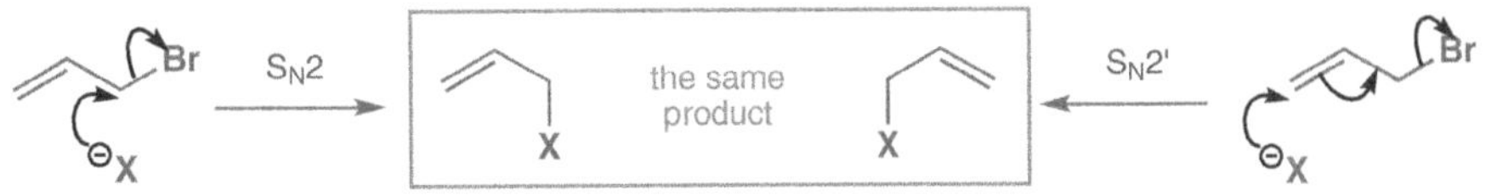

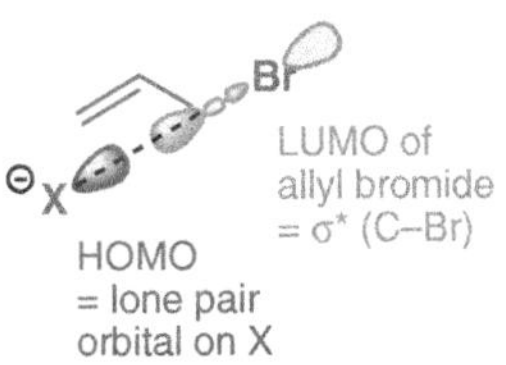

We can explain both mechanisms in a unified way if we look at the frontier orbitals involved. The nucleophile must attack an empty orbital (the LUMO), which we might expect to be simply σ^* (C–Br) for the S_N2 reaction. But this ignores the alkene. The interaction between π^* (C=C) and the adjacent σ^* (C–Br) will as usual produce two new orbitals, one higher and one lower in energy. The lower-energy orbital, $\pi^* + \sigma^*$, will now be the LUMO. To construct this orbital we must put all the atomic orbitals parallel and make the contact between $\pi^* + \sigma^*$ a bonding interaction.

If the allylic halide is unsymmetrically substituted, a question of regioselectivity arises. The products from S_N2 and S_N2' are different and the normal result is that nucleophilic attack occurs at the less hindered end of the allylic system, whether that means S_N2 or S_N2'. This important allylic bromide, known as prenyl bromide, normally reacts entirely via the S_N2 reaction.

prenyl bromide reacts like this   and not like this

The two ends of the allylic system are contrasted sterically: direct (S_N2) attack is at a primary carbon while allylic (S_N2') attack is at a tertiary carbon atom so that steric hindrance favours the S_N2 reaction. In addition, the number of substituents on the alkene product means that the S_N2 product is nearly always preferred—S_N2 gives a trisubstituted alkene while the S_N2' product has a less stable monosubstituted alkene.

An important example is the reaction of prenyl bromide with phenols. This is simply carried out with K_2CO_3 in acetone as phenols are acidic enough ($pK_a \sim 10$) to be substantially deprotonated by carbonate. The product is almost entirely from the S_N2 route, and is used in the Claisen rearrangement (Chapter 35).

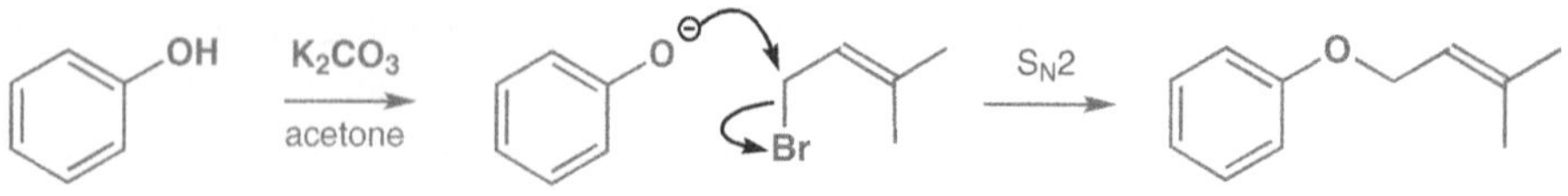

If we make the two ends of the allyl system more similar, say one end primary and one end secondary, things are more equal. We could consider the two isomeric butenyl chlorides.

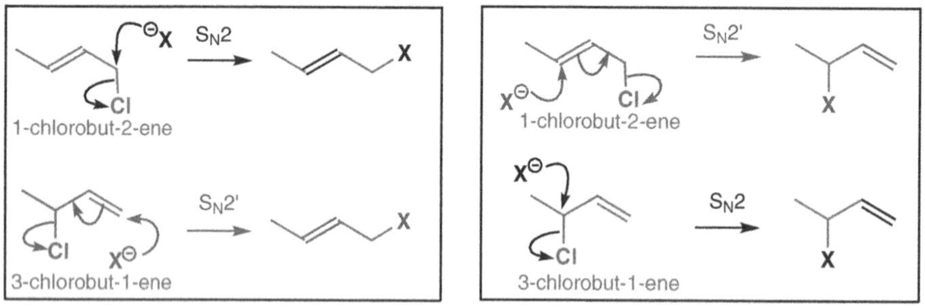

All routes look reasonable, although we might again expect faster attack at the primary carbon. The reactions in the left-hand box are preferred to those in the right-hand box. But there is no special preference for the S_N2 over the S_N2' mechanism or vice versa—the individual case decides. If we react the secondary butenyl chloride with an amine we get the S_N2' mechanism entirely.

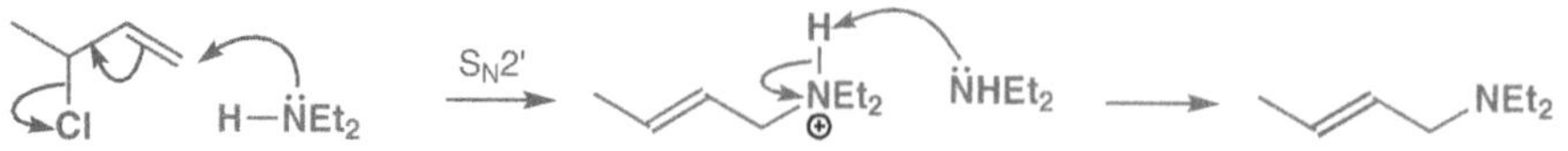

If the primary chloride is used, once again we get nucleophilic attack at the primary centre. The more stable product with the more highly substituted alkene is formed this time by the S_N2 reaction. Here is a slightly more advanced example:

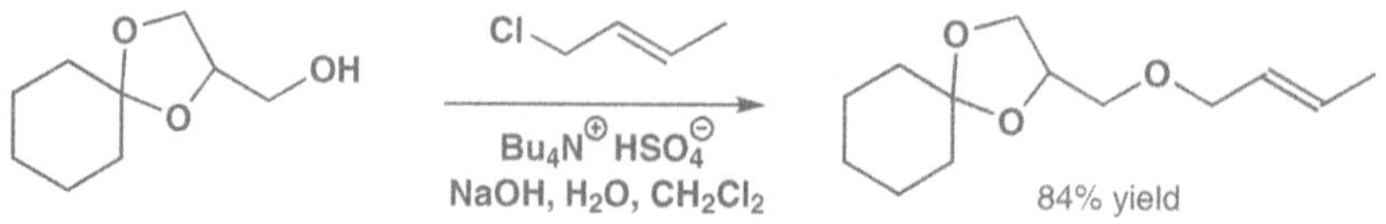

84% yield

■ So far we have used the word 'allyl' to describe these compounds. Strictly, that word applies only to specific compounds $CH_2=CH–CH_2X$ with no substituents other than hydrogen. Allyl is often used loosely to describe any compound with a functional group on the carbon atom *next* to the alkene. We shall use 'allylic' for that and 'allyl' only for the unsubstituted version.

🖱 Interactive mechanisms for various nucleophilic substitutions

🖱 Interactive mechanism for S_N2' nucleophilic substitutions

➡ We explained why adjacent double bonds assist S$_N$2 reactions on p. 341.

Notice that these reactions take place with allylic *chlorides*. We should not expect an alkyl chloride to be particularly good at S$_N$2 reactions as chloride ion is only a moderate leaving group and we should normally prefer to use alkyl bromides or iodides. *Allylic* chlorides are more reactive because of the alkene. Even if the reaction occurs by a simple S$_N$2 mechanism without rearrangement, the alkene is still making the molecule more electrophilic.

You might ask a very good question at this point. How do we know that these reactions really take place by S$_N$2 and S$_N$2' mechanisms and not by an S$_N$1 mechanism via the stable allyl cation? Well in the case of prenyl bromide, we don't! In fact, we suspect that the cation probably *is* an intermediate because prenyl bromide and its allylic isomer are in rapid equilibrium in solution at room temperature.

primary isomer
prenyl bromide
>99%

alternative representations:

tertiary isomer
<1%

The equilibrium is entirely in favour of prenyl bromide because of its more highly substituted double bond. Reactions on the tertiary allylic isomer are very likely to take place by the S$_N$1 mechanism: the cation is stable because it is tertiary and allylic and the equilibration tells us it is already there. Even if the reactions were bimolecular, no S$_N$2' mechanism would be necessary for the tertiary bromide because it can equilibrate to the primary isomer more rapidly than the S$_N$2 or S$_N$2' reaction takes place.

Even the secondary system we also considered is in rapid equilibrium when the leaving group is bromide. This time both allylic isomers are present, and the primary allylic isomer (known as crotyl bromide) is an *E/Z* mixture. The bromides can be made from either alcohol with HBr and the same ratio of products results, indicating a common intermediate in the two mechanisms. You saw at the beginning of Chapter 15 that this reaction is restricted to alcohols that can react by S$_N$1.

80–87%; ca. 4:1 *E:Z*

13–20%

same products from both
starting materials

Displacement of the bromide by cyanide ion, using the copper(I) salt as the reagent, gives a mixture of nitriles in which the more stable primary nitrile predominates even more. These can be separated by a clever device. Hydrolysis in concentrated HCl is successful with the predominant primary nitrile but the more hindered secondary nitrile does not hydrolyse. Separation of compounds having two different functional groups is easy: in this case the acid can be extracted into aqueous base, leaving the neutral nitrile in the organic layer.

80–87%

13–20%

CuCN

92%

8%

conc. HCl

70% yield

Once again, we do not know for sure whether this displacement by cyanide goes by the S$_N$1 or S$_N$2' mechanism, as the reagents equilibrate under the reaction conditions. However, the chlorides do *not* equilibrate and so, if we want a clear-cut result on a single well-defined starting material, the chlorides are the compounds to use. But you already see that regioslectivity with allylic compounds may depend on steric hindrance, rates of reaction, and stability of the product.

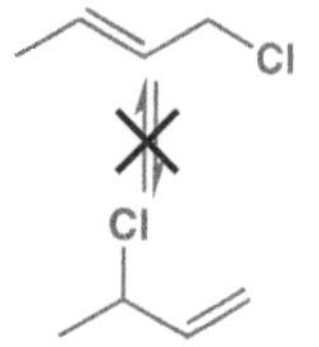

Regiospecific preparation of allylic chlorides

Allylic alcohols are good starting materials for making allylic compounds with control over where the double bond and the leaving group will be. Allylic alcohols are easily made by addition of Grignard reagents or organolithium compounds to enals or enones (Chapter 9) or by reduction of enals or enones (Chapter 23). More to the point, they do not equilibrate except in strongly acidic solution, so we know which allylic isomer we have.

allylic *chlorides*
do not equilibrate

■ By analogy with *stereospecific*, we can define *regiospecific* to mean a reaction where the regiochemistry (that is, the location of the functional groups) of the product is determined by the regiochemistry of the starting material.

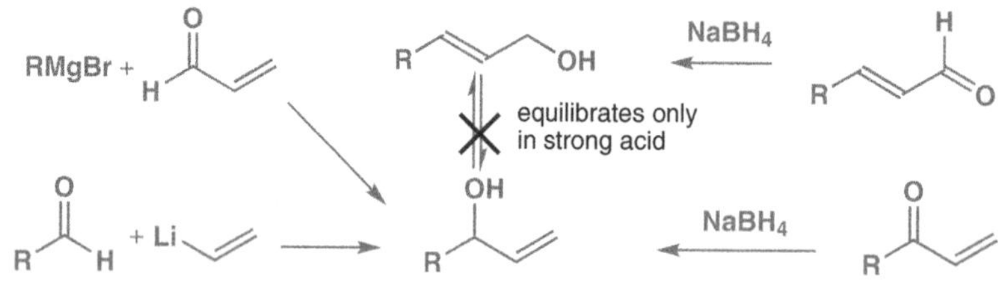

Conversion of the alcohols into the chlorides is easier with the primary than with the secondary alcohols. We need to convert OH into a leaving group and provide a source of chloride ion to act as a nucleophile. One way to do this is with methanesulfonyl chloride ($MeSO_2Cl$) and LiCl.

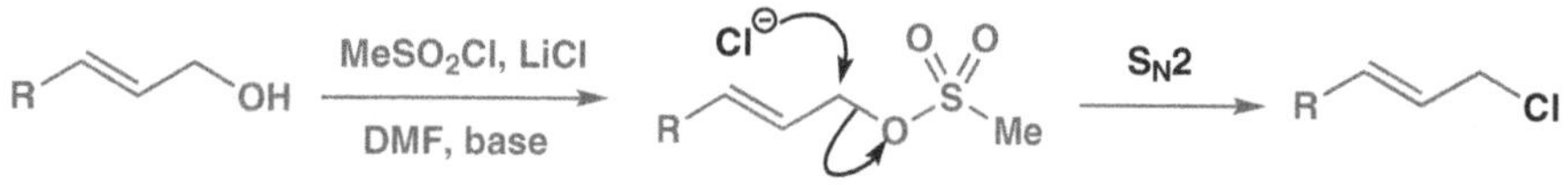

This result hardly looks worth reporting and, anyway, how do we know that equilibration or S_N1 reactions aren't happening? Well, here the mechanism must be S_N2 because the corresponding *Z*-allylic alcohol preserves its alkene configuration. If there were equilibration of any sort, the *Z*-alkene would give the *E*-alkene because *E*- and *Z*-allylic cations are not geometrically stable.

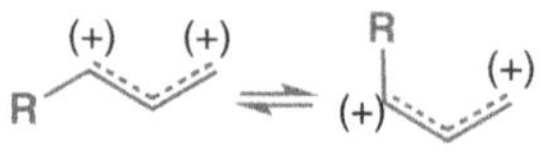

no cation can be involved because *E* and *Z* carbocations are in rapid equilibrium

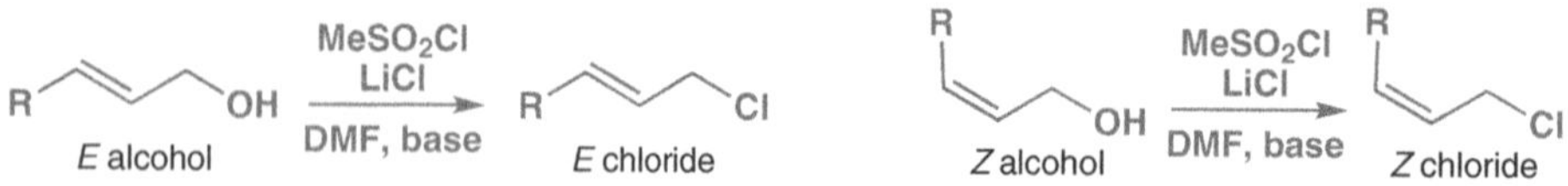

Sadly, this method fails to preserve the integrity of the secondary allylic alcohol, which gives a mixture of allylic chlorides.

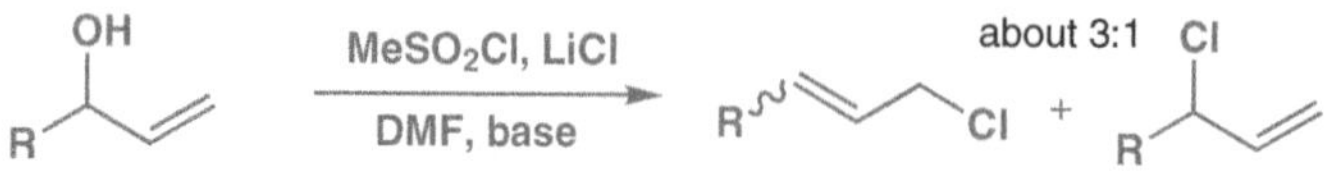

Reliable clean S_N2 reactions with secondary allylic alcohols can be achieved only with Mitsunobu chemistry. Here is a well-behaved example with a *Z*-alkene. The reagents have changed since your last encounter with a Mitsunobu-type reaction: instead of DEAD and a carboxylic acid we have hexachloroacetone, with, of course, triphenylphosphine.

➡ The Mitsunobu reaction was discussed in Chapter 15, p. 349. Mitsunobu chemistry involves using a phosphorus atom to remove the OH group, after the style of PBr_3 as a reagent to make alkyl bromides from alcohols.

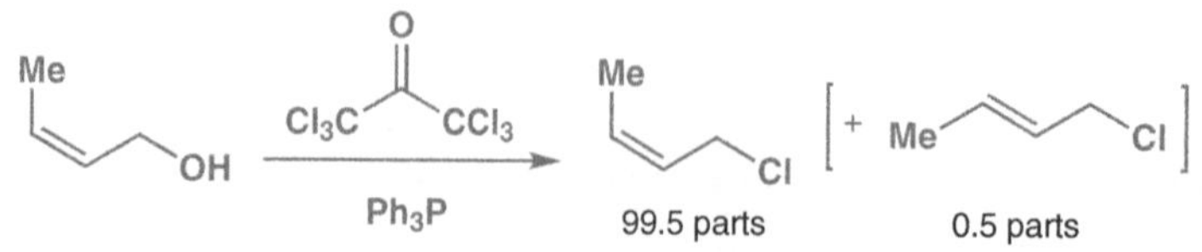

The first thing that happens is that the lone pair on phosphorus attacks one of the chlorine atoms in the chloroketone. The leaving group in this S_N2 reaction at chlorine is an enolate, which is a basic species and can remove the proton from the OH group in the allylic alcohol.

■ Phosphorus doing a substitution at a C–Cl bond the wrong way round? But P is soft, so it cares little about the polarization of the bond, only about the energy of the C–Cl σ*. The energy is the same whichever end of the bond is attacked. You may see similar reactions of PPh_3 with CBr_4 or CCl_4: all produce stabilized carbanions.

Now the alkoxide anion can attack the positively charged phosphorus atom. This is a good reaction in two ways. First, there is the obvious neutralization of charge and, second, the P–O bond is very strong.

■ We looked at the converse— 'loose' S_N2 transition states with considerable S_N1 character—in the reactions of bromonium ions and protonated epoxides in Chapter 17.

The next step is a true S_N2 reaction at carbon as the very good leaving group is displaced. The already strong P–O single bond becomes an even stronger P=O double bond to compensate for the loss of the strong C–O single bond. There is obviously no S_N1 component in this displacement (otherwise the Z-alkene would have partly isomerized to the E-alkene) and very little S_N2' presumably as only 0.5% of the rearrangement product is formed. These displacements of Ph_3P=O are often the 'tightest' of S_N2 reactions.

Now for the really impressive result. Even if the alcohol is secondary, and the rearranged product would be thermodynamically more stable, very little of it is formed and almost all the reaction is clean S_N2.

S_N2 preferred to S_N2'

There is a bit more rearrangement than there was with the other isomer but that is only to be expected. The very high proportion of direct S_N2 product shows that there is a real preference for the S_N2 over the S_N2' reaction in this displacement.

Now that we know how to make allylic chlorides of known structure—whether primary or secondary—we need to discover how to replace the chlorine with a nucleophile with predictable regioselectivity. We have said little so far about carbon nucleophiles (except cyanide ion) so we shall concentrate on simple carbon nucleophiles in the S_N2' reaction of allylic chlorides.

The S_N2' reaction of carbon nucleophiles on allylic chlorides

Ordinary carbon nucleophiles such as cyanide or Grignard reagents or organolithium compounds fit the patterns we have described already. They usually give the more stable product by S_N2 or S_N2' reactions depending on the starting material. If we use copper compounds, there is a tendency—no more than that—to favour the S_N2' reaction. You will recall that copper(I) was the metal we used to ensure conjugate addition to enones (Chapter 22) and its use in S_N2' reactions is obviously related. Simple alkyl copper reagents (RCu, known as Gilman reagents) generally favour the S_N2' reaction but we can do much better by using RCu complexed with BF_3.

➡ The nature of metal–alkene complexes is discussed in Chapter 40.

The copper must complex to the alkene and then transfer the alkyl group to the S_N2' position as it gathers in the chloride. This might well be the mechanism, although it is often difficult to draw precise mechanisms for organometallic reactions.

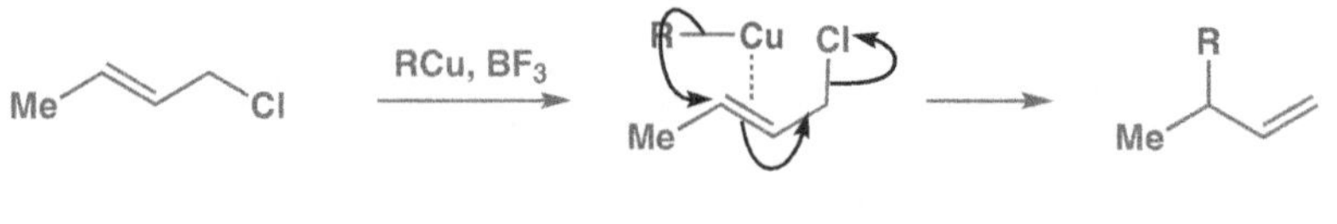

The secondary allylic isomer also gives almost entirely the rearranged product. This is perhaps less surprising, as the major product is the more stable isomer, but it means that either product can be formed in high yield simply by choosing the right (or should we say *wrong*, since there is complete allylic rearrangement during the reaction) isomer. The reaction is *regiospecific*.

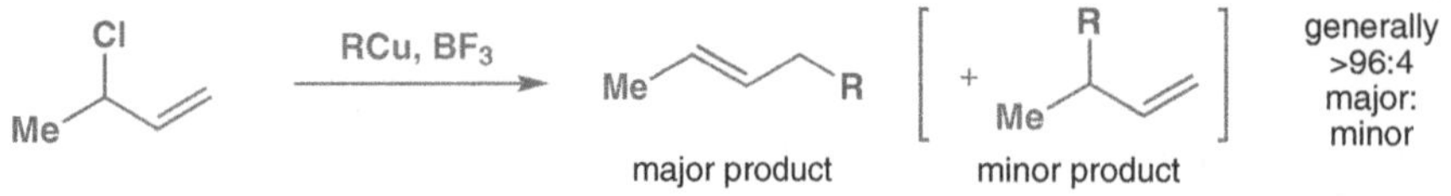

The most remarkable result of all is that prenyl chloride gives rearranged products in good yield. This is about the only way in which these compounds suffer attack at the tertiary centre by S_N2' reaction when there is the alternative of an S_N2 reaction at a primary centre.

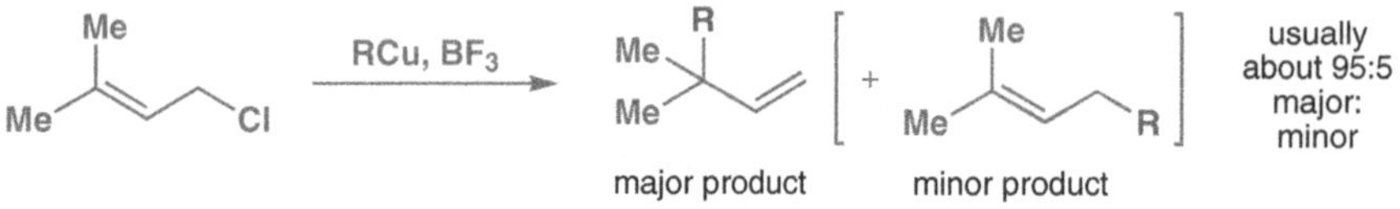

Electrophilic attack on conjugated dienes

Another way to make allylic chlorides is by treating dienes with HCl. Electrophiles attack conjugated dienes more readily than they do isolated alkenes. There was some discussion of this in Chapter 19, establishing the main point that the terminal carbon atoms are the most nucleophilic and that the initial attack produces an allylic cation. A simple example is the addition of HCl to cyclopentadiene.

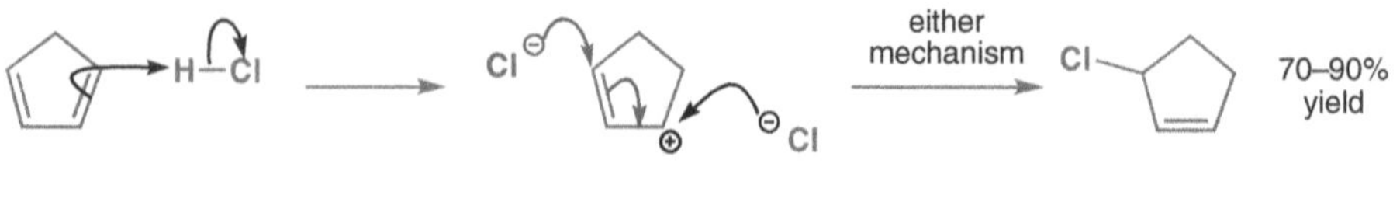

Although there is a question of regioselectivity in the initial protonation, the allylic cation is symmetrical and attack by chloride at either end produces the same product. However, if the electrophile is a halogen rather than HCl or HBr then the reaction becomes regioselective as the cationic intermediate is no longer symmetrical. What happens is this:

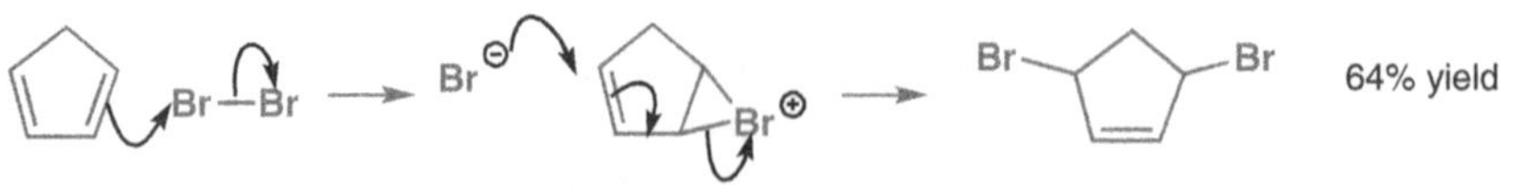

The alternative is direct attack on the bromonium ion intermediate, which we assume would occur at the allylic site (black arrows) and not at the other (green arrows). Although this 1,2-dibromide product is not observed, it is still possible that this reaction happens because the 1,2-product can rearrange by bromide shift to the observed 1,4-dibromide.

➡ By 'bromide shift' we mean the reversible isomerization of allylic bromides you saw on p. 576.

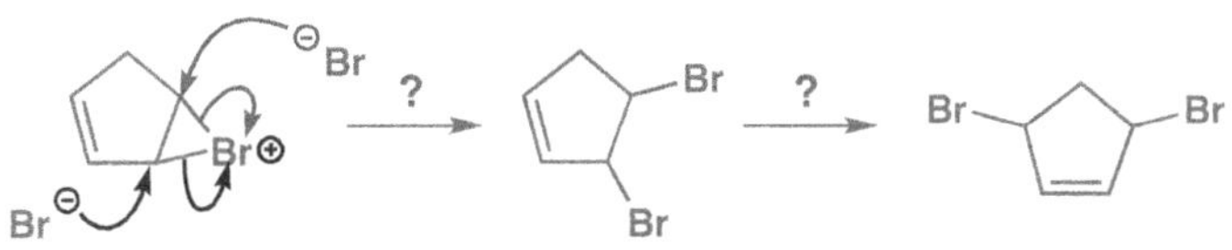

The final product of this reaction could in fact be either of two compounds as the two bromine atoms may be *cis* or *trans*. Bromination in chloroform at –20 °C gives mostly a liquid *cis* dibromide while reaction in hydrocarbon solvents gives the crystalline *trans* isomer. On standing the *cis* isomer slowly turns into the *trans*.

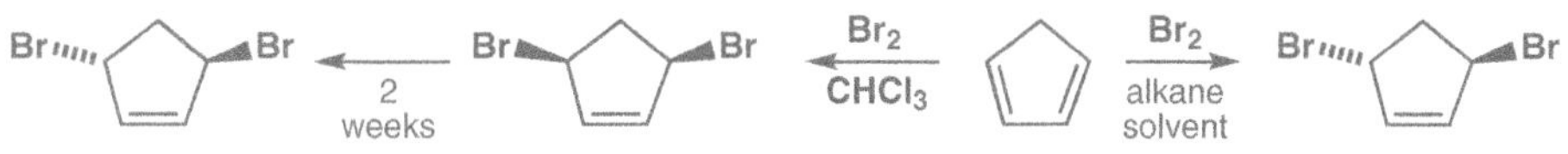

This suggests that the *cis* bromide is the kinetic product and the more stable *trans* compound is the thermodynamic product, formed by reversible loss of bromide and reformation of the bromonium ion.

Similar questions arise when nucleophilic substitution occurs on the dibromides. Reaction of either the *cis* or the *trans* dibromide with dimethylamine gives the *trans* isomer of a diamine. But look at the regioselectivity—it's not the diamine you might expect. The only explanation is one S_N2 displacement and one S_N2' displacement.

this regioisomer not formed

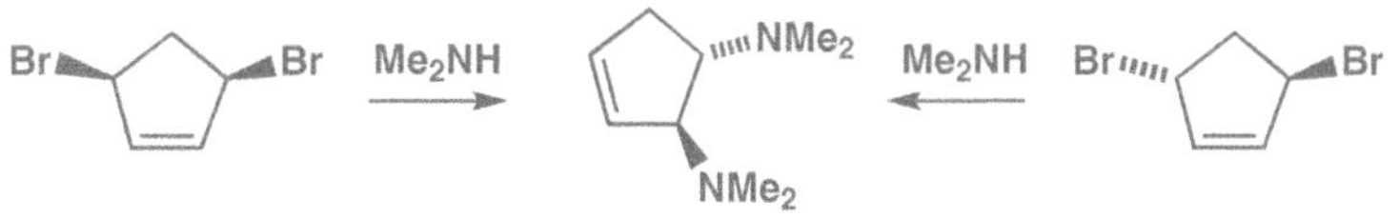

But what about the stereochemistry? Starting with the *cis* isomer, one S_N2 displacement with inversion might be followed by an intramolecular S_N2' displacement and finally another S_N2 displacement with inversion at the allylic centre.

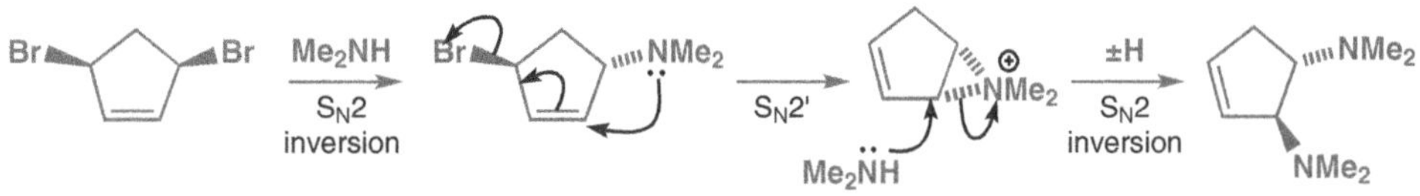

The reaction with the *trans* isomer is almost identical: the same three-membered ring is an intermediate in both sequences so the products are bound to be the same.

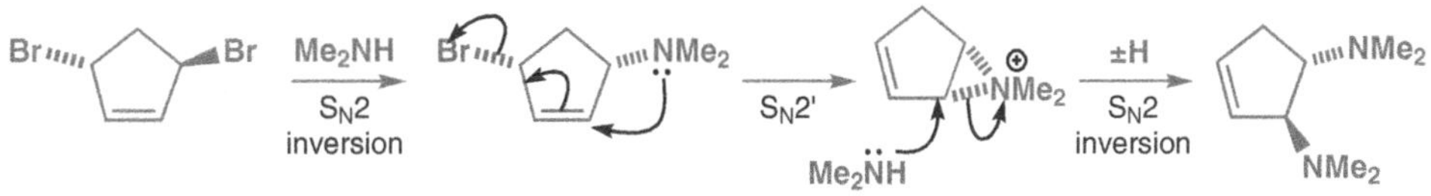

If the nucleophile is different from the electrophile we can get a bit more information about the course of the reaction. When butadiene is treated with bromine in methanol as solvent, two adducts are formed in a 15:1 ratio along with some dibromide. Methanol is a weak nucleophile and adds to the bromonium ion mainly at the allylic position (black arrow below); only a small amount of product is formed by attack at the far end of the allylic system. Note that no attack occurs at the other end of the bromonium ion (green dotted arrow).

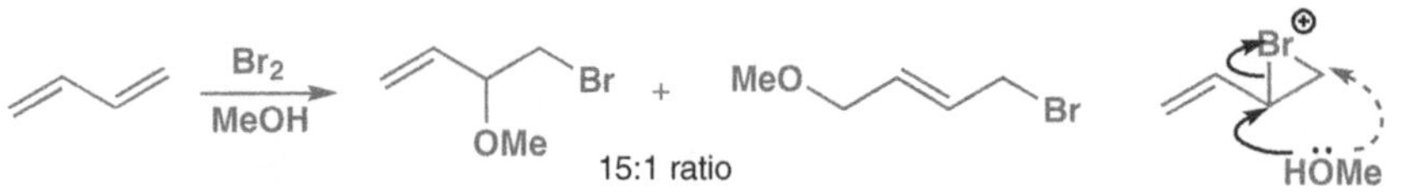

Conjugate addition

In Chapter 22 we devoted considerable space to discussing conjugate addition and the reasons why some reactions occur by direct attack on the carbonyl group of an α,β-unsaturated carbonyl compound and why others occur by conjugate addition. We shall briefly revise the regioselectivity aspects of these reactions.

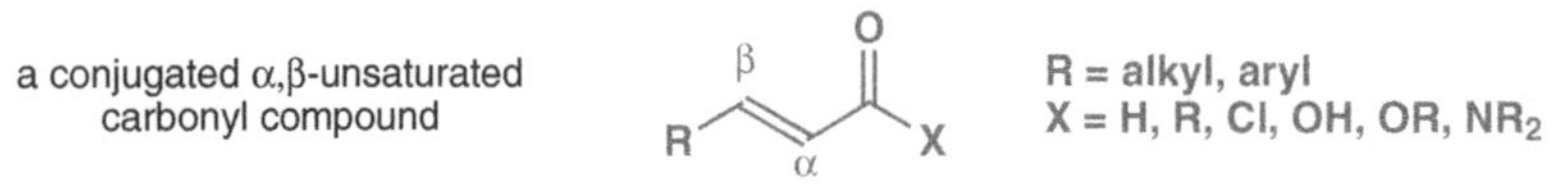

a conjugated α,β-unsaturated carbonyl compound

R = alkyl, aryl
X = H, R, Cl, OH, OR, NR$_2$

Direct (or 1,2) addition means that the nucleophile attacks the carbonyl group directly. An addition compound is formed which may lose X$^-$, if it is a leaving group, or become protonated to give an alcohol.

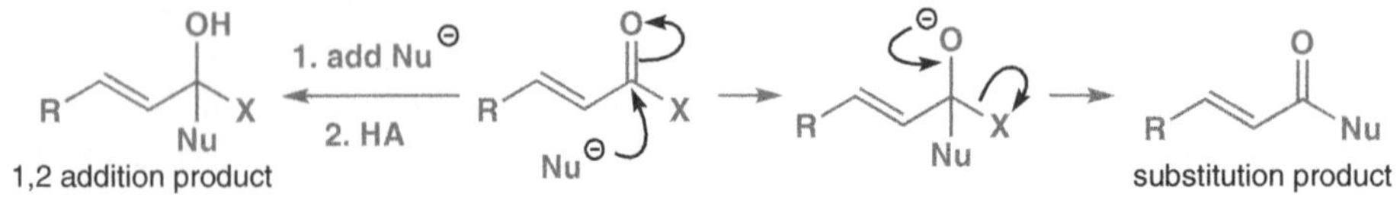

Conjugate (or 1,4) addition means that the nucleophile adds to the end of the alkene furthest from the carbonyl group. The electrons move through into the carbonyl group to produce an enolate anion that usually becomes protonated to give a ketone.

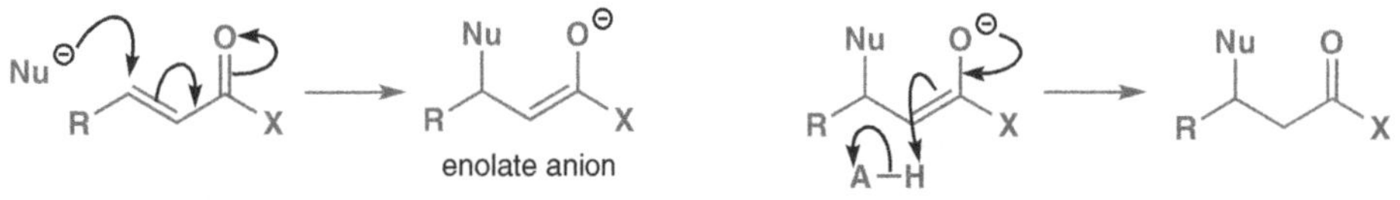

The first difference between the two routes is that the product from direct addition keeps the alkene but loses the carbonyl group while conjugate addition keeps the carbonyl group but loses the alkene. As a C=O π bond is stronger than a C=C π bond, **conjugate addition gives the thermodynamic product**. But as the carbonyl group is more electrophilic than the far end of the alkene, especially to charged, hard nucleophiles, **direct addition gives the kinetic product**. So direct addition is favoured by low temperatures and short reaction times while conjugate addition is favoured by higher temperatures and longer reaction times, provided the 1,2 addition is reversible.

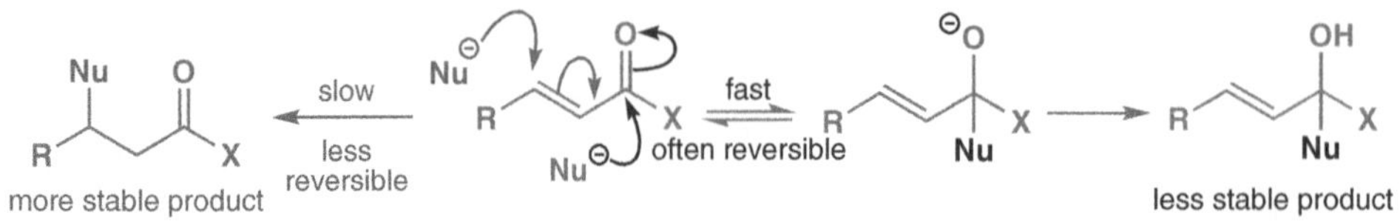

The second difference depends on how electrophilic is the α,β-unsaturated carbonyl compound. The more electrophilic such as aldehydes and acid chlorides tend to prefer direct addition while the less electrophilic such as ketones or esters tend to prefer conjugate addition.

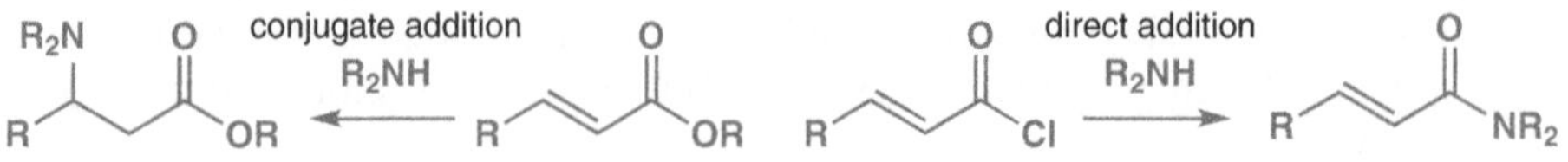

It is similar with the choice of nucleophile: more nucleophilic species, such as MeLi or Grignard reagents, prefer direct addition, particularly as they react irreversibly, while less nucleophilic species like amines and thiols prefer conjugate addition. These nucleophiles add reversibly to the C=O group, giving an opportunity for any direct addition product to revert to starting materials and react again.

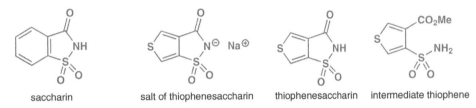

Regioselectivity in action

We finish with an example that illustrates several aspects of chemoselectivity as well as introducing the subjects of the next two chapters. The first synthetic sweetener was saccharin but newer ones such as the BASF compound thiophenesaccharin are much in demand. The sodium salt is the active sweetener but the neutral compound has to be made via the simpler intermediate thiophene.

■ Thiophene is the name for this sulfur-containing aromatic compound. There is more about it in Chapter 29.

thiophene

saccharin   salt of thiophenesaccharin   thiophenesaccharin   intermediate thiophene

The synthesis started with a conjugate addition of a thiol to an unsaturated ester. The thiol is obviously the nucleophile and regioselectively chooses conjugate addition rather than attack on either ester group.

■ The thiol *could* attack its own ester group, leading to polymerization, but it doesn't.

MeO₂C⟍SH   ⟍CO₂Me   ⟶   MeO₂C⟍S⟍CO₂Me   85% yield

In the next step the diester is treated with base and a carbonyl condensation reaction occurs of the type you will meet in Chapter 26. There is a real question of regioselectivity here: an enolate could form next to either ester (as shown by the orange circles) and would then attack the other ester as a nucleophile. There is little to choose between these alternatives but the first was wanted and was selected by careful experimentation, although only in 50% yield. This was acceptable on a large scale as the product could be separated by crystallization, the most practical of all methods.

Reactions such as this—the attack of enolates on carbon electrophiles—form the subject of the next two chapters, where we will discuss in detail the mechanism of this type of reaction.

Further reading

There is a basic introduction in S. Warren and P. Wyatt, *Organic Synthesis: the Disconnection Approach*, Wiley, Chichester, 2008, chapter 3.

Ortholithiation: P. Wyatt and S. Warren, *Organic Synthesis: Strategy and Control*, Wiley, Chichester, 2007. J. Clayden, *Organolithiums: Selectivity for Synthesis*, Pergamon, 2002.

Reduction of nitro groups: L. McMaster and A. C. Magill, *J. Am. Chem. Soc.*, 1928, **50**, 3038. Bromination of nitrobenzene: B. S. Furniss, A. J. Hannaford, P. W. G. Smith, and A. T. Tatchell, *Vogel's Textbook of Practical Organic Chemistry*, Longman, 5th edn, 1989, p. 864.

Formation of diazonium salts and conversion into aryl halides: B. S. Furniss, A. J. Hannaford, P. W. G. Smith, and A. T. Tatchell, *Vogel's Textbook of Practical Organic Chemistry*, Longman, 5th edn, 1989, pp. 933, 935. Iodolactonisation: B. S. Furniss, A. J. Hannaford, P. W. G. Smith, and A. T. Tatchell, *Vogel's Textbook of Practical Organic Chemistry*, Longman, 5th edn, 1989, p. 734.

The Wittig-style Horner-Wadworth-Emmons alkene synthesis: W. S. Wadsworth and W. D. Emmons, *Org. Synth. Coll.*, 1973, **5**, 547. P. Wyatt and S. Warren, *Organic Synthesis: Strategy and Control*, Wiley, Chichester, 2007. Synthesis of non-conjugated compounds: C. W. Whitehead, J. J. Traverso, F. J. Marshall, and D. E. Morrison, *J. Org. Chem*, 1961, **26**, 2809.

Regioslective electrophilic attack on dienes: R. B. Moffett, *Org. Synth. Coll.*, 1963, **4**, 238. K. Nakayama, S. Yamada, H. Takayama, Y. Nawata, and Y. Itaka, *J. Org. Chem.*, 1984, **49**, 1537.

Buffered epoxidation to avoid rearrangement of product: M. Imuta and H. Ziffer, *J. Org. Chem.*, 1979, **44**, 1351. Mono- and di-epoxidation of dienes: M. A. Hashem, E. Manteuffel, and P. Weyerstahl, *Chem. Ber.*, 1985, **118**, 1267.

Regioselective bromination of dienes: A. T. Blomquist and W. G. Mayes, *J. Org. Chem.*, 1945, **10**, 134. Regioselective nucleophilic substitution on allylic bromides: A. C. Cope, L. L. Estes, J. R. Emery, and A. C. Haven, *J. Am. Chem. Soc.*, 1951, **73**, 1199. V. H. Heasley and P. H. Chamberlain, *J. Org. Chem.*, 1970, **35**, 539. But ignore the theoretical part especially the three 'different' intermediates.

Check your understanding

 To check that you have mastered the concepts presented in this chapter, attempt the problems that are available in the book's Online Resource Centre at http://www.oxfordtextbooks.co.uk/orc/clayden2e/

Suggested solutions for Chapter 24

24

PROBLEM 1

Two routes are proposed for the preparation of this amino-alcohol. Which do you think is more likely to succeed and why?

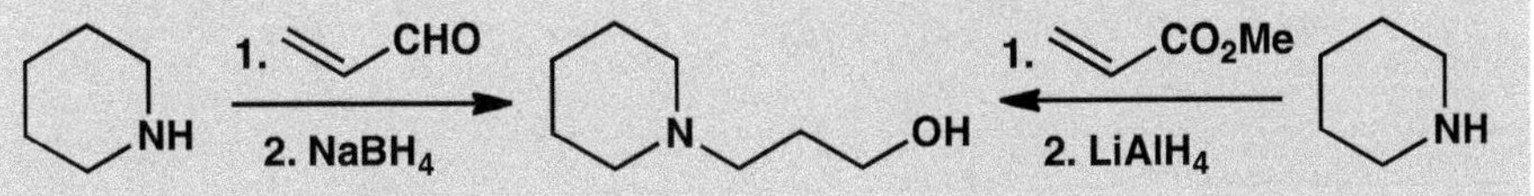

Purpose of the problem

Practical application of the choice of reagent to ensure the correct regioselectivity in a conjugate addition.

Suggested solution

Either route might give the product but enals are more likely to undergo direct addition to the carbonyl group rather than conjugate addition while conjugated esters are better at conjugate addition. So the ester is probably better.

■ See pages 505–510 and 581–582 in the textbook.

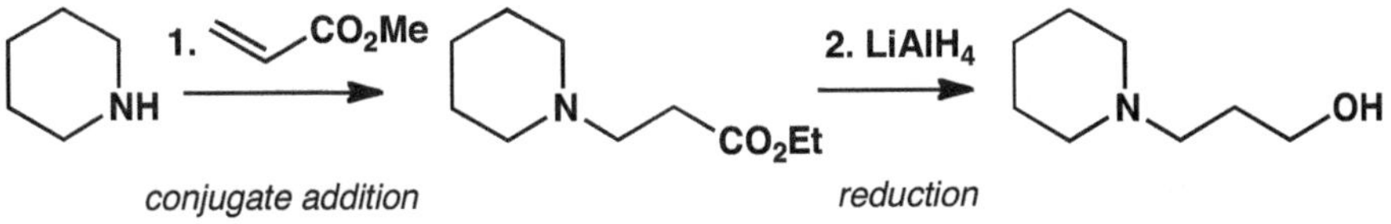

conjugate addition   reduction

PROBLEM 2

Predict the products of these reactions.

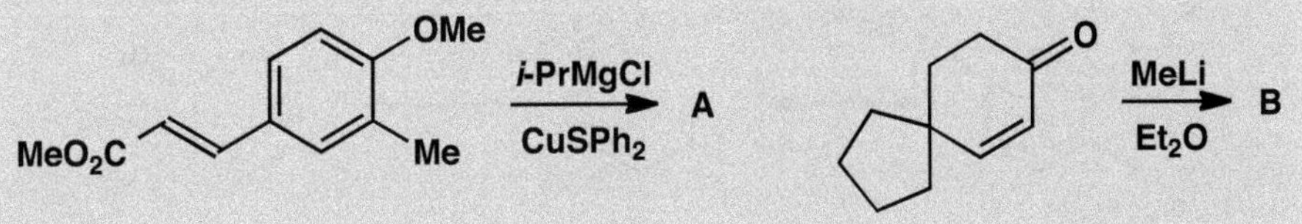

Purpose of the problem

Practice at predicting the regioselectivity of a direct or conjugate addition.

Suggested solution

Both reactions involve addition of organometallic compounds to unsaturated carbonyl compounds. The key difference is the metal. With Cu(I) as catalyst, the Grignard reagent will give conjugate addition in the first case. MeLi will give direct addition in the second.

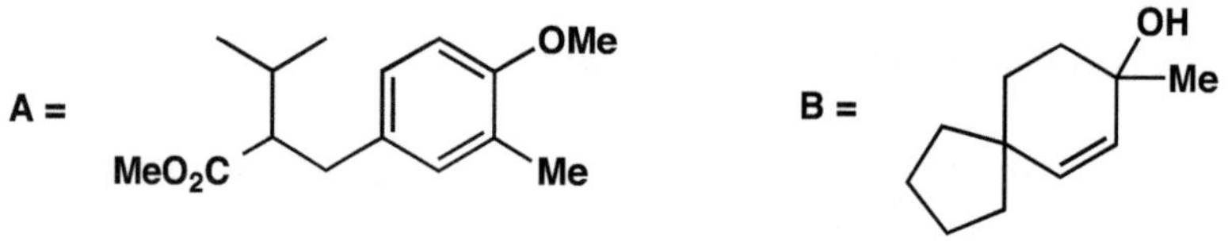

PROBLEM 3

Explain the different regioselectivity in these two brominations of 1,2-dimethylbenzene.

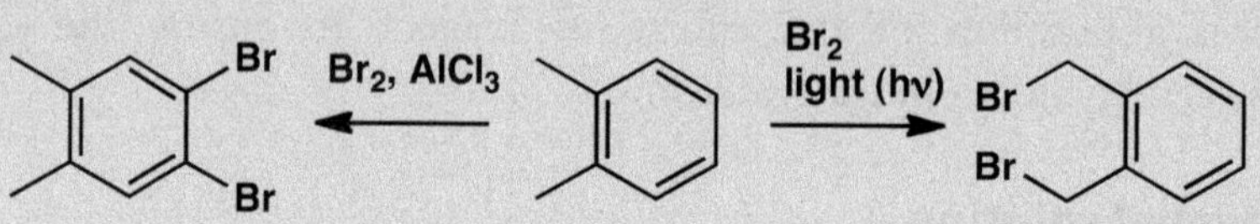

Purpose of the problem

Regioselectivity in electrophilic aromatic substitution and in radical substitution.

Suggested solution

■ Make sure you understand *why* methyl groups direct *ortho, para*: if you need reminding, see p. 484 of the textbook.

$AlCl_3$ is a commonly used Lewis acid in electrophilic aromatic substitution reactions. Here it activates the bromine to form the electrophile 'Br^+', which attacks the aromatic ring. Methyl groups are *ortho,para* directors, so any of the four unsubstituted positions could be attacked, but steric hindrance directs the first bromine to go to one of the positions that does not lead to a 1,2,3-trisubstituted ring.

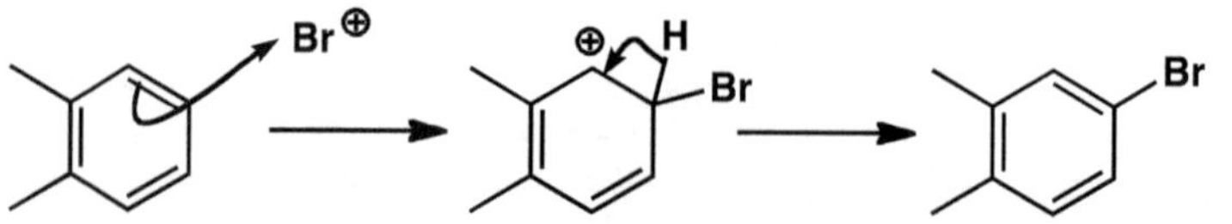

Now we have three *ortho,para* directors, and bromine (with its lone pairs) is the strongest, so the next bromine will go *ortho* to the bromine in the less sterically hindered of the two possibilities.

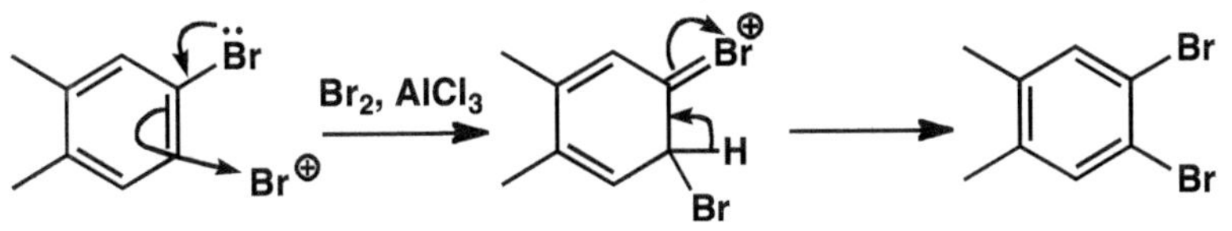

In the presence of light, bromine's weak Br–Br bond undergoes homolysis, and Br• radicals are formed. One of these can abstract a hydrogen atom, breaking the weakest C–H bond. The methyl groups' C–H bonds are weaker than those of the phenyl ring because the benzyl radical that forms is delocalized into the aromatic ring. The benzyl radical attacks another molecule of bromine, and the cycle continues.

■ A similar argument was used on p. 573 of the textbook to explain why Br• abstracts H from an allylic, rather than a vinylic, position.

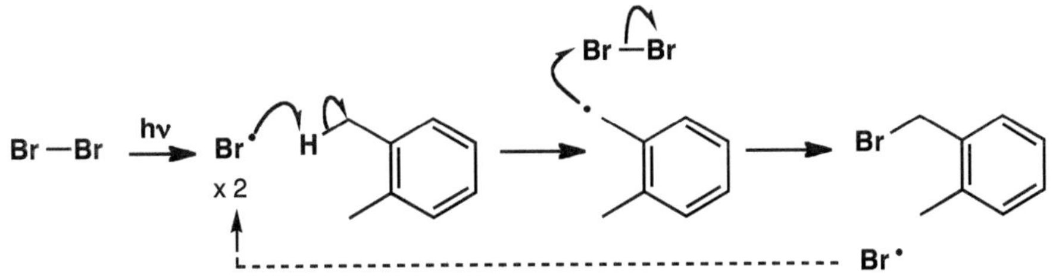

The mechanism is shown here for the first bromination; the same thing can happen on the other methyl group.

PROBLEM 4

The nitro compound below was needed for the synthesis of an anti-emetic drug. It was proposed to make it by nitration of the hydrocarbon shown. How successful do you think this would be?

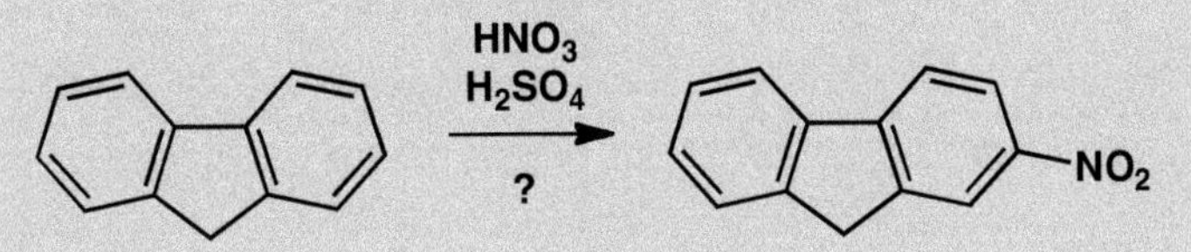

Purpose of the problem

Predicting regioselectivity in electrophilic aromatic substitution where directing effects are more subtle.

Suggested solution

The standard conditions for nitration generate the electrophile NO_2^+, and to get the product shown here, this species has to attack the ring as shown below. The intermediate cation looks quite all right, since the positive charge can be delocalized even into the other ring.

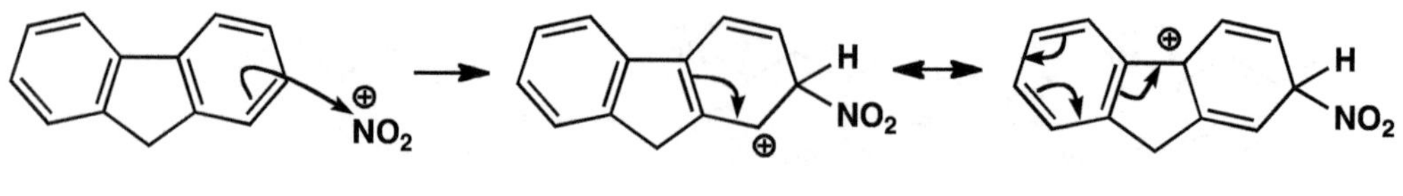

What about the alternatives? A similar cation is formed if the electrophile attacks the position labelled '1', but the nitro group is in a more hindered position here, so we don't expect this to contribute much to the product mixture. Position '2' gives the cation shown below, which although perfectly feasible as an intermediate, does not benefit from the same degree of stabilization as the one in the reaction we want (it can't be delocalized into the other ring). Position 4 is similar but more hindered. Overall we can reasonably expect the reaction to give the product we want.

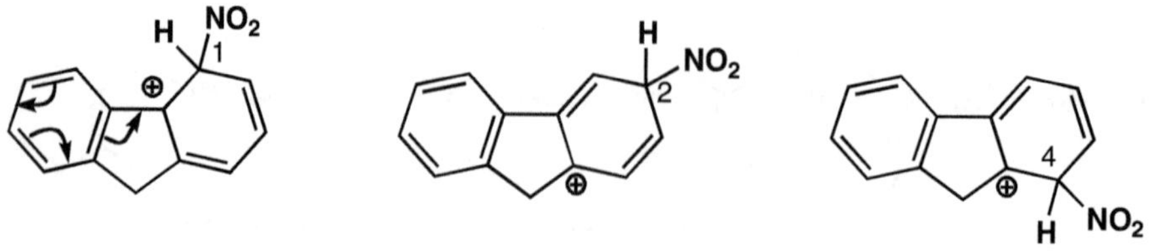

PROBLEM 5

Comment on the regioselectivity and chemoselectivity of the reactions shown below.

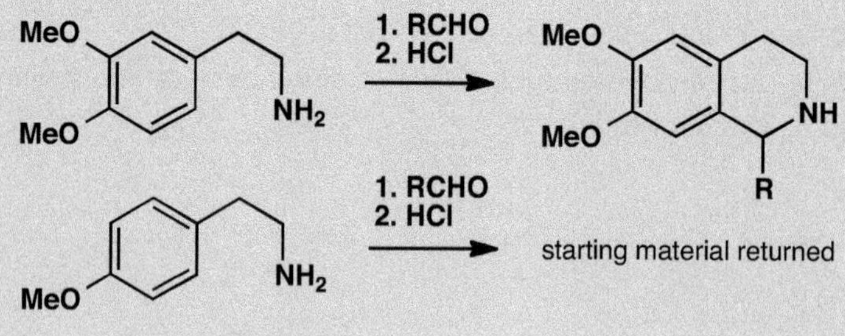

Purpose of the problem

Regioselectivity in electrophilic aromatic substitution with an intramolecular electrophile: an important reaction for making heterocycles.

Suggested solution

The reaction of an aldehyde with an amine gives an imine, and in acid (HCl), protonation gives an iminium ion, the electrophile that attacks the aromatic ring. The iminium ion is tethered to the ring, so it has only two choices of reaction site, since it can't reach any further than the positions *ortho* to the tether. The one it chooses is the less hindered. It is also *para* to an electron-donating methoxy group, so the reaction works well.

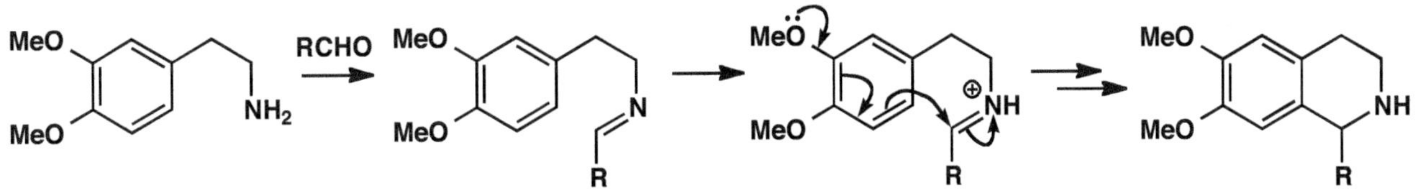

In the second case, there is only one methoxy group, and both the positions *ortho* to the tether are *meta* to it, where it can't activate substitution. The positions *ortho* to itself, where it can activate, are too far away for the iminium to reach, so no substitution takes place. Presumably the iminium ion forms, but it is just hydrolysed back to the aldehyde.

■ This reaction is a useful way of making some important alkaloid natural products (and indeed it mimics the way nature makes them). It is sometimes known as the 'Pictet-Spengler reaction'.

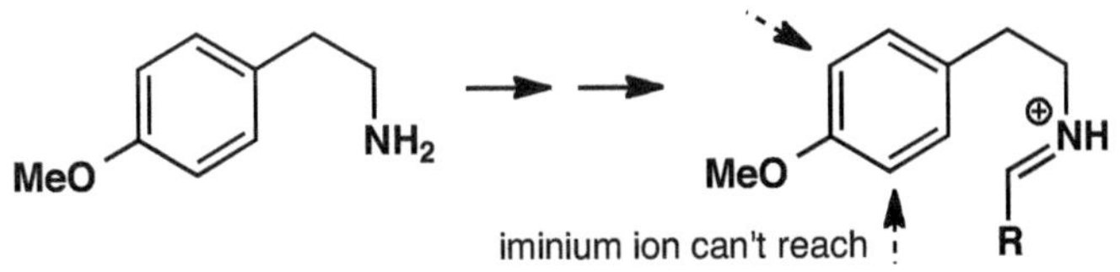

iminium ion can't reach

PROBLEM 6

Identify **A** and **B** and account for the selectivity displayed in this sequence of reactions.

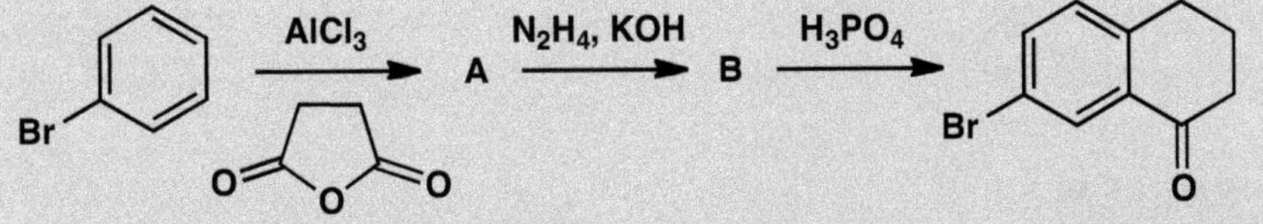

Purpose of the problem

Analysing selectivity in a useful ring-forming sequence.

Suggested solution

The Friedel-Crafts acylation in the first step is controlled by the bromo substituent, which is an *ortho,para* director: here we get *para* selectivity as usual for steric reasons. Work through the mechanism and you find a ketoacid as the product **A**.

■ You have seen this sort of thing in the textbook on pp. 49–4

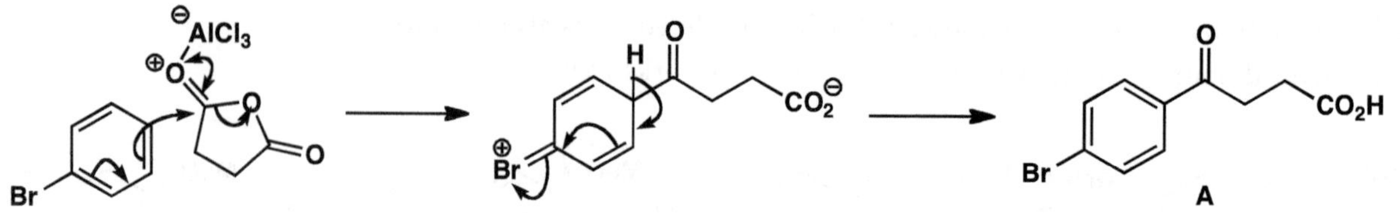

The next step is the reduction we introduced on p. 540 of the textbook, the 'Wolff-Kishner' reduction. The mechanism is there so we need not repeat it here; the product is the acid **B** (or, rather, its potassium salt). Now adding acid forms a ring in another Friedel-Crafts acylation. The electrophile must be the acylium ion: usually Friedel-Crafts acylations need more than just strong acid, but this one is fast because it is intramolecular. What about regioselectivity? Well, the only positions the electrophile can reach are *ortho* to the carbon chain, so it must react there (they are both the same) even though that means it has to attack *meta* to the Br group. It's still *ortho* to the alkyl chain though, which is *ortho,para* directing.

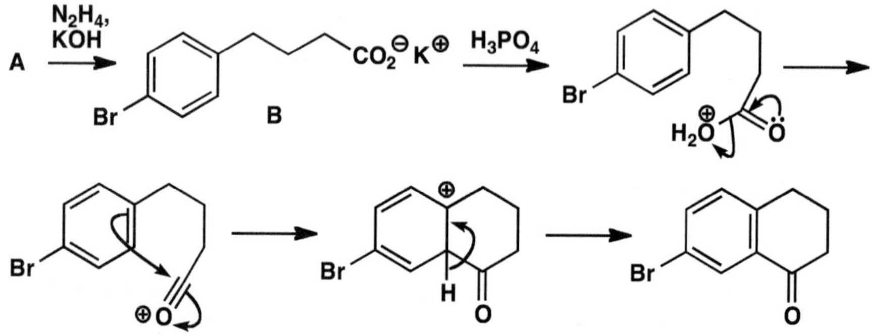

PROBLEM 7

The sequence of reactions below shows the preparation of a compound needed for the synthesis of a powerful anti-cancer compound. Explain the regioselectivity of the reactions. Why do you think two equivalents of BuLi are needed in the second step?

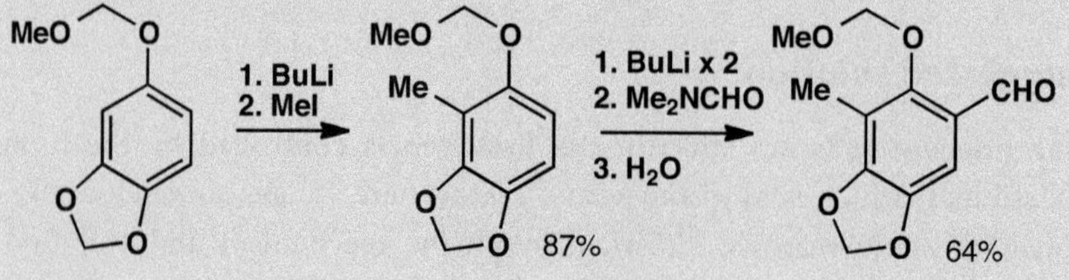

Purpose of the problem

Explaining regioselectivity in ortholithiation reactions.

Suggested solution

Both reactions involve ortholithiation—deprotonation of the aromatic ring to form an intermediate aryllithium. As we explain on pp. 563–4 of the textbook, the deprotonation occurs where the BuLi can be 'guided in' by coordinating oxygen atoms. The methoxymethyl acetal, with its two oxygen atoms, is very good at doing this, so we expect deprotonation at one of the two positions *ortho* to this group. The other acetal is also a complexing group, so the deprotonation happens in between the two oxygen atoms.

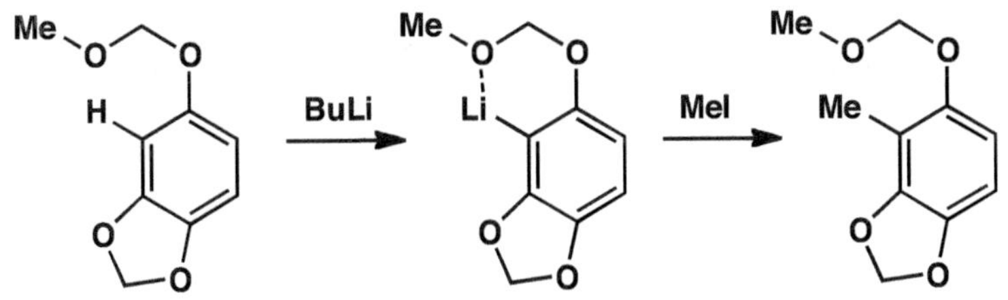

In the second step, deprotonation can again take place next to the methoxymethyl group. Two equivalents of BuLi are needed because the most acidic proton is in fact one of the protons of the methyl group: a benzyllithium forms first, and then a more reactive aryllithium. When the electrophile (DMF) is added, it reacts only with the last formed, more basic anion.

■ Phenyllithiums are more basic than benzyllithiums, because in benzyllithiums the 'anion' is conjugated with the ring; in phenyllithiums the 'anion' is perpendicular to the π system (like the lone pair in pyridine).

■ Selectivity in the reactions of dianions is described on p. 547 of the textbook.

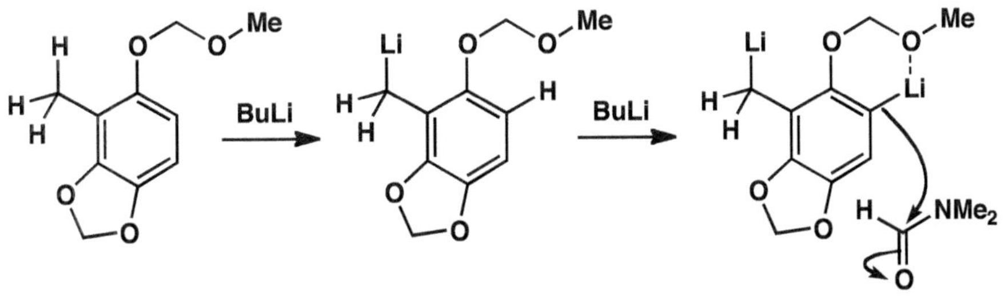

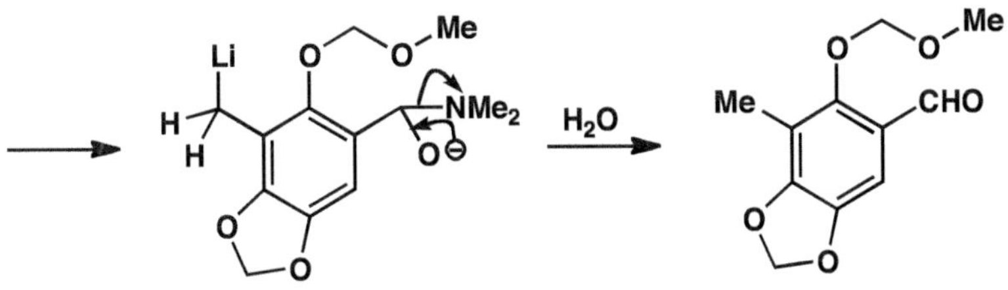

■ This chemistry is from Corey's synthesis of ecteinascidin: E. J. Corey, D. Y. Gin and R. S. Kania, *J. Am. Chem. Soc.* 1996, **118**, 9202; E. J. Martinez and E. J. Corey, *Org. Lett.* 2000, **2**, 993.

PROBLEM 8

Comment on the regioselectivity and chemoselectivity of the reactions in the sequence below.

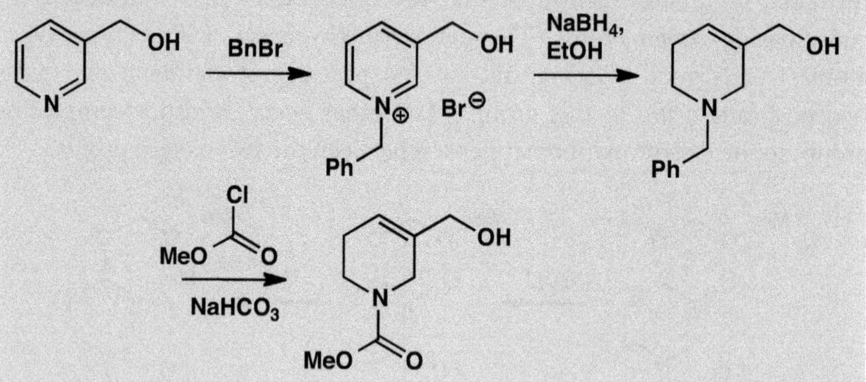

Purpose of the problem

Practice using simple principles of reactivity to explain why nucleophiles and electrophiles choose to react at particular sites, and using your knowledge of reactivity to deduce mechanisms for some unfamiliar reactions.

Suggested solution

Benzyl bromide is a good electrophile and it reacts well with alkoxides to make ethers. With neutral alcohols however the substitution is very slow, so only the more nucleophilic (and more basic) pyridine nitrogen is attacked, to make a pyridinium salt.

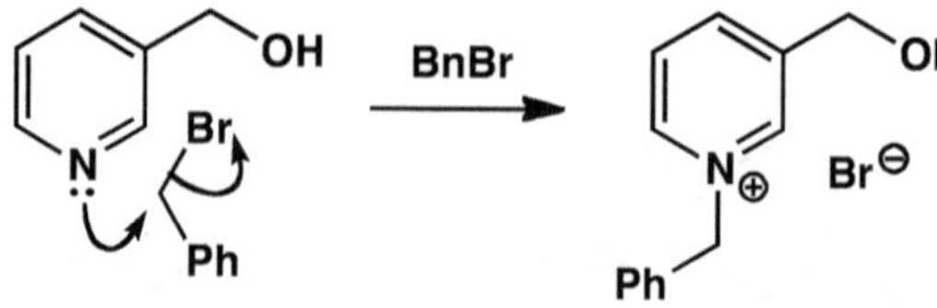

The pyridinium salt is a bit like an iminium ion, so sodium borohydride attacks it at the C=N⁺ bond to make a neutral enamine. Looking at the product, you can see that another reduction must take place as well, but enamines are nucleophilic, so the borohydride can't atttack directly. What must happen instead is that the enamine is protonated to make another iminium, which can then be reduced. The final double bond is safe from attack, since it is an isolated, electron-rich alkene.

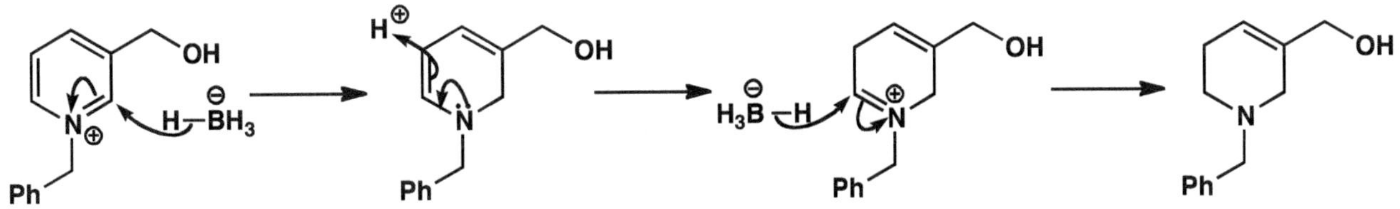

In the final step, another electrophile is added: it's an acid chloride, though a slightly unusual one, usually called 'methyl chloroformate'. As in the first step, the most nucleophilic atom is the pyridine N, so we use its lone pair to attack the carbonyl group and displace chloride. Now we have to lose the benzyl group, but the only reagent we have to help us is the chloride we just lost. Chloride is a nucleophile—a weak one, but powerful enough to attack the cationic species we have just generated. Which is the site most susceptible to nucleophilic substitution? The benzylic carbon, quite reasonably, because of the accelerating effect of the adjacent π system. Nucleophilic substitution here gives the final product.

■ See p. 431 of the textbook for a reminder about the reactivity of benzylic electrophiles.

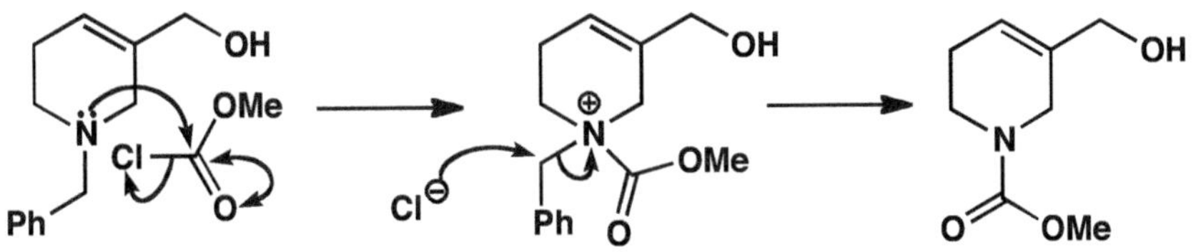

PROBLEM 9

Explain the regioselectivity displayed in this synthesis of the drug tanomastat.

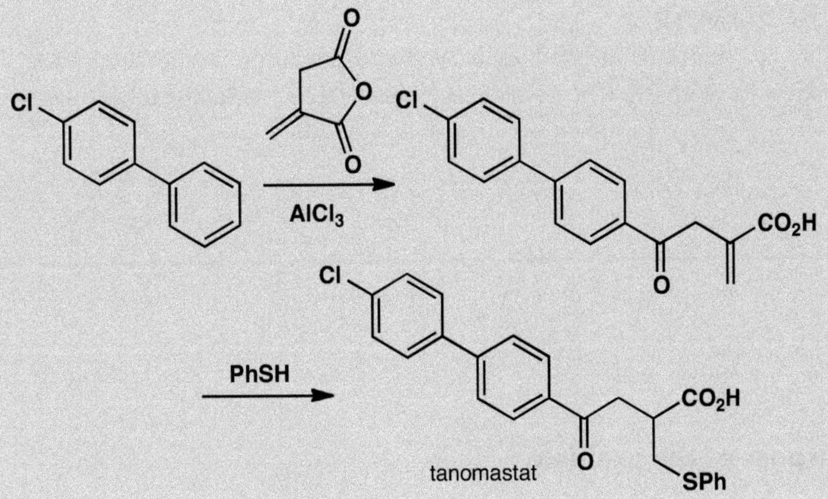

Purpose of the problem

Explaining why moderately complex molecules choose to react selectively.

Suggested solution

The first reaction is a Friedel-Crafts acylation. There are two rings and two carbonyl groups, so we must explain first of all the choice between each of these pairs. One ring is chlorinated: chlorine has a deactivating effect on electrophilic aromatic substitution, so the non-chlorinated ring reacts. The two carbonyls differ in that the top one is (a) less hindered and (b) not conjugated, both of which contribute to its greater reactivity. There is also the question of regioselectivity in the way that the acylation occurs at the *para* position of the non-chlorinated ring. Aryl substituents are *ortho,para* directing, because they can delocalize the positive charge formed from attack in this way; steric factors favour the *para* over the *ortho* positions.

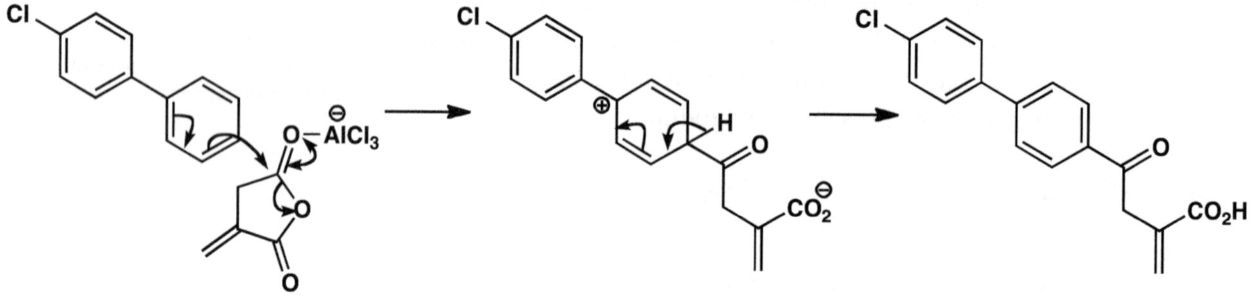

- The effect of halogens on the reactivity of aromatic rings is described on pp. 489–490 of the textbook.

- The synthesis of tanomastat, a protease inhibitor, is from US patent 5,789,434.

In the second step, thiophenol gives the conjugate addition, rather than the direct addition product to either carbonyl group. Sulfur nucleophiles are soft, and this is typical behaviour for thiols.

PROBLEM 10

This compound is needed as a synthetic precursor to the drug etalocib. Suggest a synthesis. *Hint*: consider using nucleophilic aromatic substitution.

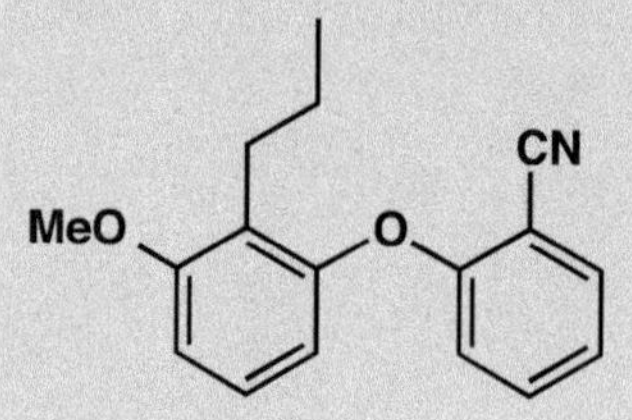

Purpose of the problem

Thinking about regioselectivity and reactivity in the synthesis of a moderately complex aromatic compound.

Suggested solution

There are lots of *ortho* relationships in this compound! And somehow we have to join the two aromatic rings together to make an ether. This can only really be done by nucleophilic aromatic substitution, so we need to look for an electron-withdrawing group to help us. The nitrile is in the right place, provided we have a leaving group (such as fluoride) *ortho* to it. So our last step can be as shown here:

■ See pp. 514–520 of the textbook for a reminder of nucleophilic aromatic substitution.

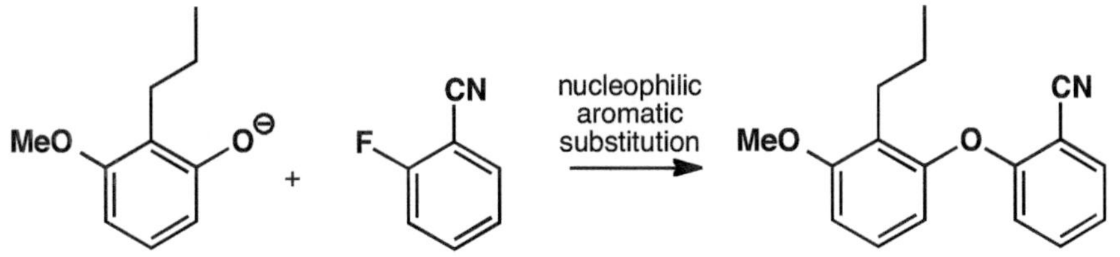

To make the left hand ring we have to consider what methods are available to introduce the three substituents we have. It's always easier to add C-substituents than O-substituents, so we might consider how to alkylate the phenol below. Both the OH and OMe groups are *ortho,para* directing, so we could consider a Friedel-Crafts reaction, but there are problems with this approach. One is the usual problem with primary alkyl groups—we would have to do an acylation and then reduce. Another is more serious: the less hindered positions the other side of the OMe or OH groups are also activated, so we will have a regioselectivity problem. The solution used by the chemists making this compound for the first time was to use ortholithiation, making the dianion with two equivalents of BuLi and making use of the fact that two O substituents guide the BuLi in to deprotonate the position between them.

■ The work is described by Sawyer *et al., J. Med. Chem.* 1993, **38**, 4411.

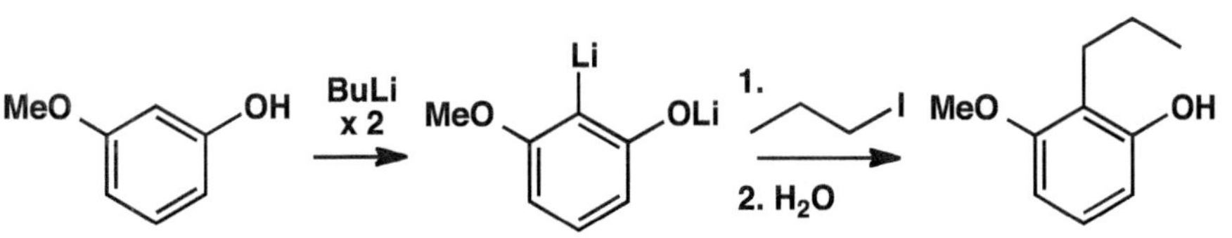

Aromatic heterocycles 1: reactions

29

Connections

⇒ **Building on**	⇒ **Arriving at**	⇒ **Looking forward to**
• Aromaticity ch7	• Aromatic systems conceptually derived from benzene: replacing CH with N to get pyridine	• Synthesis of aromatic heterocycles ch30
• Enols and enolates ch20		• Saturated heterocycles ch31
• Electrophilic aromatic substitution ch21	• Replacing CH=CH with N to get pyrrole	• Biological chemistry ch42
• Nucleophilic attack on aromatic rings ch22	• How pyridine reacts	
• Reactions of enols and enolates ch25 & ch26	• How pyridine derivatives can be used to extend pyridine's reactivity	
	• How pyrrole reacts	
	• How furan and thiophene compare with pyrrole	
	• Putting more nitrogens in five- and six-membered rings	
	• Fused rings: indole, quinoline, isoquinoline, and indolizine	
	• Rings with nitrogen and another heteroatom: oxygen or sulfur	

Introduction

Benzene is aromatic because it has six electrons in a cyclic conjugated system. We know it is aromatic because it is exceptionally stable, it has a ring current and hence large chemical shifts in the proton NMR spectrum, and it has special chemistry involving substitution rather than addition with electrophiles. This chapter and the next are about the very large number of other aromatic systems in which one or more atoms in the benzene ring are replaced by heteroatoms such as N, O, and S. There are thousands of these systems with five- and six-membered rings, and we will examine just a few.

⇒ The rather precise chemical definition of 'aromatic' is explained in Chapter 7. You will find the reactions of benzene and its aromatic derivatives described in Chapters 21 and 22: those two chapters are essential reading before you tackle this one.

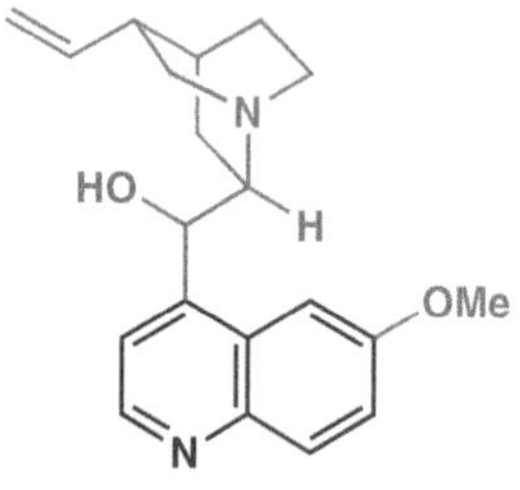

quinine

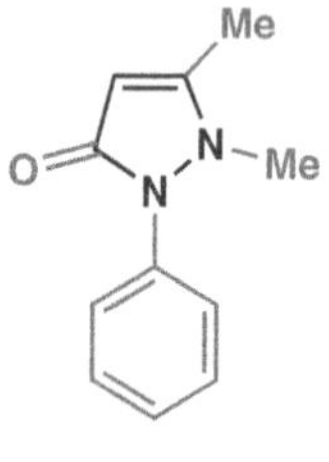

antipyrine

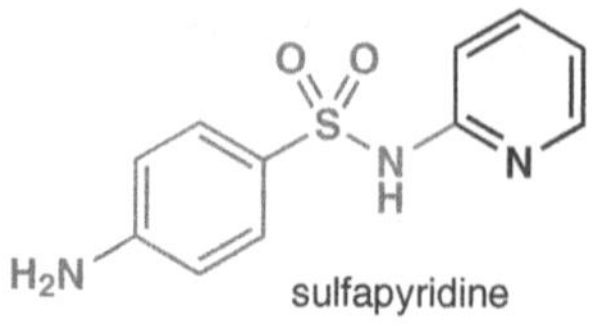

sulfapyridine

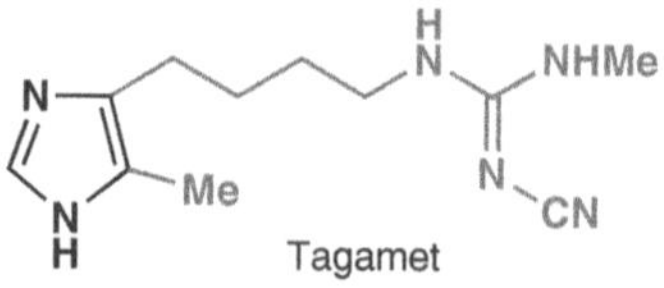

Tagamet

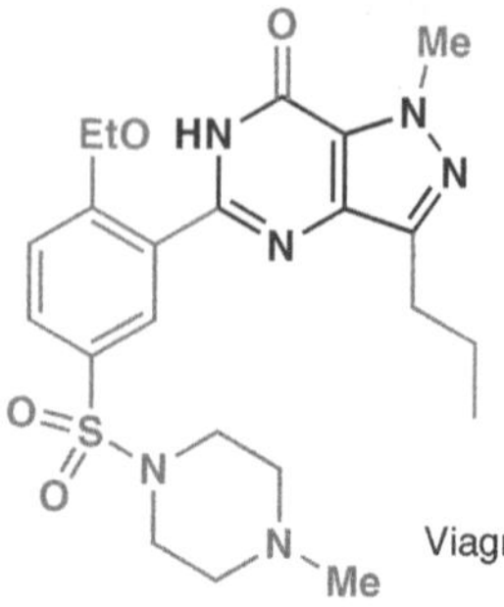

Viagra

Online support. The icon 🖰 in the margin indicates that accompanying interactive resources are provided online to help your understanding: just type **www.chemtube3d.com/clayden/123** into your browser, replacing **123** with the number of the page where you see the icon. For pages linking to more than one resource, type **123-1**, **123-2** etc. (replacing **123** with the page number) for access to successive links.

Our subject is aromatic heterocycles and it is important that we treat it seriously because most—probably about two-thirds of—organic compounds belong to this class, and they number among them some of the most significant compounds for human beings. If we think only of drugs we can define the history of medicine by heterocycles. Even in the sixteenth century quinine was used to prevent and treat malaria, although the structure of the drug was not known. The first synthetic drug was antipyrine (1887) for the reduction of fevers. The first effective antibiotic was sulfapyridine (1938). The first multi-million pound drug (1970s) was Tagamet, the anti-ulcer drug, and among the most topical of current drugs is Viagra (1997) for treatment of male impotence.

All these compounds have heterocyclic aromatic rings shown in black. Three have single rings, five- or six-membered, two have five- or six-membered rings fused together. The number of nitrogens in the rings varies from one to four. We will start by looking at the simple six-membered ring with one nitrogen atom: pyridine.

Aromaticity survives when parts of benzene's ring are replaced by nitrogen atoms

There is no doubt that benzene is aromatic. Now we must ask: how can we insert a heteroatom into the ring and retain aromaticity? What kind of atom is needed? If we want to replace one of the carbon atoms of benzene with a heteroatom, we need an atom that can be trigonal to keep the flat hexagonal ring, and that has a p orbital to keep the six delocalized electrons. Nitrogen fits all of these requirements. This is what happens if we replace a CH group in benzene with a nitrogen atom.

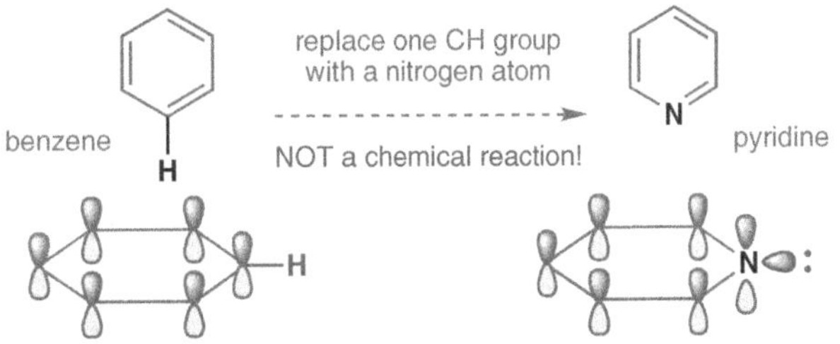

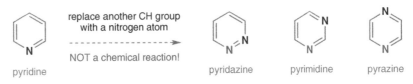

Interactive structure of pyridine

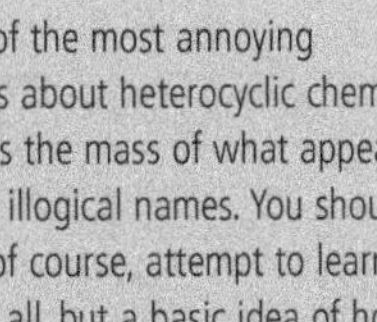

δ_H 7.5

H δ_H 7.1

H δ_H 8.5

^{1}H NMR spectrum of pyridine

The orbitals in the ring have not changed in position or shape and we still have the six electrons from the three double bonds. One obvious difference is that nitrogen is trivalent and thus there is no NH bond. Instead, a lone pair of electrons occupies the space of the C–H bond in benzene.

In theory then, pyridine is aromatic. But is it in real life? The most important evidence comes from the proton NMR spectrum. The six protons of benzene resonate at 7.27 ppm, some 2 ppm downfield from the alkene region, clear evidence for a ring current (Chapter 13). Pyridine is not as symmetrical as benzene but the three types of proton all resonate in the same region. As we will see, pyridine is also very stable and, by any reasonable assessment, pyridine is aromatic.

We could continue the process of replacing, on paper, more CH groups with nitrogen atoms, and would find three new aromatic heterocycles: pyridazine, pyrimidine, and pyrazine:

Nomenclature

One of the most annoying things about heterocyclic chemistry is the mass of what appear to be illogical names. You should not, of course, attempt to learn them all, but a basic idea of how they are designed will help you. We will give you a guide on which names to learn shortly. For the moment accept that 'amine' ends in '-ine' and any heterocyclic compound whose name ends in '-ine' is a nitrogen heterocycle. The syllable 'azo-' also implies nitrogen and 'pyr-' (usually) implies a six-membered ring (except in pyrrole!).

There is another way in which we might transform benzene into a heterocycle. Instead of using just one electron from N to replace an electron in the π system, we could use nitrogen's lone pair of electrons to replace two electrons in the π system. We can substitute a CH=CH unit in benzene with a nitrogen atom providing that we can use the lone pair in the delocalized system. This means putting it into a p orbital. We still have the four electrons from the remaining double bonds and, with the two electrons of the lone pair on nitrogen, that makes six in all. The nitrogen atom must still be trigonal with the lone pair in a p orbital so the N–H bond is in the plane of the five-membered ring.

replace a CH=CH unit
with a nitrogen atom

NOT a chemical reaction!

benzene

pyrrole

The ^{1}H NMR spectrum of pyrrole is slightly less convincing as the two types of proton on the ring resonate at higher field (6.5 and 6.2 ppm) than those of benzene or pyridine but they still fall in the aromatic rather than the alkene region. Pyrrole is also more reactive towards electrophiles than benzene or pyridine, but it does the usual aromatic substitution reactions (Friedel–Crafts, nitration, halogenation) rather than addition reactions: pyrrole is also aromatic.

Inventing heterocycles by further replacement of CH groups by nitrogen in pyrrole leads to two compounds, pyrazole and imidazole, after one replacement, to two triazoles after two replacements, and to a single tetrazole after three.

H δ_H 6.2

H δ_H 6.5

δ_H~10

replace one
CH group
with a
nitrogen atom

NOT a
chemical
reaction!

pyrazole

replace a second
CH group
with a
nitrogen atom

NOT a
chemical
reaction!

imidazole

1,2,3-triazole

replace a third
CH group
with a
nitrogen atom

NOT a
chemical
reaction!

1,2,4-triazole

N—N

tetrazole

All of these compounds are generally accepted as aromatic too as they broadly have the NMR spectra and reactivities expected for aromatic compounds. As you may expect, introducing heteroatoms into the aromatic ring and, even more, changing the ring size actually affect the chemistry a great deal. We must now return to pyridine and work our way more slowly through the chemistry of these important heterocycles to establish the principles that govern their behaviour.

More nomenclature

The ending '-ole' is systematic and refers to a five-membered heterocyclic ring. All the five-membered aromatic heterocycles with nitrogen in the ring are sometimes called 'the azoles'. Oxazole and thiazole are used for the oxygen and sulfur analogues of imidazole.

oxazole   thiazole

Pyridine is a very unreactive aromatic imine

The nitrogen atom in the pyridine ring is planar and trigonal with the lone pair in the plane of the ring. This makes it an imine. Most of the imines you have met before (in Chapter 11, for example), have been unstable intermediates in carbonyl group reactions, but in pyridine we have a stable imine—stable because of its aromaticity. All imines are more weakly basic than saturated amines and pyridine is a weak base with a pK_a (for its conjugate acid) of 5.5. This means that the pyridinium ion is about as strong an acid as a carboxylic acid.

■ Pyridine is also toxic and has a foul smell—so there are disadvantages in using pyridine as a solvent. But it is cheap and remains a popular solvent in spite of the problems.

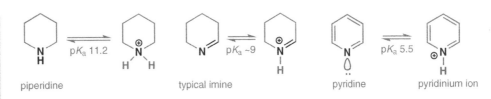

piperidine   typical imine   pyridine   pyridinium ion

Pyridine is a reasonable nucleophile for carbonyl groups and is often used as a nucleophilic catalyst in acylation reactions. Esters are often made in pyridine solution from alcohols and acid chlorides (the full mechanism is on p. 199 of Chapter 10).

🖱 Interactive mechanism for pyridine nucleophilic catalysis

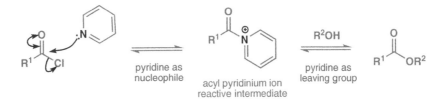

pyridine as nucleophile

acyl pyridinium ion reactive intermediate

pyridine as leaving group

DMAP

One particular amino-pyridine has a special role as a more effective acylation catalyst than pyridine itself. This is DMAP (*N,N*-dimethylaminopyridine) in which the amino group is placed to reinforce the nucleophilic nature of the nitrogen atom. Whereas acylations 'catalysed' by pyridine are normally carried out in solution in pyridine, only small amounts of DMAP in other solvents are needed to do the same job.

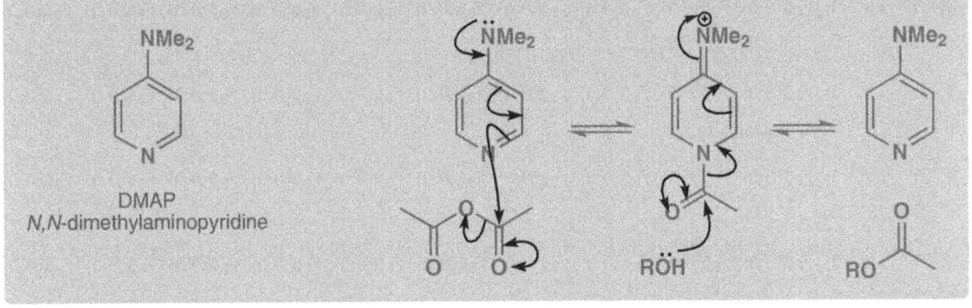

NMe₂

DMAP
N,N-dimethylaminopyridine

Pyridine is nucleophilic at the nitrogen atom because *the lone pair of electrons on nitrogen cannot be delocalized around the ring*. They are in an sp² orbital orthogonal to the p orbitals in the ring and there is no interaction between orthogonal orbitals. Try it for yourself, drawing arrows. All attempts to delocalize the electrons lead to impossible results!

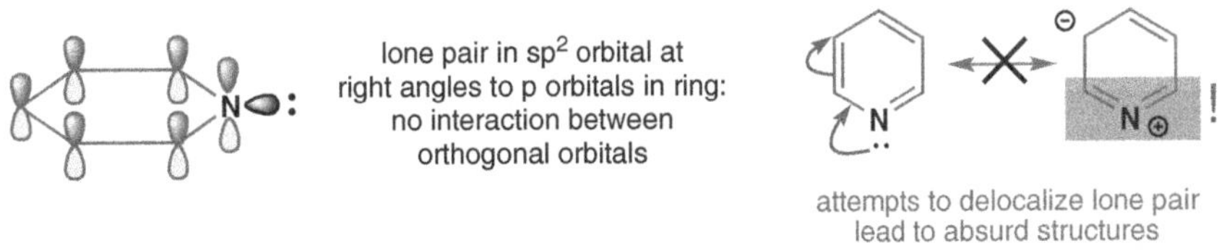

lone pair in sp² orbital at right angles to p orbitals in ring: no interaction between orthogonal orbitals

attempts to delocalize lone pair lead to absurd structures

● **The lone pair of pyridine's nitrogen atom is not delocalized.**

Our main question about the reactivity of pyridine must be this: what does the nitrogen atom do to the rest of the ring? The important orbitals—the p orbitals of the aromatic system—are superficially the same as in benzene, but the more electronegative nitrogen atom will lower the energy of all the orbitals. Lower-energy filled orbitals mean a *less* reactive nucleophile but a lower-energy LUMO means a *more* reactive electrophile. This is a good guide to the chemistry of pyridine. It is less reactive than benzene in electrophilic aromatic substitution reactions, but nucleophilic substitution, which is difficult for benzene, comes easily to pyridine.

➡ Electrophilic substitution in benzene is discussed in Chapter 21.

Pyridine is bad at electrophilic aromatic substitution

The lower energy of the orbitals of pyridine's π system means that electrophilic attack on the ring is difficult. Another way to look at this is to see that the nitrogen atom destabilizes the cationic would-be intermediate, especially when it can be delocalized onto nitrogen.

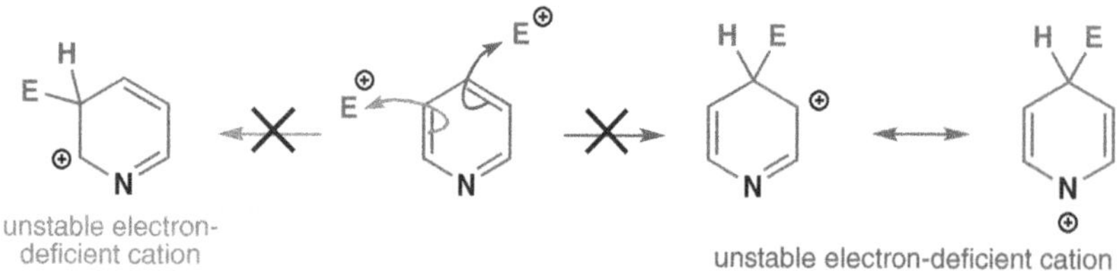

unstable electron-deficient cation

unstable electron-deficient cation

■ Contrast the unstable electron-deficient cationic intermediate with the stable pyridinium ion. The nitrogen lone pair is used to make the pyridinium ion but is not involved in the unstable intermediate. Note that reaction at the 3-position is the best option but still doesn't occur. Reaction at the 2- and 4-positions is worse.

An equally serious problem is that the nitrogen lone pair is basic and a reasonably good nucleophile—this is the basis for its role as a nucleophilic catalyst in acylations. The normal reagents for electrophilic substitution reactions, such as nitration, are acidic. Treatment of pyridine with the usual mixture of HNO_3 and H_2SO_4 merely protonates the nitrogen atom. Pyridine itself is not very reactive towards electrophiles: the pyridinium ion is totally unreactive.

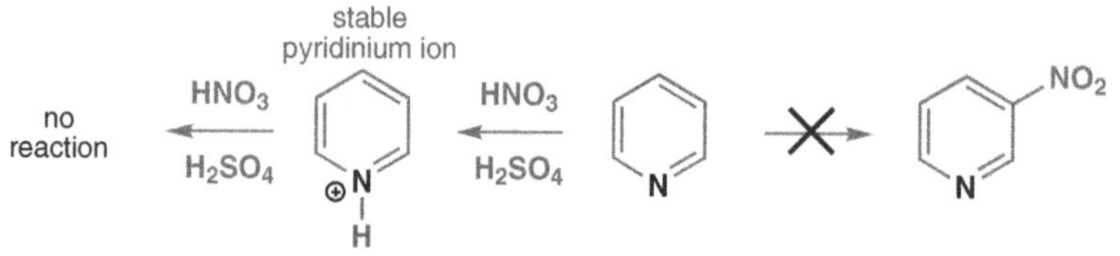

Other reactions, such as Friedel–Crafts acylations, require Lewis acids and these too react at nitrogen. Pyridine is a good ligand for metals such as Al(III) or Sn(IV) and, once again, the complex with its cationic nitrogen is completely unreactive towards electrophiles.

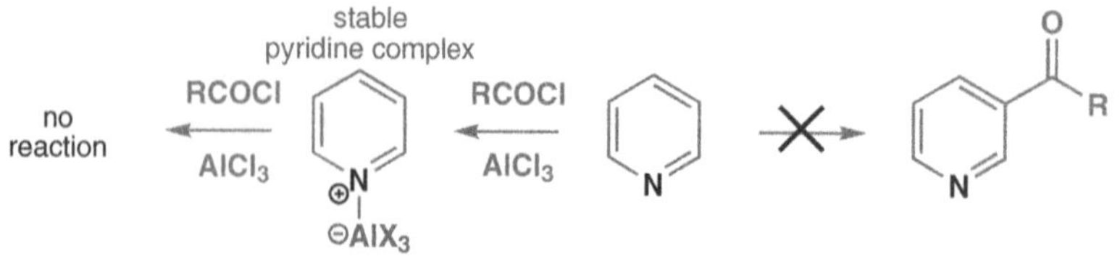

● Pyridine does not undergo electrophilic substitution

Aromatic electrophilic substitution on pyridine is not a useful reaction. The ring is unreactive and the electrophilic reagents attack nitrogen, making the ring even less reactive. Avoid nitration, sulfonation, halogenation, and Friedel–Crafts reactions on simple pyridines.

Nucleophilic substitution is easy with pyridines

By contrast, the nitrogen atom makes pyridines *more* reactive towards nucleophilic substitution, particularly at the 2- and 4-positions, by lowering the LUMO energy of the π system of pyridine. You can see this effect in action in the ease of replacement of halogens in these positions by nucleophiles.

➡ Nucleophilic substitution in benzene is discussed in Chapter 22.

Interactive mechanism for nucleophilic substitution on pyridines

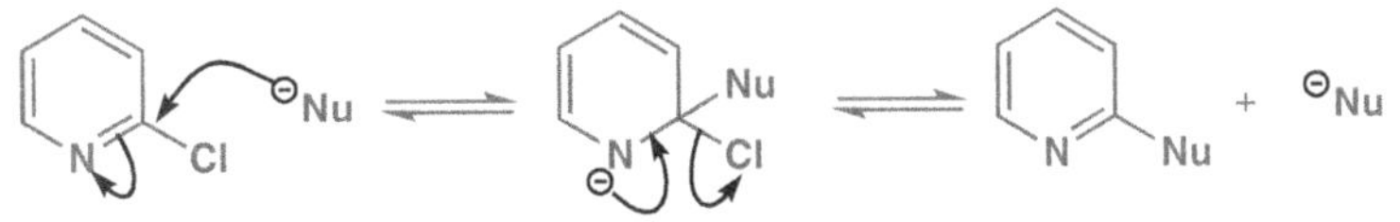

The intermediate anion is stabilized by electronegative nitrogen and by delocalization round the ring. These reactions have some similarity to nucleophilic aromatic substitution (Chapter 22) but are more similar to carbonyl reactions. The intermediate anion is a tetrahedral intermediate that loses the best leaving group to regenerate the stable aromatic system. Nucleophiles such as amines or thiolate anions work well in these reactions.

Note the similarity to nucleophilic substitution on the carbonyl group (Chapter 10).

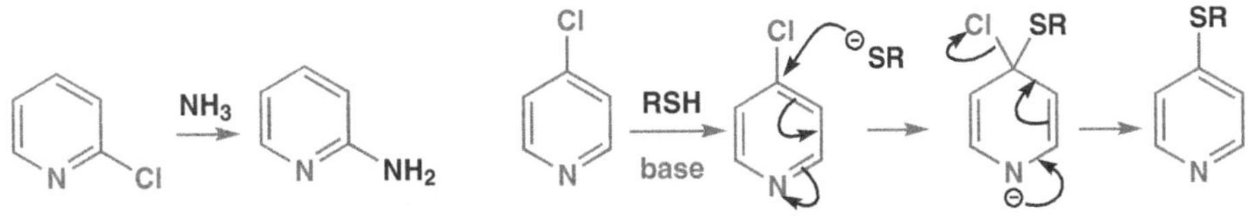

The leaving group does not have to be as good as chloride in these reactions. Continuing the analogy with carbonyl reactions, 2- and 4-chloropyridines are rather like acid chlorides but we need only use less reactive pyridyl ethers, which react like esters, to make amides. Substitution of a 2-methoxypyridine allows the synthesis of flupirtine.

You will see more of this synthesis later in the chapter.

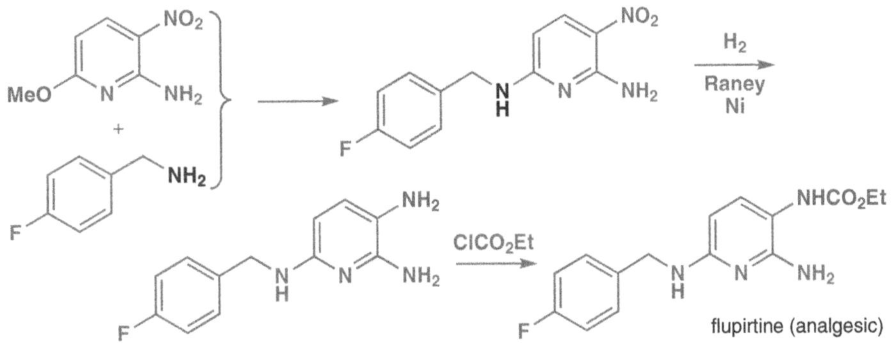

The first step is a nucleophilic aromatic substitution. In the second step the nitro group is reduced to an amino group without any effect on the pyridine ring—another piece of evidence for its aromaticity. Finally, the one amino group whose lone pair is not delocalized onto the pyridine N is acylated in the presence of two others.

Pyridones are good substrates for nucleophilic substitution

The starting materials for these nucleophilic substitutions (2- and 4-chloro- or methoxypyridines) are themselves made by nucleophilic substitution on *pyridones*. If you were asked to propose how 2-methoxypyridine might be made, you would probably suggest, by analogy with the corresponding benzene compound, alkylation of a phenol. Let us look at this in detail.

The starting material for this reaction is a 2-hydroxypyridine that can tautomerize to an amide-like structure known as a pyridone by the shift of the acidic proton from oxygen to nitrogen. In the phenol series there is no doubt about which structure will be stable as the ketone is not aromatic; for the pyridine both structures are aromatic.

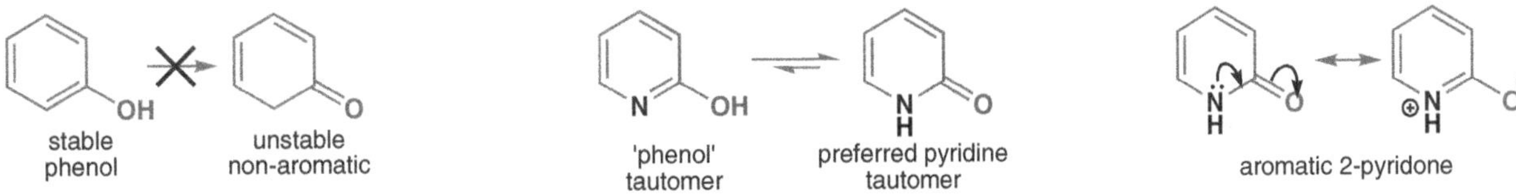

stable phenol

unstable non-aromatic

'phenol' tautomer

preferred pyridine tautomer

aromatic 2-pyridone

In fact, 2-hydroxypyridine prefers to exist as the 'amide' because that has the advantage of a strong C=O bond and is still aromatic. There are two electrons in each of the C=C double bonds and two also in the lone pair of electrons on the trigonal nitrogen atom of the amide. Delocalization of the lone pair in typical amide style makes the point clearer.

Interactive tautomerism between 2-hydroxypyridine and pyridone

Pyridones are easy to prepare (see Chapter 30) and can be alkylated on oxygen as predicted by their structure. A more important reaction is the direct conversion to chloropyridines with $POCl_3$. The reaction starts by attack of the oxygen atom at phosphorus to create a leaving group, followed by aromatic nucleophilic substitution. The overall effect is very similar to acyl chloride formation from a carboxylic acid (Chapter 10).

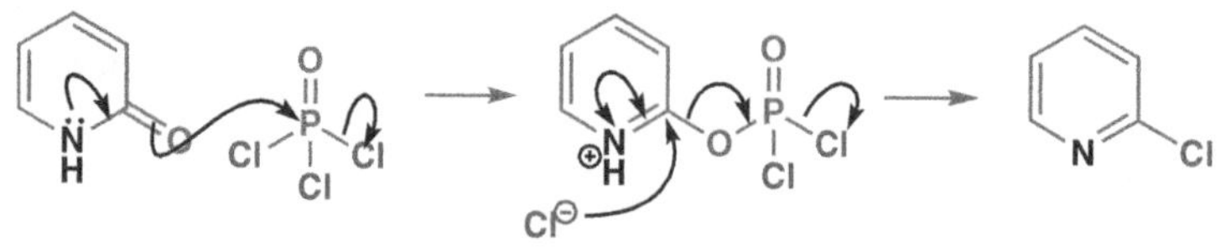

The same reaction occurs with 4-pyridone, which is also delocalized in the same way and exists in the 'amide' form, but not with 3-hydroxypyridine, which exists in the 'phenol' form. Its only tautomer is a zwitterion but the pyridine nitrogen is too weak to remove a proton from the hydroxyl group.

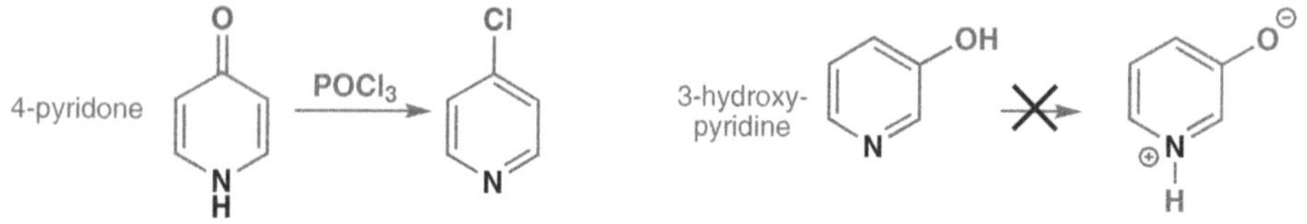

• Pyridines undergo nucleophilic substitution

Pyridines can undergo *electrophilic* substitution only if they are activated by electron-donating substituents (see next section) but they readily undergo *nucleophilic* substitution without any activation other than the ring nitrogen atom.

Activated pyridines will do electrophilic aromatic substitution

Useful electrophilic substitutions occur only on pyridines having electron-donating substituents such as NH_2 or OMe. These activate benzene rings too (Chapter 21) but here their help is vital. They supply a non-bonding pair of electrons that raises the energy of the HOMO and carries out the reaction. Simple amino- or methoxypyridines react reasonably well *ortho* and *para* to the activating group. These reactions happen in spite of the molecule being a pyridine, not because of it.

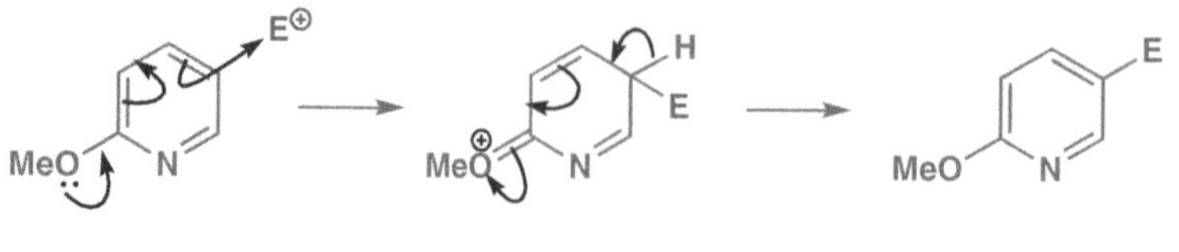

A practical example occurs in the manufacture of the analgesic flupirtine where a doubly activated pyridine having both MeO and NH_2 groups is nitrated just as if it were a benzene ring. The nitro group goes in *ortho* to the amino group and *para* to the methoxy group. The activation is evidently enough to compensate for the molecule being almost entirely protonated under the conditions of the reaction.

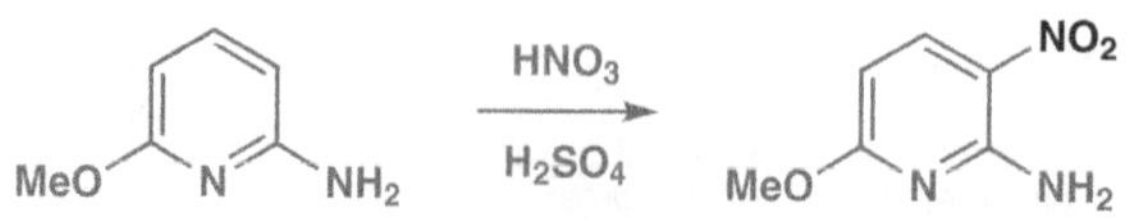

This is the starting material for the flupirtine synthesis on p. 728.

Pyridine *N*-oxides are reactive towards both electrophilic and nucleophilic substitution

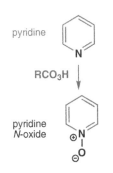

pyridine

RCO₃H

pyridine
N-oxide

🖱 Interactive structure of pyridine *N*-oxide

This is all very well if the molecule has such activating groups, but supposing it doesn't? How are we to nitrate pyridine itself? The answer involves an ingenious trick. We need to activate the ring with an electron-rich substituent that can later be removed and we also need to stop the nitrogen atom reacting with the electrophile. All of this can be done with a single atom!

Because the nitrogen atom is nucleophilic, pyridine can be oxidized to pyridine *N*-oxide with reagents such as *m*-CPBA or just H₂O₂ in acetic acid. These *N*-oxides are stable dipolar species with the electrons on oxygen delocalized round the pyridine ring, raising the HOMO of the molecule. Reaction with electrophiles occurs at the 2- (*ortho*) and 4- (*para*) positions, chiefly at the 4-position to keep away from positively charged nitrogen.

Now the oxide must be removed and this is best done with trivalent phosphorus compounds such as (MeO)₃P or PCl₃. The phosphorus atom detaches the oxygen atom in a single step to form the very stable P=O double bond. In this reaction the phosphorus atom is acting as both a nucleophile and an electrophile, but mainly as an electrophile since PCl₃ is more reactive here than (MeO)₃P.

phosphorus donates its lone pair while accepting electrons into its d orbitals

The same activation that allowed simple electrophilic substitution—oxidation to the *N*-oxide—can also allow a useful *nucleophilic* substitution. The positive nitrogen atom encourages nucleophilic attack and the oxygen atom can be turned into a leaving group with PCl₃. Our example is nicotinic acid, whose biological importance we will discuss in Chapter 42.

nicotinic acid

The *N*-oxide reacts with PCl₃ through oxygen and the chloride ion released in this reaction adds to the most electrophilic position between the two electron-withdrawing groups. Now a simple elimination restores aromaticity and gives a product looking as though it results from chlorination rather than nucleophilic attack.

🖱 Interactive mechanism for nucleophilic substitution on pyridine *N*-oxide

The reagent PCl₃ also converts the carboxylic acid to the acyl chloride, which is hydrolysed back again in the last step. This is a useful sequence because the chlorine atom has been introduced into the 2-position, from which it may in turn be displaced by, for example, amines.

nifluminic acid
(an analgesic)

● **Pyridine *N*-oxides**
Pyridine *N*-oxides are useful for both electrophilic and nucleophilic substitutions on the same carbon atoms (2-, 4-, and 6-) in the ring.

Nucleophilic addition at an even more distant site is possible on reaction with acid anhydrides if there is an alkyl group in the 2-position. Acylation occurs on oxygen as in the last reaction but then a proton is lost from the side chain to give an uncharged intermediate.

This compound rearranges with migration of the acetate group to the side chain and the restoration of aromaticity. This may be an ionic reaction or a type of rearrangement that you will learn to call a [3,3]-sigmatropic rearrangement (Chapter 35).

Pyridine as a catalyst and reagent

Since pyridine is abundant and cheap and has an extremely rich chemistry, it is not surprising that it has many applications. One of the simplest ways to brominate benzenes is not to bother with the Lewis acid catalysts recommended in Chapter 21 but just to add liquid bromine to the aromatic compound in the presence of a small amount of pyridine. Only about one mole per cent is needed and even then the reaction has to be cooled to stop it getting out of hand.

As we have seen, pyridine attacks electrophiles through its nitrogen atom. This produces the reactive species, the *N*-bromo-pyridinium ion, which is attacked by the benzene. Pyridine is a better nucleophile than benzene and a better leaving group than bromide. This is another example of **nucleophilic catalysis**.

➡ Nucleophilic catalysis is discussed on p. 200.

Another way to use pyridine in brominations is to make a stable crystalline compound to replace the dangerous liquid bromine. This compound, known by names such as pyridinium tribromide, is simply a salt of pyridine with the anion Br_3^-. It can be used to brominate reactive compounds such as alkenes (Chapter 19).

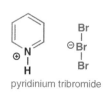

pyridinium tribromide

Both of these methods depend on the lack of reactivity of pyridine's π system towards electrophiles such as bromine. Notice that, in the first case, both benzene and pyridine are present together. The pyridine attacks bromine only through nitrogen (and reversibly at that) and never through carbon.

Oxidation of alcohols is normally carried out with Cr(VI) reagents (Chapter 23) but these, like the Jones' reagent ($Na_2Cr_2O_7$ in sulfuric acid), are usually acidic. Some pyridine complexes of Cr(VI) compounds solve this problem by having the pyridinium ion (pK_a 5) as the only acid. The two most famous are PDC (pyridinium dichromate) and PCC (pyridinium chloro-chromate). Pyridine forms a complex with CrO_3 but this is liable to burst into flames. Treatment with HCl gives PCC, which is much less dangerous. PCC is particularly useful in the oxidation of primary alcohols to aldehydes as over-oxidation is avoided in the only slightly acidic conditions (Chapter 23).

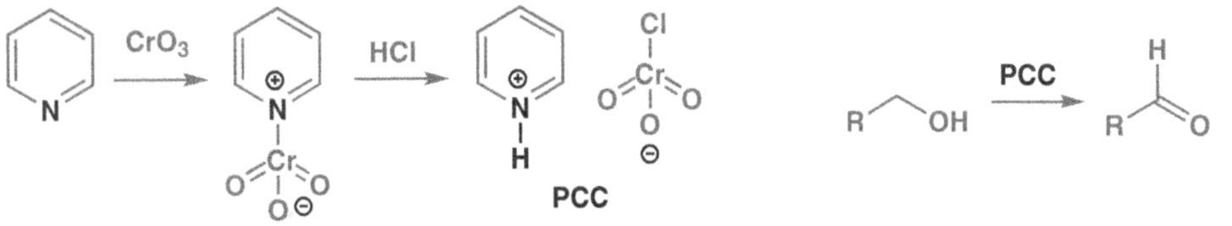

Bipyridyl (bipy)

The ability of pyridine to form metal complexes is greatly enhanced in a dimer—the famous ligand 'bipy' or 2,2'-bipyridyl. It is bidentate and because of its 'bite' it is a good ligand for many transition metals, with a partiality for Fe(II).

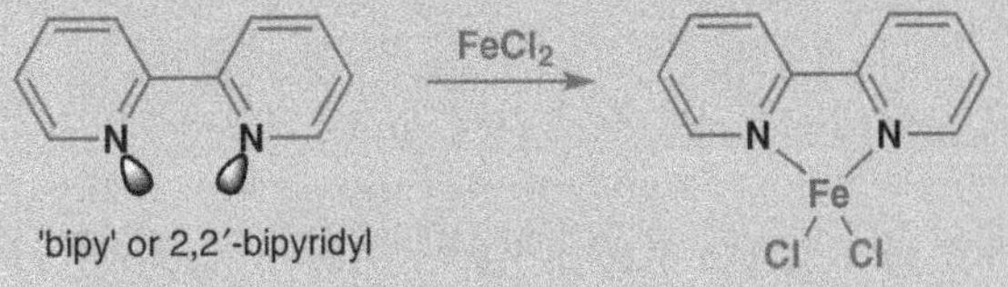

'bipy' or 2,2'-bipyridyl

It looks like a rather difficult job to persuade two pyridine rings to join together in this way to form bipy. It is indeed very difficult unless you make things easier by using a reagent that favours the product. And what better than Fe(II) to do the job? Bipy is manufactured by treating pyridine with $FeCl_2 \cdot 4H_2O$ at high temperatures and high pressures. Only a small proportion of the pyridine is converted to the Fe(II) complex of bipy (about 5%) but the remaining pyridine goes back in the next reaction. This is probably a radical process (Chapter 37) within the coordination sphere of Fe(II).

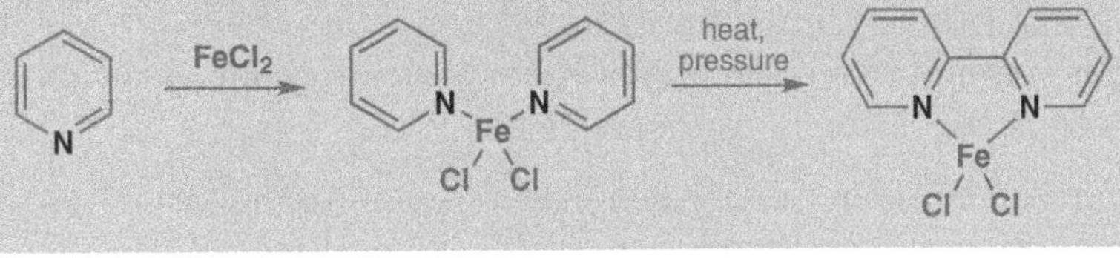

Six-membered aromatic heterocycles can have oxygen in the ring

Although pyridine is overwhelmingly the most important of the six-membered aromatic het-erocycles, there are oxygen heterocycles, pyrones, that resemble the pyridones. The pyrones are aromatic, although α-pyrone is rather unstable.

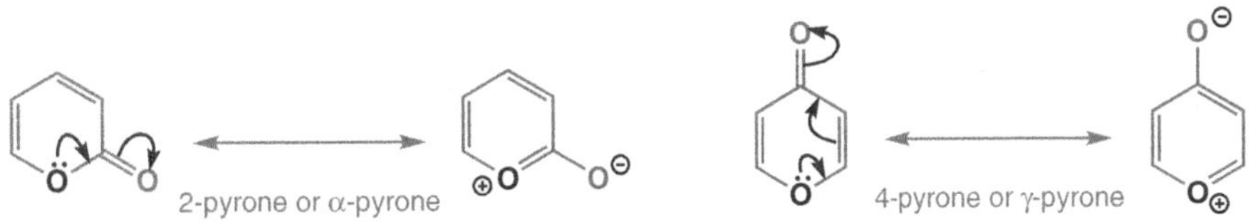

2-pyrone or α-pyrone   4-pyrone or γ-pyrone

The pyrylium salts are stable aromatic cations and are responsible as metal complexes for some flower colours. Heterocycles with six-membered rings based on other elements (for example, P) do exist but they are outside the scope of this book.

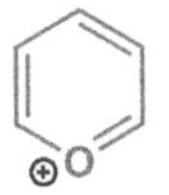

the pyrylium cation

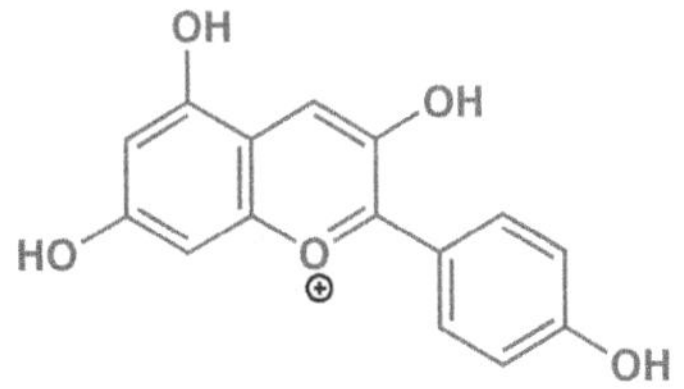

a red pyrylium flower pigment

Five-membered aromatic heterocycles are good at electrophilic substitution

Just about everything is the other way round with pyrrole. Electrophilic substitution is much easier than it is with benzene—almost too easy in fact—while nucleophilic substitution is more difficult. Pyrrole is not a base nor can it be converted to an N-oxide. We need to find out why this is. The big difference is that the nitrogen lone pair is delocalized round the ring. The NMR spectrum suggests that all the positions in the ring are about equally electron-rich with chemical shifts about 1 ppm smaller than those of benzene. The ring is flat and the bond lengths are very similar, although the bond opposite the nitrogen atom is a bit longer than the others.

The delocalization of the lone pair can be drawn equally well to any ring atom because of the five-membered ring and we shall soon see the consequences of this. All the delocalization pushes electrons from the nitrogen atom into the ring and we expect the ring to be electron-rich at the expense of the nitrogen atom. The HOMO should go up in energy and the ring become more nucleophilic.

An obvious consequence of this delocalization is the decreased basicity of the nitrogen atom and the increased acidity of the NH group. In fact, the pK_a of pyrrole acting as a base is about −4, and protonation occurs at carbon below pH −4. By contrast, the NH proton (pK_a 16.5) can be removed by much weaker bases than those that can remove protons on normal secondary amines. The nucleophilic nature of the ring means that pyrrole is attacked readily by electrophiles. Reaction with bromine requires no Lewis acid and leads to substitution (confirming the aromaticity of pyrrole) at all four free positions. Contrast pyridine's reactivity with bromine (p. 731): it reacts just once, at nitrogen.

This is a fine reaction in its way, but we don't usually want four bromine atoms in a molecule so one problem with pyrrole is to control the reaction to give only monosubstitution. Another problem is that strong acids cannot be used. Although protonation does not occur at nitrogen, it does occur at carbon and the protonated pyrrole then adds another molecule like this.

● Pyrrole polymerizes!

Strong acids, those such as H_2SO_4 with a pK_a of less than −4, cannot be used without polymerization of pyrrole.

Some reactions can be controlled to give good yields of monosubstituted products. One is the Vilsmeier reaction, in which a combination of an N,N-dimethylamide and $POCl_3$ is used to make a carbon electrophile in the absence of strong acid or Lewis acid. It is a substitute for the Friedel–Crafts acylation, and works with aromatic compounds at the more reactive end of the scale (where pyrrole is).

Interactive structure of pyrrole

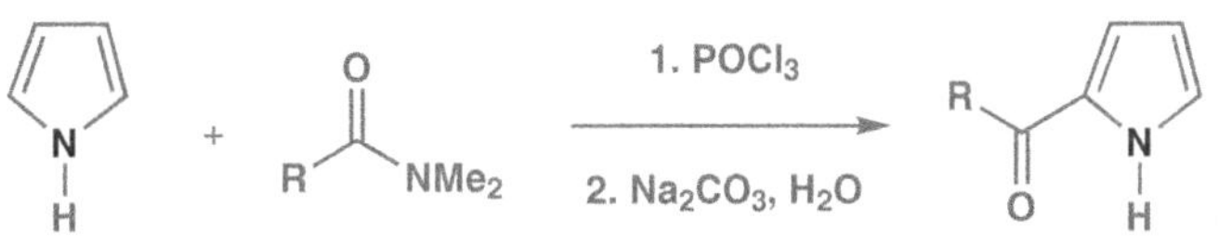

In the first step, the amide reacts with $POCl_3$, which makes off with the amide oxygen atom and replaces it with chlorine. This process would be very unfavourable but for the formation of the strong P–O bond, and is the direct analogy of the chloropyridine-forming reaction you have just seen.

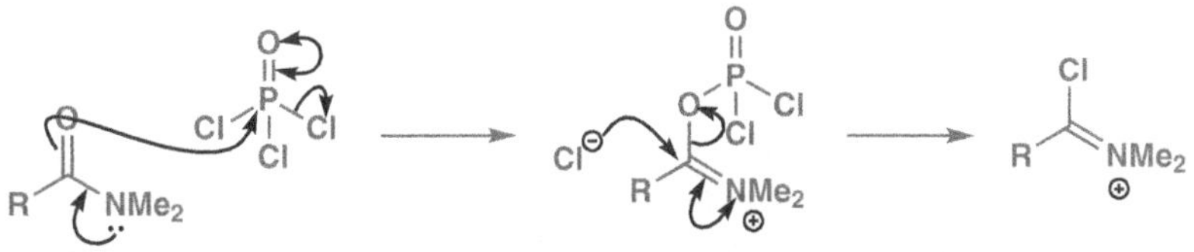

The product from this first step is an iminium cation that reacts with pyrrole to give a more stable iminium salt. The extra stability comes from the conjugation between the pyrrole nitrogen and the iminium group. The work-up with aqueous Na_2CO_3 hydrolyses the imine salt and removes any acid formed. This method is particularly useful because it works well with Me_2NCHO (DMF) to add a formyl (CHO) group. This is difficult to do with a conventional Friedel–Crafts reaction.

🖱 Interactive mechanism for Vilsmeier reaction of pyrrole

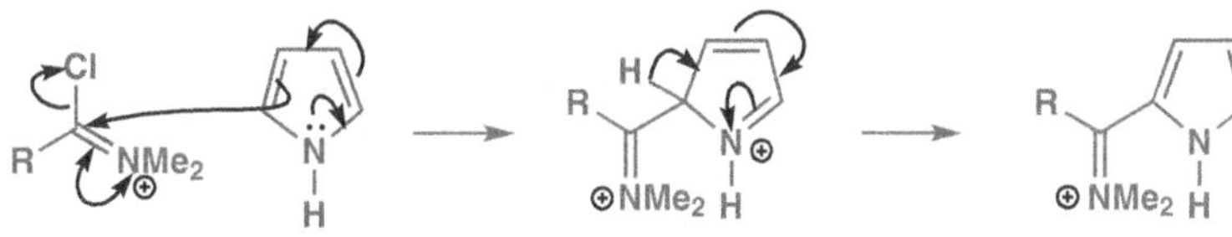

You may have noticed that the reaction occurred only at the 2-position on pyrrole. Although all positions react with reagents like bromine, most reagents go for the 2- (or 5-) position and attack the 3- (or 4-) position only if the 2- and 5-positions are blocked. A good example is the Mannich reaction. In these two examples N-methylpyrrole reacts cleanly at the 2-position while the other pyrrole with both 2- and 5-positions blocked by methyl groups reacts cleanly at the 3-position. These reactions are used in the manufacture of the non-steroidal anti-inflammatory compounds tolmetin and clopirac.

➡ Remind yourself of the Mannich reaction on p. 621 of Chapter 26.

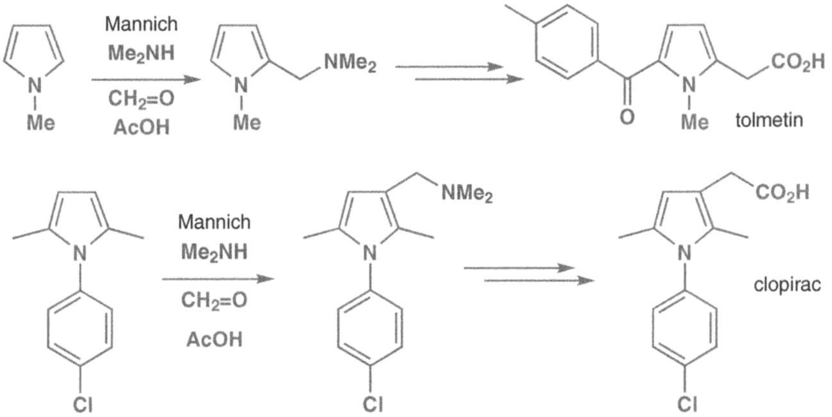

🖱 Interactive mechanism for the Mannich reaction on pyrrole

Now we need an explanation. The mechanisms for both 2- and 3-substitutions look good and we will draw both, using a generalized E^+ as the electrophile. Both mechanisms can occur very readily. Reaction in the 2-position is somewhat better than in the 3-position but the difference is small. Substitution is favoured at *all* positions. Calculations show that the HOMO of pyrrole does indeed have a larger coefficient in the 2-position, and one way to explain this result is to look at the structure of the intermediates. The intermediate from attack at the

2-position has a linear conjugated system. In both intermediates the two double bonds are, of course, conjugated with each other, but only in the first intermediate are both double bonds conjugated with N^+. The second intermediate is 'cross-conjugated', while the first has a more stable linear conjugated system.

reaction with electrophiles in the 2-position

reaction with electrophiles in the 3-position

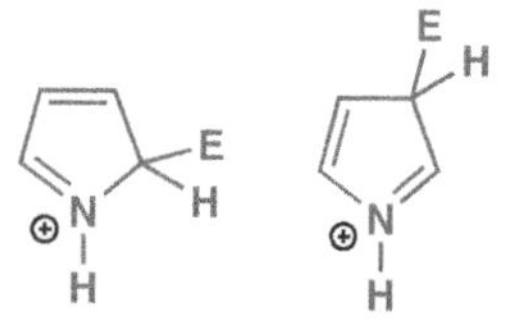

more stable   less stable

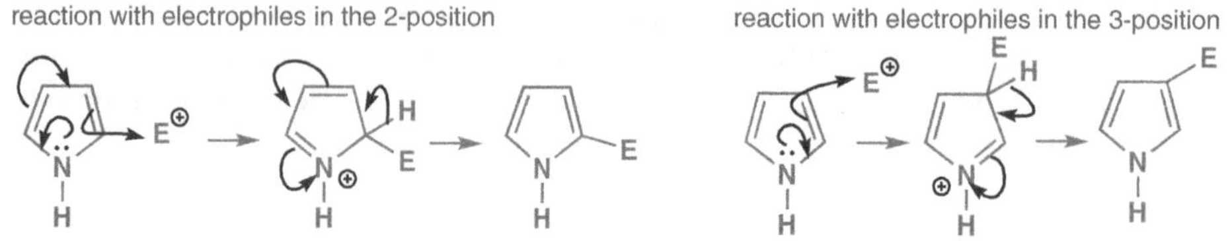

Since electrophilic substitution on pyrroles occurs so easily, it can be useful to block substitution with a removable substituent. This is usually done with an ester group. Hydrolysis of the ester (this is particularly easy with *t*-butyl esters—see Chapter 23) releases the carboxylic acid, which decarboxylates on heating. There is no doubt that the final electrophilic substitution must occur at C2.

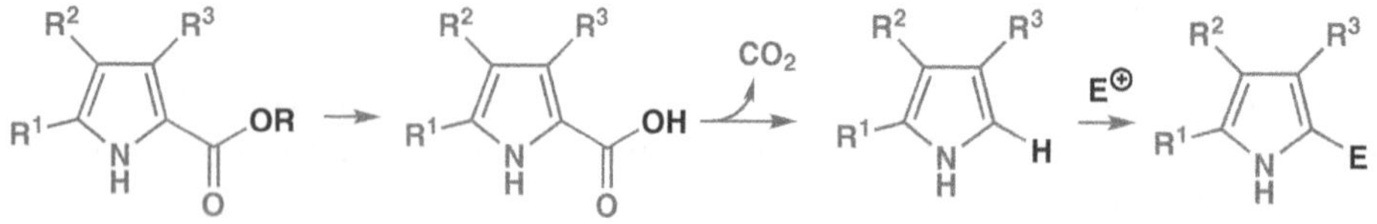

The decarboxylation is a general reaction of pyrroles: it's a kind of reverse Friedel–Crafts reaction in which the electrophile is a proton (provided by the carboxylic acid itself) and the leaving group is carbon dioxide. The protonation may occur anywhere but it leads to reaction only if it occurs where there is a CO_2H group.

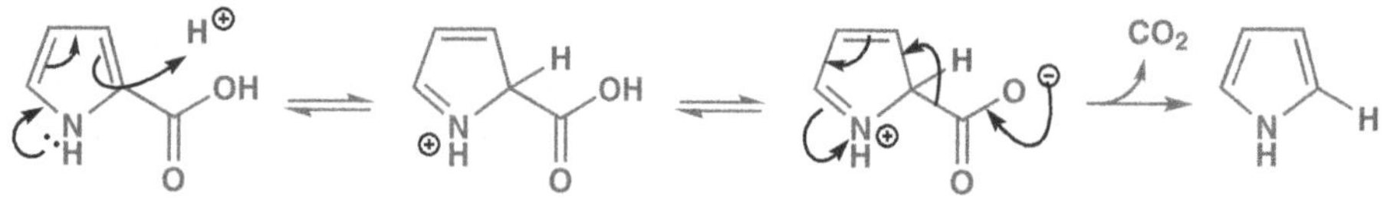

Furan and thiophene are oxygen and sulfur analogues of pyrrole

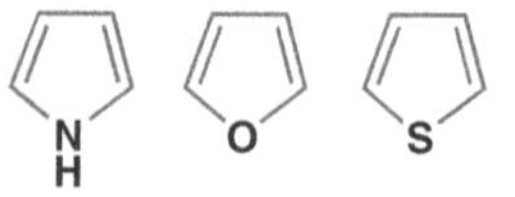

The other simple five-membered heterocycles are furan, with an oxygen atom instead of nitrogen, and thiophene, with a sulfur atom. They also undergo electrophilic aromatic substitution very readily, although not so readily as pyrrole. Nitrogen is the most powerful electron donor of the three, oxygen the next, and sulfur the least. Thiophene is very similar to benzene in reactivity.

Thiophene is the least reactive of the three because the p orbital of the lone pair of electrons on sulfur that conjugates with the ring is a 3p orbital rather than the 2p orbital of N or O, so overlap with the 2p orbitals on carbon is less good. Both furan and thiophene undergo more or less normal Friedel–Crafts reactions, although the less reactive anhydrides (here acetic anhydride, Ac_2O) are used instead of acid chlorides, and weaker Lewis acids than $AlCl_3$ are preferred.

pyrrole   furan   thiophene

Notice that the regioselectivity is the same as it was with pyrrole—the 2-position is more reactive than the 3-position in both cases. The product ketones are less reactive towards electrophiles than the starting heterocycles and deactivated furans can even be nitrated with the reagents used for benzene derivatives. Notice that reaction has occurred at the 5-position in spite of the presence of the ketone. The preference for 2- and 5-substitution is quite marked.

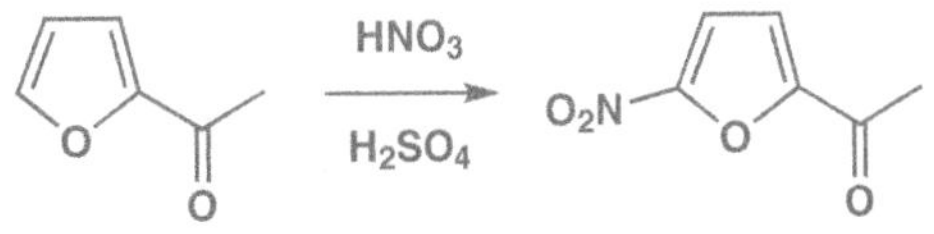

Electrophilic addition may be preferred to substitution with furan

So far, thiophenes and furans look much the same as pyrrole but there are other reactions in which they behave quite differently and we shall now concentrate on those. Furan is less aromatic than pyrrole, and if there is the prospect of forming stable bonds such as C–O single bonds by addition, this may be preferred to substitution. A famous example is the reaction of furan with bromine in methanol. In non-hydroxylic solvents, polybromination occurs as expected, but in MeOH no bromine is added at all!

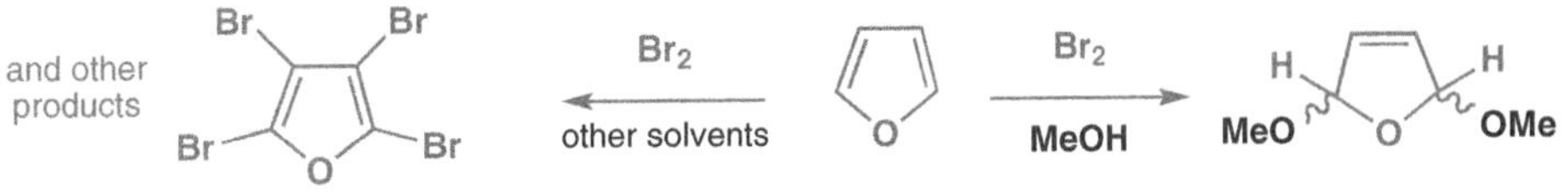

Bromination must start in the usual way, but a molecule of methanol captures the first formed cation in a 1,4-addition to furan.

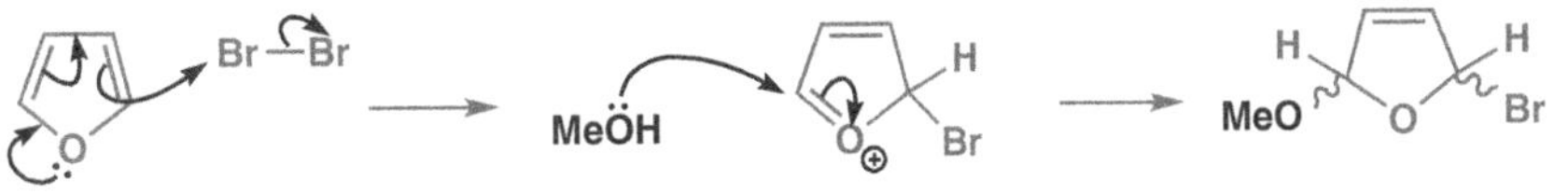

The bromine atom that was originally added is now pushed out by the furan oxygen atom to make a relatively stable conjugated oxonium ion, which adds a second molecule of methanol.

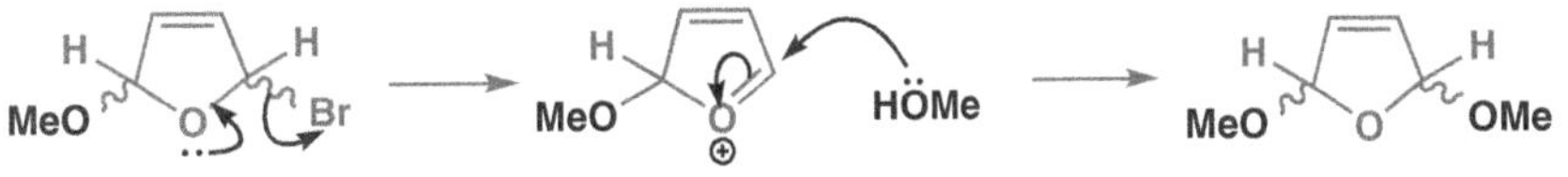

This product conceals an interesting molecule. At each side of the ring we have an acetal, and if we were to hydrolyse the acetals, we would have 'maleic dialdehyde' (*cis*-butenedial)—a molecule that is too unstable to be isolated. The furan derivative may be used in its place.

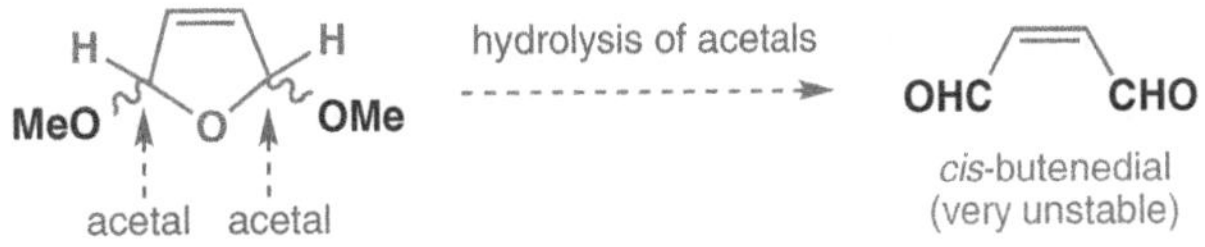

The same 1,4-dialdehyde can be made by oxidizing furan with the mild oxidizing agent dimethyldioxirane, which you met on p. 432. In this sequence, it is trapped in a Wittig reaction to give an *E,Z*-diene, which is easily isomerized to *E,E*.

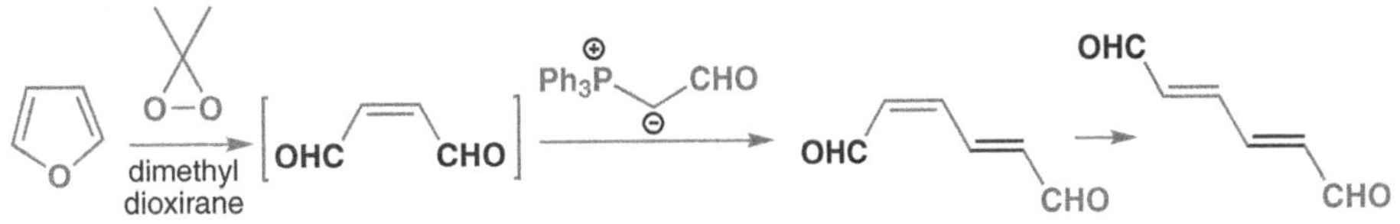

We can extend this idea of furan being the origin of 1,4-dicarbonyl compounds if we consider that furan is, in fact, an enol ether on both sides of the ring. If these enol ethers were hydrolysed we would get a 1,4-diketone.

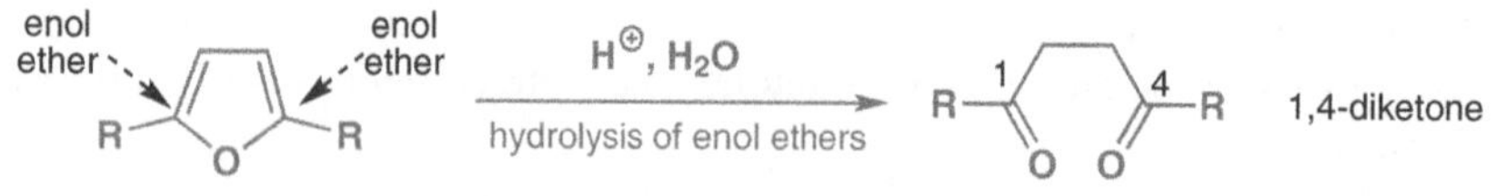

1,4-diketone

This time the arrow is solid, not dotted, because this reaction really happens. You will discover in the next chapter that furans can also be made from 1,4-diketones so this whole process is reversible. The example we are choosing has other features worth noting. The cheapest starting material containing a furan is furan-2-aldehyde or 'furfural', a by-product of breakfast cereal manufacture. Here it reacts in a typical Wittig process with a stabilized ylid.

Now comes the interesting step: treatment of this furan with acidic methanol gives a white crystalline compound having two 1,4-dicarbonyl relationships. You might like to try and draw a mechanism for this reaction.

> ➡ We explained some of the challenges in making 1,4-difunctionalized compounds in Chapter 28.

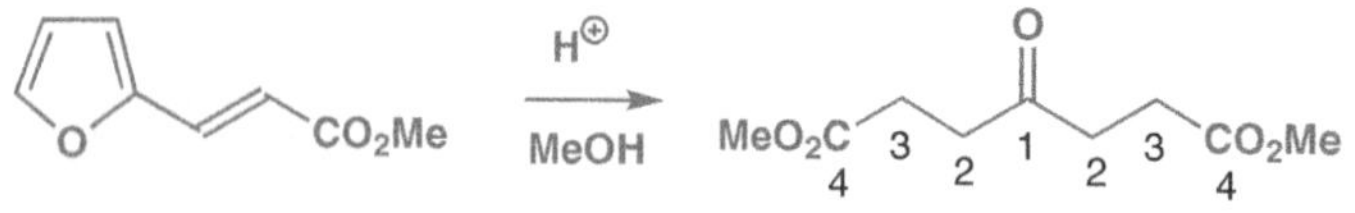

The thiophene ring can also be opened up, but in a very different way. Reductive removal of the sulfur atom with Raney nickel reduces not only the C–S bonds but also the double bonds in the ring and the four carbons in the ring form a saturated alkyl chain. If the reduction follows two Friedel–Crafts reactions on thiophene the product is a 1,6-diketone instead of the 1,4-diketones from furan. Thiophene is well behaved in Friedel–Crafts acylations, and reaction occurs at the 2- and 5-positions unless these are blocked.

> ➡ Raney nickel was introduced in Chapter 23, p. 537.

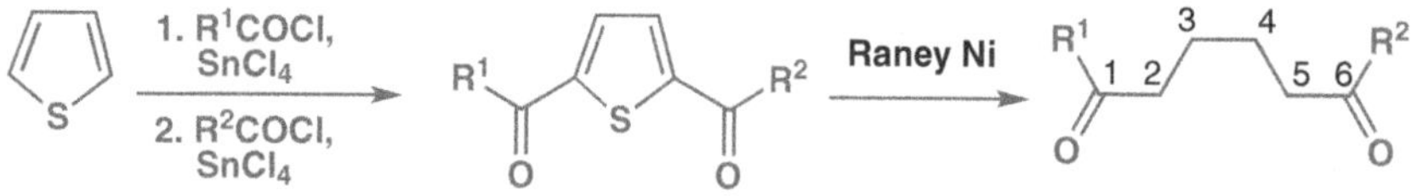

Lithiation of thiophenes and furans

A reaction that furans and thiophenes do particularly well and that fits well with these last two reactions is metallation, particularly lithiation, of a C–H group next to the heteroatom. Metallation of benzene rings (Chapter 24) is carried out by lithium–halogen (Br or I) exchange—a method that works well for heterocycles too as we will see later with pyridine—or by directed (*ortho*) lithiation of a C–H group next to an activating group such as OMe. With thiophene and furan, the heteroatom in the ring provides the necessary activation.

Activation is by coordination of O or S to Li followed by proton removal by the butyl group—the by-product is gaseous butane. These lithium compounds have a carbon–lithium σ bond and are soluble in organic solvents. We shall represent them very simply, but in fact they are typically dimers or more complex aggregates, with the coordination sphere of Li completed by THF molecules.

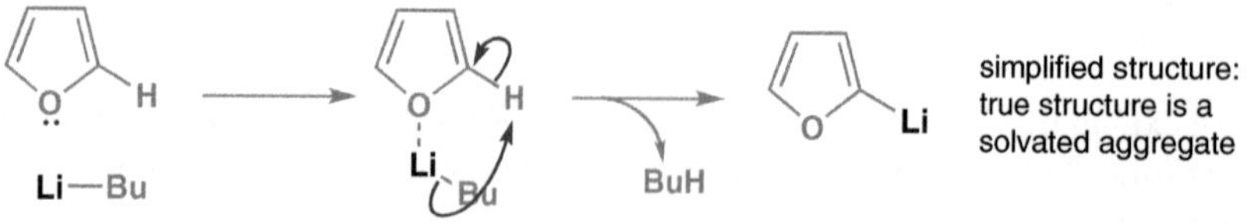

simplified structure: true structure is a solvated aggregate

These lithium compounds are very reactive and will combine with most electrophiles—in this example the organolithium is alkylated by a benzylic halide. Treatment with aqueous acid gives the 1,4-diketone by hydrolysis of the two enol ethers.

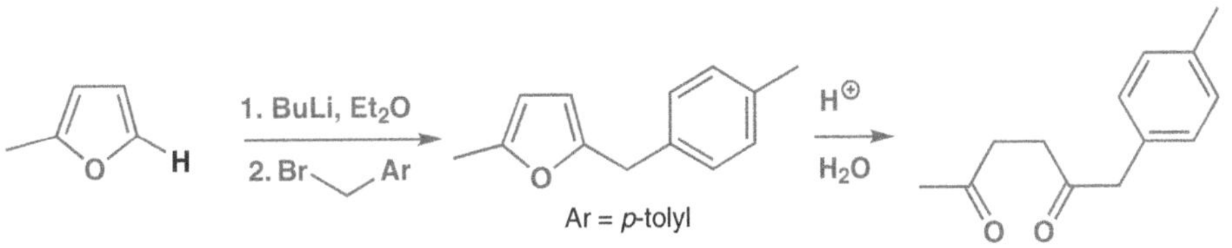

Treatment of this diketone with *anhydrous* acid would cause recyclization to the same furan (see Chapter 30) but it can alternatively be cyclized in base by an intramolecular aldol reaction (Chapter 26) to give a cyclopentenone.

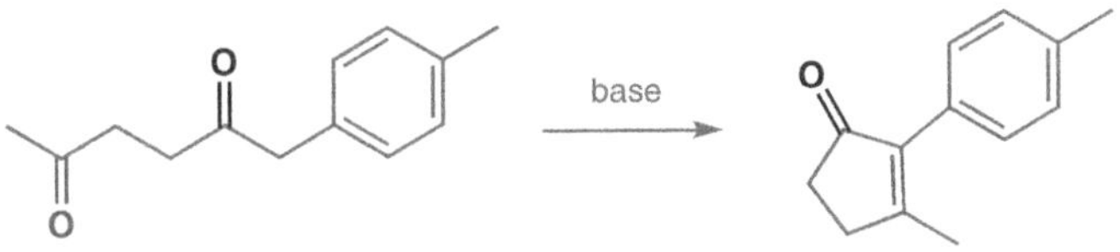

This completes our exploration of chemistry special to thiophene and furan, and we now return to all three heterocycles (pyrrole in particular) and look at *nucleophilic* substitution.

More reactions of five-membered heterocycles

Nucleophilic substitution requires an activating group

Nucleophilic substitution is a relatively rare reaction with pyrrole, thiophene, or furan and requires an activating group such as nitro, carbonyl, or sulfonyl, just as it does with benzene (Chapter 22). This intramolecular example is used to make the painkiller ketorolac.

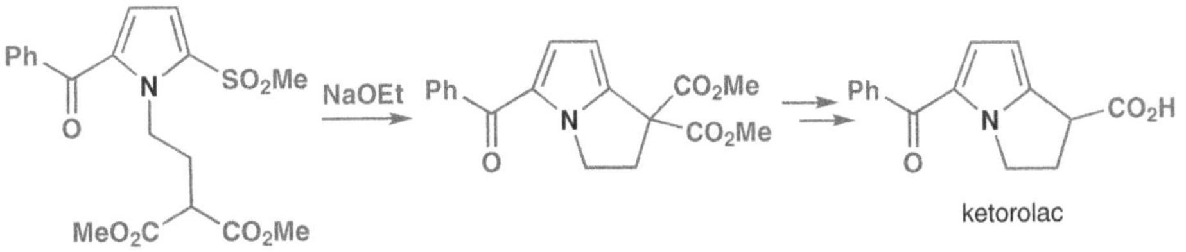

The nucleophile is a stable enolate and the leaving group is a sulfinate anion. An intermediate must be formed in which the negative charge is delocalized onto the carbonyl group on the ring, just as you saw in the benzene ring examples in Chapter 22. Attack occurs at the 2-position because the leaving group is there and because the negative charge can be delocalized onto the ketone from that position.

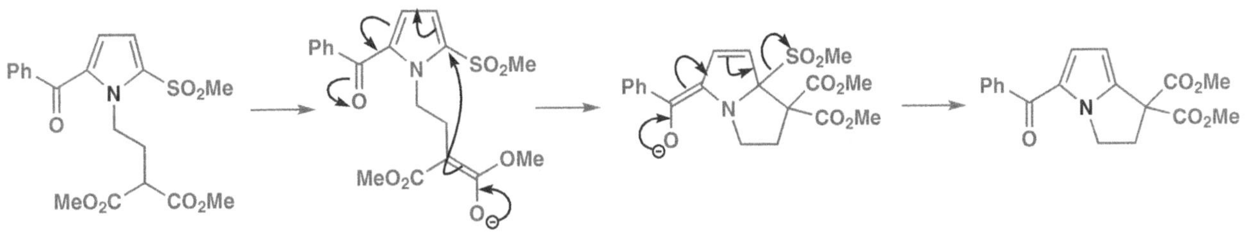

Five-membered heterocycles act as dienes in Diels–Alder reactions

All of the reactions of pyrrole, furan, and thiophene we have discussed so far have been variations on reactions of benzene. But heterocycles also do reactions totally unlike those of benzene and we are now going to explore two of them.

The first is a reaction you will meet in detail in Chapter 34. It is known as the Diels–Alder reaction, and although it has a number of subtleties we will not discuss here, it has a simple cyclic mechanism in which six electrons (three curly arrows) move around to form a new six-membered ring.

Here is an example with the Boc derivative of pyrrole. The electron-deficient Boc group makes pyrrole less nucleophilic and promotes the Diels–Alder reaction with an alkynyl sulfone. Benzene, and even many other heterocycles, will not do this sort of reaction.

➡ The Boc protecting group is discussed in Chapter 23, p. 558.

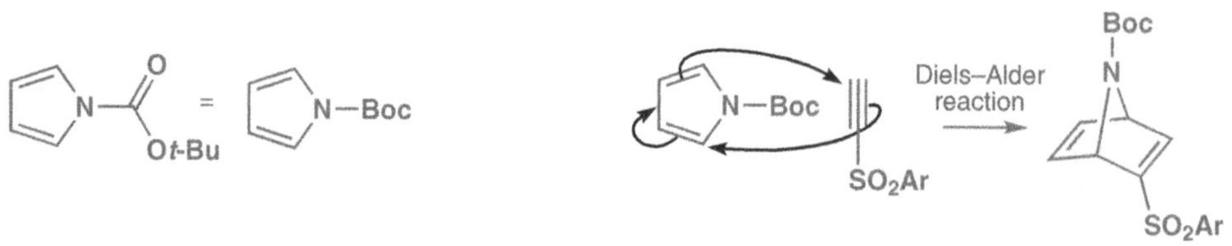

The product is a useful intermediate in the synthesis of the analgesic epibatidine. Selective reduction of the non-conjugated double bond is followed by addition of a pyridine nucleophile (a lithium derivative can be prepared from a bromopyridine) to the vinyl sulfone.

Epibatidine was discovered in the skin of Ecuadoran frogs in 1992. It is an exceptionally powerful analgesic and works by a different mechanism from that of morphine so there is hope that it will not be addictive. The compound can now be synthesized so there is no need to kill the frogs to get it—indeed, they are a protected species.

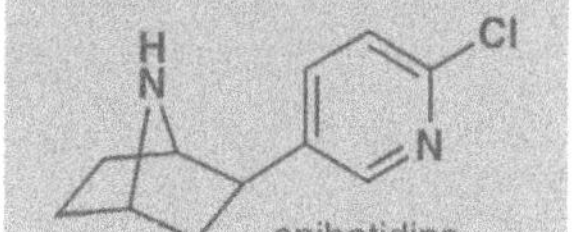

epibatidine

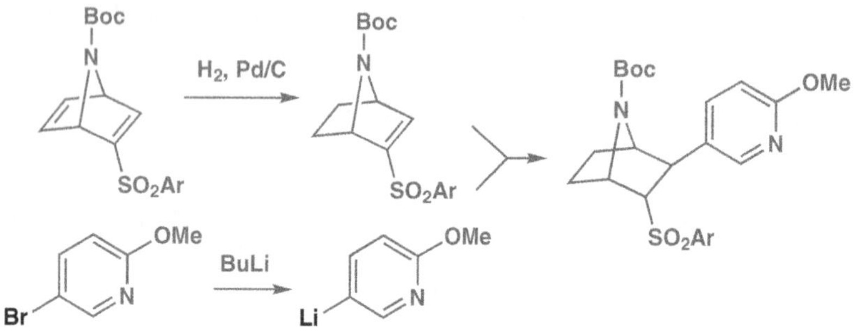

Furan is particularly good at Diels–Alder reactions but it gives the thermodynamic product, the *exo* adduct, because with this aromatic diene the reaction is reversible.

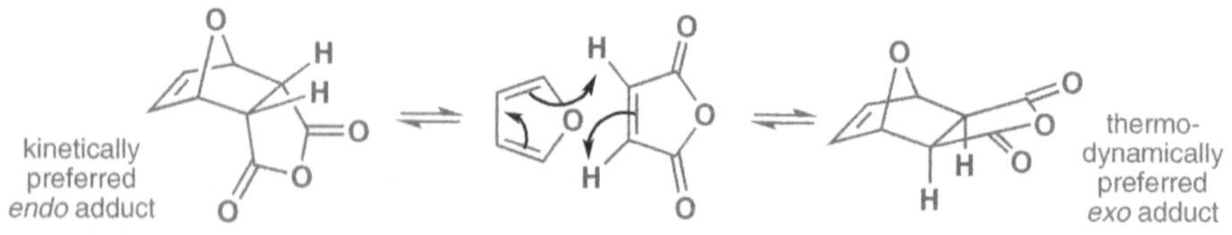

kinetically preferred *endo* adduct

thermo-dynamically preferred *exo* adduct

➡ *Endo* and *exo* Diels–Alder adducts are explained in Chapter 34.

Aromaticity prevents thiophene taking part in Diels–Alder reactions, but oxidation to the sulfone destroys the aromaticity because both lone pairs become involved in bonds to oxygen. The sulfone is unstable and reacts with itself but will also do Diels–Alder reactions. With an alkyne, loss of SO_2 gives a substituted benzene derivative.

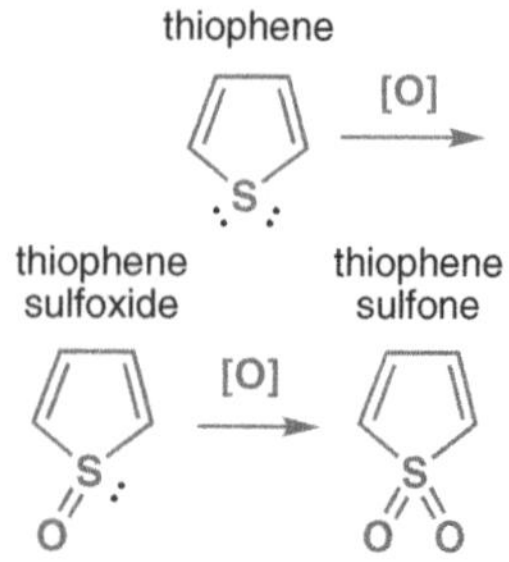

thiophene

[O]

thiophene sulfoxide   thiophene sulfone

[O]

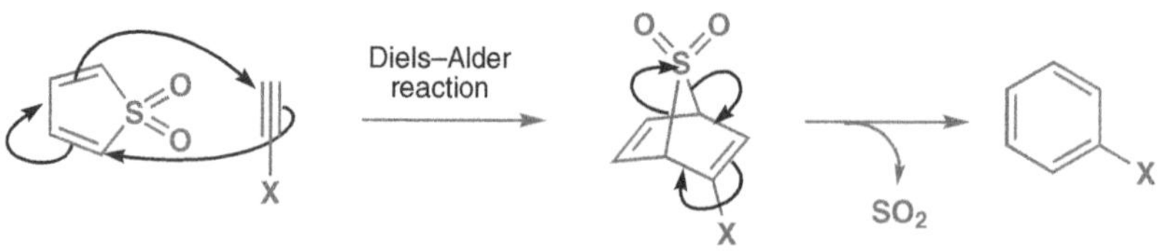

Diels–Alder reaction

SO_2

Similar reactions occur with α-pyrones. These are also rather unstable and barely aromatic and they react with alkynes by Diels–Alder reactions followed by reverse Diels–Alder reactions to give benzene derivatives with the loss of CO_2.

Nitrogen anions can be easily made from pyrrole

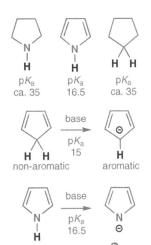

Pyrrole is much more acidic than comparable saturated amines. The pK_a of pyrrolidine is about 35, but pyrrole has a pK_a of 16.5, making it some 10^{23} times more acidic! Pyrrole is about as acidic as a typical alcohol so bases stronger than alkoxides will convert it to its anion. We should not be too surprised at this as the corresponding hydrocarbon, cyclopentadiene, is also extremely acidic, with a pK_a of 15. The reason is that the anions are aromatic with six delocalized π electrons. The effect is much greater for cyclopentadiene because the hydrocarbon is not aromatic and much less for pyrrole because it is already aromatic and has less to gain.

In all of the reactions of pyrrole that we have so far seen, new groups have added to the carbon atoms of the ring. The anion of pyrrole is useful because it reacts at nitrogen. The nitrogen atom has two lone pairs of electrons in the anion: one is delocalized around the ring but the other is localized in an sp^2 orbital on nitrogen. This high-energy pair is the new HOMO and this is where the molecule reacts. *N*-acylated derivatives in general can be made in this way. A commonly used base is sodium hydride (NaH) but weaker bases produce enough anion for reaction to occur.

- **Anions of pyrroles react with electrophiles at the *nitrogen* atom.**

This is how the *N*-Boc pyrrole was made for use in the synthesis of epibatidine on p. 739. The base used was the pyridine derivative DMAP, which you met earlier in the chapter (p. 726). Its conjugate acid has a pK_a of 9.7 and so produces small, equilibrating amounts of the anion as well as acting as a nucleophilic catalyst. 'Boc anhydride' is used as the acylating agent.

➡ DMAP's pK_a of 9.7 is between those of pyridine (5.5) and tertiary alkyl amines (ca. 10) but is much closer to the latter.

'Boc anhydride'   DMAP   *N*-Boc-pyrrole

Anion formation is important in the next main section of this chapter, which is about what happens when we insert more nitrogen atoms into the pyrrole ring.

Five-membered rings with two or more nitrogen atoms

Imidazole

At the beginning of this chapter we imagined adding more nitrogen atoms to the pyrrole ring and noticed then that there were two compounds with two nitrogen atoms: pyrazole and imidazole.

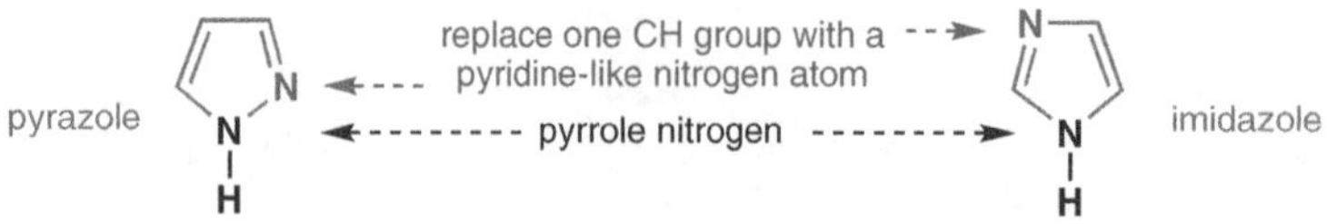

Only one nitrogen atom in a five-membered ring can contribute two electrons to the aromatic sextet. The other replaces a CH group, has no hydrogen, and is like the nitrogen atom in pyridine. The black nitrogens are the pyrrole-like nitrogens; the green ones are pyridine-like. The lone pairs on the black nitrogens are delocalized round the ring; those on the green nitrogens are localized in sp² orbitals on nitrogen. We can expect these compounds to have properties intermediate between those of pyrrole and pyridine. Imidazole is a stronger base than either pyrrole or pyridine—the imidazolium ion has a pK_a of almost exactly 7, meaning that it is 50% protonated in neutral water. Imidazole is also more acidic than pyrrole, with a pK_a of 14.5.

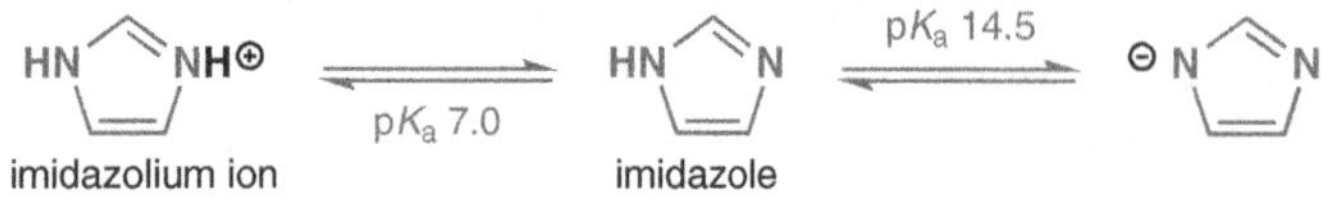

These curious results are a consequence of the 1,3 relationship between the two nitrogen atoms. Both the (protonated) cation and the (deprotonated) anion share the charge equally between the two nitrogen atoms—they are perfectly symmetrical and unusually stable. Another way to look at the basicity of imidazole would be to say that both nitrogen atoms can act at once on the proton being attacked. It has to be the pyridine-like nitrogen that actually captures the proton but the pyrrole nitrogen can help by using its delocalized electrons like this:

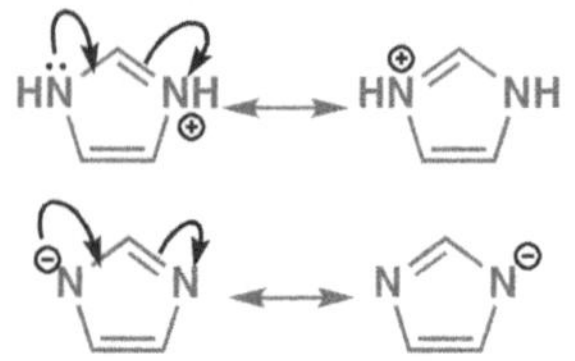

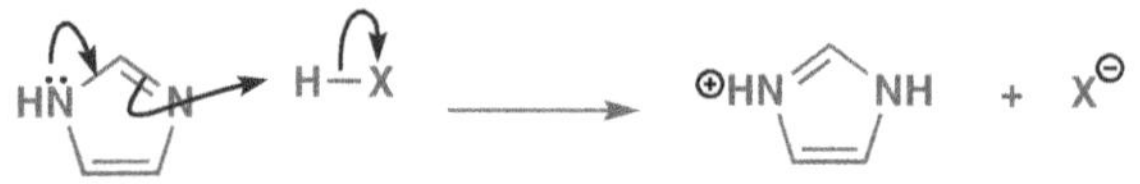

➡ A similar effect accounts for the basicity of DBU, see p. 175.

Nature makes use of this property by having imidazole groups attached to proteins in the form of the amino acid histidine and using them as nucleophilic, basic, and acidic catalytic groups in enzyme reactions (this will be discussed in Chapter 42). We use this property in the same way when we add a silyl group to an alcohol. Imidazole is a popular catalyst for these reactions.

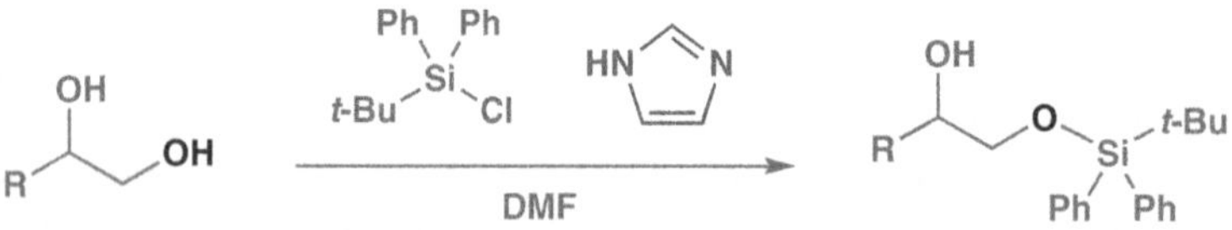

A weakly basic catalyst is needed here because we want to discriminate between the primary and secondary alcohols in the diol. Imidazole is too weak a base to remove protons from an alcohol (pK_a ~ 16) but it can remove a proton after the OH group has attacked the silicon atom.

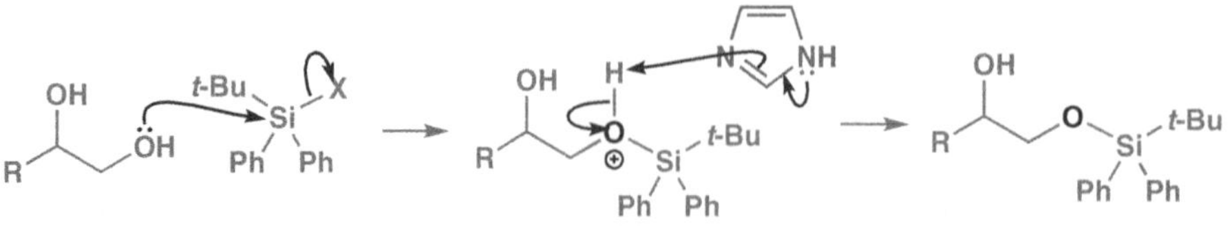

In fact, the imidazole is also a nucleophilic catalyst of this reaction, and the first step is substitution of Cl by imidazole—that is why the leaving group in the last scheme was shown as 'X'. The reaction starts off like this:

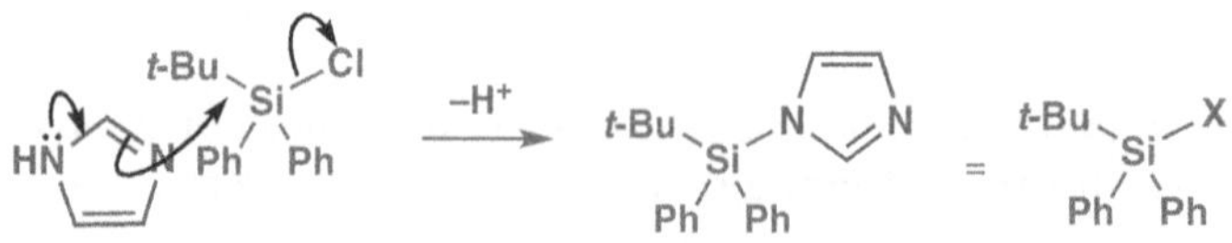

The same idea leads to the use of carbonyl diimidazole (CDI) as a double electrophile when we want to link two nucleophiles together by a carbonyl group. Phosgene (COCl$_2$) has been used for this but it is appallingly toxic (it was used in the First World War as a poison gas with dreadful effects). CDI is safer and more controlled. In these reactions imidazole acts (twice) as a leaving group.

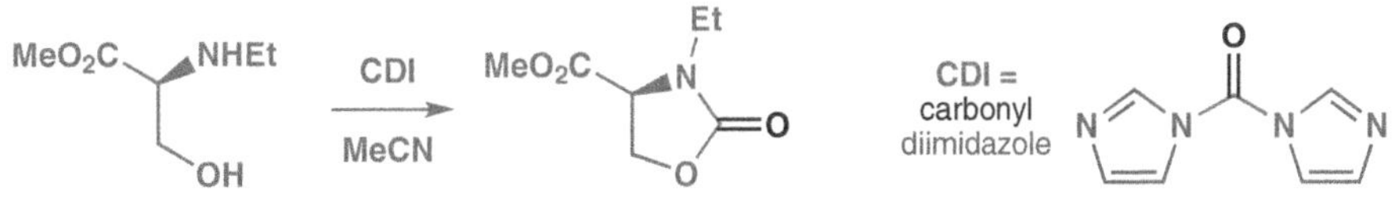

The amino group probably attacks first to displace one imidazole anion, which returns to deprotonate the ammonium salt. The alcohol can then attack intramolecularly, displacing the second imidazole anion, which deprotonates the OH group in its turn. The other product is just two molecules of imidazole.

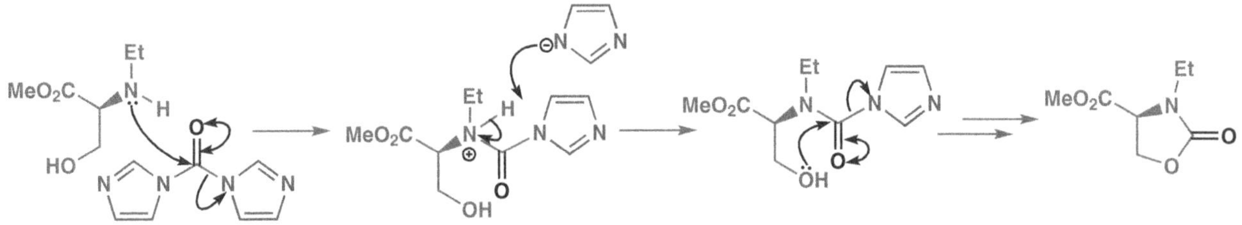

The relationship between the delocalized imidazole anion and imidazole itself is rather like that between an enolate anion and an enol. It will come as no surprise therefore that, like an enol, imidazole tautomerizes rapidly at room temperature in solution. For the parent compound the two tautomers are the same, but with unsymmetrical imidazoles the tautomerism is more interesting. We will explore this question alongside electrophilic aromatic substitution of imidazoles. Imidazoles with a substituent between the two nitrogen atoms (position 2) can be nitrated with the usual reagents and the product consists of a mixture of tautomers.

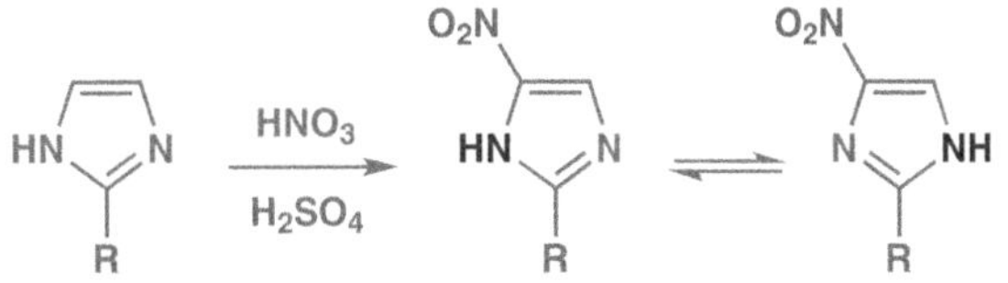

two different tautomers of a nitro-imidazole

The initial nitration may occur at either of the remaining sites on the ring with the electrons coming from the pyrrole-like nitrogen atom. Tautomerism after nitration gives the mixture. Tautomerism is rapid and the tautomers cannot be separated.

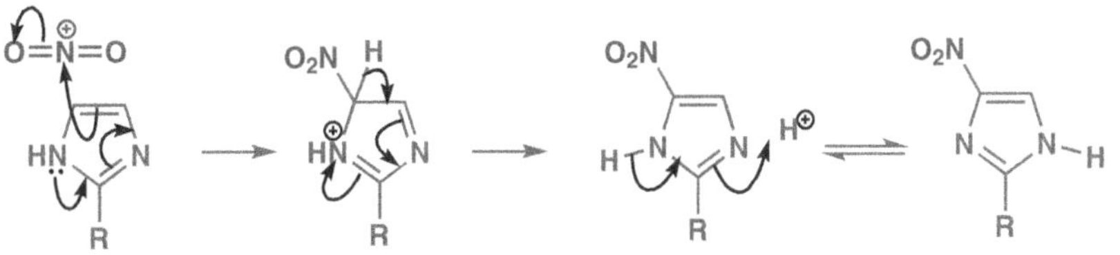

The tautomerism can be stopped by alkylation at one of the nitrogen atoms. If this is done in basic solution, the anion is an intermediate and the alkyl group adds to the nitrogen atom next to the nitro group. Again, it does not matter from which tautomer the anion is derived— there is only one anion delocalized over both nitrogen atoms and the nitro group. One reason for the formation of this isomer is that it has the linear conjugated system between the pyrrole-like nitrogen and the nitro group (see p. 734).

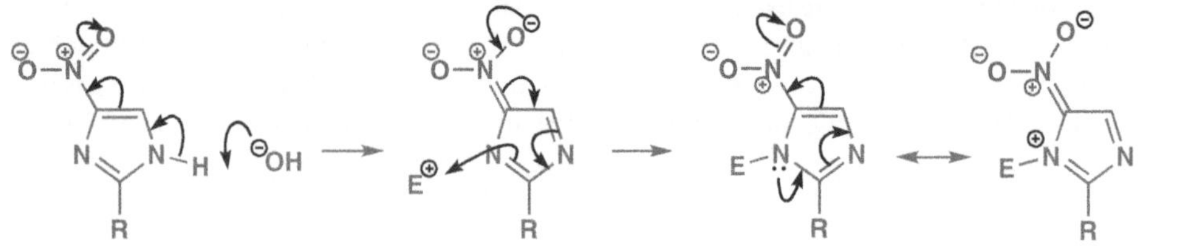

Important medicinal compounds are made in this way. The antiparasitic metronidazole comes from 2-methyl imidazole by nitration and alkylation with an epoxide in base.

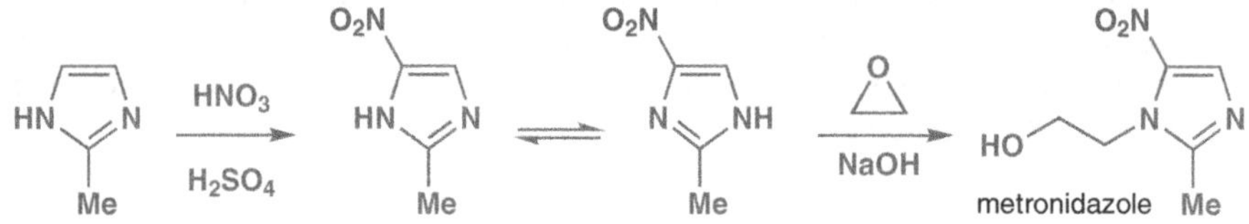

The triazoles

There are two triazoles, and each has one pyrrole-like nitrogen and two pyridine-like nitrogens. Both triazoles have the possibility of tautomerism (in 1,2,3-triazole the tautomers are identical) and both give rise to a single anion.

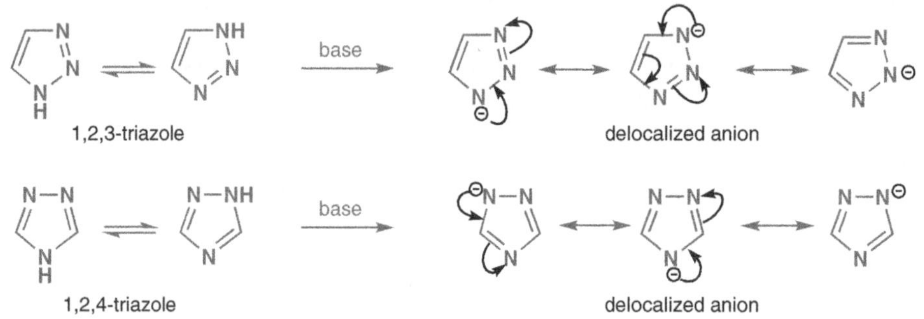

The 1,2,4-triazole is more important because it is the basis of the best modern agricultural fungicides as well as drugs for fungal diseases in humans. The extra nitrogen atom, inevitably of the pyridine type, makes it more weakly basic than imidazole, but it increases its acidity so that the anion is now easy to make.

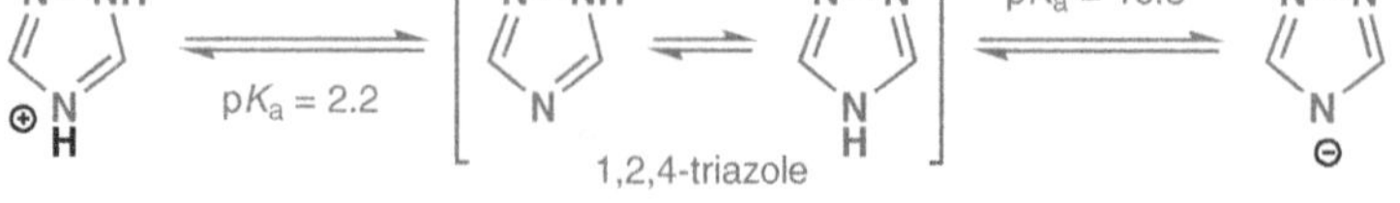

The fungicides are usually made by the addition of the triazole anion to an epoxide or other carbon electrophile. The anion normally reacts at one of the two linked nitrogen atoms (it does not matter which—the product is the same).

A modern example of an agent used against human fungal infections is Pfizer's fluconazole, which actually contains two triazoles. The first is added as the anion to an α-chloroketone and the second is added to an epoxide made with the sulfur ylid chemistry you met in Chapter 27. Note that weak bases were used to catalyse both of these reactions. Triazole is acidic enough for even $NaHCO_3$ to produce a small amount of the anion.

manufacture of Pfizer's fluconazole

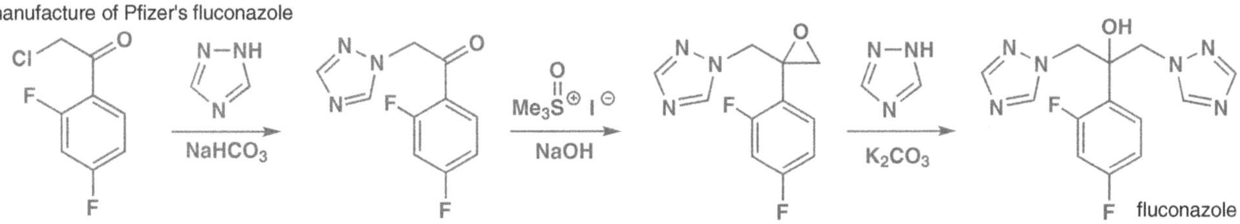

Tetrazole

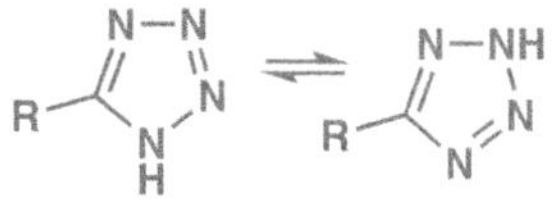

two tautomers of a tetrazole

There is only one isomer of tetrazole or of *C*-substituted tetrazoles, as there is only one carbon atom in the ring, although there may be several tautomers. The main interest in tetrazoles is that they are rather acidic: the pK_a for the loss of the NH proton to form an anion is about 5, essentially the same as that of a carboxylic acid. The anion is delocalized over all four nitrogen atoms (as well as the one carbon atom), and four nitrogen atoms do the work of two oxygen atoms.

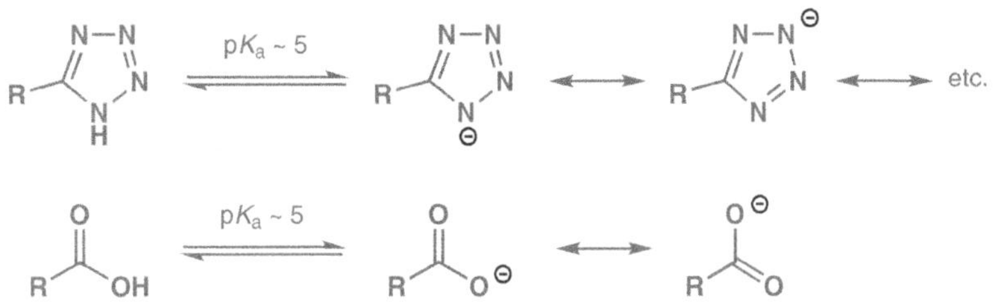

Because tetrazoles have similar acidities to those of carboxylic acids, they have been used in drugs as replacements for the CO_2H unit when the carboxylic acid has unsatisfactory properties for human medicine. A simple example is the anti-arthritis drug indomethacin whose carboxylic acid group may be replaced by a tetrazole with no loss of activity.

indomethacin

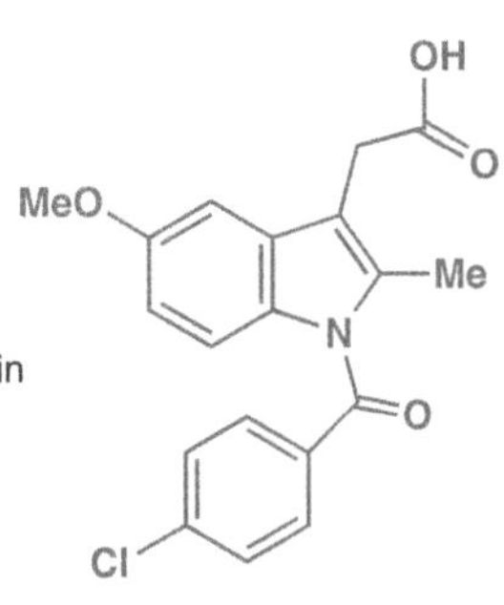

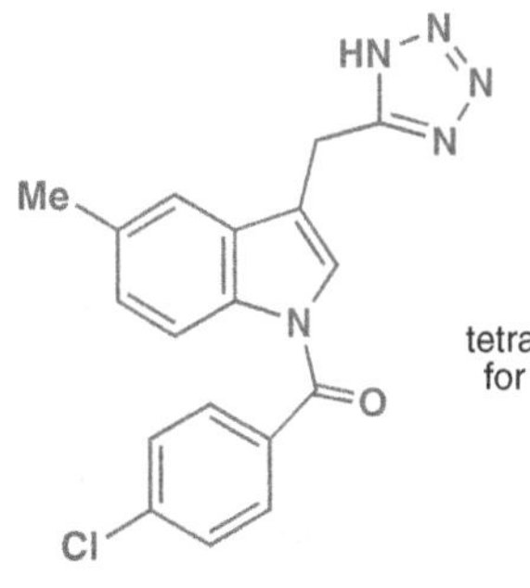

tetrazole substitute for indomethacin

Nitrogen atoms and explosions

Compounds with even two or three nitrogen atoms joined together, such as diazomethane (CH_2N_2) or azides (RN_3), are potentially explosive because they can suddenly give off stable gaseous nitrogen. Compounds with more nitrogen atoms, such as tetrazoles, are likely to be more dangerous and few people have attempted to prepare pentazoles. The limit is reached with diazotetrazole, with the amazing formula CN_6! It is made by diazotization of 5-aminotetrazole, which first gives a diazonium salt.

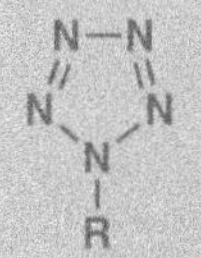

a pentazole
highly explosive!

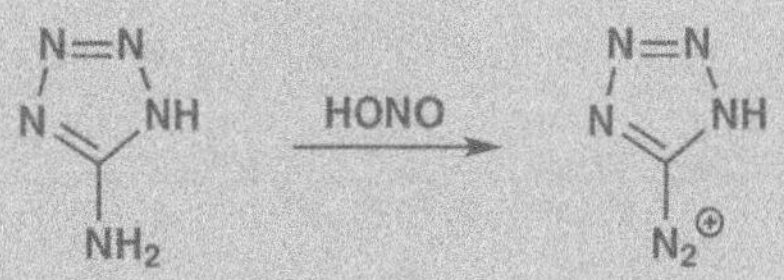

5-amino-(1H)-tetrazole

the diazonium salt
highly explosive!

The diazonium salt is extremely dangerous: 'It should be emphasised that [the diazonium salt] is extremely explosive and should be handled with great care. We recommend that no more than 0.75 mmol be isolated at one time. Ethereal solutions are somewhat more stable but explosions have occurred after standing at −70 °C for 1 hr.' So much for that, but what about the diazo compound? It is extremely unstable and decomposes to a carbene with loss of one molecule of nitrogen and then loses two more to give...

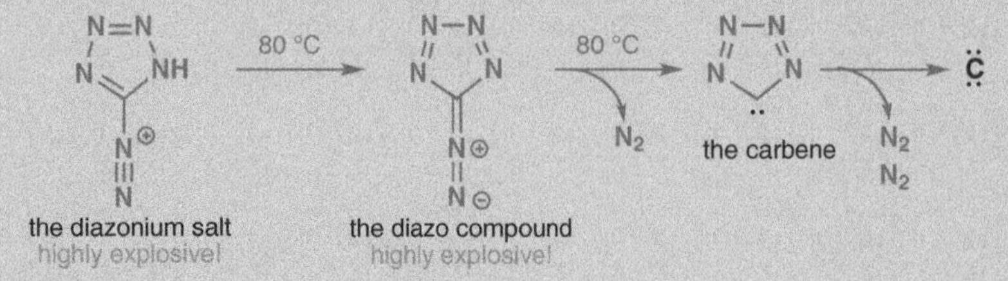

All that is left is a carbon atom and this is one of very few ways to make carbon atoms chemically. The carbon atoms have remarkable reactions and these have been studied briefly, but the hazardous preparation of the starting materials discourages too much research. However, you will see in the next chapter that 1-amino tetrazole is a useful starting material for making an anti-allergic drug.

Benzo-fused heterocycles

Indoles are benzo-fused pyrroles

Indomethacin and its tetrazole analogue contain pyrrole rings with benzene rings fused to the side. Such bicyclic heterocyclic structures are called **indoles** and are our next topic. Indole itself has a benzene ring and a pyrrole ring sharing one double bond, or, if you prefer to look at it this way, it is an aromatic system with 10 electrons—eight from four double bonds and the lone pair from the nitrogen atom.

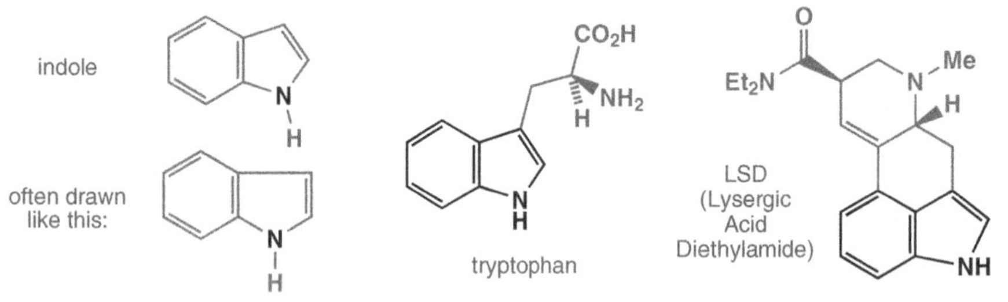

Indole is an important heterocyclic system because it is built into proteins in the form of the amino acid tryptophan (Chapter 42) because it is the basis of important drugs such as indomethacin, and because it provides the skeleton of the **indole alkaloids**—biologically active compounds from plants including strychnine and LSD (alkaloids are discussed in Chapter 42).

In many ways the chemistry of indole is that of a reactive pyrrole ring with a relatively unreactive benzene ring standing on one side—electrophilic substitution almost always occurs on the pyrrole ring, for example. But indole and pyrrole differ in one important respect. In indole, electrophilic substitution is preferred in the 3-position with almost all reagents whereas it occurs in the 2-position with pyrrole. Halogenation, nitration, sulfonation, Friedel−Crafts acylation, and alkylation all occur cleanly at that position.

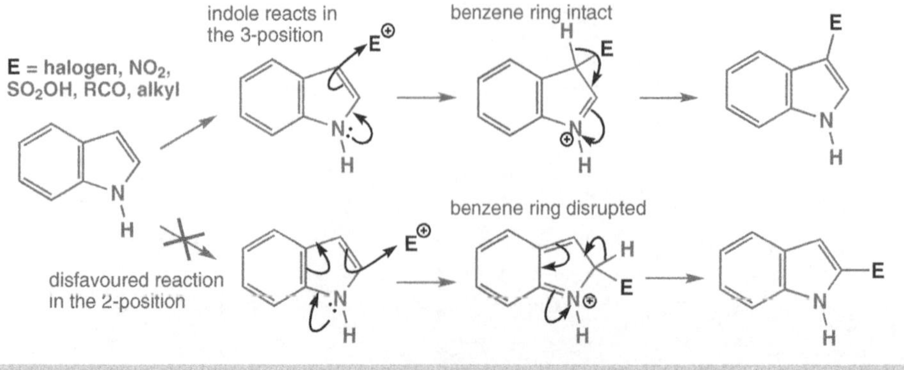

This is, of course, the reverse of what happens with pyrrole. Why should this be? A simple explanation is that reaction at the 3-position simply involves the rather isolated enamine system in the five-membered ring and does not disturb the aromaticity of the benzene ring. The positive charge in the intermediate is, of course, delocalized round the benzene ring, but it gets its main stabilization from the nitrogen atom. It is not possible to get reaction in the 2-position without seriously disturbing the aromaticity of the benzene ring.

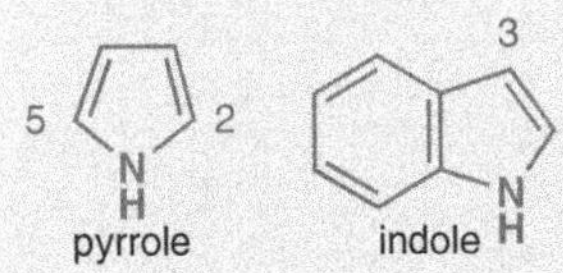

● Electrophilic substitution on pyrrole and indole

Pyrrole reacts with electrophiles at all positions but prefers the 2- and 5-positions, while indole much prefers the 3-position.

A simple example of electrophilic substitution is the Vilsmeier formylation with DMF and POCl$_3$, showing that indole has similar reactivity, if different regioselectivity, to pyrrole. If the 3-position is blocked, reaction occurs at the 2-position and this at first seems to suggest that it is all right after all to take the electrons the 'wrong way' round the five-membered ring. This intramolecular Friedel–Crafts alkylation is an example.

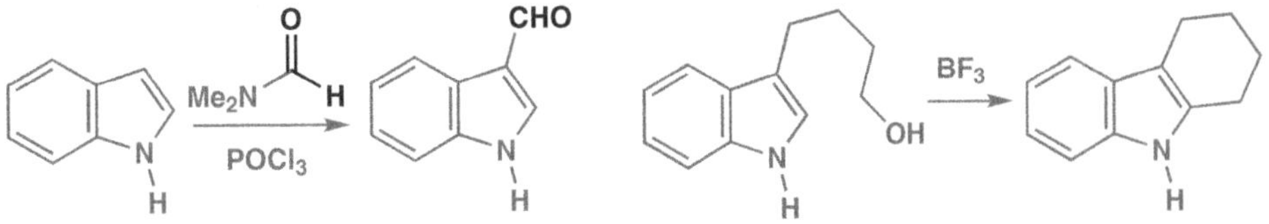

An ingenious experiment showed that this cyclization is not as simple as it seems. If the starting material is labelled with tritium (radioactive ^{3}H) next to the ring, the product shows exactly 50% of the label where it is expected and 50% where it is not.

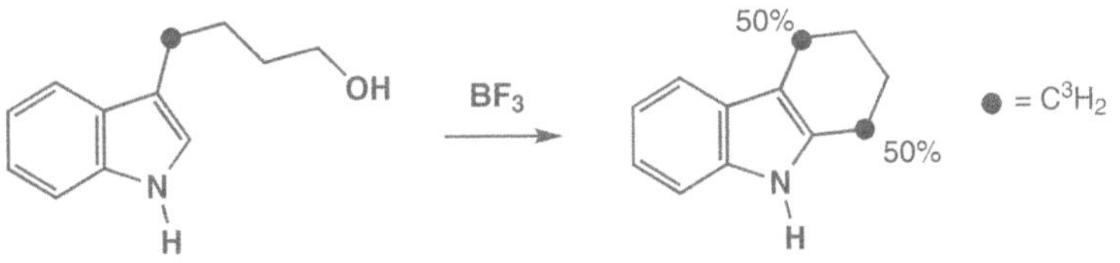

To give this result, the reaction must have a symmetrical intermediate and the obvious candidate arises from attack at the 3-position. The product is formed from the intermediate *spiro* compound, which has the five-membered ring at right angles to the indole ring—each CH$_2$ group has an exactly equal chance of migrating.

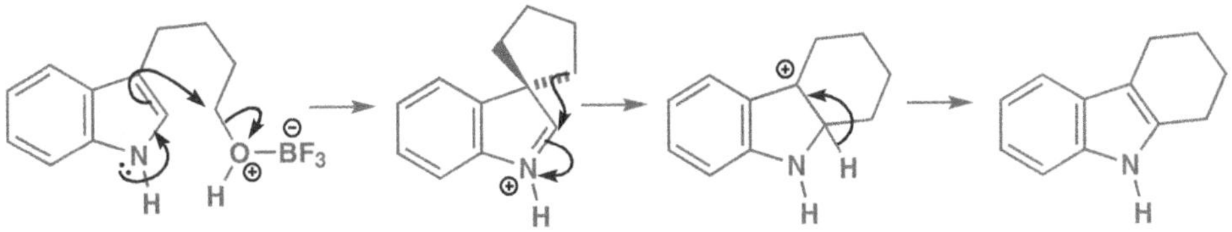

➡ The migration is a pinacol-like rearrangement similar to those in Chapter 36.

It is now thought that most substitutions in the 2-position go by this migration route but that some go by direct attack with disruption of the benzene ring. A good example of indole's 3-position preference is the Mannich reaction, which works as well with indole as it does with pyrrole or furan.

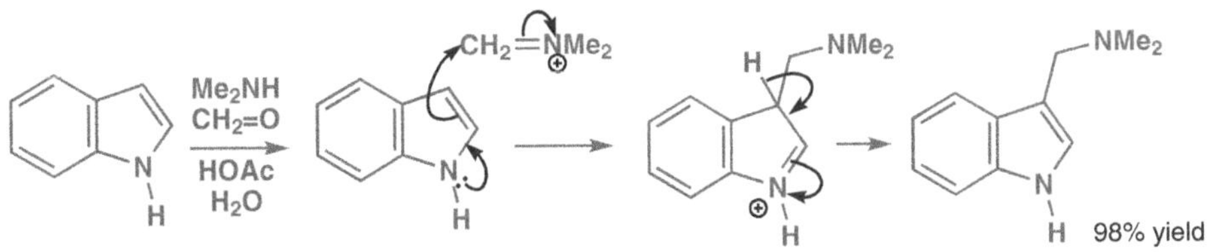

The electron-donating power of the indole and pyrrole nitrogens is never better demonstrated than in the use to which these Mannich bases (the products of the reaction) are put. You may remember that normal Mannich bases can be converted to other compounds by alkylation and elimination (see p. 621). No alkylation is needed here as the indole nitrogen can even expel the Me₂N group when NaCN is around as a base and nucleophile. The reaction is slow and the yield not wonderful but it is amazing that it happens at all. The reaction is even easier with pyrrole derivatives.

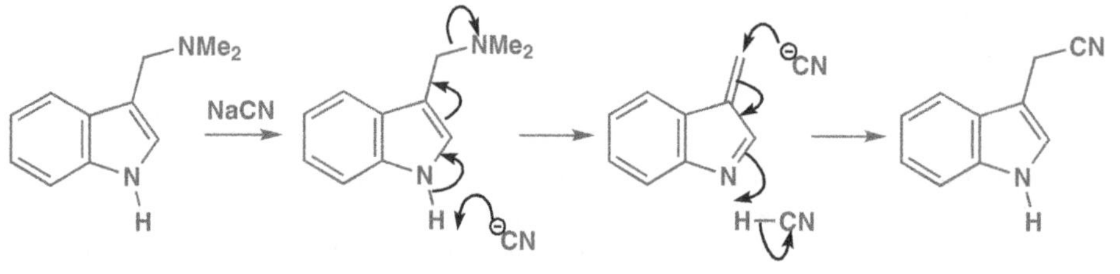

All of the five-membered rings we have looked at have their benzo-derivatives but we will concentrate on just one, 1-hydroxybenzotriazole, both because it is an important compound and because we have said little about simple 1,2,3-triazoles.

HOBt is an important reagent in peptide synthesis

1-Hydroxybenzotriazole (HOBt) is a friend in need in the lives of biochemists. It is added to many reactions where an activated ester of one amino acid is combined with the free amino group of another (see Chapter 23 for some examples). It was first made in the nineteenth century by a remarkably simple reaction.

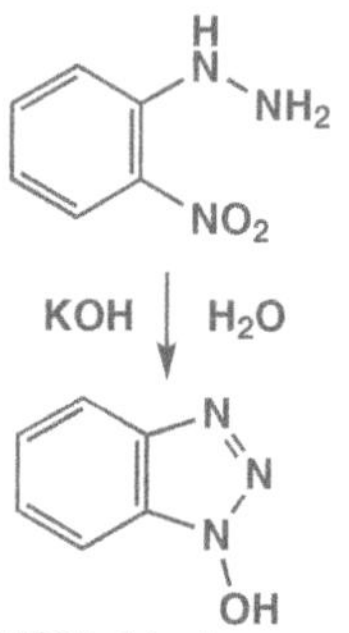

The structure of HOBt appears quite straightforward, except for the unstable N–O single bond, but we can easily draw some other tautomers in which the proton on oxygen—the only one in the heterocyclic ring—can be placed on some of the nitrogen atoms. These structures are all aromatic, the second and third are nitrones, and the third structure looks less good than the other two.

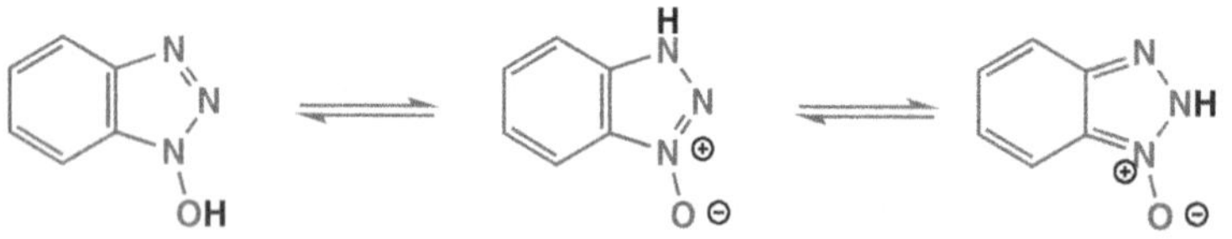

HOBt, 1-hydroxy-benzotriazole

➡ You will meet some nitrone chemistry in Chapter 34.

HOBt comes into play when amino acids are being coupled together in the laboratory. The reaction is an amide formation, but in Chapter 23 we mentioned that amino-acyl chlorides cannot be used to make polypeptides—they are too reactive and they lead to side reactions. Instead, activated amino-esters (with good RO⁻ leaving groups) are used, such as the phenyl esters of Chapter 23. It is even more common to form the activated ester in the coupling reaction, using a coupling reagent, the most common being DCC, dicyclohexylcarbodiimide. DCC reacts with carboxylic acids like this:

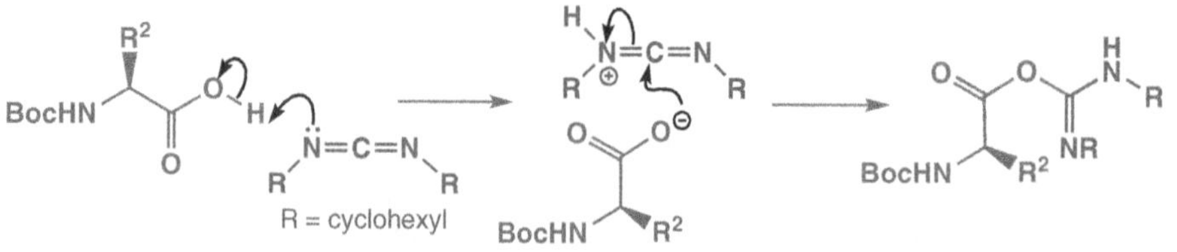

The product ester is activated because substitution with any nucleophile expels this very stable urea as a leaving group.

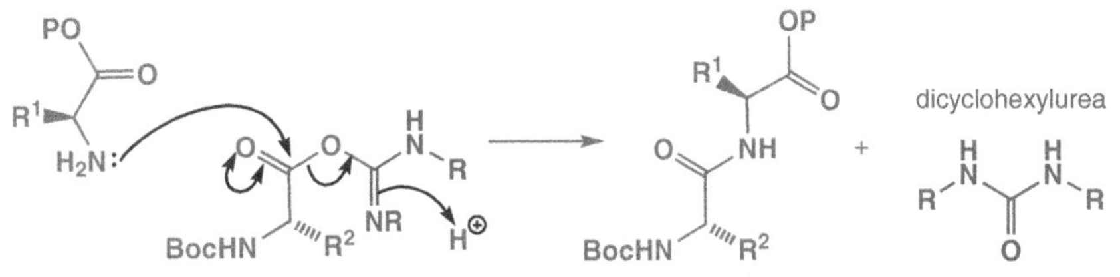

The problem with attacking this ester directly with the amino group of the second amino acid is that some racemization of the active ester is often found. A better method is to have plenty of HOBt around. It intercepts the activated ester first and the new intermediate does not racemize, mostly because the reaction is highly accelerated by the addition of HOBt. The second amino acid, protected on the carboxyl group, attacks the HOBt ester and gives the dipeptide in a very fast reaction without racemization.

> ➡ You saw in Chapter 26 that the most electrophilic carboxylic acid derivatives are also the most enolizable.

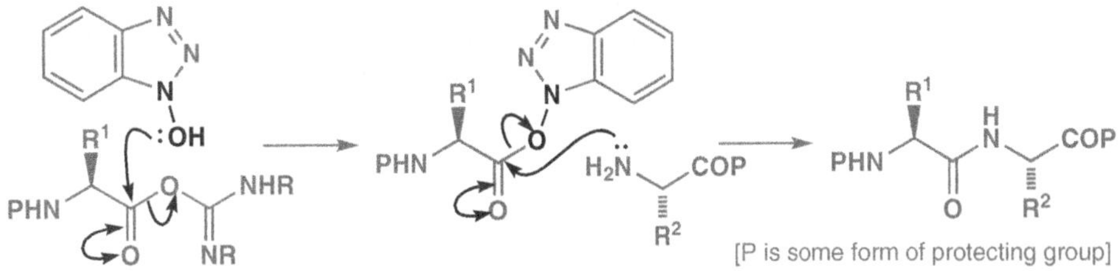

[P is some form of protecting group]

Putting more nitrogen atoms in a six-membered ring

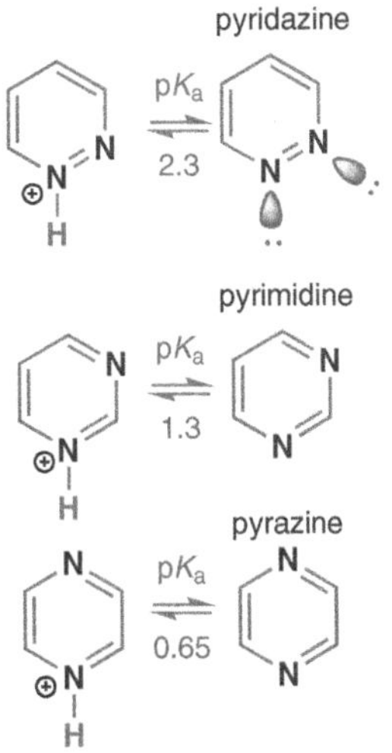

At the beginning of the chapter we mentioned the three six-membered aromatic heterocycles with two nitrogen atoms—pyridazine, pyrimidine, and pyrazine. In these compounds both nitrogen atoms must be of the pyridine sort, with lone pair electrons not delocalized round the ring.

We are going to look at these compounds briefly here. Pyrimidine is more important than either of the others because of its involvement in DNA and RNA—you will find this in Chapter 42. All three compounds are very weak bases—hardly basic at all in fact. Pyridazine is slightly more basic than the other two because the two adjacent lone pairs repel each other and make the molecule more nucleophilic (the α effect again: see p. 513). The chemistry of these very electron-deficient rings mostly concerns nucleophilic attack and displacement of leaving groups such as Cl by nucleophiles such as alcohols and amines. To introduce this subject we need to take one heterocyclic synthesis at this point, although these are properly the subject of the next chapter. The compound maleic hydrazide has been known for some time because it is easily formed when hydrazine is acylated twice by maleic anhydride.

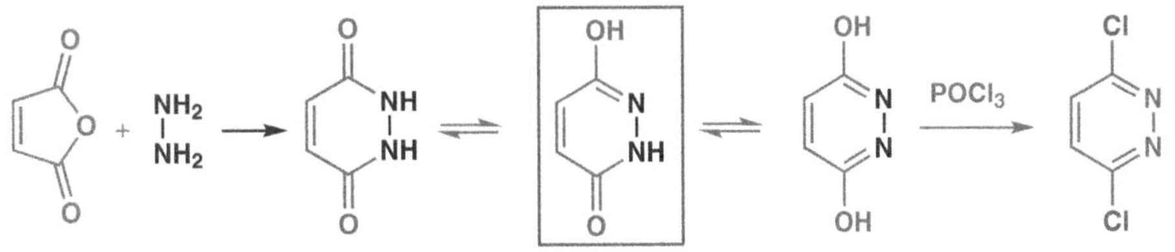

The compound actually prefers to exist as the second tautomer (in the green frame). Reaction with $POCl_3$ in the way we have seen for pyridine gives the undoubtedly aromatic pyridazine dichloride. Now we come to the point. Each of these chlorides can be displaced in turn with an oxygen or nitrogen nucleophile. Only one chloride is displaced in the first reaction, if that is required, and then the second can be displaced with a different nucleophile.

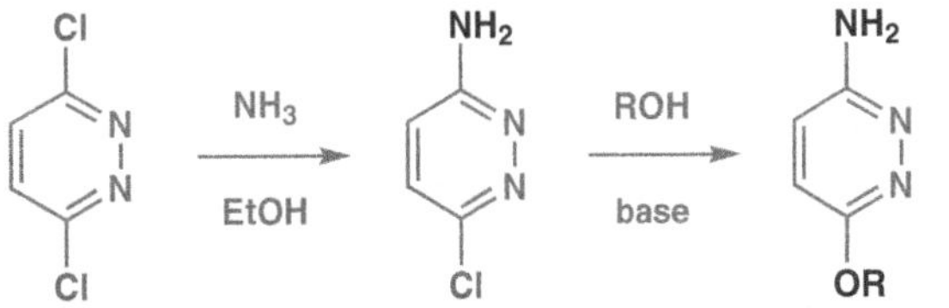

How is this possible? The mechanism of the reactions is addition to the pyridazine ring followed by loss of the leaving group. When the second nucleophile attacks it is forced to attack a less electrophilic ring. An electron-withdrawing group (Cl) has been replaced by a strongly electron-donating group (NH_2) so the rate-determining step, the addition of the nucleophile, is slower.

The same principle applies to other easily made symmetrical dichloro derivatives of these rings and their benzo analogues. The nitrogen atoms can be related 1,2, 1,3, or 1,4, as in these examples. The first two are used to link the quinine-derived ligands required for the Sharpless asymmetric dihydroxylation, which will be described in Chapter 41.

Fusing rings to pyridines: quinolines and isoquinolines

quinoline

isoquinoline

A benzene ring can be fused on to the pyridine ring in two ways, giving the important heterocycles quinoline, with the nitrogen atom next to the benzene ring, and isoquinoline, with the nitrogen atom in the other possible position. Quinoline forms part of quinine (structure at the head of this chapter) and isoquinoline forms the central skeleton of the isoquinoline alkaloids, which we will discuss in Chapter 42. In this chapter we need not say much about quinoline because it behaves rather as you would expect—its chemistry is a mixture of that of benzene and pyridine. Electrophilic substitution favours the benzene ring and nucleophilic substitution favours the pyridine ring. So nitration of quinoline gives two products—the 5-nitroquinolines and the 8-nitroquinolines—in about equal quantities (although you will realize that the reaction really occurs on protonated quinoline).

This is obviously rather unsatisfactory but nitration is actually one of the better behaved reactions. Chlorination gives ten products (at least!), of which no fewer than five are chlorinated quinolines of various structures. The nitration of isoquinoline is rather better behaved, giving 72% of one isomer (5-nitroisoquinoline) at 0 °C.

To get reaction on the pyridine ring, the *N*-oxide can be used—as with pyridine itself. A good example is acridine, with two benzene rings, which gives four nitration products, all on the benzene rings. Its *N*-oxide, on the other hand, gives just one product in good yield—nitration takes place at the only remaining position on the pyridine ring.

In general, these reactions are not of much use and most substituents are put into quinolines during ring synthesis from simple precursors, as we will explain in the next chapter. There are a couple of quinoline reactions that are unusual and interesting. Vigorous oxidation goes for the more electron-rich ring, the benzene ring, and destroys it leaving pyridine rings with carbonyl groups in the 2- and 3-positions.

A particularly interesting nucleophilic substitution occurs when quinoline *N*-oxide is treated with acylating agents in the presence of nucleophiles. These two examples show that nucleophilic substitution occurs in the 2-position and you may compare these reactions with those of pyridine *N*-oxide. The mechanism is similar.

In considering quinolines and indoles with their fused rings we kept the benzene and heterocyclic rings separate. Yet there is a way in which they can be combined more intimately, and that is to have a nitrogen atom at a ring junction.

A nitrogen atom can be at a ring junction

indolizine

It has to be a pyrrole-type nitrogen as it must have three σ bonds, so the lone pair must be in a p orbital. This means that one of the rings must be five-membered and the simplest member of this interesting class is called indolizine—it has pyridine and pyrrole rings fused together along a C–N bond. If you examine this structure you will see that there is definitely a pyrrole ring but that the pyridine ring is not all there. Of course, the lone pair and the π electrons are all delocalized but this system, unlike indole and quinoline, is much better regarded as a ten-electron outer ring than as two six-electron rings joined together. Indolizidine reacts with electrophiles on the five-membered rings by substitution reactions as expected.

Fused rings with more than one nitrogen

It is easily possible to continue to insert nitrogen atoms into fused ring systems and some important compounds belong to these groups. The purines are part of DNA and RNA, one example is adenine and another is guanine in the box below, but simple purines play an important part in our lives. Coffee and tea owe their stimulant properties to caffeine, a simple trimethyl purine derivative. It has an imidazole ring fused to a pyrimidine ring and is aromatic in spite of the two carbonyl groups.

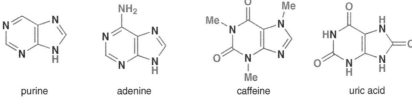

purine   adenine   caffeine   uric acid

Uric acid, gout, and allopurinol

Another purine, uric acid, occurs widely in nature—it is used by birds, and to some extent by humans, as a way to excrete excess nitrogen—but it causes much distress in humans when crystalline uric acid is deposited in joints. We call the pain 'gout'. The solution is a specific inhibitor of the enzyme producing uric acid and it is no surprise that a compound closely resembling uric acid, allopurinol, is the best.

allopurinol

Two of the carbonyl groups have gone and the imidazole ring has been replaced by a pyrazole ring. Purines from DNA are degraded in the body to xanthine, which is oxidized to uric acid. Allopurinol binds to the enzyme xanthine oxidase but inactivates it by not reacting. In fact it imitates not uric acid but the true substrate xanthine in a competitive fashion. This enzyme plays a minor part in human metabolism so inhibiting it is not serious—it just prevents over-production of uric acid.

guanine   xanthine   *xanthine oxidase* — *inhibited by allopurinol*   uric acid

Other fused heterocycles have very attractive flavour and odour properties. Pyrazines, in general, are important in many strong food flavours: a fused pyrazine with a ring junction nitrogen atom is one of the most important components in the smell of roast meat. You can read about the simple pyrazine that provides green peppers with their flavour in the box on the next page.

Finally, the compounds in the margin form a medicinally important group of molecules, which includes antitumour compounds for humans and anthelmintics (compounds that get rid of parasitic worms) for animals. They are derived from a 6/5 fused aromatic ring system that resembles the ten-electron system of the indolizine ring system but has three nitrogen atoms.

All this multiple heteroatom insertion is possible only with nitrogen and we need to look briefly at what happens when we combine nitrogen with oxygen or in heterocycles.

smell of roast meat

useful medicinal compounds

Aromatic heterocycles can have many nitrogens but only one sulfur or oxygen in any ring

A neutral oxygen or sulfur atom can have only two bonds and so it can never be like the nitrogen atom in pyridine—it can only be like the nitrogen atom in pyrrole. We can put as many pyridine-like nitrogens as we like in an aromatic ring, but never more than one pyrrole-like nitrogen. Similarly, we can put only one oxygen or sulfur atom in an aromatic ring. The simplest examples are oxazoles and thiazoles, and their less stable isomers isoxazoles and isothiazoles.

The instability of the 'iso-' compounds comes from the weak O–N or S–N bond. These bonds can be cleaved by reducing agents, which then usually reduce the remaining functional groups further. The first product from reduction of the N–O bond is an unstable imino-enol. The enol tautomerizes to the ketone and the imine may be reduced further to the amine.

Such heterocycles with even more nitrogen atoms exist but are relatively unimportant and we shall mention just one, the 1,2,5-thiadiazole, because it is part of a drug, timolol.

oxazole   thiazole

isoxazole   isothiazole

■ Timolol is a β-blocker that blocks one action of adrenaline (epinephrine) and keeps heart disease at bay by counteracting high blood pressure.

1,2,5-thiadiazole

Timolol–a beta-blocker

The flavour of green peppers

The discovery of the compound responsible for the flavour of green peppers provides us with a chance to review some spectroscopy. This powerful compound was isolated from the oil of the green pepper (*Capsicum annuum var. grossum*). The oil makes up about 0.0001% of the mass of the peppers and the main pepper flavour comes from one compound which is 30% of the oil. It had an even molecular ion at 166 and looks like a compound without nitrogen, perhaps $C_{11}H_{18}O$. But a high-resolution mass spectrum revealed that M^+ was actually 166.1102, which corresponds almost exactly to $C_9H_{14}N_2O$ (166.1106).

The IR had no OH, NH, or C=O peaks, and the proton NMR looked like this.

δ_H, ppm	Integral	Shape	J, Hz	Comments
0.91	6H	d	6.7	Me_2CH^-
1.1–2.4	1H	m	?	
2.61	2H	d	7.0	CH_2CH^-
3.91	3H	s	—	$-OMe$?
7.80	1H	d	2.4	aromatic
7.93	1H	d	2.4	aromatic

The 'CH' feature in the Me_2CH and CH_2CH signals must be the same CH and it must be the signal at 1.1–2.4 ppm described as a 'multiplet' as it is the only one showing enough coupling. It will be a septuplet of triplets, that is, 21 lines. We can easily reconstruct the aliphatic part of the molecule because it has two methyl groups and a CH_2 group joined to the same CH group.

side chain of green pepper flavour compound

We also have an OMe group (only oxygen is electronegative enough to take a methyl group to nearly 4 ppm). This adds up to $C_5H_{12}O$. What is left? Only $C_4H_2N_2$—and no clue yet as to the nitrogen functionality. We also have an aromatic ring that must have nitrogen in it (because there are only five carbon atoms—not enough for a benzene ring!) and the coupling constant between the two aromatic hydrogens is 2.4 Hz. So could we perhaps have a pyrrole ring? Well, no, and for two reasons. If we try and construct such a molecule, we can't fit in the last nitrogen! If we put it on the end of the dotted line, it would have to be an NH_2 group, and there isn't one.

A better reason is that the chemical shifts are all wrong. The protons on an electron-rich pyrrole ring come at around 6–6.5 ppm, upfield from benzene (7.27 ppm). But these protons are at 7.8–8.0 ppm, downfield from benzene. We have a deshielded (electron-poor) ring, not a shielded (electron-rich) ring. From what you now know of heterocyclic chemistry, the ring must be a six-membered one, and we must put both nitrogen atoms in the ring. There are three ways to do this.

pyridazine (1,2-nitrogens)

pyrimidine (1,3-nitrogens)

pyrazine (1,4-nitrogens)

The small coupling constant really fits the pyrazine alone and the chemical shifts are about right for that molecule too, although not as far downfield. But we have a MeO group on the ring feeding electrons into the aromatic system and that will increase the shielding slightly and move the protons upfield. This gives us a unique structure.

correct structure
of the main flavour
compound from
green peppers

There is only one way to be sure and that is to make this compound and see if it is the same as the natural product in all respects, including biological activity. The investigators did this but then wished that they hadn't! The structure was indeed correct but the biological activity—the smell of green peppers—was so intense that they had to seal up the laboratory where the work was done as no one would work there. Human beings can detect 2 parts in 10^{12} of this compound in water.

There are thousands more heterocycles out there

But we're not going to discuss them and we hope you're grateful. In fact, it's about time to stop, and we shall leave you with a hint of the complexity that is possible. If pyrrole is combined with benzaldehyde a good yield of a highly coloured crystalline compound is formed: a porphyrin. Now, what about this ring system—is it aromatic? It's certainly highly delocalized and your answer to the question clearly depends on whether you include the nitrogen electrons or not. In fact, if you ignore the pyrrole-like nitrogen atoms but include the pyridine-like nitrogens and weave round the periphery, you have nine double bonds and hence 18 electrons—a $4n + 2$ number. Most people agree that these compounds are aromatic.

Some heterocycles are simple, some very complex, but we cannot live without them. We shall end this chapter with a wonderful story of heterocyclic chemistry at work. Folic acid is much in the news today as a vitamin that is particularly important for pregnant women, but is involved in the metabolism of all living things. Folic acid is built up in nature from three pieces: a heterocyclic starting material (red), *p*-aminobenzoic acid (black), and the amino acid glutamic acid (green). Here you see the precursor, dihydrofolic acid.

a porphyrin
with the 18 π-electron
system in green

action of dihydropteroate synthase (enzyme)

enzymatic acylation
with glutamic acid

dihydrofolic acid

Although folic acid is vital for human health, we don't have the enzymes to make it: it's a vitamin, which means we must take it in our diet or we die. Bacteria, on the other hand, do make folic acid. This is very useful because it means that if we inhibit the enzymes of folic acid synthesis we can kill bacteria but we cannot possibly harm ourselves as we don't have those enzymes. The sulfa drugs, such as sulfamethoxypyridazine or sulfamethoxazole, imitate *p*-aminobenzoic acid and inhibit the enzyme dihydropteroate synthase. Each has a new heterocyclic system added to the sulfonamide part of the drug.

sulfa-
methoxy-
pyridazine

sulfa-
meth-
oxazole

The next step in folic acid synthesis is the reduction of dihydrofolate to tetrahydrofolate. This can be done by both humans and bacteria, and although it looks like a rather trivial reaction (see black portion of molecules), it can only be done by the very important enzyme **dihydrofolate reductase.**

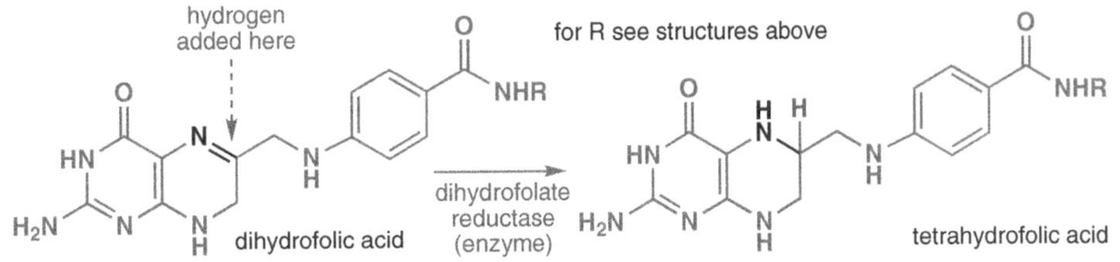

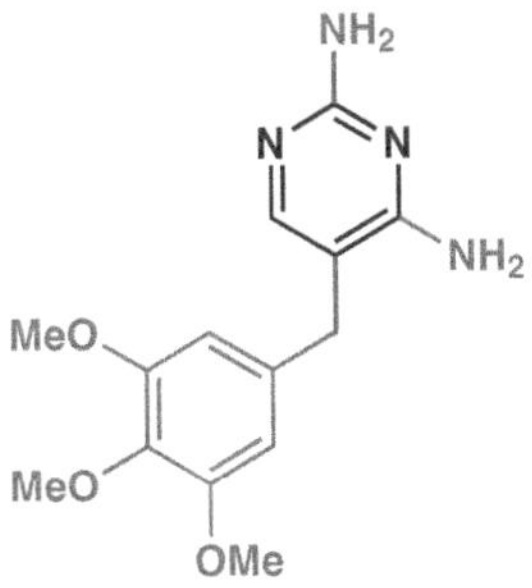

Although both bacteria and humans have this enzyme, the bacterial version is different enough for us to attack it with specific drugs. An example is trimethoprim—yet another heterocyclic compound with a pyrimidine core (black on diagram). These two types of drugs that attack the folic acid metabolism of bacteria are often used together.

We will see in the next chapter how to make these heterocyclic systems and, in Chapter 42, other examples of how important they are in living things.

trimethoprim

Which heterocyclic structures should you learn?

This is, of course, nearly a matter of personal choice. Every chemist really must know the names of the simplest heterocycles and we give those below along with a menu of suggestions. First of all, those every chemist must know:

Now the table gives a suggested list of five ring systems that have important roles in the chemistry of life and in human medicine—many drugs are based on these five structures.

1 Imidazole

the most important five-membered ring with two nitrogen atoms

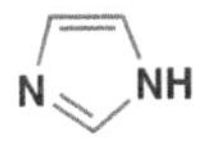

part of the amino acid histidine, occurs in proteins and is important in enzyme mechanisms

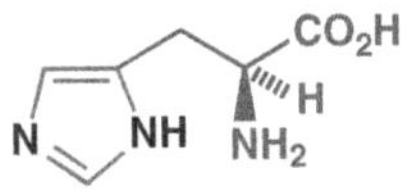

a substituted imidazole is an essential part of the anti-ulcer drug cimetidine

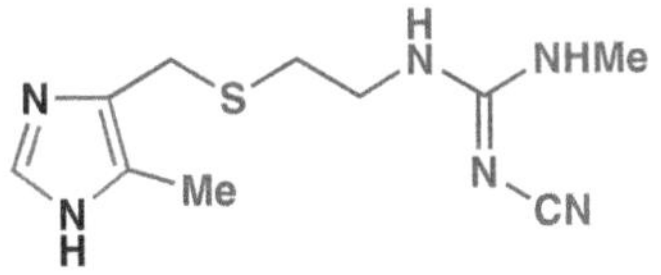

imidazole

the amino acid histidine

the anti-ulcer drug cimetidine (Tagamet) a selective histidine mimic

2 Pyrimidine

the most important six-membered ring with two nitrogen atoms

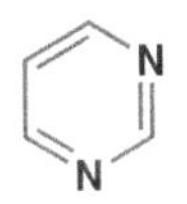

three functionalized pyrimidines are part of DNA and RNA structure, e.g. uracil

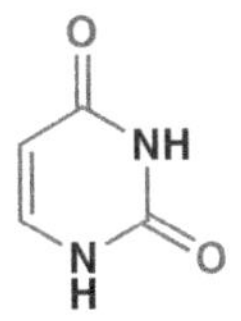

many antiviral drugs, particularly anti-HIV drugs, are modified pieces of DNA and contain pyrimidines

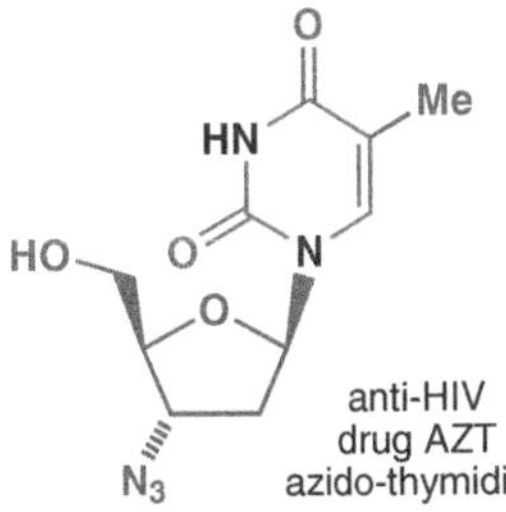

pyrimidine

uracil

anti-HIV drug AZT azido-thymidine

3 Quinoline

one of two benzo-pyri-dines with many applications

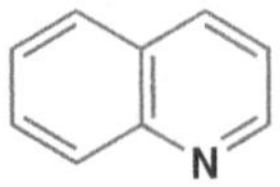

quinoline

occurs naturally in the important antimalar-ial drug quinine

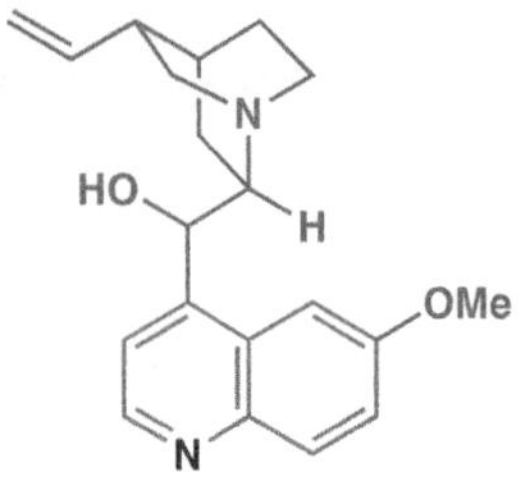

quinine

'cyanine' dyestuffs used as sensitizers for particular light wavelengths in colour photography

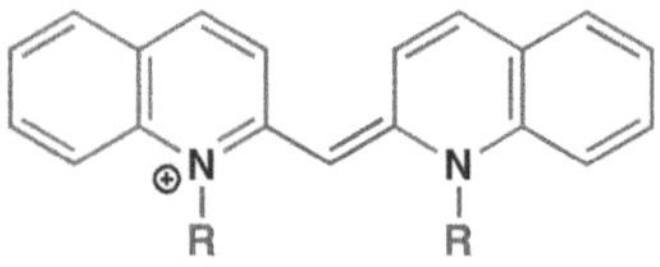

a 'cyanine' dyestuff

4 Isoquinoline

the other benzo-pyridine with many applications

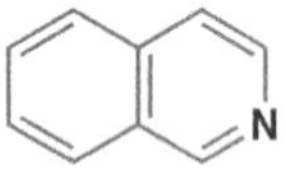

isoquinoline

occurs naturally in the benzyl isoquinoline alkaloids like papaverine

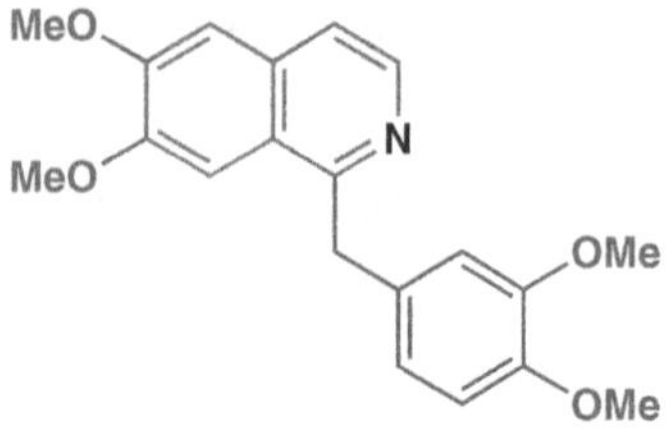

papaverine—a benzyl isoquinoline alkaloid

5 Indole

the more important benzo-pyrrole

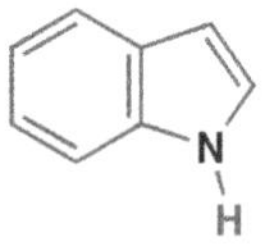

indole

occurs in proteins as tryptophan and in the brain as the neurotransmitter serotonin (5-hydroxytryptamine)

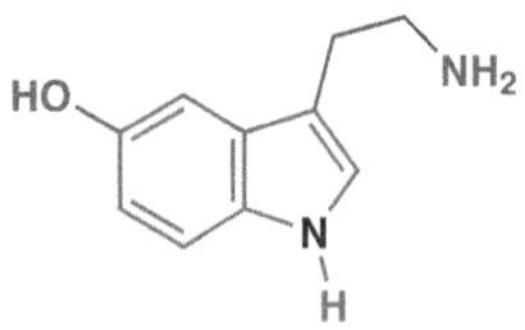

serotonin—a neurotransmitter

important modern drugs are based on serotonin, including sumatriptan for migraine and ondansetron, an antiemetic for cancer chemotherapy

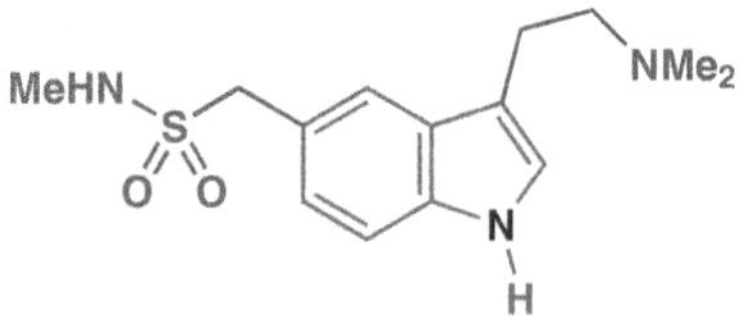

sumatriptan: for treatment of migraine

Further reading

Basic introduction: *Aromatic Heterocyclic Chemistry*, D. T. Davies, Oxford Primer, OUP, 1992. The best general text on heterocycles is J. A. Joule and K. Mills, *Heterocyclic Chemistry*, 4th edn, Chapman and Hall, London, 2010. S. Warren and P. Wyatt, *Workbook for Organic Synthesis: the Disconnection Approach*, Wiley, Chichester, 2009, chapters 32, 34, and 35.

Reference for the green pepper compound: R. G. Buttery, R.M. Seifert, R. E. Lundin, D. G. Guadagni, and L. C. Ling, *Chemistry and Industry (London)*, 1969, 490.

Diazotetrazole was made by P. B. Shevlin, *J. Am. Chem. Soc.* 1972, 94, 1379 and used in a reaction with buckminsterfullerene by R. M. Strongin and group, *J. Org. Chem.*, 1998, 63, 3522.

Check your understanding

 To check that you have mastered the concepts presented in this chapter, attempt the problems that are available in the book's Online Resource Centre at http://www.oxfordtextbooks.co.uk/orc/clayden2e/

Suggested solutions for Chapter 29

29

For each of the following reactions (a) state what kind of substitution is suggested and (b) suggest what product might be formed if monosubstitution occured.

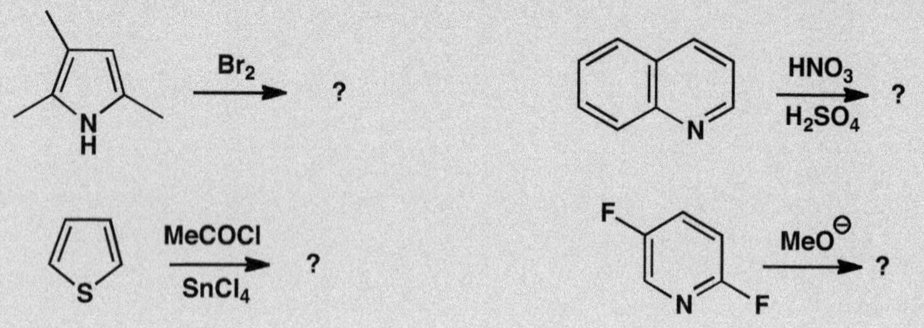

Purpose of the problem

A simple exercise in aromatic substitution on heterocycles.

Suggested solution

The first three reactions are all electrophilic substitutions: a bromination of a pyrrole, the nitration of quinoline, and a Friedel-Crafts reaction of thiophene. Bromination of the pyrrole occurs at the only remaining site. Nitration of quinoline occurs on the benzene rather than the pyridine ring (actually giving a mixture of 5- and 8-nitroquinolines) and the acylation occurs next to sulfur.

■ You needn't be concerned with the mixture of 5- and 8-nitroquinoline here, but p. 749 of the textbook has more detail.

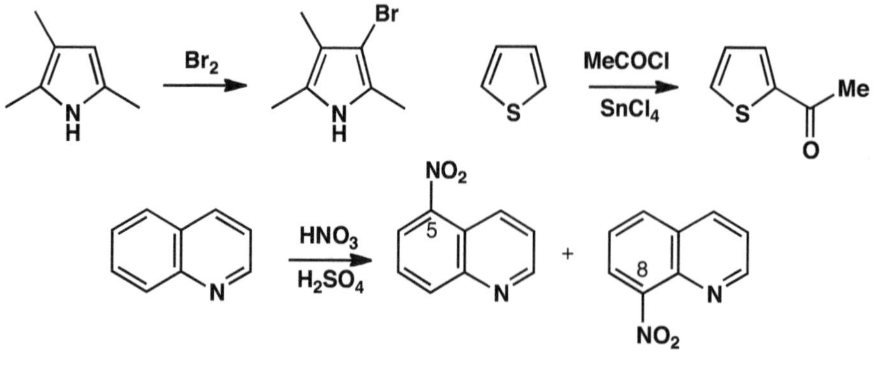

The last reaction is a nucleophilic aromatic substitution on a pyridine. It occurs only at the site where the negative change in the intermediate can be delocalized onto the nitrogen.

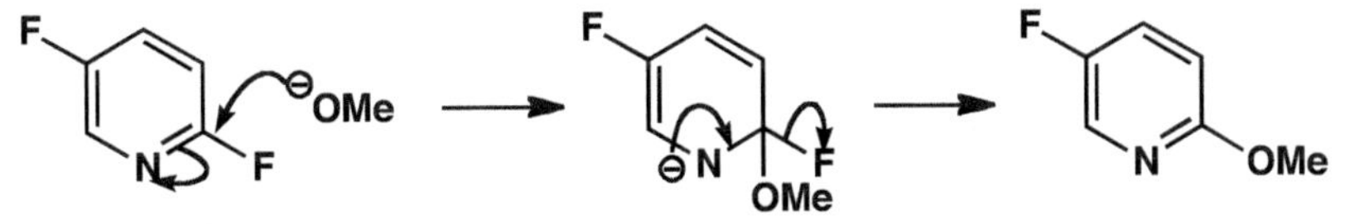

PROBLEM 2

Give a mechanism for this side-chain extension of a pyridine.

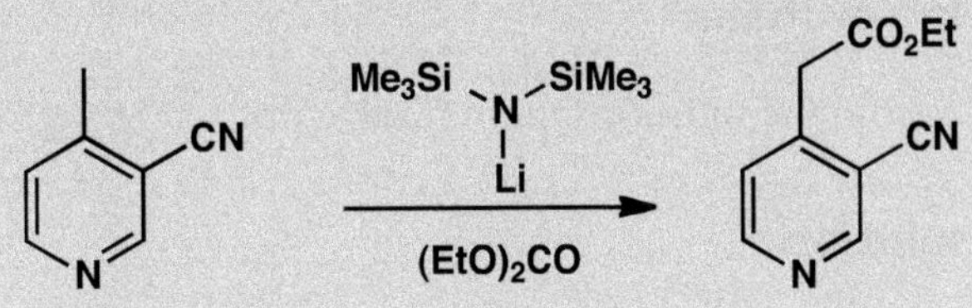

Purpose of the problem

An exercise in thinking about the reactivity of alkylated pyridines.

Suggested solution

The strong base (LHMDS, lithium hexamethyldisilazide) removes a proton from the methyl group so that the anion is stabilized both by the nitrile and the pyridine nitrogen atom. Acylation occurs outside the ring to preserve the aromaticity. If you drew the lithium atom covalently bound to nitrogen, your answer is better than ours.

■ This sort of chemistry was introduced by D. J. Sheffield and K. R. H. Wooldridge, *J.Chem. Soc., Perlin 1*, 1972, 2506 and by A. S. Kende and T. P. Demuth, *Tetrahedron Lett.*, 1980, **21**, 715 in a synthesis of the antileukaemic sesbanine.

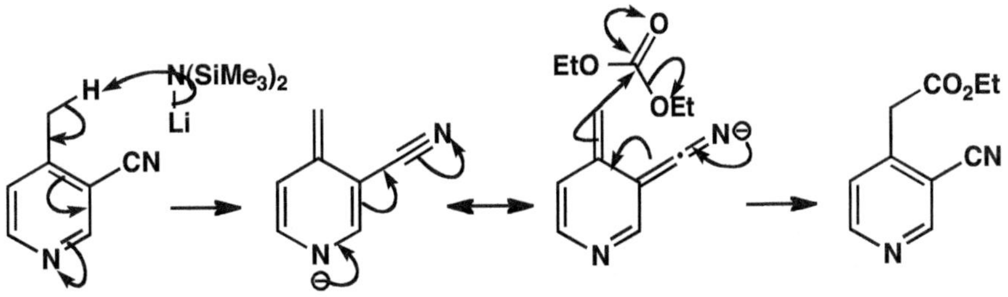

PROBLEM 3

Give a mechanism for this reaction, commenting on the position in the furan ring that reacts.

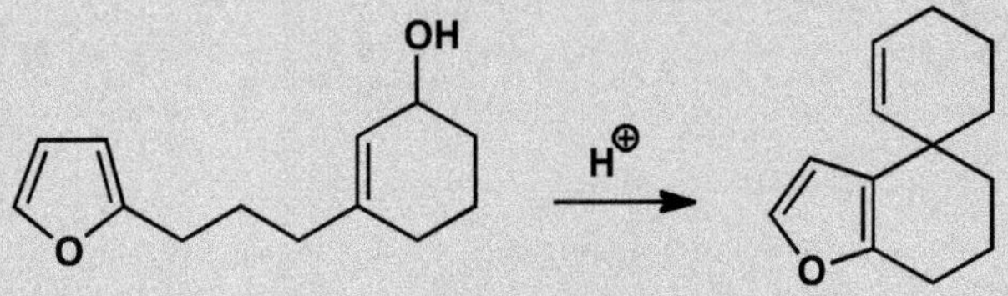

Purpose of the problem

An unusual electrophilic substitution on furan with interesting selectivity.

Suggested solution

Furans normally prefer substitution at the α-positions (2 or 5) but one α-position is already blocked and the other is too far away to reach the allyl cation. Attack at the other end of the allylic system would give an eight-membered ring with a *trans* alkene in it. This would theoretically be possible but closure of a six-membered ring is much faster. In other words, the electrophile and nucleophile are *tethered*.

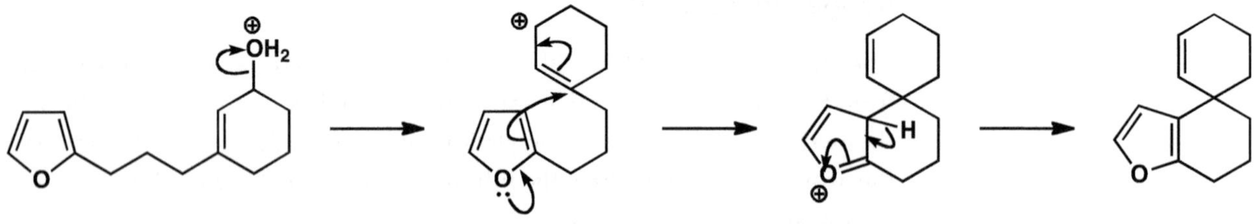

PROBLEM 4

Suggest which product might be formed in these reactions and justify your choice.

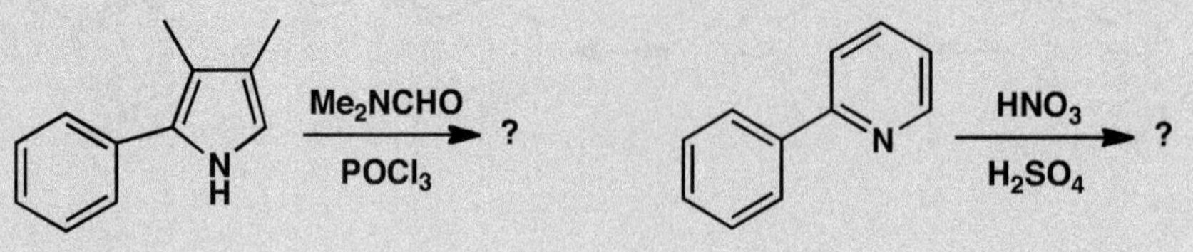

Purpose of the problem

Regioselectivity test with contrasted electrophilic aromatic substitution.

Suggested solution

In each case we have a choice between reaction on a benzene ring or an aromatic heterocycle. The pyrrole is more reactive than the benzene and the pyridine less so. The pyrrole does a Vilsmeier reaction (p. 734 of the textbook) in the remaining free position while nitration occurs on the benzene. Pyridine acts as an electron-withdrawing and deactivating substituent, and therefore directs *meta*.

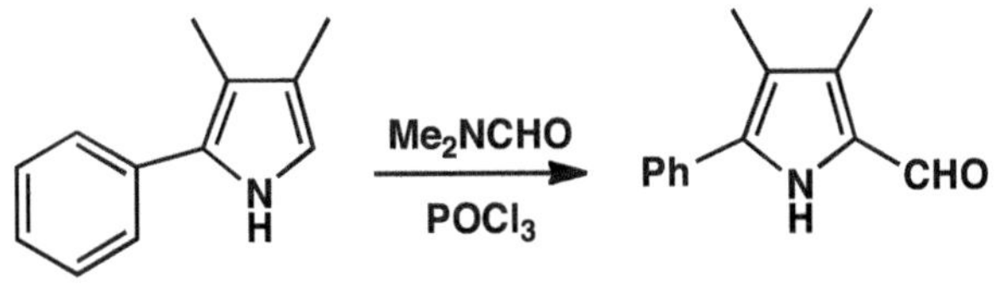

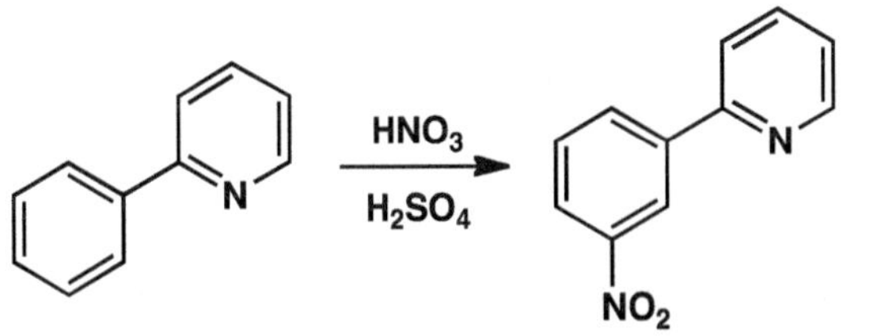

PROBLEM 5

Explain the formation of the product in this Friedel-Crafts alkylation of an indole.

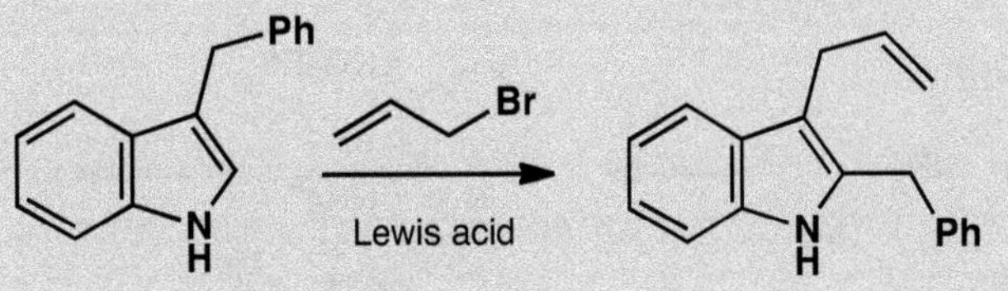

Purpose of the problem

Checking up on your understanding of indole chemistry.

Suggested solution

■ The textbook, pp. 745–6, explains why it is the 3- and not the the 2-position that is attacked.

The Lewis acid combines with allyl bromide to give either the allyl cation or the complex we show here. In either case, electrophilic attack occurs at the 3-position of the indole. The benzyl group migrates to the 2-position where there is a proton that can be lost to restore aromaticity.

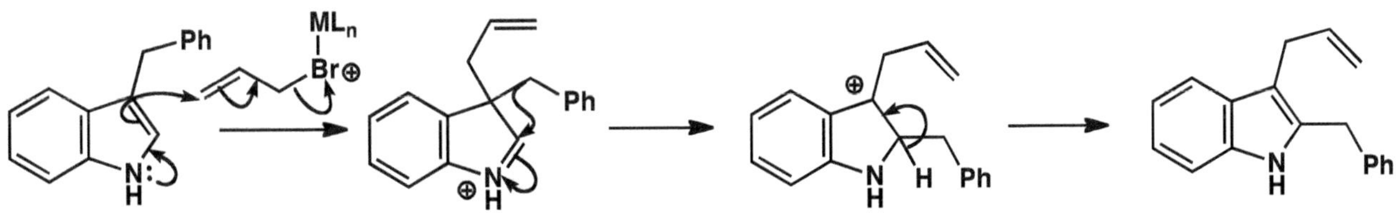

PROBLEM 6

Suggest what the products of these nucleophilic substitutions might be.

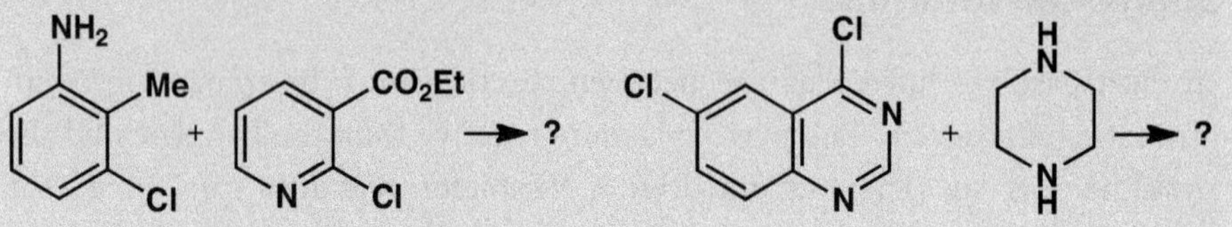

Purpose of the problem

Checking your understanding of nucleophilic aromatic substitution involving decisions on chemoselectivity and regioselectivity.

Suggested solution

Each compound has potential nucleophilic and electrophilic sites. In the first case the benzene ring is not activated towards nucleophilic substitution but the pyridine is, both by the pyridine nitrogen atom and by the ester group. The NH_2 on the benzene ring is much more nucleophilic than the pyridine nitrogen atom.

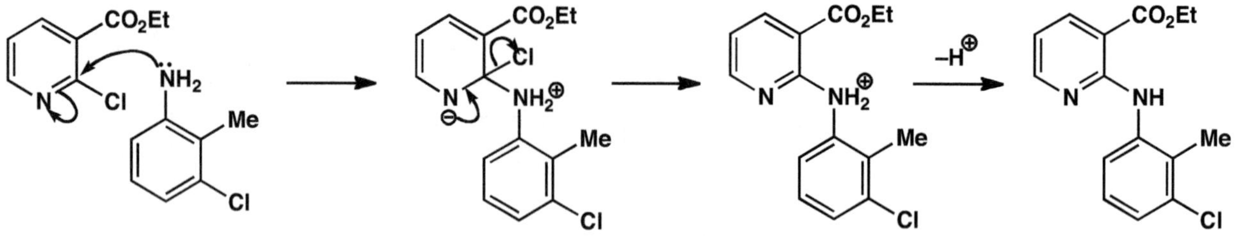

In the second case, the chlorine on the heterocyclic ring is much more reactive towards nucleophilic substitution as the intermediate is stabilized by two nitrogen atoms and the benzene ring is not disturbed. The saturated heterocycle (piperazine) can be made to react once only as the product under the reaction conditions is strictly the hydrochloride of the unreacted amino group. This is much more basic than the one that has reacted as that lone pair is conjugated with the heterocyclic ring.

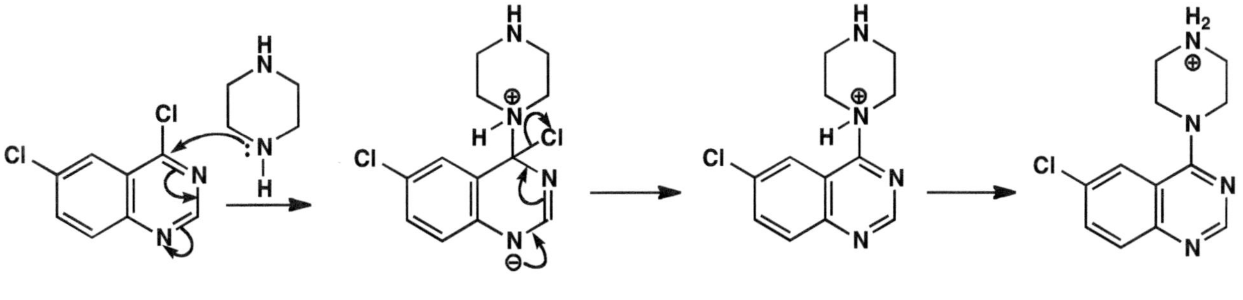

PROBLEM 7

Suggest how 2-pyridone might be converted into the amine shown. This amine undergoes nitration to give compound **A** with the NMR spectrum given. What is the structure of **A**? Why is this isomer formed?

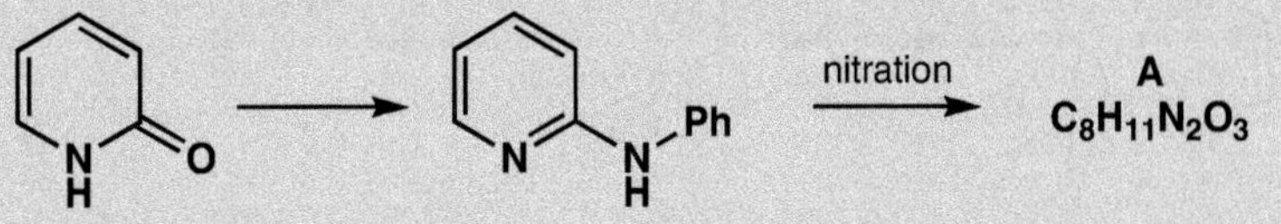

NMR of **A**: δ_H 1.0 (3H, t, *J* 7 Hz), 1.7 (2H, sextet, *J* 7 Hz), 3.3 (2H, t, *J* 7 Hz), 5.9 (1H, broad s), 6.4 (1H, d, *J* 8 Hz), 8.1 (1H, dd, *J* 8 and 2 Hz), and 8.9 (1H, d, *J* 2 Hz). Compound **A** was needed for conversion into the enzyme inhibitor below. How might this be achieved?

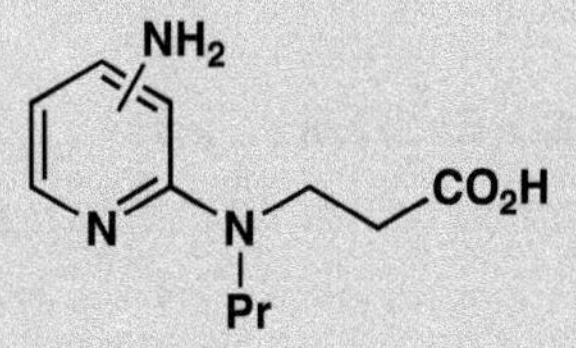

Purpose of the problem

Revision of proof of structure together with electrophilic and nucleophilic substitution on pyridines and a bit of synthesis.

Suggested solution

The first step requires nucleophilic substitution so we could convert the pyridine into 2-chloropyridine and displace the chlorine with the amine.

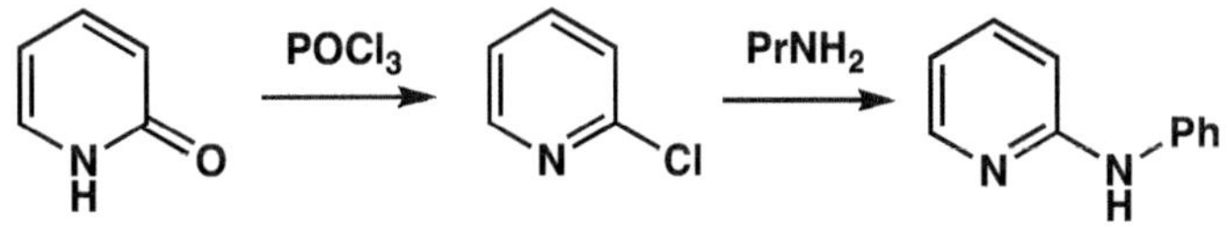

The nitration occurs only because this pyridine is activated by the extra amino group so you could start by predicting which compound might be made. Alternatively you could work out the structure from the NMR. The key points are (i) **A** has only three aromatic protons so nitration has occurred on the ring, (ii) there is only one coupling large enough to be between *ortho* hydrogens (8 Hz), and (iii) there is a proton that has only *meta* coupling (2 Hz) a long way downfield (at large chemical shift). The pyridine nitrogen causes large downfield shifts at positions 2, 4, and 6, the nitro group causes large downfield *ortho* shifts, and the amino group causes upward *ortho* shifts (to smaller δ). All this fits the structure and mechanism shown. The amino group directs *ortho, para* and *para* is preferred sterically.

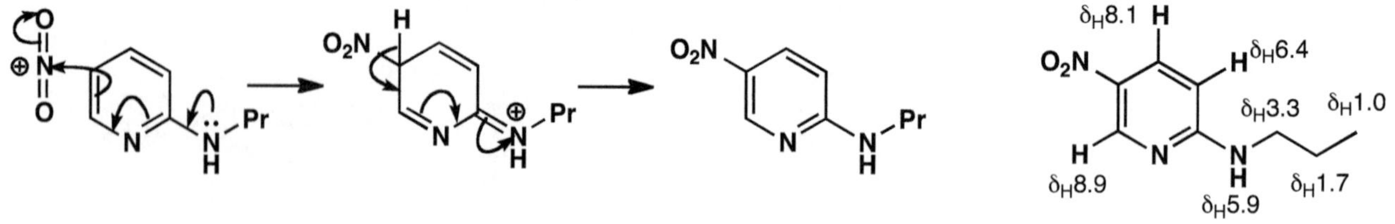

To get the enzyme inhibitor we need to reduce the nitro group to an amine and add the new chain to the other amine. This conjugate addition is best done first while there is only one nucleophilic amine. The ester is probably the best derivative to use, but you may have chosen something else.

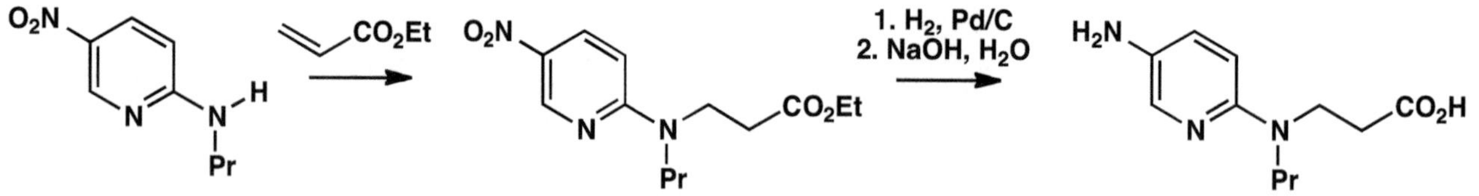

PROBLEM 8

The reactions outlined in the chart below were the early stages in a synthesis of an antiviral drug by the Parke-Davis company. Consider how the reactivity of imidazoles is illustrated in these reactions, which involve not only the skeleton of the molecule but also the reagent **D**. You will need to draw mechanisms for the reactions and explain how they are influenced by the heterocycles.

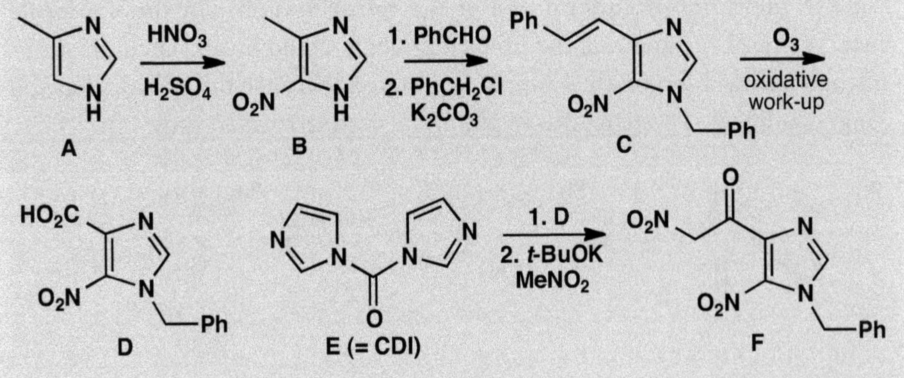

Purpose of the problem

An exploration of the chemistry of imidazole beyond that considered in chapter 29.

Suggested solution

The first reaction is the nitration of an imidazole in one of only two free positions. The position next to one nitrogen is more nucleophilic than the one between the two nitrogens. Imidazole has one pyridine-like and one pyrrole-like nitrogen so it is more nucleophilic than pyridine but less so than pyrrole.

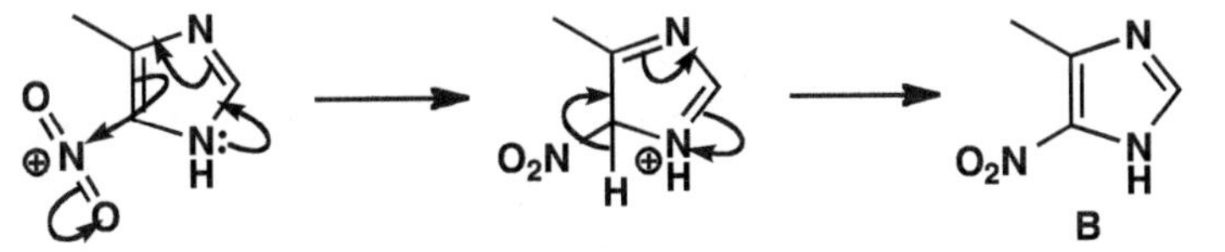

The second reaction is like an aldol condensation between the methyl group on the ring and the benzaldehyde as the electrophile. The nitro group provides some stabilization for the 'enolate' but that would not be enough without the imidazole—*ortho*-nitro toluene would not do this reaction. The elimination is E1cB-like, going through a similar 'enol' intermediate.

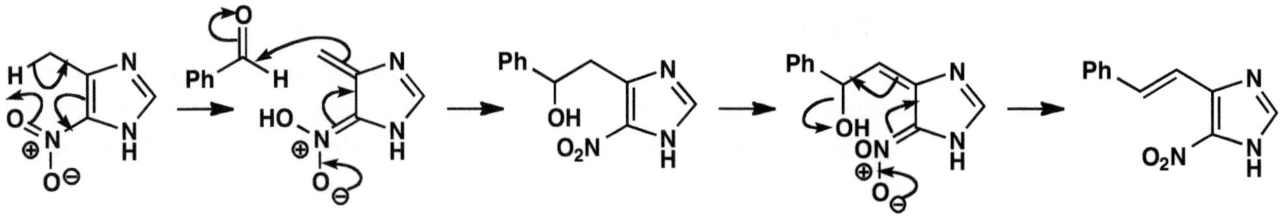

Next, alkylation occurs on one of the nitrogen atoms in the imidazole ring. We need the anion of the imidazole which could be alkylated on either nitrogen. Alkylation on the lower N is preferred because the product has the longer conjugated system—we've put in the curly arrows to show it.

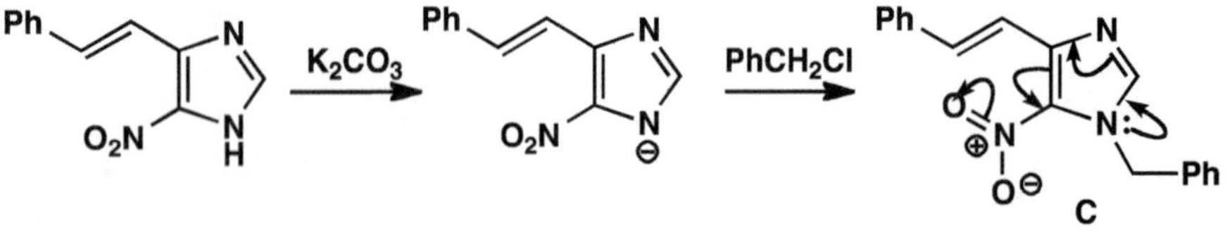

Ozonolysis of the alkene of **C** frees the carboxylic acid of **D** which reacts with carbonyl diimidazole **E** (CDI) in a nucleophilic substitution at the carbonyl group, with the relatively stable imidazole anion as the leaving group. The product is an 'activated ester', like an anhydride, from which the anion of nitromethane displaces the second molecule of imidazole to give the product **F**.

■ See pp. 622–3 of the textbook for some chemistry of the anion of nitromethane.

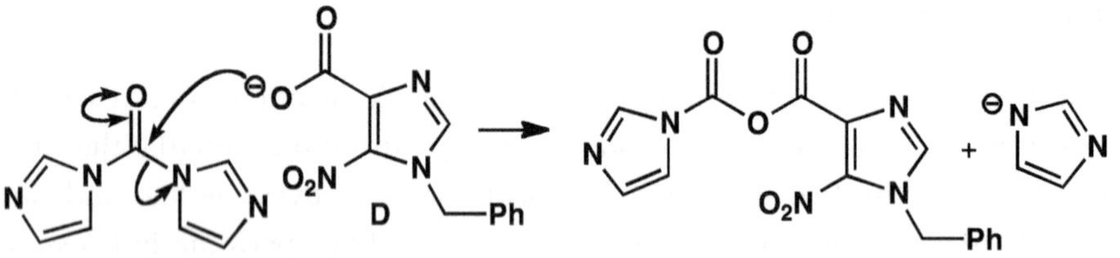

E = carbonyl diimidazole (CDI)

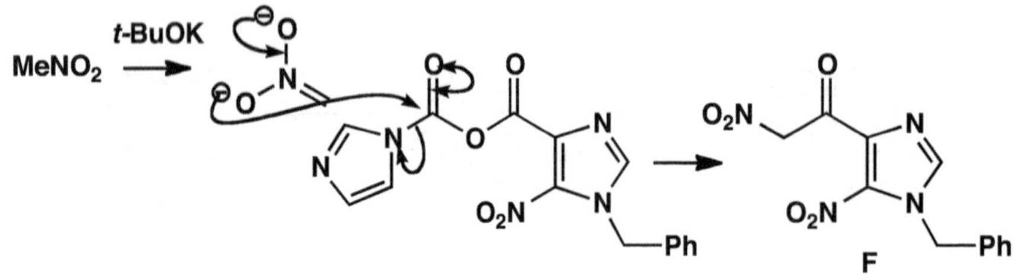

PROBLEM 9

What aromatic system might be based on this ring system? What sort of reactivity might it display?

Purpose of the problem

A chance for you to think creatively about aromatic heterocycles.

Suggested solution

The aromatic system has the poetic name 'pyrrocoline' and you will have found it by trial and error. One ring looks like a pyridine and one like a pyrrole but counting the electrons should have made you realize that you need the lone pair on nitrogen to give a ten electron system. The nitrogen is therefore pyrrole-like and so if you predicted that this compound would react well in electrophilic substitutions on the five-membered ring you would be right: that is exactly what it does. The easiest pyrrocolines to make have alkyl groups at position 3 and these compounds are nitrated to give the 4-nitro compounds. Friedel-Crafts reactions happen at the same atom.

pyrrocoline

PROBLEM 10

Explain the order of events and the choice of bases in this sequence.

Purpose of the problem

The use of selective lithiation in furan chemistry.

Suggested solution

The allylic group evidently goes into the 2-position so deprotonation of the starting material by LDA must occur there, directed by both the oxygen and bromine atoms. The second electrophile (MeI) talkes the place of the Br atom, so BuLi must lead to bromine-lithium exchange rather than deprotonation. The alternative order of events would require selective lithiation adjacent to the methyl group—not something you would expect to work reliably.

■ Ortholithiation is introduced on pp. 563–4 of the textbook, and the lithiation of furan is on p. 737. Halogen-metal exchange is on p. 188.

■ The product is related to a constituent of the perfume of roses and was made by N. D. Ly and M. Schlosser, *Helv. Chim. Acta* 1977, **60**, 2085.

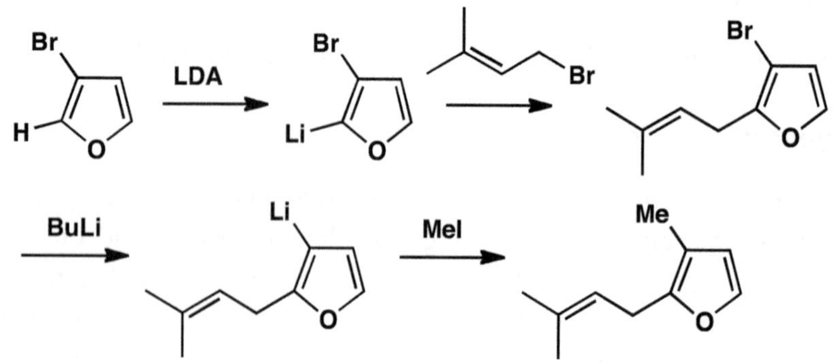

Aromatic heterocycles 2: synthesis

30

Connections

Building on

- Aromaticity ch7
- Enols and enolates ch20
- Michael additions of enolates ch25
- Aldol reactions and acylation of enolates ch26
- Retrosynthetic analysis ch28
- Reactions of heterocycles ch29

Arriving at

- Thermodynamics is on our side
- Disconnecting the carbon–heteroatom bonds first
- How to make pyrroles, thiophenes, and furans from 1,4-dicarbonyl compounds
- How to make pyridines and pyridones
- How to make pyridazines and pyrazoles
- How to make pyrimidines from 1,3-dicarbonyl compounds and amidines
- How to make thiazoles
- How to make isoxazoles and tetrazoles by 1,3-dipolar cycloadditions
- The Fischer indole synthesis
- Making drugs: Viagra, sumatriptan, ondansetron, indomethacin
- How to make quinolines and isoquinolines

Looking forward to

- Cycloadditions ch34
- Biological chemistry ch42

In this chapter you will revisit the heterocyclic systems you have just met and find out how to make them. You'll also meet some new heterocyclic systems and find out how to make those. With so many heterocycles to consider, you'd be forgiven for feeling rather daunted by this prospect, but do not be alarmed. Making heterocycles is easy—that's precisely why there are so many of them. Just reflect...

- Making C–O, C–N, and C–S bonds is easy.
- Intramolecular reactions are preferred to intermolecular reactions.
- Forming five- and six-membered rings is easy.
- We are talking about aromatic, that is, very stable molecules.

If we are to use these bullet points to our advantage we must think strategy before we start. When we were making benzene compounds we usually started with a preformed simple benzene derivative—toluene, phenol, aniline—and added side chains by electrophilic substitution. In this chapter our strategy will usually be to build the heterocyclic ring with most of its substituents already in place and add just a few others, perhaps by electrophilic substitution, but mostly by nucleophilic substitution.

We will usually make the rings by cyclization reactions with the heteroatom (O, N, S) as a nucleophile and a suitably functionalized carbon atom as the electrophile. This electrophile

Online support. The icon in the margin indicates that accompanying interactive resources are provided online to help your understanding: just type **www.chemtube3d.com/clayden/123** into your browser, replacing **123** with the number of the page where you see the icon. For pages linking to more than one resource, type **123-1, 123-2** etc. (replacing **123** with the page number) for access to successive links.

will almost always be a carbonyl compound of some sort and this chapter will help you revise your carbonyl chemistry from Chapters 10, 11, 20, 25, and 26 as well as the approach to synthesis described in Chapter 28.

Thermodynamics is on our side

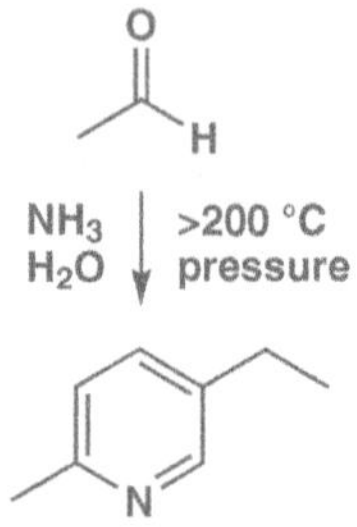

Some of the syntheses we will meet will be quite surprisingly simple! It sometimes seems that we can just mix a few things together with about the right number of atoms and let thermodynamics do the rest. A commercial synthesis of pyridines combines acetaldehyde and ammonia under pressure to give a simple pyridine.

The yield is only about 50%, but what does that matter in such a simple process? By counting atoms we can guess that four molecules of aldehyde and one of ammonia react, but exactly how is a triumph of thermodynamics over mechanism. Much more complex molecules can sometimes be made very easily too. Take allopurinol, for example, which you met in the last chapter. It is not too difficult to work out where the atoms go—the hydrazine obviously gives rise to the pair of adjacent nitrogen atoms in the pyrazole ring and the ester group must be the origin of the carbonyl group (the colours and numbers illustrate this)—but would you have planned this synthesis?

➡ Allopurinol was discussed in Chapter 29, p. 751.

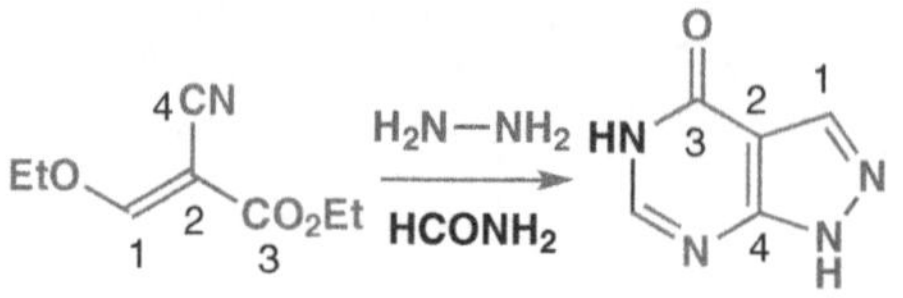

We will see that this sort of 'witch's brew' approach to heterocyclic synthesis is restricted to a few basic ring systems and that, in general, careful planning is just as important here as elsewhere. The difference is that the synthesis of aromatic heterocycles is very forgiving—it often 'goes right' instead of going wrong. We'll now look seriously at planning the synthesis of aromatic heterocycles.

Disconnect the carbon–heteroatom bonds first

The simplest synthesis for a heterocycle emerges when we remove the heteroatom and see what electrophile we need. We shall use pyrroles as examples. The nitrogen forms an enamine on each side of the ring and we know that enamines are made from carbonyl compounds and amines.

➡ These arrows are the retrosynthetic arrows you met in Chapter 28.

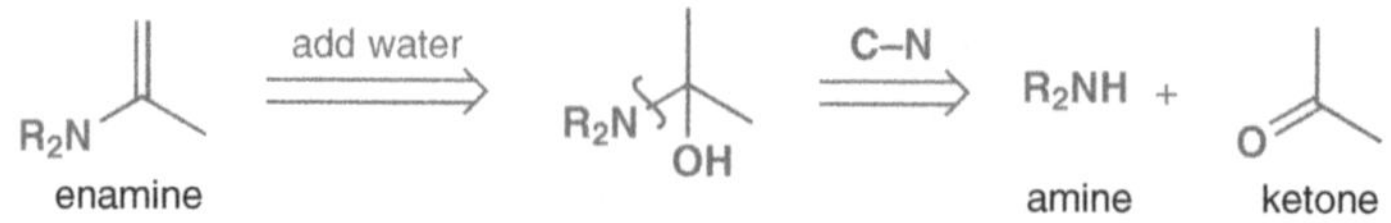

If we do the same disconnection with a pyrrole, omitting the intermediate stage, we can repeat the C–N disconnection on the other side too:

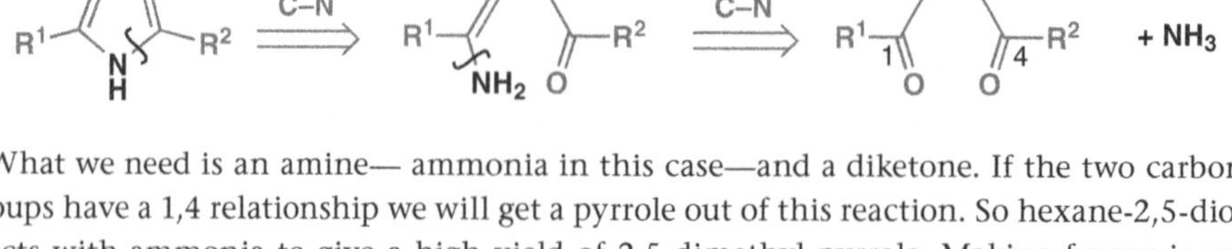

What we need is an amine— ammonia in this case—and a diketone. If the two carbonyl groups have a 1,4 relationship we will get a pyrrole out of this reaction. So hexane-2,5-dione reacts with ammonia to give a high yield of 2,5-dimethyl pyrrole. Making furans is even easier because the heteroatom (oxygen) is already there. All we have to do is to dehydrate the 1,4-diketone instead of making enamines from it. Heating with acid is enough.

Avoiding the aldol product

1,4-Diketones also self-condense rather easily in an intramolecular aldol reaction (see Chapter 26, p. 636) to give a cyclopentenone with an all-carbon five-membered ring. This too is a useful reaction but we need to know how to control it. The usual rule is:

- Base gives the cyclopentenone.
- Acid gives the furan.

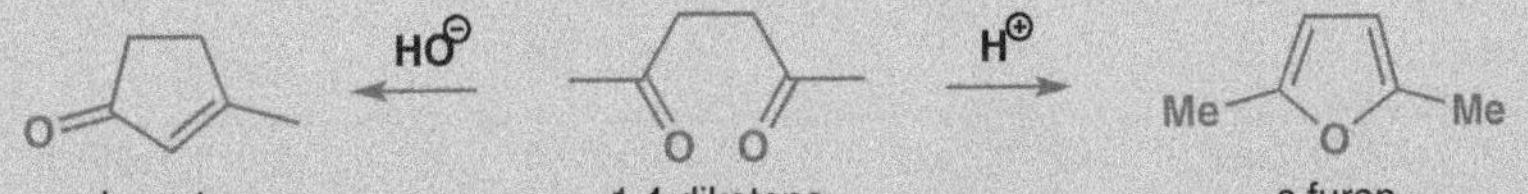

For thiophenes we could in theory use H_2S or some other sulfur nucleophile but, in practice, an electrophilic reagent is usually used to convert the two C=O bonds to C=S bonds. Thioketones are much less stable than ketones and cyclization is swift. Reagents such as P_2S_5 or Lawesson's reagent are the usual choice here.

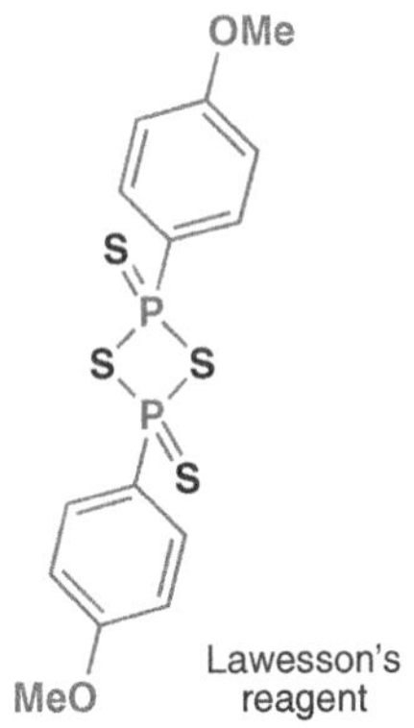

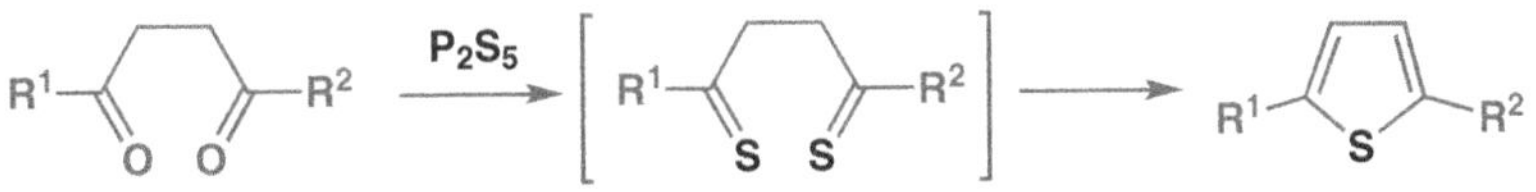

● Making five-membered heterocycles

Cyclization of 1,4-dicarbonyl compounds with nitrogen, sulfur, or oxygen nucleophiles gives the five-membered aromatic heterocycles pyrrole, thiophene, and furan.

It seems a logical extension to use a 1,5-diketone to make substituted pyridines but there is a slight problem here as we will introduce only two of the required three double bonds when the two enamines are formed. To get the pyridine by enamine formation we should need a double bond somewhere in the chain between the two carbonyl groups. But here another difficulty arises—it will have to be a *cis* (*Z*) double bond or cyclization would be impossible.

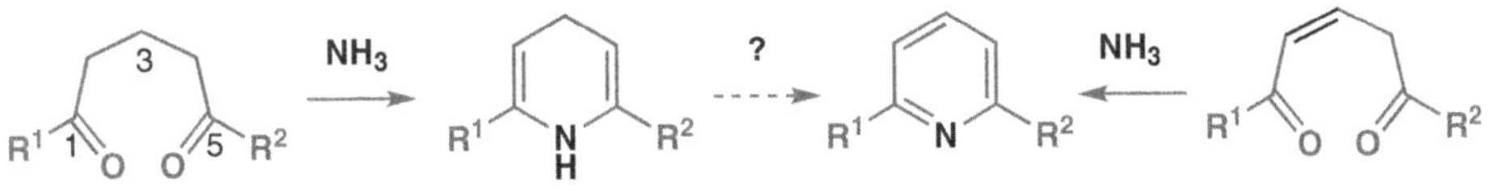

On the whole it is easier to use the saturated 1,5-diketone and oxidize the product to the pyridine. As we are going from a non-aromatic to an aromatic compound, oxidation is easy and we can replace the question mark above with almost any simple oxidizing agent, as we shall soon see.

● Making six-membered heterocycles

Cyclization of 1,5-dicarbonyl compounds with nitrogen nucleophiles leads to the six-membered aromatic heterocycle pyridine.

Heterocycles with two nitrogen atoms come from the same strategy

Reacting a 1,4-diketone with hydrazine (NH_2NH_2) makes a double enamine again and this is only an oxidation step away from a pyridazine. This is also a good synthesis.

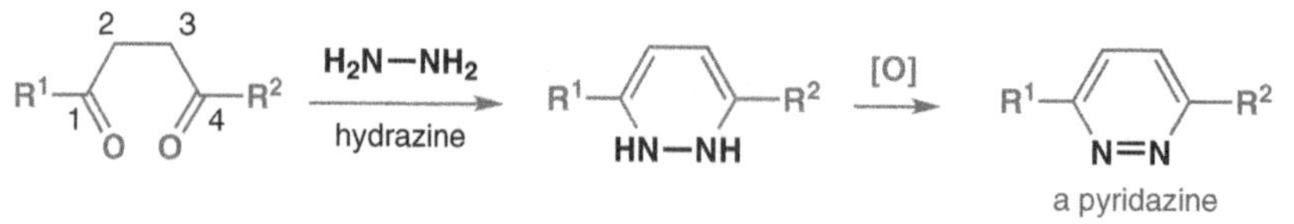

If we use a 1,3-diketone instead we will get a five-membered heterocycle and the imine and enamine formed are enough to give aromaticity without any need for oxidation. The product is a pyrazole. The two heteroatoms do not, of course, need to be joined together for this strategy to work. If an amidine is combined with the same 1,3-diketone we get a six-membered heterocycle. As the nucleophile contains one double bond already, an aromatic pyrimidine is formed directly.

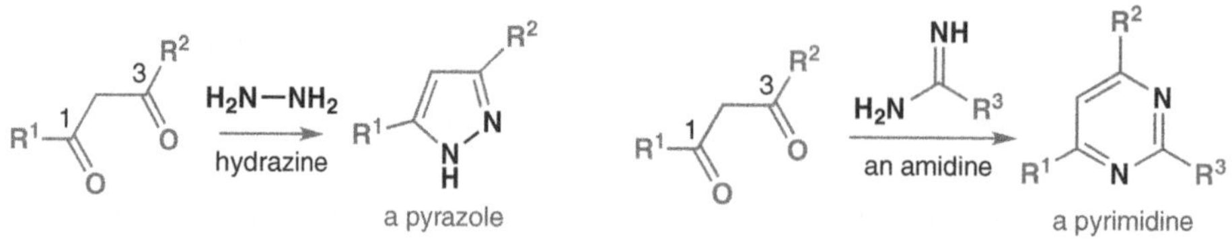

Since diketones and other dicarbonyl compounds are easily made by enolate chemistry (Chapters 25, 26, and 28) this strategy has been very popular and we will look at some detailed examples before moving on to more specialized reactions for the different classes of aromatic heterocycles.

Pyrroles, thiophenes, and furans from 1,4-dicarbonyl compounds

We need to make the point that pyrrole synthesis can be done with primary amines as well as with ammonia and a good example is the pyrrole needed for clopirac, a drug we discussed in Chapter 29. The synthesis is very easy.

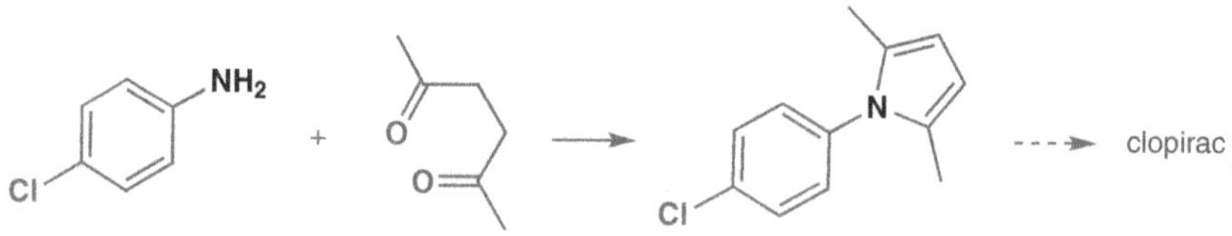

For an example of furan synthesis we choose menthofuran, which contributes to the flavour of mint. It has a second ring, but that is no problem if we simply disconnect the enol ethers as we have been doing so far.

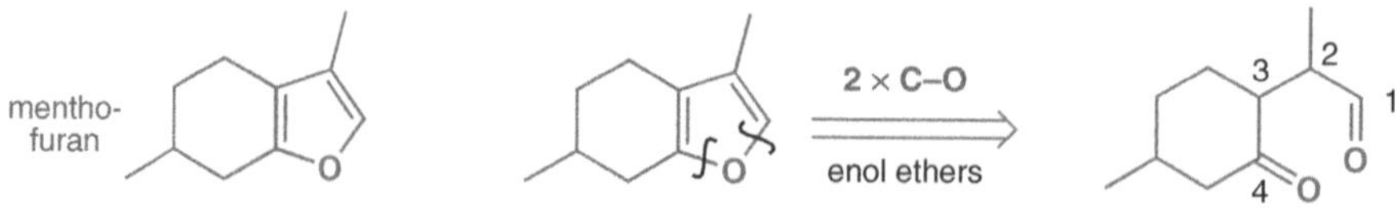

The starting material is again a 1,4-dicarbonyl compound but as there was no substituent at C1 of the furan, that atom is an aldehyde rather than a ketone. This might lead to problems in the synthesis so a few changes (using the notation you met in Chapter 28) are made to the intermediate before further disconnection.

■ Halo aldehydes are unstable and should be avoided.

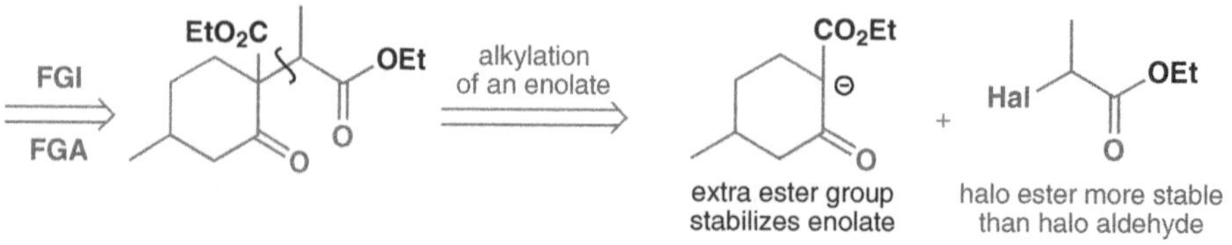

Notice in particular that we have 'oxidized' the aldehyde to an ester to make it more stable—in the synthesis reduction will be needed. Here is the alkylation step of the synthesis, which does indeed go very well with the α-iodo-ester.

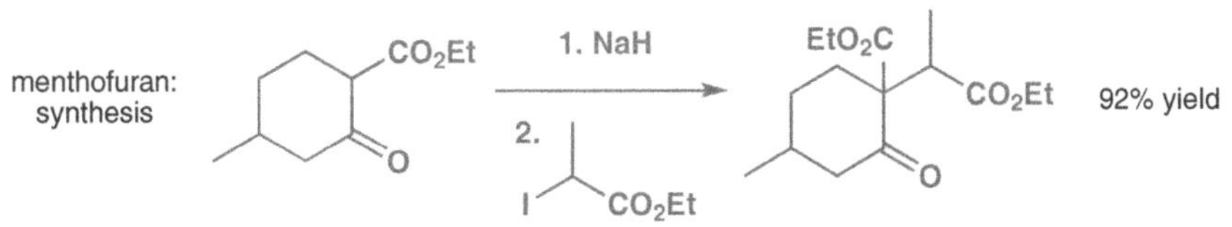

menthofuran:
synthesis

1. NaH

2.

92% yield

Cyclization with acid now causes a lot to happen. The 1,4-dicarbonyl compound cyclizes to a lactone, not to a furan, and the redundant ester group is lost by hydrolysis and decarboxylation. Notice that the double bond moves into conjugation with the lactone carbonyl group. Finally, the reduction gives the furan. No special precautions are necessary—as soon as the ester is partly reduced, it loses water to give the furan whose aromaticity prevents further reduction even with $LiAlH_4$.

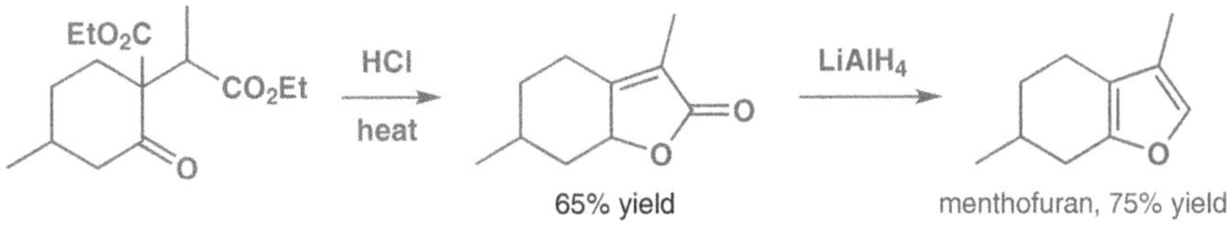

HCl
heat

65% yield

$LiAlH_4$

menthofuran, 75% yield

● **A reminder**

Cyclization of 1,4-dicarbonyl compounds with nitrogen, sulfur, or oxygen nucleophiles gives the five-membered aromatic heterocycles pyrrole, thiophene, and furan.

Now we need to take these ideas further and discuss an important pyrrole synthesis that follows this strategy but includes a cunning twist. It all starts with the porphyrin found in blood. In Chapter 29 we gave the structure of porphyrin and showed that it contains four pyrrole rings joined in a macrocycle. We are going to look at one of those pyrroles.

Porphyrins can be made by joining together the various pyrroles in the right order and what is needed for this one (and also, in fact, for another—the one in the north-east corner of the porphyrin) is a pyrrole with the correct substituents in positions 3 and 4, a methyl group in position 5, and a hydrogen atom at position 2. Position 2 must be free. Here is the molecule drawn somewhat more conveniently, together with the disconnection we have been using so far.

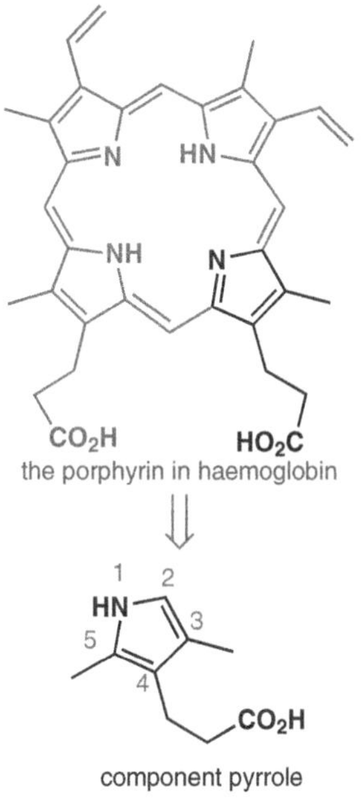

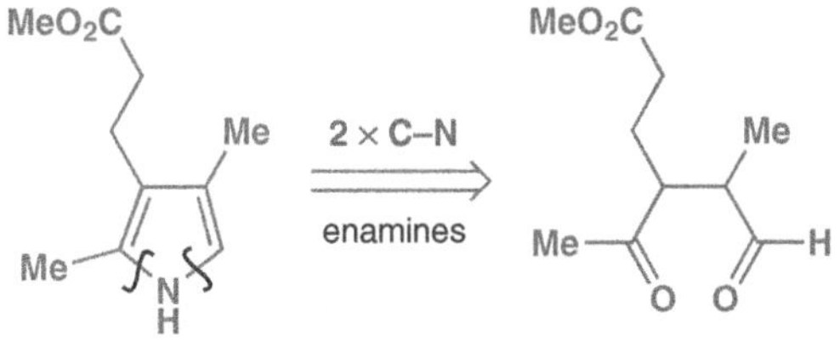

2 × C–N

enamines

No doubt such a synthesis could be carried out but it is worth looking for alternatives for a number of reasons. We would prefer not to make a pyrrole with a free position at C2 as that would be very reactive and we know from Chapter 29 that we can reversibly block such a position with a *t*-butyl ester group. This gives us a very difficult starting material with four different carbonyl groups.

2 × C–N

enamines

the porphyrin in haemoglobin

component pyrrole

➡ See p. 733 for a discussion of how to control pyrrole's reactivity.

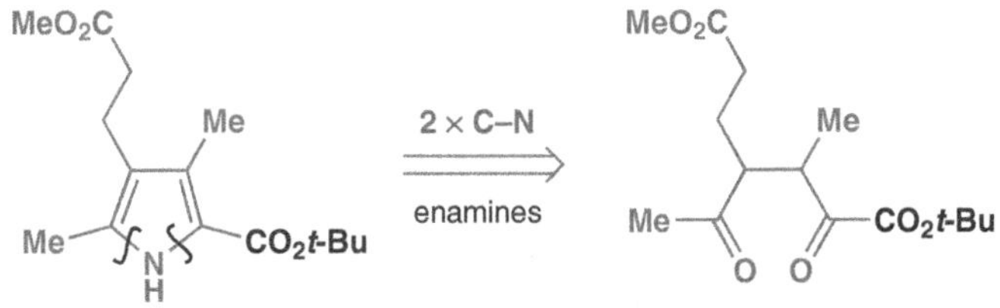

We have made a problem for ourselves by having two carbonyl groups next to each other. Could we escape from that by replacing one of them with an amine? We should then have an ester of an α amino acid, a much more attractive starting material, and this corresponds to disconnecting just one of the C–N bonds.

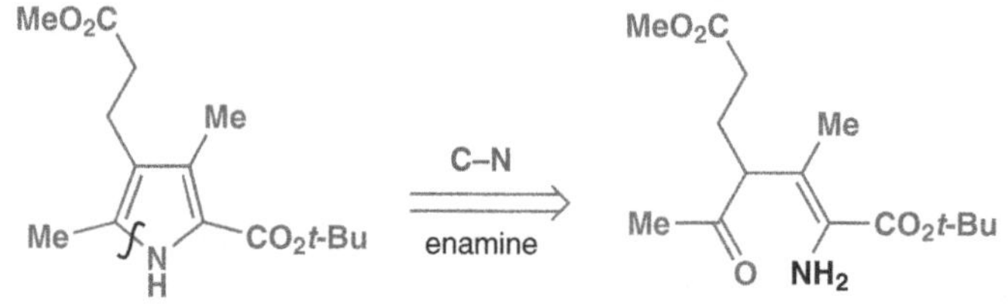

At first we seem to have made no progress but just see what happens when we move the double bond round the ring into conjugation with the ketone. After all, it doesn't matter where the double bond starts out—we will always get the aromatic product.

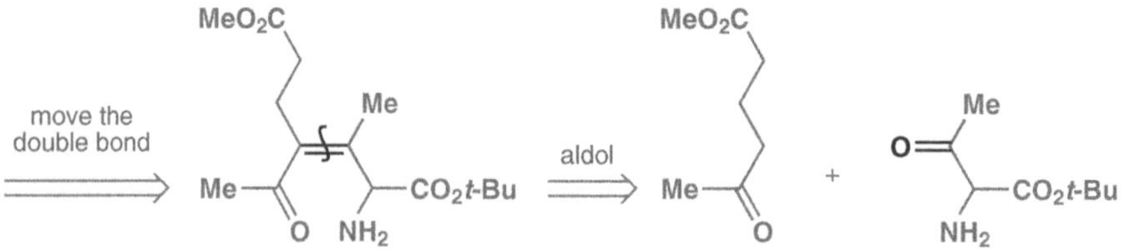

■ Conjugate additions with 1,3-dicarbonyl compounds were discussed in Chapter 26. If you have read Chapter 28 then you should be aware that such reactions are an excellent way of making 1,5-difunctionalized compounds.

Each of our two much simpler starting materials needs to be made. The keto-ester is a 1,5-dicarbonyl compound so it can be made by a conjugate addition of an enolate, a process greatly assisted by the addition of a second ester group.

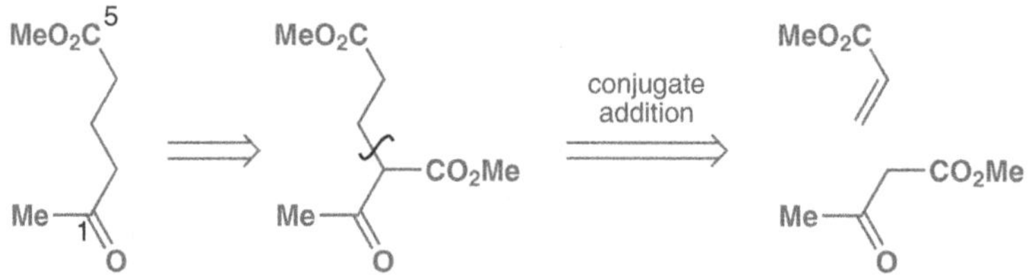

The other compound is an amino-keto-ester and will certainly react with itself if we try to prepare it as a pure compound. The answer is to release it directly into the reaction mixture and this can be done by nitrosation and reduction (Chapter 20) of another stable enolate.

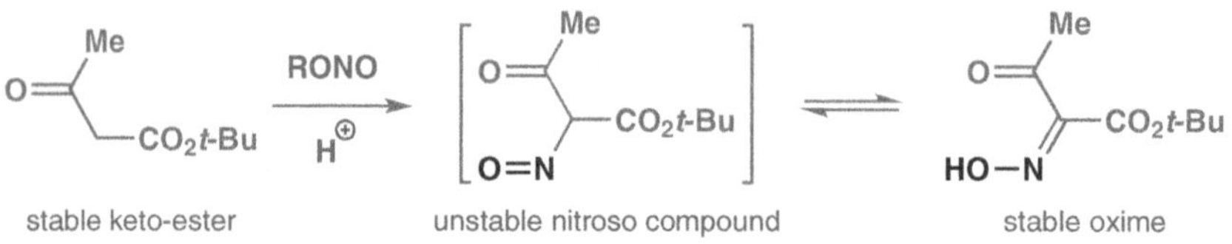

Zinc in acetic acid (Chapter 23) reduces the oxime to the amine and we can start the synthesis by doing the conjugate addition and then reducing the oxime in the presence of the keto-diester.

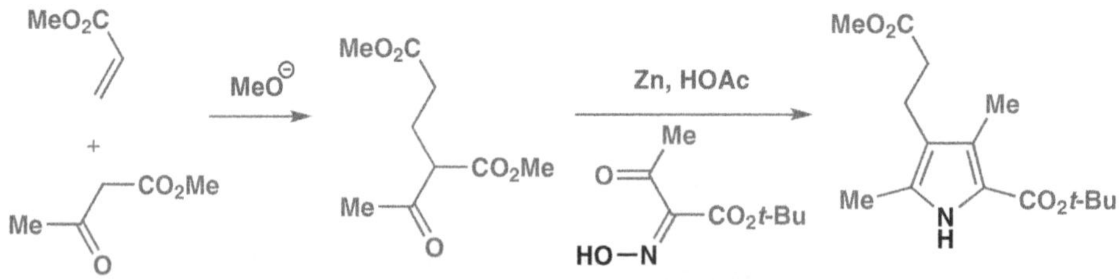

This reaction forms the required pyrrole in one step! First, the oxime is reduced to an amine, then the amino group forms an imine with the most reactive carbonyl group (the ketone) in the ketodiester. Finally, the very easily formed enamine cyclizes onto the other ketone.

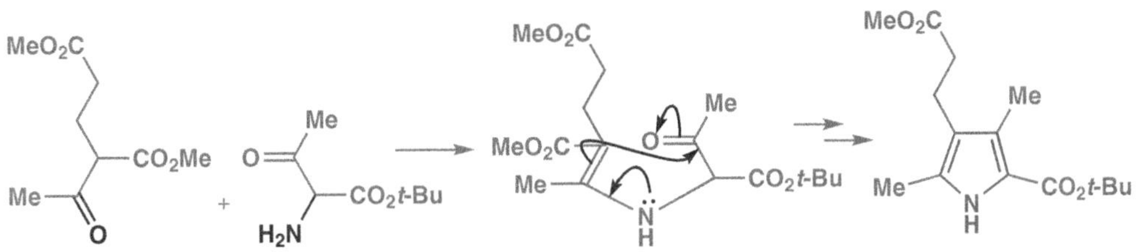

This pyrrole synthesis is important enough to be given the name of its inventor—it is the Knorr pyrrole synthesis. Knorr himself made a rather simpler pyrrole in a remarkably efficient reaction. See if you can work out what is happening here.

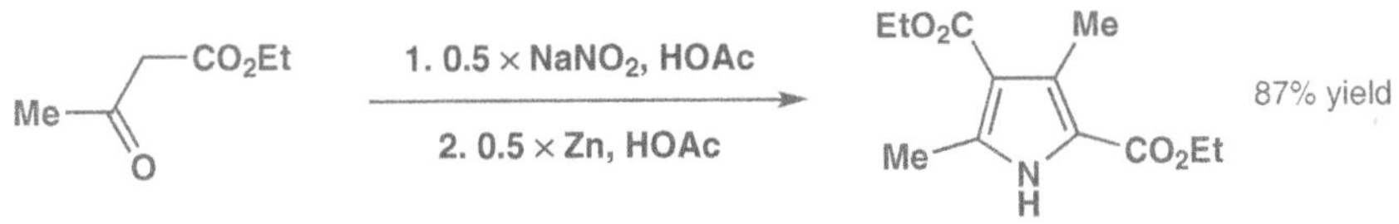

Names for heterocyclic syntheses

Standard heterocyclic syntheses tend to have a name associated with them and it is simply not worthwhile learning these names. Few chemists use any but the most famous of them: we will mention the Knorr pyrrole synthesis, the Hantzsch pyridine synthesis, and the Fischer and Reissert indole syntheses. We did not mention that the synthesis of furans from 1,4-dicarbonyl compounds is known as the Feist–Benary synthesis, and there are many more like this. If you are really interested in these other names we suggest you consult a specialist book on heterocyclic chemistry.

How to make pyridines: the Hantzsch pyridine synthesis

The idea of coupling two keto-esters together with a nitrogen atom also works for pyridines except that an extra carbon atom is needed. This is provided as an aldehyde and another important difference is that the nitrogen atom is added as a nucleophile rather than an electrophile. These are features of the **Hantzsch pyridine synthesis**. This is a four-component reaction from simple starting materials.

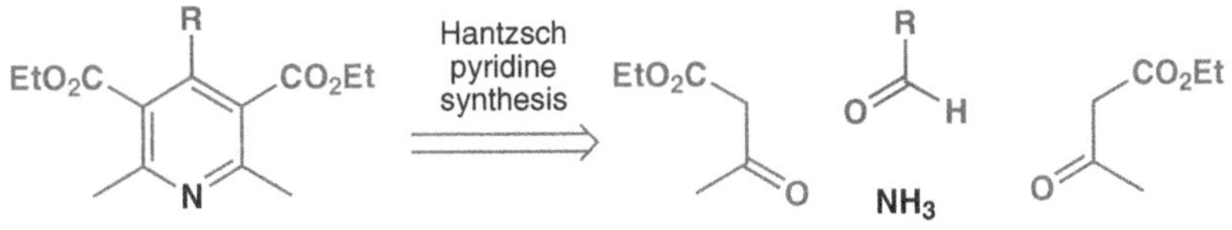

You are hardly likely to understand the rationale behind this reaction from that diagram so let's explore the details. The product of the reaction is actually the dihydropyridine, which has to be oxidized to the pyridine by a reagent such as HNO_3, Ce(IV), or a quinone.

The reaction is very simply carried out by mixing the components in the right proportions in ethanol. The presence of water does not spoil the reaction and the ammonia, or some added amine, ensures the slightly alkaline pH necessary. Any aldehyde can be used, even formaldehyde, and yields of the crystalline dihydropyridine are usually very good.

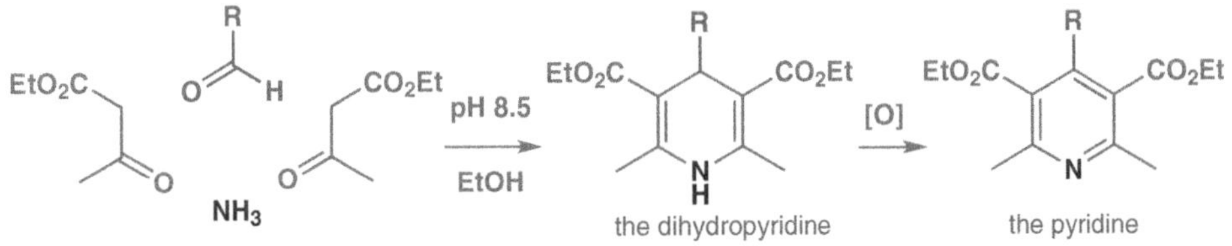

Arthur Hantzsch, 1857–1935, the 'fiery stereochemist' of Leipzig, was most famous for the work he did with Werner at the ETH in Zurich where in 1890 he suggested that oximes could exist in *cis* and *trans* forms.

the dihydropyridine   the pyridine

This reaction is an impressive piece of molecular recognition by small molecules and writing a detailed mechanism is a bold venture. We can see that certain events have to happen, but which order they happen in is a matter of conjecture. The ammonia has to attack the ketone groups, but it would prefer to attack the more electrophilic aldehyde so this is probably not the first step. The enol or enolate of the keto-ester has to attack the aldehyde (twice!) so let us start there.

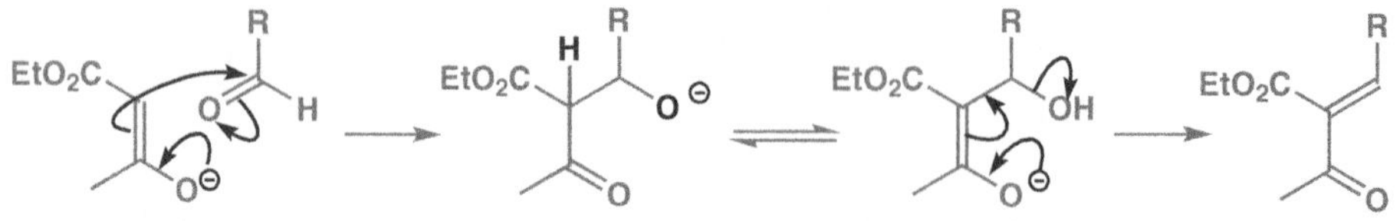

Interactive mechanism for Hantzsch pyridine synthesis

This adduct is in equilibrium with the stable enolate from the keto-ester and elimination now gives an unsaturated carbonyl compound. Such chemistry is associated with the aldol reactions we discussed in Chapter 26. The new enone has two carbonyl groups at one end of the double bond and is therefore a very good Michael acceptor (Chapter 25). A second molecule of enolate does a conjugate addition to complete the carbon skeleton of the molecule. Now the ammonia attacks either of the ketones and cyclizes on to the other. As ketones are more electrophilic than esters it is to be expected that ammonia will prefer to react there.

the dihydropyridine

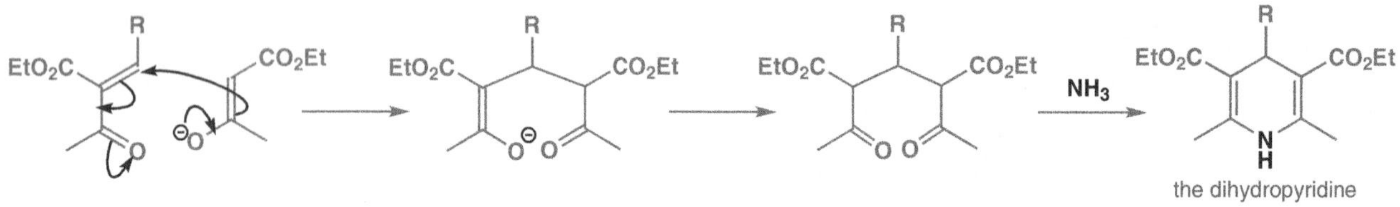

■ We will show in Chapter 42 that Nature uses related dihydropyridines as reducing agents in living things.

The necessary oxidation is easy both because the product is aromatic and because the nitrogen atom can help to expel the hydrogen atom and its pair of electrons from the 4-position. If we use a quinone as oxidizing agent, both compounds become aromatic in the same step.

Interactive mechanism for quinone oxidation of dihydropyridines

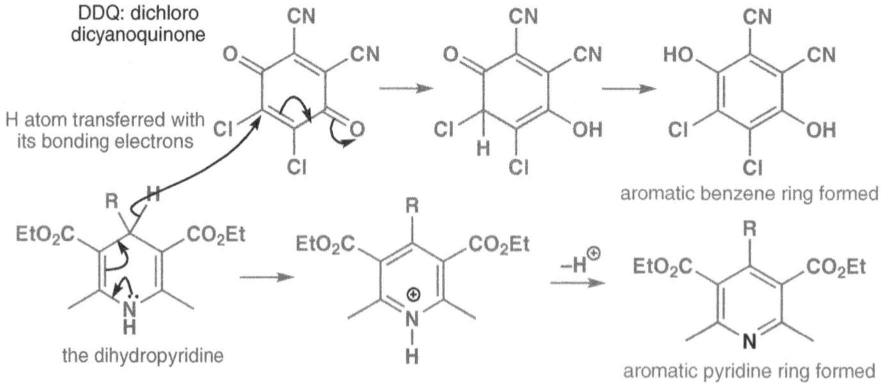

To recap this mechanism, the essentials are:

• aldol reaction between the aldehyde and the keto-ester
• Michael (conjugate) addition to the enone
• addition of ammonia to one ketone
• cyclization of the imine or enamine on to the other ketone

although several of the steps could happen in a different order.

The Hantzsch pyridine synthesis is an old discovery (1882) which sprang into prominence in the 1980s with the discovery that the dihydropyridine intermediates prepared from aromatic aldehydes are calcium channel-blocking agents and therefore valuable drugs for heart disease with useful effects on angina and hypertension.

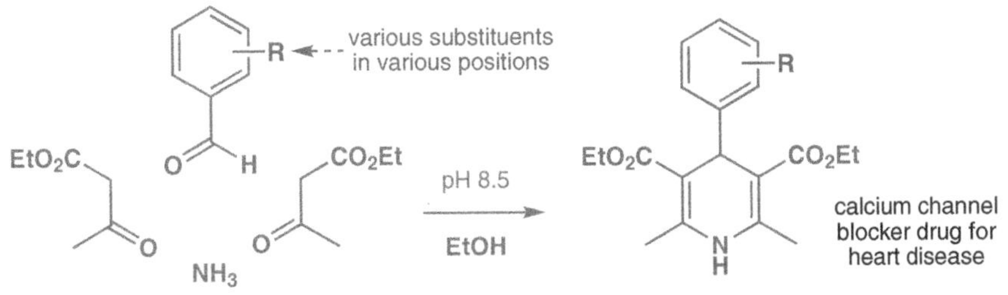

So far, so good. But it also became clear that the best drugs were unsymmetrical—some in a trivial way such as felodipine but some more seriously such as Pfizer's amlodipine. At first sight it looks as though the very simple and convenient Hantzsch synthesis cannot be used for these compounds.

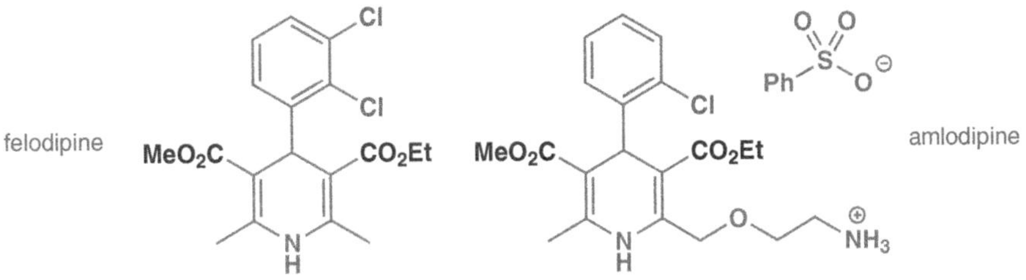

These drugs inhibit Ca^{2+} ion transport across cell membranes and relax muscle tissues selectively without affecting the working of the heart. They allow high blood pressure to be reduced. Pfizer's amlodipine (Istin™ or Norvasc™) is a very important drug.

Clearly, a modification is needed in which half of the molecule is assembled first. The solution lies in early work by Robinson, who made the very first enamines from keto-esters and amines. One half of the molecule is made from an enamine and the other half from a separately synthesized enone. We can use felodipine as a simple example.

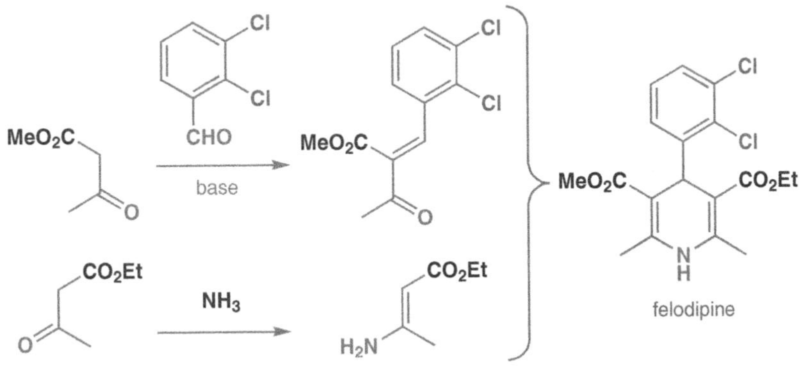

Other syntheses of pyridines

The Hantzsch synthesis produces a reduced pyridine but there are many syntheses that go directly to pyridines. One of the simplest is to use hydroxylamine (NH_2OH) instead of ammonia as the nucleophile. Reaction with a 1,5-diketone gives a dihydropyridine but then water is lost and no oxidation is needed.

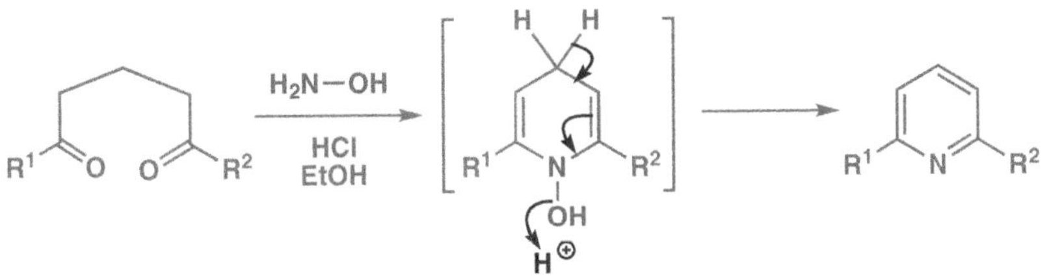

The example below shows how these 1,5-diketones may be quickly made by the Mannich (Chapter 26) and Michael (Chapter 25) reactions. Our pyridine has a phenyl substituent and a fused saturated ring. First we must disconnect to the 1,5-diketone.

Further disconnection reveals a ketone and an enone. There is a choice here and both alternatives would work well.

It is convenient to use the amine products of Mannich reactions ('Mannich bases') instead of the very reactive unsaturated ketones and we will continue with disconnection 'a'.

The synthesis is extraordinarily easy. The stable Mannich base is simply heated with the other ketone to give a high yield of the 1,5-diketone. Treatment of that with the HCl salt of NH_2OH in EtOH gives the pyridine directly, also in good yield.

nicotinamide

Another direct route leads, as we shall now demonstrate, to pyridones. These useful compounds are the basis for nucleophilic substitutions on the ring (Chapter 29). We choose an example that puts a nitrile in the 3-position. This is significant because the role of nicotinamide in living things (Chapter 42) makes such products interesting to make. Aldol disconnection of a 3-cyano pyridone starts us on the right path. If we now disconnect the C–N bond forming the enamine on the other side of the ring we will expose the true starting materials. This approach is unusual in that the nitrogen atom that is to be the pyridine nitrogen is not added as ammonia but is already present in a molecule of cyanoacetamide.

The keto-aldehyde can be made by a simple Claisen ester condensation (Chapter 26) using the enolate of the methyl ketone with ethyl formate (HCO_2Et) as the electrophile. It actually exists as a stable enol, like so many 1,3-dicarbonyl compounds (Chapter 20).

In the synthesis, the product of the Claisen ester condensation is actually the enolate anion of the keto-aldehyde and this can be combined directly without isolation with cyano-acetamide to give the pyridone in the same flask.

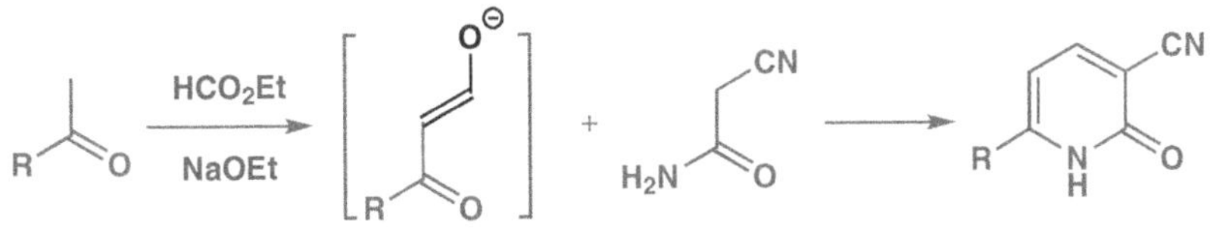

What happens here is that the two compounds must exchange protons (or switch enolates if you prefer) before the aldol reaction occurs. Cyclization probably occurs next through C–N bond formation and, finally, dehydration is forced to give the Z alkene.

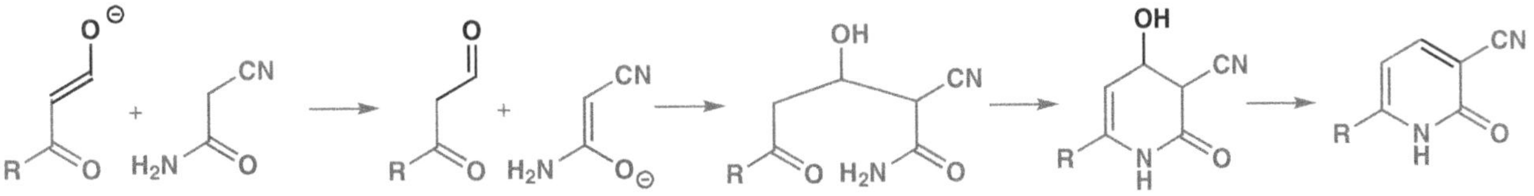

In planning the synthesis of a pyrrole or a pyridine from a dicarbonyl compound, considerable variation in oxidation state is possible. The oxidation state is chosen to make further disconnection of the carbon skeleton as easy as possible. We can now see how these same principles can be applied to pyrazoles and pyridazines.

Pyrazoles and pyridazines from hydrazine and dicarbonyl compounds

Disconnection of pyridazines reveals a molecule of hydrazine and a 1,4-diketone with the proviso that, just as with pyridines, the product will be a dihydropyrazine and oxidation will be needed to give the aromatic compound. As with pyridines, we prefer to avoid the *cis* double bond problem.

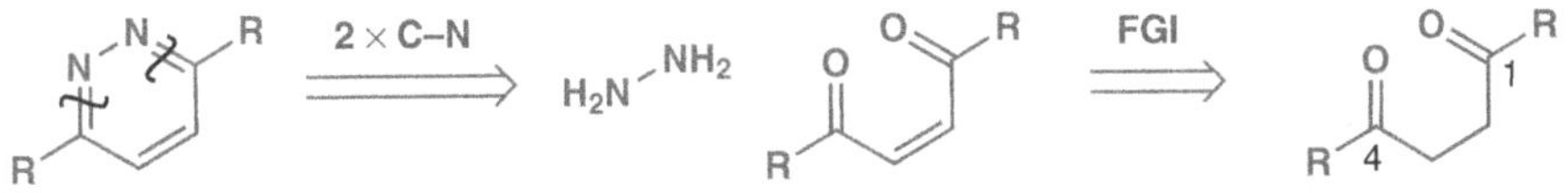

As an example we can take the cotton herbicide made by Cyanamid. Direct removal of hydrazine would require a problematic *cis* double bond in the starting material.

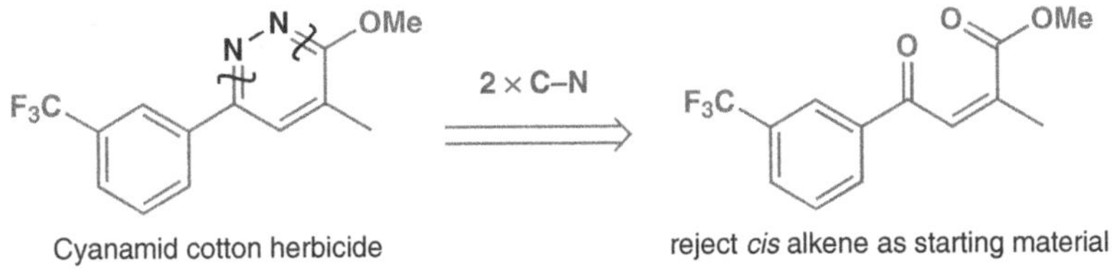

If we remove the double bond first, a much simpler compound emerges. Note that this is a ketoester rather than a diketone.

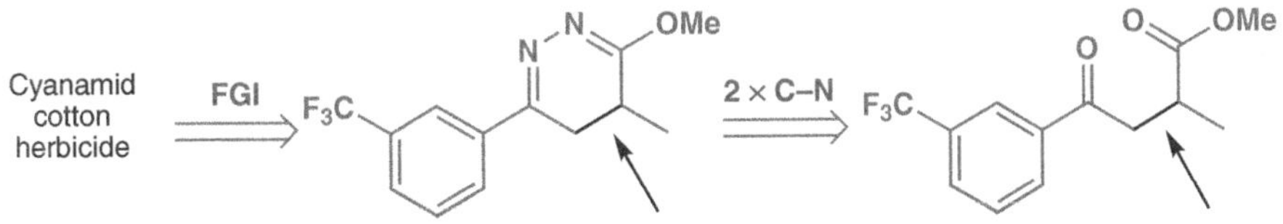

When hydrazine is added to the keto-ester an imine is formed with the ketone but acylation occurs at the ester end to give an amide rather than the imino-ester we had designed.

> ■ If dehydration occurred first, only the Z alkene could cyclize and the major product, the E alkene, would be wasted.

> ■ The herbicide kills weeds in cotton crops rather than the cotton plant itself!

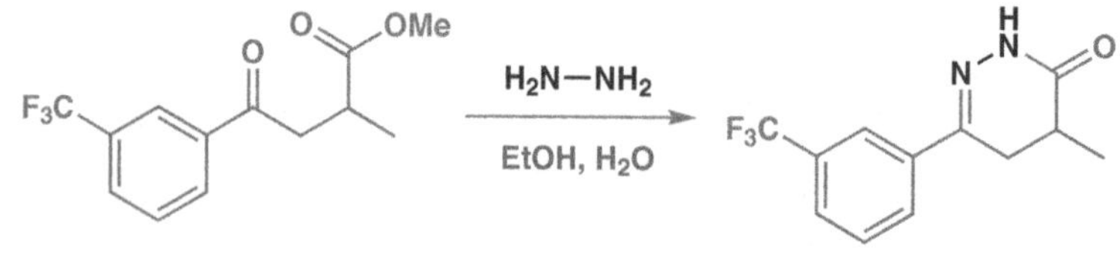

Aromatization with bromine gives the aromatic pyridazolone by bromination and dehydro-bromination, and now we invoke the nucleophilic substitution reactions introduced in Chapter 29. First we make the chloride with POCl$_3$ and then displace with methanol.

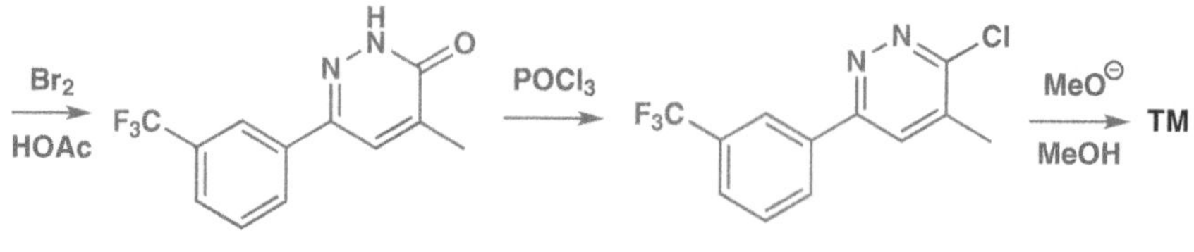

The five-membered ring pyrazoles are even simpler as the starting material is a 1,3-dicarbonyl compound available from the aldol or Claisen ester condensations.

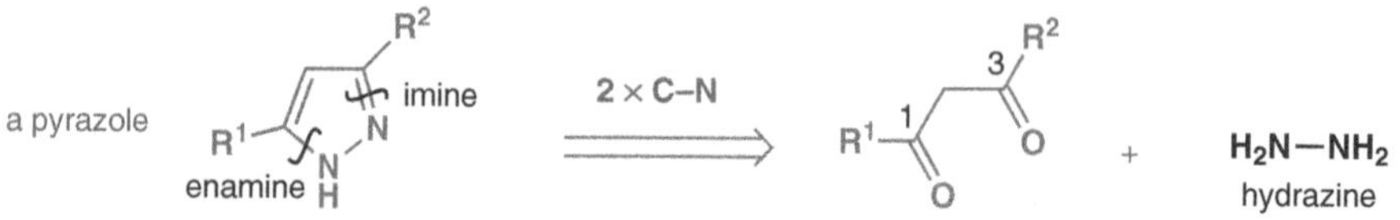

Chemistry hits the headlines—Viagra

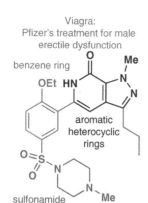

Viagra:
Pfizer's treatment for male erectile dysfunction

benzene ring

OEt HN

aromatic heterocyclic rings

sulfonamide

In 1998 chemistry suddenly appeared in the media in an exceptional way. Normally not a favourite of TV or the newspapers, chemistry produced a story with all the right ingredients—sex, romance, human ingenuity—and all because of a pyrazole. In the search for a heart drug, Pfizer uncovered a compound that allowed impotent men to have active sex lives. They called it Viagra. The molecule contains a sulfonamide and a benzene ring as well as the part that interests us most—a bicyclic aromatic heterocyclic system of a pyrazole fused to a pyrimidine. We shall discuss in detail how Pfizer made this part of the molecule and just sketch in the rest. The sulfonamide can be made from the sulfonic acid that can be added to the benzene ring by electrophilic aromatic sulfonation (Chapter 21).

Inspection of what remains reveals that the carbon atom in the heterocycles next to the benzene ring (marked with an orange blob) is at the oxidation level of a carboxylic acid. If, therefore, we disconnect both C–N bonds to this atom we will have two much simpler starting materials.

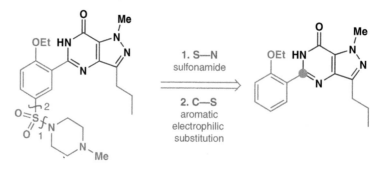

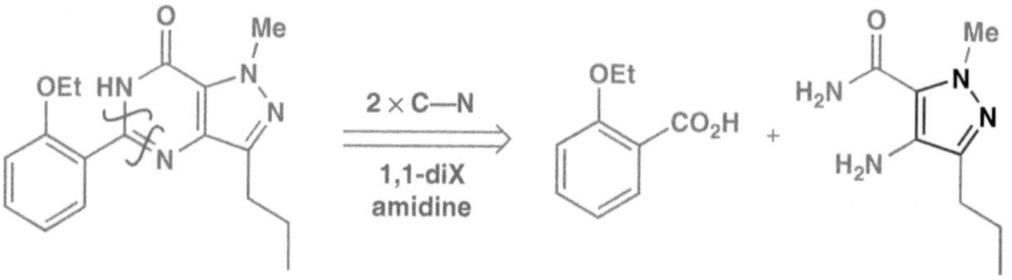

The aromatic acid is available and we need consider only the pyrazole (the core pyrazole ring in black in the diagram). The aromatic amino group can be put in by nitration and reduction, and the amide can be made from the corresponding ester. This leaves a carbon skeleton, which must be made by ring synthesis.

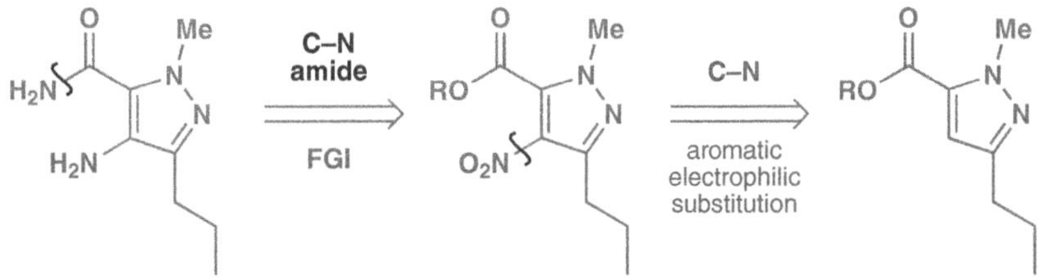

Following the methods we have established so far in this chapter, we can remove the hydrazine portion to reveal a 1,3-dicarbonyl compound. In fact, this is a tricarbonyl compound, a diketo-ester, because of the ester already present and it contains 1,2-, 1,3-, and 1,4-dicarbonyl relationships. The simplest synthesis is by a Claisen ester condensation and we choose the disconnection so that the electrophile is a reactive (oxalate) diester that cannot enolize. The only control needed will then be in the enolization of the ketone.

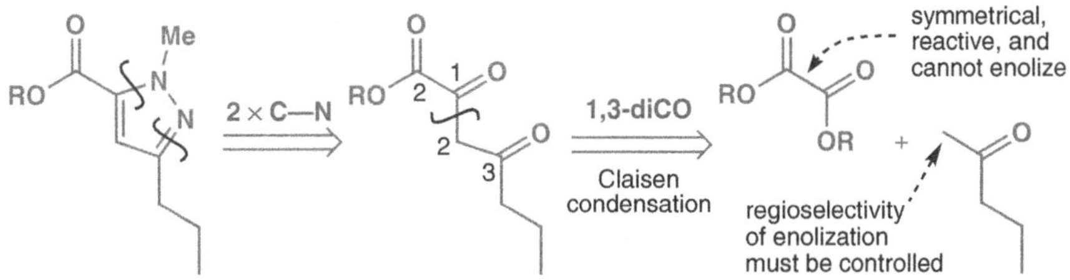

The Claisen ester condensation gives the right product just by treatment with base. The reasons for this are discussed in Chapter 26. We had then planned to treat the keto-diester with methylhydrazine but there is a doubt about the regioselectivity of this reaction—the ketones are more electrophilic than the ester all right, but which ketone will be attacked by which nitrogen atom?

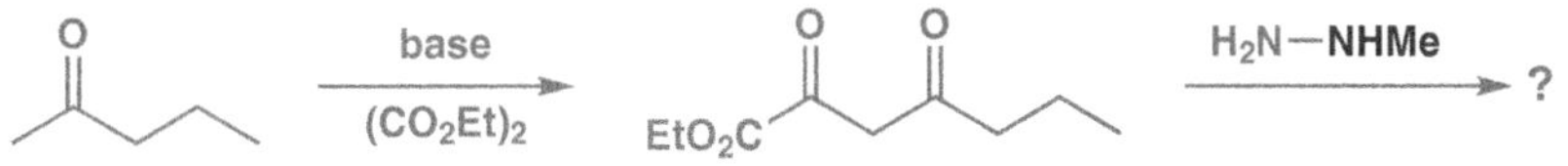

We have already seen the solution to this problem in Chapter 29. If we use symmetrical hydrazine, we can deal with the selectivity problem by alkylation. Dimethyl sulfate turns out to be the best reagent.

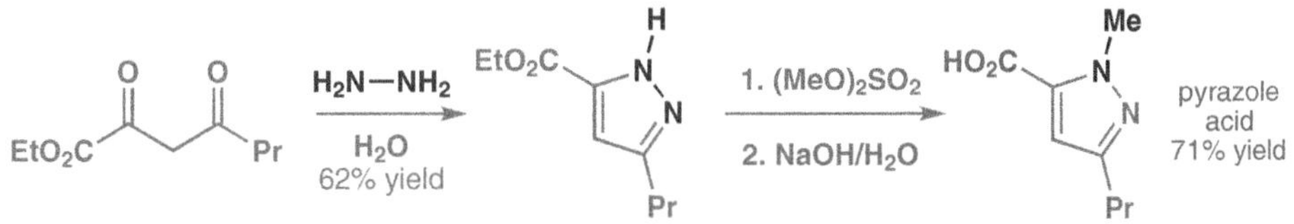

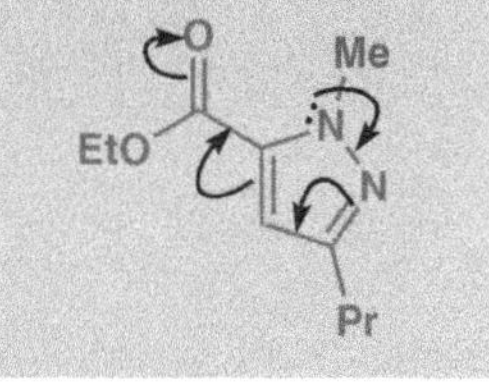

The stable pyrazole acid from the hydrolysis of this ester is a key intermediate in Viagra production. Nitration can occur only at the one remaining free position and then amide formation and reduction complete the synthesis of the amino pyrazole amide ready for assembly into Viagra.

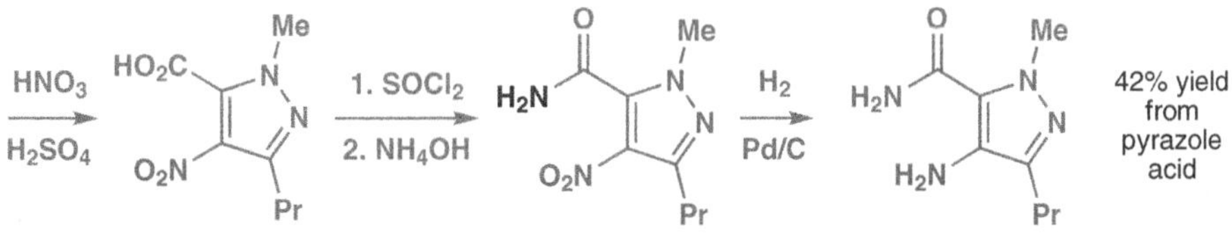

> ■ The alkylation is regioselective because the methylated nitrogen must become the pyrrole-like nitrogen atom and the molecule prefers the longest conjugated system involving that nitrogen and the ester.
>
> lone pair delocalized into ester carbonyl group
>

The rest of the synthesis can be summarized very briefly as it mostly concerns material outside the scope of this chapter. You might like to notice how easy the construction of the second heterocyclic ring is—the nucleophilic attack of the nitrogen atom of one amide on to the carbonyl of another would surely not occur unless the product were an aromatic heterocycle.

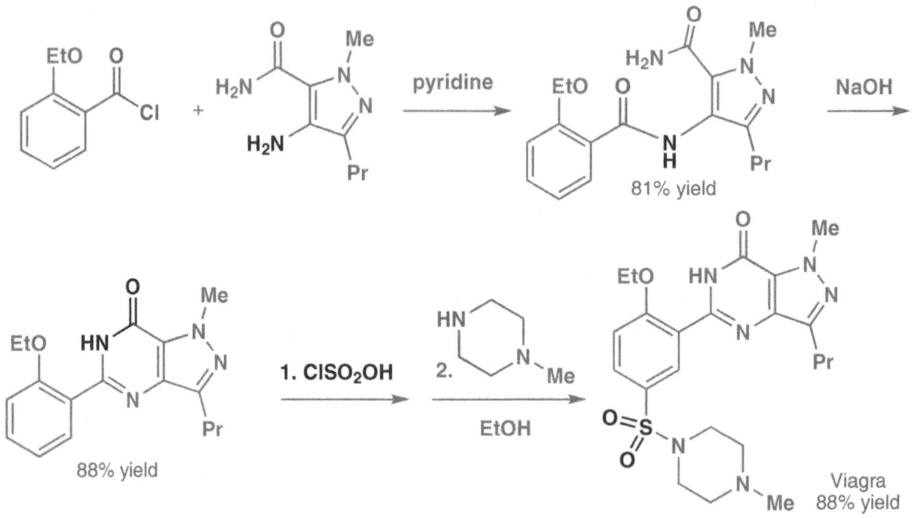

Pyrimidines can be made from 1,3-dicarbonyl compounds and amidines

In Chapter 29 we met some compounds that interfere in folic acid metabolism and are used as antibacterial agents. One of them was trimethoprim and it contains a pyrimidine ring (black on the diagram). We are going to look at its synthesis briefly because the strategy used is the opposite of that used with the pyrimidine ring in Viagra. Here we disconnect a molecule of guanidine from a 1,3-dicarbonyl compound.

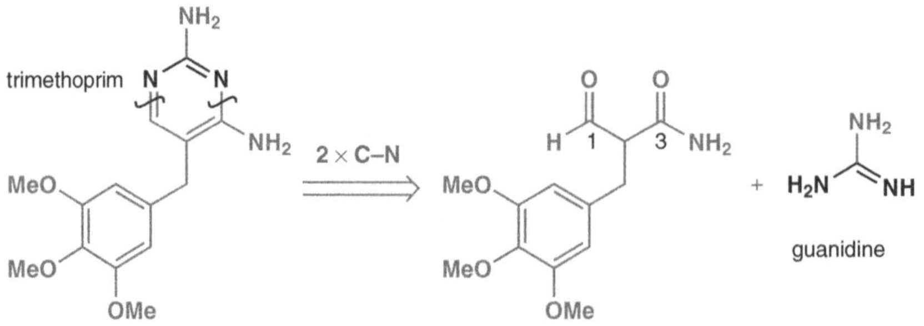

The 1,3-dicarbonyl compound is a combination of an aldehyde and an amide but is very similar to a malonic ester so we might think of making this compound by alkylation of that stable enolate (Chapter 25) with the convenient benzylic bromide.

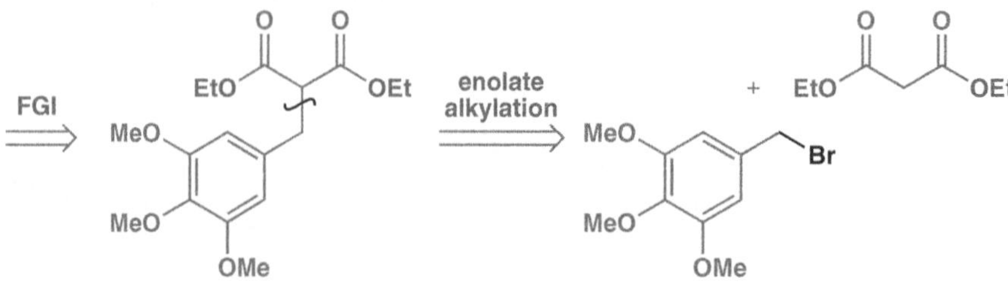

The alkylation works fine but it turns out to be better to add the aldehyde as an electrophile (cf. the pyridone synthesis on p. 766) rather than try to reduce an ester to an aldehyde. The other ester is already at the right oxidation level.

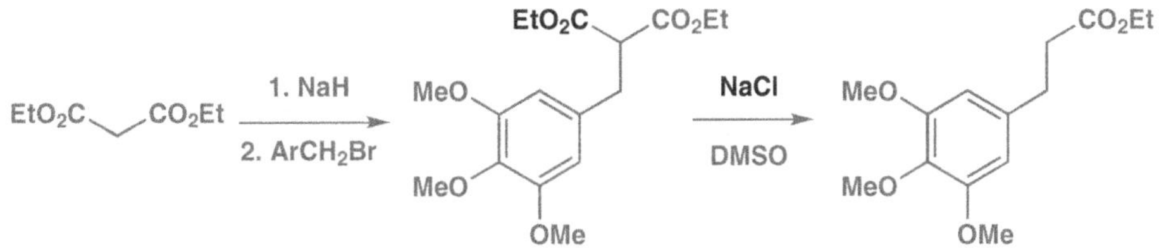

■ Notice the use of the NaCl method of decarboxylation (Chapter 25, p. 597).

Condensation with ethyl formate (HCO$_2$Et) and cyclization with guanidine gives the pyrimidine ring system but with an OH instead of the required amino group. Aromatic nucleophilic substitution the pyridone style from Chapter 29 gives trimethoprim.

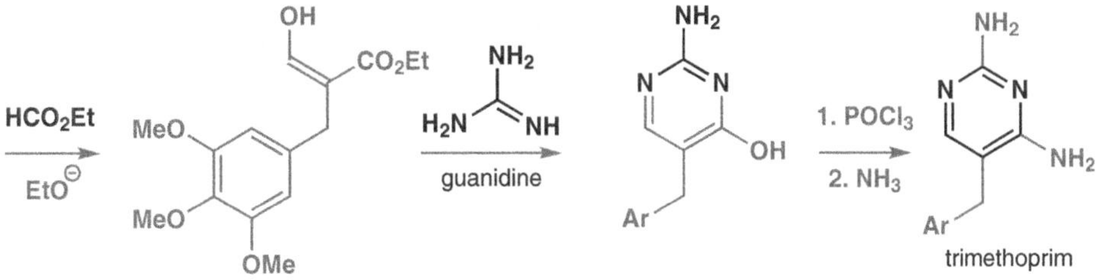

Unsymmetrical nucleophiles lead to selectivity questions

The synthesis of thiazoles is particularly interesting because of a regioselectivity problem. If we try out the two strategies we have just used for pyrimidines, the first requires the reaction of a carboxylic acid derivative with a most peculiar enamine that is also a thioenol. This does not look like a stable compound.

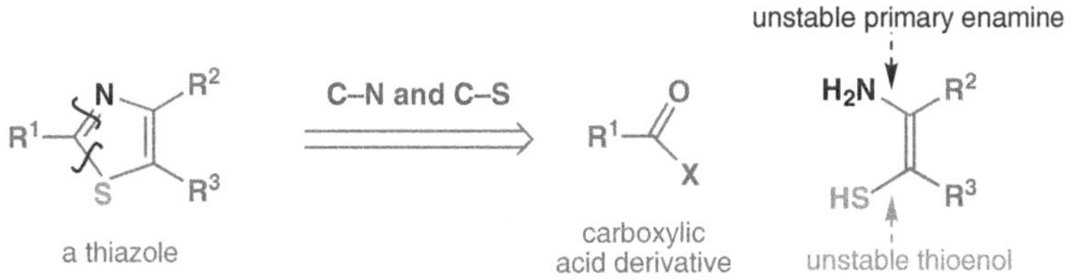

The alternative is to disconnect the C–N and C–S bonds on the other side of the hetero-atoms. Here we must be careful what we are about or we will get the oxidation state wrong. We shall do it step by step to make sure. We can rehydrate the double bond in two ways. We can first try putting the OH group next to nitrogen.

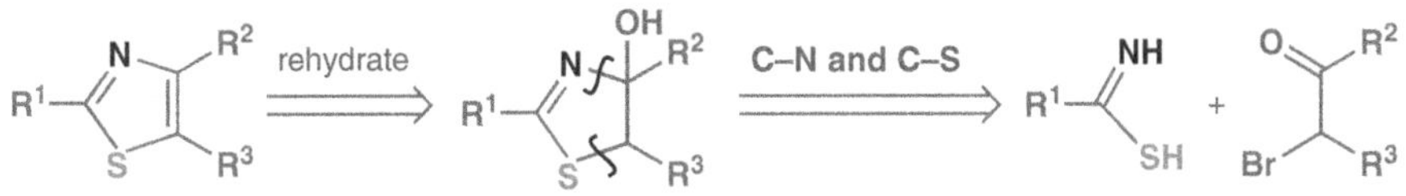

Or we can rehydrate it the other way round, putting the OH group next to the sulfur atom, and disconnect in the same way. In both cases we require an electrophilic carbon atom at the alcohol oxidation level and one at the aldehyde or ketone oxidation level. In other words we need an α-haloketone.

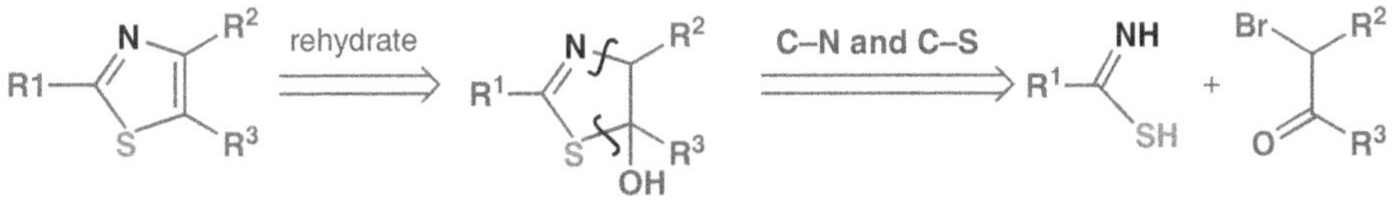

➡ The structure of Lawesson's reagent is on p. 759.

The nucleophile is the same in both cases and it is an odd-looking molecule. That is, until we realize that it is just a tautomer of a thioamide. Far from being odd, thioamides are among the few stable thiocarbonyl derivatives and can be easily made from ordinary amides with P_2S_5 or Lawesson's reagent.

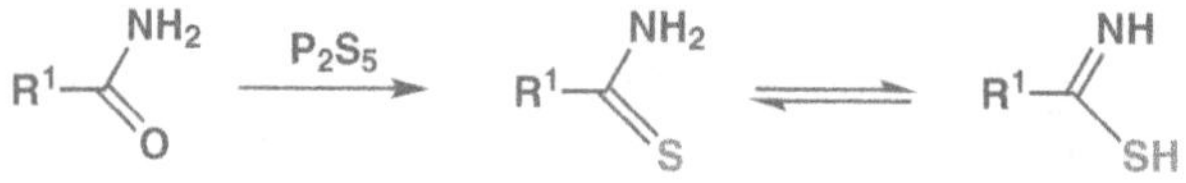

So the only remaining question is: when thioamides combine with α-haloketones, which nucleophilic atom (N or S) attacks the ketone, and which atom (N or S) attacks the alkyl halide? Carbonyl groups are 'hard' electrophiles—their reactions are mainly under charge control and so they react best with basic nucleophiles (Chapter 10). Alkyl halides are 'soft' electrophiles—their reactions are mainly under frontier orbital control and they react best with large uncharged nucleophiles from the lower rows of the periodic table. The ketone reacts with nitrogen and the alkyl halide with sulfur.

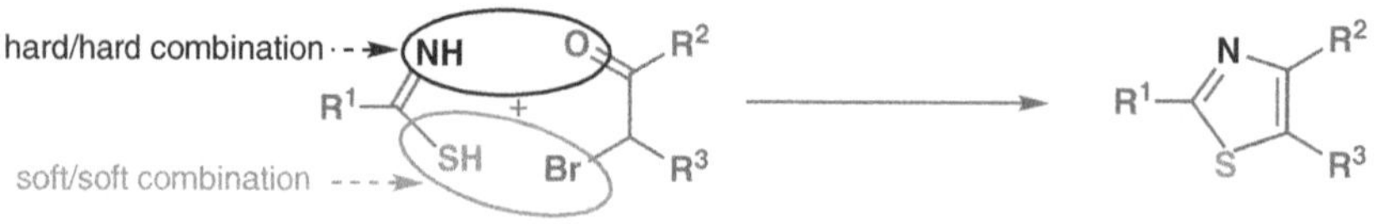

Fentiazac, a non-steroidal anti-inflammatory drug, is a simple example. Disconnection shows that we need thiobenzamide and an easily made α-haloketone (easily made because the ketone can enolize on this side only—see Chapter 20).

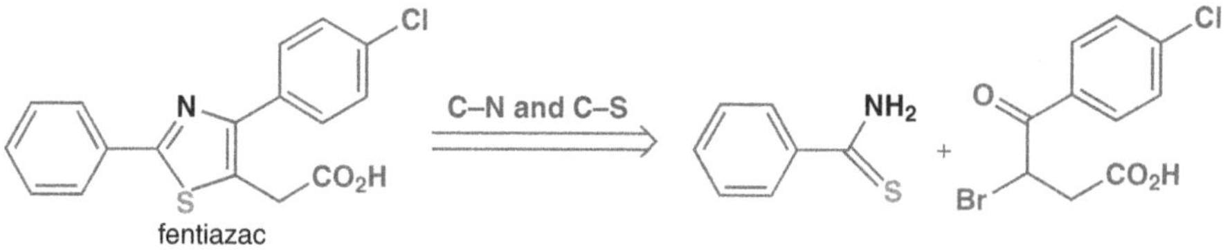

The synthesis involves heating these two compounds together and the correct thiazole forms easily with the double bonds finding their right positions in the product—the only positions for a stable aromatic heterocycle.

Isoxazoles are made from hydroxylamine or by cycloaddition

The two main routes for the synthesis of isoxazoles are (a) the attack of hydroxylamine (NH_2OH) on diketones and (b) a reaction of nitrile oxides called a 1,3-dipolar cycloaddition. They thus form a link between the strategy we have been discussing (cyclization of a nucleophile with two heteroatoms and a compound with two electrophilic carbon atoms) and the next strategy—cycloaddition reactions.

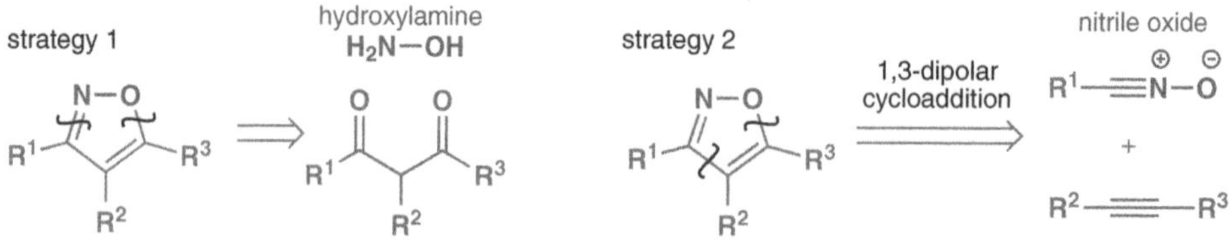

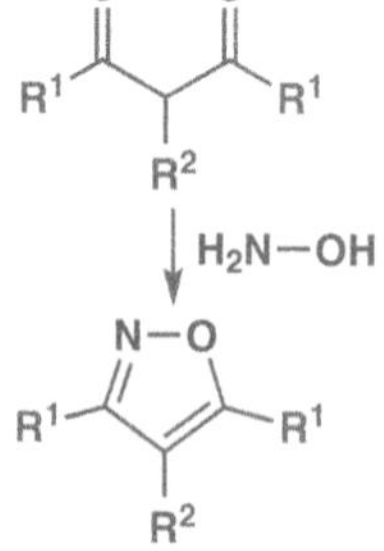

■ We will cover cycloadditions in detail in Chapter 34—in fact you have met an example already in the last chapter: the Diels–Alder reaction (p. 739) is also a cycloaddition. The cycloadditions here involve a 1,3-dipole and form a five-membered ring, by a cyclic mechanism involving six electrons.

Simple symmetrical isoxazoles are easily made by the hydroxylamine route. If $R^1 = R^3$, we have a symmetrical and easily prepared 1,3-diketone as starting material. The central R^2 group can be inserted by alkylation of the stable enolate of the diketone (Chapter 25).

When $R^1 \neq R^3$, we have an unsymmetrical dicarbonyl compound and we must be sure that we know which way round the reaction will proceed. The more nucleophilic end of NH_2OH will attack the more electrophilic carbonyl group. It seems obvious that the more nucleophilic end of NH_2OH will be the nitrogen atom but that depends on the pH of the solution. Normally, hydroxylamine is supplied as the crystalline hydrochloride salt and a base of some kind added to give the nucleophile. The relevant pK_as are shown in the margin. Bases such as pyridine or sodium acetate produce some of the reactive neutral NH_2OH in the presence of the less reactive cation, but bases such as NaOEt produce the anion. Reactions of keto-aldehydes with acetate-buffered hydroxylamine usually give the isoxazole from nitrogen attack on the aldehyde as expected.

state of hydroxylamine changes with pH: the more nucleophilic atom is marked in black

$$\overset{\oplus}{H_3N}-OH \quad \text{unreactive}$$

$\updownarrow$ pK_a 6

$$H_2N-OH \quad \text{reactive}$$

$\updownarrow$ pK_a 13

$$H_2N-O^{\ominus} \quad \text{very reactive}$$

Modification of the electrophile may also be successful. Reaction of hydroxylamine with 1,2,4-diketo-esters usually gives the isoxazole from attack of nitrogen at the more reactive keto group next to the ester.

A clear demonstration of selectivity comes from the reactions of bromoenones. It is not immediately clear which end of the electrophile is more reactive but the reactions tell us the answer.

The alternative approach to isoxazoles relies on the reaction of nitrile oxides with alkynes. We shall see in Chapter 34 that there are two good routes to these reactive compounds, the γ-elimination of chlorooximes or the dehydration of nitroalkanes.

Interactive mechanism for nitrile oxide formation

A few nitrile oxides are stable enough to be isolated (those with electron-withdrawing or highly conjugating substituents, for example) but most are prepared in the presence of the alkyne by one of these methods because otherwise they dimerize rapidly. Both methods of forming nitrile oxides are compatible with their rapid reactions with alkynes. With aryl alkynes the reaction is usually clean and regioselective.

Interactive mechanism for nitrile oxide cycloaddition

The reaction forms the five-membered ring in a single step: it is a cycloaddition, in which the alkyne is using its HOMO to attack the LUMO of the nitrile oxide (see Chapter 34 for an explanation). If the alkyne has an electron-withdrawing group, mixtures of isomers are usually formed as the HOMO of the nitrile oxide also attacks the LUMO of the alkyne. Intramolecular reactions are usually clean regardless of the preferred electronic orientation if the tether is too short to allow any cyclization except one. In this example, even the more favourable orientation looks very bad because of the linear nature of the reacting species, but only one isomer is formed.

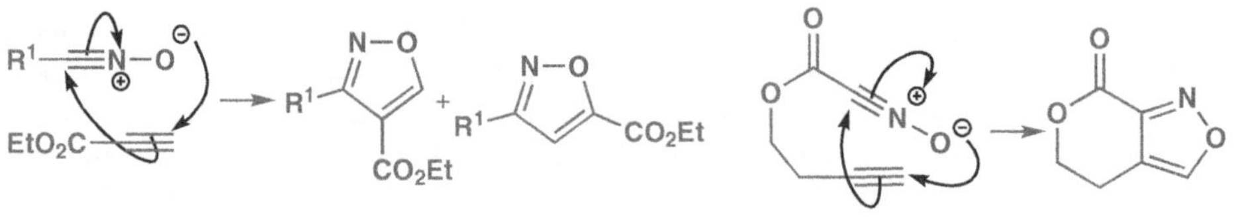

Tetrazoles and triazoles are also made by cycloadditions

Disconnection of tetrazoles with a 1,3-dipolar cycloaddition in mind is easy to see once we realize that a nitrile (RCN) is going to be one of the components. It can be done in two ways: disconnection of the neutral compound would require the dangerous hydrazoic acid (HN_3) as the dipole but the anion disconnects directly to azide ion.

■ You saw in Chapter 29 that tetrazoles are about as acidic as carboxylic acids.

Unpromising though this reaction may look, it actually works well if an ammonium chloride-buffered mixture of sodium azide and the nitrile is heated in DMF. The reagent is really ammonium azide and the reaction occurs faster with electron-withdrawing substituents in R. In the reaction mixture, the anion of the tetrazole is formed but neutralization with acid gives the free tetrazole.

🖱 Interactive mechanism for tetrazole formation

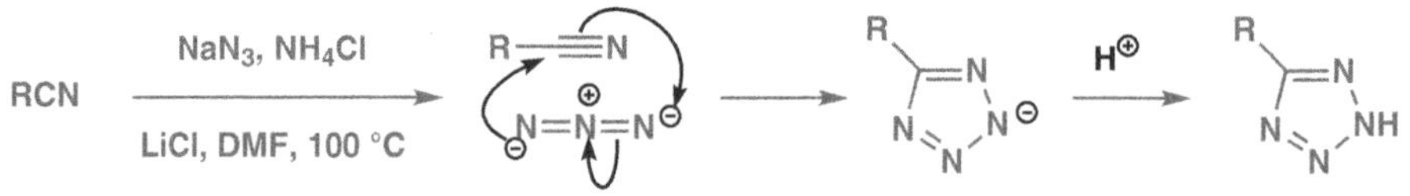

As nitriles are generally readily available this is the main route to simple tetrazoles. More complicated ones are made by alkylation of the product of a cycloaddition. The tetrazole substitute for indomethacin that we mentioned in Chapter 29 is made by this approach. First, the nitrile is prepared from the indole. The 1,3-dipolar cycloaddition works well by the azide route we have just discussed, even though this nitrile will form an 'enol' rather easily. Finally, the indole nitrogen atom must be acylated. The tetrazole is more acidic so it is necessary to form a dianion to get reaction at the right place. The usual rule is followed (see Chapter 23)—the second anion formed is less stable and so it reacts first.

➡ The synthesis of indoles with this substitution pattern by the Fischer indole synthesis is described on p. 777.

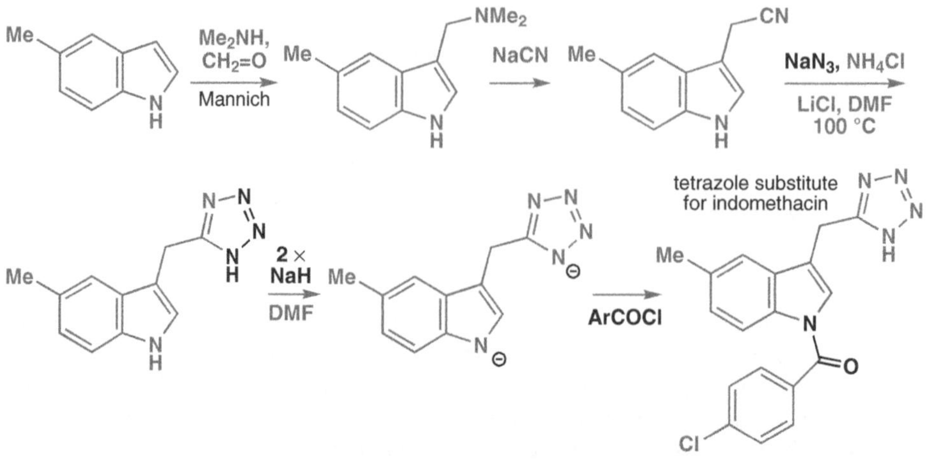

The synthesis of the anti-inflammatory drug broperamole illustrates modification of a tetrazole using its anion. The tetrazole is again constructed from the nitrile—it's an aromatic nitrile with an electron-withdrawing substituent so this will be a good reaction.

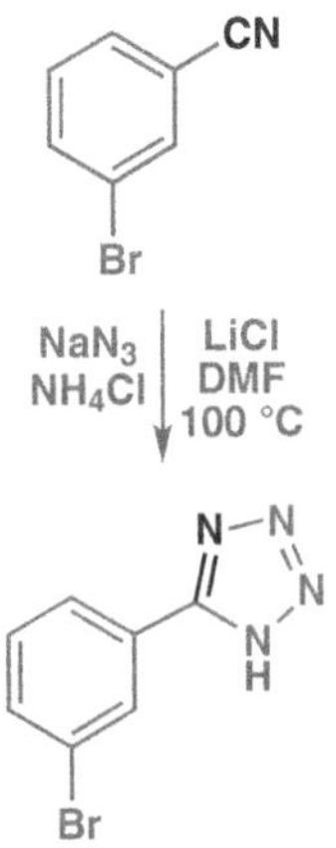

Conjugate addition to acrylic acid (Chapter 22) occurs to give the other tautomer to the one we have drawn. The anion intermediate is, of course, delocalized and can react at any of the nitrogen atoms. Amide formation completes the synthesis of broperamole.

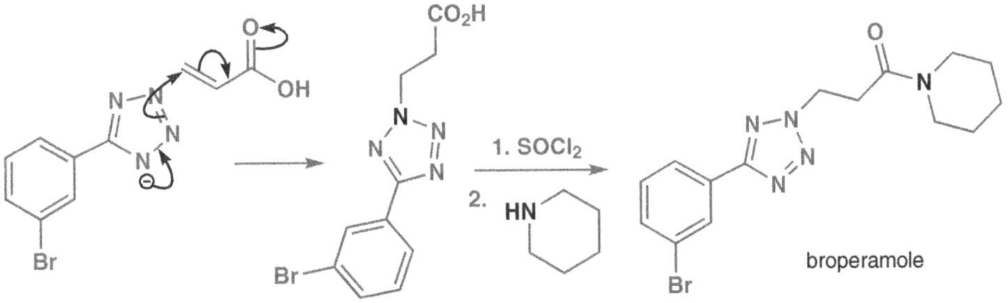

One of the best reactions of all is the related cycloaddition of a substituted azide to an alkyne. Just mixing together and heating an azide and an alkyne will give a triazole, but often as a mixture of two regioisomers.

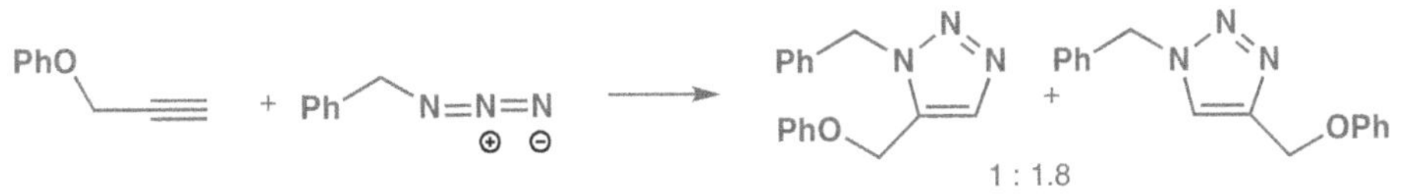

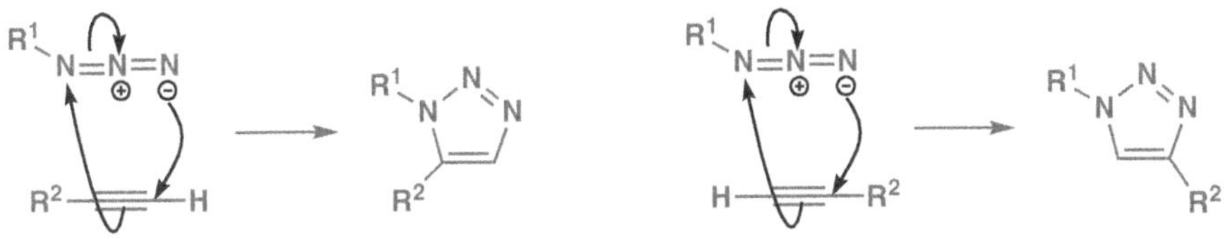

Interactive mechanism for triazole formation

However, a simple addition to the reaction mixture improves the situation hugely: a catalytic amount of Cu(I), often made *in situ* by adding $CuSO_4$ and a mild reducing agent, makes the reaction much faster and gives the 1,4-disubstituted triazole selectively. The work of Sharpless has turned this reaction into not only a very powerful way of making triazoles, but also a very simple way of linking two otherwise relatively unreactive molecules together—the reaction even works in water.

■ 1,2,4-Triazoles are usually made from the reaction of the unsubstituted 1,2,4-triazole anion with electrophiles, as described in Chapter 29.

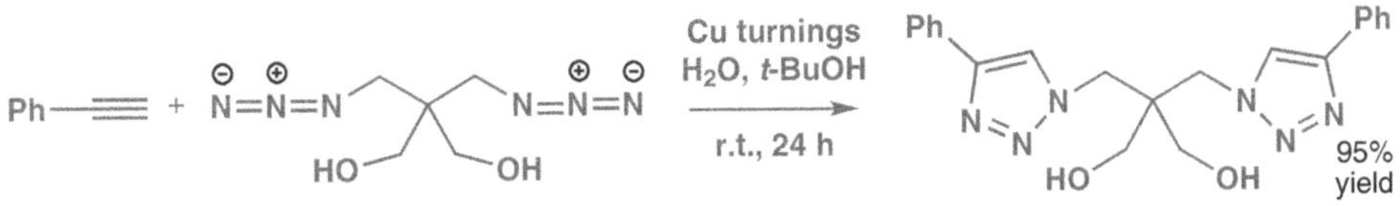

■ It may look as though the more nucleophilic end of the azide has attacked the wrong end of the alkyne but you will see in Chapter 34 that (1) it is very difficult to predict which is the more nucleophilic end of a 1,3-dipole and (2) it may be either HOMO (dipole) and LUMO (alkyne), or LUMO (dipole) and HOMO (alkyne) that dominate the reaction.

The Fischer indole synthesis

You are about to see one of the great inventions of organic chemistry. It is a remarkable reaction, amazing in its mechanism, and it was discovered in 1883 by one of the greatest organic chemists of all, Emil Fischer. Fischer had earlier discovered phenylhydrazine ($PhNHNH_2$) and, in its simplest form, the Fischer indole synthesis occurs when phenylhydrazine is heated in acidic solution with an aldehyde or ketone.

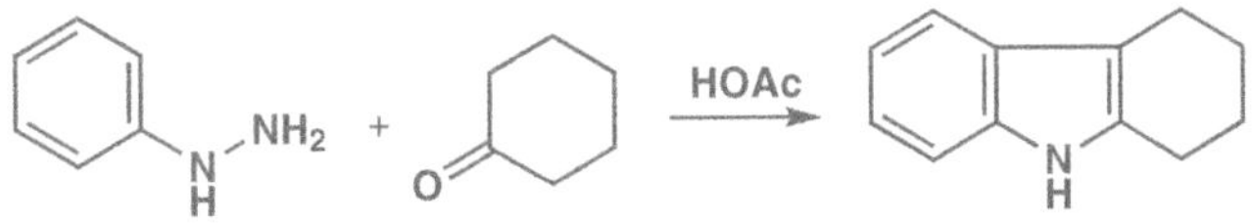

■ This step is a [3,3]-sigmatropic rearrangement, as you will discover in Chapter 35: the new single bond (C–C) bears a 3,3 relationship to the old single bond (N–N).

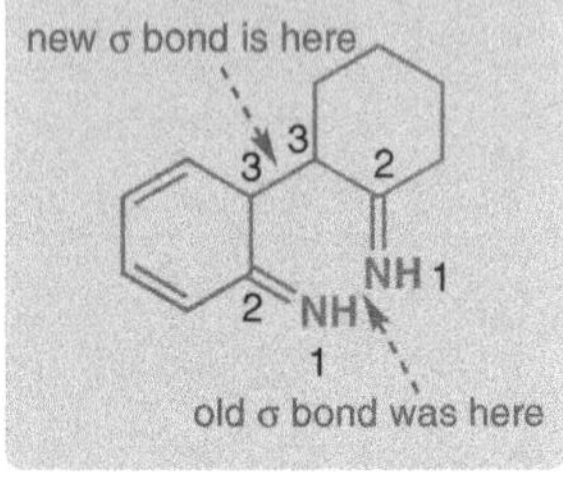

new σ bond is here

old σ bond was here

🖱 Interactive mechanism for Fischer indole synthesis

The first step in the mechanism is formation of the phenylhydrazone (the imine) of the ketone. This can be isolated as a stable compound (Chapter 11).

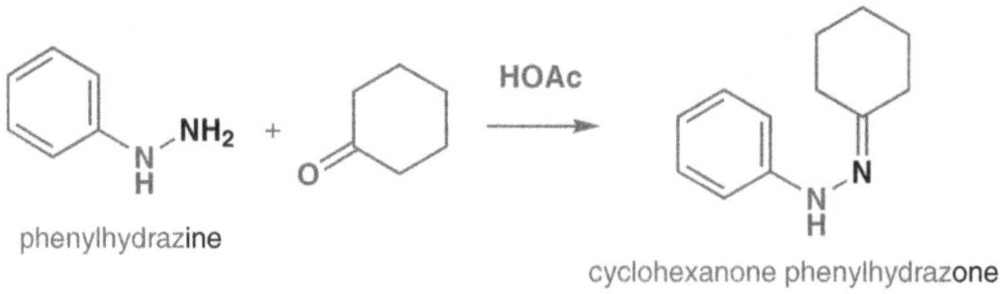

phenylhydraz**ine**

cyclohexanone phenylhydraz**one**

The hydrazone then needs to tautomerize to the enamine, and now comes the key step in the reaction. The enamine can rearrange with formation of a strong C–C bond and cleavage of the weak N–N single bond by moving electrons round a six-membered ring.

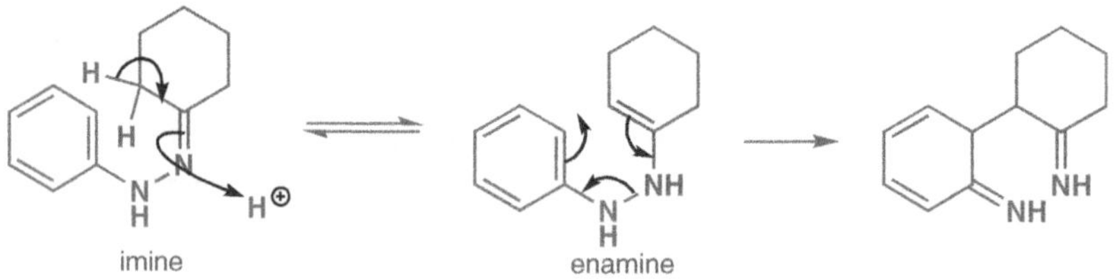

imine   enamine

Next, re-aromatization of the benzene ring (by proton transfer from carbon to nitrogen) creates an aromatic amine that immediately attacks the other imine. This gives an aminal, the nitrogen equivalent of an acetal.

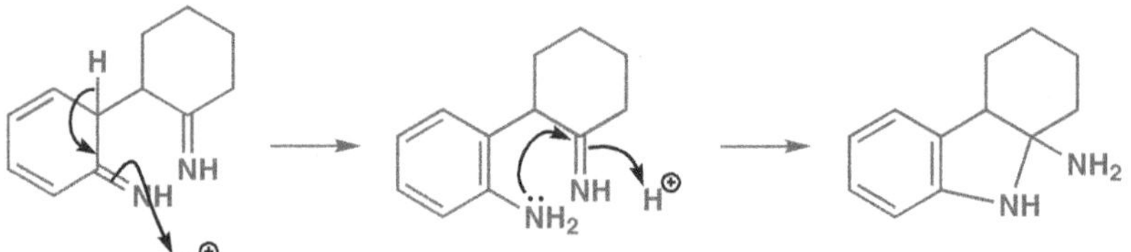

Finally, acid-catalysed decomposition of the aminal in acetal fashion with expulsion of ammonia allows the loss of a proton and the formation of the aromatic indole.

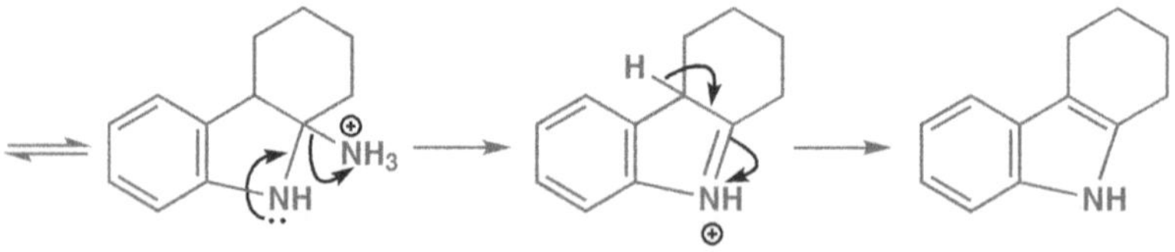

This is admittedly a complicated mechanism but if you remember the central step—the rearrangement of the enamine—the rest should fall into place. The key point is that the C–C bond is established at the expense of a weak N–N bond. Naturally, Fischer had no idea of any of the steps in the mechanism. He was sharp enough to see that something remarkable had happened and skilful enough to find out what it was.

The Fischer method is the main way of making indoles, but it is not suitable for them all. We need now to consider its applicability to various substitution patterns. If the carbonyl compound can enolize on one side only, as is the case with an aldehyde, then the obvious product is formed.

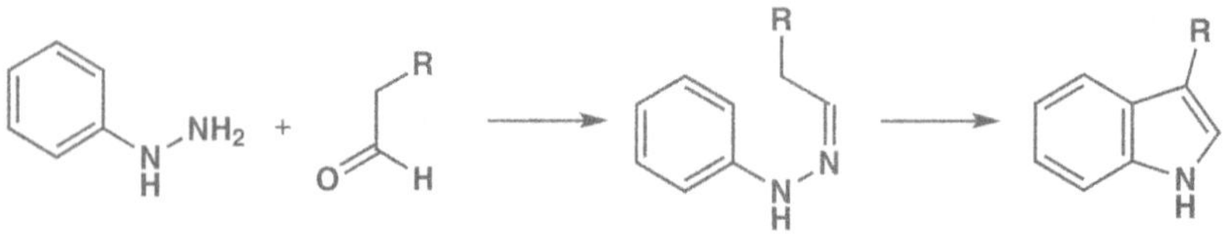

If the benzene ring has only one *ortho* position, then again cyclization must occur to that position. Other substituents on the ring are irrelevant.

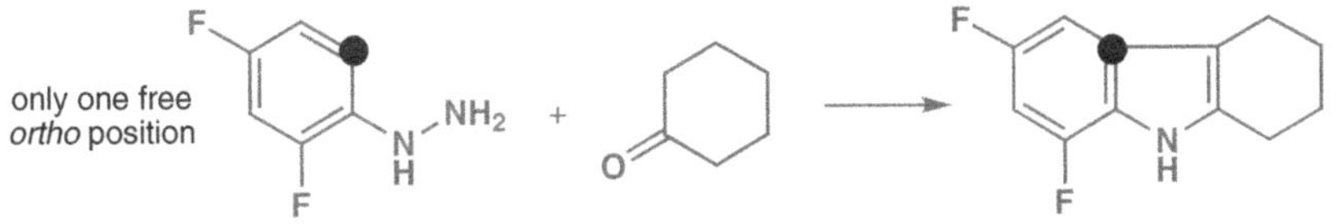

■ At this point we shall stop drawing the intermediate phenylhydrazone.

Another way to secure a single indole as product from the Fischer indole synthesis is to make sure the reagents are symmetrical. These two examples should make plain the types of indole available from symmetrical starting materials.

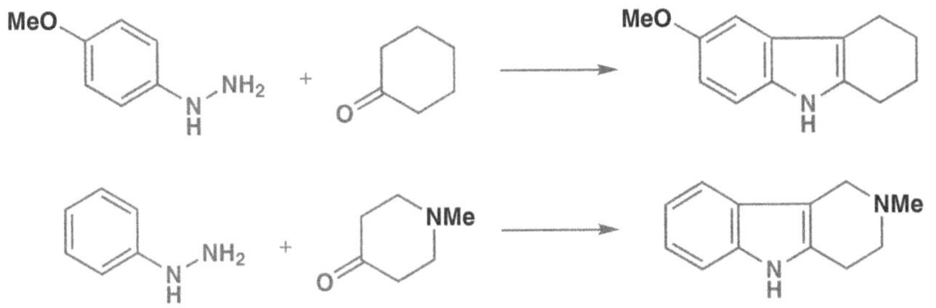

The substitution pattern of the first example is particularly important as the neurotransmitter serotonin is an indole with a hydroxyl group in the 5-position, and many important drugs follow that pattern. Sumatriptan (marketed as Imigran, the migraine treatment) is an analogue of serotonin, whose synthesis starts with the formation of a diazonium salt (Chapter 22) from the aniline shown below. Nitrosation gives the diazonium salt, and reduction with SnCl$_2$ and HCl returns the salt of the phenylhydrazine.

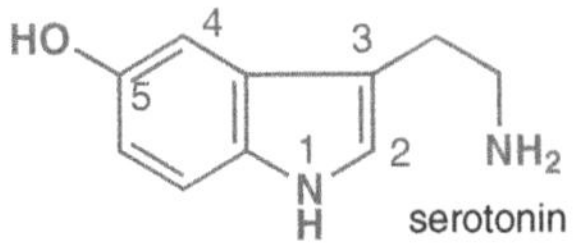

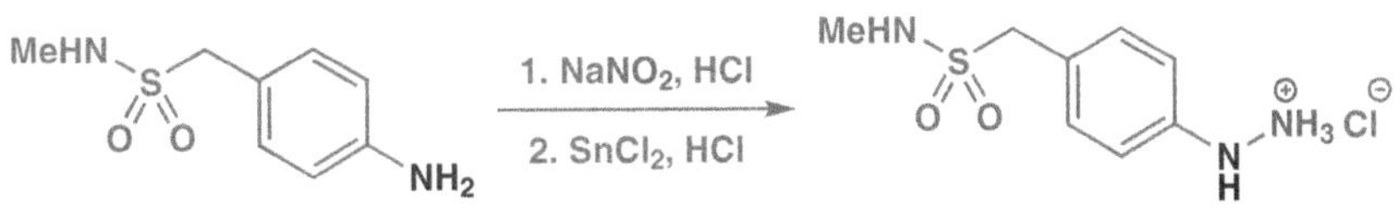

The required aldehyde (3-cyanopropanal) is added as an acetal to prevent self-condensation. The acidic conditions release the aldehyde, which forms the phenylhydrazone, ready for the next step.

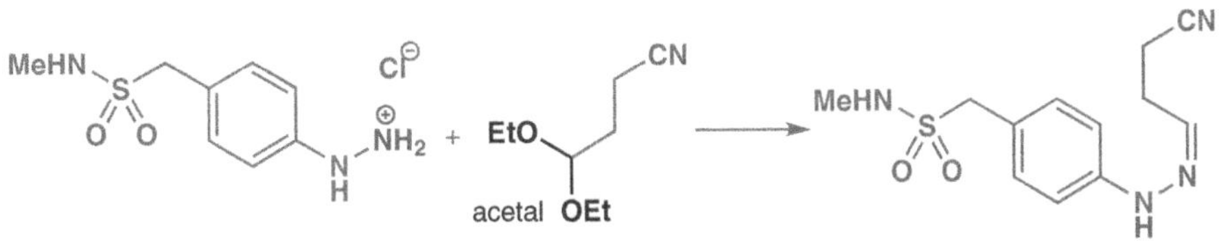

The Fischer indole synthesis itself is catalysed in this case by polyphosphoric acid (PPA), a sticky gum based on phosphoric acid (H$_3$PO$_4$) but dehydrated so that it contains some oligomers. It is often used as a catalyst in organic reactions and residues are easily removed in water.

All that remains is to introduce the dimethylamino group. The nitrile is reduced by hydrogenation and the two methyl groups added by reductive amination with formaldehyde. The reducing agent is formic acid, and the reaction works by sequential formation and reduction of an imine, followed by an iminium formation and reduction to introduce the second methyl group.

■ The dimethylation of primary amines (or methylation of secondary amines) by this method is sometimes called the Eschweiler–Clarke method, and was also mentioned in Chapter 28, p. 716.

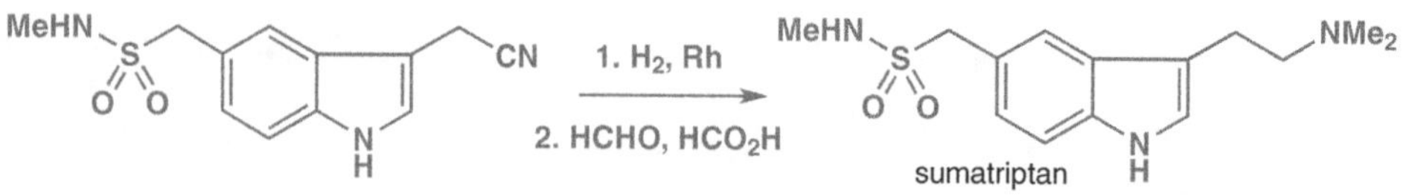

For some indoles it is necessary to control regioselectivity with unsymmetrical carbonyl compounds. Ondansetron, the anti-nausea compound that is used to help cancer patients take larger doses of antitumour compounds than was previously possible, is an example. It contains an indole and an imidazole ring.

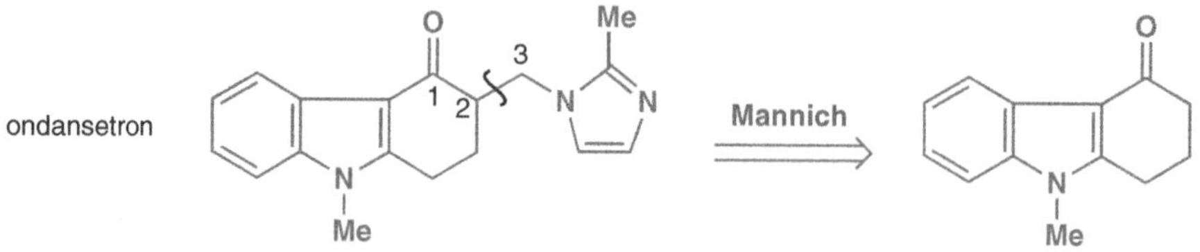

The 1,3 relationship between C–N and C–O suggests a Mannich reaction to add the imidazole ring (Chapters 26 and 28), and that disconnection reveals an indole with an unsymmetrical right-hand side, having an extra ketone group. Fischer disconnection will reveal a diketone as partner for phenylhydrazine. We shall leave aside for the moment when to add the methyl group to the indole nitrogen.

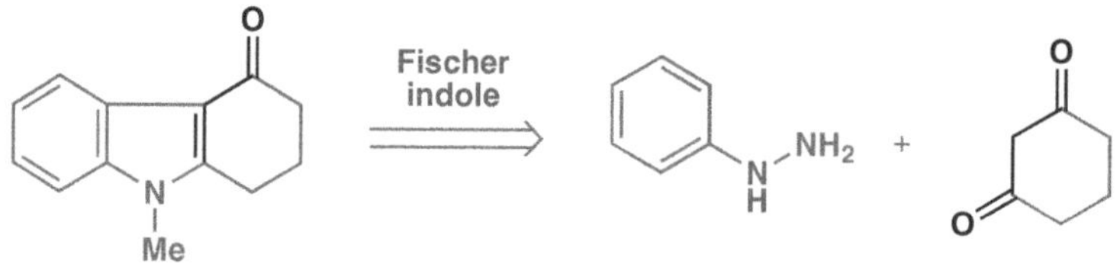

The diketone has two identical carbonyl groups and will enolize (or form an enamine) exclusively towards the other ketone. The phenylhydrazone therefore forms only the enamine we want.

In this case, the Fischer indole reaction was catalysed by a Lewis acid, $ZnCl_2$, and base-catalysed methylation followed. The final stages are summarized below.

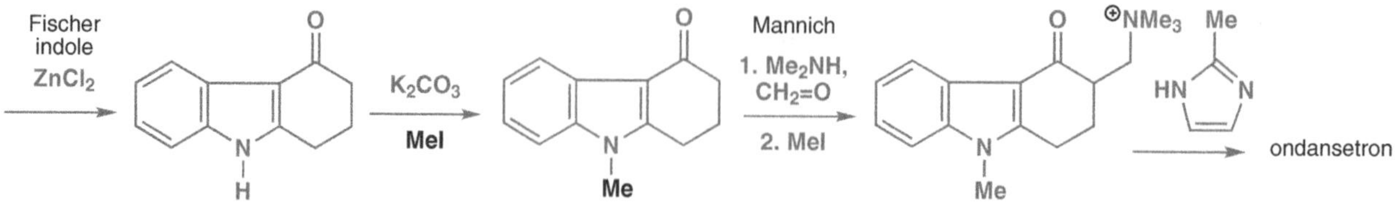

In the worst case, there is no such simple distinction between the two sites for enamine formation and we must rely on other methods of control. The non-steroidal anti-inflammatory drug indomethacin is a good example. Removing the N-acyl group reveals an indole with substituents in both halves of the molecule.

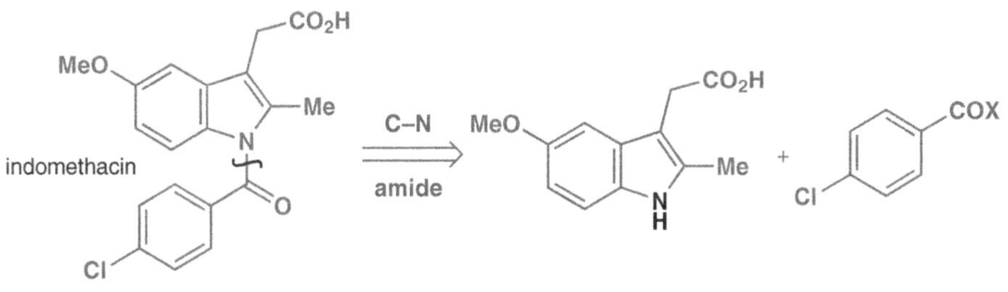

The benzene ring portion is symmetrical and is ideal for the Fischer synthesis but the right-hand half must come from an unsymmetrical open-chain keto-acid. Is it possible to control such a synthesis?

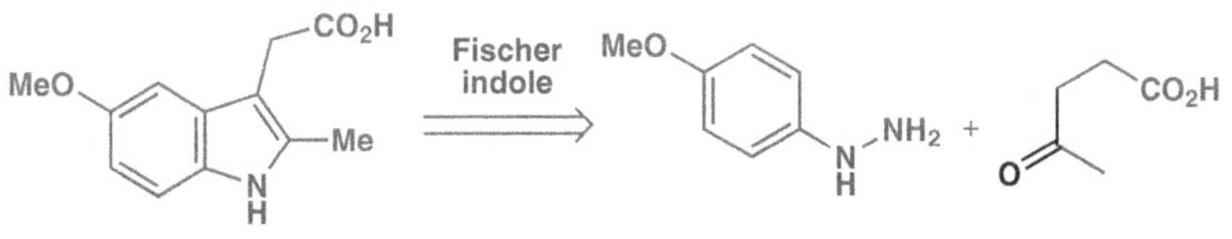

The Fischer indole is acid-catalysed so we must ask: on what side of the ketone is enolization (and therefore enamine formation) expected in acid solution? The answer (see Chapter 20) is away from the methyl group and into the alkyl chain. This is what we want and the reaction does indeed go this way. In fact, the *tert*-butyl ester is used instead of the free acid.

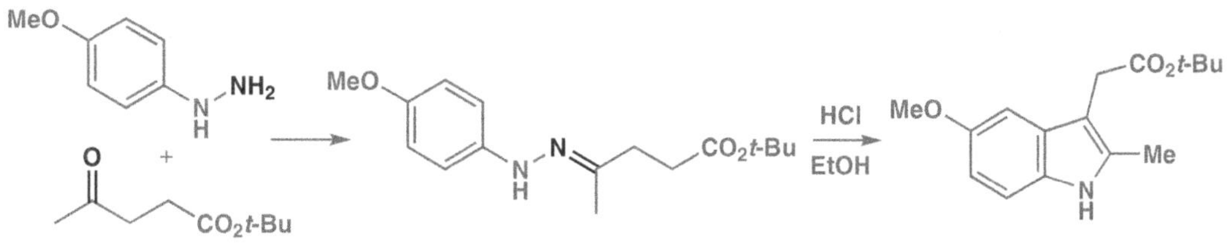

Acylation at the indole nitrogen atom is achieved with acid chloride in base and removal of the *t*-butyl ester gives free indomethacin.

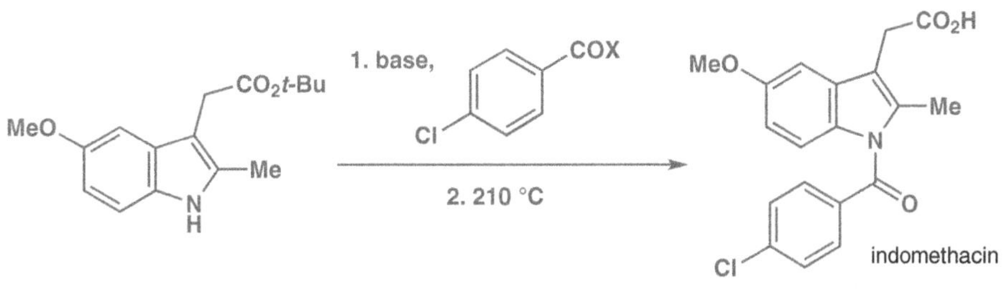

There are many other indole syntheses but we will give a brief mention to only one other, which allows the synthesis of indoles with a different substitution pattern in the benzene ring. If you like names, you may call it the Reissert synthesis, and this is the basic reaction.

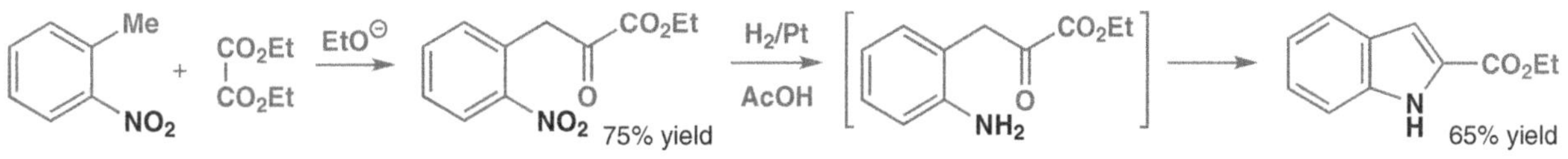

Ethoxide is a strong enough base to remove a proton from the methyl group, delocalizing the negative charge into the nitro group. The anion then attacks the reactive diester (diethyl oxalate) and is acylated by it.

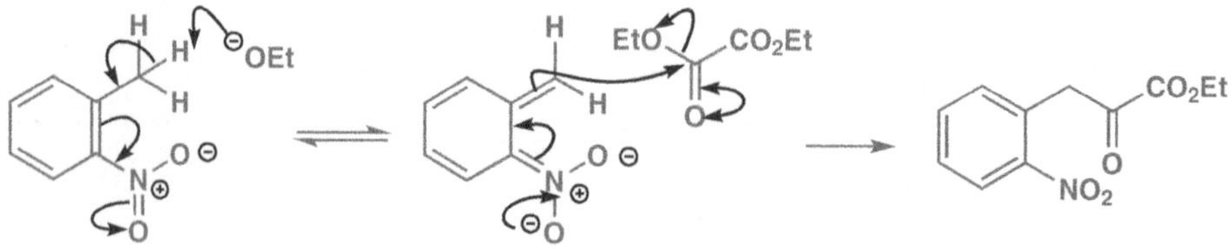

The rest of the synthesis is more straightforward: the nitro group can be reduced to an amine, which immediately forms an enamine by intramolecular attack on the more reactive carbonyl group (the ketone) to give the aromatic indole. Since the nitro compound is made by nitration of a benzene ring, the preferred symmetry is very different from that needed for the Fischer synthesis. Nitration of *para*-xylene (1,4-dimethylbenzene) is a good example.

Fischer vs Reissert

We can contrast the types of indole made by the Fischer and Reissert syntheses by the different ideal positions for substituents. These are, of course, not the only possible substitution patterns.

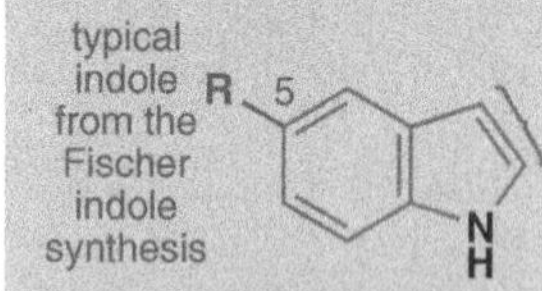

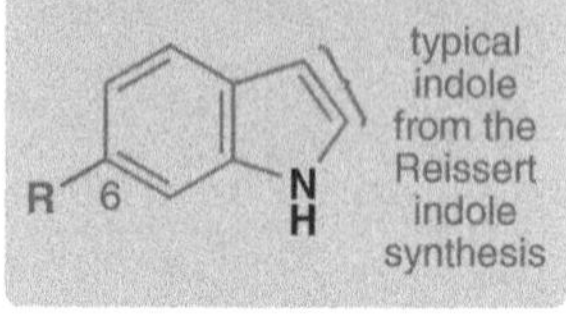

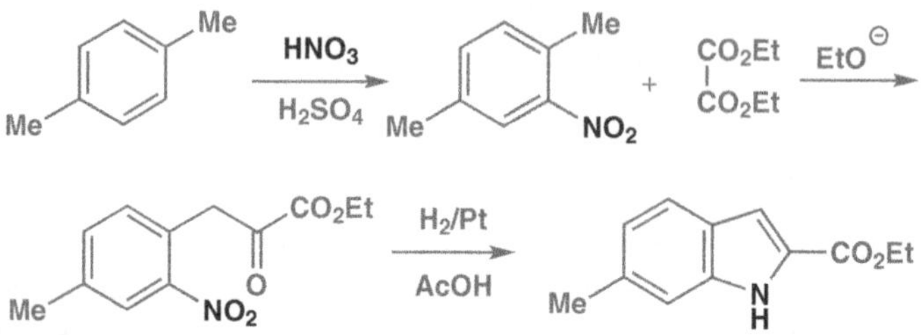

The ester products we have been using so far can be hydrolysed and decarboxylated by the mechanism described in the last chapter if a free indole is required. In any case, it is not necessary to use diethyl oxalate as the electrophilic carbonyl compound. The synthesis below, using the acetal of DMF as the electrophile, forms part of the synthesis of the strange antibiotic chuangxinmycin.

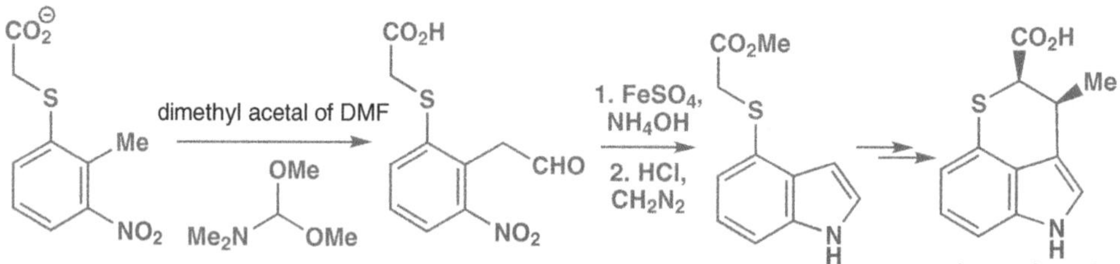

chuangxinmycin

Quinolines and isoquinolines

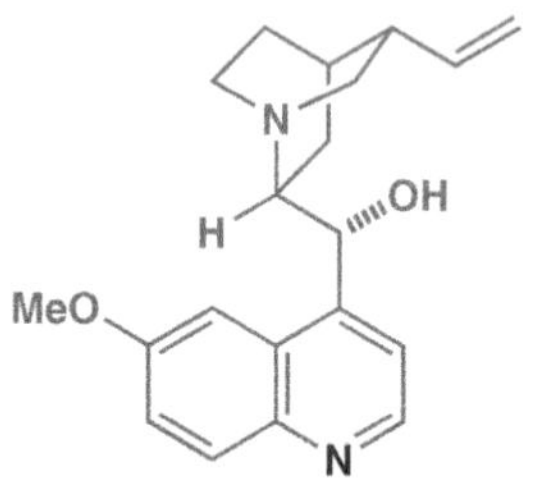

quinine

Quinoline forms part of the structure of quinine, the malaria remedy found in cinchona bark and known since the time of the Incas. The quinoline in quinine has a 6-MeO substituent and a side chain attached to C4. In discussing the synthesis of quinolines, we will be particularly interested in this pattern. This is because the search for anti-malarial compounds continues and other quinolines with similar structures are among the available anti-malarial drugs.

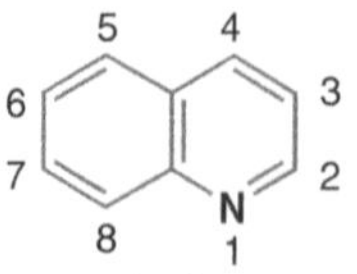

quinoline

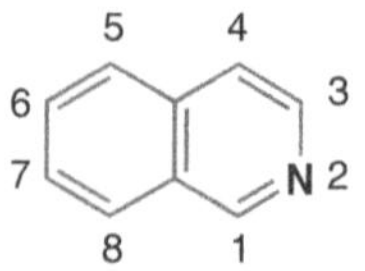

isoquinoline

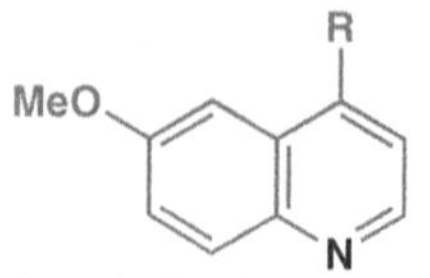

the quinoline in quinine

We shall also be very interested in quinolones, analogous to pyridones, with carbonyl groups at positions 2 and 4, as these are useful antibiotics. A simple example is pefloxacin, which has typical 6-F and 7-piperazine substituents.

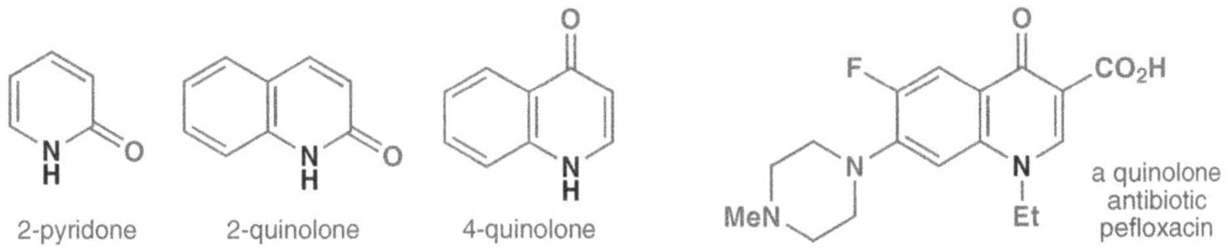

2-pyridone   2-quinolone   4-quinolone   a quinolone antibiotic pefloxacin

When we consider the synthesis of a quinoline, the obvious disconnections are, first, the C–N bond in the pyridine ring and, then, the C–C bond that joins the side chain to the benzene ring. We will need a three-carbon (C_3) synthon, electrophilic at both ends, which will yield two double bonds after incorporation. The obvious choice is a 1,3-dicarbonyl compound.

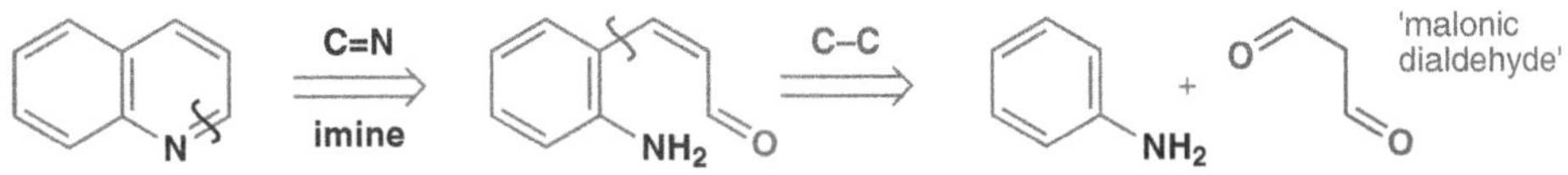

The choice of an aromatic amine is a good one as the NH_2 group reacts well with carbonyl compounds and it activates the *ortho* position to electrophilic attack. However, the dialdehyde is malonic dialdehyde, a compound that does not exist, so some alternative must be found. If the quinoline is substituted in the 2- and 4-positions this approach looks better.

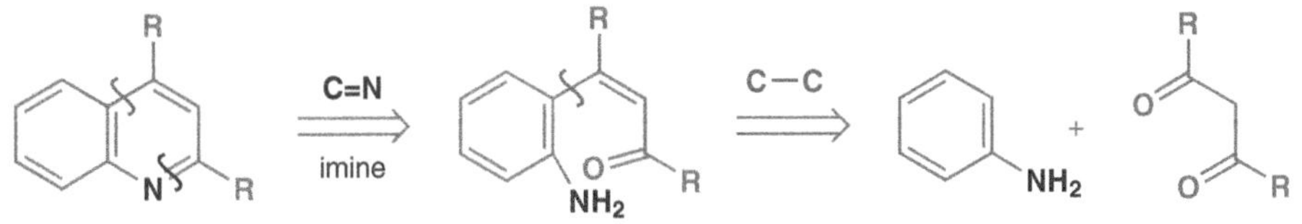

The initially formed imine will tautomerize to a conjugated enamine and cyclization now occurs by electrophilic aromatic substitution. The enamine will normally prefer to adopt the first configuration shown in which cyclization is not possible, and (perhaps for this reason or perhaps because it is difficult to predict which quinoline will be formed from an unsymmetrical 1,3-dicarbonyl compound) this has not proved a very important quinoline synthesis. However, the synthetic plan is sound, and we shall describe two important variants on this theme, one for quinolines and one for quinolones.

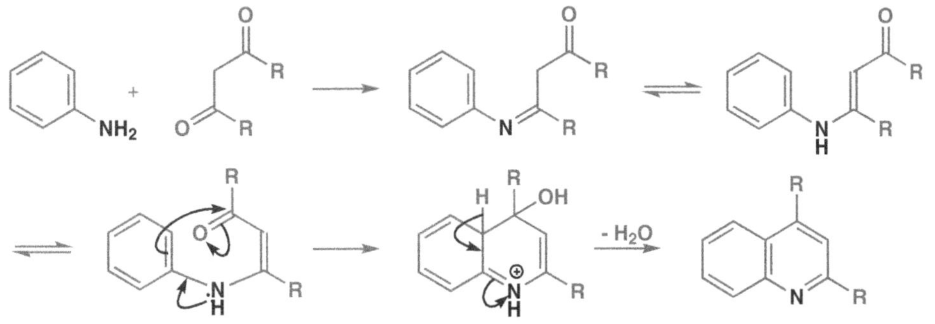

In the synthesis of pyridines it proved advantageous to make a dihydropyridine and oxidize it to a pyridine afterwards. The same idea works well in probably the most famous quinoline synthesis, the Skraup reaction. The diketone is replaced by an unsaturated carbonyl compound so that the quinoline is formed regiospecifically. The first step is conjugate addition of the amine. Under acid catalysis the ketone now cyclizes in the way we have just described to give a dihydroquinoline after dehydration. Oxidation to the aromatic quinoline is an easy step accomplished by many possible oxidants.

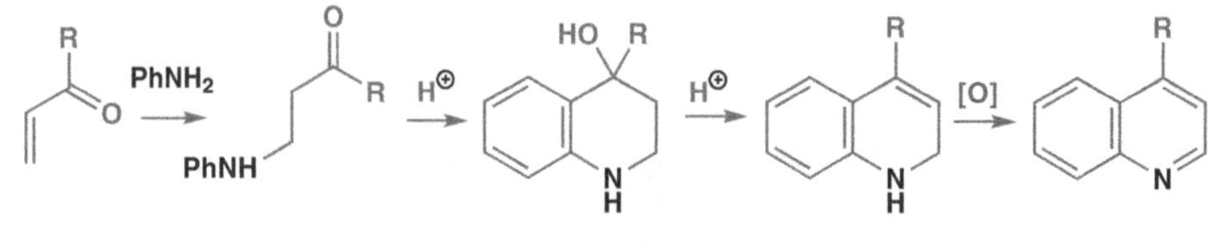

Traditionally, the Skraup reaction was carried out by mixing everything together and letting it rip. A typical mixture to make a quinoline without substituents on the pyridine ring would be the aromatic amine, concentrated sulfuric acid, glycerol, and nitrobenzene all heated up in a large flask at over 100 °C with a wide condenser.

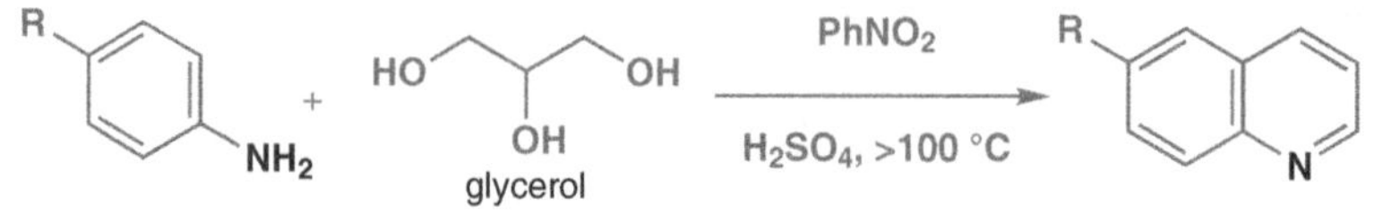

The ugly name of the Skraup reaction appropriately applies to the worst 'witch's brew' of all the heterocyclic syntheses. Some workers have added strange oxidizing agents such as arsenic acid, iron (III) salts, tin (IV) salts, nitrobenzenes of various substitution patterns, or iodine to make it 'go better'.

The glycerol was to provide acrolein ($CH_2{=}CH{\cdot}CHO$) by dehydration, the nitrobenzene was to act as oxidant, and the wide condenser…? All too often Skraup reactions did let rip—with destructive results. A safer approach is to prepare the conjugate adduct first, cyclize it in acid solution, and then oxidize it with one of the reagents we described for pyridine synthesis, particularly quinones such as DDQ.

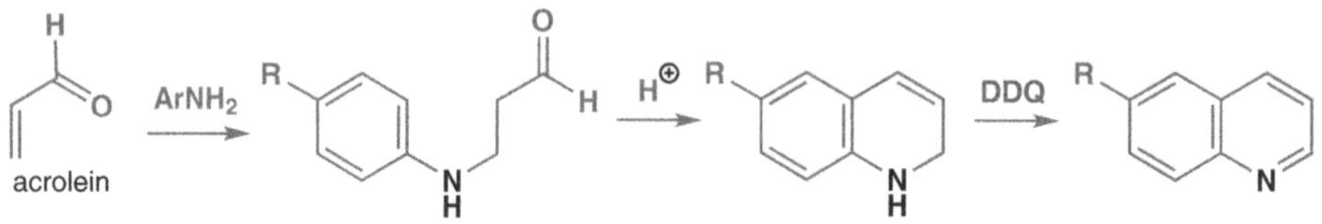

The more modern style of Skraup synthesis is used to make 8-quinolinol or 'oxine'. *ortho*-Aminophenol has only one free position *ortho* to the amino group and is very nucleophilic, so acrolein can be used in weak acid with only a trace of strong acid. Iron(III) is the oxidant, with a bit of boric acid for luck, and the yield is excellent.

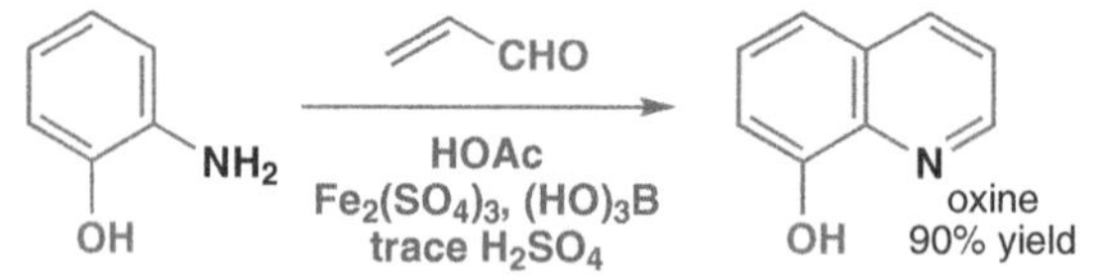

Oxine

This compound is important because it forms unusually stable metal complexes with metal ions such as Mg(II) or Al(III). It is also used as a corrosion inhibitor on copper because it forms a stable layer of the Cu(II) complex that prevents oxidation of the interior.

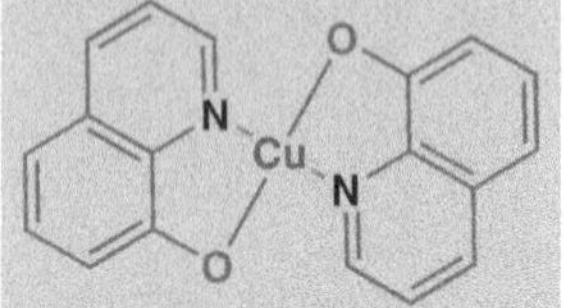

oxine complex of copper

Quinolones also come from anilines by cyclization to an *ortho* position

The usual method for making quinolone antibiotics is possible because they all have a carboxylic acid in the 3-position. The disconnection we used for quinoline suggests a rather unstable malonic ester derivative as starting material.

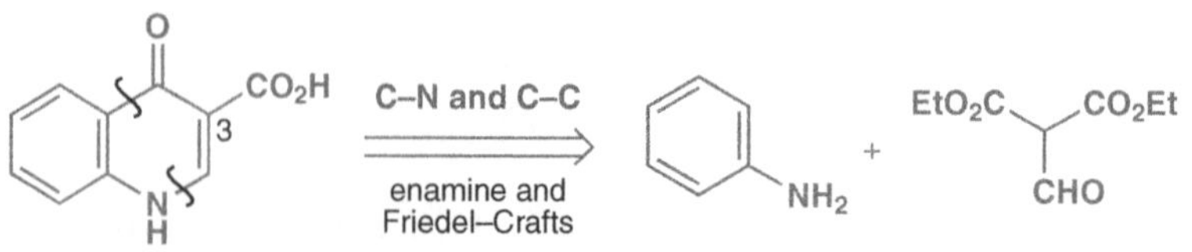

In fact, the enol ether of this compound is easily made from diethyl malonate and ethyl orthoformate [$HC(OEt)_3$]. The aromatic amine reacts with this compound by an addition–elimination sequence, giving an enamine that cyclizes on heating. This time the geometry of the enamine is not a concern.

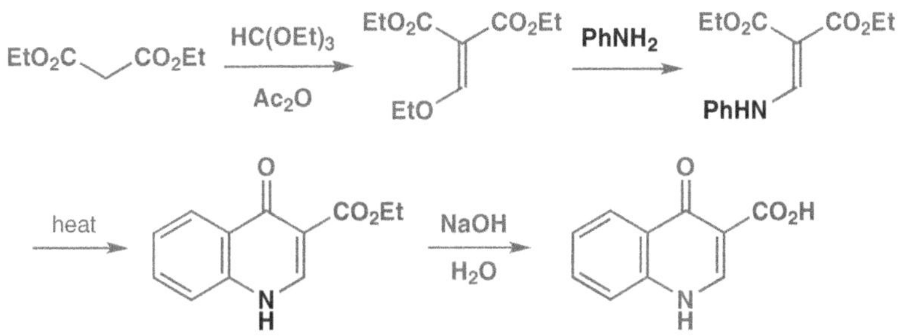

For examples of quinolone antibiotics we can choose ofloxacin, whose synthesis was discussed in detail in Chapter 22, and rosoxacin, whose synthesis is discussed below. Both molecules contain the same quinolone carboxylic acid framework, outlined in black, with another heterocyclic system at position 7 and various other substituents here and there.

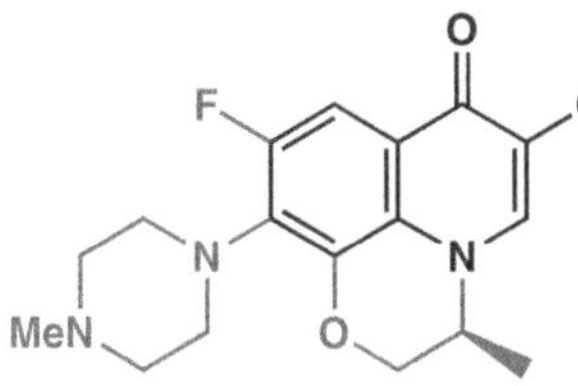

To make rosoxacin two heterocyclic systems must be constructed. Workers at the pharmaceutical company Sterling decided to build the pyridine in an ingenious version of the Hantzsch synthesis using acetylenic esters on 3-nitrobenzaldehyde. The ammonia was added as ammonium acetate. Oxidation with nitric acid made the pyridine, hydrolysis of the esters and decarboxylation removed the acid groups, and reduction with Fe(II) and HCl converted the nitro group into the amino group required for the quinolone synthesis.

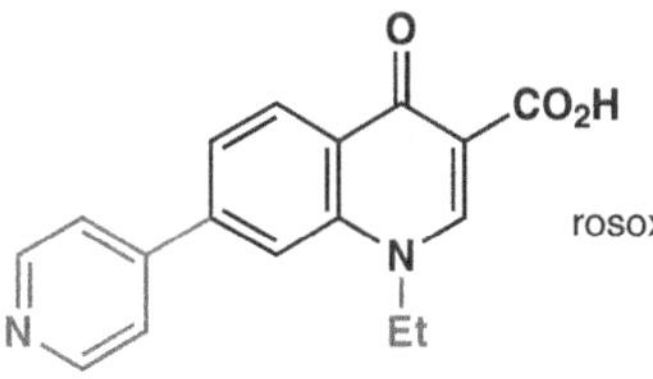

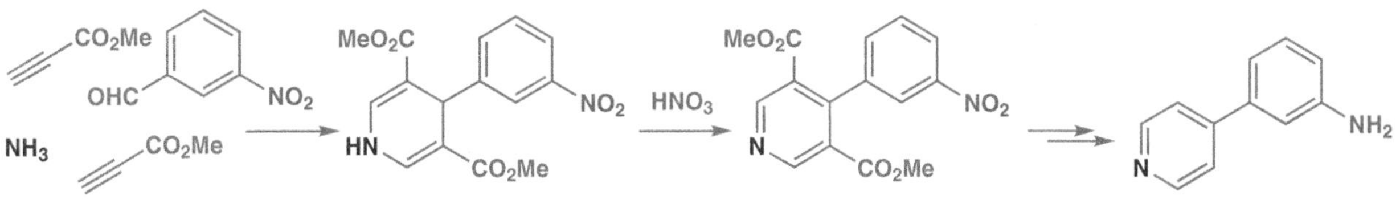

Now the quinolone synthesis can be executed with the same reagents we used before and all that remains is ester hydrolysis and alkylation at nitrogen. Notice that the quinolone cyclization could in theory have occurred in two ways as the two positions *ortho* to the amino group are different. In practice cyclization occurs away from the pyridine ring as the alternative quinolone would be impossibly crowded.

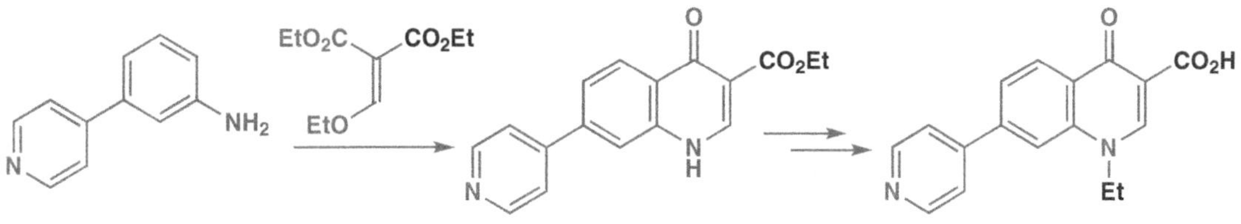

Since quinolones, like pyridones, can be converted into chloro-compounds with $POCl_3$, they can be used in nucleophilic substitution reactions to build up more complex quinolines.

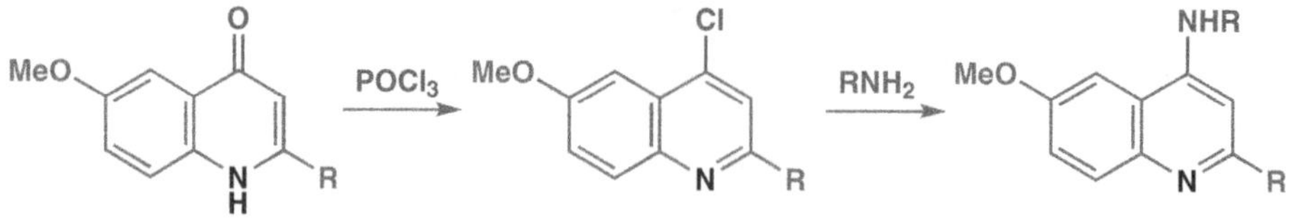

■ The Vilsmeier reaction with DMF is on p. 734.

We will give just one important synthesis of isoquinolines here. It is a synthesis of a dihydroisoquinoline by what amounts to an intramolecular Vilsmeier reaction in which the electrophile is made from an amide and POCl₃. Oxidation (in this case a dehydrogenation with Pd(0)) gives the quinoline.

■ The reaction with Pd is simply the reverse of a Pd-catalysed hydrogenation.

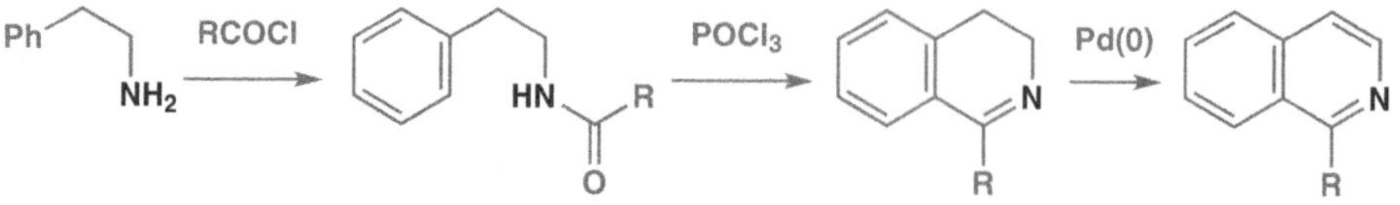

More heteroatoms in fused rings mean more choice in synthesis

The imidazo-pyridazine ring system forms the basis for a number of drugs in human and animal medicine. The synthesis of this system uses the chemistry discussed in Chapter 29 to build the pyridazine ring. There we established that it was easy to make dichloropyridazines and to displace the chlorine atoms one by one with different nucleophiles. Now we will move on from these intermediates to the bicyclic system.

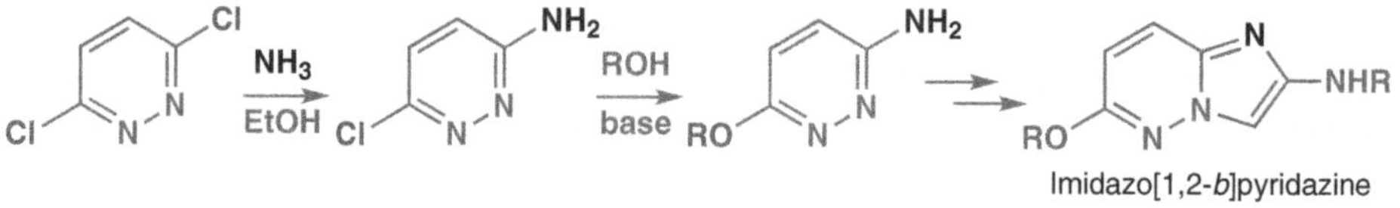

Imidazo[1,2-b]pyridazine

A 2-bromo-acid derivative is the vital reagent. It reacts at the amino nitrogen atom with the carbonyl group and at the pyridazine ring nitrogen atom with the alkyl halide. This is the only way the molecule can organize itself into a ten-electron aromatic system.

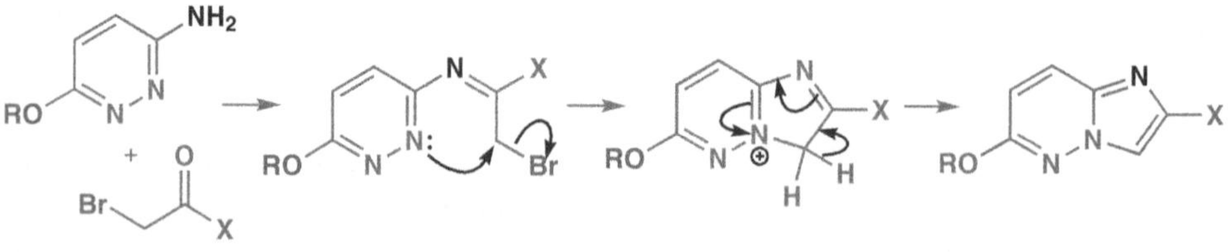

In Chapter 29 we also gave the structure of timolol, a thiadiazole-based β-blocker drug for reduction of high blood pressure. This compound has an aromatic 1,2,5-thiadiazole ring system and a saturated morpholine as well as an aliphatic side chain. Its synthesis relies on ring formation by rather a curious method followed by selective nucleophilic substitution, rather in the style of the last synthesis. The aromatic ring is made by the action of S_2Cl_2 on 'cyanamide'.

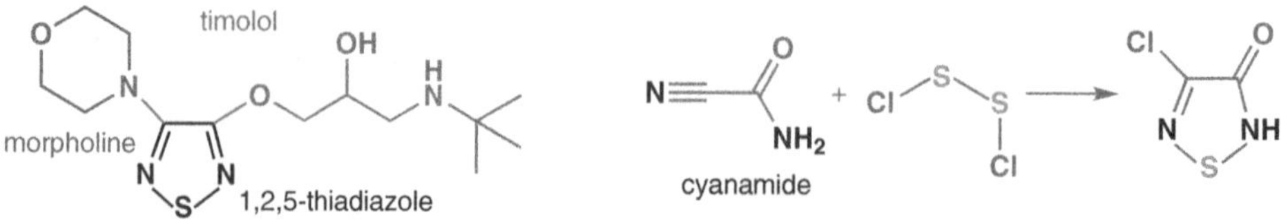

This reaction must start by attack of the amide nitrogen on the electrophilic sulfur atom. Cyclization cannot occur while the linear nitrile is in place so chloride ion (from disproportionation of ClS⁻) must first attack CN. Thereafter cyclization is easy.

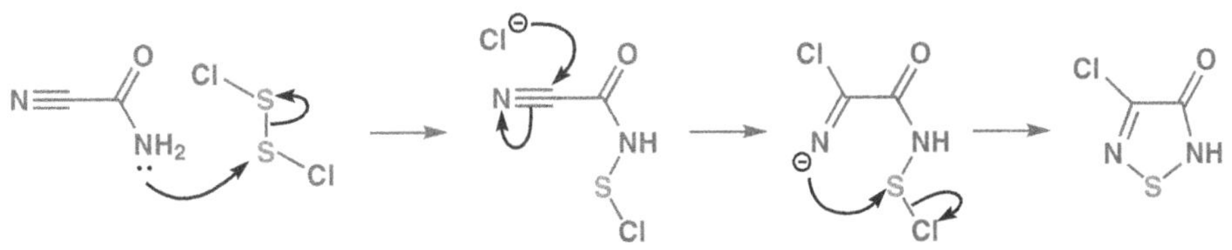

Reaction with epichlorohydrin followed by amine displacement puts in one of the side chains and nucleophilic substitution with morpholine on the ring completes the synthesis.

➡ We used epichlorohydrin a lot in Chapter 28.

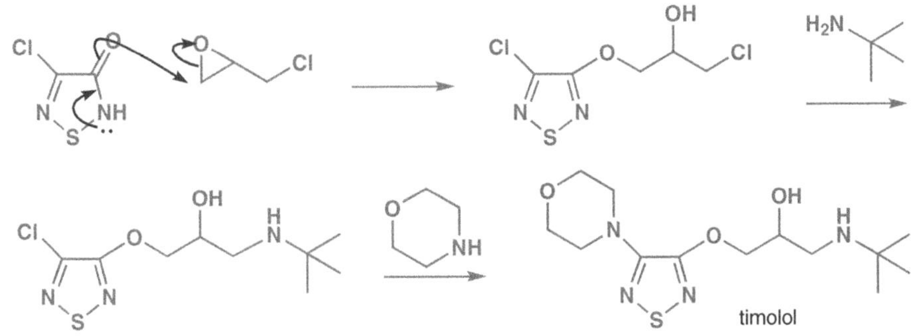

Summary: the three major approaches to the synthesis of aromatic heterocycles

We end this chapter with summaries of the three major strategies in the synthesis of heterocycles:

- ring construction by ionic reactions
- ring construction by cycloadditions
- modification of existing rings by electrophilic or nucleophilic aromatic substitution or by lithiation and reaction with electrophiles.

We will summarize the different applications of these strategies, and also suggest cases for which each strategy is not suitable. This section revises material from Chapter 29 as well since most of the ring modifications appear there.

■ This is only a summary. There are more details in the relevant sections of Chapters 29 and 30. There are also many, many more ways of making all these heterocycles. These methods are just where we suggest you *start*.

Ring construction by ionic cyclization

The first strategy you should try out when faced with the synthesis of an aromatic heterocyclic ring is the disconnection of bonds between the heteroatom or atoms and carbon, with the idea of using the heteroatoms as nucleophiles and the carbon fragment as a double electrophile.

Heterocycles with one heteroatom

five-membered rings

- pyrroles, thiophenes, and furans ideally made by this strategy from 1,4-dicarbonyl compounds

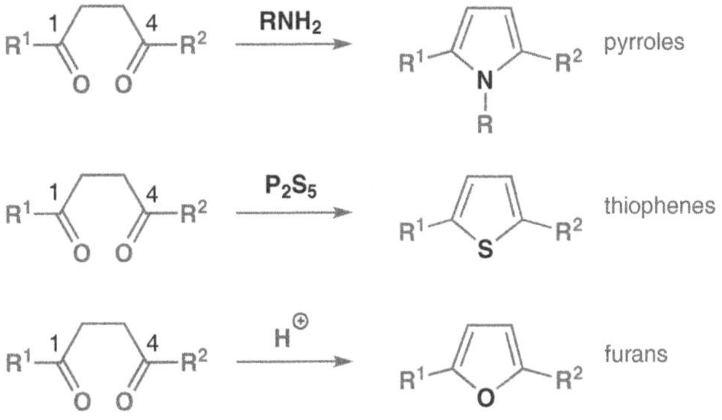

six-membered rings
- pyridines made by this strategy from 1,5-dicarbonyl compounds with oxidation

Heterocycles with two adjacent heteroatoms

five-membered rings
- pyrazoles and isoxazoles ideally made by this strategy from 1,3-dicarbonyl compounds

Note. This strategy is *not* suitable for isothiazoles as 'thiolamine' does not exist

six-membered rings
- pyridazines ideally made by this strategy from 1,4-dicarbonyl compounds with oxidation

Heterocycles with two non-adjacent heteroatoms

five-membered rings
- imidazoles and thiazoles ideally made by this strategy from α-halocarbonyl compound

Note. This strategy is *not* suitable for oxazoles as amides are not usually reactive enough: cyclization of acylated carbonyl compounds is usually preferred

six-membered rings
- pyrimidines ideally made by this strategy from 1,3-dicarbonyl compounds

Ring construction by cycloadditions

1,3-dipolar cycloaddition reactions

- ideal for the construction of isoxazoles, 1,2,3-triazoles, and tetrazoles

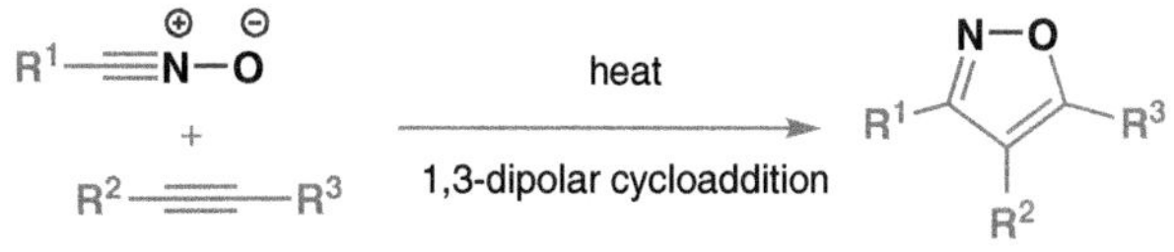

...or sigmatropic rearrangements

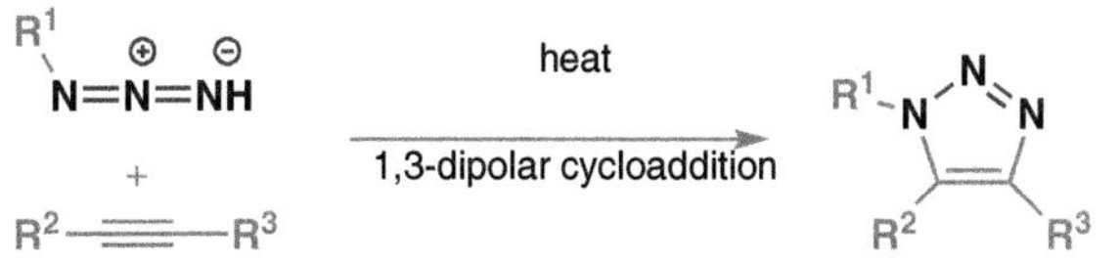

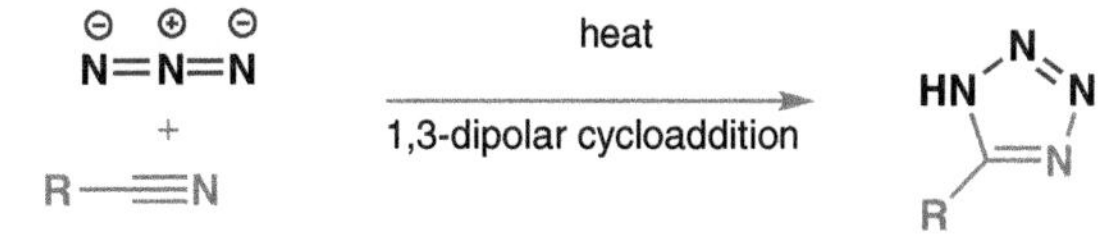

- a special reaction that is the vital step of the Fischer indole synthesis

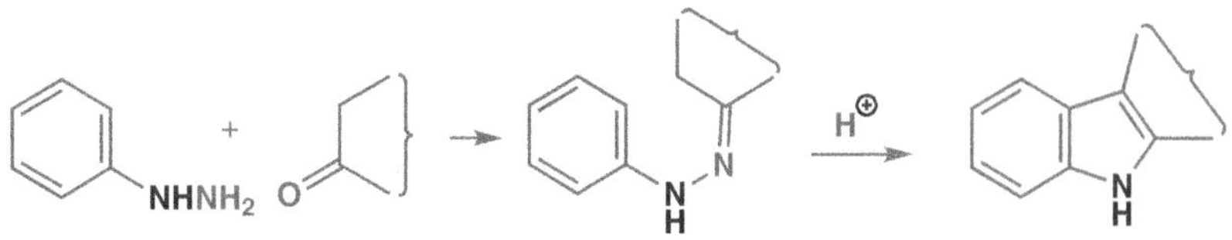

phenylhydrazine   a phenylhydrazone   an indole

Ring modification

Electrophilic aromatic substitution

- works very well on pyrroles, thiophenes, and furans, where it occurs best in the 2- and 5-positions and nearly as well in the 3- and 4-positions
- often best to block positions where substitution not wanted

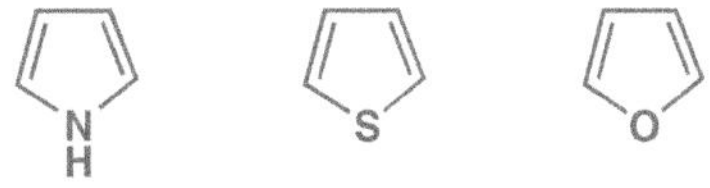

pyrrole   thiophene   furan

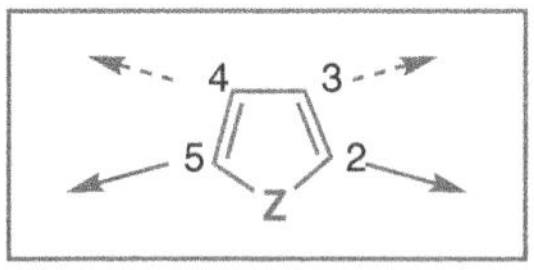

- works well for indole—occurs only in the 3-position but the electrophile may migrate to the 2-position

indole

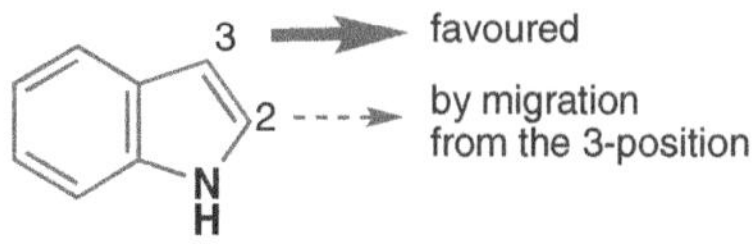

favoured

by migration from the 3-position

- works well for five-membered rings with a sulfur, oxygen, or pyrrole-like nitrogen atom and occurs anywhere that is not blocked (see earlier sections)

Note. Not recommended for pyridine, quinoline, or isoquinoline

Nucleophilic aromatic substitution

- works particularly well for pyridine and quinoline where the charge in the intermediate can rest on nitrogen

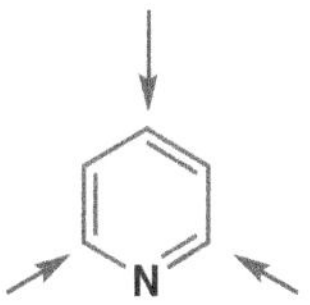

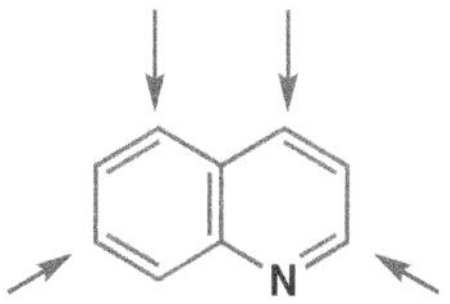

621

- especially important for pyridones and quinolones with conversion to the chloro-compound and displacement of chlorine by nucleophiles and, for quinolines, displacement of fluorine atoms on the benzene ring

- works well for the six-membered rings with two nitrogens (pyridazines, pyrimidines, and piperazines) in all positions

Lithiation and reaction with electrophiles

- works well for pyrrole (if NH blocked), thiophene, or furan next to the heteroatom. Exchange of Br or I for Li works well for most electrophiles providing any acidic hydrogens (including the NH in the ring) are blocked

Further reading

The best general text on heterocycles is J. A. Joule and K. Mills, *Heterocyclic Chemistry* 4th edn, Chapman and Hall, London, 2010.

S. Warren and P. Wyatt, *Workbook for Organic Synthesis: the Disconnection Approach*, Wiley, Chichester, 2009, chapters 34–35.

Check your understanding

To check that you have mastered the concepts presented in this chapter, attempt the problems that are available in the book's Online Resource Centre at http://www.oxfordtextbooks.co.uk/orc/clayden2e/

Suggested solutions for Chapter 30

30

PROBLEM 1

Suggest a mechanism for this synthesis of a tricyclic aromatic heterocycle.

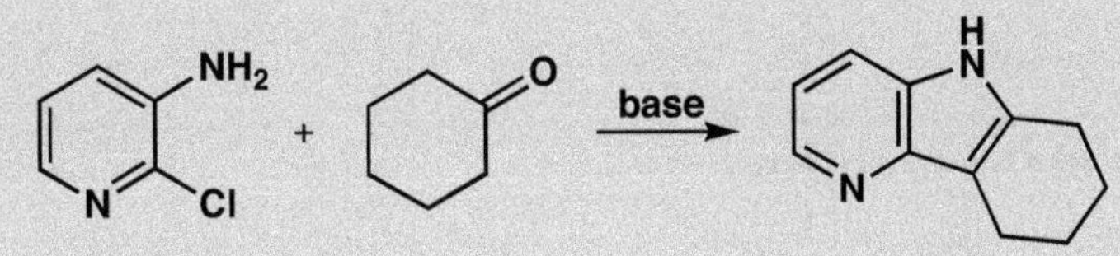

Purpose of the problem

A simple exercise in the synthesis of a pyridine fused to a pyrrole (or an indole with an extra nitrogen atom).

Suggested solution

The first step must be the formation of an enamine between the primary amine and the ketone. Now, because we have a pyridine and not a benzene ring, nucleophilic aromatic substitution can occur. These 'aza-indoles' are more easily formed than indoles.

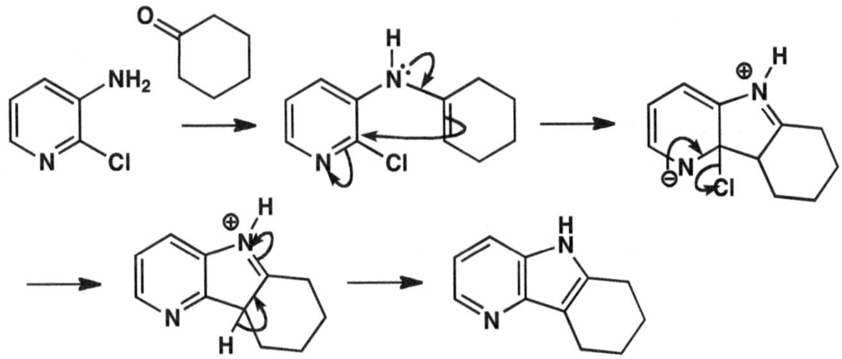

PROBLEM 2

Is the heterocyclic ring created in this reaction aromatic? How does the reaction proceed? Comment on the regioselectivity of this cyclization.

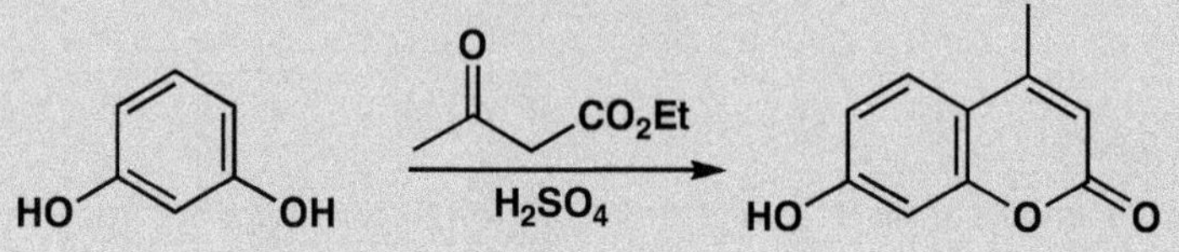

Purpose of the problem

Exploring the synthesis and aromaticity of an unfamiliar heterocycle.

Suggested solution

The left-hand ring is obviously aromatic as it is a benzene ring. The right-hand ring has four electrons from the double bonds and can have two from a lone pair on oxygen, making six in all. This is more obvious in a delocalized form. Alternatively the whole system can be considered as a 10-electron molecule. Strangely enough, this is easier to see in the other Kekulé form.

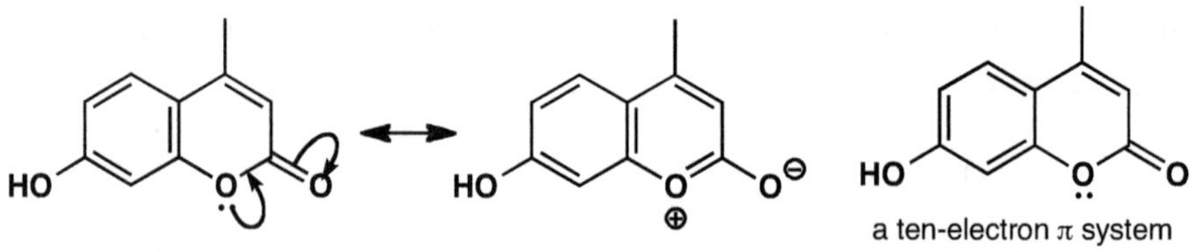

a ten-electron π system

■ This is a very old reaction discovered by H. von Pechmann and C. Duisberg, *Ber.*, 1883, 2119.

■ See p. 483 of the textbook for more on the selectivity between the *ortho* and *para* positions.

The first step in the reaction is a transesterification and cyclization then occurs in the *ortho* position, *para* to the other hydroxyl group. Cyclization might have happened to the position in between the two substituents, as the other OH is *ortho*, *para*-directing, but the position chosen is more reactive for both steric and electronic reasons.

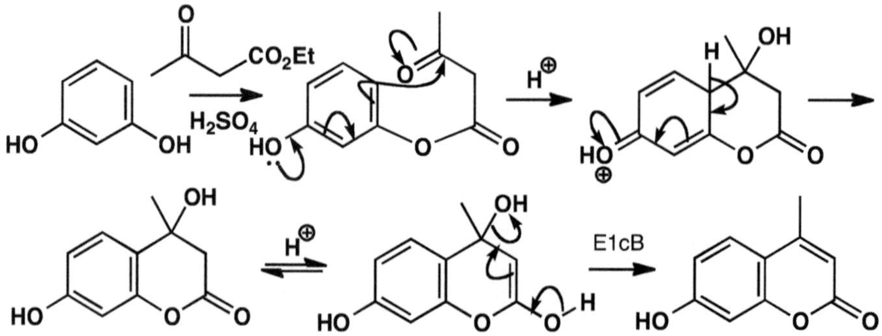

PROBLEM 3

Suggest mechanisms for this unusual indole synthesis. How does the second mechanism relate to electrophilic substitution on indoles (p. 746) ?

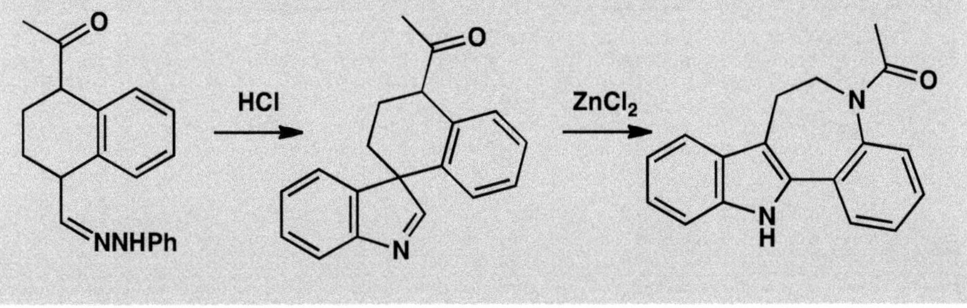

Purpose of the problem

A combination of a Fischer indole synthesis with revision of a bit of indole chemistry from the last chapter.

Suggested solution

The first step starts off as a normal Fischer indole synthesis (we have omitted the first step); you just have to draw the molecules carefully to show the *spiro* ring system, and you have to stop before an indole is formed as the quaternary centre prevents aromatization.

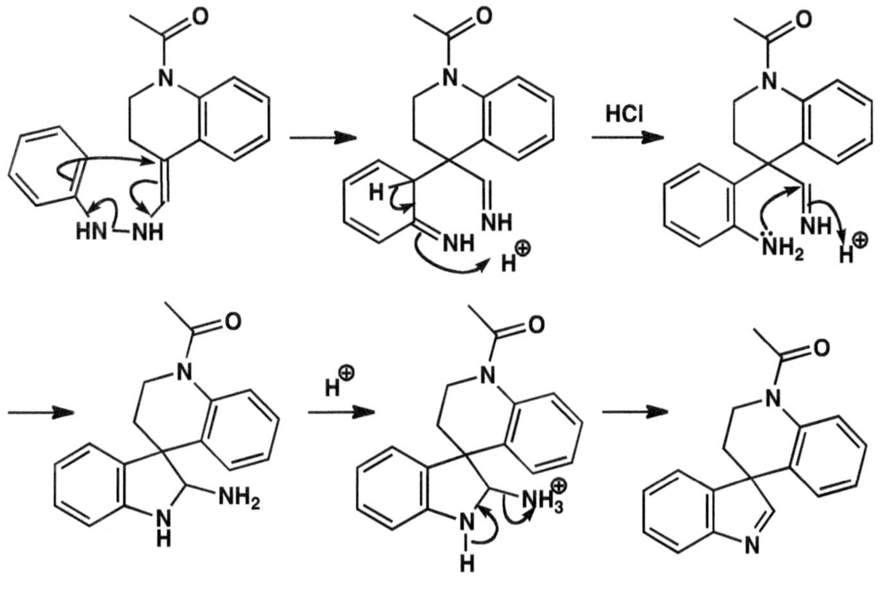

Treatment with a Lewis acid initiates a rearrangement very like those occurring when 3-substituted indoles are attacked by electrophiles (p. 746 of the textbook). The aromatic ring is a better migrating group than the primary alkyl alternative and an indole can finally be formed.

■ The new seven-membered heterocycle (an azepine) is found in some tranquilizers: see T. S. T. Wang, *Tetrahedron Lett.*, 1975, 1637.

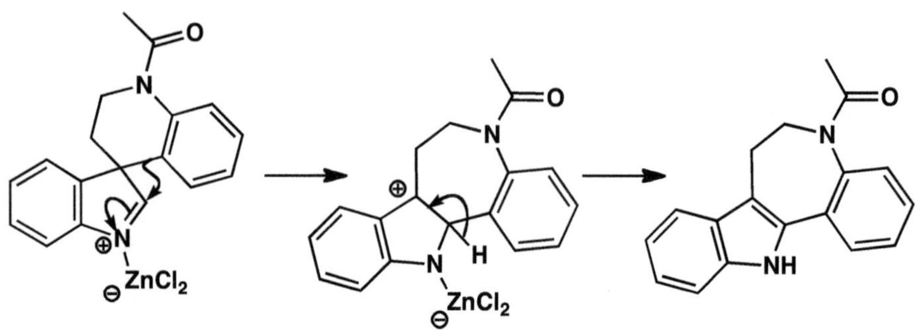

PROBLEM 4

Explain the reactions in this partial synthesis of methoxatin, the coenzyme of bacteria living on methanol.

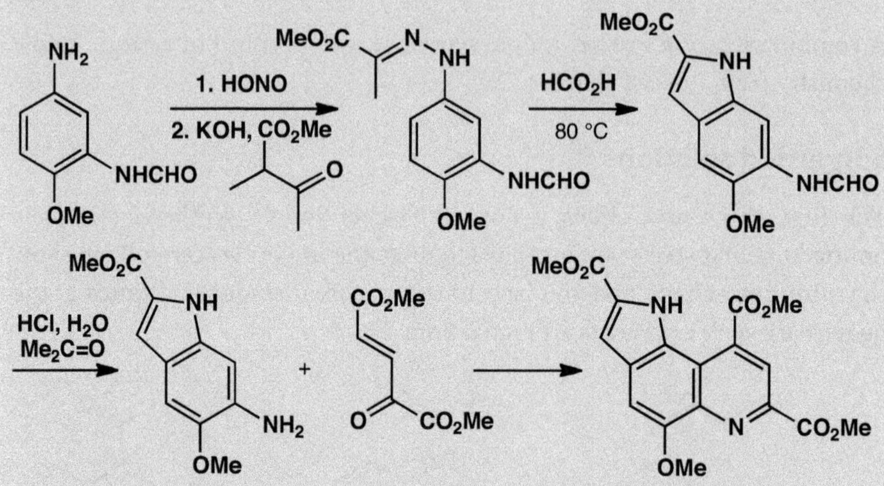

Purpose of the problem

A combination of Fischer indole synthesis with revision of indole chemistry from chapter 29.

Suggested solution

■ Diazotization: see p. 521 of the textbook.

There is clearly a Fischer indole synthesis in the second step but the first step makes the usual hydrazone in a most unusual way. The first reaction is a diazotization so we have to combine the diazonium salt with the enolate of the keto-ester. That creates a quaternary centre and the KOH deacylates it to give the aryl hydrazone needed for the next step.

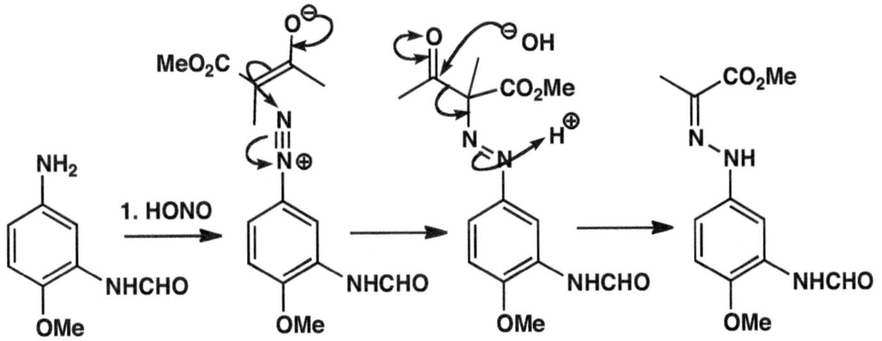

Now that we have the hydrazone, the Fischer indole step is straightforward and gives the indole-2-carboxylic acid derivative. There is only one site for an enamine and the indole is formed on the side of the benzene ring away from the other substituents.

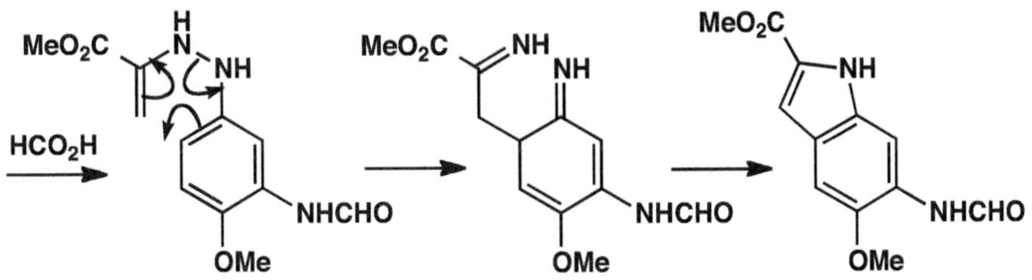

The next stage must involve the primary amine as nucleophile and the conjugated keto-diester as electrophile. You may have expected direct addition of the amine to the ketone as that gives the product by a reasonable mechanism. In fact, conjugate addition must occur first as the tertiary alcohol **A** can be isolated. The dehydration is obviously acid-catalysed and the oxidation by air [or Ce(IV)] is also acid-catalysed.

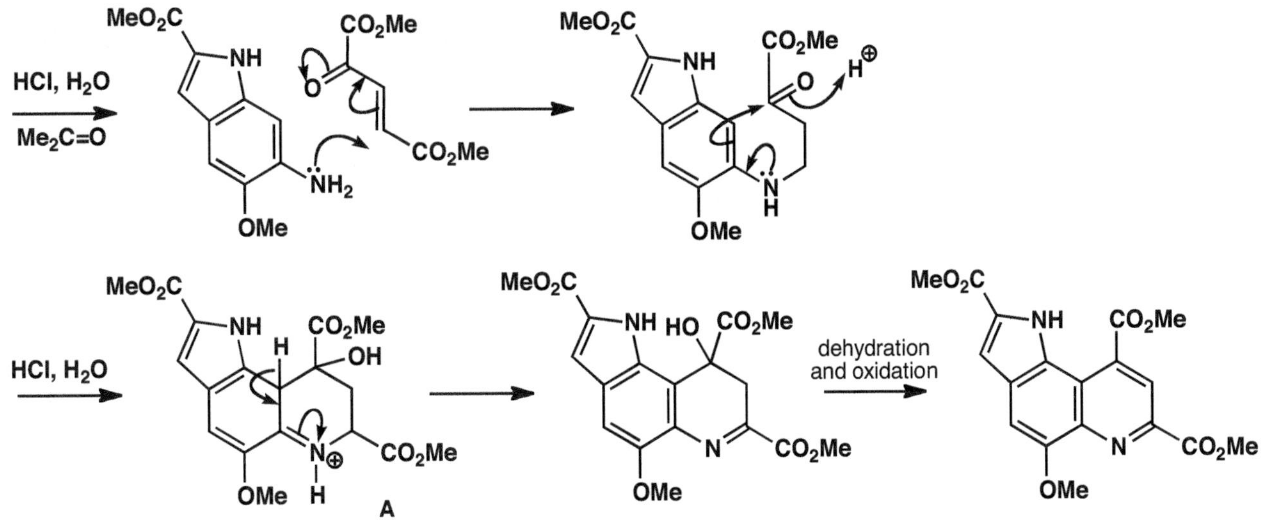

PROBLEM 5

Explain why these two quinoline syntheses from the same starting materials give (mainly) different products.

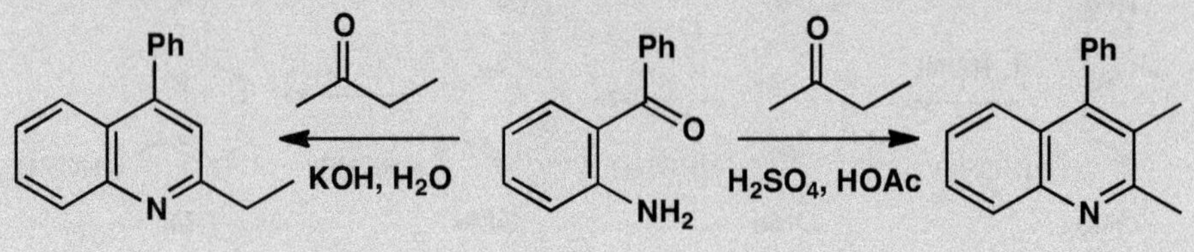

Purpose of the problem

An exercise in regioselectivity in a heterocyclic synthesis controlled by pH.

Suggested solution

■ This selective route to quinolines by the Friedländer synthesis was discovered by E. A. Fehnel, *J. Org. Chem.*, 1966, **31**, 2899.

You have a choice here: either you first form an enol(ate) from butanone and do an aldol reaction with the aromatic ketone or you first make an imine and then form enamines from that. In either case, you would expect enol or enamine formation on the more substituted side in acid but the less substituted side in base.

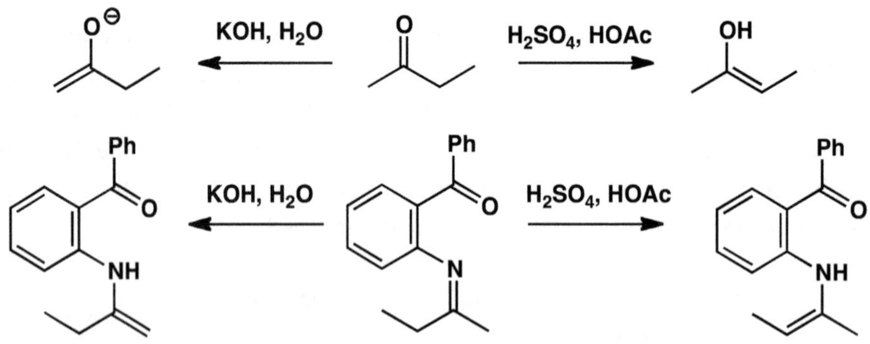

PROBLEM 6

Give mechanisms for these reactions used to prepare a fused pyridine. Why is it necessary to use a protecting group?

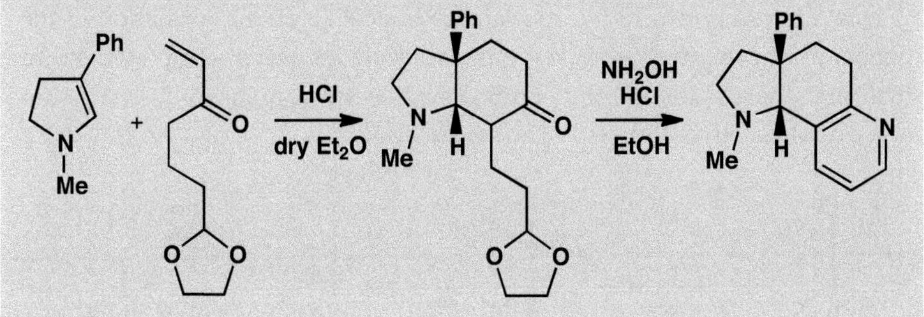

Purpose of the problem

Saturated and aromatic heterocycles combined with stereochemistry make an interesting synthesis for you to explore.

Suggested solution

The first starting material is a stable cyclic enamine and conjugate addition is what we should expect with an enone. Of course, if the aldehyde were unprotected, direct addition might occur there as well as carbonyl condensations. The product is in equilibrium with both its enols, one of which can cyclize to form the new six-membered ring.

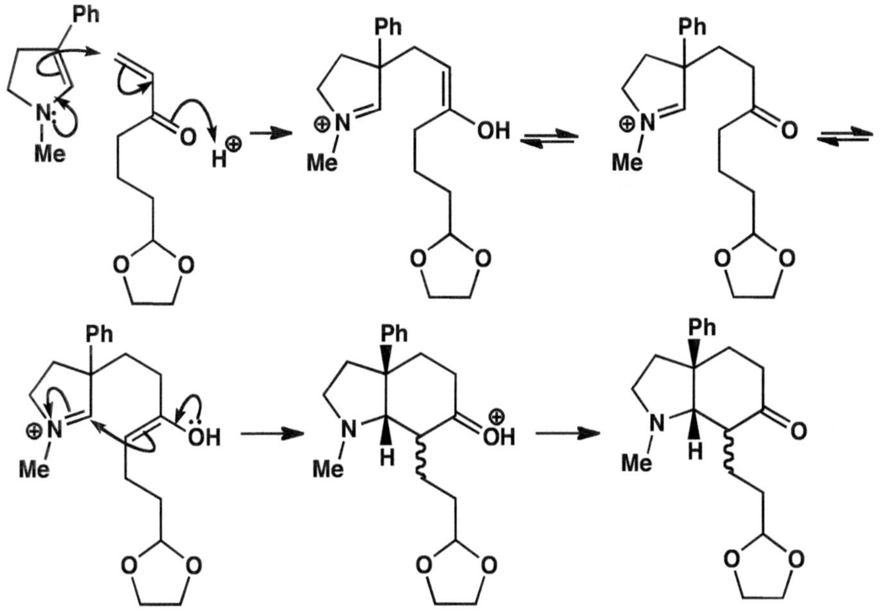

The enol must attack the five-membered ring in a *cis* fashion as the tether is too short to reach the other side. There is no control over one stereogenic centre (represented with a wiggly line) but that is unimportant as it is soon to disappear.

Now the reaction with hydroxylamine in acid solution. Formation of the oxime of the ketone produces one molecule of water—just enough to hydrolyse the acetal—and the pyridine synthesis is completed by cyclization and a double dehydration (p. 765 of the textbook).

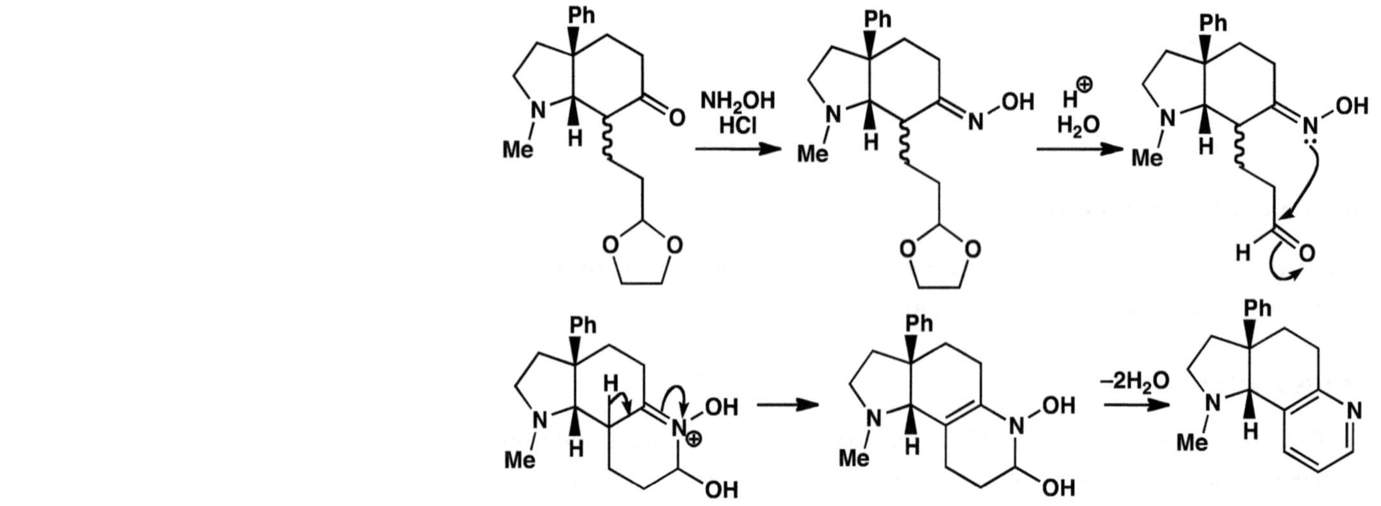

PROBLEM 7

Identify the intermediates and give mechanisms for the steps in this synthesis of a triazole.

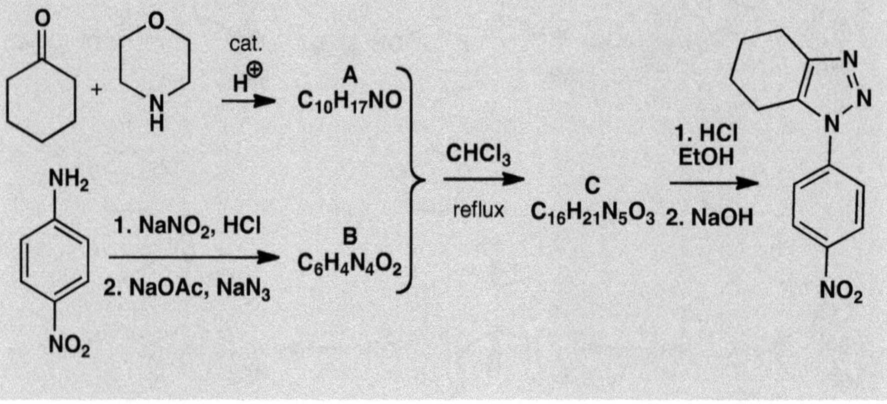

Purpose of the problem

Revision of aromatic nucleophilic substitution and a chance to unravel an interesting mechanism.

Suggested solution

The first reaction forms **A**, just the enamine from the ketone and the secondary amine (morpholine). Below we have diazotization of an aromatic amine and replacement by azide to give **B**. This nucleophilic substitution could occur by the addition-elimination mechanism activated by the nitro group or by the S_N1 mechanism (chapter 22).

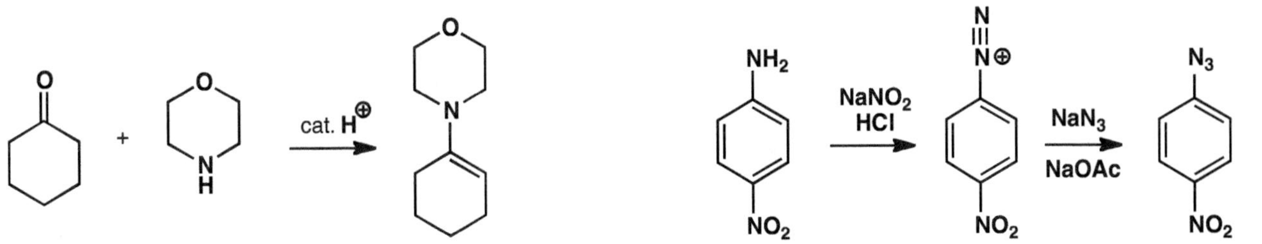

A
$C_{10}H_{17}NO$

B
$C_6H_4N_4O_2$

Now comes the interesting bit. The two reagents **A** and **B** combine without losing anything—it is evident that the enamine must be the nucleophile and so the azide must be the electrophile. We can see from the final product that the enamine attacks one end or the other of the azide. Trial and error takes over! Here is one possible solution with some side chains in the intermediate abbreviated for clarity. This product **C** can be isolated but its stereochemistry is not known.

■ An alternative is a 1,3-dipolar cycloaddition, see chapter 34.

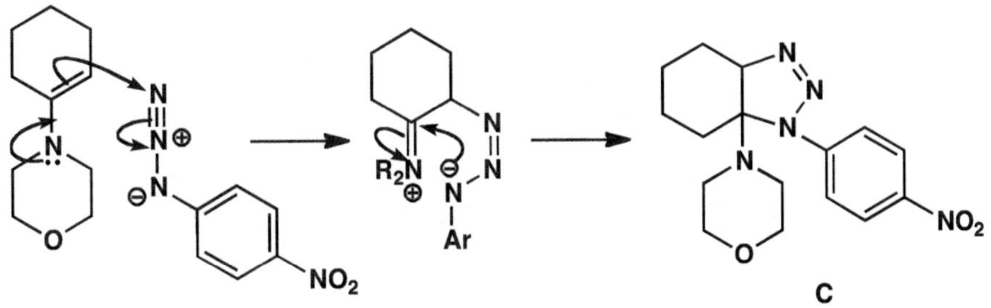

C

■ This synthesis was discovered in Milan during a mechanistic study of the reactions between enamines and azides: R. Fusco *et al.*, *Gazz. Chim. Ital.*, 1961, **91**, 849.

Finally, the new aromatic system (a triazole) is formed by elimination of the aminal. Protonation of the most basic nitrogen is followed by expulsion of morpholine and aromatization by deprotonation.

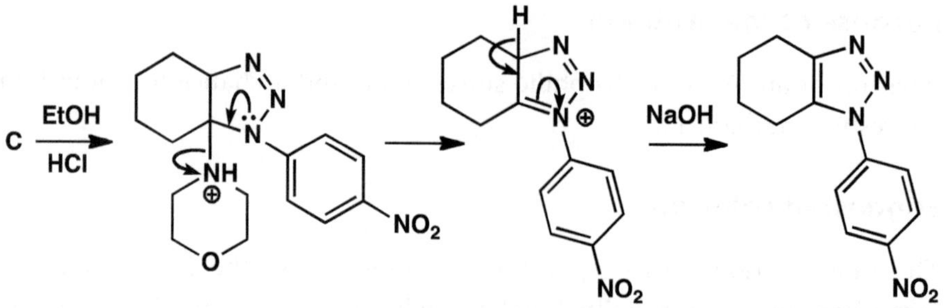

PROBLEM 8

Give detailed mechanisms for this pyridine synthesis.

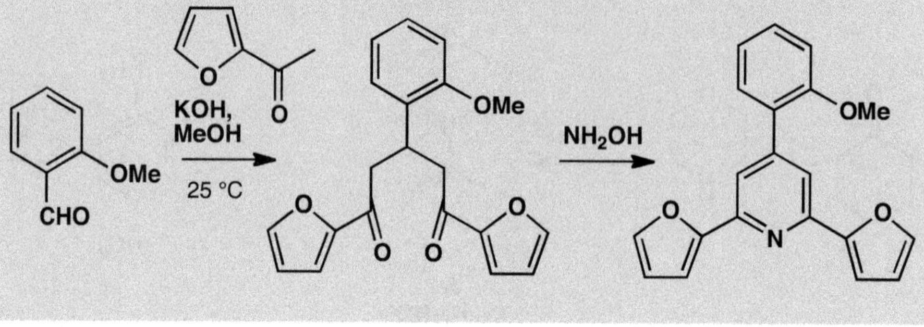

Purpose of the problem

Revision of aldol and conjugate addition reactions of enol(ate)s and a synthesis involving two furans and one pyridine.

Suggested solution

The first reactions are an aldol condensation and a conjugate addition. We have shown just the first steps, but make sure that you can draw full mechanisms for both. The last step is a standard pyridine synthesis.

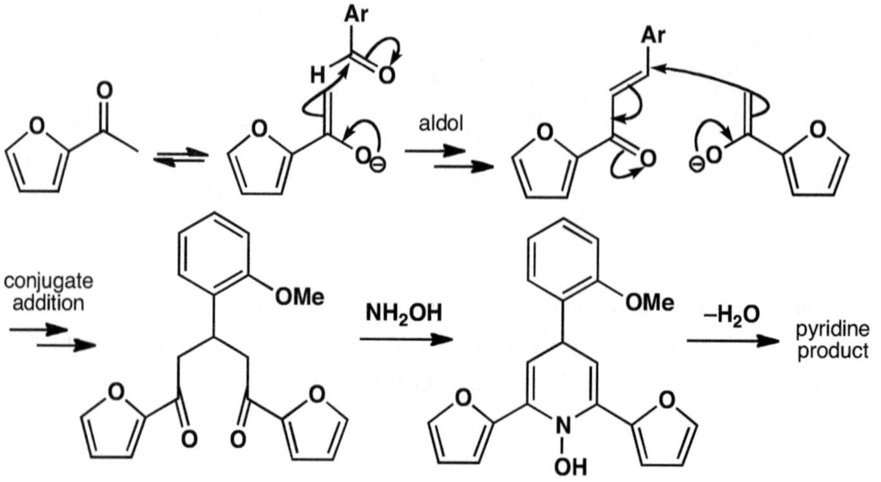

PROBLEM 9

Suggest a synthesis for this compound.

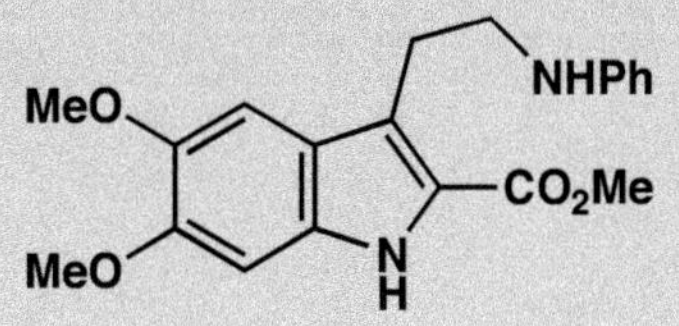

Purpose of the problem

The synthesis of an indole with a slight twist.

Suggested solution

This looks very much like a perfect subject for the Fischer indole synthesis. Let's see.

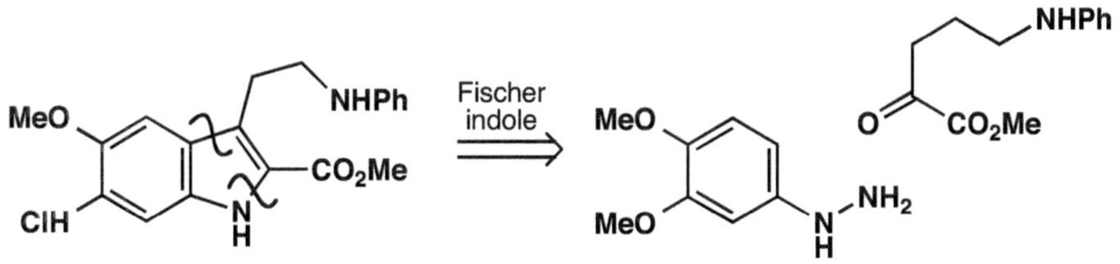

This looks fine, though we may wonder how we are going to have an amino group in that position on the keto ester. Surely it will cyclize onto the ester to form a lactam? One solution would be to protect it with something like a Boc group, but the solution found by the Sterling drug company was partly motivated by a desire to make a variety of compounds with different amine substituents. They chose hydroxyl as an easily replaceable group and accepted that the starting material would exist as a lactone. They made it like this:

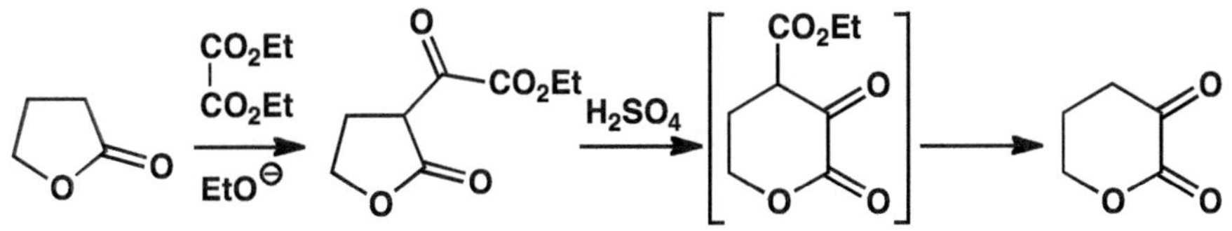

The first step is a typical Claisen ester condensation and the second is an acid-catalysed thermodynamically controlled transesterification (the lactone and ethyl ester exchange alcohol partners) to give the more stable six-membered lactone, followed by decarboxylation. Now the Fischer indole synthesis works well and work-up with dry HCl in methanol gave the alkyl

■ This chemistry is in the patent literature but see S. Archer, *Chem. Abstr.*, 1971, **78**, 29442..

chloride that could be displaced with amines to give a series of anti-depressants.

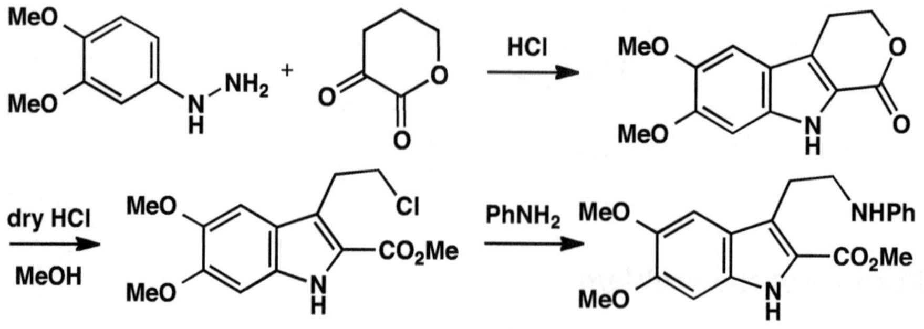

PROBLEM 10

How would you synthesize these aromatic heterocycles?

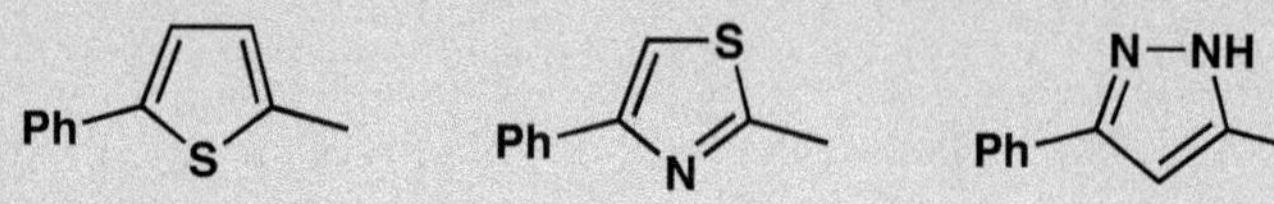

Purpose of the problem

A chance to devise syntheses for five-membered aromatic heterocycles with one or two heteroatoms.

Suggested solution

These compounds all look much the same but the strategies needed for each are rather different. Removing the heteroatom from the thiophene reveals a 1,4-diketone to be made by one of the methods in chapter 28. We have chosen to propose an enamine and an α-bromoketone though there are many other good choices.

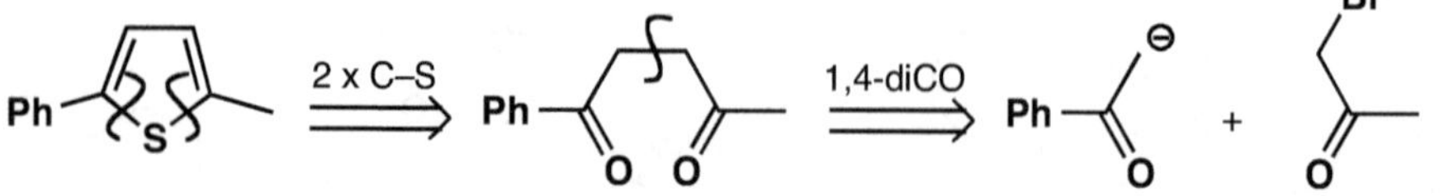

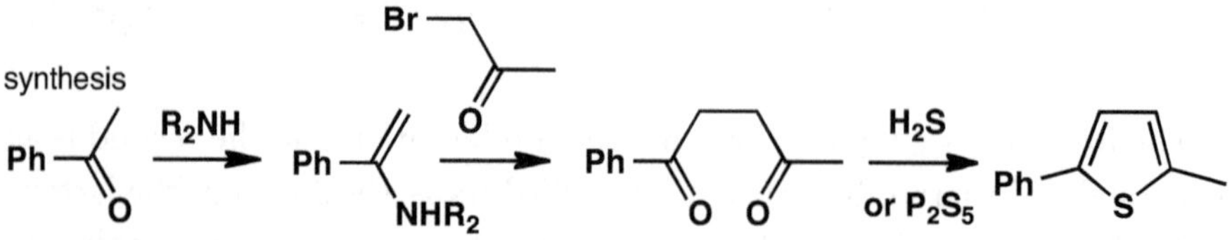

The second compound is a thiazole and we want to use a thioamide to make it (see p. 771 of the textbook). We should disconnect C–N and C–S bonds to give the thioamide and another α-bromoketone remembering to let the nucleophiles exercise their natural preferences: sulfur attacking saturated carbon and nitrogen attacking the carbonyl group.

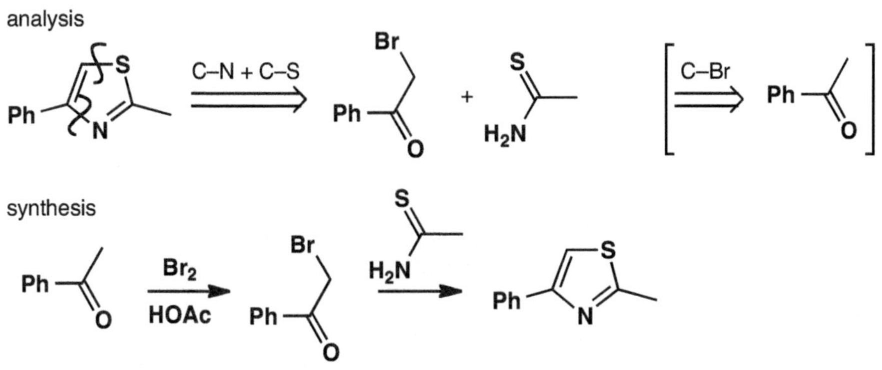

analysis

synthesis

The third compound has the two heteroatoms joined together so we should keep them that way. We disconnect both C–N bonds revealing the hidden molecule of hydrazine (NH₂NH₂). We then need a 1,3-diketone so we need Claisen ester chemistry (chapter 26).

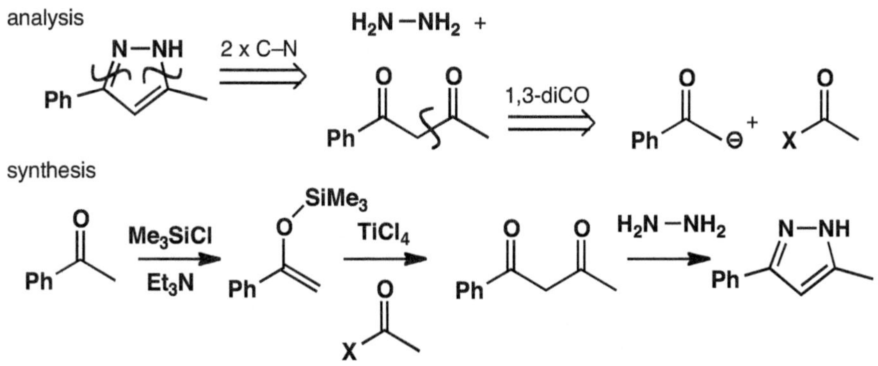

analysis

synthesis

Acronyms and abbreviations

Note: Abbreviations for amino acids are given in Appendix 1

5-HT5	hydroxytryptamine (serotonin)
7-ACA7	aminocephalosporinic acid
6-APA6	aminopenicillanic acid
ACE	angiotensin-converting enzyme
ACh	acetylcholine
AChE	acetylcholinesterase
ACP	acyl carrier protein
ACT	artemisinin combination therapy
ADAPT	antibody-directed abzyme prodrug therapy
ADEPT	antibody-directed enzyme prodrug therapy
ADH	alcohol dehydrogenase
ADME	absorption, distribution, metabolism, excretion
ADP	adenosine 5′-diphosphate
AGO	argonaute protein
AIC	5-aminoimidazole-4-carboxamide
AIDS	acquired immune deficiency syndrome
Akt	protein kinase B
ALK	anaplastic lymphoma kinase
AME	aminoglycoside modifying enzyme
AML	acute myeloid leukaemia
AMP	adenosine 5′-monophosphate
AT	angiotensin
ATP	adenosine 5′-triphosphate
AUC	area under the curve
BiTE	bi-specific T-cell engager
BuChE	butyrlcholinesterase
BTK	Bruton's tyrosine kinase
cAMP	cyclic AMP
β-CCE	carboline-3-carboxylate
CCK	cholecystokinin
CDKs	cyclin-dependent kinases
CETP	cholesteryl ester transfer protein
cGMP	cyclic GMP
CHO cells	Chinese hamster ovarian cells
CKIs	cyclin-dependent kinase inhibitors
c-KIT	mast/stem cell growth factor receptor
Clog *P*	calculated logarithm of the partition coefficient
c-MET receptor	hepatocyte growth factor receptor
CML	chronic myeloid leukaemia
CMV	cytomegalovirus

CNS	central nervous system
CoA	coenzyme A
CoMFA	comparative molecular field analysis
COMT	catechol O-methyltransferase
COPD	chronic obstructive pulmonary disease
COX	cyclooxygenase
CSD	Cambridge Structural Database
CYP	enzymes that constitute the cytochrome P450 family
D-receptor	dopamine receptor
dATP	deoxyadenosine triphosphate
DCC	dicyclohexylcarbodiimide
dCTP	deoxycytosine triphosphate
DG	diacylglycerol
dGTP	deoxyguanosine triphosphate
DHFR	dihydrofolate reductase
Dhh	desert hedgehog
DMAP	dimethlaminopyridine
DNA	deoxyribonucleic acid
DOR	delta opioid receptor
dsDNA	double-stranded DNA
dsRNA	double-stranded RNA
dTMP	deoxythymidylate monophosphate
dTTP	deoxythymidylate triphosphate
dUMP	deoxyuridylate monophosphate
EC_{50}	concentration of drug required to produce 50% of the maximum possible effect
E_s	Taft's steric factor
EGF	epidermal growth factor
EGFR	epidermal growth factor receptor
EMEA	European Agency for the Evaluation of Medicinal Products
EPC	European Patent Convention
EPO	European Patent Office
EPO	erythropoietin
ErbB	epidermal growth factor receptor
ERK	see MAPK
ET	endothelin
FDA	US Food and Drug Administration
FdUMP	fluorodeoxyuracil monophosphate
FGF	fibroblast growth factor
FGFR	fibroblast growth factor receptor

FH$_4$	tetrahydrofolate
F	oral bioavailability
F	inductive effect of an aromatic substituent in QSAR
F-SPE	fluorous solid-phase extraction
FLOG	Flexible Ligands Orientated on Grid
FPGS	folylpolyglutamate synthetase
FPP	farnesyl diphosphate
FT	farnesyl transferase
FTI	farnesyl transferase inhibitor
G-protein	guanine nucleotide binding protein
GABA	γ-aminobutyric acid
GAP	GTPase activating protein
GCP	Good Clinical Practice
GDEPT	gene-directed enzyme prodrug therapy
GDP	guanosine diphosphate
GEF	guanine nucleotide exchange factors
GGTase	geranylgeranyltransferase
GH	growth hormone
GIT	gastrointestinal tract
GLP	Good Laboratory Practice
GMC	General Medical Council
GMP	Good Manufacturing Practice
GMP	guanosine monophosphate
GnRH	gonadotrophin-releasing hormone
gp	glycoprotein
GRB2	growth factor receptor bound protein 2
gt	genotype
GTP	guanosine triphosphate
h-PEPT	human intestinal proton-dependent oligopeptide transporter
H-receptor	histamine receptor
HA	haemagglutinin
HAART	highly active antiretroviral therapy
HAMA	human anti-mouse antibodies
HBA	hydrogen bond acceptor
HBD	hydrogen bond donor
HCV	hepatitis C virus
HDL	high density lipoprotein
HERG	human ether-a-go-go related gene
HER	human epidermal growth factor receptor
HGFR	hepatocyte growth factor receptor
HIF	hypoxia-inducible factor
HIV	human immunodeficiency virus

HMG-SCoA	3-hydroxy-3-methylglutaryl-coenzyme A
HMGR	3-hydroxy-3-methylglutaryl-coenzyme A reductase
HOMO	highest occupied molecular orbital
HPLC	high-performance liquid chromatography
HPMA	*N*-(2-hydroxypropyl)methacrylamide
HPT	human intestinal di-/tripeptide transporter
HRV	human rhinoviruses
HSV	herpes simplex virus
HTS	high-throughput screening
IC$_{50}$	concentration of drug required to inhibit a target by 50%
ICMT1	isoprenylcysteine carboxylmethyltransferase
If	funny ion channels
IGF-1R	insulin growth factor 1 receptor
Ihh	Indian hedgehog
IND	Investigational exemption to a New Drug application
IP$_3$	inositol triphosphate
IPER	International Preliminary Examination Report
IRB	Institutional Review Board
ISR	International Search Report
ITC	isothermal titration calorimetry
IUPAC	International Union of Pure and Applied Chemistry
IV	intravenous
JAK	Janus kinase
K_D	dissociation binding constant
K_i	inhibition constant
K_M	Michaelis constant
KOR	kappa opioid receptor
LAAM	L-α-acetylmethadol
LD$_{50}$	lethal dose required to kill 50% of a test sample of animals
LDH	lactate dehydrogenase
LDL	low density lipoprotein
LH	luteinizing hormone
LHRH	luteinizing hormone-releasing hormones
LipE	lipophilic efficiency
log *P*	logarithm of the partition coefficient
LDL	low density lipoprotein
LUMO	lowest unoccupied molecular orbital
M-receptor	muscarinic receptor
MAA	Marketing Authorization Application

MAB	monoclonal antibody
MAO	monoamine oxidase
MAOI	monoamine oxidase inhibitor
MAOS	microwave assisted organic synthesis
MAP	mitogen-activated protein
MAPK	mitogen-activated protein kinases
MCHR	melanin-concentrating hormone receptor
MDR	multidrug resistance
MDRTB	multidrug-resistant tuberculosis
MEP	molecular electrostatic potential
miRNA	micro RNA
miRNP	micro RNA protein
MMAE	monomethyl auristatin E (vedotin)
MMP	matrix metalloproteinase
MMPI	matrix metalloproteinase inhibitor
MOR	mu opioid receptor
MR	molar refractivity
mRNA	messenger RNA
MRSA	methicillin-resistant *Staphylococcus aureus*
mRTKI	multi-receptor tyrosine kinase inhibitors
MTP	microsomal triglyceride transfer protein
MTDD	multi-target drug discovery
mTOR	mechanistic or mammalian target of rapamycin
mTORC	mechanistic or mammalian target of rapamycin complex
mTRKI	multi-tyrosine receptor kinase inhibitor
MWt	molecular weight
N-receptor	nicotinic receptor
NA	neuraminidase or noradrenaline
NAD$^+$ / NADH	nicotinamide adenine dinucleotide
NADP$^+$ / NADPH	nicotinamide adenine dinucleotide phosphate
NAG	*N*-acetylglucosamine
NAM	*N*-acetylmuramic acid
NCE	new chemical entity
NDA	new drug application
NEP	neutral endopeptidase
NHS	National Health Service
NICE	National Institute for Health and Clinical Excellence
NMDA	*N*-methyl-D-aspartate
NME	new molecular entity
NMR	nuclear magnetic resonance

NNRTI	non-nucleoside reverse transcriptase inhibitor
NO	nitric oxide
NOR	nociceptin opioid receptor
NOS	nitric oxide synthase
NRTI	nucleoside reverse transcriptase inhibitor
NS	non-structural
NSAID	non-steroidal anti-inflammatory drug
NSCLC	non-small-cell lung carcinoma
NVOC	nitroveratryloxycarbonyl
ORL1	opioid receptor-like receptor
P	partition coefficient
P$_2$Y receptor	purinergic G-protein-coupled receptor
PABA	*p*-aminobenzoic acid
PAR	protease activated receptor
PARP	poly ADP ribose polymerase
PBP	penicillin binding protein
PCP	phencyclidine, otherwise known as 'angel dust'
PCT	patent cooperation treaty
PD-1 receptor	programmed cell death 1 receptor
PDB	protein data bank
PDE	phosphodiesterase
PDGF	platelet-derived growth factor
PDGFR	platelet-derived growth factor receptor
PDK1	phosphoinositide dependent kinase 1
PDT	photodynamic therapy
PEG	polyethylene glycol
PGE	prostaglandin E
PGF	prostaglandin F
PGI$_2$	prostacyclin
PH	Pleckstrin homology
PI3K	phosphoinositide 3-kinases
PIP$_2$	phosphatidylinositol diphosphate
PIP$_3$	phosphatidylinositol (3,4,5)-triphosphate
PI	protease inhibitor
piRNA	piwi-interacting RNA
PKA	protein kinase A
PKB	protein kinase B
PKC	protein kinase C
PLC	phospholipase C
PLS	partial least squares
PPAR	peroxisome proliferator-activated receptor
PPBI	protein–protein binding inhibitor
PPI	proton pump inhibitor

PPts	pyridinium 4-toluenesulphonate		SPA	scintillation proximity assay
PTase	palmitoyl transferase		SPE	solid-phase extraction
PTCH	patched receptor		SPOS	solution phase organic synthesis
QSAR	quantitative structure–activity relationships		SPR	surface plasmon resonance
r	regression or correlation coefficient		ssDNA	single-stranded DNA
R	resonance effect of an aromatic substituent in QSAR		SSRI	selective serotonin reuptake inhibitor
			ssRNA	single-stranded RNA
RAAS	renin–angiotensin–aldosterone system		STAT	signal transducer and activator of transcription
RANK	receptor activator of nuclear factor-kappa B		TB	tuberculosis
RCE1	ras converting enzyme 1		TCA	tricyclic antidepressants
RES	reticuloendothelial system		TFA	trifluoroacetic acid
RET	rearranged during transcription		TGF-α	transforming growth factor α
RFC	reduced folate carrier		TGF-β	transforming growth factor β
RISC	RNA induced silencing complex		THF	tetrahydrofuran
RMSD	root mean square distance		TM	transmembrane
rRNA	ribosomal RNA		TNF	tumour necrosis factor
RNA	ribonucleic acid		TNFR	tumour necrosis factor receptor
RNAi	RNA interference		TNT	trinitrotoluene
s	standard error of estimate or standard deviation		TRAIL	TNF-related apoptosis-inducing ligand
			TRIPS	trade related aspects of intellectual property rights
SAR	structure–activity relationships			
SCAL	safety-catch acid-labile linker		tRNA	transfer RNA
SCF	stem cell factor		T-VEC	talimogene laherparepvec
SCFR	mast/stem cell growth factor receptor		UTI	urinary tract infection
SCID	severe combined immunodeficiency disease		vdW	van der Waals
sGC	soluble guanylate cyclase		VEGF	vascular endothelial growth factor
SH	src homology		VEGFR	vascular endothelial growth factor receptor
Shh	sonic hedgehog		VIP	vasoactive intestinal peptide
siRNA	small interfering RNA		VOC-Cl	vinyloxycarbonyl chloride
SKF	Smith-Kline and French		VRE	vancomycin-resistant enterococci
Smo	Smoothened receptor		VRSA	vancomycin-resistant *Staphylococci aureus*
SNRI	selective noradrenaline reuptake inhibitors		VZV	varicella-zoster viruses
siRNA	Small inhibitory RNA		WHO	World Health Organization
snRNA	Small nuclear RNA		WTO	World Trade Organization
SOP	standard operating procedure			
SOS	son of sevenless protein			

1 Drugs and drug targets: an overview

1.1 What is a drug?

Medicinal chemistry involves the design and synthesis of a pharmaceutical agent that has a desired biological effect on the human body or some other living system. Such a compound could also be called a 'drug', but this is a word that many scientists dislike because of the way it is viewed by society. With media headlines such as 'Drugs Menace' or 'Drug Addiction Sweeps City Streets', this is hardly surprising. However, it suggests that a distinction can be drawn between drugs that are used in medicine and drugs that are abused. But is this really true? Can we draw a neat line between 'good drugs' like penicillin and 'bad drugs' like heroin? If so, how do we define what is meant by a good or a bad drug in the first place? Where would we place a so-called social drug like cannabis in this divide? What about nicotine, or alcohol?

The answers we get depend on who we ask. As far as the law is concerned, the dividing line is defined in black and white. As far as the party-going teenager is concerned, the law is an ass. As far as we are concerned, the questions are irrelevant. Trying to divide drugs into two categories—safe or unsafe, good or bad—is futile and could even be dangerous.

First, let us consider the so-called 'good' drugs used in medicines. How 'good' are they? If a drug is to be truly 'good' it would have to do what it is meant to do, have no toxic or unwanted side effects, and be easy to take.

How many drugs fit these criteria?

The short answer is 'none'. There is no pharmaceutical compound on the market today that can completely satisfy all these conditions. Admittedly, some come quite close to the ideal. **Penicillin**, for example, has been one of the safest and most effective antibacterial agents ever discovered. Yet it too has drawbacks. It cannot treat all known bacterial infections, and, as the years have gone by, more and more bacterial strains have become resistant. Moreover, some individuals can experience severe allergic reactions to the compound.

Penicillin is a relatively safe drug, but there are some drugs that are distinctly dangerous. **Morphine** is one such example. It is an excellent analgesic, yet it suffers from serious side effects such as tolerance, respiratory depression, and addiction. It can even kill if taken in excess. **Barbiturates** are also known to be dangerous. At Pearl Harbor, American casualties were given barbiturates as general anaesthetics before surgery. However, a poor understanding of how barbiturates are stored in the body led to many patients receiving a fatal overdose. In fact, it is thought that more casualties died at the hands of the anaesthetists at Pearl Harbor than died of their wounds.

To conclude, the 'good' drugs are not as perfect as one might think.

What about the 'bad' drugs then? Is there anything good that can be said about them? Surely there is nothing we can say in defence of the highly addictive drug **heroin**?

Well, let us look at the facts about heroin. It is one of the best painkillers known to medicine. In fact, it was named heroin at the end of the nineteenth century because it was thought to be the 'heroic' drug that would banish pain for good. Heroin went on the market in 1898, but had to be withdrawn from general distribution 5 years later when its addictive properties became evident. However, heroin is still used in medicine today—under strict control, of course. The drug is called **diamorphine** and it is the drug of choice for treating patients dying of cancer. Not only does diamorphine reduce pain to acceptable levels, it also produces a euphoric effect that helps to counter the depression faced by patients close to death. Can we really condemn such a drug as being all 'bad'?

By now, it should be evident that the division between 'good' and 'bad' drugs is a woolly one and is not really relevant to our discussion of medicinal chemistry. All drugs have their good and bad points. Some have more good points than bad and vice versa, but, like people, they all have their own individual characteristics. So how are we to define a drug in general?

One definition could be to classify drugs as 'compounds which interact with a biological system to produce a biological response'. This definition covers all the drugs we have discussed so far, but it goes further. There are chemicals which we take every day and which have a biological effect on us. What are these everyday drugs?

One is contained in the cups of tea, coffee, and cocoa that we consume. All of these beverages contain the stimulant **caffeine**. Whenever you take a cup of coffee, you are a drug user. We could go further. Whenever you crave a cup of coffee, you are a drug addict. Even children are not immune. They get their caffeine 'shot' from Coke or Pepsi. Whether you like it or not, caffeine is a drug. When you take it, you experience a change of mood or feeling.

So too, if you are a worshipper of the 'nicotine stick'. The biological effect is different. In this case you crave sedation or a calming influence, and it is the **nicotine** in the cigarette smoke which induces that effect. **Alcohol** is another example of a 'social' drug and, as such, causes society more problems than all other drugs put together. One only has to study road accident statistics to appreciate that fact. If alcohol was discovered today, it would probably be restricted in exactly the same way as **cocaine**. Considered in a purely scientific way, alcohol is a most unsatisfactory drug. As many will testify, it is notoriously difficult to judge the correct dose required to gain the beneficial effect of 'happiness' without drifting into the higher dose levels that produce unwanted side effects such as staggering down the street. Alcohol is also unpredictable in its biological effects. Either happiness or depression may result, depending on the user's state of mind. On a more serious note, **addiction** and **tolerance** in certain individuals have ruined the lives of addicts and relatives alike.

Our definition of a drug can also be used to include less obvious compounds; for example poisons and toxins. They too interact with a biological system and produce a biological response—a bit extreme perhaps, but a response all the same. The idea of poisons acting as drugs may not appear so strange if we consider penicillin. We have no problem in thinking of penicillin as a drug, but if we were to look closely at how penicillin works, then it acts as a poison. It interacts with bacteria (the biological system) and kills them (the biological response). Fortunately for us, penicillin has no such effect on human cells.

Even those drugs which do not act as poisons have the potential to become poisons—usually if they are taken in excess. We have already seen this with morphine. At low doses it is a painkiller. At high doses, it is a poison which kills by suppressing breathing. Therefore, it is important that we treat all medicines as potential poisons and treat them with respect.

There is a term used in medicinal chemistry known as the **therapeutic index**, which indicates how safe a particular drug is. The therapeutic index is a measure of the drug's beneficial effects at a low dose, versus its harmful effects at a high dose. To be more precise, the therapeutic index compares the dose level required to produce toxic effects in 50% of patients to the dose level required to produce the maximum therapeutic effects in 50% of patients. A high therapeutic index means that there is a large safety margin between beneficial and toxic doses. The values for cannabis and alcohol are 1000 and 10 respectively, which might imply that cannabis is safer and more predictable than alcohol. Indeed, a cannabis preparation (**nabiximols**) has now been approved to relieve the symptoms of multiple sclerosis. However, this does not suddenly make cannabis safe. For example, the favourable therapeutic index of cannabis does not indicate its potential toxicity if it is taken over a long period of time (chronic use). For example, the various side effects of cannabis include panic attacks, paranoid delusions, and hallucinations. Clearly, the safety of drugs is a complex matter and it is not helped by media sensationalism.

If useful drugs can be poisons at high doses or over long periods of use, does the opposite hold true? Can a poison be a medicine at low doses? In certain cases, this is found to be so.

Arsenic is well known as a poison, but arsenic-derived compounds are used as antiprotozoal and anticancer agents. **Curare** is a deadly poison which was used by the native people of South America to tip their arrows such that a minor arrow wound would be fatal, yet compounds based on the **tubocurarine** structure (the active principle of curare) are used in surgical operations to relax muscles. Under proper control and in the correct dosage, a lethal poison may well have an important medical role. Alternatively, lethal poisons can be the starting point for the development of useful drugs. For example, ACE inhibitors are important cardiovascular drugs that were developed, in part, from the structure of a snake venom.

Since our definition covers any chemical that interacts with any biological system, we can include all the pesticides used in agriculture as drugs. They interact with the biological systems of harmful bacteria, fungi, and insects to produce a toxic effect that protects plants.

Even food can act like a drug. Junk foods and fizzy drinks have been blamed for causing hyperactivity in children. It is believed that junk foods have high concentrations of certain amino acids which can be converted in the body to neurotransmitters—chemicals that pass messages between nerves. In excess, these chemical messengers overstimulate the nervous system, leading to the disruptive behaviour observed in susceptible individuals. Allergies due to food additives and preservatives are also well recorded.

Some foods even contain toxic chemicals. Broccoli, cabbage, and cauliflower all contain high levels of a chemical that can cause reproductive abnormalities in rats. Peanuts and maize sometimes contain fungal toxins, and it is thought that fungal toxins in food were responsible for one of the biblical plagues. Basil contains over 50 compounds that are potentially carcinogenic, and other herbs contain some of the most potent carcinogens known. Carcinogenic compounds have also been identified in radishes, brown mustard, apricots, cherries, and plums. Such unpalatable facts might put you off your dinner, but take comfort—these chemicals are present in such small quantities that the risk is insignificant. Therein lies a great truth, which was recognized as long ago as the fifteenth century when it was stated that 'Everything is a poison, nothing is a poison. It is the dose that makes the poison'.

Almost anything taken in excess will be toxic. You can make yourself seriously ill by taking 100 aspirin tablets or a bottle of whisky or 9 kg of spinach. The choice is yours!

To conclude, drugs can be viewed as actual or potential poisons. An important principle is that of **selective toxicity**. Many drugs are effective because they are toxic to 'problem cells', but not normal cells. For example, antibacterial, antifungal, and antiprotozoal drugs are useful in medicine when they show a selective toxicity to microbial cells, rather than mammalian cells. Clinically effective anticancer agents show a selective toxicity for cancer cells over normal cells. Similarly, effective antiviral agents are toxic to viruses rather than normal cells.

Having discussed what drugs are, we shall now consider why, where, and how they act.

KEY POINTS

- Drugs are compounds that interact with a biological system to produce a biological response.

- No drug is totally safe. Drugs vary in the side effects they might have.

- The dose level of a compound determines whether it will act as a medicine or as a poison.

- The therapeutic index is a measure of a drug's beneficial effect at a low dose versus its harmful effects at a higher dose. A high therapeutic index indicates a large safety margin between beneficial and toxic doses.

- The principle of selective toxicity means that useful drugs show toxicity against foreign or abnormal cells, but not against normal host cells.

1.2 Drug targets

Why should chemicals, some of which have remarkably simple structures, have such an important effect on such a complicated and large structure as a human being? The answer lies in the way that the human body operates. If we could see inside our bodies to the molecular level, we would see a magnificent array of chemical reactions taking place, keeping the body healthy and functioning.

Drugs may be mere chemicals, but they are entering a world of chemical reactions with which they interact. Therefore, there should be nothing odd in the fact that they can have an effect. The surprising thing might be that they can have such *specific* effects. This is more a result of *where* they act in the body—the drug targets.

1.2.1　Cell structure

Since life is made up of cells, then quite clearly drugs must act on cells. The structure of a typical mammalian cell is shown in Fig. 1.1. All cells in the human body contain a boundary wall called the **cell membrane** which encloses the contents of the cell—the **cytoplasm**. The cell membrane seen under the electron microscope consists of two identifiable layers, each of which is made up of an ordered row of phosphoglyceride molecules such as **phosphatidylcholine (lecithin)** (Fig. 1.2). The outer layer of the membrane is made up of phosphatidylcholine whereas the inner layer is made up of phosphatidylethanolamine, phosphatidylserine, and phosphatidylinositol. Each phosphoglyceride molecule consists of a small polar head-group, and two long hydrophobic (water-hating) chains.

In the cell membrane, the two layers of phospholipids are arranged such that the hydrophobic tails point towards each other and form a fatty, hydrophobic centre, while the ionic head-groups are placed at the inner and outer surfaces of the cell membrane (Fig. 1.3). This is a stable structure because the ionic, hydrophilic head-groups

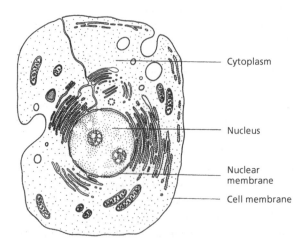

FIGURE 1.1　A typical mammalian cell. Taken from J. Mann, *Murder, magic, and medicine*, Oxford University Press (1992), with permission.

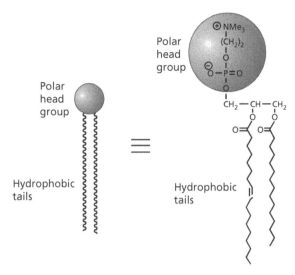

FIGURE 1.2 Phosphoglyceride structure.

interact with the aqueous media inside and outside the cell, whereas the hydrophobic tails maximize hydrophobic interactions with each other and are kept away from the aqueous environments. The overall result of this structure is to construct a fatty barrier between the cell's interior and its surroundings.

The membrane is not just made up of phospholipids, however. There are a large variety of proteins situated in the cell membrane (Fig. 1.3). Some proteins lie attached to the inner or the outer surface of the membrane. Others are embedded in the membrane with part of their structure exposed to one surface or both. The extent to which these proteins are embedded within the cell membrane structure depends on the types of amino acid present. Portions of protein that are embedded in the cell membrane have a large number of hydrophobic amino acids, whereas those portions that stick out from the surface have a large number of hydrophilic amino acids. Many surface proteins also have short chains of carbohydrates attached to them and are thus classed as **glycoproteins**.

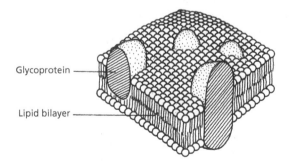

FIGURE 1.3 Cell membrane. Taken from J. Mann, *Murder, magic, and medicine*, Oxford University Press (1992), with permission.

These carbohydrate segments are important to cell–cell recognition (section 10.7).

Within the cytoplasm there are several structures, one of which is the **nucleus**. This acts as the 'control centre' for the cell. The nucleus contains the genetic code—the DNA—which acts as the blueprint for the construction of all the cell's proteins. There are many other structures within a cell, such as the mitochondria, the Golgi apparatus, and the endoplasmic reticulum, but it is not the purpose of this book to look at the structure and function of these organelles. Suffice it to say that different drugs act on molecular targets at different locations in the cell.

1.2.2 Drug targets at the molecular level

We shall now move to the molecular level, because it is here that we can truly appreciate how drugs work. The main molecular targets for drugs are proteins (enzymes, receptors, and transport proteins), and nucleic acids (DNA and RNA). These are large molecules (**macromolecules**) having molecular weights measured in the order of several thousand atomic mass units. They are much bigger than a typical drug, which has a molecular weight in the order of a few hundred atomic mass units.

The interaction of a drug with a macromolecular target involves a process known as binding. There is usually a specific area of the macromolecule where this takes place, and this is known as the **binding site** (Fig. 1.4). Typically, this takes the form of a hollow or canyon on the surface of the macromolecule allowing the drug to sink into the body of the larger molecule. Some drugs react with the binding site and become permanently attached via a covalent bond that has a bond strength of 200–400 kJ mol^{-1}. However, most drugs interact through weaker forms of interaction known as **intermolecular bonds**. These include electrostatic or ionic bonds, hydrogen bonds, van der Waals interactions, dipole–dipole interactions, and hydrophobic interactions. (It is also possible for these interactions to take place *within* a molecule, in which case they are called **intramolecular bonds**; see for example protein structure, sections 2.2 and 2.3). None of these bonds is as strong as the covalent bonds that make up the skeleton of a molecule, and so they can be formed, then broken again. This means that an equilibrium takes place between the drug being bound and unbound to its target. The binding forces are strong enough to hold the drug for a certain period of time to let it have an effect on the target, but weak enough to allow it to depart once it has done its job. The length of time the drug remains at its target will depend on the number of intermolecular bonds involved in holding it there. Drugs having a large number of interactions are likely to remain bound longer than those that have only a few. The relative strength of the different intermolecular binding forces

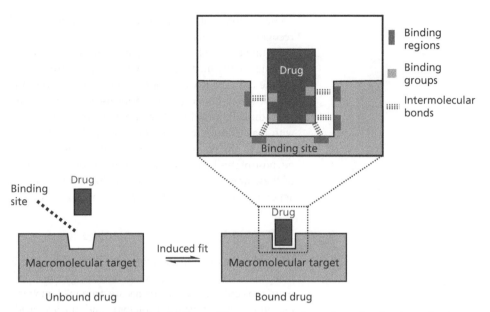

FIGURE 1.4 The equilibrium of a drug being bound and unbound to its target.

is also an important factor. Functional groups present in the drug can be important in forming intermolecular bonds with the target binding site. If they do so, they are called **binding groups**. However, the carbon skeleton of the drug also plays an important role in binding the drug to its target through van der Waals interactions. As far as the target binding site is concerned, it too contains functional groups and carbon skeletons which can form intermolecular bonds with 'visiting' drugs. The specific regions where this takes place are known as **binding regions**. The study of how drugs interact with their targets through binding interactions and produce a pharmacological effect is known as **pharmacodynamics**. Let us now consider the types of intermolecular bond that are possible.

1.3 Intermolecular bonding forces

There are several types of intermolecular bonding interactions, which differ in their bond strengths. The number and types of these interactions depend on the structure of the drug and the functional groups that are present (section 13.1 and Appendix 7). Thus, each drug may use one or more of the following interactions, but not necessarily all of them.

1.3.1 Electrostatic or ionic bonds

An ionic or electrostatic bond is the strongest of the intermolecular bonds (20–40 kJ mol^{-1}) and takes place between groups having opposite charges such as a carboxylate ion and an aminium ion (Fig. 1.5). The strength of the interaction is inversely proportional to the distance between the two charged atoms, and it is also dependent on the nature of the environment, being stronger in hydrophobic environments than in polar environments. Usually, the binding sites of macromolecules are more hydrophobic in nature than the surface, and so this enhances the effect of an ionic interaction. The drop-off in ionic bonding strength with separation is less than in other intermolecular interactions, so if an ionic interaction is possible, it is likely to be the most important initial interaction as the drug enters the binding site.

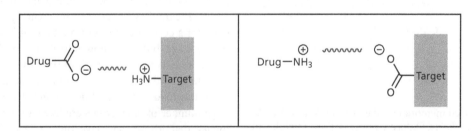

FIGURE 1.5 Electrostatic (ionic) interactions between a drug and the binding site.

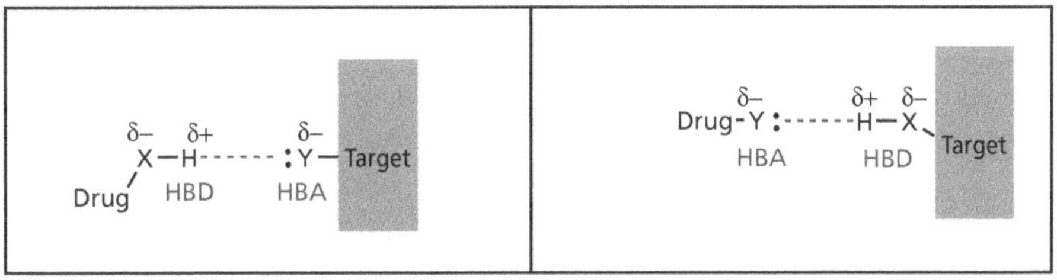

FIGURE 1.6 Hydrogen bonding shown by a dashed line between a drug and a binding site (X, Y = oxygen or nitrogen; HBD = hydrogen bond donor, HBA = hydrogen bond acceptor).

1.3.2 Hydrogen bonds

A hydrogen bond can vary substantially in strength, and normally takes place between an electron-rich heteroatom and an electron-deficient hydrogen (Fig. 1.6). The electron-rich heteroatom has to have a lone pair of electrons and is usually oxygen or nitrogen.

The electron-deficient hydrogen is usually linked by a covalent bond to an electronegative atom, such as oxygen or nitrogen. As the electronegative atom (X) has a greater attraction for electrons, the electron distribution in the covalent bond (X–H) is weighted towards the more electronegative atom, and so the hydrogen gains a slight positive charge. Such a hydrogen atom can act as a **hydrogen bond donor (HBD)**. The electron-rich heteroatom that receives the hydrogen bond is known as the **hydrogen bond acceptor (HBA)**. Some functional groups can provide both hydrogen bond donors and hydrogen bond acceptors (e.g. OH, NH_2). When such a group is present in a binding site, it is possible that it might bind to one ligand as a hydrogen bond donor and to another as a hydrogen bond acceptor. This characteristic is given the term **hydrogen bond flip-flop**.

Hydrogen bonds have been viewed as a weak form of electrostatic interaction, because the heteroatom is slightly negative and the hydrogen is slightly positive. However, there is more to hydrogen bonding than an attraction between partial charges. Unlike other intermolecular interactions, an interaction of orbitals takes place between the two molecules (Fig. 1.7). The orbital containing the lone pair of electrons on heteroatom Y interacts with the atomic orbitals normally involved in the covalent bond between X and H. This results in a weak form of sigma (σ) bonding and has an important directional consequence

that is not evident in electrostatic bonds. The optimum orientation is where the X–H bond points directly to the lone pair on Y, such that the angle formed between X, H, and Y is 180°. This is observed in very strong hydrogen bonds. However, the angle can vary between 130° and 180° for moderately strong hydrogen bonds, and can be as low as 90° for weak hydrogen bonds. The lone pair orbital of Y also has a directional property, depending on its hybridization. For example, the nitrogen of a pyridine ring is sp^2 hybridized and so the lone pair points directly away from the ring, and in the same plane (Fig. 1.8). The best location for a hydrogen bond donor would be the region of space indicated in the figure.

The strength of a hydrogen bond can vary widely, but most hydrogen bonds in drug–target interactions are moderate in strength, varying from 16 to 60 kJ mol^{-1}—approximately 10 times less than a covalent bond. The bond distance reflects this, and hydrogen bonds are typically 1.5–2.2 Å compared with 1.0–1.5 Å for a covalent bond. The strength of a hydrogen bond depends on the strengths of the hydrogen bond acceptor and the hydrogen bond donor. A good hydrogen bond acceptor has to be electronegative and have a lone pair of electrons. Nitrogen and oxygen are the most common atoms involved as hydrogen bond acceptors in biological systems. Nitrogen has one lone pair of electrons and can act as an acceptor for one hydrogen bond; oxygen has two lone pairs of electrons and can act as an acceptor for two hydrogen bonds (Fig. 1.9).

Several drugs and macromolecular targets contain a sulphur atom, which is also electronegative. However, sulphur is a weak hydrogen bond acceptor because its lone pairs are in third-shell orbitals, which are larger and more diffuse than second-shell orbitals. This means that

FIGURE 1.7 Orbital overlap in a hydrogen bond.

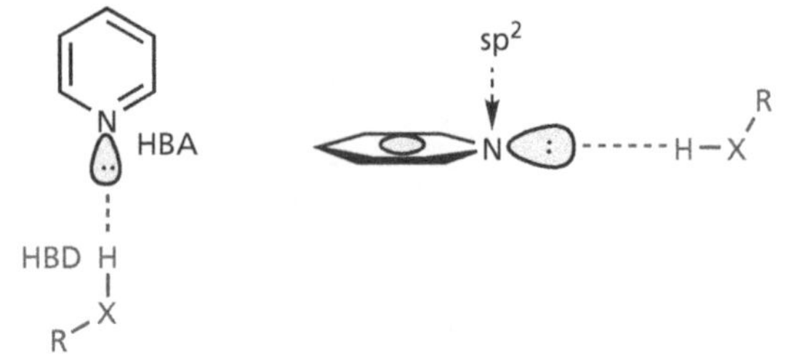

FIGURE 1.8 Directional influence of hybridization on hydrogen bonding.

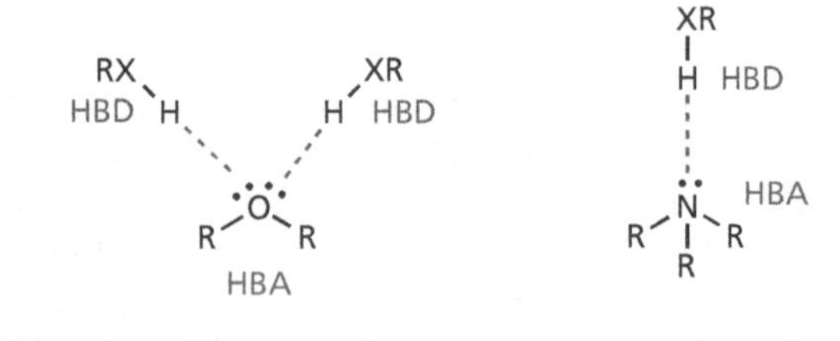

FIGURE 1.9 Oxygen and nitrogen acting as hydrogen bond acceptors (HBD = hydrogen bond donor, HBA = hydrogen bond acceptor).

the orbitals concerned interact less efficiently with the small 1s orbital of a hydrogen atom.

Fluorine, which is present in several drugs, is more electronegative than either oxygen or nitrogen. It also has three lone pairs of electrons, and this might suggest that it would make a good hydrogen bond acceptor. In fact, it is rather a weak hydrogen bond acceptor. It has been suggested that fluorine is so electronegative that it clings on tightly to its

lone pairs of electrons, making them incapable of hydrogen bond interactions. This is in contrast to a fluoride ion which is a very strong hydrogen bond acceptor.

Any feature that affects the electron density of the hydrogen bond acceptor is likely to affect its ability to act as a hydrogen bond acceptor; the greater the electron density of the heteroatom the greater its strength as a hydrogen bond acceptor. For example, the oxygen of a negatively charged carboxylate ion is a stronger hydrogen bond acceptor than the oxygen of the uncharged carboxylic acid (Fig. 1.10). Phosphate ions can also act as good hydrogen bond acceptors. Most hydrogen bond acceptors present in drugs and binding sites are neutral functional groups such as ethers, alcohols, phenols, amides, amines, and ketones. These groups will form moderately strong hydrogen bonds.

It has been proposed that the pi (π) systems present in alkynes and aromatic rings are regions of high electron density and can act as hydrogen bond acceptors. However, the electron density in these systems is diffuse, and so the hydrogen bonding interaction is much weaker than those involving oxygen or nitrogen. As a result, aromatic rings and alkynes are only likely to be significant hydrogen bond acceptors if they interact with a strong hydrogen bond donor such as an alkylammonium ion (NHR_3^+).

More subtle effects can influence whether an atom is a good hydrogen bond acceptor or not. For example, the nitrogen atom of an aliphatic tertiary amine is a better hydrogen bond acceptor than the nitrogen of an amide or an aniline (Fig. 1.11). In the latter functional groups,

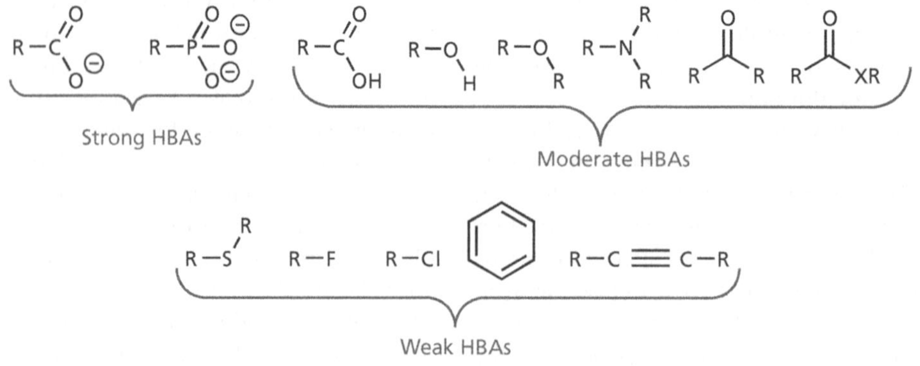

FIGURE 1.10 Relative strengths of hydrogen bond acceptors (HBAs).

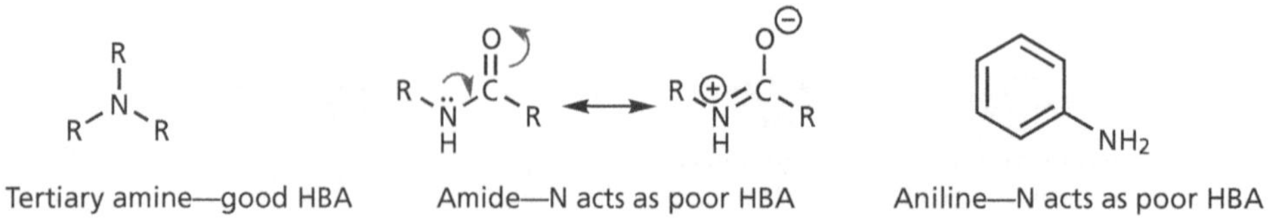

FIGURE 1.11 Comparison of different nitrogen-containing functional groups as hydrogen bond acceptors (HBAs).

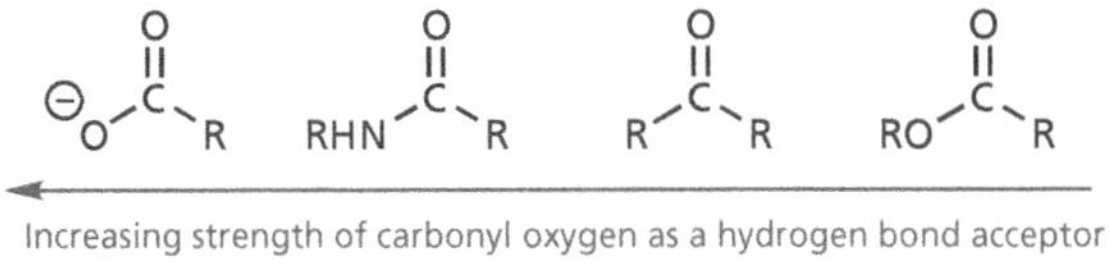

O=C(O⁻)R   O=C(RHN)R   O=C(R)R   O=C(RO)R

← *Increasing strength of carbonyl oxygen as a hydrogen bond acceptor*

FIGURE 1.12 Comparison of carbonyl oxygens as hydrogen bond acceptors.

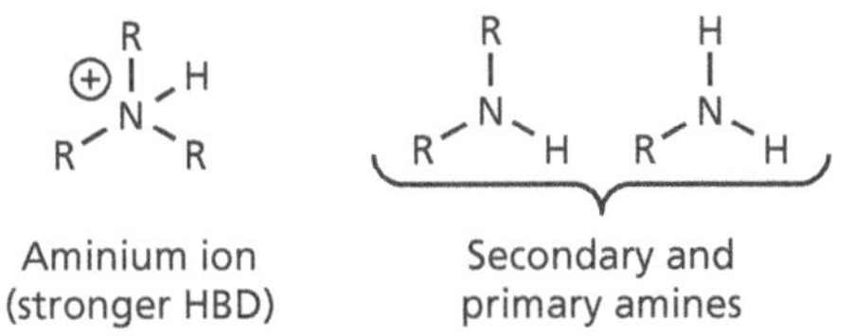

Aminium ion
(stronger HBD)

Secondary and
primary amines

FIGURE 1.13 Comparison of hydrogen bond donors (HBDs).

the lone pair of the nitrogen can interact with neighbouring pi systems to form various resonance structures. As a result, it is less likely to take part in a hydrogen bond.

Similarly, the ability of a carbonyl group to act as a hydrogen bond acceptor varies depending on the functional group involved (Fig. 1.12).

It has also been observed that an sp³ hybridized oxygen atom linked to an sp² carbon atom rarely acts as an HBA. This includes the alkoxy oxygen of esters, and the oxygen atom present in aromatic ethers or furans.

Good hydrogen bond donors contain an electron-deficient proton linked to oxygen or nitrogen. The more electron-deficient the proton, the better it will act as a hydrogen bond donor. For example, a proton attached to a positively charged nitrogen atom acts as a stronger hydrogen bond donor than the proton of a primary or secondary amine (Fig. 1.13). Because the nitrogen is positively charged, it has a greater pull on the electrons surrounding it, making attached protons even more electron-deficient.

1.3.3 Van der Waals interactions

Van der Waals interactions are very weak interactions that are typically 2–4 kJ mol⁻¹ in strength. They involve interactions between hydrophobic regions of different molecules, such as aliphatic substituents or the overall carbon skeleton. The electronic distribution in neutral, non-polar regions is never totally even or symmetrical, and there are always transient areas of high and low electron densities leading to temporary dipoles. The dipoles in one molecule can induce dipoles in a neighbouring molecule, leading to weak interactions between the two molecules (Fig. 1.14). Thus, an area of high electron density on one molecule can have an attraction for an area of low electron density on another molecule. The strength of these interactions falls off rapidly the further the two molecules are apart, decreasing to the seventh power of the separation. Therefore, the drug has to be close to the target binding site before the interactions become important. Van der Waals interactions are also referred to as **London forces**. Although the interactions are individually weak, there may be many such interactions between a drug and its target, and so the overall contribution of van der Waals interactions is often crucial to binding. Hydrophobic forces are also important when the non-polar regions of molecules interact (section 1.3.6).

1.3.4 Dipole–dipole and ion–dipole interactions

Many molecules have a permanent dipole moment resulting from the different electronegativities of the atoms and functional groups present. For example, a ketone has a dipole moment due to the different electronegativities of the carbon and oxygen making up the carbonyl bond. The binding site also contains functional groups, so it is inevitable that it too will have various local dipole moments. It is possible for the dipole moments of the drug and the binding site to interact as a drug approaches, aligning the drug such that the dipole moments are parallel and in opposite directions (Fig. 1.15). If this positions the drug such that other intermolecular interactions can take place between the drug and the binding site, then the alignment is beneficial to both binding and activity. If not, then binding and activity may be weakened. An example of such an effect can be found in anti-ulcer drugs (section 25.2.8.3).

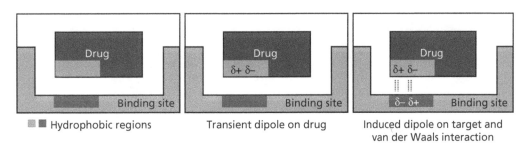

Hydrophobic regions   Transient dipole on drug   Induced dipole on target and van der Waals interaction

FIGURE 1.14 Van der Waals interactions between hydrophobic regions of a drug and a binding site.

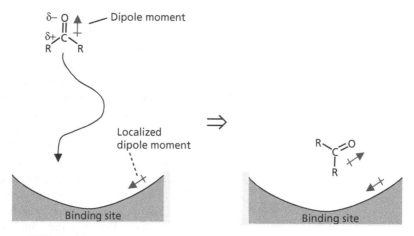

FIGURE 1.15 Dipole–dipole interactions between a drug and a binding site.

The strength of dipole–dipole interactions reduces with the cube of the distance between the two dipoles. This means that dipole–dipole interactions fall away more quickly with distance than electrostatic interactions, but less quickly than van der Waals interactions.

An ion–dipole interaction is where a charged or ionic group in one molecule interacts with a dipole in a second molecule (Fig. 1.16). This is stronger than a dipole–dipole interaction, and falls off less rapidly with separation (decreasing relative to the square of the separation).

Interactions involving an induced dipole moment have been proposed. There is evidence that an aromatic ring can interact with an ionic group such as a quaternary ammonium ion. Such an interaction is feasible if the positive charge of the quaternary ammonium group distorts the π electron cloud of the aromatic ring to produce a dipole moment, where the face of the aromatic ring is electron-rich and the edges are electron-deficient (Fig. 1.17). This is also called a **cation–pi interaction**. An important neurotransmitter called **acetylcholine** forms this type of interaction with its binding site (section 22.5).

1.3.5 Repulsive interactions

So far we have concentrated on attractive forces, which increase in strength the closer the molecules approach each other. Repulsive interactions are also important. Otherwise, there would be nothing to stop molecules

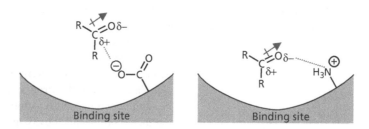

FIGURE 1.16 Ion–dipole interactions between a drug and a binding site.

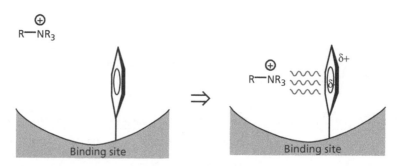

FIGURE 1.17 Induced dipole interaction between an alkylammonium ion and an aromatic ring.

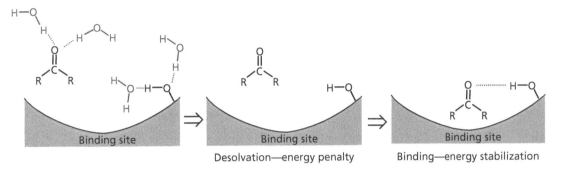

FIGURE 1.18 Desolvation of a drug and its target binding site prior to binding.

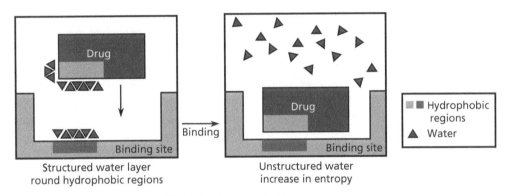

FIGURE 1.19 Hydrophobic interactions.

trying to merge with each other! If molecules come too close, their molecular orbitals start to overlap and this results in repulsion. Other forms of repulsion are related to the types of groups present in both molecules. For example, two charged groups of identical charge are repelled.

1.3.6 The role of water and hydrophobic interactions

A crucial feature that is often overlooked when considering the interaction of a drug with its target is the role of water. The macromolecular targets in the body exist in an aqueous environment, and the drug has to travel through that environment in order to reach its target. Therefore, both the drug and the macromolecule are solvated with water molecules before they meet each other. The water molecules surrounding the drug and the target binding site have to be stripped away before the interactions described above can take place (Fig. 1.18). This requires energy, and if the energy required to desolvate both the drug and the binding site is greater than the stabilization energy gained by the binding interactions, then the drug may be ineffective. In certain cases, it has even proved beneficial to remove a polar binding group from a drug in order to lower its energy of desolvation. For example, a polar binding group was removed during the development of the antiviral drug **ritonavir** (section 20.7.4.4).

Sometimes polar groups are added to a drug to increase its water solubility. If this is the case, it is important that such groups are positioned in such a way that they protrude from the binding site when the drug binds; in other words they are solvent-accessible or solvent-exposed. In this way, the water that solvates this highly polar group does not have to be stripped away, and there is no energy penalty when the drug binds to its target. Examples of this can be seen in sections 21.6.2.1, 26.9.1.2, and Case study 5.

It is not possible for water to solvate the non-polar or hydrophobic regions of a drug or its target binding site. Instead, the surrounding water molecules form stronger than usual interactions with each other, resulting in an ordered layer of water next to the non-polar surface. This represents a negative entropy due to the increase in order. When the hydrophobic region of a drug interacts with a hydrophobic region of a binding site, these water molecules are freed and become less ordered (Fig. 1.19). This leads to an increase in entropy and a gain in binding energy.[1] The interactions involved are small, at 0.1–0.2 kJ mol⁻¹ for each square angstrom of hydrophobic surface, but overall they can be substantial. Sometimes, a hydrophobic region in the drug may not be sufficiently close to a

---

[1]The free energy gained by binding (ΔG) is related to the change in entropy (ΔS) by the equation $\Delta G = \Delta H - T\Delta S$. If entropy increases, ΔS is positive which makes ΔG more negative. The more negative the value of ΔG, the more likely binding will take place.

hydrophobic region in the binding site, and water may be trapped between the two surfaces. The entropy increase is not so substantial in that case, and there is a benefit in designing a better drug that fits more snugly.

1.4 Pharmacokinetic issues and medicines

Pharmacodynamics is the study of how a drug binds to its target binding site and produces a pharmacological effect. However, a drug capable of binding to a particular target is not necessarily going to be useful as a clinical agent or medicine. For that to be the case, the drug not only has to bind to its target, it has to reach it in the first place. For an orally administered drug, that involves a long journey with many hazards to be overcome. The drug has to survive stomach acids, then digestive enzymes in the intestine. It has to be absorbed from the gut into the blood supply, then it has to survive the liver where enzymes try to destroy it (drug metabolism). It has to be distributed round the body and not get mopped up by fat tissue. It should not be excreted too rapidly or else frequent doses will be required to maintain activity. On the other hand, it should not be excreted too slowly or its effects could linger on longer than required. The study of how a drug is absorbed, distributed, metabolized, and excreted (known as ADME in the pharmaceutical industry) is called **pharmacokinetics**. Pharmacokinetics has sometimes been described as 'what the body does to the drug' as opposed to pharmacodynamics—'what the drug does to the body'.

There are many ways in which medicinal chemists can design a drug to improve its pharmacokinetic properties, but the methods by which a drug is formulated and administered are just as important. Medicines are not just composed of the active pharmaceutical agent. For example, a pill contains a whole range of chemicals which are present to give structure and stability to the pill, and also to aid the delivery and breakdown of the pill at the desired part of the gastrointestinal tract.

KEY POINTS

- Drugs act on molecular targets located in the cell membrane of cells or within the cells themselves.

- Drug targets are macromolecules that have a binding site into which the drug fits and binds.

- Most drugs bind to their targets by means of intermolecular bonds.

- Pharmacodynamics is the study of how drugs interact with their targets and produce a pharmacological effect.

- Electrostatic or ionic interactions occur between groups of opposite charge.

- Hydrogen bonds occur between an electron-rich heteroatom and an electron-deficient hydrogen.

- The hydrogen involved in a hydrogen bond is called the hydrogen bond donor. The electronegative atom that interacts with the hydrogen in a hydrogen bond is called the hydrogen bond acceptor.

- Van der Waals interactions take place between non-polar regions of molecules and are caused by transient dipole–dipole interactions.

- Ion–dipole and dipole–dipole interactions are a weak form of electrostatic interaction.

- Hydrophobic interactions involve the displacement of ordered layers of water molecules which surround hydrophobic regions of molecules. The resulting increase in entropy contributes to the overall binding energy.

- Polar groups have to be desolvated before intermolecular interactions take place. This results in an energy penalty.

- The pharmacokinetics of a drug relate to its absorption, distribution, metabolism, and excretion in the body.

1.5 Classification of drugs

There are four main ways in which drugs might be classified or grouped.

By pharmacological effect. Drugs can be classified depending on the biological or pharmacological effect that they have; for example analgesics, antipsychotics, antihypertensives, anti-asthmatics, and antibiotics. This is useful if one wishes to know the full scope of drugs available for a certain ailment, but it means that the drugs included are numerous and highly varied in structure. This is because there are a large variety of targets at which drugs could act in order to produce the desired effect. It is, therefore, not possible to compare different painkillers and expect them to look alike or to have some common mechanism of action.

The chapters on antibacterial, antiviral, anticancer, anti-ulcer, and cardiovascular drugs (Chapters 19, 20, 21, 25, and 26) illustrate the variety of drug structures and mechanisms of action that are possible when drugs are classified according to their pharmacological effect.

By chemical structure. Many drugs which have a common skeleton are grouped together; for example penicillins, barbiturates, opiates, steroids, and catecholamines. In some cases, this is a useful classification since the biological activity and mechanism of action is the same for the structures involved; for example, the antibiotic activity of penicillins. However, not all compounds with similar chemical structure have the same biological action. For example, steroids share a similar tetracyclic structure, but they have very different effects in the body. In this text, various groups of structurally related drugs are discussed; for example, penicillins, cephalosporins, sulphonamides, opioids, and glucocorticoids (sections 19.4–19.5, Chapter 24, and Case study 6). These are examples of compounds with a similar structure and similar mechanism of action. However, there are exceptions. Most sulphonamides are used as antibacterial agents, but there are a few which have totally different medical applications.

By target system. Drugs can be classified according to whether they affect a certain target system in the body. An example of a target system is where a neurotransmitter is synthesized, released from its neuron, interacts with a protein target, and is either metabolized or reabsorbed into the neuron. This classification is a bit more specific than classifying drugs by their overall pharmacological effect. However, there are still several different targets with which drugs could interact in order to interfere with the system, and so the drugs included in this category are likely to be quite varied in structure due to the different mechanisms of action that are involved. In Chapters 22 and 23, we look at drugs that act on target systems—the cholinergic and the adrenergic system respectively.

By target molecule. Some drugs are classified according to the molecular target with which they interact. For example, anticholinesterases (sections 22.12–22.15) are drugs which act by inhibiting the enzyme acetylcholinesterase. This is a more specific classification since we have now identified the precise target at which the drugs act. In this situation, we might expect some structural similarity between the agents involved and a common mechanism of action, although this is not an inviolable assumption. However, it is easy to lose the wood for the trees and to lose sight of why it is useful to have drugs which switch off a particular enzyme or receptor. For example, it is not intuitively obvious why an anticholinesterase agent could be useful in treating Alzheimer's disease or glaucoma.

1.6 Naming of drugs and medicines

The vast majority of chemicals that are synthesized in medicinal chemistry research never make it to the market place and it would be impractical to name them all. Instead, research groups label them with a code which usually consists of letters and numbers. The letters are specific to the research group undertaking the work, and the number is specific for the compound. Thus, Ro31-8959, ABT-538, and MK-639 were compounds prepared by Roche, Abbott, and Merck pharmaceuticals respectively. If the compounds concerned show promise as therapeutic drugs, they are taken forward to pre-clinical trials then clinical studies, by which time they are often named. For example, the above compounds showed promise as anti-HIV drugs and were named **saquinavir**, **ritonavir**, and **indinavir** respectively. Finally, if the drugs prove successful and are marketed as medicines, they are given a proprietary,

brand, or trade name which only the company can use. For example, the above compounds were marketed as **Fortovase®**, **Norvir®**, and **Crixivan®** respectively (note that brand names always start with a capital letter and have the symbol R or TM to indicate that they are registered brand names). The proprietary names are also specific for the preparation or formulation of the drug. For example, Fortovase® (or Fortovase™) is a preparation containing 200 mg of saquinavir in a gel-filled, beige-coloured capsule. If the formulation is changed, then a different name is used. For example, Roche sell a different preparation of saquinavir called **Invirase®** which consists of a brown/green capsule containing 200 mg of saquinavir as the mesylate salt. When a drug's patent has expired, it is possible for any pharmaceutical company to produce and sell that drug as a generic medicine. However, they are not allowed to use the trade name used by the company that originally invented it. European law requires that generic medicines are given a **recommended International Non-proprietary Name (rINN)** which is usually identical to the name of the drug. In Britain, such drugs were given a **British Approved Name** (BAN), but these have now been modified to fall in line with rINNs. rINNs generally have a suffix which indicates the therapeutic area for the named drug. For example, saquinavir, ritonavir, and indinavir all end with the suffix -vir indicating that they are antiviral agents.

Since the naming of drugs is progressive, early research papers in the literature may only use the original letter/number code since the name of the drug had not been allocated at the time of publication.

Throughout this text, the names of the active constituents are used rather than the trade names, although the trade name may be indicated if it is particularly well known. For example, it is indicated that **sildenafil** is **Viagra®** and that **paclitaxel** is **Taxol®**. If you wish to find out the trade name for a particular drug, these are listed in the index. If you wish to 'go the other way', Appendix 6 contains trade names and directs you to the relevant compound name. Only those drugs covered in the text are included and if you cannot find the drug you are looking for, you should refer to other textbooks or formularies such as the British National Formulary (see General further reading).

KEY POINTS

- Drugs can be classified by their pharmacological effect, their chemical structure, their effect on a target system, or their effect on a target structure.

- Clinically useful drugs have a trade (or brand) name, as well as a recommended international non-proprietary name.

- Most structures produced during the development of a new drug are not considered for the clinic. They are identified by simple codes that are specific to each research group.

QUESTIONS

1. The hormone adrenaline interacts with proteins located on the surface of cells and does not cross the cell membrane. However, larger steroid molecules such as estrone cross cell membranes and interact with proteins located in the cell nucleus. Why is a large steroid molecule able to cross the cell membrane when a smaller molecule such as adrenaline cannot?

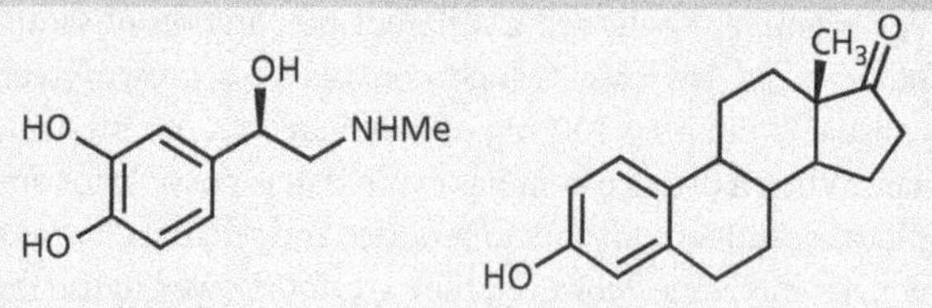

Adrenaline   Estrone

Structure I

2. Valinomycin is an antibiotic which is able to transport ions across cell membranes and disrupt the ionic balance of the cell. Find out the structure of valinomycin and explain why it is able to carry out this task.

3. Archaea are microorganisms which can survive in extreme environments such as high temperature, low pH, or high salt concentration. It is observed that the cell membrane phospholipids in these organisms (see structure I) are markedly different from those in eukaryotic cell membranes. What differences are present and what function might they serve?

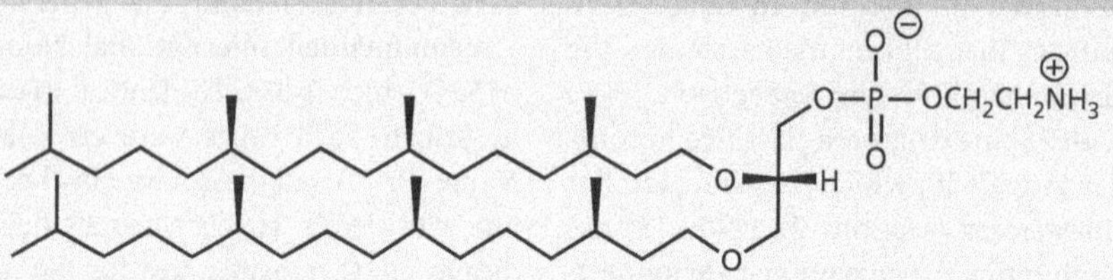

4. Teicoplanin is an antibiotic which 'caps' the building blocks used in the construction of the bacterial cell wall, such that they cannot be linked up. The cell wall is a barrier surrounding the bacterial cell membrane, and the building blocks are anchored to the outside of this cell membrane prior to their incorporation into the cell wall. Teicoplanin contains a very long alkyl substituent which plays no role in the capping mechanism. However, if this substituent is absent, activity drops. What role do you think this alkyl substituent might serve?

5. The Ras protein is an important protein in signalling processes within the cell. It exists freely in the cell cytoplasm, but must become anchored to the inner surface of the cell membrane in order to carry out its function. What kind of modification to the protein might take place to allow this to happen?

6. Cholesterol is an important constituent of eukaryotic cell membranes and affects the fluidity of the membrane. Consider the structure of cholesterol (shown below) and suggest how it might be orientated in the membrane.

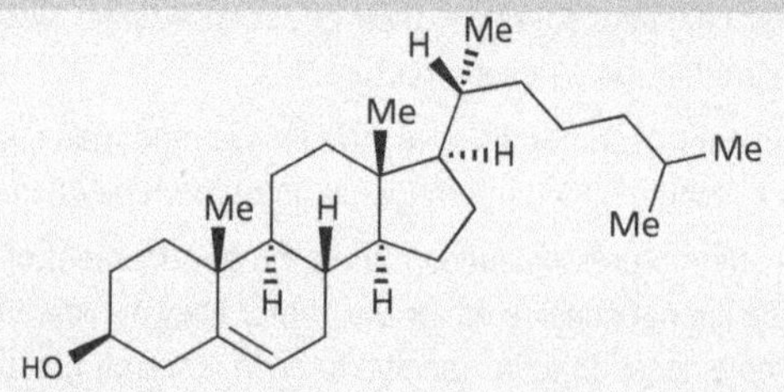

Cholesterol

7. Most unsaturated alkyl chains in phospholipids are *cis* rather than *trans*. Consider the *cis*-unsaturated alkyl chain in the phospholipid shown in Fig. 1.2. Redraw this chain to give a better representation of its shape and compare it with the shape of its *trans*-isomer. What conclusions can you make regarding the packing of such chains in the cell membrane, and the effect on membrane fluidity?

8. The relative strength of carbonyl oxygens as hydrogen bond acceptors is shown in Fig. 1.12. Suggest why the order is as shown.

9. Consider the structures of adrenaline, estrone, and cholesterol and suggest what kind of intermolecular interactions are possible for these molecules and where they occur.

10. Using the index and Appendix 8 (on the website), identify the structures and trade names for the following drugs—amoxicillin, ranitidine, gefitinib, atracurium.

Multiple-choice questions are available on the Online Resource Centre at www.oxfordtextbooks.co.uk/orc/patrick6e/

FURTHER READING

Kubinyi, H. (2001) Hydrogen bonding: The last mystery in drug design? in Testa, B., van de Waterbeemd, H., Folkers, G., and Guy, R. (eds), *Pharmacokinetic optimization in drug research*. Wiley-VCH, Weinheim.

Mann, J. (1992) *Murder, magic, and medicine*, Chapter 1. Oxford University Press, Oxford.

Page, C., Curtis, M., Sutter, M., Walker, M., and Hoffman, B. (2002) Drug names and drug classification systems. in *Integrated pharmacology 2nd edn*, Chapter 2. Mosby, Elsevier, Maryland Heights, MO.

Sneader, W. (2005) *Drug discovery: a history.* John Wiley and Sons, Chichester.

WEBSITES

International non-proprietary names, World Health Organization. www.who.int/medicines/services/inn/en/

Brand names of some commonly used drugs. www.mwrckmanuals.com/professional/appendices/brand-names-of-some-commonly-used-drugs?starting with=a

Titles for general further reading are listed on p.845.

11 Pharmacokinetics and related topics

11.1 The three phases of drug action

There are three phases involved in drug action. The first of these is the **pharmaceutical phase**. For an orally administered drug, this includes the disintegration of a pill or capsule in the gastrointestinal tract (GIT), the release of the drug, and its dissolution. The pharmaceutical phase is followed by the **pharmacokinetic phase**, which includes absorption from the GIT into the blood supply, and the various factors that affect a drug's survival and progress as it travels to its molecular target. The final **pharmacodynamic phase** involves the mechanism by which a drug interacts with its molecular target and the resulting pharmacological effect.

In previous chapters, we have focused on drug targets and drug design, where the emphasis is on the pharmacodynamic aspects of drug action; for example, optimizing the binding interactions of a drug with its target. However, the compound with the best binding interactions for a target is not necessarily the best drug to use in medicine. This is because a drug has to reach its target in the first place if it is to be effective. Therefore, when carrying out a drug design programme, it is important to study pharmacokinetics alongside pharmacodynamics. The four main topics to consider in pharmacokinetics are absorption, distribution, metabolism, and excretion (often abbreviated to ADME).

11.2 A typical journey for an orally active drug

The preferred method of drug administration is the oral route, and so we shall consider some of the hurdles and hazards faced by such a drug in order to reach its eventual target. When a drug is swallowed, it enters the **gastrointestinal tract** (GIT), which comprises the mouth, throat, stomach, and the upper and lower intestines. A certain amount of the drug may be absorbed through the mucosal membranes of the mouth, but most passes down into the stomach where it encounters gastric juices and hydrochloric acid. These chemicals aid in the digestion of food and will treat a drug in a similar fashion if it is susceptible to breakdown and is not protected within an acid-resistant pill or capsule. For example, the first clinically useful penicillin was broken down in the stomach and had to be administered by injection. Other acid-labile drugs include the **local anaesthetics** and **insulin**. If the drug *does* survive the stomach, it enters the upper intestine where it encounters digestive enzymes that serve to break down food. Assuming the drug survives this attack, it then has to pass through the cells lining the gut wall. This means that it has to pass through a cell membrane on two occasions, first to enter the cell and then to exit it on the other side. Once the drug has passed through the cells of the gut wall, it can enter the blood supply relatively easily as the cells lining the blood vessels are loose fitting and there are pores through which most drugs can pass. In other words, drugs enter blood vessels by passing between cells rather than through them.

The drug is now transported in the blood to the body's 'customs office'—the liver. The liver contains enzymes which are ready and waiting to intercept foreign chemicals, and modify them such that they are more easily excreted—a process called drug metabolism (section 11.5). Following this, the drug has to be carried by the blood supply round the body to reach its eventual target, which may require crossing further cell membranes—always assuming that it is neither excreted before it gets there, nor diverted to parts of the body where it is not needed.

It can be seen that stringent demands are made on any orally administered drug. It must be stable to both chemical and enzymatic attack. It must also have the correct physicochemical properties to allow it to reach its target in therapeutic concentrations. This includes efficient absorption, effective distribution to target tissues, and an acceptable rate of excretion. We will now look more closely at the various stages.

11.3 **Drug absorption**

In order to be absorbed efficiently from the GIT, a drug must have the correct balance of water versus fat solubility. If the drug is too polar (hydrophilic), it will fail to pass through the fatty cell membranes of the gut wall (section 1.2.1). On the other hand, if the drug is too fatty (hydrophobic), it will be poorly soluble in the gut and will dissolve in fat globules. This means that there will be poor surface contact with the gut wall, resulting in poor absorption.

It is noticeable how many drugs contain an amine functional group. There are good reasons for this. Amines are often involved in a drug's binding interactions with its target. However, they are also an answer to the problem of balancing the dual requirements of water and fat solubility. Amines are weak bases, and it is found that many of the most effective drugs contain amine groups having a pK_a value in the range 6–8. In other words, they are partially ionized at the slightly acidic and alkaline pHs present in the intestine and blood respectively, and can easily equilibrate between their ionized and non-ionized forms. This allows them to cross cell membranes in the non-ionized form, while the presence of the ionized form gives the drug good water solubility and permits good binding interactions with its target binding site (Fig. 11.1).

The extent of ionization at a particular pH can be determined by the **Henderson–Hasselbalch equation**:

$$pH = pK_a + \log\frac{[RNH_2]}{[RNH_3^+]}$$

where $[RNH_2]$ is the concentration of the free base and $[RNH_3^+]$ is the concentration of the ionized amine. K_a is the equilibrium constant for the equilibrium shown in Fig. 11.1, and the Henderson–Hasselbalch equation can be derived from the equilibrium constant:

$$K_a = \frac{[H^+][RNH_2]}{[RNH_3^+]}$$

Therefore $pK_a = -\log\frac{[H^+][RNH_2]}{[RNH_3^+]}$

$$= -\log[H^+] - \log\frac{[RNH_2]}{[RNH_3^+]}$$

$$= pH - \log\frac{[RNH_2]}{[RNH_3^+]}$$

Therefore $pH = pK_a + \log\frac{[RNH_2]}{[RNH_3^+]}$

Ionized amine

Non-ionized amine (free base)

Receptor interaction and water solubility

Crosses membranes

FIGURE 11.1 Equilibrium between the ionized and non-ionized form of an amine.

Note that when the concentration of the ionized and unionized amines are identical (i.e. when $[RNH_2]$ = $[RNH_3^+]$), the ratio $[RNH_2]/[RNH_3^+]$ is 1. Since $\log 1 = 0$, the Henderson–Hasselbalch equation will simplify to pH = pK_a. In other words, when the amine is 50% ionized, pH = pK_a. Therefore, drugs with a pK_a of 6–8 are approximately 50% ionized at blood pH (7.4) or the slightly acidic pH of the intestines.

The hydrophilic/hydrophobic character of the drug is the crucial factor affecting absorption through the gut wall; in theory the molecular weight of the drug should be irrelevant. For example, **ciclosporin** is successfully absorbed through cell membranes, although it has a molecular weight of about 1200. In practice, however, larger molecules tend to be poorly absorbed because they are likely to contain a large number of polar functional groups. As a rule of thumb, orally absorbed drugs tend to obey what is known as Lipinski's **rule of five**. The rule of five was derived from an analysis of compounds from the World Drugs Index database, aimed at identifying features that were important in making a drug orally active. It was found that the factors concerned involved numbers that are multiples of 5:

- a molecular weight less than 500
- no more than 5 hydrogen bond donor (HBD) groups
- no more than 10 hydrogen bond acceptor (HBA) groups
- a calculated **log *P*** value less than +5 (log *P* is a measure of a drug's hydrophobicity—section 14.1).

The rule of five has been an extremely useful rule of thumb for many years, but it is neither quantitative nor foolproof. For example, orally active drugs such as **atorvastatin**, **rosuvastatin**, **ciclosporin**, and **vinorelbine** do not obey the rule of five. It has also been demonstrated that a high molecular weight does not in itself cause poor oral bioavailability. Another source of debate concerns the calculation of the number of HBAs. In Lipinski's original paper, the number of HBAs corresponded to the total number of oxygen and nitrogen atoms present in a structure. This was done for simplicity's sake, but most medicinal chemists would discount weak HBAs such as amide nitrogens (see also section 1.3.2 and Appendix 7). Therefore, it is better to view Lipinski's rules as a set of guidelines rather than rules. Lipinski himself stated that a compound was likely to be orally active as long as it did not break more than one of his 'rules'.

Further research has been carried out to find guidelines that are independent of molecular weight. Work carried out by Veber et al. in 2002 demonstrated the rather surprising finding that molecular flexibility plays an important role in oral bioavailability; the more flexible the molecule, the less likely it is to be orally active. In order to measure flexibility, one can count the number of freely rotatable bonds that result in significantly different conformations. Bonds to simple substituents such as methyl or alcohol groups are not included in this analysis as their rotation does not result in significantly different conformations.

Veber's studies also demonstrated that the polar surface area of the molecule could be used as a factor instead of the number of hydrogen bonding groups. These findings led to the following parameters for predicting acceptable oral activity. Either:

- a polar surface area ≤ 140 Å and ≤ 10 rotatable bonds;

or

- ≤ 12 hydrogen bond donors and acceptors in total and ≤ 10 rotatable bonds.

Some researchers set the limit of rotatable bonds to ≤ 7 since the analysis shows a marked improvement in oral bioavailability for such molecules.

These rules are independent of molecular weight and open the way to studying larger structures that have been 'shelved' up to now. Unfortunately, structures having a molecular weight larger than 500 are quite likely to have more than 10 rotatable bonds. However, the new rules suggest that rigidifying the structures to reduce the number of rotatable bonds would be beneficial. Rigidification tactics are described in section 13.3.9 as a strategy to improve a drug's pharmacodynamic properties, but these same tactics could also be used to improve pharmacokinetic properties. Appendix 9 (available on the website) provides information on MWt, log P, HBDs, HBAs, rotatable bonds, and polar surface area for several of the drugs covered in this text.

Polar drugs that break the above rules are usually poorly absorbed and have to be administered by injection.

Nevertheless, some highly polar drugs *are* absorbed from the digestive system as they are able to 'hijack' **transport proteins** present in the membranes of cells lining the gut wall (sections 2.7.2 and 10.1). These transport proteins normally transport the highly polar building blocks required for various biosynthetic pathways (e.g. amino acids and nucleic acid bases) across cell membranes. If the drug bears a structural resemblance to one of these building blocks, then it, too, may be smuggled across. For example, **levodopa** is transported by the transport protein for the amino acid phenylalanine, while **fluorouracil** is transported by transport proteins for the nucleic acid bases thymine and uracil. The antihypertensive agent **lisinopril** is transported by transport proteins for dipeptides. The anticancer agent **methotrexate** and the antibiotic **erythromycin** are also absorbed by means of transport proteins.

Other highly polar drugs can be absorbed into the blood supply if they have a low molecular weight (less than 200), as they can then pass through small pores between the cells lining the gut wall.

Occasionally, polar drugs with high molecular weight can cross the cells of the gut wall without actually passing through the membrane. This involves a process known as **pinocytosis** where the drug is engulfed by the cell membrane and a membrane-bound vesicle is pinched off to carry the drug across the cell (Fig. 11.2). The vesicle then fuses with the membrane to release the drug on the other side of the cell.

Sometimes, drugs are deliberately designed to be highly polar so that they are *not* absorbed from the GIT. These are usually antibacterial agents targeted against gut infections. Making them highly polar ensures that the drug reaches the site of infection in higher concentration (Box 19.2).

Finally, it should be noted that the absorption of some drugs can be affected adversely by interactions with food or other drugs in the gut (section 11.7.1).

Other drug administration routes may involve an absorption process and this is discussed in section 11.7.

For additional information see Web article 25: Looking at medicinal chemistry post Lipinski on the Online Resource Centre at www.oxfordtextbooks.co.uk/orc/patrick6e/

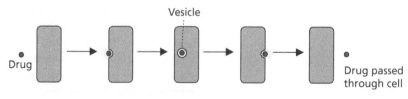

FIGURE 11.2 Pinocytosis.

11.4 Drug distribution

Once a drug has been absorbed, it is rapidly distributed around the blood supply, then more slowly distributed to the various tissues and organs. The rate and extent of distribution depends on various factors, including the physical properties of the drug itself.

11.4.1 Distribution round the blood supply

The vessels carrying blood round the body are called **arteries**, **veins**, and **capillaries** (section 26.2). The heart is the pump that drives the blood through these vessels. The major artery carrying blood from the heart is called the **aorta**, and as it moves further from the heart, it divides into smaller and smaller arteries—similar to the limbs and branches radiating from the trunk of a tree. Eventually, the blood vessels divide to such an extent that they become extremely narrow—equivalent to the twigs of a tree. These blood vessels are called capillaries, and it is from these vessels that oxygen, nutrients, and drugs can escape in order to reach the tissues and organs of the body. At the same time, waste products such as cell breakdown products and carbon dioxide are transferred from the tissues into the capillaries to be carried away and disposed of. The capillaries now start uniting into bigger and bigger vessels, resulting in the formation of veins which return the blood to the heart.

Once a drug has been absorbed into the blood supply, it is rapidly and evenly distributed throughout the blood supply within a minute—the time taken for the blood volume to complete one circulation. However, this does not mean that the drug is evenly distributed around the body—the blood supply is richer to some areas of the body than to others.

11.4.2 Distribution to tissues

Drugs do not stay confined to the blood supply. If they did, they would be of little use since their targets are the cells of various organs and tissues. The drug has to leave the blood supply in order to reach those targets. The body has an estimated 10 billion capillaries with a total surface area of 200 m². They probe every part of the body, such that no cell is more than 20–30 μm away from a capillary. Each capillary is very narrow, not much wider than the red blood cells that pass through it. Its walls are made up of a thin single layer of cells packed tightly together. However, there are pores between the cells which are 90–150 Å in diameter—large enough to allow most drug-sized molecules to pass though, but not large enough to allow the **plasma proteins** present in blood to escape. Therefore, drugs do not have to cross cell membranes in

order to leave the blood system and can be freely and rapidly distributed into the aqueous fluid surrounding the various tissues and organs of the body. Having said that, some drugs bind to plasma proteins in the blood. Since the plasma proteins cannot leave the capillaries, the proportion of drug bound to these proteins is also confined to the capillaries and cannot reach its target.

11.4.3 Distribution to cells

Once a drug has reached the tissues, it can immediately be effective if its target site is a receptor situated in a cell membrane. However, there are many drugs that have to enter the individual cells of tissues in order to reach their target. These include local anaesthetics, enzyme inhibitors, and drugs which act on nucleic acids or intracellular receptors. Such drugs must be hydrophobic enough to pass through the cell membrane, unless they are smuggled through by carrier proteins or taken in by pinocytosis. Since many drugs contain an amine functional group, the same principles described in section 11.3 apply. The drug must pass through the cell membrane as the free base, but, once it is inside the cell, the amine may become protonated to allow a strong interaction with the target binding site. Experiments have demonstrated this with a number of drugs such as **verapamil** (section 26.6.3).

11.4.4 Other distribution factors

The concentration levels of free drug circulating in the blood supply rapidly fall away after administration as a result of the distribution patterns described above, but there are other factors at work. Drugs that are excessively hydrophobic are often absorbed into fatty tissues and removed from the blood supply. This fat solubility can lead to problems. For example, obese patients undergoing surgery require a larger than normal volume of general anaesthetic because the gases used are particularly fat soluble. Unfortunately, once surgery is over and the patient has regained consciousness, the anaesthetics stored in the fat tissues will be released and may render the patient unconscious again. **Barbiturates** were once seen as potential intravenous anaesthetics which could replace the anaesthetic gases. Unfortunately, they, too, are fat soluble and it is extremely difficult to estimate a sustained safe dosage. The initial dose can be estimated to allow for the amount of barbiturate taken up by fat cells, but further doses eventually lead to saturation of the fat depot and result in a sudden, and perhaps fatal, increase of barbiturate levels in the blood supply.

Ionized drugs may become bound to various macromolecules and be removed from the blood supply. Drugs may also be bound reversibly to blood plasma proteins such as **albumin**, thus lowering the level of free drug.

Therefore, only a small proportion of the drug that has been administered may actually reach the desired target.

11.4.5 Blood–brain barrier

The blood–brain barrier is an important barrier that drugs have to negotiate if they are to enter the brain. The blood capillaries feeding the brain are lined with tight-fitting cells which do not contain pores (unlike capillaries elsewhere in the body). Moreover, the capillaries are coated with a fatty layer formed from nearby cells, providing an extra fatty barrier through which drugs have to cross. Therefore, drugs entering the brain have to dissolve through the cell membranes of the capillaries and also through the fatty cells coating the capillaries. As a result, polar drugs such as **penicillin** do not easily enter the brain.

The existence of the blood–brain barrier makes it possible to design drugs which will act at various parts of the body (e.g. the heart) and have no activity in the brain, thus reducing any central nervous system (CNS) side effects. This is done by increasing the polarity of the drug such that it does not cross the blood–brain barrier. On the other hand, drugs that are intended to act in the brain must be designed such that they *are* able to cross the blood–brain barrier. This means that they are limited in the number of polar groups that they can have. Alternatively, polar groups may have to be temporarily masked in order to allow passage through the blood–brain barrier (see prodrugs; section 14.6). Having said that, some polar drugs can cross the blood–brain barrier with the aid of carrier proteins, while others (e.g. **insulin**) can cross by the process of pinocytosis previously described. The ability to cross the blood–brain barrier has an important bearing on the analgesic activity of opioids (section 24.5). Research is also being carried out to find ways of increasing the permeability of the blood–brain barrier using techniques such as ultrasound or drugs such as **sildenafil**.

11.4.6 Placental barrier

The placental membranes separate a mother's blood from the blood of her fetus. The mother's blood provides the fetus with essential nutrients and carries away waste products, but these chemicals must pass through the placental barrier. As food and waste products can pass through the placental barrier, it is perfectly feasible for drugs to pass through as well. Drugs such as **alcohol**, **nicotine**, and **cocaine** can all pass into the fetal blood supply. Fat-soluble drugs will cross the barrier most easily, and drugs such as **barbiturates** will reach the same levels in fetal blood as in maternal blood. Such levels may have unpredictable effects on fetal

development. They may also prove hazardous once the baby is born. Before birth, drugs and other toxins can be removed from fetal blood by the maternal blood and detoxified. Once the baby is born, it may have the same levels of drugs in its blood as the mother, but it does not have the same ability to detoxify or eliminate them. As a result, drugs will have a longer lifetime and may have fatal effects.

11.4.7 Drug–drug interactions

Drugs such as **warfarin** and **methotrexate** are bound to albumin and plasma proteins in the blood, and are unavailable to interact with their targets. When another drug is taken which can compete for plasma protein binding (e.g. **sulphonamides**), then a certain percentage of previously bound drug is released, increasing the concentration of the drug and its effect.

KEY POINTS

- Pharmacodynamics is the study of how drugs interact with a molecular target to produce a pharmacological effect, whereas pharmacokinetics is the study of how a drug reaches its target in the body and how it is affected on that journey.

- The four main issues in pharmacokinetics are absorption, distribution, metabolism, and excretion.

- Orally taken drugs have to be chemically stable to survive the acidic conditions of the stomach, and metabolically stable to survive digestive and metabolic enzymes.

- Orally taken drugs must be sufficiently polar to dissolve in the GIT and blood supply, but sufficiently fatty to pass through cell membranes.

- Most orally taken drugs obey Lipinski's rule of five and have no more than seven rotatable bonds.

- Highly polar drugs can be orally active if they are small enough to pass between the cells of the gut wall, are recognized by carrier proteins, or are taken across the gut wall by pinocytosis.

- Distribution round the blood supply is rapid. Distribution to the interstitial fluid surrounding tissues and organs is also rapid if the drug is not bound to plasma proteins.

- Some drugs have to enter cells in order to reach their target.

- A certain percentage of a drug may be absorbed into fatty tissue and/or bound to macromolecules.

- Drugs entering the CNS have to cross the blood–brain barrier. Polar drugs are unable to cross this barrier unless they make use of carrier proteins or are taken across by pinocytosis.

- Some drugs cross the placental barrier into the fetus and may harm development or prove toxic in newborn babies.

11.5 Drug metabolism

When drugs enter the body, they are subject to attack from a range of metabolic enzymes. The role of these enzymes is to degrade or modify the foreign structure, such that it can be more easily excreted. As a result, most drugs undergo some form of metabolic reaction, resulting in structures known as **metabolites**. Very often these metabolites lose the activity of the original drug, but, in some cases, they may retain a certain level of activity. In exceptional cases, the metabolite may even be more active than the parent drug. Some metabolites can possess a different activity from the parent drugs, resulting in side effects or toxicity. A knowledge of drug metabolism and its possible consequences can aid the medicinal chemist in designing new drugs which do not form unacceptable metabolites. Equally, it is possible to take advantage of drug metabolism to activate drugs in the body. This is known as a prodrug strategy (see section 14.6). It is now a requirement to identify all the metabolites of a new drug before it can be approved. The structure and stereochemistry of each metabolite has to be determined and each metabolite must be tested for biological activity (section 15.1.2).

11.5.1 Phase I and phase II metabolism

The body treats drugs as foreign substances and has methods of getting rid of such chemical invaders. If the drug is polar, it will be quickly excreted by the kidneys (section 11.6). However, non-polar drugs are not easily excreted and the purpose of drug metabolism is to convert such compounds into more polar molecules that *can* be easily excreted.

Non-specific enzymes (particularly **cytochrome P450 enzymes** in the liver) are able to add polar functional groups to a wide variety of drugs. Once the polar functional group has been added, the overall drug is more polar and water soluble, and is more likely to be excreted when it passes through the kidneys. An alternative set of enzymatic reactions can reveal masked polar functional groups which might already be present in a drug. For example, there are enzymes which can demethylate a methyl ether to reveal a more polar hydroxyl group. Once again, the more polar product (metabolite) is excreted more efficiently.

These reactions are classed as phase I reactions and generally involve oxidation, reduction, and hydrolysis (see Figs. 11.3–11.9). Most of these reactions occur in the liver, but some (such as the hydrolysis of esters and amides) can also occur in the gut wall, blood plasma, and other tissues. Some of the structures most prone to oxidation are *N*-methyl groups, aromatic rings, the terminal positions of alkyl chains, and the least hindered positions of alicyclic rings. Nitro, azo, and carbonyl groups are prone to reduction by **reductases**, while amides and esters are prone to hydrolysis by **peptidases** and **esterases** respectively. For many drugs, two or more metabolic reactions might occur, resulting in different metabolites; other drugs may not be metabolized at all. A knowledge of the metabolic reactions that are possible for different functional groups allows the medicinal chemist to predict the likely metabolic products for any given drug, but only drug metabolism studies will establish whether these metabolites are really formed.

Drug metabolism has important implications when it comes to using chiral drugs, especially if the drug is to be used as a racemate. The enzymes involved in catalysing metabolic reactions will often distinguish between the two enantiomers of a chiral drug, such that one enantiomer undergoes different metabolic reactions from the other. As a result, both enantiomers of a chiral drug have to be tested separately to see what metabolites are formed. In practice, it is usually preferable to use a single enantiomer in medicine, or design the drug such that it is not asymmetric (section 13.3.8).

A series of metabolic reactions classed as phase II reactions also occur, mainly in the liver (see Figs. 11.10–11.16). Most of these reactions are **conjugation reactions**, whereby a polar molecule is attached to a suitable polar 'handle' that is already present on the drug or has been introduced by a phase I reaction. The resulting conjugate has greatly increased polarity, thus increasing its excretion rate in urine or bile even further.

Both phase I and phase II reactions can be species specific, which has implications for *in vivo* metabolic studies. In other words, the metabolites formed in an experimental animal may not necessarily be those formed in humans. A good knowledge of how metabolic reactions differ from species to species is important in determining which test animals are relevant for drug metabolism tests. Both sets of reactions can also be regioselective and stereoselective. This means that metabolic enzymes can distinguish between identical functional groups or alkyl groups located at different parts of the molecule (regioselectivity) as well as between different stereoisomers of chiral molecules (stereoselectivity).

11.5.2 Phase I transformations catalysed by cytochrome P450 enzymes

The enzymes that constitute the cytochrome P450 family are the most important metabolic enzymes and are located in liver cells. They are **haemoproteins** (containing haem and iron) and they catalyse a reaction that splits molecular oxygen, such that one of the oxygen atoms is introduced into the drug and the other ends up in water (Fig. 11.3). As a result they belong to a general class of enzymes called the **monooxygenases**.

$$\text{Drug} - \text{H} + \text{O}_2 + \text{NADPH} + \text{H}^+ \xrightarrow[\text{enzymes}]{\text{Cytochrome P450}} \text{Drug} - \text{OH} + \text{NADP}^+ + \text{H}_2\text{O}$$

FIGURE 11.3 Oxidation by cytochrome P450 enzymes.

There are at least 33 different cytochrome P450 enzymes, grouped into four main families, CYP1–CYP4. Within each family there are various subfamilies designated by a letter, and each enzyme within that subfamily is designated by a number. For example, CYP3A4 is enzyme 4 in the sub family A of the main family 3. Most drugs in current use are metabolized by five primary CYP enzymes (CYP3A, CYP2D6, CYP2C9, CYP1A2, and CYP2E1). The isozyme CYP3A4 is particularly important in drug metabolism and is responsible for the metabolism of most drugs. The reactions catalysed by cytochrome P450

enzymes are shown in Figs. 11.4 and 11.5 and can involve the oxidation of carbon, nitrogen, phosphorus, sulphur, and other atoms.

Oxidation of carbon atoms can occur if the carbon atom is either exposed (i.e. easily accessible to the enzyme) or activated (Fig. 11.4). For example, methyl substituents on the carbon skeleton of a drug are often easily accessible and are oxidized to form alcohols, which may be oxidized further to carboxylic acids. In the case of longer-chain substituents, the terminal carbon and the penultimate carbon are the most exposed carbons in

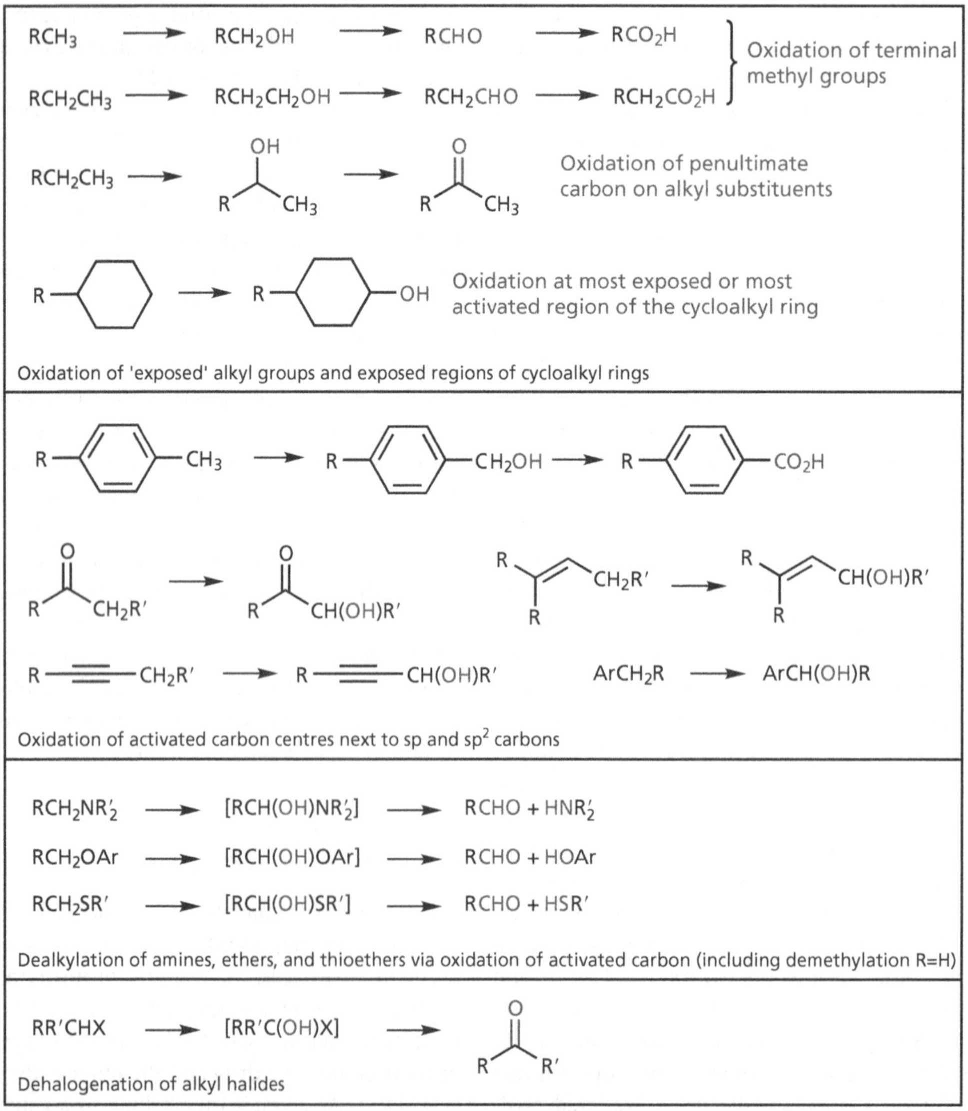

FIGURE 11.4 Oxidative reactions catalysed by cytochrome P450 enzymes on saturated carbon centres.

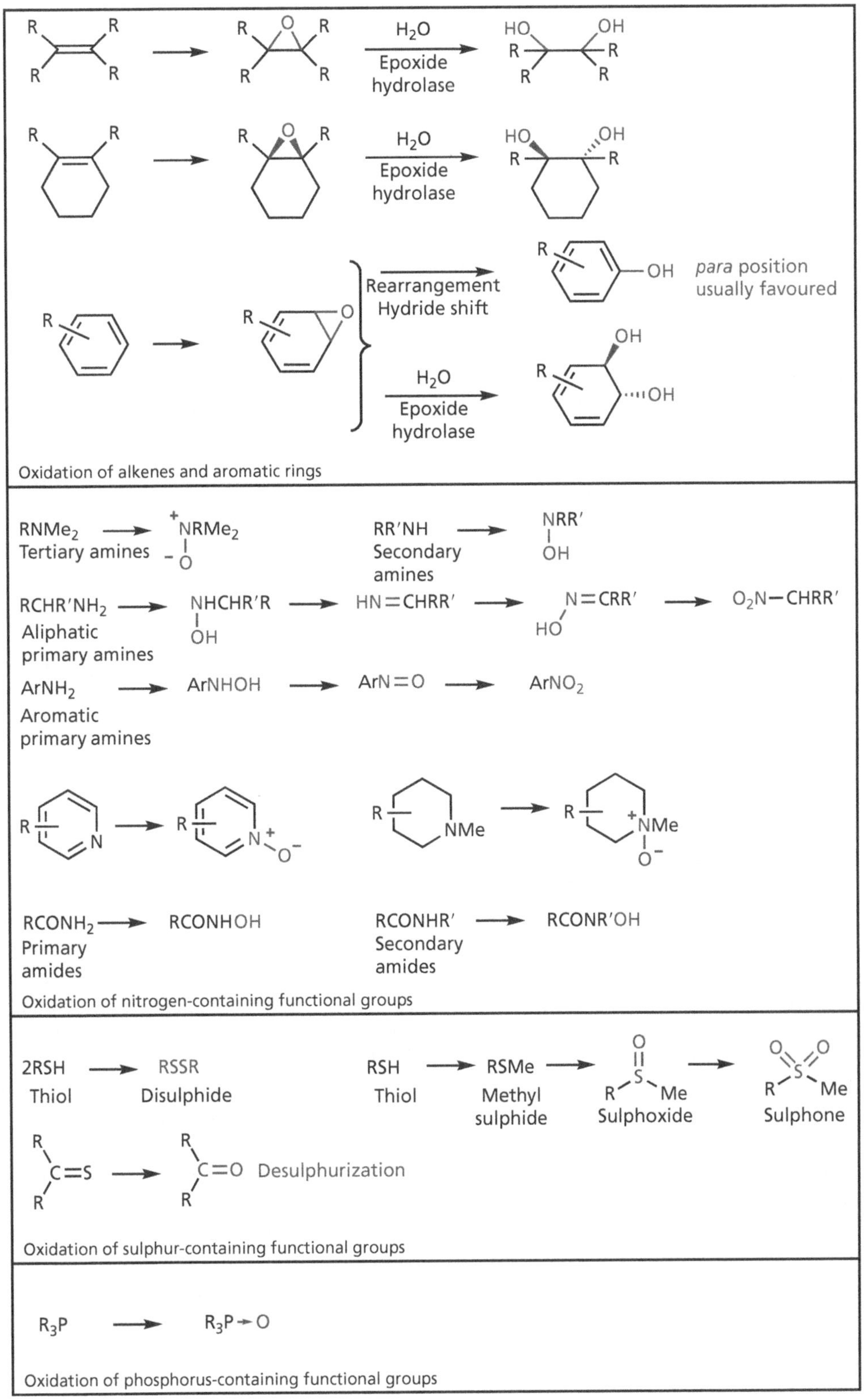

FIGURE 11.5 Oxidative reactions catalysed by cytochrome P450 enzymes on heteroatoms and unsaturated carbon centres.

the chain and are both susceptible to oxidation. If an aliphatic ring is present, the most exposed region is the part most likely to be oxidized.

Activated carbon atoms next to an sp^2 carbon centre (i.e. allylic or benzylic positions) or an sp carbon centre (i.e. a propynylic position) are more likely to be oxidized than exposed carbon atoms (Fig. 11.4). Carbon atoms which are alpha to a heteroatom are also activated and prone to oxidation. In this case, hydroxylation results in an unstable metabolite that is immediately hydrolysed resulting in the dealkylation of amines, ethers, and thioethers, or the dehalogenation of alkyl halides. The aldehydes which are formed from these reactions generally undergo further oxidation to carboxylic acids by aldehyde dehydrogenases (section 11.5.4). Tertiary amines are found to be more reactive to oxidative dealkylation than secondary amines because of their greater basicity, while O-demethylation of aromatic ethers is faster than O-dealkylation of larger alkyl groups. O-Demethylation is important to the analgesic activity of **codeine** (section 24.5).

Cytochrome P450 enzymes can catalyse the oxidation of unsaturated sp^2 and sp carbon centres present in alkenes, alkynes, and aromatic rings (Fig. 11.5). In the case of alkenes, a reactive epoxide is formed which is deactivated by the enzyme **epoxide hydrolase** to form a diol. In some cases, the epoxide may evade the enzyme. If this happens, it can act as an alkylating agent and react with nucleophilic groups present in proteins or nucleic acids, leading to toxicity. The oxidation of an aromatic ring results in a similarly reactive epoxide intermediate which can have several possible fates. It may undergo a rearrangement reaction involving a hydride transfer to form a phenol, normally at the *para* position. Alternatively, it may be deactivated by epoxide hydrolase to form a diol, or react with **glutathione S-transferase** to form a conjugate (section 11.5.5). If the epoxide intermediate evades these enzymes, it may act as an alkylating agent and prove toxic. Electron-rich aromatic rings are likely to be epoxidized more quickly than those with electron-withdrawing substituents, and this has consequences for drug design.

Tertiary amines are oxidized to N-oxides as long as the alkyl groups are not sterically demanding. Primary and secondary amines are also oxidized to N-oxides, but these are rapidly converted to hydroxylamines and beyond. Aromatic primary amines are also oxidized in stages to aromatic nitro groups—a process which is related to the toxicity of aromatic amines, as highly electrophilic intermediates are formed which can alkylate proteins or nucleic acids. Aromatic primary amines can also be methylated in a phase II reaction (section 11.5.5) to a secondary amine which can then undergo phase I oxidation to produce formaldehyde and primary hydroxylamines. Primary and secondary amides can be

oxidized to hydroxylamides. These functional groups have also been linked with toxicity and carcinogenicity. Thiols can be oxidized to disulphides. There is evidence that thiols can be methylated to methyl sulphides, which are then oxidized to sulphides and sulphones.

For additional information see Web article 5: The design of a serotonin antagonist as a possible anxiolytic agent on the Online Resource Centre at www.oxfordtextbooks.co.uk/orc/patrick6e/

11.5.3 Phase I transformations catalysed by flavin-containing monooxygenases

Another group of metabolic enzymes present in the endoplasmic reticulum of liver cells consists of the **flavin-containing monooxygenases.** These enzymes are chiefly responsible for metabolic reactions involving oxidation at nucleophilic nitrogen, sulphur, and phosphorus atoms, rather than at carbon atoms. Several examples are given in Fig. 11.6. Many of these reactions are also catalysed by cytochrome P450 enzymes.

11.5.4 Phase I transformations catalysed by other enzymes

There are several oxidative enzymes in various tissues around the body that are involved in the metabolism of endogenous compounds, but can also play a role in drug metabolism (Fig. 11.7). For example, **monoamine oxidases** are involved in the deamination of catecholamines (section 23.5), but have been observed to oxidize some drugs. Other important oxidative enzymes include alcohol dehydrogenases and aldehyde dehydrogenases. The aldehydes formed by the action of alcohol dehydrogenases on primary alcohols are usually not observed, as they are converted to carboxylic acids by aldehyde dehydrogenases.

Reductive phase I reactions are less common than oxidative reactions, but reductions of aldehyde, ketone, azo, and nitro functional groups have been observed in specific drugs (Fig. 11.8). Many of the oxidation reactions described for heteroatoms in Figs. 11.5–11.7 are reversible, and are catalysed by reductase enzymes. Cytochrome P450 enzymes are involved in catalysing some of these reactions. Remember: enzymes can catalyse a reaction in both directions, depending on the nature of the substrate. So although cytochrome P450 enzymes are predominantly oxidative enzymes, it is possible for them to catalyse some reductions.

The hydrolysis of esters and amides is a common metabolic reaction, catalysed by **esterases** and **peptidases** respectively (Fig. 11.9). These enzymes are present in various organs of the body including the liver. Amides

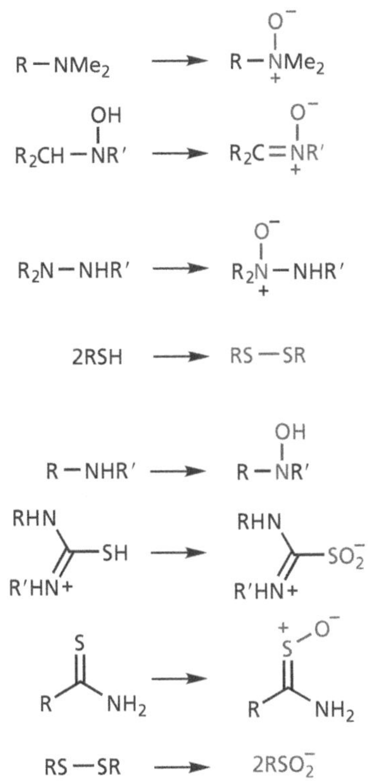

FIGURE 11.6 Phase I reactions catalysed by flavin monooxygenases.

tend to be hydrolysed more slowly than esters. The presence of electron-withdrawing groups can increase the susceptibility of both amides and esters to hydrolysis.

11.5.5 Phase II transformations

Most phase II reactions are **conjugation reactions** catalysed by transferase enzymes. The resulting conjugates are usually inactive, but there are exceptions to this rule. Glucuronic acid conjugation is the most common of these reactions. Phenols, alcohols, hydroxylamines, and carboxylic acids form **O-glucuronides** by reaction with **UDFP-glucuronate** such that a highly polar glucuronic acid molecule is attached to the drug (Fig. 11.10). The resulting conjugate is excreted in the urine, but may also be excreted in the bile if the molecular weight is over 300.

A variety of other functional groups such as sulphonamides, amides, amines, and thiols (Fig. 11.11) can react to form *N*- or *S*-glucuronides. *C*-Glucuronides are also possible in situations where there is an activated carbon centre next to carbonyl groups.

Another form of conjugation is sulphate conjugation (Fig. 11.12). This is less common than glucuronidation and is restricted mainly to phenols, alcohols, arylamines, and *N*-hydroxy compounds. The reaction is catalysed by **sulphotransferases** using the cofactor **3′-phosphoadenosine 5′-phosphosulphate** as the sulphate source. Primary and secondary amines, secondary alcohols, and phenols form stable conjugates, whereas primary alcohols form reactive sulphates which can act as toxic alkylating agents. Aromatic hydroxylamines and hydroxylamides also form unstable sulphate conjugates that can be toxic.

Drugs bearing a carboxylic acid group can become conjugated to amino acids by the formation of a peptide link. In most animals, glycine conjugates are generally formed, but L-glutamine is the most common amino acid used for conjugation in primates. The carboxylic acid present in the drug is first activated by formation of a coenzyme A thioester which is then linked to the amino acid (Fig. 11.13).

Electrophilic functional groups such as epoxides, alkyl halides, sulphonates, disulphides, and radical species can react with the nucleophilic thiol group of the tripeptide **glutathione** to give glutathione conjugates which can be subsequently transformed to **mercapturic acids** (Fig. 11.14). The glutathione conjugation reaction can take place in most cells, especially those in the liver and kidney, and is catalysed by **glutathione transferase.** This conjugation reaction is important in detoxifying potentially dangerous environmental toxins or electrophilic alkylating agents formed by phase I reactions (Fig. 11.15). Glutathione conjugates are often excreted in the bile, but are more usually converted to mercapturic acid conjugates before excretion.

Not all phase II reactions result in increased polarity. Methylation and acetylation are important phase II reactions which usually *decrease* the polarity of the drug (Fig. 11.16). An important exception is the methylation of pyridine rings, which leads to polar quaternary salts. The functional groups that are susceptible to

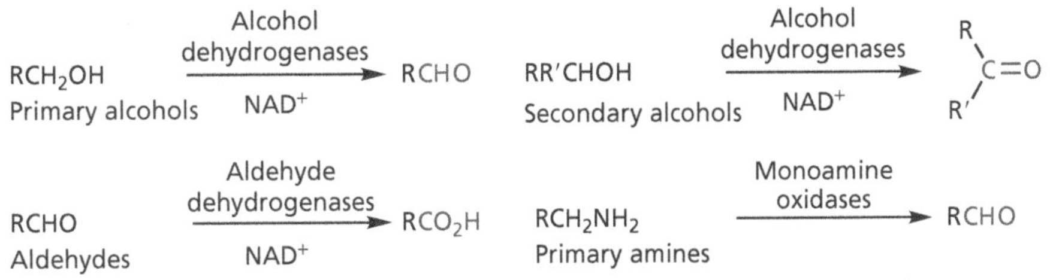

FIGURE 11.7 Phase I oxidative reactions catalysed by miscellaneous enzymes.

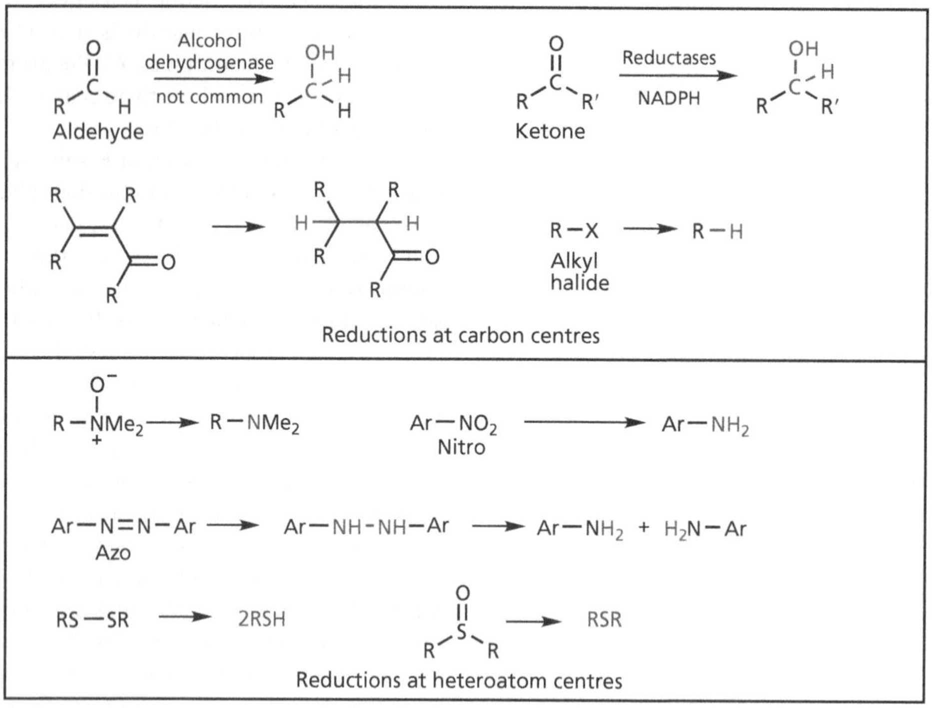

FIGURE 11.8 Phase I reductive reactions.

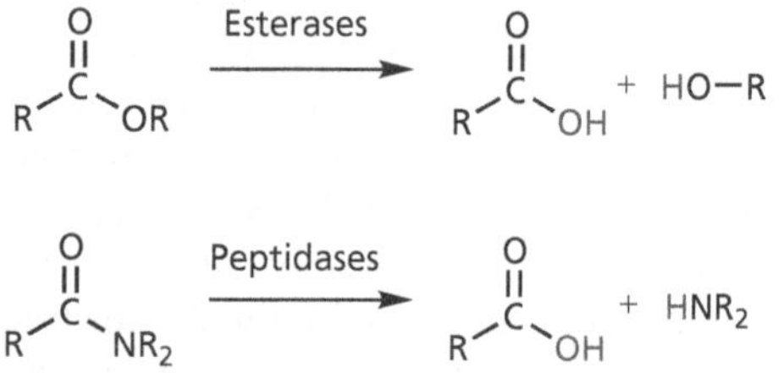

FIGURE 11.9 Hydrolysis of esters and amides.

meta position of catechols (section 23.5). It should be pointed out, however, that methylation occurs less frequently than other conjugation reactions and is more important in biosynthetic pathways or the metabolism of endogenous compounds.

It is possible for drugs bearing carboxylic acids to become conjugated with **cholesterol.** Cholesterol conjugates can also be formed with drugs bearing an ester group by means of a transesterification reaction. Some drugs with an alcohol functional group form conjugates with fatty acids by means of an ester link.

methylation are phenols, amines, and thiols. Primary amines are also susceptible to acetylation. The enzyme cofactors involved in contributing the methyl group or acetyl group are **S-adenosyl methionine** and **acetyl SCoA** respectively. Several methyltransferase enzymes are involved in the methylation reactions. The most important enzyme for *O*-methylations is **catechol *O*-methyltransferase**, which preferentially methylates the

11.5.6 Metabolic stability

Ideally, a drug should be resistant to drug metabolism because the production of metabolites complicates drug therapy (see Box 11.1). For example, the metabolites formed will usually have different properties from the

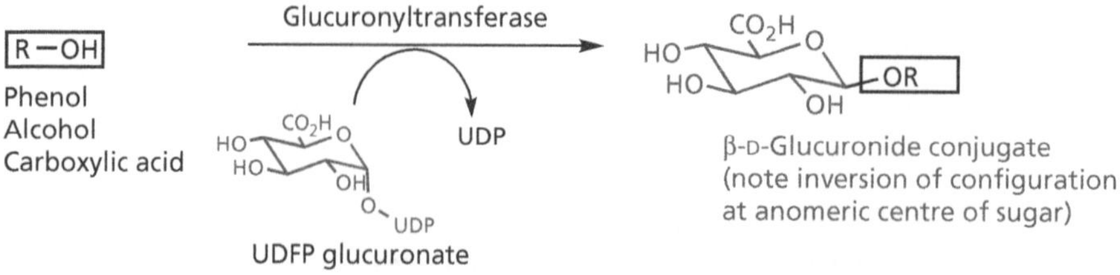

FIGURE 11.10 Glucuronidation of alcohols, phenols, and carboxylic acids.

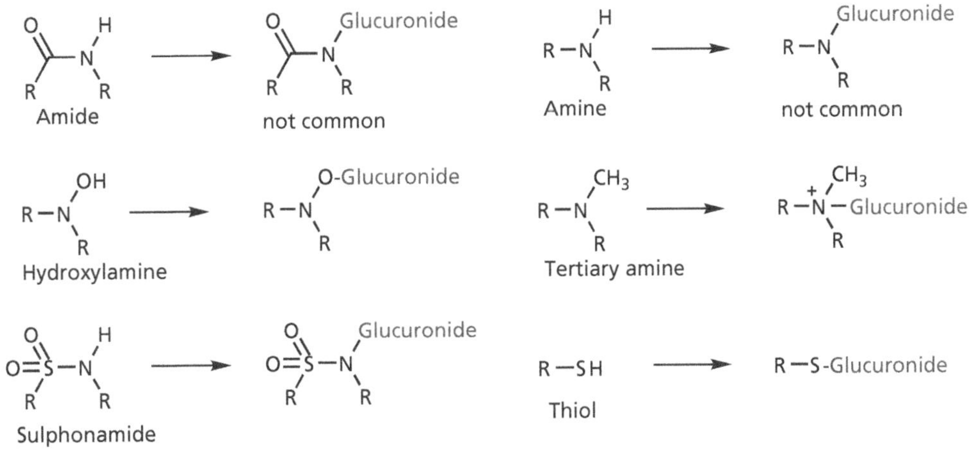

FIGURE 11.11 Glucuronidation of miscellaneous functional groups.

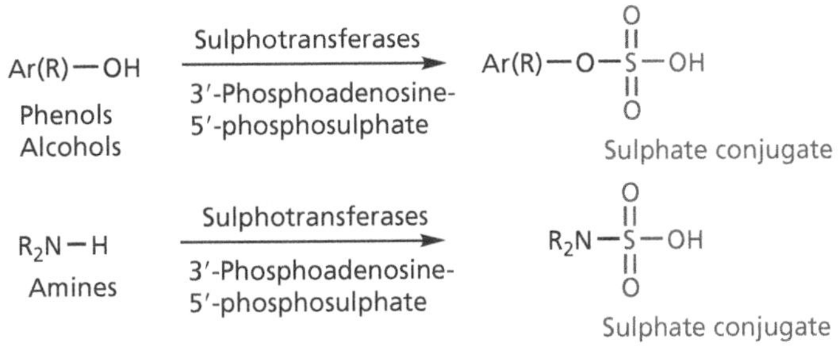

FIGURE 11.12 Examples of sulphoconjugation phase II reactions.

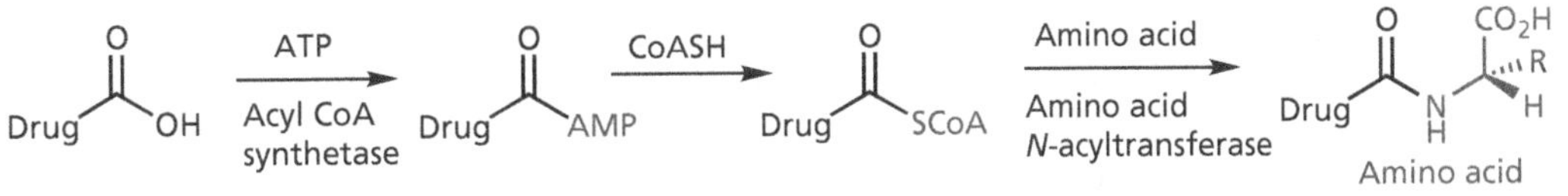

FIGURE 11.13 Formation of amino acid conjugates.

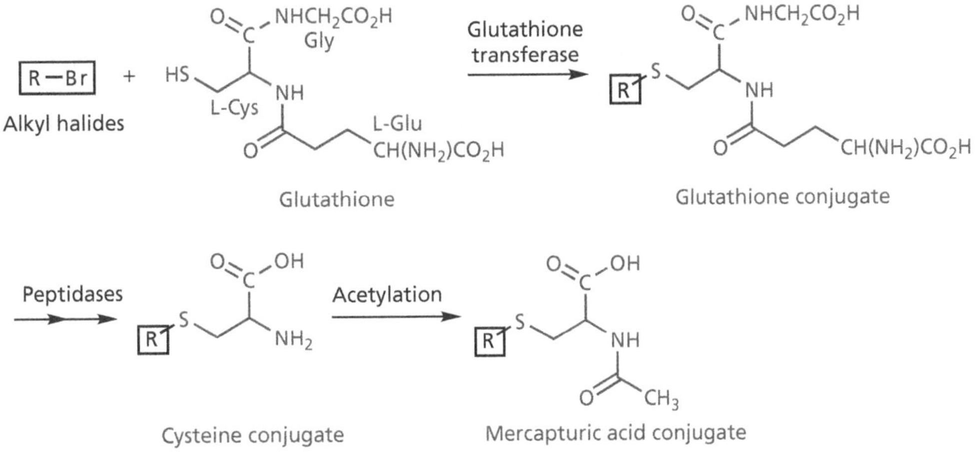

FIGURE 11.14 Formation of glutathione and mercapturic acid conjugates from an alkyl halide.

665

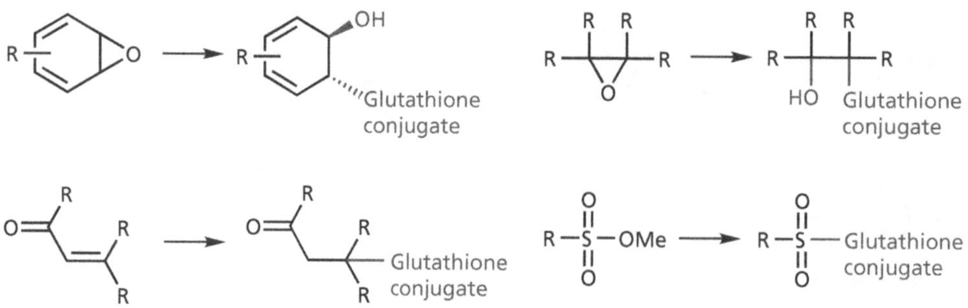

FIGURE 11.15 Formation of glutathione conjugates (Glu-Cys-Gly) with electrophilic groups.

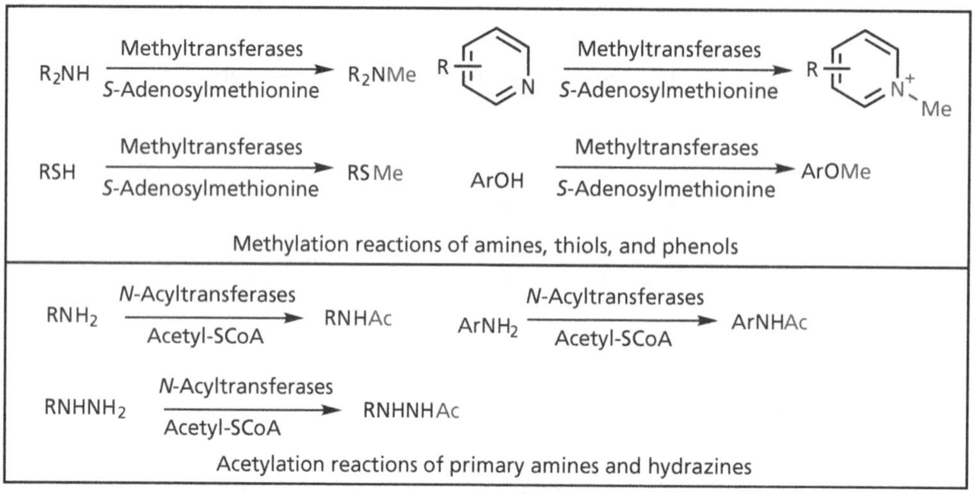

FIGURE 11.16 Methylation and acetylation.

original drug. In some cases, activity may be lost. In others, the metabolite may prove to be toxic. For example, the metabolites of **paracetamol** cause liver toxicity, and the carcinogenic properties of some polycyclic hydrocarbons are due to the formation of epoxides.

Another problem arises from the fact that the activity of metabolic enzymes varies from individual to individual. This is especially true of the cytochrome P450 enzymes with at least a 10-fold variability for the most important isoform CYP3A4. Individuals may even lack particular isoforms. For example, 8% of Americans lack the CYP2D6 isoform, which means that drugs normally metabolized by this enzyme can rise to toxic levels. Examples of drugs that are normally metabolized by this isozyme are **desipramine**, **haloperidol**, and **tramadol**. Some prodrugs require metabolism by CYP2D6 in order to be effective. For example, the analgesic effects of **codeine** are due to its metabolism by CYP2D6 to morphine. Therefore, codeine is ineffective in patients lacking this isozyme. The profile of these enzymes in different patients can vary, resulting in a difference in the way a drug is metabolized. As a result, the amount of drug that can be safely administered also varies.

Differences across populations can be quite significant, resulting in different countries having different recommended dose levels for particular drugs. For example, the rate at which the antibacterial agent **isoniazid** is acetylated and deactivated varies amongst populations. Asian populations acylate the drug at a fast rate, whereas 45–65% of Europeans and North Americans have a slow rate of acylation. **Pharmacogenomics** is the study of genetic variations between individuals and the effect that has on individual responses to drugs. In the future, it is possible that 'fingerprints' of an individual's genome may allow better prediction of which drugs would be suitable for that individual, and which drugs might produce unacceptable side effects—an example of **personalized medicine**. This in turn may avoid drugs having to be withdrawn from the market as a result of rare toxic side effects.

Another complication involving drug metabolism and drug therapy relates to the fact that cytochrome P450 activity can be affected by other chemicals. For example, certain foods have an influence. Brussels sprouts and cigarette smoke enhance activity, whereas grapefruit juice inhibits activity. This can have a significant effect on the activity of drugs metabolized by cytochrome P450

BOX 11.1 Metabolism of an antiviral agent

Indinavir is an antiviral agent used in the treatment of HIV and is prone to metabolism, resulting in seven different metabolites (Fig. 1). Studies have shown that the CYP3A subfamily of cytochrome P450 enzymes is responsible for six of these metabolites. The metabolites concerned arise from N-dealkylation of the piperazine ring, N-oxidation of the pyridine ring, para-hydroxylation of the phenyl ring, and hydroxylation of the indane ring. The seventh metabolite is a glucuronide conjugate of the pyridine ring. All these reactions occur individually to produce five separate metabolites. The remaining two metabolites arise from two or more metabolic reactions taking place on the same molecule.

The major metabolites are those resulting from N-dealkylation. As a result, research has been carried out to try and design indinavir analogues that are resistant to this reaction.

For example, structures having two methyl substituents on the activated carbon next to pyridine have been effective in blocking N-dealkylation (Fig. 2).

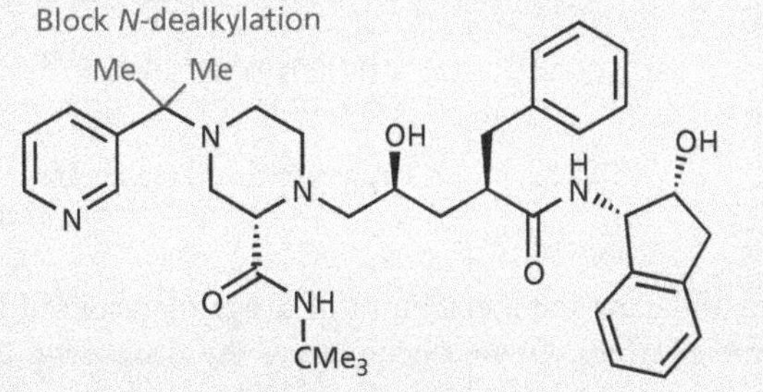

FIGURE 2 Analogue of indinavir resistant to N-dealkylation.

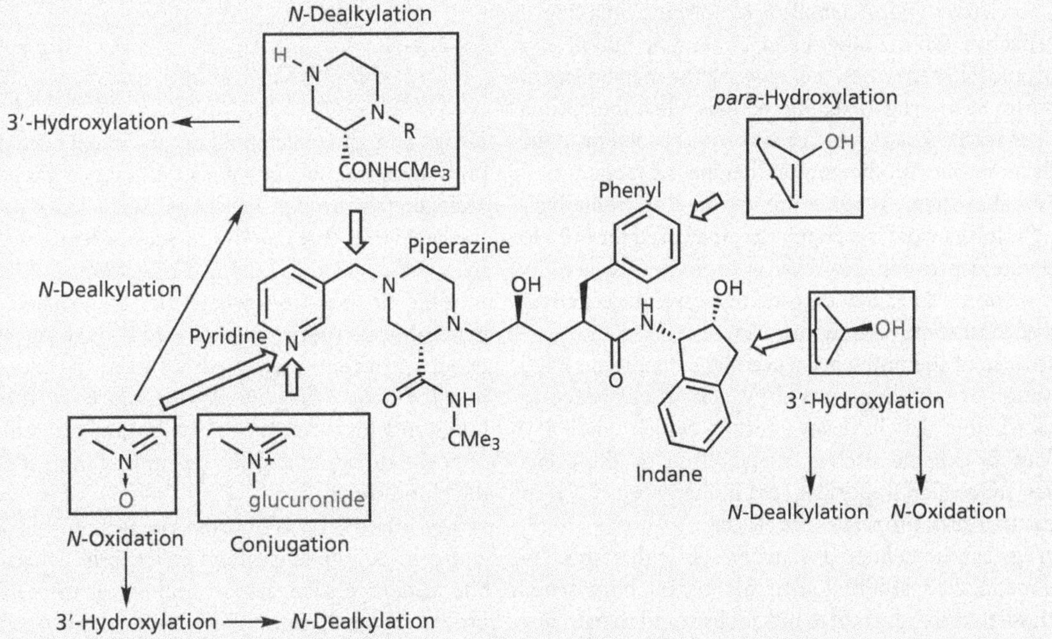

FIGURE 1 Metabolism of indinavir.

enzymes. For example, the immunosuppressant drug **ciclosporin** and the dihydropyridine hypotensive agents are more efficient when taken with grapefruit juice, because their metabolism is reduced. However, serious toxic effects can arise if the antihistamine agent **terfenadine** is taken with grapefruit juice. Terfenadine is actually a prodrug and is metabolized to the active agent **fexofenadine** (Fig. 11.17). If metabolism is inhibited by grapefruit juice, terfenadine persists in the body and can cause serious cardiac toxicity. As a result, fexofenadine itself is now favoured over terfenadine and is marketed as **Allegra**.

Certain drugs are also capable of inhibiting or promoting cytochrome P450 enzymes, leading to a phenomenon known as **drug–drug interactions** where the presence of one drug affects the activity of another. For example, several antibiotics can act as cytochrome P450 inhibitors

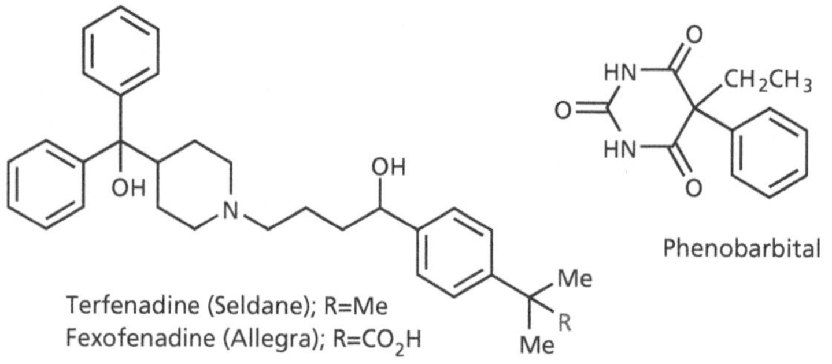

FIGURE 11.17 Drugs which are metabolized by cytochrome P450 enzymes or affect the activity of cytochrome P450 enzymes.

and will slow the metabolism of drugs metabolized by these enzymes. Other examples are the drug–drug interactions that occur between the anticoagulant **warfarin** (section 26.9.1.1) and the barbiturate **phenobarbital** (Fig. 11.17), or between warfarin and the anti-ulcer drug **cimetidine** (section 25.2.7.3).

Phenobarbital stimulates cytochrome P450 enzymes and accelerates the metabolism of warfarin, making it less effective. On the other hand, cimetidine inhibits cytochrome P450 enzymes, thus slowing the metabolism of warfarin. Such drug–drug interactions affect the plasma levels of warfarin and could cause serious problems if the levels move outwith the normal therapeutic range.

Herbal medicine is not immune from this problem either. **St. John's wort** is a popular remedy used for mild to moderate depression. However, it promotes the activity of cytochrome P450 enzymes and decreases the effectiveness of contraceptives and warfarin.

Because of the problems caused by cytochrome P450 activation or inhibition, new drugs are usually tested to check whether they have any effect on cytochrome P450 activity, or are, themselves, metabolized by these enzymes. Indeed, an important goal in many projects is to ensure that such properties are lacking.

Drugs can be defined as hard or soft with respect to their metabolic susceptibility. In this context, **hard drugs** are those that are resistant to metabolism and remain unchanged in the body. **Soft drugs** are designed to have a predictable, controlled metabolism where they are inactivated to non-toxic metabolites and excreted. A group is normally incorporated which is susceptible to metabolism, but will survive sufficiently long to allow the drug to achieve what it is meant to do before it is metabolized and excreted. Drugs such as these are also called **antedrugs**.

11.5.7 The first pass effect

Drugs that are taken orally pass directly to the liver once they enter the blood supply. Here, they are exposed to drug metabolism before they are distributed around the rest of the body, and so a certain percentage of the drug

is transformed before it has the chance to reach its target. This is known as the first pass effect. Drugs that are administered in a different fashion (e.g. injection or inhalation) avoid the first pass effect and are distributed around the body before reaching the liver. Indeed, a certain proportion of the drug may not pass through the liver at all, but may be taken up in other tissues and organs en route.

11.6 Drug excretion

Drugs and their metabolites can be excreted from the body by a number of routes. Volatile or gaseous drugs are excreted through the lungs. Such drugs pass out of the capillaries that line the air sacs (alveoli) of the lungs, then diffuse through the cell membranes of the alveoli into the air sacs, from where they are exhaled. Gaseous **general anaesthetics** are excreted in this way and move down a concentration gradient from the blood supply into the lungs. They are also administered through the lungs, in which case the concentration gradient is in the opposite direction and the gas moves from the lungs to the blood supply.

The **bile duct** travels from the liver to the intestines and carries a greenish fluid called **bile** which contains bile acids and salts that are important to the digestion process. A small number of drugs are diverted from the blood supply back into the intestines by this route. Since this happens from the liver, any drug eliminated in this way has not been distributed round the body. Therefore, the amount of drug distributed is less than that absorbed. However, once the drug has entered the intestine, it can be reabsorbed, so it has another chance.

It is possible for as much as 10–15% of a drug to be lost through the skin in sweat. Drugs can also be excreted through saliva and breast milk, but these are minor excretion routes compared with the kidneys. There are concerns, however, that mothers may be passing on drugs such as **nicotine** to their baby through breast milk.

The **kidneys** are the principal route by which drugs and their metabolites are excreted (Fig. 11.18). The kidneys

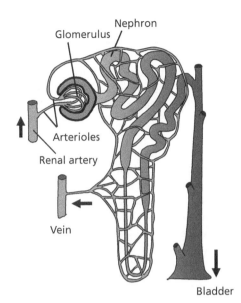

FIGURE 11.18 Excretion by the kidneys.

filter the blood of waste chemicals and these chemicals are subsequently removed in the urine. Drugs and their metabolites are excreted by the same mechanism.

Blood enters the kidneys by means of the **renal artery**. This divides into a large number of capillaries, each one of which forms a knotted structure called a **glomerulus** that fits into the opening of a duct called a **nephron**. The blood entering these glomeruli is under pressure, and so plasma is forced through the pores in the capillary walls into the nephron, carrying with it any drugs and metabolites that might be present. Any compounds that are too big to pass through the pores, such as plasma proteins and red blood cells, remain in the capillaries with the remaining plasma. Note that this is a filtration process, so it does not matter whether the drug is polar or hydrophobic: all drugs and drug metabolites will be passed equally efficiently into the nephron. However, this does not mean that every compound will be *excreted* equally efficiently, because there is more to the process than simple filtration.

The filtered plasma and chemicals now pass through the nephron on their route to the bladder. However, only a small proportion of what starts that journey actually finishes it. This is because the nephron is surrounded by a rich network of blood vessels carrying the filtered blood away from the glomerulus, permitting much of the contents of the nephron to be reabsorbed into the blood supply. Most of the water that was filtered into the nephron is quickly reabsorbed through pores in the nephron cell membrane which are specific for water molecules, and bar the passage of ions or other molecules. These pores are made up of protein molecules called **aquaporins**. As water is reabsorbed, drugs and other agents are concentrated in the nephron and a concentration gradient

is set up. There is now a driving force for compounds to move back into the blood supply down the concentration gradient. However, this can only happen if the drug is sufficiently hydrophobic to pass through the cell membranes of the nephron. This means that hydrophobic compounds are efficiently reabsorbed back into the blood, whereas polar compounds remain in the nephron and are excreted. This process of excretion explains the importance of drug metabolism to drug excretion. Drug metabolism creates polar metabolites which are less likely to be reabsorbed from the nephrons.

Some drugs are actively transported from blood vessels into the nephrons. This process is called **facilitated transport**, and is important in the excretion of penicillins (section 19.5.1.9).

KEY POINTS

- Drugs are exposed to enzyme-catalysed reactions which modify their structure. This is called drug metabolism and can take place in various tissues. However, most reactions occur in the liver.

- Orally taken drugs are subject to the first pass effect.

- Drugs administered by methods other than the oral route avoid the first pass effect.

- Phase I metabolic reactions typically involve the addition or exposure of a polar functional group. Cytochrome P450 enzymes present in the liver carry out important phase I oxidation reactions. The types of cytochrome P450 enzymes present vary between individuals, leading to varying rates of drug metabolism.

- The activity of cytochrome P450 enzymes can be affected by food, chemicals, and drugs, resulting in drug–drug interactions and possible side effects.

- Phase II metabolic reactions involve the addition of a highly polar molecule to a functional group. The resulting conjugates are more easily excreted.

- Drug excretion can take place through sweat, exhaled air, or bile, but most excretion takes place through the kidneys.

- The kidneys filter blood such that drugs and their metabolites enter nephrons. Non-polar substances are reabsorbed into the blood supply, but polar substances are retained in the nephrons and excreted in the urine.

11.7 Drug administration

There are a large variety of ways in which drugs can be administered and many of these avoid some of the problems associated with oral administration. The main routes are: oral, sublingual, rectal, epithelial, inhalation, and injection. The method chosen will depend on the target organ and the pharmacokinetics of the drug.

11.7.1 Oral administration

Orally administered drugs are taken by mouth. This is the preferred option for most patients, so there is more chance that the patient will comply with the drug regime and complete the course. However, the oral route places the greatest demands on the chemical and physical properties of the drug, as described in the previous sections of this chapter.

Drugs given orally can be taken as pills, capsules, or solutions. Drugs taken in solution are absorbed more quickly and a certain percentage may even be absorbed through the stomach wall. For example, approximately 25–33% of **alcohol** is absorbed into the blood supply from the stomach; the rest is absorbed from the upper intestine. Drugs taken as pills or capsules are mostly absorbed in the upper intestine. The rate of absorption is partly determined by the rate at which the pills and capsules dissolve. This, in turn, depends on such factors as particle size and crystal form. In general, about 75% of an orally administered drug is absorbed into the body within 1–3 hours. Specially designed pills and capsules can remain intact in the stomach to help protect acid-labile drugs from stomach acids. The containers then degrade once they reach the intestine.

Care has to be taken if drugs interact with food. For example, **tetracycline** binds strongly to calcium ions, which inhibits absorption, so foods such as milk should be avoided. Some drugs bind other drugs and prevent absorption. For example, **colestyramine** (used to lower cholesterol levels) binds to **warfarin** and also to the thyroid drug **levothyroxine sodium**, so these drugs should be taken separately.

11.7.2 Absorption through mucous membranes

Some drugs can be absorbed through the mucous membranes of the mouth or nose, thus avoiding the digestive and metabolic enzymes encountered during oral administration. For example, heart patients take **glyceryl trinitrate** (Fig. 11.19) by placing it under the tongue (sublingual administration). The opiate analgesic **fentanyl** (Fig. 11.19) has been given to children in the form of a lollipop, and is absorbed through the mucous membranes of the mouth. The Incas absorbed **cocaine** sublingually by chewing coca leaves.

Nasal decongestants are absorbed through the mucous membranes of the nose. Cocaine powder is absorbed in this way when it is sniffed, as is **nicotine** in the form of snuff. Nasal sprays have been used to administer analogues of peptide hormones such as **antidiuretic hormone**. These drugs would be quickly degraded if taken orally.

Eye drops are used to administer drugs directly to the eye and thus reduce the possibility of side effects elsewhere in the body. For example, the eye condition known as glaucoma is treated in this way. Nevertheless, some absorption into the blood supply can still occur and some asthmatic patients suffer bronchospasms when taking **timolol** eye drops.

11.7.3 Rectal administration

Some drugs are administered rectally as **suppositories**, especially if the patient is unconscious, vomiting, or unable to swallow. However, there are several problems associated with rectal administration: the patient may suffer membrane irritation, and, although drug absorption is efficient, it can be unpredictable. It is not the most popular of methods with patients either!

11.7.4 Topical administration

Topical drugs are those which are applied to the skin. For example, steroids are applied topically to treat local skin irritations. It is also possible for some of the drug to be absorbed through the skin (**transdermal absorption**) and to enter the blood supply, especially if the drug is lipophilic. **Nicotine patches** work in this fashion, as do hormone replacement therapies for **estrogen**. Drugs are absorbed by this method at a steady rate, and avoid the acidity of the stomach, or the enzymes in the gut or gut wall. Other drugs that have been applied in this way include the analgesic **fentanyl** and the antihypertensive agent **clonidine**. Once applied, the drug is slowly released from the patch and absorbed through the skin into the blood supply over several days. As a result, the level of drug remains relatively constant over that period.

A technique known as **iontophoresis** is being investigated as a means of topical administration. Two miniature electrode patches are applied to the skin and linked

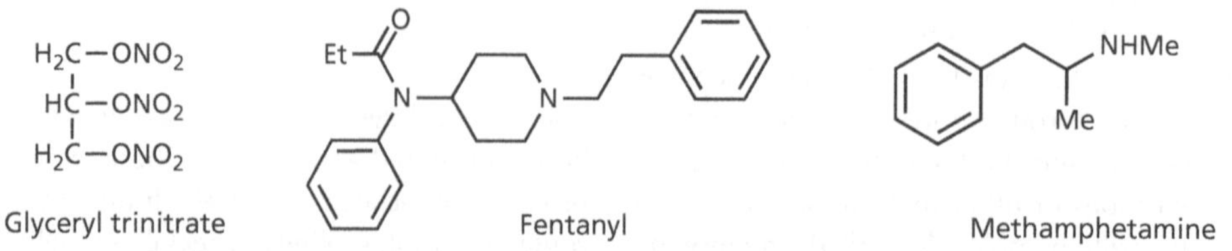

FIGURE 11.19 Glyceryl trinitrate, fentanyl, and methamphetamine.

to a reservoir of the drug. A painless pulse of electricity is applied, which has the effect of making the skin more permeable to drug absorption. By timing the electrical pulses correctly, the drug can be administered such that fluctuations in blood levels are kept to a minimum. Similar devices are being investigated which use ultrasound to increase skin permeability.

11.7.5 Inhalation

Drugs administered by inhalation avoid the digestive and metabolic enzymes of the GIT or liver. Once inhaled, the drugs are absorbed through the cell linings of the respiratory tract into the blood supply. Assuming the drug is able to pass through the hydrophobic cell membranes, absorption is rapid and efficient because the blood supply is in close contact with the cell membranes of the lungs. For example, **general anaesthetic gases** are small, highly lipid-soluble molecules which are absorbed almost as fast as they are inhaled.

Non-gaseous drugs can be administered as **aerosols**. This is how anti-asthmatic drugs are administered and it allows them to be delivered to the lungs in far greater quantities than if they were given orally or by injection. In the case of anti-asthmatics, the drug is made sufficiently polar that it is poorly absorbed into the bloodstream. This localizes it in the airways and lowers the possibility of side effects elsewhere in the body (e.g. action on the heart). However, a certain percentage of an inhaled drug is inevitably swallowed and can reach the blood supply by the oral route. This may lead to side effects. For example, tremor is a side effect of the anti-asthmatic **salbutamol** as a result of the drug reaching the blood supply.

Several drugs of abuse are absorbed through inhalation or smoking (e.g. **nicotine**, **cocaine**, **marijuana**, **heroin**, and **methamphetamine** (Fig. 11.19)). Smoking is a particularly hazardous method of taking drugs. A normal cigarette is like a mini-furnace producing a complex mixture of potentially carcinogenic compounds, especially from the tars present in tobacco. These are not absorbed into the blood supply but coat the lung tissue, leading to long-term problems such as lung cancer. The tars in cannabis are considerably more dangerous than those in tobacco. If cannabis is to be used in medicine, safer methods of administration are desirable (i.e. inhalers).

11.7.6 Injection

Drugs can be introduced into the body by intravenous, intramuscular, subcutaneous, or intrathecal injection. Injection of a drug produces a much faster response than oral administration because the drug reaches the blood supply more quickly. The levels of drug administered are also more accurate because absorption by the oral route

has a level of unpredictability due to the first pass effect. Injecting a drug, however, is potentially more hazardous. For example, some patients may have an unexpected reaction to a drug and there is little that can be done to reduce the levels once the drug has been injected. Such side effects would be more gradual and treatable if the drug was given orally. Furthermore, sterile techniques are essential when giving injections to avoid the risks of bacterial infection or of transmitting hepatitis or AIDS from a previous patient. Finally, there is a greater risk of receiving an overdose when injecting a drug.

The **intravenous** route involves injecting a solution of the drug directly into a vein. This method of administration is not particularly popular with patients, but it is a highly effective method of administering drugs in accurate doses and it is the fastest of the injection methods. However, it is also the most hazardous method of injection. Since its effects are rapid, the onset of any serious side effects or allergies is also rapid. It is important, therefore, to administer the drug as slowly as possible and to monitor the patient closely. An intravenous drip allows the drug to be administered in a controlled manner such that there is a steady level of drug in the system. The local anaesthetic **lidocaine** is given by intravenous injection. Drugs that are dissolved in oily liquids cannot be given by intravenous injection as this may result in the formation of blood clots.

The **intramuscular** route involves injecting drugs directly into muscle, usually in the arm, thigh, or buttocks. Drugs administered in this way do not pass round the body as rapidly as they would if given by intravenous injection, but they are still absorbed faster than by oral administration. The rate of absorption depends on various factors such as the diffusion of the drug, blood supply to the muscle, the solubility of the drug, and the volume of the injection. Local blood flow can be reduced by adding adrenaline to constrict blood vessels. Diffusion can be slowed by using a poorly absorbed salt, ester, or complex of the drug (see also section 14.6.2). The advantage of slowing down absorption is in prolonging activity. For example, oily suspensions of steroid hormone esters are used to slow absorption. Drugs are often administered by intramuscular injection when they are unsuitable for intravenous injection, and so it is important to avoid injecting into a vein.

Subcutaneous injection involves injecting the drug under the surface of the skin. Absorption depends on factors such as how fast the drug diffuses, the level of blood supply to the skin, and the ability of the drug to enter the blood vessels. Absorption can be slowed by the same methods described for intramuscular injection. Drugs which can act as irritants should not be administered in this way as they can cause severe pain and may damage local tissues.

Intrathecal injection means that the drug is injected into the spinal cord. Antibacterial agents that do not normally cross the blood–brain barrier are often administered in this way. Intrathecal injections are also used to administer **methotrexate** in the treatment of childhood leukaemia in order to prevent relapse in the CNS.

Intraperitoneal injection involves injecting drugs directly into the abdominal cavity. This is very rarely used in medicine, but it is a method of injecting drugs into animals during preclinical tests.

11.7.7 Implants

Continuous osmotically driven minipumps for **insulin** have been developed which are implanted under the skin. The pumps monitor the level of insulin in the blood and release the hormone as required to keep levels constant. This avoids the problem of large fluctuations in insulin levels associated with regular injections.

Gliadel is a wafer that has been implanted into the brain to administer anticancer drugs direct to brain tumours, thus avoiding the blood–brain barrier.

Polymer-coated, drug-releasing stents have been used to keep blood vessels open after a clot clearing procedure called angioplasty.

Investigations are underway into the use of implantable microchips which could detect chemical signals in the body and release drugs in response to these signals.

KEY POINTS

- Oral administration is the preferred method of administering drugs, but it is also the most demanding on the drug.

- Drugs administered by methods other than the oral route avoid the first pass effect.

- Drugs can be administered such that they are absorbed through the mucous membranes of the mouth, nose, or eyes.

- Some drugs are administered rectally as suppositories.

- Topically administered drugs are applied to the skin. Some drugs are absorbed through the skin into the blood supply.

- Inhaled drugs are administered as gases or aerosols to act directly on the respiratory system. Some inhaled drugs are absorbed into the blood supply to act systemically.

- Polar drugs which are unable to cross cell membranes are given by injection.

- Injection is the most efficient method of administering a drug, but it is also the most hazardous. Injection can be intravenous, intramuscular, subcutaneous, or intrathecal.

- Implants have been useful in providing controlled drug release such that blood concentrations of the drug remain as level as possible.

11.8 Drug dosing

Because of the number of pharmacokinetic variables involved, it can be difficult to estimate the correct dose regimen for a drug (i.e. the amount of drug used for each dose and the frequency of administration). There are other issues to consider as well. Ideally, the blood levels of any drug should be constant and controlled, but this would require a continuous, intravenous drip which is clearly impractical for most drugs. Therefore, drugs are usually taken at regular time intervals. This means that the doses taken have to be such that blood levels of the drug are within a maximum and minimum level, where they are not too high to be toxic, yet not too low to be ineffective. In general, the concentration of free drug in the blood (i.e. not bound to plasma protein) is a good indication of the availability of that drug at its target site. This does not mean that blood concentration levels are the same as the concentration levels at the target site. However, any variations in blood concentration will result in similar fluctuations at the target site. Thus, blood concentration levels can be used to determine therapeutic and safe dosing levels for a drug.

Figure 11.20 shows two dose regimens. Dose regimen A quickly reaches the therapeutic level but continues to rise to a steady state which is toxic. Dose regimen B involves half the amount of drug provided with the same frequency. The time taken to reach the therapeutic level is certainly longer, but the steady state levels of the drug remain between the therapeutic and toxic levels—the **therapeutic window**.

Dose regimens involving regular administration of a drug work well in most cases, especially if the size of each dose is less than 200 mg and doses are taken once or twice a day. However, there are certain situations where timed doses are not suitable. The treatment of diabetes with **insulin** is a case in point. Insulin is normally secreted continuously by the pancreas, so the injection of insulin at timed intervals is unnatural and can lead to a whole range of physiological complications.

Other dosing complications include differences of age, sex, and race. Diet, environment, and altitude also have an influence. Obese people present a particular problem, as it can be very difficult to estimate how much of a drug will be stored in fat tissue and how much will remain free in the blood supply. The precise time when drugs are taken may be important, because metabolic reaction rates can vary throughout the day.

Drugs can interact with other drugs. For example, some drugs used for diabetes are bound by plasma protein in the blood supply and are therefore not free to react with their targets. However, drugs such as **aspirin** may displace them from plasma protein, leading

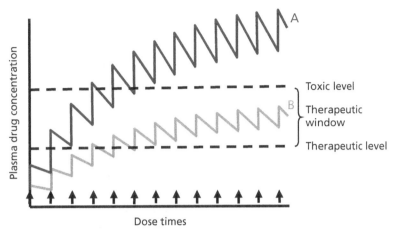

FIGURE 11.20 Dosing regimens.

to a drug overdose. Aspirin has this same effect on anticoagulants.

Problems can also occur if a drug inhibits a metabolic reaction and is taken with a drug normally metabolized by that reaction. The latter is more slowly metabolized than normal, increasing the risk of an overdose. For example, the antidepressant drug **phenelzine** inhibits the metabolism of amines and should not be taken with drugs such as **amphetamines** or **pethidine**. Even amine-rich foods can lead to adverse effects, implying that cheese and wine parties are hardly the way to cheer the victim of depression. Other examples were described in section 11.5.6.

When one considers all these complications, it is hardly surprising that individual variability to drugs can vary by as much as a factor of 10.

11.8.1 Drug half-life

The half-life ($t_{1/2}$) of a drug is the time taken for the concentration of the drug in blood to fall by half. The removal or elimination of a drug takes place through both excretion and drug metabolism, and is not linear with time. Therefore, drugs can linger in the body for a significant period of time. For example, if a drug has a half-life of 1 hour, then there is 50% of it left after 1 hour. After 2 hours, there is 25% of the original dose left, and after 3 hours, 12.5% remains. It takes 7 hours for the level to fall below 1% of the original dose. Some drugs, such as the

opioid analgesic **fentanyl**, have short half-lives (45 minutes), whereas others such as **diazepam** (**Valium**) have a half-life measured in days. In the latter case, recovery from the drug may take a week or more.

11.8.2 Steady state concentration

Drugs are metabolized and eliminated as soon as they are administered, so it is necessary to provide regular doses in order to maintain therapeutic levels in the body. Therefore, it is important to know the half-life of the drug in order to calculate the frequency of dosing required to reach and maintain these levels. In general, the time taken to reach a **steady state concentration** is six times the drug's half-life. For example, the concentration levels of a drug with a half-life of 4 hours, supplied at 4-hourly intervals, are shown in Table 11.1 and Fig. 11.21.

Note that there is a fluctuation in level in the period between each dose. The level is at a maximum after each dose, and falls to a minimum before the next dose is provided. It is important to ensure that the level does not drop below the therapeutic level, but does not rise to such a level that side effects are induced. The time taken to reach steady state concentration is not dependent on the size of the dose, but the blood level achieved at steady state is. Therefore, the levels of drug present at steady state concentration depend on the size of each dose given, as well as the frequency of dosing. During

TABLE 11.1 Fluctuation of drug concentration levels on regular dosing

Time of dosing (h)	0	4	8	12	16	20	24
Max level (µg/ml)	1.0	1.5	1.75	1.87	1.94	1.97	1.98
Min level (µg/ml)	0.5	0.75	0.87	0.94	0.97	0.98	0.99

673

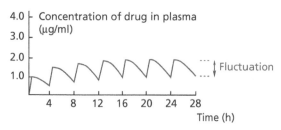

FIGURE 11.21 Graphical representation of fluctuation of drug concentration levels on regular dosing.

clinical trials, blood samples are taken from patients at regular time intervals to determine the concentration of the drug in the blood. This helps determine the proper dosing regimen in order to get the ideal blood levels.

The **area under the plasma drug concentration curve** (AUC) represents the total amount of drug that is available in the blood supply during the dosing regimen.

11.8.3 Drug tolerance

With certain drugs, it is found that the effect of the drug diminishes after repeated doses, and it is necessary to increase the size of the dose in order to achieve the same results. This is known as drug tolerance. There are several mechanisms by which drug tolerance can occur. For example, the drug can induce the synthesis of metabolic enzymes which result in increased metabolism of the drug. **Pentobarbital** (Fig. 11.22) is a barbiturate sedative which induces enzymes in this fashion.

Alternatively, the target may adapt to the presence of a drug. Occupancy of a target receptor by an antagonist may induce cellular effects which result in the synthesis of more receptor (section 8.7). As a result, more drug will be needed in the next dose to antagonize all the receptors.

Physical dependence is usually associated with drug tolerance. Physical dependence is a state in which a patient becomes dependent on the drug in order to feel normal. If the drug is withdrawn, uncomfortable **withdrawal symptoms** may arise which can only be alleviated by re-taking the drug. These effects can be explained in part by the effects which lead to drug tolerance. For example, if cells have synthesized more receptors to counteract the presence of an antagonist, the removal of the antagonist means that the body will have too many receptors. This results in a 'kickback' effect where the cell

becomes oversensitive to the normal neurotransmitter or hormone, and this is what produces withdrawal symptoms. These will continue until the excess receptors have been broken down by normal cellular mechanisms—a process that may take several days or weeks (see also sections 8.6–8.7).

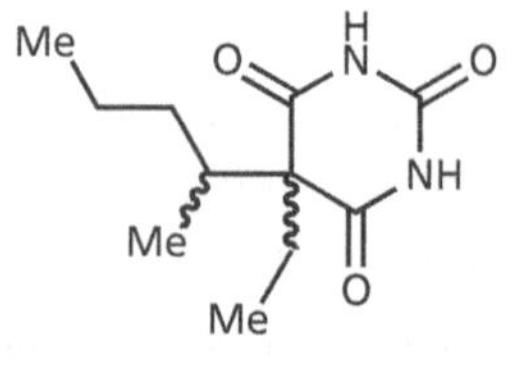

11.8.4 Bioavailability

Bioavailability refers to how quickly and how much of a particular drug reaches the blood supply once all the problems associated with absorption, distribution, metabolism, and excretion have been taken into account. **Oral bioavailability (F)** is the fraction of the ingested dose that survives to reach the blood supply. This is an important property when it comes to designing new drugs and should be considered alongside the pharmacodynamics of the drug (i.e. how effectively the drug interacts with its target).

11.9 Formulation

The way a drug is formulated can avoid some of the problems associated with oral administration. Drugs are normally taken orally as tablets or capsules. A tablet is usually a compressed preparation that contains 5–10% of the drug, 80% of fillers, disintegrants, lubricants, glidants, and binders, and 10% of compounds which ensure easy disintegration, disaggregation, and dissolution of the tablet in the stomach or the intestine—a process which is defined as the **pharmaceutical phase** of drug action. The disintegration time can be modified for a rapid effect or for sustained release. Special coatings can make the tablet resistant to the stomach acids such that it only disintegrates in the duodenum as a result of enzyme action or pH. Pills can also be coated with sugar, varnish, or wax to disguise taste. Some tablets are designed with an osmotically active bilayer core surrounded by a semi-permeable membrane with one or more laser drilled pores in it. The osmotic pressure of water entering the tablet pushes the drug through the pores at a constant rate as the tablet moves through the digestive tract. Therefore, the rate of release is independent of varying pH or gastric motility. Several drugs such as **hydromorphone, albuterol**, and **nifedipine** have been administered in this way.

A capsule is a gelatinous envelope enclosing the active substance. Capsules can be designed to remain intact for some hours after ingestion in order to delay absorption. They may also contain a mixture of slow-release and fast-release particles to produce rapid and sustained absorption in the same dose.

The drug itself needs to dissolve in aqueous solution at a controlled rate. Such factors as particle size and crystal

FIGURE 11.22 Pentobarbital.

form can significantly affect dissolution. Fast dissolution is not always ideal. For example, slow dissolution rates can prolong the duration of action or avoid initially high plasma levels.

Formulation can also play an important role in preventing drugs being abused. For example, a tablet preparation (**Oxecta**) of the opioid analgesic **oxycodone** was approved in 2011 as an orally active opioid analgesic and includes deterrents to abuse. For example, chemicals are present that prevent the drug being dissolved in solvent and injected. Other chemicals cause a burning sensation in the nose, which discourages drug abusers crushing the tablets and snorting the powder. Finally, other chemicals are present which produce non-toxic, but very unpleasant, effects if too many pills are taken orally.

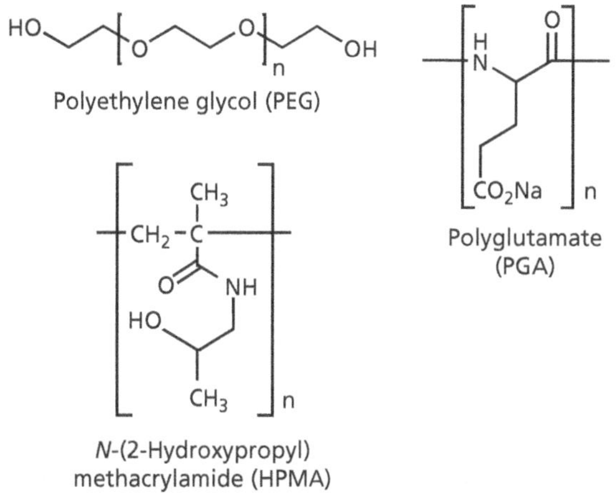

FIGURE 11.23 Synthetic polymers used for polymer–drug conjugates.

11.10 Drug delivery

The various aspects of drug delivery could fill a textbook in themselves, so any attempt to cover the topic in a single section is merely tickling the surface, let alone scratching it! However, it is worth appreciating that there are various methods by which drugs can be physically protected from degradation and/or targeted to treat particular diseases such as cancer and inflammation. One approach is to use a prodrug strategy (section 14.6), which involves chemical modifications to the drug. Another approach covered in this section is the use of water-soluble macromolecules to help the drug reach its target. The macromolecules concerned are many and varied and include synthetic polymers, proteins, liposomes, and antibodies. The drug itself may be covalently linked to the macromolecule or encapsulated within it. The following are some illustrations of drug delivery systems.

Antibodies were described in section 10.7.2 and have long been seen as a method of targeting drugs to cancer cells. Methods have been devised for linking anticancer drugs to antibodies to form **antibody–drug conjugates** that remain stable on their journey through the body, but release the drug at the target cell. A lot of research has been carried out on these conjugates, and this is discussed in detail in section 21.10.2. However, there are problems associated with antibodies. The amount of drug that can be linked to the protein is quite limited and there is the risk of an immune reaction where the body identifies the antibody as foreign and tries to reject it.

A similar approach is to link the drugs to synthetic polymers such as polyethylene glycol (PEG), polyglutamate, or *N*-(2-hydroxypropyl)methacrylamide (HPMA) to form polymer–drug conjugates (Fig. 11.23). Again the amount of drug that can be linked is limited, but a variety of anticancer–polymer conjugates are currently undergoing clinical trials. Such conjugates help to protect the

lifetime of the drug by decreasing the rates of metabolism and excretion. **Pegaptanib** is a preparation that was approved for treating a vascular disease in the eye and consists of an oligonucleotide drug linked to PEG (section 10.5). Pegylation has also been used to design a peripherally acting opioid that is unable to cross the blood–brain barrier (section 24.9.4)

Protein-based polymers are being developed as drug delivery systems for the controlled release of ionized drugs. For example, the cationic drugs **Leu-enkephalin** or **naltrexone** could be delivered using polymers with anionic carboxylate groups. Ionic interactions between the drug and the protein result in folding and assembly of the protein polymer to form a protein–drug complex and the drug is then released at a slow and constant rate. The amount of drug carried could be predetermined by the density of carboxylate binding sites present and the accessible surface area of the vehicle. The rate of release could be controlled by varying the number of hydrophobic amino acids present. The greater the number of hydrophobic amino acids present, the weaker the affinity between the carboxylate binding groups and the drug. Once the drug is released, the protein carrier would be metabolized like any normal protein.

A physical method of protecting drugs from metabolic enzymes in the bloodstream and allowing a steady slow release of the drug is to encapsulate the drug within small vesicles called **liposomes**, and then inject them into the blood supply (Fig. 11.24). These vesicles or globules consist of a bilayer of fatty phospholipid molecules (similar to a cell membrane) and will travel round the circulation, slowly leaking their contents. Liposomes are known to be concentrated in malignant tumours and this provides a possible method of delivering antitumour drugs to these cells. It is also found that liposomes can fuse with the plasma membranes of a variety of cells, allowing the delivery of drugs or DNA into these cells. As a result, they

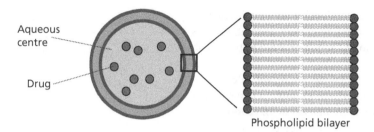

FIGURE 11.24 Liposome containing a drug.

may be useful for gene therapy. The liposomes can be formed by sonicating a suspension of a phospholipid (e.g. phosphatidylcholine) in an aqueous solution of the drug.

Another future possibility for targeting liposomes is to incorporate antibodies into the liposome surface such that specific tissue antigens are recognized. Liposomes have a high drug-carrying capacity, but it can prove difficult to control the release of drug at the required rate. Slow leakage is a problem if the liposome is carrying a toxic anticancer drug such as **doxorubicin**. The liposomes can also be trapped by the **reticuloendothelial system** (RES) and removed from the blood supply. The RES is a network of cells which can be viewed as a kind of filter. One answer to this problem has been to attach PEG polymers to the liposome (see also section 14.8.2). The tails of the PEG polymers project out from the liposome surface and act as a polar outer shell which protects and shields the liposome from both destructive enzymes and the reticuloendothelial system. This significantly increases its lifetime and reduces leakage of its passenger drug. **DOXIL** is a PEGylated liposome containing doxorubicin which is used successfully in anticancer therapy as a once-monthly infusion.

The use of injectable **microspheres** has been approved for the delivery of human growth hormone. The microspheres containing the drug are made up of a biologically degradable polymer and slowly release the hormone over a 4-week period.

A large number of important drugs have to be administered by injection because they are either susceptible to digestive enzymes or cannot cross the gut wall. This includes the ever-growing number of therapeutically useful peptides and proteins being generated by biotechnology companies using recombinant DNA technology. Drug delivery systems which could deliver these drugs orally would prove a huge step forward in medicine. For example, liposomes are currently being studied as possible oral delivery systems. Another approach currently being investigated is to link a therapeutic protein to a hydrophobic polymer such that it is more likely to be absorbed. However, it is important that the conjugate breaks up before the drug enters the blood supply or else

it would have to be treated as a new drug and undergo expensive preclinical and clinical trials. **Hexyl-insulin monoconjugate 2** consists of a polymer linked to a lysine residue of insulin. It is currently being investigated as an oral delivery system for **insulin**.

Biologically erodible microspheres have also been designed to stick to the gut wall such that absorption of the drug within the sphere through the gut wall is increased. This has still to be used clinically, but has proved effective in enhancing the absorption of insulin and **plasmid DNA** in test animals. In a similar vein, drugs have been coated with bioadhesive polymers designed to adhere to the gut wall so that the drug has more chance of being absorbed. The use of anhydride polymers has the added advantage that these polymers are capable of crossing the gut wall and entering the bloodstream, taking their passenger drug with them. **Emisphere Technologies Inc.** have developed derivatives of amino acids and shown that they can enhance the absorption of specific proteins. It is thought that the amino acid derivatives interact with the protein and make it more lipophilic so that it can cross cell membranes directly.

Drug delivery systems are being investigated which will carry oligonucleotides such as DNA, antisense molecules, and siRNAs (section 9.7.2). For example, nucleic acid–lipid particles are being investigated as a means of delivering oligonucleotides into liver cells. Such particles are designed to have a positive charge on their exterior since this encourages adsorption to the negatively charged cell membranes of target cells. Another method of carrying and delivering oligonucleotides is to incorporate them into viruses that are capable of infecting cells. However, there are risks associated with this approach and there have been instances of fatalities during clinical trials. Therefore, nanotechnology is being used to construct artificial viruses which will do the job more safely. Clinical trials have demonstrated that it is possible to use engineered viruses to target drugs to tumour cells.

Other areas of research include studies of crown ethers, nanoparticles, nanospheres, nanowires, nanomagnets, biofuel cells, hydrogel polymers, and superhydrophobic materials as methods of delivering drugs.

KEY POINTS

- Drugs should be administered at the correct dose levels and frequency to ensure that blood concentrations remain within the therapeutic window.

- The half-life of a drug is the time taken for the blood concentration of the drug to fall by half. A knowledge of the half-life is required to calculate how frequently doses should be given to ensure a steady state concentration.

- Drug tolerance is where the effect of a drug diminishes after repeated doses. In physical dependence a patient becomes dependent on a drug and suffers withdrawal symptoms on stopping the treatment.

- Formulation refers to the method by which drugs are prepared for administration, whether by solution, pill, capsule, liposome, or microsphere. Suitable formulations can protect drugs from particular pharmacokinetic problems.

QUESTIONS

1. Benzene used to be a common solvent in organic chemistry, but is no longer used because it is a suspect carcinogen. Benzene undergoes metabolic oxidation by cytochrome P450 enzymes to form an electrophilic epoxide which can alkylate proteins and DNA. Toluene is now used as a solvent in place of benzene. Toluene is also oxidized by cytochrome P450 enzymes, but the metabolite is less toxic and is rapidly excreted. Suggest what the metabolite might be, and why the metabolism of toluene is different from that of benzene.

2. The prodrug of the antipsychotic drug **fluphenazine** shown below has a prolonged period of action when it is given by intramuscular injection, but not when it is given by intravenous injection. Suggest why this is the case.

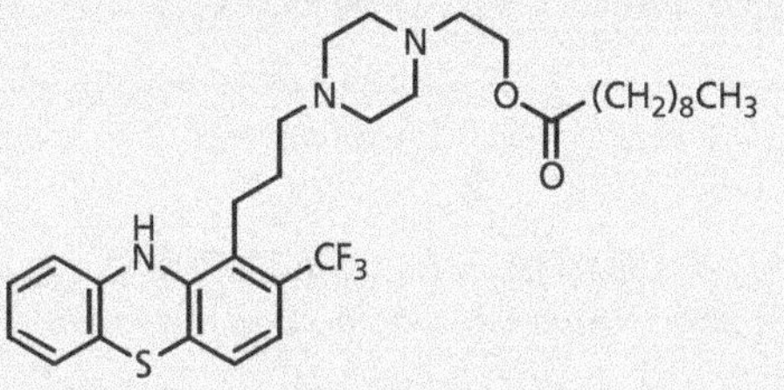

Fluphenazine prodrug

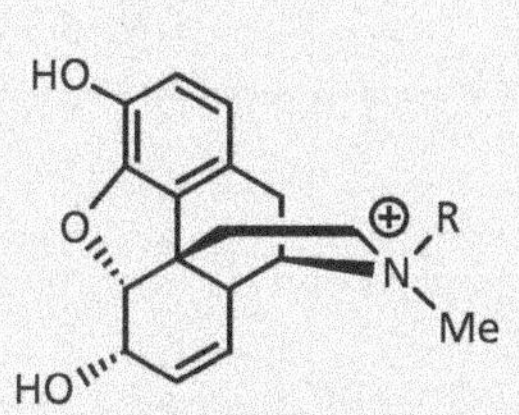

Morphine; R = H
Quaternary salt; R = Me

3. Morphine binds strongly to opioid receptors in the brain to produce analgesia. *In vitro* studies on opioid receptors show that the quaternary salt of morphine also binds strongly. However, the compound is inactive *in vivo* when injected intravenously. Explain this apparent contradiction.

4. The phenol group of morphine is important in binding morphine to opioid receptors and causing analgesia. Codeine

has the same structure as morphine but the phenol group is masked as a methyl ether. As a result, codeine binds poorly to opioid receptors and should show no analgesic activity. However, when it is taken *in vivo*, it shows useful analgesic properties. Explain how this might occur.

5. The pK_a of histamine is 5.74. What is the ratio of ionized to unionized histamine (a) at pH 5.74 (b) at pH 7.4?

6. A drug contains an ionized carboxylate group and shows good activity against its target in *in vitro* tests. When *in vivo* tests were carried out, the drug showed poor activity when it was administered orally, but good activity when it was administered by intravenous injection. The same drug was converted to an ester, but proved inactive *in vitro*. Despite that, it proved to be active *in vivo* when it was administered orally. Explain these observations.

7. **Atomoxetine** and **methylphenidate** are used in the treatment of attention deficit hyperactivity disorder. Suggest possible metabolites for these structures.

8. Suggest metabolites for the proton pump inhibitor **omeprazole**.

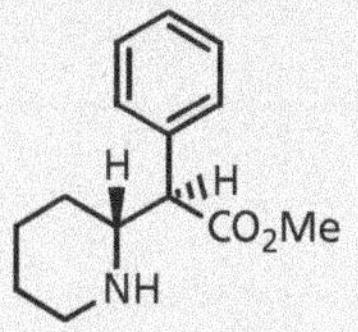

Methylphenidate 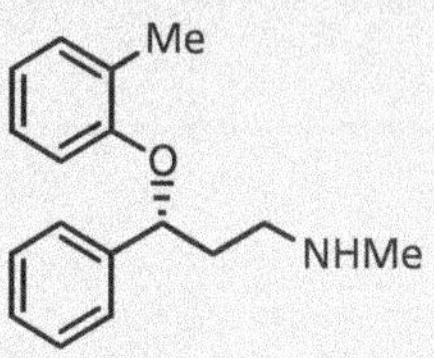   Atomoxetine

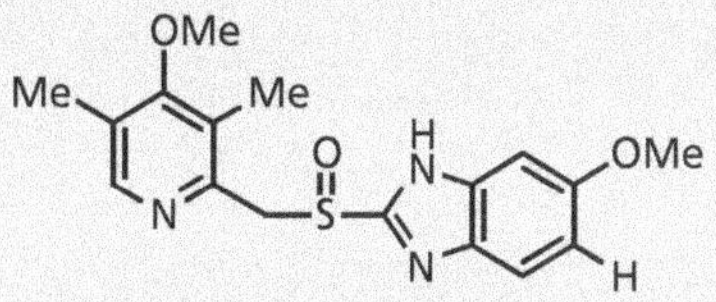

Omeprazole

9. A drug has a half-life of 4 hours. How much of the drug remains after 24 hours?

10. Salicylic acid is absorbed more effectively from the stomach than from the intestines, whereas quinine is absorbed more effectively from the intestines than from the stomach. Explain these observations.

Multiple-choice questions are available on the Online Resource Centre at www.oxfordtextbooks.co.uk/orc/patrick6e/

FURTHER READING

Cairns, D. (2012) *Essentials of pharmaceutical chemistry*, 4th edn. Pharmaceutical Press, London.

Duncan, R. (2003) The dawning era of polymer therapeutics. *Nature Reviews Drug Discovery*, **2**, 347–60.

Goldberg, M., and Gomez-Orellana, I. (2003) Challenges for the oral delivery of macromolecules. *Nature Reviews Drug Discovery*, **2**, 257–8.

Guengerich, F. P. (2002) Cytochrome P450 enzymes in the generation of commercial products. *Nature Reviews Drug Discovery*, **1**, 359–66.

King, A. (2011) Breaking through the barrier. *Chemistry World*, June, 36–9.

Langer, R. (2003) Where a pill won't reach. *Scientific American*, April, 32–9.

LaVan, D. A., Lynn, D. M., and Langer, R. (2002) Moving smaller in drug discovery and delivery. *Nature Reviews Drug Discovery*, **1**, 77–84.

Lindpaintner, K. (2002) The impact of pharmacogenetics and pharmacogenomics on drug discovery. *Nature Reviews Drug Discovery*, **1**, 463–9.

Lipinski, C. A., Lombardo, F., Dominy, B. W., and Feeney, P. J. (1997) Experimental and computational approaches to estimate solubility and permeability in drug discovery and development settings. *Advanced Drug Delivery Reviews*, **23**, 3–25 (rule of five).

Mastrobattista, E., et al. (2006) Artificial viruses: a nanotechnological approach to gene delivery. *Nature Reviews Drug Discovery*, **5**, 115–21.

Nicholson, J. K., and Wilson, I. D. (2003) Understanding global systems biology: metabonomics and the continuum of metabolism. *Nature Reviews Drug Discovery*, **2**, 668–76.

Pardridge, W. M. (2002) Drug and gene targeting to the brain with molecular Trojan horses. *Nature Reviews Drug Discovery*, **1**, 131–9.

Roden, D. M., and George, A. L. (2002) The genetic basis of variability in drug responses. *Nature Reviews Drug Discovery*, **1**, 37–44.

Roses, A. D. (2002) Genome-based pharmacogenetics and the pharmaceutical industry. *Nature Reviews Drug Discovery*, **1**, 541–9.

Saltzman, W. M., and Olbricht, W. L. (2002) Building drug delivery into tissue engineering. *Nature Reviews Drug Discovery*, **1**, 177–86.

Stevenson, R. (2003) Going with the flow. *Chemistry in Britain*, November, 18–20 (aquaporins).

Veber, D. F., et al. (2002) Molecular properties that influence the oral bioavailability of drug candidates. *Journal of Medicinal Chemistry*, **45**, 2615–23.

Willson, T. M., and Kliewer, S. A. (2002) PXR, CAR and drug metabolism. *Nature Reviews Drug Discovery*, **1**, 259–66.

Titles for general further reading are listed on p.845.

12 Drug discovery: finding a lead

In this chapter, we shall look at what happens when a pharmaceutical company or university research group initiates a new medicinal chemistry project through to the identification of a lead compound.

12.1 Choosing a disease

How does a pharmaceutical company decide which disease to target when designing a new drug? Clearly, it would make sense to concentrate on diseases where there is a need for new drugs. However, pharmaceutical companies have to consider economic factors as well as medical ones. A huge investment has to be made towards the research and development of a new drug. Therefore, companies must ensure that they get a good financial return for their investment. As a result, research projects tend to focus on diseases that are important in the developed world, because this is the market best able to afford new drugs. A great deal of research is carried out on ailments such as migraine, depression, ulcers, obesity, flu, cancer, and cardiovascular disease. Less is carried out on the tropical diseases of the developing world. Only when such diseases start to make an impact on western society do the pharmaceutical companies sit up and take notice. For example, there has been a noticeable increase in antimalarial research as a result of the increase in tourism to more exotic countries, and the spread of malaria into the southern states of the USA (see also Case study 3). Moreover, pharmaceutical companies are becoming more involved in partnerships with governments and philanthropic organizations such as the **Wellcome Trust**, the **Bill and Melinda Gates Foundation,** and **Medicines for Malaria Venture** in order to study diseases such as tuberculosis, malaria, and dengue.

Choosing which disease to tackle is usually a matter for a company's market strategists. The science becomes important at the next stage.

12.2 Choosing a drug target

12.2.1 Drug targets

Once a therapeutic area has been identified, the next stage is to identify a suitable drug target (e.g. receptor, enzyme, or nucleic acid). An understanding of which biomacromolecules are involved in a particular disease state is clearly important (see Box 12.1). This allows the medicinal research team to identify whether agonists or antagonists should be designed for a particular receptor, or whether inhibitors should be designed for a particular enzyme. For example, agonists of serotonin receptors are useful for the treatment of migraine, while antagonists of dopamine receptors are useful as antidepressants. Sometimes it is not known for certain whether a particular target will be suitable or not. For example, **tricyclic antidepressants,** such as **desipramine** (Fig. 12.1), are known to inhibit the uptake of the neurotransmitter **noradrenaline** from nerve synapses by inhibiting the carrier protein for noradrenaline (section 23.12.4). However, these drugs also inhibit uptake of a separate neurotransmitter called **serotonin,** and the possibility arose that inhibiting serotonin uptake might also be beneficial. A search for **selective serotonin uptake inhibitors** was initiated, which led to the discovery of the best-selling antidepressant drug **fluoxetine (Prozac)** (Fig. 12.1), but when this project was initiated it was not known for certain whether serotonin uptake inhibitors would be effective or not.

12.2.2 Discovering drug targets

If a drug or a poison produces a biological effect, there must be a molecular target for that agent in the body. In the past, the discovery of drug targets depended on finding the drug first. Many early drugs such as the analgesic **morphine** are natural products derived from plants, and just happen to interact with a molecular target in the human body. As this involves coincidence more than

BOX 12.1 Recently discovered targets: the caspases

The **caspases** are examples of recently discovered enzymes which may prove useful as drug targets. They are a family of protease enzymes that catalyse the hydrolysis of important cellular proteins, and which have been found to play a role in inflammation and cell death. Cell death is a natural occurrence in the body, and cells are regularly recycled. Therefore, caspases should not necessarily be seen as 'bad' or 'undesirable' enzymes. Without them, cells could be more prone to unregulated growth, resulting in diseases such as cancer.

The caspases catalyse the hydrolysis of particular target proteins such as those involved in DNA repair and the regulation of cell cycles. By understanding how these enzymes operate, there is the possibility of producing new therapies for a variety of diseases. For example, agents which promote the activity of caspases and lead to more rapid cell death might be useful in the treatment of diseases such

as cancer, autoimmune disease, and viral infections. For example, **carboplatin** is an anticancer agent that promotes caspase activity. Alternatively, agents which inhibit caspases and reduce the prevalence of cell death could provide novel treatments for trauma, neurodegenerative disease, and strokes. It is already known that the active site of caspases contains two amino acids that are crucial to the mechanism of hydrolysis—cysteine, which acts as a nucleophile, and histidine, which acts as an acid–base catalyst. The mechanism is similar to that used by acetylcholinesterase (section 22.12.3.2).

Caspases recognize an aspartate residue within protein substrates and cleave the peptide link next to it. Selective inhibitors have been developed which include aspartate or a mimic of it, but it remains to be seen whether such inhibitors have a clinical role.

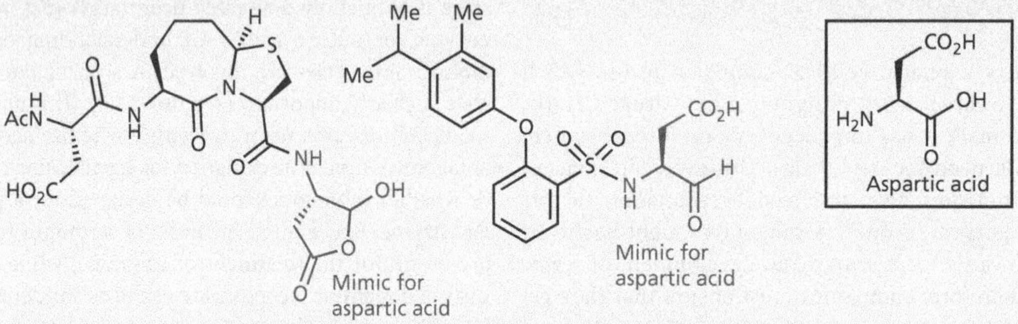

FIGURE 1 Selective caspase inhibitors.

design, the detection of drug targets was very much a hit and miss affair. Later, the body's own chemical messengers started to be discovered and pointed the finger at further targets. For example, since the 1970s a variety of peptides and proteins have been discovered which act as the body's own analgesics (enkephalins and endorphins). Another example is the rather surprising discovery that

nitric oxide acts as a chemical messenger (Box 3.1 and sections 22.3.2 and 26.5.1). Despite this, relatively few of the body's messengers were identified, either because they were present in such small quantity or because they were too short lived to be isolated. Indeed, many chemical messengers still remain undiscovered today. This, in turn, means that many of the body's potential drug

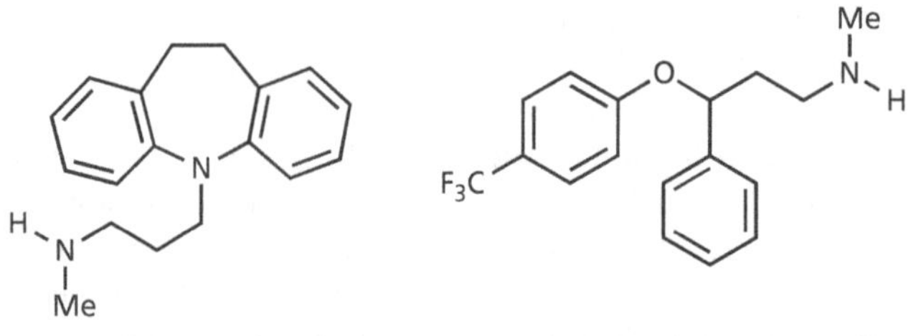

Desipramine Fluoxetine (Prozac)

FIGURE 12.1 Antidepressant drugs.

targets remain hidden. Or at least it did! The advances in genomics and proteomics have changed all that. The various genome projects which have mapped the DNA of humans and other life forms, along with the newer field of proteomics (section 2.6), are revealing an ever increasing number of new proteins which are potential drug targets for the future. In many cases, the natural chemical messengers for these proteins are unknown, and medicinal chemistry is faced with new targets, but with no lead compounds to interact with them. Such targets have been defined as **orphan receptors**. The challenge is now to find a chemical which will interact with each of these targets in order to find out what their function is and whether they will be suitable as drug targets. This was one of the main driving forces behind the development of **combinatorial** and **parallel synthesis** (Chapter 16).

12.2.3 Target specificity and selectivity between species

Target specificity and selectivity is a crucial factor in modern medicinal chemistry research. The more selective a drug is for its target, the less chance there is that it will interact with different targets and have undesirable side effects.

In the field of antimicrobial agents, the best targets to choose are those that are unique to the microbe and are not present in humans. For example, **penicillin** targets an enzyme involved in bacterial cell wall biosynthesis. Mammalian cells do not have a cell wall, so this enzyme is absent in human cells and penicillin has minimal side effects (section 19.5). In a similar vein, sulphonamides inhibit a bacterial enzyme not present in human cells (section 19.4.1.5), while several agents used to treat AIDS inhibit an enzyme called **retroviral reverse transcriptase** which is unique to the infectious agent HIV (section 20.7.3).

Other cellular features that are unique to microorganisms could also be targeted. For example, the microorganisms which cause sleeping sickness in Africa are propelled by means of a tail-like structure called a **flagellum**. This feature is not present in mammalian cells, so designing drugs that bind to the proteins making up the flagellum to prevent it working could be potentially useful in treating that disease.

Having said all that, it is still possible to design drugs against targets which are present both in humans and microbes, as long as the drugs show selectivity against the microbial target. Fortunately, this is perfectly feasible. An enzyme which catalyses a reaction in a bacterial cell differs significantly from the equivalent enzyme in a human cell. The enzymes may have been derived from an ancient common ancestor, but several million years of evolution have

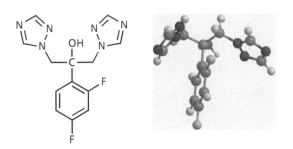

FIGURE 12.2 Fluconazole.

resulted in significant structural differences. For example, the antifungal agent **fluconazole** (Fig. 12.2) inhibits a fungal demethylase enzyme involved in steroid biosynthesis. This enzyme is also present in humans, but the structural differences between the two enzymes are significant enough that the antifungal agent is highly selective for the fungal enzyme. Other examples of bacterial or viral enzymes which are sufficiently different from their human equivalents are **dihydrofolate reductase** (section 19.4.2) and **viral DNA polymerase** (section 20.6.1).

12.2.4 Target specificity and selectivity within the body

Selectivity is also important for drugs acting on targets within the body. Enzyme inhibitors should only inhibit the target enzyme and not some other enzyme. Receptor agonists/antagonists should ideally interact with a specific kind of receptor (e.g. the adrenergic receptor) rather than a variety of different receptors. However, nowadays medicinal chemists aim for even higher standards of target selectivity. Ideally, enzyme inhibitors should show selectivity between the various isozymes of an enzyme (isozymes are the structural variants of an enzyme that result from different amino acid sequences or quaternary structure—section 3.7). For example, there are three different isoforms of **nitric oxide synthase** (NOS)—the enzyme responsible for generating the chemical messenger **nitric oxide** (Box 3.1). Selective inhibitors for one of these isoforms (nNOS) could potentially be useful in treating cerebral palsy and other neurodegenerative diseases.

Receptor agonists and antagonists should not only show selectivity for a particular receptor (e.g. an adrenergic receptor) or even a particular receptor type (e.g. the β-adrenergic receptor), but also for a particular receptor subtype (e.g. the β$_2$-adrenergic receptor). One of the current areas of research is to find antipsychotic agents with fewer side effects. Traditional antipsychotic agents act as antagonists of dopamine receptors. However, it has been found that there are five dopamine receptor subtypes and that traditional antipsychotic agents antagonize two of these (D$_3$ and D$_2$). There is good evidence that the D$_2$ receptor is responsible for the undesirable Parkinsonian type side effects of current drugs, and so research is now underway to find a selective D_3 antagonist.

12.2.5 Targeting drugs to specific organs and tissues

Targeting drugs against specific receptor subtypes often allows drugs to be targeted to specific organs or to specific areas of the brain. This is because the various receptor subtypes are not uniformly distributed around the body, but are often concentrated in particular tissues. For example, the β-adrenergic receptors in the heart are predominantly β_1, whereas those in the lungs are β_2. This makes it feasible to design drugs that will work on the lungs with a minimal side effect on the heart, and vice versa.

Attaining subtype selectivity is particularly important for drugs that are intended to mimic neurotransmitters. Neurotransmitters are released close to their target receptors and once they have passed on their message, they are quickly deactivated and do not have the opportunity to 'switch on' more distant receptors. Therefore, only those receptors which are fed by 'live' nerves are switched on.

In many diseases, there is a 'transmission fault' to a particular tissue or in a particular region of the brain. For example, in Parkinson's disease, **dopamine** transmission is deficient in certain regions of the brain, although it is functioning normally elsewhere. A drug could be given to mimic dopamine in the brain. However, such a drug acts like a hormone rather than as a neurotransmitter because it has to travel round the body in order to reach its target. This means that the drug could potentially 'switch on' all the dopamine receptors around the body and not just the ones that are suffering the dopamine deficit. Such drugs would have a large number of side effects, so it is important to make the drug as selective as possible for the particular type or subtype of dopamine receptor affected in the brain. This would target the drug more effectively to the affected area and reduce side effects elsewhere in the body.

12.2.6 Pitfalls

A word of caution! It is possible to identify whether a particular enzyme or receptor plays a role in a particular ailment. However, the body is a highly complex system. For any given function, there are usually several messengers, receptors, and enzymes involved in the process. For example, there is no one simple cause for hypertension (high blood pressure). This is illustrated by the variety of receptors and enzymes which can be targeted in its treatment. These include β_1-adrenoceptors, calcium ion channels, angiotensin-converting enzyme (ACE), potassium ion channels, and angiotensin II receptors (Chapters 23 and 26).

As a result, more than one target may need to be addressed for a particular ailment (Box 12.2). For example, most of the current therapies for asthma involve a combination of a bronchodilator (β_2-agonist) and an anti-inflammatory agent such as a corticosteroid.

BOX 12.2 Pitfalls in choosing particular targets

Drugs are designed to interact with a particular target because that target is believed to be important to a particular disease process. Occasionally, though, a particular target may not be as important to a disease as was first thought. For example, the dopamine D_2 receptor was thought to be involved in causing nausea. Therefore, the D_2 receptor antagonist **metoclopramide** was developed as an anti-emetic agent. However, it was found that more potent D_2 antagonists were less effective, implying that a different receptor might be more important in producing nausea. Metoclopramide also antagonizes the 5-hydroxytryptamine (5-HT$_3$) receptor, so antagonists for this receptor were studied, which led to the development of the anti-emetic drugs **granisetron** and **ondansetron**.

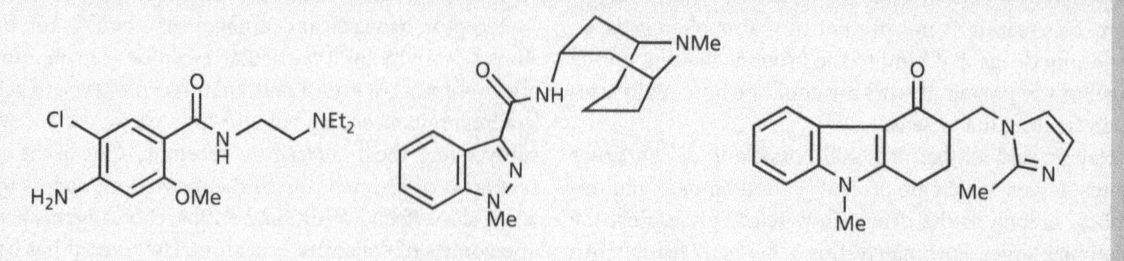

FIGURE 1 Anti-emetic agents.

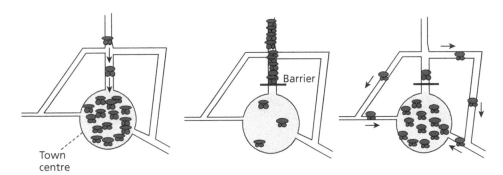

FIGURE 12.3 Avoiding the jam.

Sometimes, drugs designed against a specific target become less effective over time. Because cells have a highly complex system of signalling mechanisms, it is possible that the blockade of one part of that system could be bypassed. This could be compared to blocking the main road into town to try and prevent congestion in the town centre. To begin with the policy works, but in a day or two commuters discover alternative routes, and congestion in the centre becomes as bad as ever (Fig. 12.3).

12.2.7 Multi-target drugs

In certain diseases and afflictions, there can be an advantage in 'hitting' a number of different targets selectively, as this can be more beneficial than hitting just one. Combination therapy is normally used to achieve this by administering two or more drugs showing selectivity against the different targets. This is particularly the case in the treatment of cancer (Chapter 21) and HIV infection (Box 20.2). However, combination therapies are also used in a variety of other situations (sections 19.4.2.1, 19.5.4, and 20.10). The disadvantage of combination therapies is the number of different medications and the associated dose regimens. Therefore, there are benefits in designing a single drug that can act selectively at different targets in a controlled manner—a **multi-target-directed ligand**. Many research projects now set out to discover new drugs with a defined profile of activity against a range of specific targets. For example, a research team may set out to find a drug that has agonist activity for one receptor subtype and antagonist activity at another. A further requirement may be that the drug neither acts as a substrate for cytochrome p450 enzymes, nor affects the activity of those metabolic enzymes (section 11.5). Tests are also carried out to ensure that the drug does not interact with targets that could lead to toxicity (Box 12.3). A current area of research is in designing dual-action drugs to treat depression (section 23.12.4 and Case study 7). Dual and triple action drugs are also being studied for the treatment of Alzheimer's disease (section 22.15).

BOX 12.3 Early tests for potential toxicity

In vivo and *in vitro* tests are often carried out at an early stage to find out whether lead compounds or candidate drugs are likely to have certain types of toxicity. One such test is to see whether compounds inhibit HERG potassium ion channels in the heart. HERG stands for the gene that codes for this protein, the so called **Human Ether-a-go-go Related Gene**! Who makes up these names? Several promising drugs have had to be withdrawn at a very late stage in their development because they were found to inhibit the HERG potassium ion channels. Inhibition can result in disruption in the normal rhythm of the heart, leading to fibrillation, heart failure, and death. The gastric agent **cisapride** (Fig. 1) and the antihistamine **terfenadine** (section 11.5.6) both had to be withdrawn from the market because of this problem. A large variety

FIGURE 1 Cisapride.

of other structures have been found to have this unwanted effect, and so tests to detect this property are best done as

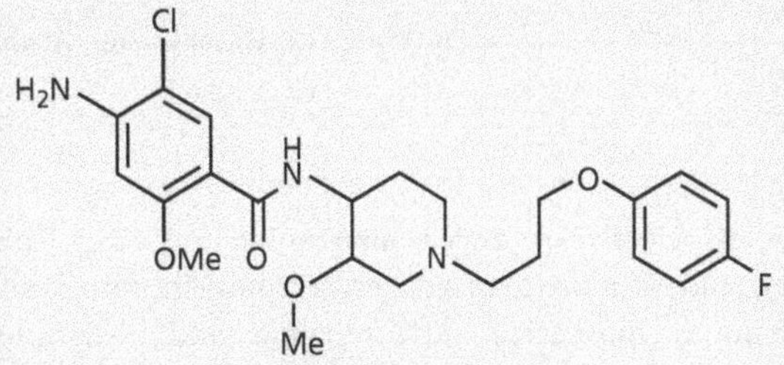

(Continued)

BOX 12.3 Early tests for potential toxicity (*Continued*)

early as possible in order to remove this property as part of the drug optimization process.

The **Ames test** is another early test that is worth carrying out in order to detect potential mutagenicity or carcinogenicity in new compounds. It involves the use of a mutated bacterial strain of *Salmonella typhimurium* that lacks the ability to synthesize histidine and can only grow in a medium containing that amino acid. The test involves growing the mutant strain in a medium that contains a small amount of histidine, as well as the test compound. Since there is only a small amount of histidine present, the mutant bacteria will soon stop growing and dividing. However, some of the mutant bacteria will 'back mutate' to the original wild-type strain. These cells are now able to synthesize their own histidine and will keep growing. The bacterial colonies that are present on the plate are subcultured onto plates lacking histidine to detect the presence of the wild-type strains, allowing a measure of the mutation rate. Any mutagenic or carcinogenic drug that is present in the original medium will increase the mutation rate, relative to a reference culture containing no drug.

Many research groups now concentrate on 'taming' Ames and HERG liabilities at an early stage of drug development. For example, structure I (Fig. 2) is an antagonist for the **melanin-concentrating hormone receptor** (MCHR)—a receptor that has been identified as an important target for novel anti-obesity drugs. Unfortunately, structure I blocks HERG ion channels and has Ames liability (i.e. it has mutagenic properties). A library of analogues was prepared by parallel synthesis (Chapter 16), which identified structure II as a potent antagonist having no Ames liability. Further work led to structure III, which lacked the Ames liability and has a greatly reduced capacity to block HERG ion channels.

Another example where studies were carried out to avoid interactions with the HERG ion channels was in the development of the antiviral agent **maraviroc** (section 20.7.5).

Microbioassay tests are also been developed to test for drug toxicity. These involve the use of microfluidic systems on microchips. Cells from different organs are grown in microchannels on the microchip and then tiny volumes of drug solution are passed through the microchip to see what effect they have.

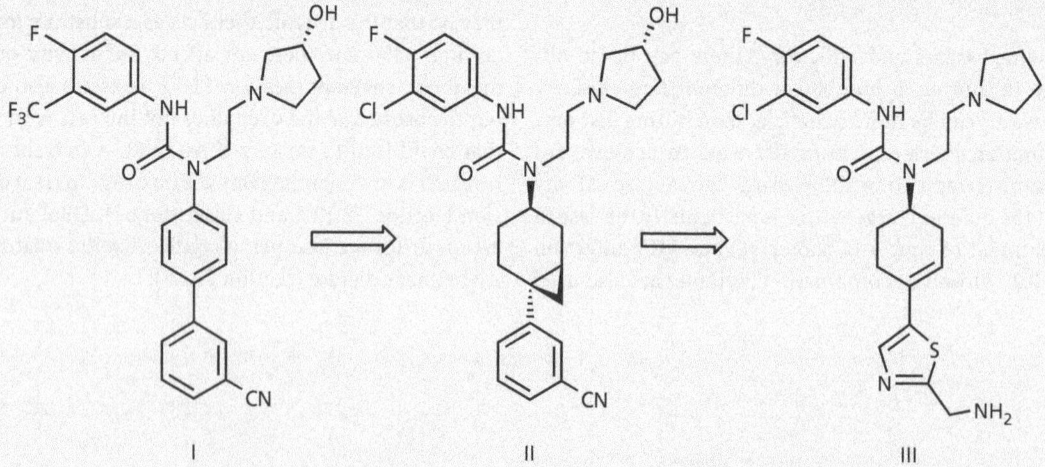

FIGURE 2 The development of agents to remove undesirable properties.

A less selective example is **olanzapine** (Fig. 12.4). This drug binds to more than a dozen receptors for serotonin, dopamine, muscarine, noradrenaline, and histamine. This kind of profile would normally be unacceptable, but olanzaprine has been highly effective in the treatment of schizophrenia, probably because it blocks both serotonin and dopamine receptors. Drugs which interact with a large range of targets are called **promiscuous ligands** or **dirty drugs**. Such drugs can act as lead compounds for the development of more selective multi-targeted ligands (see also section 22.15.3).

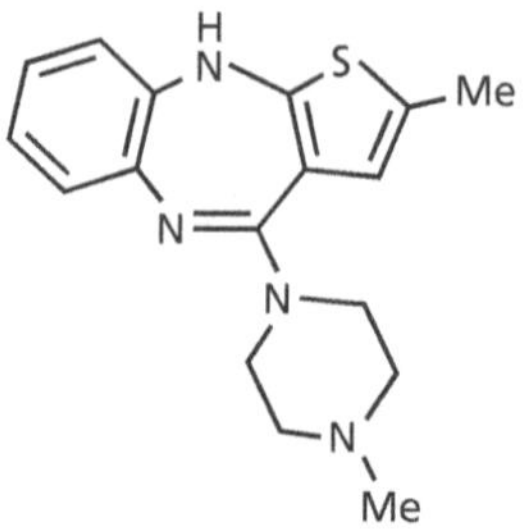

FIGURE 12.4 Olanzapine.

12.3 Identifying a bioassay

12.3.1 Choice of bioassay

Choosing the right bioassay or test system is crucial to the success of a drug research programme. The test should be simple, quick, and relevant, as there are usually a large number of compounds to be analysed. Human testing is not possible at such an early stage, so the test has to be done *in vitro* (i.e. on isolated cells, tissues, enzymes, or receptors) or *in vivo* (on animals). In general, *in vitro* tests are preferred over *in vivo* tests because they are cheaper, easier to carry out, less controversial, and they can be automated. However, *in vivo* tests are often needed to check whether drugs have the desired pharmacological activity, and whether they have acceptable pharmacokinetic properties. In modern medicinal chemistry, a variety of tests are usually carried out both *in vitro* and *in vivo* to determine not only whether the candidate drugs are acting at the desired target, but also whether they have activity at other undesired targets (Box 12.3). The direction taken by projects is then determined by finding drugs that have the best balance of good activity at the desired target and minimal activity at other targets. In this way, there is less likelihood of millions of dollars being wasted developing a drug that will either fail clinical trials or be withdrawn from the market with all the associated litigation that involves—a '**fail fast, fail cheap**' strategy.

12.3.2 *In vitro* tests

In vitro tests do not involve live animals. Instead, specific tissues, cells, or enzymes are used. Enzyme inhibitors can be tested on the pure enzyme in solution. In the past, it could be a major problem to isolate and purify sufficient enzyme to test, but nowadays genetic engineering can be used to incorporate the gene for a particular enzyme into fast-growing cells such as yeast or bacteria. These then produce the enzyme in larger quantities, making isolation easier. For example, **HIV protease** (section 20.7.4.1) has been cloned and expressed in the bacterium *Escherichia coli*. A variety of experiments can be carried out on this enzyme to determine whether an enzyme inhibitor is competitive or non-competitive, and to determine IC_{50} values (section 7.8).

Receptor agonists and antagonists can be tested on isolated tissues or cells which express the target receptor on their surface. Sometimes these tissues can be used to test drugs for physiological effects. For example, bronchodilator activity can be tested by observing how well compounds inhibit contraction of isolated tracheal smooth muscle. Alternatively, the affinity of drugs for receptors (how strongly they bind) can be measured by radioligand studies (section 8.9). Many *in vitro* tests have been designed by genetic engineering where the gene coding for a specific receptor is identified, cloned, and expressed in fast-dividing cells such as bacterial, yeast, or tumour cells. For example, **Chinese Hamster Ovarian cells** (CHO cells) are commonly used for this purpose, as they express a large amount of the cloned receptor on their cell surface. *In vitro* studies on whole cells are useful because there are none of the complications of *in vivo* studies where the drug has to survive metabolic enzymes or cross barriers such as the gut wall. The environment surrounding the cells can be easily controlled and both intracellular and intercellular events can be monitored, allowing measurement of efficacy and potency (section 8.9). Primary cell cultures (i.e. cells that have not been modified) can be produced from embryonic tissues; transformed cell lines are derived from tumour tissue. Cells grown in this fashion are all identical.

Antibacterial drugs are tested *in vitro* by measuring how effectively they inhibit or kill bacterial cells in culture. It may seem strange to describe this as an *in vitro* test, as bacterial cells are living microorganisms. However, *in vivo* antibacterial tests are defined as those that are carried out on animals or humans to test whether antibacterial agents combat infection.

The lack of a suitable *in vitro* test can actually prevent progress being made in a particular field of medicinal chemistry. For example, research into finding novel antiviral agents for the treatment of hepatitis C was hindered for many years because of the lack of a suitable *in vitro* test. An *in vitro* test has now been developed, but the time needed to achieve that goal was one of the main reasons why novel antiviral agents for the treatment of hepatitis C have only appeared on the market since 2011, whereas antiviral drugs for the treatment of HIV appeared in the 1980s.

In vitro tests are also used to test for the pharmacokinetic properties of compounds. For example, the **Caco-2 cell monolayer absorption** model is used to assess how well a drug is likely to be absorbed from the gastrointestinal tract. Microsomes and hepatocytes extracted from liver cells contain cytochrome P450 enzymes, and can be used to assess the likely metabolism of drug candidates, as well as identifying possible drug–drug interactions. Another *in vitro* assay using artificial membranes has been developed as a simple and rapid measure of how effectively drugs will cross the blood–brain barrier.

12.3.3 *In vivo* tests

In vivo tests on animals often involve inducing a clinical condition in the animal to produce observable symptoms. The animal is then treated to see whether the drug alleviates the problem by eliminating the observable symptoms.

For example, the development of non-steroidal inflammatory drugs was carried out by inducing inflammation on test animals, then testing drugs to see whether they relieved the inflammation.

Transgenic animals are often used in *in vivo* testing. These are animals whose genetic code has been altered. For example, it is possible to replace some mouse genes with human genes. The mouse produces the human receptor or enzyme and this allows *in vivo* testing against that target. Alternatively, the mouse's genes could be altered such that the animal becomes susceptible to a particular disease (e.g. breast cancer). Drugs can then be tested to see how well they prevent that disease.

There are several problems associated with *in vivo* testing. It is slow and expensive, and it also causes animal suffering. There are the many problems of pharmacokinetics (Chapter 11), and so the results obtained may be misleading and difficult to rationalize if *in vivo* tests are carried out in isolation. For example, how can one tell whether a negative result is due to the drug failing to bind to its target or not reaching the target in the first place? Thus, *in vitro* tests are usually carried out first to determine whether a drug interacts with its target, and *in vivo* tests are then carried out to test pharmacokinetic properties.

Certain *in vivo* tests might turn out to be invalid. It is possible that the observed symptoms might be caused by a different physiological mechanism than the one intended. For example, many promising anti-ulcer drugs which proved effective in animal testing were ineffective in clinical trials. Finally, different results may be obtained in different animal species. For example, **penicillin methyl ester prodrugs** (Box 19.7) are hydrolysed in mice or rats to produce active penicillins, but are not hydrolysed in rabbits, dogs, or humans. Another example involves **thalidomide** which is teratogenic in rabbits and humans, but has no such effect in mice.

Despite these issues, *in vivo* testing is still crucial in identifying the particular problems that might be associated with using a drug *in vivo* and which cannot be picked up by *in vitro* tests.

12.3.4 Test validity

Sometimes the validity of testing procedures is easy and clear-cut. For example, an antibacterial agent can be tested *in vitro* by measuring how effectively it kills bacterial cells. A local anaesthetic can be tested *in vitro* on how well it blocks action potentials in isolated nerve tissue. In other cases, the testing procedure is more difficult. For example, how do you test a new antipsychotic drug? There is no animal model for this condition and so a simple *in vivo* test is not possible. One way round this problem is to propose which receptor or receptors might be involved in a

medical condition, and to carry out *in vitro* tests against these in the expectation that the drug will have the desired activity when it comes to clinical trials. One problem with this approach is that it is not always clear-cut whether a specific receptor or enzyme is as important as one might think to the targeted disease (see Box 12.2).

12.3.5 High-throughput screening

Robotics and the miniaturization of *in vitro* tests on genetically modified cells has led to a process called high-throughput screening (HTS) which is particularly effective in identifying potential new lead compounds. This involves the automated testing of large numbers of compounds versus a large number of targets; typically, several thousand compounds can be tested at once in 30–50 biochemical tests. It is important that the test should produce an easily measurable effect which can be detected and measured automatically. This effect could be cell growth, an enzyme-catalysed reaction which produces a colour change, or displacement of radioactively labelled ligands from receptors.

Receptor antagonists can be studied using modified cells which contain the target receptor in their cell membrane. Detection is possible by observing how effectively the test compounds inhibit the binding of a radiolabelled ligand. Another approach is to use yeast cells which have been modified such that activation of a target receptor results in the activation of an enzyme which, when supplied with a suitable substrate, catalyses the release of a dye. This produces an easily identifiable colour change.

In general, positive hits are compounds which have an activity in the range 30 µM–1 nM. Unfortunately, HTS can generate many false-positive hits, and there is a high failure rate between the number of hits and those compounds which are eventually identified as authentic lead compounds. One of the main causes of false hits is what are known as **promiscuous inhibitors**. These are agents which appear to inhibit a range of different target proteins and show very poor selectivity. It is believed that agents working in this manner come together in solution to form molecular aggregates which adsorb target proteins onto their surface, resulting in the inhibition observed. The effect is more pronounced if mixtures of compounds are being tested in solution, such as those prepared by combinatorial syntheses. This kind of inhibition is of no use to drug design and it is important to eliminate these agents early on as potential lead compounds, such that time is not wasted resynthesizing and investigating them. One way of finding out whether promiscuous inhibition is taking place is to add a detergent to the test solution. This reverses and prevents the phenomenon.

Other false hits include agents which are chemically reactive and carry out a chemical reaction with the target protein, such as the alkylation or acylation of a susceptible nucleophilic group. This results in an irreversible inhibition of the protein since the agent becomes covalently linked to the target. Although there are important drugs which act as irreversible inhibitors, the emphasis in HTS is to find reversible inhibitors which interact with their targets through intermolecular binding interactions. For that reason, known alkylating or acylating agents should not be included in HTS, or, if they are, they should not be considered as potential lead compounds. Examples of reactive groups include alkyl halides, acid chlorides, epoxides, aldehydes, α-chloroketones, and trifluoromethyl ketones.

12.3.6 Screening by NMR

Nuclear magnetic resonance (NMR) spectroscopy is an analytical tool which has been used for many years to determine the molecular structure of compounds. More recently, it has been used to detect whether a compound binds to a protein target. In NMR spectroscopy, a compound is radiated with a short pulse of energy which excites the nuclei of specific atoms such as hydrogen, carbon, or nitrogen. Once the pulse of radiation has stopped, the excited nuclei slowly relax back to the ground state giving off energy as they do so. The time taken by different nuclei to give off this energy is called the **relaxation time** and this varies depending on the environment or position of each atom in the molecule. Therefore, a different signal will be obtained for each atom in the molecule, and a spectrum is obtained which can be used to determine the structure.

The size of the molecule also plays an important role in the length of the relaxation time. Drugs are generally small molecules and have long relaxation times, whereas large molecules such as proteins have short relaxation times. Therefore it is possible to delay the measurement of energy emission such that only small molecules are detected. This is the key to the detection of binding interactions between a protein and a test compound.

First of all, the NMR spectrum of the drug is taken, then the protein is added and the spectrum is re-run, introducing a delay in the measurement such that the protein signals are not detected. If the drug fails to bind to the protein, then its NMR spectrum will still be detected. If the drug binds to the protein, it essentially becomes part of the protein. As a result, its nuclei will have a shorter relaxation time and no NMR spectrum will be detected.

This screening method can also be applied to a mixture of compounds arising from a natural extract or from a combinatorial synthesis. If any of the compounds present bind to the protein, its relaxation time is shortened and so signals due to that compound will disappear from the spectrum. This will show that a component of the mixture is active and determine whether it is worthwhile separating the mixture or not.

There are several advantages in using NMR as a detection system:

- It is possible to screen 1000 small-molecular-weight compounds a day with one machine.
- The method can detect weak binding which would be missed by conventional screening methods.
- It can identify the binding of small molecules to different regions of the binding site (section 12.4.10).
- It is complimentary to HTS. The latter may give false-positive results, but these can be checked by NMR to ensure that the compounds concerned are binding in the correct binding site (section 12.4.10).
- The identification of small molecules which bind weakly to part of the binding site allows the possibility of using them as building blocks for the construction of larger molecules that bind more strongly (section 12.4.10).
- Screening can be done on a new protein without needing to know its function.

Disadvantages include the need to purify the protein and to obtain it in a significant quantity (at least 200 mg).

12.3.7 Affinity screening

A nice method of screening mixtures of compounds for active constituents is to take advantage of the binding affinity of compounds for the target. This not only detects the presence of such agents, but picks them out from the mixture. For example, the vancomycin family of antibacterial agents has a strong binding affinity for the dipeptide D-Ala-D-Ala (section 19.5.5.2). D-Ala-D-Ala was linked to sepharose resin, and the resin was mixed with extracts from various microbes which were known to have antibacterial activity. If an extract lost antibacterial activity as a result of this operation, it indicated that active compounds had bound to the resin. The resin could then be filtered off and, by changing the pH, the compounds could be released from the resin for identification.

12.3.8 Surface plasmon resonance

Surface plasmon resonance (SPR) is an optical method of detecting when a ligand binds to its target. The procedure is patented by Pharmacia Biosensor as **BIAcore** and makes use of a dextran-coated, gold-surfaced glass chip (Fig. 12.5). A ligand that is known to bind to the target is immobilized by linking it covalently to the dextran matrix, which is in a flow of buffer solution. Monochromatic, plane-polarized light is shone at an angle of incidence (α)

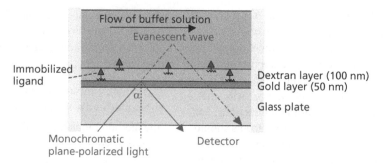

FIGURE 12.5 Surface plasmon resonance. The word evanescent means 'passing out of sight'.

from below the glass plate and is reflected back at the interface between the dense gold-coated glass and the less dense buffer solution. However, a component of the light called the **evanescent wave** penetrates a distance of about one wavelength into the buffer/dextran matrix. Normally, all of the light including the evanescent wave is reflected back, but if the gold film is very thin (a fraction of the evanescent wavelength), and the angle of incidence is exactly right, the evanescent wave interacts with free oscillating electrons called **plasmons** in the metal film. This is the surface plasmon resonance. Energy from the incident light is then lost to the gold film. As a result, there is a decrease in the reflected light intensity, which can be measured.

The angle of incidence when SPR occurs depends crucially on the refractive index of the buffer solution close to the metal film surface. This means that if the refractive index of the buffer changes, the angle of incidence at which SPR takes place changes as well.

If the macromolecular target for the immobilized ligand is now introduced into the buffer flow, some of it will be bound by the immobilized ligand. This leads to a change of refractive index in the buffer solution close to the metal-coated surface, which can be detected by measuring the change in the angle of incidence required to get SPR. The technique allows the detection of ligand–target binding and can also be used to measure rate and equilibrium binding constants.

Suppose, now, we want to test whether a novel compound is binding to the target. This can be tested by introducing the novel compound into the buffer flow along with the target. If the test compound *does* bind to the target, less target will be available to bind to the immobilized ligands, so there will be a different change in both the refractive index and the angle of incidence.

12.3.9 Scintillation proximity assay

Scintillation proximity assay (SPA) is a visual method of detecting whether a ligand binds to a target. It involves the immobilization of the target by linking it covalently to beads which are coated with a scintillant. A solution of a known ligand labelled with iodine-125 is then added to the beads. When the labelled ligand binds to the immobilized target, the ^{125}I acts as an energy donor and the scintillant-coated bead acts as an energy acceptor, resulting in an emission of light that can be detected. In order to find out whether a novel compound interacts with the target, the compound is added to the solution of the labelled ligand and the mixture is added to the beads. Successful binding by the novel compound will mean that less of the labelled ligand will bind, resulting in a reduction in the emission of light.

12.3.10 Isothermal titration calorimetry

Isothermal titration calorimetry (ITC) is a technique that is used to determine the thermodynamic properties of binding between a drug and its protein target, in particular the binding affinity and enthalpy change. Two identical glass cells are used which are filled with buffer solution. One of the cells acts as the reference cell, while the other acts as the sample cell and contains the protein target in solution. The reference cell is heated slightly to a constant temperature. The sample cell is heated to the same temperature through an automatic feedback system, whereby any temperature difference between the two cells is detected and power is applied to the sample cell to equalize the temperature. Once the apparatus has stabilized, a constant level of power is being used to maintain the two cells at the same constant temperature.

The drug is now added to the sample cell and binds to the protein target. If the binding interaction is exothermic, heat energy is generated within the sample cell and so less external power is needed to maintain the cell temperature. If the interaction is endothermic, the opposite holds true and more external power has to be applied to maintain the temperature. The external power required to maintain the temperature of the sample cell is measured with respect to time, with power 'spikes' occurring every time the drug is injected into the cell. Measurement of these spikes allows the determination of the thermodynamic properties of binding.

12.3.11 Virtual screening

Virtual screening involves the use of computer programs to assess whether known compounds are likely to be lead compounds for a particular target. There is no guarantee that 'positive hits' from a virtual screening will, in fact, be active, and the compounds still have to be screened experimentally, but the results from a virtual screening can be used to make experimental screening methods more efficient. In other words, if there are several thousand compounds available for testing, virtual screening can be used to identify those compounds which are most likely to be active, and so those are the structures which would be given priority for actual screening. Virtual screening can involve a search for pharmacophores known to be required for activity. Alternatively, compounds can be docked into target binding sites (sections 17.11–17.13).

KEY POINTS

- Pharmaceutical companies tend to concentrate on developing drugs for diseases which are prevalent in developed countries, and aim to produce compounds with better properties than existing drugs.

- A molecular target is chosen which is believed to influence a particular disease when affected by a drug. The greater the selectivity that can be achieved, the less chance of side effects.

- A suitable bioassay must be devised which will demonstrate whether a drug has activity against a particular target. Bioassays can be carried out *in vitro* or *in vivo*, and usually a combination of tests is used.

- HTS involves the miniaturization and automation of *in vitro* tests such that a large number of tests can be carried out in a short period of time.

- Compounds can be tested by NMR spectroscopy for their affinity to a macromolecular target. The relaxation times of ligands bound to a macromolecule are shorter than when they are unbound.

- SPR, SPA, and ITC are three visual methods of detecting whether ligands bind to macromolecular targets.

- Virtual screening can be used to identify the compounds most likely to be active in experimental screening.

12.4 Finding a lead compound

Once a target and a testing system have been chosen, the next stage is to find a lead compound—a compound which shows the desired pharmacological activity. The level of activity may not be very great and there may be undesirable side effects, but the lead compound provides a start for the drug design and development process. There are various ways in which a lead compound might be discovered as described in the following sections.

12.4.1 Screening of natural products

Natural products are a rich source of biologically active compounds. Many of today's medicines are either obtained directly from a natural source or were developed from a lead compound originally obtained from a natural source. Usually, the natural source has some form of biological activity, and the compound responsible for that activity is known as the **active principle**. Such a structure can act as a lead compound. Most biologically active natural products are **secondary metabolites** with quite complex structures and several chiral centres. This has an advantage in that they are extremely novel compounds. Unfortunately, this complexity also makes their synthesis difficult and the compounds usually have to be extracted from their natural source—a slow, expensive, and inefficient process. As a result, there is usually an advantage in designing simpler analogues (section 13.3.8).

Many natural products have radically new chemical structures which no chemist would dream of synthesizing. For example, the antimalarial drug **artemisinin** (Fig. 12.6) is a natural product with an extremely unstable looking trioxane ring—one of the most unlikely structures to have appeared in recent years (see also Case study 3).

The study of medicines derived from natural sources is known as **pharmacognosy**, and includes both crude extracts and purified active principles.

12.4.1.1 The plant kingdom

Plants have always been a rich source of lead compounds (e.g. **morphine**, **cocaine**, **digitalis**, **quinine**, **tubocurarine**, **nicotine**, and **muscarine**). Many of these lead compounds are useful drugs in themselves (e.g. morphine and quinine), and others have been the basis for synthetic drugs (e.g. local anaesthetics developed from cocaine). Plants still remain a promising source of new drugs and will continue to be so. Clinically useful drugs which have recently been isolated from plants include the anticancer agent **paclitaxel (Taxol)** from the yew tree, the antimalarial agent artemisinin from a Chinese plant (Fig. 12.6), and the Alzheimer's drug **galantamine** from daffodils (section 22.15.1).

Plants provide a bank of rich, complex, and highly varied structures which are unlikely to be discovered from other sources. Furthermore, evolution has already carried out a screening process that favours compounds which provide plants with an 'edge' when it comes to survival. For example, biologically potent compounds

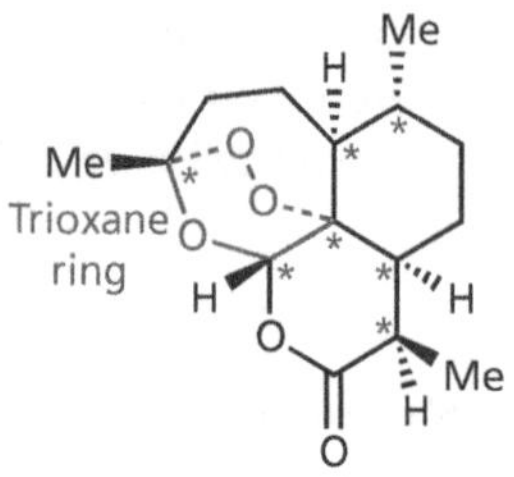

Artemisinin

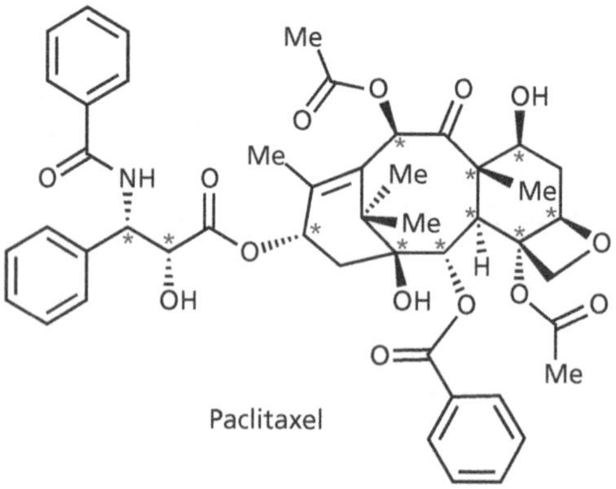

Paclitaxel

FIGURE 12.6 Plant natural products as drugs (the asterisks indicate chiral centres).

can deter animals or insects from eating the plants that contain them. Considering the debt medicinal chemistry owes to the natural world, it is sobering to think that very few plants have been fully studied and the vast majority have not been studied at all. The rainforests of the world are particularly rich in plant species which have still to be discovered, let alone studied. Who knows how many exciting new lead compounds await discovery for the fight against cancer, AIDS, or any of the other myriad of human afflictions? This is one reason why the destruction of rainforests and other ecosystems is so tragic; once these ecosystems are destroyed, unique plant species are lost to medicine for ever. For example, **silphion**—a plant that was cultivated near Cyrene in North Africa and was famed as a contraceptive agent in ancient Greece—is now extinct. It is certain that many more useful plants have become extinct without medicine ever being aware of them.

12.4.1.2 Microorganisms

Microorganisms such as bacteria and fungi have also provided rich pickings for drugs and lead compounds. These organisms produce a large variety of antimicrobial agents which have evolved to give their hosts an advantage over their competitors in the microbiological world. The screening of microorganisms became highly popular

after the discovery of **penicillin**. Soil and water samples were collected from all round the world in order to study new fungal or bacterial strains, leading to an impressive arsenal of antibacterial agents such as the **cephalosporins**, **tetracyclines**, **aminoglycosides**, **rifamycins**, **chloramphenicol**, and **vancomycin** (Chapter 19). Although most of the drugs derived from microorganisms are used in antibacterial therapy, some microbial metabolites have provided lead compounds in other fields of medicine. For example, **asperlicin**—isolated from *Aspergillus alliaceus*—is a novel antagonist of a peptide hormone called **cholecystokinin** (CCK) which is involved in the control of appetite. CCK also acts as a neurotransmitter in the brain and is thought to be involved in panic attacks. Analogues of asperlicin may therefore have potential in treating anxiety (see also Box 13.2).

Other examples include the fungal metabolite **lovastatin**, which was the first of the clinically useful statins found to lower cholesterol levels (Case study 1), and another fungal metabolite called **ciclosporin** (Fig. 12.7) which is used to suppress the immune response after organ transplants. **Lipstatin** (Fig. 12.7) is a natural product which was isolated from *Streptomyces toxytricini*. It inhibits pancreatic lipase, and was the lead compound for the anti-obesity compound **orlistat** (Box 7.2). Finally, a fungal metabolite called **rasfonin** (isolated from a fungus in New Zealand) promotes cell death (apoptosis) in

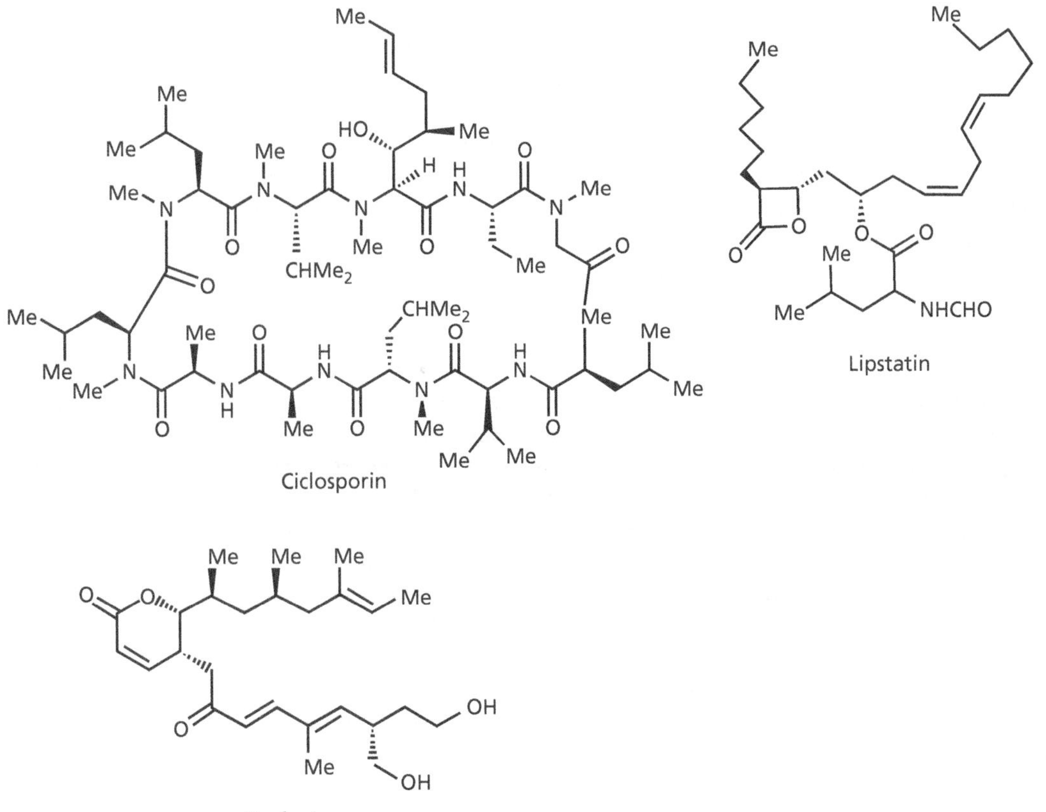

Ciclosporin

Lipstatin

Rasfonin

FIGURE 12.7 Lead compounds from microbiological sources.

cancer cells, but not normal cells. It represents a promising lead compound for novel anticancer agents.

12.4.1.3 Marine sources

In recent years, there has been great interest in finding lead compounds from marine sources. Coral, sponges, fish, and marine microorganisms have a wealth of biologically potent chemicals with interesting inflammatory, antiviral, and anticancer activity. For example, **curacin A** (Fig. 12.8) is obtained from a marine cyanobacterium, and shows potent antitumour activity. Other antitumour agents derived from marine sources include **eleutherobin, bryostatins, dolastatins, cephalostatins,** and **halichondrin B** (sections 21.5.2 and 21.9.2). In 2010, a simplified analogue of halichondrin B was approved for the treatment of breast cancer.

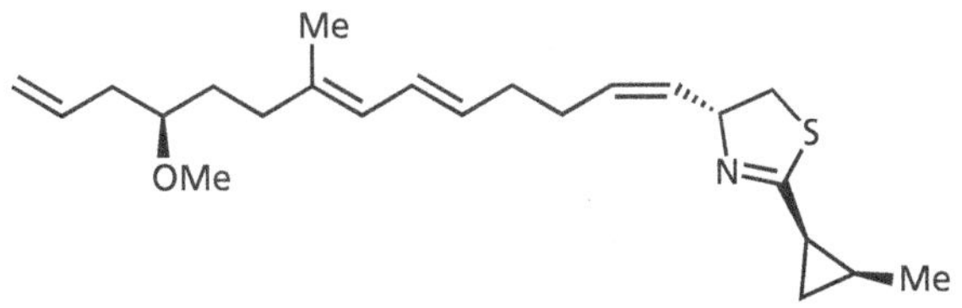

FIGURE 12.8 Curacin A.

12.4.1.4 Animal sources

Animals can sometimes be a source of new lead compounds. For example, a series of antibiotic polypeptides known as the **magainins** were extracted from the skin of the African clawed frog *Xenopus laevis*. These agents protect the frog from infection and may provide clues to the development of novel antibacterial and antifungal agents in human medicine. Another example is a potent analgesic compound called **epibatidine** (Fig. 12.9), obtained from the skin extracts of the Ecuadorian poison frog.

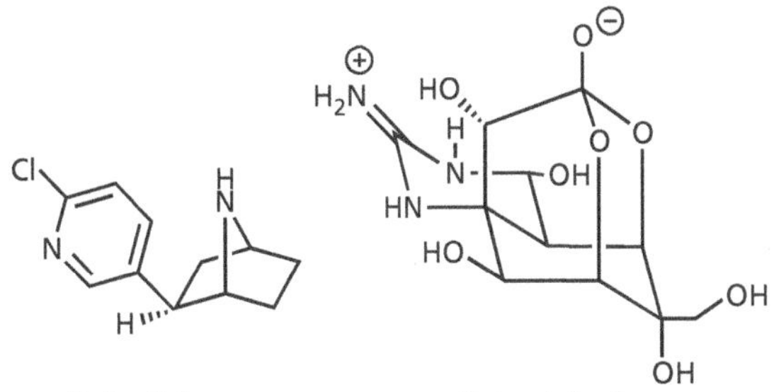

Epibatidine

Tetrodotoxin

FIGURE 12.9 Natural products as drugs.

12.4.1.5 Venoms and toxins

Venoms and toxins from animals, plants, snakes, spiders, scorpions, insects, and microorganisms are extremely potent because they often have very specific interactions with a macromolecular target in the body. As a result, they have proved important tools in studying receptors, ion channels, and enzymes. Many of these toxins are polypeptides (e.g. α-**bungarotoxin** from cobras). However, non-peptide toxins such as **tetrodotoxin** from the puffer fish (Fig. 12.9) are also extremely potent.

Venoms and toxins have been used as lead compounds in the development of novel drugs. For example, **teprotide**, a peptide isolated from the venom of the Brazilian viper, was a lead compound for the development of the antihypertensive agents **cilazapril** and **captopril** (Case study 2).

The neurotoxins from *Clostridium botulinum* are responsible for serious food poisoning (**botulism**), but they have a clinical use as well. They can be injected into specific muscles (such as those controlling the eyelid) to prevent muscle spasm. These toxins prevent cholinergic transmission (Chapter 22) and could well prove a lead for the development of novel anticholinergic drugs.

Finally, **conotoxin** is a peptide toxin derived from the marine cone snail, and has very powerful analgesic properties in humans. A synthetic form of conotoxin called **ziconotide** was approved in 2004 for the treatment of chronic pain.

12.4.2 Medical folklore

In the past, ancient civilizations depended greatly on local flora and fauna for their survival. They would experiment with various berries, leaves, and roots to find out what effects they had. As a result, many brews were claimed by the local healer or shaman to have some medicinal use. More often than not, these concoctions were useless or downright dangerous, and if they worked at all, it was because the patient willed them to work—a **placebo effect**. However, some of these extracts may indeed have a real and beneficial effect, and a study of medical folklore can give clues as to which plants might be worth studying in more detail. **Rhubarb** root has been used as a purgative for many centuries. In China, it was called 'The General' because of its 'galloping charge'! The most significant chemicals in rhubarb root are anthraquinones, which were used as the lead compounds in the design of the laxative—**dantron** (Fig. 12.10).

The ancient records of Chinese medicine also provided the clue to the novel antimalarial drug **artemisinin** mentioned in section 12.4.1 (see also Case study 3). The therapeutic properties of the opium poppy (active principle **morphine**) were known in Ancient Egypt, as were those of the *Solanaceae* plants in ancient Greece (active principles **atropine** and **hyoscine**; section 22.9.2). The snakeroot plant was well regarded in India (active principle **reserpine**; Fig. 12.10), and herbalists in medieval England used extracts from the willow tree (active principle **salicin**; Fig. 12.10) and foxglove (active principle **digitalis**—a mixture of compounds such as digitoxin, digitonin, and digitalin). The Aztec and Mayan cultures of South America used extracts from a variety of bushes and trees including the ipecacuanha root (active principle **emetine**; Fig. 12.10), coca bush (active principle **cocaine**), and cinchona bark (active principle **quinine**).

12.4.3 Screening synthetic compound 'libraries'

The thousands of compounds which have been synthesized by the pharmaceutical companies over the years

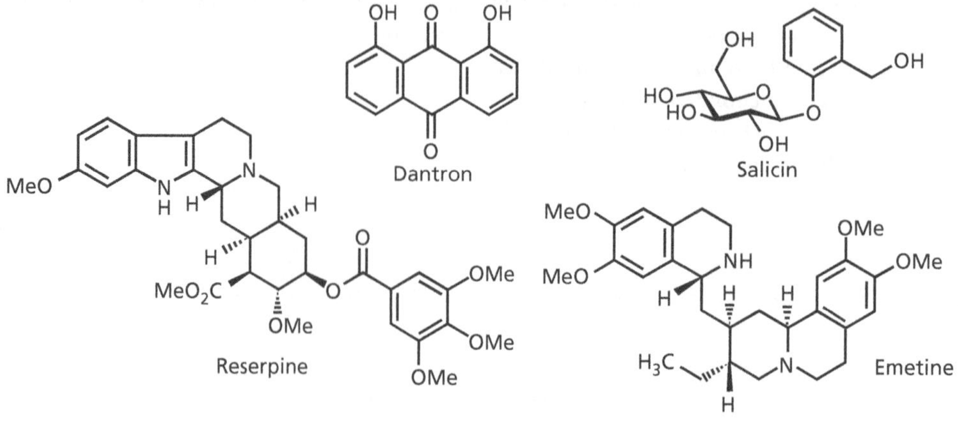

FIGURE 12.10 Active compounds resulting from studies of herbs and potions.

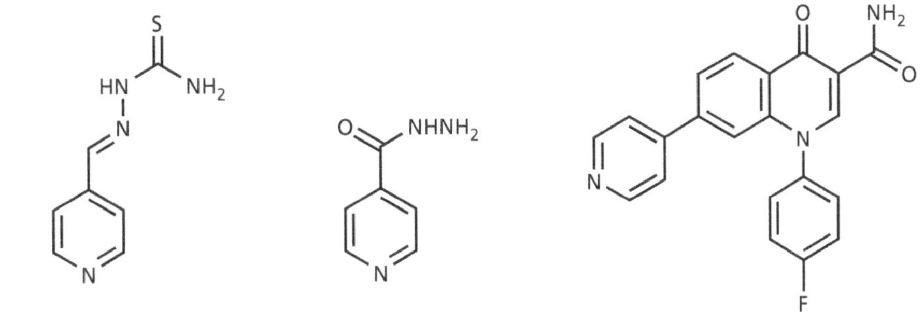

FIGURE 12.11 Pharmaceutically active compounds discovered from synthetic intermediates.

are another source of lead compounds. The vast majority of these compounds have never made the market place, but they have been stored in compound 'libraries' and are still available for testing. Pharmaceutical companies often screen their library of compounds whenever they study a new target. However, it has to be said that the vast majority of these compounds are merely variations on a theme; for example 1000 or so different penicillin structures. This reduces the chances of finding a novel lead compound.

Pharmaceutical companies often try to diversify their range of structures by purchasing novel compounds prepared by research groups elsewhere—a useful source of revenue for hard-pressed university departments! These compounds may never have been synthesized with medicinal chemistry in mind, but there is always the chance that they may have useful biological activity.

It can also be worth testing synthetic intermediates. For example, a series of thiosemicarbazones was synthesized and tested as antitubercular agents in the 1950s. This included **isonicotinaldehyde thiosemicarbazone**, the synthesis of which involved the hydrazide structure **isoniazid** (Fig. 12.11) as a synthetic intermediate. It was subsequently found that isoniazid had greater activity than the target structure. Similarly, a series of **quinoline-3-carboxamide** intermediates (Fig. 12.11) were found to have antiviral activity.

12.4.4 Existing drugs

12.4.4.1 'Me too' and 'me better' drugs

Many companies use established drugs from their competitors as lead compounds in order to design a drug that gives them a foothold in the same market area. The aim is to modify the structure sufficiently such that it avoids patent restrictions, retains activity, and ideally has improved therapeutic properties. For example, the antihypertensive drug **captopril** was used as a lead compound by various

companies to produce their own antihypertensive agents (Fig. 12.12, see also Case study 2).

Although often disparaged as 'me too' drugs, they can often offer improvements over the original drug ('me better' drugs). For example, modern penicillins are more selective, more potent, and more stable than the original penicillins. Newer statins that lower cholesterol levels also have improved properties over older ones (Case study 1). It should also be noted that it is not unusual for companies to be working on similar looking structures for a particular disease at the same time. The first of these drugs to reach the market gets all the kudos, but it is rather unfair to call the drugs that follow it as 'me too' drugs, since they were designed and developed independently.

12.4.4.2 Enhancing a side effect

An existing drug usually has a minor property or an undesirable side effect which could be of use in another area of medicine. As such, the drug could act as a lead compound on the basis of its side effects. The aim would then be to enhance the desired side effect and to eliminate the major biological activity. This has been described as the SOSA approach—**selective optimization of side activities**. Choosing a known drug as the lead compound for a side effect has the advantage that the compound is already 'drug-like' and it should be more feasible to develop a clinically useful drug with the required pharmacodynamic and pharmacokinetic properties. Many of the 'hits' obtained from HTS do not have a 'drug-like' structure and it may require far more effort to optimize them. Indeed, it has been argued that modifications of known drug structures should provide lead compounds in several areas of medicinal chemistry. Many research groups are now screening compounds that are either in clinical use or have reached late-stage clinical trials to see whether they have side effects that would make them suitable lead compounds. The John Hopkins Clinical Compound Library is one such source of these compounds.

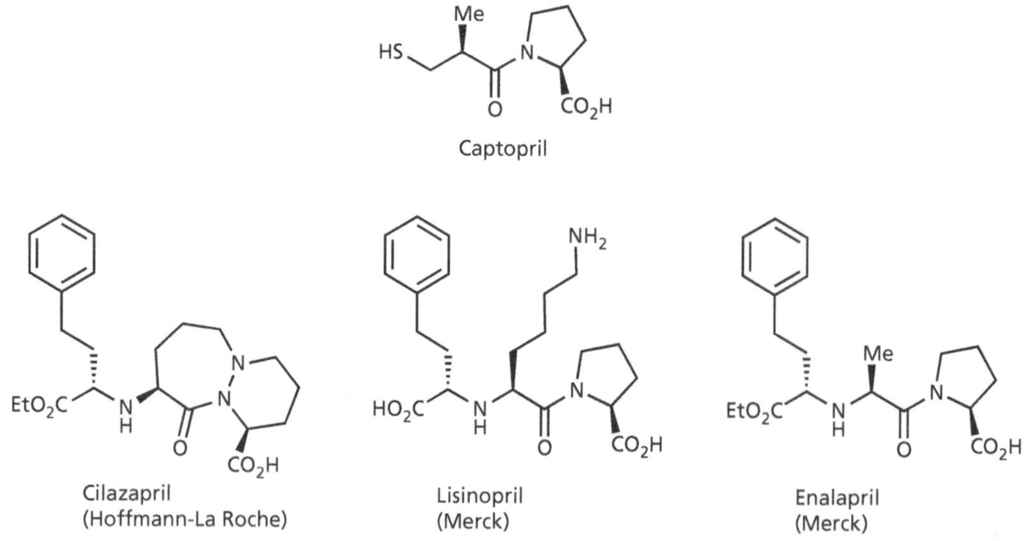

FIGURE 12.12 Captopril and 'me too' drugs.

For example, most sulphonamides have been used as antibacterial agents. However, some sulphonamides with antibacterial activity could not be used clinically because they had convulsive side effects brought on by **hypoglycaemia** (lowered glucose levels in the blood). Clearly, this is an undesirable side effect for an antibacterial agent, but the ability to lower blood glucose levels would be useful in the treatment of diabetes. Therefore, structural alterations were made to the sulphonamides concerned in order to eliminate the antibacterial activity and to enhance the hypoglycaemic activity. This led to the antidiabetic agent **tolbutamide** (Fig. 12.13). Another example was the discovery that the anticoagulant **warfarin** is also a weak inhibitor of a viral enzyme that is important in the life cycle of HIV. Warfarin was used as the lead compound in the development of an anti-HIV drug called **tipranavir** (section 20.7.4.10).

In some cases, the side effect may be strong enough that the drug can be used without modification. For example, the anti-impotence drug **sildenafil (Viagra)** (Fig. 12.13) was originally designed as a vasodilator to treat angina and hypertension (section 26.5.2). During clinical trials, it was found that it acted as a vasodilator more effectively in the penis than in the heart, resulting in increased erectile function. The drug is now used to treat erectile dysfunction and sexual impotence. Another example is the antidepressant drug **bupropion**. Patients taking this drug reported that it helped them give up smoking, and so the drug is now marketed as an antismoking aid (**Zyban**) (section 23.12.4). **Astemizole** (Fig. 12.13) is a medication used in the treatment of allergy, but has been found to be a potent antimalarial agent.

The moral of the story is that a drug used in one field of medicinal chemistry could be the lead compound

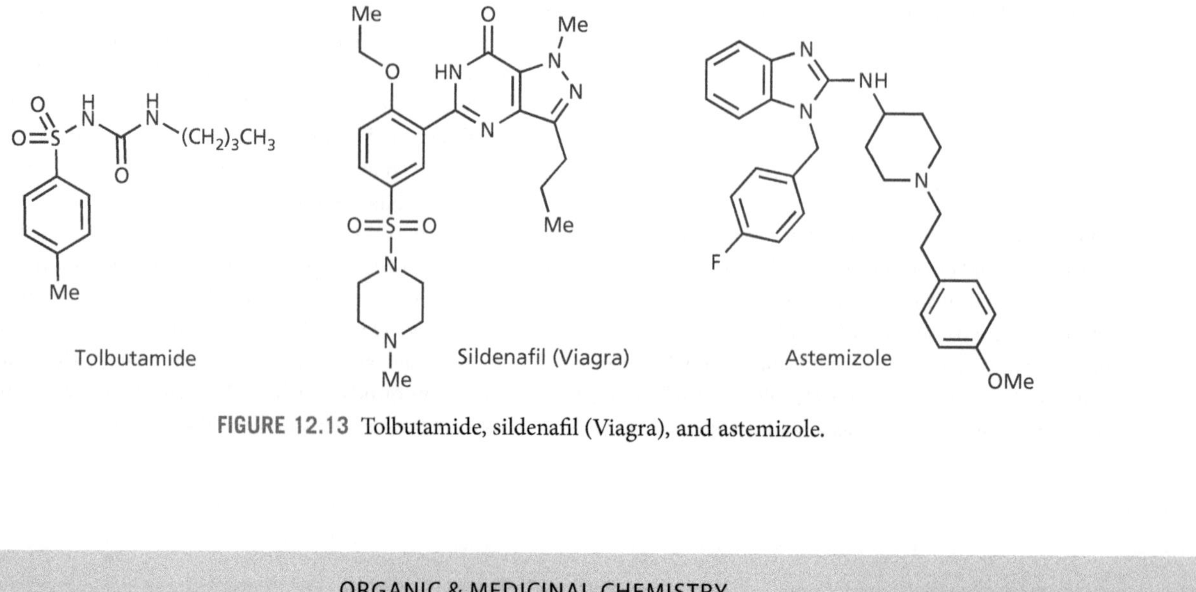

FIGURE 12.13 Tolbutamide, sildenafil (Viagra), and astemizole.

BOX 12.4 Selective optimization of side activities (SOSA)

Several drugs have been developed by enhancing the side effect of another drug (Fig. 1). **Chlorpromazine** is used as a neuroleptic agent in psychiatry, but was developed from the antihistamine agent **promethazine**. This might appear an odd thing to do, but it is known that promethazine has sedative side effects, and so medicinal chemists modified the structure to enhance the sedative effects at the expense of antihistamine activity. Similarly, the development of sulphonamide diuretics such as **chlorothiazide** arose from the observation that **sulphanilamide** has a diuretic effect in large doses (due to its action on an enzyme called **carbonic anhydrase**).

Sometimes, slight changes to a structure can result in significant changes in pharmacological activity. For example,

minaprine (Fig. 2) is an antidepressant agent that acts as a serotonin agonist. Adding a phenolic substituent resulted in **4-hydroxyminaprine**, which is a potent dopamine agonist, whereas adding a cyano substituent gave **bazinaprine**, which is a potent inhibitor of the enzyme **monoamine oxidase-A**. Minaprine also binds weakly to muscarinic receptors, and modifications were successfully carried out to give structure I, having potent activity for the muscarinic receptor and negligible activity for dopamine and serotonin receptors. Minaprine also has weak affinity for the cholinesterase enzyme, and modifications led to structure II with over 1000-fold increased affinity.

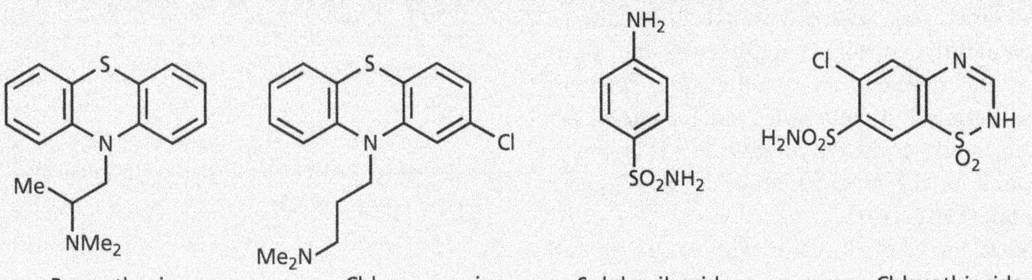

FIGURE 1 Drugs developed by enhancing a side effect.

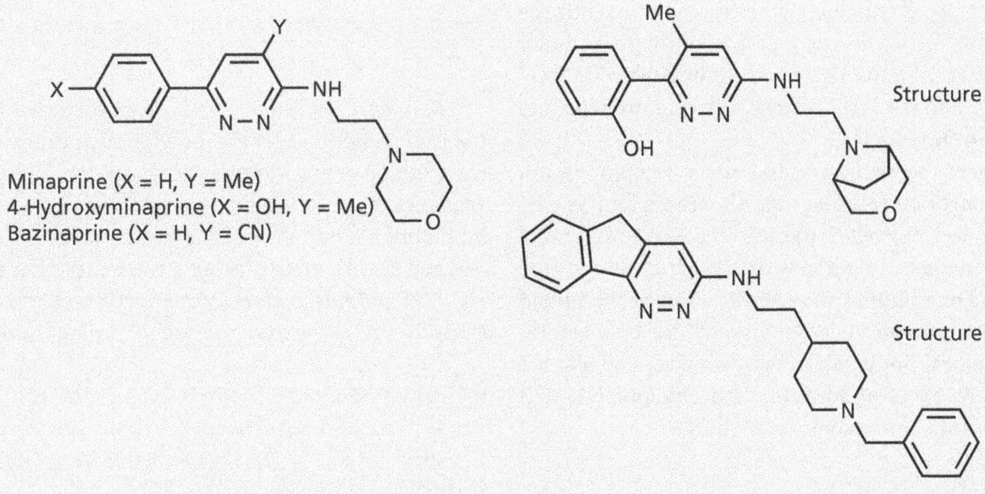

FIGURE 2 Structures with different pharmacological properties derived from the lead compound minaprine.

in another field (Box 12.4). Furthermore, one can fall into the trap of thinking that a structural group of compounds all have the same type of biological activity. The

sulphonamides are generally thought of as antibacterial agents, but we have seen that they can also have other properties.

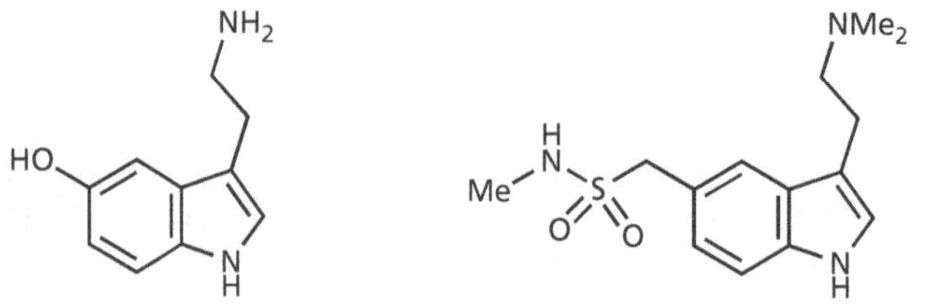

5-Hydroxytryptamine   Sumatriptan (5-HT$_1$ agonist)

FIGURE 12.14 5-Hydroxytryptamine and sumatriptan.

12.4.5 Starting from the natural ligand or modulator

12.4.5.1 Natural ligands for receptors

The natural ligand of a target receptor has sometimes been used as a lead compound. The natural ligands **adrenaline** and **noradrenaline** were the starting points for the development of adrenergic β-agonists such as **salbutamol**, **dobutamine**, and **xamoterol** (section 23.10), and 5-hydroxytryptamine (5-HT) was the starting point for the development of the 5-HT$_1$ agonist **sumatriptan** (Fig. 12.14).

The natural ligand of a receptor can also be used as the lead compound in the design of an antagonist. For example, **histamine** was used as the original lead compound in the development of the H$_2$ histamine antagonist **cimetidine** (section 25.2). Turning an agonist into an antagonist is frequently achieved by adding extra binding groups to the lead structure. Other examples include the development of the adrenergic antagonist **pronethalol** (section 23.11.3.1), the H$_2$ antagonist **burimamide** (section 25.2.4), and the 5-HT$_3$ antagonists **ondansetron** and **granisetron** (Box 12.2)

Sometimes the natural ligand for a receptor is not known (an **orphan receptor**) and the search for it can be a major project in itself. If the search is successful, however, it opens up a brand-new area of drug design (see Box 12.5). For example, the identification of the opioid receptors for **morphine** led to a search for endogenous opioids (natural body painkillers), which eventually led to the discovery of **endorphins** and **enkephalins**, and their use as lead compounds (section 24.8).

12.4.5.2 Natural substrates for enzymes

The natural substrate for an enzyme can be used as the lead compound in the design of an enzyme inhibitor. For example, **enkephalins** have been used as lead compounds for the design of enkephalinase inhibitors. **Enkephalinases** are enzymes which metabolize enkephalins, and their inhibition should prolong the activity of enkephalins (section 24.8.4).

> **BOX 12.5** Natural ligands as lead compounds
>
> The discovery of **cannabinoid** receptors in the early 1990s led to the discovery of two endogenous cannabinoid messengers—**arachidonylethanolamine (anandamide)** and **2-arachidonyl glycerol**. These have now been used as lead compounds to develop agents that interact with cannabinoid receptors. Such agents may prove useful in suppressing nausea during chemotherapy or in stimulating appetite in patients with AIDS. 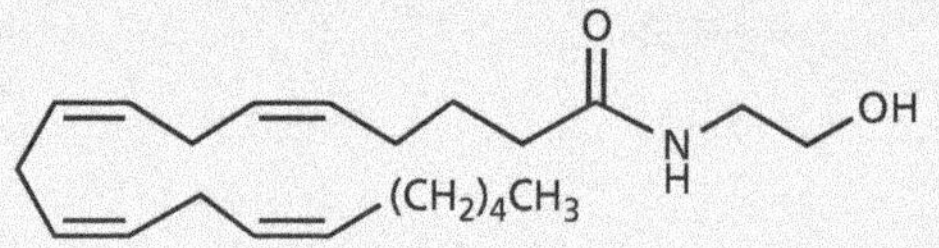
>
>
> **FIGURE 1** Anandamide.

The natural substrate for HIV protease was used as the lead compound for the development of the first protease inhibitor used to treat HIV (section 20.7.4). Other examples of substrates being used as lead compounds for inhibitors include the substrates for farnesyl transferase (section 21.6.1), matrix metalloproteinase (section 21.7.1), HCV NS3-4A protease (section 20.10.1), renin (Case study 8), and 17β-hydroxysteroid dehydrogenase type 1.

For additional material see Web article 1: Steroids as novel anticancer agents (*web article on 17β-HSD type 1*) on the Online Resource Centre at www.oxfordtextbooks.co.uk/orc/patrick6e/

12.4.5.3 Enzyme products as lead compounds

It should be remembered that enzymes catalyse a reaction in both directions, and so the product of an enzyme-catalysed reaction can also be used as a lead compound for an enzyme inhibitor. For example, the design of the carboxypeptidase inhibitor **L-benzylsuccinic acid** was based on the products arising from the carboxypeptidase-catalysed hydrolysis of peptides (see Case study 2). Similarly, the development of some antiviral agents for the treatment of hepatitis C began with a lead compound that was the product of the enzyme-catalysed reaction (Case study 10).

12.4.5.4 Natural modulators as lead compounds

Many receptors and enzymes are under allosteric control (sections 3.6 and 8.3.2). The natural or endogenous chemicals that exert this control (modulators) could also serve as lead compounds. For example, ATP is a natural antagonist for a platelet receptor called the P2Y$_{12}$ receptor. It was used as the lead compound in the development of the antiplatelet agent **cangrelor** (section 26.9.2.3).

In some cases, a modulator for an enzyme or receptor is suspected but has not yet been found. For example, the **benzodiazepines** are synthetic compounds that modulate the receptor for γ-**aminobutyric acid** (GABA) by binding to an allosteric binding site. The natural modulators for this allosteric site were not known at the time benzodiazepines were synthesized, but endogenous peptides called **endozepines** have since been discovered which bind to the same allosteric binding site, and which may serve as lead compounds for novel drugs having the same activity as the benzodiazepines.

12.4.6 Combinatorial and parallel synthesis

The growing number of potentially new drug targets arising from genomic and proteomic projects has meant that there is an urgent need to find new lead compounds to interact with them. Unfortunately, the traditional sources of lead compounds have not managed to keep pace and in the last decade or so, research groups have invested greatly in combinatorial and parallel synthesis in order to tackle this problem. Combinatorial synthesis is an automated solid-phase procedure aimed at producing as many different structures as possible in as short a time as possible. The reactions are carried out on very small scale, often in a way that will produce mixtures of compounds in each reaction vial. In a sense, combinatorial synthesis aims to mimic what plants do, i.e. produce a pool of chemicals, one of which may prove to be a useful lead compound. Combinatorial synthesis has developed so swiftly that it is almost a branch of chemistry in itself and a separate chapter is devoted to it (Chapter 16). Parallel synthesis involves the small scale synthesis of large numbers of compounds at the same time using specialist miniaturized equipment. The synthesis can be carried out in solution or solid phase, and each reaction vial contains a distinct product (Chapter 16). Nowadays, parallel synthesis is generally preferred over combinatorial synthesis in order to produce smaller, more focused compound libraries.

12.4.7 Computer-aided design of lead compounds

A detailed knowledge of a target binding site significantly aids in the design of novel lead compounds intended to bind with that target. In cases where enzymes or receptors can be crystallized, it is possible to determine the structure of the protein and its binding site by **X-ray crystallography**. Molecular modelling software programs can then be used to study the binding site, and to design molecules which will fit and bind to the site—*de novo* drug design (section 17.15).

In some cases, the enzyme or receptor cannot be crystallized and so X-ray crystallography cannot be carried out. However, if the structure of an analogous protein has been determined, this can be used as the basis for generating a computer model of the protein. This is covered in more detail in section 17.14. NMR spectroscopy has also been effective in determining the structure of proteins and can be applied to proteins that cannot be studied by X-ray crystallography.

12.4.8 Serendipity and the prepared mind

Frequently, lead compounds are found as a result of serendipity (i.e. chance). However, it still needs someone with an inquisitive nature or a prepared mind to recognize the significance of chance discoveries and to take advantage of these events. The discovery of **cisplatin** (section 9.3.4) and **penicillin** (section 19.5.1.1) are two such examples, but there are many more (see Box 12.6).

Sometimes, the research carried out to improve a drug can have unexpected and beneficial spin offs. For example, **propranolol** (Fig. 12.15) and its analogues are effective β-blockers (antagonists of β-adrenergic receptors) (section 23.11.3). However, they are also lipophilic, which means that they can cross the blood–brain barrier and cause CNS side effects. To counteract this, more hydrophilic analogues were designed by decreasing the size of the aromatic ring system and adding a hydrophilic amide group. One of the compounds made was **practolol**. As expected, this compound had fewer CNS side effects, but, more importantly, it was found to be a selective antagonist for the β-receptors of the heart over β-receptors in other organs—a result that was highly desirable, but not the one that was being looked for at the time.

Frequently, new lead compounds have arisen from research projects carried out in a totally different field of medicinal chemistry. This emphasizes the importance of keeping an open mind, especially when testing for biological activity. For example, we have already described the development of the antidiabetic drug **tolbutamide** (section 12.4.4.2), based on the observation that some antibacterial sulphonamides could lower blood glucose levels.

BOX 12.6 Examples of serendipity

During the Second World War, an American ship carrying **mustard gas** exploded in an Italian harbour. It was observed that many of the survivors who had inhaled the gas lost their natural defences against microbes. Further study showed that their white blood cells had been destroyed. It is perhaps hard to see how a drug that weakens the immune system could be useful. However, there is one disease where this *is* the case—leukaemia. Leukaemia is a form of cancer which results in the excess proliferation of white blood cells, so a drug that kills these cells is potentially useful. As a result, a series of mustard-like drugs were developed based on the structure of the original mustard gas (sections 9.3.1 and 21.2.3.1).

Another example involved the explosives industry, where it was quite common for workers to suffer severe headaches. These headaches resulted from dilatation of blood vessels in the brain, caused by handling **trinitroglycerine.** Once again, it is hard to see how such a drug could be useful. Certainly, the dilatation of blood vessels in the brain may not be particularly beneficial, but dilating the blood vessels in the heart is useful in cardiovascular medicine. As a result, trinitroglycerine (or **glyceryl trinitrate** as it is called in medical circles) is used as a spray or sublingual tablet for the prophylaxis and treatment of angina. The agent acts as a prodrug for the generation of **nitric oxide**, which causes vasodilation (see also section 26.5.1).

Workers in the rubber industry found that they often acquired a distaste for **alcohol**! This was caused by an antioxidant used in the rubber manufacturing process which found its way into workers' bodies and prevented the normal oxidation of alcohol in the liver. As a result, there was a build-up of **acetaldehyde**, which was so unpleasant that workers preferred not to drink. The antioxidant became the lead compound for the development of **disulfiram (Antabuse)**—used for the treatment of chronic alcoholism.

The following are further examples of lead compounds arising as a result of serendipity:

• **Clonidine** was originally designed to be a nasal vasoconstrictor for use in nasal drops and shaving soaps. Clinical trials revealed that it caused a marked fall in blood pressure, and so it became an important antihypertensive instead.

• **Imipramine** was synthesized as an analogue of chlorpromazine (Box 12.4), and was initially to be used as an antipsychotic. However, it was found to alleviate depression and this led to the development of a series of compounds classified as the **tricyclic antidepressants** (section 23.12.4).

• **Aminoglutethimide** was prepared as a potential anti-epileptic drug, but is now used as an anticancer agent (section 21.4.5).

• The anti-impotence drug **sildenafil (Viagra)** (Fig. 12.13) was discovered by chance from a project aimed at developing a new heart drug (see also section 26.5.2).

• **Isoniazid** (Fig. 12.11) was originally developed as an antituberculosis agent. Patients taking it proved remarkably cheerful and this led to the drug becoming the lead compound for a series of antidepressant drugs known as the **monoamine oxidase inhibitors** (MAOIs) (section 23.12.5).

• **Chlorpromazine** (Box 12.4) was synthesized as an antihistamine for possible use in preventing surgical shock, and was found to make patients relaxed and unconcerned. This led to the drug being tested in people with manic depression, where it was found to have tranquillizing effects. As a result, it was marketed as the first of the neuroleptic drugs (major tranquillizers) used for schizophrenia.

• **Ciclosporin A** (Fig. 12.7) suppresses the immune system and is used during organ and bone marrow transplants to prevent the immune response rejecting the donor organs. The compound was isolated from a soil sample as part of a study aimed at finding new antibiotics. Fortunately, the compounds were more generally screened and the immunosuppressant properties of ciclosporin A were identified.

• In a similar vein, the anticancer alkaloids **vincristine** and **vinblastine** (section 10.2.2) were discovered by chance when searching for compounds that could lower blood sugar levels. Vincristine is used in the treatment of Hodgkin's disease.

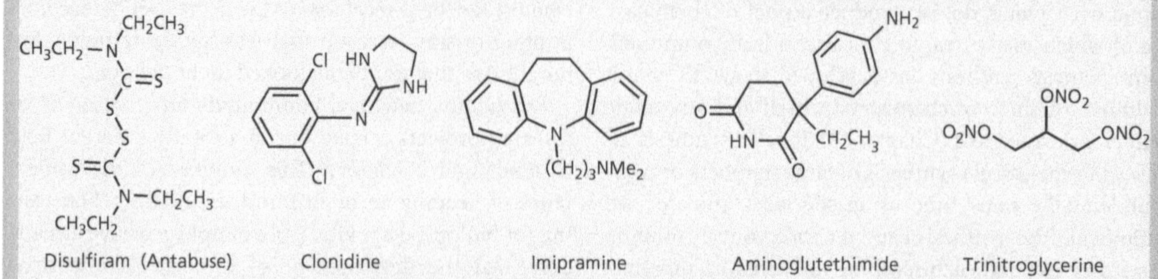

FIGURE 1 Drugs discovered by serendipity.

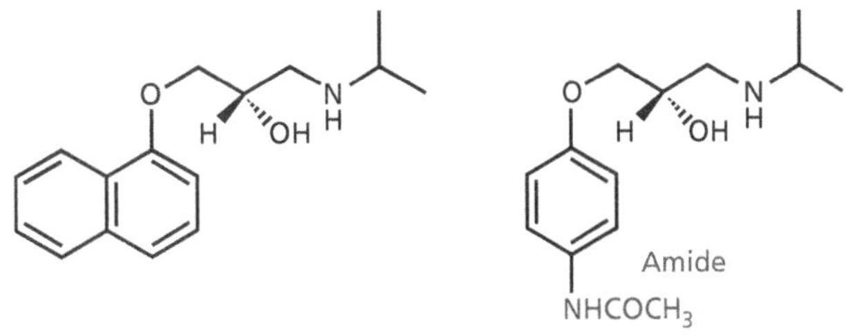

FIGURE 12.15 Propranolol and practolol.

12.4.9 Computerized searching of structural databases

New lead compounds can be found by carrying out computerized searches of structural databases. In order to carry out such a search, it is necessary to know the desired **pharmacophore** (sections 13.2 and 17.11). Alternatively, docking experiments can be carried out if the structure of the target binding site is known (section 17.12). This type of database searching is also known as **database mining** and is described in section 17.13.

12.4.10 Fragment-based lead discovery

So far we have described methods by which a lead compound can be discovered from a natural or synthetic source, but all these methods rely on an active compound being present. Unfortunately, there is no guarantee that this will be the case. Recently, NMR spectroscopy has been used to *design* a lead compound rather than to discover one (see Box 12.7). In essence, the method sets out to find small molecules (**epitopes**) which will bind to specific, but different, regions of a protein's binding site.

BOX 12.7 The use of NMR spectroscopy in finding lead compounds

NMR spectroscopy was used in the design of high-affinity ligands for the FK506 binding protein—a protein involved in the suppression of the immune response. Two optimized epitopes (A and B) were discovered, which bound to different regions of the binding site. Structure C was then synthesized, where the two epitopes were linked by a propyl link. This compound had higher affinity than either of the individual epitopes and represents a lead compound for further development.

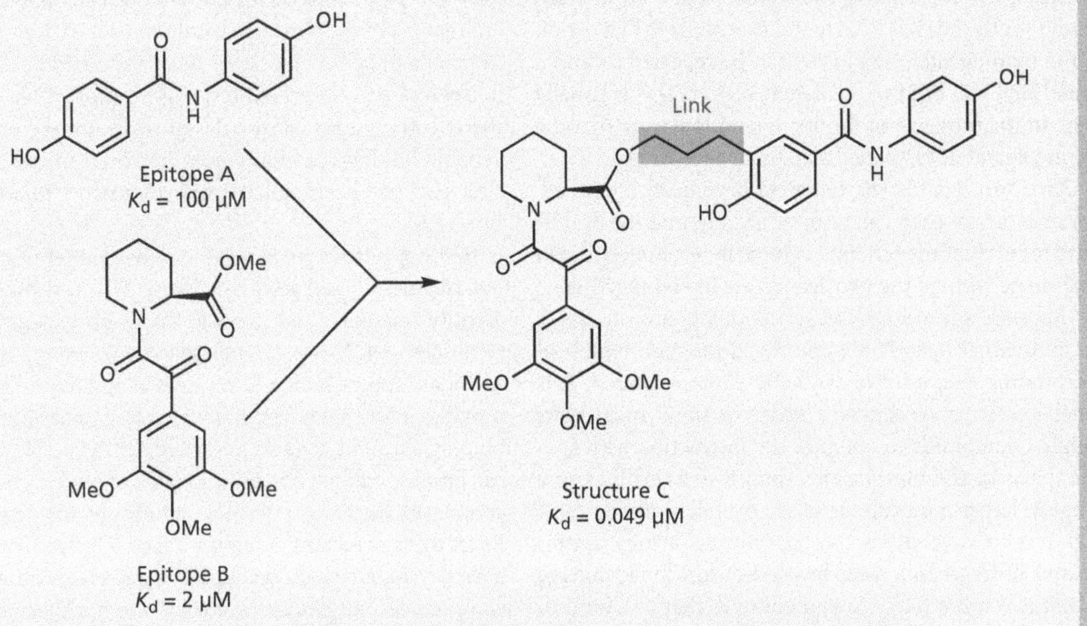

FIGURE 1 Design of a ligand for the FK506 binding protein. (K_d is defined in section 8.9.)

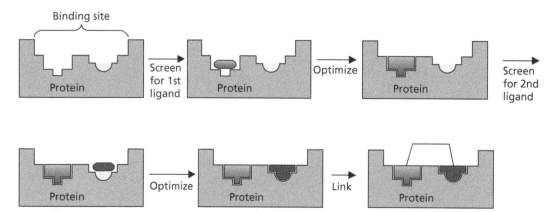

FIGURE 12.16 Epitope mapping.

These molecules will have no activity in themselves since they only bind to one part of the binding site, but if a larger molecule is designed which links these epitopes together, then a lead compound may be created which *is* active and which binds to the whole of the binding site (Fig. 12.16).

Lead discovery by NMR is also known as SAR by NMR (SAR = structure–activity relationships) and can be applied to proteins of known structure which are labelled with ^{15}N or ^{13}C, such that each amide bond in the protein has an identifiable peak.

A range of low-molecular-weight compounds is screened to see whether any of them bind to a specific region of the binding site. Binding can be detected by observing a shift in any of the amide signals, which will not only show that binding is taking place, but will also reveal which part of the binding site is occupied. Once a compound (or ligand) has been found that binds to one region of the binding site, the process can be repeated to find a ligand that will bind to a different region. This is usually done in the presence of the first ligand to ensure that the second ligand does, in fact, bind to a distinct region.

Once two ligands (or epitopes) have been identified, the structure of each can be optimized to find the best ligand for each of the binding regions, then a molecule can be designed where the two ligands are linked together.

There are several advantages to this approach. Since the individual ligands are optimized for each region of the binding site, a lot of synthetic effort is spared. It is much easier to synthesize a series of small-molecular-weight compounds to optimize the interaction with specific parts of the binding site, than it is to synthesize a range of larger molecules to fit the overall binding site. A high level of diversity is also possible, as various combinations of fragments could be used. A further advantage is that it is more likely to find epitopes that will bind to a particular region of a binding site, than to find a lead compound that will bind to the overall binding site.

Moreover, fragments are more likely to be efficient binders, having a high binding energy per unit molecular mass. Finally, some studies have demonstrated a 'super-additivity' effect where the binding affinity of the two linked fragments is much greater than one might have expected from the binding affinities of the two independent fragments.

The method described above involves the linking of fragments. Another strategy is to 'grow' a lead compound from a single fragment—a process called **fragment evolution**. This involves the identification of a single fragment that binds to part of the binding site, then finding larger and larger molecules which contain that fragment, but which bind to other parts of the binding site as well.

A third strategy is known as **fragment self-assembly** and is a form of dynamic combinatorial chemistry (section 16.6.3). Fragments are chosen that can bind to different regions of the binding site, then react with each other to form a linked molecule *in situ*. This could be a reversible reaction as described in section 16.6.3. Alternatively, the two fragments can be designed to undergo an irreversible linking reaction when they bind to the binding site. This has been called '**click chemistry** *in situ*' (see Box 12.8).

NMR spectroscopy is not the only method of carrying out fragment-based lead discovery. It is also possible to identify fragments that bind to target proteins using the techniques of X-ray crystallography, *in vitro* bioassays, and mass spectrometry. X-ray crystallography, like NMR, provides information about how the fragment binds to the binding site, and does so in far greater detail. However, it can be quite difficult obtaining crystals of protein–fragment complexes because of the low affinity of the fragments. Recently, a screening method called **CrystalLEAD** has been developed which can quickly screen large numbers of compounds, and detect ligands by monitoring changes in the electron density map of protein–fragment complexes, relative to the unbound protein.

A femtomolar inhibitor for the acetylcholinesterase enzyme was obtained by fragment self-assembly within the active site of the enzyme. One of the molecular fragments contained an azide group while the other contained an alkyne group. In the presence of the enzyme, both fragments were bound to the active site, and were positioned close enough to each other for an irreversible 1,3 dipolar cycloaddition to take place, forming the inhibitor *in situ* (Fig. 1). This type of reaction has been called 'click chemistry *in situ*'.

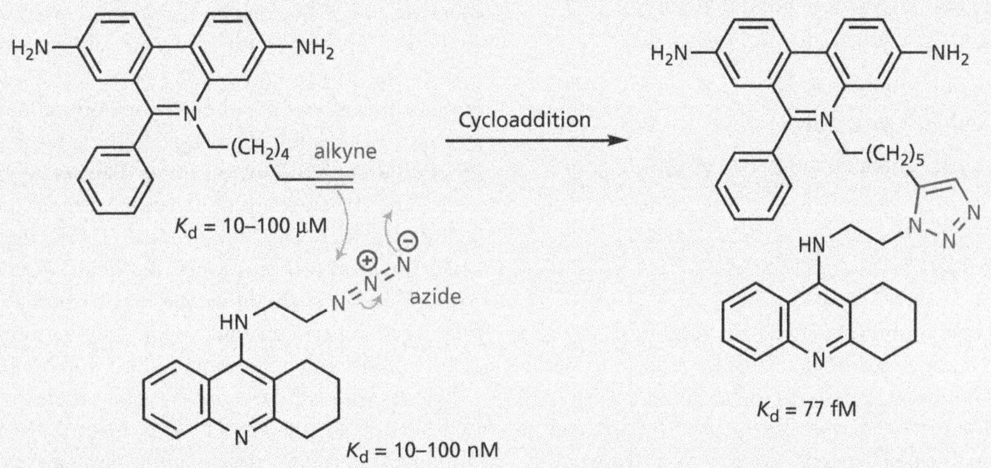

FIGURE 1 'Click' chemistry by means of a cycloaddition reaction. (K_d is defined in section 8.9.)

Finally, it is possible to use fragment-based strategies as a method of optimizing lead compounds that may have been obtained by other means. The strategy is to identify distinct fragments within the lead compound and then to optimize these fragments by the procedures already described. Once the ideal fragments have been identified, the full structure is synthesized incorporating the optimized fragments. This can be a much quicker method of optimization than synthesizing analogues of the larger lead compound.

For additional material see Web article 17: Fragment-based drug discovery on the Online Resource Centre at www.oxfordtextbooks.co.uk/orc/patrick6e/

12.4.11 Properties of lead compounds

Some of the lead compounds that have been isolated from natural sources have sufficient activity to be used directly in medicine without serious side effects; for example morphine, quinine, and paclitaxel. However, most lead compounds have low activity and/or unacceptable side effects, which means that a significant amount of structural modification is required (see Chapters 13 and 14). If the aim of the research is to develop an orally active compound, certain properties of the lead compound should be taken into account. Most orally active drugs obey the rules laid down in Lipinski's rule of five or Veber's parameters (section 11.3). A study of known orally active drugs and the lead compounds from which they were derived demonstrated that the equivalent rules for a lead compound should be more stringent. This is because the structure of the lead compound almost certainly has to be modified and increased, both in terms of size and hydrophobicity. The suggested properties for a lead compound are that it should have a molecular weight of 100–350 amu and a Clog *P* value of 1–3. (Clog *P* is a measure of how hydrophobic a compound is; section 14.1). In general, there is an average increase in molecular weight of 80 amu, and an increase of 1 in Clog *P* when going from a lead compound to the final drug. Studies also show that a lead compound generally has fewer aromatic rings and hydrogen bond acceptors compared to the final drug. Such considerations can be taken into account when deciding which lead compound to use for a research project if several such structures are available. Another approach in making this decision is to calculate the **binding** or **ligand 'efficiency'** of each potential lead compound. This can be done by dividing the free energy of binding for each molecule by the number of non-hydrogen atoms present in the structure. The better the ligand efficiency, the lower the molecular weight of the final optimized structure is likely to be. Moreover, if you have a choice of lead compounds, the most suitable one is not necessarily the most potent.

For fragment-based lead discovery (section 12.4.10), a rule of three has been suggested for the fragments used:

- a molecular weight less than 300;
- no more than 3 hydrogen bond donors;
- no more than 3 hydrogen bond acceptors;
- a Clog *P* of no more than 3;
- no more than 3 rotatable bonds;
- a polar surface area no more than 60 Å².

12.5 Isolation and purification

If the lead compound (or **active principle**) is present in a mixture of compounds from a natural source or a combinatorial synthesis (Chapter 16), it has to be isolated and purified. The ease with which the active principle can be isolated and purified depends very much on the structure, stability, and quantity of the compound. For example, Fleming recognized the antibiotic qualities of **penicillin** and its remarkable non-toxic nature to humans, but he disregarded it as a clinically useful drug because he was unable to purify it. He could isolate it in aqueous solution, but whenever he tried to remove the water, the drug was destroyed. It was not until the development of new experimental procedures such as freeze-drying and chromatography that the successful isolation and purification of penicillin and other natural products became feasible. A detailed description of the experimental techniques involved in the isolation and purification of compounds is outwith the scope of this textbook, and can be obtained from textbooks covering the practical aspects of chemistry.

12.6 Structure determination

It is sometimes hard for present-day chemists to appreciate how difficult structure determinations were before the days of NMR and IR spectroscopy. A novel structure, which may now take a week's work to determine, would have provided two or three decades of work in the past. For example, the microanalysis of **cholesterol** was carried out in 1888 to get its molecular formula, but its chemical structure was not fully established until an X-ray crystallographic study was carried out in 1932.

In the past, structures had to be degraded to simpler compounds, which were further degraded to recognizable fragments. From these scraps of evidence, a possible structure was proposed, but the only sure way of proving the proposal was to synthesize the structure and to compare its chemical and physical properties with those of the natural compound.

Today, structure determination is a relatively straightforward process and it is only when the natural product is obtained in minute quantities that a full synthesis is required to establish its structure. The most useful analytical techniques are **X-ray crystallography** and **NMR spectroscopy**. The former technique comes closest to giving a 'snapshot' of the molecule, but requires a suitable crystal of the sample. The latter technique is used more commonly as it can be carried out on any sample, whether it be a solid, oil, or liquid. There are a large variety of different NMR experiments that can be used to establish the structures of quite complex molecules. These include various two-dimensional NMR experiments which involve a comparison of signals from different types of nuclei in the molecule (e.g. carbon and hydrogen). Such experiments allow the chemist to build up a picture of the molecule atom by atom, and bond by bond.

In cases where there is not enough sample for an NMR analysis, mass spectrometry can be helpful. The fragmentation pattern can give useful clues about a structure, but does not prove it. A full synthesis is still required as final proof.

12.7 Herbal medicine

We have described how useful drugs and lead compounds can be isolated from natural sources, so where does this place herbal medicine? Are there any advantages or disadvantages in using herbal medicines instead of the drugs developed from their active principles? There are no simple answers to this. Herbal medicines contain a large variety of different compounds, several of which may have biological activity, so there is a significant risk of side effects and even toxicity. The active principle is also present in small quantity, so the herbal medicine may be expected to be less active than the pure compound. Herbal medicines such as **St. John's wort** can also interact with prescribed medicines (section 11.5.6), and, in general, there is a lack of regulation or control over their use. Another example is **Ginkgo** which is often used to treat memory problems. However, it also has anticoagulant properties and should not be used alongside other drugs having similar properties; for example warfarin, aspirin, or ibuprofen. Having said all that, several of the issues identified above may actually be advantageous. If the herbal extract contains the active principle in small quantities, there is an inbuilt safety limit to the dose levels received. Different compounds within the extract may also have roles to play in the medicinal properties of the plant and enhance the effect of the active principle—a phenomenon known as **synergy.** Alternatively, some plant extracts have a wide variety of different active principles which act together to produce a beneficial effect. The **aloe plant** (the 'wand of heaven') is an example of this. It is a cactus-like plant found in the deserts of Africa and Arizona and has long been revered for its curative properties. Supporters of herbal medicine have proposed the use of aloe preparations to treat burns, irritable bowel syndrome, rheumatoid arthritis, asthma, chronic leg ulcers, itching, eczema, psoriasis, and acne, thus avoiding the undesirable side effects of long-term steroid use. The preparations are claimed to contain analgesic, anti-inflammatory, antimicrobial, and many other agents, which all contribute to the overall effect. Trying to isolate each active principle would detract from this. On the other hand, critics have stated that many of the beneficial effects claimed for aloe preparations have not been proven and that, although the effects may be useful in some ailments, they are not very effective.

KEY POINTS

- A lead compound is a structure which shows a useful pharmacological activity and can act as the starting point for drug design.

- Natural products are a rich source of lead compounds. The agent responsible for the biological activity of a natural extract is known as the active principle.

- Lead compounds have been isolated from plants, trees, microorganisms, animals, venoms, and toxins. A study of medical folklore indicates plants and herbs which may contain novel lead compounds.

- Lead compounds can be found by screening synthetic compounds obtained from combinatorial syntheses and other sources.

- Existing drugs can be used as lead compounds for the design of novel structures in the same therapeutic area. Alternatively, the side effects of an existing drug can be enhanced to design novel drugs in a different therapeutic area.

- The natural ligand, substrate, product, or modulator for a particular target can act as a lead compound.

- The ability to crystallize a molecular target allows the use of X-ray crystallography and molecular modelling to design lead compounds which will fit the relevant binding site.

- Serendipity has played a role in the discovery of new lead compounds.

- A knowledge of an existing drug's pharmacophore allows the computerized searching of structural databases to identify possible new lead compounds which share that pharmacophore. Docking experiments are also used to identify potential lead compounds.

- NMR spectroscopy can be used to identify whether small molecules (epitopes) bind to specific regions of a binding site. Epitopes can be optimized then linked together to give a lead compound.

- If a lead compound is present in a natural extract or a combinatorial synthetic mixture, it has to be isolated and purified such that its structure can be determined. X-ray crystallography and NMR spectroscopy are particularly important in structure determination.

- Herbal medicines contain different active principles that may combine to produce a beneficial effect. However, toxic side effects and adverse interactions may occur when taken in combination with prescribed medicines.

QUESTIONS

1. What is meant by target specificity and selectivity? Why is it important?

2. What are the advantages and disadvantages of natural products as lead compounds?

3. Fungi have been a richer source of antibacterial agents than bacteria. Suggest why this might be so.

4. Scuba divers and snorkelers are advised not to touch coral. Why do you think this might be? Why might it be of interest to medicinal chemists?

5. You are employed as a medicinal chemist and have been asked to initiate a research programme aimed at finding a drug which will prevent a novel tyrosine kinase receptor from functioning. There are no known lead compounds that have this property. What approaches can you make to establish a lead compound? (Consult section 4.8 to find out more about protein kinase receptors.)

6. A study was set up to look for agents that would inhibit the kinase active site of the epidermal growth factor receptor (section 4.8). Three assay methods were used: an assay

carried out on a genetically engineered form of the protein that was water soluble and contained the kinase active site; a cell assay that measured total tyrosine phosphorylation in the presence of epidermal growth factor; and an *in vivo* study on mice that had tumours grafted onto their backs. How do you think these assays were carried out to measure the effect of an inhibitor? Why do you think three assays were necessary? What sort of information did they provide?

Multiple-choice questions are available on the Online Resource Centre at www.oxfordtextbooks.co.uk/orc/patrick6e/

FURTHER READING

Abad-Zapatero, C., and Metz, J. T. (2005) Ligand efficiency indices as guideposts for drug discovery. *Drug Discovery Today*, **10**, 464–9.

Bleicher, K. H., et al. (2003) Hit and lead generation: beyond high-throughput screening. *Nature Reviews Drug Discovery*, **2**, 369–78.

Blundell, T. L., Jhoti, H., and Abell, C. (2002) High-throughput crystallography for lead discovery in drug design. *Nature Reviews Drug Discovery*, **1**, 45–54.

Bolognesi, M. L., et al. (2009) Alzheimer's disease: new approaches to drug discovery. *Current Opinion in Chemical Biology*, **13**, 303–8.

Cavalli, A., et al. (2008) Multi-target-directed ligands to combat neurodegenerative diseases. *Journal of Medicinal Chemistry*, **51**, 347–72.

Clardy, J., and Walsh, C. (2004) Lessons from natural molecules. *Nature*, **432**, 829–37.

Di, L., et al. (2003) High throughput artificial membrane permeability assay for blood–brain barrier. *European Journal of Medicinal Chemistry*, **38**, 223–32.

Engel, L. W., and Straus, S. E. (2002) Development of therapeutics: opportunities within complementary and alternative medicine. *Nature Reviews Drug Discovery*, **1**, 229–37.

Gershell, L. J., and Atkins, J. H. (2003) A brief history of novel drug discovery technologies. *Nature Reviews Drug Discovery*, **2**, 321–7.

Honma, T. (2003) Recent advances in *de novo* design strategy for practical lead identification. *Medicinal Research Reviews*, **23**, 606–32.

Hopkins, A. L., and Groom, C. R. (2002) The druggable genome. *Nature Reviews Drug Discovery*, **1**, 727–30.

Keseru, G. M., Erlanson, D. A., Ferenczy, G. G., et al. (2016) Design principle for fragment libraries: maximising the value of learnings from Pharma fragment-based drug discovery (FBDD) programs for use in academia. *Journal of Medicinal Chemistry*, DOI: 10.1021/acs.jmedchem.6b00197.

Lewis, R. J., and Garcia, M. L. (2003) Therapeutic potential of venom peptides. *Nature Reviews Drug Discovery*, **2**, 790–802.

Lindsay, M. A. (2003) Target discovery. *Nature Reviews Drug Discovery*, **2**, 831–8.

Lipinski, C., and Hopkins, A. (2004) Navigating chemical space for biology and medicine. *Nature*, **432**, 855–61.

Lowe, D. (2009) In the pipeline. *Chemistry World*, Nov., 20 (screening assays).

Megget, K. (2011) Of mice and men. *Chemistry World*, April, 42–5.

Pellecchia, M., Sem, D. S., and Wuthrich, K. (2002) NMR in drug discovery. *Nature Reviews Drug Discovery*, **1**, 211–9.

Perks, B. (2011) Extreme potential. *Chemistry World*, June, 48–51.

Phillipson, J. D. (2007) Phytochemistry and pharmacognosy. *Phytochemistry*, **68**, 2960–72.

Rees, D. C., et al. (2004) Fragment-based lead discovery. *Nature Reviews Drug Discovery*, **3**, 660–72.

Rishton, G. B. (2003) Nonleadlikeness and leadlikeness in biochemical screening. *Discovering Drugs Today*, **8**, 86–96.

Sauter, G., Simon, R., and Hillan, K. (2003) Tissue microarrays in drug discovery. *Nature Reviews Drug Discovery*, **2**, 962–72.

Shuker, S. B., Hajduk, P. J., Meadows, R. P., and Fesik, S. W. (1996) Discovering high-affinity ligands for proteins: SAR by NMR. *Science*, **274**, 1531–4.

Srivastava, A. S., et al. (2005) Plant-based anticancer molecules. *Bioorganic Medicinal Chemistry*, **13**, 5892–908.

Stockwell, B. R. (2004) Exploring biology with small organic molecules. *Nature*, **432**, 846–54.

Su, J., et al. (2007) SAR study of bicyclo[4.1.0]heptanes as melanin-concentrating hormone receptor R1 antagonists: taming hERG. *Bioorganic and Medicinal Chemistry*, **15**, 5369–85.

Walters, W. P., and Namchuk, M. (2003) Designing screen: how to make your hits a hit. *Nature Reviews Drug Discovery*, **2**, 259–66.

Wermuth, C. G. (2006) Selective optimization of side activities: the SOSA approach. *Drug Discovery Today*, **11**, 160–4.

Titles for general further reading are listed on p.845.

13 Drug design: optimizing target interactions

In Chapter 12, we looked at the various methods of discovering a lead compound. Once it *has* been discovered, the lead compound can be used as the starting point for drug design. There are various aims in drug design. The eventual drug should have a good selectivity and level of activity for its target, and have minimal side effects. It should be easily synthesized and chemically stable. Finally, it should be non-toxic and have acceptable pharmacokinetic properties. In this chapter, we concentrate on design strategies that can be used to optimize the interaction of the drug with its target in order to produce the desired pharmacological effect; in other words its **pharmacodynamic** properties. In Chapter 14, we look at the design strategies that can improve the drug's ability to reach its target and have an acceptable lifetime—in other words its **pharmacokinetic** properties. Although these topics are in separate chapters, it would be wrong to think that they are tackled separately during drug optimization. For example, it would be foolish to spend months or years perfecting a drug that interacts perfectly with its target, but has no chance of reaching that target because of adverse pharmacokinetic properties. Pharmacodynamics and pharmacokinetics should have equal priority in influencing drug design strategies and determining which analogues are synthesized.

13.1 Structure–activity relationships

Once the structure of a lead compound is known, the medicinal chemist moves on to study its structure–activity relationships (SAR). The aim is to identify those parts of the molecule that are important to biological activity and those that are not. If it is possible to crystallize the target with the lead compound bound to the binding site, the crystal structure of the complex could be solved by X-ray crystallography, then studied with molecular modelling software to identify important binding interactions.

However, this is not possible if the target structure has not been identified or cannot be crystallized. It is then necessary to revert to the traditional method of synthesizing a selected number of compounds that vary slightly from the original structure, then studying what effect that has on the biological activity.

One can imagine the drug as a chemical knight going into battle with an affliction. The drug is armed with a variety of weapons and armour, but it may not be obvious which weapons are important to the drug's activity, or which armour is essential to its survival. We can only find this out by removing some of the weapons and armour to see if the drug is still effective. The weapons and armour involved are the various structural features in the drug that can either act as binding groups with the target binding site (section 1.3), or assist and protect the drug on its journey through the body (Chapter 14). Recognizing functional groups and the sort of intermolecular bonds that they can form is important in understanding how a drug might bind to its target.

Let us imagine that we have isolated a natural product with the structure shown in Fig. 13.1. We shall name it glipine. There are a variety of functional groups present in the structure, and the diagram shows the potential binding interactions that are possible with a target binding site.

It is unlikely that all of these interactions take place, so we have to identify those that do. By synthesizing analogues (such as the examples shown in Fig. 13.2) where one particular functional group of the molecule is removed or altered, it is possible to find out which groups are essential and which are not. This involves testing all the analogues for biological activity and comparing them with the original compound. If an analogue shows a significantly lowered activity, then the group that has been modified must have been important. If the activity remains similar, then the group is not essential.

The ease with which this task is carried out depends on how easily we can synthesize the necessary analogues. It may be possible to modify some lead compounds directly

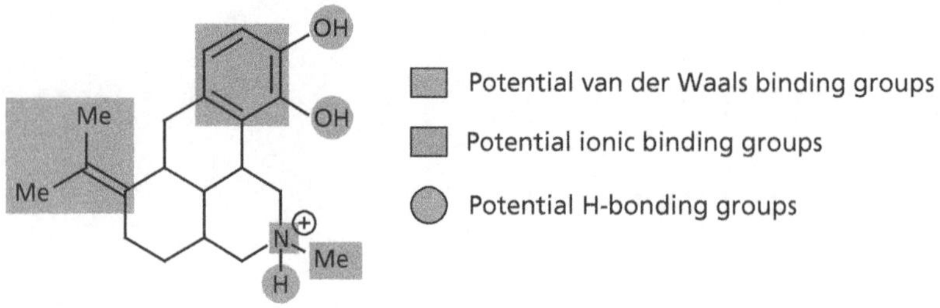

FIGURE 13.1 Glipine.

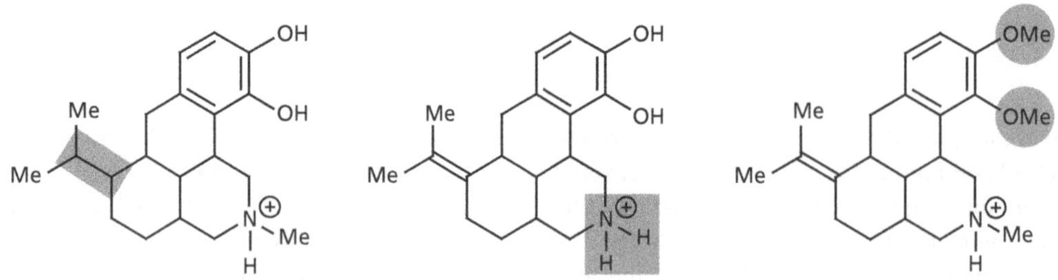

FIGURE 13.2 Modifications of glipine.

to the required analogues, whereas the analogues of other lead compounds may best be prepared by total synthesis. Let us consider the binding interactions that are possible for different functional groups, and the analogues that could be synthesized to establish whether they are involved in binding or not (see also section 1.3 and Appendix 7).

13.1.1 Binding role of alcohols and phenols

Alcohols and phenols are functional groups which are commonly present in drugs and are often involved in hydrogen bonding. The oxygen can act as a hydrogen bond acceptor, and the hydrogen can act as a hydrogen bond donor (Fig. 13.3). The directional preference for hydrogen bonding is indicated by the arrows in the figure, but it is important to realize that slight deviations are possible (section 1.3.2). One or all of these interactions may be important in binding the drug to the binding site. Synthesizing a methyl ether or an ester analogue would be relevant in testing this, as it is highly likely that

the hydrogen bonding would be disrupted in either analogue. Let us consider the methyl ether first.

There are two reasons why the ether might hinder or prevent the hydrogen bonding of the original alcohol or phenol. The obvious explanation is that the proton of the original hydroxyl group is involved as a hydrogen bond donor and, by removing it, the hydrogen bond is lost (Frames 1 and 2 in Fig. 13.4). However, suppose the oxygen atom is acting as a hydrogen bond acceptor (Frame 3, Fig. 13.4)? The oxygen is still present in the ether analogue, so could it still take part in hydrogen bonding? Well, it may, but possibly not to the same extent. The extra bulk of the methyl group should hinder the close approach that was previously attainable and is likely to disrupt hydrogen bonding (Frame 4, Fig. 13.4). The hydrogen bonding may not be completely prevented, but we could reasonably expect it to be weakened.

An ester analogue cannot act as a hydrogen bond donor either. There is still the possibility of it acting as a hydrogen bond acceptor, but the extra bulk of the acyl

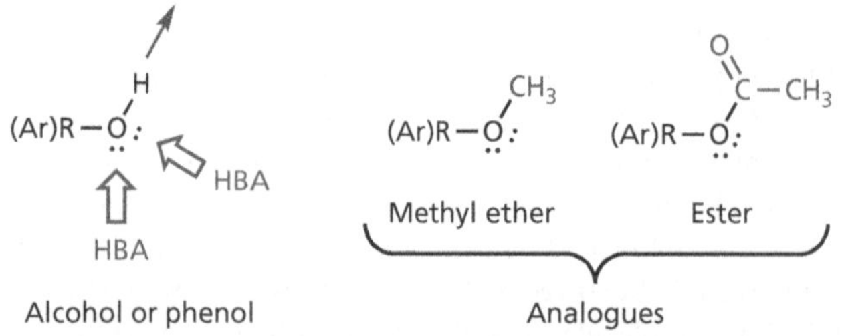

FIGURE 13.3 Possible hydrogen bonding interactions for an alcohol or phenol.

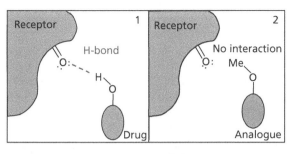

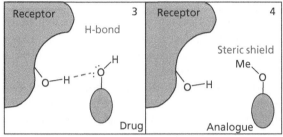

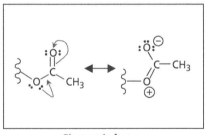

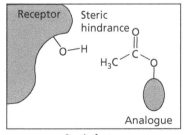

FIGURE 13.4 Possible hydrogen bond interactions for an alcohol/phenol in comparison with an ether analogue.

group is even greater than the methyl group of the ether, and this too should hinder the original hydrogen bonding interaction. There is also a difference between the electronic properties of an ester and an alcohol. The carboxyl group has a weak pull on the electrons from the neighbouring oxygen, giving the resonance structure shown in Fig. 13.5. Because the lone pair is involved in such an interaction, it will be less effective as a hydrogen bond acceptor. Of course, one could then argue that the

carbonyl oxygen is potentially a more effective hydrogen bond acceptor; however, it is in a different position relative to the rest of the molecule and may be poorly positioned to form an effective hydrogen bond interaction with the target binding region.

It is relatively easy to acetylate alcohols and phenols to their corresponding esters, and this was one of the early reactions that was carried out on natural products such as morphine (sections 24.3 and 24.5). Alcohols and phenols can also be converted easily to ethers.

In this section, we considered the OH group of alcohols and phenols. It should be remembered that the OH group of a phenol is linked to an aromatic ring, which can also be involved in intermolecular interactions (section 13.1.2).

13.1.2 Binding role of aromatic rings

Aromatic rings are planar, hydrophobic structures, commonly involved in van der Waals interactions with flat hydrophobic regions of the binding site. An analogue containing a cyclohexane ring in place of the aromatic ring is less likely to bind so well, as the ring is no longer flat. The axial protons can interact weakly, but they also serve as buffers to keep the rest of the cyclohexane ring at a distance (Fig. 13.6). The binding region for the aromatic ring may also be a narrow slot rather than a planar surface. In that scenario, the cyclohexane ring would be incapable of fitting into it, because it is a bulkier structure.

Although there are methods of converting aromatic rings to cyclohexane rings, they are unlikely to be

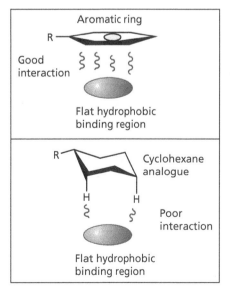

FIGURE 13.6 Binding comparison of an aromatic ring with a cyclohexyl ring.

Electronic factor

Steric factor

FIGURE 13.5 Factors by which an ester group can disrupt the hydrogen bonding of the original hydroxyl group.

successful with most lead compounds, and so such analogues would normally be prepared using a full synthesis.

Aromatic rings could also interact with an aminium or quaternary ammonium ion through induced dipole interactions or hydrogen bonding (sections 1.3.4 and 1.3.2). Such interactions would not be possible for the cyclohexyl analogue.

13.1.3 Binding role of alkenes

Like aromatic rings, alkenes are planar and hydrophobic, so they too can interact with hydrophobic regions of the binding site through van der Waals interactions. The activity of the equivalent saturated analogue would be worth testing, since the saturated alkyl region is bulkier and cannot approach the relevant region of the binding site so closely (Fig. 13.7). Alkenes are generally easier to reduce than aromatic rings, so it may be possible to prepare the saturated analogue directly from the lead compound.

13.1.4 The binding role of ketones and aldehydes

A ketone group is not uncommon in many of the structures studied in medicinal chemistry. It is a planar group that can interact with a binding site through hydrogen bonding where the carbonyl oxygen acts as a hydrogen bond acceptor (Fig. 13.8). Two such interactions are possible, as two lone pairs of electrons are available on the carbonyl oxygen. The lone pairs are in sp^2 hybridized orbitals which are in the same plane as the functional group. The carbonyl group also has a significant dipole

moment and so a dipole–dipole interaction with the binding site is also possible.

It is relatively easy to reduce a ketone to an alcohol and it may be possible to carry out this reaction directly on the lead compound. This significantly changes the geometry of the functional group from planar to tetrahedral. Such an alteration in geometry may well weaken any existing hydrogen bonding interactions and will certainly weaken any dipole–dipole interactions, as both the magnitude and orientation of the dipole moment will be altered (Fig. 13.9). If it was suspected that the oxygen present in the alcohol analogue might still be acting as a hydrogen bond acceptor, then the ether or ester analogues could be studied, as described in section 13.1.1. Reactions are available that can reduce a ketone completely to an alkane and remove the oxygen, but they are unlikely to be practical for many of the lead compounds studied in medicinal chemistry.

Aldehydes are less common in drugs because they are more reactive and are susceptible to metabolic oxidation to carboxylic acids. However, they could interact in the same way as ketones, and similar analogues could be studied.

13.1.5 Binding role of amines

Amines are extremely important functional groups in medicinal chemistry and are present in many drugs. They may be involved in hydrogen bonding, either as a hydrogen bond acceptor or a hydrogen bond donor (Fig. 13.10). The nitrogen atom has one lone pair of electrons and can act as a hydrogen bond acceptor for one

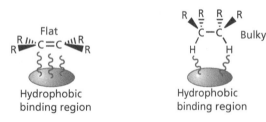

FIGURE 13.7 Binding comparison of an alkene with an alkane.

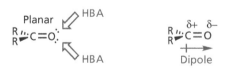

FIGURE 13.8 Binding interactions that are possible for a carbonyl group.

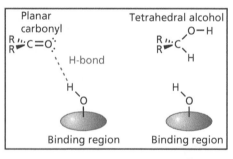

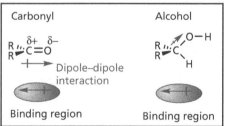

FIGURE 13.9 Effect on binding interactions following the reduction of a ketone or aldehyde.

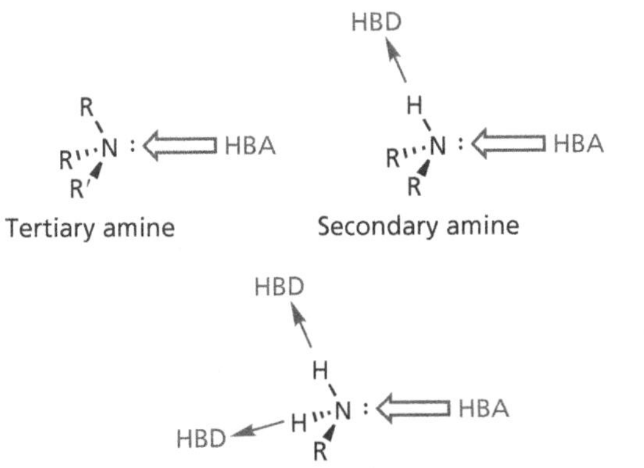

FIGURE 13.10 Possible binding interactions for amines.

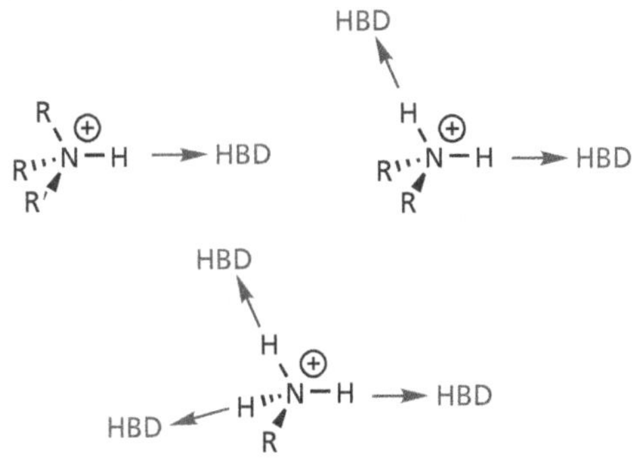

FIGURE 13.11 Possible hydrogen bonding interactions for ionized amines.

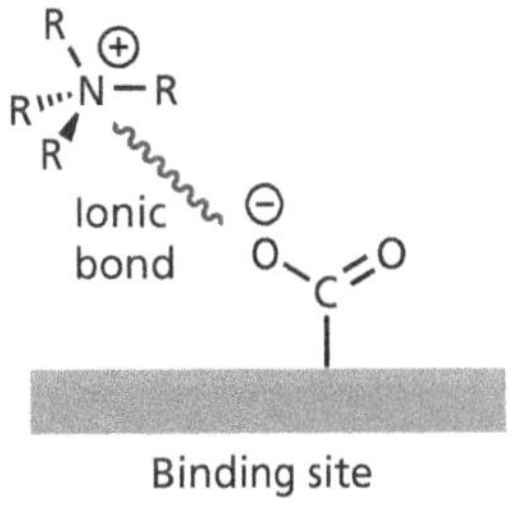

FIGURE 13.12 Ionic interaction between an ionized amine and a carboxylate ion (R = H, alkyl, or aryl).

hydrogen bond. Primary and secondary amines have N–H groups and can act as hydrogen bond donors. Aromatic and heteroaromatic amines act only as hydrogen bond donors, because the lone pair interacts with the aromatic or heteroaromatic ring.

In many cases, the amine may be protonated when it interacts with its target binding site, which means that it is ionized and cannot act as a hydrogen bond acceptor. However, it can still act as a hydrogen bond donor and will form stronger hydrogen bonds than if it was not ionized (Fig. 13.11). Alternatively, a strong ionic interaction may take place with a carboxylate ion in the binding site (Fig. 13.12).

To test whether ionic or hydrogen bonding interactions are taking place, an amide analogue could be studied. This will prevent the nitrogen acting as a hydrogen bond acceptor, as the nitrogen's lone pair will interact with the neighbouring carbonyl group (Fig. 13.13). This interaction also prevents protonation of the nitrogen and rules out the possibility of ionic interactions. You might argue that the right-hand structure in Fig. 13.13a has a positive charge on the nitrogen and could still take part in an ionic interaction. However, this resonance structure represents one extreme and is never present as a distinct entity. The amide group as a whole is neutral, and so lacks the net positive charge required for ionic bonding.

It is relatively easy to form secondary and tertiary amides from primary and secondary amines respectively, and it may be possible to carry out this reaction directly on the lead compound. A tertiary amide lacks the N–H group of the original secondary amine and would test whether this is involved as a hydrogen bond donor. The secondary amide formed from a primary amine still has an N–H group present, but the steric bulk of the acyl group should hinder it acting as a hydrogen bond donor.

Tertiary amines cannot be converted directly to amides, but if one of the alkyl groups is a methyl group, it is often possible to remove it with vinyloxycarbonyl chloride (VOC-Cl) to form a secondary amine, which could then be converted to the amide (Fig. 13.14). This demethylation reaction is extremely useful and has been used to good effect in the synthesis of morphine analogues (see Box 24.2 for the reaction mechanism).

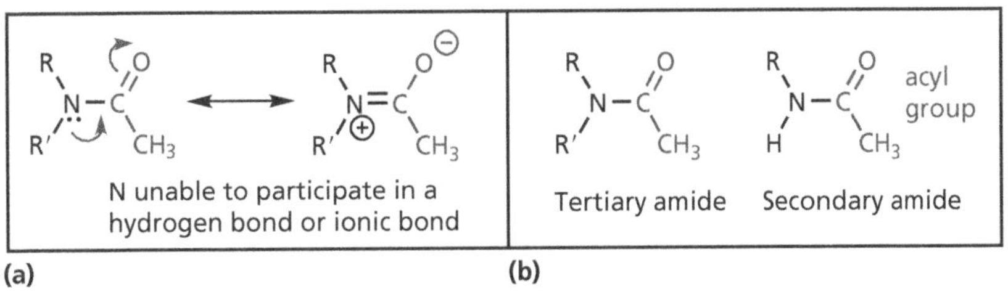

FIGURE 13.13 (a) Interaction of the nitrogen lone pair with the neighbouring carbonyl group in amides. (b) Secondary and tertiary amides.

709

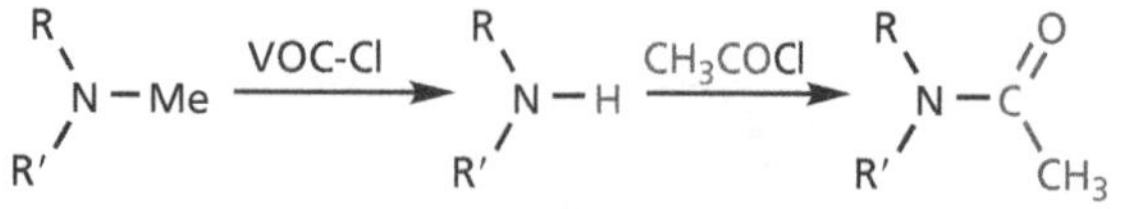

FIGURE 13.14 Demethylation of a tertiary amine and formation of a secondary amide.

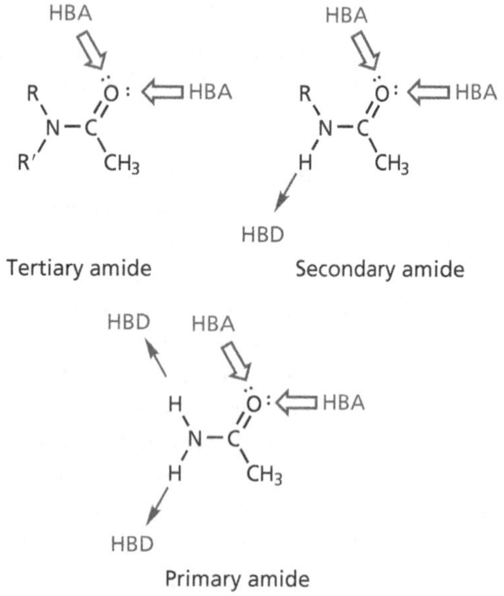

FIGURE 13.15 Possible hydrogen bonding interactions for amides.

13.1.6 Binding role of amides

Many of the lead compounds currently studied in medicinal chemistry are peptides or polypeptides consisting of amino acids linked together by peptide or amide bonds (section 2.1). Amides are likely to interact with binding sites through hydrogen bonding (Fig. 13.15). The carbonyl oxygen atom can act as a hydrogen bond acceptor and has the potential to form two hydrogen bonds. Both the lone pairs involved are in sp^2 hybridized orbitals which are located in the same plane as the amide group. The

nitrogen cannot act as a hydrogen bond acceptor because the lone pair interacts with the neighbouring carbonyl group (Fig. 13.13a). Primary and secondary amides have an N–H group, which allows the possibility of this group acting as a hydrogen bond donor.

The most common type of amide in peptide lead compounds is the secondary amide. Suitable analogues that could be prepared to test out possible binding interactions are shown in Fig. 13.16. All the analogues, apart from the primary and secondary amines, could be used to check whether the amide is acting as a hydrogen bond donor. The alkenes and amines could be tested to see whether the amide is acting as a hydrogen bond acceptor. However, there are traps for the unwary. The amide group is planar and does not rotate because of its partial double bond character. The ketone, the secondary amine and the tertiary amine analogues have a single bond at the equivalent position which *can* rotate. This would alter the relative positions of any binding groups on either side of the amide group and lead to a loss of binding, even if the amide itself was not involved in binding. Therefore, a loss of activity would not necessarily mean that the amide is important as a binding group. With these groups, it would only be safe to say that the amide group is not essential if activity is retained. Similarly, the primary amine and carboxylic acid may be found to have no activity, but this might be due to the loss of important binding groups in one half of the molecule. These particular analogues would only be worth considering if the amide group is peripheral to the molecule (e.g. R–NHCOMe or R–CONHMe) and not part of the main skeleton.

The alkene would be a particularly useful analogue to test because it is planar, cannot rotate, and cannot act as a hydrogen bond donor or hydrogen bond acceptor. However, the synthesis of this analogue may not be simple. In fact, it is likely that all the analogues described would have to be prepared using a full synthesis. Amides are relatively stable functional groups and, although several of the analogues described might be attainable directly from the lead compound, it is more likely that the lead

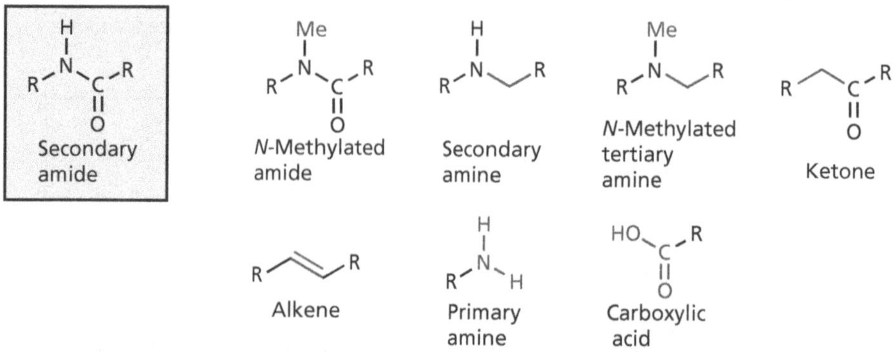

FIGURE 13.16 Possible analogues to test the binding interactions of a secondary amide.

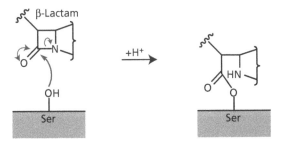

FIGURE 13.17 β-Lactam ring acting as an acylating agent.

compound would not survive the forcing conditions required.

Amides which are within a ring system are called lactams. They, too, can form intermolecular hydrogen bonds as described above. However, if the ring is small and suffers ring strain, the lactam can undergo a chemical reaction with the target leading to the formation of a covalent bond. The best examples of this are the penicillins which contain a four-membered β-lactam ring. This acts as an acylating agent and irreversibly inhibits a bacterial enzyme by acylating a serine residue in the active site (Fig. 13.17) (section 19.5.1.4).

13.1.7 Binding role of quaternary ammonium salts

Quaternary ammonium salts are ionized and can interact with carboxylate groups by ionic interactions (Fig. 13.18). Another possibility is an **induced dipole interaction** between the quaternary ammonium ion and any aromatic rings in the binding site. The positively charged nitrogen can distort the π electrons of

the aromatic ring such that a dipole is induced, whereby the face of the ring is slightly negative and the edges are slightly positive. This allows an interaction between the slightly negative faces of the aromatic rings and the positive charge of the quaternary ammonium ion. This is also known as a **π-cation interaction**.

The importance of these interactions could be tested by synthesizing an analogue that has a tertiary amine group rather than the quaternary ammonium group. Of course, it is possible that such a group could ionize by becoming protonated, then interact in the same way. Converting the amine to an amide would prevent this possibility. The neurotransmitter **acetylcholine** has a quaternary ammonium group which is thought to bind to the binding site of its target receptor by ionic bonding and/or induced dipole interactions (section 22.5).

13.1.8 Binding role of carboxylic acids

The carboxylic acid group is reasonably common in drugs. It can act as a hydrogen bond acceptor or as a hydrogen bond donor (Fig. 13.19). Alternatively, it may exist as the carboxylate ion. This allows the possibility of an ionic interaction and/or a strong hydrogen bond where the carboxylate ion acts as a hydrogen bond acceptor. The carboxylate ion is also a good ligand for metal ion cofactors present in several enzymes; for example zinc metalloproteinases (section 21.7.1 and Case study 2).

In order to test the possibility of such interactions, analogues such as esters, primary amides, primary alcohols, and ketones could be synthesized and tested (Fig. 13.20). None of these functional groups can ionize, so a loss of activity could imply that an ionic bond is important. The primary alcohol could shed light on whether the carbonyl oxygen is involved in hydrogen bonding, whereas the ester and ketone could indicate whether the hydroxyl group of the carboxylic acid is involved in hydrogen

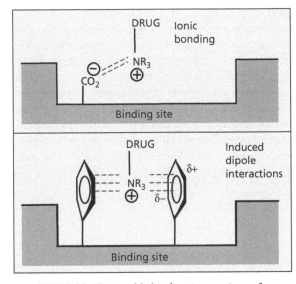

FIGURE 13.18 Possible binding interactions of a quaternary ammonium ion.

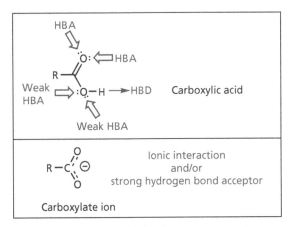

FIGURE 13.19 Possible binding interactions for a carboxylic acid and carboxylate ion.

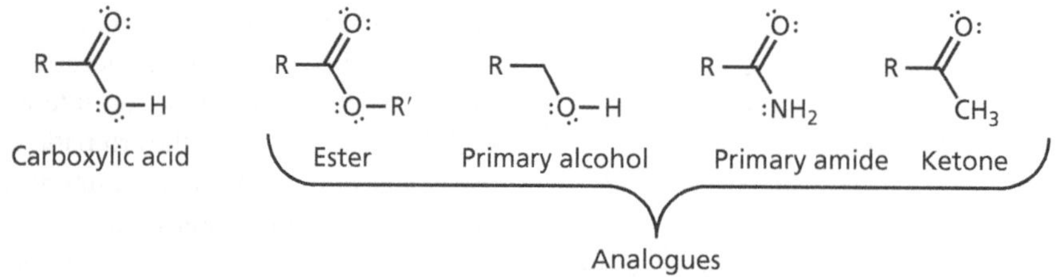

FIGURE 13.20 Analogues to test the binding interactions for a carboxylic acid.

bonding. It may be possible to synthesize the ester and amide analogues directly from the lead compound, but the reduction of a carboxylic acid to a primary alcohol requires harsher conditions and this sort of analogue would normally be prepared by a full synthesis. The ketone would also have to be prepared by a full synthesis.

13.1.9 Binding role of esters

An ester functional group has the potential to interact with a binding site as a hydrogen bond acceptor only (Fig. 13.21). The carbonyl oxygen is more likely to act as the hydrogen bond acceptor than the alkoxy oxygen (section 1.3.2), as it is sterically less hindered and has a greater electron density. The importance or otherwise of the carbonyl group could be judged by testing an equivalent ether, which would require a full synthesis.

Esters are susceptible to hydrolysis *in vivo* by metabolic enzymes called **esterases**. This may pose a problem if the lead compound contains an ester that is important to binding, as it means the drug might have a short lifetime *in vivo*. Having said that, there are several drugs that *do* contain esters and are relatively stable to metabolism,

thanks to electronic factors that stabilize the ester or steric factors that protect it.

Esters that are susceptible to metabolic hydrolysis are sometimes used deliberately to mask a polar functional group such as a carboxylic acid, alcohol, or phenol in order to achieve better absorption from the gastrointestinal tract. Once in the blood supply, the ester is hydrolysed to release the active drug. This is known as a **prodrug** strategy (section 14.6).

Special mention should be made of the ester group in aspirin. Aspirin has an anti-inflammatory action resulting from its ability to inhibit an enzyme called **cyclooxygenase** (COX) which is required for **prostaglandin** synthesis. It is often stated that aspirin acts as an acylating agent, and that its acetyl group is covalently attached to a serine residue in the active site of COX (Fig. 13.22). However, this theory has been disputed and it is stated that aspirin acts, instead, as a prodrug to generate salicylic acid, which then inhibits the enzyme through non-covalent interactions.

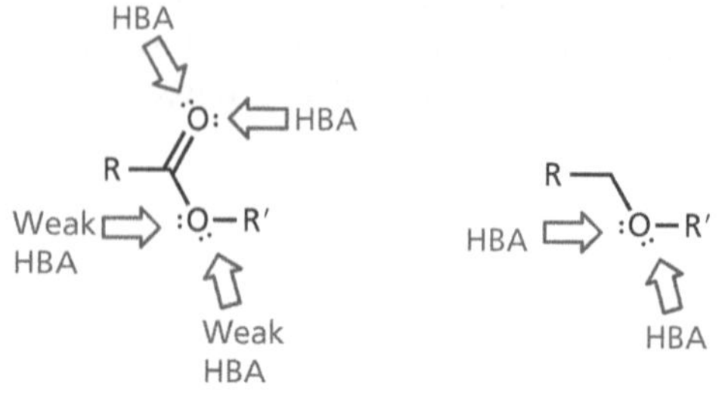

13.1.10 Binding role of alkyl and aryl halides

Alkyl halides involving chlorine, bromine, or iodine tend to be chemically reactive, since the halide ion is a good leaving group. As a result, a drug containing an alkyl halide is likely to react with any nucleophilic group that it encounters and become permanently linked to that group by a covalent bond—an alkylation reaction (Fig. 13.23). This poses a problem, as the drug is likely to alkylate a large variety of macromolecules which have nucleophilic groups, especially amine groups in proteins and nucleic acids. It is possible to moderate the reactivity to some extent, but selectivity is still a problem and leads to severe side effects. These drugs are, therefore, reserved

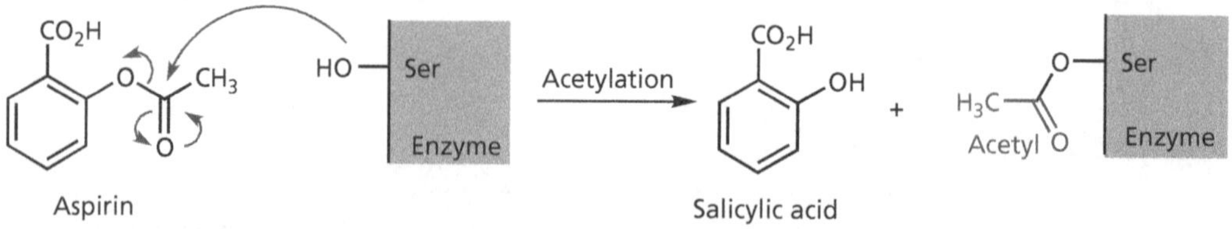

FIGURE 13.21 Possible binding interactions for an ester and an ether.

FIGURE 13.22 The disputed theory of aspirin acting as an acylating agent.

FIGURE 13.23 Alkylation of macromolecular targets by alkyl halides.

for life-threatening diseases such as cancer (sections 9.3 and 21.2.3). Alkyl fluorides, on the other hand, are not alkylating agents, because the strong C–F bond is not easily broken. Fluorine is commonly used to replace a proton as it is approximately the same size, but has different electronic properties. It may also protect the molecule from metabolism (sections 13.3.7 and 14.2.4).

Aryl halides do not act as alkylating agents and pose less of a problem in that respect. As the halogen substituents are electron-withdrawing groups, they affect the electron density of the aromatic ring and this may have an influence on the binding of the aromatic ring. The halogen substituents chlorine and bromine are hydrophobic in nature and may interact favourably with hydrophobic pockets in a binding site. Hydrogen bonding is likely to be weak if it occurs at all. Although halide ions are strong hydrogen bond acceptors, halogen substituents are generally poor hydrogen bond acceptors. Having said that, an ion–dipole interaction involving an aryl fluoride is important in the binding of some statins to their target binding site (Case study 1).

Aliphatic and aromatic analogues lacking the halogen substituent could be prepared by a full synthesis to test whether the halogen has any importance towards the activity of the lead compound.

13.1.11 Binding role of thiols and ethers

The thiol group (S–H) is known to be a good ligand for d-block metal ions and has been incorporated into several drugs designed to inhibit enzymes containing a zinc cofactor; for example, the zinc metalloproteinases (sections 21.7.1 and 26.3.3 and Case study 2). If the lead compound has a thiol group, the corresponding alcohol could be tested as a comparison. This would have a far weaker interaction with zinc.

An ether group (R′OR) might act as a hydrogen bond acceptor through the oxygen atom (Fig. 13.21). This could be tested by increasing the size of the neighbouring alkyl group to see whether it diminishes the ability of the group to take part in hydrogen bonding. Analogues where the oxygen is replaced with a methylene (CH$_2$) isostere should show significantly decreased binding affinity.

The oxygen atom of an aromatic ether is generally a poor hydrogen bond acceptor (section 1.3.2).

13.1.12 Binding role of other functional groups

In some drugs, sulphonamides can have a binding role where the oxygen atoms act as hydrogen bond acceptors and the NH proton acts as a hydrogen bond donor. In certain circumstances, the sulphonamide group may be ionized and participate in an ionic interaction with the binding site (section 26.4.2)

Lead compounds may contain a wide variety of other functional groups that have no direct binding role, but could be important in other respects. Some may influence the electronic properties of the molecule (e.g. nitro groups or nitriles). Others may restrict the shape or conformation of a molecule (e.g. alkynes) (Box 13.3). Functional groups may also act as metabolic blockers (e.g. aryl halides) (section 14.2.4).

13.1.13 Binding role of alkyl groups and the carbon skeleton

The alkyl substituents and carbon skeleton of a lead compound are hydrophobic and may bind with hydrophobic regions of the binding site through van der Waals interactions. The relevance of an alkyl substituent to binding can be determined by synthesizing an analogue which lacks the substituent. Such analogues generally have to be synthesized using a full synthesis if the substituents are attached to the carbon skeleton of the molecule. However, if the alkyl group is attached to nitrogen or oxygen, it may be possible to remove the group from the lead compound, as shown in Fig. 13.24. The analogues obtained may then be expected to have less activity if the alkyl group was involved in important hydrophobic interactions.

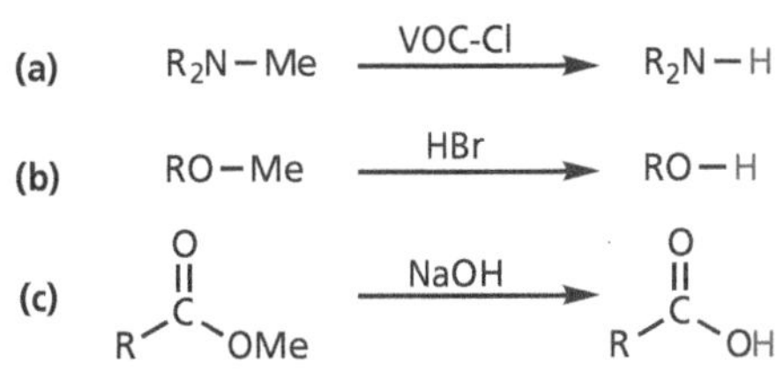

FIGURE 13.24 (a) *N*-Demethylation of a tertiary amine with vinyloxycarbonyl chloride (see Box 24.2 for mechanism). (b) Demethylation of a methyl ether using hydrogen bromide where nucleophilic substitution leads to an alcohol (or phenol) plus bromomethane. (c) Hydrolysis of an ester using sodium hydroxide where OH replaces OMe.

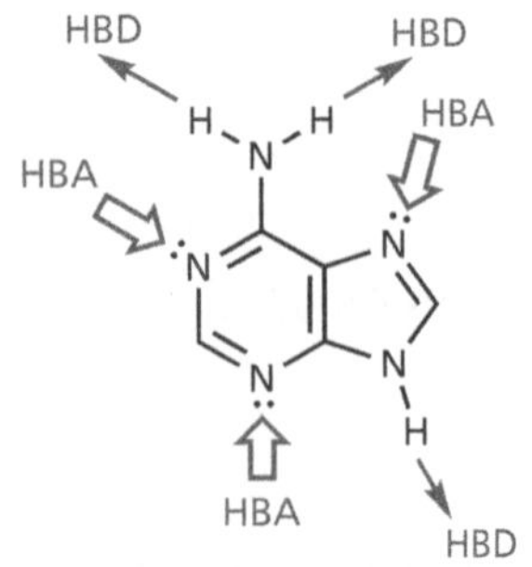

FIGURE 13.25 Possible hydrogen bonding interactions for adenine.

13.1.14 Binding role of heterocycles

A large diversity of heterocycles are found in lead compounds. Heterocycles are cyclic structures that contain one or more heteroatoms such as oxygen, nitrogen, or sulphur. Nitrogen-containing heterocycles are particularly prevalent. The heterocycles can be aliphatic or aromatic in character, and have the potential to interact with binding sites through a variety of bonding forces. For example, the overall heterocycle can interact through van der Waals and pi–pi interactions, while the individual heteroatoms present in the structure could interact by hydrogen bonding or ionic bonding.

As far as hydrogen bonding is concerned, there is an important directional aspect. The position of the heteroatom in the ring and the orientation of the ring in the binding site can be crucial in determining whether or not a good interaction takes place. For example, adenine can take part in six hydrogen bonding interactions, three as a hydrogen bond donor and three as a hydrogen bond acceptor. The ideal directions for these interactions are

shown in Fig. 13.25. Van der Waals or pi–pi interactions are also possible to regions of the binding site above and below the plane of the ring system.

Heterocycles can be involved in quite intricate hydrogen bonding networks within a binding site. For example, the anticancer drug **methotrexate** contains a diaminopteridine ring system that interacts with its binding site as shown in Fig. 13.26.

If the lead compound contains a heterocyclic ring, it is worth synthesizing analogues containing a benzene ring or different heterocyclic rings to explore whether all the heteroatoms present are really necessary.

A complication with heterocycles is the possibility of **tautomers**. This played an important role in determining the structure of DNA (section 6.1.2). The structure of DNA consists of a double helix with base pairing between two sets of heterocyclic nucleic acid bases. Base pairing involves three hydrogen bonds between the base pair guanine and cytosine, and two hydrogen bonds between the base pair adenine and thymine (Fig. 13.27). The rings involved in the base pairing are coplanar, allowing the optimum orientation for the hydrogen bond donors and hydrogen bond acceptors. This in turn means that the base pairs are stacked above each other, allowing van der Waals interactions between the faces of each base pair. However, when Watson and Crick originally tried to devise a model for DNA, they incorrectly assumed that the preferred tautomers for the nucleic acid bases were as shown in the right-hand part of Fig. 13.27. With these tautomers, the required hydrogen bonding is not possible and would not explain the base pairing observed in the structure of DNA.

In a similar vein, knowing the preferred tautomers of heterocycles can be important in understanding how

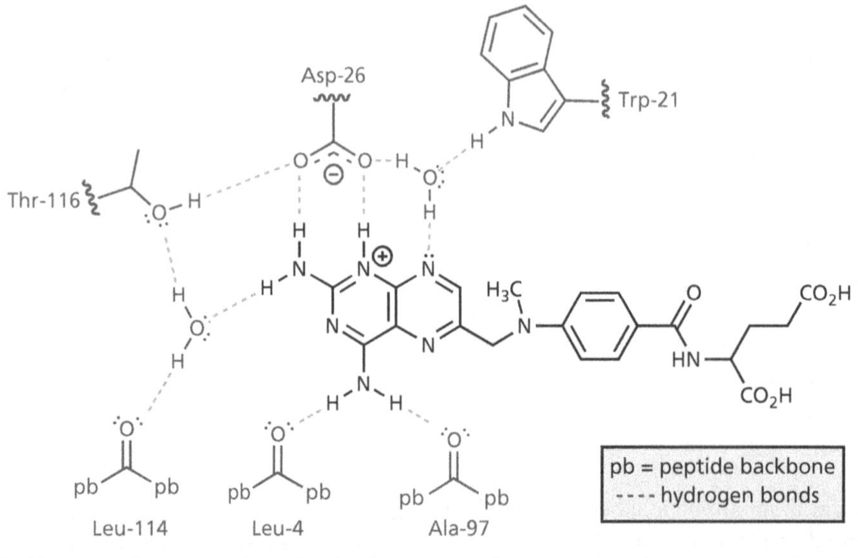

FIGURE 13.26 Binding interactions for the diaminopteridine ring of methotrexate in its binding site.

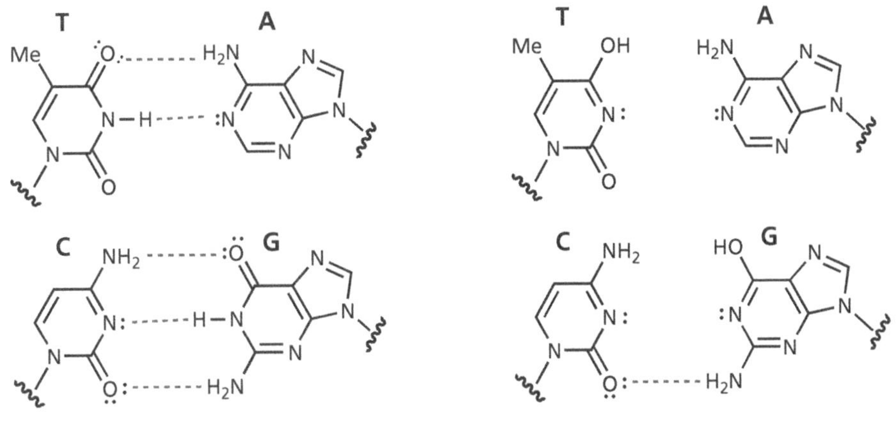

Correct tautomers for base-pairing   Tautomers resulting in weak base-pairing

FIGURE 13.27 Base pairing in DNA and the importance of tautomers.

drugs interact with their binding sites. This is amply illustrated in the design of the anti-ulcer agent **cimetidine** (section 25.2).

With heterocyclic compounds, it is possible for a hydrogen bond donor and a hydrogen bond acceptor to be part of a conjugated system. Polarization of the electrons in the conjugated system permits **π-bond cooperativity**, where the strength of the hydrogen bond donor is enhanced by the hydrogen bond acceptor and vice versa. This has also been called **resonance-assisted hydrogen bonding**. This type of hydrogen bonding is possible for the hydrogen bond donors and acceptors for the nucleic acid base pairs (Fig. 13.28).

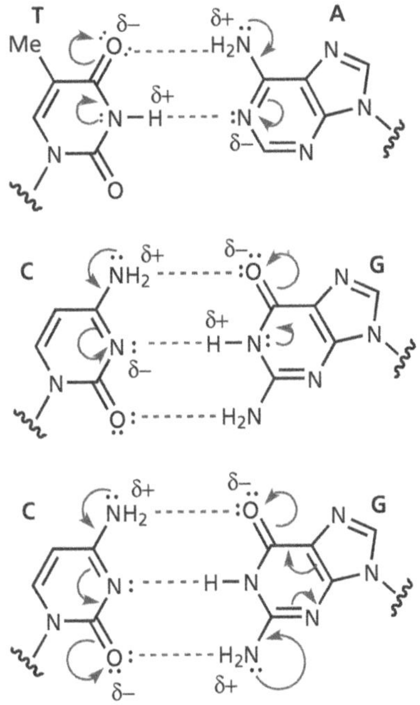

FIGURE 13.28 π-Bond cooperativity in hydrogen bonding.

Note that not all heteroatoms in heterocyclic systems are able to act as good hydrogen bond acceptors. If a heteroatom's lone pair of electrons is part of an aromatic sextet of electrons, it is not available to form a hydrogen bond.

13.1.15 Isosteres

Isosteres are atoms or groups of atoms which share the same valency and which have chemical or physical similarities (Fig. 13.29).

For example, SH, NH_2, and CH_3 are isosteres of OH, whereas S, NH, and CH_2 are isosteres of O. Isosteres can be used to determine whether a particular group is an important binding group or not, by altering the character of the molecule in as controlled a way as possible. Replacing O with CH_2, for example, makes little difference to the size of the analogue, but will have a marked effect on its polarity, electronic distribution, and bonding. Replacing OH with the larger SH may not have such an influence on the electronic character, but steric factors become more significant.

Isosteric groups could be used to determine whether a particular group is involved in hydrogen bonding. For example, replacing OH with CH_3 would completely eliminate hydrogen bonding, whereas replacing OH with NH_2 would not.

The β-blocker **propranolol** has an ether linkage (Fig. 13.30). Replacement of the OCH_2 segment with the isosteres CH=CH, SCH_2, or CH_2CH_2 eliminates activity, whereas replacement with $NHCH_2$ retains activity (though reduced). These results show that the ether oxygen is important to the activity of the drug and suggests that it is involved in hydrogen bonding with the receptor.

The use of isosteres in drug design is described in section 13.3.7.

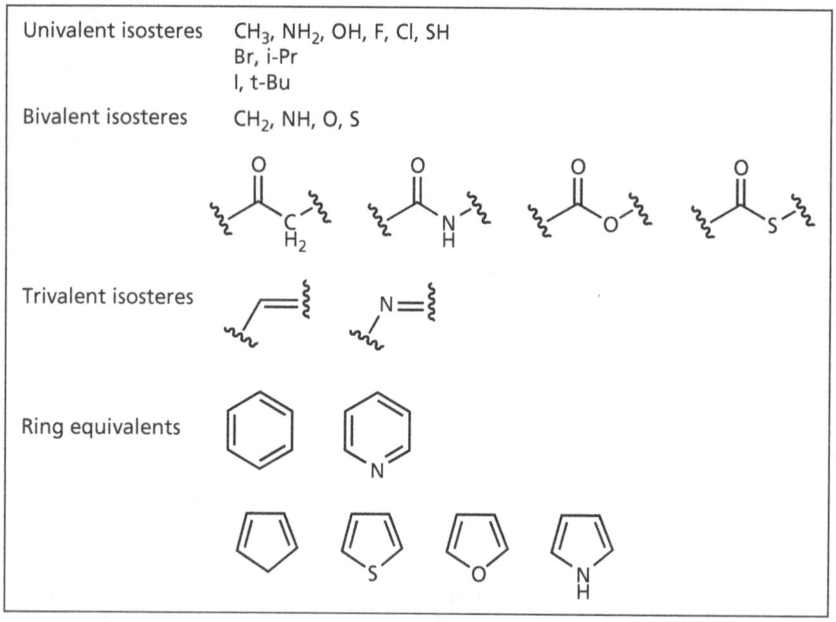

FIGURE 13.29 Examples of classic isosteres.

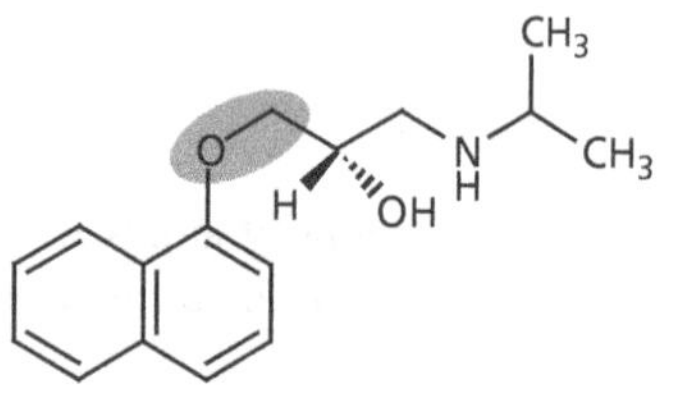

FIGURE 13.30 Propranolol.

13.1.16 Testing procedures

When investigating structure–activity relationships for drug–target binding interactions, biological testing should involve *in vitro* tests; for example inhibition studies on isolated enzymes or binding studies on membrane-bound receptors in whole cells. The results then show conclusively which binding groups are important in drug–target interactions. If *in vivo* testing is carried out, the results are less clear-cut because loss of activity may be due to the inability of the drug to reach its target rather than reduced drug–target interactions. However, *in vivo* testing may reveal functional groups that are important in protecting or assisting the drug in its passage through the body. This would not be revealed by *in vitro* testing.

NMR spectroscopy can also be used to test structure–activity relationships, as described in section 12.4.10.

As mentioned in the introduction, there is little point in designing a drug that has optimum interactions with its target if it has undesirable pharmacokinetic properties. Calculating a structure's hydrophobicity can provide an indication as to whether it is likely to suffer from pharmacokinetic problems. This is because hydrophobic drugs have been found to be more prone to adverse pharmacokinetic properties. For example, they are more likely to interact with other protein targets, resulting in unwanted side effects. They are generally less soluble, show poor permeability, and are more likely to produce toxic metabolites. The hydrophobic nature of a drug can be calculated by its Clog D value (section 14.1). In recent years, several research groups have optimized drugs by optimizing **lipophilic efficiency** (LipE), where LipE = pK_i (or pIC$_{50}$) − Clog D. Drugs with a good level of lipophilic efficiency will have high activity (pK_i or pIC$_{50}$) and low hydrophobic character. Optimizing LipE involves a parallel optimization of potency and hydrophobicity. This quantitative method of optimizing both the pharmacodynamic and pharmacokinetic properties has been called **property-based drug design** and was used in the structure-based drug design of **crizotinib** (Box 13.4).

13.1.17 SAR in drug optimization

In this section, we have focused on SAR studies aimed at identifying important binding groups in a lead compound. SAR studies are also used in drug optimization, where the aim is to find analogues with better activity and selectivity. This involves further modifications of the lead compound to identify whether these are beneficial or detrimental to activity. This is covered in section 13.3 where the different strategies of optimizing drugs are discussed.

13.2 Identification of a pharmacophore

Once it is established which groups are important for a drug's activity, it is possible to move on to the next stage—the identification of the **pharmacophore**. The pharmacophore summarizes the important binding groups which are required for activity, and their relative positions in space with respect to each other. For example, if we discover that the important binding groups for our hypothetical drug glipine are the two phenol groups, the aromatic ring, and the nitrogen atom, then the pharmacophore is as shown in Fig. 13.31. Structure I shows the two-dimensional (2D) pharmacophore and structure II shows the three-dimensional (3D) pharmacophore. The latter specifies the relative positions of the important groups in space. In this case, the nitrogen atom is 5.063 Å from the centre of the phenolic ring and lies at an angle of 18° from the plane of the ring. Note that it is not necessary to show the specific skeleton connecting the important groups. Indeed, there are benefits in not doing so, as it is easier to compare the 3D pharmacophores from different structural classes of compound to see if they are similar. Three-dimensional pharmacophores can be defined using molecular modelling (section 17.11), which allows the definition of 'dummy bonds', such as the one in Fig. 13.31 between nitrogen and the centre of the aromatic ring. The centre of the ring can be defined by a dummy atom called a **centroid** (not shown).

An even more general type of 3D pharmacophore is the one shown as structure III (Fig. 13.31)—a bonding-type pharmacophore. Here the bonding characteristics of each functional group are defined, rather than the group itself. Note also that the groups are defined as points in space. This includes the aromatic ring, which is defined by the centroid. All the points are connected by pharmacophoric triangles to define their positions. This allows the comparison of molecules which may have the same pharmacophore and binding interactions, but which use different functional groups to achieve these interactions. In this case, the phenol groups can act as hydrogen bond donors or acceptors, the aromatic ring can participate in van der Waals interactions, and the amine can act as a hydrogen bond acceptor (or as an ionic centre if it is protonated). We shall return to the concept and use of 3D pharmacophores in sections 17.11 and 18.10.

Identifying 3D pharmacophores is relatively easy for rigid cyclic structures such as the hypothetical glipine. With more flexible structures, it is not so straightforward because the molecule can adopt a large number of shapes or conformations which place the important binding groups in different positions relative to each other. Normally, only one of these conformations is recognized and bound by the binding site. This conformation is known as the **active conformation**. In order to identify the 3D pharmacophore, it is necessary to know the active conformation. There are various ways in which this might be done. Rigid analogues of the flexible compound could be synthesized and tested to see whether activity is retained (section 13.3.9). Alternatively, it may be possible to crystallize the target with the compound bound to the binding site. X-ray crystallography could then be used to identify the structure of the complex as well as the active conformation of the bound ligand (section 17.10). Finally, progress has been made in using NMR spectroscopy to solve the active conformation of isotopically labelled molecules bound to their binding sites (see also Case study 10).

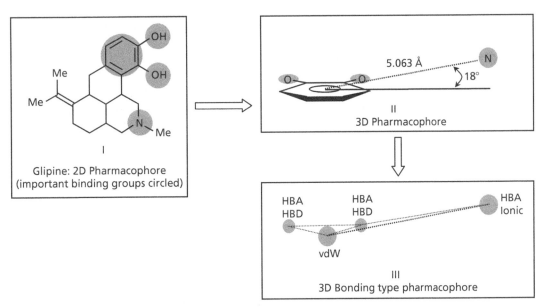

FIGURE 13.31 Pharmacophore for the fictitious structure glipine.

We finish this section with a warning! A drawback with pharmacophores is their unavoidable emphasis on functional groups as the crucial binding groups. In many situations this is certainly true, but in other situations, it is not. It is not uncommon to find compounds that have the correct pharmacophore, but show disappointing activity and poor binding. It is important to realize that the overall skeleton of the molecule is involved in interactions with the binding site through van der Waals and hydrophobic interactions. The strength of these interactions can sometimes be crucial in whether a drug binds effectively or not, and the 3D pharmacophore does not take this into account. The pharmacophore also does not take into account the size of a molecule and whether it will fit the binding site. Finally, a functional group that is part of the pharmacophore may not be so crucial if an agent can form an alternative binding interaction with the binding site. For example, the phenol group is an important part of the analgesic pharmacophore for **morphine** and closely related analogues, but is less important for analgesics such as the **oripavines**. Other analgesics such as **pethidine** and **methadone** lack the phenol group entirely (Chapter 24).

KEY POINTS

- SARs define the functional groups or regions of a lead compound which are important to its biological activity.

- Functional groups such as alcohols, amines, esters, amides, carboxylic acids, phenols, and ketones can interact with binding sites by means of hydrogen bonding.

- Functional groups such as aminium ions, quaternary ammonium salts, and carboxylate groups can interact with binding sites by ionic bonding. In some cases, a sulphonamide group can become ionized and form an ionic bond.

- Functional groups such as alkenes and aromatic rings can interact with binding sites by means of van der Waals interactions. π-π interactions are also possible with aromatic rings.

- Alkyl substituents and the carbon skeleton of the lead compound can interact with hydrophobic regions of binding sites by means of van der Waals interactions.

- Interactions involving dipole moments or induced dipole moments may play a role in binding a lead compound to a binding site.

- Reactive functional groups such as alkyl halides may lead to irreversible covalent bonds being formed between a lead compound and its target.

- The relevance of a functional group to binding can be determined by preparing analogues where the functional group is modified or removed in order to see whether activity is affected by such a change.

- Some functional groups can be important to the activity of a lead compound for reasons other than target binding. They

may play a role in the electronic or stereochemical properties of the compound, or they may have an important pharmacokinetic role.

- Replacing a group in the lead compound with an isostere (a group having the same valency) makes it easier to determine whether a particular property such as hydrogen bonding is important.

- *In vitro* testing procedures should be used to determine the SAR for target binding.

- The pharmacophore summarizes the groups which are important in binding a lead compound to its target, as well as their relative positions in three dimensions.

13.3 Drug optimization: strategies in drug design

Once the important binding groups and pharmacophore of the lead compound have been identified, it is possible to synthesize analogues that contain the same pharmacophore. But why is this necessary? If the lead compound has useful biological activity, why bother making analogues? The answer is that very few lead compounds are ideal. Most are likely to have low activity, poor selectivity, and significant side effects. They may also be difficult to synthesize, so there is an advantage in finding analogues with improved properties. We look now at strategies that can be used to optimize the interactions of a drug with its target in order to gain better activity and selectivity.

13.3.1 Variation of substituents

Varying easily accessible substituents is a common method of fine tuning the binding interactions of a drug.

13.3.1.1 Alkyl substituents

Certain alkyl substituents can be varied more easily than others. For example, the alkyl substituents of ethers, amines, esters, and amides are easily varied as shown in Fig. 13.32. In these cases, the alkyl substituent already present can be removed and replaced by another substituent. Alkyl substituents which are part of the carbon skeleton of the molecule are not easily removed, and it is usually necessary to carry out a full synthesis in order to vary them.

If alkyl groups are interacting with a hydrophobic pocket in the binding site, then varying the length and bulk of the alkyl group (e.g. methyl, ethyl, propyl, butyl, isopropyl, isobutyl, or *t*-butyl) allows one to probe the depth and width of the pocket. Choosing a substituent that will fill the pocket will then increase the binding interactions (Fig. 13.33).

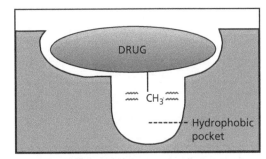

FIGURE 13.32 Methods of modifying an alkyl group.

Larger alkyl groups may also confer selectivity on the drug. For example, in the case of a compound that interacts with two different receptors, a bulkier alkyl substituent may prevent the drug from binding to one of those receptors and so cut down side effects (Fig. 13.34). For example, **isoprenaline** is an analogue of **adrenaline** where a methyl group was replaced by an isopropyl group, resulting in selectivity for adrenergic α-receptors over adrenergic β-receptors (section 23.11.3).

13.3.1.2 Substituents on aromatic or heteroaromatic rings

If a drug contains an aromatic or heteroaromatic ring, the position of substituents can be varied to find better binding interactions, resulting in increased activity (Fig. 13.35).

For example, the best anti-arrhythmic activity for a series of benzopyrans was found when the sulphonamide substituent was at position 7 of the aromatic ring (Fig. 13.36).

Changing the position of one substituent may have an important effect on another. For example, an

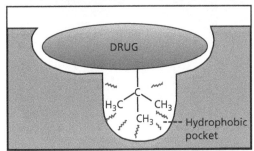

FIGURE 13.33 Variation of an alkyl substituent to fill a hydrophobic pocket.

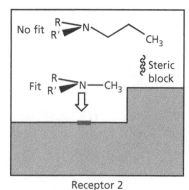

Receptor 1

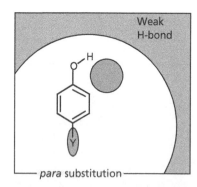

Receptor 2

■■■ Binding region for N

FIGURE 13.34 Use of a larger alkyl group to confer selectivity on a drug.

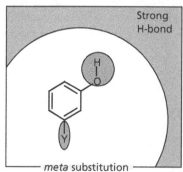

FIGURE 13.35 Varying the position of a substituent on an aromatic ring.

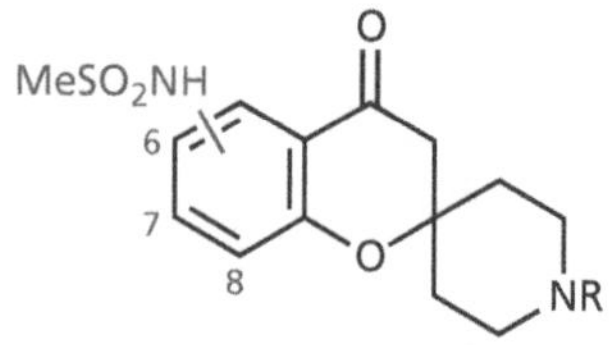

FIGURE 13.36 Benzopyrans.

electron-withdrawing nitro group will affect the basicity of an aromatic amine more significantly if it is at the *para* position rather than the *meta* position (Fig. 13.37). At the *para* position, the nitro group will make the amine a weaker base and less liable to protonate. This would decrease the amine's ability to interact with ionic binding groups in the binding site, and decrease activity.

If the substitution pattern is ideal, then we can try varying the substituents themselves. Substituents have different steric, hydrophobic, and electronic properties, and so varying these properties may have an effect on binding and activity. For example, activity might be improved by having a more electron-withdrawing substituent, in which case a chloro substituent might be tried in place of a methyl substituent.

The chemistry involved in these procedures is usually straightforward, so these analogues are made as a matter of course whenever a novel drug structure is developed. Furthermore, the variation of substituents is open to **quantitative structure–activity relationship** (QSAR) studies, as described in Chapter 18.

13.3.1.3 Synergistic effects

Finally, a warning! When varying substituents, it is normal to study analogues where only one substituent is

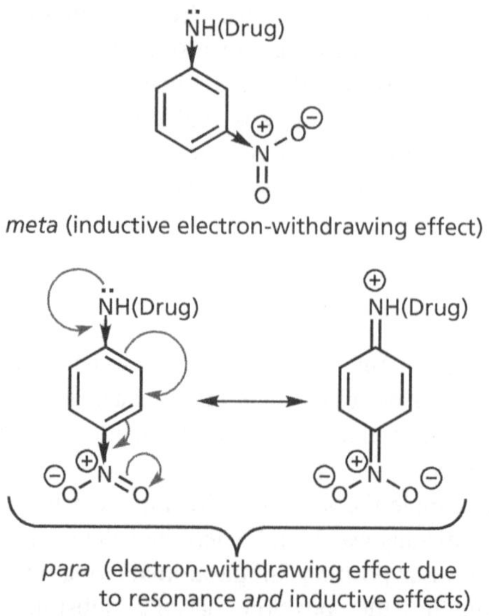

meta (inductive electron-withdrawing effect)

para (electron-withdrawing effect due to resonance *and* inductive effects)

FIGURE 13.37 Electronic effects of different substitution patterns on an aromatic ring.

added or altered at a time. In that way, one can identify those substituents that are good for activity and those that are not. However, it does not take into account the synergistic effect that two or more substituents may have on activity. For example, two substituents that are individually bad for activity may actually be beneficial for activity when they are both present. The design of the anticancer drug **sorafenib** provides an illustration of this effect (Box 21.10).

13.3.2 Extension of the structure

The strategy of extension involves the addition of another functional group or substituent to the lead compound in order to probe for extra binding interactions with the target. Lead compounds are capable of fitting the binding site and have the necessary functional groups to interact with some of the important binding regions present. However, it is possible that they do not interact with all the binding regions available. For example, a lead compound may bind to three binding regions in the binding site but fail to use a fourth (Fig. 13.38). Therefore, why not add extra functional groups to probe for that fourth region?

Extension tactics are often used to find extra hydrophobic regions in a binding site by adding various alkyl or arylalkyl groups. These groups can be added to functional groups such as alcohols, phenols, amines, and carboxylic acids should they be present in the drug, as long as this does not disrupt important binding interactions that are already present. Alternatively, they could be built into the building blocks used in the synthesis of various analogues. By the same token, substituents containing polar functional groups could be added to probe for extra hydrogen bonding or ionic interactions. A good example of the use of extension tactics to increase binding interactions involves the design of the ACE inhibitor **enalaprilat** from the lead compound **succinyl proline**; see Case study 2, Figs CS2.8–2.9.

Extension strategies are used to strengthen the binding interactions and activity of a receptor agonist or an enzyme inhibitor, but they can also be used to convert an agonist into an antagonist. This will happen if the extra binding interaction results in a different induced fit from that required to activate the receptor. As a result, the antagonist binds to an inactive conformation of the receptor and blocks access to the endogenous agonist. The strategy has also been used to alter an enzyme substrate into an inhibitor (Box 13.1).

The extension tactic has been used successfully to produce more active analogues of morphine (sections 24.6.2 and 24.6.4) and more active adrenergic agents (sections 23.9–23.11). It was also used to improve the activity and selectivity of the protein kinase inhibitor **imatinib** (section 21.6.2.2). Other examples of the extension strategy can be found in Case studies 2, 5, 6, and 7, and Box 17.6, as well as sections 19.7.7, 20.7.4, and 26.4.2.

An unusual example of an extension strategy is where a substituent was added to an enzyme substrate such that extra binding interactions took place with a neighbouring cofactor in the binding site. This resulted in the analogue acting as an inhibitor, rather than a substrate (Box 13.1).

13.3.3 Chain extension/contraction

Some drugs have two important binding groups linked together by a chain, in which case it is possible that the chain length is not ideal for the best interaction. Therefore, shortening or lengthening the chain length is a useful tactic to try (Fig. 13.39, see also Box 13.1, section 24.6.2, and Case study 2).

13.3.4 Ring expansion/contraction

If a drug has one or more rings that are important binding groups, it is generally worth synthesizing analogues where one of these rings is expanded or contracted. The principle behind this approach is much the same

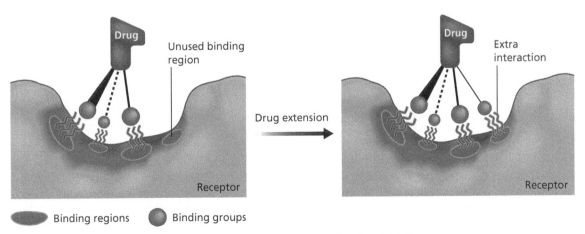

FIGURE 13.38 Extension of a drug to provide a fourth binding group.

BOX 13.1 Converting an enzyme substrate to an inhibitor by extension tactics

The enzyme **17β-hydroxysteroid dehydrogenase type 1** catalyses the conversion of **estrone** to the female steroid hormone **estradiol**, with the cofactor **NADH** acting as the reducing agent for the reaction (Fig. 1; see also Chapter 3, Figs. 3.11–3.12). Inhibition of this enzyme may prove useful in the treatment of estradiol-dependent tumours since the levels of estradiol present in the body would be lowered.

The cofactor NADH is bound next to estrone in the active site, and so it was reasoned that a direct bonding interaction between an estrone analogue and NADH would lock the analogue into the active site and block access to estrone itself. Therefore, the analogue would act as an enzyme inhibitor. Various substituents were added at position 16 to achieve this goal since crystallographic and molecular modelling studies had shown that such substituents would

be ideally placed for an interaction with the cofactor. This led to a structure (Fig. 2) which showed promising activity as an inhibitor. The amide group interacts with the primary amide of NADH by hydrogen bonding, while the pyridine ring interacts with the phosphate groups of the cofactor. A more conventional extension strategy was to add an ethyl group at C-2, which allowed additional van der Waals interactions with a small hydrophobic pocket in the active site. It was also observed that two protons acted as steric blockers and prevented NADH reducing the ketone group of the analogue.

For additional material see Web article 1: Steroids as novel anticancer agents on the Online Resource Centre at www.oxfordtextbooks.co.uk/orc/patrick6e/

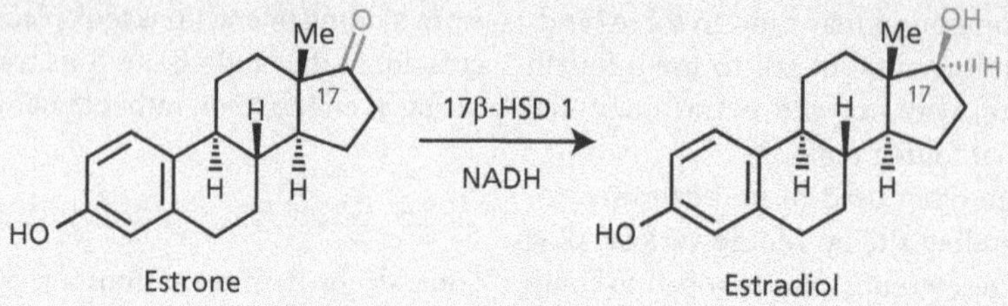

FIGURE 1 The enzyme-catalysed conversion of estrone to estradiol.

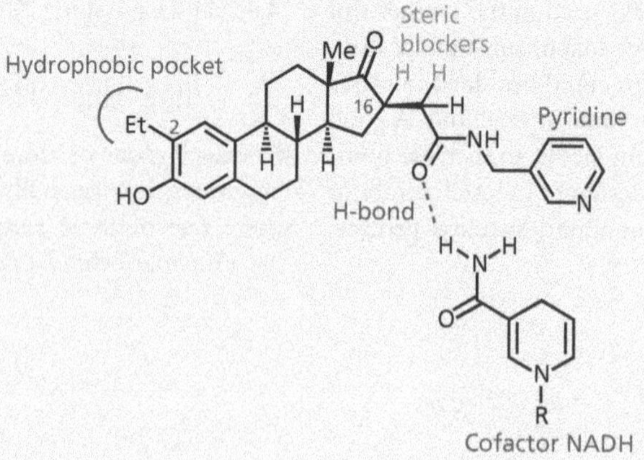

FIGURE 2 Extra binding interactions resulting from the extension strategy.

as varying the substitution pattern of an aromatic ring. Expanding or contracting a ring may put other rings in different positions relative to each other, and may lead to better interactions with specific regions in the binding site (Fig. 13.40).

Varying the size of a ring can also bring substituents into a good position for binding. For example, during the development of the antihypertensive agent **cilazaprilat** (another ACE inhibitor), the bicyclic structure I showed promising activity (Fig. 13.41). The important binding

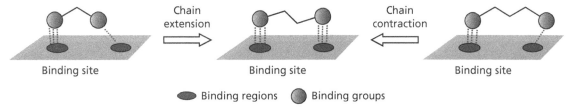

FIGURE 13.39 Chain contraction and chain extension.

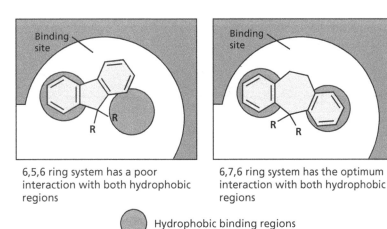

6,5,6 ring system has a poor interaction with both hydrophobic regions

6,7,6 ring system has the optimum interaction with both hydrophobic regions

Hydrophobic binding regions

FIGURE 13.40 Ring expansion.

groups were the two carboxylate groups and the amide group. By carrying out various ring contractions and expansions, cilazaprilat was identified as the structure having the best interaction with the binding site.

Another example of a ring expansion can be seen in the design and development of the cardiovascular agent **ivabradine** (section 26.7). An example of a beneficial ring contraction can be seen in the design of the antiviral agent **simeprevir**, where a 15-membered macrocycle was contracted to a 14-membered macrocycle (Case study 10).

13.3.5 Ring variations

A popular strategy used for compounds containing an aromatic or heteroaromatic ring is to replace the original ring with a range of other heteroaromatic rings of different ring size and heteroatom positions. For example, several non-steroidal anti-inflammatory agents (NSAIDs) have been reported, all consisting of a central ring with 1,2-biaryl substitution. Different pharmaceutical companies have varied the central ring to produce a range of active compounds (Fig. 13.42).

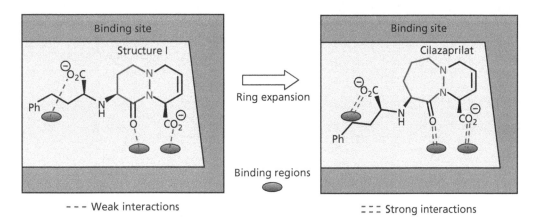

– – – Weak interactions   ‗‗‗ Strong interactions

FIGURE 13.41 Development of cilazaprilat.

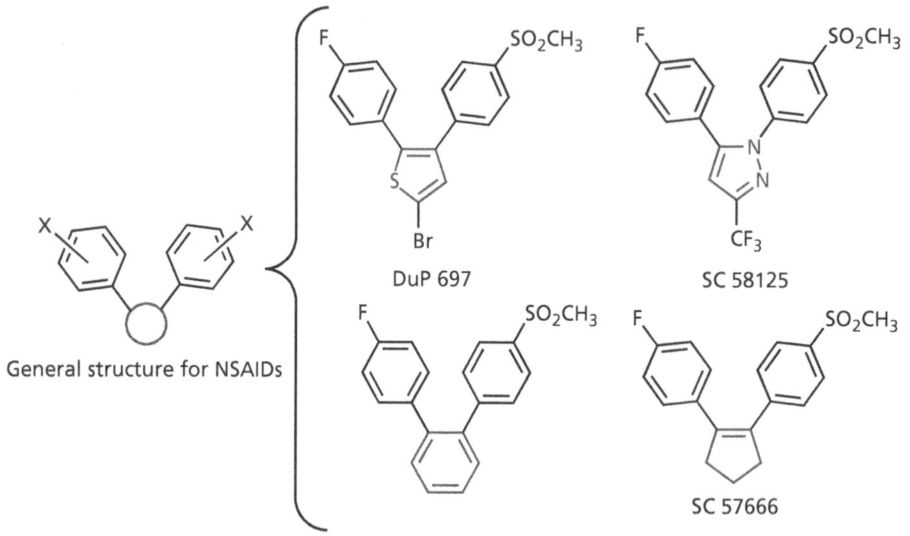

FIGURE 13.42 Non-steroidal anti-inflammatory drugs (NSAIDS).

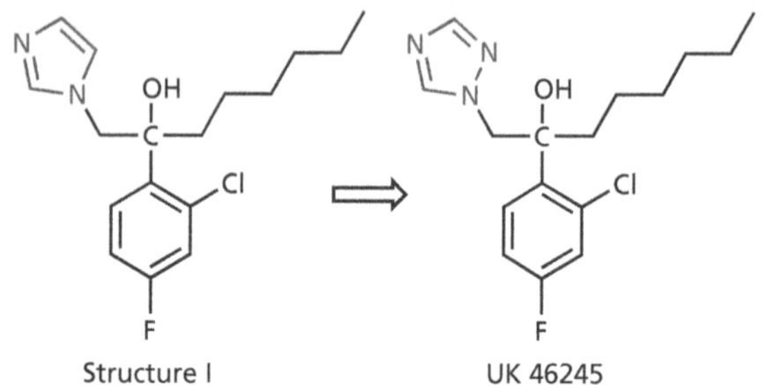

FIGURE 13.43 Development of UK 46245.

Admittedly, a lot of these changes are merely ways of avoiding patent restrictions ('**me too' drugs**), but there can often be significant improvements in activity, as well as increased selectivity and reduced side effects ('**me better' drugs**). For example, the antifungal agent (I) (Fig. 13.43) acts against an enzyme present in both fungal and human cells. Replacing the imidazole ring of structure (I) with a 1,2,4-triazole ring to give UK 46245 resulted in better selectivity against the fungal form of the enzyme.

One advantage of altering an aromatic ring to a heteroaromatic ring is that it introduces the possibility of an extra hydrogen bonding interaction with the binding site, should a suitable binding region be available (*extension strategy*). For example, structure I (Fig. 13.44) was the lead compound for a project looking into novel antiviral agents. Replacing the aromatic ring with a pyridine ring resulted in an additional binding interaction with the target enzyme. Further development led eventually to the antiviral agent **nevirapine** (Fig. 13.44 and section 20.7.3.2).

13.3.6 Ring fusions

Extending a ring by ring fusion can sometimes result in increased interactions or increased selectivity. One of the major advances in the development of the selective β-blockers was the replacement of the aromatic ring in **adrenaline** with a naphthalene ring system (**pronethalol**) (Fig. 13.45). This resulted in a compound that was able to distinguish between two very similar receptors—the α- and β-receptors for adrenaline. One possible explanation

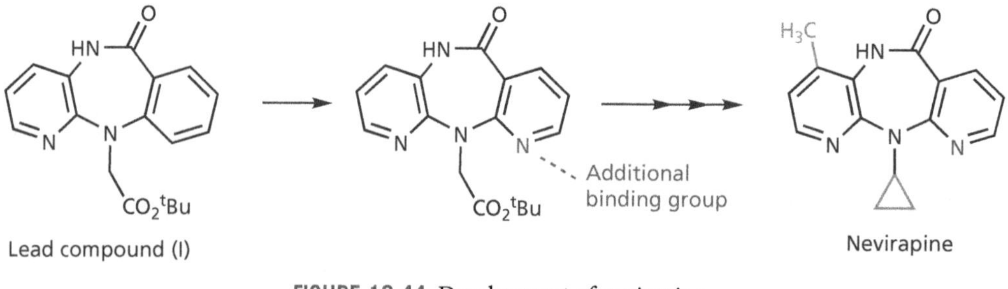

FIGURE 13.44 Development of nevirapine.

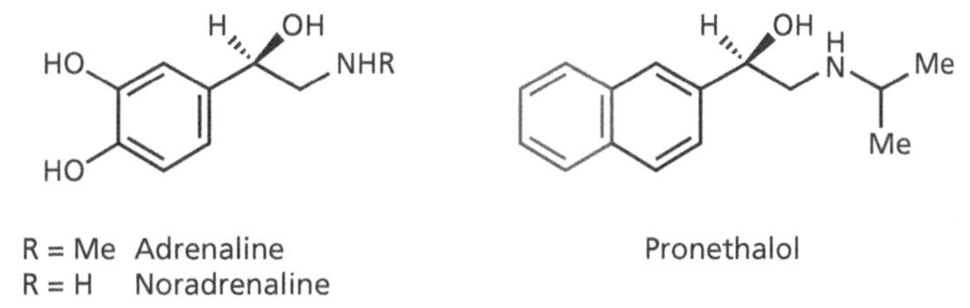

FIGURE 13.45 Structures of adrenaline, noradrenaline, and pronethalol.

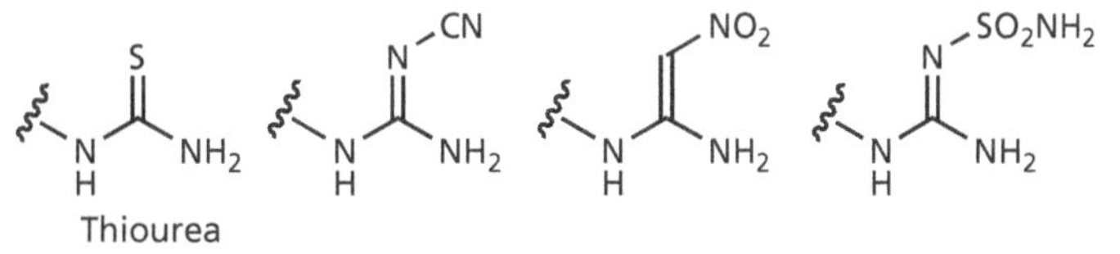

FIGURE 13.46 Non-classical isosteres for a thiourea group.

for this could be that the β-receptor has a larger van der Waals binding area for the aromatic system than the α-receptor, and can interact more strongly with pronethalol than with adrenaline. Another possible explanation is that the naphthalene ring system is sterically too big for the α-receptor, but is just right for the β-receptor.

13.3.7 Isosteres and bio-isosteres

Isosteres (section 13.1.15) have often been used in drug design to vary the character of the molecule in a rational way with respect to features such as size, polarity, electronic distribution, and bonding. Some isosteres can be used to determine the importance of size towards activity, whereas others can be used to determine the importance of electronic factors. For example, fluorine is often used as an isostere of hydrogen since it is virtually the same size. However, it is more electronegative and can be used to vary the electronic properties of the drug without having any steric effect.

The presence of fluorine in place of an enzymatically labile hydrogen can also disrupt an enzymatic reaction, as C–F bonds are not easily broken. For example, the antitumour drug **5-fluorouracil** described in section 21.3.2 is accepted by its target enzyme because it appears little different from the normal substrate—**uracil**. However, the mechanism of the enzyme-catalysed reaction is totally disrupted, as the fluorine has replaced a hydrogen which is normally lost during the enzyme mechanism.

Several non-classical isosteres have been used in drug design as replacements for particular functional groups. Non-classical isosteres are groups which do not obey the steric and electronic rules used to define classical isosteres, but which have similar physical and chemical properties. For example, the structures shown in Fig. 13.46 are non-classical isosteres for a thiourea group. They are all planar groups of similar size and basicity.

The term **bio-isostere** is used in drug design and includes both classical and non-classical isosteres. A bio-isostere is a group that can be used to replace another group while retaining the desired biological activity. For example, a cyclopropyl group has been used as a bio-isostere for an alkene group in prodrugs (section 14.6.1.1) and opioid antagonists (section 24.6.2). Bio-isosteres are often used to replace a functional group that is important for target binding, but is problematic in one way or another. For example, the thiourea group was present as an important binding group in early histamine antagonists, but was responsible for toxic side effects. Replacing it with bio-isosteres allowed the important binding interactions to be retained for histamine antagonism, but avoided the toxicity problems (section 25.2.6). Further examples of the use of bio-isosteres are given in sections 14.1.6, 14.2.2, 20.7.4, and 20.10.3. It is important to realize that bio-isosteres are specific for a particular group of compounds and their target. Replacing a functional group with a bio-isostere is not guaranteed to retain activity for every drug at every target.

As stated above, bio-isosteres are commonly used in drug design to replace a problematic group while

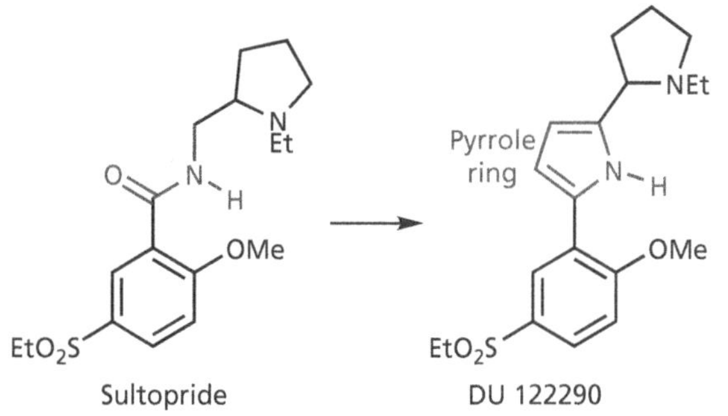

FIGURE 13.47 Introducing a pyrrole ring as a bio-isostere for an amide group.

retaining activity. In some situations, the use of a bio-isostere can actually increase target interactions and/or selectivity. For example, a pyrrole ring has frequently been used as a bio-isostere for an amide. Carrying out this replacement on the dopamine antagonist **sultopride** led to increased activity and selectivity towards the dopamine D_3-receptor over the dopamine D_2-receptor (Fig. 13.47). Such agents show promise as antipsychotic agents which lack the side effects associated with the D_2-receptor.

Introducing a bio-isostere to replace a problematic group often involves introducing further functional groups that might form extra binding interactions with the target binding site (section 13.3.2). For example, a ten-fold increase in activity was observed for an antiviral agent when an *N*-acylsulphonamide was used as a bio-isostere for a carboxylic acid (Fig. 13.48). The *N*-acylsulphonamide group introduces the possibility of further hydrogen bonding and van der Waals interactions with the binding site (see also Case study 10).

Transition-state isosteres are a special type of isostere used in the design of transition-state analogues. These are drugs that are used to inhibit enzymes (section 7.4). During an enzymatic reaction, a substrate goes through a transition state before it becomes product. It is proposed that the transition state is bound more strongly than either the substrate or the product, so it makes sense to design drugs based on the structure of the transition state rather than the structure of the substrate or the product. However, the transition state is inherently unstable and so transition-state isosteres are moieties that are used to mimic the crucial features of the transition state, but which are stable to the enzyme-catalysed reaction. For example, the transition state of an amide hydrolysis is

thought to resemble the tetrahedral reaction intermediate shown in Fig. 13.49. This is a geminal diol, which is inherently unstable. The hydroxyethylene moiety shown is a transition-state isostere because it shares the same tetrahedral geometry, retains one of the hydroxyl groups, and is stable to hydrolysis. Further examples of the use of transition-state isosteres are given in sections 20.7.4, 20.8.3, and 21.3.4, and Case studies 1 and 2.

13.3.8 Simplification of the structure

Simplification is a strategy which is commonly used on the often complex lead compounds arising from natural sources (see Box 13.2). Once the essential groups of such a drug have been identified by SAR, it is often possible to discard the non-essential parts of the structure without losing activity. Consideration is given to removing functional groups which are not part of the pharmacophore, simplifying the carbon skeleton (for example removing rings), and removing asymmetric centres.

This strategy is best carried out in small stages. For example, consider our hypothetical natural product glipine (Fig. 13.50). The essential groups have been highlighted, and we might aim to synthesize simplified compounds in the order shown. These still retain the essential groups making up the pharmacophore.

Chiral drugs pose a particular problem. The easiest and cheapest method of synthesizing a chiral drug is to make the racemate. However, both enantiomers then have to be tested for their activity and side effects, doubling the number of tests that have to be carried out. This is because different enantiomers can have different activities. For example, compound **UH-301** (Fig. 13.51) is inactive as a racemate, whereas its enantiomers have opposing agonist

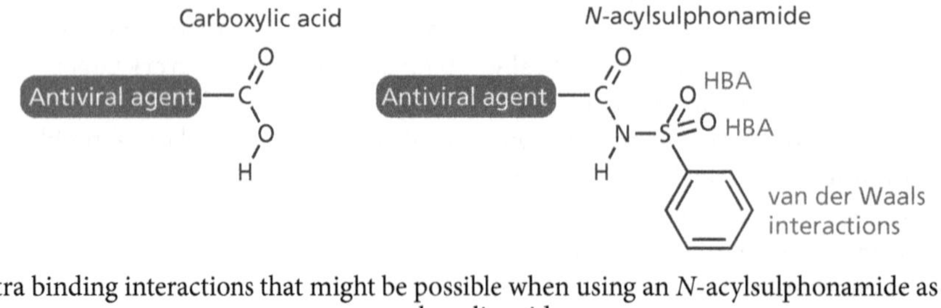

FIGURE 13.48 Extra binding interactions that might be possible when using an *N*-acylsulphonamide as a bio-isostere for a carboxylic acid.

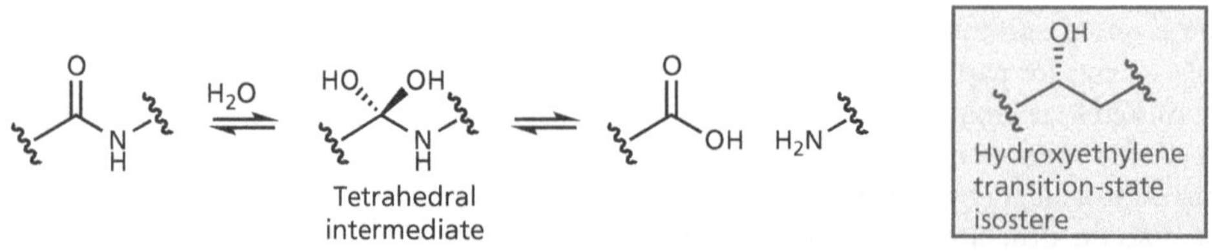

FIGURE 13.49 Example of a transition-state isostere designed to resemble a tetrahedral intermediate formed during an enzyme-catalysed reaction. The transition state is believed to resemble the tetrahedral intermediate.

BOX 13.2 Simplification

Simplification tactics have been used successfully with the alkaloid **cocaine**. Cocaine has local anaesthetic properties, and its simplification led to the development of local anaesthetics which could be easily synthesized in the laboratory. One of the earliest was **procaine** (**Novocaine**), discovered in 1909 (Fig. 1). Simplification tactics have also proved effective in the design of simpler morphine analogues (section 24.6.3).

Simplification tactics were also used in the development of **devazepide** from the microbial metabolite **asperlicin**. The benzodiazepine and indole skeletons inherent in asperlicin are important to activity and have been retained. Both asperlicin and devazepide act as antagonists of a neuropeptide chemical messenger called **cholecystokinin** (CCK) which has been implicated in causing panic attacks. Therefore, antagonists may be of use in treating such attacks.

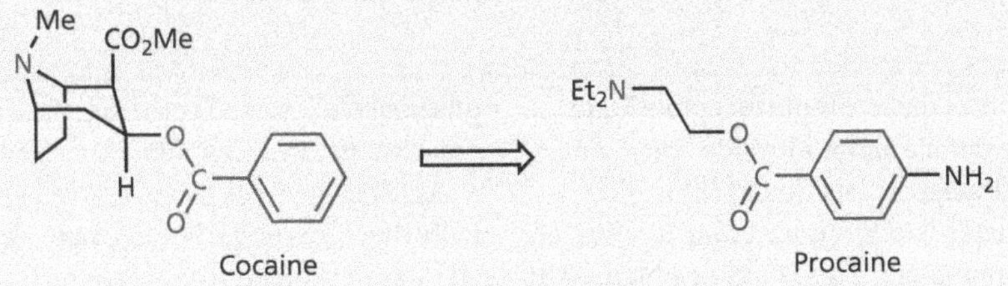

FIGURE 1 Simplification of cocaine (pharmacophore shown in colour).

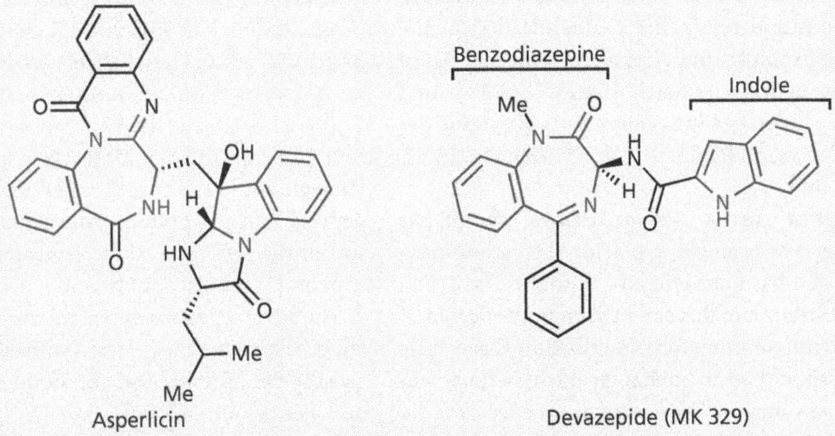

FIGURE 2 Simplification of asperlicin.

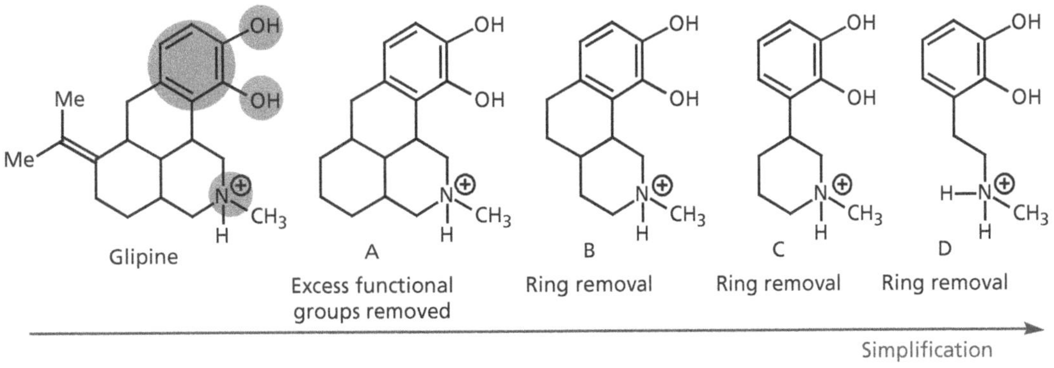

FIGURE 13.50 Glipine analogues.

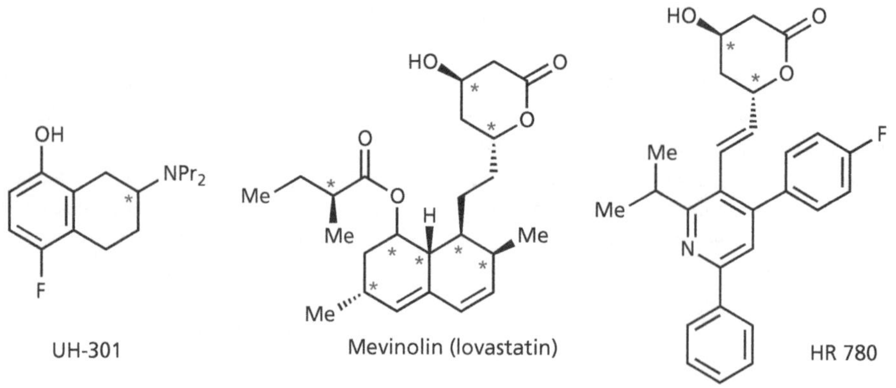

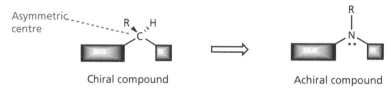

FIGURE 13.51 UH-301, mevinolin, and HR 780.

and antagonist activity at the serotonin receptor (5-HT$_{1A}$). Another notorious example is **thalidomide** where one of the enantiomers is teratogenic (section 21.9.1).

The use of racemates is discouraged and it is preferable to use a pure enantiomer. This could be obtained by separating the enantiomers of the racemic drug, or carrying out an asymmetric synthesis. Both options inevitably add to the cost of the synthesis, and so designing a structure that lacks some or all of the asymmetric centres can be advantageous and represents a simplification of the structure. For example, the cholesterol-lowering agent **mevinolin** has eight asymmetric centres, but a second generation of cholesterol-lowering agents has been developed which contain far fewer (e.g. **HR 780**; Fig. 13.51, see also Case study 1).

Various tactics can be used to remove asymmetric carbon centres. For example, replacing the carbon centre with nitrogen has been effective in many cases (Fig. 13.52). An illustration of this can be seen in the design of thymidylate synthase inhibitors described in Case study 5. However, it should be noted that the introduction of an amine in this way may well have significant effects on the

pharmacokinetics of the drug in terms of log *P*, basicity, polarity, etc.; see Chapters 11 and 14.

Another tactic is to introduce symmetry where originally there was none. For example, the muscarinic agonist (II) was developed from (I) in order to remove asymmetry (Fig. 13.53). Both structures have the same activity.

Simplification strategies have been applied extensively in many areas of medicinal chemistry, some of which are described in this text; for example, antiprotozoal agents (Case studies 3 and 4), local anaesthetics (section 17.9, Box 13.2), antibacterial agents (section 19.5.5.2), antiviral agents (section 20.7.4.8), anticancer agents (sections 21.2.1, 21.2.3.3, and 21.5.2), muscarinic antagonists (section 22.9.2.3), and opioids (section 24.6.3). Simplification strategies are also crucial when developing drugs from peptide lead compounds, where the aim is to reduce the size of the structure to the equivalent of a dipeptide or tripeptide (see Case study 10).

The advantage of simpler structures is that they are easier, quicker, and cheaper to synthesize in the laboratory. Usually the complex lead compounds obtained from natural sources are impractical to synthesize and have to be

FIGURE 13.52 Replacing an asymmetric carbon with nitrogen.

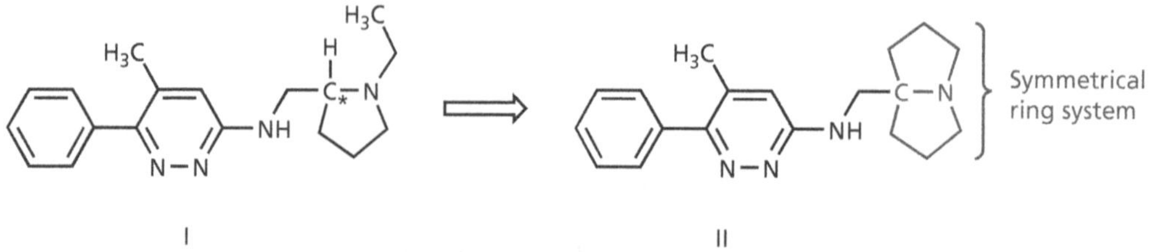

FIGURE 13.53 Introducing symmetry.

extracted from the source material—a slow, tedious, and expensive business. Removing unnecessary functional groups can also be advantageous in removing side effects if these groups interact with other targets, or are chemically reactive. There are, however, potential disadvantages in oversimplifying molecules. Simpler molecules are often more flexible and can sometimes bind differently to their targets compared to the original lead compound, resulting in different effects. It is best to simplify in small stages, checking that the desired activity is retained at each stage. Oversimplification may also result in reduced activity, reduced selectivity, and increased side effects. We shall see why in the next section (section 13.3.9).

13.3.9 Rigidification of the structure

Rigidification has often been used to increase the activity of a drug or to reduce its side effects. In order to understand why this tactic can work, let us consider again our hypothetical neurotransmitter from Chapter 5 (Fig. 13.54). This is quite a simple, flexible molecule with several rotatable bonds that can lead to a large number of conformations or shapes. One of these conformations is recognized by the receptor and is known as the **active conformation**. The other conformations are unable to interact efficiently with the receptor and are inactive conformations. However, it is possible that a different receptor exists which *is* capable of binding one of these alternative conformations. If this is the case, then our model neurotransmitter could switch on two different receptors

and give two different biological responses, one which is desired and one which is not.

The body's own neurotransmitters are highly flexible molecules (section 4.2), but, fortunately, the body is efficient at releasing them close to their target receptors, then quickly inactivating them so that they do not make the journey to other receptors. This is not the case for drugs. They have to be sturdy enough to travel throughout the body and will interact with all the receptors that are prepared to accept them. The more flexible a drug molecule is, the more likely it will interact with more than one receptor and produce other biological responses (side effects). Too much flexibility is also bad for oral bioavailability (section 11.3).

The strategy of rigidification is to make the molecule more rigid, such that the active conformation is retained and the number of other possible conformations is decreased. This should reduce the possibility of other receptor interactions and side effects. This same strategy should also increase activity. By making the drug more rigid, it is more likely to be in the active conformation when it approaches the target binding site and should bind more readily. This is also important when it comes to the thermodynamics of binding. A flexible molecule has to adopt a single active conformation in order to bind to its target, which means that it has to become more ordered. This results in a decrease in entropy and, since the free energy of binding is related to entropy by the equation $\Delta G = \Delta H - T\Delta S$, any decrease in entropy will adversely affect ΔG. This in turn lowers the binding affinity (K_i), which is related to ΔG by the equation $\Delta G = -RT\ln K_i$. A totally rigid molecule, on the other hand, is already in its active conformation, and there is no loss of entropy involved in binding to the target. If the binding interactions (ΔH) are exactly the same as for the more flexible molecule, the rigid molecule will have the better overall binding affinity.

Incorporating the skeleton of a flexible drug into a ring is the usual way of locking a conformation. For our model compound, the analogue shown in Fig. 13.55 would be suitably rigid.

A ring was used to rigidify the acyclic pentapeptide shown in Fig. 13.56. This is a highly flexible molecule that acts as an inhibitor of a proteolytic enzyme. It was decided to rigidify the structure by linking the asparagine

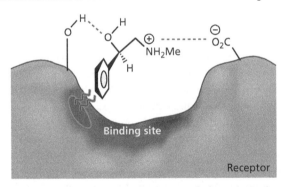

FIGURE 13.54 Active conformation of a hypothetical neurotransmitter.

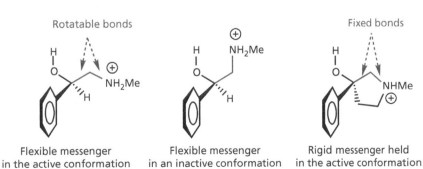

FIGURE 13.55 Rigidification of a molecule by locking rotatable bonds within a ring.

residue with the aromatic ring of the phenylalanine residue to form a macrocyclic ring. The resulting structure showed a 400-fold increase in activity. Macrocyclic rings have an advantage over smaller rings because they rigidify the molecule, but still allow a level of flexibility that increases the chances of the active conformation being adopted. Another example of macrocycles being used in rigidification can be found in Case study 10.

Similar rigidification tactics have been useful in the development of the antihypertensive agent **cilazapril** (Fig. 12.12) from **captopril**, and the development of the sedative **etorphine** (section 24.6.4). Other examples of rigidification can be seen in sections 25.2.8.1, 26.3.3, and 26.7.

Locking a rotatable bond into a ring is not the only way a structure can be rigidified. A flexible side chain can be partially rigidified by incorporating a rigid functional group such as a double bond, alkyne, amide, or aromatic ring (see Box 13.3).

Rigidification also has potential disadvantages. Rigidified structures may be more complicated to synthesize. There is also no guarantee that rigidification will retain the active conformation; it is perfectly possible that rigidification will lock the compound into an inactive conformation. Another disadvantage involves drugs acting on targets which are prone to mutation. If a mutation alters the shape of the binding site, then the drug may no longer be able to bind, whereas a more flexible drug may adopt a different conformation that *could* bind.

For additional material see Web article 5: The design of a serotonin antagonist as a possible anxiolytic agent on the Online Resource Centre at www.oxfordtextbooks.co.uk/orc/patrick6e/

13.3.10 Conformational blockers

We have seen how rigidification tactics can restrict the number of possible conformations for a compound. Another tactic that has the same effect is the use of conformational blockers. In certain situations, a quite simple substituent can hinder the free rotation of a single bond. For example, introducing a methyl substituent to the dopamine (D_3) antagonist (I in Fig. 13.57) gives structure II and results in a dramatic reduction in affinity. The explanation lies in a bad steric clash between the new methyl group and an *ortho* proton on the neighbouring ring which prevents both rings being in the same plane. Free rotation around the bond between the two rings is no longer possible and so the structure adopts a conformation where the two rings are at an angle to each other. In structure I, free rotation around the connecting bond allows the molecule to adopt a conformation where the aromatic rings are coplanar—the active conformation for the receptor. In this case, a conformational blocker

'rejects' the active conformation. Examples of a conformational blocker favouring the active conformation can be seen with 4-methylhistamine (section 25.2.2.2), the design of a serotonin antagonist (see Web article 5), and the development of the anticancer agent imatinib (section 21.6.2.2). In the last case, conformational restraint not only increased activity, but also introduced selectivity between two similar target binding sites.

Rigidification is also possible through intramolecular hydrogen bonding, which may help to stabilize particular conformations (Fig. 13.58).

13.3.11 Structure-based drug design and molecular modelling

So far we have discussed the traditional strategies of drug design. These were frequently carried out with no knowledge of the target structure, and the results obtained were useful in providing information about the target binding site. Clearly, if a drug has an important binding group, there must be a complementary binding region present in the binding site of the receptor or enzyme.

If the macromolecular target can be isolated and crystallized, then it may be possible to determine the structure using X-ray crystallography. Unfortunately, this does not reveal where the binding site is, and so it is better to crystallize the protein with a known inhibitor or antagonist (ligand) bound to the binding site. X-ray crystallography can then be used to determine the structure of the complex and this can be downloaded to a computer. Molecular modelling software is then used to identify where the ligand is and thus identify the binding site. Moreover, by measuring the distances between the atoms of the ligand and neighbouring atoms in the binding site, it is possible to identify important binding interactions between the ligand and the binding site. Once this has been done, the ligand can be removed from the binding site *in silico*, and novel lead compounds can be inserted *in silico* to see how well they fit. (The term *in silico* indicates that the virtual process concerned is being carried out on a computer using molecular modelling software.) Regions in the binding site which are not occupied by the lead compound can be identified and used to guide the medicinal chemist as to what modifications and additions can be made to design a new drug that occupies more of the available space and binds more strongly. The drug can then be synthesized and tested for activity. If it proves active, the target protein can be crystallized with the new drug bound to the binding site, and then X-ray crystallography and molecular modelling can be used again to identify the structure of the complex to see if binding took place as expected. This approach is known as structure-based drug design. Examples of the use of structure-based drug design can be found in Case studies

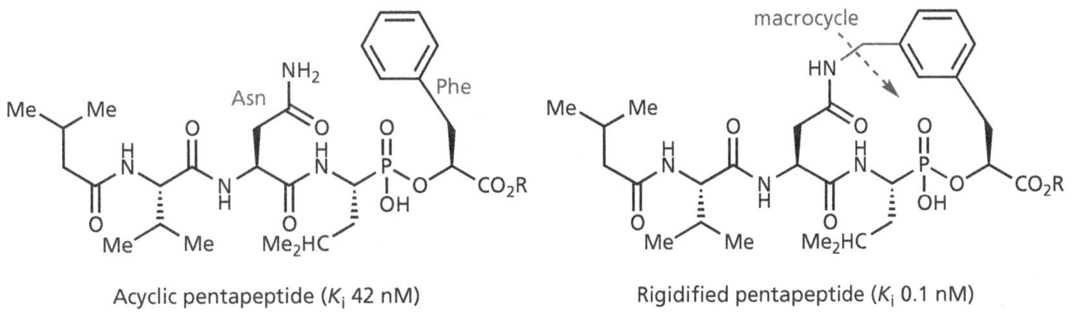

Acyclic pentapeptide (K_i 42 nM)   Rigidified pentapeptide (K_i 0.1 nM)

FIGURE 13.56 Rigidification of an acyclic pentapeptide.

2 and 5, Box 13.4, and Web article 5, as well as sections 14.9.1, 20.7.3.2, 20.7.4, 20.9, and 21.6.2.

A related process is known as *de novo* drug design (section 17.15). This involves the design of a novel drug structure, based on a knowledge of the binding site alone.

This is quite a demanding exercise, but there are examples where *de novo* design has successfully led to a novel lead compound which can then be the starting point for structure-based drug design (see Case study 5 and section 20.7.4.4).

BOX 13.3 Rigidification tactics in drug design

The diazepine (I) is an inhibitor of platelet aggregation and binds to its target receptor by means of a guanidine functional group and a diazepine ring system. These binding groups are linked together by a highly flexible chain. Structures (II) and (III) are examples of active compounds where the connecting chain between the guanidine group and the bicyclic system has been partially rigidified by the introduction of rigid functional groups.

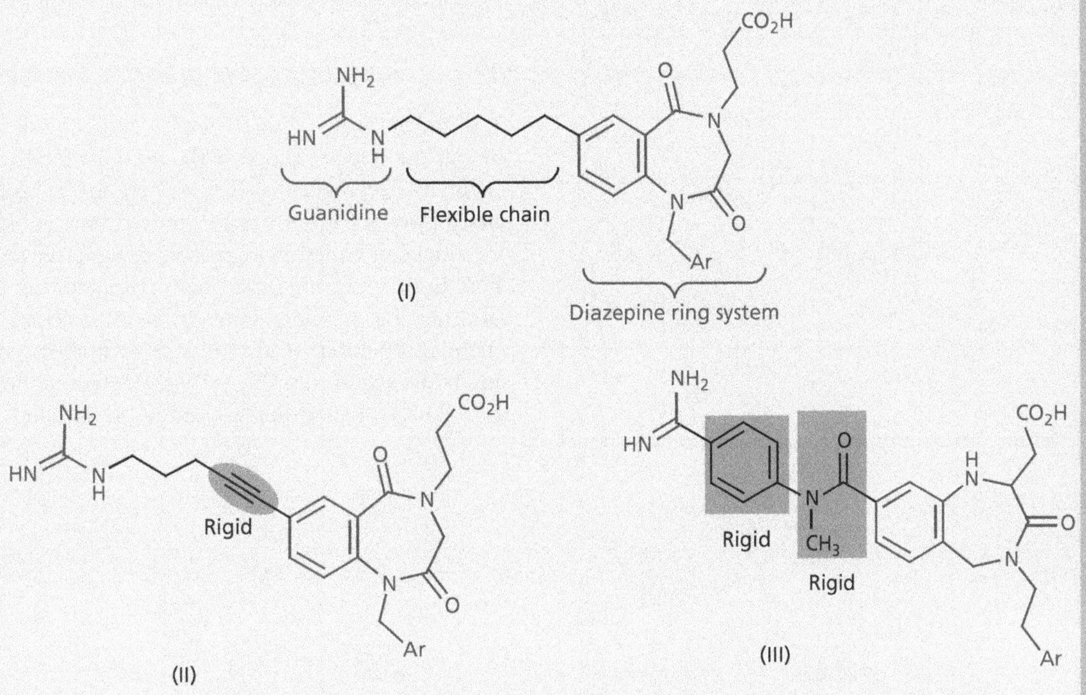

FIGURE 1 Rigidification of flexible chains.

Structure-based drug design cannot be used in all cases. Sometimes the target for a lead compound may not have been identified and, even if it has, it may not be possible to crystallize it. This is particularly true for membrane-bound proteins. One way round this is to identify a protein which is thought to be similar to the target protein, and which *has* been crystallized and studied by X-ray crystallography. The structural and mechanistic information obtained from that analogous protein can then be used to design drugs for the target protein (see Case studies 2 and 5).

Molecular modelling can also be used to study different compounds which are thought to interact with the same target. The structures can be compared and the important pharmacophore identified (section 17.11), allowing the design of novel structures containing the same pharmacophore. Compound databanks can be searched for those pharmacophores to identify novel lead compounds (section 17.13).

There are many other applications of molecular modelling in medicinal chemistry, some of which are described in Chapter 17. However, a point of caution is worth making at this stage. Molecular modelling studies tackle only one part of a much bigger problem—the design of an effective drug. True, one might design a

compound that binds perfectly to a particular enzyme or receptor *in silico*, but that becomes pointless if the drug cannot be synthesized or never reaches the target protein in the body.

There have also been various examples where a binding site has altered shape in an unpredictable way to accommodate ligands that would not normally be expected to bind. Examples include binding sites for the **statins** (Case study 1) and an anti-inflammatory steroid (Box 8.1). Another example involves the dimeric structure of **galantamine** which has been studied as an inhibitor of the enzyme **acetylcholinesterase** (section 22.15.2).

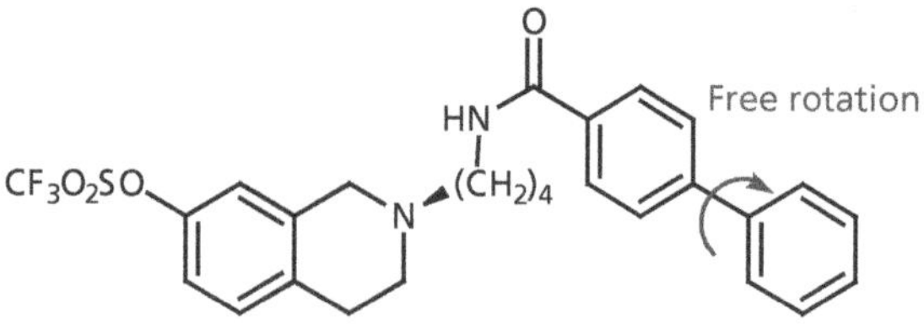

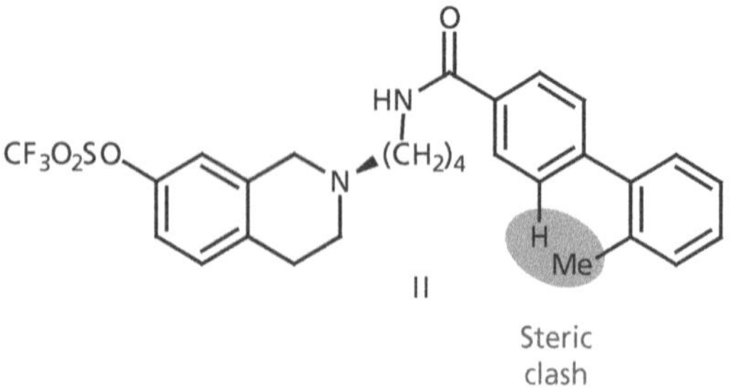

13.3.12 Drug design by NMR spectroscopy

The use of NMR spectroscopy in designing lead compounds has already been discussed in section 12.4.10. This can also be seen as a method of drug design since the focus is not only on designing a lead compound, but in designing a *potent* lead compound. Usually, drug design aims to optimize a lead compound once it has been discovered. In the NMR method, the component parts (**epitopes**) are optimized first to maximize binding interactions, then linked together to produce the final compound.

NMR is also being increasingly used to identify the structure of target proteins that cannot be crystallized and studied by X-ray crystallography. Once the structure has been identified, molecular modelling techniques can be used for drug design as described in section 13.3.11.

13.3.13 The elements of luck and inspiration

It is true to say that drug design has become more rational, but the role of chance or the need for hard-working, mentally alert medicinal chemists has not yet been eliminated. Most of the drugs currently on the market were developed by a mixture of rational design, trial and error, hard graft, and pure luck. There are a growing number of drugs that were developed by rational design such as the ACE inhibitors (Case study 2), thymidylate synthase inhibitors (Case study 5), HIV protease inhibitors (section 20.7.4), neuraminidase inhibitors (section 20.8.3),

FIGURE 13.57 Introducing rigidity by conformational blocking.

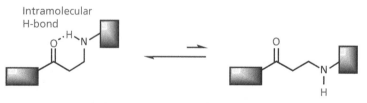

FIGURE 13.58 Rigidification involving an intramolecular hydrogen bond.

Structure-based drug design is normally used to observe the binding interactions of a ligand, then to identify modifications that will result in better interactions and greater activity. This approach was used in the design of a recently approved anticancer agent called crizotinib, and included a substantial modification which totally altered the scaffold of the molecule. **PHA-665752** was the starting point for this research, and had been obtained from structure-based drug design of a previous lead compound. However, it had a large molecular weight and was too hydrophobic to be orally active. The structure was co-crystallized with the target enzyme and the crucial binding interactions were identified. These

included the dihydroindolone ring system which formed two important hydrogen bonds (HBD and HBA), as well as the dichloroaromatic ring. As a result of this study, it was noted that much of the scaffold connecting these binding groups was redundant, and so a much simpler, less hydrophobic skeleton was designed which would position the important binding groups in a similar but more efficient manner. The thought process behind this design involved a ring fusion, ring cleavage, and chain contraction. When the novel structures were synthesized, they were found to bind as predicted, and further structure-based drug design was used in the optimization process leading to crizotinib.

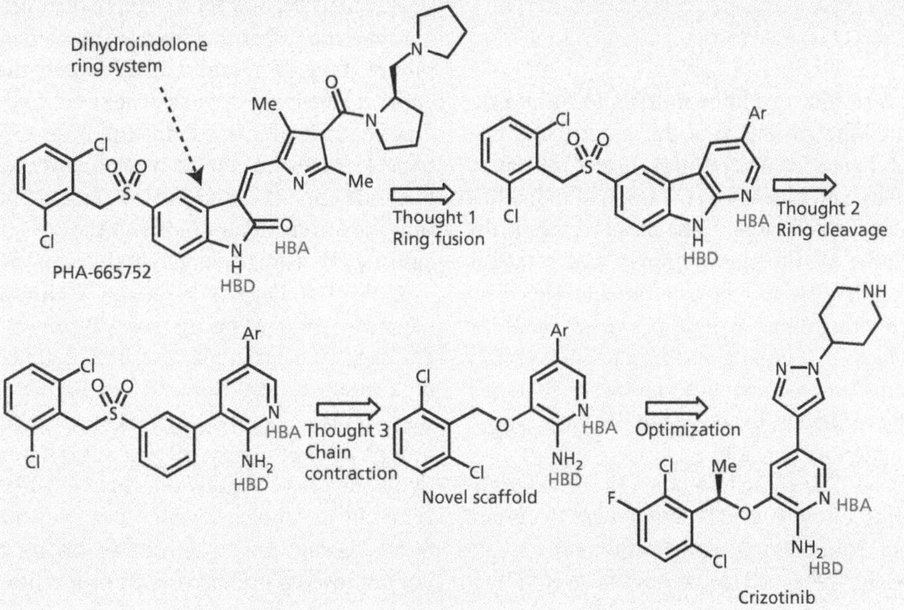

FIGURE 1 Design process in the development of crizotinib.

pralidoxime (section 22.14), and **cimetidine** (section 25.2), but they are still in the minority.

Frequently, the development of drugs is helped by watching the literature to see what works on related compounds and what doesn't, then trying out similar alterations to one's own work. It is often a case of groping in the dark, with the chemist asking whether the addition of a group at a certain position will have a steric, electronic, or interactive effect. Even when drug design is carried out on rational lines, good fortune often has a role to play—for example the discovery of the β-blocker **propranolol** (section 23.11.3).

Finally, there are some cases where the use of logical step-by-step modifications to a structure fails to result in significantly improved activity. In such cases, there may be some advantage in synthesizing a large range of structures with different substituents or modifications in the hope of striking lucky. This is illustrated in the development of the anticancer agent **sorafenib** (Box 21.9). The breakthrough here was the discovery of an active structure which contained two substituents that were known to be bad for activity when only one or other was present. When both were present, however, there was a beneficial synergistic effect.

13.3.14 Designing drugs to interact with more than one target

Many diseases require a cocktail of drugs interacting with different targets to provide suitable treatments. A better approach would be to design agents that interact with two or more targets in a controlled fashion in order to reduce the number of drugs that have to be taken. This is known as **multi-target drug discovery** (MTDD) (section 12.2.7). There have been two approaches to designing such multi-target-directed ligands. One is to design agents from known drugs and pharmacophores such that the new agent has the combined properties of the drugs involved. The other approach is to start from a lead compound which has activity against a wide range of targets, and then modify the structure to try and narrow the activity down to the desired targets.

13.3.14.1 Agents designed from known drugs

One strategy is to link two known drugs to form a dimeric structure. The advantage of this approach is that there is a good chance that the resulting dimer will have a similar selectivity and potency to the original individual drugs for both intended targets. The disadvantage is the increased number of functional groups and rotatable bonds that result, as this may have detrimental effects on whether the resulting dimer is orally active or not. There is also the problem that linking one drug to another may block each individual component binding to its target binding site. Nevertheless, the design of dimers has been successful in a number of fields.

Dimers can be defined as homodimeric or heterodimeric depending on whether the component drugs are the same or not. Homodimeric and heterodimeric opioid ligands have been synthesized to take advantage of the fact

that opioid receptors form homodimeric and heterodimeric arrays in certain tissues of the body (section 24.9.2).

There is also a potential use for dimers in the treatment of Alzheimer's disease. The **acetylcholinesterase** enzyme has an active site and a peripheral binding site, both of which play a role in the symptoms of the disease. Dimers have been designed that can interact with both of these sites and act as **dual-action agents** (section 22.15.2). Research is also being carried out to design triple-action agents that will interact with the two binding sites in the acetylcholinesterase enzyme plus a totally different target that is also involved in the symptoms or development of the disease.

Enzyme inhibitors have also been designed that contain structural components of the substrate and cofactor of **17β-hydroxysteroid dehydrogenase type 1** (see Web article 1).

A second strategy for designing dual-action drugs is to consider the pharmacophores of two different drugs, and to then design a hybrid structure where the two pharmacophores are merged. Such drugs are called **hybrid drugs**. One example of this is **ladostigil** (Fig. 13.59), which is a hybrid structure of the acetylcholinesterase inhibitor **rivastigmine** and the monoamine oxidase inhibitor **rasagiline**. The feature in blue indicates the structural features of ladostigil that are present in both component drugs.

A third strategy is to design a **chimeric drug** that contains key pharmacophore features from two different drugs. For example, a structure containing features of **2-methoxyestradiol** and **colchicine** has been synthesized as a potential anticancer agent (Fig. 13.60). Although both of the parent structures have anticancer activity, they have serious drawbacks. 2-Methoxyestradiol is rapidly metabolized, while colchicine has toxic side effects. The chimeric structure also has anticancer activity, but improved pharmacokinetic properties.

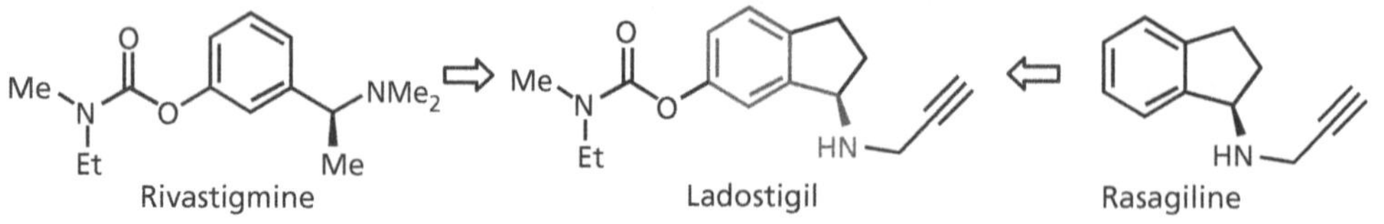

FIGURE 13.59 Design of the hybrid drug ladostigil.

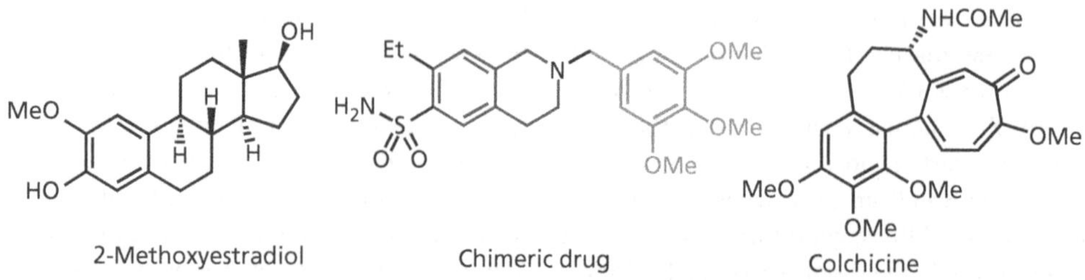

FIGURE 13.60 Design of a chimeric drug.

13.3.14.2 Agents designed from non-selective lead compounds

The second approach to designing multi-target drugs is to identify a lead compound that already shows the ability to interact with a wide variety of targets. Such an agent is termed a **promiscuous ligand** or a **dirty drug**. Linear polyamines have been suggested as ideal lead compounds in this approach as they have several amine groups that can act as good binding groups to protein targets. Moreover, the flexibility of the structure means that an active conformation is likely to exist for a large number of protein targets. The challenge is then to modify the structure such that it shows selectivity towards the desired targets. This approach has been used in the design of an agent which shows activity both as an acetylcholinesterase inhibitor and a muscarinic antagonist (section 22.15.3). Such agents may be useful in the treatment of Alzheimer's disease. Multi-tyrosine receptor kinase inhibitors have also been developed as anticancer agents (section 21.6.2.10).

KEY POINTS

- Drug optimization aims to maximize the interactions of a drug with its target binding site in order to improve activity and selectivity, and to minimize side effects. Designing a drug that can be synthesized efficiently and cheaply is another priority.

- The length and size of alkyl substituents can be modified to fill up hydrophobic pockets in the binding site or to introduce selectivity for one target over another. Alkyl groups attached to heteroatoms are most easily modified.

- Aromatic substituents can be varied in character and/or ring position.

- Extension is a strategy where extra functional groups are added to the lead compound such that they interact with extra binding regions in the binding site.

- Chains connecting two important binding groups can be modified in length in order to maximize the interactions of each group with the corresponding binding regions.

- Ring systems can be modified to maximize binding interactions through strategies such as expansion, contraction, variation, or fusion with other rings.

- Classical and non-classical isosteres are frequently used in drug optimization.

- Simplification involves removing functional groups from the lead compound that are not part of the pharmacophore. Unnecessary parts of the carbon skeleton or asymmetric centres can also be removed in order to design drugs that are easier and cheaper to synthesize. Oversimplification can result in molecules that are too flexible, resulting in decreased activity and selectivity.

- Rigidification is applicable to flexible lead compounds. The aim is to reduce the number of conformations available while retaining the active conformation. Locking rotatable rings into ring structures or introducing rigid functional groups are common methods of rigidification.

- Conformational blockers are groups which are introduced into a lead compound to reduce the number of conformations that the molecule can adopt.

- Structure-based drug design makes use of X-ray crystallography and computer-based molecular modelling to study how a lead compound and its analogues bind to a target binding site.

- NMR studies can be used to determine protein structure and to design novel drugs.

- Serendipity plays a role in drug design and optimization.

- Multi-target-directed ligands can be designed by linking or merging established drugs, or by modifying a lead compound that interacts with a large number of targets.

QUESTIONS

1. DU 122290 was developed from sultopride (Fig. 13.47) and shows improved activity and selectivity. Suggest possible reasons for this.

2. Methotrexate inhibits the enzyme dihydrofolate reductase. The pteridine ring system of methotrexate binds to the binding site as shown in Fig. 13.26. Suggest how dihydrofolate (the natural substrate for the enzyme) might bind.

Dihydrofolate

3. A lead compound containing a methyl ester was hydrolysed to give a carboxylic acid. An *in vivo* bioassay suggested that the ester was active and the acid was inactive. However, an *in vitro* bioassay suggested that the ester was inactive and the acid was active. Explain these contradictory results.

4. A lead compound contains an aromatic ring. The following structures were made as analogues. Structures I and II were similar in activity to the lead compound, whereas structure III showed a marked increase in activity. Explain these results and describe the strategies involved.

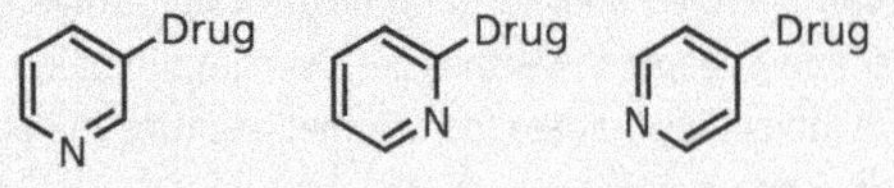

5. The pharmacophore of cocaine is shown in Box 13.2. Identify possible cyclic analogues which are simpler than cocaine and which would be expected to retain activity.

6. Procaine (Box 13.2) has been a highly successful local anaesthetic and yet there are three bonds between the important ester and amine binding groups, compared to four in cocaine. This might suggest that these groups are too close together in procaine. In fact, this is not the case. Suggest why not.

7. The aromatic amine on procaine is not present in cocaine. Comment on its possible role.

8. Explain how you would apply the principles of rigidification to structure IV below in order to improve its pharmacological properties. Give two specific examples of rigidified structures.

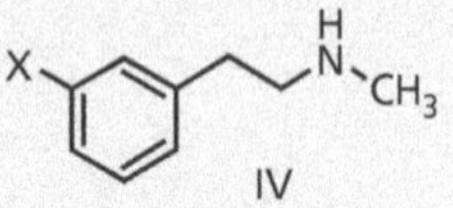

9. Combretastatin is an anticancer agent discovered from an African plant. Analogue V is more active than combretastatin whereas analogue VI is less active. What strategy was used in designing analogues V and VI? Why is analogue V more active and analogue VI less active than combretastatin?

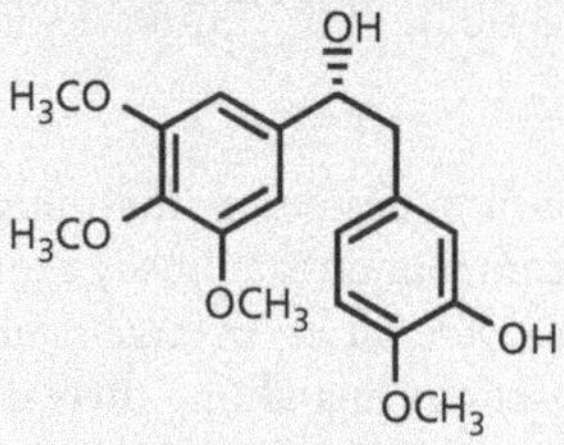

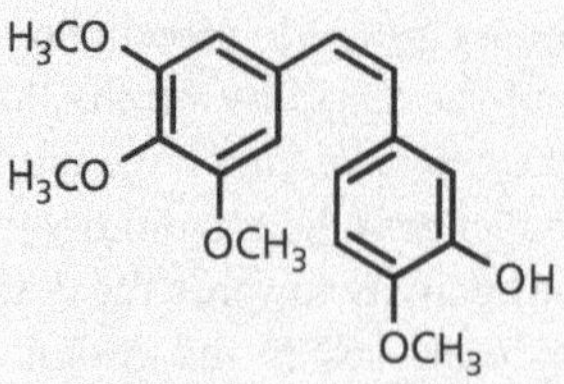

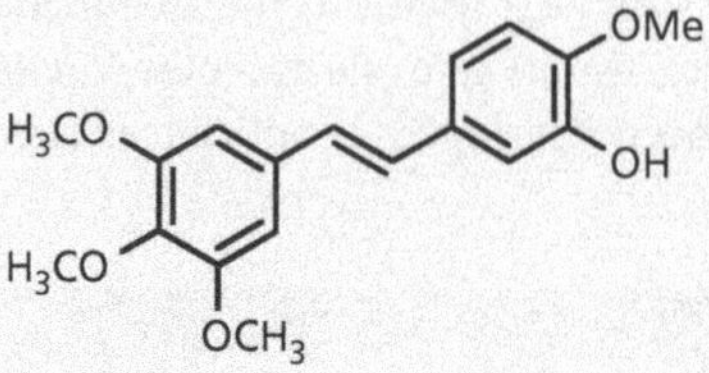

10. Structure VII is a serotonin antagonist. A methyl group has been introduced into analogue VIII, resulting in increased activity. What role does the methyl group play and what is the term used for such a group? Explain why increased activity arises.

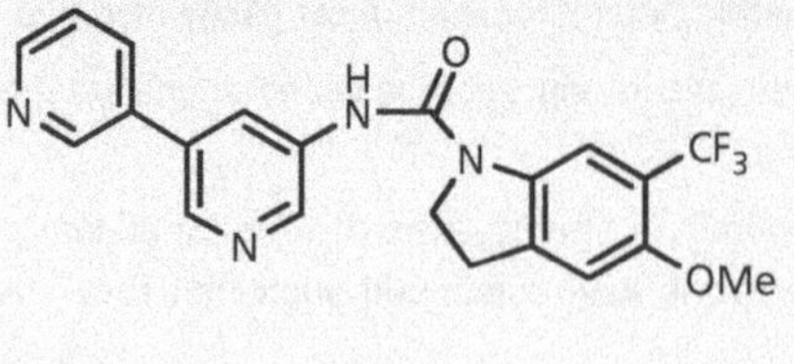

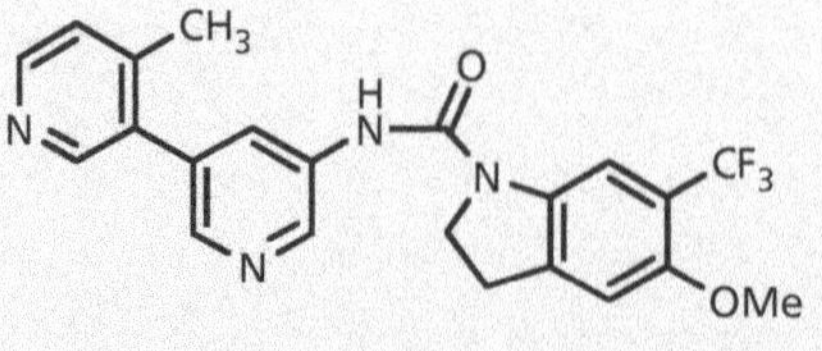

11. Explain what kind of drug design strategies were carried out in the design of enalaprilat (Case study 2).

12. Salicylamides are inhibitors for an enzyme called scytalone dehydratase. SAR shows that there are three important hydrogen bonding interactions. Explain whether you think quinazolines could act as a bio-isostere for salicylamides.

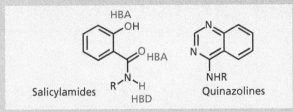

Salicylamides   Quinazolines

13. Structure IX (X = NH) is an inhibitor of a metalloenzyme called thermolysin and forms interactions as shown. Explain why the analogue (X = O) has reduced binding affinity by a factor of 1000, and why the analogue (X = CH_2) has roughly the same binding affinity.

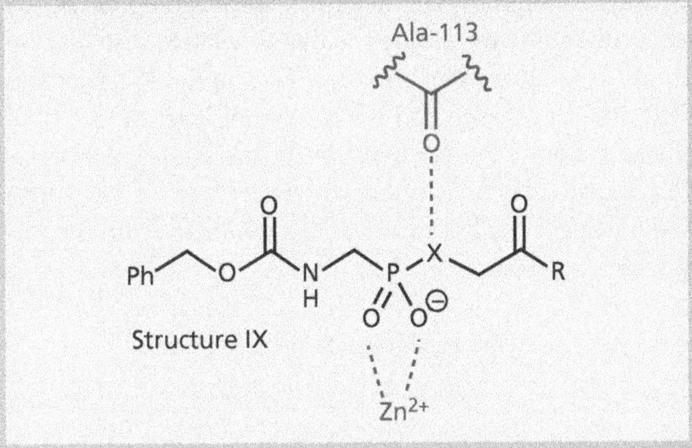

Ala-113

Structure IX

Zn^{2+}

14. Suggest why the oxygen atoms in the following structures are poor hydrogen bond acceptors.

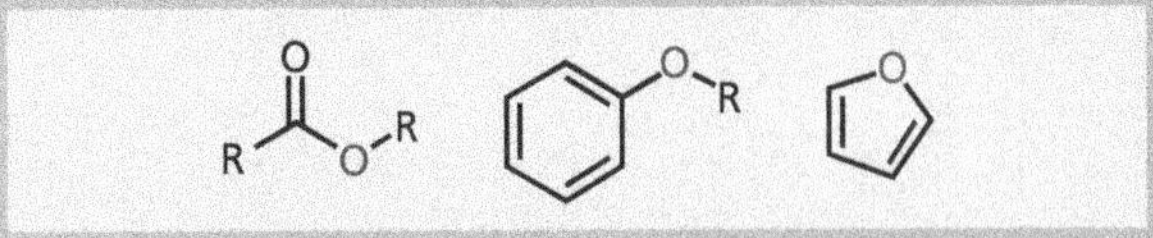

15. Compare the ability of the nitrogen atoms in the following structures to act as hydrogen bond acceptors.

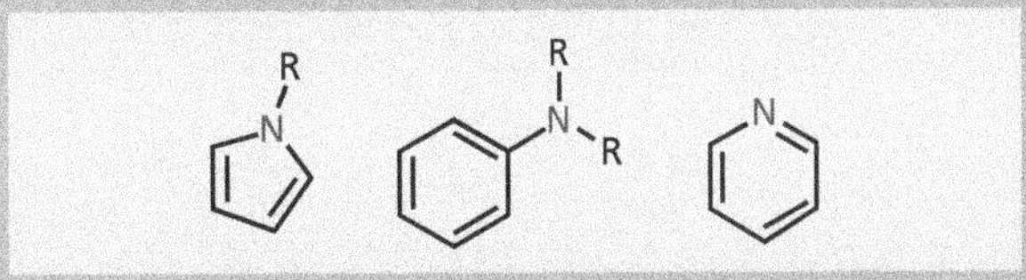

16. Explain why a benzimidazole group could be considered a bio-isostere for an *N*-phenyl amide, and why using a benzimidazole group might increase binding affinity.

Multiple-choice questions are available on the Online Resource Centre at www.oxfordtextbooks.co.uk/orc/patrick6e/

FURTHER READING

Acharya, K. R., et al. (2003) ACE revisited: a new target for structure-based drug design. *Nature Reviews Drug Discovery*, **2**, 891–902.

Cavalli, A., et al. (2008) Multi-target-directed ligands to combat neurodegenerative diseases. *Journal of Medicinal Chemistry*, **51**, 347–72.

Cui, J. J., et al. (2011) Structure-based drug design of crizotinib. *Journal of Medicinal Chemistry*, **54**, 6342–63.

Hruby, V. J. (2002) Designing peptide receptor agonists and antagonists. *Nature Reviews Drug Discovery*, **1**, 847–58.

Kubinyi, H. (2001) Hydrogen bonding: the last mystery in drug design? in Testa, B., van de Waterbeemd, H., Folkers, G., and Guy, R. (eds), *Pharmacokinetic optimization in drug research*. Wiley-VCH, Weinheim.

Luca, S., et al. (2003) The conformation of neurotensin bound to its G-protein-coupled receptor. *Proceedings of the National Academy of Sciences of the USA*, **100**, 10706–11 (active conformation by NMR).

Morphy, R., et al. (2004) From magic bullets to designed multiple ligands. *Drug Discovery Today*, **9**, 641–51.

Morphy, R., and Rankovic, Z. (2005) Designed multiple ligands. An emerging drug discovery paradigm. *Journal of Medicinal Chemistry*, **48**, 6523–43.

Pellecchia, M., Sem, D. S., and Wuthrich, K. (2002) NMR in drug discovery. *Nature Reviews Drug Discovery*, **1**, 211–9.

Rees, D. C., et al. (2004) Fragment-based lead discovery. *Nature Reviews Drug Discovery*, **3**, 660–72.

Titles for general further reading are listed on p.845.

14 Drug design: optimizing access to the target

In Chapter 13, we looked at drug design strategies aimed at optimizing the binding interactions of a drug with its target. However, the compound with the best binding interactions is not necessarily the best drug to use in medicine. The drug needs to overcome many barriers if it is to reach its target in the body (Chapter 11). In this chapter, we shall study design strategies which can be used to counter such barriers, and which involve modification of the drug itself. There are other methods of aiding a drug in reaching its target, which include linking the drug to polymers or antibodies, or encapsulating it within a polymeric carrier. These topics are discussed in sections 11.10 and 21.10. In general, the aim is to design drugs that will be absorbed into the blood supply, will reach their target efficiently, be stable enough to survive the journey, and will be eliminated in a reasonable period of time. This all comes under the banner of a drug's pharmacokinetics.

14.1 Optimizing hydrophilic/hydrophobic properties

The relative hydrophilic/hydrophobic properties of a drug are crucial in influencing its solubility, absorption, distribution, metabolism, and excretion (ADME). Drugs which are too polar or hydrophilic do not easily cross the cell membranes of the gut wall. One way round this is to inject them, but they cannot be used against intracellular targets since they will not cross cell membranes. They are also likely to have polar functional groups which will make them prone to plasma protein binding, metabolic phase II conjugation reactions, and rapid excretion (Chapter 11). Very hydrophobic drugs fare no better. If they are administered orally, they are likely to be dissolved in fat globules in the gut and will be poorly absorbed. If they are injected, they are poorly soluble in blood and are likely to be taken up by fat tissue, resulting in low circulating levels. It has also been observed that toxic metabolites are more likely to be formed from hydrophobic drugs.

The hydrophobic character of a drug can be measured experimentally by testing the drug's relative distribution in an *n*-octanol/water mixture. Hydrophobic molecules will prefer to dissolve in the *n*-octanol layer of this two-phase system, whereas hydrophilic molecules will prefer the aqueous layer. The relative distribution is known as the partition coefficient (*P*) and is obtained from the following equation:

$$P = \frac{\text{Concentration of drug in octanol}}{\text{Concentration of drug in aqueous solution}}$$

Hydrophobic compounds have a high *P* value, whereas hydrophilic compounds have a low *P* value. In fact, log *P* values are normally used as a measure of hydrophobicity. Other experimental procedures to determine log *P* include high-performance liquid chromatography (HPLC) and automated potentiometric titration procedures. It is also possible to calculate log *P* values for a given structure using suitable software programs. Such estimates are referred to as **Clog *P*** values to distinguish them from experimentally derived log *P* values.

Many drugs can exist as an equilibrium between an ionized and an un-ionized form. However, log *P* measures only the relative distribution of the un-ionized species between water and octanol. The relative distribution of all species (both ionized and un-ionized) is given by **log *D***.

In general, the hydrophilic/hydrophobic balance of a drug can be altered by changing easily accessible substituents. Such changes are particularly open to a quantitative approach known as QSAR (quantitative structure–activity relationships), discussed in Chapter 18.

As a postscript, the hydrophilic/hydrophobic properties of a drug are not the only factors that influence drug absorption and oral bioavailability. Molecular flexibility

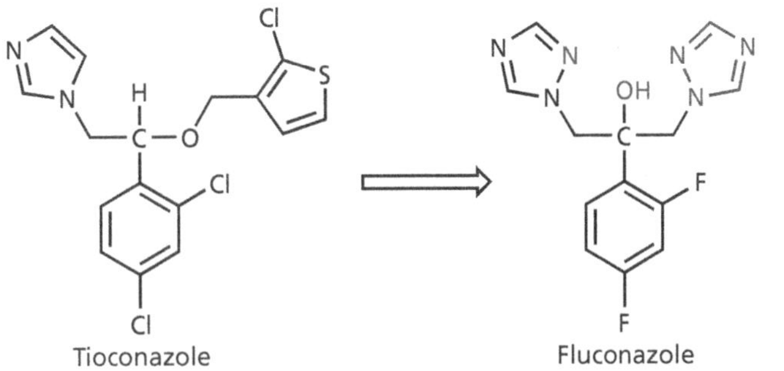

FIGURE 14.1 Increasing polarity in antifungal agents.

also has an important role in oral bioavailability (section 11.3), and so the tactics of rigidification described in section 13.3.9 can be useful in improving drug absorption.

14.1.1 Masking polar functional groups to decrease polarity

Molecules can be made less polar by masking a polar functional group with an alkyl or acyl group. For example, an alcohol or a phenol can be converted to an ether or ester, a carboxylic acid can be converted to an ester or amide, and primary and secondary amines can be converted to amides or to secondary and tertiary amines. Polarity is decreased not only by masking the polar group, but by the addition of an extra hydrophobic alkyl group—larger alkyl groups having a greater hydrophobic effect. One has to be careful in masking polar groups, though, as they may be important in binding the drug to its target. Masking such groups would decrease binding interactions and lower activity. If this is the case, it is often useful to mask the polar group temporarily such that the mask is removed once the drug is absorbed (section 14.6).

14.1.2 Adding or removing polar functional groups to vary polarity

A polar functional group could be added to a drug to increase its polarity. For example, the antifungal agent **tioconazole** is only used for skin infections because it is non-polar and poorly soluble in blood. Introducing a polar hydroxyl group and more polar heterocyclic rings led to the orally active antifungal agent **fluconazole** with improved solubility and enhanced activity against systemic infection (i.e. in the blood supply) (Fig. 14.1). Another example can be found in Case study 1 where a polar sulphonamide group was added to **rosuvastatin** to make it more hydrophilic and more tissue selective. Finally, nitrogen-containing heterocycles (e.g. morpholine or pyridine) are often added to drugs in order to increase their polarity and water solubility. This is because the

nitrogen is basic in character, and it is possible to form water-soluble salts. Examples of this tactic can be seen in the design of **gefitinib** (section 21.6.2.1) and a thymidylate synthase inhibitor (Case study 5). If a polar group is added in order to increase water solubility, it is preferable to add it to the molecule in such a way that it is still exposed to surrounding water when the drug is bound to the target binding site. This means that energy does not have to be expended in desolvation (section 1.3.6).

The polarity of an excessively polar drug can be lowered by removing polar functional groups. This strategy has been particularly successful with lead compounds derived from natural sources (e.g. alkaloids or endogenous peptides). It is important, though, not to remove functional groups which are important to the drug's binding interactions with its target. In some cases, a drug may have too many essential polar groups. For example, the antibacterial agent shown in Fig. 14.2 has good *in vitro* activity but poor *in vivo* activity because of the large number of polar groups. Some of these groups can be removed or masked, but most of them are required for activity. As a result, the drug cannot be used clinically.

14.1.3 Varying hydrophobic substituents to vary polarity

Polarity can be varied by the addition, removal, or variation of suitable hydrophobic substituents. For

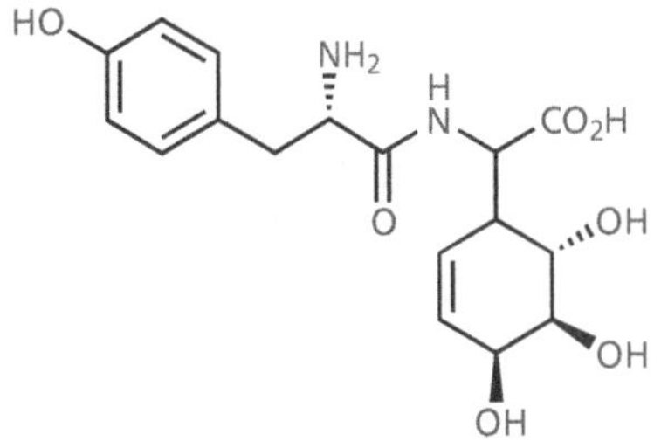

FIGURE 14.2 Excess polarity (coloured) in a drug.

example, extra alkyl groups could be included within the carbon skeleton of the molecule to increase hydrophobicity if the synthetic route permits. Alternatively, alkyl groups already present might be replaced with larger groups. If the molecule is not sufficiently polar, then the opposite strategy can be used (i.e. replacing large alkyl groups with smaller alkyl groups, or removing them entirely). Sometimes there is a benefit in increasing the size of one alkyl group and decreasing the size of another. This is called a **methylene shuffle** and has been found to modify the hydrophobicity of a compound. The addition of halogen substituents also increases hydrophobicity. Chloro or fluoro substituents are commonly used, and, less commonly, a bromo substituent.

14.1.4 Variation of *N*-alkyl substituents to vary p*K*$_a$

Drugs with a p*K*$_a$ outside the range 6–9 tend to be too strongly ionized and are poorly absorbed through cell membranes (section 11.3). The p*K*$_a$ can often be altered to bring it into the preferred range. For example, this can be done by varying any *N*-alkyl substituents that are present. However, it is sometimes difficult to predict how such variations will affect the p*K*$_a$. Extra *N*-alkyl groups or larger *N*-alkyl groups have an increased electron-donating effect which should increase basicity, but increasing the size or number of alkyl groups increases the steric bulk around the nitrogen atom. This hinders water molecules from solvating the ionized form of the base and prevents stabilization of the ion. This in turn decreases the basicity of the amine. Therefore, there are two different effects acting against each other. Nevertheless, varying alkyl substituents is a useful tactic to try.

A variation of this tactic is to 'wrap up' a basic nitrogen within a ring. For example, the benzamidine structure (I in Fig. 14.3) has antithrombotic activity, but the amidine group present is too basic for effective absorption. Incorporating the group into an isoquinoline ring system (**PRO 3112**) reduced basicity and increased absorption.

14.1.5 Variation of aromatic substituents to vary p*K*$_a$

The p*K*$_a$ of an aromatic amine or carboxylic acid can be varied by adding electron-donating or electron-withdrawing substituents to the ring. The position of the substituent relative to the amine or carboxylic acid is important if the substituent interacts with the ring through resonance (section 18.2.2). An illustration of this can be seen in the development of **oxamniquine** (Case study 4).

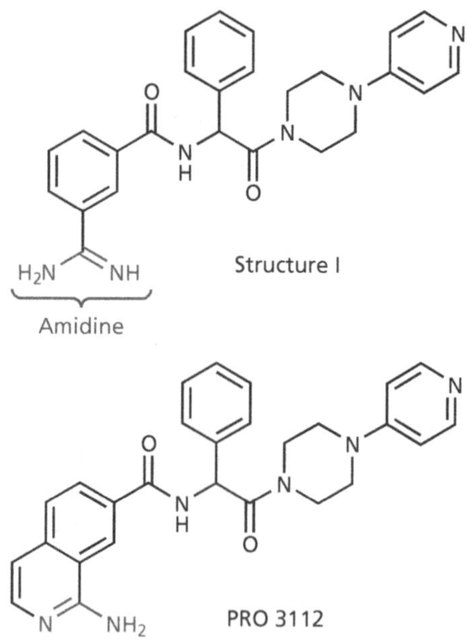

FIGURE 14.3 Varying basicity in antithrombotic agents.

14.1.6 Bio-isosteres for polar groups

The use of bio-isosteres has already been described in section 13.3.7 in the design of compounds with improved target interactions. Bio-isosteres have also been used as substitutes for important functional groups that are required for target interactions, but which pose pharmacokinetic problems. For example, a carboxylic acid is a highly polar group which can ionize and hinder absorption of any drug containing it. One way of getting round this problem is to mask it as an ester prodrug (section 14.6.1.1). Another strategy is to replace it with a bio-isostere which has similar physicochemical properties, but which offers some advantage over the original carboxylic acid. Several bio-isosteres have been used for carboxylic acids, but among the most popular is a 5-substituted tetrazole ring (Fig. 14.4). Like carboxylic acids, tetrazoles contain an acidic proton and are ionized at pH 7.4. They are also planar in structure. However, they have an advantage in that the tetrazole anion is 10 times more lipophilic than a carboxylate anion. Drug absorption is enhanced as a result (see Box 14.1). They are also resistant to many of the metabolic reactions that occur on carboxylic acids. *N*-Acylsulphonamides have

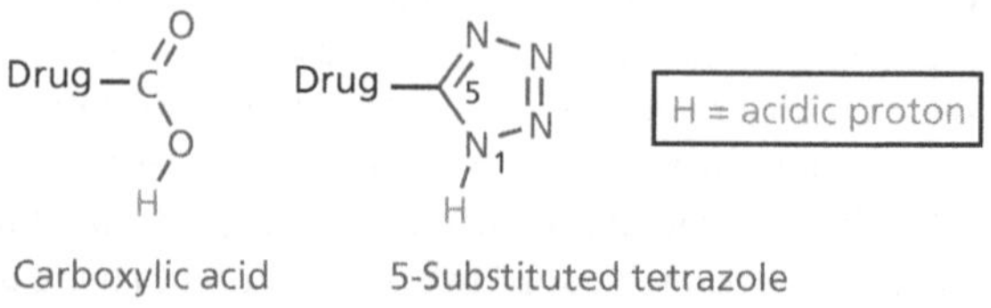

FIGURE 14.4 5-Substituted tetrazole ring as a bio-isostere for a carboxylic acid.

BOX 14.1 The use of bio-isosteres to increase absorption

The biphenyl structure (Structure I) was shown by Du Pont to inhibit the receptor for angiotensin II and had potential as an antihypertensive agent. However, the drug had to be injected as it showed poor absorption through the gut wall. Replacing the carboxylic acid with a tetrazole ring led to **losartan**, which was launched in 1994 (section 26.3.4).

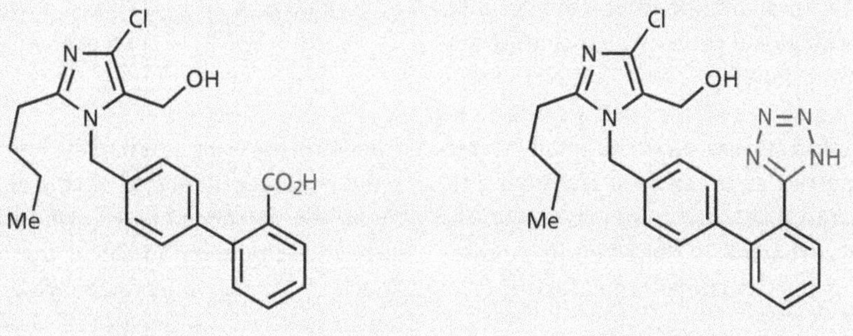

FIGURE 1 Development of losartan.

also been used as bio-isosteres for carboxylic acids (section 13.3.7).

Phenol groups are commonly present in drugs but are susceptible to metabolic conjugation reactions. Various bio-isosteres involving amides, sulphonamides, or heterocyclic rings have been used where an N–H group mimics the phenol O–H group.

14.2 Making drugs more resistant to chemical and enzymatic degradation

There are various strategies that can be used to make drugs more resistant to hydrolysis and drug metabolism, and thus prolong their activity.

14.2.1 Steric shields

Some functional groups are more susceptible to chemical and enzymatic degradation than others. For example, esters and amides are particularly prone to hydrolysis. A common strategy that is used to protect such groups is to add steric shields, designed to hinder the approach of a nucleophile or an enzyme to the susceptible group. These usually involve the addition of a bulky alkyl group close to the functional group. For example, the *t*-butyl group in the antirheumatic agent **D 1927** serves as a

steric shield and blocks hydrolysis of the terminal peptide bond (Fig. 14.5). Steric shields have also been used to protect penicillins from lactamases (section 19.5.1.8), and to prevent drugs interacting with cytochrome P450 enzymes (section 22.7.1).

For additional material see Web article 5: The design of a serotonin antagonist as a possible anxiolytic agent on the Online Resource Centre at www.oxfordtextbooks.co.uk/orc/patrick6e/

14.2.2 Electronic effects of bio-isosteres

Another popular tactic used to protect a labile functional group is to stabilize the group electronically using a bio-isostere. Isosteres and non-classical isosteres are frequently used as bio-isosteres (see also sections 13.1.15, 13.3.7, and 14.1.6). For example, replacing

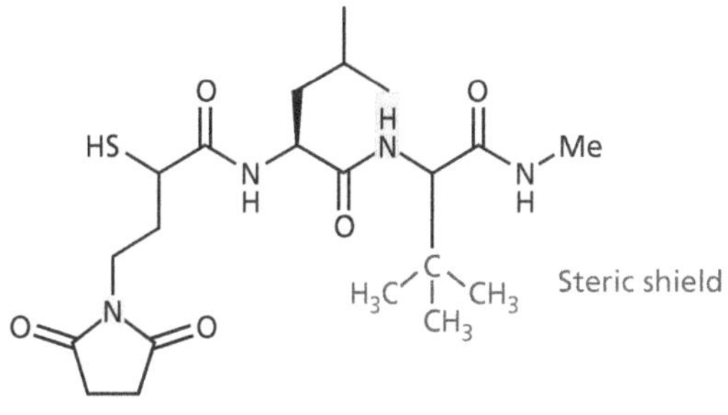

FIGURE 14.5 The use of a steric shield to protect the antirheumatic agent D 1927.

FIGURE 14.6 Isosteric replacement of a methyl group with an amino group.

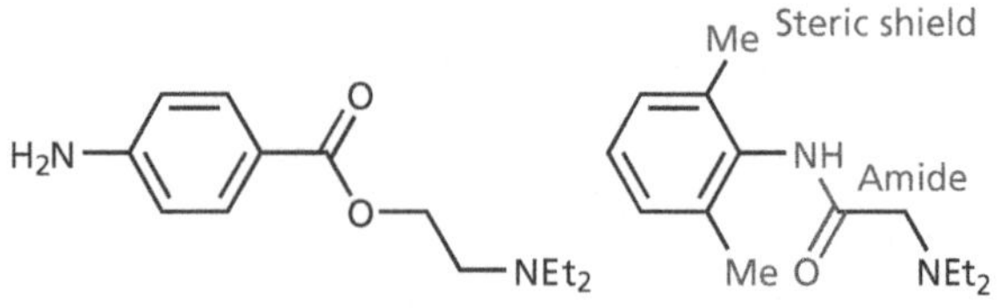

FIGURE 14.7 Steric and electronic modifications which make lidocaine a longer lasting local anaesthetic compared to procaine.

the methyl group of an ethanoate ester with NH_2 results in a urethane functional group which is more stable than the original ester (Fig. 14.6). The NH_2 group is the same valency and size as the methyl group and so it has no steric effect. However, it has totally different electronic properties as it can feed electrons into the carboxyl group and stabilize it from hydrolysis. The cholinergic agonist **carbachol** is stabilized in this way (section 22.7.2), as is the cephalosporin **cefoxitin** (section 19.5.2.4).

Alternatively, a labile ester group could be replaced by an amide group (NH replacing O). Amides are more resistant to chemical hydrolysis, due again to the lone pair of the nitrogen feeding its electrons into the carboxyl group and making it less electrophilic.

It is important to realize that bio-isosteres are often specific to a particular area of medicinal chemistry. Replacing an ester with a urethane or an amide may work in one category of drugs, but not another. One must also appreciate that bio-isosteres are different from isosteres. It is the retention of important biological activity that determines whether a group is a bio-isostere, not the valency. Therefore, non-isosteric groups can be used as bio-isosteres. For example, a pyrrole ring was used as a bio-isostere for an amide bond in the development of the dopamine antagonist **Du 122290** from **sultopride** (section 13.3.7). Similarly, thiazolyl rings were used as bio-isosteres for pyridine rings in the development of **ritonavir** (section 20.7.4.4).

One is not confined to the use of bio-isosteres to increase stability. Groups or substituents having an inductive electronic effect have frequently been incorporated into molecules to increase the stability of a labile functional group. For example, electron-withdrawing groups were incorporated into the side chain of penicillins to increase their resistance to acid hydrolysis (section 19.5.1.8). The inductive effects of groups can also determine the ease with which ester prodrugs are hydrolysed (Box 14.4).

14.2.3 Steric and electronic modifications

Steric hindrance and electronic stabilization have often been used together to stabilize labile groups. For example, **procaine** (Fig. 14.7) is a good, but short-lasting, local anaesthetic because its ester group is quickly hydrolysed.

Changing the ester group to the less reactive amide group reduces susceptibility to chemical hydrolysis. Furthermore, the presence of two *ortho*-methyl groups on the aromatic ring helps to shield the carbonyl group from attack by nucleophiles or enzymes. This results in the longer acting local anaesthetic **lidocaine**. Further successful examples of steric and electronic modifications are demonstrated by **oxacillin** (Box 19.5) and **bethanechol** (section 22.7.3).

14.2.4 Metabolic blockers

Some drugs are metabolized by the introduction of polar groups at particular positions in their skeleton. For example, steroids can be oxidized at position 6 of the tetracyclic skeleton to introduce a polar hydroxyl group (Fig. 14.8). The introduction of this group allows the formation of polar conjugates which can be quickly eliminated from the system. By introducing a methyl group at position 6, metabolism is blocked and the activity of the steroid is prolonged. The oral contraceptive **megestrol acetate** is an agent which contains a 6-methyl blocking group.

On the same lines, a popular method of protecting aromatic rings from metabolism at the *para*-position is to introduce a fluoro substituent. For example, **CGP 52411** (Fig. 14.9) is an enzyme inhibitor which acts on the kinase active site of the epidermal growth factor receptor

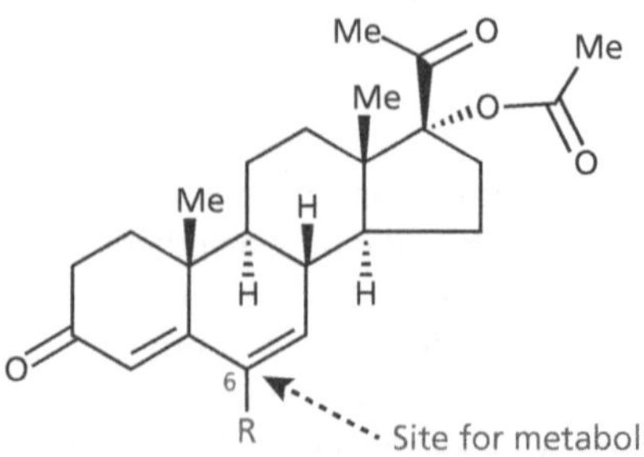

FIGURE 14.8 Metabolically susceptible steroid (R = H), metabolite (R = OH), and megestrol acetate (R = Me).

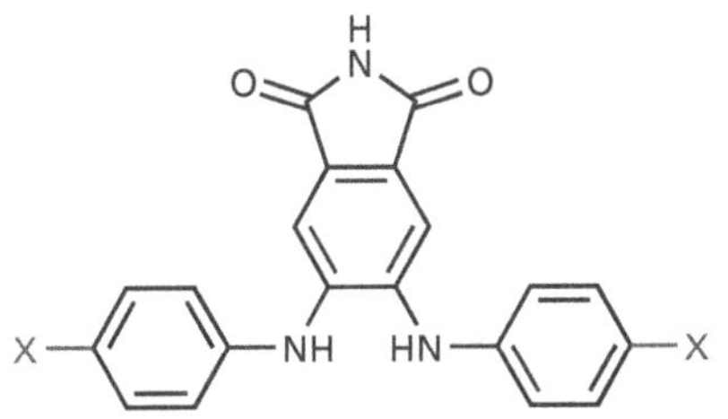

FIGURE 14.9 The use of fluorine substituents as metabolic blockers. X = H, CGP 52411; X = OH, metabolite; X = F, CGP 53353.

(section 4.8). It went forward for clinical trials as an anti-cancer agent and was found to undergo oxidative metabolism at the *para*-position of the aromatic rings. Fluoro-substituents were successfully added in the analogue **CGP 53353** to block this metabolism. This tactic was also applied successfully in the design of **gefitinib** (section 21.6.2.1). Fluorine has now been used extensively to block metabolism in a variety of structural situations.

Another approach which is actively being explored is to replace a hydrogen atom with a deuterium isotope. The covalent bond between carbon and deuterium is twice as strong as that between carbon and hydrogen, and this might help to block metabolic mechanisms.

For additional information see Web article 28: Use of fluorine in medicinal chemistry on the Online Resource Centre at www.oxfordtextbooks.co.uk/orc/patrick6e/

14.2.5 Removal or replacement of susceptible metabolic groups

Certain substituents are particularly susceptible to metabolic enzymes. For example, methyl groups on aromatic rings are often oxidized to carboxylic acids (section 11.5.2). These acids can then be quickly eliminated from the body. Other common metabolic reactions include aliphatic and aromatic *C*-hydroxylations, *N*- and *S*-oxidations, *O*- and *S*-dealkylations, and deaminations (section 11.5).

Susceptible groups can sometimes be removed or replaced by groups that are stable to oxidation, in order to prolong the lifetime of the drug. For example, the aromatic methyl substituent of the antidiabetic **tolbutamide** was

replaced by a chloro substituent to give **chlorpropamide**, which is much longer lasting (Fig. 14.10). This tactic was also used in the design of **gefitinib** (section 21.6.2.1). An alternative strategy which is often tried is to replace the susceptible methyl group with CF_3, CHF_2, or CH_2F. The fluorine atoms alter the oxidation potential of the methyl group and make it more resistant to oxidation.

Another example where a susceptible metabolic group is replaced is seen in section 19.5.2.3 where a susceptible ester in cephalosporins is replaced with metabolically stable groups to give **cephaloridine** and **cefalexin**.

14.2.6 Group shifts

Removing or replacing a metabolically vulnerable group is feasible if the group concerned is not involved in important binding interactions with the binding site. If the group *is* important, then we have to use a different strategy.

There are two possible solutions. We can either mask the vulnerable group on a temporary basis by using a prodrug (section 14.6) or we can try shifting the vulnerable group within the molecular skeleton. The latter tactic was used in the development of **salbutamol** (Fig. 14.11). Salbutamol was introduced in 1969 for the treatment of asthma, and is an analogue of the neurotransmitter **noradrenaline**—a catechol structure containing two *ortho*-phenolic groups.

One of the problems faced by catechol compounds is metabolic methylation of one of the phenolic groups. Since both phenol groups are involved in hydrogen bonds to the receptor, methylation of one of the phenol groups disrupts the hydrogen bonding and makes the compound inactive. For example, the noradrenaline analogue (I in Fig. 14.12) has useful anti-asthmatic activity, but the effect is of short duration because the compound is rapidly metabolized to the inactive methyl ether (II in Fig. 14.12).

Removing the OH or replacing it with a methyl group prevents metabolism, but also prevents the important hydrogen bonding interactions with the binding site. So

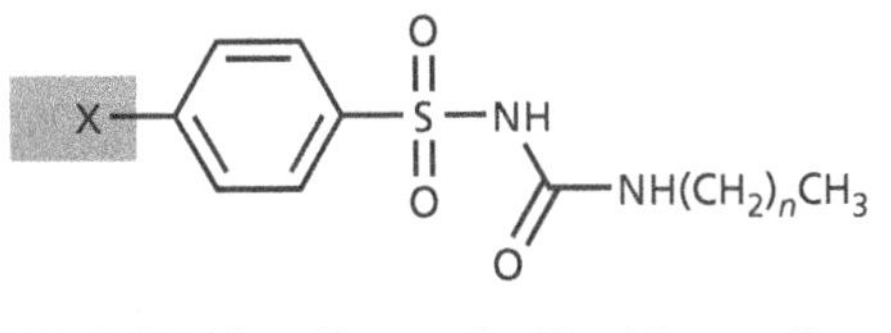

FIGURE 14.10 Tolbutamide (X = Me; *n* = 3) and chlorpropamide (X = Cl; *n* = 2).

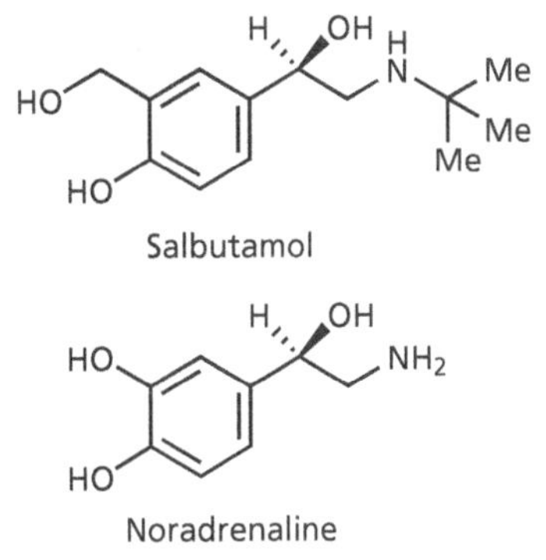

Salbutamol

Noradrenaline

FIGURE 14.11 Salbutamol and noradrenaline.

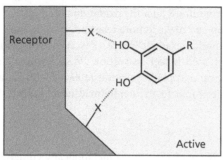

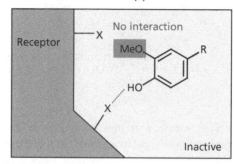

FIGURE 14.12 Metabolic methylation of a noradrenaline analogue. X denotes an electronegative atom.

how can this problem be solved? The answer was to move the vulnerable hydroxyl group out from the ring by one carbon unit. This was enough to make the compound unrecognizable to the metabolic enzyme, but not to the receptor binding site.

Fortunately, the receptor appears to be quite lenient over the position of this hydrogen bonding group and it is interesting to note that a hydroxyethyl group is also acceptable (Fig. 14.13). Beyond that, activity is lost because the OH group is out of range, or the substituent is too large to fit. These results demonstrate that it is better to consider a binding region within the receptor binding site as an available volume, rather than imagining it as being fixed at one spot. A drug can then be designed such that the relevant binding group is positioned in any part of that available volume. Another example of a successful group shift strategy can be seen in Case study 7.

Shifting an important binding group that is metabolically susceptible cannot be guaranteed to work in every situation. It may well make the molecule unrecognizable both to its target and to the metabolic enzyme.

14.2.7 Ring variation and ring substituents

Certain ring systems may be susceptible to metabolism, and so varying the ring might improve metabolic stability. This can be done by adding a nitrogen into the ring to lower the electron density of the ring system. For example, the imidazole ring of the antifungal agent **tioconazole** mentioned in section 14.1.2 is susceptible to metabolism, but replacement with a 1,2,4-triazole ring as in **fluconazole** results in improved stability (Fig. 14.1).

Electron-rich aromatic rings such as phenyl groups are particularly prone to oxidative metabolism, but can be stabilized by replacing them with nitrogen-containing heterocyclic rings such as pyridine or pyrimidine. Alternatively, electron-withdrawing substituents could be added to the aromatic ring to lower the electron density (see Web article 5).

Ring variation can also help to stabilize metabolically susceptible aromatic or heteroaromatic methyl substituents. Such substituents could be replaced with more stable substituents as described in section 14.2.5, but sometimes the methyl substituent has to be retained for good

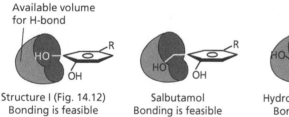

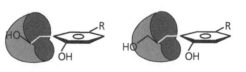

FIGURE 14.13 Viewing a binding region as an available volume.

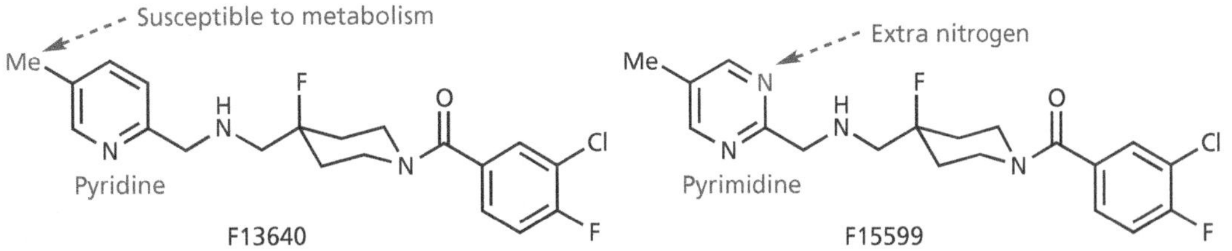

FIGURE 14.14 Stabilizing an aromatic or heteroaromatic methyl substituent by adding a nitrogen to the ring.

activity. In such cases, introducing a nitrogen atom into the aromatic/heteroaromatic ring can be beneficial, since lowering the electron density in the ring also helps to make the methyl substituent more resistant to metabolism. For example, **F13640** underwent phase II clinical trials as an analgesic (Fig. 14.14). The methyl substituent on the pyridine ring is susceptible to oxidation and is converted to a carboxylic acid which is inactive. The methyl group plays an important binding role and has to be present. Therefore, the pyridine ring was changed to a pyrimidine ring resulting in a compound (**F15599**) that has increased metabolic stability without affecting binding affinity.

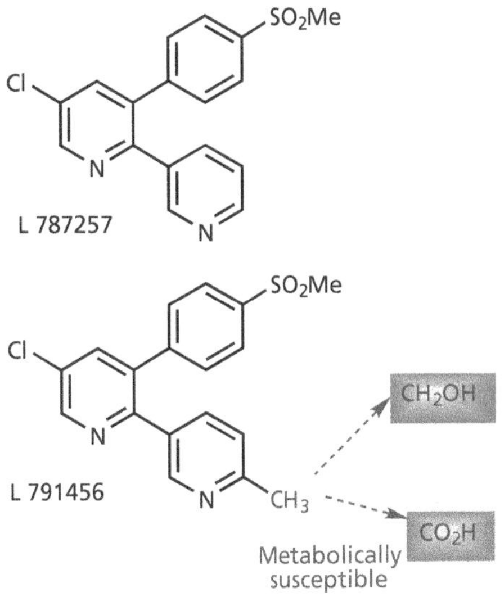

14.3 Making drugs less resistant to drug metabolism

So far, we have looked at how the activity of drugs can be prolonged by inhibiting their metabolism. However, a drug that is extremely stable to metabolism and is very slowly excreted can pose just as many problems as one that is susceptible to metabolism. It is usually desirable to have a drug that does what it is meant to do, then stops doing it within a reasonable time. If not, the effects of the drug could last too long and cause toxicity and lingering side effects. Therefore, designing drugs with decreased chemical and metabolic stability can sometimes be useful.

14.3.1 Introducing metabolically susceptible groups

Introducing groups that are susceptible to metabolism is a good way of shortening the lifetime of a drug (see Box 14.2). For example, a methyl group was introduced to the anti-arthritic agent **L 787257** to shorten its lifetime. The methyl group of the resulting compound (**L 791456**) was metabolically oxidized to a polar alcohol as well as to a carboxylic acid (Fig. 14.15).

FIGURE 14.15 Adding a metabolically labile methyl group to shorten a drug's lifetime.

Another example is the analgesic **remifentanil** (section 24.6.3.4), where ester groups were incorporated to make it a short-lasting agent. The beta-blocker **esmolol** was also designed to be a short-acting agent by introducing an ester group (section 23.11.3.4).

14.3.2 Self-destruct drugs

A self-destruct drug is one which is chemically stable under one set of conditions, but becomes unstable and spontaneously degrades under another set of conditions. The advantage of a self-destruct drug is that inactivation does not depend on the activity of metabolic enzymes, which could vary from patient to patient. The best example of a self-destruct drug is the neuromuscular blocking agent **atracurium**, which is stable at acid pH but self-destructs when it meets the slightly alkaline conditions of the blood (section 22.10.2.4). This means that the drug has a short duration of action, allowing anaesthetists to control its blood levels during surgery by providing it as a continuous, intravenous drip.

BOX 14.2 Shortening the lifetime of a drug

Anti-asthmatic drugs are usually taken by inhalation to reduce the chances of side effects elsewhere in the body. However, a significant amount is swallowed and can be absorbed into the blood supply from the gastrointestinal tract. Therefore, it is desirable to have an anti-asthmatic drug which is potent and stable in the lungs, but which is rapidly metabolized in the blood supply. **Cromakalim** has useful anti-asthmatic properties, but has cardiovascular side effects if

it gets into the blood supply. Structures **UK 143220** and **UK 157147** were developed from cromakalim so that they would be quickly metabolized. UK 143220 contains an ester which is quickly hydrolysed by esterases in the blood to produce an inactive carboxylic acid, while UK 157147 contains a phenol group which is quickly conjugated by metabolic conjugation enzymes and eliminated. Both these compounds were considered as clinical candidates.

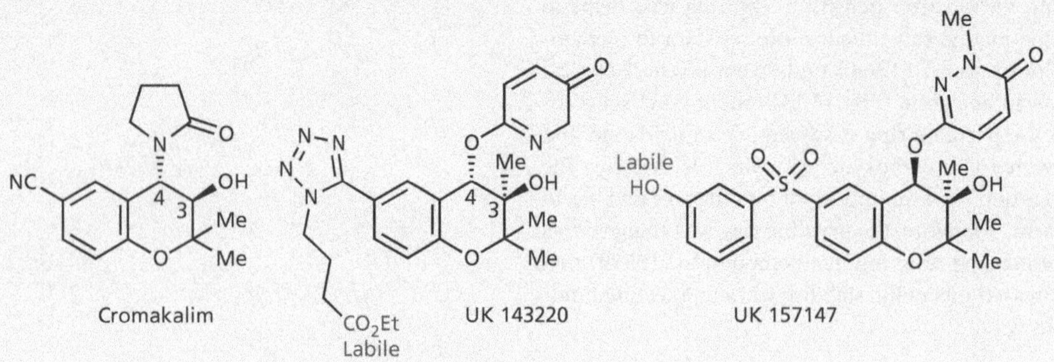

FIGURE 1 Metabolically labile analogues of cromakalim.

KEY POINTS

- The polarity or pK_a of a lead compound can be altered by varying alkyl substituents or functional groups, allowing the drug to be absorbed more easily.

- Drugs can be made more resistant to metabolism by introducing steric shields to protect susceptible functional groups. It may also be possible to modify the functional group itself to make it more stable as a result of electronic factors.

- Metabolically stable groups can be added to block metabolism at certain positions.

- Groups which are susceptible to metabolism may be modified or removed to prolong activity, as long as the group is not required for drug–target interactions.

- Metabolically susceptible groups which are necessary for drug–target interactions can be shifted in order to make them unrecognizable by metabolic enzymes, as long as they are still recognizable to the target.

- Varying a heterocyclic ring in the lead compound can sometimes improve metabolic stability.

- Drugs which are slowly metabolized may linger too long in the body and cause side effects.

- Groups which are susceptible to metabolic or chemical change can be incorporated to reduce a drug's lifetime.

14.4 Targeting drugs

One of the major goals in drug design is to find ways of targeting drugs to the exact locations in the body where they are most needed. The principle of targeting drugs can be traced back to Paul Ehrlich who developed antimicrobial drugs that were selectively toxic for microbial cells over human cells. Drugs can also be made more selective to distinguish between different targets within the body, as discussed in Chapter 13. Here, we discuss other tactics related to the targeting of drugs.

14.4.1 Targeting tumour cells: 'search and destroy' drugs

A major goal in cancer chemotherapy is to target drugs efficiently against tumour cells rather than normal cells. One method of achieving this is to design drugs which make use of specific molecular transport systems. The idea is to attach the active drug to an important 'building block' molecule that is needed in large amounts by the rapidly dividing tumour cells. This could be an amino acid or a nucleic acid base (e.g. uracil mustard; section 21.2.3.1). Of course, normal cells require these building blocks as well, but tumour cells often grow more quickly than normal cells and require the building blocks more urgently. Therefore, the uptake is greater in tumour cells.

A more recent idea has been to attach the active drug (or a poison such as **ricin**) to **monoclonal antibodies** which can recognize antigens unique to the tumour cell. Once the antibody binds to the antigen, the drug or poison is released to kill the cell. The difficulties in this approach include the identification of suitable antigens and the production of antibodies in significant quantity. Nevertheless, the approach has great promise for the future and is covered in more detail in section 21.10.2. Another tactic which has been used to target anticancer drugs is to administer an enzyme–antibody conjugate where the enzyme serves to activate an anticancer prodrug, and the antibody directs the enzyme to the tumour. This is a strategy known as **ADEPT** and is covered in more detail in section 21.10.3. Other targeting strategies include **ADAPT** and **GDEPT** covered in sections 21.10.4 and 21.10.5 respectively. Antibodies are also being studied as a means of targeting viruses (section 20.11.5).

14.4.2 Targeting gastrointestinal infections

If a drug is to be targeted against an infection of the gastrointestinal tract, it must be prevented from being absorbed into the blood supply. This can be done by using a fully ionized drug which is incapable of crossing cell membranes. For example, highly ionized sulphonamides are used against gastrointestinal infections (Box 19.2).

14.4.3 Targeting peripheral regions rather than the central nervous system

It is often possible to target drugs such that they act peripherally and not in the central nervous system. By increasing the polarity of drugs, they are less likely to cross the blood–brain barrier (section 11.4.5), and this means they are less likely to have central nervous system side effects. Achieving selectivity for the central nervous system over the peripheral regions of the body is not so straightforward. In order to achieve that, the drug would have to be designed to cross the blood–brain barrier efficiently, whilst being metabolized rapidly to inactive metabolites in the peripheral system.

14.4.4 Targeting with membrane tethers

Several drug targets are associated with cell membranes, and one way of targeting drugs to these targets is to attach membrane tethers to the drug such that the molecule is anchored in the membrane close to the target. The antibacterial agent **teicoplanin** is one such example and is discussed in section 19.5.5.2. Another membrane-tethered drug has been designed to inhibit the enzyme β-**secretase**, with the ultimate aim of treating Alzheimer's disease (AD). This enzyme generates the proteins that are responsible for the toxic protein aggregates found in the brains of AD sufferers, and does so mainly in cellular organelles called **endosomes**. A peptide transition-state inhibitor has been linked to a sterol such that it is taken into endosomes by endocytosis. The sterol then acts as the membrane tether to lock the drug in position, such that it targets β-secretase in endosomes rather than β-secretase in other locations. Potential agents for AD treatment are also being targeted to mitochondria where AD leads to the generation of radicals and oxidation reactions that are damaging to the cell. **MitoQ** (Fig. 14.16) is an agent undergoing clinical trials which contains an antioxidant prodrug linked to a hydrophobic triphenylphosphine moiety. The latter group aids the drug's entry into mitochondria, then tethers it to the phospholipid bilayers of the mitochondria membrane. The quinone ring system is rapidly reduced to the active quinol form which can then act as an antioxidant to neutralize free radicals. A different approach for targeting antioxidant drugs to mitochondria has been to modify known antibacterial agents (e.g. **gramicidin S**) such that they act as antioxidants rather than antibacterial agents. The

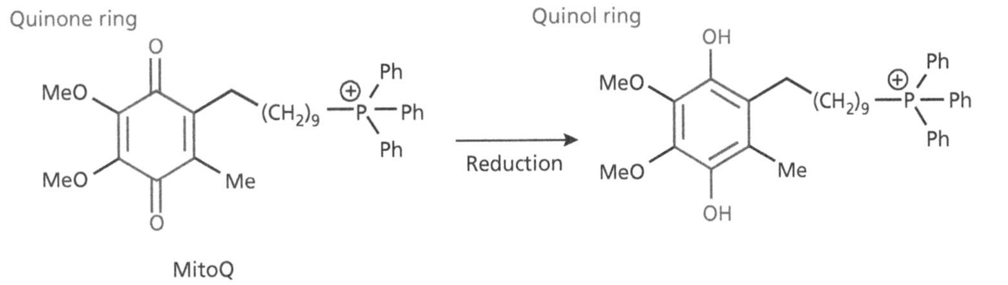

FIGURE 14.16 MitoQ acting as a prodrug.

rationale here is that the mitochondrial membrane is similar in nature to bacterial cell membranes, and so antibacterial agents may show selectivity for mitochondrial membranes over cell membranes.

14.5 Reducing toxicity

It is often found that a drug fails clinical trials because of toxic side effects. This may be due to toxic metabolites, in which case the drug should be made more resistant to metabolism as described earlier (section 14.2). It is also worth checking to see whether there are any functional groups present which are particularly prone to producing toxic metabolites. For example, it is known that functional groups such as aromatic nitro groups, aromatic amines, bromoarenes, hydrazines, hydroxylamines, or polyhalogenated groups are often metabolized to toxic products (see section 11.5 for typical metabolic reactions; see also Box 14.3).

Side effects might also be reduced or eliminated by varying apparently harmless substituents. For example, the halogen substituents of the antifungal agent **UK 47265** were varied in order to find a compound that was less toxic to the liver. This led to the successful antifungal agent **fluconazole** (Fig. 14.17).

Varying the position of substituents can sometimes reduce or eliminate side effects. For example, the dopamine antagonist **SB 269652** inhibits cytochrome P450 enzymes as a side effect. Placing the cyano group at a different position prevented this inhibition (Fig. 14.18).

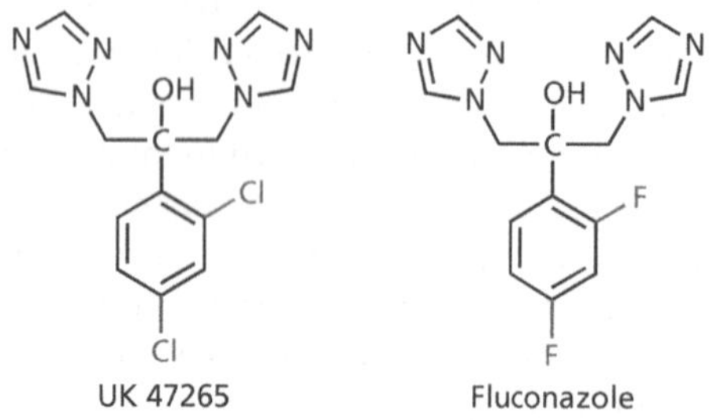

FIGURE 14.17 Varying aromatic substituents to reduce toxicity.

KEY POINTS

- Strategies designed to target drugs to particular cells or tissues are likely to lead to safer drugs with fewer side effects.
- Drugs can be linked to amino acids or nucleic acid bases to target them against fast-growing and rapidly dividing cells.
- Drugs can be targeted to the gastrointestinal tract by making them ionized or highly polar such that they cannot cross the gut wall.
- The central nervous system side effects of peripherally acting drugs can be eliminated by making the drugs more polar so that they do not cross the blood–brain barrier.
- Drugs with toxic side effects can sometimes be made less toxic by varying the nature or position of substituents, or by preventing their metabolism to a toxic metabolite.

14.6 Prodrugs

Prodrugs are compounds which are inactive in themselves, but which are converted in the body to the active drug. They have been useful in tackling problems such as acid sensitivity, poor membrane permeability, drug toxicity, bad taste, and short duration of action. Usually, a metabolic enzyme is involved in converting the prodrug to the active drug, and so a good knowledge of drug metabolism and the enzymes involved allows the medicinal chemist to design a suitable prodrug which turns drug metabolism into an advantage rather than a problem. Prodrugs have been designed to be activated by a variety of metabolic enzymes. Ester prodrugs which are hydrolysed by esterase enzymes are particularly common, but prodrugs have also been designed which are activated by *N*-demethylation, decarboxylation, and the hydrolysis of amides and phosphates. Not all prodrugs are activated by metabolic enzymes, however. For example, photodynamic therapy involves the use of an external light source to activate prodrugs. When designing prodrugs, it is important to ensure that the prodrug is effectively converted to the active drug once it has been absorbed into the blood supply, but it is also important to ensure that any groups that are cleaved from the molecule are non-toxic.

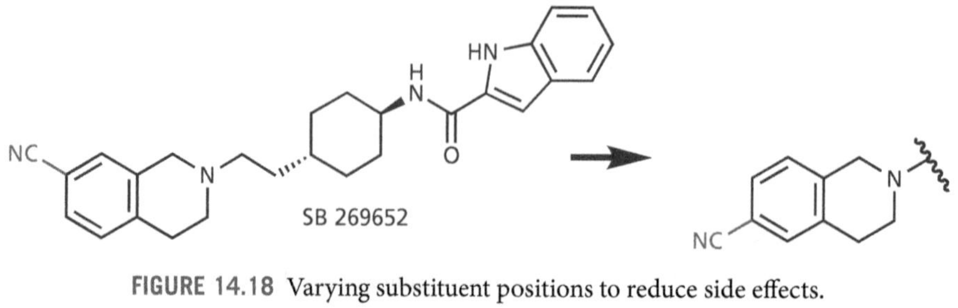

FIGURE 14.18 Varying substituent positions to reduce side effects.

BOX 14.3 Identifying and replacing potentially toxic groups

Replacing potentially toxic functional groups is often carried out at an early stage in drug development, even if there is no direct evidence of actual toxicity. For example, the presence of a 1,4-diamino-substituted aromatic ring is seen as a significant risk factor for toxicity. The presence of the two amino groups activates the aromatic ring and makes it electron rich such that it can undergo metabolism to form a 1,4-diiminoquinone structure. Diiminoquinones can act as electrophilic agents and react with nucleophilic groups such as lysine in proteins, or guanine in nucleic acids.

One strategy that can be used to combat this problem is to make the ring less electron rich by introducing a nitrogen atom. In other words, the benzene ring is replaced with a pyridine ring. This strategy was used in the development of the anticancer agent **sonidegib** (section 21.6.3). Another strategy that was used in the development of the anticancer agent **ceritinib** was to carry out a 'ring reversal'. This involved reversing a piperidine ring in the lead compound such that the nitrogen was no longer linked to the aromatic ring. In addition, a methyl substituent was added to act as a metabolic blocker.

1,4-Diamino-substituted aromatic ring

1,4-Diiminoquinone ring

Ring reversal & metabolic blocker

14.6.1 Prodrugs to improve membrane permeability

14.6.1.1 Esters as prodrugs

Prodrugs have proved very useful in temporarily masking an 'awkward' functional group which is important to target binding, but which hinders the drug from crossing the cell membranes of the gut wall. For example, a carboxylic acid functional group may have an important role to play in binding a drug to its binding site via ionic or hydrogen bonding. However, the very fact that it is an ionizable group may prevent it from crossing a fatty cell membrane. The answer is to protect the acid function as an ester. The less polar ester can cross fatty cell membranes and, once it is in the bloodstream, it is hydrolysed

back to the free acid by esterases in the blood. Examples of ester prodrugs used to aid membrane permeability include many of the ACE inhibitors (section 26.3.3 and Case study 2), **sacubitril** (section 26.5.3), and **pivampicillin**, which is a penicillin prodrug (Box 19.7).

Not all esters are hydrolysed equally efficiently, and a range of esters may need to be tried to find the best one (Box 14.4). It is possible to make esters more susceptible to hydrolysis by introducing electron-withdrawing groups to the alcohol moiety (e.g. OCH_2CF_3, OCH_2CO_2R, $OCONR_2$, OAr). The inductive effect of these groups aids the hydrolytic mechanism by stabilizing the alkoxide leaving group (Fig. 14.19). Care has to be taken, however, not to make the ester too reactive in case it becomes chemically unstable and is hydrolysed by the acid conditions of

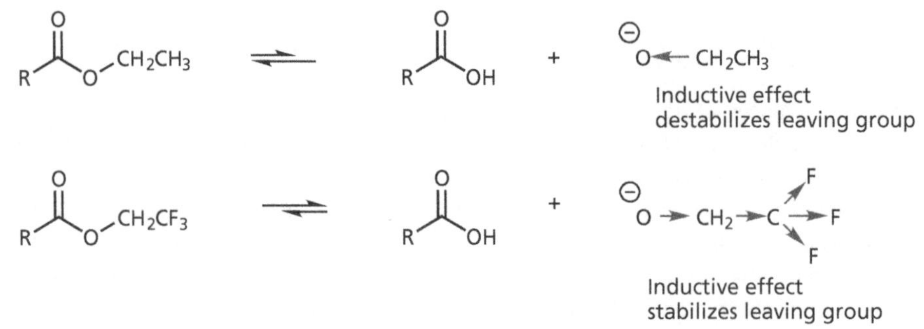

FIGURE 14.19 Inductive effects on the stability of leaving groups.

the stomach or the more alkaline conditions of the intestine before it reaches the blood supply. To that end, it may be necessary to make the ester more stable. For example, cyclopropanecarboxylic acid esters have been studied as potential prodrugs because the cyclopropane ring has the ability to stabilize the carbonyl group of a neighbouring ester (Fig. 14.20). In this respect, it is acting as a bio-isostere for a double bond (see also section 13.3.7). A conjugated double bond stabilizes a neighbouring carbonyl group due to interaction of the π-systems involved. It is proposed that the σ bonds of a cyclopropane ring are correctly orientated to allow a hyperconjugative interaction that has a similar stabilizing effect on a neighbouring carbonyl group. The interaction proposed involves hyperconjugative donation to the antibonding π orbital of the carbonyl group.

There are some instances where esterases fail to hydrolyse an ester prodrug because the ester is shielded by a bulky group. For example, this is often the case if the ester is linked to a multicylic ring system. In such cases, an **extended ester** can be the answer. This involves esterifying the drug with a group that contains a second ester or carbonate group, which will be positioned further away from the ring system. Consequently, it will be more accessible to esterase enzymes. Enzyme-catalysed hydrolysis of the more accessible ester then leads to a product which is designed to be chemically unstable and will spontaneously degrade to give the active drug without the need for enzyme intervention. Examples of extended esters as prodrugs can

be found in some penicillins (Box 19.7) and antihypertensive agents (section 26.3.4).

14.6.1.2 *N*-Methylated prodrugs

N-Demethylation is a common metabolic reaction in the liver, so polar amines can be *N*-methylated to reduce polarity and improve membrane permeability. Several hypnotics and anti-epileptics take advantage of this reaction, for example **hexobarbitone** (Fig. 14.21).

14.6.1.3 Trojan horse approach for transport proteins

Another way round the problem of membrane permeability is to design a prodrug which can take advantage of transport proteins in the cell membrane, such as the ones responsible for carrying amino acids into a cell (section 2.7.2). A well-known example of such a prodrug is **levodopa** (Fig. 14.22). Levodopa is a prodrug for the neurotransmitter **dopamine** and has been used in the treatment of Parkinson's disease—a condition due primarily to a deficiency of that neurotransmitter in the brain. Dopamine itself cannot be used, since it is too polar to cross the blood–brain barrier. Levodopa is even more polar and seems an unlikely prodrug, but it is also an amino acid, and so it is recognized by the transport proteins for amino acids which carry it across the cell membrane. Once in the brain, a decarboxylase enzyme removes the acid group and generates dopamine.

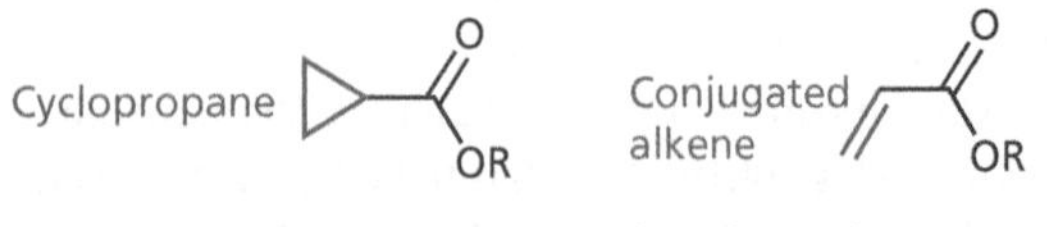

FIGURE 14.20 Cyclopropane carboxylic acid esters as prodrugs and bio-isosteres for α,β-unsaturated esters.

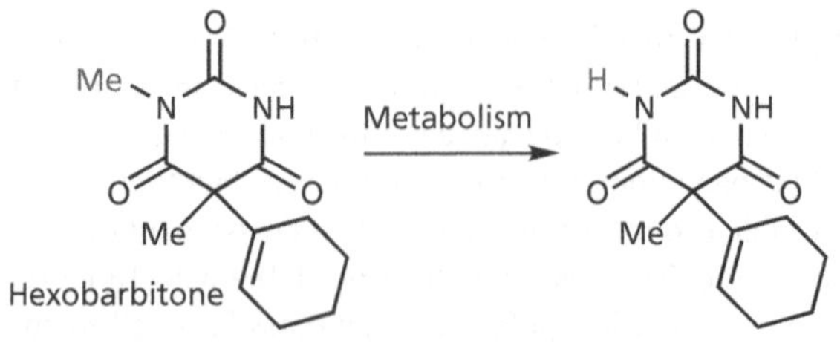

FIGURE 14.21 *N*-Demethylation of hexobarbitone.

BOX 14.4 Varying esters in prodrugs

The protease inhibitor **candoxatrilat** has to be given intravenously because it is too polar to be absorbed from the gastrointestinal tract. Different esters were tried as prodrugs to get round this problem. It was found that an ethyl ester was absorbed but was inefficiently hydrolysed. A more activated ester was required, and a 5-indanyl ester proved to be the best. The 5-indanol released on hydrolysis is non-toxic (Fig. 1).

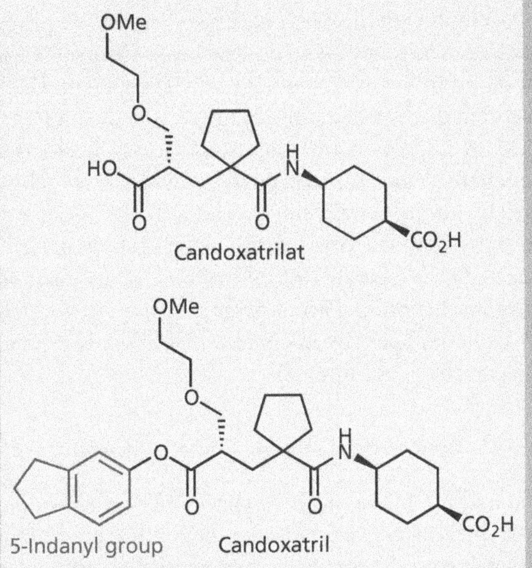

FIGURE 1 Protease inhibitors.

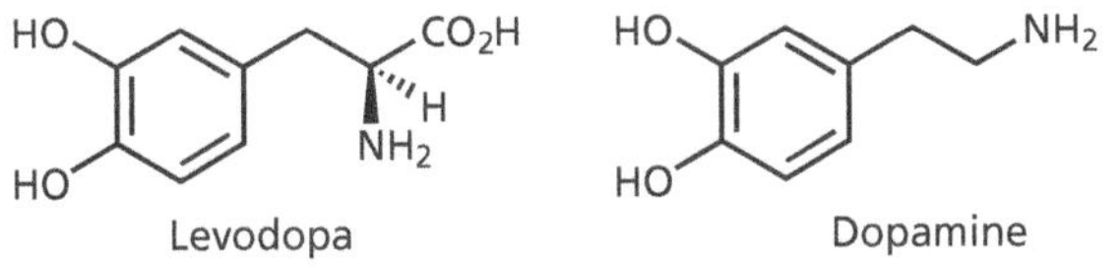

FIGURE 14.22 Levodopa and dopamine.

14.6.2 Prodrugs to prolong drug activity

Sometimes prodrugs are designed to be converted slowly to the active drug, thus prolonging a drug's activity. For example, **6-mercaptopurine** (Fig. 14.23) suppresses the body's immune response and is, therefore, useful in protecting donor grafts. Unfortunately, the drug tends to be eliminated from the body too quickly. The prodrug **azathioprine** has the advantage that it is slowly converted to 6-mercaptopurine by being attacked by **glutathione** (section 11.5.5), allowing a more sustained activity. The rate of conversion can be altered, depending on the electron-withdrawing ability of the heterocyclic group. The greater the electron-withdrawing power, the faster the breakdown. The NO_2 group is therefore present to ensure an efficient conversion to 6-mercaptopurine, since it is strongly electron-withdrawing on the heterocyclic ring.

There is a belief that the well-known sedatives **Valium** (Fig. 14.24) and **Librium** might be prodrugs and are active because they are metabolized by *N*-demethylation to **nordazepams**. Nordazepam itself has been used as a sedative, but loses activity quite quickly as a result of metabolism and excretion. Valium, if it is a prodrug for nordazepam, demonstrates again how a prodrug can be used to produce a more sustained action.

Another approach to maintaining a sustained level of drug over long periods is to deliberately associate a very lipophilic group to the drug. This means that most of the drug is stored in fat tissue, from where it is steadily and slowly released into the bloodstream. The antimalarial agent **cycloguanil pamoate** (Fig. 14.25) is one such agent. The active drug is bound ionically to an anion containing a large lipophilic group, and is only released into the blood supply following slow dissociation of the ion complex.

Similarly, lipophilic esters of the antipsychotic drug **fluphenazine** are used to prolong its action. The prodrug is given by intramuscular injection and slowly diffuses from fat tissue into the blood supply where it is rapidly hydrolysed (Fig. 14.26).

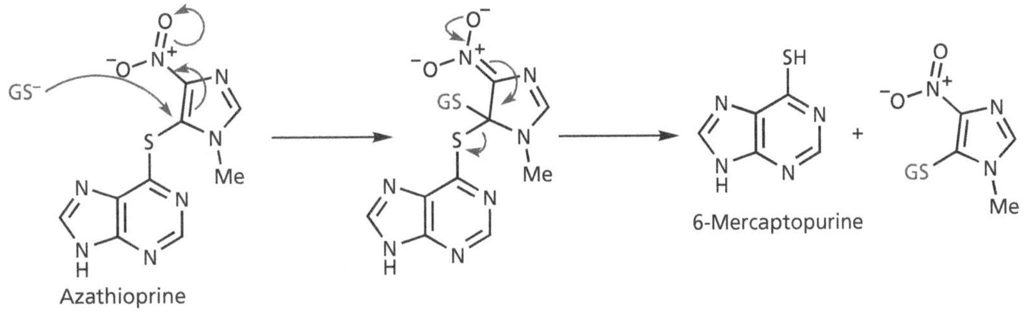

FIGURE 14.23 Azathioprine acts as a prodrug for 6-mercaptopurine (GS = glutathione).

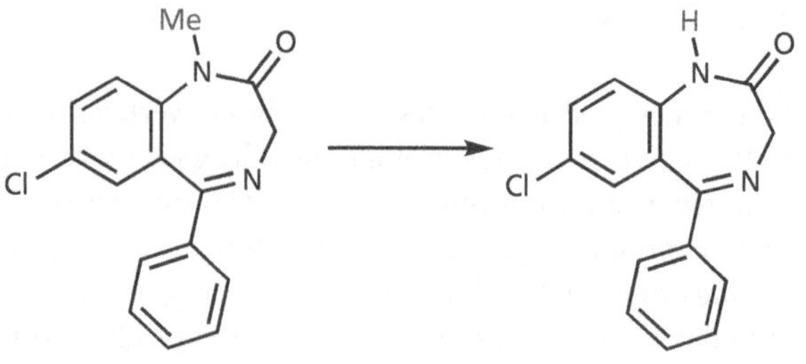

FIGURE 14.24 Valium (diazepam) as a possible prodrug for nordazepam.

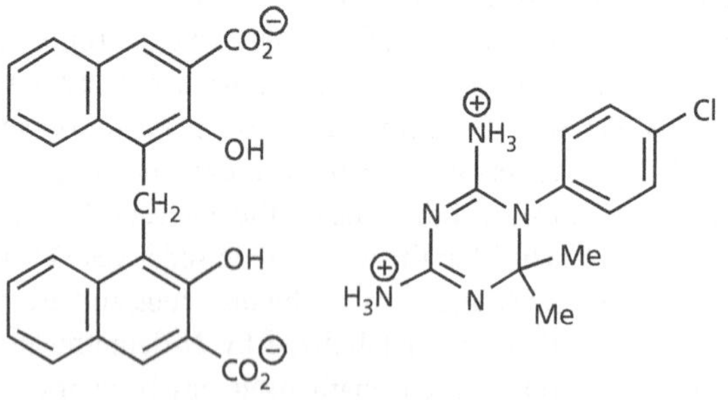

FIGURE 14.25 Cycloguanil pamoate.

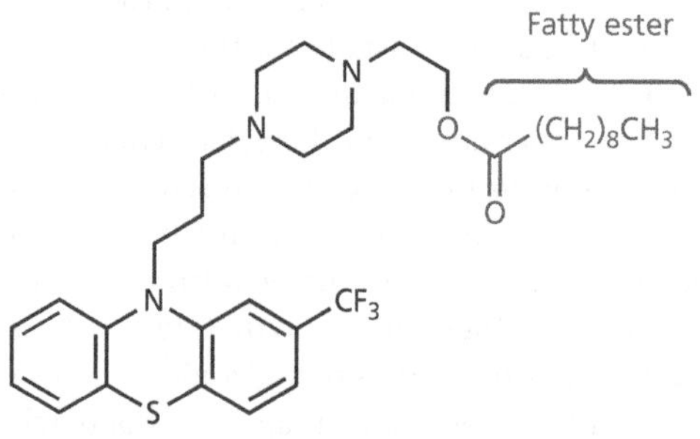

FIGURE 14.26 Fluphenazine decanoate.

14.6.3 Prodrugs masking drug toxicity and side effects

Prodrugs can be used to mask the side effects and toxicity of drugs (Box 14.5). For example, **salicylic acid** is a good painkiller, but causes gastric bleeding due to the free phenolic group. This is overcome by masking the phenol as an ester (**aspirin**) (Fig. 14.27). The ester is later hydrolysed to free the active drug.

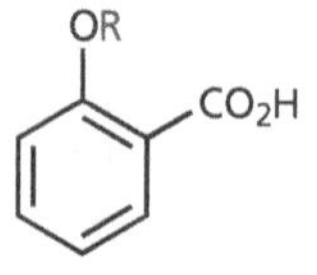

FIGURE 14.27 Aspirin (R = COCH$_3$) and salicylic acid (R = H).

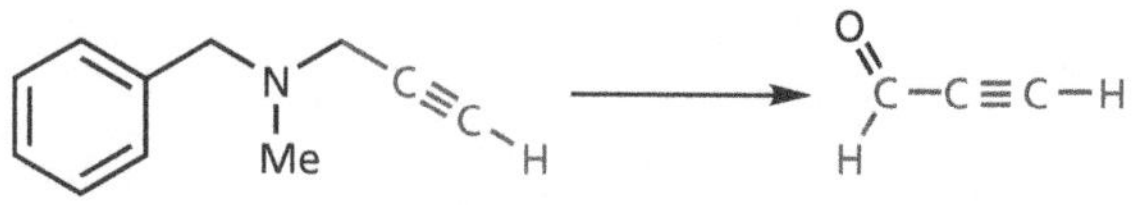

FIGURE 14.28 Pargyline as a prodrug for propiolaldehyde.

Prodrugs can be used to give a slow release of drugs that would be too toxic to give directly. **Propiolaldehyde** is useful in the aversion therapy of alcohol, but is not used itself because it is an irritant. The prodrug **pargyline** can be converted to propiolaldehyde by enzymes in the liver (Fig. 14.28).

Cyclophosphamide is a successful, non-toxic prodrug which can be safely taken orally. Once absorbed, it is metabolized in the liver to a toxic alkylating agent which is useful in the treatment of cancer (section 21.2.3.1).

Many important antiviral drugs such as **aciclovir** and **penciclovir** are non-toxic prodrugs which show selective toxicity towards virally infected cells. This is because they are activated by a viral enzyme which is only present in infected cells (sections 9.5 and 20.6.1). In a similar vein, the antischistosomal agent **oxamniquine** is converted to an alkylating agent by an enzyme which is only present in the parasite (Case study 4).

14.6.4 Prodrugs to lower water solubility

Some drugs have a revolting taste! One way to avoid this problem is to reduce their water solubility to prevent them dissolving on the tongue. For example, the bitter taste of the antibiotic **chloramphenicol** can be avoided by using the palmitate ester (Fig. 14.29). This is more hydrophobic because of the masked alcohol and the long chain fatty group that is present. It does not dissolve easily on the tongue and is quickly hydrolysed once swallowed.

14.6.5 Prodrugs to improve water solubility

Prodrugs have been used to increase the water solubility of drugs (Box 14.6). This is particularly useful for drugs which are given intravenously, as it means that higher

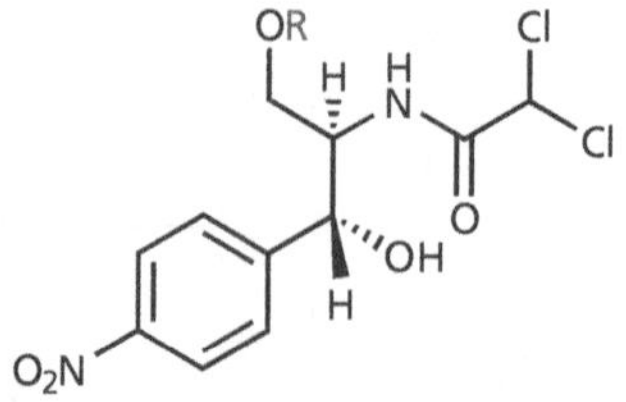

FIGURE 14.29 Chloramphenicol (R = H) and chloramphenicol prodrugs: chloramphenicol palmitate (R = CO(CH$_2$)$_{14}$CH$_3$); chloramphenicol succinate (R = CO(CH$_2$)$_2$CO$_2$H).

BOX 14.5 Prodrugs masking toxicity and side effects

LDZ is an example of a diazepam prodrug which avoids the drowsiness side effects associated with **diazepam**. These side effects are associated with the high initial plasma levels of diazepam following administration. The use of a prodrug avoids this problem. An aminopeptidase enzyme hydrolyses the prodrug to release a non-toxic lysine moiety, and the resulting amine spontaneously cyclizes to the diazepam (as shown below).

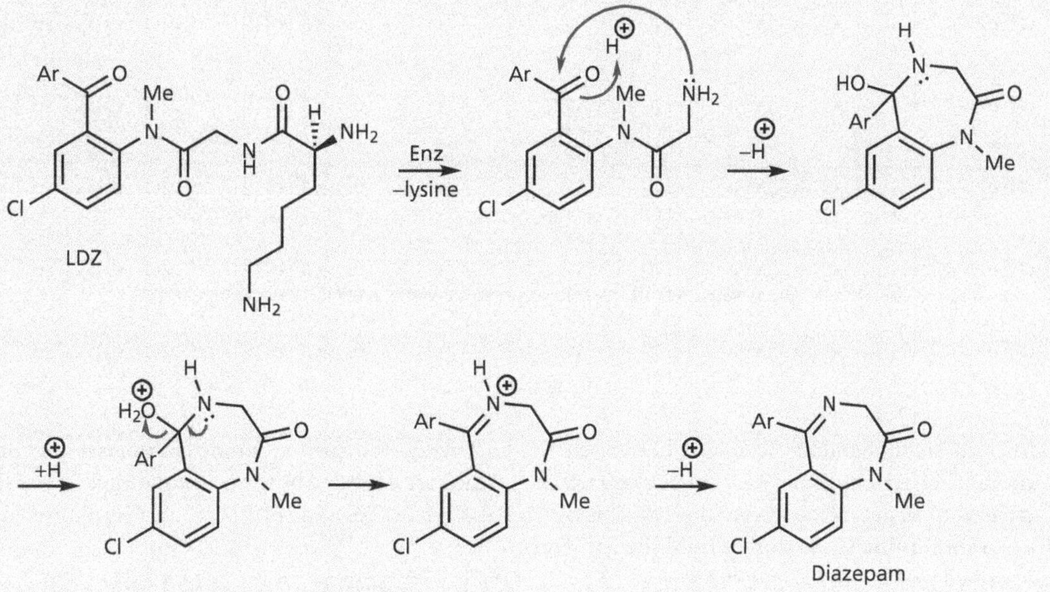

FIGURE 1 LDZ as a diazepam prodrug.

concentrations and smaller volumes can be used. For example, the succinate ester of **chloramphenicol** (Fig. 14.29) increases the latter's water solubility due to the extra carboxylic acid that is present. Once the ester is hydrolysed, chloramphenicol is released along with succinic acid, which is naturally present in the body.

Prodrugs designed to increase water solubility have proved useful in preventing the pain associated with some injections, which is caused by the poor solubility of the drug at the site of injection. For example, the antibacterial agent **clindamycin** is painful when injected, but this is avoided by using a phosphate ester prodrug which has much better solubility because of the ionic phosphate group (Fig. 14.30).

14.6.6 Prodrugs used in the targeting of drugs

Methenamine (Fig. 14.31) is a stable, inactive compound when the pH is more than 5. At a more acidic pH, however, the compound spontaneously degrades to generate **formaldehyde,** which has antibacterial properties. This

is useful in the treatment of some urinary tract infections. The normal pH of blood is slightly alkaline (7.4) and so methenamine passes round the body unchanged. However, once it is excreted into the infected urinary tract, it encounters urine which can be acidic as a result of certain bacterial infections. Consequently, methenamine degrades to generate formaldehyde just where it is needed.

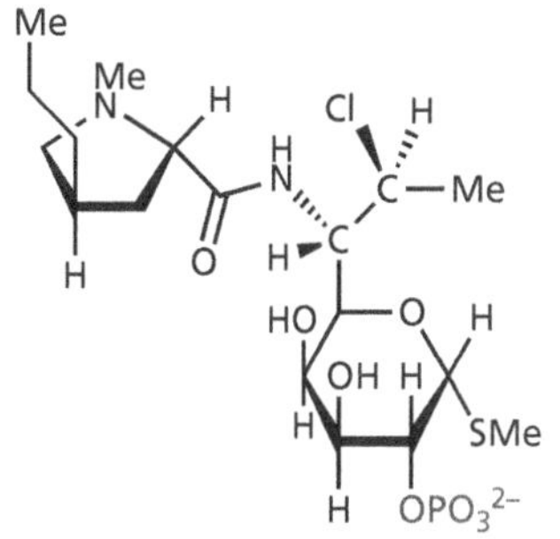

FIGURE 14.30 Clindamycin phosphate.

Polar prodrugs have been used to improve the absorption of non-polar drugs from the gut. Drugs have to have some water solubility if they are to be absorbed, otherwise they dissolve in fatty globules and fail to interact effectively with the gut wall. The steroid **estrone** is one such drug. By using a lysine ester prodrug, water solubility and absorption are increased. Hydrolysis of the prodrug releases the active drug, and the amino acid lysine as a non-toxic by-product.

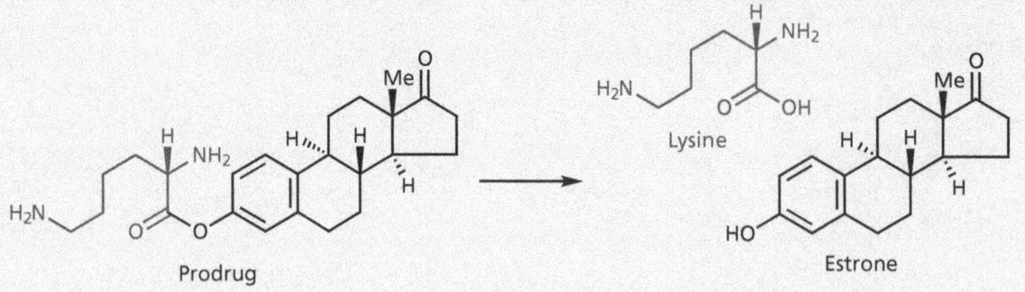

FIGURE 1   The lysine ester of estrone to improve water solubility and absorption.

Prodrugs of sulphonamides have also been used to target intestinal infections (Box 19.2). Other examples of prodrugs used to target infections are the antischistoso-mal drug **oxamniquine** (Case study 4) and the antiviral drugs described in sections 9.5 and 20.6.1.

The targeting of prodrugs to tumour cells by antibody-related strategies was mentioned in section 14.4.1 and is described in more detail in section 21.10. Antibody–drug conjugates can also be viewed as prodrugs and are described in that section.

Finally, the **proton pump inhibitors** are prodrugs which are activated by the acid conditions of the stomach (section 25.3).

14.6.7 Prodrugs to increase chemical stability

The antibacterial agent **ampicillin** decomposes in con-centrated aqueous solution as a result of intramolecular attack of the side chain amino group on the lactam ring (section 19.5.1.8). **Hetacillin** (Fig. 14.32) is a prodrug which locks up the offending nitrogen in a ring and pre-vents this reaction. Once the prodrug has been adminis-tered, hetacillin slowly decomposes to release ampicillin and acetone.

In the field of antiviral agents, cyclopropane carboxylic acid esters (section 14.6.1.1) are being studied as poten-tial prodrugs for aciclovir, in order to prolong chemical stability in solution.

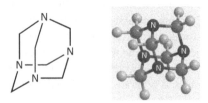

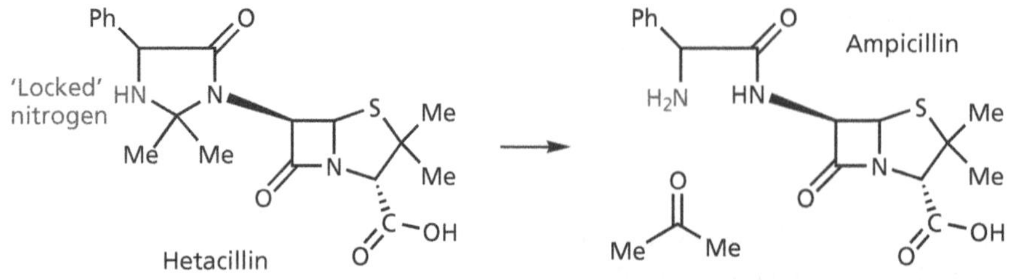

FIGURE 14.31   Methenamine.

FIGURE 14.32   Hetacillin and ampicillin.

14.6.8 Prodrugs activated by external influence (sleeping agents)

Conventional prodrugs are inactive compounds which are normally metabolized in the body to the active form. A variation of the prodrug approach is the concept of a 'sleeping agent'. This is an inactive compound which is only converted to the active drug by some form of external influence. The best example of this approach is the use of photosensitizing agents such as **porphyrins** or **chlorins** in cancer treatment—a strategy known as **photodynamic therapy**. Given intravenously, these agents accumulate within cells and have some selectivity for tumour cells. By themselves, the agents have little effect, but if the cancer cells are irradiated with light, the porphyrins are converted to an excited state and react with molecular oxygen to produce highly toxic singlet oxygen. This is covered in section 21.11.

KEY POINTS

- Prodrugs are inactive compounds which are converted to active drugs in the body—usually by drug metabolism.

- Esters are commonly used as prodrugs to make a drug less polar, allowing it to cross cell membranes more easily. The nature of the ester can be altered to vary the rate of hydrolysis.

- Extended esters can be used if a simple ester is shielded by a bulky group.

- Introducing a metabolically susceptible N-methyl group can sometimes be advantageous in reducing polarity.

- Prodrugs with a similarity to important biosynthetic building blocks may be capable of crossing cell membranes with the aid of transport proteins.

- The activity of a drug can be prolonged by using a prodrug which is converted slowly to the active drug.

- The toxic nature of a drug can be reduced by using a prodrug which is slowly converted to the active compound, preferably at the site of action.

- Prodrugs which contain metabolically susceptible polar groups are useful in improving water solubility. They are particularly useful for drugs which have to be injected, or for drugs which are too hydrophobic for effective absorption from the gut.

- Prodrugs which are susceptible to pH or chemical degradation can be effective in targeting drugs or increasing stability in solution prior to injection.

- Prodrugs which are activated by light are the basis for photodynamic therapy.

14.7 Drug alliances

Some drugs are found to affect the activity or pharmacokinetic properties of other drugs, and this can be put to good use. The following are some examples.

14.7.1 'Sentry' drugs

In this approach, a second drug is administered with the principal drug in order to guard or assist it. Usually, the second drug inhibits an enzyme that metabolizes the principal drug. For example, **clavulanic acid** inhibits the enzyme β-**lactamase** and is therefore able to protect penicillins from that particular enzyme (sections 7.5 and 19.5.4.1).

The antiviral preparation **Kaletra**, used in the treatment of AIDS, is a combination of two drugs called **ritonavir** and **lopinavir**. Although the former has antiviral activity, it is principally present to protect lopinavir, which is metabolized by the metabolic cytochrome P450 enzyme (CYP3A4). Ritonavir is a strong inhibitor of this enzyme and so the metabolism of lopinavir is decreased allowing lower doses to be used for therapeutic plasma levels (section 20.7.4.4).

Another example is to be found in the drug therapy of Parkinson's disease. The use of **levodopa** as a prodrug for **dopamine** has already been described (section 14.6.1.3). To be effective, however, large doses of levodopa (3–8 g per day) are required, and over a period of time these dose levels lead to side effects such as nausea and vomiting. Levodopa is susceptible to the enzyme **dopa decarboxylase** and, as a result, much of the levodopa administered is decarboxylated to dopamine before it reaches the central nervous system (Fig. 14.33). This build-up of dopamine in the peripheral blood supply leads to the observed nausea and vomiting.

The drug **carbidopa** has been used successfully as an inhibitor of dopa decarboxylase and allows smaller doses of levodopa to be used. Furthermore, since it is a highly polar compound containing two phenolic groups, a hydrazine moiety, and an acidic group, it is unable to cross the blood–brain barrier, and so cannot prevent the

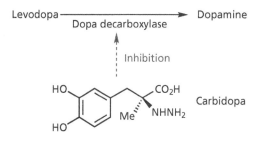

FIGURE 14.33 Inhibition of levodopa decarboxylation.

conversion of levodopa to dopamine in the brain. Carbidopa is marketed as a mixture with levodopa and is called **co-careldopa**.

Several important peptides and proteins could be used as drugs if it were not for the fact that they are quickly broken down by **protease** enzymes. One way round this problem is to inhibit the protease enzymes. **Candoxatril** (Box 14.4) is a protease inhibitor which has some potential in this respect and is under clinical evaluation.

Further examples of enzyme inhibitors that are used to block the metabolism of other drugs include **cilastatin** (Box 19.11), **tipiracil** (section 21.3.2), and **cobicistat** (section 20.7.5).

Finally, the action of penicillins can be prolonged if they are administered alongside **probenecid** (section 19.5.1.9). This agent slows the rate at which penicillins are excreted in the kidneys.

14.7.2 Localizing a drug's area of activity

Adrenaline is an example of a drug which has been used to localize the area of activity for another drug. When injected with the local anaesthetic **procaine**, adrenaline constricts the blood vessels in the vicinity of the injection, and so prevents procaine being rapidly removed from the area by the blood supply.

14.7.3 Increasing absorption

Metoclopramide (Fig. 14.34) is administered alongside analgesics in the treatment of migraine. Its function is to increase gastric motility, leading to faster absorption of the analgesic and quicker pain relief.

KEY POINTS

- A sentry drug is a drug which is administered alongside another drug to enhance the latter's activity.

- Many sentry drugs protect their partner drug by inhibiting an enzyme which acts on the latter.

- Other drugs have been used to localize the site of action of local anaesthetics, and to increase the absorption of drugs from the gastrointestinal tract.

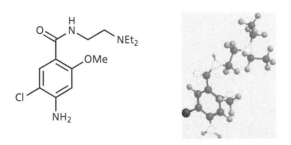

FIGURE 14.34 Metoclopramide.

14.8 Endogenous compounds as drugs

Endogenous compounds are molecules which occur naturally in the body. Many of these could be extremely useful in medicine. For example, the body's hormones are natural chemical messengers, so why not use them as medicines instead of synthetic drugs that are foreign to the body? In this section, we look at important molecules such as neurotransmitters, hormones, peptides, proteins, and antibodies, to see how feasible it is to use them as drugs.

14.8.1 Neurotransmitters

Many non-peptide neurotransmitters are simple molecules which can easily be prepared in the laboratory, so why are these not used commonly as drugs? For example, if there is a shortage of dopamine in the brain, why not administer more dopamine to make up the balance?

Unfortunately, this is not possible for a number of reasons. Many neurotransmitters are not stable enough to survive the acid of the stomach, and would have to be injected. Even if they were injected, there is little chance that they would survive to reach their target receptors. The body has efficient mechanisms which inactivate neurotransmitters as soon as they have passed on their message from nerve to target cell. Therefore, any neurotransmitter injected into the blood supply would be swiftly inactivated by enzymes, or taken up by cells via transport proteins. Even if they were not inactivated or removed, they would be poor drugs indeed, leading to many undesirable side effects. For example, the shortage of neurotransmitter may only be at one small area in the brain; the situation may be normal elsewhere. If we gave the natural neurotransmitter, how would we stop it producing an overdose of transmitter at these other sites? Of course this is a problem with all drugs, but it has been discovered that the receptors for a specific neurotransmitter are not all identical. There are different types and subtypes of a particular receptor, and their distribution around the body is not uniform. One subtype of receptor may be common in one tissue, whereas a different subtype is common in another tissue. The medicinal chemist can design synthetic drugs which take advantage of that difference, ignoring receptor subtypes which the natural neurotransmitter would not. In this respect, the medicinal chemist has actually improved on nature.

We cannot even assume that the body's own neurotransmitters are perfectly safe, and free from the horrors of tolerance and addiction associated with drugs such as **heroin**. It is quite possible to be addicted to one's own neurotransmitters and hormones. Some people are addicted to exercise, and are compelled to exercise long hours each day in order to feel good. The very process of exercise leads to the release of hormones and neurotransmitters which can produce a 'high', and this drives susceptible people to exercise more and more. If they stop exercising, they suffer withdrawal symptoms such as deep depression. The same phenomenon probably drives mountaineers into attempting feats which they know might well lead to their death. The thrill of danger produces hormones and neurotransmitters which in turn produce a 'high'. This may also explain why some individuals choose to become mercenaries and risk their lives travelling the globe in search of wars to fight.

To conclude, many of the body's own neurotransmitters are known and can be easily synthesized, but they cannot be effectively used as medicines.

14.8.2 Natural hormones, peptides, and proteins as drugs

Unlike neurotransmitters, natural hormones have potential in drug therapy as they normally circulate round the body and behave like drugs. Indeed, **adrenaline** is commonly used in medicine to treat (among other things) severe allergic reactions (section 23.10.1). Most hormones are peptides and proteins, and some naturally occurring peptide and protein hormones are already used in medicine. These include **insulin, calcitonin, erythropoietin, human growth factor, interferons**, and **colony stimulating factors**.

The availability of many protein hormones owes a great deal to genetic engineering (section 6.4). It is extremely tedious and expensive to obtain substantial quantities of these proteins by other means. For example, isolating and purifying a hormone from blood samples is impractical because of the tiny quantities of hormone present. It is far more practical to use **recombinant DNA techniques**, whereby the human genes for the protein are cloned and then incorporated into the DNA of fast-growing bacterial, yeast, or mammalian cells. These cells then produce sufficient quantities of the protein.

Using these techniques, it is possible to produce 'cut down' versions of important body proteins and polypeptides which can also be used therapeutically. For example, **teriparatide** is a polypeptide which has been approved for the treatment of osteoporosis, and was produced by recombinant DNA technology using a genetically modified strain of the bacterium *Escherichia coli*. It consists of 34 amino acids that represent the *N*-terminal end of **human parathyroid hormone** (consisting of 84 amino acids). Another recombinant protein that has been approved is **etanercept**, which is used for the treatment of rheumatoid arthritis. More than 80 polypeptide drugs

have reached the market as a result of the biotechnology revolution, with more to come. Another example is **abatacept** which was approved in 2005 for the treatment of rheumatoid arthritis. This disease is caused by T-cells binding and interacting with susceptible cells to cause cell damage and inflammation. The binding process involves a protein–protein interaction between a T-cell protein and a protein in the membrane of the susceptible cell. Abatacept is an agent which mimics the T-cell protein and binds to the susceptible cell before the T-cell does, thus preventing the damage and inflammation that would result from such an interaction. Abatacept was prepared by taking the extracellular portion of the T-cell protein and linking it to part of an antibody. Therefore, it is classed as a **fusion protein**. **Belatacept** is a very similar fusion protein that was approved in 2011 as an immunosuppressant used for extending graft survival.

Recombinant enzymes have also been produced. For example, **glucarpidase** is a carboxypeptidase enzyme which was approved in 2012. It is administered to cancer patients with failed kidneys when they are taking the anticancer drug **methotrexate**. The enzyme serves to metabolize methotrexate and prevent it from reaching toxic levels. Another recombinant enzyme that has been recently approved for the treatment of cancer patients is **rasburicase**. This is a recombinant version of the enzyme urate oxidase which catalyses the conversion of uric acid to allantoin. Uric acid can build up to toxic levels as a result of cell death from chemotherapy—a condition called **tumour lysis syndrome**. If untreated, this can result in kidney failure. The agent has also been investigated as a treatment for gout.

Other examples of recombinant enzymes used in the clinic include **imiglucerase, taliglucerase alfa**, and **velaglucerase alfa** for the treatment of **Gaucher's disease**, which is a hereditary disease caused by a deficiency in the enzyme glucocerebrosidase, and **agalsidase beta** for the treatment of **Fabry disease**. **Alglucosidase alfa** is a recombinant enzyme that is used to treat **Pompe disease**, where the patient suffers a deficiency in alpha-glucosidase, while **elsosulfase alfa** was approved in 2014 for the treatment of **Morquio syndrome**, where the patient suffers a deficiency in *N*-acetylgalactosamine-6-sulphatase.

Collagenase clostridium histolyticum is a bacterial enzyme produced by *Clostridium histolyticum* that catalyses the degradation of collagen. It has been approved for the treatment of a condition where patients cannot straighten their fingers due to a build-up of collagen in the palms of the hand. **Ocriplasmin** was given approval in 2012 for the treatment of vitreomacular adhesion. This is a condition where there is abnormal adhesion between the vitreous gel and retina of the eye. Ocriplasmin is a truncated version of plasmin—a serine protease enzyme that dissolves the proteins responsible for this condition.

In 2015, **sebelipase alfa** was approved as an orphan drug for patients deficient in the enzyme lysosomal acid lipase (**Wolman disease**). This deficiency means that fats are not properly metabolized, leading to a build-up of fat in the digestive system, the internal organs, and blood vessels. Infants suffering from this enzyme deficiency rarely survive beyond their first year due to poor absorption through the fat-coated digestive system. It is estimated that less than 0.002% of the population in Europe suffers from this deficiency, and sebelipase alfa is the first licensed treatment. The recombinant enzyme is extracted from the egg white produced by genetically modified chickens and is the first recombinant protein to be produced in that way for human medicine.

Asfatase alfa is another recombinant enzyme that was approved in 2015 to treat a rare condition where patients have a deficit of the enzyme tissue-non-specific alkaline phosphatase—an enzyme that plays an important role in bone mineralization. Such patients suffer from bone diseases such as rickets.

Despite all these successes, many endogenous peptides and proteins have proved ineffective. This is because peptides and proteins suffer serious drawbacks such as susceptibility to digestive and metabolic enzymes, poor absorption from the gut, and rapid clearance from the body. Furthermore, proteins are large molecules which could possibly induce an adverse immunological response where the body produces antibodies against the therapeutic agent.

Solutions to some of these problems are appearing, though. It has been found that linking the polymer **polyethylene glycol** (PEG) to a protein can increase the latter's solubility and stability, as well as decreasing the likelihood of an immune response (Fig. 14.35). PEGylation, as it is called, also prevents the removal of small proteins from the blood supply by the kidneys or the reticuloendothelial system. The increased size of the PEGylated protein means that it is not filtered into the kidney nephrons and remains in the blood supply.

The PEG molecules surrounding the protein can be viewed as a kind of hydrophilic, polymeric shield which both protects and disguises the protein. The PEG polymer has the added advantage that it shows little toxicity. The enzymes L-asparaginase and **adenosine deaminase** have

been treated in this way to give protein–PEG conjugates called **pegaspargase** and **pegademase**, which have been used for the treatment of leukaemia and **severe combined immunodeficiency** (SCID) syndrome respectively. SCID is an immunological defect associated with a lack of **adenosine deaminase**. The conjugates have longer plasma half-lives than the enzymes alone and are less likely to produce an immune response. **Interferon** has similarly been PEGylated to give a preparation called **peginterferon α2b** which is used for the treatment of hepatitis C.

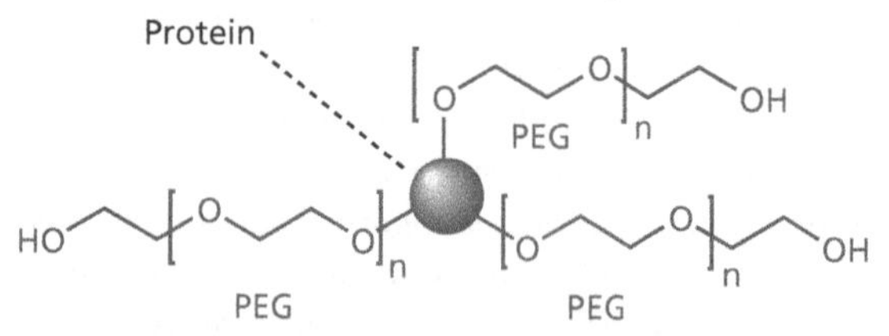

Pegvisomant is the PEGylated form of **human growth hormone antagonist** and is used for the treatment of a condition known as acromegaly which results in abnormal enlargement of the skull, jaw, hands, and feet due to the excessive production of growth hormone. **Pegfilgrastim** is the PEGylated form of **filgrastim** (**recombinant human granulocyte-colony stimulating factor**), and is used as an anticancer agent. **Pegloticase** is a recombinant porcine-like uricase that has been PEGylated and was approved in 2010 for the treatment of gout. It metabolizes uric acid to allantoin. Compared to rasburicase (see above), PEGylation increases the half-life from 8 hours to 10 or 12 days and decreases the immunogenicity of the protein.

PEGylation has also been used to protect liposomes for drug delivery (section 11.10).

14.8.3 Antibodies as drugs

Biotechnology companies are producing an ever increasing number of antibodies and antibody-based drugs with the aid of genetic engineering and monoclonal antibody technology.

Because antibodies can recognize the chemical signature of a particular cell or macromolecule, they have great potential in targeting cancer cells or viruses. Alternatively, they could be used to carry drugs or poisons to specific targets (see sections 14.4.1, 20.11.5 and 21.10). Antibodies that recognize a particular antigen are generated by exposing a mouse to the antigen so that the mouse produces the desired antibodies (known as **murine antibodies**). However, the antibodies themselves are not isolated. Antibodies are produced by cells called **B lymphocytes**, and it is a mixture of B lymphocytes that is isolated from the mouse. The next task is to find the B lymphocyte responsible for producing the desired antibody. This is done by fusing the mixture with immortal (cancerous) human B lymphocytes to produce cells called **hybridomas**. These are then separated and cultured. The culture that produces the desired antibody can be identified by its ability to bind to the antigen, and is then used to produce antibody on a large scale. Since all the cells in this culture are identical, the antibodies produced are also identical and are called **monoclonal antibodies**.

FIGURE 14.35 PEGylated protein.

There was great excitement when this technology appeared in the 1980s, which spawned an expectation that antibodies would be the magic bullet to tackle many diseases. Unfortunately, the early antibodies failed to reach the clinic, because they triggered an immune response in patients which resulted in antibodies being generated against the antibodies! In hindsight, this is not surprising; the antibodies were mouse-like in character and were identified as 'foreign' by the human immune system, resulting in the production of human anti-mouse antibodies (the **HAMA response**).

In order to tackle this problem, **chimeric antibodies** have been produced which are part human (66%) and part mouse in origin, to make them less 'foreign'. Genetic engineering has also been used to generate **humanized antibodies** which are 90% human in nature. In another approach, genetic engineering has been used to insert the human genes responsible for antibodies into mice, such that the mice (transgenic mice) produce human antibodies rather than murine antibodies when they are exposed to the antigen. As a result of these efforts, a variety of antibodies have reached the clinic and are being used as antiviral and anticancer agents (sections 20.11.5 and 21.10.1), as well as lipid-lowering agents (section 26.8.6) and drug antidotes (section 26.9.1.2). Others are being used as immunosuppressants, For example, **omalizumab** is an example of a recombinant humanized monoclonal antibody which targets **immunoglobulin E** (IgE) and was approved in 2003 for the treatment of allergic asthmatic disease. It is known that exposure to allergens results in increased levels of IgE, which triggers the release of many of the chemicals responsible for the symptoms of asthma. Omalizumab works by binding to IgE and prevents it from acting in this way. More recently, **reslizumab** was approved in 2016 for the treatment of severe asthma. Another example of an immunosuppressant is **belimumab**, which was approved in 2011 for the treatment of lupus—an autoimmune disease. The antibody downgrades the immune response by inhibiting B-cell activating factor.

A number of antibodies have been marketed that target the cytokines and cytokine receptors that play an important role in inflammation. **Adalimumab** was the first fully humanized antibody to reach the market, and was approved for the treatment of rheumatoid arthritis. It works by binding to **tumour necrosis factor** (TNF-α) which is overproduced in arthritis and causes chronic inflammation. By binding to the cytokine, the antibody prevents it interacting with its receptor. The antibody can also tag cells that are producing the chemical messenger, leading to the cell's destruction by the body's immune system. **Infliximab** is another monoclonal antibody that targets TNF-α, but this is a chimeric monoclonal antibody and there is greater chance of the body developing an immune response against it during long-term use. **Tocilizumab** is a humanized monoclonal antibody that was approved in 2010, and targets a cytokine receptor rather than a cytokine itself. To be specific, it targets the interleukin-6 receptor and prevents interleukin 6 from binding. **Secukinumab** was approved in 2015 for the treatment of psoriasis and acts by targeting the cytokine interleukin 17, while **mepolizumab** was approved in 2015 for the treatment of eosinophilic asthma and targets interleukin 5.

Vedolizumab was approved in 2014 for the treatment of ulcerative colitis and Crohn's disease. It binds to an integrin and blocks it from promoting inflammation. **Natalizumab** was approved in 2004 for the treatment of multiple sclerosis.

Ranibizumab is a fragment of the monoclonal antibody **bevacizumab** used in cancer therapy (section 21.10.1), and is used for the treatment of a condition that results in age-related vision loss. **Denosumab** is a fully humanized monoclonal antibody that was approved in 2009 for the treatment of osteoporosis, while **raxibacumab** is a human monoclonal antibody that was approved in 2012 for the treatment of inhaled anthrax.

Work on the large-scale production of antibodies has also been continuing. They have traditionally been produced using hybridoma cells in bioreactors, but more recently companies have been looking at the possibility of using transgenic animals in order to collect antibodies in milk. Another possibility is to harvest transgenic plants which produce the antibody in their leaves or seeds.

A different approach to try and prevent antibodies producing an immune response has been to treat them with PEG (section 14.8.2). Unfortunately, this tends to be counterproductive, as it prevents the antibody acting out its role as a targeting molecule. However, controlling the PEGylation such that it only occurs on the thiol group of cysteine residues could be beneficial, as it would limit the number of PEG molecules attached and make it more likely that the antibody remains functional.

An interesting idea under investigation involves coating the inside of nanotubes with antibodies that can recognize infectious agents such as viruses. It is hoped that such nanotubes could be administered to trap and remove viruses from the blood supply.

14.9 Peptides and peptidomimetics in drug design

Endogenous peptides and proteins serve as highly important lead compounds for the design of novel drugs. Current examples include renin inhibitors (Case study 8), protease inhibitors (section 20.7.4), LHRH agonists (section 21.4.2), matrix metalloproteinase

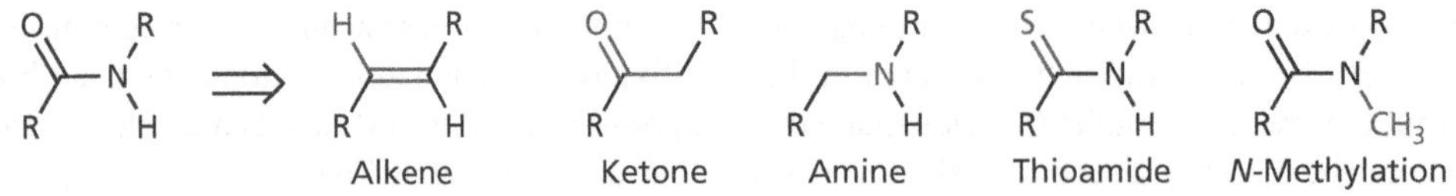

FIGURE 14.36 Examples of functional groups that might be used to replace a peptide bond.

inhibitors (section 21.7.1), and enkephalin analogues (section 24.8.2). Peptides will continue to be important lead compounds because many of the new targets in medicinal chemistry involve peptides as receptor ligands or as enzyme substrates; for example the protein kinases. Consequently, drugs which are designed from these lead compounds are commonly peptide-like in nature. The pharmacokinetic properties of these 'first-generation' drugs are often unsatisfactory, and so various strategies have been developed to try and improve bioavailability and attain more acceptable levels in the blood supply. This usually involves strategies aimed at disguising or reducing the peptide nature of the lead compound to generate a structure which is more easily absorbed from the gastrointestinal tract, and is more resistant to digestive and metabolic enzymes. Such analogues are known as **peptidomimetics**.

14.9.1 Peptidomimetics

One approach that is used to increase bioavailability is to replace a chemically or enzymatically susceptible peptide bond with a functional group that is either more stable to hydrolytic attack by peptidase enzymes, or binds less readily to the relevant active sites. For example, a peptide bond might be replaced by an alkene (Fig. 14.36). If the compound retains activity, then the alkene represents a **bio-isostere** for the peptide link. An alkene has the advantage that it mimics the double bond nature of a peptide bond and is not a substrate for peptidases. However, the peptide bonds in lead compounds are often involved in hydrogen bond interactions with the target binding site, where the NH acts as a hydrogen bond donor and the carbonyl C=O acts as a hydrogen bond acceptor. Replacing both of these groups may result in a significant drop in binding strength. Therefore, an alternative approach might be to replace the amide with a ketone or an amine, such that only one possible interaction is lost. The problem now is that the double bond nature of the original amide group is lost, resulting in greater chain flexibility and a possible drop in binding affinity (see section 13.3.9). A thioamide group is another option. This group retains the planar shape of the amide, and the NH moiety can still act as a hydrogen bond donor. The sulphur is a poor hydrogen bond acceptor, but this could be advantageous if the original

carbonyl oxygen forms a hydrogen bond to the active site of peptidase enzymes.

A different approach is to retain the amide, but to protect or disguise it. One strategy that has been used successfully is to methylate the nitrogen of the amide group. The methyl group may help to protect the amide from hydrolysis by acting as a steric shield, or prevent an important hydrogen bonding interaction taking place between the NH of the original amide and the active site of the peptidase enzyme that would normally hydrolyse it.

A second strategy is to replace an L-amino acid with the corresponding D-enantiomer (Fig. 14.37). Such a move alters the relative orientation of the side chain with respect to the rest of the molecule and can make the molecule unrecognizable to digestive or metabolic enzymes, especially if the side chain is involved in binding interactions. The drawback to this strategy is that the resulting peptidomimetic may become unrecognizable to the desired target as well.

A third strategy is to replace natural amino acid residues with unnatural ones. This is a tactic that has worked successfully in structure-based drug design where the binding interactions of the peptidomimetic and a protein target are studied by X-ray crystallography and molecular modelling. The idea is to identify binding subsites in the target binding site into which various amino acid side chains fit and bind. The residues are then replaced by groups which are designed to fit the subsites better, but which are not found on natural amino acids. This increases the binding affinity of the peptidomimetic to the target binding site, and, at the same time, makes it less recognizable to digestive and metabolic enzymes.

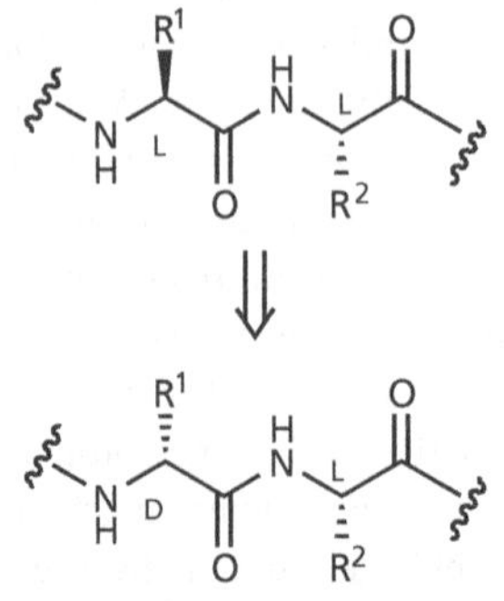

FIGURE 14.37 Replacing an L-amino acid with a D-amino acid. The common L-amino acids have the *R*-configuration except for L-cysteine which has the *S*-configuration.

For example, the lead compound for the antiviral drug **saquinavir** contained an L-proline residue that occupied a hydrophobic subsite of a viral protease enzyme. The proline residue was replaced by a decahydroisoquinoline ring which filled the hydrophobic subsite more fully, resulting in better binding interactions (Fig. 14.38; see also section 20.7.4.3).

It is even possible to design extended groups which fill two different subsites (Fig. 14.39). This means that the peptidomimetic can be pruned to a smaller molecule. The resulting decrease in molecular weight often leads to better absorption (see also Case study 8 and sections 20.7.4.6 and 20.7.4.7).

Peptidomimetics are often hydrophobic in nature, and this can pose a problem because poor water solubility may result in poor oral absorption. Water solubility can be increased by increasing the polarity of residues. For example, an aromatic ring could be replaced by a pyridine ring. However, it is important that this group is not involved in any binding interactions with the target and remains exposed to the surrounding water medium when the peptidomimetic is bound (Fig. 14.40). Otherwise, it would have to be desolvated and this would carry an energy penalty that would result in a decreased binding affinity.

Another potential problem with peptide lead compounds is that they are invariably flexible molecules with a large number of freely rotatable bonds. Flexibility has been shown to be detrimental to oral bioavailability (section 11.3), and so rigidification tactics (section 13.3.9) may well be beneficial.

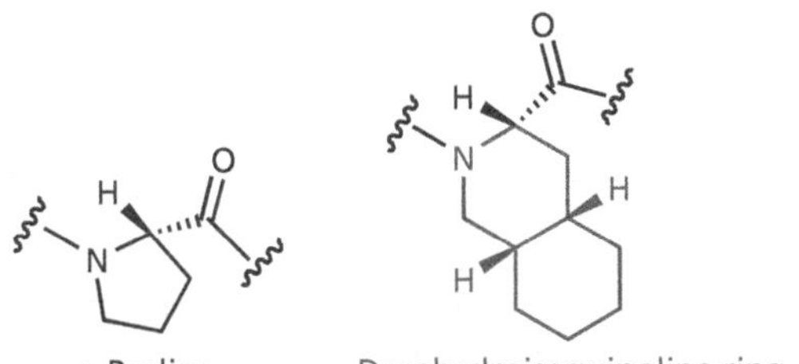

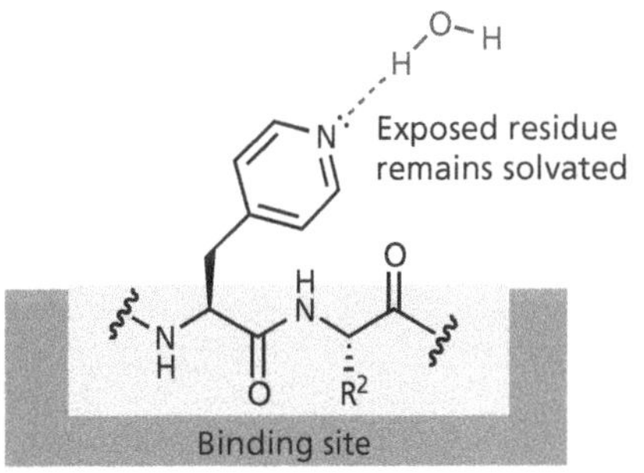

FIGURE 14.40 Altering exposed residues to increase water solubility.

The structure-based design of various peptidomimetic enzyme inhibitors is described in sections 20.7.4 and 20.10.1, as well as Case study 10. These examples illustrate many of the principles described above. Crystal structures of peptide lead compounds bound to their target binding site are invaluable in identifying what kind of modifications are likely to be successful in achieving a particular strategy. Crystal structures also allow docking studies to be carried out to see whether planned analogues are likely to bind or not. It is even possible to use docking alongside virtual screening to modify peptide lead compounds (section 17.13).

Finally, there is current research into designing structures which mimic particular features of protein secondary structure such as α-helices, β-sheets and β-turns (section 10.5). The goal here is to design a stable molecular scaffold that contains substituents capable of mimicking the side chains of amino acids as they would be positioned in common protein features such as α-helices. This might be useful in designing peptidomimetics that mimic peptide neurotransmitters or peptide hormones. For example, it is found that such messengers adopt a helical conformation when they bind to their receptor. 1,1,6-Trisubstituted indanes have been designed to mimic three consecutive amino acid side chains in an α-helix (Fig. 14.41).

FIGURE 14.38 Replacing a natural residue with an unnatural one.

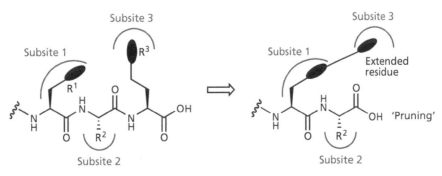

FIGURE 14.39 Extended residues.

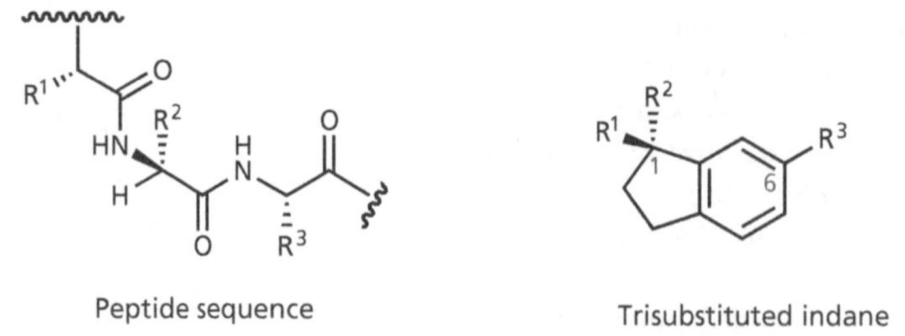

FIGURE 14.41 Trisubstituted indanes as a peptidomimetic for a tripeptide sequence in an α-helix.

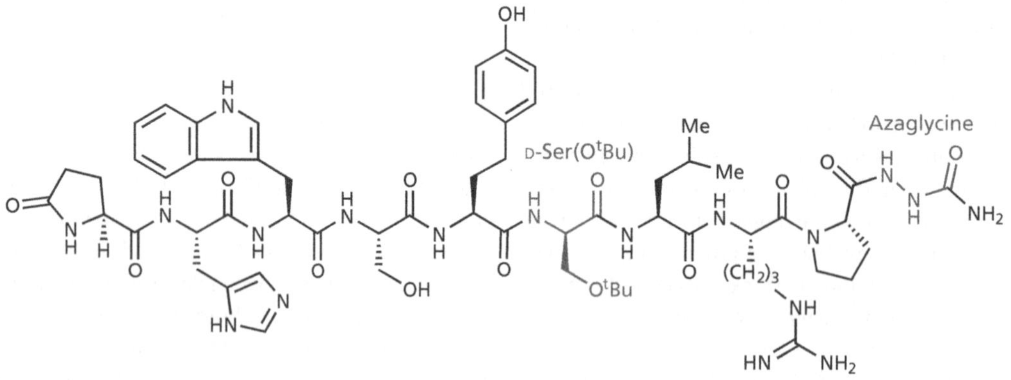

FIGURE 14.42 Goserelin (Zoladex). Moieties in blue increase metabolic resistance and receptor affinity.

14.9.2 Peptide drugs

As stated above, there is often a reluctance to use peptides as drugs because of the many pharmacokinetic difficulties that can be encountered, but this does not mean that peptide drugs have no role to play in medicinal chemistry. For example, the immunosuppressant **ciclosporin** can be administered orally (section 11.3). Another important peptide drug is **goserelin** (Fig. 14.42), which is administered as a subcutaneous implant and is used against breast and prostate cancers, earning $700 million dollars a year for its maker (section 21.4.2). In 2003, **enfuvirtide (Fuzeon)** was approved as the first of a new class of anti-HIV drugs (section 20.7.5). It is a polypeptide of 36 amino acids which is injected subcutaneously and offers another weapon in the combination therapies used against HIV. **Teriparatide**, which was mentioned in section 14.8.2, is also administered by subcutaneous injection. Peptide drugs can be useful if one chooses the right disease and method of administration.

14.10 Oligonucleotides as drugs

Oligonucleotides are being studied as **antisense drugs** and **aptamers**. The rationale and therapeutic potential of these agents are described in sections 9.7.2 and 10.5. However, there are disadvantages to the use of oligonucleotides as drugs, as they are rapidly degraded by enzymes called **nucleases**. They are also large and highly charged, and are not easily absorbed through cell membranes. Attempts to stabilize these molecules, and to reduce their polarity, have involved modifying the phosphate linkages in the sugar–phosphate backbone. For example, phosphorothioates and methylphosphonates have been extensively studied, and oligonucleotides containing these linkages show promise as therapeutic agents (Fig. 14.43). An antisense oligonucleotide with such a modified backbone has been approved as an antiviral drug (section 20.6.3), as has one for the treatment of high cholesterol levels (section 26.8.4). Alterations to the sugar moiety have also been tried. For example, placing a methoxy group at position 2′, or using the α-anomer of a deoxyribose sugar, increases resistance to nucleases. Bases have also been modified to improve and increase the number of hydrogen bonding interactions with target nucleic acids.

The biopharmaceutical company Genta has developed an antisense drug called **oblimersen** which consists of 18 deoxynucleotides linked by a phosphorothioate backbone. It binds to the initiation codon of the messenger RNA molecule carrying the genetic instructions for **Bcl-2**. Bcl-2 is a protein which suppresses cell death

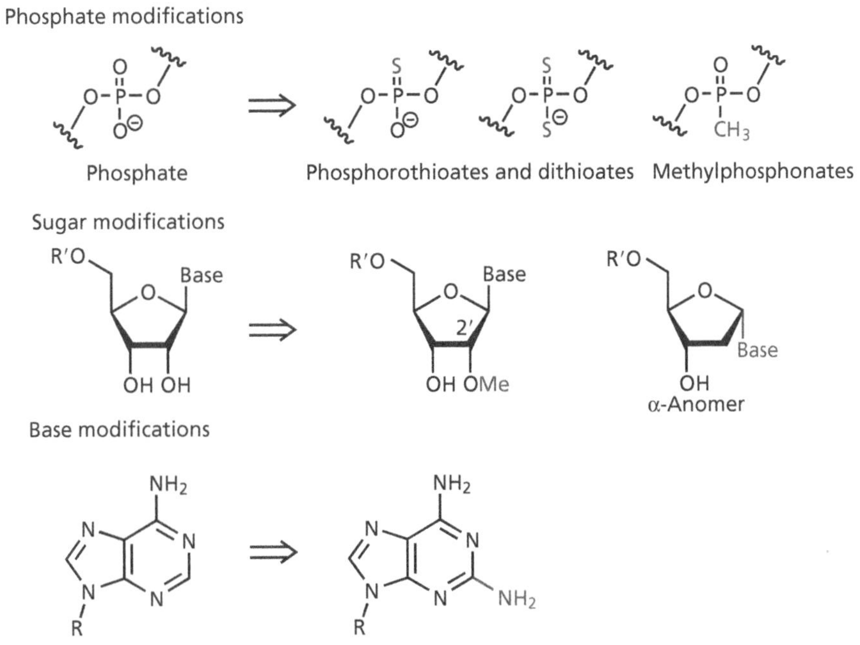

Phosphate modifications

Phosphate   Phosphorothioates and dithioates   Methylphosphonates

Sugar modifications

Base modifications

FIGURE 14.43 Modifications on oligonucleotides.

(apoptosis) (section 21.1.7), and so suppressing its synthesis will increase the chances of apoptosis taking place when chemotherapy or radiotherapy is being used for the treatment of cancer. The drug is currently undergoing Phase III clinical trials in combination with the anticancer drugs **docetaxel** and **irenotecan**.

Phosphorothioate oligonucleotides are also being investigated which will target the genetic instructions for **Raf** and **PKCγ**, two proteins which are involved in signal transduction pathways (section 5.4). These oligonucleotides also have potential as anticancer drugs.

KEY POINTS

- Neurotransmitters are not effective as drugs as they have a short lifetime in the body, and have poor selectivity for the various types and subtypes of a particular target.

- Hormones are more suitable as drugs and several are used clinically. Others are susceptible to digestive or metabolic enzymes, and show poor absorption when taken orally. Adverse immune reactions are possible.

- Peptides and proteins generally suffer from poor absorption or metabolic susceptibility. Peptidomimetics are compounds that are derived from peptide lead compounds, but have been altered to disguise their peptide character.

- Many of the body's hormones are peptides and proteins and can be produced by recombinant DNA techniques. However, there are several disadvantages in using such compounds as drugs.

- Antibodies are proteins which are important to the body's immune response, and can identify foreign cells or macromolecules, marking them out for destruction. They have been used therapeutically and can also be used to carry drugs to specific targets.

- Oligonucleotides are susceptible to metabolic degradation, but can be stabilized by modifying the sugar–phosphate backbone so that they are no longer recognized by relevant enzymes.

- Antisense molecules have been designed to inhibit the mRNA molecules that code for the proteins which suppress cell death.

QUESTIONS

1. Suggest a mechanism by which methenamine (Fig. 14.31) is converted to formaldehyde under acid conditions.

2. Suggest a mechanism by which ampicillin (Fig. 14.32) decomposes in concentrated solution.

3. Carbidopa (Fig. 14.34) protects levodopa from decarboxylation in the peripheral blood supply, but is too polar to cross the blood–brain barrier into the central nervous system. Carbidopa is reasonably similar in structure to levodopa, so why can it not mimic levodopa and cross the blood–brain barrier by means of a transport protein?

4. Acetylcholine (Fig. 4.3) is a neurotransmitter that is susceptible to chemical and enzymatic hydrolysis. Suggest strategies that could be used to stabilize the ester group of acetylcholine, and show the sort of analogues which might have better stability.

5. Decamethonium is a neuromuscular blocking agent which requires both positively charged nitrogen groups to be present. Unfortunately, it is slowly metabolized and lasts too long in the body. Suggest analogues which might be expected to be metabolized more quickly and lead to inactive metabolites.

6. Miotine has been used in the treatment of a muscle-wasting disease, but there are side effects because a certain amount of the drug enters the brain. Suggest how one might modify the structure of miotine to eliminate this side effect.

7. The oral bioavailability of the antiviral drug aciclovir is only 15–30%. Suggest why this may be the case, and how one might increase the bioavailability of this drug.

Decamethonium

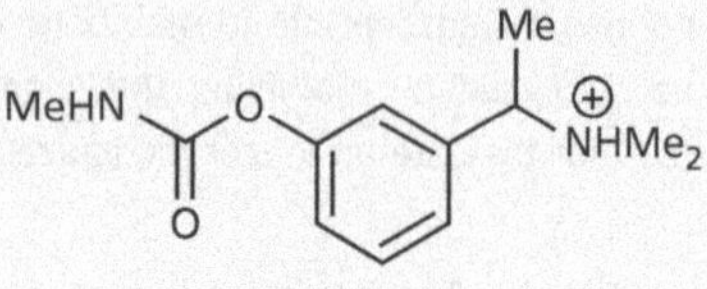

Miotine

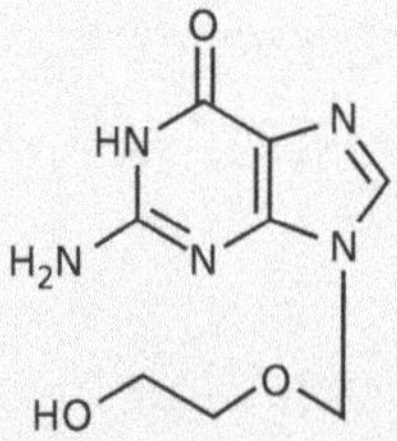

Aciclovir

8. CGP 52411 is a useful inhibitor of a protein kinase enzyme. Studies on structure–activity relationships demonstrate that substituents on the aromatic rings such as Cl, Me, or OH are bad for activity. Drug metabolism studies also show that *para*-hydroxylation occurs to produce inactive metabolites. How would you modify the structure to protect it from metabolism?

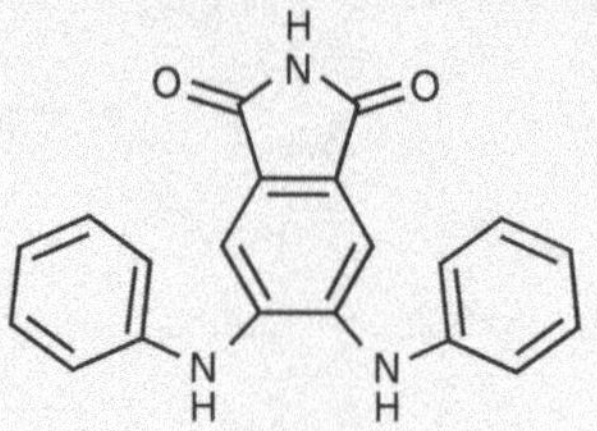

CGP52411

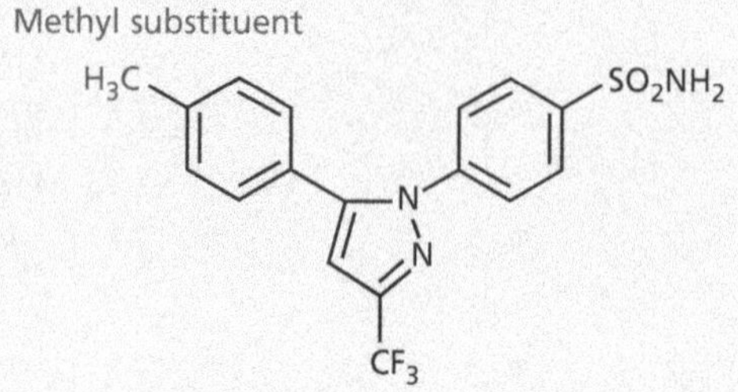

Methyl substituent

Celecoxib

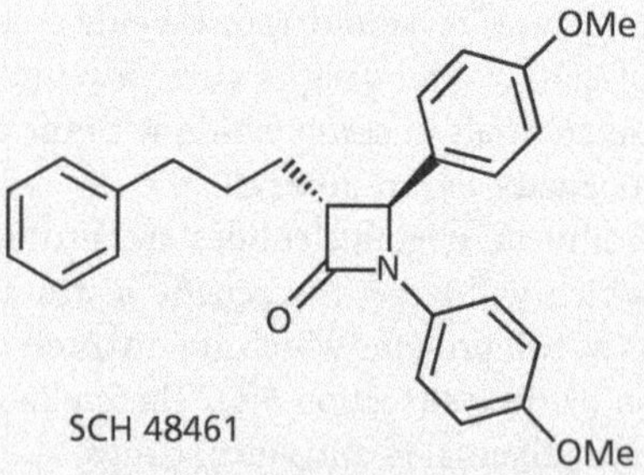

SCH 48461

9. Celecoxib is a COX-2 inhibitor and contains a methyl substituent on the phenyl ring. It is known that inhibitory activity increases if this methyl substituent is not present, or if it is replaced with a chloro substituent. However, neither of these analogues were used clinically. Why not?

10. SCH 48461 has been found to lower cholesterol levels by inhibiting cholesterol absorption. Unfortunately, it is susceptible to metabolism. Identify the likely metabolic reactions which this molecule might undergo, and what modifications could be made to reduce metabolic susceptibility.

Multiple-choice questions are available on the Online Resource Centre at www.oxfordtextbooks.co.uk/orc/patrick6e/

FURTHER READING

Berg, C., Neumeyer, K., and Kirkpatrick, P. (2003) Teriparatide. *Nature Reviews Drug Discovery*, **2**, 257–8.

Bolgnesi, M. L., et al. (2009) Alzheimer's disease: new approaches to drug discovery. *Current Opinions in Chemical Biology*, **13**, 303–8.

Burke, M. (2002) Pharmas market. *Chemistry in Britain*, June, 30–2 (antibodies).

Duncan, R. (2003) The dawning era of polymer therapeutics. *Nature Reviews Drug Discovery*, **2**, 347–60.

Ezzell, C. (2001) Magic bullets fly again. *Scientific American*, October, 28–35 (antibodies).

Harris, J. M., and Chess, R. B. (2003) Effect of pegylation on pharmaceuticals. *Nature Reviews Drug Discovery*, **2**, 214–21.

Herr, R. J. (2002) 5-Substituted-1H-tetrazoles as carboxylic acid isosteres: medicinal chemistry and synthetic methods. *Bioorganic and Medicinal Chemistry*, **10**, 3379–93.

Matthews, T., et al. (2004) Enfuvirtide: the first therapy to inhibit the entry of HIV-1 into host CD4 lymphocytes. *Nature Reviews Drug Discovery*, **3**, 215–25.

Moreland, L., Bate, G., and Kirkpatrick, P. (2006) Abatacept. *Nature Reviews Drug Discovery*, **5**, 185–6.

Opalinska, J. B., and Gewirtz, A. M. (2002) Nucleic-acid therapeutics: basic principles and recent applications. *Nature Reviews Drug Discovery*, **1**, 503–14.

Pardridge, W. M. (2002) Drug and gene targeting to the brain with molecular Trojan horses. *Nature Reviews Drug Discovery*, **1**, 131–9.

Reichert, J. M., and Dewitz, M. C. (2006) Anti-infective monoclonal antibodies: perils and promise of development. *Nature Reviews Drug Discovery*, **5**, 191–5.

Rotella, D. P. (2002) Phosphodiesterase 5 inhibitors: current status and potential applications. *Nature Reviews Drug Discovery*, **1**, 674–82.

Titles for general further reading are listed on p.845.

18 Quantitative structure–activity relationships (QSAR)

In Chapters 13 and 14 we studied the various strategies which can be used in the design of drugs. Several of these strategies involved a change in shape such that the new drug had a better 'fit' for its target binding site. Other strategies involved a change in functional groups or substituents such that the drug's pharmacokinetics or binding site interactions were improved. These latter strategies often involved the synthesis of analogues containing a range of substituents on aromatic or heteroaromatic rings or accessible functional groups. The number of possible analogues that could be made is infinite if we were to try and synthesize analogues with every substituent and combination of substituents possible. Therefore, it is clearly advantageous if a rational approach can be followed in deciding which substituents to use. The quantitative structure–activity relationship (QSAR) approach has proved extremely useful in tackling this problem.

The QSAR approach attempts to identify and quantify the physicochemical properties of a drug and to see whether any of these properties has an effect on the drug's biological activity. If such a relationship holds true, an equation can be drawn up which quantifies the relationship and allows the medicinal chemist to say with some confidence which property (or properties) has an important role in the pharmacokinetics or mechanism of action of the drug. It also allows the medicinal chemist some level of prediction. By quantifying physicochemical properties, it should be possible to calculate in advance what the biological activity of a novel analogue might be. There are two advantages to this. Firstly, it allows the medicinal chemist to target efforts on analogues which should have improved activity and, thus, cut down the number of analogues that have to be made. Secondly, if an analogue is discovered which does not fit the equation, it implies that some other feature is important and provides a lead for further development.

What are these physicochemical features that we have mentioned?

Essentially, they could be any structural, physical, or chemical property of a drug. Clearly, any drug will have a large number of such properties and it would be a Herculean task to quantify and relate them all to biological activity at the same time. A simple, more practical approach is to consider one or two physicochemical properties of the drug and to vary these while attempting to keep other properties constant. This is not as simple as it sounds, as it is not always possible to vary one property without affecting another. Nevertheless, there have been numerous examples where the approach has worked.

It is important that the QSAR method is used properly and in relevant situations. Firstly, the compounds studied must be structurally related, act at the same target, and have the same mechanism of action. Secondly, it is crucial that the correct testing procedures are used. *In vitro* tests carried out on isolated enzymes are relevant for a QSAR study since the activities measured for different inhibitors are related directly to how each compound binds to the active site. *In vivo* tests carried out to measure the physiological effects of enzyme inhibitors are not valid, however, since both pharmacodynamic and pharmacokinetic factors come into play. This makes it impossible to derive a sensible QSAR equation.

18.1 Graphs and equations

In the simplest situation, a range of compounds is synthesized in order to vary one physicochemical property (e.g. log P) and to test how this affects the biological activity (log $1/C$) (we will come to the meaning of log $1/C$ and log P in due course). A graph is then drawn to plot the biological activity on the y-axis versus the physicochemical feature on the x-axis (Fig. 18.1).

It is then necessary to draw the best possible line through the data points on the graph. This is done by a procedure known as '**linear regression analysis by the least squares method**'. This is quite a mouthful and can produce a glazed expression on any chemist who is not

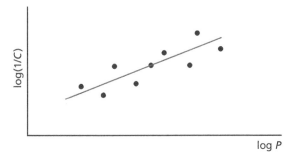

FIGURE 18.1 Biological activity versus log P.

mathematically orientated. In fact, the principle is quite straightforward. If we draw a line through a set of data points, most of the points will be scattered on either side of the line. The best line will be the one closest to the data points. To measure how close the data points are, vertical lines are drawn from each point (Fig. 18.2). These verticals are measured and then squared in order to eliminate the negative values. The squares are then added up to give a total (the sum of the squares). The best line through the points will be the line where this total is a minimum. The equation of the straight line will be $y = k_1 x + k_2$ where k_1 and k_2 are constants. By varying k_1 and k_2, different equations are obtained until the best line is obtained. This whole process can be speedily done using relevant software.

The next stage in the process is to see whether the relationship is meaningful. As any good politician knows, numbers can be used to 'prove' whatever you want them to prove. Therefore, a proper statistical analysis has to be carried out to assess the validity of any QSAR equation and quantify the goodness of fit for any plot. The **regression** or **correlation coefficient** (r) is a measure of how well the physicochemical parameters present in the equation explains the observed variance in activity. An explanation of how r is derived is given in Appendix 3. For a perfect fit, $r = 1$, in which case the observed activities would be the same as those calculated by the equation. Such perfection is impossible with biological data and so r values greater than 0.9 are considered

acceptable. The regression coefficient is often quoted as r^2, in which case values over 0.8 are considered a good fit. If r^2 is multiplied by 100 it indicates the percentage variation in biological activity that is accounted for by the physicochemical parameters used in the equation. Thus, an r^2 value of 0.85 signifies that 85% of the variation in biological activity is accounted for by the parameters used. There are dangers in putting too much reliance on r, as the value obtained takes no account of the number of compounds (n) involved in the study and it is possible to obtain higher values of r by increasing the number of compounds tested.

Therefore, another statistical measure for the goodness of fit should be quoted alongside r. This is the **standard error of estimate** or the **standard deviation** (s). Ideally, s should be zero, but this would assume there were no experimental errors in the experimental data or the physicochemical parameters. In reality, s should be small, but not smaller than the standard deviation of the experimental data. It is therefore necessary to know the latter to assess whether the value of s is acceptably low. Appendix 3 shows how s is obtained and demonstrates that the number of compounds (n) in the study influences the value of s.

Statistical tests called **Fisher's F-tests** are often quoted (Appendix 3). These tests are used to assess the significance of the coefficients k for each parameter in the QSAR equation. Normally p values (derived from the F-test) should be less than or equal to 0.05 if the parameter is significant. If this is not the case, the parameter should not be included in the QSAR equation.

18.2 Physicochemical properties

Many physical, structural, and chemical properties have been studied by the QSAR approach, but the most common are hydrophobic, electronic, and steric properties. This is because it is possible to quantify these effects. Hydrophobic properties can be easily quantified for complete molecules or for individual substituents. On the other hand, it is more difficult to quantify electronic and steric properties for complete molecules, and this is only really feasible for individual substituents.

Consequently, QSAR studies on a variety of totally different structures are relatively rare and are limited to studies on hydrophobicity. It is more common to find QSAR studies being carried out on compounds of the same general structure, where substituents on aromatic rings or accessible functional groups are varied. The QSAR study then considers how the hydrophobic, electronic, and steric properties of the substituents affect biological activity. The three most studied physicochemical properties are now considered in some detail.

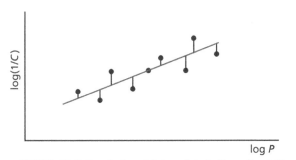

FIGURE 18.2 Proximity of data points to line of best fit.

Test your understanding and practise your molecular modelling with Exercise 18.1 on the Online Resource Centre at www.oxfordtextbooks.co.uk/orc/patrick6e/. You might also find Exercises 25.3 and 25.4 useful at this point.

18.2.1 Hydrophobicity

The hydrophobic character of a drug is crucial to how easily it crosses cell membranes (section 11.3) and may also be important in receptor interactions. Changing substituents on a drug may well have significant effects on its hydrophobic character and, hence, its biological activity. Therefore, it is important to have a means of predicting this quantitatively.

18.2.1.1 The partition coefficient (P)

The hydrophobic character of a drug can be measured experimentally by testing the drug's relative distribution in an n-octanol/water mixture. Hydrophobic molecules will prefer to dissolve in the n-octanol layer of this two-phase system, whereas hydrophilic molecules will prefer the aqueous layer. The relative distribution is known as the partition coefficient (P) and is obtained from the following equation:

$$P = \frac{\text{Concentration of drug in octanol}}{\text{Concentration of drug in aqueous solution}}$$

Hydrophobic compounds have a high P value, whereas hydrophilic compounds have a low P value.

Varying substituents on the lead compound will produce a series of analogues having different hydrophobicities and, therefore, different P values. By plotting these P values against the biological activity of these drugs, it is possible to see if there is any relationship between the two properties. The biological activity is normally expressed as $1/C$, where C is the concentration of drug required to achieve a defined level of biological activity. The reciprocal of the concentration ($1/C$) is used, since more active drugs will achieve a defined biological activity at lower concentration.

The graph is drawn by plotting $\log(1/C)$ versus $\log P$. In studies where the range of the $\log P$ values is restricted to a small range (e.g. $\log P = 1$–4), a straight-line graph is obtained (Fig. 18.1) showing that there is a relationship between hydrophobicity and biological activity. Such a line would have the following equation:

$$\log\left(\frac{1}{C}\right) = -k_1 \log P + k_2$$

For example, the binding of drugs to serum albumin is determined by their hydrophobicity and a study of 42 compounds resulted in the following equation:

$$\log\left(\frac{1}{C}\right) = 0.75 \log P + 230 \quad (n = 42, r = 0.960, s = 0.159)$$

The equation shows that serum albumin binding increases as $\log P$ increases. In other words, hydrophobic drugs bind more strongly to serum albumin than hydrophilic drugs. Knowing how strongly a drug binds to serum albumin can be important in estimating effective dose levels for that drug. When bound to serum albumin, the drug cannot bind to its receptor, and so the dose levels for the drug should be based on the amount of unbound drug present in the circulation. The equation above allows us to calculate how strongly drugs of similar structure will bind to serum albumin, and gives an indication of how 'available' they will be for receptor interactions. The r value of 0.96 is close to 1, which shows that the line resulting from the equation is a good fit. The value of r^2 is 92%, which indicates that 92% of the variation in serum albumin binding can be accounted for by the different hydrophobicities of the drugs tested. This means that 8% of the variation is unaccounted for, partly as a result of the experimental errors involved in the measurements.

Despite such factors as serum albumin binding, it is generally found that increasing the hydrophobicity of a lead compound results in an increase in biological activity. This reflects the fact that drugs have to cross hydrophobic barriers such as cell membranes in order to reach their target. Even if no barriers are to be crossed (e.g. in *in vitro* studies), the drug has to interact with a target system, such as an enzyme or receptor, where the binding site is more hydrophobic than the surface. Therefore, increasing hydrophobicity also aids the drug in binding to its target site.

This might imply that increasing $\log P$ should increase the biological activity ad infinitum. In fact, this does not happen. There are several reasons for this. For example, the drug may become so hydrophobic that it is poorly soluble in the aqueous phase. Alternatively, it may be 'trapped' in fat depots and never reach the intended site. Finally, hydrophobic drugs are often more susceptible to metabolism and subsequent elimination.

A straight-line relationship between $\log P$ and biological activity is observed in many QSAR studies because the range of $\log P$ values studied is often relatively narrow. For example, the study carried out on serum albumin binding was restricted to compounds having $\log P$ values in the range 0.78–3.82. If these studies were to be extended to include compounds with very high $\log P$ values, then we would see a different picture. The graph would be parabolic, as shown in Fig. 18.3. Here, the biological activity increases as $\log P$ increases until a maximum value is obtained. The value of $\log P$ at the maximum ($\log P^0$) represents the optimum partition coefficient for

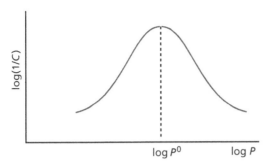

FIGURE 18.3 Parabolic curve of log(1/C) versus log P.

biological activity. Beyond that point, an increase in log P results in a decrease in biological activity.

If the partition coefficient is the only factor influencing biological activity, the parabolic curve can be expressed by the equation:

$$\log\left(\frac{1}{C}\right) = -k_1 (\log P)^2 + k_2 \log P + k_3$$

Note that the $(\log P)^2$ term has a minus sign in front of it. When P is small, the $(\log P)^2$ term is very small and the equation is dominated by the log P term. This represents the first part of the graph where activity increases with increasing P. When P is large, the $(\log P)^2$ term becomes more significant and eventually 'overwhelms' the log P term. This represents the last part of the graph where activity drops with increasing P. k_1, k_2, and k_3 are constants and can be determined by a suitable software program.

There are relatively few drugs where activity is related to the log P factor alone. Such drugs tend to operate in cell membranes where hydrophobicity is the dominant feature controlling their action. The best examples of drugs which operate in cell membranes are the general anaesthetics. Although they also bind to GABA$_A$ receptors, general anaesthetics are thought to function by entering the central nervous system and 'dissolving' into cell membranes where they affect membrane structure and nerve function. In such a scenario, there are no specific drug–receptor interactions and the mechanism of the drug is controlled purely by its ability to enter cell membranes (i.e. its hydrophobic character). The general anaesthetic activity of a range of ethers was found to fit the following parabolic equation:

$$\log\left(\frac{1}{C}\right) = -0.22 \, (\log P)^2 + 1.04 \log P + 2.16$$

According to this equation, anaesthetic activity increases with increasing hydrophobicity (P), as determined by the log P factor. The negative $(\log P)^2$ factor shows that the relationship is parabolic and that there is an optimum

value for log P (log P^0) beyond which increasing hydrophobicity causes a decrease in anaesthetic activity. With this equation, it is now possible to predict the anaesthetic activity of other compounds, given their partition coefficients. However, there are limitations. The equation is derived purely for anaesthetic ethers and is not applicable to other structural types of anaesthetics. This is generally true in QSAR studies. The procedure works best if it is applied to a series of compounds which have the same general structure.

However, QSAR studies *have* been carried out on other structural types of general anaesthetics, and a parabolic curve has been obtained in each case. Although, the constants for each equation are different, it is significant that the optimum hydrophobicity (represented by log P^0) for anaesthetic activity is close to 2.3, regardless of the class of anaesthetic being studied. This finding suggests that all general anaesthetics are operating in a similar fashion, controlled by the hydrophobicity of the structure.

Because different anaesthetics have similar log P^0 values, the log P value of any compound can give some idea of its potential potency as an anaesthetic. For example, the log P values of the gaseous anaesthetics **ether, chloroform**, and **halothane** are 0.98, 1.97, and 2.3 respectively. Their anaesthetic activity increases in the same order.

As general anaesthetics have a simple mechanism of action based on the efficiency with which they enter the central nervous system, it implies that log P values should give an indication of how easily any compound can enter the central nervous system. In other words, compounds having a log P value close to 2 should be capable of entering the central nervous system efficiently. This is generally found to be true. For example, the most potent barbiturates for sedative and hypnotic activity are found to have log P values close to 2.

As a rule of thumb, drugs which are to be targeted for the central nervous system should have a log P value of approximately 2. Conversely, drugs which are designed to act elsewhere in the body should have log P values significantly different from 2 in order to avoid possible central nervous system side effects (e.g. drowsiness) (see Box 18.1).

18.2.1.2 The substituent hydrophobicity constant (π)

We have seen how the hydrophobicity of a compound can be quantified using the partition coefficient P. In order to get P, we have to measure it experimentally and that means that we have to synthesize the compounds. It would be much better if we could calculate P theoretically and decide in advance whether the compound is worth synthesizing. QSAR would then allow us to target the most promising-looking structures. For example, if we were planning to synthesize a range of barbiturate

BOX 18.1 Altering log *P* to remove central nervous system side effects

The cardiotonic agent (I) was found to produce 'bright visions' in some patients, which implied that it was entering the central nervous system. This was supported by the fact that the log *P* value of the drug was 2.59. In order to prevent the drug entering the central nervous system, the 4-OMe group was replaced by a 4-S(O)Me group. This particular group is approximately the same size as the methoxy group, but more hydrophilic. The log *P* value of the new drug (**sulmazole**) was found to be 1.17. The drug was now too hydrophilic to enter the central nervous system and was free of central nervous system side effects.

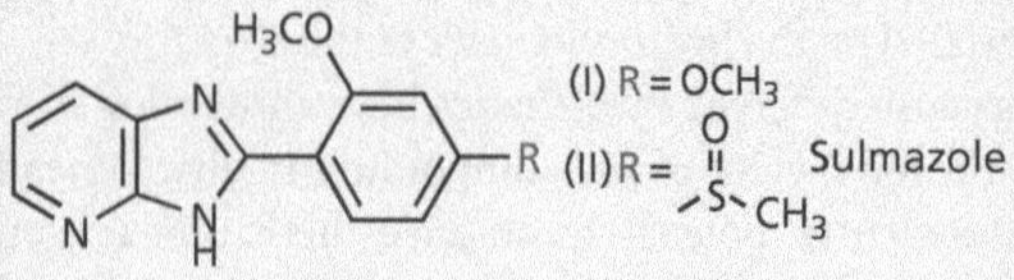

FIGURE 1 Cardiotonic agents.

structures, we could calculate log *P* values for them all and concentrate on the structures which had log *P* values closest to the optimum log P^0 value for barbiturates.

Partition coefficients can be calculated by knowing the contribution that various substituents make to hydrophobicity. This contribution is known as the **substituent hydrophobicity constant** (π) and is a measure of how hydrophobic a substituent is relative to hydrogen. The value can be obtained as follows. Partition coefficients are measured experimentally for a standard compound such as benzene, with and without a substituent (X). The hydrophobicity constant (π_X) for the substituent (X) is then obtained using the following equation:

$$\pi_X = \log P_X - \log P_H$$

where P_H is the partition coefficient for the standard compound and P_X is the partition coefficient for the standard compound with the substituent.

A positive value of π indicates that the substituent is more hydrophobic than hydrogen; a negative value indicates that the substituent is less hydrophobic. The π values for a range of substituents are shown in Table 18.1. These π values are characteristic for the substituent and can be used to calculate how the partition coefficient of a drug would be affected if these substituents

were present. The *P* value for the lead compound would have to be measured experimentally, but, once that is known, the *P* value for analogues can be calculated quite simply.

As an example, consider the log *P* values for benzene (log *P* = 2.13), chlorobenzene (log *P* = 2.84), and benzamide (log *P* = 0.64) (Fig. 18.4). Benzene is the parent compound, and the substituent constants for Cl and CONH₂ are 0.71 and –1.49 respectively. Having obtained these values, it is now possible to calculate the theoretical log *P* value for *meta*-chlorobenzamide:

$$\log P_{(chlorobenzamide)} = \log P_{(benzene)} + \pi_{Cl} + \pi_{CONH_2}$$
$$= 2.13 + 0.71 + (-1.49)$$
$$= 1.35$$

The observed log *P* value for this compound is 1.51.

It should be noted that π values for aromatic substituents are different from those used for aliphatic substituents. Furthermore, neither of these sets of π values are in fact true constants, and they are accurate only for the structures from which they were derived. They can be used as good approximations when studying other structures, but it is possible that the values will have to be adjusted in order to get accurate results.

In order to distinguish calculated log *P* values from experimental ones, the former are referred to as **Clog *P*** values. There are also software programs which will calculate Clog *P* values for a given structure.

Test your understanding and practise your molecular modelling with Exercise 18.2 on the Online Resource Centre at www.oxfordtextbooks.co.uk/orc/patrick6e/

18.2.1.3 *P* versus π

QSAR equations relating biological activity to the partition coefficient *P* have already been described, but there is no reason why the substituent hydrophobicity constant π cannot be used in place of *P* if only the substituents are being varied. The equation obtained would be just as relevant as a study of how hydrophobicity affects biological activity. That is not to say that *P* and π are exactly equivalent—different equations would be obtained with different constants. Apart from that, the two factors have different emphases. The partition coefficient *P* is a measure of the drug's overall hydrophobicity and is, therefore, an important measure of how efficiently a drug is transported to its target and bound to its binding site. The π factor measures the hydrophobicity of a specific region on the drug's skeleton and, if it is present in the QSAR equation, it could emphasize important hydrophobic interactions involving that region of the molecule with the binding site.

TABLE 18.1 Values of π for a range of substituents on aliphatic and aromatic scaffolds

Group	CH_3	*t*-Bu	OH	OCH_3	CF_3	Cl	Br	F
π (aliphatic scaffolds)	0.50	1.68	−1.16	0.47	1.07	0.39	0.60	−0.17
π (aromatic scaffolds)	0.52	1.68	−0.67	−0.02	1.16	0.71	0.86	0.14

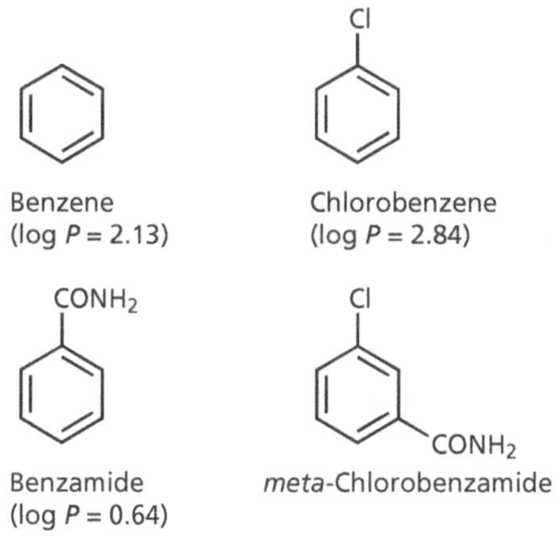
Cl
Benzene
(log P = 2.13)
Chlorobenzene
(log P = 2.84)
CONH2
Cl
CONH2
Benzamide
(log P = 0.64)
meta-Chlorobenzamide

FIGURE 18.4 Values for log *P*.

Most QSAR equations have a contribution from *P* or from π, but there are examples of drugs for which they have only a slight contribution. For example, a study on antimalarial drugs showed very little relationship between antimalarial activity and hydrophobic character. This finding supports the theory that these drugs act in red blood cells, since previous research has shown that the ease with which drugs enter red blood cells is not related to their hydrophobicity.

18.2.2 Electronic effects

The electronic effects of various substituents will clearly have an effect on a drug's ionization or polarity. This, in turn, may have an effect on how easily a drug can pass through cell membranes or how strongly it can interact with a binding site. It is, therefore, useful to measure the electronic effect of a substituent.

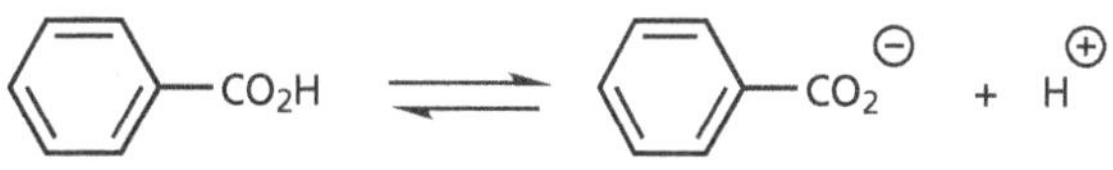
CO2H
CO2
+ H

FIGURE 18.5 Ionization of benzoic acid in water.

As far as substituents on an aromatic ring are concerned, the measure used is known as the **Hammett substituent constant** (σ). This is a measure of the electron-withdrawing or electron-donating ability of a substituent, and has been determined by measuring the dissociation of a series of substituted benzoic acids compared with the dissociation of benzoic acid itself.

Benzoic acid is a weak acid and only partially ionizes in water (Fig. 18.5). An equilibrium is set up between the ionized and non-ionized forms, where the relative proportion of these species is known as the **equilibrium** or **dissociation constant** K_H (the subscript H signifies that there are no substituents on the aromatic ring).

$$K_H = \frac{[PhCO_2^-]}{[PhCO_2H]}$$

When a substituent is present on the aromatic ring, this equilibrium is affected. Electron-withdrawing groups, such as a nitro group, result in the aromatic ring having a stronger electron-withdrawing and stabilizing influence on the carboxylate anion, and so the equilibrium will shift to the ionized form. Therefore, the substituted benzoic acid is a stronger acid and has a larger K_X value (X represents the substituent on the aromatic ring) (Fig. 18.6).

If the substituent X is an electron-donating group such as an alkyl group, then the aromatic ring is less able to

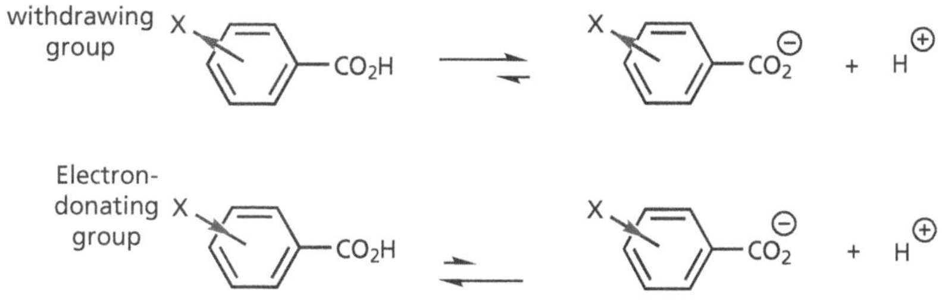
Electron-
withdrawing
group
X
CO2H
X
CO2
+ H
Electron-
donating
group
X
CO2H
X
CO2
+ H

FIGURE 18.6 Position of equilibrium dependent on substituent group X.

stabilize the carboxylate ion. The equilibrium shifts to the left indicating a weaker acid with a smaller K_x value (Fig. 18.6).

The Hammett substituent constant (σ_x) for a particular substituent (X) is defined by the following equation:

$$\sigma_x = \log \frac{K_x}{K_H} = \log K_x - \log K_H$$

Benzoic acids containing electron-withdrawing substituents will have larger K_x values than benzoic acid itself (K_H) and, therefore, the value of σ_x for an electron-withdrawing substituent will be positive. Substituents such as Cl, CN, or CF_3 have positive σ values.

Benzoic acids containing electron-donating substituents will have smaller K_x values than benzoic acid itself and, hence, the value of σ_x for an electron-donating substituent will be negative. Substituents such as Me, Et, and *t*-Bu have negative values of σ. The Hammett substituent constant for H is zero.

The Hammett substituent constant takes into account both resonance and inductive effects. Therefore, the value of σ for a particular substituent will depend on whether the substituent is *meta* or *para*. This is indicated by the subscript *m* or *p* after the σ symbol. For example, the nitro substituent has $\sigma_p = 0.78$ and $\sigma_m = 0.71$. In the *meta* position, the electron-withdrawing power is due to the inductive influence of the substituent, whereas at the *para* position inductive and resonance effects both play a part and so the σ_p value is greater (Fig. 18.7).

For the hydroxyl group, $\sigma_m = 0.12$ and $\sigma_p = -0.37$. At the *meta* position, the influence is inductive and electron-withdrawing. At the *para* position, the electron-donating

influence due to resonance is more significant than the electron-withdrawing influence due to induction (Fig. 18.8).

Most QSAR studies start off by considering σ, and, if there is more than one substituent present, the σ values can be summed ($\Sigma\sigma$). However, as more compounds are synthesized, it is possible to fine-tune the QSAR equation. As mentioned above, σ is a measure of a substituent's inductive and resonance electronic effects. With more detailed studies, the inductive and resonance effects can be considered separately. Tables of constants are available which quantify a substituent's inductive effect (*F*) and its resonance effect (*R*). In some cases, it might be found that a substituent's effect on activity is due to *F* rather than *R*, and vice versa. It might also be found that a substituent has a more significant effect at a particular position on the ring and this can also be included in the equation.

There are limitations to the electronic constants described so far. For example, Hammett substituent constants cannot be measured for *ortho* substituents as such substituents have an important steric, as well as electronic, effect.

There are very few drugs whose activities are solely influenced by a substituent's electronic effect, and hydrophobicity usually has to be considered as well. Those that do are generally operating by a mechanism whereby they do not have to cross any cell membranes (see Box 18.2). Alternatively, *in vitro* studies on isolated enzymes may result in QSAR equations lacking the hydrophobicity factor, as there are no cell membranes to be considered.

The constants σ, *R*, and *F* can only be used for aromatic substituents and are, therefore, only suitable for drugs containing aromatic rings. However, a series of aliphatic

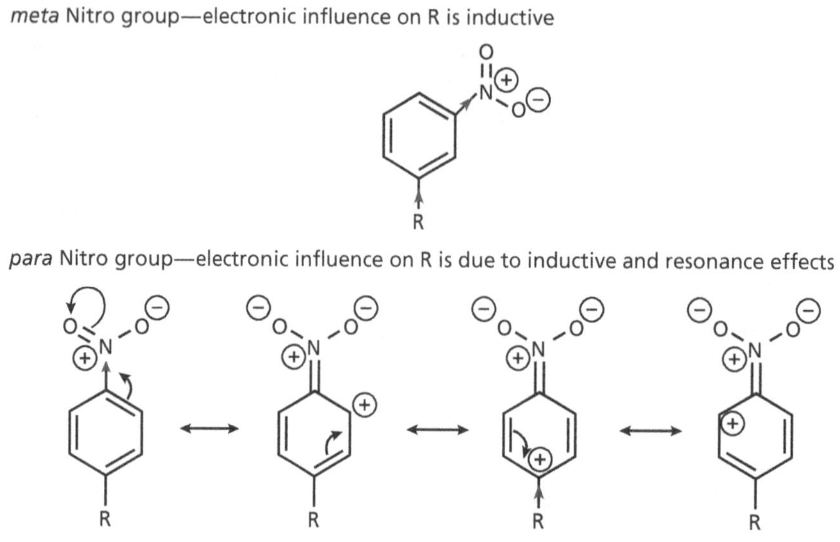

meta Nitro group—electronic influence on R is inductive

para Nitro group—electronic influence on R is due to inductive and resonance effects

FIGURE 18.7 Substituent effects of a nitro group at the *meta* and *para* positions.

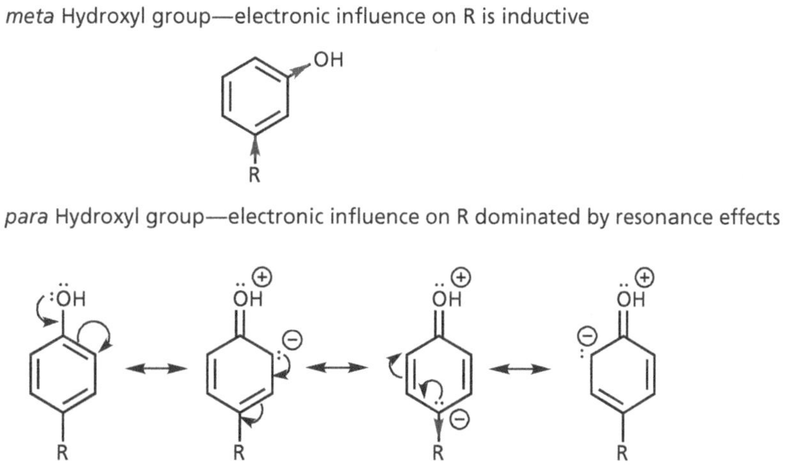

meta Hydroxyl group—electronic influence on R is inductive

para Hydroxyl group—electronic influence on R dominated by resonance effects

FIGURE 18.8 Substituent effects of a phenol at the *meta* and *para* positions.

electronic substituent constants are available. These were obtained by measuring the rates of hydrolysis for a series of aliphatic esters (Fig. 18.9). Methyl ethanoate is the parent ester and it is found that the rate of hydrolysis is affected by the substituent X. The extent to which the

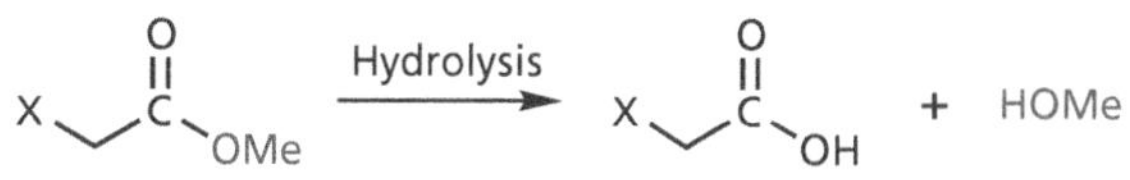

FIGURE 18.9 Hydrolysis of an aliphatic ester.

rate of hydrolysis is affected is a measure of the substituent's electronic effect at the site of reaction (i.e. the ester group). The electronic effect is purely inductive and is given the symbol σ_I. Electron-donating groups reduce the rate of hydrolysis and, therefore, have negative values. For example, σ_I values for methyl, ethyl, and propyl are −0.04, −0.07, and −0.36 respectively. Electron-withdrawing groups increase the rate of hydrolysis and have positive values. The σ_I values for NMe_3^+ and CN are 0.93 and 0.53 respectively.

It should be noted that the inductive effect is not the only factor affecting the rate of hydrolysis. The substituent may also have a steric effect. For example, a bulky substituent may shield the ester from attack and lower the rate of hydrolysis. It is, therefore, necessary to separate out these two effects. This can be done by measuring hydrolysis rates under both basic and acidic conditions. Under basic conditions, steric and electronic factors are important, whereas under acidic conditions only steric factors are important. By comparing the rates, values for the electronic effect (σ_I) and the steric effect (E_s) (see section 18.2.3.1) can be determined.

Test your understanding and practise your molecular modelling with Exercise 18.3 on the Online Resource Centre at www.oxfordtextbooks.co.uk/orc/patrick6e/

18.2.3 Steric factors

The bulk, size, and shape of a drug will influence how easily it can approach and interact with a binding site.

BOX 18.2 Insecticidal activity of diethyl phenyl phosphates

The insecticidal activity of diethyl phenyl phosphates is one of the few examples where activity is related to electronic factors alone:

$$\log\left(\frac{1}{C}\right) = 2.282\sigma - 0.348. \quad (r^2\ 0.952,\ r\ 0.976,\ s\ 0.286)$$

The equation reveals that substituents with a positive value for σ (i.e. electron-withdrawing groups) will increase activity. The fact that a hydrophobic parameter is not present is a good indication that the drugs do not have to pass into, or through, a cell membrane to have activity. In fact, these drugs are known to act on an enzyme called **acetylcholinesterase** which is situated on the outside of cell membranes (section 22.12).

The value of *r* is close to 1, which demonstrates that the line is a good fit, and the value of r^2 demonstrates that 95% of the data is accounted for by the σ parameter.

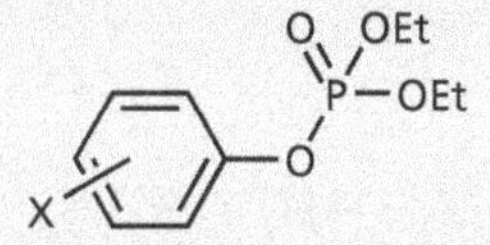

FIGURE 1 Diethyl phenyl phosphates.

A bulky substituent may act like a shield and hinder the ideal interaction between a drug and its binding site. Alternatively, a bulky substituent may help to orientate a drug properly for maximum binding and increase activity. Steric properties are more difficult to quantify than hydrophobic or electronic properties. Several methods have been tried, of which three are described here. It is highly unlikely that a drug's biological activity will be affected by steric factors alone, but these factors are frequently found in Hansch equations (section 18.3).

18.2.3.1 Taft's steric factor (E_s)

Attempts have been made to quantify the steric features of substituents by using Taft's steric factor (E_s). The value for E_s can be obtained by comparing the rates of hydrolysis of substituted aliphatic esters against a standard ester under acidic conditions (Fig. 18.9). Thus,

$$E_s = \log k_x - \log k_0$$

where k_x represents the rate of hydrolysis of an aliphatic ester bearing the substituent X, and k_0 represents the rate of hydrolysis of the reference ester.

The substituents that can be studied by this method are restricted to those which interact sterically with the tetrahedral transition state of the reaction, and not by resonance or internal hydrogen bonding. For example, unsaturated substituents which are conjugated to the ester cannot be measured by this procedure. Examples of E_s values are shown in Table 18.2. Note that the reference ester is X = Me. Substituents such as H and F are smaller than a methyl group and result in a faster rate of hydrolysis ($k_x > k_0$), making E_s positive. Substituents which are larger than methyl reduce the rate of hydrolysis ($k_x < k_0$), making E_s negative. A disadvantage of E_s values is that they are a measure of an *intramolecular* steric effect, whereas drugs interact with target binding sites in an *intermolecular* manner. For example, consider the E_s values for *i*-Pr, *n*-Pr, and *n*-Bu. The E_s value for the branched isopropyl group is significantly greater than that for the linear *n*-propyl group since the bulk of the substituent is closer to the reaction centre. Extending the alkyl chain from *n*-propyl to *n*-butyl has little effect on E_s. The larger *n*-butyl group is extended away from the reaction centre, and so it has little additional steric effect on the rate of

hydrolysis. As a result, the E_s value for the *n*-butyl group undervalues the steric effect which this group might have if it was present on a drug approaching a binding site.

18.2.3.2 Molar refractivity

Another measure of the steric factor is provided by a parameter known as **molar refractivity** (**MR**). This is a measure of the volume occupied by an atom or a group of atoms. The MR is obtained from the following equation:

$$MR = \frac{(n^2 - 1)}{(n^2 + 2)} \times \frac{MW}{d}$$

where n is the index of refraction, MW is the molecular weight, and d is the density. The term MW/d defines a volume, and the $(n^2 - 1)/(n^2 + 2)$ term provides a correction factor by defining how easily the substituent can be polarized. This is particularly significant if the substituent has π electrons or lone pairs of electrons.

18.2.3.3 Verloop steric parameter

Another approach to measuring the steric factor involves a computer program called **Sterimol**, which calculates steric substituent values (**Verloop steric parameters**) from standard bond angles, van der Waals radii, bond lengths, and possible conformations for the substituent. Unlike E_s, the Verloop steric parameters can be measured for any substituent. For example, the Verloop steric parameters for a carboxylic acid group are demonstrated in Fig. 18.10. L is the length of the substituent and B_1–B_4 are the radii of the group in different dimensions.

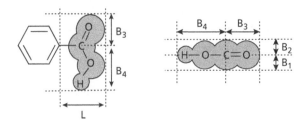

FIGURE 18.10 Verloop parameters for a carboxylic acid group.

TABLE 18.2 Values of E_s for various substituents

Substituent	H	F	Me	Et	*n*-Pr	*n*-Bu	*i*-Pr	*i*-Bu	Cyclopentyl
E_s	1.24	0.78	0	−0.07	−0.36	−0.39	−0.47	−0.93	−0.51

18.2.4 Other physicochemical parameters

The physicochemical properties most commonly studied by the QSAR approach have been described above, but other properties have been studied including dipole moments, hydrogen bonding, conformations, and interatomic distances. Difficulties in quantifying these properties limit the use of these parameters, however. Several QSAR formulae have been developed based on the highest occupied and/or the lowest unoccupied molecular orbitals of the test compounds. The energy calculations of these orbitals can be carried out using semi-empirical quantum mechanical methods (section 17.7.3). Indicator variables for different substituents can also be used. These are described in section 18.7.

18.3 Hansch equation

In section 18.2, we looked at the physicochemical properties commonly used in QSAR studies and how it is possible to quantify them. In a situation where biological activity is related to only one such property, a simple equation can be drawn up. The biological activity of most drugs, however, is related to a combination of physicochemical properties. In such cases, simple equations involving only one parameter are relevant only if the other parameters are kept constant. In reality, this is not easy to achieve and equations which relate biological activity to a number of different parameters are more common (Box 18.3). These equations are known as Hansch equations and they usually relate biological activity to the most commonly used physicochemical properties (log P, π, σ, and a steric factor). If the range of hydrophobicity values is limited to a small range then the equation will be linear, as follows:

$$\log\left(\frac{1}{C}\right) = k_1 \log P + k_2\sigma + k_3 E_s + k_4$$

If the log P values are spread over a large range, then the equation will be parabolic for the same reasons described in section 18.2.1.

$$\log\left(\frac{1}{C}\right) = -k_1(\log P)^2 + k_2 \log P + k_3\sigma + k_4 E_s + k_5$$

The constants k_1–k_5 are determined by computer software in order to get the best fitting equation. Not all the parameters will necessarily be significant. For example, the adrenergic blocking activity of β-halo-arylamines (Fig. 18.11) was related to π and σ and did not include a steric factor. This equation tells us that biological activity increases if the substituents have a positive π value and a negative σ value. In other words, the substituents should be hydrophobic and electron donating.

When carrying out a Hansch analysis, it is important to choose the substituents carefully to ensure that the change in biological activity can be attributed to a particular parameter. There are plenty of traps for the unwary. Take, for example, drugs which contain an amine group. One of the studies most frequently carried out on amines is to synthesize analogues containing a homologous series of alkyl substituents on the nitrogen atom (i.e. Me, Et, n-Pr, n-Bu). If activity increases with the chain length of the substituent, is it due to increasing hydrophobicity, increasing size, or both? If we look at the π and MR values of these substituents, we find that both sets of values increase in a similar fashion across the series and we would not be able to distinguish between them (Table 18.3). In this example, a series of substituents would have to be chosen where π and MR are not correlated. The substituents H, Me, OMe, NHCOCH$_2$, I, and CN would be more suitable.

> Test your understanding and practise your molecular modelling with Exercise 18.4 on the Online Resource Centre at www.oxfordtextbooks.co.uk/orc/patrick6e/

18.4 The Craig plot

Although tables of π and σ factors are readily available for a large range of substituents, it is often easier to visualize the relative properties of different substituents by considering a plot where the y-axis is the value of the σ factor and the x-axis is the value of the π factor. Such a plot is known as a Craig plot. The example shown in Fig. 18.12 is the Craig plot for the σ and π factors of *para*-aromatic substituents. There are several advantages to the use of such a Craig plot.

- The plot shows clearly that there is no overall relationship between π and σ. The various substituents are scattered around all four quadrants of the plot.

- It is possible to tell at a glance which substituents have positive π and σ parameters, which substituents have negative π and σ parameters, and which substituents have one positive and one negative parameter.

- It is easy to see which substituents have similar π values. For example, the ethyl, bromo, trifluoromethyl, and trifluoromethylsulphonyl groups are all approximately on the same vertical line on the plot. In theory, these groups could be interchangeable on drugs where the principal factor affecting biological activity

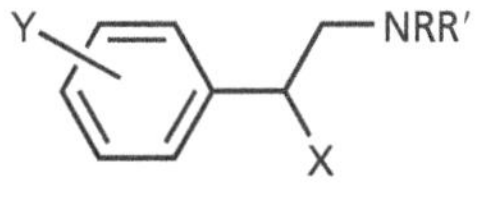

$$\log\left(\frac{1}{C}\right) = 1.22\pi - 1.59\sigma + 7.89$$

$$(n = 22,\ r^2 = 0.841,\ s = 0.238)$$

β-Halo-arylamines

FIGURE 18.11 QSAR equation for β-halo-arylamines.

BOX 18.3 Hansch equation for a series of antimalarial compounds

A series of 102 phenanthrene aminocarbinols was tested for antimalarial activity. In the structure shown, X represents up to four substituents on the left-hand ring while Y represents up to four substituents on the right-hand ring. Experimental log *P* values for the structures were not available and equations were derived which compared the activity with some or all of the following terms:

π_{sum} — The π constants for *all* the substituents in the molecule (i.e. all the X and Y substituents, as well as the amino substituents R and R'). This term was used in place of log *P* to represent the overall hydrophobicity for the molecule.

σ_{sum} — The σ constants for *all* the substituents in the molecule.

$\Sigma\pi_X$ — The sum of the π constants for all the substituents X in the left-hand ring.

$\Sigma\pi_Y$ — The sum of the π constants for all the substituents Y in the right-hand ring.

$\Sigma\pi_{X+Y}$ — The sum of the π constants for all the substituents X and Y in both the left- and right-hand rings

$\Sigma\sigma_{X+Y}$ — The sum of the σ constants for all the substituents X and Y in both the left- and right-hand rings.

$\Sigma\sigma_X$ — The sum of the σ constants for all the substituents X in the left-hand ring.

$\Sigma\sigma_Y$ — The sum of the σ constants for all the substituents Y in the right-hand ring.

Equations such as equations 1–3 were derived which matched activity against one of the above terms, but none of them had an acceptable value of r^2.

A variety of other equations were derived which included two of the above terms but these were not satisfactory either. Finally, an equation was derived which contained 6 terms and proved satisfactory:

$$\log\left(\frac{1}{C}\right) = -0.015(\pi_{sum})^2 + 0.14\pi_{sum} + 0.27\Sigma\pi_X$$
$$+ 0.40\Sigma\pi_Y + 0.65\Sigma\sigma_X + 0.88\Sigma\sigma_Y + 2.34$$
$$(n = 102, r = 0.913, r^2 = 0.834, s = 0.258)$$

The equation shows that antimalarial activity increases very slightly as the overall hydrophobicity of the molecule (π_{sum}) increases (the constant 0.14 is low). The $(\pi_{sum})^2$ term shows that there is an optimum overall hydrophobicity for activity and this is found to be 4.44. Activity increases if hydrophobic substituents are present on ring X and, in particular, on ring Y. This could be taken to imply that some form of hydrophobic interaction is involved near both rings. Electron-withdrawing substituents on both rings are also beneficial to activity, more so on ring Y than ring X. The r^2 value is 0.834 which is above the minimum acceptable value of 0.8.

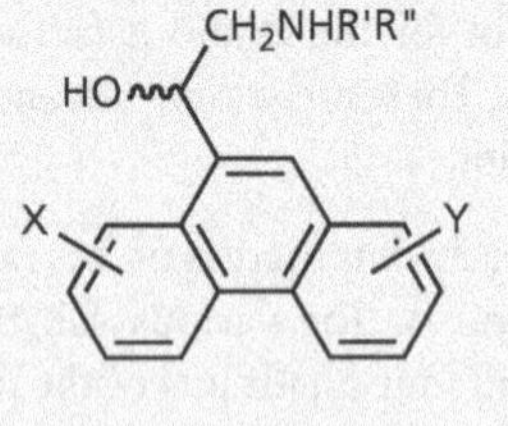

1) $\log\left(\frac{1}{C}\right) = 0.557\Sigma\pi_{x+y} + 2.699$ ($n = 102, r = 0.768, r^2 = 0.590, s = 0.395$)

2) $\log\left(\frac{1}{C}\right) = 0.017\pi_{sum} + 3.324$ ($n = 102, r = 0.069, r^2 = 0.005, s = 0.616$)

3) $\log\left(\frac{1}{C}\right) = 1.218\sigma_{sum} + 2.721$ ($n = 102, r = 0.814, r^2 = 0.663, s = 0.359$)

FIGURE 1 Phenanthrene aminocarbinols.

is the π factor. Similarly, groups which form a horizontal line can be identified as being isoelectronic or having similar σ values (e.g. CO_2H, Cl, Br, I).

- The Craig plot is useful in planning which substituents should be used in a QSAR study. In order to derive the most accurate equation involving π and σ, analogues should be synthesized with substituents from each quadrant. For example, halogen substituents are useful representatives of substituents with increased hydrophobicity and electron-withdrawing properties (positive

TABLE 18.3 Values for π and *MR* for a series of substituents.

Substituent	H	Me	Et	*n*-Pr	*n*-Bu	OMe	NHCONH$_2$	I	CN
π	0.00	0.56	1.02	1.50	2.13	−0.02	−1.30	1.12	−0.57
MR	0.10	0.56	1.03	1.55	1.96	0.79	1.37	1.39	0.63

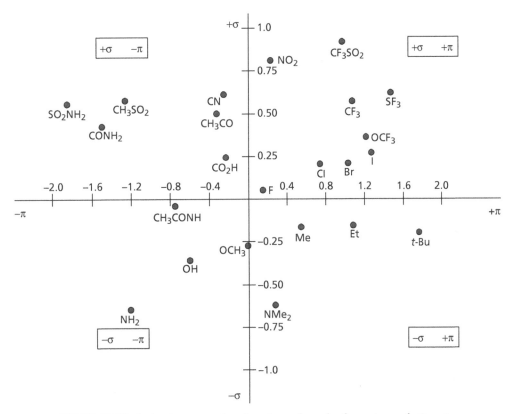

FIGURE 18.12 Craig plot comparing the values of σ and π for various substituents.

π and positive σ), whereas an OH substituent has more hydrophilic and electron-donating properties (negative π and negative σ). Alkyl groups are examples of substituents with positive π and negative σ values, whereas acyl groups have negative π and positive σ values.

- Once the Hansch equation has been derived, it will show whether π or σ should be negative or positive in order to get good biological activity. Further developments would then concentrate on substituents from the relevant quadrant. For example, if the equation shows that positive π and positive σ values are necessary, then further substituents should only be taken from the top right quadrant.

Craig plots can also be drawn up to compare other sets of physicochemical parameters, such as hydrophobicity and *MR*.

18.5 The Topliss scheme

In certain situations, it might not be feasible to make the large range of structures required for a Hansch equation. For example, the synthetic route involved might be so difficult that only a few structures can be made in a limited time. In these circumstances, it would be useful to test compounds for biological activity as they are synthesized and to use these results to determine the next analogue to be synthesized.

A Topliss scheme is a 'flow diagram' which allows such a procedure to be followed. There are two Topliss schemes, one for substituents on an aromatic ring (Fig. 18.13) and one for substituents on aliphatic moieties (Fig. 18.14). The schemes were drawn up by considering the hydrophobicity and electronic factors of various substituents, and are designed such that the optimum substituent can be found as efficiently as possible. They are not meant to be a replacement for a full Hansch analysis, however. Such an analysis would be carried out in due course, once a suitable number of structures have been synthesized.

The Topliss scheme for substituents on an aromatic ring (Fig. 18.13) assumes that the lead compound has been tested for biological activity and contains a monosubstituted aromatic ring. The first analogue in the scheme is the 4-chloro derivative, as this derivative is usually easy to synthesize. The chloro substituent is more hydrophobic and electron-withdrawing than hydrogen and, therefore, π and σ are positive.

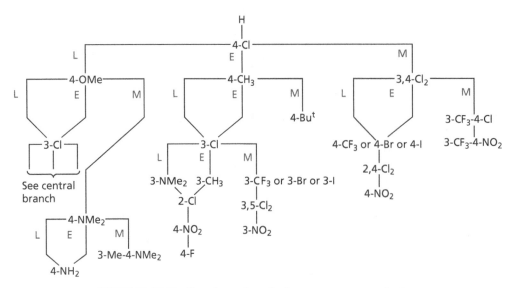

FIGURE 18.13 Topliss scheme for substituents on an aromatic ring.

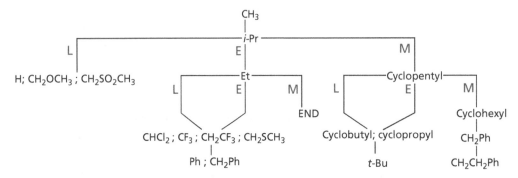

FIGURE 18.14 Topliss scheme for substituents on an aliphatic moiety.

Once the chloro analogue has been synthesized, the biological activity is measured. There are three possibilities. The analogue will have less activity (L), equal activity (E), or more activity (M). The type of activity observed will determine which branch of the Topliss scheme is followed next.

If the biological activity increases, the (M) branch is followed and the next analogue to be synthesized is the 3,4-dichloro-substituted analogue. If, on the other hand, the activity stays the same, then the (E) branch is followed and the 4-methyl analogue is synthesized. Finally, if activity drops, the (L) branch is followed and the next analogue is the 4-methoxy analogue. Biological results from the second analogue now determine the next branch to be followed in the scheme.

What is the rationale behind this?

Let us consider the situation where the 4-chloro derivative increases in biological activity. The chloro substituent has positive π and σ values, which implies that one, or both, of these properties are important to biological activity. If both are important, then adding a second chloro group should increase biological activity yet further. If it does, substituents are varied to increase the π and σ values even further. If it does not, then an unfavourable steric interaction or excessive hydrophobicity is indicated. Further modifications then test the relative importance of π and steric factors.

Now consider the situation where the 4-chloro analogue drops in activity. This suggests that either negative π and/or σ values are important to activity or a *para* substituent is sterically unfavourable. It is assumed that an unfavourable σ effect is the most likely reason for the reduced activity and so the next substituent is one with a negative σ factor (i.e. 4-OMe). If activity improves, further changes are suggested to test the relative importance of the σ and π factors. On the other hand, if the 4-OMe group does not improve activity, it is assumed that an unfavourable steric factor is at work and the next substituent is a 3-chloro group. Modifications of this group would then be carried out in the same way as shown in the centre branch of Fig. 18.13.

The last scenario is where the activity of the 4-chloro analogue is little changed from the lead compound. This could arise from the drug requiring a positive π value and a negative σ value. As both values for the chloro group are positive, the beneficial effect of the positive π value might be cancelled out by the detrimental effects of a positive σ value. The next substituent to try in that case is the 4-methyl group which has the necessary positive π value and negative σ value. If this still has no beneficial effect, then it is assumed that there is an unfavourable steric interaction at the *para* position and the 3-chloro substituent is chosen next. Further changes continue to vary the relative values of the π and σ factors.

The validity of the Topliss scheme was tested by looking at structure–activity results for various drugs which had been reported in the literature. For example, the biological activities of 19 substituted benzenesulphonamides (Fig. 18.15) have been reported. The second most active compound was the nitro-substituted analogue, which would have been the fifth compound synthesized if the Topliss scheme had been followed.

Another example comes from the anti-inflammatory activities of substituted aryltetrazolylalkanoic acids (Fig. 18.16), of which 28 were synthesized. Using the Topliss scheme, 3 out of the 4 most active structures would have been prepared from the first 8 compounds synthesized.

The Topliss scheme for substituents on aliphatic moieties (Fig. 18.14) was set up following a similar rationale to the aromatic scheme and is used in the same way for substituents attached to a carbonyl, amino, amide, or similar functional group. The scheme attempts to differentiate only between the hydrophobic and electronic effects of substituents, and not their steric properties. Thus, the substituents involved have been chosen to try to minimize any steric differences. It is assumed that the lead compound has a methyl group. The first analogue suggested is the isopropyl analogue. This has an increased π value and, in most cases, would be expected to increase activity. It has been found from experience that the hydrophobicity of most lead compounds is less than optimum.

Let us concentrate first of all on the situation where activity increases. Following this branch, a cyclopentyl group is now used. A cyclic structure is used since it has a larger π value, but keeps any increase in steric factor to a minimum. If activity rises again, more hydrophobic substituents are tried. If activity does not rise, then there could be two explanations. Either the optimum hydrophobicity has been passed or there is an electronic effect (σ_I) at work. Further substituents are then used to determine which is the correct explanation.

Let us now look at the situation where the activity of the isopropyl analogue stays much the same. The most likely explanation is that the methyl group and the isopropyl group are on either side of the hydrophobic optimum. Therefore, an ethyl group is used next, since it has an intermediate π value. If this does not lead to an improvement, it is possible that there is an unfavourable electronic effect. The groups used have been electron-donating, and so electron-withdrawing groups with similar π values are now suggested.

Finally, we shall look at the case where activity drops for the isopropyl group. In this case, hydrophobic and/

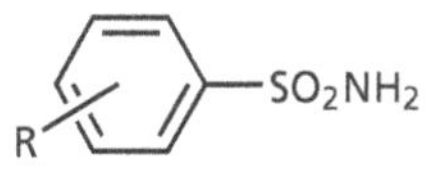

Order of synthesis	R	Biological activity	High potency
1	H	–	
2	4-Cl	More	
3	3,4-Cl$_2$	Less	
4	4-Br	Equal	
5	4-NO$_2$	More	*

FIGURE 18.15 The order of benzenesulphonamide synthesis as directed by the Topliss scheme.

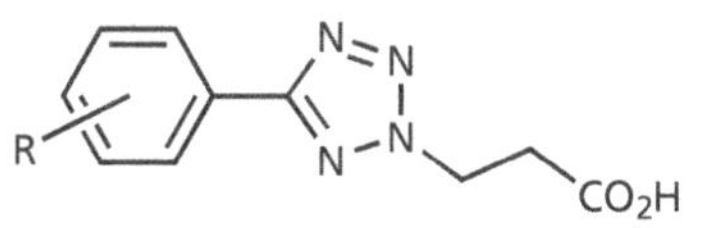

Order of synthesis	R	Biological activity	High potency
1	H	–	
2	4-Cl	Less	
3	4-OMe	Less	
4	3-Cl	More	*
5	3-CF$_3$	Less	
6	3-Br	More	*
7	3-I	Less	
8	3,5-Cl$_2$	More	*

FIGURE 18.16 The order of synthesis for substituted aryltetrazolylalkanoic acids as directed by the Topliss scheme.

or electron-donating groups could be bad for activity and the groups suggested are suitable choices for further development.

18.6 Bio-isosteres

Tables of substituent constants are available for various physicochemical properties. A knowledge of these constants allows the medicinal chemist to identify substituents which may be potential bio-isosteres. Thus, the substituents CN, NO_2, and COMe have similar hydrophobic, electronic, and steric factors, and might be interchangeable. Such interchangeability was observed in the development of **cimetidine** and its analogues (sections 25.2.6 and 25.2.8). The important thing to note is that groups can be bio-isosteric in some situations, but not others. Consider for example the table shown in Fig. 18.17.

This table shows physicochemical parameters for six different substituents. If the most important physicochemical parameter for biological activity is σ_p, then the $COCH_3$ group (0.50) would be a reasonable bio-isostere for the $SOCH_3$ group (0.49). If, on the other hand, the dominant parameter is π, then a more suitable bio-isostere for $SOCH_3$ (−1.58) would be SO_2CH_3 (−1.63).

18.7 The Free–Wilson approach

In the Free–Wilson approach to QSAR, the biological activity of a parent structure is measured then compared with the activities of a range of substituted analogues. An equation is then derived which relates biological activity to the presence or otherwise of particular substituents $(X_1–X_n)$.

$$\text{Activity} = k_1X_1 + k_2X_2 + k_3X_3 + \ldots\ldots + k_nX_n + Z$$

In this equation, X_n is defined as an **indicator variable** and is given the value 1 or 0, depending on whether the substituent (n) is present or not. The contribution that each substituent makes to the activity is determined by

the value of k_n. Z is a constant representing the average activity of the structures studied.

Since the approach considers the overall effect of a substituent on biological activity rather than its various physicochemical properties, there is no need for physicochemical constants and tables, and the method only requires experimental measurements of biological activity. This is particularly useful when trying to quantify the effect of unusual substituents that are not listed in the tables, or when quantifying specific molecular features which cannot be tabulated.

The disadvantage in the approach is the large number of analogues which have to be synthesized and tested to make the equation meaningful. For example, each of the terms k_nX_n refers to a specific substituent at a specific position in the parent structure. Therefore, analogues would not only have to have different substituents, but also have them at different positions of the skeleton.

Another disadvantage is the difficulty in rationalizing the results and explaining why a substituent at a particular position is good or bad for activity. Finally, the effects of different substituents may not be additive. There may be intramolecular interactions which affect activity.

Nevertheless, indicator variables can be useful in certain situations and they can also be used as part of a Hansch equation. An example of this can be seen in the later case study (section 18.9).

18.8 Planning a QSAR study

When starting a QSAR study it is important to decide which physicochemical parameters are going to be studied and to plan the analogues such that the parameters under study are suitably varied. For example, it would be pointless to synthesize analogues where the hydrophobicity and steric volume of the substituents are correlated if these two parameters are to go into the equation.

It is also important to make enough structures to make the results statistically meaningful. As a rule of thumb, five structures should be made for every parameter studied. Typically, the initial QSAR study would involve the

Substituent	$\underset{CH_3}{\overset{O}{\overset{\|}{C}}}$	$\underset{CH_3}{\overset{NC\diagdown \diagup CN}{\overset{C}{\overset{\|}{C}}}}$	$\underset{CH_3}{\overset{O}{\overset{\|}{S}}}$	$\overset{O}{\underset{O}{\overset{\|}{S}}}{-}CH_3$	$\overset{O}{\underset{O}{\overset{\|}{S}}}{-}NHCH_3$	$\underset{NMe_2}{\overset{O}{\overset{\|}{C}}}$
π	−0.55	0.40	−1.58	−1.63	−1.82	−1.51
σ_p	0.50	0.84	0.49	0.72	0.57	0.36
σ_m	0.38	0.66	0.52	0.60	0.46	0.35
MR	11.2	21.5	13.7	13.5	16.9	19.2

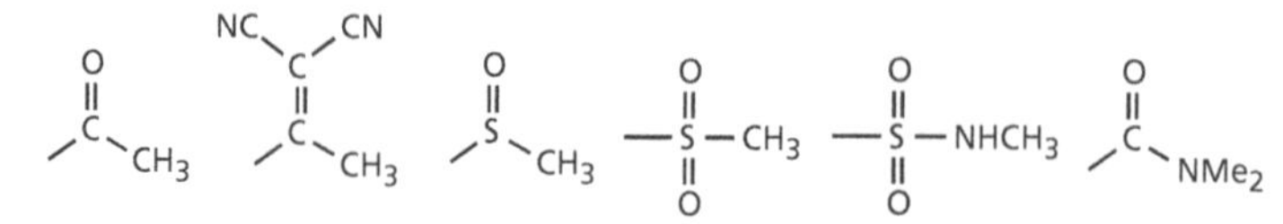

FIGURE 18.17 Physicochemical parameters for six substituents.

two parameters π and σ, and possibly E_s. Craig plots could be used in order to choose suitable substituents.

Certain substituents are worth avoiding in the initial study, as they may have properties other than those being studied. For example, it is best to avoid substituents that might ionize (CO_2H, NH_2, SO_3H) and groups that might be easily metabolized (e.g. esters or nitro groups).

If there are two or more substituents, then the initial equation usually considers the total π and σ contribution.

As more analogues are made, it is often possible to consider the hydrophobic and electronic effect of substituents at specific positions of the molecule. Furthermore, the electronic parameter σ can be split into its inductive and resonance components (F and R). Such detailed equations may show up a particular localized requirement for activity. For example, a hydrophobic substituent may be favoured in one part of the skeleton, while an electron-withdrawing substituent is favoured at another. This, in turn, gives clues about the binding interactions involved between drug and receptor.

18.9 **Case study**

An example of how a QSAR equation can become more specific as a study develops is demonstrated from work carried out on the anti-allergic activity of a series of pyranenamines (Fig. 18.18). In this study, substituents were varied on the aromatic ring, while the remainder of the molecule was kept constant. Nineteen compounds were synthesized and the first QSAR equation was obtained by considering π and σ:

$$\log\left(\frac{1}{C}\right) = -0.14\sum\pi - 1.35(\sum\sigma)^2 - 0.72$$

$$(n\ 19,\ r^2\ 0.48,\ s\ 0.47,\ F_{2,16}\ 7.3)$$

where $\sum\pi$ and $\sum\sigma$ are the total π and σ values for all the substituents present.

The negative coefficient for the π term shows that activity is inversely proportional to hydrophobicity, which is quite unusual. The $(\sum\sigma)^2$ term is also quite unusual. It was chosen because there was no simple relationship between activity and σ. In fact, it was observed

that activity decreased if the substituent was electron-withdrawing *or* electron-donating. Activity was best with neutral substituents. To take account of this, the $(\sum\sigma)^2$ term was introduced. As the coefficient in the equation is negative, activity is lowered if σ is anything other than zero.

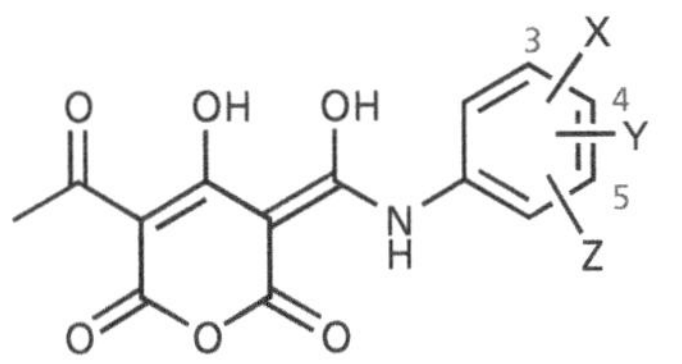

A further range of compounds was synthesized with hydrophilic substituents to test this equation, making a total of 61 structures. This resulted in the following inconsistencies.

- The activities for the substituents 3-NHCOMe, 3-NHCOEt, and 3-NHCOPr were all similar, but according to the equation, these activities should have dropped as the alkyl group got larger as a result of increasing hydrophobicity.

- Activity was greater than expected if there was a substituent such as OH, SH, NH_2, or NHCOR at position 3, 4, or 5.

- The substituent $NHSO_2R$ was bad for activity.

- The substituents 3,5-$(CF_3)_2$ and 3,5-$(NHCOMe)_2$ had much greater activity than expected.

- An acyloxy group at the 4-position resulted in an activity five times greater than predicted by the equation.

These results implied that the initial equation was too simple and that properties other than π and σ were important to activity. At this stage, the following theories were proposed to explain the above results.

- The similar activities for 3-NHCOMe, 3-NHCOEt, and 3-NHCOPr could be due to a steric factor. The substituents had increasing hydrophobicity, which is bad for activity, but they were also increasing in size and it was proposed that this was good for activity. The most likely explanation is that the size of the substituent is forcing the drug into the correct orientation for optimum receptor interaction.

- The substituents which unexpectedly increased activity when they were at positions 3, 4, or 5 are all capable of hydrogen bonding. This suggests an important hydrogen bonding interaction with the receptor. For some reason, the $NHSO_2R$ group is an exception, which implies there is some other unfavourable steric or electronic factor peculiar to this group.

- The increased activity for 4-acyloxy groups was explained by suggesting that these analogues are acting as prodrugs. The acyloxy group is less polar than the hydroxyl group and so these analogues would be expected to cross cell membranes and reach the receptor more efficiently than analogues bearing a free hydroxyl group. Hydrolysis of the ester group would reveal the hydroxyl group which would then take part in hydrogen bonding with the receptor.

FIGURE 18.18 Structure of pyranenamines.

- The structures having substituents 3,5-$(CF_3)_2$ and 3,5-$(NHCOMe)_2$ are the only disubstituted structures where a substituent at position 5 has an electron-withdrawing effect, so this feature was also introduced into the next equation.

The revised QSAR equation was as follows:

$$\log\left(\frac{1}{C}\right) = -0.30\sum\pi - 1.5(\sum\sigma)^2 + 2.0(F\text{-}5)$$
$$+ 0.39(345\text{-}HBD) - 0.63(NHSO_2)$$
$$+ 0.78(M\text{-}V) + 0.72(4\text{-}OCO) - 0.75$$
$$(n\ 61,\ r^2\ 0.77,\ s\ 0.40,\ F_{7,53}\ 25.1)$$

The π and σ parameters are still present, but a number of new parameters have now been introduced.

- The *F*-5 term represents the inductive effect of a substituent at position 5. The coefficient is positive and large, showing that an electron-withdrawing group substantially increases activity. However, only 2 compounds of the 61 synthesized had a 5-substituent, so there might be quite an error in this result.
- The *M-V* term represents the volume of any *meta* substituent. The coefficient is positive, indicating that substituents with a large volume at the *meta* position increase activity.
- The advantage of having hydrogen bonding substituents at position 3, 4, or 5 is accounted for by including a hydrogen bonding term (*345-HBD*). The value of this term depends on the number of hydrogen bonding substituents present. If one such group is present, the *345-HBD* term is 1. If two such groups are present, the parameter is 2. Therefore, for each hydrogen bonding substituent present at positions 3, 4, or 5, log (1/*C*) increases by 0.39. This sort of term is known as an **indicator variable**, which is the basis of the **Free–Wilson approach** described earlier. There is no tabulated value one can use for a hydrogen bonding substituent and so the contribution that this term makes to the biological activity is determined by the value of *k*, and whether the relevant group is present or not. Indicator variables were also used for the following terms.
- The *NHSO₂* term was introduced because this group was bad for activity despite being capable of hydrogen bonding. The negative coefficient indicates the drop in activity. A figure of 1 is used for any $NHSO_2R$ substituent present, resulting in a drop of activity by 0.63.
- The *4-OCO* term is 1 if an acyloxy group is present at position 4, and so log (1/*C*) is increased by 0.72 if this is the case.

A further 37 structures were synthesized to test steric and *F*-5 parameters as well as exploring further groups capable of hydrogen bonding. Since hydrophilic substituents were good for activity, a range of very hydrophilic substituents were also tested to see if there was an optimum value for hydrophilicity. The results obtained highlighted one more anomaly, in that two hydrogen bonding groups *ortho* to each other were bad for activity. This was attributed to the groups hydrogen bonding with each other rather than to the receptor. A revised equation was obtained as follows:

$$\log\left(\frac{1}{C}\right) = -0.034(\sum\pi)^2 - 0.33(\sum\pi) + 4.3(F\text{-}5)$$
$$+ 1.3(R\text{-}5) - 1.7(\sum\sigma)^2 + 0.73(345\text{-}HBD)$$
$$- 0.86(HB\text{-}INTRA) - 0.69(NHSO_2)$$
$$+ 0.72(4\text{-}OCO) - 0.59$$
$$(n\ 98,\ r^2\ 0.75,\ s\ 0.48,\ F_{9,88}\ 28.7)$$

The main points of interest from this equation are as follows.

- Increasing the hydrophilicity of substituents allowed the identification of an optimum value for hydrophobicity ($\sum\pi = -5$) and introduced the $(\sum\pi)^2$ parameter into the equation. The value of −5 is remarkably low and indicates that the region of the binding site occupied by the drug's aromatic ring is hydrophilic.
- As far as electronic effects are concerned, it is revealed that the resonance effects of substituents at the 5-position also have an influence on activity.
- The unfavourable situation where two hydrogen bonding groups are *ortho* to each other is represented by the *HB-INTRA* parameter. This parameter is given the value 1 if such an interaction is possible, and the negative constant (−0.86) shows that such interactions decrease activity.
- It is interesting to note that the steric parameter is no longer significant and has been removed from the equation.

The compound having the greatest activity has two $NHCOCH(OH)CH_2OH$ substituents at the 3- and 5-positions and is 1000 times more active than the original lead compound. The substituents are very polar and are not ones that would normally be used. They satisfy all the requirements determined by the QSAR study. They are highly polar groups which can take part in hydrogen bonding. They are *meta* with respect to each other, rather than *ortho*, to avoid undesirable intramolecular hydrogen bonding. One of the groups is at the 5-position and has a favourable *F*-5 parameter. Together, the two groups have a negligible $(\sum\sigma)^2$ value. Such an analogue would certainly not have been obtained by trial and error and this example demonstrates the strengths of the QSAR approach.

All the evidence from this study suggests that the aromatic ring of this series of compounds is fitting into a hydrophilic pocket in the target binding site which contains polar groups capable of hydrogen bonding.

It is further proposed that a positively charged residue such as arginine, lysine, or histidine might be present in the pocket which could interact with an electronegative substituent at position 5 of the aromatic ring (Fig. 18.19).

This example demonstrates that QSAR studies and computers are powerful tools in medicinal chemistry. However, it also shows that the QSAR approach is a long way from replacing the human factor. One cannot put a series of facts and figures into a computer and expect it to magically produce an instant explanation of how a drug works. The medicinal chemist still has to interpret results, propose theories, and test those theories by incorporating the correct parameters into the QSAR equation. Imagination and experience still count for a great deal.

KEY POINTS

- QSAR relates the physicochemical properties of a series of drugs to their biological activity by means of a mathematical equation.

- The commonly studied physicochemical properties are hydrophobicity, electronic factors, and steric factors.

- The partition coefficient is a measure of a drug's overall hydrophobicity. Values of log P are used in QSAR equations, with larger values indicating greater hydrophobicity.

- The substituent hydrophobicity constant is a measure of the hydrophobic character of individual substituents. The values are different for substituents attached to aliphatic and aromatic systems and are only directly relevant to the class of structures from which the values were derived. Positive values represent substituents more hydrophobic than hydrogen; negative values represent substituents more hydrophilic than hydrogen.

- The Hammett substituent constant is a measure of how electron-withdrawing or electron-donating a substituent is. It is measured experimentally and is dependent on the relative position of the substituent on an aromatic ring. The value takes into account both inductive and resonance effects.

- The parameters F and R are constants quantifying the inductive and resonance effects of a substituent on an aromatic ring.

- The inductive effect of substituents on aliphatic moieties can be measured experimentally and tabulated.

- Steric factors can be measured experimentally or calculated using physical parameters or computer software.

- The Hansch equation is a mathematical equation which relates a variety of physicochemical parameters to biological activity for a series of related structures.

- The Craig plot is a visual comparison of two physicochemical properties for a variety of substituents. It facilitates the choice of substituents for a QSAR study such that the values of each property are not correlated.

- The Topliss scheme is used when structures can only be synthesized and tested one at a time. The scheme is a guide to which analogue should be synthesized next in order to get good activity. There are different schemes for substituents on aromatic and aliphatic systems.

- Indicator variables are used when there are no tabulated or experimental values for a particular property or substituent. The Free–Wilson approach to QSAR only uses indicator variables, whereas the Hansch approach can use a mixture of indicator variables and physicochemical parameters.

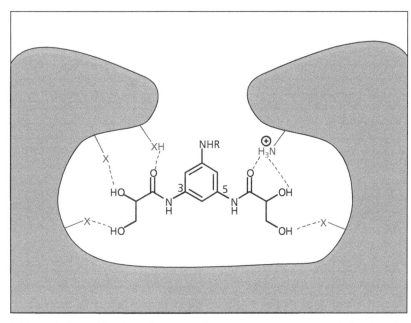

FIGURE 18.19 Hypothetical binding interactions between a pyranenamine and the target binding site.

18.10 3D QSAR

In recent years, a method known as 3D QSAR has been developed in which the three-dimensional properties of a molecule are considered as a whole rather than by considering individual substituents or moieties. This has proved remarkably useful in the design of new drugs. Moreover, the necessary software and hardware are readily affordable and relatively easy to use. The philosophy of 3D QSAR revolves around the assumption that the most important features about a molecule are its overall size and shape, and its electronic properties.

If these features can be defined, then it is possible to study how they affect biological properties. There are several approaches to 3D QSAR, but the method which has gained ascendancy was developed by the company Tripos and is known as **CoMFA (Comparative Molecular Field Analysis)**. CoMFA methodology is based on the assumption that drug–receptor interactions are non-covalent and that changes in biological activity correlate with the changes in the steric and/or electrostatic fields of the drug molecules.

18.10.1 Defining steric and electrostatic fields

The steric and electrostatic fields surrounding a molecule can be measured and defined using the grid and probe method described in section 17.7.5. This can be repeated for all the molecules in the 3D QSAR study, but it is crucial that the molecules are all in their **active conformation**, and that they are all positioned within the grid in exactly the same way. In other words, they must all be correctly aligned. Identifying a **pharmacophore** (section 13.2) that is common to all the molecules can assist in this process (Fig. 18.20).

The pharmacophore is placed into the grid and its position is kept constant such that it acts as a reference point when positioning each molecule into the lattice. For each molecule studied, the active conformation and pharmacophore is identified and then the molecule is placed into the lattice such that its pharmacophore matches the reference pharmacophore (Fig. 18.21). Once a molecule has been placed into the lattice, the steric and electrostatic fields around it are measured as described in section 17.7.5.

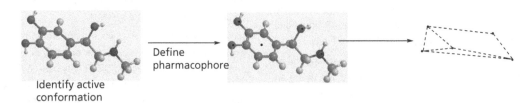

Identify active conformation

Define pharmacophore

FIGURE 18.20 Identification of the active conformation and pharmacophore.

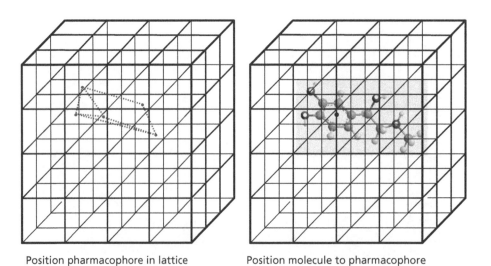

Position pharmacophore in lattice

Position molecule to pharmacophore

FIGURE 18.21 Positioning a pharmacophore and molecules into a lattice.

18.10.2 Relating shape and electronic distribution to biological activity

Defining the steric and electrostatic fields of a series of molecules is relatively straightforward and is carried out automatically by the software program. The next stage is to relate these properties to the biological activity of the molecules. This is less straightforward and differs significantly from traditional QSAR. In traditional QSAR, there are relatively few variables involved. For example, if we consider log P, π, σ, and a steric factor for each molecule, then we have four variables per molecule to compare against biological activity. With 100 molecules in the study, there are far more molecules than variables, and it is possible to come up with an equation relating variables to biological activity as previously described.

In 3D QSAR, the variables for each molecule are the calculated steric and electronic interactions at a couple of thousand lattice points. With 100 molecules in the study, the number of variables now far outweighs the number of structures, and it is not possible to relate these to biological potency by the standard multiple linear regression analysis described in section 18.1. A different statistical procedure has to be followed using a technique called **partial least squares** (PLS). Essentially, it is an analytical computing process which is repeated over and over again (iterated) to try to find the best formula relating biological potency against the various variables. As part of the process, the number of variables is reduced as the software filters out those which are clearly unrelated to biological activity.

An important feature of the analysis is that a structure is deliberately left out as the computer strives to form some form of relationship. Once a formula has been defined, the formula is tested against the structure which was left out. This is called **cross-validation** and tests how well the formula predicts the biological property for the molecule which was left out. The results of this are fed back into another round of calculations, but now the structure which was left out is included in the calculations, and a different structure is left out. This leads to a new improved formula which is once again tested against the compound that was left out, and so the process continues until cross-validation has been carried out against all the structures.

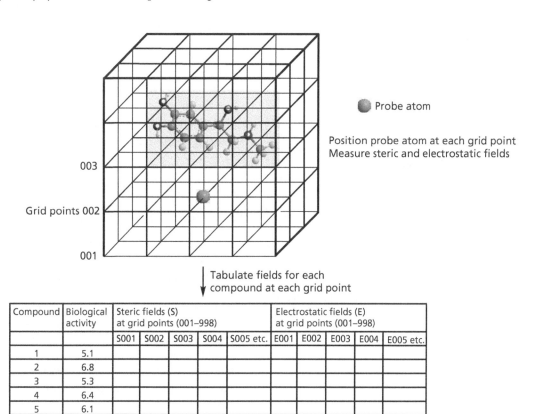

QSAR equation   Activity = aS001 + bS002 +........mS998 + nE001 +.......+ yE998 + z

FIGURE 18.22 Measuring steric and electronic fields.

At the end of the process, the final formula is obtained (Fig. 18.22). The predictability of this final equation is quantified by the **cross-validated correlation coefficient** r^2, which is usually referred to as q^2. In contrast to normal QSAR, where r^2 should be greater than 0.8, values of q^2 greater than 0.3 are considered significant. It is more useful, though, to give a graphical representation showing which regions around the molecule are important to biological activity on steric or electronic grounds. Therefore, a steric map shows a series of coloured contours indicating beneficial and detrimental steric interactions around a representative molecule from the set of molecules tested (Fig. 18.23). A similar contour map is created to illustrate electrostatic interactions.

An example of a 3D QSAR study is described in the case study in section 18.10.6.

For additional material see Web article 5: The design of a serotonin antagonist as a possible anxiolytic agent on the Online Resource Centre at www.oxfordtext-books.co.uk/orc/patrick6e/

18.10.3 Advantages of CoMFA over traditional QSAR

Some of the problems involved with a traditional QSAR study include the following:

- Only molecules of similar structure can be studied.
- The validity of the numerical descriptors is open to doubt. These descriptors are obtained by measuring reaction rates and equilibrium constants in model reactions and are listed in tables. However, separating one property from another is not always possible in experimental measurement. For example, the Taft steric factor is not purely a measure of the steric factor. This is because the measured reaction rates used to define it are also affected by electronic factors. Also, the n-octanol/water partition coefficients which are used to measure log P are known to be affected by the hydrogen bonding character of molecules.
- The tabulated descriptors may not include entries for unusual substituents.

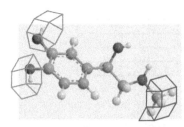

FIGURE 18.23 Definition of favourable and unfavourable interactions around a representative molecule.

- It is necessary to synthesize a range of molecules where substituents are varied in order to test a particular property (e.g. hydrophobicity). However, synthesizing such a range of compounds may not be straightforward or feasible.
- Traditional QSAR equations do not directly suggest new compounds to synthesize.

These problems are avoided with CoMFA which has the following advantages:

- Favourable and unfavourable interactions are represented graphically by 3D contours around a representative molecule. A graphical picture such as this is easier to visualize than a mathematical formula.
- In CoMFA, the properties of the test molecules are calculated individually by the computer program. There is no reliance on experimental or tabulated factors. There is no need to confine the study to molecules of similar structure. As long as one is confident that all the compounds in the study share the same pharmacophore and interact in the same way with the target, they can all be analysed in a CoMFA study.
- The graphical representation of beneficial and non-beneficial interactions allows medicinal chemists to design new structures. For example, if a contour map shows a favourable steric effect at one particular location, this implies that the target binding site has space for further extension at that location. This may lead to further favourable receptor–drug interactions.
- Both traditional and 3D QSAR can be used without needing to know the structure of the biological target.

18.10.4 Potential problems of CoMFA

There are several potential problems in using CoMFA:

- It is important to know the active conformation for each of the molecules in the study. Identifying the active conformation is easy for rigid structures such as steroids, but it is more difficult for flexible molecules that are capable of several bond rotations. Therefore, it is useful to have a conformationally restrained analogue which is biologically active and which can act as a guide to the likely active conformation. More flexible molecules can then be constructed on the computer with the conformation most closely matching that of the more rigid analogue. If the structure of the target binding site is known, this can be useful in deciding the likely active conformations of molecules.
- Each molecule in the study must be correctly positioned in the grid so that it is properly aligned with respect to all the other molecules. A common phar-

macophore can be used to aid this process as described earlier. However, it may be difficult to identify the pharmacophore in some molecules. In that case, a pharmacophore mapping exercise could be carried out (section 17.11). This is likely to be successful if there are some rigid active compounds amongst the compounds being studied. An alternative method of alignment is to align the molecules based on their structural similarity. This can be done automatically using what is known as 'topomer' methodology.

- One has to be careful to ensure that all the compounds in the study interact with the target in similar ways. For example, a 3D QSAR study on all possible **acetylcholinesterase** inhibitors is doomed to failure. In the first place, the great diversity of structures involved makes it impossible to align these structures in an unbiased way or to generate a 3D pharmacophore. Secondly, the various inhibitors do not interact with the target enzyme in the same way. X-ray crystallographic studies of enzyme–inhibitor complexes show that the inhibitors **tacrine**, **edrophonium**, and **decamethonium** all have different binding orientations in the active site.

- 3D QSAR provides a summary of how structural changes in a drug affect biological activity, but it is dangerous to assume too much. For example, a 3D QSAR model may show that increasing the bulk of the molecule at a particular location increases activity. This might suggest that there is an accessible hydrophobic pocket allowing extra binding interactions. On the other hand, it is possible that the extra steric bulk causes the molecule to bind in a different orientation from the other molecules in the analysis, and that this is the reason for the increased activity.

- It has been found that slightly different orientations of the grid from one study to another can produce different results for the same set of compounds.

18.10.5 Other 3D QSAR methods

CoMFA continues to be the most popular program for studies into 3D QSAR, but it does suffer a number of disadvantages as described above. Two of the more serious problems are the spurious results that can be obtained if compounds are not properly aligned, or if the orientation of the grid box is slightly different between studies. Users of the program can also choose different spacings between the grid points, and this can give poor results if the grid is too coarse or too fine. The method is also computationally expensive, requiring a lot of calculations for each molecule in the study, and so powerful computers are needed to cope with the huge memory requirement.

Other 3D QSAR programs have been developed in an attempt to address some of these issues. Examples include **HINT** which can be used alongside CoMFA to measure a hydrophobic field, **CoMSIA** which includes hydrogen bonding and hydrophobic fields as well as steric and electrostatic fields, and **CoMASA** which uses fewer calculations.

Some 3D QSAR programs use the intrinsic molecular properties of compounds rather than using a probe to measure the property fields surrounding them. Four examples are **SOMFA**, **HASL**, **CoMMA**, and **MS-Whim**. Other programs are used to model hypothetical pseudoreceptors. These include **Quasar**, **WeP**, and **GRIND**.

KEY POINTS

- CoMFA is an example of a 3D QSAR program which measures steric and electrostatic fields round a series of structures and relates these to biological activity.
- A comparison of the steric and electrostatic fields for different molecules against their biological activity allows the definition of steric and electrostatic interactions which are favourable and unfavourable for activity. These can be displayed visually as contour lines.
- It is necessary to define the active conformation and pharmacophore for each molecule in a CoMFA study. Alignment of the molecules is crucial.
- Unlike conventional QSAR studies, molecules of different structural classes can be compared if they share the same pharmacophore.
- 3D QSAR does not depend on experimentally measured parameters.
- A variety of different 3D QSAR programs have been developed.

18.10.6 Case study: inhibitors of tubulin polymerization

Colchicine (Fig. 18.24) is a lead compound for agents which act as inhibitors of tubulin polymerization (section 2.7.1) and might be useful in the treatment of arthritis. Other lead compounds have been discovered which bind to tubulin at the same binding site, and so a study

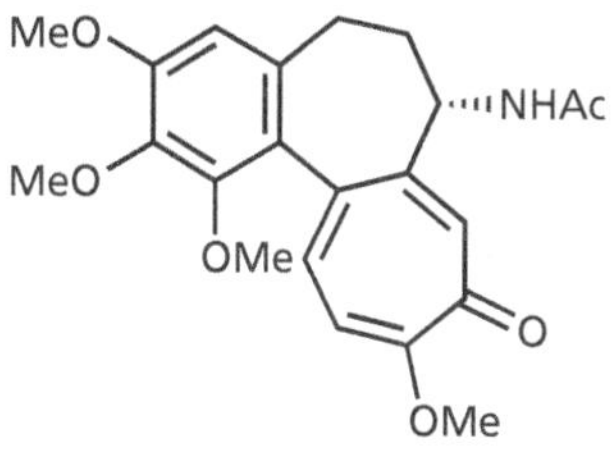

FIGURE 18.24 Colchicine.

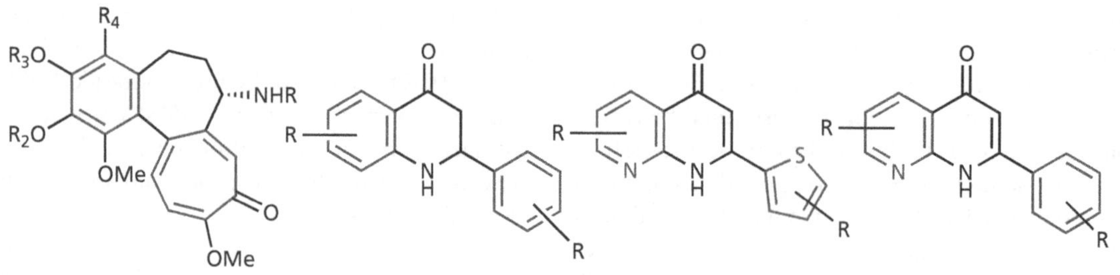

FIGURE 18.25 Structural classes used in the 3D QSAR study.

was carried out to compare the various structural classes interacting in this way. In this 3D QSAR study, 104 such agents were tested, belonging to four distinct families of compounds (Fig. 18.25); 51 compounds were used as a 'training set' for the analysis itself and 53 were used as a 'testing set' to test the predictive value of the results. Both sets contained a mixture of structural classes having both low and high activity.

The first task was to work out how to align these different classes of molecule. Colchicine is the most rigid of the four and also has a high affinity for tubulin. Therefore, it was chosen as the template on to which the other structures would be aligned. The relevant pharmacophore in colchicine was identified as the two aromatic rings. Molecular modelling was now carried out on each of the remaining structures to generate various conformations. Each conformation was compared with colchicine to find the one that would allow the pharmacophores in each structure to be aligned. This was then identified as the active conformation.

Once the active conformations for each structure had been identified, they were fitted into the lattice of grids previously described such that each structure was properly aligned. The steric and electrostatic fields round each molecule were calculated using a probe atom, then the 3D QSAR analysis was carried out to relate the fields to the measured biological activity.

The results of the 3D QSAR analysis are summarized as contour lines round a representative molecule

(Fig. 18.26). For the steric interactions, solid contours represent fields that are favourable for activity, and the dashed lines show fields that are unfavourable. For the electrostatic interactions, solid lines are regions where positively charged species improve affinity, and dashed lines indicate regions where negatively charged groups are favourable.

The results revealed that introducing steric bulk round the aromatic ring is more crucial to activity than introducing steric bulk round the bicyclic system. Based on this evidence, the structure shown in Fig. 18.27 was synthesized. The predicted value of pIC_{50} for the compound was 5.62. The actual value was in close agreement at 6.04 ($pIC_{50} = -\log [IC_{50}]$ where IC_{50} is the concentration of inhibitor required to produce 50% enzyme inhibition).

The steric fields of the Tripos CoMFA analysis (Fig. 18.26) were subsequently placed into a model of the binding site. It was found that the bad steric regions were in the same regions as the peptide backbone, whereas the favourable steric areas were in empty spaces.

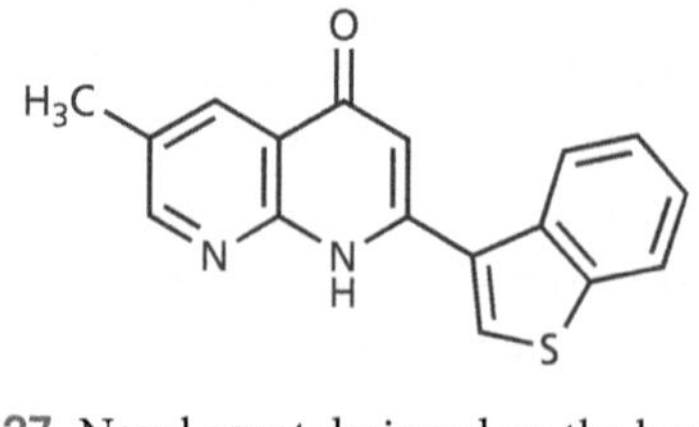

FIGURE 18.27 Novel agent designed on the basis of the 3D QSAR study.

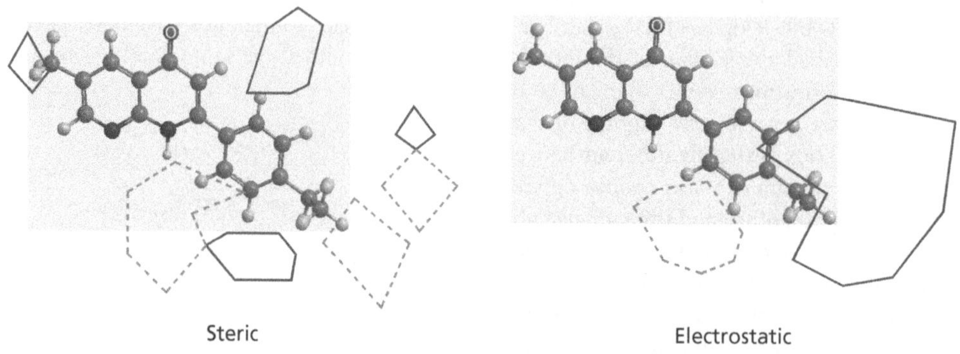

Steric   Electrostatic

FIGURE 18.26 Results of the 3D QSAR analysis ($q^2 = 0.637$).

QUESTIONS

1. Using values from Table 18.1, calculate the log P value for structure (I) (log P for benzene = 2.13).

2. Several analogues of a drug are to be prepared for a QSAR study which will consider the effect of various aromatic substituents on biological activity. You are asked whether the substituents (SO_2NH_2, CF_3, CN, CH_3SO_2, SF_3, $CONH_2$, OCF_3, CO_2H, Br, I) are relevant to the study. What are your thoughts?

3. A lead compound has a monosubstituted aromatic ring present as part of its structure. An analogue was synthesized containing a *para*-chloro substituent which had approximately the same activity. It was decided to synthesize an analogue bearing a methyl group at the *para* position. This showed increased activity. What analogue would you prepare next, and why?

4. The following QSAR equation was derived for the pesticide activity of structure (II). Explain what the various terms mean and whether the equation is a valid one. Identify what kind of substituents would be best for activity.

$$\log 1/C = 1.08\pi_x + 2.41F_x + 1.40\ R_x - 0.072\ MR_x + 5.25$$
$$(n = 16,\ r^2 = 0.840,\ s = 0.59)$$

5. A QSAR equation for the anticonvulsant (III) was derived as follows:

$$\log 1/C = 0.92\ \pi_x - 0.34\ \pi_x^2 + 3.18 \quad (n = 15,\ r^2 = 0.902,\ s = 0.09,\ \pi_o = 1.35)$$

What conclusions can you draw from this equation? Would you expect activity to be greater if X = CF_3 rather than H or CH_3?

6. The following QSAR equation is related to the mutagenic activity of a series of nitrosoamines; $\log 1/C = 0.92\ \pi + 2.08\ \sigma - 3.26$ ($n = 12$, $r^2 = 0.794$, $s = 0.314$). What sort of substituent is likely to result in high mutagenic activity?

@ Multiple-choice questions are available on the Online Resource Centre at www.oxfordtextbooks.co.uk/orc/patrick6e/

FURTHER READING

Craig, P. N. (1971) Interdependence between physical parameters and selection of substituent groups for correlation studies. *Journal of Medicinal Chemistry*, **14**, 680–4.

Cramer, R. D., et al. (1979) Application of quantitative structure–activity relationships in the development of the antiallergenic pyranenamines. *Journal of Medicinal Chemistry*, **22**, 714–25.

Cramer, R. D., Patterson, D. E., and Bunce, J. D. (1988) Comparative field analysis (CoMFA). *Journal of the American Chemical Society*, **110**, 5959–67.

Cramer, R. D. (2003) Topomer CoMFA: A design methodology for rapid lead optimization. *Journal of Medicinal Chemistry*, **46**, 374–88.

Hansch, C., and Leo, A. (1995) *Exploring QSAR*. American Chemical Society, Washington, DC.

Kellogg, G. E., and Abraham, D. J. (2000) Hydrophobicity: is Log P$_{o/w}$ more than the sum of its parts? *European Journal of Medicinal Chemistry*, **35**, 651–61.

Kotani, T., and Higashiura, K. (2004) Comparative molecular active site analysis (CoMASA). 1. An approach to rapid evaluation of 3D QSAR. *Journal of Medicinal Chemistry*, **47**, 2732–42.

Kubini, H., Folkers, G., and Martin, Y. C. (eds) (1998) *3D QSAR in drug design*. Kluwer/Escom, Dordrecht.

Martin, Y. C., and Dunn, W. J. (1973) Examination of the utility of the Topliss schemes by analog synthesis. *Journal of Medicinal Chemistry*, **16**, 578–9.

Sutherland, J. J., O'Brien, L. A., and Weaver, D. F. (2004) A comparison of methods for modeling quantitative structure–activity relationships. *Journal of Medicinal Chemistry*, **47**, 5541–54.

Taft, R. W. (1956) Separation of polar, steric and resonance effects in reactivity. In Newman, M. S. (ed.), *Steric effects in organic chemistry*. Chapter 13, John Wiley and Sons, New York.

Topliss, J. G. (1972) Utilization of operational schemes for analog synthesis in drug design. *Journal of Medicinal Chemistry*, **15**, 1006–11.

Verloop, A., Hoogenstraaten, W., and Tipker, J. (1976) Development and application of new steric substituent parameters in drug design. *Medicinal Chemistry*, **11**, 165–207.

van de Waterbeemd, H., Testa, B., and Folkers, G. (eds) (1997) *Computer-assisted lead finding and optimization*. Wiley–VCH, New York.

Zhang, S.-X., et al. (2000) Antitumor agents. 199. Three-dimensional quantitative structure–activity relationship study of the colchicine binding site ligands using comparative molecular field analysis. *Journal of Medicinal Chemistry*, **43**, 167–76.

Titles for general further reading are listed on p.845.

Periodic table of the elements

	1 I	2 II		3 III	4 IV	5 V	6 VI	7 VII	8	9 VIII
s	3 **Li** RAM: 6.941 P: 0.98	4 **Be** RAM: 9.012182 P: 1.57								
2	Lithium	Beryllium								
	11 **Na** RAM: 22.98977 P: 0.93	12 **Mg** RAM: 24.305 P: 1.31								
3	Sodium	Magnesium								
	19 **K** RAM: 39.0983 P: 0.82	20 **Ca** RAM: 40.078 P: 1	**d**	21 **Sc** RAM: 44.95591 P: 1.36	22 **Ti** RAM: 47.88 P: 1.54	23 **V** RAM: 50.9415 P: 1.63	24 **Cr** RAM: 51.9961 P: 1.66	25 **Mn** RAM: 54.93805 P: 1.55	26 **Fe** RAM: 55.847 P: 1.83	27 **Co** RAM: 58.9332 P: 1.88
4	Potassium	Calcium		Scandium	Titanium	Vanadium	Chromium	Manganese	Iron	Cobalt
	37 **Rb** RAM: 85.4678 P: 0.82	38 **Sr** RAM: 87.62 P: 0.95		39 **Y** RAM: 88.90585 P: 1.22	40 **Zr** RAM: 91.224 P: 1.33	41 **Nb** RAM: 92.90638 P: 1.6	42 **Mo** RAM: 95.94 P: 2.16	43 **Tc** RAM: 98 P: 1.9	44 **Ru** RAM: 101.07 P: 2.2	45 **Rh** RAM: 102.9055 P: 2.28
5	Rubidium	Strontium		Yttrium	Zirconium	Niobium	Molybdenum	Technetium	Ruthenium	Rhodium
	55 **Cs** RAM: 132.9054 P: 0.79	56 **Ba** RAM: 137.327 P: 0.89		71 **Lu** RAM: 174.967 P: 1.27	72 **Hf** RAM: 178.49 P: 1.3	73 **Ta** RAM: 180.9479 P: 1.5	74 **W** RAM: 183.85 P: 2.36	75 **Re** RAM: 186.207 P: 1.9	76 **Os** RAM: 190.2 P: 2.2	77 **Ir** RAM: 192.22 P: 2.2
6	Cesium	Barium		Lutetium	Hafnium	Tantalum	Tungsten	Rhenium	Osmium	Iridium
	87 **Fr** RAM: 223 P: 0.7	88 **Ra** RAM: 226.0254 P: 0.9		103 **Lr** RAM: 260 P:	104 **Rf** RAM: 261 P:	105 **Db** RAM: 262 P:	106 **Sg** RAM: 263 P:	107 **Bh** RAM: 262 P:	108 **Hs** RAM: 265 P:	109 **Mt** RAM: 266 P:
7	Francium	Radium		Lawrencium	Rutherfordium	Dubnium	Seaborgium	Bohrium	Hassium	Meitnerium

f

57 **La** RAM: 138.9055 P: 1.1	58 **Ce** RAM: 140.115 P: 1.12	59 **Pr** RAM: 140.9077 P: 1.13	60 **Nd** RAM: 144.24 P: 1.14	61 **Pm** RAM: 145 P: 1.13	62 **Sm** RAM: 150.36 P: 1.17	63 **Eu** RAM: 151.965 P: 1.2
Lanthanum	Cerium	Praseodymium	Neodymium	Promethium	Samarium	Europium
89 **Ac** RAM: 227 P: 1.1	90 **Th** RAM: 232.0381 P: 1.3	91 **Pa** RAM: 213.0359 P: 1.5	92 **U** RAM: 238.0289 P: 1.38	93 **Np** RAM: 237.0482 P: 1.36	94 **Pu** RAM: 244 P: 1.28	95 **Am** RAM: 243 P: 1.3
Actinium	Thorium	Protactinium	Uranium	Neptunium	Plutonium	Americium

Key

Symbol

Atomic number .....00 **Xx**
Relative Atomic Mass ..... RAM: 0.000
Electronegativity (Pauling) ..... P: 0.0
Element .... Name

III

1s

1 **H**	2 **He**
RAM: 1.00794	RAM: 4.002602
P: 2.2	P: 0
Hydrogen	Helium

10	11 I	12 II	13 III	14 IV	15 V	16 VI	17 VII	18 VIII

p

5 **B**	6 **C**	7 **N**	8 **O**	9 **F**	10 **Ne**
RAM: 10.811	RAM: 12.011	RAM: 14.00674	RAM: 15.9994	RAM: 18.9984	RAM: 20.1797
P: 2.04	P: 2.55	P: 3.04	P: 3.44	P: 3.98	P: 0
Boron	Carbon	Nitrogen	Oxygen	Fluorine	Neon

13 **Al**	14 **Si**	15 **P**	16 **S**	17 **Cl**	18 **Ar**
RAM: 26.98154	RAM: 28.0855	RAM: 30.97376	RAM: 32.066	RAM: 35.4527	RAM: 39.948
P: 1.61	P: 1.9	P: 2.19	P: 2.58	P: 3.16	P: 0
Aluminium	Silicon	Phosphorus	Sulfur	Chlorine	Argon

28 **Ni**	29 **Cu**	30 **Zn**	31 **Ga**	32 **Ge**	33 **As**	34 **Se**	35 **Br**	36 **Kr**
RAM: 58.6934	RAM: 63.546	RAM: 65.39	RAM: 69.723	RAM: 72.61	RAM: 74.92159	RAM: 78.96	RAM: 79.904	RAM: 83.8
P: 1.91	P: 1.9	P: 1.65	P: 1.81	P: 2.01	P: 2.18	P: 2.55	P: 2.96	P: 0
Nickel	Copper	Zinc	Gallium	Germanium	Arsenic	Selenium	Bromine	Krypton

46 **Pd**	47 **Ag**	48 **Cd**	49 **In**	50 **Sn**	51 **Sb**	52 **Te**	53 **I**	54 **Xe**
RAM: 106.42	RAM: 107.8682	RAM: 112.411	RAM: 114.82	RAM: 118.71	RAM: 121.757	RAM: 127.6	RAM: 126.9045	RAM: 131.29
P: 2.2	P: 1.93	P: 1.69	P: 1.78	P: 1.96	P: 2.05	P: 2.1	P: 2.66	P: 0
Palladium	Silver	Cadmium	Indium	Tin	Antimony	Tellurium	Iodine	Xenon

78 **Pt**	79 **Au**	80 **Hg**	81 **Tl**	82 **Pb**	83 **Bi**	84 **Po**	85 **At**	86 **Rn**
RAM: 195.08	RAM: 196.9665	RAM: 200.59	RAM: 204.3833	RAM: 207.2	RAM: 208.9804	RAM: 209	RAM: 210	RAM: 222
P: 2.28	P: 2.54	P: 2	P: 2.04	P: 2.33	P: 2.02	P: 2	P: 2.2	P: 0
Platinum	Gold	Mercury	Thallium	Lead	Bismuth	Polonium	Astatine	Radon

Other artificially-produced elements have
been isolated, but are of no practical interest to
organic chemists.

64 **Gd**	65 **Tb**	66 **Dy**	67 **Ho**	68 **Er**	69 **Tm**	70 **Yb**	Lanthanides
RAM: 157.25	RAM: 158.9253	RAM: 162.5	RAM: 164.9303	RAM: 167.26	RAM: 168.9342	RAM: 173.04	
P: 1.2	P: 1.2	P: 1.22	P: 1.23	P: 1.24	P: 1.25	P: 1.1	
Gadolinium	Terbium	Dysprosium	Holmium	Erbium	Thulium	Ytterbium	

96 **Cm**	97 **Bk**	98 **Cf**	99 **Es**	100 **Fm**	101 **Md**	102 **No**	Actinides
RAM: 247	RAM: 247	RAM: 251	RAM: 252	RAM: 257	RAM: 258	RAM: 259	
P: 1.3	P: 1.3	P: 1.3	P: 1.3	P: 1.3	P: 1.3	P: 1.3	
Curium	Berkelium	Californium	Einsteinium	Fermium	Mendelevium	Nobelium	